Springer Monographs in Mathematics

This series publishes advanced monographs giving well-written presentations of the "state-of-the-art" in fields of mathematical research that have acquired the maturity needed for such a treatment. They are sufficiently self-contained to be accessible to more than just the intimate specialists of the subject, and sufficiently comprehensive to remain valuable references for many years. Besides the current state of knowledge in its field, an SMM volume should ideally describe its relevance to and interaction with neighbouring fields of mathematics, and give pointers to future directions of research.

More information about this series at http://www.springer.com/series/3733

Ali Baklouti • Hidenori Fujiwara • Jean Ludwig

Representation Theory of Solvable Lie Groups and Related Topics

Springer

Ali Baklouti
Department of Mathematics
University of Sfax
Sfax, Tunisia

Hidenori Fujiwara
Faculté de Science et Technologie pour l'Humanité
Université de Kinki
Iizuka, Japan

Jean Ludwig
Institut Élie Cartan de Lorraine
Université de Lorraine
Metz, France

ISSN 1439-7382 ISSN 2196-9922 (electronic)
Springer Monographs in Mathematics
ISBN 978-3-030-82043-5 ISBN 978-3-030-82044-2 (eBook)
https://doi.org/10.1007/978-3-030-82044-2

This Springer imprint is published by the registered company Springer Nature Switzerland AG.
The registered company address is: Gewerbestrasse 11, 6330 Cham, Switzerland

Preface

Representation theory of Lie groups and noncommutative harmonic analysis on Lie groups and homogeneous spaces have witnessed a significant growth starting from the 1940s when research work in these areas began. Taken jointly, they constitute a pivotal and compelling field within mathematics, due to the tight relationships with so many other areas—think of number theory, algebraic geometry, differential geometry, operator algebras, PDEs, to name but a few—and also with farther fields like physics.

Within the setting of exponential solvable Lie groups, the so-called orbit method is a fundamental tool to associate through an elegant way unitary duals to the space of coadjoint orbits. Still, the harmonic analysis on the corresponding homogeneous spaces remains a difficult subject.

The purpose of this book is to discuss the latest advances in this area, prove novel results on noncommutative harmonic analysis on solvable homogeneous spaces, and provide many applications. The text offers the most recent solutions to a number of open questions posed over the last decades, presents the newest research results on the matter, and provides an alluring platform for progressing in this research area.

Throughout the text, unless otherwise explicitly stated, G will always denote a connected and simply connected exponential solvable Lie group with Lie algebra $\mathfrak{g}$. This means the exponential mapping $\exp : \mathfrak{g} \longrightarrow G$ is a diffeomorphism. The first chapter aims to build the branching laws of *mixed representations* (or representations of mixed type), showing that they obey the orbital spectrum formula. Let A and H be two closed connected subgroups of G, and $\pi and \sigma$ two unitary irreducible representations of G and H, respectively. Mixed representations are unitary representations of the form $(\mathrm{ind}_H^G \sigma)|_A$ (up-down representation) or $\mathrm{ind}_H^G(\pi_{|H})$ (down-up representation). We shall provide explicit formulas for the multiplicity function occurring in the disintegration of such representations, and detailed information regarding its behavior. We shall study many situations where the multiplicity function is uniformly infinite or finite and bounded, and discuss along the way necessary and sufficient conditions for finiteness. One major motivation for this work goes back to the theory of ergodic Lie group actions. Another is the need to build smooth disintegrations of the space $L^2(G)$ on a Lie group G from a representation-theoretical point of view.

Now, one of the most important and difficult problems in the disintegration theory of group representations is to write down explicit unitary operators that intertwine unitary representations and their decomposition into irreducibles. This is, as a matter of fact, the basic aim of Chap. 3. The answer to this question can help to solve several important problems, for example regarding the study of the solvability of invariant differential operators on homogeneous spaces, or the Frobenius reciprocity problem related to the corresponding multiplicities, the disintegration of generalized invariant vectors, etc. We shall construct intertwining operators of induced representations (sometimes concerning discrete subgroups) and restricted representations of arbitrary nilpotent Lie groups. This will elicit the analysis of the tensor product of two unitary representations. Even in the case of exponential solvable Lie groups these problems are still not fully solved, and we shall only discuss the case of induced representations from normal subgroups.

With the above in mind, we establish the Bonnet-Plancherel formula associated with a monomial representation of a nilpotent Lie group and, more generally, of an exponential solvable Lie group when the inducing subgroup is normal. In Chap. 4, we will also discuss a variant of the Penney-Plancherel formula for induced representations and restrictions.

The orbit method of Chap. 2, which goes back to the early 1970s, asserts that two monomial representations $\pi_i = \mathrm{ind}_{H_i}^G \chi_f (H_i = \exp(\mathfrak{h}_i), 1 \leq i \leq 2)$ of G are irreducible and mutually equivalent. Here $\mathfrak{h}_1$ and $\mathfrak{h}_2$ are two polarizations of $\mathfrak{g}$ at $f \in \mathfrak{g}^*$ that satisfy the Pukanszky condition. We construct an explicit intertwining operator between these representations in the absence of a complete proof of the fact that the product $H_1 H_2$ is a closed subset of G. A third Vergne polarization will come into play and give rise to a unitary intertwining operator. We shall prove a composition theorem involving the Maslov index of the three polarizations. The fundamental results of Auslander-Kostant [3] and Pukanszky [144] about holomorphically induced representations of solvable Lie groups, which hold without any knowledge of real polarizations for a given linear form in the dual Lie algebra, live outside the world of exponential Lie groups. There clearly remains a massive number of open questions to be solved. Chapter 6 will provide the occasion to study certain intertwining operators and real polarizations in the aforementioned rather general setting.

The aim of Chap. 5 is to present two open conjectures concerning certain algebras of invariant differential operators on the space of induced and restricted representations of nilpotent Lie groups. The claim is that whenever these representations have finite multiplicities, the corresponding algebras are isomorphic to the algebra of invariant polynomial functions on some related geometric object (manifold). We will furnish solutions in many special cases and lay out the latest advances in this direction.

Chapter 7 is a sort of generalization of Chaps. 2 and 4 to the setting of exponential groups G. We shall study in every detail the finite and infinite multiplicity of discrete-type monomial representations. Our interest lies in the Penney-Plancherel formula for monomial representations of discrete type, and in a question of Michel Duflo (cf. [54]) regarding the commutativity of the algebra of G-invariant differential operators on line bundles associated with the representation in question. We shall emphasize the result whereby such algebra turns out to be commutative in the case of monomial representations of discrete type and hence gives a counterexample to Duflo's question. The orbit method (cf. [70]) will be a fundamental technique in the study.

The last chapter (Chap. 8) is an application of earlier results of the book. We will characterize bounded, topologically irreducible representations of an exponential solvable Lie group G on Banach spaces by using a triple involving topologically irreducible weighted representations. We shall determine the simple modules of the group algebra $L^1(G)$. These are essentially obtained in the same way as for nilpotent groups, except that one has to generalize the induced representations. This is no longer true for topologically irreducible representations, as shown in [117]. We will show that if the group G is not symmetric, a new type of bounded, irreducible representation of G on Banach spaces crops up. The corresponding representation of $L^1(G)$ is fundamentally different from the induced representation. The construction of such representations is related to the invariant-subspace problem.

Sfax, Tunisia
Iizuka, Japan
Metz, France

Ali Baklouti
Hidenori Fujiwara
Jean Ludwig

Nomenclature

$\rho(G, H, A, \sigma)$	up-down representation
$\mathscr{E}(G, K)$	the space of continuous functions ξ on G with compact support modulo K
$(\mathscr{H}_\rho^{-\infty})^{K,\lambda}$	space of K-semi-invariant generalized vectors
β_ϕ	Penney distribution
$\exp : \mathfrak{g} \longrightarrow G$	The exponential mapping
λ_G	left regular representation
$\mathbb{C}(\Omega)^K$	field of invariant rational functions
$\mathbb{C}(\Gamma_\tau)^H$	field of invariant rational functions
$\mathbb{C}[\Omega(\pi)]^K$	algebra of invariant polynomial functions
$\mathbb{C}[\Gamma_\tau]^H$	algebra of invariant polynomial functions
$\mathfrak{b}$	real kernel part
$\mathfrak{d}$	real part
$\mathfrak{e}$	complex part
$\mathfrak{g}^*/G$	space of coadjoint orbits
$\mathscr{S}(N)$	the Schwartz-algebra of N
$\mathscr{C}$	composition sequence
$\mathscr{H}(f, \mathfrak{h}, N)$	space of holomorphically induced representation
$\mathscr{H}(f, \eta_f, \mathfrak{h}_i, G)$	space of holomorphically induced representation
$\mathscr{S}$	good sequence of subalgebras
$\mathscr{U}(\mathfrak{g}, \tau)$	infinitesimal invariant operator
ω_U	fundamental weight
$\pi \times \pi'$	outer tensor product of two representations
$\pi_{\vert H}$	a restriction of a representation to a subgroup
$\rho(f, \mathfrak{h}, N)$	holomorphically induced representation
$\rho(f, \eta_f, \mathfrak{h}, G)$	holomorphically induced representation
$\rho(G, H, \pi)$	down-up representation
$\mathrm{ind}_H^G \nu$	induced representation
Δ_G	modular function
Θ	algebra homomorphism
Θ_G	the Kirillov-Bernat mapping

$\widehat{A}^{\text{top}}$	equivalence classes of irreducible Banach space modules of A
$a(\mathfrak{h}_i, \mathfrak{h}_j, \mathfrak{h}_k)$	$= e^{\frac{i\pi}{4}\tau_{ijk}}$
a_ℓ	pointwise Penney distribution
a_π^k	Penney's distribution
$c(i, j, k)$	Maslov index
$D_\pi(G)^K$	algebra of invariant differential operator
$D_\tau(G/H)$	algebra of invariant differential operators
$G \cdot l$	coadjoint orbit
$H \diagdown G \diagup B$	a double coset
$H(x, y)$	Hermitian form
$H_f(x, y)$	Hermitian form
$I_{\mathfrak{h}_2\mathfrak{h}_1}$	formal intertwining integral
J	complex structure
$K(s, m)$	integral
$P^+ = P^+(f, \mathfrak{n})$	set of positive polarizations
P_X	image function
$R_{a,G/P,\ell}$	retract
$S(e)$	set of jump indices
$S(x, y)$	symmetric form
$Simple(A)$	equivalence classes of simple modules
$T(e)$	set of non-jump indices
$T(f, \eta_f, \mathfrak{h}, G)$	induced representation
$T_{\mathfrak{h}_2\mathfrak{h}_1}$	intertwining operator
U_e	layer
X^{fin}	finite-rank subspace
$\delta_\ell(s) = \delta_{\ell,\mathscr{S}}(s)$	weight factor
$\mathfrak{a}_\tau$	kernel
$\nu_{\mathfrak{h}}(s)$	action of an automorphism
τ_{ijk}	Maslov index
$\theta_{\mathfrak{h}}(s)$	density
$\Lambda_{\mathfrak{h}}$	intertwining isomorphism

Contents

Chapter 1
Branching Laws and the Multiplicity Function of Unitary Representations of Exponential Solvable Lie Groups

1.1 Introduction

Throughout the chapter G denotes an exponential solvable Lie group. Let A and H be two closed connected subgroups of G and π and σ unitary and irreducible representations of G and H respectively. Branching laws and explicit formulas of the disintegration into irreducibles of the representations $\mathrm{ind}_H^G\sigma$ and $\pi_{|A}$ in terms of the representations' spectrum and multiplicities are well known (see for instance [63, 64] and [104] and the references therein), so we shall provide a quick reminder. We will then introduce so-called mixed representations, or representations of mixed type. These are unitary representations of the form $(\mathrm{ind}_H^G\sigma)_{|A}$ and $\mathrm{ind}_H^G(\pi_{|H})$.

One motivation is that the study of mixed representations is of great interest since their decompositions into irreducibles, as well as the corresponding intertwining operators, provide us with new smooth disintegrations of $L^2(G)$ for exponential groups (cf. Sect. 3.3.4 for instance). Other reasons of interest go back to the theory of ergodic actions as explained by R. L. Lipsman [107], where the spectrum and the multiplicities of particular types of mixed representations are studied. More specifically, the representation $\mathrm{ind}_H^G 1_{|A}$, where H is a closed subgroup of G and A an Abelian subgroup of G. Lipsman determined the concrete spectrum and the explicit multiplicity formula using the orbit method.

The first aim of the chapter is to show that mixed representations (up-down and down-up representations) of exponential solvable Lie groups obey the orbital spectrum formula. Explicit formulas of the multiplicity function occurring in the disintegration of such representations will be given. We also prove that for completely solvable Lie groups the multiplicities of such representations are uniformly infinite or finite and bounded. Necessary and sufficient conditions for the finiteness will be provided. These conditions clearly pass to the subgroups H and A and their actions by using semi-pseudo-algebraic geometry.

A. Baklouti et al., *Representation Theory of Solvable Lie Groups and Related Topics*, Springer Monographs in Mathematics,
https://doi.org/10.1007/978-3-030-82044-2_1

For exponential solvable Lie groups the situation turns out to be more complicated. In fact, the non-unipotency of the group actions greatly complicates the manipulation of the multiplicity functions. However, we show that the multiplicity results do hold when H is a normal subgroup. We also produce counterexamples showing that these results fail in general.

1.2 Generalities and Notations

We begin this section by reviewing some facts about induced and restricted representations of a solvable Lie group. One says that G is an exponential solvable Lie group if the exponential mapping $\exp : \mathfrak{g} \longrightarrow G$ is a diffeomorphism, where $\mathfrak{g}$ designates the Lie algebra of G. Let $\mathfrak{g}^*$ be the dual space of $\mathfrak{g}$. The Lie algebra $\mathfrak{g}$ acts on $\mathfrak{g}$ by the adjoint representation $\mathrm{ad}_{\mathfrak{g}}$, that is

$$\mathrm{ad}_{\mathfrak{g}}(X)(Y) = \mathrm{ad}(X)(Y) = [X, Y],\ X, Y \in \mathfrak{g}.$$

The group G acts on $\mathfrak{g}$ by the adjoint representation $\mathrm{Ad}_{\mathfrak{g}}$, that is

$$\mathrm{Ad}_{\mathfrak{g}}(g)Y = \mathrm{Ad}(g)Y = e^{\mathrm{ad}(X)}Y,\ g \in G,\ Y \in \mathfrak{g}$$

and on $\mathfrak{g}^*$ by the coadjoint representation $\mathrm{Ad}^*_{\mathfrak{g}}$, i.e.

$$\mathrm{Ad}^*_{\mathfrak{g}}(g)l(X) = g \cdot l(X) = l(\mathrm{Ad}(g^{-1})X),\quad g \in G,\ l \in \mathfrak{g}^*\ X \in \mathfrak{g}.$$

The coadjoint orbit $G \cdot l$ (or Gl for short) of l is the set

$$G \cdot l = \{g \cdot l, g \in G\}. \tag{1.2.1}$$

The space of coadjoint orbits is denoted by $\mathfrak{g}^*/G$.

1.2.1 Coexponential Bases

Let $\mathfrak{k}$ be a subalgebra of $\mathfrak{g}$. A subset $\{X_1, X_2, \ldots, X_d\}$ of elements of $\mathfrak{g}$ is called a *coexponential basis* for $\mathfrak{k}$ in $\mathfrak{g}$ if the mapping

$$((x_1, x_2, \ldots, x_d), Y) \longmapsto \left(\prod_{j=1}^{d} \exp\ (x_j X_j)\right) \exp\ Y$$

is a diffeomorphism of $\mathbb{R}^d \times \mathfrak{k}$ on the group G. Such a basis always exists for connected subgroups of G [24, chap. I].

1.2.2 Modular Functions and Quotient Measures

Let G be an exponential solvable Lie group with Lie algebra $\mathfrak{g}$. Let dg be a left Haar measure of G and Δ_G the modular function of G, which is defined by the relation

$$\int_G F(gx^{-1})\, dg = \Delta_G(x) \int_G F(g)\, dg \ (x \in G)$$

for any F in the set $C_c(G)$ of continuous functions on G with compact support. It is well known that

$$\Delta_G(x) = |\det \mathrm{Ad}\, x|^{-1} \ (x \in G).$$

Define for a subgroup K in G the function $\Delta_{K,G}(k)$ on K by

$$\Delta_{K,G}(k) = \frac{\Delta_K(k)}{\Delta_G(k)}, \ (k \in K).$$

Let $\mathscr{E}(G, K)$ be the space of continuous functions ξ on G with compact support modulo K which satisfy

$$\xi(gk) = \Delta_{K,G}(k)\xi(g) \quad (g \in G,\ k \in K).$$

Then G acts on $\mathscr{E}(G, K)$ by left translations. It is known [24] that, up to a positive scalar factor, there exists a unique G-invariant positive linear form on $\mathscr{E}(G, K)$. We denote it by $\mu_{G,K}$ and write it as

$$\mu_{G,K}(\xi) = \oint_{G/K} \xi(g)\ \mu_{G,K}(g).$$

If $\Delta_K = \Delta_G$ on K (this is for example the case if K is a normal subgroup of G), then $\mu_{G,K}$ turns out to be a G-invariant measure on the homogeneous space G/K.

Lemma 1.2.1 (Transitivity Formula (See [24])) *Let $K \subset H$ be two closed connected subgroups of G, and ξ a $\mu_{G,H}$-integrable function on G. Then the set of $g \in G$ such that the function on H defined by*

$$h \mapsto \xi(gh)\Delta_{H,G}^{-1}(h)$$

is not $\mu_{G,H}$-integrable, but $\mu_{H,K}$-negligible. The function on G

$$g \mapsto \oint_{H/K} \xi(gh)\Delta_{H,G}^{-1}(h) d\mu_{H,K}(h)$$

is $\mu_{G,H}$-integrable, and we have

$$\oint_{G/K} \xi(x) d\mu_{G,K}(x) = \oint_{G/H} d\mu_{G,H}(g) \oint_{H/K} \xi(gh)\Delta_{H,G}^{-1}(h) d\mu_{H,K}(h).$$

To describe these measures we need suitable Malcev bases or coexponential bases. Let $(\mathfrak{a}_i)_{i=n}^1$ be a good sequence of subalgebras of $\mathfrak{g}$. This means that

$$\mathfrak{g} = \mathfrak{a}_n \supset \ldots \supset \mathfrak{a}_1 \supset \mathfrak{a}_0 = (0),$$

with $\dim \mathfrak{a}_i/\mathfrak{a}_{i-1} = 1$, for $i = n, n-1, \ldots, 1$, and also that if $\mathfrak{a}_i$ is not an ideal in $\mathfrak{g}$, then $\mathfrak{a}_{i+1}$ and $\mathfrak{a}_{i-1}$ are ideals of $\mathfrak{g}$ and the $\mathfrak{g}$-module $\mathfrak{a}_{i+1}/\mathfrak{a}_{i-1}$ is irreducible. Let $\mathfrak{k}$ be a subalgebra of $\mathfrak{g}$ and $K = \exp \mathfrak{k}$. We set

$$I^K = \{i \in \{n, \ldots, 1\};\ \mathfrak{a}_i \cap \mathfrak{k} \neq \mathfrak{a}_{i-1} \cap \mathfrak{k}\}.$$

Now we choose for $i \notin I^K$ an element $Z_i \in \mathfrak{a}_i \setminus \mathfrak{a}_{i-1}$, and for $i \in I^K$ a vector $Z_i \in \mathfrak{k} \cap \mathfrak{a}_i \setminus \mathfrak{a}_{i-1} \cap \mathfrak{k}$. We obtain this way a strong Malcev basis $\mathscr{Z} = \{Z_1, \ldots, Z_n\}$ of $\mathfrak{g}$, which contains a strong Malcev basis $\mathscr{Z}_{\mathfrak{k}} = \{Z_i,\ i \in I^K\}$ of $\mathfrak{k}$. The mapping

$$E_{\mathscr{Z}} : \mathbb{R}^n \to G;\ E_{\mathscr{Z}}(T) = \exp(t_n Z_n) \cdots \exp(t_1 Z_1),\ T = (t_n, \ldots, t_1) \in \mathbb{R}^n,$$

is then a diffeomorphism. Similarly, the family

$$\mathscr{X} = \{X_k, \ldots, X_1\} = \{Z_{i_k}, \ldots, Z_{i_1}\} = \{Z_i,\ i = 1, \ldots, n,\ i \notin I^K\}$$

is a coexponential basis of $\mathfrak{g}$ relative to $\mathfrak{k}$, and the mapping

$$E_{\mathscr{X}} : \mathbb{R}^k \to G/K;\ E_{\mathscr{X}}(T) = \exp(t_k X_k) \cdots \exp(t_1 X_1) K,\ T = (t_k, \ldots, t_1) \in \mathbb{R}^k,$$

is a diffeomorphism. We can now write the Haar measure dx of G explicitly as follows:

$$\int_G \varphi(x) dx = \int_{\mathbb{R}^n} \varphi(E_{\mathscr{Z}}(T)) dT, \quad \varphi \in C_c(G),$$

and

$$\oint_{G/K} \psi(s) d\mu_{G,K}(s) = \int_{\mathbb{R}^k} \psi(E_{\mathscr{X}}(T)) dT, \quad \psi \in C_c(G/K).$$

1.2.3 Induced Representations

Let ν be a unitary representation of H acting on a Hilbert space $\mathscr{H}_\nu$. We define the unitary induced representation $\tau(\nu) = \mathrm{ind}_H^G \nu$ of G, i.e. the representation of G induced by the representation ν of H, by using a left translation on the completion of the space of continuous function ξ on G with values in $\mathscr{H}_\nu$, satisfying

$$\xi(gh) = \Delta_{H,G}(h^{-1})\nu(h^{-1})(\xi(g)), \ g \in G, \ h \in H, \tag{1.2.2}$$

and having a compact support modulo H. This space is endowed with the norm

$$||\xi|| = \Big(\oint_{G/H} ||\xi(g)||^2_{\mathscr{H}_\nu} d\mu_{G,H}(g)\Big)^{\frac{1}{2}}.$$

1.2.4 Polarizations

For any f in $\mathfrak{g}^*$ we define the skew-symmetric bilinear form B_f on $\mathfrak{g}$ by the formula $B_f(X, Y) = \langle f, [X, Y]\rangle$. If $\mathfrak{h}$ is a Lie subalgebra of $\mathfrak{g}$, we write $\mathfrak{h}(f) = \{X \in \mathfrak{g} : B_f(X, \mathfrak{h}) = \{0\}\}$ and $\mathfrak{h}^\perp = \{l \in \mathfrak{g}^* : l_{|\mathfrak{h}} = 0\}$, where $l_{|\mathfrak{h}}$ stands for the restriction of l to $\mathfrak{h}$. If $\mathfrak{h} \subset \mathfrak{h}(f)$, then $\mathfrak{h}$ is said to be a totally isotropic subalgebra for B_f. We denote by $i(f, \mathfrak{g})$ the set of maximal, totally isotropic subspaces for B_f, and by $S(f, \mathfrak{g})$ the set of all isotropic subalgebras for B_f. If $\mathfrak{h}$ is in $S(f, \mathfrak{g})$, we define the unitary character χ_f of the group $H = \exp \mathfrak{h}$ by

$$\chi_f(\exp Y) = e^{-2i\pi\langle f,Y\rangle}, \ Y \in \mathfrak{h}. \tag{1.2.3}$$

Let $M(f, \mathfrak{g})$ be the set of elements of $S(f, \mathfrak{g})$ of maximal dimension. An element of $M(f, \mathfrak{g})$ will be called (real) polarization at the linear form f. We also call $I(f, \mathfrak{g})$ the subset of $S(f, \mathfrak{g})$ of subalgebras $\mathfrak{b}$ such that $\tau(f) = \tau(\chi_f)$ is irreducible. We shall simply write τ instead of $\tau(f)$ when no confusion arises.

Theorem 1.2.2 ([141], [24, chap. VI]) *Let $\mathfrak{b}$ be an element of $S(f, \mathfrak{g})$ and $B =$* $\exp \mathfrak{b}$. *The following assertions are equivalent:*

(i) $B \cdot f = f + \mathfrak{b}^\perp$;
(ii) $\mathfrak{b}$ *is in* $M(f, \mathfrak{g})$ *and* $f + \mathfrak{b}^\perp \subset G \cdot f$;
(iii) *For all* l *in* $\mathfrak{b}^\perp$, $\mathfrak{b}$ *belongs to* $M(f + l, \mathfrak{g})$;
(iv) $\mathfrak{b}$ *is in* $I(f, \mathfrak{g})$.

Definition 1.2.3 If $\mathfrak{b}$ meets one of the above conditions, we say that $\mathfrak{b}$ is a polarization of l in $\mathfrak{g}$ which meets the Pukanszky condition (or simply, a Pukanszky polarization).

There is a standard process due to M. Vergne to construct an element of $I(f, \mathfrak{g})$. Let $\mathfrak{s} = \left(\mathfrak{a}_j\right)_{j=0}^n$ be a good sequence of subalgebras of $\mathfrak{g}$. Let f_j be the restriction $f_{|\mathfrak{a}_j}$ of f to $\mathfrak{a}_j$, for $1 \leq j \leq n$. The following theorem is proved in [24]:

Theorem 1.2.4 *The subspace $\mathfrak{b} = \sum_{1\leq j\leq n} \mathfrak{a}_j(f_j)$ is a polarization of f (called Vergne polarization) that satisfies the Pukanszky condition.*

1.2.5 *Orbit Theory*

For $l \in \mathfrak{g}^*$, we take a real polarization $\mathfrak{b}$ of l meeting the Pukanszky condition. For such a polarization, define

$$\pi_l = \pi_{l,\mathfrak{b}} = \mathrm{ind}_B^G \chi_l, \quad B = \exp \mathfrak{b}.$$

Then π is a unitary and irreducible representation of G, and its equivalence class $[\pi_{l,\mathfrak{b}}]$ only depends on the coadjoint orbit of l as of (1.2.1). Moreover, every irreducible representation π is equivalent to an induced representation $\pi_{l,\mathfrak{b}}$ for some $l \in \mathfrak{g}^*$ and some polarization $\mathfrak{b}$ of l with the character χ_l. The map

$$\Theta_G : \mathfrak{g}^* \longrightarrow \hat{G}; l \longmapsto [\pi_{l,\mathfrak{b}}]$$

is called the Kirillov-Bernat mapping, and quotients to a homeomorphism

$$K : \mathfrak{g}^*/G \longrightarrow \hat{G}; G \cdot l \longmapsto \pi_{l,\mathfrak{b}},$$

when the spaces are endowed with suitable topologies (cf. [24]). The symbols K and Θ_G may be used interchangeably when no confusion arises.

1.2.6 *Branching Laws: Induced Representations*

Let $p_{\mathfrak{h}} : \mathfrak{g}^* \to \mathfrak{h}^*$ be the canonical projection of $\mathfrak{h}$, and Ω_σ^H the coadjoint orbit associated to the representation σ by the Kirillov-Bernat map $\Theta_H : \mathfrak{h}^* \to \hat{H}$, (where $\hat{H}$ is the unitary dual of H). The natural measure on $p_{\mathfrak{h}}^{-1}(\Omega_\sigma^H)$ is the fiber measure that has H-invariant measure on the base Ω_σ^H and Lebesgue measure on the affine fiber $\mathfrak{h}^\perp$. The representation $\tau(\sigma)$ obeys the orbital spectrum formula (cf. [63])

$$\tau(\sigma) \simeq \int_{G\cdot p_{\mathfrak{h}}^{-1}(\Omega_\sigma^H)/G}^{\oplus} n_\phi^\sigma \pi_\phi d\nu_{G,H}^\sigma(\phi), \tag{1.2.4}$$

where $\nu_{G,H}^\sigma$ is the push-forward of the natural measure on $p_{\mathfrak{h}}^{-1}(\Omega_\sigma^H) \subset \mathfrak{g}^*$ under the mapping $p_{\mathfrak{h}}^{-1}(\Omega_\sigma^H) \to G \cdot p_{\mathfrak{h}}^{-1}(\Omega_\sigma^H)/G$, and the value of the multiplicity function

n_ϕ^σ is the number of H-orbits in $p_\mathfrak{h}^{-1}(\Omega_\sigma^H) \cap G \cdot \phi$. We denote this number by $\#[p_\mathfrak{h}^{-1}(\Omega_\sigma^H) \cap G \cdot \phi/H]$ (the symbol $\#E$ denotes the cardinality of the set E).

If G is a completely solvable Lie group , the multiplicity function $\phi \longmapsto n_\phi^\sigma$ is either uniformly infinite or uniformly finite and bounded. In particular, the multiplicities are finite if and only if

$$\dim G \cdot \phi - 2 \dim H \cdot \phi + \dim \Omega_\sigma^H = 0, \tag{1.2.5}$$

for a generic $\phi \in p_\mathfrak{h}^{-1}(\Omega_\sigma^H)$ (see [104]).

1.2.7 Restrictions

Let π be a unitary and irreducible representation associated to the coadjoint orbit $\Omega_\pi^G \subset \mathfrak{g}^*$. Let $p_\mathfrak{a} : \mathfrak{g}^* \to \mathfrak{a}^*$ be the canonical projection of $\mathfrak{a}$, and let $\pi_{|A}$ be the restriction of π to the closed connected subgroup A. The restriction $\pi_{|A}$ obeys the orbital spectrum formula

$$\pi_{|A} = \int_{p_\mathfrak{a}(\Omega_\pi^G)/A}^{\oplus} n_\pi^\psi \sigma_\psi d\mu_{A,G}^\pi(\psi) \tag{1.2.6}$$

where $n_\pi^\psi = \#[(\Omega_\pi^G \cap p_\mathfrak{a}^{-1}(A \cdot \psi))]/A$, and $\mu_{A,G}^\pi$ is the push-forward of the canonical measure on Ω_π^G under $\Omega_\pi^G \longrightarrow p_\mathfrak{a}(\Omega_\pi^G)/A$ (see [64]).

In the completely solvable case, as before it is known that the multiplicity function $\psi \longmapsto n_\pi^\psi$ is either uniformly infinite or uniformly finite and bounded. These multiplicities are finite if and only if

$$\dim \Omega_\pi^G - 2 \dim A \cdot \phi + \dim A \cdot p_\mathfrak{a}(\phi) = 0 \tag{1.2.7}$$

generically on Ω_π^G (see [104]).

1.3 Pseudo-Algebraic Geometry

1.3.1 Pseudo-Algebraic Sets

Definition 1.3.1 Let P be a polynomial function with real coefficients on $\mathbb{R}^{m+k}$ and $\lambda_i : \mathbb{R}^m \to \mathbb{R}$, $1 \le i \le k$ real linear functionals. The function

$$F_P = F : \mathbb{R}^m \longrightarrow \mathbb{R},\ x \longmapsto P(x, e^{\lambda_1(x)}, \dots, e^{\lambda_k(x)}) \tag{1.3.1}$$

is called a *pseudo-algebraic* function associated to the polynomial P.

Definition 1.3.2 A subset V of $\mathbb{R}^m$ is called *pseudo-algebraic* if it admits some representation of the form

$$V = \Big\{x \in \mathbb{R}^m : F_1(x) = \ldots = F_r(x) = 0\Big\}, \tag{1.3.2}$$

where F_i, $1 \le i \le r$ are pseudo-algebraic functions on $\mathbb{R}^m$.

Definition 1.3.3 A subset A of $\mathbb{R}^m$ is called *semi-pseudo-algebraic* if it admits some representation of the form

$$A = \Big\{x \in \mathbb{R}^m : F_1(x) = \ldots = F_r(x) = 0,\ G_1(x) > 0, \ldots, G_l(x) > 0\Big\}, \tag{1.3.3}$$

where $F_i = F_{P_i}, 1 \le i \le r$ and $G_j = G_{Q_j}, 1 \le j \le l$ are pseudo-algebraic functions on $\mathbb{R}^m$ associated to the polynomials P_i and Q_j respectively.

Then it is clear that any semi-pseudo-algebraic set of $\mathbb{R}^m$ is the intersection between the closed set of common zeroes of a finite number of pseudo-algebraic functions on $\mathbb{R}^m$ and an open set of $\mathbb{R}^m$.

Remark 1.3.4 When the linear forms $\lambda_i, i = 1, \ldots, k$ are null, the notion of pseudo-algebraic set coincides with the familiar notion of algebraic set. Let us recall the definition of semi-algebraic sets.

Definition 1.3.5 A subset of $\mathbb{R}^n$ is said to be a semi-algebraic set if it admits a representation of the form

$$\{x \in \mathbb{R}^n : h_1(x) = \cdots = h_p(x) = 0,\ g_1(x) > 0, \ldots,\ g_q(x) > 0\}$$

where $h_1, \ldots, h_p, g_1, \ldots, g_q$ are polynomial functions on $\mathbb{R}^n$.

The following theorem plays an important role in the sequel, since the analysis of the multiplicity function of mixed representations naturally involves pseudo-algebraic geometry (see the next section for details).

Theorem 1.3.6 *Let A be a semi-pseudo-algebraic set of $\mathbb{R}^m$ as of Definition 1.3.3. Then the number of connected components of A in the Euclidean topology is finite and bounded by a scalar that depends only on m, r, l, k, $\deg P_i$, and $\deg Q_j$, but not on the coefficients of either the polynomials or the linear functionals.*

Proof We argue as in Proposition 4.4.5 of [20]. Let $\pi : \mathbb{R}^{m+1} \to \mathbb{R}^m$ be the natural continuous projection $(x_1, \ldots, x_{m+1}) \mapsto (x_1, \ldots, x_m)$ and let $A_2 \subset \mathbb{R}^{m+1}$ be the pseudo-algebraic set defined by

$$A_2 = \Big\{(x_1, \ldots, x_{m+1}) = (x, x_{m+1}) \in \mathbb{R}^{m+1} : F_1(x) = \cdots = F_r(x) = 0$$
$$x_{m+1}G_1(x) \cdots G_l(x) = 1\Big\}.$$

Then the set $A_1 = \pi(A_2)$ is defined by

$$A_1 = \left\{x \in \mathbb{R}^m : F_1(x) = \cdots = F_r(x) = 0, G_1(x) \cdots G_l(x) \neq 0\right\}.$$

Indeed, suppose $(x, x_{m+1}) \in A_2$, so that

$$F_1(x) = \cdots = F_r(x) = 0 \quad \text{and} \quad G_1(x) \cdots G_l(x) \neq 0,$$

and then $x \in A_1$ and $\pi(A_2) \subset A_1$. Conversely, let $x \in A_1$. As $G_1(x) \cdots G_l(x) \neq 0$, we take $x_{m+1} = (G_1(x) \cdots G_l(x))^{-1}$. Then $(x, x_{m+1}) \in A_2$, so it follows that $x \in \pi(A_2)$ and $A_1 \subset \pi(A_2)$. On the other hand, the subset $\{x \in \mathbb{R}^m : G_i(x) > 0\}$ is open and closed in $\{x \in \mathbb{R}^m : G_i(x) \neq 0\}$, for all $i = 1, \ldots, l$. Hence A is open and closed in A_1 and each connected component of A is also a connected component of A_1. Finally, the number of connected components of A is less than or equal to the number of connected components of A_1, which is in turn does not exceed the number of connected components of A_2, as $A_1 = \pi(A_2)$. The following result, which can be found in [104], enables us to conclude. Every pseudo-algebraic subset V of $\mathbb{R}^m$ as per Definition 1.3.2 has only finitely many components in the Euclidean topology. In fact, the number of the connected components of V is bounded by a scalar which depends on m, r, l, k, deg P_i and deg Q_j, but neither on the coefficients of the polynomials nor the linear functionals. □

Corollary 1.3.7 *Let $S \subset \mathbb{R}^m$ be a semi-pseudo-algebraic set defined as in (3.3). Let $F(x_1, \ldots, x_m, y_1, \ldots, y_k)$ be a polynomial function with real coefficients. Suppose $\gamma_i : \mathbb{R}^m \to \mathbb{R}$, $1 \leq i \leq k$ are real linear functionals. Then there is a number N depending only on m, r, l, k,* deg P_i, deg Q_j, *and* deg F *such that either*

(1) *$F(x, e^{\gamma_1(x)}, \ldots, e^{\gamma_k(x)}) = 0$ has an infinite number of solutions in S,*
or
(2) *the number of solutions in S is bounded by N.*

Proof Suppose that the number of solutions in S of equation

$$F(x, e^{\gamma_1(x)}, \ldots, e^{\gamma_k(x)}) = 0 \tag{1.3.4}$$

is finite. It must equal the number of connected components of the semi-pseudo-algebraic set $V \cap S$ where

$$V = \left\{x \in \mathbb{R}^m : \ F(x, e^{\gamma_1(x)}, \ldots, e^{\gamma_k(x)}) = 0\right\},$$

and the claim follows from the previous theorem. □

1.3.2 Semi-analytic Sets

We now recall some properties of semi-analytic sets in $\mathbb{R}^n$. The material is quite standard (Cf. [88, 109]). Given an open neighbourhood $U \subset \mathbb{R}^n$, let $C^\omega(U, \mathbb{R})$ be the set of real analytic functions on U. Denote by $B(C^\omega(U, \mathbb{R}))$ the Boolean algebra generated by sets of the form

$$\{x \in U : f_1(x) = \cdots = f_p(x) = 0,\ g_1(x) > 0, \ldots, g_q(x) > 0\}, \tag{1.3.5}$$

where $f_1, \ldots, f_p, g_1, \ldots, g_q$ are in $C^\omega(U, \mathbb{R})$.

Definition 1.3.8 A subset M of $\mathbb{R}^n$ is *semi-analytic* if, for every $x \in \mathbb{R}^n$, there exists an open neighbourhood U of x such that $M \cap U \in B(C^\omega(U, \mathbb{R}))$.

It is well-known that set-complements, finite intersections and locally finite unions of semi-analytic sets are semi-analytic. The pre-image of a semi-analytic by any analytic map is semi-analytic. The closure, interior and boundary of any semi-analytic set are semi-analytic.

Let M be a semi-analytic set in $\mathbb{R}^n$ and p a positive integer. A point $x \in M$ is said to be p-regular if there is an open neighbourhood U of x such that $M \cap U$ is an analytic submanifold of dimension p of U; x is 0-regular if it is an isolated point of M. The set of regular points of M (ie, p-regular points for some p) is dense in M. The dimension $\dim M$ of M is less than or equal to p if there do not exist q-regular points of M with $q > p$; $\dim M = p$ if $\dim M \leq p$ but $\dim M \not\leq p - 1$. Let $\dim M = p$; then $\dim \overline{M} = p$ and $\dim(\overline{M} \setminus M) < p$ ($\overline{M}$ denotes the closure of M).

1.3.3 Structure of Coadjoint Orbits

The following theorem describes the structure of the space of coadjoint orbits of an exponential solvable Lie group, and is proved in [45].

Theorem 1.3.9 *Let G be a connected and simply connected exponential solvable Lie group with Lie algebra $\mathfrak{g}$. There exists a finite partition $\wp$ of $\mathfrak{g}^*$ such that:*

(1) each $U \in \wp$ is G-invariant,
(2) for a given $U \in \wp$, the dimension of coadjoint orbits in U is constant,
(3) there is a total ordering $U_1 < U_2 < \cdots < U_u$ on $\wp$ such that for each U, $\cup\{U' : U' \leq U\}$ is Zariski open in $\mathfrak{g}^*$.
Given $U \in \wp$, there exist a subspace V of $\mathfrak{g}^*$ and index sets ι and φ such that for each $j \in \iota$ we have a complex-valued rational map p_j on $\mathfrak{g}^*$ and for each $j \in \varphi$ a complex-valued rational map q_j on $\mathfrak{g}^*$ satisfying the following:
(4) the set $\Sigma = \{l \in V \cap U : p_j(l) = 0,\ j \in \iota,\ |q_j(l)|^2 = 1,\ j \in \varphi\}$ is a cross-section of the coadjoint orbits in U.

Moreover, there exists a G-invariant analytic function $P : U \to U$ such that $P(U) = \Sigma$.

If G is a completely solvable Lie group, the set $\wp$ is empty and the p_j are real-valued for each $j \in \iota$ (see [47]).

Corollary 1.3.10 *Let G be a connected and simply connected exponential solvable Lie group with Lie algebra $\mathfrak{g}$. Every coadjoint orbit Ω (lying in some U of the above partition) is closed in U and is a semi-analytic set in $\mathfrak{g}^*$. In particular Ω is locally closed in $\mathfrak{g}^*$.*

Proof Let U be a layer in $\wp$ such that $\Omega \subset U$ and let $P : U \to U$ be the analytic G-invariant function such that $P(U) = \Sigma$, where Σ is a cross-section of the coadjoint orbits in U, as in Theorem 1.3.9. Then Ω meets Σ in a single point. Let $\{l\} = \Sigma \cap \Omega$. The subset $P^{-1}(\{l\})$ is G-invariant and every orbit $\Omega' \subset P^{-1}(\{l\})$ intersects Ω, so $\Omega = P^{-1}(\{l\})$. This shows that Ω is closed in U, and semi-analytic as P is analytic. Finally, as U is a semi-algebraic set in $\mathfrak{g}^*$, Ω is a semi-analytic set in $\mathfrak{g}^*$. □

Saturated Orbits with Respect to an Ideal of Codimension One

We now recall a well-known result, which will be used on many subsequent occasions. Let $\mathfrak{g}_0$ be an ideal of $\mathfrak{g}$ such that $\dim \mathfrak{g}/\mathfrak{g}_0 = 1$ and $G_0 = \exp \mathfrak{g}_0$. Let $p : \mathfrak{g}^\star \to \mathfrak{g}_0^\star$ be the canonical projection. With respect to $\mathfrak{g}_0$, coadjoint orbits of G, and so also unitary irreducible representations of G, fall into two categories. We call $G(f)$ the stabilizer of $f \in \mathfrak{g}^\star$ in G, whose Lie algebra is written $\mathfrak{g}(f)$. In the following we say that the orbit Ω of f is saturated (or not saturated) with respect to $\mathfrak{g}_0$ if $\mathfrak{g}(f) \subset \mathfrak{g}_0$ (respectively, $\mathfrak{g}(f) \subsetneq \mathfrak{g}_0$).

Theorem 1.3.11 (cf. [39]) *Let Ω be a coadjoint orbit of G, $f \in \Omega$ and $\pi(\Omega)$ the unitary and irreducible representation of G associated to Ω. We have the following alternative:*

(i) if Ω is saturated with respect to $\mathfrak{g}_0$, then $p(\Omega)$ is a union of G_0-orbits ω_t parametrized by $t \in \mathbb{R}$, $\omega_t \exp(tX) \cdot \omega_0$ with $X \notin \mathfrak{g}_0$ and $\Omega = p^{-1}(p(\Omega))$. On the other hand, $\pi(\Omega)_{|G_0} \simeq \int_{\mathbb{R}}^{\oplus} \pi(\omega_t)dt$ while $\mathrm{ind}_{G_0}^G \pi(\omega_t)$ is irreducible.

(ii) If Ω is not saturated respect to $\mathfrak{g}_0$, then $p(\Omega)$ is a G_0-orbit, say ω, and $p^{-1}(\omega)$ is a union of G-orbits Ω_t parametrized by $t \in \mathbb{R}$. On the other hand, $\pi(\Omega_t)_{|G_0}$ is irreducible and $\mathrm{ind}_{G_0}^G \pi(\omega) \simeq \int_{\mathbb{R}}^{\oplus} \pi(\Omega_t)dt$.

1.4 Up-Down Representations of Exponential Solvable Lie Groups

Let G be a real Lie group, A and H closed connected subgroups of G and σ a unitary representation of H. The representation

$$\rho(G, H, A, \sigma) = \mathrm{ind}_H^G \sigma|_A, \tag{1.4.1}$$

of A is called an *up-down representation.*

When the unitary representation σ of H is replaced by a unitary character χ_f, the representation $\rho(G, H, A, \sigma)$ will be simply denoted by $\rho(G, H, A, f)$. It is clear that $\rho(G, H, G, \sigma) = \tau(\sigma)$ and $\rho(G, G, A, \sigma) = \sigma_{|A}$. On the other hand, if the product AH is a closed subgroup of G, then Mackey's subgroup theorem allows us to say that

$$\rho(AH, H, A, \sigma) \simeq \rho(A, A \cap H, A, \sigma_{|A\cap H}) \simeq \mathrm{ind}_{A\cap H}^{A}(\sigma_{A\cap H}).$$

If G is exponential solvable and σ is a unitary irreducible representation, then σ is a monomial representation, which means that it is induced by a unitary character. We write $\sigma \simeq \mathrm{ind}_B^H \chi_l$ where $l \in \mathfrak{g}^\star$, $B = \exp(\mathfrak{b})$ and $\mathfrak{b} \in I(l_\mathfrak{h}, \mathfrak{h})$. Then it is immediate that

$$\rho(G, H, A, \sigma) \simeq \rho(G, B, A, l). \tag{1.4.2}$$

Thus, the study of the mixed representation (1.4.1) when $\sigma \in \widehat{H}$ can be reduced to mixed representations of type $\rho(G, H, A, f)$. This is exactly what many authors usually do. Some of our results will be proved in full generality for mixed-type representations (1.4.1). Relationship (1.4.2) is sometimes very useful for transferring known results on mixed representations of type $\rho(G, H, A, f)$ to general mixed representations. When $A = H$, the representation $\rho(G, H, H, \sigma)$ will be simply denoted by $\rho(G, H, \sigma)$.

Remark 1.4.1 Mackey studied the disintegration of such representations when A and H are regularly related in the context of arbitrary locally compact groups. At the same time, Lemma 4.2 in [115] proves that if H and A are normal subgroups of a locally compact group G and $A \subset H$, the support of the mixed representation (1.4.1) is of the form $(\sigma_{|A})^g$, $g \in G$ where $(\sigma_{|A})^g(a) = \sigma(g \cdot a \cdot g^{-1}), \in A$.

1.4.1 Disintegration of Up-Down Representations

In this section we describe the orbital spectrum formula of the representation $\rho(G, H, A, \sigma)$ for exponential solvable Lie groups. We disintegrate the representation $\rho(G, H, A, \sigma)$, giving an explicit description of the disintegration space, the measures and the irreducible unitary multiplicity representations which appear in this decomposition. We prove the following:

Theorem 1.4.2 *Let G be an exponential solvable Lie group, H and A two closed connected subgroups of G and σ an irreducible unitary representation of H. Then the mixed representation $\rho(G, H, A, \sigma)$ satisfies the orbital spectrum formula, that is*

$$\rho(G, H, A, \sigma) = \int_{[p(G\cdot q^{-1}(\Omega_\sigma^H))]/A}^{\oplus} m_\sigma(\psi)\rho_\psi d\mu_{G,H,A}^\sigma(\psi),$$

where $\mu^{\sigma}_{G,H,A}$ *is the image measure of* $(dg \times \widetilde{\mu})$ *on* $G \times q^{-1}(\Omega^H_\sigma)$ *under*

$$G \times q^{-1}(\Omega^H_\sigma) \to G \cdot q^{-1}(\Omega^H_\sigma) \to p(G \cdot q^{-1}(\Omega^H_\sigma)) \to \big[p(G \cdot q^{-1}(\Omega^H_\sigma))\big]/A$$

and

$$m_\sigma(\psi) = \sum_{\Omega \in [G \cdot q^{-1}(\Omega^H_\sigma) \cap G \cdot p^{-1}(\psi)]/G} n^\sigma_\Omega n^\psi_\Omega \tag{1.4.3}$$

where $n^\sigma_\Omega = \#[\Omega \cap q^{-1}(\Omega^H_\sigma)]/H$ *and* $n^\psi_\Omega = \#[\Omega \cap p^{-1}(A \cdot \psi)]/A$.

Remark 1.4.3 If G is nilpotent, A is Abelian and $\sigma = 1_H$, [107, Theorem 3.1] shows that the representation $\rho(G, H, A) = \mathrm{ind}_H^G 1_{|A}$ satisfies the orbital spectrum formula:

$$\rho(G, H, A) = \int^{\oplus}_{p(G \cdot \mathfrak{h}^\perp)} n_\psi \chi_\psi d\psi \quad \text{and} \quad n_\psi = \sum_{\Omega \in [G \cdot \mathfrak{h}^\perp \cap G \cdot p^{-1}(\psi)]/G} n^\sigma_\Omega n^\psi_\Omega .$$

Proof The proof of this theorem is based on the disintegrations of the induced unitary map $\tau(\sigma)$ and the restrictions $\pi_{\phi|A}$, where π_ϕ is the representation corresponding to the linear form $\phi \in q^{-1}(\Omega^H_\sigma)$. From formula (1.2.4) we have

$$\tau(\sigma) = \int^{\oplus}_{G \cdot q^{-1}(\Omega^H_\sigma)/G} n^\sigma_\phi \pi_\phi d\nu^\sigma_{G,H}(\phi).$$

It follows that

$$\begin{aligned}\rho(G, H, A, \sigma) &= \Big(\int^{\oplus}_{G \cdot q^{-1}(\Omega^H_\sigma)/G} n^\sigma_\phi \pi_\phi d\nu^\sigma_{G,H}(\phi)\Big)_{|A} \\ &= \int^{\oplus}_{G \cdot q^{-1}(\Omega^H_\sigma)/G} n^\sigma_\phi \pi_\phi d\nu^\sigma_{G,H}(\phi)\end{aligned}$$

On the other hand, from formula (1.2.6)

$$\pi_{\phi|A} = \int^{\oplus}_{p(G \cdot \phi)/A} n^\psi_\phi \gamma_\psi d\mu^{\pi_\phi}_{G,A}(\psi).$$

Then

$$\rho(G,H,A,\sigma)=\int_{G\cdot q^{-1}(\Omega_\sigma^H)/G}^{\oplus} n_\phi^\sigma \int_{p(G\cdot\phi)/A}^{\oplus} n_\phi^\psi \gamma_\psi d\mu_{G,A}^{\pi_\phi}(\psi)d\nu_{G,H}^\sigma(\phi)$$
$$=\int_{G\cdot q^{-1}(\Omega_\sigma^H)/G}^{\oplus} n_\phi^\sigma \int_{p(G\cdot\phi)/A}^{\oplus} n_\phi^\sigma n_\phi^\psi \gamma_\psi d\mu_{G,A}^{\pi_\phi}(\psi)d\nu_{G,H}^\sigma(\phi) \tag{1.4.4}$$
$$=\int_S^{\oplus} m(\psi,\phi)\gamma_\psi d\mu_{G,A}^{\pi_\phi}(\psi)\otimes d\nu_{G,H}^\sigma(\phi),$$

where $S=\bigcup_{G\cdot\phi\in[G\cdot q^{-1}(\Omega_\sigma^H)]/G} p(G\cdot\phi)/A$.

Observe first that the projection p is $\mathrm{Ad}^*(A)$-equivariant, i.e. $p\circ\mathrm{Ad}^*(a)=\mathrm{Ad}^*(a)\circ p$. As $G\cdot q^{-1}(\Omega_\sigma^H)$ is G-invariant, $p(G\cdot q^{-1}(\Omega_\sigma^H))$ is A-invariant. Then

$$S=[p(G\cdot q^{-1}(\Omega_\sigma^H))]/A. \tag{1.4.5}$$

We denote by $\widetilde{\mu}_{G,H,A}^\sigma$ the measure on $p(G\cdot q^{-1}(\Omega_\sigma^H))$ obtained by the image of $(dg\times\widetilde{\nu})$ on $G\times q^{-1}(\Omega_\sigma^H)$ through the maps $G\times q^{-1}(\Omega_\sigma^H)\to G\cdot q^{-1}(\Omega_\sigma^H)\to p(G\cdot q^{-1}(\Omega_\sigma^H))$. Apart from transporting the measure $\widetilde{\mu}_{G,H,A}^\sigma$ on $p(G\cdot q^{-1}(\Omega_\sigma^H))$ by the map $p(G\cdot q^{-1}(\Omega_\sigma^H))\to p(G\cdot q^{-1}(\Omega_\sigma^H))/A$, we obtained the measure $\mu_{G,H,A}^\sigma$ on $[p(G\cdot q^{-1}(\Omega_\sigma^H))]/A$ described in the theorem. On the other hand, from formulas (4) and (5) the measure $\nu_{G,H}^\sigma$ is the image of $(dg\times\widetilde{\mu})$ under $G\times q^{-1}(\Omega_\sigma^H)\to G\cdot q^{-1}(\Omega_\sigma^H)$, and for $G\cdot\phi\in(G\cdot q^{-1}(\Omega_\sigma^H))/G$ the measure $\mu_{G,A}^{\pi_\phi}$ is the image of the canonical measure on $G\cdot\phi$ under $G\cdot\phi\to p(G\cdot\phi)/A$. As the class of the canonical measure on the orbit $G\cdot\phi$ is the image of the Haar measure dg under $G\to G\cdot\phi$, it follows that $G\to G\cdot\phi\to p(G\cdot\phi)/A$ maps dg to $\mu_{G,A}^{\pi_\phi}$. Hence by (1.4.5), if $G\cdot\phi$ varies in $(G\cdot q^{-1}(\Omega_\sigma^H))/G$ the tensor product of $\mu_{G,A}^{\pi_\phi}$ and $\nu_{G,H}^\sigma$ also gives the measure $\mu_{G,H,A}^\sigma$ on $[p(G\cdot q^{-1}(\Omega_\sigma^H))]/A$. From all of this, and formula (1.4.4), we get

$$\rho(G,H,A,\sigma)=\int_{p(G\cdot q^{-1}(\Omega_\sigma^H))/A}^{\oplus} m(\psi)\rho_\psi d\mu_{G,H,A}^\sigma(\psi)$$

for some multiplicity functions $m(\psi)$, which we shall prove coincide with the map in (1.4.3). For $\psi\in\mathfrak{a}^\star$, let

$$M_\psi(\sigma)=\{\phi\in G\cdot q^{-1}(\Omega_\sigma^H);\ G\cdot\phi\cap p^{-1}\{A\cdot\psi\}\neq\emptyset\}.$$

As $p^{-1}\{A \cdot \psi\} = A \cdot p^{-1}(\psi)$, it is clear that $M_\psi(\sigma)$ is G-invariant and $M_\psi(\sigma) = G \cdot q^{-1}(\Omega_\sigma^H) \cap G \cdot p^{-1}(\psi)$. Hence the set $M_\psi(\sigma)$ only depends on the orbit $A \cdot \psi$. We then find the multiplicity function

$$m(\psi) = \sum_{G\cdot\phi\in[M_\psi(\sigma)]/G} n_{G\cdot\phi}^{\sigma} n_{G\cdot\phi}^{\psi} = \sum_{\Omega\in[M_\psi(\sigma)]/G} n_\Omega^\sigma n_\Omega^\psi.$$

From the definition of the multiplicity function n_Ω^ψ, we can see that m is constant on each A-orbit. From this the claimed disintegration formula follows. □

As an immediate consequence of Theorem 1.4.2 we deduce the central disintegration of the mixed representation $\rho(G, H, A, \sigma)$ in irreducible unitary representations. Let $M'_\psi(\sigma)$ be the dense open subset of $M_\psi(\sigma)$ formed by elements with G-orbits of maximal dimension. When $\sigma = \chi_f$, we write $M_\psi(\sigma) = M_\psi$ and $M'_\psi(\sigma) = M_\psi$.

Corollary 1.4.4 *Let G be an exponential solvable Lie group, H and A two connected closed subgroups of G and σ an irreducible unitary representation of H. Let β be the image of the measure $\mu_{G,H,A}^\sigma$ on $[p(G \cdot q^{-1(\Omega_\sigma^H)})]/A$ under the Kirillov-Bernat map $K_A : \mathfrak{a}^\star/A \to \widehat{A}$. Then*

$$\rho(G, H, A, \sigma) \simeq \int_{\widehat{A}}^{\oplus} m(\gamma)\gamma d\beta(\gamma). \tag{1.4.6}$$

A representation $\gamma = \gamma_\psi$, $\psi \in \mathfrak{a}^\star$ appears in decomposition (1.4.6) if and only if $\psi \in p(G \cdot q^{-1}(\Omega_\sigma^H))$. The multiplicities $m(\gamma_\psi)$ of $\rho(G, H, A, \sigma)$, as per (1.4.3), are β-almost everywhere finite if and only if $\tau(\sigma)$, $\pi_{\phi|A}$ have finite multiplicities for almost every $\phi \in q^{-1}(\Omega_\sigma^H)$ and $[M'_\psi(\sigma)]/G$, and they have finite cardinality for almost every $\psi \in p(G \cdot q^{-1}(\Omega_\sigma^H))$.

1.4.2 The Multiplicity Function of Up-Down Representations

Assume now that G is a completely solvable Lie group with Lie algebra $\mathfrak{g}$. Let χ_f be a unitary character of H. For $\psi \in p_{\mathfrak{a}}(G \cdot \Gamma_f)$, let

$$\mathscr{M}_\psi = G \cdot \Gamma_f \cap G \cdot p_{\mathfrak{a}}^{-1}(\psi), \tag{1.4.7}$$

where

$$\Gamma_f := f + \mathfrak{h}^\perp. \tag{1.4.8}$$

Let e be the largest index in $\{1, \ldots, u\}$ such that $U = U_e$, defined as in Theorem 1.3.9, meets Γ_f in a non-empty Zariski-open subset of Γ_f. Let

$$\Gamma_0 = U \cap \Gamma_f. \tag{1.4.9}$$

For any $\psi \in p_{\mathfrak{a}}(G \cdot \Gamma_0)$, let

$$\mathscr{M}'_\psi = G \cdot \Gamma_0 \cap G \cdot p_{\mathfrak{a}}^{-1}(\psi) = \mathscr{M}_\psi \cap U. \tag{1.4.10}$$

Let d_G be the maximal dimension of G-orbits in $G \cdot \Gamma_f$, d_A the maximal dimension of A-orbits in $G \cdot \Gamma_f$, d'_A the maximal dimension of A-orbits in $p_{\mathfrak{a}}(G \cdot \Gamma_f)$ and d_H the maximal dimension of H-orbits in $G \cdot \Gamma_f$. Then the set of linear forms ϕ such that $\dim G \cdot \phi = d_G$, $\dim H \cdot \phi = d_H$ and $\dim A \cdot \phi = d_A$ is an open and dense co-null set in $G \cdot \Gamma_f$. Likewise, the set of $\psi \in p_{\mathfrak{a}}(G \cdot \Gamma_f)$ such that $\dim A \cdot \psi = d'_A$ is open, dense and co-null in $p_{\mathfrak{a}}(G \cdot \Gamma_f)$.

For $\psi \in p_{\mathfrak{a}}(G \cdot \Gamma_0)$, let

$$\mathscr{B}_\psi = \Gamma_f \cap G \cdot p_{\mathfrak{a}}^{-1}(\psi) \text{ and } \mathscr{B}'_\psi = \mathscr{B}_\psi \cap \Gamma_0$$

The set of generic orbits in $\mathscr{M}_\psi$ is then the set

$$\mathscr{M}'_\psi = G \cdot \mathscr{B}'_\psi.$$

We now prove the main result of this section.

Theorem 1.4.5 *Let G be a completely solvable Lie group, A, H analytic subgroups of G and χ_f a unitary character of H. Then*

1. *the multiplicity function of the representation $\rho(G, H, A, f)$ is either uniformly infinite or uniformly finite and bounded;*
2. *the multiplicities of the representation $\rho(G, H, A, f)$ are uniformly finite if and only if $\mathscr{M}'_\psi$ is a semi-analytic subset of $\mathfrak{g}^*$ and the equalities*

$$\dim \mathscr{M}'_\psi = 2d_H = 2d_A - d'_A$$

hold generically on $p_{\mathfrak{a}}(G \cdot \Gamma_f)$.

Proof The proof of the theorem is based on the following series of lemmas.

Lemma 1.4.6 *We keep the same notation and hypotheses as above. The integer $m(\psi)$ is finite if and only if $\mathscr{M}'_\psi$ is a semi-analytic subset of $\mathfrak{g}^*$ and $\dim \mathscr{M}'_\psi = 2d_A - d'_A = 2d_H$.*

Proof If $m(\psi)$ is finite, then $n(\psi) = \#[\mathscr{M}'_\psi]/G < \infty$, and for every G-orbit $\Omega \subset \mathscr{M}'_\psi$, we have

$$n_\Omega^f < \infty \text{ and } n_\Omega^\psi < \infty. \tag{1.4.11}$$

Write

$$\mathcal{M}'_\psi = \bigcup_{i=1}^{k} \Omega_i,$$

where the Ω_i, $i = 1, \ldots, k$ are distinct G-orbits with common dimension. It follows from Corollary 1.3.10 that $\mathcal{M}'_\psi$ is a semi-analytic set as it is a finite union of semi-analytic sets. Moreover, the dimension of $\mathcal{M}'_\psi$ equals $\dim \Omega_i$ for all $i = 1, \ldots, k$. Hypothesis (1.4.11) implies that $2d_H = \dim \Omega_i$ by means of (1.2.5) and $2d_A - d'_A = \dim \Omega_i$ by formula (1.2.7) for $i = 1, \ldots, k$. Hence

$$\dim \mathcal{M}'_\psi = 2d_H = 2d_A - d'_A.$$

Conversely, write again

$$\mathcal{M}'_\psi = \bigcup_{i \in I} \Omega_i.$$

In this case

$$\dim \Omega_i \leq \dim \mathcal{M}'_\psi = 2d_H \leq \dim \Omega_i,$$

for any $i \in I$. Hence $\dim \mathcal{M}'_\psi = \dim \Omega_i$, which means that the G-orbit Ω_i is open in $\mathcal{M}'_\psi$ and, therefore, closed in $\mathcal{M}'_\psi$. As coadjoint orbits are connected, the orbits Ω_i, $i \in I$ turn out to be the connected components of $\mathcal{M}'_\psi$. Let $\Phi : G \times \mathfrak{g}^* \longrightarrow \mathfrak{g}^*$ be defined by $\Phi(g, l) = \mathrm{Ad}^* g\ l$. Since G is completely solvable, Φ is a pseudo-algebraic. The set

$$\mathcal{D} = \{(g, l) \in G \times p_{\mathfrak{a}}^{-1}(\psi) : \ \Phi(g, l) \in \Gamma_0\}$$

is therefore a semi-pseudo-algebraic set. It is clear that $\mathcal{B}'_\psi = \Phi(\mathcal{D})$. By Theorem 1.3.6, the set $\mathcal{D}$ has a finite number of connected components. Due to the continuity of Φ, therefore, $\mathcal{B}'_\psi$ has a finite number of connected components. Finally, the number of connected components of $\mathcal{M}'_\psi = G \cdot \mathcal{B}'_\psi$ is less than or equal to the number of connected components of $\mathcal{B}'_\psi$, that being the image of $G \times \mathcal{B}'_\psi$ under Φ and because of $\#[\mathcal{M}'_\psi]/G < \infty$. The multiplicity functions $n^f_{\Omega_i}$ and $n^\psi_{\Omega_i}$ are then finite by (1.2.5) and (1.2.7). □

Lemma 1.4.7 *Let G be a completely solvable Lie group, and suppose H, A are closed, connected subgroups of G. Then $\#[\mathcal{M}'_\psi]/G$ is either uniformly infinite or uniformly finite and bounded on $p_{\mathfrak{a}}(G \cdot \Gamma_f)$.*

Proof Let Σ be a cross-section of the G-orbits in U and

$$\mathcal{M} = \mathcal{M}'_\psi \cap \Sigma.$$

Let

$$\mathcal{N} = \{(g_1, g_2, l_1, l_2, l_3) \in G \times G \times \Gamma_f \times p_{\mathfrak{a}}^{-1}(\psi) \times \Sigma : \Phi(g_1, l_1) = l_3,\ \Phi(g_2, l_2) = l_3\},$$

where Φ is defined as in the proof of the previous lemma. As Σ is an algebraic set of U, $\mathcal{N}$ is a semi-pseudo-algebraic set. On the other hand, $\mathcal{M}$ is a cross-section for the G-orbits in $\mathcal{M}'_\psi$ and $\mathcal{M} = q(\mathcal{N})$, where $q : G \times G \times \Gamma_f \times p_{\mathfrak{a}}^{-1}(\psi) \times \Sigma \longrightarrow \Sigma$ is the canonical projection. If the number $n(\psi) = \#[\mathcal{M}'_\psi]/G$ is finite then $n(\psi) = \#[q(\mathcal{N})]$ equals the number of connected components of $q(\mathcal{N})$. This is less than or equal to the number of connected components of $\mathcal{N}$. It follows from Theorem 1.3.6 that the number of connected components of $\mathcal{N}$ is bounded by a number that does not depend on ψ over $p_{\mathfrak{a}}(G \cdot \Gamma_f)$, since ψ intervenes only in the expression of the coefficients of the pseudo-algebraic functions. Hence $n(\psi)$ is either uniformly infinite or uniformly finite and bounded by a number that is independent of ψ on $p_{\mathfrak{a}}(G \cdot \Gamma_f)$. This completes the proof. □

Lemma 1.4.8 *The multiplicity function of $\rho(G, H, A, f)$ is either uniformly infinite or uniformly finite and bounded on $p_{\mathfrak{a}}(G \cdot \Gamma_f)$.*

Proof We recall that the set for which the linear form ϕ satisfies $\dim G \cdot \phi = d_G$, $\dim H \cdot \phi = d_H$ and $\dim A \cdot \phi = d_A$, is an open dense co-null set in $G \cdot \Gamma_f$, on which $\dim G \cdot \phi - 2 \dim H \cdot \phi$ and $\dim G \cdot \phi - 2 \dim A \cdot \phi$ are constant. It follows from the above lemmas that the multiplicity function $m(\psi)$ is either uniformly infinite or uniformly finite. There remains to prove that if $m(\psi)$ is uniformly finite, then it is bounded. But the multiplicity n_Ω^ψ equals the number of connected components of the pseudo-algebraic set $\Omega \cap p_{\mathfrak{a}}^{-1}(A \cdot \psi)$, which does not depend on ψ as shown above (see Theorem 1.3.6). An argument similar to that of the nilpotent situation allows to conclude (see [12], page 190). □

All in all, this settles Theorem 1.4.5. □

We now study the multiplicities of mixed representations for exponential solvable groups. We have the following proposition:

Proposition 1.4.9 *Let G be an exponential solvable Lie group, H a closed connected subgroup of G and $\sigma \in \widehat{H}$. Let $\mathfrak{a}$ be a G-invariant polarization common to all generic elements of $q^{-1}(\Omega_\sigma^H)$ and $A = \exp \mathfrak{a}$. Then the multiplicity of a mixed representation $\rho(G, H, A, \sigma)$ coincides with the multiplicity of $\tau(\sigma)$, and its spectrum contains only characters. If G is completely solvable, these multiplicities are uniformly bounded if and only if $\dim G \cdot \phi = 2 \dim H \cdot \phi - \dim \Omega_\sigma^H$ generically.*

Proof Let $\mathcal{O}_1$ be the dense open set of elements of $q^{-1}(\Omega_\sigma^H)$ whose G-orbits are of maximal dimension. Let $\widetilde{\mathcal{O}_1}$ be the set of generic elements in $G \cdot q^{-1}(\Omega_\sigma^H)$, that is to say

$$\widetilde{\mathcal{O}_1} = \{\phi \in G \cdot q^{-1}(\Omega_\sigma^H) / \phi \in G \cdot \phi',\ \phi' \in \mathcal{O}_1\} = G \cdot \mathcal{O}_1.$$

The set $\widetilde{\mathscr{O}_1}$ is open in $G \cdot q^{-1}(\Omega_\sigma^H)$. Let U be a non-empty open of $G \cdot q^{-1}(\Omega_\sigma^H)$. Then $G \cdot U$ is a non-empty open of $G \cdot q^{-1}(\Omega_\sigma^H)$, as

$$G \cdot [G \cdot U \cap q^{-1}(\Omega_\sigma^H)] = G \cdot U \cap G \cdot q^{-1}(\Omega_\sigma^H) = G \cdot U \neq \emptyset,$$

and $G \cdot U \cap q^{-1}(\Omega_\sigma^H)$ is not empty and open in $q^{-1}(\Omega_\sigma^H)$. Hence $G \cdot U \cap q^{-1}(\Omega_\sigma^H) \cap \mathscr{O}_1 \neq \emptyset$, which means $G \cdot U \cap \mathscr{O}_1 \neq \emptyset$, and so $U \cap G \cdot \mathscr{O}_1 \neq \emptyset$, which entails $\widetilde{\mathscr{O}_1}$ is dense and open in $G \cdot q^{-1}(\Omega_\sigma^H)$. Let now $\mathscr{O} = \{p(\phi);\ \dim G \cdot \phi$ is maximal in $G \cdot q^{-1}(\Omega_\sigma^H)\} = p(\widetilde{\mathscr{O}_1}) = p(G \cdot \mathscr{O}_1)$. By definition of image measure,

$$\mu^\sigma_{G,H,A}(C^{\mathscr{O}/A}_{p(G \cdot q^{-1}(\Omega_\sigma^H))/A}) = (dg \otimes \widetilde{\mu})(G \times C^{\mathscr{O}_1}_{q^{-1}(\Omega_\sigma^H)}).$$

As $\mathscr{O}_1$ is a $\widetilde{\mu}$-negligible complementary open, $\mathscr{O}/A$ is $\mu^\sigma_{G,H,A}$-negligible and complementary in $p(G \cdot q^{-1}(\Omega_\sigma^H))/A$.

Let $\psi = p(\phi) \in \mathscr{O}$. We can then define an extension of ψ on any $\mathfrak{g}^\star$ that we will still call ψ. In view of $A \cdot \phi = \phi + \mathrm{ad}^\star \mathfrak{a} \cdot \phi$, we see that the A-orbit is closed in $\mathfrak{g}^\star$, and then $\mathfrak{a}$ is a Pukanszky polarization. Hence

$$A \cdot \phi = \phi + \mathfrak{a}^\perp = \psi + \mathfrak{a}^\perp = p(\phi) + \mathfrak{a}^\perp. \tag{1.4.12}$$

Since $p^{-1}(\psi) = \psi + \mathfrak{a}^\perp$ we deduce from relation (1.4.12) that $G \cdot \phi = G \cdot p^{-1}(\{\psi\})$. Then for all $\psi = p(\phi) \in \mathscr{O}$ we have $G \cdot p^{-1}(\psi) = G \cdot \phi$.

Then $M_\psi(\sigma) = G \cdot \phi$ and $\#[G \cdot q^{-1}(\Omega_\sigma^H) \cap G \cdot p^{-1}(\{\psi\})]/G = 1$, which implies that $m_\sigma(\psi) = n_\Omega^\sigma n_\Omega^\psi$. On the other hand we know that $\Omega \cap p^{-1}(A \cdot \psi) = \Omega \cap A \cdot p^{-1}(\{\psi\})$. It follows that $\Omega \cap p^{-1}(A \cdot \psi) = \Omega \cap (A \cdot \phi) = A \cdot \phi$. Finally, $n_\Omega^\psi = \#[(\Omega \cap p^{-1}(A \cdot \psi))]/A = 1$. Then $m_\sigma(\psi) = n_\Omega$. On the other hand, for $\psi \in \mathscr{O}$ we have $\psi([\mathfrak{a}, \mathfrak{a}]) = 0$, so the spectrum of $\rho(G, H, A, \sigma)$ only involves characters. When G is completely solvable, the multiplicities of $\tau(\sigma)$ are uniformly bounded if condition $\dim \mathscr{M}'_\psi = 2d_H = 2d_A - d'_A$ is satisfied, otherwise they are uniformly infinite. □

Corollary 1.4.10 *Suppose G is nilpotent and that $\tau(f)$ is has almost everywhere finite multiplicities. Suppose that* $\mathrm{Ad}_G([\mathfrak{a}, \mathfrak{a}]) \subset \mathfrak{h} \cap \ker(f)$ *and that* $\dim(\mathfrak{g}/\mathfrak{a}) = \frac{1}{2} d_G$. *Then the multiplicities of $\rho(G, H, A, f)$ are finite almost everywhere.*

Note that the first condition is trivial when $\mathfrak{a}$ is Abelian. If so we recover Theorem 1.4.2.

Proof Let ϕ be a generic element in Γ_f, so $\phi = f + \phi_1$ with $\phi_1 \in \mathfrak{h}^\perp$ and $g \cdot \phi([\mathfrak{a}, \mathfrak{a}]) = 0$ for all $g \in G$. On the other hand:

$$\begin{aligned}\dim \mathfrak{a} &= \dim \mathfrak{g} - \frac{1}{2} \dim G \cdot \phi \\ &= \dim \mathfrak{g} - \frac{1}{2}(\dim \mathfrak{g} - \dim \mathfrak{g}(\phi)) \\ &= \frac{1}{2}(\dim \mathfrak{g} + \dim \mathfrak{g}(\phi)).\end{aligned}$$

Thus $\mathfrak{a}$ is a real polarization for any element of the orbit $G \cdot \phi$. As $\mathfrak{g}$ is nilpotent, $\mathfrak{a}$ is a Pukanszky polarization. The rest of the proof follows from the final part of the previous proposition, without the fact that $\mathfrak{a}$ is an ideal of $\mathfrak{g}$. We get that the multiplicities of the representation $\rho(G, H, A, f)$ coincide with those of the representation $\tau(f)$, which are almost everywhere finite by hypothesis. □

Example 1.4.11 Let G be the nilpotent, connected and simply connected Lie group with Lie algebra $\mathfrak{g}$ generated by $(e_1, \ldots, e_6)$ and nonzero brackets:

$$[e_1, e_5] = -e_4,$$

$$[e_6, e_4] = -e_3,$$

$$[e_6, e_3] = -e_2,$$

$$[e_5, e_2] = -e_1,$$

$$[e_4, e_3] = -e_1.$$

Let $\mathfrak{h} = \mathbb{R}\text{-span}(e_2, e_3, e_6)$ and $\mathfrak{a} = \mathbb{R}\text{-span}(e_1, e_2, e_3, e_6)$. Then $\mathfrak{h}$ and $\mathfrak{a}$ are subalgebras of $\mathfrak{g}$. Write $f = \lambda e_3^\star$. The generic orbits in Γ_τ are of dimension $d_G = 4$. Indeed, a simple calculation shows that for $x \neq 0$, the orbit $G \cdot f_x$ of $f_x = \lambda e_3^\star + \lambda e_1^\star$ is:

$$\Omega_x = \{x e_1^\star + t_2 e_2^\star + t_3 e_3^\star + t_4 e_4^\star + t_5 e_5^\star + \frac{1}{2x}(t_3^2 - 2t_2t_4 - \lambda^2)e_6^\star;\ t_i \in \mathbb{R}\}.$$

Then $\tau \simeq \int_{\mathbb{R}}^{\oplus} \pi_x dx$ where π_x is the representation associated to the orbit Ω_x. The hypothesis that n_Ω is finite, for all Ω generic in $G \cdot \Gamma_\tau$, holds.
On the other hand, we have:

$$\pi_{x|A} \simeq \int_{\mathbb{R}}^{\oplus} \rho_{x,y} dy$$

where $\rho_{x,y}$ is the representation in $\widehat{A}$ associated to the orbit $A \cdot (x e_1^\star + y e_2^\star)$. Then

$$\tau_{|A} \simeq \int_{\mathbb{R}^2}^{\oplus} \pi_{x,y}^{\circ} dx dy.$$

On the other hand the subgroup A is of codimension $2 = \frac{1}{2} d_G$, but the assumption $\mathrm{Ad}_G([\mathfrak{a}, \mathfrak{a}]) \subset \mathfrak{h} \cap \ker(f)$ of Corollary 1.4.10 does not hold, while $m(\psi) = 1$ for all generic ψ in $p(G \cdot \Gamma_\tau)$.

The Case of Normal Subgroups

In this section we assume that G is an exponential solvable Lie group, H and A are closed, connected and normal subgroups of G. For $\psi \in \mathfrak{a}^\star$, let $\mathbf{B}_\psi = \Gamma_f \cap G \cdot p^{-1}(\psi)$, then it is clear that $\mathbf{B}_\psi$ is an analytic set (i.e. the common zeroes of a finite family of analytic functions) and that $M_\psi = G \cdot \mathbf{B}_\psi$.

As in the previous section, let d_H, d_A, d'_A indicate the maximum dimensions of H-orbits in $G\cdot\Gamma_f$, A-orbits in $G\cdot\Gamma_f$ and A-orbits in $p(G\cdot\Gamma_f)$. In accordance with the nilpotent case, generically in Γ_f, the dimensions of G-orbits, H-orbits, A-orbits in $G\cdot\Gamma_f$ are maximal, as is the dimension of A-orbits in $p(G\cdot\Gamma_f)$.

Let $\phi \in \Gamma_f$, $g \in G$ and $\psi = (g\cdot\phi) \in p(G\cdot\Gamma_f)$. We denote

$$\mathfrak{k} = (\mathfrak{a}\cap\mathfrak{h})^\phi = (\mathfrak{a}\cap\mathfrak{h})^f = \{X \in \mathfrak{g} \,:\, f[X, \mathfrak{a}\cap\mathfrak{h}] = 0\},$$

and

$$K = \exp(\mathfrak{k}) = \{g \in G \,:\, g\cdot\phi_{|\mathfrak{a}\cap\mathfrak{h}} = \phi_{|\mathfrak{a}\cap\mathfrak{h}}\}.$$

Define subspaces $\mathfrak{m}$ of $\mathfrak{g}$, $\mathfrak{u}$ of $\mathfrak{a}$ and $\mathfrak{n}$ of $\mathfrak{h}$ such that

$$\mathfrak{g} = \mathfrak{m}\oplus(\mathfrak{a}+\mathfrak{h}),\ \mathfrak{h} = \mathfrak{n}\oplus(\mathfrak{a}\cap\mathfrak{h})$$

and

$$\mathfrak{a} = \mathfrak{u}\oplus(\mathfrak{a}\cap\mathfrak{h}).$$

We begin by proving the following lemma:

Lemma 1.4.12

$$\mathbf{B}_\psi = \widetilde{\phi} + p(K\cdot\phi) + (\mathfrak{a}+\mathfrak{h})^\perp,$$

where $\widetilde{\phi}$ is the linear form on $\mathfrak{g}$ defined by $\widetilde{\phi} = f$ on $\mathfrak{n}$ and $\widetilde{\phi} = 0$ otherwise. In particular, the sets $\mathbf{B}_\psi$ and M_ψ are connected.

Proof Let $\psi = p(g\cdot\phi) \in p(G\cdot\Gamma_f)$ for some $g \in G$ and $\phi \in \Gamma_f$. We remark first that $\mathbf{B}_\psi = (\phi + \mathfrak{h}^\perp) \cap G\cdot(\phi\mathfrak{a}^\perp) \neq 0$. An element l of $\mathfrak{g}^\star$ is in $\mathbf{B}_\psi$ if and only if l is can be written as $l = \phi + \alpha = k\cdot\phi + \beta$ for some linear forms $\alpha \in \mathfrak{h}^\perp$, $\beta \in \mathfrak{a}^\perp$ and an element $k \in G$. As $k\cdot\phi = \phi$ in $\mathfrak{h}\cap\mathfrak{a}$, then $k \in K$. Thus $\alpha = k\cdot\phi - \phi$ on $\mathfrak{a}$ and $\beta = \phi - k\cdot\phi$ on $\mathfrak{h}$. Let F be the map: $K\times(\mathfrak{a}+\mathfrak{h})^\perp \to \mathfrak{g}^\star$ defined by $(k,\phi) \mapsto k\cdot\phi + \beta(k) + \varphi$, with $\beta(k) \in \mathfrak{a}^\perp$ defined by $\beta(k) = f - k\cdot f$ on $\mathfrak{n}$ and $\beta(k) = 0$ on $\mathfrak{m}$. Then $\mathbf{B}_\psi$ is contained in the image of $K\times(\mathfrak{a}+\mathfrak{h})^\perp$ under the map F. Conversely, let l be an element of this image, $l = k\cdot\phi + \beta(k) + \varphi$. Then we can write l in the form $l = f + \alpha$ with $\alpha \in \mathfrak{h}^\perp$ defined by $\alpha = k\cdot\phi$ on $\mathfrak{u}$ and $\alpha = \varphi + k\cdot\phi - f$ on $\mathfrak{m}$. Hence l is an element of $f + \mathfrak{h}^\perp$. Likewise, we see that

$l = k \cdot \phi + \gamma$ with $\gamma \in \mathfrak{a}^{\perp}$ defined by $\gamma = \beta(k)$ on $\mathfrak{n}$ and $\gamma = \varphi$ on $\mathfrak{m}$, so l is an element of $G \cdot \phi + \mathfrak{a}^{\perp}$. This shows $\mathbf{B}_{\psi}$ is the image of $K \times (\mathfrak{a} + \mathfrak{h})^{\perp}$ under F. Hence $\mathbf{B}_{\psi} = \tilde{\phi} + p(k \cdot \phi) + (\mathfrak{a} + \mathfrak{h})^{\perp}$, where we consider elements of $p(K \cdot \phi)$ as elements of $\mathfrak{g}^{\star}$ by extending trivially. □

We are now ready to prove the main result of this section.

Theorem 1.4.13 *Let G be an exponential solvable Lie group, H and A two closed connected normal subgroups of G, and χ_f a unitary character of H. Then the multiplicities of the mixed representation $\rho(G, H, A, f)$ are uniformly infinite or uniformly equal to 1. A necessary and sufficient condition for $\rho(G, H, A, f)$ to be multiplicity-free is that* $\dim \mathbf{B}_{\psi} = d_H = d_A - \frac{1}{2}d'_A$ *for almost every $\psi \in p(G \cdot \Gamma_f)$.*

Proof According to formula (1.4.3), $m(\psi) = 1$ or $+\infty$. Indeed, if $m(\psi) < \infty$, then $\#[M'_{\psi}]/G < \infty$ and for all Ω generic in $G \cdot \Gamma_f$, $n_{\Omega} = n_{\Omega}^{\psi} = 1$. Let then $M'_{\psi} = \cup_{i=1}^{m} G \cdot \phi_i$, where $G \cdot \phi$ are generic orbits. This gives

$$\mathbf{B}'_{\psi} = \bigcup_{i=1}^{m} G \cdot \phi_i \cap \Gamma_f = \bigcup_{i=1}^{m} H \cdot \phi_i = \mathbf{B}_{\psi}$$

since for all linear forms $\phi \in \Gamma_f$, $\dim H \cdot \phi = \dim H \cdot f$. We have $H \cdot \phi_i = \phi_i + \mathrm{ad}^{\star}\mathfrak{h} \cdot \phi_i = \phi_i + \mathrm{ad}^{\star}\mathfrak{h} \cdot f$, so $H \cdot \phi_i$ is open and closed in $\mathbf{B}_{\psi}$, which is connected. Hence $m = 1$ and $\mathbf{B}_{\psi} = H \cdot \phi$ for some ϕ in Γ_f, which implies $M_{\psi} = G \cdot \phi$ and $m(\psi) = 1$. We claim

$$m(\psi) = 1 \Leftrightarrow \dim \mathbf{B}_{\psi} = \dim H \cdot \phi = \dim A \cdot \phi - \frac{1}{2} \dim A \cdot \psi. \tag{1.4.13}$$

Indeed, suppose first that $m(\psi) = 1$. In this case we have $M_{\psi} = M'_{\psi} = G \cdot \phi$, $n_{G \cdot \phi} = 1$ and $n_{G \cdot \phi}^{\psi} = 1$. Then $\mathbf{B}_{\psi} = H \cdot \phi$, and this entails that $\dim \mathbf{B}_{\psi} = \dim H \cdot \phi = \dim A \cdot \phi - \frac{1}{2} \dim A \cdot p(\phi)$. Reciprocally, if $\dim \mathbf{B}_{\psi} = \dim H \cdot \phi$, then $H \cdot \phi$ is open in the connected manifold $\mathbf{B}_{\psi}$. As $H \cdot \phi$ is closed, we have $\mathbf{B}_{\psi} = H \cdot \phi$. Moreover $M_{\psi} = G \cdot \phi$ and $\mathbf{B}_{\psi} = M_{\psi} \cap \Gamma_f = \Gamma_f \cap G \cdot \phi = H \cdot \phi$. This leads to $n_{G \cdot \phi} = 1$, hence $\dim G \cdot \phi = 2H \cdot \phi$, and the fact that $\dim H \cdot \phi = \dim A \cdot \phi - \frac{1}{2} \dim A \cdot \psi$ allows us to deduce that $n_{G \cdot \phi}^{\psi} = 1$. This concludes the proof of (1.4.13). Now we observe that $\dim \mathbf{B}_{\psi} = \dim p(k \cdot \phi) + \dim(\mathfrak{a} + \mathfrak{h})^{\perp} = \dim K \cdot p(\phi) + \dim(\mathfrak{a} + \mathfrak{h})^{\perp}$, which is generically constant on Γ_f. Therefore the multiplicities of $\rho(G, H, A, f)$ are uniformly 1 or infinite. □

Corollary 1.4.14 *Under the same hypotheses of the above theorem, the representation $\rho(G, H, A, f)$ is multiplicity-free if and only if, for almost every $\psi \in p(G \cdot \Gamma_f)$, M_{ψ} is a manifold of dimension*

$$\dim M_{\psi} = 2d_H = 2d_A - d'_A.$$

Proof Suppose M_ψ is a manifold of dimension $\dim M_\psi = 2d_H = 2d_A - d'_A$. Let $M_\psi = \cup_{i\in I} G\cdot\phi_i$. Then for all $i \in I$ we have $\dim G\cdot\phi_i \le \dim M_\psi = 2\dim H\cdot\phi_i \le \dim G\cdot\phi_i$. Hence $\dim M_\psi = \dim G\cdot\phi_i$, and every orbit is open and hence closed in M_ψ connected. We therefore deduce that I has one element and $M_\psi = G\cdot\phi$, so $\dim \mathbf{B}_\psi = \dim H\cdot\phi$. The assumption $\dim H\cdot\phi = d_A - \frac{1}{2}d'_A$ and Theorem 1.4.13 enable us to deduce that the representation $\rho(G, H, A, f)$ is multiplicity-free.

Conversely, if the multiplicities of $\rho(G, H, A, f)$ are uniformly finite, then for almost all $\psi \in p(G\cdot\Gamma_f)$, $m(\psi) = 1$. Then $M_\psi = G\cdot\phi$ is a manifold, with ϕ a generic element in Γ_f satisfying $\psi = p(G\cdot\phi)$. Also $n^\psi_{G\cdot\phi} = n_{G\cdot\phi} = 1$ and then $\dim M_\psi = \dim G\cdot\phi = 2d_H = 2d_A - d'_A$. □

Corollary 1.4.15 *Let G be an exponential solvable Lie group; H and A two closed connected normal subgroups of G and χ_f a unitary character of H. We suppose that $[\mathfrak{a},\mathfrak{a}] \subset \mathfrak{h}\cap\ker(f)$ and $\dim\mathfrak{g}/\mathfrak{a} = \frac{1}{2}d_G$. If τ is of finite multiplicities, then $\rho(G, H, A, f)$ is multiplicity-free.*

Proof The hypothesis implies $\mathfrak{a}$ is a common polarization at all elements whose G-orbits are maximal in Γ_f. Theorems 1.4.2 and 1.4.13 allow to conclude. □

Now we set out to study the particular case where $A = H$.

Proposition 1.4.16 *Let G be an exponential solvable Lie group, H a closed connected normal subgroup of G and χ_f a character of H. Then:*

$$\rho(G, H, f) \simeq m(f,\mathfrak{h}) \int^{\oplus}_{p(G\cdot f)} \chi_\psi d\mu^{\pi_f}_{G,H}(\psi) \simeq m(f,\mathfrak{h})\pi_{f|H},$$

where $m(f,\mathfrak{h}) = 1$ or ∞ according to whether $\mathfrak{h} \in M(f, g)$ or not.

Proof Let

$$K = H_f = \{g \in G : g\cdot(f_{|\mathfrak{h}}) = f_{|\mathfrak{h}}\}.$$

The representation space of $\rho(G, H, f)$ is $L^2(G/H, f)$. For $\xi \in L^2(G/H, f)$, we have

$$\rho(h)\xi(g) = \chi_{g\cdot f}(h)\xi(g),\ h \in H,\ g \in G,$$

and

$$L^2(G/H, f) = L^2(G/K)\otimes L^2(K/H, f) \simeq \int^{\oplus}_{\Omega} L^2(K/H,\varphi)d\mu^f_{G,H}(\varphi),$$

where $\Omega = G\cdot f_{|\mathfrak{h}} = G/K$. Hence if $H = K$ then H is a Pukanszky polarization at f, and

$$\rho \simeq \int^{\oplus}_{\Omega} \chi_\varphi d\mu^f_{G,H}(\varphi) \simeq \pi_{f|\mathfrak{h}}.$$

If $H \neq K$, then

$$\rho \simeq \int_{\Omega}^{\oplus} \infty \chi_{\varphi} d\mu_{G,H}^{f}(\varphi) \simeq \infty \pi_{f|H}.$$

□

Example 1.4.17 Let G be an exponential solvable Lie group with Lie algebra $\mathfrak{g}$. We know that $(\mathrm{ind}_{\{e\}}^{G} 1)$ is the left-regular representation λ_G on $L^2(G)$ defined by $\lambda_G(g)\xi(x) = \xi(g^{-1}x)$ for all $g, x \in G$ and $\xi \in L^2(G)$. Here $f = 0$ and $H = \{e\}$. Then $\mathfrak{h}$ is not a polarization in f, and $\rho(G, \{e\}, 0) = \lambda_{G|\{e\}} \simeq \infty \cdot 1$.

Let now H be a normal subgroup of codimension 1 in G such that $\mathfrak{h}$ is subordinate to f, $p : \mathfrak{g}^{\star} \to \mathfrak{h}^{\star}$ the canonical projection and write $\mathfrak{g} = \mathfrak{h} \oplus \mathbb{R}X$. If $\mathfrak{g}(f) \subset \mathfrak{h}$, then $\mathfrak{h}$ is a polarization at f and we have

$$\rho(G, H, f) \simeq \int_{\mathbb{R}}^{\oplus} \chi_p(\exp tX \cdot f)dt.$$

If $\mathfrak{g}(f) \not\subset \mathfrak{h}$, then $\mathfrak{h}$ is not a polarization at f and we have $\rho(G, H, f) \simeq \infty\chi_p(f)$. Finally, suppose G is the Boidol group, whose Lie algebra $\mathfrak{g}$ is generated by vectors A, X, Y, Z such that $[A, X] = X$, $[A, Y] = -Y$, $[X, Y] = Z$. Take $f_\alpha = \alpha Z^{\star}$, $\alpha \neq 0$. If $H = \exp(\mathbb{R}Z)$, then $\mathfrak{h}$ is not a polarization at f_α, $m(f, \mathfrak{h}) = +\infty$ and $\rho(G, H, f, \alpha) \simeq \infty\chi_{\alpha Z^{\star}}$. If $H = \exp(\mathbb{R}Z + \mathbb{R}X)$ then $\mathfrak{h}$ is not a polarization at f, $\mathfrak{h}$ is not maximal and $m(f, \mathfrak{h}) = +\infty$. We get $\rho(G, H, f, \alpha) \simeq \int_{\mathbb{R}}^{\oplus} \infty\chi_{(\alpha Z^{\star} + \beta X^{\star})} d\beta$.

1.5 Down-Up Representations

Definition 1.5.1 Let G be a Lie group, H a closed subgroup of G and π a unitary representation of G. The unitary representation $\rho(G, H, \pi) = \mathrm{ind}_H^G(\pi_{|H})$ of G is called a *down-up representation*.

If G is an exponential solvable Lie group, then the following holds [96].

Theorem 1.5.2 *Let G be an exponential solvable Lie group and let H be a proper, connected, closed subgroup of G. If π is a unitary representation of G, then the down-up representation $\rho(G, H, \pi)$ is reducible.*

Remark 1.5.3 Let $G = \exp \mathfrak{g}$ be an exponential solvable Lie group, let $H = \exp \mathfrak{h}$ be an analytic subgroup, and let π be an irreducible unitary representation of G.

(1) It follows from Theorem 1.5.2 that a down-up representation $\rho(G, H, \pi)$ is irreducible if and only if $\rho(G, H, \pi) \simeq \pi$, and this can occur only if $G = H$.
(2) By Kirillov's theory there exist a linear form f in $\mathfrak{g}^{\star}$ and a Pukanszky polarization $B = \exp(\mathfrak{b})$ for f such that $\pi = \mathrm{ind}_B^G \chi_f$. Suppose that $K = BH$ is a closed subgroup of G and that $\pi_{|K}$ is irreducible. In this case $\pi_{|K} = \mathrm{ind}_B^K \chi_f$.

In fact, since $\pi = \mathrm{ind}_B^G \chi_f = \mathrm{ind}_K^G(\mathrm{ind}_B^K \chi_f)$, it follows that $\mathrm{ind}_B^K \chi_f$ is irreducible, and hence equivalent to $\pi_{|K}$. Therefore $K = G$. Moreover,

$$\rho(G, H, \pi) = \mathrm{ind}_H^G(\pi|H) = \mathrm{ind}_H^G(\mathrm{ind}_B^G \chi_f|H) = \mathrm{ind}_H^G(\mathrm{ind}_{H\cap B}^H \chi_f) = \mathrm{ind}_{H\cap B}^G \chi_f,$$

which follows from the Mackey subgroup theorem. In this situation we must only apply the results concerning $\mathrm{ind}_{H\cap B}^G \chi_f$.

(3) Suppose that $\mathfrak{h}$ is a Pukanszky polarization common to elements of the orbit $G \cdot f = \Omega(\pi)$. Then one can readily see that

$$\pi_{|H} \simeq \int_{\Omega(\pi)}^{\oplus} \chi_l d\Omega(l),$$

where $d\Omega$ is the canonical measure on the orbit $\Omega(\pi)$. It follows that

$$\rho(G, H, \pi) \simeq \int_{\Omega(\pi)}^{\oplus} \mathrm{ind}_H^G \chi_l d\Omega(l) \simeq \infty \cdot \pi.$$

1.5.1 The Down-Up Formula

In this section we prove the orbital spectrum formula for down-up representations of exponential solvable Lie groups.

Theorem 1.5.4 *Let G be an exponential solvable Lie group, H be a closed connected subgroup of G with Lie algebra $\mathfrak{h}$, and π be an irreducible unitary representation of G. Let $d\Omega$ be a canonical measure on the orbit $\Omega(\pi)$, and let λ be the Lebesgue measure on $\mathfrak{h}^\perp$. Let $\mu_\pi^{G,H}$ be the push-forward of the measure $(d\Omega \times \lambda)$ on $\Omega(\pi) \times \mathfrak{h}^\perp$ under the map*

$$\Omega(\pi)\times\mathfrak{h}^\perp \to p(\Omega(\pi))\times\mathfrak{h}^\perp \to p(\Omega(\pi))+\mathfrak{h}^\perp \to G\cdot(p(\Omega(\pi))+\mathfrak{h}^\perp)/G = G\cdot(\Omega(\pi)+\mathfrak{h}^\perp)/G.$$

The down-up orbital spectrum formula reads

$$\rho(G, H, \pi) \simeq \int_{G\cdot(\Omega(\pi)+\mathfrak{h}^\perp)/G}^{\oplus} m_\pi(\phi)\pi_\phi d\mu_\pi^{G,H}(\phi), \tag{1.5.1}$$

where

$$m_\pi(\phi) = \sum_{\Omega^H\in[p(\Omega(\pi))\cap p(\Omega(\pi_\phi))]/H} n_\pi^{\Omega^H} n_{\pi_\phi}^{\Omega^H} \tag{1.5.2}$$

with $n_\pi^{\Omega^H} = \#[\Omega(\pi)\cap p^{-1}(\Omega^H)]/H$ and $n_{\pi_\phi}^{\Omega^H} = \#[\Omega(\pi_\phi)\cap p^{-1}(\Omega^H)]/H$.

Proof The proof of this theorem is based on the decompositions of the restriction $\pi_{|H}$ and of the induced representation $\mathrm{ind}_H^G\sigma_\psi$, where ψ represents an H-orbit in $p(\Omega(\pi))$ (p is the canonical projection of $\mathfrak{h}$). In fact, by formula (1.5.2) we see that

$$\pi_{|H} \simeq \int_{p(\Omega(\pi))/H}^{\oplus} n_\pi^\psi \sigma_\psi d\mu_{H,G}^\pi(\psi),$$

and hence

$$\rho(G,H,\pi) \simeq \mathrm{ind}_H^G\Big(\int_{p(\Omega(\pi))/H}^{\oplus} n_\pi^\psi \sigma_\psi d\mu_{H,G}^\pi(\psi)\Big) \simeq \int_{p(\Omega(\pi))/H}^{\oplus} n_\pi^\psi \mathrm{ind}_H^G \sigma_\psi d\mu_{H,G}^\pi(\psi).$$

On the other hand, formula (2) implies

$$\mathrm{ind}_H^G\sigma_\psi \simeq \int_{G\cdot p^{-1}(H\cdot\psi)/G}^{\oplus} n_\phi^\psi \pi_\phi d\nu_{G,H}^{\sigma_\psi}(\phi).$$

It follows that

$$\rho(G,H,\pi) \simeq \int_{p(\Omega(\pi))/H}^{\oplus} n_\pi^\psi \int_{G\cdot p^{-1}(H\cdot\psi)/G}^{\oplus} n_\phi^\psi \pi_\phi d\nu_{G,H}^{\sigma_\psi}(\phi) d\mu_{H,G}^\pi(\psi) \tag{1.5.3}$$

$$\simeq \int_{G\cdot(\Omega(\pi)+\mathfrak{h}^\perp)/G} m(\phi)\pi_\phi d\mu_\pi^{G,H}(\phi). \tag{1.5.4}$$

We claim that (1.5.3) and (1.5.4) are equivalent. To show this, it suffices to prove that the two integral representations in (1.5.3) and (1.5.4) have the same spectrum, the same multiplicity function and equivalent measures.

Spectrum We need only show that

$$\bigcup_{H\cdot\psi\in[p(\Omega(\pi))]/H} [G\cdot p^{-1}(H\cdot\psi)]/G = [G\cdot(\Omega(\pi)+\mathfrak{h}^\perp)]/G.$$

Let $H\cdot\psi$ lie in an H-orbit in $p(\Omega(\pi))/H$. In this case $p^{-1}(H\cdot\psi) = H\cdot\psi+\mathfrak{h}^\perp = H\cdot(\psi+\mathfrak{h}^\perp)$. Hence

$$[G\cdot(\Omega(\pi)+\mathfrak{h}^\perp)]/G \subset \bigcup_{H\cdot\psi\in p(\Omega(\pi))/H} [G\cdot p^{-1}(H\cdot\psi)]/G,$$

where $p(f)$ is regarded as a linear form on $\mathfrak{g}$ that can be taken to be an arbitrary extension of $p(f)\in\mathfrak{h}^\star$.

Multiplicity For any linear form ϕ in $\Omega(\pi)+\mathfrak{h}^{\perp}$

$$m_\pi(\phi)=\sum\{n_\pi^\psi n_\phi^\psi,\ \psi \text{ represents an } H\text{-orbit in } p(\Omega(\pi)) : G\cdot\phi \text{ meets } p^{-1}(H\cdot\psi)\}.$$

The set

$$\mathscr{A}_\phi=\{\psi\in p(\Omega(\pi)) : G\cdot\phi\cap p^{-1}(\psi)\neq\emptyset\}$$

is non-empty because ϕ is in $\Omega(\pi)+\mathfrak{h}^{\perp}$. On the other hand, one can readily see that $\mathscr{A}_\phi$ is H-invariant. We claim that $H\cdot\psi\in[\mathscr{A}_\phi]/H$ if and only if $H\cdot\psi$ belongs to $[p(\Omega(\pi))\cap p(G\cdot\phi)]/H$. In fact, let $H\cdot\psi\in[\mathscr{A}_\phi]/H$. In this case $G\cdot\phi\cap p^{-1}(H\cdot\psi)\neq\emptyset$, and hence there exists a linear form f such that $f\in G\cdot\phi$ and $f\in p^{-1}(H\cdot\psi)$. Moreover, it follows from $f\in G\cdot\phi$ that $p(f)\in p(G\cdot\phi)$, which is H-invariant, and therefore $H\cdot p(f)\subset p(G\cdot\phi)$. On the other hand, the second condition $f\in p^{-1}(H\cdot\psi)$ is equivalent to the fact that $p(f)\in H\cdot\psi$, or, equivalently, $H\cdot p(f)=H\cdot\psi$, and therefore $H\cdot\psi\subset p(G\cdot\phi)$. Hence $[\mathscr{A}_\phi]/H\subset[p(\Omega(\pi))\cap p(G\cdot\phi)]/H$. For the other inclusion it suffices to show that $\psi\in[p(\Omega(\pi))\cap p(G\cdot\phi)]$ implies $p^{-1}(\psi)\cap G\cdot\phi\neq\emptyset$. But

$$[\mathscr{A}_\phi]/H=[p(\Omega(\pi))\cap p(G\cdot\phi)]/H,$$

and

$$m_\pi(\phi)=\sum_{\Omega^H\in[\mathscr{A}_\phi]/H} n_\pi^{\Omega^H} n_{\pi_\phi}^{\Omega^H}.$$

Measures Recall that

$$\bigcup_{\Omega^H\in p(\Omega(\pi))/H}[G\cdot p^{-1}(\Omega^H)]/G=G\cdot[\bigcup_{\Omega^H\in p(\Omega(\pi))/H}p^{-1}(\Omega^H)]/G.$$

The measure $\mu_{H,G}^{\pi}$ is the push-forward of the canonical measure on $\Omega(\pi)$ under the mapping $\Omega(\pi)\to p(\Omega(\pi))/H$, and hence, for each $\Omega^H\in p(\Omega(\pi))/H$, the measure $\nu_{G,H}^{\Omega^H}$ is the push-forward of the natural measure on $p^{-1}(\Omega^H)$ under $p^{-1}(\Omega^H)\to G\cdot p^{-1}(\Omega^H)/G$, where the natural measure is the direct product of the canonical measure on Ω^H and the Lebesgue measure on $\mathfrak{h}^{\perp}$. This completes the proof. □

Remark 1.5.5 Denote by $d\nu_\pi^{G,H}$ the push-forward of the measure $d\mu_\pi^{G,H}$ on $G\cdot(\Omega(\pi)+\mathfrak{h}^{\perp})/G$ under the canonical injection $G\cdot(\Omega(\pi)+\mathfrak{h}^{\perp})/G\to\mathfrak{g}^\star/G$. In this case

$$\rho(G,H,\pi)\simeq\int_{\mathfrak{g}^\star/G}^{\oplus}m_\pi(\Omega)\rho_\Omega d\nu_\pi^{G,H}(\Omega),$$

where ρ_Ω is the irreducible unitary representation associated with the orbit Ω by the Kirillov-Bernat homeomorphism, and $m_\pi(\Omega)$ is as in (1.5.2).

As an immediate corollary to Theorem 1.5.2 we obtain the central decomposition formula into irreducible unitary representations of any down-up representation $\rho(G, H, \pi)$.

Corollary 1.5.6 *Let the assumptions of Theorem 1.5.2 be satisfied. Let β be the push-forward of the measure $d\mu_\pi^{G,H}$ under the Kirillov-Bernat homeomorphism $K_G : \mathfrak{g}^\star/G \to \widehat{G}$. Then*

$$\rho(G, H, \pi) \simeq \int_{\widehat{G}}^{\oplus} m_\pi(\gamma)\gamma d\beta(\gamma), \tag{1.5.5}$$

where the multiplicity function $m_\pi(\gamma)$ is given by (1.5.2). A representation $\gamma = \gamma_\phi$, $\phi \in \mathfrak{g}^\star$, occurs in decomposition (1.5.5) if and only if $\phi \in G \cdot (p^{-1}p(\Omega(\pi))) = G \cdot (\Omega(\pi) + \mathfrak{h}^\perp)$.

1.5.2 The Down-Up Multiplicity Formula

Let $V_1 < V_2 < \cdots < V_s$ be the stratification of $\mathfrak{h}^*$ of Theorem 1.3.9, and call e the largest index in $\{1, \ldots, s\}$ such that V_e meets $p_\mathfrak{h}(\Omega_\pi^G)$. The set $(\Omega_\pi^G + \mathfrak{h}^\perp) \cap p_\mathfrak{h}^{-1}(V_e)$ is a non-empty, open and co-null in $(\Omega_\pi^G + \mathfrak{h}^\perp)$. For any $\phi \in (\Omega_\pi^G + \mathfrak{h}^\perp) \cap p_\mathfrak{h}^{-1}(V_e)$, let

$$\mathscr{A}_\phi = p_\mathfrak{h}(\Omega_\pi^G) \cap p_\mathfrak{h}(G \cdot \phi) \text{ and } \mathscr{A}_\phi' = V_e \cap \mathscr{A}_\phi. \tag{1.5.6}$$

It is clear that the set $\mathscr{A}_\phi$ is H-invariant. The set $\mathscr{A}_\phi'$ is an open dense and H-invariant subset of $\mathscr{A}_\phi$ as $\mathscr{A}_\phi' = (\cup_{e' \leq e} V_{e'}) \cap \mathscr{A}_\phi$ and $(\cup_{e' \leq e} V_{e'})$ is a Zariski open set in $\mathfrak{h}^*$ (see property (3) of Theorem 1.3.9). It is clear that $\mathscr{A}_\phi'$ is the set of H-orbits of maximal dimension in $\mathscr{A}_\phi$. Recall the notation $\Omega_\pi^G = G \cdot f$.

Let ρ be a unitary and irreducible representation of G. For ϕ in Ω_ρ^G, write

$$d_\rho(\phi) = \dim H \cdot \phi, \quad d_\rho = \max_{\phi \in \Omega_\rho^G} d_\rho(\phi),$$

and

$$d_G(\rho) = 2d_\rho - \dim \Omega_\rho^G.$$

The set $Z_\rho = \{\phi \in \Omega_\rho^G : \ d(\phi) = d_\rho\}$ is a non-empty co-null set in Ω_ρ^G.

Our main result of this section is the following:

Theorem 1.5.7 *Let G be a completely solvable Lie group, $H = \exp \mathfrak{h}$ an analytic subgroup of G and π an irreducible unitary representation of G. Then*

1. *the multiplicity function of the representation $\rho(G, H, \pi)$ is either uniformly infinite or uniformly finite and bounded;*
2. *the multiplicities of $\rho(G, H, \pi)$ are finite if and only if $\mathscr{A}'_\phi$ is a semi-analytic subset of $\mathfrak{h}^*$ and the two equalities*

$$\dim \mathscr{A}'_\phi = d_G(\pi_\phi) = d_G(\pi),$$

hold generically on $\Omega_\pi^G + \mathfrak{h}^\perp$.

Proof For starters, we shall prove that for ϕ in $(\Omega_\pi^G + \mathfrak{h}^\perp) \cap p_{\mathfrak{h}}^{-1}(V_e)$, the set $\mathscr{A}'_\phi$ has finitely many connected components in the Euclidean topology. In fact, let $\{X_1, \ldots, X_n\}$ be a Jordan-Hölder basis of $\mathfrak{g}$ (i.e. the subspace $\mathfrak{g}_i = \mathbb{R} - \text{span}\{X_1, \ldots, X_i\}$ is an ideal of $\mathfrak{g}$ for every $i = 1, \ldots, n$). Since G is completely solvable, the function

$$F : \mathbb{R}^n \times \mathfrak{g}^* \longrightarrow \mathfrak{g}^*$$
$$(x, l) = ((x_1, \ldots, x_n), l) \longmapsto \exp x_n X_n \cdots \exp x_1 X_1 \cdot l$$

is pseudo-algebraic, and for $l \in \mathfrak{g}^*$ one has $G \cdot l = \{F(x, l),\ x \in \mathbb{R}^n\}$. We view an element l in $\mathfrak{h}^*$ as an element of $\mathfrak{g}^*$ with trivial extension on $\mathfrak{g}$. The set

$$\mathscr{A} = \{(x, y, l) \in \mathbb{R}^n \times \mathbb{R}^n \times \mathfrak{h}^* : l \in V_e, F(x, f) - l \in \mathfrak{h}^\perp,\ F(y, \phi) - l \in \mathfrak{h}^\perp\},$$

is semi-pseudo-algebraic in $\mathbb{R}^n \times \mathbb{R}^n \times \mathfrak{h}^*$. Let $q : \mathbb{R}^n \times \mathbb{R}^n \times \mathfrak{h}^* \to \mathfrak{h}^*$ be the canonical projection, so clearly $\mathscr{A}'_\phi = q(\mathscr{A})$. Therefore the number of connected components of $\mathscr{A}'_\phi$ is not larger than the number of connected components of $\mathscr{A}$, and Theorem 1.3.6 allows us to conclude. With the above in mind, the number of connected components of $\mathscr{A}'_\phi$ is bounded by a scalar that does not depend on $\phi \in \Omega_\pi^G + \mathfrak{h}^\perp$.

Let now $\phi \in \Omega_\pi^G + \mathfrak{h}^\perp$ and π_ϕ its irreducible and unitary representation. We show that $m_\pi(\phi)$ is finite if and only if $\mathscr{A}'_\phi$ is a semi-analytic set and $\dim \mathscr{A}'_\phi = d_G(\pi_\phi) = d_G(\pi)$. Assume first that $m_\pi(\phi) < \infty$, then $\#[\mathscr{A}'{}_\phi]/H < \infty$ and for all $H \cdot \psi$ in $[\mathscr{A}'_\phi]/H$, we have that $n_\pi^{H\cdot\psi} = n_\pi^\psi < \infty$ and $n_{\pi_\phi}^{H\cdot\psi} = n_{\pi_\phi}^\psi < \infty$. Hence $d_G(\pi) = \dim H \cdot \psi$ and $d_G(\pi_\phi) = \dim H \cdot \psi$. Write

$$\mathscr{A}'_\phi = \bigcup_{i=1}^{k} H \cdot \psi_i,$$

where the H-orbits, $H \cdot \psi_i, i = 1, \ldots, k$ have the same dimension. It follows from Corollary 1.3.10 that $\mathscr{A}'_\phi$ is a finite union of semi-analytic sets, which in turn is a semi-analytic set. Moreover, the dimension of $\mathscr{A}'_\phi$ is equal to $\dim H \cdot \psi_i$ for all $i = 1, \ldots, k$. Hence $\dim \mathscr{A}'_\phi = d_G(\pi_\phi) = d_G(\pi)$. On the other hand, let

$$\mathscr{A}'_\phi = \bigcup_{i \in I} H \cdot \psi_i.$$

Then for all $i \in I$, we have:

$$\dim H \cdot \psi_i \geq d_G(\pi_\phi) = d_G(\pi) = \dim \mathscr{A}'_\phi \geq \dim H \cdot \psi_i. \tag{1.5.7}$$

Therefore $\dim \mathscr{A}'_\phi = \dim H \cdot \psi_i$, which means that $H \cdot \psi_i$ is open in $\mathscr{A}'_\phi$ and therefore closed in $\mathscr{A}'_\phi$. Since H is connected, the orbits $H \cdot \psi_i$, $i \in I$ are the connected components of $\mathscr{A}'_\phi$, which implies I is finite. Finally, by (1.5.7), (1.2.5) and (1.2.7) we conclude that the multiplicity functions $n_\pi^{\psi_i}$ and $n_{\pi_\phi}^{\psi_i}$ are finite.

Let Σ_e be the cross-section of H-orbits in V_e. For f in Ω_π^G let

$$\mathbf{U} = \{(x, y, l) \in \mathbb{R}^n \times \mathbb{R}^n \times \mathfrak{h}^* : l \in \Sigma_e, F(x, f) - l \in \mathfrak{h}^\perp, \ F(y, \phi) - l \in \mathfrak{h}^\perp\}.$$

As Σ_e is algebraic, $\mathbf{U}$ is pseudo-algebraic. Moreover,

$$\#[\mathscr{A}'_\phi]/H = \#q(\mathbf{U}).$$

Corollary 1.3.7 tells that the set $\#[\mathscr{A}'_\phi]/H$ is either uniformly infinite or uniformly finite and bounded on $\Omega_\pi^G + \mathfrak{h}^\perp$, thus ending the proof. □

1.5.3 Examples

Let $\mathfrak{g}$ be the 4-dimensional Lie algebra spanned by A, X, Y, Z with nonzero Lie brackets

$$[A, X] = X, \quad [A, Y] = -Y, \quad [X, Y] = Z.$$

Let $\{A^*, X^*, Y^*, Z^*\}$ be the dual basis of $\mathfrak{g}^*$ and $\varphi = \varphi_{(\alpha,\beta,\gamma,\delta)} = \alpha A^* + \beta X^* + \gamma Y^* + \delta Z^* \in \mathfrak{g}^*$. We denote by $\Omega^G_{(\alpha,\beta,\gamma,\delta)}$ the G-orbit of $\varphi_{(\alpha,\beta,\gamma,\delta)}$. A routine computation gives

$$\mathrm{Ad}^*\left((\exp aA \exp xX \exp yY \exp zZ)^{-1}\right)\varphi_{(\alpha,\beta,\gamma,\delta)} = \varphi_{(\alpha - \beta e^a x + \gamma e^{-a} y + \delta xy, \beta e^a - \delta y, \gamma e^{-a} + \delta x, \delta)}.$$

(1) $\mathfrak{h} = \mathbb{R}\text{-span}\{A, X, Z\}$, $\mathfrak{a} = \mathbb{R}\text{-span}\{A, Z\}$ and $f = \lambda A^*$. Then $\Gamma_f = f + \mathfrak{h}^\perp = \lambda A^* \oplus \mathbb{R}Y^*$ and

$$G \cdot \Gamma_f = \{(\lambda + xy)A^* + yY^*,\ x, y \in \mathbb{R}\}.$$

Therefore, $p_{\mathfrak{a}}(G \cdot \Gamma_f) = \mathbb{R}A^*$. Moreover, for $\psi = \alpha A^* \in p_{\mathfrak{a}}(G \cdot \Gamma_f)$, one has

$$\mathscr{M}_\psi = \begin{cases} \mathbb{R}A^* \oplus \mathbb{R}^* Y^* & \text{if } \alpha \neq \lambda, \\ G \cdot \Gamma_f & \text{otherwise,} \end{cases}$$

and $\mathscr{M}'_\psi = \Omega^G_{(0,0,1,0)} \cup \Omega^G_{(0,0,-1,0)} = \Omega_+ \cup \Omega_-$. Note that

$$\dim \mathscr{M}'_\psi = 2, \quad d_H = 1, \quad d_A = 1, \quad d'_A = 0,$$

and for $\Omega \in [\mathscr{M}'_\psi]/G$,

$$n^f_\Omega = 1 \text{ and } n^\psi_\Omega = 1.$$

It follows that $m(\psi) = 2$ and

$$\rho(G, H, A, f) \simeq 2 \int_{\mathbb{R}}^{\oplus} \chi_{tA^*} dt.$$

(2) $\mathfrak{h} = \mathbb{R}\text{-span}\{A, X\}$, $\mathfrak{a} = \mathbb{R}\text{-span}\{A, Y\}$ and $f = \lambda A^*$. We have in this case $\Gamma_f = \lambda A^* \oplus \mathbb{R}Y^* \oplus \mathbb{R}Z^*$ and

$$\begin{aligned} G \cdot \Gamma_f = \{\lambda A^*\} \cup \{aA^* + yY^* :\ a \in \mathbb{R}, y \in \mathbb{R}^*\} \cup \\ \{(\lambda - \frac{xy}{z})A^* + xX^* + yY^* + zZ^* :\ x, y \in \mathbb{R}, z \in \mathbb{R}^*\}. \end{aligned}$$

So, $p_{\mathfrak{a}}(G \cdot \Gamma_f) = \{\lambda A^*\} \cup \{xA^* + yY^* :\ x \in \mathbb{R}, y \in \mathbb{R}^*\}$. A-orbits in general position in $p_{\mathfrak{a}}(G \cdot \Gamma_f)$ are parametrized by the forms $\psi = \pm Y^*$. Moreover, for $\psi = \pm Y^* \in p_{\mathfrak{a}}(G \cdot \Gamma_f)$, we have

$$\mathscr{M}_\psi = \{aA^* + yY^* :\ a \in \mathbb{R}, y \in \mathbb{R}^*\} \cup \{(\lambda - \frac{xy}{z})A^* + xX^* + yY^* + zZ^* :\ x, y \in \mathbb{R}, z \in \mathbb{R}^*\},$$

$$\mathscr{M}'_\psi = \{(\lambda - \frac{xy}{z})A^* + xX^* + yY^* + zZ^* :\ x, y \in \mathbb{R}, z \in \mathbb{R}^*\},$$

and therefore $\dim \mathscr{M}'_\psi = 3$. It follows that for $\psi \in p_{\mathfrak{a}}(G \cdot \Gamma_f)$, $m(\psi) = \infty$ and

$$\rho(G, H, A, f) \simeq \infty\sigma_- \oplus \infty\sigma_+,$$

where $\sigma_\pm$ is the irreducible unitary representation of A associated to the form $\psi = \pm Y^*$.

(3) Suppose $\mathfrak{h} = \mathbb{R}\text{-span}\{A, Y, Z\}$ and let $\varphi^\circ_{(\alpha,\gamma,\delta)} = \alpha A^* + \gamma Y^* + \delta Z^* \in \mathfrak{h}^*$. As above,

$$\mathrm{Ad}^*((\exp aA \exp yY \exp zZ)^{-1})\varphi^\circ_{(\alpha,\gamma,\delta)} = \varphi^\circ_{(\alpha+\gamma e^{-a}y,\gamma e^{-a},\delta)}.$$

Let $f = \beta X^*,\ \beta > 0$. Then

$$\Omega^G_\pi = G \cdot f = \{aA^* + xX^*,\ \ a \in \mathbb{R}, x \in \mathbb{R}^*_+\}$$

and

$$G \cdot (\Omega^G_\pi + \mathfrak{h}^\perp) = \mathbb{R}A^* \oplus \mathbb{R}X^*.$$

The generic orbits in $G \cdot (\Omega^G_\pi + \mathfrak{h}^\perp)$ are associated to the linear forms $\pm X^*$. For $\phi = \pm X^*$, we have $\mathscr{A}'_\phi = \mathscr{A}_\phi = \mathbb{R}A^*$ and $\#[\mathscr{A}_\phi]/H = \infty$. Hence

$$\rho(G, H, \pi_f) \simeq \infty\pi_{-X^*} \oplus \infty\pi_{X^*}.$$

Moreover, we have $\dim \mathscr{A}_\phi = 1$ and $d_G(\pi) = 0$.

(4) Finally, consider the case where $\mathfrak{h} = \mathbb{R}\text{-span}\{A, X, Z\}$. If $\varphi^\circ_{(\alpha,\beta,\delta)} = \alpha A^* + \beta X^* + \delta Z^* \in \mathfrak{h}^*$,

$$\mathrm{Ad}^*((\exp aA \exp xX \exp zZ)^{-1})\varphi^\circ_{(\alpha,\beta,\delta)} = \varphi^\circ_{(\alpha-\beta e^a x,\beta e^a,\delta)}.$$

Let $f = X^*$, so

$$\Omega^G_\pi = G \cdot f = \{\alpha A^* + \beta X^*,\ \ \alpha \in \mathbb{R},\ \beta \in \mathbb{R}^*_+\},$$

$$\Omega^G_\pi + \mathfrak{h}^\perp = \{\alpha A^* + \beta X^* + \gamma Y^*,\ \ \alpha, \gamma \in \mathbb{R},\ \beta \in \mathbb{R}^*_+\},$$

and

$$G \cdot (\Omega^G_\pi + \mathfrak{h}^\perp) = \Omega^G_\pi + \mathfrak{h}^\perp.$$

Let ϕ be in $\Omega^G_\pi + \mathfrak{h}^\perp$ and write $\phi = \alpha A^* + \beta X^* + \gamma Y^*$. Then

$$G \cdot \phi = \{(\alpha - \beta e^a x + \gamma e^{-a} y)A^* + \beta e^a X^* + \gamma e^{-a} Y^*,\ \ a, x, y \in \mathbb{R}\},$$

and

$$H \cdot \phi = \{(\alpha - \beta e^a x)A^* + \beta e^a X^* + \gamma e^{-a} Y^*,\ \ a, x \in \mathbb{R}\},$$

It follows that $\mathscr{A}_\phi = p_{\mathfrak{h}}(\Omega_\pi^G) \cap p_{\mathfrak{h}}(G \cdot \phi) = \mathbb{R}A^* + \mathbb{R}_+^* X^*$, and

$$\dim \mathscr{A}_\phi = d_G(\pi) = d_G(\pi_\phi) = 2.$$

Moreover, for $\varphi^\circ = X^* \in \mathfrak{h}^*$ we have $H \cdot \varphi^\circ = \mathscr{A}_\phi$. Hence $\mathscr{A}'_\phi = \mathscr{A}_\phi$ is a semi-analytic subset of $\mathfrak{h}^*$ and $\#[\mathscr{A}_\phi]/H = 1$. Let $\Omega^H_{\sigma(\varphi^\circ)} = H \cdot \varphi^\circ$ where $\sigma(\varphi^\circ)$ is the unitary and irreducible representation associated to φ° by the Kirillov-Bernat mapping Θ_H. It is easy to check that $\Omega_\pi^G \cap p_{\mathfrak{h}}^{-1}(\Omega^H_{\sigma(\varphi^\circ)}) = H \cdot \varphi_{(0,1,0,0)}$ and $\Omega^G_{\pi_\phi} \cap p_{\mathfrak{h}}^{-1}(\Omega^H_{\sigma(\varphi^\circ)}) = H \cdot \varphi_{(\alpha,1,\gamma,0)}$, so

$$n_\pi^{\Omega^H_{\sigma(\varphi^\circ)}} = n_{\pi_\phi}^{\Omega^H_{\sigma(\varphi^\circ)}} = 1.$$

Finally, we have $m_\pi(\phi) = 1$.

1.5.4 The Case of Exponential Solvable Groups

In this section we denote by $G = \exp \mathfrak{g}$ an exponential solvable Lie group, by $H = \exp(\mathfrak{h})$ a closed connected normal subgroup of G, and by $\pi = \pi_f$ an irreducible unitary representation of G. Let us study the multiplicity function of the down-up representation $\rho(G, H, \pi)$. We prove a result similar to that in Theorem 4.1, which deals with the nilpotent situation. Let $p : \mathfrak{g}^\star \to \mathfrak{h}^\star$ be the canonical projection, recall $\Gamma_f = f + \mathfrak{h}^\perp$, and set $d_G = 2 \dim H \cdot f - \dim G \cdot f$.

Theorem 1.5.8 *Let G be an exponential solvable Lie group, H a closed connected normal subgroup of G with Lie algebra* $\mathfrak{h}$, *and* $\pi = \pi_f$ *an irreducible unitary representation of G. Then:*

(1) *the multiplicity function of the down-up representation* $\rho(G, H, \pi)$ *is either uniformly infinite or uniformly finite and free on* Γ_f.
(2) *The down-up representation is multiplicity-free if and only if*

$$\dim p(G \cdot f) = 2 \dim H \cdot \phi - \dim G \cdot \phi = dG$$

generically on Γ_f.

Proof The proof of the theorem will be split over several into several lemmas. First of all, let us describe in more detail the disintegration of the representation $\rho(G, H, \pi)$ when H is normal in G.

Lemma 1.5.9 *The down-up orbital spectrum formula of* $\rho(G, H, \pi_f)$ *is*

$$\rho(G, H, \pi_f) \simeq \int^{\oplus}_{G \cdot (f + \mathfrak{h}^\perp)/G} m_\pi(\phi) \pi_\phi d\mu_\pi^{G,H}(\phi),$$

where $m_\pi(\phi) = \#[p(G\cdot f)/H]n_{\pi_f}^{p(f)}n_{\pi_\phi}^{p(f)}$*, and* $\mu_\pi^{G,H}$ *is defined as in Theorem 1.5.4.*

Proof Let us first recall (see Theorem 1.5.4) that the spectrum of the representation $\rho(G, H, \pi_f)$ is $[G\cdot(G\cdot f+\mathfrak{h}^\perp)/G]$. Since $\mathfrak{h}$ is an ideal of $\mathfrak{g}$, it follows that

$$G\cdot(G\cdot f+\mathfrak{h}^\perp)/G = G\cdot(f+\mathfrak{h}^\perp)/G.$$

Let us now examine the multiplicity function $m_\pi(\phi)$, which is given for all linear forms ϕ in Γ_f by $m_\pi(\phi) = \sum_{H\cdot\psi\in[\mathscr{A}_\phi]/H} n_{\pi_f}^{\psi}n_{\pi_\phi}^{\psi}$, where $\mathscr{A}_\phi = p(G\cdot f)\cap p(G\cdot\phi)$. It is clear that

$$p(G\cdot\phi) = p(G\cdot p(\phi)) = p(G\cdot p(f)) = p(G\cdot f),$$

and finally $\mathscr{A}_\phi = p(G\cdot f)$. Concerning the multiplicity function $n_{\pi_f}^{\psi}$ for $\psi\in p(G\cdot f)$, we obtain

$$n_{\pi_f}^{\psi} = \#[G\cdot f\cap p^{-1}(H\cdot\psi)]/H = \#[G\cdot f\cap H\cdot(\psi+\mathfrak{h}^\perp)]/H = \#[G\cdot f\cap H\cdot(p(g\cdot f)+\mathfrak{h}^\perp)]/H$$

for $\psi = p(g\cdot f)$. Continuing the calculation,

$$n_{\pi_f}^{\psi} = \#[G\cdot f\cap H\cdot(g\cdot f+\mathfrak{h}^\perp)]/H = \#[G\cdot f\cap Hg\cdot(f+\mathfrak{h}^\perp)]/H = \#[G\cdot f\cap H\cdot(f+\mathfrak{h}^\perp)]/H.$$

(The map $[G\cdot f\cap Hg\cdot(f+\mathfrak{h}^\perp)/H]\to[G\cdot f\cap H\cdot(f+\mathfrak{h}^\perp)/H]$ given by $Hg\cdot\phi\mapsto H\cdot\phi$ is bijective). Therefore, $n_{\pi_f}^{\psi} = n_{\pi_f}^{p(f)}$. We may similarly show that $n_{\pi_\phi}^{\psi} = n_{\pi_\phi}^{p(f)}$. Thus the lemma is proved. □

It follows from formula (1.2.4) that

$$\mathrm{ind}_H^G\sigma_{p(f)} \simeq \int_{G\cdot(f+\mathfrak{h}^\perp)/G}^{\oplus} n_{\pi_\phi}^{p(f)}\pi_\phi d\nu_{G,H}^{\sigma_{p(f)}}(\phi).$$

Hence $\rho(G, H, \pi_f)\simeq m\cdot\mathrm{ind}_H^G(\sigma_{p(f)})$, where $m = \#[p(G\cdot f)/H]n_{\pi_f}^{p(f)}$. Indeed,

$$\begin{aligned}\rho(G, H, \pi_f) &\simeq \int_{G\cdot(f+\mathfrak{h}^\perp)/G}^{\oplus} \#[p(G\cdot f)/H]n_{\pi_f}^{p(f)}n_{\pi_\phi}^{p(f)}\pi_\phi d\mu_\pi^{G,H}(\phi)\\ &\simeq \#[p(G\cdot f)/H]n_{\pi_f}^{p(f)}\int_{G\cdot(f+\mathfrak{h}^\perp)/G}^{\oplus} n_{\pi_\phi}^{p(f)}\pi_\phi d\mu_\pi^{G,H}(\phi)\\ &\simeq \#[p(G\cdot f)/H]n_{\pi_f}^{p(f)}\mathrm{ind}_H^G\sigma_{p(f)}.\end{aligned}$$

Since the multiplicity function of $\tau(\sigma_{p(f)})$ is either uniformly infinite or uniformly finite and bounded, $\rho(G, H, \pi_f)$ obeys the same alternative.

Lemma 1.5.10 *Let G be an exponential solvable Lie group, H a closed connected normal subgroup of G with Lie algebra $\mathfrak{h}$ and $\pi = \pi_f$ an irreducible unitary representation of G. Then for ϕ in Γ_f, the number $m_\pi(\phi)$ is equal to 1 if and only if* $\dim p(G \cdot f) = 2 \dim H \cdot \phi - \dim G \cdot \phi = d_G$.

Proof Let $m_\pi(\phi)$ be finite. This means $\#[p(G \cdot f)/H] < \infty$, $n^{\omega}_{\pi_f} = 1$, and $n^{\omega}_{\pi_\phi} = 1$ for any H-orbit ω in $[p(G \cdot f)/H]$. Set

$$p(G \cdot f) = \bigcup_{i=1}^{k} H \cdot p(g_i \cdot f)$$

for some $g_i \in G$ and $i = 1, \ldots, k$. Then

$$\dim H \cdot p(g_i \cdot f) = \dim g_i \cdot (H \cdot p(f)) = \dim H \cdot p(f). \tag{1.5.8}$$

It follows that all H-orbits $H \cdot p(g_i \cdot f)$ are of the same dimension. We now consider the action of G on $\mathfrak{h}^\star$ and the variety $p(G \cdot f) = G \cdot p(f)$. As was proved above, we have $\dim p(G \cdot f) = \dim H \cdot p(f)$. Then every H-orbit is open, and therefore closed in $p(G \cdot f)$ because $p(G \cdot f)$ is connected. Finally, $p(G \cdot f) = H \cdot p(f)$. We conclude that $m_\pi(\phi) = 1$ and

$$\dim p(G \cdot f) = 2 \dim H \cdot \phi - \dim G \cdot \phi = d_G. \tag{1.5.9}$$

Conversely, assume (1.5.9) holds, so

$$\dim H \cdot p(f) \leq \dim p(G \cdot f) \leq \dim H \cdot p(f).$$

Hence $\dim p(G \cdot f) = \dim H \cdot p(f)$, $2 \dim H \cdot \phi - \dim G \cdot \phi = \dim H \cdot p(f) = d_G$. The last relations enable us to infer $n^{p(f)}_{\pi_f} = 1$ and $n^{p(f)}_{\pi_\phi} = 1$. Let us now prove that $\#[p(G \cdot f)/H] = 1$. We write

$$p(G \cdot f) = \bigcup_{i \in I} H \cdot p(g_i \cdot f).$$

It follows from $\dim p(G \cdot f) = \dim H \cdot p(f)$ and from (1.5.8) that every H-orbit is open in $p(G \cdot f)$, and, using the same arguments as above, we see $p(G \cdot f) = H \cdot p(f)$. □

Using the above lemmas, we obtain the proof of Theorem 1.5.8. □

Example 1.5.11 Let $G = \exp \mathfrak{g}$ be an exponential solvable Lie group, let $H = \exp(\mathfrak{h})$ be a codimension-one subgroup, and let $\pi = \pi_l$ be an irreducible unitary representation of G. Suppose first that the orbit of π is saturated with respect to $\mathfrak{h}$. Let

$$\pi_{|H} \simeq \int_{\mathbb{R}}^{\oplus} \pi_t dt,$$

where $\pi_t = \exp(tX)\pi,\ X \in \mathfrak{g} \setminus \mathfrak{h}$. It follows that

$$\rho(G, H, \pi) \simeq \int_{\mathbb{R}}^{\oplus} \mathrm{ind}_H^G \pi_t dt \simeq \infty \cdot \pi.$$

If the orbit of π is saturated with respect to $\mathfrak{h}$, the representation $\pi_{|H}$ is irreducible and equivalent to $\sigma_{l_0} \in \widehat{H},\ l_0 = l_{|\mathfrak{h}}$. Hence

$$\rho(G, H, \pi) = \mathrm{ind}_H^G \sigma_{l_0} \simeq \int_{\mathbb{R}}^{\oplus} \pi_{l_0 + sX^\star} ds.$$

Example 1.5.12 Let $\mathfrak{g}_\alpha = \mathbb{R}\text{-span}(A, X, Y)$ be a Lie algebra with non-vanishing brackets

$$[A, X] = X - \alpha Y, \quad [A, Y] = \alpha X + Y.$$

Then $G_\alpha = \exp \mathfrak{g}_\alpha$ is an exponential solvable Lie group which is not completely solvable. Let $\{A^\star, X^\star, Y^\star\}$ be the dual basis of $\{A, X, Y\}$ in $\mathfrak{g}_\alpha^\star$. The generic orbits of G_α are the orbits $\mathscr{O}_\theta$ of the form f_θ. A routine calculation shows that

$$\mathscr{O}_\theta = \{sA^\star + e^{-t}\cos(\theta - \alpha t)X^\star + e^{-t}\sin(\theta - \alpha t)Y^\star,\ s, t \in \mathbb{R}\}.$$

Suppose that $\mathfrak{h} = \mathbb{R}X \oplus \mathbb{R}Y$. By Example 1.5.11 we have $\rho(G, H, \pi_\theta) \simeq \infty \cdot \pi_\theta$. On the other hand,

$$2\dim H \cdot l - \dim G \cdot l = 2\dim H \cdot f_\theta - \dim G \cdot f_\theta = 0$$

for all $l \in f_\theta + \mathbb{R}A^\star$. Moreover, $\dim p(\mathscr{O}_\theta) = 1$.

Example 1.5.13 Suppose that $\mathfrak{h} = \mathbb{R}X$. It is immediate to see that the representation $\pi_{\theta|H}$ has infinite multiplicities, and hence is of the form $\rho(G, H, \pi_\theta)$. More precisely,

$$\rho(G, H, \pi_\theta) \simeq \int_{\mathbb{R}^2}^{\oplus} \pi e^{-t}\cos(\theta - \alpha t)X^\star + sY^\star dtds.$$

On the other hand, the dimension of the intersection $\mathscr{A}_l = p(\mathscr{O}_\theta) \cap p(G \cdot l)$ is generically zero on $\mathscr{O}_\theta + \mathfrak{k}^\perp$, and hence it equals $2\dim H \cdot f_\theta - \dim G \cdot f_\theta$. Here $p : \mathfrak{g}^\star \to \mathfrak{h}^\star$ is the canonical projection. Hence the equations in Theorem 1.5.8 holds. This example shows that Theorem 1.5.8 fails for arbitrary exponential solvable groups.

1.6 The Multiplicity Function of Monomial Representations

We return to the general setting. Let $H = \exp \mathfrak{h}$ be an analytic subgroup of G and let us build, as we did earlier, the monomial representation

$$\tau = \operatorname{ind}_H^G \chi_f \tag{1.6.1}$$

induced by the unitary character χ_f of H where $f \in \mathfrak{g}^*$, the linear dual space of $\mathfrak{g}$, and $\langle f, [\mathfrak{h}, \mathfrak{h}] \rangle = \{0\}$. As above, the representation τ obeys the orbital spectrum formula:

$$\tau \simeq \int_{\hat{G}}^{\oplus} m(\pi)\pi \, d\mu(\pi), \tag{1.6.2}$$

where μ denotes a finite measure that is equivalent to the image, under the Kirillov-Bernat map $\Theta = \Theta_G : \mathfrak{g}^\star \longrightarrow \widehat{G}$, of the Lebesgue measure on Γ_f of (1.4.8), and $m(\pi)$ denotes the number of H-orbits lying inside the intersection of Γ_f with the coadjoint orbit $\Omega(\pi)$ of π. (This is also the number of connected components in $\Gamma_f \cap \Omega(\pi)$, when finite). Here, $\Omega(\pi) \subset \mathfrak{g}^*$ designates the coadjoint orbit associated to π. The affine space Γ_f is sometimes called $\Gamma_{\mathfrak{h},f}$ or Γ_τ.

On the basis of disintegration (1.6.2), several results have appeared in the literature regarding harmonic analysis on exponential homogenous spaces. The most notable are those about connected and simply connected nilpotent Lie groups of differential operators (cf. [5, 10] and [6]), the Plancherel-Bonnet formula (see [28, 33, 65, 76, 78, 85, 86, 101, 103, 105, 108] and references). These results relate the aforementioned objects to the behaviour of the multiplicities $m(\pi)$, $\pi \in \hat{G}$, yet sometimes fail to encompass the general exponential case.

As shown above, and in the setup of completely solvable Lie groups, the representation τ either has uniformly infinite multiplicities or it has finite, bounded multiplicities. Even more, these multiplicities are uniformly bounded if and only if

$$\dim G \cdot \phi = 2 \dim H \cdot \phi \quad \text{for generic } \phi \in \Gamma_f. \tag{1.6.3}$$

Beyond this case and other particular situations (for example when H is normal or maximal and G is exponential), the problem is considerably more difficult. First of all, the last result fails to hold for exponential solvable Lie groups. A counterexample may be found in [44] and [99]. Secondly, it is not known whether these multiplicities might be uniformly finite or bounded, which greatly complicates the study, since both algebraic and pseudo-algebraic geometry cannot be used. We shall, nevertheless, provide a solution to this problem for so-called *adapted* data (G, H, f). The latter is the situation where $\mathfrak{h} + \mathfrak{g}(\phi)$ is a polarizing subalgebra of a generic $\phi \in \Gamma_f$, possibly without the Pukanszky condition. In this case we shall establish a necessary and sufficient condition to have finite multiplicities.

Definition 1.6.1 Let G be a connected and simply connected exponential solvable Lie group with Lie algebra $\mathfrak{g}$, $H = \exp \mathfrak{h}$ a closed connected subgroup of G and $f \in \mathfrak{g}^*$ such that $\langle f, [\mathfrak{h}, \mathfrak{h}] \rangle = \{0\}$. We will prove below that the set $\{\phi \in \Gamma_f : \mathfrak{g}(\phi) + \mathfrak{h}$ is a polarization$\}$ either contains a non-empty Zariski open set, or it is negligible with respect to the Lebesgue measure on Γ_f. In the first situation we say that the triple (G, H, f) is *adapted.*

We start by presenting some examples of adapted triples.

1. If there exists a good sequence $\mathfrak{s} = (\mathfrak{a}_j)_{j=0}^n$ of subalgebras of $\mathfrak{g}$ containing $\mathfrak{h}$, i.e. there exists $i_0 \in \{1, \ldots, n\}$ such that $\mathfrak{a}_{i_0} = \mathfrak{h}$ (this is the case when $\mathfrak{h}$ is an ideal of $\mathfrak{g}$), then the triple (G, H, f) is adapted if and only if τ has finite multiplicities.
2. If $\mathfrak{h}$ is a subalgebra of codimension 1, then (G, H, f) is adapted.
3. Let $\mathfrak{g}$ be the 2-dimensional Lie algebra spanned by A, X with nonzero bracket

$$[A, X] = X.$$

Let $f = \beta A^* + \alpha X^*$ ($\alpha \neq 0$) and $\mathfrak{h} = \mathbb{R}A$. For $\phi = \beta A^* + x X^* \in \Gamma_f$ ($x \neq 0$), we have $\mathfrak{g}(\phi) = \{0\}$, hence (G, H, f) is adapted.

4. Let $\alpha \in \mathbb{R}^*$ and $\mathfrak{g}_\alpha$ be the Lie algebra spanned by A, X, Y such that

$$[A, X] = X - \alpha Y, \qquad [A, Y] = \alpha X + Y.$$

Let $\{A^*, X^*, Y^*\}$ be the dual basis of $\{A, X, Y\}$ in $\mathfrak{g}_\alpha^*$. The non-trivial coadjoint orbits of G_α are the orbits Ω_θ associated to a linear form $f_\theta = \cos\theta X^* + \sin\theta Y^*$ for some $0 < \theta \leq 2\pi$. It is not hard to see that

$$\Omega_\theta = \{sA^* + e^{-t}\cos(\theta - \alpha t)X^* + e^{-t}\sin(\theta - \alpha t)Y^*, s, t \in \mathbb{R}\}.$$

Let $\mathfrak{h} = \mathbb{R}X$ and let $f = 0$. For a generic $\phi = \lambda Y^* + \alpha A^*$ ($\lambda \neq 0$) in $\mathfrak{h}^\perp = \mathbb{R}A^* + \mathbb{R}Y^*$, we have $\mathfrak{g}(\phi) = \mathbb{R}(X + \alpha Y)$, which implies that $\mathfrak{h} + \mathfrak{g}(\phi) = \mathbb{R}X + \mathbb{R}Y$. So, the triple (G, H, f) is adapted.

5. Let $\mathfrak{g}$ be the four-dimensional real Lie algebra spanned by vectors A, X, Y, Z with the following non-vanishing brackets

$$[X, Y] = Z, \qquad [A, X] = X, \qquad [A, Y] = -Y.$$

Let $\{A^*, X^*, Y^*, Z^*\}$ be the dual basis of $\{A, X, Y, Z\}$ in $\mathfrak{g}^*$. Let $\mathfrak{h} = \mathbb{R}A + \mathbb{R}Y$ and $f = \lambda A^*$ ($\lambda \neq 0$). The generic G-orbits in $\Gamma_f = \lambda A^* + \mathbb{R}X^* + \mathbb{R}Z^*$ are associated to the linear forms

$$\phi = \lambda A^* + xX^* + zZ^*$$

such that $x^2 + z^2 \neq 0$. It is clear that for each ϕ $\mathfrak{g}(\phi) = \mathbb{R}(zA + xY) \oplus \mathbb{R}Z$, and $\mathfrak{h} + \mathfrak{g}(\phi) = \mathbb{R}A + \mathbb{R}Y + \mathbb{R}Z$ is a Vergne polarization of ϕ. Hence the triple (G, H, f) is adapted.

6. Let G be the simply connected four-dimensional nilpotent Lie group whose Lie algebra $\mathfrak{g}$ has generators $X_1, \ldots, X_4$ satisfying the bracket relations

$$[X_4, X_2] = X_1, \quad [X_4, X_3] = X_2.$$

Let $f = \alpha X_4^*$ ($\alpha \in \mathbb{R}$) and $\mathfrak{h} = \mathbb{R}X_4$. We have $\Gamma_f = \alpha X_4^* + \mathbb{R}X_1^* + \mathbb{R}X_2^* + \mathbb{R}X_3^*$, and

$$\tau = \int_{\mathbb{R}^2}^{\oplus} \pi_{\alpha X_4^* + xX_1^* + yX_3^*} dxdy.$$

Moreover, for $\phi = \alpha X_4^* + xX_1^* + yX_3^* \in \Gamma_f$ ($x \neq 0$) we have $\mathfrak{g}(\phi) = \mathbb{R}X_1 \oplus \mathbb{R}X_3$. It is clear that the triple (G, H, f) is not adapted.

We shall prove a necessary and sufficient condition for the multiplicity function appearing in the direct-integral decomposition of the induced representation τ to be finite for adapted triples. The section's main theorem is

Theorem 1.6.2 *Let G be an exponential solvable Lie group. Let H be a closed connected subgroup of G with Lie algebra $\mathfrak{h}$, and let χ_f be a unitary character of H. If the triple (G, H, f) is adapted then:*

1. *generically, on Γ_f, the multiplicity $m(\pi_\phi)$ is finite if and only if the variety $G \cdot \phi \cap \Gamma_f$ is a semi-algebraic set.*
2. *If the multiplicity function of τ is uniformly finite, then it is uniformly bounded.*

Proof We shall first show that the set $\{\phi \in \Gamma_f : \mathfrak{h} + \mathfrak{g}(\phi) \text{ is a polarization}\}$ is either Lebesgue-negligible or it contains a non-empty Zariski open set. Let v be the smallest index in $\{1, \ldots, u\}$ such that $U = U_v$ (defined as in Theorem 1.3.9) meets Γ_f in a Zariski open and dense subset $\Gamma_0 = U \cap \Gamma_f$ of Γ_f. The function $d_G(\phi) = \dim G \cdot \phi$ is then constant in Γ_0, equal to d_G. Let also $d_H = \max_{\phi \in \Gamma_0} \dim H \cdot \phi$. It is known (see [63]) that, for fixed ϕ, if the multiplicity $m(\pi_\phi)$ is finite then the subspace $\mathfrak{h} + \mathfrak{g}(\phi)$ belongs to $i(\phi, \mathfrak{g})$.

Lemma 1.6.3 *The subset*

$$\Gamma_1 = \{\phi \in \Gamma_0 : \mathfrak{h} + \mathfrak{g}(\phi) \in i(\phi, \mathfrak{g})\},$$

is a Zariski-open, H-invariant subset of Γ_f.

Proof It is clear that Γ_1 is H-invariant. Let $\phi \in \Gamma_0$, then

$$\begin{aligned} \mathfrak{h} + \mathfrak{g}(\phi) \in i(\phi, \mathfrak{g}) &\iff \dim\big(\mathfrak{h} + \mathfrak{g}(\phi)\big) = \frac{1}{2}\big(\dim \mathfrak{g} + \dim \mathfrak{g}(\phi)\big) \\ &\iff \dim \mathfrak{h} - \dim \mathfrak{h} \cap \mathfrak{g}(\phi) = \frac{1}{2}\big(\dim \mathfrak{g} - \dim \mathfrak{g}(\phi)\big) \\ &\iff \dim H \cdot \phi = \frac{1}{2} \dim G \cdot \phi. \end{aligned}$$

Hence

$$\Gamma_1 = \{\phi \in \Gamma_0 : \dim H \cdot \phi = \frac{d_G}{2}\}.$$

Moreover

$$\dim H \cdot \phi = \operatorname{rank}\Big(\langle \phi, [Y_i, X_j]\rangle, \ 1 \leq i \leq m, \ 1 \leq j \leq n\Big),$$

where $\{Y_1, \dots, Y_m\}$ is a basis of $\mathfrak{h}$, and $\{X_1, \dots, X_n\}$ is a basis of $\mathfrak{g}$. As $\dim H \cdot \phi \leq \frac{d_G}{2}$ for all ϕ in Γ_0 (see [99], Lemma 1.4), Γ_1 turns out to be a Zariski open set of Γ_0, and therefore Γ_1 is Zariski open in Γ_f, on which we have $\dim H \cdot \phi = d_H = \frac{d_G}{2}$. □

We remark by the way that if Γ_1 is empty, the multiplicities $\{m(\pi_\phi) : \phi \in \Gamma_f\}$ of the induced representation τ are generically infinite.

Lemma 1.6.4 *Let Γ_2 be the H-invariant subset of Γ_f defined as*

$$\Gamma_2 = \{\phi \in \Gamma_1 : \mathfrak{h} + \mathfrak{g}(\phi) \text{ is a subalgebra of } \mathfrak{g}\}.$$

Then there are two mutually exclusive possibilities:

1. *Γ_2 is a negligible set in Γ_f;*
2. *Γ_2 contains a non-empty Zariski open set. In particular Γ_2 is co-negligible in Γ_f for the Lebesgue measure.*

Proof We suppose that Γ_1 is a non-empty set. Let $\phi \in \Gamma_1$, so $\mathfrak{b}_\phi = \mathfrak{h} + \mathfrak{g}(\phi)$ is a subalgebra of $\mathfrak{g}$ if and only if $[\mathfrak{h}, \mathfrak{g}(\phi)] \subset \mathfrak{b}_\phi$, which is in turn equivalent to the fact that $\langle l, [\mathfrak{h}, \mathfrak{g}(\phi)]\rangle = \{0\}$ for all $l \in \mathfrak{b}_\phi^\perp$. Taking bases of $\mathfrak{b}_\phi$ and $\mathfrak{b}_\phi^\perp$ that vary rationally with ϕ (possibly, when restricted to a small non-empty Zariski open set of Γ_1), one easily finds a real polynomial Q such that

$$\Gamma_2 = \{\phi \in \Gamma_1 : Q(\phi) = 0\}.$$

The result follows immediately. □

Back to Theorem 1.6.2, let us assume that the triple (G, H, f) is adapted. Obviously, then, Γ_2 is a non-empty Zariski open set of Γ_f. Take ϕ in Γ_2 and suppose that $m(\pi_\phi)$ is finite. As $G \cdot \phi \cap \Gamma_f$ is H-invariant, we write

$$G \cdot \phi \cap \Gamma_f = \bigcup_{i=1}^{m(\pi_\phi)} H \cdot \phi_i. \tag{1.6.4}$$

Observe that here all H-orbits in the intersection evidently have the same dimension, and there are no lower-dimensional H-orbits, which obviously comes from the definition of Γ_1. We then obtain that

$$\dim G \cdot \phi \cap \Gamma_f = \dim H \cdot \phi_i,$$

for all $i \in \{1, \ldots, m(\pi_\phi)\}$, which merely implies that the H-orbits $H \cdot \phi_i$ are open and therefore closed in $G \cdot \phi \cap \Gamma_f$. Moreover, for any $i = 1, \ldots, m(\pi_\phi)$, the subspace $\mathfrak{b}_{\phi_i} = \mathfrak{h} + \mathfrak{g}(\phi_i)$ is a polarization of ϕ_i in $\mathfrak{g}$ according to our hypothesis. Let $B_{\phi_i} = \exp \mathfrak{b}_{\phi_i}$. It is known ([24]) that $B_{\phi_i} \cdot \phi_i$ is open in $\phi + \mathfrak{b}_{\phi_i}^{\perp}$, whence $H \cdot \phi_i = B_{\phi_i} \cdot \phi_i$ is open in $\phi_i + \mathfrak{b}_{\phi_i}^{\perp}$. We deduce therefore that for all $i = 1, \ldots, m(\pi_\phi)$

$$H \cdot \phi_i \subset (\phi_i + \mathfrak{b}_{\phi_i}^{\perp}) \cap U \subset \Gamma_0. \tag{1.6.5}$$

Hence $H \cdot \phi_i$ is an open subset of $(\phi_i + \mathfrak{b}_{\phi_i}^{\perp}) \cap U$. We now prove that $H \cdot \phi_i$ is closed in $(\phi_i + \mathfrak{b}_{\phi_i}^{\perp}) \cap U$. We show first that for every $\phi \in U$, $\Omega = G \cdot \phi$ is closed in U. In fact, let $P : U \to U$ be the analytic G-invariant function such that $P(U) = \Sigma$, where Σ is a cross-section of the coadjoint orbits in U, as in Theorem 1.3.9. Thus $\{\phi\} = \Sigma \cap \Omega$. The subset $P^{-1}(\{\phi\})$ is G-invariant and every G-orbit $\Omega' \subset P^{-1}(\{\phi\})$ intersects Ω, so $\Omega = P^{-1}(\{\phi\})$, giving the result. Hence $G \cdot \phi \cap \Gamma_f$ is closed in $U \cap \Gamma_f = \Gamma_0$. As $H \cdot \phi_i$ is closed in $G \cdot \phi \cap \Gamma_f$, then $H \cdot \phi_i$ is closed in Γ_0. Hence it follows from (1.6.5) that $H \cdot \phi_i$ is closed in $(\phi_i + \mathfrak{b}_{\phi_i}^{\perp}) \cap U$. Finally, $H \cdot \phi_i$ is both closed and open in $(\phi_i + \mathfrak{b}_{\phi_i}^{\perp}) \cap U$, for all $i = 1, \ldots, m(\pi_\phi)$. Moreover, $H \cdot \phi_i$ is connected, which implies that $H \cdot \phi_i$ is a connected component of the semi-algebraic set $(\phi_i + \mathfrak{b}_{\phi_i}^{\perp}) \cap U$. It is known (see [20] and [25]) that any connected component of a semi-algebraic set is also semi-algebraic. Hence $G \cdot \phi \cap \Gamma_f$ is semi-algebraic by (1.6.4).

Let, conversely, $\phi \in \Gamma_0$ be such that $G \cdot \phi \cap \Gamma_f$ is semi-algebraic. Write

$$G \cdot \phi \cap \Gamma_f = \cup_{i \in I} H \cdot \phi_i.$$

We know, according to [99], that generically on Γ_f the set $G \cdot \phi \cap \Gamma_f$ is a manifold of dimension $\dim G \cdot \phi - \dim H \cdot \phi$. As the triple (G, H, f) is adapted, we get

$$\dim(\mathfrak{h} + \mathfrak{g}(\phi)) = \frac{1}{2}(\dim \mathfrak{g} + \dim \mathfrak{g}(\phi))$$

which implies that $\dim \mathfrak{h} - \dim \mathfrak{h} \cap \mathfrak{g}(\phi) = \frac{1}{2}(\dim \mathfrak{g} - \dim \mathfrak{g}(\phi))$. We then obtain that $\dim G \cdot \phi = 2 \dim H \cdot \phi$ and so $d_H = \dim G \cdot \phi \cap \Gamma_f$. Thus the orbits $H \cdot \phi_i, i \in I$ are the connected components of $G \cdot \phi \cap \Gamma_f$, being open and closed connected sets in $G \cdot \phi \cap \Gamma_f$. As previously mentioned, the number of connected components of a semi-algebraic set is finite and bounded by an integer which does not depend on ϕ. The result follows immediately.

Suppose finally that the multiplicity function $m(\pi_\phi)$ is uniformly finite, so $m(\pi_\phi)$ coincides with the number of connected components of the semi-algebraic set $G \cdot \phi \cap \Gamma_f$, which is uniformly finite by the above. This completes the proof of the theorem. □

Remark 1.6.5 Let $G = \exp \mathfrak{g}$ be a nilpotent Lie group, $H = \exp \mathfrak{h}$ a closed connected subgroup of G (or G exponential solvable and H normal in G). Take f in $\mathfrak{g}^*$ such that $\langle f, [\mathfrak{h}, \mathfrak{h}] \rangle = \{0\}$. Suppose that $\mathfrak{g}(\phi) \subset \mathfrak{h}$ generically on Γ_f. Then the representation τ is either irreducible or it has infinite multiplicities, uniformly. Assume in fact that the multiplicities of τ are finite. Then by [63] the subspace $\mathfrak{h} + \mathfrak{g}(\phi)$ is in $i(\phi, \mathfrak{g})$. It follows that $\mathfrak{h} + \mathfrak{g}(\phi) = \mathfrak{h}$ is a polarization of ϕ. It is then fairly easy to check that $\mathfrak{h}$ is a Pukanszky polarization of ϕ. Hence $\tau = \tau(\phi)$ is irreducible.

Chapter 2
Intertwining Operators for Irreducible Representations of an Exponential Solvable Lie Group

2.1 Introduction

When we study problems in representation theory, it is often very useful to construct an intertwining operator between two equivalent unitary representations. A prototype arises in the following situation. Let $G = \exp \mathfrak{g}$ be an exponential solvable Lie group with Lie algebra $\mathfrak{g}$ and $\mathfrak{h}_1, \mathfrak{h}_2$ two (real) polarizations of $\mathfrak{g}$ at $f \in \mathfrak{g}^*$ verifying the Pukanszky condition. We define the unitary characters χ_f of $H_j = \exp(\mathfrak{h}_j)(j = 1, 2)$ by $\chi_f(\exp X) = e^{if(X)}, X \in \mathfrak{h}_j$ and the induced monomial representations $\tau_j = \mathrm{ind}_{H_j}^G \chi_f (j = 1, 2)$ of G. Then it is well known that τ_1, τ_2 are equivalent. One main issue is how to explicitly construct an intertwining operator between them. When G is nilpotent, this problem was solved by G. Lion [97]. His work suggests using a formal intertwining integral in the general exponential case. We shall see that the essential obstacle is the integral's convergence (cf. [17]). We also propose a way to overcome this obstacle by introducing a third Vergne polarization.

2.2 A Trace Relation

Let $G = \exp \mathfrak{g}$ be an exponential solvable Lie group with Lie algebra $\mathfrak{g}$. If $\mathfrak{h}_1$ and $\mathfrak{h}_2$ are two polarizations of $\mathfrak{g}$ at $f \in \mathfrak{g}^*$ which satisfy the Pukanszky condition, the orbit method asserts that the monomial representations $\pi_i = \mathrm{ind}_{H_i}^G \chi_f (H_i = \exp(\mathfrak{h}_i), 1 \leq i \leq 2)$ of G are irreducible and mutually equivalent. We are interested in constructing an explicit intertwining operator between these representations. Let us sketch the idea, which goes back to Vergne [151]. If we have

$$\mathrm{Tr}\, \mathrm{ad}_{\mathfrak{h}_1/(\mathfrak{h}_1 \cap \mathfrak{h}_2)} X + \mathrm{Tr}\, \mathrm{ad}_{\mathfrak{h}_2/(\mathfrak{h}_1 \cap \mathfrak{h}_2)} X = 0$$

A. Baklouti et al., *Representation Theory of Solvable Lie Groups and Related Topics*, Springer Monographs in Mathematics,
https://doi.org/10.1007/978-3-030-82044-2_2

for any $X \in \mathfrak{h}_1 \cap \mathfrak{h}_2$, then

$$\Delta_{H_1,G}(h) = \Delta_{H_2,G}(h)\Delta_{H_1\cap H_2,H_2}(h)^2$$

for all $h \in H_1 \cap H_2$, and, for ϕ in the Hilbert space $\mathscr{H}_{\pi_1}$ of π_1 and $g \in G$, the function Φ_g on H_2 given by

$$\Phi_g(h) = \phi(gh)\chi_f(h)\Delta_{H_2,G}^{-1/2}(h)$$

verifies the relation

$$\Phi_g(hx) = \Delta_{H_1\cap H_2,H_2}(x)\Phi_g(h)\ (h \in H_2, x \in H_1 \cap H_2).$$

Thus we can, at least formally, consider the integral

$$(I_{\mathfrak{h}_2\mathfrak{h}_1}\phi)(g) = \oint_{H_2/(H_1\cap H_2)} \phi(gh)\chi_f(h)\Delta_{H_2,G}^{-1/2}(h)d\nu_{H_2,H_1\cap H_2}(h) \tag{2.2.1}$$

for $\phi \in \mathscr{H}_{\pi_1}$ and $g \in G$. If this integral converges for any $g \in G$, it is clear that $I_{\mathfrak{h}_2\mathfrak{h}_1}\phi$ verifies the covariance relation required on the elements of the space $\mathscr{H}_{\pi_2}$, and $I_{\mathfrak{h}_2\mathfrak{h}_1}$ commutes with the action of G by left translations. In fact, Vergne proved the following proposition. An ideal of $\mathfrak{g}$ is said to be minimal non-central if it is minimal among all non-central ideals of $\mathfrak{g}$. For an ideal $\mathfrak{a}$ of $\mathfrak{g}$, we put

$$\mathfrak{a}^f = \{X \in \mathfrak{g};\ f([X, \mathfrak{a}]) = \{0\}\}.$$

Proposition 2.2.1 ([24, Chap. VI], [151]) *Take $\mathfrak{h} \in I(f, \mathfrak{g})$ and let $\mathfrak{a}$ be a minimal non-central ideal of $\mathfrak{g}$. Put $\mathfrak{h}_0 = \mathfrak{h} \cap \mathfrak{a}^f$, $\mathfrak{h}' = \mathfrak{h}_0 + \mathfrak{a}$, $\mathfrak{k} = \mathfrak{h} + \mathfrak{a}$, $\tilde{f} = f|_{\mathfrak{k}} \in \mathfrak{k}^*$ and $H = \exp\mathfrak{h}$, $K = \exp\mathfrak{k}$. Then*

(i) $\mathfrak{a}$ is Abelian, $\dim \mathfrak{a}/(\mathfrak{z}\cap\mathfrak{a}) \leq 2$, where $\mathfrak{z}$ denotes the center of $\mathfrak{g}$, and $\mathfrak{h} = \mathfrak{h}_0+\mathfrak{b}$, where $\mathfrak{b} = \{X \in \mathfrak{h};\ [X, \mathfrak{a}] \subset \mathfrak{h} \cap \mathfrak{a}\}$.

*(ii) By passing to the quotient, B_f induces a pairing **B** between $\mathfrak{h}/\mathfrak{h}_0$ and $\mathfrak{a}/(\mathfrak{h}\cap\mathfrak{a})$, and also an isomorphism of $\mathfrak{h}_0$-modules from $\mathfrak{h}/\mathfrak{h}_0$ onto $(\mathfrak{a}/(\mathfrak{h} \cap \mathfrak{a}))^*$.*

(iii) We have

$$Tr\ ad_{\mathfrak{h}/\mathfrak{h}_0}X + Tr\ ad_{\mathfrak{h}'/\mathfrak{h}_0}X = 0$$

for any $X \in \mathfrak{h}_0$.

(iv) Let $\sigma = ind_H^K\chi_f$ and $\sigma' = ind_{H'}^K\chi_f$ with $H' = \exp(\mathfrak{h}')$. The operator $I_{\mathfrak{h}'\mathfrak{h}} : \mathscr{H}_\sigma \to \mathscr{H}_{\sigma'}$ formally defined by (2.2.1) is interpreted as a Fourier transform, and consequently gives an intertwining operator from $\pi = ind_H^G\chi_f$ to $\pi' = ind_{H'}^G\chi_f$.

(v) $\mathfrak{h}' \in I(f, \mathfrak{g})$.

We begin with the following theorem which generalizes equation (iii) in the above proposition.

Theorem 2.2.2 *For any* $X \in \mathfrak{h}_1 \cap \mathfrak{h}_2$,

$$Tr\ ad_{\mathfrak{h}_1/(\mathfrak{h}_1\cap\mathfrak{h}_2)}X + Tr\ ad_{\mathfrak{h}_2/(\mathfrak{h}_1\cap\mathfrak{h}_2)}X = 0.$$

Proof The theorem being trivial if $\dim \mathfrak{g} = 1$, we proceed by induction on $\dim \mathfrak{g}$. We examine different cases.

I. We suppose that there is a non-trivial ideal $\mathfrak{a}$ of $\mathfrak{g}$ on which f vanishes. Then $\mathfrak{a} \subset \mathfrak{g}(f) \subset \mathfrak{h}_1 \cap \mathfrak{h}_2$. Let $p : \mathfrak{g} \to \tilde{\mathfrak{g}} = \mathfrak{g}/\mathfrak{a}$ be the canonical mapping, $\tilde{\mathfrak{h}}_i = p(\mathfrak{h}_i)(1 \le i \le 2)$ and $\tilde{f}$ the image of f in $(\tilde{\mathfrak{g}})^*$. We have $\tilde{\mathfrak{h}}_i \in I(\tilde{f}, \tilde{\mathfrak{g}})(1 \le i \le 2)$, so we can descend to $\tilde{\mathfrak{g}}$ where the induction hypothesis applies.
II. We suppose that there is no non-trivial ideal of $\mathfrak{g}$ on which f vanishes. Let $\mathfrak{z}$ be the center of $\mathfrak{g}$ and $\mathfrak{a}$ a minimal non-central ideal of $\mathfrak{g}$. Then $\dim \mathfrak{z} \le 1$ and $\mathfrak{a}^f \subsetneq \mathfrak{g}$. Put $f' = f|_{\mathfrak{a}^f} \in (\mathfrak{a}^f)^*$.

(II-1) If $\mathfrak{h}_1$ and $\mathfrak{h}_2$ contain $\mathfrak{a}$, then $\mathfrak{h}_i \subset \mathfrak{a}^f$ and $\mathfrak{h}_i \in I(f', \mathfrak{a}^f)$ for $1 \le i \le 2$. Hence the theorem immediately follows from the induction hypothesis. We are in this situation when $\mathfrak{a}$ is minimal among non-trivial ideals of $\mathfrak{g}$ (cf. [24]).

From now on we suppose that $\dim \mathfrak{z} = 1, \mathfrak{z} \subset \mathfrak{a}$. We therefore have $\dim \mathfrak{a} = 2$ or 3. Put $\mathfrak{h}_i^0 = \mathfrak{h}_i \cap \mathfrak{a}^f$, $\mathfrak{h}_i' = \mathfrak{h}_i^0 + \mathfrak{a} \in I(f, \mathfrak{g})(1 \le i \le 2)$.

(II-2) We suppose that $\mathfrak{a} \subset \mathfrak{h}_i$ and $\mathfrak{a} \not\subset \mathfrak{h}_j$ for $\{i, j\} = \{1, 2\}$. For example, suppose that $\mathfrak{a} \subset \mathfrak{h}_1$ and $\mathfrak{a} \not\subset \mathfrak{h}_2$. Then $\mathfrak{h}_1 \cap \mathfrak{h}_2' = \mathfrak{h}_1 \cap \mathfrak{h}_2 + \mathfrak{a}$ and $\mathfrak{h}_1, \mathfrak{h}_2'$ belong to $I(f', \mathfrak{a}^f)$. From the induction hypothesis and Proposition 2.2.1 we have, for all $X \in \mathfrak{h}_1 \cap \mathfrak{h}_2$,

$$\begin{aligned}
0 &= \mathrm{Tr\ ad}_{\mathfrak{h}_1/(\mathfrak{h}_1\cap\mathfrak{h}_2')}X + \mathrm{Tr\ ad}_{\mathfrak{h}_2'/(\mathfrak{h}_1\cap\mathfrak{h}_2')}X \\
&= \mathrm{Tr\ ad}_{\mathfrak{h}_1/(\mathfrak{h}_1\cap\mathfrak{h}_2)}X - \mathrm{Tr\ ad}_{\mathfrak{a}/(\mathfrak{h}_2\cap\mathfrak{a})}X + \mathrm{Tr\ ad}_{\mathfrak{h}_2'/(\mathfrak{h}_1\cap\mathfrak{h}_2)}X - \mathrm{Tr\ ad}_{\mathfrak{a}/(\mathfrak{h}_2\cap\mathfrak{a})}X \\
&= \mathrm{Tr\ ad}_{\mathfrak{h}_1/(\mathfrak{h}_1\cap\mathfrak{h}_2)}X + \mathrm{Tr\ ad}_{\mathfrak{h}_2^0/(\mathfrak{h}_1\cap\mathfrak{h}_2)}X - \mathrm{Tr\ ad}_{\mathfrak{a}/(\mathfrak{h}_2\cap\mathfrak{a})}X \\
&= \mathrm{Tr\ ad}_{\mathfrak{h}_1/(\mathfrak{h}_1\cap\mathfrak{h}_2)}X + \mathrm{Tr\ ad}_{\mathfrak{h}_2/(\mathfrak{h}_1\cap\mathfrak{h}_2)}X.
\end{aligned}$$

(II-3) We suppose that $\mathfrak{a} \not\subset \mathfrak{h}_i\,(i = 1, 2)$. Let

$$\mathfrak{g}_0 = \{X \in \mathfrak{g};\ [X, \mathfrak{a}] \subset \mathfrak{z}\}.$$

Since $\mathfrak{g}$ is exponential solvable, $\mathfrak{g}_0$ is an ideal of $\mathfrak{g}$ such that $\dim(\mathfrak{g}/\mathfrak{g}_0) \le 1$.

Lemma 2.2.3 *In* $\mathfrak{h}_1 \cap \mathfrak{h}_2$ *we can take a complementary subspace* $\mathfrak{t}$ *of* $\mathfrak{h}_1^0 \cap \mathfrak{h}_2^0$ *in such a manner that* $\mathfrak{t}$ *is contained in* $\mathfrak{g}_0$.

Proof The form B_f defines by passage to the quotient a bilinear form

$$\mathbf{B} : (\mathfrak{h}_1 \cap \mathfrak{h}_2)/(\mathfrak{h}_1^0 \cap \mathfrak{h}_2^0) \times \mathfrak{a}/((\mathfrak{h}_1 + \mathfrak{h}_2) \cap \mathfrak{a}) \to \mathbb{R}.$$

It gives a pairing of $(\mathfrak{h}_1 \cap \mathfrak{h}_2)/(\mathfrak{h}_1^0 \cap \mathfrak{h}_2^0)$ and $\mathfrak{a}/((\mathfrak{h}_1 + \mathfrak{h}_2) \cap \mathfrak{a})$, and induces an isomorphism of $(\mathfrak{h}_1^0\cap\mathfrak{h}_2^0)$-modules from $(\mathfrak{h}_1\cap\mathfrak{h}_2)/(\mathfrak{h}_1^0\cap\mathfrak{h}_2^0)$ onto $(\mathfrak{a}/((\mathfrak{h}_1+\mathfrak{h}_2)\cap\mathfrak{a}))^*$.

Now, the adjoint representation of $\mathfrak{h}_1 \cap \mathfrak{h}_2$ on $\mathfrak{a}/(\mathfrak{a} \cap \mathfrak{z})$ is either trivial or irreducible. Hence it suffices to consider the case where $\mathfrak{h}_1 \cap \mathfrak{h}_2 \not\subset \mathfrak{g}_0$ and $(\mathfrak{h}_1 + \mathfrak{h}_2) \cap \mathfrak{a} = \mathfrak{z}$.

We take a complementary subspace $\mathfrak{t}_0$ of $\mathfrak{h}_1^0 \cap \mathfrak{h}_2^0$ in $\mathfrak{h}_1 \cap \mathfrak{h}_2$. If $\mathfrak{h}_1^0 \cap \mathfrak{h}_2^0 \not\subset \mathfrak{g}_0$, then $\mathfrak{t} = [X, \mathfrak{t}_0]$ will do, where X is an element of $\mathfrak{h}_1^0 \cap \mathfrak{h}_2^0$ not belonging to $\mathfrak{g}_0$ (cf. Proposition 2.2.1).

Suppose now $[\mathfrak{h}_1^0 \cap \mathfrak{h}_2^0, \mathfrak{a}] = \{0\}$, $\mathfrak{h}_1 \cap \mathfrak{h}_2 \not\subset \mathfrak{g}_0$. There exist $T \in \mathfrak{t}_0$ and $Y \in \mathfrak{a}$ satisfying $[T, Y] \notin \mathfrak{z}$, $[\mathfrak{t}_0 \cap \mathfrak{g}_0, Y] = \{0\}$ and $B_f(T, Y) \neq 0$. Then we can choose $l \in \mathfrak{a}^*$ so that $l|_\mathfrak{z} = f|_\mathfrak{z}$, $l([T, Y]) = 0$. Finally, let $f_0 \in \mathfrak{g}^*$ be such that $f_0|_{\mathfrak{h}_1+\mathfrak{h}_2} = f|_{\mathfrak{h}_1+\mathfrak{h}_2}$ and $f_0|_\mathfrak{a} = l$. We therefore obtain $B_{f_0}(Y, \mathfrak{h}_1 \cap \mathfrak{h}_2) = \{0\}$. But $\mathfrak{h}_1, \mathfrak{h}_2 \in M(f_0, \mathfrak{g})$ from the Pukanszky condition, so it follows that $Y \in \mathfrak{h}_1 + \mathfrak{h}_2$, which is absurd. □

Let us go back to the proof of the theorem.

A. Suppose first $\mathfrak{a} \subset \mathfrak{h}_1 + \mathfrak{h}_2$. Then $\mathfrak{h}_1 \cap \mathfrak{h}_2 \subset \mathfrak{a}^f$. We easily see $\mathfrak{h}_1' \cap \mathfrak{h}_2' = (\mathfrak{h}_1 \cap \mathfrak{h}_2') + \mathfrak{a} = (\mathfrak{h}_1' \cap \mathfrak{h}_2) + \mathfrak{a}$. As $(\mathfrak{h}_1 \cap \mathfrak{h}_2)$-module, $\mathfrak{h}_i'/(\mathfrak{h}_1' \cap \mathfrak{h}_2')$ is isomorphic to $\mathfrak{h}_i^0/(\mathfrak{h}_i \cap \mathfrak{h}_j')$, which gives

$$\begin{aligned}\mathrm{Tr}\,\mathrm{ad}_{\mathfrak{h}_i/(\mathfrak{h}_1\cap\mathfrak{h}_2)} X &= \mathrm{Tr}\,\mathrm{ad}_{\mathfrak{h}_i/\mathfrak{h}_i^0} X + \mathrm{Tr}\,\mathrm{ad}_{\mathfrak{h}_i^0/(\mathfrak{h}_i\cap\mathfrak{h}_j')} X + \mathrm{Tr}\,\mathrm{ad}_{(\mathfrak{h}_i\cap\mathfrak{h}_j')/(\mathfrak{h}_1\cap\mathfrak{h}_2)} X \\ &= \mathrm{Tr}\,\mathrm{ad}_{\mathfrak{h}_i/\mathfrak{h}_i^0} X + \mathrm{Tr}\,\mathrm{ad}_{\mathfrak{h}_i'/(\mathfrak{h}_1'\cap\mathfrak{h}_2')} X + \mathrm{Tr}\,\mathrm{ad}_{(\mathfrak{h}_i\cap\mathfrak{h}_j')/(\mathfrak{h}_1\cap\mathfrak{h}_2)} X\end{aligned}$$

for all $X \in \mathfrak{h}_1 \cap \mathfrak{h}_2$, where $\{i, j\} = \{1, 2\}$.

If $\mathfrak{h}_1 \cap \mathfrak{h}_2 = \mathfrak{h}_1^0 \cap \mathfrak{h}_2^0 \subset \mathfrak{g}_0$, $[X, \mathfrak{h}_i] \subset \mathfrak{h}_i^0$ and $[X, \mathfrak{h}_i \cap \mathfrak{h}_j'] \subset \mathfrak{h}_1 \cap \mathfrak{h}_2$. Hence

$$\mathrm{Tr}\,\mathrm{ad}_{\mathfrak{h}_1/(\mathfrak{h}_1\cap\mathfrak{h}_2)} X + \mathrm{Tr}\,\mathrm{ad}_{\mathfrak{h}_2/(\mathfrak{h}_1\cap\mathfrak{h}_2)} X = \mathrm{Tr}\,\mathrm{ad}_{\mathfrak{h}_1'/(\mathfrak{h}_1'\cap\mathfrak{h}_2')} X + \mathrm{Tr}\,\mathrm{ad}_{\mathfrak{h}_2'/(\mathfrak{h}_1'\cap\mathfrak{h}_2')} X = 0$$

because $\mathfrak{h}_1', \mathfrak{h}_2' \subset \mathfrak{a}^f$.

Therefore we may suppose $\mathfrak{h}_1 \cap \mathfrak{a} = \mathfrak{h}_2 \cap \mathfrak{a} = \mathfrak{z}$ and that $(\mathfrak{h}_1^0 + \mathfrak{h}_2^0) \cap \mathfrak{a}$ is equal to either $\mathfrak{z}$ or $\mathfrak{a}$, because the adjoint representation of $\mathfrak{h}_1\cap\mathfrak{h}_2$ on $\mathfrak{a}/\mathfrak{z}$ is either trivial or irreducible. But the condition $(\mathfrak{h}_1^0 + \mathfrak{h}_2^0) \cap \mathfrak{a} = \mathfrak{z}$ implies $\mathfrak{h}_1 \cap \mathfrak{h}_2 \subset \mathfrak{g}_0$. Indeed, as $(\mathfrak{h}_1 \cap \mathfrak{h}_2)$-module, $\mathfrak{a}/\mathfrak{z}$ is isomorphic to $\mathfrak{h}_i/\mathfrak{h}_i^0 (i = 1, 2)$ and then to $(\mathfrak{a}/\mathfrak{z})^*$ by Proposition 2.2.1. We have $\mathfrak{h}_1 \cap \mathfrak{h}_2 \subset \mathfrak{g}_0$ because $\mathfrak{g}$ is exponential solvable.

There remains to examine the case where $\mathfrak{h}_1\cap\mathfrak{a} = \mathfrak{h}_2\cap\mathfrak{a} = \mathfrak{z}$, $(\mathfrak{h}_1^0+\mathfrak{h}_2^0)\cap\mathfrak{a} = \mathfrak{a}$. By quotienting, the form B_f defines a bilinear form $\mathbf{B} : \mathfrak{h}_i/\mathfrak{h}_i^0\times(\mathfrak{h}_i'\cap\mathfrak{h}_j)/(\mathfrak{h}_1\cap\mathfrak{h}_2) \to \mathbb{R}$ which pairs $\mathfrak{h}_i/\mathfrak{h}_i^0$ and $(\mathfrak{h}_i' \cap \mathfrak{h}_j)/(\mathfrak{h}_1 \cap \mathfrak{h}_2)$ and further induces an isomorphism of $(\mathfrak{h}_1 \cap \mathfrak{h}_2)$-modules from $\mathfrak{h}_i/\mathfrak{h}_i^0$ onto $((\mathfrak{h}'_i \cap \mathfrak{h}_j)/(\mathfrak{h}_1 \cap \mathfrak{h}_2))^*$, where $\{i, j\} = \{1, 2\}$. This proves

$$\mathrm{Tr}\ \mathrm{ad}_{\mathfrak{h}_i/\mathfrak{h}_i^0} X + \mathrm{Tr}\ \mathrm{ad}_{(\mathfrak{h}_i \cap \mathfrak{h}'_j)/(\mathfrak{h}_1 \cap \mathfrak{h}_2)} X = 0$$

with $\{i, j\} = \{1, 2\}$, and therefore

$$\mathrm{Tr}\ \mathrm{ad}_{\mathfrak{h}_1/(\mathfrak{h}_1 \cap \mathfrak{h}_2)} X + \mathrm{Tr}\ \mathrm{ad}_{\mathfrak{h}_2/(\mathfrak{h}_1 \cap \mathfrak{h}_2)} X = \mathrm{Tr}\ \mathrm{ad}_{\mathfrak{h}'_1/(\mathfrak{h}'_1 \cap \mathfrak{h}'_2)} X + \mathrm{Tr}\ \mathrm{ad}_{\mathfrak{h}'_2/(\mathfrak{h}'_1 \cap \mathfrak{h}'_2)} X = 0$$

for all $X \in \mathfrak{h}_1 \cap \mathfrak{h}_2$.

B. Suppose now $\mathfrak{a} \not\subset \mathfrak{h}_1 + \mathfrak{h}_2$. Then $\mathfrak{h}_1 \cap \mathfrak{h}_2 \not\subset \mathfrak{a}^f$. As $f|_{\mathfrak{z}} \neq 0$, let $Z \in \mathfrak{z}$ satisfy $f(Z) = 1$.

(i) The case where $\dim \mathfrak{a} = 2$. Let $Y \in \mathfrak{a} \cap \ker(f)$ and $X \in \mathfrak{h}_1 \cap \mathfrak{h}_2$ verify $[X, Y] = Z$ (by Lemma 2.2.3 such X exists). Then $\mathfrak{a}^f$ being of codimension 1 in $\mathfrak{g}$, we have

$$\mathfrak{h}_i = \mathbb{R}X \oplus \mathfrak{h}_i^0 (i = 1, 2), \mathfrak{h}_1 \cap \mathfrak{h}_2 = \mathbb{R}X \oplus (\mathfrak{h}_1^0 \cap \mathfrak{h}_2^0), \mathfrak{h}'_1 \cap \mathfrak{h}'_2 = (\mathfrak{h}_1^0 \cap \mathfrak{h}_2^0) \oplus \mathbb{R}Y.$$

For any $X_0 \in \mathfrak{h}_1^0 \cap \mathfrak{h}_2^0$,

$$\begin{aligned}\mathrm{Tr}\ \mathrm{ad}_{\mathfrak{h}_i/(\mathfrak{h}_1 \cap \mathfrak{h}_2)} X_0 &= \mathrm{Tr}\ \mathrm{ad}_{(\mathfrak{h}_i^0 \oplus \mathbb{R}X)/((\mathfrak{h}_1^0 \cap \mathfrak{h}_2^0) \oplus \mathbb{R}X)} X_0 \\ &= \mathrm{Tr}\ \mathrm{ad}_{\mathfrak{h}_i^0/(\mathfrak{h}_1^0 \cap \mathfrak{h}_2^0)} X_0 = \mathrm{Tr}\ \mathrm{ad}_{\mathfrak{h}'_i/(\mathfrak{h}'_1 \cap \mathfrak{h}'_2)} X_0,\end{aligned}$$

from which

$$\mathrm{Tr}\ \mathrm{ad}_{\mathfrak{h}_1/(\mathfrak{h}_1 \cap \mathfrak{h}_2)} X_0 + \mathrm{Tr}\ \mathrm{ad}_{\mathfrak{h}_2/(\mathfrak{h}_1 \cap \mathfrak{h}_2)} X_0 = 0. \tag{2.2.2}$$

Next we show

$$\mathrm{Tr}\ \mathrm{ad}_{\mathfrak{h}_1/(\mathfrak{h}_1 \cap \mathfrak{h}_2)} X + \mathrm{Tr}\ \mathrm{ad}_{\mathfrak{h}_2/(\mathfrak{h}_1 \cap \mathfrak{h}_2)} X = 0.$$

If $\mathfrak{h}_1 \cap \mathfrak{h}_2 \not\subset \mathfrak{g}_0$, then $\mathfrak{h}_1^0 \cap \mathfrak{h}_2^0 \not\subset \mathfrak{g}_0$ by Lemma 2.2.3, and from Proposition 2.2.1 we can choose X in $[\mathfrak{g}, \mathfrak{g}]$. Suppose henceforth $\mathfrak{h}_1 \cap \mathfrak{h}_2 \subset \mathfrak{g}_0$.

Set $\mathfrak{g}_X = \{T \in \mathfrak{g}; B_f(X, T) = 0\}$. Let $\mathfrak{m}_i (i = 1, 2)$ be a complementary subspace of $\mathfrak{h}_1 \cap \mathfrak{h}_2$ in $\mathfrak{h}_i$, take a basis $\{\hat{e}_j; 1 \leq j \leq n\}$ of $\mathfrak{m}_2$ and $\{e_i; 1 \leq i \leq n\}$ the dual basis of $\mathfrak{m}_1$ with respect to B_f. We write

$$[e_i, \hat{e}_j] \equiv \beta_{ji} Y + \delta_{ij} Z \ (\mathrm{mod}\ \mathfrak{p} = \mathfrak{g}_X \cap \ker(f))$$

with $\beta_{ij} \in \mathbb{R}(1 \le i, j \le n)$, where δ_{ij} is the Kronecker delta. Then

$$\operatorname{Tr} \operatorname{ad}_{\mathfrak{h}_1/(\mathfrak{h}_1\cap\mathfrak{h}_2)} X + \operatorname{Tr} \operatorname{ad}_{\mathfrak{h}_2/(\mathfrak{h}_1\cap\mathfrak{h}_2)} X = \sum_{i=1}^{n} \beta_{ii}.$$

Therefore it is a matter to check that $\beta = \sum_{i=1}^{n} \beta_{ii} = 0$. If we denote by $f_c, c \in \mathbb{R}$, the linear form on $\mathfrak{g}$ defined by the conditions $f_c \in f + (\mathfrak{p} \oplus \mathfrak{z})^{\perp}$ and $f_c(Y) = c$, then $f_c \in f + (\mathfrak{h}_1 + \mathfrak{h}_2)^{\perp}$. From the Pukanszky condition, $\mathfrak{h}_i \in M(f_c, \mathfrak{g})(i = 1, 2)$ and the form B_{f_c} gives a pairing of $\mathfrak{m}_1$ and $\mathfrak{m}_2$. Let $\{e'_k; 1 \le k \le n\}$ be the basis of $\mathfrak{m}_1$ dual to $\{\hat{e}_j; 1 \le j \le n\}$ with respect to the form B_{f_c}. If we write $e'_k = \sum_{j=1}^{n} \lambda_{kj}(c) e_j$ with $\lambda_{kj}(c) \in \mathbb{R}(1 \le j, k \le n)$, we have

$$[e'_k, \hat{e}_i] = \sum_{j=1}^{n} \lambda_{kj}(c)[e_j, \hat{e}_i] \equiv \sum_{j=1}^{n} \lambda_{kj}(c)\beta_{ij} Y + \lambda_{ki}(c) Z \pmod{\mathfrak{p}}.$$

Hence

$$c(\sum_{j=1}^{n} \beta_{ij}\lambda_{kj}(c)) + \lambda_{ki}(c) = \delta_{ki} \ (1 \le i, k \le n). \tag{2.2.3}$$

Taking relation (2.2.2) into account, $B_{f_c}(\mathfrak{h}_1 \cap \mathfrak{h}_2^0, \sum_{k=1}^{n}[e'_k, \hat{e}_k]) = \{0\}$. We deduce from this

$$\sum_{k=1}^{n}[e'_k, \hat{e}_k] \in \mathfrak{h}_1 + \mathfrak{h}_2 + \mathbb{R}Y$$

because $\mathfrak{h}_1 \cap \mathfrak{h}_2 \subset \mathfrak{g}_0$. Hence

$$\beta = B_{f_c}(X, \sum_{k=1}^{n}[e'_k, \hat{e}_k]) = \sum_{i,j=1}^{n} \beta_{ij}\lambda_{ij}(c). \tag{2.2.4}$$

In order to have $\lambda_{ij}(c) \in \mathbb{R}(1 \le i, j \le n)$ verifying equalities (2.2.3), (2.2.4) we must have

$$\begin{vmatrix} B_c & 0 & \cdots & 0 & \mathbf{v}_1 \\ 0 & B_c & \ddots & \vdots & \mathbf{v}_2 \\ \vdots & \ddots & \ddots & 0 & \vdots \\ 0 & \cdots & 0 & B_c & \mathbf{v}_n \\ \mathbf{b}_1 & \mathbf{b}_2 & \cdots & \mathbf{b}_n & \beta \end{vmatrix} = 0, \tag{2.2.5}$$

where

$$\mathbf{b}_j = (\beta_{j1}, \beta_{j2}, \ldots, \beta_{jn}), \mathbf{v}_j = \begin{pmatrix} 0 \\ \vdots \\ 0 \\ 1 \\ 0 \\ \vdots \\ 0 \end{pmatrix} \ (1 \le j \le n),$$

and the 1 is in the jth entry, and

$$B_c = \begin{pmatrix} c\beta_{11}+1 & c\beta_{12} & \cdots & c\beta_{1n} \\ c\beta_{21} & c\beta_{22}+1 & \cdots & c\beta_{2n} \\ \vdots & \vdots & \cdots & \vdots \\ c\beta_{n1} & c\beta_{n2} & \cdots & c\beta_{nn}+1 \end{pmatrix} \in M_n(\mathbb{R}).$$

Consequently, equality (2.2.5) is an identity for $c \in \mathbb{R}$. Letting $c \to \infty$ after dividing by c each row of the matrix in (2.2.5), we obtain $\beta(\det B)^n = 0$, where $B = (\beta_{ij}) \in M_n(\mathbb{R})$.

If $\beta \neq 0$, then $\det B = 0$ and we can suppose $\beta_{j1} = 0 (1 \le j \le n)$, possibly after basis change. Then (2.2.5) reads

$$\begin{vmatrix} \tilde{B}_c^1 & 0 & \cdots & 0 & \mathbf{v}_1 \\ 0 & \tilde{B}_c^1 & \ddots & \vdots & \mathbf{v}_2 \\ \vdots & \ddots & \ddots & \vdots & \vdots \\ 0 & \cdots & 0 & \tilde{B}_c^1 & \mathbf{v}_n \\ \tilde{\mathbf{b}}_1^1 & \tilde{\mathbf{b}}_2^1 & \cdots & \tilde{\mathbf{b}}_n^1 & \beta \end{vmatrix} = 0,$$

where $\tilde{\mathbf{b}}_j^1 = (0, \beta_{j2}, \ldots, \beta_{jn}) (1 \le j \le n)$, and

$$\tilde{B}_c^1 = \begin{pmatrix} 1 & c\beta_{12} & \cdots & c\beta_{1n} \\ 0 & c\beta_{22}+1 & \cdots & c\beta_{2n} \\ \vdots & \vdots & \cdots & \vdots \\ 0 & c\beta_{n2} & \cdots & c\beta_{nn}+1 \end{pmatrix}.$$

It follows that

$$\begin{vmatrix} B_c^1 & 0 & \cdots & 0 & \mathbf{v}_1^1 \\ 0 & B_c^1 & \ddots & \vdots & \mathbf{v}_2^1 \\ \vdots & \ddots & \ddots & \vdots & \vdots \\ 0 & \cdots & 0 & B_c^1 & \mathbf{v}_n^1 \\ \mathbf{b}_1^1 & \mathbf{b}_2^1 & \cdots & \mathbf{b}_n^1 & \beta \end{vmatrix} = 0,$$

where

$$\mathbf{b}_j^1 = (\beta_{j2}, \beta_{j3}, \cdots, \beta_{jn})(1 \le j \le n), \mathbf{v}_k^1 = \begin{pmatrix} 0 \\ \vdots \\ 0 \\ 1 \\ 0 \\ \vdots \\ 0 \end{pmatrix} (2 \le k \le n),$$

where the 1 in $\mathbf{v}_k^1$ appears in the $(k-1)$th entry, and

$$B_c^1 = \begin{pmatrix} c\beta_{22}+1 & c\beta_{23} & \cdots & c\beta_{2n} \\ c\beta_{32} & c\beta_{33}+1 & \cdots & c\beta_{3n} \\ \vdots & \vdots & \cdots & \vdots \\ c\beta_{n2} & c\beta_{n3} & \cdots & c\beta_{nn}+1 \end{pmatrix} \in M_{n-1}(\mathbb{R})$$

Exactly as before, we find $\beta(\det B^1)^n = 0$ where

$$B^1 = \begin{pmatrix} \beta_{22} & \cdots & \beta_{2n} \\ \vdots & \cdots & \vdots \\ \beta_{n2} & \cdots & \beta_{nn} \end{pmatrix},$$

and we can suppose $\beta_{j2} = 0 (2 \le j \le n)$ if $\beta \neq 0$.

Iterating this process will eventually show that $\beta = 0$.

(ii) The case where $\dim \mathfrak{a} = 3$. If $\dim(\mathfrak{h}_i \cap \mathfrak{a}) = 2 (i = 1, 2)$, then $\mathfrak{h}_1, \mathfrak{h}_2 \subset \mathfrak{g}_0 \subsetneq \mathfrak{g}$ and it suffices to apply the induction hypothesis to $\mathfrak{g}_0$.

(a) Suppose $\dim(\mathfrak{h}_i \cap \mathfrak{a}) = 1$ and $\dim(\mathfrak{h}_j \cap \mathfrak{a}) = 2$, where $\{i, j\} = \{1, 2\}$. We have $\mathfrak{h}_1 \cap \mathfrak{h}_2 \subset \mathfrak{g}_0$. Let, for example, $\dim(\mathfrak{h}_1 \cap \mathfrak{a}) = 1$ and $\dim(\mathfrak{h}_2 \cap \mathfrak{a}) = 2$. From Lemma 2.2.3, there exist two elements $X_1 \in \mathfrak{h}_1 \cap \mathfrak{h}_2, X_2 \in \mathfrak{h}_1$

and a basis $\{Y_1, Y_2\}$ of $\mathfrak{a} \cap \ker(f)$ such that

$$Y_2 \in \mathfrak{h}_2, \mathfrak{h}_1 = \mathbb{R}X_1 \oplus \mathbb{R}X_2 \oplus \mathfrak{h}_1^0, [X_i, Y_j] = \delta_{ij} Z (1 \le i, j \le 2)$$

(such an element X_2 exists by Proposition 2.2.1). We then have

$$\mathfrak{h}_1 \cap \mathfrak{h}_2 = \mathbb{R}X_1 \oplus (\mathfrak{h}_1^0 \cap \mathfrak{h}_2^0), \mathfrak{h}_1 \cap \mathfrak{h}_2' = \mathfrak{h}_1^0 \cap \mathfrak{h}_2^0.$$

For $X \in \mathfrak{h}_1^0 \cap \mathfrak{h}_2^0$,

$$\mathrm{Tr}\, \mathrm{ad}_{\mathfrak{h}_i/(\mathfrak{h}_1 \cap \mathfrak{h}_2)} X = \mathrm{Tr}\, \mathrm{ad}_{\mathfrak{h}_i/(\mathfrak{h}_1 \cap \mathfrak{h}_2')} X \ (i = 1, 2)$$

and

$$\mathrm{Tr}\, \mathrm{ad}_{\mathfrak{h}_2'/(\mathfrak{h}_1 \cap \mathfrak{h}_2')} X = \mathrm{Tr}\, \mathrm{ad}_{\mathfrak{h}_2^0/(\mathfrak{h}_1 \cap \mathfrak{h}_2')} X = \mathrm{Tr}\, \mathrm{ad}_{\mathfrak{h}_2/(\mathfrak{h}_1 \cap \mathfrak{h}_2')} X,$$

which leads us directly to the case $\mathfrak{a} \subset \mathfrak{h}_2$, and we conclude

$$\mathrm{Tr}\, \mathrm{ad}_{\mathfrak{h}_1/(\mathfrak{h}_1 \cap \mathfrak{h}_2)} X + \mathrm{Tr}\, \mathrm{ad}_{\mathfrak{h}_2/(\mathfrak{h}_1 \cap \mathfrak{h}_2)} X = 0$$

for all $X \in \mathfrak{h}_1^0 \cap \mathfrak{h}_2^0$.

Finally, the relation

$$\mathrm{Tr}\, \mathrm{ad}_{\mathfrak{h}_1/(\mathfrak{h}_1 \cap \mathfrak{h}_2)} X + \mathrm{Tr}\, \mathrm{ad}_{\mathfrak{h}_2/(\mathfrak{h}_1 \cap \mathfrak{h}_2)} X = 0$$

is found using the same reasoning of case (i).

(b) Suppose $\mathfrak{h}_1 \cap \mathfrak{a} = \mathfrak{h}_2 \cap \mathfrak{a} = \mathfrak{z}$. If $\mathfrak{h}_1^0 \cap \mathfrak{h}_2^0$ is of codimension 2 in $\mathfrak{h}_1 \cap \mathfrak{h}_2$, we may prove the desired relation just as in the case (i). Suppose from now on that $\mathfrak{h}_1^0 \cap \mathfrak{h}_2^0$ is of codimension 1 in $\mathfrak{h}_1 \cap \mathfrak{h}_2$. We have $\mathfrak{h}_1 \cap \mathfrak{h}_2 \subset \mathfrak{g}_0$ from Lemma 2.2.3 and we can choose an element $X_0 \in \mathfrak{h}_1 \cap \mathfrak{h}_2$ and a basis $\{Y_1, Y_2\}$ of $\mathfrak{a} \cap \ker(f)$ in such a way that

$$\mathfrak{h}_1 \cap \mathfrak{h}_2 = \mathbb{R}X_0 \oplus (\mathfrak{h}_1^0 \cap \mathfrak{h}_2^0), [X_0, Y_i] = \delta_{1i} Z (1 \le i \le 2).$$

Then $Y_2 \in \mathfrak{h}_1 + \mathfrak{h}_2$.

(α) First suppose $Y_2 \in \mathfrak{h}_1^0 + \mathfrak{h}_2^0$. We write $Y_2 = X_2 - X_1$ with $X_i \in \mathfrak{h}_i^0 (i = 1, 2)$. We may verify without difficulty

$$\mathfrak{h}_1' \cap \mathfrak{h}_2 = (\mathfrak{h}_1^0 \cap \mathfrak{h}_2^0) \oplus \mathbb{R}X_2, \mathfrak{h}_1 \cap \mathfrak{h}_2' = (\mathfrak{h}_1^0 \cap \mathfrak{h}_2^0) \oplus \mathbb{R}X_1$$

and

$$\mathfrak{h}_1' \cap \mathfrak{h}_2' = (\mathfrak{h}_1^0 \cap \mathfrak{h}_2^0) \oplus \mathbb{R}X_1 \oplus \mathbb{R}X_2 \oplus \mathbb{R}Y_1.$$

As $\mathfrak{h}_1 \cap \mathfrak{h}_2 \subset \mathfrak{g}_0$, we see that, for all $X \in \mathfrak{h}_1^0 \cap \mathfrak{h}_2^0$,

$$\mathrm{Tr}\ \mathrm{ad}_{\mathfrak{h}_i'/(\mathfrak{h}_1' \cap \mathfrak{h}_2')} X = \mathrm{Tr}\ \mathrm{ad}_{\mathfrak{h}_i'/((\mathfrak{h}_1^0 \cap \mathfrak{h}_2^0) \oplus \mathbb{R}X_1 \oplus \mathbb{R}X_2 \oplus \mathbb{R}Y_1)} X$$

$$= \mathrm{Tr}\ \mathrm{ad}_{\mathfrak{h}_i^0/((\mathfrak{h}_1^0 \cap \mathfrak{h}_2^0) \oplus \mathbb{R}X_i)} X = \mathrm{Tr}\ \mathrm{ad}_{\mathfrak{h}_i^0/(\mathfrak{h}_1^0 \cap \mathfrak{h}_2^0)} X = \mathrm{Tr}\ \mathrm{ad}_{\mathfrak{h}_i/(\mathfrak{h}_1 \cap \mathfrak{h}_2)} X \ (i = 1, 2),$$

which gives

$$\mathrm{Tr}\ \mathrm{ad}_{\mathfrak{h}_1/(\mathfrak{h}_1 \cap \mathfrak{h}_2)} X + \mathrm{Tr}\ \mathrm{ad}_{\mathfrak{h}_2/(\mathfrak{h}_1 \cap \mathfrak{h}_2)} X = 0.$$

We then deduce the required relation for X_0 as in case (i).

(β) Suppose $Y_2 \notin \mathfrak{h}_1^0 + \mathfrak{h}_2^0$. There exist two elements $X_1 \in \mathfrak{h}_1$ and $X_2 \in \mathfrak{h}_2$ such that

$$Y_2 = a(X_2 - X_1)(0 \neq a \in \mathbb{R}), [X_i, Y_j] = \delta_{2j} Z, \mathfrak{h}_i = \mathbb{R}X_i \oplus \mathbb{R}X_0 \oplus \mathfrak{h}_i^0 (1 \leq i, j \leq 2),$$

because $\mathfrak{h}_1^0 \cap \mathfrak{h}_2^0$ is of codimension 1 in $\mathfrak{h}_1 \cap \mathfrak{h}_2$. It immediately follows that $\mathfrak{h}_1' \cap \mathfrak{h}_2 = \mathfrak{h}_1 \cap \mathfrak{h}_2' = \mathfrak{h}_1^0 \cap \mathfrak{h}_2^0$. As $\mathfrak{h}_1 \cap \mathfrak{h}_2 \subset \mathfrak{g}_0$, we have, for any $X \in \mathfrak{h}_1^0 \cap \mathfrak{h}_2^0$,

$$\mathrm{Tr}\ \mathrm{ad}_{\mathfrak{h}_i/(\mathfrak{h}_1 \cap \mathfrak{h}_2)} X = \mathrm{Tr}\ \mathrm{ad}_{\mathfrak{h}_i^0/(\mathfrak{h}_1^0 \cap \mathfrak{h}_2^0)} X \ (i = 1, 2),$$

and again

$$\mathrm{Tr}\ \mathrm{ad}_{\mathfrak{h}_1'/(\mathfrak{h}_1' \cap \mathfrak{h}_2)} X = \mathrm{Tr}\ \mathrm{ad}_{\mathfrak{h}_1^0/(\mathfrak{h}_1^0 \cap \mathfrak{h}_2^0)} X, \mathrm{Tr}\ \mathrm{ad}_{\mathfrak{h}_2/(\mathfrak{h}_1' \cap \mathfrak{h}_2)} X = \mathrm{Tr}\ \mathrm{ad}_{\mathfrak{h}_2/(\mathfrak{h}_1^0 \cap \mathfrak{h}_2^0)} X.$$

In this way we are back in the case where $\mathfrak{a} \subset \mathfrak{h}_1$. The relation for X_0 is proved as in the case (i), and the theorem's proof ends. □

2.3 Relations Between Two Polarizations

We keep the notations.

Remark 2.3.1 Without assuming the Pukanzsy condition Theorem 2.2.2 might fail, and we cannot write the integral of (2.2.1) even if $\pi_1 \simeq \pi_2$. Indeed, let $\mathfrak{g}$ be the completely solvable Lie algebra of dimension 4 with basis (T, P, Q, Z) satisfying

$$[T, P] = \frac{1}{2}P, [T, Q] = \frac{1}{2}Q, [T, Z] = [P, Q] = Z.$$

Let $f = Z^* \in \mathfrak{g}^*$. Then $\mathfrak{g}(f) = \{0\}$ and

$$\mathfrak{h}_1 = \mathbb{R}T + \mathbb{R}P + \mathbb{R}Z, \mathfrak{h}_2 = \mathbb{R}T + \mathbb{R}Q + \mathbb{R}Z$$

belong to $M(f, \mathfrak{g})$, but neither $\mathfrak{h}_1$ nor $\mathfrak{h}_2$ verifies the Pukanszky condition.

There exist two open coadjoint orbits $O_\pm$ of $G = \exp \mathfrak{g}$:

$$O_+ = \{l \in \mathfrak{g}^*; l(Z) > 0\}, \ O_- = \{l \in \mathfrak{g}^*; l(Z) < 0\}.$$

Denoting by $\rho(\Omega)$ the irreducible unitary representation of G corresponding to the orbit $\Omega \in \mathfrak{g}^*/G$, we know that $\pi_1 \simeq \pi_2 \simeq \rho(O_+) \oplus \rho(O_-)$ (cf. [151]). However, it is evident that $\mathrm{Tr}\ \mathrm{ad}_{\mathfrak{h}_1/(\mathfrak{h}_1 \cap \mathfrak{h}_2)} T = \mathrm{Tr}\ \mathrm{ad}_{\mathfrak{h}_2/(\mathfrak{h}_1 \cap \mathfrak{h}_2)} T = 1/2$.

When $G = \exp \mathfrak{g}$ is nilpotent and K_1, K_2 are analytic subgroups, the product $K_1 K_2 = \{k_1 k_2; k_1 \in K_1, k_2 \in K_2\}$ is always a closed subset of G (cf. [97]). We denote by $\mathcal{K}(f, \mathfrak{h}_1, G)$ the dense subspace of $\mathcal{H}_{\pi_1}$ consisting of continuous functions with compact support modulo H_1. It follows that the integral (2.2.1) converges for $\phi \in \mathcal{K}(f, \mathfrak{h}_1, G)$. In fact, Lion proved that (2.2.1) gives an intertwining operator between π_1 and π_2. On the other hand, when we pass to the exponential case $K_1 K_2$ might not be closed, thus making the convergence of integral (2.2.1) a serious issue.

Example 2.3.2 Take

$$G_2 = \exp \mathfrak{g}_2 = \left\{ \begin{pmatrix} a & b \\ 0 & 1 \end{pmatrix} \in M_2(\mathbb{R}); a > 0 \right\} (ax + b \text{ group}).$$

The elements

$$e_1 = \begin{pmatrix} 1 & 0 \\ 0 & 0 \end{pmatrix}, e_2 = \begin{pmatrix} 0 & 1 \\ 0 & 0 \end{pmatrix}$$

form a basis of $\mathfrak{g}_2$ such that $[e_1, e_2] = e_2$. Let $f = e_2^* \in \mathfrak{g}_2^*$, $\mathfrak{k}_1 = \mathbb{R}e_1$, $\mathfrak{k}_2 = \mathbb{R}(e_1 + e_2)$ and $K_i = \exp \mathfrak{k}_i (i = 1, 2)$. Then $\mathfrak{k}_i \in M(f, \mathfrak{g}_2)$ but $\mathfrak{k}_i \notin I(f, \mathfrak{g}_2) (i = 1, 2)$. We immediately see

$$K_2 K_1 = \left\{ g = \begin{pmatrix} a & b \\ 0 & 1 \end{pmatrix} \in G_2; b > -1 \right\}$$

and $\mathfrak{k}_2 = g_0 \cdot \mathfrak{k}_1$ with $g_0 = \begin{pmatrix} 1 & -1 \\ 0 & 1 \end{pmatrix} \in G_2$.

Let $G = \exp \mathfrak{g}$ still be an exponential solvable Lie group with Lie algebra $\mathfrak{g}$, $f \in \mathfrak{g}^*$, $\mathfrak{h}_i \in I(f, \mathfrak{g})$ and $H_i = \exp(\mathfrak{h}_i) (i = 1, 2)$. If $H_2 H_1$ is closed in G, the canonical mapping $p : H_2/(H_1 \cap H_2) \to G/H_1$ becomes a homeomorphism onto its closed image $H_2 H_1/H_1$. Then integral (2.2.1) converges for $\phi \in \mathcal{K}(f, \mathfrak{h}_1, G)$.

Example 2.3.3 In case $\dim \mathfrak{g} \leq 5$ one can show that H_2H_1 is always closed, and hence (2.2.1) converges for $\phi \in \mathscr{K}(f, \mathfrak{h}_1, G)$.

We do not know whether H_2H_1 is closed or not in general. Nevertheless, we shall see a few properties of H_2H_1 in particular cases. It is clear that H_2H_1 is a union of countable closed subsets of G. Moreover,

$$H_2H_1 = \{g \in G; g \cdot (f + (\mathfrak{h}_1)^\perp) \cap (f + (\mathfrak{h}_2)^\perp) \neq \emptyset\}$$

from the Pukanszky condition.

Proposition 2.3.4 *The subset H_2H_1 is locally closed in G and H_2H_1/H_1 is homeomorphic to the homogeneous space $H_2/(H_1 \cap H_2)$.*

Proof It suffices to prove that H_2H_1 is locally closed in G. Let

$$F = \{g \in G; \dim(g \cdot (\mathfrak{h}_1)^\perp + (\mathfrak{h}_2)^\perp) \leq \dim((\mathfrak{h}_1)^\perp + (\mathfrak{h}_2)^\perp)\}$$

and

$$O = \{g \in G; \dim(g \cdot (\mathfrak{h}_1)^\perp + (\mathfrak{h}_2)^\perp) \geq \dim((\mathfrak{h}_1)^\perp + (\mathfrak{h}_2)^\perp)\}.$$

Then F is closed and O is open in G. Furthermore, $H_2H_1 \subset F \cap O$ and H_2H_1 is closed in $F \cap O$ because

$$H_2H_1 = \{g \in G; g{\cdot}f - f \in g \cdot (\mathfrak{h}_1)^\perp + (\mathfrak{h}_2)^\perp\}.$$

Hence H_2H_1 is locally closed in G. □

Proposition 2.3.5 *Assume $\mathfrak{g} = \mathfrak{h}_1 + \mathfrak{h}_2$. Then H_2H_1 is open and everywhere dense in G.*

Proof It is evident that H_2H_1 is open in G. Put

$$G_0 = \{g \in G; \dim(g \cdot (\mathfrak{h}_1)^\perp + (\mathfrak{h}_2)^\perp) \leq \dim((\mathfrak{h}_1)^\perp + (\mathfrak{h}_2)^\perp)\}.$$

Then G_0 can be expressed as $G_0 = \{g \in G; F_1(g) = 0\}$ for a certain real analytic function F_1. Since G_0 contains H_2H_1, which is open, $F_1 \equiv 0$ and $G_0 = G$. Hence if we set

$$O = \{g \in G; \dim(g \cdot (\mathfrak{h}_1)^\perp + (\mathfrak{h}_2)^\perp) = \dim((\mathfrak{h}_1)^\perp + (\mathfrak{h}_2)^\perp)\},$$

O reads $O = \{g \in G; F_2(g) \neq 0\}$ for some real analytic function F_2. Since $O \supset H_2H_1$,

$$H_2H_1 = \{g \in O; g{\cdot}f - f \in g \cdot (\mathfrak{h}_1)^\perp + (\mathfrak{h}_2)^\perp\},$$

which is of the form $O \cap \{g \in G; F_3(g) = 0\}$ for some real analytic function F_3. Because H_2H_1 is open, $F_3 \equiv 0$ and $H_2H_1 = O$, which is evidently everywhere dense in G. □

Let $\mathfrak{k}$ be a subalgebra of $\mathfrak{g}$. A ordered set $(X_1, X_2, \ldots, X_d)$ of elements of $\mathfrak{g}$ is called a coexponential basis of $\mathfrak{k}$ in $\mathfrak{g}$ if the mapping

$$((x_1, x_2, \ldots, x_d), Y) \mapsto \Big(\prod_{j=1}^{d} \exp(x_j X_j)\Big) \exp Y$$

is a diffeomorphism from $\mathbb{R}^d \times \mathfrak{k}$ onto G.

Proposition 2.3.6 *Suppose that $\mathfrak{g}$ is completely solvable and $\mathfrak{g} = \mathfrak{h}_1 + \mathfrak{h}_2$. Then $G = H_2H_1$ and the integral in (2.2.1) converges for any $\phi \in \mathscr{K}(f, \mathfrak{h}_1, G)$.*

Proof We find an increasing sequence $(\mathfrak{g}_i)_{0 \le i \le r}$ of subalgebras of $\mathfrak{g}$ such that $\mathfrak{g}_0 = \mathfrak{h}_1$, $\mathfrak{g}_r = \mathfrak{g}$ and $\dim(\mathfrak{g}_i/\mathfrak{g}_{i-1}) = 1 (1 \le i \le r)$. Put $G_i = \exp(\mathfrak{g}_i)(0 \le i \le r)$. Supposing $G_{i-1} \subset H_2H_1$, we shall show $G_i \subset H_2H_1$.

Let $\mathfrak{h}_2^i = \mathfrak{h}_2 \cap \mathfrak{g}_i$ and $H_2^i = \exp(\mathfrak{h}_2^i)$. As $H_2^iG_{i-1} = (H_2H_1) \cap G_i$ is locally closed in G, the quotient space $H_2^iG_{i-1}/G_{i-1}$ is homeomorphic to the connected homogeneous space $H_2^i/(G_{i-1} \cap H_2^i)$. Let X_i be a coexponential basis of $\mathfrak{g}_{i-1}$ in $\mathfrak{g}_i$. Taking the proof of Proposition 2.3.5 into account, there exists on $\mathbb{R}$ a real analytic function $F \not\equiv 0$ such that $H_2^iG_{i-1} = \{\exp(tX_i); F(t) \neq 0\}G_{i-1}$. As the set $\{t \in \mathbb{R}; F(t) \neq 0\}$ must be connected, $F(t) \neq 0$ for any $t \in \mathbb{R}$ and the claim is proved. □

Corollary 2.3.7 *In the set-up of Proposition 2.3.6, assume $\mathfrak{h}_1 \neq \mathfrak{h}_2$. Then $\mathfrak{h}_1$ and $\mathfrak{h}_2$ are never conjugate under inner automorphisms.*

Proposition 2.3.8 *Let $\mathfrak{h} \in I(f, \mathfrak{g})$. We suppose $\mathfrak{g}$ non-Abelian and denote by $\mathfrak{z}$ the center of $\mathfrak{g}$. Then $\mathfrak{z} \subsetneq g \cdot \mathfrak{h} \cap \mathfrak{h}$ for any $g \in G$.*

Proof If we had $\mathfrak{z} = g \cdot \mathfrak{h} \cap \mathfrak{h} (g \in G)$, we would have $g \cdot f - f \in \mathfrak{z}^{\perp} = g \cdot \mathfrak{h}^{\perp} + \mathfrak{h}^{\perp}$, in other words $g \cdot (f + \mathfrak{h}^{\perp}) \cap (f + \mathfrak{h}^{\perp}) \neq \emptyset$. Then $g \in H = \exp \mathfrak{h}$ by the Pukanszky condition, which is a contradiction. □

Corollary 2.3.9 *Let $\mathfrak{h}_i \in I(f, \mathfrak{g})(i = 1, 2)$. Suppose $\mathfrak{g}$ non-Abelian and denote by $\mathfrak{z}$ the center of $\mathfrak{g}$. If $\mathfrak{z} = \mathfrak{h}_1 \cap \mathfrak{h}_2$, then $\mathfrak{h}_1$ and $\mathfrak{h}_2$ are never conjugate under inner automorphisms.*

As Example 2.3.2 shows, these properties are no longer valid without the Pukanszky condition.

Let $\mathfrak{a}$ be an ideal of $\mathfrak{g}$ and $f' = f|_{\mathfrak{a}} \in \mathfrak{a}^*$. A polarization $\mathfrak{h} \in M(f, \mathfrak{g})$ is said to be admissible for $\mathfrak{a}$ if $\mathfrak{h} \cap \mathfrak{a} \in M(f', \mathfrak{a})$.

Proposition 2.3.10 (Lion) *Suppose $\mathfrak{h}_1, \mathfrak{h}_2 \in M(f, \mathfrak{g})$ are admissible for a nilpotent ideal $\mathfrak{n}$ of $\mathfrak{g}$ containing $[\mathfrak{g}, \mathfrak{g}]$ (Such polarizations belong to $I(f, \mathfrak{g})$). Then H_2H_1 is closed in G.*

Proof If we put $\mathfrak{h}_i' = \mathfrak{h}_i \cap \mathfrak{n}$ and $\mathfrak{h}_i'' = \mathfrak{h}_i \cap \mathfrak{n}^f$, we have $\mathfrak{h}_i = \mathfrak{h}_i' + \mathfrak{h}_i''$, $\mathfrak{h}_i' \in M(f', \mathfrak{n})$ and $\mathfrak{h}_i'' \in M(f'', \mathfrak{n}^f)(i = 1, 2)$ with $f' = f|_{\mathfrak{n}} \in \mathfrak{n}^*$ and $f'' = f|_{\mathfrak{n}^f} \in (\mathfrak{n}^f)^*$ (cf. [3]). We set $N = \exp \mathfrak{n}$ and $H_i' = \exp(\mathfrak{h}_i')(i = 1, 2)$. Considering the fact that

$$\mathfrak{n}^f \cap \mathfrak{n} = \mathfrak{n}^{f'} = \{X \in \mathfrak{n};\ B_{f'}(X, \mathfrak{n}) = \{0\}\} \subset \mathfrak{h}_1' \cap \mathfrak{h}_1'' \cap \mathfrak{h}_2' \cap \mathfrak{h}_2'',$$

we denote by $\mathfrak{p}$ (resp. $\mathfrak{p}_1, \mathfrak{p}_2$) a complementary subspace of $\mathfrak{n}^{f'}$ (resp. $\mathfrak{h}_1'' \cap \mathfrak{h}_2''$) in $\mathfrak{h}_1'' \cap \mathfrak{h}_2''$ (resp. $\mathfrak{h}_1'', \mathfrak{h}_2''$). Finally, we take a complementary subspace $\mathfrak{q}$ of $\mathfrak{h}_1'' + \mathfrak{h}_2''$ in $\mathfrak{n}^f$ in such a way that

$$\mathfrak{h}_i'' = \mathfrak{n}^{f'} \oplus \mathfrak{p} \oplus \mathfrak{p}_i \ (i = 1, 2)$$

and

$$\mathfrak{n}^f = \mathfrak{n}^{f'} \oplus \mathfrak{p} \oplus \mathfrak{p}_1 \oplus \mathfrak{p}_2 \oplus \mathfrak{q}.$$

We also have

$$\mathfrak{n} + \mathfrak{n}^f = \mathfrak{n} \oplus \mathfrak{p} \oplus \mathfrak{p}_1 \oplus \mathfrak{p}_2 \oplus \mathfrak{q}.$$

Choose bases $(P_1, \ldots, P_l)$ of $\mathfrak{p}$, $(P_1^i, \ldots, P_k^i)$ of $\mathfrak{p}_i (i = 1, 2)$ and $(Q_1, \ldots, Q_m)$ of $\mathfrak{q}$. Since $\mathfrak{n} \supset [\mathfrak{g}, \mathfrak{g}]$, $(P_1, \ldots, P_l)$, say, is a coexponential basis for $\mathfrak{n}$ in $\mathfrak{n} + \mathfrak{p}$. The mapping

$$\begin{aligned}
&\theta : (u_1, \ldots, u_k, v_1, \ldots, v_l, n, x_1, \ldots, x_k, y_1, \ldots, y_m) \\
&\to \exp(u_1 P_1^1) \cdots \exp(u_k P_k^1) \exp(v_1 P_1) \cdots \exp(v_l P_l) n \exp(x_1 P_1^2) \cdots \exp(x_k P_k^2) \\
&\quad \times \exp(y_1 Q_1) \cdots \exp(y_m Q_m)
\end{aligned}$$

is a diffeomorphism from $\mathbb{R}^k \times \mathbb{R}^l \times N \times \mathbb{R}^k \times \mathbb{R}^m$ onto the analytic subgroup $G_0 = \exp(\mathfrak{n} + \mathfrak{n}^f)$. As $\exp X \cdot H_i' = H_i' \cdot \exp X (i = 1, 2)$ for any $X \in \mathfrak{p}$, we have

$$H_2 H_1 = \theta(\mathbb{R}^k \times \mathbb{R}^l \times H_2' H_1' \times \mathbb{R}^k \times \{0\}).$$

Since $H_2' H_1'$ is closed in N, it follows that $H_2 H_1$ is closed in G. □

We consider three Lagrangian subspaces $W_j (1 \leq j \leq 3)$ of $\mathfrak{g}$ for the bilinear form B_f and define, following Kashiwara, a quadratic form Q on $W_1 \oplus W_2 \oplus W_3$ by the formula

$$Q(X_1, X_2, X_3) = f([X_1, X_2]) + f([X_2, X_3]) + f([X_3, X_1]).$$

The index of Q is called the Maslov index of the spaces $W_j (1 \leq j \leq 3)$ and is denoted by $\tau(W_1, W_2, W_3)$. Recall the main properties of this index.

Lemma 2.3.11 (cf. [98]) *We agree to write* τ_{ijk} *instead of* $\tau(W_i, W_j, W_k)$. *Then*

(a) $\tau_{123} = -\tau_{213} = -\tau_{132}$.
(b) $\tau_{234} - \tau_{134} + \tau_{124} - \tau_{123} = 0$.
(c) *If* $\mathfrak{p}$ *is an isotropic subspace of* $\mathfrak{g}$ *(with respect to* B_f*) containing* $\mathfrak{g}(f)$ *and* W *is a Lagrangian subspace of* $\mathfrak{g}$, *then* $W^{\mathfrak{p}} = (W \cap \mathfrak{p}^{\perp}) + \mathfrak{p}$ *is Lagrangian, where* $\mathfrak{p}^{\perp}$ *denotes the* B_f*-orthogonal of* $\mathfrak{p}$. *Moreover, if* $\mathfrak{p}$ *is contained in* $W_1 \cap W_2 + W_2 \cap W_3 + W_3 \cap W_1$, *we have* $\tau_{123} = \tau(W_1^{\mathfrak{p}}, W_2^{\mathfrak{p}}, W_3^{\mathfrak{p}})$.

Let us recap our set-up and briefly review some facts that will be needed in the sequel. Let $G = \exp \mathfrak{g}$ be an exponential solvable Lie group with Lie algebra $\mathfrak{g}$, $f \in \mathfrak{g}^*$, $\mathfrak{h}_i \in I(f, \mathfrak{g})$, $H_i = \exp(\mathfrak{h}_i)$, $\chi_f : H_i \to \mathbb{T}$ given by $\chi_f(\exp X) = e^{if(X)}(X \in \mathfrak{h}_i)$ and $\pi_i = \mathrm{ind}_{H_i}^{G} \chi_f (i = 1, 2)$. Then we have a natural candidate (2.2.1) of an intertwining operator from π_1 to π_2. The formal operator $I_{\mathfrak{h}_2\mathfrak{h}_1}$ will be called $I^{G}_{\mathfrak{h}_2\mathfrak{h}_1}$ if necessary. When $I_{\mathfrak{h}_2\mathfrak{h}_1}$ defines a true intertwining operator, we write it as $T_{\mathfrak{h}_2\mathfrak{h}_1}$.

Now, we know from the Pukanszky condition that

$$H_2H_1 = \{g \in G; g \cdot (f + \mathfrak{h}_1^{\perp}) \cap (f + \mathfrak{h}_2^{\perp}) \neq \emptyset\}.$$

Let $\{l_j; 1 \leq j \leq d\}$ be a basis of $(\mathfrak{h}_1)^{\perp} \subset \mathfrak{g}^*$ and $\{X_i; 1 \leq i \leq m\}$ a basis of $\mathfrak{h}_2$. The condition $g \cdot (f + \mathfrak{h}_1^{\perp}) \cap (f + \mathfrak{h}_2^{\perp}) \neq \emptyset$ says that the system of linear equations

$$g \cdot (f + \sum_{j=1}^{d} x_j l_j)(X_i) = f(X_i) \; (1 \leq i \leq m) \tag{2.3.1}$$

has a solution. Let us introduce the coefficient matrix $A(g) = (a_{ij}(g))$, where $a_{ij}(g) = g{\cdot}l_j(X_i)(1 \leq i \leq m, 1 \leq j \leq d)$, and the matrix $B(g) = (A(g)b(g))$ obtained from $A(g)$ by appending the row

$$b(g) = \begin{pmatrix} b_1(g) \\ \vdots \\ b_m(g) \end{pmatrix}, \quad b_i(g) = (f - g{\cdot}f)(X_i).$$

We point out the following facts. First, the functions $a_{ij}(g)$, $b_i(g)$ are analytic in the variable g. The rank of $A(g)$ depends only on the double coset H_2gH_1. The relation $g \in H_2H_1$ is equivalent to $\mathrm{rank}A(g) = \mathrm{rank}B(g)$, or again to

$$\mathrm{rank}A(g) = \mathrm{rank}B(g) = \mathrm{rank}A(e) = \dim(\mathfrak{h}_2/(\mathfrak{h}_1 \cap \mathfrak{h}_2)),$$

where e denotes the neutral element of G.

Take an open neighbourhood V of e in G such that $\mathrm{rank}A(g) \geq \mathrm{rank}A(e)$ on V. Then

$$H_2H_1 \cap VH_1 = \{g \in VH_1;\ \mathrm{rank}B(g) \leq \mathrm{rank}A(e)\}$$

is closed in VH_1. Proposition 2.3.4 implies H_2H_1 is locally closed in G.

Let us examine in detail the set-up of Proposition 2.2.1. Let $\mathfrak{z}$ be the center of $\mathfrak{g}$, $\mathfrak{a}$ a minimal non-central ideal of $\mathfrak{g}$ and $\mathfrak{h} \in I(f, \mathfrak{g})$ such that $\mathfrak{a} \not\subset \mathfrak{h}$. We know $\mathfrak{a} \cap \mathfrak{z} \neq \{0\}$ (cf. [24, Chap. VI]). We put

$$\mathfrak{h}' = (\mathfrak{h} \cap \mathfrak{a}^f) + \mathfrak{a},\ \ \mathfrak{k} = \mathfrak{h} + \mathfrak{a} = \mathfrak{h} + \mathfrak{h}',\ \ \tilde{f} = f|_{\mathfrak{k}} \in \mathfrak{k}^*.$$

Then $\mathfrak{h}' \in I(f, \mathfrak{g})$. Put $H = \exp \mathfrak{h}$, $H' = \exp(\mathfrak{h}')$ and $K = \exp \mathfrak{k}$. The representations $\sigma = \mathrm{ind}_H^K \chi_f$, $\sigma' = \mathrm{ind}_{H'}^K \chi_f$ of K are evidently irreducible. We consider the spaces $\mathscr{H}_\sigma^\infty$, $\mathscr{H}_{\sigma'}^\infty$ of C^∞-vectors of σ, σ'. We can choose (cf. [24, Chap. VI]) Y (resp. Y_1, Y_2) in $\mathfrak{a}\backslash(\mathfrak{a}\cap\mathfrak{h})$ such that $Y + (\mathfrak{a}\cap\mathfrak{h})$ (resp. $\{Y_j + (\mathfrak{a}\cap\mathfrak{h});\ 1 \leq j \leq 2\}$) forms a basis of $\mathfrak{a}/(\mathfrak{a}\cap\mathfrak{h})$ and X (resp. X_1, X_2) in $\mathfrak{h} \setminus (\mathfrak{a}^f \cap \mathfrak{h})$ in such a manner that $[X, \mathfrak{a}] \subset \mathfrak{z}$ (resp. $[X_j, \mathfrak{a}] \subset \mathfrak{z}$, $1 \leq j \leq 2$). The spaces $\mathscr{H}_\sigma, \mathscr{H}_{\sigma'}$ are identified with $L^2(\mathbb{R})$ (resp. $L^2(\mathbb{R}^2)$) by the maps

$$\varphi \mapsto \tilde{\varphi}, \tilde{\varphi}(y) = \varphi(\exp(yY))(resp.\ \tilde{\varphi}(y_1, y_2) = \varphi(\exp(y_1Y_1 + y_2Y_2)))$$

and

$$\psi \mapsto \tilde{\psi}, \tilde{\psi}(x) = \psi(\exp(xX))(resp.\ \tilde{\psi}(x_1, x_2) = \psi(\exp(x_1X_1 + x_2X_2))).$$

Then the situation is identical to that of the nilpotent case, more precisely the case of the Heisenberg group of dimension 3 (resp. 5) and operating X, Y (resp. X_j, Y_j, $1 \leq j \leq 2$), we see that $\mathscr{H}_\sigma^\infty$ (resp. $\mathscr{H}_{\sigma'}^\infty$) is identified with the Schwartz space $\mathscr{S}(\mathbb{R})$ (resp. $\mathscr{S}(\mathbb{R}^2)$).

According to a result of Poulsen [140] on the space of C^∞-vectors of an induced representation, the rapid decay (with respect to the coordinates in question) of C^∞-vectors of representations

$$\pi = \rho(f, \mathfrak{h}, G) = \mathrm{ind}_K^G \sigma,\ \pi' = \rho(f, \mathfrak{h}', G) = \mathrm{ind}_K^G \sigma'.$$

Hence our intertwining operators

$$(T_{\mathfrak{h}'\mathfrak{h}}\varphi)(g) = (I_{\mathfrak{h}'\mathfrak{h}}\varphi)(g) = \oint_{H'/(H'\cap H)} \varphi(gh)\chi_f(h)\Delta_{H',G}^{-1/2}(h)d\nu(h)\ (g \in G)$$

and

$$(T_{\mathfrak{h}\mathfrak{h}'}\psi)(g) = (I_{\mathfrak{h}\mathfrak{h}'}\psi)(g) = \oint_{H/(H\cap H')} \psi(gh)\chi_f(h)\Delta_{H,G}^{-1/2}(h)d\nu(h)\ (g \in G)$$

are nothing but ordinary partial Fourier transforms, and therefore they converge on $\mathscr{H}_\sigma^\infty$ and $\mathscr{H}_{\sigma'}^\infty$. Finally, the Fourier inversion formula says $T_{\mathfrak{h}'\mathfrak{h}} \circ T_{\mathfrak{h}\mathfrak{h}'} = c\mathbb{I}$ and $T_{\mathfrak{h}\mathfrak{h}'} \circ T_{\mathfrak{h}'\mathfrak{h}} = c\mathbb{I}$ for some positive real number c. To sum up:

Lemma 2.3.12 *Let $\mathfrak{a}$ be a minimal non-central ideal of $\mathfrak{g}$. Given an element $\mathfrak{h}$ of $I(f, \mathfrak{g})$ such that $\mathfrak{a} \not\subset \mathfrak{h}$, we transform $\mathfrak{h}$ into the element $\mathfrak{h}' = (\mathfrak{h} \cap \mathfrak{a}^f) + \mathfrak{a}$ of $I(f, \mathfrak{g})$. Then the operators $T_{\mathfrak{h}'\mathfrak{h}}, T_{\mathfrak{h}\mathfrak{h}'}$ are ordinary partial Fourier transforms, they converge on C^∞-vectors and give intertwining operators between $\rho(f, \mathfrak{h}, G)$ and $\rho(f, \mathfrak{h}', G)$. Furthermore, we have*

$$T_{\mathfrak{h}'\mathfrak{h}} \circ T_{\mathfrak{h}\mathfrak{h}'} = \mathbb{I}, \quad T_{\mathfrak{h}\mathfrak{h}'} \circ T_{\mathfrak{h}'\mathfrak{h}} = \mathbb{I}$$

up to a normalization of the measures $d\nu_{\cdot,\cdot}$.

Suppose that $\mathfrak{h}_1, \mathfrak{h}_2$ belong to $I(f, \mathfrak{g})$ and are contained in a subalgebra $\mathfrak{g}_0$ of $\mathfrak{g}$. We put $f_0 = f|_{\mathfrak{g}_0} \in (\mathfrak{g}_0)^*$ and $G_0 = \exp(\mathfrak{g}_0)$.

Lemma 2.3.13

(i) *If $T_{\mathfrak{h}_2\mathfrak{h}_1}^{G_0}$ is an intertwining operator from $\rho(f_0, \mathfrak{h}_1, G_0)$ to $\rho(f_0, \mathfrak{h}_2, G_0)$, $T_{\mathfrak{h}_2\mathfrak{h}_1}^{G}$ is an intertwining operator from $\rho(f, \mathfrak{h}_1, G)$ to $\rho(f, \mathfrak{h}_2, G)$.*

(ii) *Suppose that $T_{\mathfrak{h}_2\mathfrak{h}_1}^{G_0}$ and $T_{\mathfrak{h}_1\mathfrak{h}_2}^{G_0}$ are intertwining operators between $\rho_1^{G^0} = \rho(f_0, \mathfrak{h}_1, G_0)$ and $\rho_2^{G^0} = \rho(f_0, \mathfrak{h}_2, G_0)$. If $T_{\mathfrak{h}_1\mathfrak{h}_2}^{G_0} \circ T_{\mathfrak{h}_2\mathfrak{h}_1}^{G_0} = c_0\mathbb{I}$ with $c_0 \in \mathbb{R}_+ = \{r \in \mathbb{R}; r > 0\}$, then $T_{\mathfrak{h}_1\mathfrak{h}_2}^{G} \circ T_{\mathfrak{h}_2\mathfrak{h}_1}^{G} = c\mathbb{I}$ with $c \in \mathbb{R}_+$.*

Proof Put $\rho_i^G = \rho(f, \mathfrak{h}_i, G)$ and call $\mathscr{H}_i^G = \mathscr{H}(f, \mathfrak{h}_i, G)$ the space of the ρ_i^G, and $\mathscr{H}_i^{G_0} = \mathscr{H}(f_0, \mathfrak{h}_i, G_0)$ $(i = 1, 2)$. Let $\pi_i = \mathrm{ind}_{G_0}^G \rho_i^{G_0}$ and set $\mathscr{L}_i$ to be the representation space of $\pi_i (i = 1, 2)$. For arbitrary $\phi \in \mathscr{H}_i^G$, let $(R_i\phi)(g)(g \in G)$ be the function on G_0 defined by

$$(R_i\phi)(g) : G_0 \ni g_0 \mapsto \phi(gg_0)\Delta_{G_0,G}^{-1/2}(g_0).$$

Then $(R_i\phi)(g) \in \mathscr{H}_i^{G_0}$ for almost all $g \in G$ and the mapping $R_i : \mathscr{H}_i^G \to \mathscr{L}_i$ is an isometry that permits us to identify ρ_i^G with $\pi_i (i = 1, 2)$.

For $\theta \in \mathscr{L}_1$, we define the function $P_{21}\theta$ on G with values in $\mathscr{H}_2^{G_0}$ by

$$(P_{21}\theta)(g) = T_{\mathfrak{h}_2\mathfrak{h}_1}^{G_0}(\theta(g))(g \in G).$$

We then see that $P_{21} : \mathscr{L}_1 \to \mathscr{L}_2$ is an intertwining operator between π_1 and π_2. We finally find an intertwining operator $\tilde{T}_{\mathfrak{h}_2\mathfrak{h}_1} : \mathscr{H}_1^G \to \mathscr{H}_2^G$ in such a way that the following diagram is commutative:

$$\begin{array}{ccc} \mathcal{H}_1^G & \xrightarrow{R_1} & \mathcal{L}_1 \\ \tilde{T}_{\mathfrak{h}_2\mathfrak{h}_1} \downarrow & & \downarrow P_{21} \\ \mathcal{H}_2^G & \xrightarrow[R_2]{} & \mathcal{L}_2 \end{array} \tag{2.3.2}$$

For $\phi \in \mathcal{K}(f, \mathfrak{h}_1, G)$,

$$(R_2 \circ \tilde{T}_{\mathfrak{h}_2\mathfrak{h}_1}\phi)(g) : G_0 \ni g_0 \mapsto (\tilde{T}_{\mathfrak{h}_2\mathfrak{h}_1}\phi)(gg_0)\Delta_{G_0,G}^{-1/2}(g_0).$$

On the other hand,

$$\begin{aligned}
&(P_{21} \circ R_1\phi)(g) : G_0 \ni g_0 \mapsto T_{\mathfrak{h}_2\mathfrak{h}_1}^{G_0}(R_1\phi(g))(g_0) \\
&= \oint_{H_2/(H_1\cap H_2)} (R_1\phi)(g)(g_0h)\chi_f(h)\Delta_{H_2,G_0}^{-1/2}(h)d\nu_{H_2,H_1\cap H_2}(h) \\
&= \oint_{H_2/(H_1\cap H_2)} \phi(gg_0h)\Delta_{G_0,G}^{-1/2}(g_0h)\chi_f(h)\Delta_{H_2,G_0}^{-1/2}(h)d\nu_{H_2,H_1\cap H_2}(h) \\
&= \Delta_{G_0,G}^{-1/2}(g_0)\oint_{H_2/(H_1\cap H_2)} \phi(gg_0h)\chi_f(h)\Delta_{H_2,G}^{-1/2}(h)d\nu_{H_2,H_1\cap H_2}(h).
\end{aligned}$$

Modifying if necessary $\tilde{T}_{\mathfrak{h}_2\mathfrak{h}_1}\phi$ on a negligible set, we find

$$(\tilde{T}_{\mathfrak{h}_2\mathfrak{h}_1}\phi)(g) = \oint_{H_2/(H_1\cap H_2)} \phi(gh)\chi_f(h)\Delta_{H_2,G}^{-1/2}(h)d\nu_{H_2,H_1\cap H_2}(h)$$

for all $g \in G$. This proves the claim. □

Poulsen's above result, together with the last lemma's proof, gives:

Lemma 2.3.14

(i) If the integral form $I_{\mathfrak{h}_2\mathfrak{h}_1}^{G_0}$ converges on C^∞-vectors, the same happens for $T_{\mathfrak{h}_2\mathfrak{h}_1}^{G}$.
(ii) If $I_{\mathfrak{h}_2\mathfrak{h}_1}^{G_0}$ extends to an intertwining operator, so does $T_{\mathfrak{h}_2\mathfrak{h}_1}^{G}$.
(iii) If the intertwining operators $T_{\mathfrak{h}_2\mathfrak{h}_1}^{G_0}, T_{\mathfrak{h}_1\mathfrak{h}_2}^{G_0}$ verify $T_{\mathfrak{h}_2\mathfrak{h}_1}^{G_0} \circ T_{\mathfrak{h}_1\mathfrak{h}_2}^{G_0} = \mathbb{I}$, so do $T_{\mathfrak{h}_2\mathfrak{h}_1}^{G}, T_{\mathfrak{h}_1\mathfrak{h}_2}^{G}$.

2.4 Vergne Polarizations

We retain the notations $G = \exp \mathfrak{g}$, $f \in \mathfrak{g}^*$, $\mathfrak{h}_j \in I(f, \mathfrak{g})$ and $H_j = \exp(\mathfrak{h}_j)$ for $j = 1, 2$.

Lemma 2.4.1 *If $\mathfrak{h}_1$ is a Vergne polarization, then there exists a coexponential basis of $\mathfrak{h}_1 \cap \mathfrak{h}_2$ in $\mathfrak{h}_2$ which is a part of a coexponential basis of $\mathfrak{h}_1$ in $\mathfrak{g}$. Likewise, there exists a coexponential basis of $\mathfrak{h}_1 \cap \mathfrak{h}_2$ in $\mathfrak{h}_1$ which is a part of a coexponential basis of $\mathfrak{h}_2$ in $\mathfrak{g}$. In particular, H_2H_1 (hence H_1H_2, too) is closed in G.*

Proof We proceed by induction on $\dim G$. If there exists a non-trivial ideal $\mathfrak{a}$ of $\mathfrak{g}$ on which f vanishes, everything passes to the quotient G/A with $A = \exp \mathfrak{a}$. This case will be excluded in what follows. If there exists a minimal ideal $\mathfrak{a}$ which is not central, we know (cf. [24, Chap. VI]) that $\mathfrak{a}$ is contained in any element of $I(f, \mathfrak{g})$. Hence for $j = 1, 2$, $\mathfrak{h}_j \subset \mathfrak{a}^f \neq \mathfrak{g}$. Suppose that the Vergne polarization $\mathfrak{h}_1$ is constructed starting from a good sequence $\mathfrak{s} = (\mathfrak{a}_j)_{0 \leq j \leq n}$ of subalgebras, namely:

$$\mathfrak{a}_{j-1} \subset \mathfrak{a}_j, \dim(\mathfrak{a}_j) = j, \mathfrak{h}_1 = \sum_{j=1}^{n} \mathfrak{a}_j(f_j), f_j = f|_{\mathfrak{a}_j} (1 \leq j \leq n).$$

Put $\mathfrak{b}_j = \mathfrak{a}_j \cap \mathfrak{a}^f$ and $f'_j = f|_{\mathfrak{b}_j} (0 \leq j \leq n)$. There exists an index $j_0 (1 \leq j_0 \leq n)$ such that $\mathfrak{b}_{j_0} = \mathfrak{b}_{j_0-1}$ and it is clear that $\mathfrak{s}' = (\mathfrak{b}_j)_{0 \leq j \leq n, j \neq j_0}$ is a good sequence of subalgebras of $\mathfrak{a}^f$. Since $\mathfrak{h}_1 = \sum_{j=1}^{n} \mathfrak{a}_j(f_j) \subset \mathfrak{a}^f$, we see $\mathfrak{a}_j(f_j) \subset \mathfrak{b}_j(f'_j)$ for every j and hence

$$\mathfrak{h}_1 \subset \sum_{j=1}^{n} \mathfrak{b}_j(f'_j). \tag{2.4.1}$$

Well, $\mathfrak{h}_1$ being a Lagrangian subspace, we necessarily have equality in (2.4.1). Now it suffices to apply the induction hypothesis to $A^f = \exp(\mathfrak{a}^f)$.

Suppose now that there exists no minimal ideal of $\mathfrak{g}$ which is not central. Take the good sequence $\mathfrak{s}$ which defines $\mathfrak{h}_1$. With our hypothesis, we necessarily have $\mathfrak{a}_1 = \mathfrak{z}$ and $f_1 \neq 0$. If $\mathfrak{a}_2$ is an ideal of $\mathfrak{g}$, the above reasoning shows that $\mathfrak{h}_1$ is a Vergne polarization of the subalgebra $(\mathfrak{a}_2)^f$. If $\mathfrak{h}_2 \subset (\mathfrak{a}_2)^f$, it suffices to apply the induction hypothesis. If not, we modify $\mathfrak{h}_2$ to $\mathfrak{h}'_2 = (\mathfrak{h}_2 \cap (\mathfrak{a}_2)^f) + \mathfrak{a}_2$ and H_2 to $H'_2 = \exp(\mathfrak{h}'_2)$. From the Pukanszky condition we can take (cf. [24, Chap. VI]) a coexponential basis $\{X\}$ of $(\mathfrak{a}_2)^f$ in $\mathfrak{g}$ in such a way that X belongs to $\mathfrak{h}_2$. Take also Y in $\mathfrak{a}_2 \setminus \mathfrak{a}_1$.

On the other hand, the induction hypothesis says that there exists a coexponential basis $\{X'_1, \dots, X'_k\}$ to $\mathfrak{h}_1$ in $(\mathfrak{a}_2)^f$ which contains a coexponential basis $\{X'_{i_1}, \dots, X'_{i_m}\}$ of $\mathfrak{h}_1 \cap \mathfrak{h}'_2$ in $\mathfrak{h}'_2$ and even in $\mathfrak{h}_2 \cap (\mathfrak{a}_2)^f$. Then $\{X, X'_1, \dots, X'_k\}$ is a coexponential basis of $\mathfrak{h}_1$ in $\mathfrak{g}$, whose part $\{X'_{i_1}, \dots, X'_{i_m}\}$ is coexponential for $\mathfrak{h}_1 \cap \mathfrak{h}_2$ in $\mathfrak{h}_2$.

Next, the induction hypothesis gives a coexponential basis $\{X_1, \ldots, X_k\}$ of $\mathfrak{h}_2'$ in $\mathfrak{g}$, whose part $\{X_{i_1}, \ldots, X_{i_l}\}$ is coexponential for $\mathfrak{h}_1 \cap \mathfrak{h}_2'$ in $\mathfrak{h}_1$. Then $\{X_1, \ldots, X_k, Y\}$ is coexponential for $\mathfrak{h}_2$ in $\mathfrak{g}$ and $\{X_{i_1}, \ldots, X_{i_l}, Y\}$ is coexponential for $\mathfrak{h}_1 \cap \mathfrak{h}_2$ in $\mathfrak{h}_1$.

If $\mathfrak{a}_2$ is not an ideal, $\mathfrak{a}_3$ is an ideal of $\mathfrak{g}$. We see as before that $\mathfrak{h}_1$ is the Vergne polarization of $(\mathfrak{a}_3)^f$, constructed from the good sequence $\mathfrak{s}' = (\mathfrak{a}_j \cap (\mathfrak{a}_3)^f)$ of subalgebras. It then suffices to apply the induction hypothesis to $(\mathfrak{a}_3)^f$ and argue as above.

□

Theorem 2.4.2 *Let $\mathfrak{h}_j \in I(f, \mathfrak{g})$ and $H_j = \exp(\mathfrak{h}_j)(j = 1, 2)$. Assume that either $\mathfrak{h}_1$ or $\mathfrak{h}_2$ is a Vergne polarization. Let T be an intertwining isometry from $\pi_1 = ind_{H_1}^G \chi_f$ to $\pi_2 = ind_{H_2}^G \chi_f$. Then there exist a constant $c \in \mathbb{C}$ of modulus 1 and a positive H_2-invariant form $\nu = \nu_{H_2, H_1 \cap H_2}$ such that*

$$(T\varphi)(g) = cT_{\mathfrak{h}_2\mathfrak{h}_1}\varphi(g) = cI_{\mathfrak{h}_2\mathfrak{h}_1}\varphi(g)$$
$$= c \oint_{H_2/(H_1 \cap H_2)} \varphi(gh)\chi_f(h)\Delta_{H_2,G}^{-1/2}(h)d\nu(h) \ (g \in G)$$

for all $\varphi \in \mathscr{K}(f, \mathfrak{h}_1, G)$. Moreover,

$$T_{\mathfrak{h}_2\mathfrak{h}_1} \circ T_{\mathfrak{h}_1\mathfrak{h}_2} = \mathbb{I}, \ T_{\mathfrak{h}_1\mathfrak{h}_2} \circ T_{\mathfrak{h}_2\mathfrak{h}_1} = \mathbb{I}.$$

Proof We first remark that, $H_2 H_1$ being closed, the integral in question converges. The proof will be carried out by induction on $\dim G$. If there exists a non-trivial ideal of $\mathfrak{g}$ on which f vanishes, everything passes to the quotient. We suppose from here on that such an ideal does not exist. If there exists a minimal ideal $\mathfrak{a}$ which is not central, then the polarizations $\mathfrak{h}_j (j = 1, 2)$ are contained in $\mathfrak{a}^f$ and it suffices to apply the induction hypothesis to $A^f = \exp(\mathfrak{a}^f)$. We now exclude this case too.

Suppose first that $\mathfrak{h}_1$ is a Vergne polarization constructed from a good sequence $\mathfrak{s} = (\mathfrak{a}_j)_{0 \le j \le n}$ of subalgebras of $\mathfrak{g}$. Then $\mathfrak{a}_1$ is of dimension 1 and equals the center $\mathfrak{z}$ of $\mathfrak{g}$. Denote by $\mathfrak{a}$ the first non-central ideal appearing in $\mathfrak{s}$, namely $\mathfrak{a} = \mathfrak{a}_2$ or $\mathfrak{a} = \mathfrak{a}_3$ according to the case.

If $\mathfrak{h}_2 \subset \mathfrak{a}^f$, everything is contained in A^f and the induction hypothesis proves the theorem. Suppose now $\mathfrak{h}_2 \not\subset \mathfrak{a}^f$ and put:

$$\mathfrak{h}_2^0 = \mathfrak{h}_2 \cap \mathfrak{a}^f, H_2^0 = \exp(\mathfrak{h}_2^0), \mathfrak{h}_2' = \mathfrak{h}_2^0 + \mathfrak{a}, H_2' = \exp(\mathfrak{h}_2').$$

Then $\mathfrak{h}_2' \in I(f, \mathfrak{g})$ and by Lemma 2.4.1 there exist X in $\mathfrak{h}_2 \setminus \mathfrak{h}_2^0$ or X_1, X_2 in $\mathfrak{h}_2 \setminus \mathfrak{h}_2^0$ such that we have the following diffeomorphic decomposition:

$$H_2 = \exp(\mathbb{R}X) \cdot H_2^0 \text{ or } H_2 = \exp(\mathbb{R}X_1) \cdot \exp(\mathbb{R}X_2) \cdot H_2^0$$

and

$$G = \exp(\mathbb{R}X) \cdot A^f \text{ or } G = \exp(\mathbb{R}X_1) \cdot \exp(\mathbb{R}X_2) \cdot A^f.$$

It follows that

$$G/H_1 \simeq \exp(\mathbb{R}X) \times A^f/H_1 \text{ or } G/H_1 \simeq \exp(\mathbb{R}X_1)\exp(\mathbb{R}X_2) \times A^f/H_1.$$

By induction hypothesis, $I_{\mathfrak{h}_2'\mathfrak{h}_1}$ is an intertwining isometry from π_1 to $\pi_2' = \operatorname{ind}_{H_2'}^G \chi_f$. Let $\varphi \in \mathscr{K}(f, \mathfrak{h}_1, G)$. It follows that the function

$$G \ni g \mapsto I_{\mathfrak{h}_2'\mathfrak{h}_1}\varphi(g) = \oint_{H_2'/(H_1\cap H_2')} \varphi(gh)\chi_f(h)\Delta_{H_2',G}^{-1/2}(h)d\nu_{H_2',H_1\cap H_2'}(h)$$

is continuous and the projection of its support on $\mathbb{R}$ or $\mathbb{R}^2$ along A^f is compact. Lemma 2.3.12 combined with the transitivity of $\nu_{\cdot,\cdot}$ gives us

$$I_{\mathfrak{h}_2\mathfrak{h}_2'} \circ I_{\mathfrak{h}_2'\mathfrak{h}_1} = I_{\mathfrak{h}_2\mathfrak{h}_1}$$

and hence $I_{\mathfrak{h}_2\mathfrak{h}_1}$ defines an intertwining operator $T_{\mathfrak{h}_2\mathfrak{h}_1}$ between π_1 and π_2.

Suppose now that $\mathfrak{h}_2$ is a Vergne polarization constructed from a good sequence $\mathfrak{s} = (\mathfrak{a}_j)_{0\le j\le n}$ of subalgebras of $\mathfrak{g}$. As before, we denote by $\mathfrak{a}$ the first non-central ideal appearing in $\mathfrak{s}$, namely $\mathfrak{a} = \mathfrak{a}_2$ or $\mathfrak{a} = \mathfrak{a}_3$ according to the case. If $\mathfrak{h}_1 \subset \mathfrak{a}^f$, we apply the induction hypothesis. If $\mathfrak{h}_1 \not\subset \mathfrak{a}^f$, put this time

$$\mathfrak{h}_1^0 = \mathfrak{h}_1 \cap \mathfrak{a}^f,\ H_1^0 = \exp(\mathfrak{h}_1^0),\ \mathfrak{h}_1' = \mathfrak{h}_1^0 + \mathfrak{a},\ H_1' = \exp(\mathfrak{h}_1').$$

Then $\mathfrak{h}_1' \in I(f, \mathfrak{g})$ and we find the following diffeomorphic decomposition: there exists $X \in \mathfrak{h}_1 \setminus \mathfrak{h}_1^0$ or X_1, X_2 in $\mathfrak{h}_1 \setminus \mathfrak{h}_1^0$ such that

$$H_1 = \exp(\mathbb{R}X)\cdot H_1^0 \text{ or } H_1 = \exp(\mathbb{R}X_1)\cdot\exp(\mathbb{R}X_2)\cdot H_1^0$$

and

$$G = \exp(\mathbb{R}X)\cdot A^f \text{ or } G = \exp(\mathbb{R}X_1)\cdot\exp(\mathbb{R}X_2)\cdot A^f.$$

It follows that $G/H_1 \simeq A^f/H_1^0$.

Let $\varphi \in \mathscr{K}(f, \mathfrak{h}_1, G)$. From the transitivity of $\nu_{\cdot,\cdot}$,

$$I_{\mathfrak{h}_2\mathfrak{h}_1}\varphi(g) = \oint_{H_2/(H_1'\cap H_2)} (I_{\mathfrak{h}_1'\mathfrak{h}_1}\varphi)(gh)\chi_f(h)\Delta_{H_2,G}^{-1/2}(h)d\nu_{H_2,H_1'\cap H_2}(h)\ (g \in G).$$

For all $g \in G$, the function $A^f \ni h \mapsto I_{\mathfrak{h}_1'\mathfrak{h}_1}\varphi(gh)$ is continuous with compact support modulo H_1'. The induction hypothesis applied to A^f tells us that

$$I_{\mathfrak{h}_2\mathfrak{h}_1} = I_{\mathfrak{h}_2\mathfrak{h}_1'} \circ I_{\mathfrak{h}_1'\mathfrak{h}_1}$$

is an intertwining operator from π_1 to π_2. □

2.5 The General Case

Take $\mathfrak{h}_j \in I(f, \mathfrak{g})$ and $H_j = \exp(\mathfrak{h}_j)(j = 1, 2)$. We shall construct an intertwining operator $T_{\mathfrak{h}_2\mathfrak{h}_1}$ so that, given $\mathfrak{h}_j \in I(f, \mathfrak{g})(1 \leq j \leq 3)$, the composition formula

$$T_{\mathfrak{h}_1\mathfrak{h}_3} \circ T_{\mathfrak{h}_3\mathfrak{h}_2} \circ T_{\mathfrak{h}_2\mathfrak{h}_1} = e^{\frac{i\pi}{4}\tau(\mathfrak{h}_3,\mathfrak{h}_2,\mathfrak{h}_1)}\mathbb{I} = a(\mathfrak{h}_3, \mathfrak{h}_2, \mathfrak{h}_1)\mathbb{I}$$

holds. In order to define this operator we use a third polarization, a Vergne polarization to be precise.

Theorem 2.5.1 *Let $\mathfrak{h}_j \in I(f, \mathfrak{g})(j = 1, 2)$ and take a Vergne polarization $\mathfrak{h}_0$ at $f \in \mathfrak{g}^*$. Then the intertwining isometry*

$$T_{\mathfrak{h}_2\mathfrak{h}_1} = a(\mathfrak{h}_1, \mathfrak{h}_0, \mathfrak{h}_2) T_{\mathfrak{h}_2\mathfrak{h}_0} \circ T_{\mathfrak{h}_0\mathfrak{h}_1},$$

where $T_{\mathfrak{h}_2\mathfrak{h}_0}, T_{\mathfrak{h}_0\mathfrak{h}_1}$ are the intertwining isometries defined in Theorem 2.4.2, does not depend on the choice of the Vergne polarization $\mathfrak{h}_0$. Moreover, $T_{\mathfrak{h}_2\mathfrak{h}_1}$ coincides with $I_{\mathfrak{h}_2\mathfrak{h}_1}$ if at least one of $\mathfrak{h}_1, \mathfrak{h}_2$ is a Vergne polarization.

Proof We proceed again by induction on dim G. As in the previous proofs, we can at once suppose that there is no minimal ideal which is not central and f does not vanish on any ideal of $\mathfrak{g}$.

Let now $\mathfrak{h}_3, \mathfrak{h}_4$ be two Vergne polarizations at f. We claim

$$a(\mathfrak{h}_1, \mathfrak{h}_3, \mathfrak{h}_2) T_{\mathfrak{h}_2\mathfrak{h}_3} \circ T_{\mathfrak{h}_3\mathfrak{h}_1} = a(\mathfrak{h}_1, \mathfrak{h}_4, \mathfrak{h}_2) T_{\mathfrak{h}_2\mathfrak{h}_4} \circ T_{\mathfrak{h}_4\mathfrak{h}_1},$$

where $a(\mathfrak{h}_i, \mathfrak{h}_j, \mathfrak{h}_k) = e^{\frac{i\pi}{4}\tau(\mathfrak{h}_i,\mathfrak{h}_j,\mathfrak{h}_k)}$.

If there is a minimal non-central ideal $\mathfrak{a}$ contained in $\mathfrak{h}_3 \cap \mathfrak{h}_4$, then $\mathfrak{h}_3 \cup \mathfrak{h}_4 \subset \mathfrak{a}^f$ and we can apply the induction hypothesis to $\mathfrak{h}_i' = \mathfrak{h}_i \cap \mathfrak{a}^f + \mathfrak{a}$, $i = 1, 2$. Thus

$$a(\mathfrak{h}_1', \mathfrak{h}_3, \mathfrak{h}_2') T_{\mathfrak{h}_2'\mathfrak{h}_3} \circ T_{\mathfrak{h}_3\mathfrak{h}_1'} = a(\mathfrak{h}_1', \mathfrak{h}_4, \mathfrak{h}_2') T_{\mathfrak{h}_2'\mathfrak{h}_4} \circ T_{\mathfrak{h}_4\mathfrak{h}_1'}.$$

We deduce from this

$$a(\mathfrak{h}_1', \mathfrak{h}_3, \mathfrak{h}_2') T_{\mathfrak{h}_2\mathfrak{h}_3} \circ T_{\mathfrak{h}_3\mathfrak{h}_1} = a(\mathfrak{h}_1', \mathfrak{h}_4, \mathfrak{h}_2') T_{\mathfrak{h}_2\mathfrak{h}_4} \circ T_{\mathfrak{h}_4\mathfrak{h}_1}.$$

Hence it suffices to show

$$a(\mathfrak{h}_1, \mathfrak{h}_3, \mathfrak{h}_2) a(\mathfrak{h}_1', \mathfrak{h}_4, \mathfrak{h}_2') = a(\mathfrak{h}_1, \mathfrak{h}_4, \mathfrak{h}_2) a(\mathfrak{h}_1', \mathfrak{h}_3, \mathfrak{h}_2').$$

But from Lemma 2.3.11 we have

$$\begin{aligned} a(\mathfrak{h}_1, \mathfrak{h}_3, \mathfrak{h}_2) &= a(\mathfrak{h}_4, \mathfrak{h}_3, \mathfrak{h}_2) a(\mathfrak{h}_1, \mathfrak{h}_4, \mathfrak{h}_2) a(\mathfrak{h}_1, \mathfrak{h}_3, \mathfrak{h}_4) \\ &= a(\mathfrak{h}_4, \mathfrak{h}_3, \mathfrak{h}_2') a(\mathfrak{h}_1, \mathfrak{h}_4, \mathfrak{h}_2) a(\mathfrak{h}_1', \mathfrak{h}_3, \mathfrak{h}_4) \end{aligned}$$

$$= a(\mathfrak{h}_1', \mathfrak{h}_3, \mathfrak{h}_2')a(\mathfrak{h}_4, \mathfrak{h}_1', \mathfrak{h}_2')a(\mathfrak{h}_4, \mathfrak{h}_3, \mathfrak{h}_1')a(\mathfrak{h}_1, \mathfrak{h}_4, \mathfrak{h}_2)a(\mathfrak{h}_1', \mathfrak{h}_3, \mathfrak{h}_4)$$
$$= a(\mathfrak{h}_1, \mathfrak{h}_4, \mathfrak{h}_2)a(\mathfrak{h}_1', \mathfrak{h}_3, \mathfrak{h}_2')a(\mathfrak{h}_4, \mathfrak{h}_1', \mathfrak{h}_2').$$

This gives us the desired result. If there is no minimal non-central ideal contained in $\mathfrak{h}_3 \cap \mathfrak{h}_4$, let $\mathfrak{a}$ (resp. $\mathfrak{b}$) be a minimal non-central ideal contained in $\mathfrak{h}_3$ (resp. $\mathfrak{h}_4$). Evidently, $[\mathfrak{a}, \mathfrak{b}] \subset \mathfrak{a} \cap \mathfrak{b} = \mathfrak{z}$. If $[\mathfrak{a}, \mathfrak{b}] = \{0\}$, we can consider a Vergne polarization $\mathfrak{h}_5$ containing $\mathfrak{a} + \mathfrak{b}$. Taking into account what we have already proved, we obtain

$$a(\mathfrak{h}_1, \mathfrak{h}_3, \mathfrak{h}_2)T_{\mathfrak{h}_2\mathfrak{h}_3} \circ T_{\mathfrak{h}_3\mathfrak{h}_1} = a(\mathfrak{h}_1, \mathfrak{h}_5, \mathfrak{h}_2)T_{\mathfrak{h}_2\mathfrak{h}_5} \circ T_{\mathfrak{h}_5\mathfrak{h}_1} = a(\mathfrak{h}_1, \mathfrak{h}_4, \mathfrak{h}_2)T_{\mathfrak{h}_2\mathfrak{h}_4} \circ T_{\mathfrak{h}_4\mathfrak{h}_1}.$$

If $[\mathfrak{a}, \mathfrak{b}] \neq \{0\}$, we immediately see $\dim \mathfrak{a} = \dim \mathfrak{b}$. Let $\mathfrak{s}$ be the good sequence of subalgebras:

$$\mathfrak{z} \subset \mathfrak{a} \subset \cdots \subset \mathfrak{k}_j \subset \cdots \subset \mathfrak{a}^f \subset \mathfrak{a}^f + \mathfrak{b} = \mathfrak{g}$$

or, if $\dim \mathfrak{a} = 3$:

$$\mathfrak{z} \subset \mathfrak{z} \oplus \mathbb{R}Y \subset \mathfrak{a} \subset \cdots \subset \mathfrak{k}_j \subset \cdots \subset \mathfrak{a}^f \subset \mathfrak{a}^f + \mathbb{R}X \subset \mathfrak{a}^f + \mathfrak{b} = \mathfrak{g}.$$

We then determine a second good sequence $\mathfrak{s}'$ of subalgebras:

$$\mathfrak{z} \subset \mathfrak{b} \subset \cdots \subset (\mathfrak{k}_j \cap \mathfrak{b}^f) + \mathfrak{b} \subset \cdots \subset (\mathfrak{a}^f \cap \mathfrak{b}^f) + \mathfrak{b} \subset \mathfrak{a}^f + \mathfrak{b} = \mathfrak{g}$$

or, if $\dim \mathfrak{a} = 3$:

$$\mathfrak{z} \subset \mathfrak{z} \oplus \mathbb{R}X \subset \mathfrak{b} \subset \cdots \subset (\mathfrak{k}_j \cap \mathfrak{b}^f) + \mathfrak{b} \subset \cdots$$
$$\subset (\mathfrak{a}^f \cap \mathfrak{b}^f) + \mathfrak{b} \subset (\mathfrak{a}^f \cap \mathfrak{b}^f) + \mathfrak{b} + \mathbb{R}Y \subset \mathfrak{a}^f + \mathfrak{b} = \mathfrak{g}.$$

Let $\mathfrak{h}_5$ (resp. $\mathfrak{h}_6$) be the Vergne polarization at f constructed from $\mathfrak{s}$ (resp. $\mathfrak{s}'$). Since $\mathfrak{a} \subset \mathfrak{h}_3 \cap \mathfrak{h}_5$ and $\mathfrak{b} \subset \mathfrak{h}_4 \cap \mathfrak{h}_6$, it suffices to show

$$a(\mathfrak{h}_1, \mathfrak{h}_5, \mathfrak{h}_2)T_{\mathfrak{h}_2\mathfrak{h}_5} \circ T_{\mathfrak{h}_5\mathfrak{h}_1} = a(\mathfrak{h}_1, \mathfrak{h}_6, \mathfrak{h}_2)T_{\mathfrak{h}_2\mathfrak{h}_6} \circ T_{\mathfrak{h}_6\mathfrak{h}_1}. \tag{2.5.1}$$

This confirms $\mathfrak{h}_5 = \mathfrak{h}_6 \cap \mathfrak{a}^f + \mathfrak{a} = \mathfrak{h}_6'$. Then we can replicate the arguments developed in [60] to find

$$T_{\mathfrak{h}_6\mathfrak{h}_1} = a(\mathfrak{h}_1, \mathfrak{h}_5, \mathfrak{h}_6)T_{\mathfrak{h}_6\mathfrak{h}_5} \circ T_{\mathfrak{h}_5\mathfrak{h}_1}.$$

For this, we put $\mathfrak{h}_1^0 = \mathfrak{h}_j \cap \mathfrak{a}^f\,(j = 1, 6)$ and examine different cases according to the position of $\mathfrak{h}_1, \mathfrak{h}_6$: $\mathfrak{a} \subset \mathfrak{h}_1$, or $\mathfrak{h}_1 \cap \mathfrak{h}_6 \not\subset \mathfrak{a}^f$, or $\mathfrak{a} \subset \mathfrak{h}_1^0 + \mathfrak{h}_6^0$, or $\mathfrak{a} \subset \mathfrak{h}_1 + \mathfrak{h}_6$ but $\mathfrak{a} \not\subset \mathfrak{h}_1^0 + \mathfrak{h}_6^0$. Only the last case needs some comments.

Assume for example that $\dim \mathfrak{a} = 2$. Write $\mathfrak{z} = \mathbb{R}Z$ with $f(Z) = 1$, $\mathfrak{a} = \mathfrak{z} + \mathbb{R}Y$ with $f(Y) = 0$. Put $\mathfrak{k} = \mathfrak{h}_1 + \mathfrak{a}$ and $\mathfrak{g}_0 = \{X \in \mathfrak{g}; [X, \mathfrak{a}] \subset \mathfrak{z}\}$. Then by Lemma 2.6.4

(see later), $\mathfrak{h}_1 \cap \mathfrak{h}_6 \subset \mathfrak{g}_0$ and there exist two elements T_1, T_2 of $\mathfrak{g}_0 \cap \ker(f)$ such that

$$\mathfrak{h}_1 = \mathbb{R}T_1 \oplus \mathfrak{h}_1^0, \mathfrak{h}_6 = \mathbb{R}T_2 \oplus \mathfrak{h}_6^0, [T_1, Y] = [T_2, Y] = Z$$

and $Y = c(T_2 - T_1)$ with $c \in \mathbb{R} \setminus \{0\}$. The inclusion $\mathfrak{h}_1 \cap \mathfrak{h}_6 \subset \mathfrak{g}_0$ implies that $\mathrm{ad}T_1, \mathrm{ad}T_2$ keep $\mathfrak{h}_1 \cap \mathfrak{h}_6$ stable. Furthermore, we have the following relationship of the traces:

$$\mathrm{Tr}\,\mathrm{ad}_{\mathfrak{h}_1/(\mathfrak{h}_1\cap\mathfrak{h}_6)} T_1 + \mathrm{Tr}\,\mathrm{ad}_{\mathfrak{h}_6/(\mathfrak{h}_1\cap\mathfrak{h}_6)} T_2 = 0. \tag{2.5.2}$$

Indeed, there is nothing to prove if Y is in $\mathfrak{h}_1 + [\mathfrak{h}_6, \mathfrak{h}_6]$. Otherwise, let $\mathfrak{m}$ be a complementary subspace of $((\mathfrak{h}_1 \cap \ker(f)) + [\mathfrak{h}_6, \mathfrak{h}_6]) \oplus \mathbb{R}Y$ in $\ker(f)$ such that

$$\mathfrak{g} = \mathfrak{z} \oplus ((\mathfrak{h}_1 \cap \ker(f)) + [\mathfrak{h}_6, \mathfrak{h}_6]) \oplus \mathbb{R}Y \oplus \mathfrak{m}.$$

For any $\lambda \in \mathbb{R}$, we denote by f_λ the linear form on $\mathfrak{g}$ such that $f_\lambda(Y) = \lambda$, which coincides with f on $(\mathfrak{h}_1 + [\mathfrak{h}_6, \mathfrak{h}_6]) \oplus \mathfrak{m}$. By Pukanszky's condition $\mathfrak{h}_1$ is a polarization of $\mathfrak{g}$ at f_λ, and so is $\mathfrak{h}_6$ as well. Hence it suffices to prove

$$\mathrm{Tr}\,\mathrm{ad}_{\mathfrak{k}/(\mathfrak{k}\cap\mathfrak{h}_6)} T_2 + \mathrm{Tr}\,\mathrm{ad}_{\mathfrak{h}_6/(\mathfrak{h}_1\cap\mathfrak{h}_6)} T_2 = 0.$$

For this, note that we can choose $\mathfrak{h}_6$ to be admissible for the maximal nilpotent ideal $\mathfrak{n}$ of $\mathfrak{g}$. Then $\mathfrak{h}_6 = (\mathfrak{h}_6 \cap \mathfrak{n}) + (\mathfrak{h}_6 \cap \mathfrak{n}^f)$ (cf. [24, Chap. IV]). Hence T_2 reads $T_2 = X_1 + X_2$ with $X_1 \in \mathfrak{h}_6 \cap \mathfrak{n}$, $X_2 \in \mathfrak{h}_6 \cap \mathfrak{n}^f$. Our relation then follows, because B_f is a pairing between $\mathfrak{k}/(\mathfrak{k} \cap \mathfrak{h}_6)$ and $\mathfrak{h}_6/(\mathfrak{h}_1 \cap \mathfrak{h}_6)$.

We show in the same way

$$T_{\mathfrak{h}_2\mathfrak{h}_6} = a(\mathfrak{h}_6, \mathfrak{h}_5, \mathfrak{h}_2) T_{\mathfrak{h}_2\mathfrak{h}_5} \circ T_{\mathfrak{h}_5\mathfrak{h}_6}$$

and then

$$T_{\mathfrak{h}_2\mathfrak{h}_6} \circ T_{\mathfrak{h}_6\mathfrak{h}_1} = a(\mathfrak{h}_6, \mathfrak{h}_5, \mathfrak{h}_2) a(\mathfrak{h}_1, \mathfrak{h}_5, \mathfrak{h}_6) T_{\mathfrak{h}_2\mathfrak{h}_5} \circ T_{\mathfrak{h}_5\mathfrak{h}_1}.$$

In order to get (2.5.1) it suffices, by way of Lemma 2.3.11, to observe

$$a(\mathfrak{h}_1, \mathfrak{h}_5, \mathfrak{h}_2) = a(\mathfrak{h}_6, \mathfrak{h}_5, \mathfrak{h}_2) a(\mathfrak{h}_1, \mathfrak{h}_6, \mathfrak{h}_2) a(\mathfrak{h}_1, \mathfrak{h}_5, \mathfrak{h}_6).$$

The case where $\dim \mathfrak{a} = 3$ is similar. □

Definition 2.5.2 Let $\mathfrak{h}_j \in I(f, \mathfrak{g})$, $H_j = \exp(\mathfrak{h}_j)$ and $\pi_j = \mathrm{ind}_{H_j}^G \chi_f (j = 1, 2)$. We define an intertwining isometry $T_{\mathfrak{h}_2\mathfrak{h}_1}$ from π_1 to π_2 by putting

$$T_{\mathfrak{h}_2\mathfrak{h}_1} = a(\mathfrak{h}_1, \mathfrak{h}_0, \mathfrak{h}_2) T_{\mathfrak{h}_2\mathfrak{h}_0} \circ T_{\mathfrak{h}_0\mathfrak{h}_1} \tag{2.5.3}$$

with the help of a third polarization $\mathfrak{h}_0$, which is a Vergne polarization. This isometry coincides with $I_{\mathfrak{h}_2\mathfrak{h}_1}$ if at least one of $\mathfrak{h}_1$, $\mathfrak{h}_2$ is a Vergne polarization. In particular, $T_{\mathfrak{h}_1\mathfrak{h}_1} = \mathbb{I}$.

Theorem 2.5.3 (Composition Formula) *Let $\mathfrak{h}_j \in I(f, \mathfrak{g})$ for $1 \leq j \leq 3$. Then*

$$T_{\mathfrak{h}_1\mathfrak{h}_3} \circ T_{\mathfrak{h}_3\mathfrak{h}_2} \circ T_{\mathfrak{h}_2\mathfrak{h}_1} = a(\mathfrak{h}_3, \mathfrak{h}_2, \mathfrak{h}_1)\mathbb{I}.$$

In particular,

$$T_{\mathfrak{h}_2\mathfrak{h}_1} \circ T_{\mathfrak{h}_1\mathfrak{h}_2} = \mathbb{I}.$$

Proof This theorem descends directly from Theorem 2.5.1 and Lemma 2.3.11. Taking a Vergne polarization $\mathfrak{h}_0$,

$$\begin{aligned} T_{\mathfrak{h}_1\mathfrak{h}_3} \circ T_{\mathfrak{h}_3\mathfrak{h}_2} \circ T_{\mathfrak{h}_2\mathfrak{h}_1} =& a(\mathfrak{h}_3, \mathfrak{h}_0, \mathfrak{h}_1) a(\mathfrak{h}_2, \mathfrak{h}_0, \mathfrak{h}_3) a(\mathfrak{h}_1, \mathfrak{h}_0, \mathfrak{h}_2) \\ & \times T_{\mathfrak{h}_1\mathfrak{h}_0} \circ T_{\mathfrak{h}_0\mathfrak{h}_3} \circ T_{\mathfrak{h}_3\mathfrak{h}_0} \circ T_{\mathfrak{h}_0\mathfrak{h}_2} \circ T_{\mathfrak{h}_2\mathfrak{h}_0} \circ T_{\mathfrak{h}_0\mathfrak{h}_1} \\ =& a(\mathfrak{h}_0, \mathfrak{h}_2, \mathfrak{h}_1) a(\mathfrak{h}_3, \mathfrak{h}_0, \mathfrak{h}_1) a(\mathfrak{h}_3, \mathfrak{h}_2, \mathfrak{h}_0)\mathbb{I} \\ =& a(\mathfrak{h}_3, \mathfrak{h}_2, \mathfrak{h}_1)\mathbb{I}. \end{aligned}$$

□

2.6 A Local Result

We denote by $\mathscr{H}_\rho$ the Hilbert space of a unitary representation ρ of G, by $\mathscr{H}_\rho^\infty$ the space of C^∞-vectors of ρ equipped with the usual topology and by $\mathscr{H}_\rho^{-\infty}$ the anti-dual space of $\mathscr{H}_\rho^\infty$. Given a closed subgroup K of G and a character λ of K, we set

$$(\mathscr{H}_\rho^{-\infty})^{K,\lambda} = \{a \in \mathscr{H}_\rho^{-\infty};\ \rho(k)a = \lambda(k)a, \forall k \in K\}.$$

Let $T_{\mathfrak{h}_2\mathfrak{h}_1}$ be the isometry defined above which intertwines π_1 and π_2. This section is devoted to the proof of the next theorem.

Theorem 2.6.1 *There exists a positive form $\nu = \nu_{H_2, H_1\cap H_2}$ such that*

$$T_{\mathfrak{h}_2\mathfrak{h}_1}\varphi(e) = \oint_{H_2/(H_1\cap H_2)} \varphi(h)\chi_f(h)\Delta_{H_2,G}^{-1/2}(h)d\nu(h)$$

for any function $\varphi \in \mathscr{H}_{\pi_1}^\infty$ whose support is sufficiently small modulo H_1.

Corollary 2.6.2 *The space* $(\mathscr{H}_{\pi_1}^{-\infty})^{H_2,\chi_f\Delta_{H_2,G}^{1/2}}$ *is equal to* $\mathbb{C}\kappa$, *where* κ *is defined by*

$$\kappa(\varphi) = \oint_{H_2/(H_1\cap H_2)} \overline{\varphi(h)\chi_f(h)\Delta_{H_2,G}^{-1/2}}(h)d\nu(h)$$

for any function $\varphi \in \mathscr{H}_{\pi_1}^{\infty}$ *whose support is sufficiently small modulo* H_1.

Proof We employ, as usual, induction on $\dim G$ and closely follow the proof of Theorem 3 in [60], where the complementary condition $\mathscr{N}$ is imposed to guarantee the convergence of the integral (2.2.1).

We first recall

$$\begin{aligned} H_2H_1 &= \{g \in G;\ g{\cdot}(f + (\mathfrak{h}_1)^{\perp}) \cap (f + (\mathfrak{h}_2)^{\perp}) \neq \emptyset\} \\ &= \{g \in G;\ \mathrm{rank}A(g) = \mathrm{rank}B(g) = \mathrm{rank}A(e) = \dim(\mathfrak{h}_2/(\mathfrak{h}_1 \cap \mathfrak{h}_2))\}, \end{aligned}$$

where $A(g)$, $B(g)$ denote the coefficient matrices of the linear system (2.3.1). Put

$$V = \{g \in G;\ \mathrm{rank}A(g) \geq \mathrm{rank}A(e)\},$$

which is a neighbourhood of H_1 such that

$$H_2H_1 \cap V = \{g \in G;\ \mathrm{rank}B(g) \leq \dim(\mathfrak{h}_2/(\mathfrak{h}_1 \cap \mathfrak{h}_2))\}$$

is closed in V. Just like $\mathrm{rank}A(g)$, moreover, V is left H_1-invariant and right H_2-invariant.

It follows that the integral

$$\oint_{H_2/(H_1\cap H_2)} \varphi(h)\chi_f(h)\Delta_{H_2,G}^{-1/2}(h)d\nu(h)$$

converges for any function $\varphi \in \mathscr{H}_{\pi_1}^{\infty}$ such that $\mathrm{supp}(\varphi) = KH_1$ with K compact in V.

Let us begin with the usual formalities. If there exists a non-trivial ideal of $\mathfrak{g}$ on which f vanishes, we can pass to the quotient and use induction. We henceforth suppose f vanishes on no non-trivial ideal. In particular, the center $\mathfrak{z}$ of $\mathfrak{g}$ is of dimension 1 and $f(\mathfrak{z}) \neq \{0\}$. Let $Z \in \mathfrak{z}$ verify $f(Z) = 1$. We can also suppose that $\mathfrak{h}_1 + \mathfrak{h}_2$ is not contained in any proper subalgebra of $\mathfrak{g}$. Under these conditions the unique minimal ideal of $\mathfrak{g}$ is $\mathfrak{z}$. Let $\mathfrak{a}$ be a minimal non-central ideal of $\mathfrak{g}$. Then $\mathfrak{a}$ contains $\mathfrak{z}$ and its dimension is equal to either 2 or 3. We choose a Vergne polarization $\mathfrak{h}_0$ of $\mathfrak{g}$ at f which contains $\mathfrak{a}$. Hence

$$T_{\mathfrak{h}_2\mathfrak{h}_1} = a(\mathfrak{h}_1, \mathfrak{h}_0, \mathfrak{h}_2)T_{\mathfrak{h}_2\mathfrak{h}_0} \circ T_{\mathfrak{h}_0\mathfrak{h}_1}.$$

Let $\mathfrak{h} \in I(f, \mathfrak{g})$ satisfy $\mathfrak{a} \not\subset \mathfrak{h}$ and $\mathfrak{h}' = \mathfrak{h} \cap \mathfrak{a}^f + \mathfrak{a}$. We have shown in the proof of Theorem 2.4.2

$$T_{\mathfrak{h}_0\mathfrak{h}} = T_{\mathfrak{h}_0\mathfrak{h}'} \circ T_{\mathfrak{h}'\mathfrak{h}}, \quad T_{\mathfrak{h}\mathfrak{h}_0} = T_{\mathfrak{h}\mathfrak{h}'} \circ T_{\mathfrak{h}'\mathfrak{h}_0}.$$

We shall often use these relations. We always use the notations $\mathfrak{h}^0 = \mathfrak{h} \cap \mathfrak{a}^f$, $H^0 = \exp(\mathfrak{h}^0)$ and $H' = \exp(\mathfrak{h}')$.

We are forced to study several cases separately. We know that $\mathfrak{a}$ is not contained in $\mathfrak{h}_1 \cap \mathfrak{h}_2$ because otherwise $\mathfrak{h}_1 + \mathfrak{h}_2$ would be contained in $\mathfrak{a}^f$. The simplest cases are the following cases (a) and (b).

(a) $\mathfrak{a} \subset \mathfrak{h}_1$ and $\mathfrak{a} \not\subset \mathfrak{h}_2$. From Lemma 2.3.11 (c) applied to $\mathfrak{a} + \mathfrak{g}(f) \subset \mathfrak{h}_1 \cap \mathfrak{h}_0$, we have $a(\mathfrak{h}_1, \mathfrak{h}_0, \mathfrak{h}_2) = a(\mathfrak{h}_1, \mathfrak{h}_0, \mathfrak{h}_2')$. Let $\varphi \in \mathscr{H}^\infty_{\pi_1}$. Then $T_{\mathfrak{h}_2'\mathfrak{h}_1}\varphi$ belongs to $\mathscr{H}^\infty_{\pi_2'}$ and Lemma 2.3.14 implies

$$\begin{aligned}(T_{\mathfrak{h}_2\mathfrak{h}_1}\varphi)(e) &= a(\mathfrak{h}_1, \mathfrak{h}_0, \mathfrak{h}_2')(T_{\mathfrak{h}_2\mathfrak{h}_2'} \circ T_{\mathfrak{h}_2'\mathfrak{h}_0} \circ T_{\mathfrak{h}_0\mathfrak{h}_1}\varphi)(e) \\ &= (T_{\mathfrak{h}_2\mathfrak{h}_2'} \circ T_{\mathfrak{h}_2'\mathfrak{h}_1}\varphi)(e) \\ &= \oint_{H_2/H_2^0} (T_{\mathfrak{h}_2'\mathfrak{h}_1}\varphi)(x)\chi_f(x)\Delta^{-1/2}_{H_2,G}(x)d\nu(x).\end{aligned}$$

If $\operatorname{supp}(\varphi)$ is sufficiently small modulo H_1, applying the induction hypothesis to $\mathfrak{a}^f$ we find

$$T_{\mathfrak{h}_2'\mathfrak{h}_1}\varphi(x) = \int_{H_2'/(H_1 \cap H_2')} \varphi(xk)\chi_f(k)\Delta^{-1/2}_{H_2',G}(k)d\nu(k) \ (x \in H_2).$$

Making use of the isomorphism $H_2'/(H_1 \cap H_2') \simeq H_2^0/(H_1 \cap H_2^0)$ and Theorem 2.2.2,

$$\begin{aligned}(T_{\mathfrak{h}_2\mathfrak{h}_1}\varphi)(e) &= \oint_{H_2/H_2^0} d\nu(x) \oint_{H_2^0/(H_1\cap H_2^0)} \varphi(xk)\chi_f(k)\Delta^{-1/2}_{H_2,G}(xk)\Delta^{-1}_{H_2^0,H_2}(k)d\nu(k) \\ &= \oint_{H_2/(H_1\cap H_2)} \varphi(h)\chi_f(h)\Delta^{-1/2}_{H_2,G}(h)d\nu(h).\end{aligned}$$

(b) $\mathfrak{a} \subset \mathfrak{h}_2$ and $\mathfrak{a} \not\subset \mathfrak{h}_1$. As in the previous case, $a(\mathfrak{h}_1, \mathfrak{h}_0, \mathfrak{h}_2) = a(\mathfrak{h}_1', \mathfrak{h}_0, \mathfrak{h}_2)$. If the support of $\varphi \in \mathscr{H}^\infty_{\pi_1}$ is sufficiently small modulo H_1, then Lemma 2.3.14 says that the function $T_{\mathfrak{h}_1'\mathfrak{h}_1}\varphi \in \mathscr{H}^\infty_{\pi_1'}$, $\pi_1' = \operatorname{ind}^G_{H_1'}\chi_f$, is given by the formula

$$(T_{\mathfrak{h}_1'\mathfrak{h}_1}\varphi)(g) = \oint_{H_1'/H_1^0} \varphi(gy)\chi_f(y)\Delta^{-1/2}_{H_1',G}(y)d\nu(y) \ (g \in G).$$

Since its support is sufficiently small in A^f modulo H_1', we can apply the induction hypothesis to $(T_{\mathfrak{h}_2\mathfrak{h}_1'}\circ T_{\mathfrak{h}_1'\mathfrak{h}_1}\varphi)(e)$. By Theorem 2.2.2 and the transitivity of $\nu_{\cdot,\cdot}$.

$$(T_{\mathfrak{h}_2\mathfrak{h}_1}\varphi)(e) = a(\mathfrak{h}_1', \mathfrak{h}_0, \mathfrak{h}_2)(T_{\mathfrak{h}_2\mathfrak{h}_0}\circ T_{\mathfrak{h}_0\mathfrak{h}_1'}\circ T_{\mathfrak{h}_1'\mathfrak{h}_1}\varphi)(e)$$

$$=(T_{\mathfrak{h}_2\mathfrak{h}_1'}\circ T_{\mathfrak{h}_1'\mathfrak{h}_1}\varphi)(e)$$

$$=\oint_{H_2/(H_1'\cap H_2)} \chi_f(k)\Delta_{H_2,G}^{-1/2}(k)d\nu(k)\oint_{H_1'/H_1^0} \varphi(ky)\chi_f(y)\Delta_{H_1',G}^{-1/2}(y)d\nu(y)$$

$$=\oint_{H_2/(H_1'\cap H_2)} d\nu(k)\oint_{(H_1'\cap H_2)/(H_1\cap H_2)} \varphi(ky)\chi_f(ky)$$

$$\times\Delta_{H_2,G}^{-1/2}(ky)\Delta_{H_1'\cap H_2,H_2}(y)d\nu(y)$$

$$=\oint_{H_2/(H_1\cap H_2)} \varphi(h)\chi_f(h)\Delta_{H_2,G}^{-1/2}(h)d\nu(h).$$

(c) $\mathfrak{a}\not\subset\mathfrak{h}_j (j = 1, 2)$. Put $\mathfrak{g}_0 = \{X \in \mathfrak{g}; [X, \mathfrak{a}] \subset \mathfrak{z}\}$. If $\mathfrak{a}/\mathfrak{z}$ is not central in $\mathfrak{g}/\mathfrak{z}$, $\mathfrak{g}_0$ is an ideal of codimension 1 of $\mathfrak{g}$ and $\mathfrak{h}_1 + \mathfrak{h}_2 \not\subset \mathfrak{g}_0$ by our hypothesis.

Now we must split the study and examine special situations.

(c.a) $\dim((\mathfrak{h}_1\cap\mathfrak{h}_2)/(\mathfrak{h}_1^0\cap\mathfrak{h}_2^0)) = \dim(\mathfrak{a}/(\mathfrak{h}_1\cap\mathfrak{a})) = \dim(\mathfrak{a}/(\mathfrak{h}_2\cap\mathfrak{a}))$. This case is divided into two subcases.

(c.a.2) $\dim\mathfrak{a} = 3$ and $\dim(\mathfrak{a}/(\mathfrak{h}_1\cap\mathfrak{a})) = \dim(\mathfrak{a}/(\mathfrak{h}_2\cap\mathfrak{a})) = 2$.

(c.a.1) $\dim(\mathfrak{h}_1\cap\mathfrak{a}) = \dim(\mathfrak{h}_2\cap\mathfrak{a}) = 1$, namely $\mathfrak{h}_1\cap\mathfrak{a} = \mathfrak{h}_2\cap\mathfrak{a} = \mathfrak{z}$.

(c.2) $\dim\mathfrak{a} = 2$. There remains to consider the case where $\dim((\mathfrak{h}_1\cap\mathfrak{h}_2)/(\mathfrak{h}_1^0\cap\mathfrak{h}_2^0)) \neq \dim(\mathfrak{a}/(\mathfrak{h}_1\cap\mathfrak{a})) = \dim(\mathfrak{a}/(\mathfrak{h}_2\cap\mathfrak{a}))$. Then necessarily $\mathfrak{h}_1\cap\mathfrak{h}_2 \subset \mathfrak{a}^f$ and hence $\mathfrak{a}\subset\mathfrak{h}_1+\mathfrak{h}_2$. We have the subcases.

(c.2.0) $\mathfrak{a}\subset\mathfrak{h}_1^0+\mathfrak{h}_2^0$.

(c.2.1) $\mathfrak{a}\not\subset\mathfrak{h}_1^0+\mathfrak{h}_2^0$ but $\mathfrak{a}\subset\mathfrak{h}_1+\mathfrak{h}_2$.

(c.3) $\dim\mathfrak{a} = 3$. There are several possibilities: $\dim((\mathfrak{h}_1\cap\mathfrak{h}_2)/(\mathfrak{h}_1^0\cap\mathfrak{h}_2^0)) \neq \dim(\mathfrak{a}/(\mathfrak{h}_1\cap\mathfrak{a})) = \dim(\mathfrak{a}/(\mathfrak{h}_2\cap\mathfrak{a}))$ or $\dim((\mathfrak{h}_1\cap\mathfrak{h}_2)/(\mathfrak{h}_1^0\cap\mathfrak{h}_2^0)) = \dim(\mathfrak{a}/(\mathfrak{h}_1\cap\mathfrak{a})) \neq \dim(\mathfrak{a}/(\mathfrak{h}_2\cap\mathfrak{a}))$ etc. This gives us a number of subcases.

(c.3.2) $\dim(\mathfrak{h}_j\cap\mathfrak{a}) = 2$, $j = 1, 2$.

(c.3.1.a) $\dim(\mathfrak{h}_1\cap\mathfrak{a}) = 1$, $\dim(\mathfrak{h}_2\cap\mathfrak{a}) = 2$, $\mathfrak{h}_1\cap\mathfrak{h}_2\not\subset\mathfrak{a}^f$.

(c.3.2.a) $\dim(\mathfrak{h}_1\cap\mathfrak{a}) = 2$, $\dim(\mathfrak{h}_2\cap\mathfrak{a}) = 1$, $\mathfrak{h}_1\cap\mathfrak{h}_2\not\subset\mathfrak{a}^f$.

(c.3.1.b) $\dim(\mathfrak{h}_1\cap\mathfrak{a}) = 1$, $\dim(\mathfrak{h}_2\cap\mathfrak{a}) = 2$, $\mathfrak{a}\subset\mathfrak{h}_1^0+\mathfrak{h}_2^0$.

(c.3.2.b) $\dim(\mathfrak{h}_1\cap\mathfrak{a}) = 2$, $\dim(\mathfrak{h}_2\cap\mathfrak{a}) = 1$, $\mathfrak{a}\subset\mathfrak{h}_1^0+\mathfrak{h}_2^0$.

(c.3.1.c) $\dim(\mathfrak{h}_1\cap\mathfrak{a}) = 1$, $\dim(\mathfrak{h}_2\cap\mathfrak{a}) = 2$, $\mathfrak{a}\subset\mathfrak{h}_1+\mathfrak{h}_2$.

(c.3.2.c) $\dim(\mathfrak{h}_1\cap\mathfrak{a}) = 2$, $\dim(\mathfrak{h}_2\cap\mathfrak{a}) = 1$, $\mathfrak{a}\subset\mathfrak{h}_1+\mathfrak{h}_2$.

(c.3.a.2) $\dim(\mathfrak{h}_1\cap\mathfrak{a}) = \dim(\mathfrak{h}_2\cap\mathfrak{a}) = 1$, $\dim(\mathfrak{h}_1\cap\mathfrak{h}_2) = \dim(\mathfrak{h}_1^0\cap\mathfrak{h}_2^0)+2$.

(c.3.b.1) $\dim(\mathfrak{h}_1\cap\mathfrak{a}) = \dim(\mathfrak{h}_2\cap\mathfrak{a}) = 1$, $\dim(\mathfrak{h}_1\cap\mathfrak{h}_2) = \dim(\mathfrak{h}_1^0\cap\mathfrak{h}_2^0)+1$, $Y'\in\mathfrak{h}_1^0+\mathfrak{h}_2^0$, where Y' is a certain non-central vector of $\mathfrak{a}$.

(c.3.b.2) $\dim(\mathfrak{h}_1 \cap \mathfrak{a}) = \dim(\mathfrak{h}_2 \cap \mathfrak{a}) = 1, \dim(\mathfrak{h}_1 \cap \mathfrak{h}_2) = \dim(\mathfrak{h}_1^0 \cap \mathfrak{h}_2^0) + 1, Y' \notin \mathfrak{h}_1^0 + \mathfrak{h}_2^0$.

(c.3.c.3) $\dim(\mathfrak{h}_1 \cap \mathfrak{h}_2) = \dim(\mathfrak{h}_1^0 \cap \mathfrak{h}_2^0), \mathfrak{a} \subset \mathfrak{h}_1^0 + \mathfrak{h}_2^0$.

(c.3.c.2) $\dim(\mathfrak{h}_1 \cap \mathfrak{h}_2) = \dim(\mathfrak{h}_1^0 \cap \mathfrak{h}_2^0), \dim(\mathfrak{a} \cap (\mathfrak{h}_1^0 + \mathfrak{h}_2^0)) = 2$.

(c.3.c.1) $\dim(\mathfrak{h}_1 \cap \mathfrak{h}_2) = \dim(\mathfrak{h}_1^0 \cap \mathfrak{h}_2^0), \dim(\mathfrak{a} \cap (\mathfrak{h}_1^0 + \mathfrak{h}_2^0)) = 1$.

We are going to treat all these cases separately.

(c.a.2) As $\dim \mathfrak{a} = 3$, the $\mathfrak{g}$-module $\mathfrak{a}/\mathfrak{z}$ is irreducible and its annihilator $\mathfrak{g}_0$ is of codimension 1 in $\mathfrak{g}$. Furthermore, $\dim(\mathfrak{a}/(\mathfrak{h}_1 \cap \mathfrak{a})) = \dim(\mathfrak{a}/(\mathfrak{h}_2 \cap \mathfrak{a})) = 1$, namely $\dim(\mathfrak{h}_1 \cap \mathfrak{a}) = \dim(\mathfrak{h}_2 \cap \mathfrak{a}) = 2$ implies $\mathfrak{h}_1 + \mathfrak{h}_2 \subset \mathfrak{g}_0$. Hence this case does not appear.

(c.a.1) We have $\mathfrak{h}_1 \cap \mathfrak{a} = \mathfrak{h}_2 \cap \mathfrak{a} = \mathfrak{z}$. Take a complement $\mathfrak{m}$ (resp. $\mathfrak{t}$) of $\mathfrak{z}$ (resp. $\mathfrak{h}_1^0 \cap \mathfrak{h}_2^0$) in $\mathfrak{a}$ (resp. $\mathfrak{h}_1 \cap \mathfrak{h}_2$) contained in $\ker(f)$ (resp. $\mathfrak{g}_0 \cap \ker(f)$), which is possible under the Pukanszky condition (see Lemma 2.2.3). Then

$$\mathfrak{h}_1 \cap \mathfrak{h}_2 = \mathfrak{t} \oplus (\mathfrak{h}_1^0 \cap \mathfrak{h}_2^0) = \mathfrak{t} \oplus (\mathfrak{h}_1 \cap \mathfrak{h}_2^0).$$

We can identify H_2/H_2^0 with $\exp \mathfrak{t}$, H_2'/H_2^0 with $\exp \mathfrak{m}$ and $\nu_{H_2,H_2^0}, \nu_{H_2',H_2^0}$ with the Lebesgue measures dX, dY on $\mathfrak{t}, \mathfrak{m}$ respectively (cf. [151]).

Since $(\mathfrak{h}_1')^{\mathfrak{t}} = \mathfrak{h}_1$, the properties of the Maslov index imply

$$a(\mathfrak{h}_1, \mathfrak{h}_0, \mathfrak{h}_2) = a(\mathfrak{h}_1', \mathfrak{h}_0, \mathfrak{h}_2) = a(\mathfrak{h}_1', \mathfrak{h}_0, \mathfrak{h}_2').$$

Well, $T_{\mathfrak{h}_2'\mathfrak{h}_1'} \circ T_{\mathfrak{h}_1'\mathfrak{h}_1} \varphi = T_{\mathfrak{h}_2'\mathfrak{h}_1} \varphi$ belongs to $\mathscr{H}_{\pi_2'}^{\infty}, \pi_2' = \mathrm{ind}_{H_2'}^G \chi_f$ for any $\varphi \in \mathscr{H}_{\pi_1}^{\infty}$ and Lemma 2.3.14 says

$$\begin{aligned}(T_{\mathfrak{h}_2\mathfrak{h}_1}\varphi)(e) &= a(\mathfrak{h}_1', \mathfrak{h}_0, \mathfrak{h}_2')(T_{\mathfrak{h}_2\mathfrak{h}_2'} \circ T_{\mathfrak{h}_2'\mathfrak{h}_0} \circ T_{\mathfrak{h}_0\mathfrak{h}_1'} \circ T_{\mathfrak{h}_1'\mathfrak{h}_1}\varphi)(e) \\ &= (T_{\mathfrak{h}_2\mathfrak{h}_2'} \circ T_{\mathfrak{h}_2'\mathfrak{h}_1'} \circ T_{\mathfrak{h}_1'\mathfrak{h}_1}\varphi)(e) \\ &= \int_{\mathfrak{t}} T_{\mathfrak{h}_2'\mathfrak{h}_1'}(\pi_1'(\exp(-X)) T_{\mathfrak{h}_1'\mathfrak{h}_1}\varphi)(e) \Delta_{H_2,G}^{-1/2}(\exp X) dX.\end{aligned}$$

We now fix a compact set K in $\mathfrak{t}$. There exists a neighbourhood V of e in G such that, for any $X \in K, \varphi \in \mathscr{H}_{\pi_1}^{\infty}$ whose support is contained in VH_1, the support of $\pi_1'(\exp(-X)) \circ T_{\mathfrak{h}_1'\mathfrak{h}_1}\varphi$ is sufficiently small in A^f modulo H_1' so that we can apply to it the formula for $T_{\mathfrak{h}_2'\mathfrak{h}_1'}$:

$$\begin{aligned}&T_{\mathfrak{h}_2'\mathfrak{h}_1'}(\pi_1'(\exp(-X)) T_{\mathfrak{h}_1'\mathfrak{h}_1}\varphi)(e) \\ &= \oint_{H_2'/(H_1' \cap H_2')} T_{\mathfrak{h}_1'\mathfrak{h}_1}\varphi(\exp X \cdot k) \chi_f(k) \Delta_{H_2',G}^{-1/2}(k) d\nu(k). \qquad (2.6.1)\end{aligned}$$

For an arbitrary $X \in \mathfrak{t}$, let $f_X = (\exp X) \cdot f$, $\mathfrak{h}_j^X = \mathrm{Ad}(\exp X)(\mathfrak{h}_j')$ and $H_j^X = \exp(\mathfrak{h}_j^X) (j = 1, 2)$. We introduce $\pi_j^X = \rho(f_X, \mathfrak{h}_j^X, G)$ and the intertwining isometry

$$R_j^X; \mathscr{H}_{\pi'_j} \to \mathscr{H}_{\pi_j^X},\ R_j^X(\varphi)(g) = c_j^X \varphi(g \exp X)\ (g \in G),$$

where c_j^X is a certain positive scalar. If the support of $\varphi \in \mathscr{H}_{\pi_1}^{\infty}$ is sufficiently small modulo H_1, we can apply (2.6.1) to the isometry $T_{\mathfrak{h}_2^X \mathfrak{h}_1^X}$ acting on $R_1^X T_{\mathfrak{h}'_1 \mathfrak{h}_1}\varphi$. It follows that there exist complex numbers $\alpha^X, \beta^X, \gamma^X (|\alpha^X| = |\beta^X| = 1, \gamma^X > 0)$ such that

$$\begin{aligned}
&(T_{\mathfrak{h}'_2 \mathfrak{h}'_1} \circ T_{\mathfrak{h}'_1 \mathfrak{h}_1}\varphi)(\exp X)\\
=&(c_2^X)^{-1}(R_2^X \circ T_{\mathfrak{h}'_2 \mathfrak{h}'_1} \circ T_{\mathfrak{h}'_1 \mathfrak{h}_1}\varphi)(e)\\
=&(c_2^X)^{-1}\alpha^X\beta^X(T_{\mathfrak{h}_2^X \mathfrak{h}_1^X} \circ R_1^X \circ T_{\mathfrak{h}'_1 \mathfrak{h}_1}\varphi)(e) \qquad (2.6.2)\\
=&(c_2^X)^{-1}\alpha^X\beta^X c_1^X \oint_{H_2^X/(H_1^X \cap H_2^X)} (T_{\mathfrak{h}'_1 \mathfrak{h}_1}\varphi)(h \exp X)\chi_f(h)\Delta_{H_2^X,G}^{-1/2}(h)d\nu(h)\\
=&(c_2^X)^{-1}\alpha^X\beta^X c_1^X \gamma^X \oint_{H'_2/(H'_1 \cap H'_2)} (T_{\mathfrak{h}'_1 \mathfrak{h}_1}\varphi)(\exp X{\cdot}h)\chi_f(h)\Delta_{H'_2,G}^{-1/2}(h)d\nu(h).
\end{aligned}$$

Taking (2.6.1) into account, we deduce from this $(c_2^X)^{-1}\alpha^X\beta^X c_1^X\gamma^X = 1$ for each $X \in \mathfrak{t}$. We obtain in this way

$$\begin{aligned}
(T_{\mathfrak{h}_2 \mathfrak{h}_1}\varphi)(e) = \int_{\mathfrak{t}} \Delta_{H_2,G}^{-1/2}(\exp X)dX \oint_{H'_2/(H'_1 \cap H'_2)} (T_{\mathfrak{h}'_1 \mathfrak{h}_1}\varphi)(\exp X{\cdot}h)\\
\times \chi_f(h)\Delta_{H'_2,G}^{-1/2}(h)d\nu(h), \qquad (2.6.3)
\end{aligned}$$

if supp(φ) is sufficiently small modulo H_1.

We further find

$$\begin{aligned}
&\int_{\mathfrak{t}} \Delta_{H_2,G}^{-1/2}(\exp X)dX \oint_{H'_2/(H'_1 \cap H'_2)} |(T_{\mathfrak{h}'_1 \mathfrak{h}_1}\varphi)(\exp X{\cdot}h)|\Delta_{H'_2,G}^{-1/2}(h)d\nu(h)\\
&= \oint_{H_2/(H'_1 \cap H_2)} |(T_{\mathfrak{h}'_1 \mathfrak{h}_1}(g)|\Delta_{H_2,G}^{-1/2}(g)d\nu(g) < \infty \qquad (2.6.4)
\end{aligned}$$

due to the transitivity of $\nu_{\cdot,\cdot}$, Lemma 2.3.14 and Theorem 2.2.2 for $X \in \mathfrak{t}$.

The convergence of (2.6.4) allows us to use the transitivity of $\nu_{\cdot,\cdot}$ in the computation of $(T_{\mathfrak{h}_2 \mathfrak{h}_1}\varphi)(e)$, and we obtain the result from Fourier's inversion formula.

(c.2.0) Suppose now $\dim \mathfrak{a} = 2$ and $\mathfrak{a} \subset \mathfrak{h}_1^0 + \mathfrak{h}_2^0$. Let $0 \neq Y \in \mathfrak{a} \cap \ker(f)$. Then there are two vectors $X_j \in \mathfrak{h}_j^0 \cap \ker(f) (j = 1, 2)$ such that $Y = X_2 - X_1$. We immediately see

$$\mathfrak{h}'_1 \cap \mathfrak{h}_2 = (\mathfrak{h}_1 \cap \mathfrak{h}_2) \oplus \mathbb{R}X_2,\ \mathfrak{h}_1 \cap \mathfrak{h}'_2 = (\mathfrak{h}_1 \cap \mathfrak{h}_2) \oplus \mathbb{R}X_1$$

and

$$\mathfrak{h}'_1 \cap \mathfrak{h}'_2 = (\mathfrak{h}_1 \cap \mathfrak{h}_2) \oplus \mathbb{R}X_1 \oplus \mathbb{R}Y.$$

Lemma 2.6.3 *The elements X_1, X_2 can be taken in $\mathfrak{g}_0$.*

Proof If $\mathfrak{h}_1 \cap \mathfrak{h}_2 \not\subset \mathfrak{g}_0$, the lemma is trivial. Assume $\mathfrak{h}_1 \cap \mathfrak{h}_2 \subset \mathfrak{g}_0$. Taking Proposition 2.2.1 into account, there exists an element $T \in \mathfrak{h}_1 \cap \ker(f)$ verifying $[T, Y] = Z$. Then $T \notin \mathfrak{h}'_1 + \mathfrak{h}_2$. Suppose $[X_2, Y] = \alpha Y$ with $\alpha \neq 0$. Let $f' \in \mathfrak{g}^*$ satisfying $f' \in f + (\mathfrak{h}'_1 + \mathfrak{h}_2)^\perp$ and $f'(T) = -1/\alpha$. The Pukanszky condition requires $\mathfrak{h}_1, \mathfrak{h}_2 \in M(f', \mathfrak{g})$. Well, $[T, \mathfrak{h}_1 \cap \mathfrak{h}_2] \subset \mathfrak{h}_1^0 \cap \ker(f)$ and $B_{f'}(T, X_2) = f'([T, X_1 + Y]) = f'(\alpha T + W + Z) = 0$, where $W \in \mathfrak{h}_1^0 \cap \ker(f)$. Consequently, $T \in \mathfrak{h}'_1 + \mathfrak{h}_2$, a contradiction. □

Since $\mathfrak{p} = \mathbb{R}X_2 \oplus \mathfrak{g}(f)$ is contained in $\mathfrak{h}'_1 \cap \mathfrak{h}_2$ and $\mathfrak{h}_1^{\mathfrak{p}} = \mathfrak{h}_1 \cap \mathfrak{p}^f + \mathfrak{p} = \mathfrak{h}'_1$, Lemma 2.3.11 (a) and (c) give us $a(\mathfrak{h}_1, \mathfrak{h}'_1, \mathfrak{h}_2) = a(\mathfrak{h}'_1, \mathfrak{h}'_1, \mathfrak{h}_2) = 1$. We now use (b) to establish

$$a(\mathfrak{h}_1, \mathfrak{h}_0, \mathfrak{h}_2) = a(\mathfrak{h}'_1, \mathfrak{h}_0, \mathfrak{h}_2)a(\mathfrak{h}_1, \mathfrak{h}'_1, \mathfrak{h}_2)a(\mathfrak{h}_1, \mathfrak{h}_0, \mathfrak{h}'_1) = a(\mathfrak{h}'_1, \mathfrak{h}_0, \mathfrak{h}'_2).$$

Further, taking Lemma 2.6.3 into account,

$$H'_2 = H_2^0 \exp(\mathbb{R}Y),\ H'_1 \cap H'_2 = (H_1 \cap H_2) \exp(\mathbb{R}X_2) \exp(\mathbb{R}Y)$$

and the space $H'_2/(H'_1 \cap H'_2)$ is isomorphic to $H_2^0/((H_1 \cap H_2)\exp(\mathbb{R}X_2))$. All these observations show, for $\varphi \in \mathscr{H}_{\pi_1}^\infty$,

$$\begin{aligned}(T_{\mathfrak{h}_2\mathfrak{h}_1}\varphi)(e) &= a(\mathfrak{h}'_1, \mathfrak{h}_0, \mathfrak{h}'_2)(T_{\mathfrak{h}_2\mathfrak{h}'_2} \circ T_{\mathfrak{h}'_2\mathfrak{h}_0} \circ T_{\mathfrak{h}_0\mathfrak{h}'_1} \circ T_{\mathfrak{h}'_1\mathfrak{h}_1}\varphi)(e)\\ &= \oint_{H_2/H_2^0} (T_{\mathfrak{h}'_2\mathfrak{h}'_1}(\pi'_1(k^{-1})T_{\mathfrak{h}'_1\mathfrak{h}_1}\varphi))(e)\chi_f(k)\Delta_{H_2,G}^{-1/2}(k)d\nu(k)\end{aligned}$$

from Lemma 2.3.14. We can require $\mathrm{supp}(\varphi)$ be sufficiently small modulo H_1 so that, when k runs over a subset of G which represents H_2/H_2^0, any $\pi'_1(k^{-1})T_{\mathfrak{h}'_1\mathfrak{h}_1}\varphi$ has sufficiently small support in A^f/H'_1, and we can apply the formula to $(T_{\mathfrak{h}'_2\mathfrak{h}'_1}(\pi'_1(k^{-1})T_{\mathfrak{h}'_1\mathfrak{h}_1}\varphi))(e)$. Consequently, making use of the trace relation and the transitivity of $\nu_{\cdot,\cdot}$, we obtain

$$\begin{aligned}(T_{\mathfrak{h}_2\mathfrak{h}_1}\varphi)(e) &= \oint_{H_2/H_2^0} \chi_f(k)\Delta_{H_2,G}^{-1/2}(k)d\nu(k) \oint_{H'_2/(H'_1\cap H'_2)} \chi_f(x)\Delta_{H'_2,G}^{-1/2}(x)d\nu(x)\\ &\quad\times \int_{\mathbb{R}} \varphi(kx\exp(yY))dy\\ &= \oint_{H_2/H_2^0} \chi_f(k)\Delta_{H_2,G}^{-1/2}(k)d\nu(k) \oint_{H'_2/(H'_1\cap H'_2)} \chi_f(x)\Delta_{H'_2,G}^{-1/2}(x)d\nu(x)\end{aligned}$$

$$\times \int_{\mathbb{R}} \varphi(kx\exp(yX_2))\Delta_{H_1,G}^{-1/2}(\exp(yX_1))dy$$

$$= \oint_{H_2/H_2^0} \chi_f(k)\Delta_{H_2,G}^{-1/2}(k)d\nu(k)$$

$$\times \oint_{H_2^0/(H_1\cap H_2)} \varphi(kx)\chi_f(x)\Delta_{H_2,G}^{-1/2}(x)\Delta_{H_2^0,H_2}^{-1/2}(x)d\nu(x)$$

$$= \int_{H_2/(H_1\cap H_2)} \varphi(h)\chi_f(h)\Delta_{H_2,G}^{-1/2}(h)d\nu(h).$$

(c.2.1) Here we suppose $\mathfrak{a} \subset \mathfrak{h}_1 + \mathfrak{h}_2$ but $\mathfrak{a} \not\subset \mathfrak{h}_1^0 + \mathfrak{h}_2^0$. There are two elements $T_1, T_2 \in \mathfrak{g} \cap \ker(f)$ such that

$$\mathfrak{h}_1 = \mathbb{R}T_1 \oplus \mathfrak{h}_1^0, \mathfrak{h}_2 = \mathbb{R}T_2 \oplus \mathfrak{h}_2^0, [T_1, Y] = [T_2, Y] = \gamma Y + Z (\gamma \in \mathbb{R})$$

and $Y = \alpha(T_2 - T_1)$ with $0 \neq \alpha \in \mathbb{R}$.

Lemma 2.6.4

(1) $\mathfrak{h}_1 \cap \mathfrak{h}_2 \subset \mathfrak{g}_0$.
(2) We can choose T_1, T_2 *in* $\mathfrak{g}_0$.

Proof We first show $\mathfrak{h}_1 \cap \mathfrak{h}_2 \subset \mathfrak{g}_0$. If not, there would exist $X \in \mathfrak{h}_1 \cap \mathfrak{h}_2$ verifying $[X, Y] = Y$. Then

$$Y - [X, \alpha(Y_2 - Y_1)] = -\alpha(T_2 - T_1) + W_1 + W_2 = -Y + W_1 + W_2$$

with $W_1 \in \mathfrak{h}_1^0$, $W_2 \in \mathfrak{h}_2^0$ (cf. Proposition 2.2.1). Hence $Y \in \mathfrak{h}_1^0 + \mathfrak{h}_2^0$, contrary to the hypothesis.

Let us prove (2). Suppose $\gamma \neq 0$. Then $\alpha[T_1, T_2] = \gamma Y + Z$. Putting $\mathfrak{b} = \mathbb{R}T_2 \oplus \mathfrak{h}_1^0$, we have $\mathfrak{h}_2 \cap \mathfrak{b} = \mathbb{R}T_2 \oplus (\mathfrak{h}_1 \cap \mathfrak{h}_2)$, $f([\mathfrak{b}, \mathfrak{b}]) = \{0\}$ and $[\mathfrak{b}, \mathfrak{b}] \subset \mathbb{R}[X_1, T_2] \oplus (\mathfrak{h}_1^0 \cap \mathfrak{g}_0)$, where X_1 is an element of $\mathfrak{h}_1^0$ satisfying $[X_1, Y] = Y$ (such an element exists by Proposition 2.2.1). Suppose $[T_1, T_2] \in \mathfrak{h}_2 + [\mathfrak{b}, \mathfrak{b}]$. We have

$$\gamma Y + Z \in \mathfrak{h}_2 + \mathbb{R}([X_1, T_1] + \alpha^{-1}Y) + \mathfrak{h}_1^0 \cap \mathfrak{g}_0.$$

Hence there exists $\beta \in \mathbb{R}$ such that

$$\gamma Y - \beta([X_1, T_1] + \alpha^{-1}Y) \in \mathfrak{h}_2 + \mathfrak{h}_1^0 \cap \mathfrak{g}_0.$$

If $\beta = \alpha\gamma$, this implies $[X_1, T_1] \in \mathfrak{h}_2 + \mathfrak{h}_1^0 \cap \mathfrak{g}_0$, which is excluded. Therefore we have $\beta \neq \alpha\gamma$ and $Y \in \mathfrak{h}_2 + \mathfrak{h}_1^0 \cap \mathfrak{g}_0$, from which the result follows.

Suppose next $[T_1, T_2] \notin \mathfrak{h}_2 + [\mathfrak{b}, \mathfrak{b}]$. Let $f' \in \mathfrak{g}^*$ belong to $f + (\mathfrak{h}_2 + [\mathfrak{b}, \mathfrak{b}])^\perp$ and verify $f'([T_1, T_2]) = 0$. From the Pukanszky condition $\mathfrak{h}_2 \in M(f', \mathfrak{g})$. On the other hand, $\mathfrak{b}$ is a totally isotropic subspace and hence maximal for the form $B_{f'}$, because $\dim \mathfrak{b} = \dim \mathfrak{h}_2$. As $f'([T_1, \mathfrak{h}_2 \cap \mathfrak{b}]) = \{0\}$, we obtain $T_1 \in \mathfrak{h}_2 + \mathfrak{b}$, contrary to the hypothesis $\mathfrak{h}_1 \cap \mathfrak{h}_2 \subset \mathfrak{a}^f$. □

Now we prove a lemma which will play a key role in what follows.

Lemma 2.6.5 *We can choose* $T_1, T_2 \in \mathfrak{g}_0$ *so that*

$$Tr\ ad_{\mathfrak{h}_1/(\mathfrak{h}_1 \cap \mathfrak{h}_2)} T_1 + Tr\ ad_{\mathfrak{h}_2/(\mathfrak{h}_1 \cap \mathfrak{h}_2)} T_2 = 0.$$

Proof If $Y \in \mathfrak{h}_1 + [\mathfrak{h}_2, \mathfrak{h}_2]$, our assertion is trivial, since we can choose $T_2 \in [\mathfrak{h}_2, \mathfrak{h}_2]$ in this case and then both vectors $T_i, i = 1, 2$, are contained in the nilradical of $\mathfrak{g}$. We suppose therefore that $Y \notin \mathfrak{h}_1 + [\mathfrak{h}_2, \mathfrak{h}_2]$, and we will adapt the proof of Theorem 2.2.2.

If $\mathfrak{h}_2 \subset \mathfrak{g}_0$, then we have that $[Y, \mathfrak{h}_2] \subset \mathfrak{h}_2$ and the subspace $\mathfrak{q}_2 = \mathbb{R}T_1 + \mathfrak{h}_2^0$ belongs to $I(f, \mathfrak{g})$ and $T_1 \in \mathfrak{h}_1 \cap \mathfrak{q}_2$. Hence according to Theorem 2.2.2 we have that

$$\mathrm{Tr}\ \mathrm{ad}_{\mathfrak{h}_1/(\mathfrak{h}_1 \cap \mathfrak{q}_2)} T_1 + \mathrm{Tr}\ \mathrm{ad}_{\mathfrak{q}_2/(\mathfrak{h}_1 \cap \mathfrak{q}_2)} T_1 = 0.$$

Our relation then follows from the fact that $\mathrm{Tr}\ \mathrm{ad}_{\mathfrak{h}_1/(\mathfrak{h}_1 \cap \mathfrak{q}_2)} T_1 = \mathrm{Tr}\ \mathrm{ad}_{\mathfrak{h}_1/(\mathfrak{h}_1 \cap \mathfrak{h}_2)} T_1$ and $\mathrm{Tr}\ \mathrm{ad}_{\mathfrak{h}_2/(\mathfrak{h}_1 \cap \mathfrak{h}_2)} T_2 = \mathrm{Tr}\ \mathrm{ad}_{\mathfrak{q}_2/(\mathfrak{h}_1 \cap \mathfrak{q}_2)} T_1$.

If $\mathfrak{h}_2 \not\subset \mathfrak{g}_0$, then we can find $X_2' \in \mathfrak{h}_2 \cap \ker(f)$ which satisfies the relation $[X_2', Y] = Y + cZ$ for some $c \in \mathbb{R}$ and then $X_2 = X_2' - cT_2$ has the property that $X_2 \in \mathfrak{h}_2^0 \cap \ker(f)$ and $[X_2, Y] = Y$. We can suppose that $X_2 \notin (\mathfrak{h}_1 + \mathfrak{h}_2) \cap \mathfrak{g}_0 + \mathbb{R}Y$, since otherwise

$$X_2 - U_2 = U_1 + sY$$

for some $s \in \mathbb{R}$, $U_1 \in \mathfrak{h}_1$ and some $U_2 \in \mathfrak{h}_2 \cap \mathfrak{g}_0$. We can apply this relation to the identity $Y = \alpha(T_2 - T_1)$ and we get $T_i' \in \mathfrak{h}_i \cap [\mathfrak{g}, \mathfrak{g}], i = 1, 2$, so that $Y = \alpha(T_2' - T_1')$, $[T_2', Y] = Z$. We can therefore assume that $Y \notin \mathbb{R}X_2 + \mathfrak{h}_1 + (\mathfrak{h}_2^0 \cap \mathfrak{g}_0)$.

Let us fix a subspace $\mathfrak{w}$ of $\mathfrak{g}$ such that $\mathfrak{w}$ contains $\mathfrak{h}_1 + (\mathfrak{h}_2^0 \cap \mathfrak{g}_0) + \mathbb{R}X_2$ and such that we have $\mathfrak{g} = \mathfrak{w} \oplus \mathbb{R}Y$. Let

$$\mathfrak{w}_0 = \{U \in \mathfrak{w};\ [T_1, U] \in \mathfrak{w},\ \langle f, [T_1, U]\rangle = 0\} \cap \ker(f).$$

Then $\mathfrak{w}_0 + \mathbb{R}Z$ contains $\mathfrak{h}_1 + (\mathfrak{h}_2^0 \cap \mathfrak{g}_0)$ and furthermore

$$\mathfrak{g} = \mathbb{R}Y \oplus \mathbb{R}X_2 \oplus \mathbb{R}Z \oplus \mathfrak{w}_0.$$

Indeed, there exist a linear functional $\gamma' \in \mathfrak{g}^*$ and a linear map $w' : \mathfrak{g} \to \mathfrak{w}$, such that for any $X \in \mathfrak{g}$, $[T_1, X] = \gamma'(X)Y + w'(X)$. Since $[T_1, X_2] \in \frac{1}{\alpha}Y + [\mathfrak{h}_2, \mathfrak{h}_2]$, we have that

$$[T_1, X - \frac{\alpha}{2}\gamma'(X)X_2 + \langle f, w'(X)\rangle Y]$$

$$= \gamma'(X)Y + w'(X) - \frac{\alpha}{2}\gamma'(X)[T_1, X_2] + \langle f, w'(X)\rangle Z$$

$$\in (\ker(f) \cap \mathfrak{w}) + (\mathfrak{h}_1 + \mathfrak{h}_2^0) \cap \ker(f) = \ker(f) \cap \mathfrak{w}.$$

Hence $X \in \mathbb{R}X_2 + \mathbb{R}Y + \mathbb{R}Z + \mathfrak{w}_0$. It is clear that the sum $\mathbb{R}X_2 + \mathbb{R}Y + \mathbb{R}Z + \mathfrak{w}_0$ is direct. We can therefore write:

$$X = \beta(X)Y + \lambda(X)Z + \alpha\gamma(X)X_2 + w_0(X), \; X \in \mathfrak{g} \tag{2.6.5}$$

for certain linear forms $\lambda, \beta, \gamma \in \mathfrak{g}^*$ and for some element $w_0(X) \in \mathfrak{w}_0$. For $c, \tau \in \mathbb{R}$, let $\ell_{c,\tau} \in \mathfrak{g}^*$ be chosen so that

$$\ell_{c,\tau} = f \text{ on } \mathfrak{w}_0 + \mathbb{R}Z, \; \ell_{c,\tau}(Y) = c \text{ and } \ell_{c,\tau}(X_2) = \frac{\tau}{\alpha}.$$

In particular we have that $\ell_{0,0} = f$. The two spaces $\mathfrak{h}_1^0/(\mathfrak{h}_1 \cap \mathfrak{h}_2)$ and $\mathfrak{h}_2^0/(\mathfrak{h}_1 \cap \mathfrak{h}_2)$ are dual under $B_{\ell_{c,\tau}}$ for all $c, \tau \in \mathbb{R}$. Let us write $\mathfrak{h}_1^0 = \mathfrak{m}_1 \oplus (\mathfrak{h}_1 \cap \mathfrak{h}_2)$ and $\mathfrak{h}_2^0 = \mathfrak{m}_2 \oplus (\mathfrak{h}_1 \cap \mathfrak{h}_2)$ for certain subspaces $\mathfrak{m}_1, \; \mathfrak{m}_2$. Let $\{\hat{e}_j; \, 1 \leq j \leq m\}$ be a basis of $\mathfrak{m}_2$ and let $\{e_i; \, 1 \leq i \leq m\}$ (resp. $\{e_i(c,\tau); \, 1 \leq i \leq m\}$) be the dual basis of $\mathfrak{m}_1$ for the bilinear forms B_ℓ (resp. $B_{\ell_{c,\tau}}, c, \tau \in \mathbb{R}$).

Writing

$$W := \sum_{j=1}^{m} [e_j, \hat{e}_j]$$

and

$$W_{c,\tau} := \sum_{j=1}^{m} [e_j(c,\tau), \hat{e}_j],$$

we arrive at

$$\operatorname{Tr} \operatorname{ad}_{\mathfrak{h}_1/(\mathfrak{h}_1 \cap \mathfrak{h}_2)} T_1 + \operatorname{Tr} \operatorname{ad}_{\mathfrak{h}_2/(\mathfrak{h}_1 \cap \mathfrak{h}_2)} T_2 = B_{\ell_{c,\tau}}(T_1, W_{c,\tau}), \, c, \tau \in \mathbb{R}. \tag{2.6.6}$$

According to (2.6.5)

$$[e_i, \hat{e}_j] = \alpha\gamma_{ji}X_2 + \beta_{ji}Y + \delta_{ji}Z + w_{ji},$$

where $\gamma_{ij}, \; \beta_{ij} \in \mathbb{R}$, $w_{ij} \in \mathfrak{w}_0 (1 \leq i, j \leq m)$, and where δ_{ij} is the Kronecker delta.

If we write $e_k(c,\tau) = \sum_{j=1}^{m} \lambda_{kj}(c,\tau)e_j$ with $\lambda_{kj}(c,\tau) \in \mathbb{R} (1 \leq j, k \leq m)$, then we get

$$[e_k(c,\tau), \hat{e}_i] = \sum_{j=1}^{m} \lambda_{kj}(c,\tau)[e_j, \hat{e}_i] = \sum_{j=1}^{m} \lambda_{kj}(c,\tau) \left(\alpha\gamma_{ij}X_2 + \beta_{ij}Y + \delta_{ij}Z + w_{ij}\right). \tag{2.6.7}$$

Hence

$$\delta_{ki} = \langle \ell_{c,\tau}, [e_k(c,\tau), \hat{e}_i] \rangle = \sum_{j=1}^{m} \lambda_{kj}(c,\tau)\left(c\beta_{ij} + \tau\gamma_{ij}\right) + \lambda_{ki}(c,\tau)(1 \leq j, k \leq m). \tag{2.6.8}$$

For the trace condition, using (2.6.6) and (2.6.7), we get

$$\begin{aligned}
\beta &= B_{\ell_{c,\tau}}(T_1, \sum_{k=1}^{m} [e_k(c,\tau), \hat{e}_k]) \\
&= \langle \ell_{c,\tau}, \sum_{k=1}^{m} [T_1, [e_k(c,\tau), \hat{e}_k]] \rangle \\
&= \langle \ell_{c,\tau}, \sum_{k=1}^{m} \sum_{j=1}^{m} [T_1, \lambda_{kj}(c,\tau)\left(\alpha\gamma_{kj} X_2 + \beta_{kj} Y + \delta_{kj} Z + w_{kj}\right)] \rangle \\
&= \sum_{k,j=1}^{m} \left(\beta_{kj} + 2c\gamma_{kj}\right) \lambda_{kj}(c,\tau).
\end{aligned} \tag{2.6.9}$$

We now interpret Eqs. (2.6.8) and (2.6.9) in the following way. Consider the $m \times m$ matrices $B := (\beta_{i,j})_{1 \leq i,j \leq m}$, $\Gamma := (\gamma_{i,j})_{1 \leq i,j \leq m}$ and $\Lambda(c,\tau) := (\lambda_{i,j}(c,\tau))_{1 \leq i,j \leq m}$. Equation (2.6.8) then reads

$$(cB + \tau\Gamma + \mathbb{I}_m)\Lambda^t(c,\tau) = \mathbb{I}_m,$$

that is,

$$\Lambda^t(c,\tau) = (cB + \tau\Gamma + \mathbb{I}_m)^{-1}$$

and

$$\beta = \mathrm{Tr}(\Lambda^t(c,\tau)(B+2c\Gamma)) = \mathrm{Tr}((cB+\tau\Gamma+\mathbb{I}_m)^{-1}(B+2c\Gamma)), c \in \mathbb{R}. \tag{2.6.10}$$

We must prove that $\beta = 0$.

Let $\mathscr{S} = \mathscr{S}^k$ be the family of all subsets S of the set $\{0, \ldots, k-1\} := E_k$, and $\mathscr{S}_i = \mathscr{S}_i^k$ the family of elements $S \in \mathscr{S}$ containing i elements. We denote by δ_S, $S \in \mathscr{S}$, its characteristic function, i.e., $\delta_S(x) = 1$, if $x \in S$ and $\delta_S(x) = 0$, if $x \notin S$. For $S \in \mathscr{S}_i$, let M_S be the $m \times m$ matrix $M_S := B^{\delta_S(0)}\Gamma^{1-\delta_S(0)} \ldots B^{\delta_S(k-1)}\Gamma^{1-\delta_S(k-1)}$ and $M_i^k := \sum_{S \in \mathscr{S}_i} M_S, 0 \leq i \leq k$.

For c and τ small enough, we can express $(cB + \tau\Gamma + \mathbb{I}_m)^{-1}$ as a convergent series in c, τ:

$$\begin{aligned}(cB + \tau\Gamma + \mathbb{I}_m)^{-1} &= \sum_{k=0}^{\infty}(-1)^k(cB + \tau\Gamma)^k \\ &= \sum_{k=0}^{\infty}(-1)^k\sum_{i=0}^{k} c^i\tau^{k-i}M_i^k \\ &= \sum_{i,j=0}^{\infty}(-1)^{i+j}c^i\tau^j M_i^{i+j}.\end{aligned}$$

If we plug this identity into relation (2.6.10), we get the identity

$$\mathrm{Tr}B = \sum_{i,j=0}^{\infty}(-1)^{i+j}c^i\tau^j\mathrm{Tr}(BM_i^{i+j}) + 2\sum_{i,j=0}^{\infty}(-1)^{i+j}c^{i+1}\tau^j\mathrm{Tr}(\Gamma M_i^{i+j}),\ c, \tau \in \mathbb{R}. \tag{2.6.11}$$

Hence

$$\mathrm{Tr}(B\Gamma^j) = 0,\ j \in \mathbb{N}^*, \tag{2.6.12}$$

and

$$\mathrm{Tr}(BM_{i+1}^{k+1}) = 2\mathrm{Tr}(\Gamma M_i^k),\ k \geq i \geq 0. \tag{2.6.13}$$

Let Σ be the group of bijections of the set E_k generated by the cyclic permutation $\sigma(i) = i + 1$ modulo $k(i \in E_k)$. This group has k elements and acts on the space $\mathscr{S}_i$ in a natural way. For $S \in \mathscr{S}_i^k$ let Σ_S be the stabilizer of S in Σ. Denote by d_S the number of elements of Σ_S, $S \in \mathscr{S}_i^k$. Let also Ω be the orbit space $\mathscr{S}_i/\Sigma$. We choose for every Σ-orbit O in $\mathscr{S}_i$ a representative S_O and we denote by $|O|$ the number of its elements. For any $S \in \mathscr{S}_i$, we have that

$$\begin{aligned}M_{\sigma S} &= B^{\delta_S(\sigma^{-1}(0))}\Gamma^{1-\delta_S(\sigma^{-1}(0))}\ldots B^{\delta_S(\sigma^{-1}(k-1))}\Gamma^{1-\delta_S(\sigma^{-1}(k-1))} \\ &= B^{\delta_S(k-1)}\Gamma^{1-\delta_S(k-1)}B^{\delta_S(0)}\Gamma^{1-\delta_S(0)}\ldots B^{\delta_S(k-2)}\Gamma^{1-\delta_S(k-2)}\end{aligned}$$

Hence

$$\mathrm{Tr}(M_S) = \mathrm{Tr}(M_{\sigma S}) = \mathrm{Tr}(M_{\sigma^l S}),\ l \in \mathbb{N}.$$

This allows us to write

$$\mathrm{Tr}(M_i^k) = \sum_{O\in\Omega} |O|\mathrm{Tr}(M_{S_O}) = k \sum_{O\in\Omega} \frac{1}{d_{S_O}}\mathrm{Tr}(M_{S_O}). \tag{2.6.14}$$

Consider the mapping $I : \mathscr{S}_i^{k-1} \to \mathscr{S}_i^k, T = (t_1 < t_2 < \cdots < t_i) \mapsto (t_1 + 1 < \cdots < t_i + 1)$. This mapping is a bijection of the set $\mathscr{S}_i^{k-1}$ onto the subset $\mathscr{S}_{i,1}^k := \{S \in \mathscr{S}_i^k, 0 \notin S\}$.

Lemma 2.6.6 *We have the formula*

$$Tr(M_i^k) = \frac{k}{k-i} Tr(\Gamma M_i^{k-1}) = \frac{k}{i} Tr(BM_{i-1}^{k-1}), 1 \le i < k. \tag{2.6.15}$$

Proof Indeed, we have that

$$\mathrm{Tr}(\Gamma M_i^{k-1}) = \sum_{T\in\mathscr{S}_i^{k-1}} \mathrm{Tr}(\Gamma M_T) = \sum_{T\in\mathscr{S}_i^{k-1}} \mathrm{Tr}(M_{I(T)}) = \sum_{S\in\mathscr{S}_{i,1}^k} \mathrm{Tr}(M_S).$$

Since $i < k$, we can take every $S_O \in \mathscr{S}_{i,1}^k$ and we write

$$\mathrm{Tr}(\Gamma M_i^{k-1}) = \sum_{O\in\Omega} |O \cap \mathscr{S}_{i,1}^k| \mathrm{Tr}(M_{S_O}). \tag{2.6.16}$$

Now it is clear that for $\tau \in \Sigma$, necessarily $\tau S \in \mathscr{S}_{i,1}^k$ if and only if $\tau = \sigma^{-t}$, for some $t \in E_k \backslash S$. On the other hand the subset $\Sigma_{S,1} := \{\tau \in \Sigma, \tau S \in \mathscr{S}_{i,1}^k\}$ is invariant under multiplication by elements of Σ_S, since S is Σ_S-invariant. Hence $|O \cap \mathscr{S}_{i,1}^k| = \frac{k-i}{d_{S_O}}$, $O \in \Omega$. Comparing (2.6.14) to (2.6.16), we obtain the desired formula. The formula for $\mathrm{Tr}(BM_i^k)$ can be proved in a similar way. □

We now apply the previous identities. Since $\mathrm{Tr}(B\Gamma^k) = 0$ for every $k \ge 1$ by (2.6.12), it follows from (2.6.13) and (2.6.15) that

$$\begin{aligned} \mathrm{Tr}(BM_{2l}^k) &= 2\mathrm{Tr}(\Gamma M_{2l-1}^{k-1}) = 2\frac{k-2l+1}{k}\mathrm{Tr}(M_{2l-1}^k) \\ &= 2\frac{k-2l+1}{k}\frac{k}{2l-1}\mathrm{Tr}(BM_{2l-2}^{k-1}), k \ge 2l+1, \end{aligned} \tag{2.6.17}$$

and so

$$0 = \mathrm{Tr}(B\Gamma^k) = \mathrm{Tr}(BM_2^{k+1}) = \cdots = \mathrm{Tr}(BM_{2s}^{k+s}) = \cdots = \mathrm{Tr}(BB^{2k}). \tag{2.6.18}$$

Hence

$$\mathrm{Tr}B = \mathrm{Tr}(B((\mathbb{I}_m + c^2B^2 + c^4B^4 + \cdots)) = \mathrm{Tr}(B(\mathbb{I}_m - c^2B^2)^{-1}), \tag{2.6.19}$$

where c is small enough. Since the mapping $\mathbb{R} \to M_m(\mathbb{R}),\, c \mapsto (\mathbb{I}_m - c^2B^2)^{-1} = (\mathbb{I}_m - cB)^{-1}(\mathbb{I}_m + cB)^{-1}$ is analytic, it follows that

$$\mathrm{Tr}B = \mathrm{Tr}(B(\mathbb{I}_m + c^2B^2 + c^4B^4 + \cdots)) = \mathrm{Tr}(B(\mathbb{I}_m - c^2B^2)^{-1}),\, c \in \mathbb{R}.$$

Since

$$\mathrm{Tr}(B(\mathbb{I}_m - c^2B^2)^{-1}) = \sum_{\lambda \in \sigma(B)} \frac{\lambda}{1 - c^2\lambda^2},\, c \in \mathbb{R},$$

we see that

$$\mathrm{Tr}(B) = \lim_{c \to \infty} \sum_{\lambda \in \sigma(B)} \frac{\lambda}{1 - c^2\lambda^2} = 0.$$

□

We continue the proof of Theorem 2.6.1. As before, we see

$$a(\mathfrak{h}_1, \mathfrak{h}_0, \mathfrak{h}_2) = a(\mathfrak{h}_1', \mathfrak{h}_0, \mathfrak{h}_2')a(\mathfrak{h}_1, \mathfrak{h}_1', \mathfrak{h}_2).$$

But the B_f-orthogonal space to $\mathfrak{h}_1^0$ in $\mathfrak{h}_2$ coincides with $\mathbb{R}T_2 + (\mathfrak{h}_1^0 \cap \mathfrak{h}_2)$ and Lemma 2.3.11 (c) says $a(\mathfrak{h}_1, \mathfrak{h}_1', \mathfrak{h}_2) = a(\mathfrak{h}_1, \mathfrak{h}_1', \mathfrak{m})$, if $\mathfrak{m} = \mathfrak{h}_1^0 + \mathbb{R}T_2$. Let U', V', W' be vectors in $\mathfrak{h}_1^0$ and put

$$U = U' + uT_1,\ V = V' + \frac{v}{\alpha}Y,\ W = W' + wT_2,$$

where $(u, v, w) \in \mathbb{R}^3$. Then $\tau(\mathfrak{h}_1, \mathfrak{h}_1', \mathfrak{m})$ is the index of the quadratic form

$$Q(U, V, W) = \frac{1}{\alpha}(uv - vw - wu).$$

Hence

$$a(\mathfrak{h}_1, \mathfrak{h}_1', \mathfrak{m}) = e^{-\frac{i\pi}{4}\mathrm{sgn}\alpha},\ a(\mathfrak{h}_1, \mathfrak{h}_0, \mathfrak{h}_2) = e^{-\frac{i\pi}{4}\mathrm{sgn}\alpha}a(\mathfrak{h}_1', \mathfrak{h}_0, \mathfrak{h}_2').$$

Now, evidently $\mathfrak{h}_1 \cap \mathfrak{h}_2 = \mathfrak{h}_1' \cap \mathfrak{h}_2$ and $[T_2, \mathfrak{h}_1 \cap \mathfrak{h}_2] \subset \mathfrak{h}_1 \cap \mathfrak{h}_2$. Let $\mathfrak{k} = (\mathfrak{h}_1 \cap \mathfrak{h}_2) + \mathbb{R}T_2$ and $K = \exp \mathfrak{k}$. This permits us to identify H_2/K with $H_2^0/(H_1 \cap H_2)$. Lemmas 2.3.12, 2.6.4 and the transitivity of $\nu_{\cdot,\cdot}$ give us the following equalities:

$$
\begin{aligned}
&\oint_{H_2/(H_1'\cap H_2)} |T_{\mathfrak{h}_1'\mathfrak{h}_1}\varphi(h)|\Delta_{H_2,G}^{-1/2}(h)d\nu(h)\\
&=\oint_{H_2^0/(H_1\cap H_2)} \Delta_{H_2,G}^{-1/2}(h)d\nu(h)\\
&\quad\times\int_{\mathbb{R}} |T_{\mathfrak{h}_1'\mathfrak{h}_1}\varphi(h\exp(tT_2))|\Delta_{H_2,G}^{-1/2}(\exp(tT_2))\Delta_{H_1\cap H_2,H_2}^{-1}(\exp(tT_2))dt\\
&=\oint_{H_2^0/(H_1\cap H_2)} \Delta_{H_2,G}^{-1/2}(h)d\nu(h)\\
&\quad\int_{\mathbb{R}}\Big|\int_{\mathbb{R}} e^{i\alpha y^2/2}\varphi(h\exp((t+\alpha y)T_2))\Delta_{H_1,G}^{-1/2}(\exp(\alpha yT_1))dy\Big|\\
&\quad\times\Delta_{H_2,G}^{-1/2}(\exp(tT_2))\Delta_{H_1\cap H_2,H_2}^{-1}(\exp(tT_2))dt\\
&=\frac{1}{|\alpha|}\int_{H_2^0/(H_1\cap H_2)} \Delta_{H_2,G}^{-1/2}(h)d\nu(h)\int_{\mathbb{R}} dt\\
&\quad\times\Big|\int_{\mathbb{R}} e^{i(t-y)^2/2\alpha}\varphi(h\exp(yT_2))\Delta_{H_2,G}^{-1/2}(\exp(yT_2))\Delta_{H_1\cap H_2,H_2}^{-1}(\exp(yT_2))dy\Big| < \infty.
\end{aligned}
$$

Having established convergence, we have

$$
\begin{aligned}
(T_{\mathfrak{h}_2\mathfrak{h}_1}\varphi)(e) &= a(\mathfrak{h}_1,\mathfrak{h}_0,\mathfrak{h}_2)(T_{\mathfrak{h}_2\mathfrak{h}_0}\circ T_{\mathfrak{h}_0\mathfrak{h}_1}\varphi)(e)\\
&= e^{-\frac{i\pi}{4}\operatorname{sgn}\alpha}(T_{\mathfrak{h}_2\mathfrak{h}_2'}\circ T_{\mathfrak{h}_2'\mathfrak{h}_1'}\circ T_{\mathfrak{h}_1'\mathfrak{h}_1}\varphi)(e)\\
&= e^{-\frac{i\pi}{4}\operatorname{sgn}\alpha}\oint_{H_2/H_2^0}(T_{\mathfrak{h}_2'\mathfrak{h}_1'}\circ T_{\mathfrak{h}_1'\mathfrak{h}_1}\varphi)(k)\chi_f(k)\Delta_{H_2,G}^{-1/2}(k)d\nu(k)\\
&= e^{-\frac{i\pi}{4}\operatorname{sgn}\alpha}\oint_{H_2/H_2^0}\chi_f(k)\Delta_{H_2,G}^{-1/2}(k)d\nu(k)\\
&\quad\times\oint_{H_2'/(H_1'\cap H_2')}(\pi_1'(k^{-1})T_{\mathfrak{h}_1'\mathfrak{h}_1}\varphi)(x)\chi_f(x)\Delta_{H_2',G}^{-1/2}(x)d\nu(x)\\
&= e^{-\frac{i\pi}{4}\operatorname{sgn}\alpha}\oint_{H_2/(H_1'\cap H_2)}(T_{\mathfrak{h}_1'\mathfrak{h}_1}\varphi)(h)\chi_f(h)\Delta_{H_2,G}^{-1/2}(h)d\nu(h)
\end{aligned}
$$

from Lemma 2.3.12, the transitivity of $\nu_{\cdot,\cdot}$ and the validity of our result for the function $\pi_1'(k^{-1})(T_{\mathfrak{h}_1'\mathfrak{h}_1}\varphi)$ whose support is sufficiently small in A^f/H_1'. Then

$$(T_{\mathfrak{h}_2\mathfrak{h}_1}\varphi)(e) = \sqrt{\frac{2\pi}{|\alpha|}} \oint_{H_2^0/(H_1\cap H_2)} \chi_f(k)\Delta_{H_2,G}^{-1/2}(k)d\nu(k)\int_{\mathbb{R}} \varphi(k\exp(tT_2))$$
$$\times \Delta_{H_2,G}^{-1/2}(\exp(tT_2))\Delta_{H_1\cap H_2,H_2}(\exp(tT_2))dt$$
$$= \oint_{H_2/(H_1\cap H_2)} \varphi(h)\chi_f(h)\Delta_{H_2,G}^{-1/2}(h)d\nu(h).$$

(c.3.2) We have $\dim\mathfrak{a} = 3$ and $\dim(\mathfrak{h}_j\cap\mathfrak{a}) = 2 (j = 1,2)$. Then necessarily $\mathfrak{h}_1+\mathfrak{h}_2 \subset \mathfrak{g}_0 \subsetneq \mathfrak{g}$. Altogether, this case does not appear.

(c.3.1.a) Suppose now $\dim(\mathfrak{h}_1\cap\mathfrak{a}) = 1$, $\dim(\mathfrak{h}_2\cap\mathfrak{a}) = 2$, $\mathfrak{h}_1\cap\mathfrak{h}_2 \not\subset \mathfrak{a}$. This means $\mathfrak{h}_2 \subset \mathfrak{g}_0$. Let $\{Y, Y'\}$ be a basis of $\mathfrak{a}\cap\ker(f)$ such that $\mathfrak{h}_2\cap\mathfrak{a} = \mathbb{R}Y\oplus\mathfrak{z}$. We know that $\mathfrak{h}_1^0\cap\mathfrak{h}_2^0$ is of codimension 1 in $\mathfrak{h}_1\cap\mathfrak{h}_2$ and

$$\mathfrak{h}_1'\cap\mathfrak{h}_2 = (\mathfrak{h}_1^0\cap\mathfrak{h}_2^0)\oplus\mathbb{R}Y,\ \mathfrak{h}_1\cap\mathfrak{h}_2' = \mathfrak{h}_1^0\cap\mathfrak{h}_2^0.$$

Let X be an element of $\mathfrak{h}_1\cap\mathfrak{h}_2\cap\ker(f)$ satisfying $[X, Y'] = Z$. Evidently, $[X, Y] = 0$. Put $\mathfrak{k} = \mathbb{R}X + (\mathfrak{h}_1'\cap\mathfrak{h}_2)$ and $K = \exp\mathfrak{k}$. We consider the isotropic subspace $\mathfrak{p}$ given by

$$\mathfrak{p} = \mathbb{R}X + \mathbb{R}Y + \mathfrak{g}(f) \subset (\mathfrak{h}_1\cap\mathfrak{h}_1') + (\mathfrak{h}_1'\cap\mathfrak{h}_2) + (\mathfrak{h}_2\cap\mathfrak{h}_1).$$

We see $(\mathfrak{h}_1)^{\mathfrak{p}} = (\mathfrak{h}_1\cap\mathfrak{p}^f) + \mathfrak{p} = \mathfrak{h}_1^0 + \mathfrak{p} = (\mathfrak{h}_1')^{\mathfrak{p}}$ and hence

$$a(\mathfrak{h}_1, \mathfrak{h}_0, \mathfrak{h}_2) = a(\mathfrak{h}_1', \mathfrak{h}_0, \mathfrak{h}_2) = a(\mathfrak{h}_1', \mathfrak{h}_0, \mathfrak{h}_2').$$

We can use (2.6.3) to find

$$(T_{\mathfrak{h}_2\mathfrak{h}_1}\varphi)(e) = \int_{\mathbb{R}} \Delta_{H_2,G}^{-1/2}(\exp(xX))dx$$
$$\times \oint_{H_2'/(H_1'\cap H_2')} (T_{\mathfrak{h}_1'\mathfrak{h}_1}\varphi)(\exp(xX)\cdot k)\chi_f(k)\Delta_{H_2',G}^{-1/2}(k)d\nu(k), \qquad (2.6.20)$$

if the support of φ is sufficiently small modulo H_1. Then

$$\int_{\mathbb{R}} \Delta_{H_2,G}^{-1/2}(\exp(xX))dx \oint_{H_2'/(H_1'\cap H_2')} |(T_{\mathfrak{h}_1'\mathfrak{h}_1}(\exp(xX)\cdot k)|\Delta_{H_2',G}^{-1/2}(k)d\nu(k)$$
$$= \oint_{H_2/K} \Delta_{H_2,G}^{-1/2}(k)d\nu(k)\int_{\mathbb{R}} dt\Big|\int_{\mathbb{R}^2} e^{-ity'}\varphi(k\exp(yY)\exp(y'Y'))dydy'\Big|,$$

which is finite for each function $\varphi \in \mathscr{H}_{\pi_1}^\infty$ in question.

This integrability allows us to apply the transitivity of $\nu_{\cdot,\cdot}$, and from (2.6.20) we obtain the result for this case:

$$(T_{\mathfrak{h}_2\mathfrak{h}_1}\varphi)(e) = 2\pi \oint_{H_2/((H_1\cap H_2)\exp(\mathbb{R}Y))} \chi_f(k)\Delta_{H_2,G}^{-1/2}(k)d\nu(k) \int_{\mathbb{R}} \varphi(k\exp(yY))dy$$
$$= \oint_{H_2/(H_1\cap H_2)} \varphi(h)\chi_f(h)\Delta_{H_2,G}^{-1/2}(h)d\nu(h).$$

(c.3.2.a) Now we have $\mathfrak{h}_1 \subset \mathfrak{g}_0$ and $\mathfrak{h}_2 \not\subset \mathfrak{g}_0$. Let $\{Y, Y'\}$ be a basis of $\mathfrak{a}\cap\ker(f)$ such that $\mathfrak{h}_2\cap\mathfrak{a} = \mathbb{R}Y \oplus \mathfrak{z}$. We know that $\mathfrak{h}_1^0\cap\mathfrak{h}_2^0$ is of codimension 1 in $\mathfrak{h}_1\cap\mathfrak{h}_2$ and

$$\mathfrak{h}_2'\cap\mathfrak{h}_1 = (\mathfrak{h}_1^0\cap\mathfrak{h}_2^0)\oplus\mathbb{R}Y,\ \mathfrak{h}_2\cap\mathfrak{h}_1' = \mathfrak{h}_1^0\cap\mathfrak{h}_2^0.$$

Let X be an element of $\mathfrak{h}_1\cap\mathfrak{h}_2\cap\ker(f)$ verifying $[X, Y'] = Z$. Evidently, $[X, Y] = 0$. Put $\mathfrak{k} = \mathbb{R}X + (\mathfrak{h}_1'\cap\mathfrak{h}_2)$ and $K = \exp\mathfrak{k}$. Take two elements T, X' in $\mathfrak{h}_2$ in such a way that

$$T \in \mathfrak{h}_2^0,\ T \notin \mathfrak{g}_0,\ X' \in [\mathfrak{h}_2, \mathfrak{h}_2], \mathfrak{h}_2 = \mathbb{R}X' + \mathbb{R}X + \mathfrak{h}_2^0$$

(we can take $X' = [T, X]$ for example). As in case (c.3.1.a), $a(\mathfrak{h}_1, \mathfrak{h}_0, \mathfrak{h}_2) = a(\mathfrak{h}_1', \mathfrak{h}_0, \mathfrak{h}_2')$. For $\varphi \in \mathscr{H}_{\pi_1}^{\infty}$ with sufficiently small support modulo H_1, we obtain

$$(T_{\mathfrak{h}_2\mathfrak{h}_1}\varphi)(e) = (T_{\mathfrak{h}_2\mathfrak{h}_2'}\circ T_{\mathfrak{h}_2'\mathfrak{h}_1'}\circ T_{\mathfrak{h}_1'\mathfrak{h}_1}\varphi)(e)$$
$$= \int_{\mathbb{R}^2} (T_{\mathfrak{h}_2'\mathfrak{h}_1'}\circ\pi_1'(\exp(-x'X'))\circ T_{\mathfrak{h}_1'\mathfrak{h}_1}\varphi)(\exp(xX))\Delta_{H_2,G}^{-1/2}(\exp(xX))dxdx'.$$

Fix $(x, x') \in \mathbb{R}^2$. We can repeat the argument used at the beginning of (c) (see (2.6.2)) to find

$$(T_{\mathfrak{h}_2'\mathfrak{h}_1'}\circ\pi_1'(\exp(-x'X'))\circ T_{\mathfrak{h}_1'\mathfrak{h}_1}\varphi)(\exp(xX))$$
$$= \oint_{H_2'/(H_1'\cap H_2')} (T_{\mathfrak{h}_1'\mathfrak{h}_1}\varphi)(\exp(x'X')\exp(xX)\cdot k)\chi_f(k)\Delta_{H_2',G}^{-1/2}(k)d\nu(k).$$

Note that here, again,

$$\int_{\mathbb{R}^2} \Delta_{H_2,G}^{-1/2}(\exp(xX))dxdx'$$
$$\times \oint_{H_2'/(H_1'\cap H_2')} |T_{\mathfrak{h}_1'\mathfrak{h}_1}\varphi(\exp(x'X')\exp(xX)\cdot k)|\Delta_{H_2',G}^{-1/2}(k)d\nu(k) < \infty. \tag{2.6.21}$$

Indeed, if dP denotes a Lebesgue measure on a complement $\mathfrak{p}$ of $\mathfrak{h}_2^0$ in $\mathfrak{h}_2$ such that $\mathfrak{p} \subset [\mathfrak{h}_2, \mathfrak{h}_2]$, the left-hand side of (2.6.21) is equal to

$$\int_{\mathfrak{p}} dP \oint_{H_2'/(H_1'\cap H_2')} |T_{\mathfrak{h}_1'\mathfrak{h}_1}\varphi(\exp P\cdot k)|\Delta_{H_2',G}^{-1/2}(k)d\nu(k)$$

$$= \oint_{H_2/K} \Delta_{H_2,G}^{-1/2}(k)d\nu(k) \int_{\mathbb{R}} \Delta_{H_1,G}^{-1/2}(\exp(tX))dt \Big| \int_{\mathbb{R}} \varphi(k\exp(tX)\exp(sY'))ds \Big|$$

$$= \oint_{H_2/K} \Delta_{H_2,G}^{-1/2}(k)d\nu(k) \int_{\mathbb{R}} dt \Big| \int_{\mathbb{R}} e^{-ist}\varphi(k\exp(sY'))ds \Big| < \infty.$$

Taking (2.6.21) into consideration, we get

$$(T_{\mathfrak{h}_2\mathfrak{h}_1}\varphi)(e) = \oint_{H_2/(H_1\cap H_2)} \chi_f(k)\Delta_{H_2,G}^{-1/2}(k)d\nu(k) \int_{\mathbb{R}} dx \int_{\mathbb{R}} e^{-ixy'}\varphi(k\exp(y'Y'))dy'$$

$$= \oint_{H_2/(H_1\cap H_2)} \varphi(h)\chi_f(h)\Delta_{H_2,G}^{-1/2}(h)d\nu(h).$$

(c.3.1.b) Suppose $\dim(\mathfrak{h}_1 \cap \mathfrak{a}) = 1$, $\dim(\mathfrak{h}_2 \cap \mathfrak{a}) = 2$, $\mathfrak{a} \subset \mathfrak{h}_1^0 + \mathfrak{h}_2^0$. Hence $\mathfrak{h}_2 \subset \mathfrak{g}_0$. Let $\{Y, Y'\}$ be a basis of $\mathfrak{a} \cap \ker(f)$ such that $\mathfrak{h}_2 \cap \mathfrak{a} = \mathbb{R}Y \oplus \mathfrak{z}$. We write $Y' = X_2 - X_1$ with $X_j \in \mathfrak{h}_j \cap \ker(f) (j = 1, 2)$. Here, X_1, X_2 belong to $\mathfrak{g}_0$,

$$\mathfrak{h}_1 \cap \mathfrak{h}_2 = \mathfrak{h}_1^0 \cap \mathfrak{h}_2^0, \ \mathfrak{h}_1' \cap \mathfrak{h}_2 = (\mathfrak{h}_1^0 \cap \mathfrak{h}_2^0) \oplus \mathbb{R}X_2 \oplus \mathbb{R}Y$$

and $\mathfrak{h}_1 \cap \mathfrak{h}_2' = (\mathfrak{h}_1^0 \cap \mathfrak{h}_2^0) \oplus \mathbb{R}X_1$. The isotropic subspace $\mathfrak{p} = \mathbb{R}X_2 + \mathbb{R}Y + \mathfrak{g}(f)$ is contained in $(\mathfrak{h}_1 \cap \mathfrak{h}_1') + (\mathfrak{h}_1' \cap \mathfrak{h}_2)$, so $a(\mathfrak{h}_1, \mathfrak{h}_1', \mathfrak{h}_2) = 1$. Therefore,

$$a(\mathfrak{h}_1, \mathfrak{h}_0, \mathfrak{h}_2) = a(\mathfrak{h}_1', \mathfrak{h}_0, \mathfrak{h}_2) = a(\mathfrak{h}_1', \mathfrak{h}_0, \mathfrak{h}_2').$$

Now it suffices to adapt the proof of case (c.2.0). Take a function φ in $\mathscr{H}_{\pi_1}^{\infty}$. If its support is sufficiently small modulo H_1, when k runs over a subset which represents H_2/H_2^0, the functions $\pi_1'(k^{-1})T_{\mathfrak{h}_1'\mathfrak{h}_1}\varphi$ have sufficiently small support in A^f/H_1'. Then if $A = \exp\mathfrak{a}$, we have

$$(T_{\mathfrak{h}_2\mathfrak{h}_1}\varphi)(e) = (T_{\mathfrak{h}_2\mathfrak{h}_2'} \circ T_{\mathfrak{h}_2'\mathfrak{h}_1'} \circ T_{\mathfrak{h}_1'\mathfrak{h}_1}\varphi)(e)$$

$$= \oint_{H_2/H_2^0} (T_{\mathfrak{h}_2'\mathfrak{h}_1'} \circ \pi_1'(k^{-1}) \circ T_{\mathfrak{h}_1'\mathfrak{h}_1}\varphi)(e)\chi_f(k)\Delta_{H_2,G}^{-1/2}(k)d\nu(k)$$

$$= \oint_{H_2/H_2^0} \chi_f(k)\Delta_{H_2,G}^{-1/2}(k)d\nu(k)$$

$$\times \oint_{H_2'/(H_1'\cap H_2')} \chi_f(x)\Delta_{H_2',G}^{-1/2}(x)d\nu(x) \int_{A/(H_1\cap A)} \varphi(kxy)d\dot{y}. \tag{2.6.22}$$

Since

$$H_2'/(H_1' \cap H_2') \simeq H_2^0/((H_1 \cap H_2)\exp(\mathbb{R}X_2)\exp(\mathbb{R}Y)),$$

formula (2.6.22) becomes

$$(T_{\mathfrak{h}_2\mathfrak{h}_1}\varphi)(e) = \oint_{H_2/H_2^0} \chi_f(k)\Delta_{H_2,G}^{-1/2}(k)d\nu(k) \oint_{H_2^0/(H_1'\cap H_2^0)} \chi_f(x)\Delta_{H_2,G}^{-1/2}(x)d\nu(x)$$

$$\times \int_{\mathbb{R}^2} \varphi(kx\exp(y'Y')\exp(yY))dydy'$$

$$= \oint_{H_2/H_2^0} \chi_f(k)\Delta_{H_2,G}^{-1/2}(k)d\nu(k) \oint_{H_2^0/(H_1\cap H_2)} \varphi(kx)\chi_f(x)\Delta_{H_2,G}^{-1/2}(x)d\nu(x)$$

$$= \oint_{H_2/(H_1\cap H_2)} \varphi(h)\chi_f(h)\Delta_{H_2,G}^{-1/2}(h)d\nu(h).$$

(c.3.2.b) Suppose $\dim(\mathfrak{h}_1 \cap \mathfrak{a}) = 2$, $\dim(\mathfrak{h}_2 \cap \mathfrak{a}) = 1$, $\mathfrak{a} \subset \mathfrak{h}_1^0 + \mathfrak{h}_2^0$. Hence $\mathfrak{h}_1 \subset \mathfrak{g}_0$. Let as above $\{Y, Y'\}$ be a basis of $\mathfrak{a} \cap \ker(f)$ such that $\mathfrak{h}_1 \cap \mathfrak{a} = \mathbb{R}Y \oplus \mathfrak{z}$ and $X_j \in \mathfrak{h}_j (j = 1, 2)$ such that $Y' = X_2 - X_1$. In this situation, X_1, X_2 belong to $\mathfrak{g}_0$,

$$\mathfrak{h}_1 \cap \mathfrak{h}_2 = \mathfrak{h}_1^0 \cap \mathfrak{h}_2^0,\ \mathfrak{h}_1 \cap \mathfrak{h}_2' = (\mathfrak{h}_1^0 \cap \mathfrak{h}_2^0) \oplus \mathbb{R}X_1 \oplus \mathbb{R}Y$$

and $\mathfrak{h}_1' \cap \mathfrak{h}_2 = (\mathfrak{h}_1^0 \cap \mathfrak{h}_2^0) \oplus \mathbb{R}X_2$.

Since

$$H_2'/(H_1' \cap H_2') \simeq H_2^0/(H_1' \cap H_2^0) = H_2^0/((H_1 \cap H_2)\exp(\mathbb{R}X_2)),$$

the above calculation leads to formula (2.6.22), which now becomes

$$(T_{\mathfrak{h}_2\mathfrak{h}_1}\varphi)(e) = \oint_{H_2/H_2^0} \chi_f(k)\Delta_{H_2,G}^{-1/2}(k)d\nu(k)$$

$$\times \oint_{H_2^0/(H_1'\cap H_2^0)} \chi_f(x)\Delta_{H_2',G}^{-1/2}(x)d\nu(x) \int_{\mathbb{R}} \varphi(kx\exp(yY')dy$$

$$= \oint_{H_2/H_2^0} \chi_f(k)\Delta_{H_2,G}^{-1/2}(k)d\nu(k)$$

$$\times \oint_{H_2^0/(H_1\cap H_2)} \varphi(kx)\chi_f(x)\Delta_{H_2,G}^{-1/2}(x)\Delta_{H_2^0,H_2}^{-1}(x)d\nu(x)$$

$$= \oint_{H_2/(H_1\cap H_2)} \varphi(h)\chi_f(h)\Delta_{H_2,G}^{-1/2}(h)d\nu(h).$$

(c.3.1.c) Suppose $\dim(\mathfrak{h}_1 \cap \mathfrak{a}) = 1$, $\dim(\mathfrak{h}_2 \cap \mathfrak{a}) = 2$, $\mathfrak{a} \subset \mathfrak{h}_1 + \mathfrak{h}_2$ but $\mathfrak{a} \not\subset \mathfrak{h}_1^0 + \mathfrak{h}_2^0$. Hence $\mathfrak{h}_2 \subset \mathfrak{g}_0$. Let $\{Y, Y'\}$ be a basis of $\mathfrak{a} \cap \ker(f)$ such that $\mathfrak{h}_2 \cap \mathfrak{a} = \mathbb{R}Y \oplus \mathfrak{z}$. It is easy to check that there exist vectors $X \in [\mathfrak{h}_1, \mathfrak{h}_1]$ and $X_j \in \mathfrak{h}_j \cap \ker(f) (j = 1, 2)$ such that

$$\mathfrak{h}_1 = \mathbb{R}X \oplus \mathbb{R}X_1 \oplus \mathfrak{h}_1^0,\ \mathfrak{h}_2 = \mathbb{R}X_2 \oplus \mathfrak{h}_2^0,\ [X, Y'] = 0,\ [X_1, Y'] = [X_2, Y'] = Z$$

and $Y' = \alpha(X_2 - X_1)$ for some $0 \neq \alpha \in \mathbb{R}$. The vectors X_1, X_2 satisfy the trace relation (cf. Lemma 2.6.5). We now follow the proof of (c.2.1) closely. We show that the Maslov index is

$$a(\mathfrak{h}_1, \mathfrak{h}_0, \mathfrak{h}_2) = e^{-\frac{i\pi}{4}\operatorname{sgn}\alpha} a(\mathfrak{h}_1', \mathfrak{h}_0, \mathfrak{h}_2').$$

For $\varphi \in \mathscr{H}_{\pi_1}^\infty$, whose support is sufficiently small modulo H_1,

$$\begin{aligned}(T_{\mathfrak{h}_2\mathfrak{h}_1}\varphi)(e) =& e^{-\frac{i\pi}{4}\operatorname{sgn}\alpha}(T_{\mathfrak{h}_2\mathfrak{h}_2'} \circ T_{\mathfrak{h}_2'\mathfrak{h}_1'} \circ T_{\mathfrak{h}_1'\mathfrak{h}_1}\varphi)(e) \\ =& e^{-\frac{i\pi}{4}\operatorname{sgn}\alpha} \int_{\mathbb{R}} \Delta_{H_2,G}^{-1/2}(\exp(tX_2))dt \\ &\times \oint_{H_2'/(H_1'\cap H_2')} (T_{\mathfrak{h}_1'\mathfrak{h}_1}\varphi)(\exp(tX_2)\cdot k)\chi_f(k)\Delta_{H_2',G}^{-1/2}(k)d\nu(k).\end{aligned} \tag{2.6.23}$$

We now introduce $\mathfrak{k} = \mathfrak{h}_1 + \mathfrak{a}$, $K = \exp \mathfrak{k}$ in order to write

$$\begin{aligned}&\int_{\mathbb{R}} \Delta_{H_2,G}^{-1/2}(\exp(tX_2))dt \oint_{H_2'/(H_1'\cap H_2')} |T_{\mathfrak{h}_1'\mathfrak{h}_1}\varphi(\exp(tX_2)\cdot k)|\Delta_{H_2',G}^{-1/2}(k)d\nu(k) \\ =& \oint_{H_2/(H_1'\cap H_2)} |T_{\mathfrak{h}_1'\mathfrak{h}_1}\varphi(x)|\Delta_{H_2,G}^{-1/2}(x)d\nu(x) \\ =& \int_{H_2/(K\cap H_2)} \Delta_{H_2,G}^{-1/2}(g)d\nu(g) \int_{\mathbb{R}} dt \Big| \int_{\mathbb{R}} e^{-ist}ds \int_{\mathbb{R}} \varphi(g\exp(sY')\exp(yY))dy \Big|.\end{aligned}$$

Then

$$\int_{\mathbb{R}} \Delta_{H_2,G}^{-1/2}(\exp(tX_2))dt \oint_{H_2'/(H_1'\cap H_2')} |T_{\mathfrak{h}_1'\mathfrak{h}_1}\varphi(\exp(tX_2)\cdot k)|\Delta_{H_2',G}^{-1/2}(k)d\nu(k) < \infty$$

for each $\varphi \in \mathscr{H}_{\pi_1}^\infty$ whose support is sufficiently small modulo H_1. Therefore we may apply Lemma 2.3.12 to (2.6.23) and obtain the desired result:

$$\begin{aligned}(T_{\mathfrak{h}_2\mathfrak{h}_1}\varphi)(e) =& e^{-\frac{i\pi}{4}\operatorname{sgn}\alpha} \oint_{H_2/(H_1'\cap H_2)} (T_{\mathfrak{h}_1'\mathfrak{h}_1}\varphi)(h)\chi_f(h)\Delta_{H_2,G}^{-1/2}(h)d\nu(h) \\ =& \frac{1}{|\alpha|}e^{-\frac{i\pi}{4}\operatorname{sgn}\alpha} \oint_{H_2/(K\cap H_2)} \chi_f(g)\Delta_{H_2,G}^{1/2}(g)d\nu(g) \int_{\mathbb{R}} dt \int_{\mathbb{R}} e^{i(t-s)^2/2\alpha} \\ &\times \Delta_{H_2,G}^{-1/2}(\exp(sX_2))\Delta_{H_1\cap H_2,H_2}^{-1}(\exp(sX_2))ds \int_{\mathbb{R}} \varphi(g\exp(sX_2)\exp(yY))dy\end{aligned}$$

$$=\sqrt{\frac{2\pi}{|\alpha|}}\oint_{H_2/(K\cap H_2)}\chi_f(g)\Delta_{H_2,G}^{-1/2}(g)d\nu(g)$$

$$\times\oint_{(K\cap H_2)/(H_1\cap H_2)}\varphi(gk)\chi_f(k)\Delta_{H_2,G}^{-1/2}(k)\Delta_{K\cap H_2,H_2}^{-1}(k)d\nu(k)$$

$$=\oint_{H_2/(H_1\cap H_2)}\varphi(h)\chi_f(h)\Delta_{H_2,G}^{-1/2}(h)d\nu(h).$$

(c.3.2.c) Suppose $\dim(\mathfrak{h}_1\cap\mathfrak{a})=2$, $\dim(\mathfrak{h}_1\cap\mathfrak{a})=1$, $\mathfrak{a}\subset\mathfrak{h}_1+\mathfrak{h}_2$ but $\mathfrak{a}\not\subset\mathfrak{h}_1^0+\mathfrak{h}_2^0$. Hence $\mathfrak{h}_1\subset\mathfrak{g}_0$. Let $\{Y,Y'\}$ be a basis of $\mathfrak{a}\cap\ker(f)$ such that $\mathfrak{h}_1\cap\mathfrak{a}=\mathbb{R}Y\oplus\mathfrak{z}$. It is easy to verify that there are vectors $X\in[\mathfrak{h}_2,\mathfrak{h}_2]$ and $X_j\in\mathfrak{h}_j\cap\ker(f)$ such that

$$\mathfrak{h}_2=\mathbb{R}X\oplus\mathbb{R}X_2\oplus\mathfrak{h}_2^0,\ \mathfrak{h}_1=\mathbb{R}X_1\oplus\mathfrak{h}_1^0,\ [X,Y']=0,\ [X_1,Y']=[X_2,Y']=Z$$

and $Y'=\alpha(X_2-X_1)$, with $0\neq\alpha\in\mathbb{R}$. As in the previous case, we obtain

$$\Delta_{H_1,G}^{-1/2}(\exp(tX_1))=\Delta_{H_2,G}^{-1/2}(\exp(tX_2))\Delta_{H_1\cap H_2,H_2}(\exp(tX_2))$$

and

$$a(\mathfrak{h}_1,\mathfrak{h}_0,\mathfrak{h}_2)=e^{-\frac{i\pi}{4}\operatorname{sgn}\alpha}a(\mathfrak{h}_1',\mathfrak{h}_0,\mathfrak{h}_2').$$

Let $\mathfrak{k}=\mathfrak{h}_1+\mathfrak{a}$, $K=\exp\mathfrak{k}$, $\mathfrak{m}=\mathbb{R}X+\mathbb{R}X_2$ and dP a Lebesgue measure on $\mathfrak{m}$. We then have $K\cap H_2=(H_1\cap H_2)\exp(\mathbb{R}X_2)$. In the same way as before, we show, for $\varphi\in\mathscr{H}_{\pi_1}^\infty$ with sufficiently small support modulo H_1,

$$(T_{\mathfrak{h}_2\mathfrak{h}_1}\varphi)(e)=e^{-\frac{i\pi}{4}\operatorname{sgn}\alpha}\int_{\mathfrak{m}}\Delta_{H_2,G}^{-1/2}(\exp P)dP$$

$$\times\oint_{H_2'/(H_1'\cap H_2')}(T_{\mathfrak{h}_1'\mathfrak{h}_1}\varphi)(\exp P\cdot k)\chi_f(k)\Delta_{H_2',G}^{-1/2}(k)d\nu(k).\qquad(2.6.24)$$

Now we must apply Lemma 2.3.12 to find

$$\int_{\mathfrak{m}}\Delta_{H_2,G}^{-1/2}(\exp P)dP\oint_{H_2'/(H_1'\cap H_2')}|T_{\mathfrak{h}_1'\mathfrak{h}_1}\varphi(\exp P\cdot k)|\Delta_{H_2',G}^{-1/2}(k)d\nu(k)$$

$$=\frac{1}{|\alpha|}\oint_{H_2/(K\cap H_2)}\Delta_{H_2,G}^{-1/2}(k)d\nu(k)\int_{\mathbb{R}}dt$$

$$\times\Big|\int_{\mathbb{R}}e^{i(t-s)^2/2\alpha}\varphi(k\exp(sX_2))\Delta_{H_2,G}^{-1/2}(\exp(sX_2))\Delta_{H_1\cap H_2,H_2}^{-1}(\exp(sX_2))ds\Big|<\infty$$

for each $\varphi\in\mathscr{H}_{\pi_1}^\infty$ in question.

If we apply Lemma 2.3.12 to (2.6.24), we attain the result:

$$(T_{\mathfrak{h}_2\mathfrak{h}_1}\varphi)(e) = e^{-\frac{i\pi}{4}\operatorname{sgn}\alpha} \oint_{H_2/(H_1'\cap H_2)} (T_{\mathfrak{h}_1'\mathfrak{h}_1}\varphi)(g)\chi_f(g)\Delta_{H_2,G}^{-1/2}(g)d\nu(g)$$

$$= \frac{1}{|\alpha|}e^{-\frac{i\pi}{4}\operatorname{sgn}\alpha} \oint_{H_2/(K\cap H_2)} \chi_f(k)\Delta_{H_2,G}^{-1/2}(k)d\nu(k)\int_{\mathbb{R}} dt$$

$$\times \int_{\mathbb{R}} e^{i(t-s)^2/2\alpha}\varphi(k\exp(sX_2))\Delta_{H_2,G}^{-1/2}(\exp(sX_2))\Delta_{H_1\cap H_2,H_2}^{-1}(\exp(sX_2))ds$$

$$= \sqrt{\frac{2\pi}{|\alpha|}} \oint_{H_2/(K\cap H_2)} \chi_f(k)\Delta_{H_2,G}^{-1/2}(k)d\nu(k)$$

$$\times \int_{\mathbb{R}} \varphi(k\exp(sX_2))\Delta_{H_2,G}^{-1/2}(\exp(sX_2))\Delta_{H_1\cap H_2,H_2}^{-1}(\exp(sX_2))ds$$

$$= \sqrt{\frac{2\pi}{|\alpha|}} \oint_{H_2/(K\cap H_2)} d\nu(k) \oint_{(K\cap H_2)/(H_1\cap H_2)} \varphi(kx)\chi_f(kx)\Delta_{H_2,G}^{-1/2}(kx)d\dot{x}$$

$$= \oint_{H_2/(H_1\cap H_2)} \varphi(h)\chi_f(h)\Delta_{H_2,G}^{-1/2}(h)d\nu(h).$$

(c.3.a.2) Suppose $\dim(\mathfrak{h}_1\cap\mathfrak{a}) = \dim(\mathfrak{h}_2\cap\mathfrak{a}) = 1$, namely $\mathfrak{h}_1\cap\mathfrak{a} = \mathfrak{h}_2\cap\mathfrak{a} = \mathfrak{z}$. In this case, $\dim(\mathfrak{h}_1\cap\mathfrak{h}_2) = \dim(\mathfrak{h}_1^0\cap\mathfrak{h}_2^0)+2$ and the assertion was proved in (c.a.1).

(c.3.b.1) We suppose $\dim(\mathfrak{h}_1\cap\mathfrak{h}_2) = \dim(\mathfrak{h}_1^0\cap\mathfrak{h}_2^0)+1$. Then $\dim((\mathfrak{h}_1+\mathfrak{h}_2)\cap\mathfrak{a}) = 2$ and hence $\mathfrak{h}_1\cap\mathfrak{h}_2$ is contained in $\mathfrak{g}_0$. Let X be an element of $\mathfrak{h}_1\cap\mathfrak{h}_2\cap\ker(f)$ and $\{Y, Y'\}$ a basis of $\mathfrak{a}\cap\ker(f)$ such that

$$\mathfrak{h}_1\cap\mathfrak{h}_2 = \mathbb{R}X + (\mathfrak{h}_1^0\cap\mathfrak{h}_2^0),\ Y' \in \mathfrak{h}_1^0+\mathfrak{h}_2^0,\ [X, Y'] = 0,\ [X, Y] = Z.$$

Then there exist two vectors $X_j \in \mathfrak{h}_j^0\cap\ker(f)(j = 1, 2)$ such that $Y' = X_2 - X_1$ and $[X_1, X] \in \mathfrak{h}_1\cap\mathfrak{h}_2$. If $X_1 \notin \mathfrak{g}_0$, $[X_1, X]$ is not linearly dependent on X modulo $\mathfrak{h}_1^0\cap\mathfrak{h}_2^0$, which contradicts our hypothesis. Hence $X_j \in \mathfrak{g}_0(j = 1, 2)$. As in the other cases,

$$\mathfrak{h}_1'\cap\mathfrak{h}_2 = (\mathfrak{h}_1^0\cap\mathfrak{h}_2^0)\oplus\mathbb{R}X_2,\quad \mathfrak{h}_1\cap\mathfrak{h}_2' = (\mathfrak{h}_1^0\cap\mathfrak{h}_2^0)\oplus\mathbb{R}X_1.$$

Since $\mathfrak{p} = \mathbb{R}X + \mathbb{R}X_2 + \mathfrak{g}(f)$ is an isotropic subspace contained in $(\mathfrak{h}_1'\cap\mathfrak{h}_2)+(\mathfrak{h}_1\cap\mathfrak{h}_2)$ and $(\mathfrak{h}_1)^{\mathfrak{p}} = (\mathfrak{h}_1')^{\mathfrak{p}} = \mathfrak{h}_1^0+\mathfrak{p}$, we see $a(\mathfrak{h}_1, \mathfrak{h}_1', \mathfrak{h}_2) = 1$. Then we deduce

$$a(\mathfrak{h}_1, \mathfrak{h}_0, \mathfrak{h}_2) = a(\mathfrak{h}_1', \mathfrak{h}_0, \mathfrak{h}_2) = a(\mathfrak{h}_1', \mathfrak{h}_0, \mathfrak{h}_2').$$

At this point we shall no longer go into details, for the rest of the proof is similar to that of (c.3.2.a).

(c.3.b.2) Suppose now $Y' \notin \mathfrak{h}_1^0 + \mathfrak{h}_2^0$. If we write $Y' = X_2' - X_1'$ with $X_j' \in \mathfrak{h}_j (j = 1, 2)$, then X_1', X_2' are in $\mathfrak{g}_0$ as in the previous case. Hence there exist two elements $X_j \in \mathfrak{h}_j \cap \ker(f)(j = 1, 2)$ such that

$$[X_j, Y'] = Z, [X_j, Y] = 0, Y' = \alpha(X_2 - X_1)$$

with $0 \neq \alpha \in \mathbb{R}$. Further, X_1, X_2 verify the trace relation (cf. Lemma 2.6.5). Evidently,

$$\mathfrak{h}_1' \cap \mathfrak{h}_2 = \mathfrak{h}_1 \cap \mathfrak{h}_2' = \mathfrak{h}_1^0 \cap \mathfrak{h}_2^0.$$

The subspace $\mathfrak{p} = \mathbb{R}X + \mathbb{R}X_2 + \mathfrak{h}_1^0$ is Lagrangian, so

$$a(\mathfrak{h}_1, \mathfrak{h}_1', \mathfrak{h}_2) = a(\mathfrak{p}, \mathfrak{h}_1', \mathfrak{h}_2) a(\mathfrak{h}_1, \mathfrak{p}, \mathfrak{h}_2) a(\mathfrak{h}_1, \mathfrak{h}_1', \mathfrak{p}).$$

But if $\mathfrak{m} = \mathbb{R}X_2 + \mathfrak{g}(f)$, $\mathfrak{m} \subset \mathfrak{p} \cap \mathfrak{h}_2$ and $(\mathfrak{h}_1)^{\mathfrak{m}} = (\mathfrak{h}_1')^{\mathfrak{m}} = \mathfrak{p}$. Hence

$$a(\mathfrak{p}, \mathfrak{h}_1', \mathfrak{h}_2) = a(\mathfrak{h}_1, \mathfrak{p}, \mathfrak{h}_2) = 1.$$

In the same way for $\mathfrak{n} = \mathbb{R}X + \mathfrak{g}(f) \subset \mathfrak{h}_1 \cap \mathfrak{p}$, we see

$$a(\mathfrak{h}_1, \mathfrak{h}_1', \mathfrak{p}) = e^{-\frac{i\pi}{4} \operatorname{sgn}\alpha} a(\mathfrak{h}_1, \mathfrak{q}, \mathfrak{p}),$$

where $\mathfrak{q} = \mathbb{R}X + \mathfrak{h}_1^0 + \mathbb{R}Y'$. As before, we show

$$a(\mathfrak{h}_1, \mathfrak{h}_0, \mathfrak{h}_2) = e^{-\frac{i\pi}{4} \operatorname{sgn}\alpha} a(\mathfrak{h}_1', \mathfrak{h}_0, \mathfrak{h}_2').$$

For $\varphi \in \mathscr{H}_{\pi_1}^{\infty}$ with sufficiently small support modulo H_1, we arrive at

$$\begin{aligned}(T_{\mathfrak{h}_2\mathfrak{h}_1}\varphi)(e) &= e^{-\frac{i\pi}{4} \operatorname{sgn}\alpha} \oint_{H_2/H_2^0} \chi_f(x) \Delta_{H_2,G}^{-1/2}(x) d\nu(x) \\ &\times \oint_{H_2'/(H_1' \cap H_2')} (T_{\mathfrak{h}_1'\mathfrak{h}_1}\varphi)(xk) \chi_f(k) \Delta_{H_2',G}^{-1/2}(k) d\nu(k). \qquad (2.6.25)\end{aligned}$$

Again, we establish an overestimation. Let $\mathfrak{k} = \mathbb{R}X_2 + (\mathfrak{h}_1 \cap \mathfrak{h}_2) = \mathbb{R}X_2 + \mathbb{R}X + (\mathfrak{h}_1^0 \cap \mathfrak{h}_2^0)$ and $K = \exp \mathfrak{k}$. Then

$$\begin{aligned}&\oint_{H_2/H_2^0} \Delta_{H_2,G}^{-1/2}(x) d\nu(x) \oint_{H_2'/(H_1' \cap H_2')} |T_{\mathfrak{h}_1'\mathfrak{h}_1}\varphi(xk)| \Delta_{H_2',G}^{-1/2}(k) d\nu(k) \\ &= \frac{1}{|\alpha|} \oint_{H_2/K} \Delta_{H_2,G}^{-1/2}(k) d\nu(k) \int_{\mathbb{R}^2} \Big| \int_{\mathbb{R}^2} e^{i(t-s)^2/2\alpha - iuy} \varphi(k \exp(sX_2) \exp(yY)) \\ &\times \Delta_{H_2,G}^{-1/2}(\exp(sX_2)) \Delta_{K,H_2}^{-1}(\exp(sX_2)) ds dy \Big| < \infty.\end{aligned}$$

Starting from (2.6.25), we eventually obtain

$$(T_{\mathfrak{h}_2\mathfrak{h}_1}\varphi)(e) = e^{-\frac{i\pi}{4}\operatorname{sgn}\alpha}\frac{1}{|\alpha|}\oint_{H_2/K}\chi_f(k)\Delta_{H_2,G}^{-1/2}(k)d\nu(k)\int_{\mathbb{R}^2}dtdu$$

$$\times\int_{\mathbb{R}^2}e^{i(t-s)^2/2\alpha-iuy}\varphi(k\exp(sX_2)\exp(yY))$$

$$\Delta_{H_2,G}^{-1/2}(\exp(sX_2))\Delta_{K,H_2}^{-1}(\exp(sX_2))dsdy$$

$$=2\pi\sqrt{\frac{2\pi}{|\alpha|}}\oint_{H_2/K}\chi_f(k)\Delta_{H_2,G}^{-1/2}(k)d\nu(k)$$

$$\times\int_{\mathbb{R}}\varphi(k\exp(sX_2))\Delta_{H_2,G}^{-1/2}(\exp(sX_2))\Delta_{K,H_2}^{-1/2}((\exp(sX_2))ds$$

$$=\oint_{H_2/K}\chi_f(k)\Delta_{H_2,G}^{-1/2}(k)d\nu(k)\oint_{K/(H_1\cap H_2)}\varphi(kx)\chi_f(x)\Delta_{H_2,G}^{-1/2}(x)\Delta_{K,H_2}^{-1}(x)d\nu(x)$$

$$=\oint_{H_2/(H_1\cap H_2)}\varphi(h)\chi_f(h)\Delta_{H_2,G}^{-1/2}(h)d\nu(h).$$

(c.3.c.3) Suppose now $\dim(\mathfrak{h}_1\cap\mathfrak{h}_2)=\dim(\mathfrak{h}_1^0\cap\mathfrak{h}_2^0)$ and $\mathfrak{a}\subset\mathfrak{h}_1^0+\mathfrak{h}_2^0$. Let $\{Y, Y'\}$ be a basis of $\mathfrak{a}\cap\ker(f)$. When we write $Y=X_2-X_1$, $Y'=X_2'-X_1'$ with $X_j, X_j'\in\mathfrak{h}_j^0 (j=1,2)$, we easily see that we can take X_j, X_j' in $\mathfrak{h}_j^0\cap\mathfrak{g}_0$. Indeed, we can assume that X_1, X_2 belong to $\mathfrak{g}_0$. Then

$$[X_2, X_1']=[X_2, X_2']\in\mathfrak{h}_2\cap\mathfrak{g}_0, [X_2, X_1']-[X_1, X_1']=\alpha Y+\beta Y'$$

for certain $\alpha,\beta\in\mathbb{R}$. If $X_j'\notin\mathfrak{g}_0$, then $\beta\neq 0$ and we replace X_1', X_2', Y' respectively by $[X_1, X_1'], [X_2, X_2'], \alpha Y+\beta Y'$. We have also

$$\mathfrak{h}_1'\cap\mathfrak{h}_2=(\mathfrak{h}_1\cap\mathfrak{h}_2)+\mathbb{R}X_2+\mathbb{R}X_2',\ \mathfrak{h}_1\cap\mathfrak{h}_2'=(\mathfrak{h}_1\cap\mathfrak{h}_2)+\mathbb{R}X_1+\mathbb{R}X_1'.$$

Now we can mimic the reasoning of case (c.2.0) to conclude.

(c.3.c.2) Suppose $\dim(\mathfrak{h}_1\cap\mathfrak{h}_2)=\dim(\mathfrak{h}_1^0\cap\mathfrak{h}_2^0)$, $\dim(\mathfrak{a}\cap(\mathfrak{h}_1^0+\mathfrak{h}_2^0))=2$. Then $\mathfrak{h}_1\cap\mathfrak{h}_2$ is contained in $\mathfrak{g}_0$. Let $\{Y, Y'\}$ be a basis of $\mathfrak{a}\cap\ker(f)$ such that $Y\in\mathfrak{h}_1^0+\mathfrak{h}_2^0$. Write $Y=X_2-X_1$, $Y'=X_2''-X_1''$ with $X_j\in\mathfrak{h}_j^0\cap\ker(f)$, $X_j''\in\mathfrak{h}_j\cap\ker(f) (j=1,2)$. Then

$$B_f(X_1, Y')=B_f(X_1, X_2'')=B_f(-Y, X_2'')=0.$$

If $X_2\notin\mathfrak{g}_0$, we can suppose $X_2''\in\mathfrak{g}_0$. Then $[X_2'', X_1]=[X_2'', X_2]\in\mathfrak{h}_2$ and

$$[X_2'', X_1]-[X_1'', X_1]=\alpha Y'+\beta Y=\alpha(X_2''-X_1'')+\beta(X_2-X_1)(\alpha,\beta\in\mathbb{R}).$$

This implies $\mathfrak{h}_1 \cap \mathfrak{h}_2$ is not contained in $\mathfrak{a}^f$, contrary to our hypothesis. Similarly, we see $X_j, X_j'' \in \mathfrak{g}_0 (j = 1, 2)$.

Modifying X_1'', X_2'', Y', we find two vectors $X_j' \in \mathfrak{h}_j \cap \ker(f) (j = 1, 2)$ such that

$$[X_1', Y] = [X_2', Y] = 0, [X_1', Y'] = [X_2', Y'] = Z, Y' = \alpha(X_2' - X_1')$$

for a certain $0 \neq \alpha \in \mathbb{R}$. Besides, X_1', X_2' satisfy the trace relation (cf. Lemma 2.6.5) and we see

$$\mathfrak{h}_1' \cap \mathfrak{h}_2 = (\mathfrak{h}_1 \cap \mathfrak{h}_2) + \mathbb{R}X_2, \ \mathfrak{h}_1 \cap \mathfrak{h}_2' = (\mathfrak{h}_1 \cap \mathfrak{h}_2) + \mathbb{R}X_1.$$

We put $\mathfrak{k} = (\mathfrak{h}_1' \cap \mathfrak{h}_2') + \mathbb{R}X_2'$, $K = \exp \mathfrak{k}$. By the same calculation as in case (c.2.1),

$$a(\mathfrak{h}_1, \mathfrak{h}_0, \mathfrak{h}_2) = e^{-\frac{i\pi}{4}\operatorname{sgn}\alpha} a(\mathfrak{h}_1', \mathfrak{h}_0, \mathfrak{h}_2').$$

Let $\varphi \in \mathscr{H}_{\pi_1}^{\infty}$. If $\operatorname{supp}(\varphi)$ is sufficiently small modulo H_1, we get

$$\oint_{H_2/(H_1' \cap H_2)} |T_{\mathfrak{h}_1'\mathfrak{h}_1}\varphi(g)| \Delta_{H_2,G}^{-1/2}(g) d\nu(g)$$
$$= \frac{1}{|\alpha|} \oint_{H_2/K} \Delta_{H_2,G}^{-1/2}(k) d\nu(k) \int_{\mathbb{R}} dt \Big| \int_{\mathbb{R}^2} e^{i(t-s)^2/2\alpha} \varphi(k \exp(sX_2') \exp(uX_2))$$
$$\times \Delta_{H_2,G}^{-1/2}(\exp(sX_2') \exp(uX_2)) \Delta_{K,H_2}^{-1}(\exp(sX_2') \exp(uX_2)) ds du \Big| < \infty.$$

Since

$$(H_2/K) \times \exp(\mathbb{R}X_2') \times \exp(\mathbb{R}X_2) \simeq H_2/(H_1 \cap H_2),$$

the function $H_2 \ni g \mapsto (T_{\mathfrak{h}_1'\mathfrak{h}_1}\varphi)(g)\chi_f(g)\Delta_{H_2,G}^{-1/2}(g)$ is integrable with respect to $\nu_{H_2, H_1' \cap H_2}$. Therefore we deduce the result for $\varphi \in \mathscr{H}_{\pi_1}^{\infty}$ with sufficiently small support modulo H_1:

$$(T_{\mathfrak{h}_2\mathfrak{h}_1}\varphi)(e) = e^{-\frac{i\pi}{4}\operatorname{sgn}\alpha} \oint_{H_2/(H_1' \cap H_2)} (T_{\mathfrak{h}_1'\mathfrak{h}_1}\varphi)(g)\chi_f(g)\Delta_{H_2,G}^{-1/2}(g) d\nu(g)$$

$$= \sqrt{\frac{2\pi}{|\alpha|}} \oint_{H_2/K} \chi_f(k) \Delta_{H_2,G}^{-1/2}(k) d\nu(k) \int_{\mathbb{R}^2} \varphi(k \exp(sX_2') \exp(uX_2))$$
$$\times \Delta_{H_2,G}^{-1/2}(\exp(sX_2') \exp(uX_2)) \Delta_{K,H_2}^{-1}(\exp(sX_2') \exp(uX_2)) ds du$$
$$= \oint_{H_2/(H_1 \cap H_2)} \varphi(h)\chi_f(h)\Delta_{H_2,G}^{-1/2}(h) d\nu(h).$$

(c.3.c.1) There only remains to examine the case where $(\mathfrak{h}_1^0 + \mathfrak{h}_2^0) \cap \mathfrak{a} = \mathfrak{z}$. Note first that $\mathfrak{a}$ is contained in the sum $(\mathfrak{h}_1 \cap \mathfrak{g}_0) + (\mathfrak{h}_2 \cap \mathfrak{g}_0)$. In fact, $\{Y, Y'\}$ being a basis of $\mathfrak{a} \cap \ker(f)$, put $Y = X_2 - X_1, Y' = X_2' - X_1'$ with $X_j, X_j' \in \mathfrak{h}_j (j = 1, 2)$. Since $\mathfrak{g}$ is exponential, we can suppose $X_2 \in \mathfrak{g}_0$ for example. Then $[X_2, X_1'] \in \mathfrak{h}_2 \cap \mathfrak{g}_0$ and $[Y, X_1'] = [X_2, X_1'] - [X_1, X_1']$ belongs to the sum $(\mathfrak{h}_1 \cap \mathfrak{g}_0) + (\mathfrak{h}_2 \cap \mathfrak{g}_0)$. And if $X_1' \notin \mathfrak{g}_0$, $\{Y, [Y, X_1'], Z\}$ is a basis of $\mathfrak{a}$, proving claim.

Let now $Y = X_2 - X_1$ with $X_j \in \mathfrak{h}_j \cap \mathfrak{g}_0 \cap \ker(f)(j = 1, 2)$. If $B_f(X_j, Y) = 0$, we take $Y' \in \mathfrak{a} \cap \ker(f)$ verifying $B_f(X_j, Y') = 1$. Writing $Y' = X_2' - X_1'$ with $X_j' \in \mathfrak{h}_j \cap \mathfrak{g}_0 \cap \ker(f)(j = 1, 2)$, we have

$$1 = B_f(X_1, Y') = B_f(X_1, X_2') = B_f(X_1', X_2) = B_f(X_1', Y) = B_f(X_2', Y)$$

and hence $B_f(X_1 + \beta X_1', Y + \beta Y') = 2\beta + \beta^2 c$ for any $\beta \in \mathbb{R}$ with $c = B_f(X_1', Y')$. Choosing β appropriately and replacing Y by $Y + \beta Y'$, we can suppose $B_f(X_j, Y) = \alpha \neq 0$ from the beginning. Taking next $Y' \in \mathfrak{a} \cap \ker(f)$ verifying $B_f(X_j, Y') = 0$ and writing $Y' = X_2' - X_1'$ with $X_j' \in \mathfrak{h}_j \cap \mathfrak{g}_0 \cap \ker(f)(j = 1, 2)$, we obtain $B_f(X_j', Y) = 0$ and hence $B_f(X_j', Y') = \alpha' \neq 0$. Replacing X_j, X_j' by $\alpha^{-1}X_j, (\alpha')^{-1}X_j'(j = 1, 2)$, we are in the following situation.

There exist a basis $\{Y, Y'\}$ of $\mathfrak{a} \cap \ker(f)$ and elements $X_j, X_j' \in \mathfrak{h}_j \cap \mathfrak{g}_0 \cap \ker(f)(j = 1, 2)$ such that

$$[X_j, Y] = [X_j', Y'] = Z, \ [X_j, Y'] = [X_j', Y] = 0$$

and

$$Y = \alpha(X_2 - X_1), \ Y' = \alpha'(X_2' - X_1')$$

for certain $\alpha, \alpha' \in \mathbb{R} \setminus \{0\}$. Furthermore, $[X_i, X_j']$ belong to $\mathfrak{h}_1 \cap \mathfrak{h}_2 \cap \mathfrak{g}_0 \cap \ker(f)(1 \le i, j \le 2)$.

In fact, we can even show $\mathfrak{h}_1 \cap \mathfrak{h}_2 \subset \mathfrak{g}_0$. Finally, we point out the trace relation for the pairs $(X_1, X_2), (X_1', X_2')$. We put $\mathfrak{k} = \mathbb{R}X_2 + \mathbb{R}X_2' + (\mathfrak{h}_1 \cap \mathfrak{h}_2)$ and $K = \exp \mathfrak{k}$. Evidently, $[\mathfrak{k}, \mathfrak{k}] \subset \mathfrak{h}_1 \cap \mathfrak{h}_2 = \mathfrak{h}_1' \cap \mathfrak{h}_2$. We verify $\mathfrak{k} = (\mathfrak{h}_1^0)^f \cap \mathfrak{h}_2$ and $a(\mathfrak{h}_1, \mathfrak{h}_1', \mathfrak{h}_2) = a(\mathfrak{h}_1, \mathfrak{h}_1', \mathfrak{p})$ if $\mathfrak{p} = \mathfrak{k} + \mathfrak{h}_1^0 = \mathbb{R}X_2 + \mathbb{R}X_2' + \mathfrak{h}_1^0$. This leads to the typical situation of the five-dimensional Heisenberg group, namely two copies of the situation encountered several times. We obtain

$$a(\mathfrak{h}_1, \mathfrak{h}_0, \mathfrak{h}_2) = e^{-\frac{i\pi}{4}(\text{sgn}\alpha + \text{sgn}\alpha')} a(\mathfrak{h}_1', \mathfrak{h}_0, \mathfrak{h}_2').$$

For $\varphi \in \mathscr{H}_{\pi_1}^{\infty}$, the usual calculation shows:

$$\oint_{H_2/(H_1'\cap H_2)} |T_{\mathfrak{h}_1'\mathfrak{h}_1}\varphi(g)|\Delta_{H_2,G}^{-1/2}(g)d\nu(g)$$
$$=\frac{1}{|\alpha\alpha'|}\oint_{H_2/K}\Delta_{H_2,G}^{-1/2}(k)d\nu(k)\int_{\mathbb{R}^2}dtdu$$
$$|\int_{\mathbb{R}^2}e^{i(t-s)^2/2\alpha+i(u-v)^2/2\alpha'}\varphi(k\exp(sX_2)\exp(vX_2'))$$
$$\times\,\Delta_{H_2,G}^{-1/2}(\exp(sX_2)\exp(vX_2'))\Delta_{K,H_2}^{-1}(\exp(sX_2)\exp(vX_2'))dsdv\Big|<\infty,$$

if supp(φ) is sufficiently small modulo H_1.

Since

$$(H_2/K)\times\exp(\mathbb{R}X_2)\times\exp(\mathbb{R}X_2')\simeq H_2/(H_1\cap H_2),$$

we conclude that the function $H_2 \ni g \mapsto (T_{\mathfrak{h}_1'\mathfrak{h}_1}\varphi)(g)\chi_f(g)\Delta_{H_2,G}^{-1/2}(g)$ is $\nu_{H_2,H_1'\cap H_2}$-integrable. This implies the result:

$$(T_{\mathfrak{h}_2\mathfrak{h}_1}\varphi)(e)=\frac{2\pi}{\sqrt{|\alpha\alpha'|}}\int_{H_2/K}\chi_f(k)\Delta_{H_2,G}^{-1/2}(k)d\nu(k)\int_{\mathbb{R}^2}\varphi(k\exp(sX_2)\exp(vX_2'))$$
$$\times\,\Delta_{H_2,G}^{-1/2}(\exp(sX_2)\exp(vX_2'))\Delta_{K,H_2}^{-1}(\exp(sX_2)\exp(vX_2'))dsdv$$
$$=\oint_{H_2/(H_1\cap H_2)}\varphi(h)\chi_f(h)\Delta_{H_2,G}^{-1/2}(h)d\nu h.$$

2.7 The Case Where $\mathfrak{h}_1 + \mathfrak{h}_2$ Is a Subalgebra

Lemma 2.7.1 *Let $\mathfrak{h}_j \in I(f,\mathfrak{g})(j = 1, 2)$ be such that $\mathfrak{k} = \mathfrak{h}_1 + \mathfrak{h}_2$ is a subalgebra of $\mathfrak{g}$. Then $K = \exp\mathfrak{k} = H_2H_1$ and there exists a coexponential basis of $\mathfrak{h}_1$ in $\mathfrak{g}$, a part of which is coexponential to $\mathfrak{h}_1 \cap \mathfrak{h}_2$ in $\mathfrak{h}_2$. In particular, H_1H_2 is closed in G.*

Proof Supposing $\mathfrak{g} = \mathfrak{h}_1 + \mathfrak{h}_2$, we shall verify $G = H_1H_2$. We already know that H_1H_2 is open in G. Consider a sequence of subalgebras

$$\mathfrak{h}_1=\mathfrak{m}_0\subset\mathfrak{m}_1\subset\cdots\subset\mathfrak{m}_k=\mathfrak{h}_1+[\mathfrak{g},\mathfrak{g}]$$

such that the adjoint action of $\mathfrak{m}_{j-1}$ on $\mathfrak{m}_j/\mathfrak{m}_{j-1}$ is irreducible for all $1 \le j \le k$. We put $M_j = \exp(\mathfrak{m}_j)$. Supposing $M_{j-1} \subset H_1H_2$, we shall prove the same inclusion for M_j. Suppose first $\dim(\mathfrak{m}_j/\mathfrak{m}_{j-1}) = 1$. We take a coexponential basis $\{X\}$ of $\mathfrak{m}_{j-1}$ in $\mathfrak{m}_j$. By writing $g \in M_j$ in the form $g = \exp(xX)\cdot g_0$, where $x \in \mathbb{R}$ and $g_0 \in M_{j-1}$, we immediately see that $g \in H_2H_1$ if and only if $\exp(xX) \in H_2H_1$.

On the other hand, the condition

$$\exp(xX) \in H_2H_1 = \{a \in G; a \cdot (f + (\mathfrak{h}_1)^\perp) \cap (f + (\mathfrak{h}_2)^\perp) \neq \emptyset\}$$

is expressed by the vanishing of a real analytic function of x. Since H_2H_1 is open in G, we have $\mathrm{rank}B(g) \leq \mathrm{rank}A(e)$ for all $g \in G$. Here, $B(g)$ and $A(e)$ denote the matrices introduced just after Lemma 2.3.11. On the other hand we have $(H_2H_1) \cap M_j = \exp(\mathfrak{h}_2 \cap \mathfrak{m}_j)H_1$, which is connected. These observations give us the desired inclusion $M_j \subset H_2H_1$.

Suppose now $\dim(\mathfrak{m}_j/\mathfrak{m}_{j-1}) = 2$. We take $\{X, X'\}$ in $[\mathfrak{g}, \mathfrak{g}]$ so that $\{X, X'\}$ is a coexponential basis of $\mathfrak{m}_{j-1}$ in $\mathfrak{m}_j$. When we write $X = Y_1 + Y_2$, $X' = Y_1' + Y_2'$ with $Y_i, Y_i' \in \mathfrak{h}_i (i = 1, 2)$, we may for instance assume $[Y_i', \mathfrak{m}_j] \subset \mathfrak{m}_{j-1} (i = 1, 2)$. If, further, $[Y_i, \mathfrak{m}_j] \subset \mathfrak{m}_{j-1} (i = 1, 2)$, then $\{Y_2, Y_2'\}$ becomes a coexponential basis of $\mathfrak{m}_{j-1}$ in $\mathfrak{m}_j$. For otherwise $[Y_i, \mathfrak{m}_j] \not\subset \mathfrak{m}_{j-1}$ and it suffices to replace X by $[Y_1, X'] - [X, Y_2'] = [Y_1, Y_1'] - [Y_2, Y_2']$ to go fall back into the previous case.

Hence $H_1 \subset M_k \subset H_2H_1$. Since $\mathfrak{m}_k \supset [\mathfrak{g}, \mathfrak{g}]$, we can finally choose elements of $\mathfrak{h}_2$ which constitute a coexponential basis of $\mathfrak{m}_k$ in $\mathfrak{g}$. □

Proposition 2.7.2 *Let $\mathfrak{h}_j \in I(f, \mathfrak{g}) (j = 1, 2)$ be such that $\mathfrak{h}_1 + \mathfrak{h}_2$ is a subalgebra of $\mathfrak{g}$. Then*

$$\left(\mathscr{H}_{\pi_1}^{-\infty}\right)^{H_2, \chi_f \Delta_{H_2 \cdot G}^{1/2}} = \mathbb{C}a,$$

where $\pi_1 = \rho(f, \mathfrak{h}_1, G)$ and a is the H_2-semi-invariant form given by

$$\langle a, \psi \rangle = \oint_{H_2/(H_1 \cap H_2)} \overline{\psi(h)\chi_f(h)\Delta_{H_2,G}^{-1/2}(h)}d\nu(h)$$

for any function $\psi \in \mathscr{H}_{\pi_1}^{\infty}$ with compact support modulo H_1.

Proof Let $\mathfrak{k} = \mathfrak{h}_1 + \mathfrak{h}_2$, $K = \exp \mathfrak{k}$ and $\pi = \mathrm{ind}_{H_1}^K \chi_f$. We know [71]

$$\dim \left(\mathscr{H}_{\pi_1}^{-\infty}\right)^{H_2, \chi_f \Delta_{H_2 \cdot G}^{1/2}} = \dim \left(\mathscr{H}_{\pi}^{-\infty}\right)^{H_2, \chi_f \Delta_{H_2 \cdot K}^{1/2}} = 1.$$

If $0 \neq b \in \left(\mathscr{H}_{\pi}^{-\infty}\right)^{H_2, \chi_f \Delta_{H_2 \cdot K}^{1/2}}$, it is clear that

$$\left(\mathscr{H}_{\pi_1}^{-\infty}\right)^{H_2, \chi_f \Delta_{H_2 \cdot G}^{1/2}} = \mathbb{C}a, \ \langle a, \psi \rangle = \langle b, (\psi|_K)\Delta_{K,G}^{-1/2} \rangle$$

for any $\psi \in \mathscr{H}_{\pi_1}^{\infty}$. From the previous lemma $K = H_2H_1$, and we can find a coexponential basis of $\mathfrak{h}_1$ in $\mathfrak{k}$ constituted by elements of $\mathfrak{h}_2$. Making use of the semi-invariance under the action of H_2, a standard argument (cf. [71]) allows to say that the form b is essentially an integral in the coordinates determined by the coexponential basis:

$$\langle b, \psi\rangle = \oint_{H_2/(H_1\cap H_2)} \overline{\psi(h)\chi_f(h)}\Delta_{H_2,K}^{-1/2}(h)d\nu(h)$$

for any $\psi \in \mathscr{H}_\pi^\infty$ with compact support modulo H_1. □

Theorem 2.7.3 *Let $\mathfrak{h}_j \in I(f, \mathfrak{g})(j = 1, 2)$ be such that $\mathfrak{k} = \mathfrak{h}_1 + \mathfrak{h}_2$ is a subalgebra of $\mathfrak{g}$. If T is an intertwining isometry from $\pi_1 = \rho(f, \mathfrak{h}_1, G)$ to $\pi_2 = \rho(f, \mathfrak{h}_2, G)$, there exists an H_2-invariant positive form $\nu = \nu_{H_2,H_1\cap H_2}$ and a constant $c \in \mathbb{C}$, $|c| = 1$, such that*

$$(T\varphi)(g) = cI_{\mathfrak{h}_2\mathfrak{h}_1}\varphi(g) = c\oint_{H_2/(H_1\cap H_2)} \varphi(gh)\chi_f(h)\Delta_{H_2,G}^{-1/2}(h)d\nu(h)\ (g \in G)$$

for any continuous function $\varphi \in \mathscr{H}_{\pi_1}$ with compact support modulo H_1.

Proof Denote by δ_2 the Dirac measure for π_2 at the neutral element e:

$$\langle \delta_2, \xi\rangle = \overline{\xi(e)},\ \xi \in \mathscr{H}_{\pi_2}^\infty.$$

It is well known [136] that δ_2 belongs to $\left(\mathscr{H}_{\pi_2}^{-\infty}\right)^{H_2,\chi_f\Delta_{H_2.G}^{1/2}}$. Hence

$$\delta_2 \circ T : \mathscr{H}_{\pi_1}^\infty \to \mathbb{C}$$

belongs to $\left(\mathscr{H}_{\pi_1}^{-\infty}\right)^{H_2,\chi_f\Delta_{H_2.G}^{1/2}}$. The previous proposition means that there exist a certain linear form ν and $c \in \mathbb{C}$, $|c| = 1$, such that

$$(T\varphi)(g) = \overline{\langle \delta_2, T\circ\pi_1(g^{-1})\varphi\rangle} = c\oint_{H_2/(H_1\cap H_2)} \varphi(gh)\chi_f(h)\Delta_{H_2,G}^{-1/2}(h)d\nu(h)\ (g \in G)$$

for $\varphi \in \mathscr{H}_{\pi_1}^\infty$ with compact support modulo H_1. Let now $\varphi \in \mathscr{H}_{\pi_1}$. Assume that φ is continuous and its support is contained in PH_1, where P is a compact subset of G. The integral

$$\oint_{H_2/(H_1\cap H_2)} \varphi(gh)\chi_f(h)\Delta_{H_2,G}^{-1/2}(h)d\nu(h)$$

is convergent for any $g \in G$, because H_2H_1 is closed in G. Fixing a compact neighbourhood V of e in G, we consider a sequence $(\varphi_n)_{n\in\mathbb{N}} \subset \mathscr{H}_{\pi_1}^\infty$ verifying

$\mathrm{supp}(\varphi_n) \subset VPH_1$ and such that $(\varphi_n)_{n\in\mathbb{N}}$ converges to φ in $\mathscr{H}_{\pi_1}$. Then it is immediate that

$$\begin{aligned}(T\varphi)(g) &= \lim_{n\to\infty}(T\varphi_n)(g)\\ &= c\lim_{n\to\infty}\oint_{H_2/(H_1\cap H_2)}\varphi_n(gh)\chi_f(h)\Delta_{H_2,G}^{-1/2}(h)d\nu(h)\\ &= c\oint_{H_2/(H_1\cap H_2)}\varphi(gh)\chi_f(h)\Delta_{H_2,G}^{-1/2}(h)d\nu(h).\end{aligned}$$

□

2.8 The Key Point Is the Convergence

We keep all previous notation. In this last section we shall see that the key point is the convergence of the integral

$$(I_{\mathfrak{h}_2\mathfrak{h}_1}^G\varphi)(g) := (I_{\mathfrak{h}_2\mathfrak{h}_1}\varphi)(g) = \oint_{H_2/(H_1\cap H_2)}\varphi(gh)\chi_f(h)\Delta_{H_2,G}^{-1/2}(h)d\nu(h)\ (g\in G). \tag{2.8.1}$$

At least formally, it is clear that the function $I_{\mathfrak{h}_2\mathfrak{h}_1}\varphi$ satisfies the necessary covariance condition for belonging in the space $\mathscr{H}_{\pi_2}$, and also that $I_{\mathfrak{h}_2\mathfrak{h}_1}$ commutes with left translations. We can therefore assert that the convergence of the integral is one major issue to deal with. We call elements of $I(f,\mathfrak{g})$ Pukanszky polarizations.

In this section we look at all pairs $(\mathfrak{h}_1,\mathfrak{h}_2)$ of Pukanszky polarizations at f which meet the following convergence property:

(CP): The integral

$$\oint_{H_2/(H_1\cap H_2)}|\varphi(gh)|\Delta_{H_2,G}^{-\frac{1}{2}}(h)d\nu(h),\ g\in G, \tag{2.8.2}$$

converges on a G-invariant and $\mathscr{U}(\mathfrak{g})$-invariant dense subspace $\mathscr{H}$ of $\mathscr{H}_{\pi_1}^\infty$.

Proposition 2.8.1 *Suppose that the pair of Pukanszky polarizations $(\mathfrak{h}_1,\mathfrak{h}_2)$ satisfies property (CP). Then for any function $\varphi\in\mathscr{K}(f,\mathfrak{h}_1,G)$ we have that*

$$\oint_{H_2/(H_1\cap H_2)}|\varphi(gh)|\Delta_{H_2,G}^{-\frac{1}{2}}(h)d\nu(h)<\infty,\ \ g\in G,$$

and the function $g\mapsto\left(I_{\mathfrak{h}_2\mathfrak{h}_1}\varphi\right)(g)$ is continuous.

Proof Choose $\psi \in \mathcal{H}$ and a continuous function $\alpha \in C_c(G/H_1)$ such that $\beta := \alpha\psi \neq 0$. Then $\beta \in \mathcal{K}(f, \mathfrak{h}_1, G)$ and for every $g \in G$:

$$\oint_{H_2/(H_1\cap H_2)} |\beta(gh)|\Delta_{H_2,G}^{-\frac{1}{2}}(h)d\nu(h)$$

$$\leq \|\alpha\|_\infty \oint_{H_2/(H_1\cap H_2)} |\psi(gh)|\Delta_{H_2,G}^{-\frac{1}{2}}(h)d\nu(h) < \infty.$$

For any $\varphi \in \mathcal{K}(f, \mathfrak{h}_1, G)$, we can find a finite subset F of G and a number $C > 0$ such that, for every $g \in G$,

$$|\varphi(g)| \leq C(\sum_{x\in F} |\beta(xg)|).$$

This shows that on the subspace $\mathcal{K}(f, \mathfrak{h}_1, G)$ the integral (2.8.2) converges. Furthermore, for any compact subset K of G, using the function $\tilde{\varphi}(g) := \sup_{k\in K} |\varphi(kg)|$, $g \in G$, we can apply Lebesgue's dominated convergence theorem to prove that the function $I_{\mathfrak{h}_2\mathfrak{h}_1}\varphi$ is continuous. □

Note that the Pukanszky condition implies that the product H_2H_1 is locally closed in G, and therefore the space H_2H_1/H_1 is homeomorphic to the homogeneous space $H_2/(H_1 \cap H_2)$. If now H_2H_1 is closed in G, which is true when G is nilpotent or one of our two polarizations is a Vergne polarization, then (2.8.1) converges for every function $\varphi \in \mathcal{K}(f, \mathfrak{h}_1, G)$ and therefore the pair fulfils property (CP). In [2] we posed the following problem, whose answer is thus far still unknown:

Question Let G be an exponential solvable Lie group. Let $f \in \mathfrak{g}^*$ and let $\mathfrak{h}_j$ be Pukanszky polarizations at f and $H_j = \exp(\mathfrak{h}_j)(j = 1, 2)$. Is the product H_2H_1 closed in G?

In the absence of an answer, we believe the following conjecture holds:

Claim *Any pair of Pukanszky polarizations satisfies property (CP).*
We shall show directly that the mapping $I_{\mathfrak{h}_2\mathfrak{h}_1}$ is an isometry if we assume (CP). Similar computations have been carried out in the previous section, so here we shall only give a full proof for completely solvable groups.

Theorem 2.8.2 *Let $\mathfrak{h}_1$ and $\mathfrak{h}_2$ be two Pukanszky polarizations at f satisfying (CP). Then for every $\varphi \in \mathcal{H}$ the function $I_{\mathfrak{h}_2\mathfrak{h}_1}(\varphi)$ belongs to $\mathcal{H}_{\pi_2}$, and for a suitable choice of the linear form $\oint_{H_2/(H_1\cap H_2)} d\nu$, we have that $\|I_{\mathfrak{h}_2\mathfrak{h}_1}(\varphi)\|_2 = \|\varphi\|_1$ for every $\varphi \in \mathcal{H}$. Put in other words, the mapping $I_{\mathfrak{h}_2\mathfrak{h}_1}$ continuously extends to a unitary intertwining operator for the representations π_1 and π_2.*

Proof We proceed by induction on the dimension of $\mathfrak{g}$. We retain the notation used in the previous section. For $\mathfrak{h} \in S(f, \mathfrak{g})$, let $H = \exp \mathfrak{h}$ and $\pi := \pi_{f,\mathfrak{h}} := \mathrm{ind}_H^G \chi_f$. We shall denote also by $L^2(G/H, \chi_f)$ the space $\mathcal{H}_\pi$.

Suppose there exists a proper subalgebra $\mathfrak{b}$ of $\mathfrak{g}$ containing $\mathfrak{h}_1+\mathfrak{h}_2$. Let $B = \exp \mathfrak{b}$ and denote by $I^G_{\mathfrak{h}_2\mathfrak{h}_1}$ and $I^B_{\mathfrak{h}_2\mathfrak{h}_1}$ the operators defined as in formula (2.8.1) at the level of G and B respectively. For $u \in G$ and $b \in B$, we have that

$$I^G_{\mathfrak{h}_2\mathfrak{h}_1}\varphi(ub) = \Delta^{\frac{1}{2}}_{B,G}(b) I^B_{\mathfrak{h}_2\mathfrak{h}_1}(\Delta^{-\frac{1}{2}}_{B,G}\varphi_u)(b),$$

where $\varphi_u(b) = \varphi(ub)$. The induction hypothesis applied to $\mathfrak{b}$ tells us that for every $\varphi \in \mathscr{H}$

$$\begin{aligned}\oint_{G/H_2} |I^G_{\mathfrak{h}_2\mathfrak{h}_1}\varphi(u)|^2 d\nu(u) &= \oint_{G/B}\oint_{B/H_2} |I^G_{\mathfrak{h}_2\mathfrak{h}_1}\varphi(ub)|^2 \Delta^{-1}_{B,G}(b) d\nu_{B,H_2}(b) d\nu_{G,B}(u)\\ &= \oint_{G/B}\oint_{B/H_2} |I^B_{\mathfrak{h}_2\mathfrak{h}_1}(\Delta^{-\frac{1}{2}}_{B,G}\varphi_u)(b)|^2 d\nu_{B,H_2}(b) d\nu_{G,B}(u)\\ &= \oint_{G/B}\oint_{B/H_1} \Delta^{-1}_{B,G}(b)|\varphi(ub)|^2 d\nu_{B,H_1}(b) d\nu_{G,B}(u)\\ &= \oint_{G/H_1} |\varphi(g)|^2 d\nu(g).\end{aligned}$$

Assume now that there exists a non-trivial ideal $\mathfrak{a}$ of $\mathfrak{g}$ on which f vanishes. It follows that $\mathfrak{a} \subset \mathfrak{h}_1 \cap \mathfrak{h}_2$. Take $\tilde{\mathfrak{g}} = \mathfrak{g}/\mathfrak{a}$, $\tilde{G} = \exp \tilde{\mathfrak{g}}$, the canonical projection $p : \mathfrak{g} \to \tilde{\mathfrak{g}}$ and $\tilde{f} \in (\tilde{\mathfrak{g}})^*$ such that $\tilde{f} \circ p = f$, $\tilde{\mathfrak{h}}_j = p(\mathfrak{h}_j)$ and $\tilde{H}_j = \exp(\tilde{\mathfrak{h}}_j)$ for $j = 1, 2$. We also denote by $p : G \to G/A$, $A = \exp \mathfrak{a}$, the canonical projection. Then for $j \in \{1, 2\}$, it is clear that $L^2(\tilde{G}/\tilde{H}_j, \chi_{\tilde{f}}) \circ p = L^2(G/H_j, \chi_f)$ and $\pi_{\tilde{f},\tilde{\mathfrak{h}}_j} \circ p = \pi_{f,\mathfrak{h}_j}$. So, $\varphi \in \mathscr{H}$ is written as $\varphi = \tilde{\varphi} \circ p$ for some $\tilde{\varphi} \in L^2(\tilde{G}/\tilde{H}_j, \chi_{\tilde{f}})$ and we immediately see that $I^{\tilde{G}}_{\tilde{\mathfrak{h}}_2\tilde{\mathfrak{h}}_1}\tilde{\varphi}(\tilde{g}) = I_{\mathfrak{h}_2\mathfrak{h}_1}\varphi(g)$ for any $g \in G$ and $\tilde{g} = p(g)$.

From now on we assume f does not vanish on any non-trivial ideal of $\mathfrak{g}$. We take a minimal non-central ideal $\mathfrak{a}$ of $\mathfrak{g}$ contained in $[\mathfrak{g}, \mathfrak{g}]$. We further suppose $\mathfrak{z} = \mathbb{R}Z$ is the one-dimensional center of $\mathfrak{g}$ such that $f(Z) = 1$, and $\mathfrak{a} \supset \mathfrak{z}$ has dimension 2 since $\mathfrak{g}$ was taken completely solvable. We put

$$\mathfrak{g}^0 := \{U \in \mathfrak{g}; \langle f, [U, \mathfrak{a}]\rangle = \{0\}\} = \mathfrak{a}^f \neq \mathfrak{g}$$

and $\mathfrak{h}_i^0 := \mathfrak{h}_i \cap \mathfrak{g}^0$, $i = 1, 2$. We have by assumption that $\mathfrak{h}_1 + \mathfrak{h}_2 \not\subset \mathfrak{g}^0$. We have to treat several situations.

Case 1 Suppose first that $\mathfrak{h}_1 \not\subset \mathfrak{g}^0$ and $\mathfrak{h}_2 \subset \mathfrak{g}^0$. Let $\mathfrak{h}_1' := \mathfrak{h}_1^0 + \mathfrak{a}$. Choose a subspace $\mathfrak{y}$ in $\mathfrak{a}$ such that $\mathfrak{a} = \mathfrak{y} \oplus (\mathfrak{h}_1 \cap \mathfrak{a})$ and $\mathfrak{y} \subset \ker(f)$. Then $\mathfrak{h}_2/(\mathfrak{h}_1 \cap \mathfrak{h}_2) \simeq \mathfrak{h}_2/(\mathfrak{h}_1' \cap \mathfrak{h}_2) \oplus \mathfrak{y}$. For simplicity, let us now write (and also in subsequent cases) $I_{2,1} := I_{\mathfrak{h}_2\mathfrak{h}_1}$, $I_{2,1'} := I_{\mathfrak{h}_2\mathfrak{h}_1'}$, $I_{1,1'} = I_{\mathfrak{h}_1\mathfrak{h}_1'}$ and so on. We know that $I_{1',1}$ is in fact a Fourier transform. The following computations show that the operator $I_{2,1'}$ converges absolutely on the space $I_{1,1'}(\mathscr{H})$ and that the composition formula $I_{2,1'} \circ$

$I_{1',1} = I_{2,1}$ holds on $\mathscr{H}$. Indeed,

$$\begin{aligned}
&I_{2,1'} \circ I_{1',1}(\varphi)(g)\\
&= \oint_{H_2/(H_2\cap H_1')} \chi_f(h)\Delta_{H_2,G}^{-\frac{1}{2}}(h) \int_{\mathfrak{y}} \chi_f(\exp Y)\varphi(gh\exp Y)dYd\nu(h)\\
&= \oint_{H_2/(H_1'\cap H_2')} \int_{\mathfrak{y}} \chi_f(h\exp Y)\Delta_{H_2,G}^{-\frac{1}{2}}(h\exp Y)\varphi(gh\exp Y)dYd\nu(h)\\
&= \oint_{H_2/(H_2\cap H_1)} \chi_f(h)\Delta_{H_2,G}^{-\frac{1}{2}}(h)\varphi(gh)d\nu(h)\\
&=I_{2,1}\varphi(g).
\end{aligned}$$

Hence $I_{2,1}$ is an isometry.

Case 2 If now $\mathfrak{h}_2 \not\subset \mathfrak{g}^0$ and $\mathfrak{h}_1 \subset \mathfrak{g}^0$, let $\mathfrak{h}_2' := \mathfrak{h}_2^0 + \mathfrak{a}$. Choose a subspace $\mathfrak{x}$ in $\mathfrak{h}_2$ such that $\mathfrak{h}_2 = \mathfrak{x} \oplus \mathfrak{h}_2^0$ and $\mathfrak{x} \subset \ker(f)$. Then

$$\mathfrak{h}_2/(\mathfrak{h}_1 \cap \mathfrak{h}_2) \simeq \mathfrak{x} \oplus (\mathfrak{h}_2'/(\mathfrak{h}_2' \cap \mathfrak{h}_1)).$$

The following details show that the operator $I_{2',1}$ converges absolutely on $\mathscr{H}$ and that the composition formula $I_{2,2'}\circ I_{2',1} = I_{2,1}$ holds on $\mathscr{H}$. Indeed, for $\varphi \in \mathscr{H}$:

$$\begin{aligned}
I_{2,2'}\circ I_{2',1}(\varphi)(g) &= \int_{\mathfrak{x}} \Delta_{H_2,G}^{-\frac{1}{2}}(\exp X)\\
&\quad\times \oint_{H_2'/(H_2'\cap H_1)} \chi_f(h)\Delta_{H_2',G}^{-\frac{1}{2}}(h)\varphi(g\exp X\cdot h)d\nu(h)dX\\
&= \int_{\mathfrak{x}} \oint_{H_2'/(H_2'\cap H_1} \chi_f(\exp X\cdot h)\Delta_{H_2,G}^{-\frac{1}{2}}(\exp X\cdot h)\varphi(g\exp X\cdot h)\\
&\quad\times\Delta_{H_2^0,H_2}^{-1}(h)dXd\nu(h)\\
&= \oint_{H_2/(H_2\cap H_1)} \chi_f(h)\Delta_{H_2,G}^{-\frac{1}{2}}(h)\varphi(gh)d\nu(h)\\
&= I_{2,1}\varphi(g),
\end{aligned}$$

since $\Delta_{H_2',G}^{-\frac{1}{2}}(h) = \Delta_{H_2,G}^{-\frac{1}{2}}(h)\Delta_{H_2^0,H_2}^{-1}(h), h \in H_2^0$. Hence $I_{2,1}$ is an isometry in this case as well.

Case 3 We now come to the case where $\mathfrak{h}_1 \not\subset \mathfrak{g}^0$ and $\mathfrak{h}_2 \not\subset \mathfrak{g}^0$. Note that there exist a character $\alpha : \mathfrak{g} \to \mathbb{R}$ and a linear functional $0 \neq \beta : \mathfrak{g} \to \mathbb{R}$ such that $[U, Y] = \alpha(U)Y + \beta(U)Z$ for all $U \in \mathfrak{g}$. In particular $\mathfrak{g}^0 = \ker(\beta)$. We put $\mathfrak{g}_0 = \ker(\alpha)$.

Subcase 3.1 Suppose that $\mathfrak{h}_1 \cap \mathfrak{h}_2 \not\subset \mathfrak{g}^0$. Since $\mathfrak{h}_1$ and $\mathfrak{h}_2$ are Pukanszky polarizations, we have that $\mathrm{Ad}^*(H_1 \cap H_2)\cdot f = f + (\mathfrak{h}_1 + \mathfrak{h}_2)^{\perp}$. This shows that the restrictions of α and β to $\mathfrak{h}_1 \cap \mathfrak{h}_2$ cannot be proportional, since $Y \notin \mathfrak{h}_1 + \mathfrak{h}_2$. We can therefore choose $X \in (\ker(f) \cap \mathfrak{h}_1 \cap \mathfrak{h}_2) \setminus \mathfrak{g}^0$ so that $[X, Y] = Z$. We have $\mathfrak{h}_1 \cap \mathfrak{a} = \mathfrak{h}_2 \cap \mathfrak{a} = \mathfrak{z}$. Let $\mathfrak{h}_i' := \mathfrak{h}_i^0 + \mathfrak{a}$, $i = 1, 2$. Then

$$\mathfrak{h}_2/(\mathfrak{h}_1 \cap \mathfrak{h}_2) \simeq \mathfrak{h}_2^0/(\mathfrak{h}_1^0 \cap \mathfrak{h}_2^0) \simeq \mathfrak{h}_2^0/(\mathfrak{h}_1 \cap \mathfrak{h}_2') \simeq \mathfrak{h}_2'/(\mathfrak{h}_1' \cap \mathfrak{h}_2').$$

As previously, it is sufficient to prove that the following holds. Let $\mathcal{H}' \subset L^2(G/H_1', \chi_f)$ be such that $I_{1,1'}(\mathcal{H}') = \mathcal{H}$. Then for $\varphi \in \mathcal{H}'$ and $g \in G$, we get

$$\begin{aligned}
I_{2,1'}\varphi(g) &= \oint_{H_2/(H_2 \cap H_1')} \varphi(gh)\Delta_{H_2,G}^{-\frac{1}{2}}(h)\chi_f(h)d\nu(h) \\
&= \oint_{H_2/(H_1 \cap H_2)} \int_{\mathbb{R}} \varphi(gh\exp(xX))\Delta_{H_1\cap H_2, H_2}^{-1}(\exp(xX)) \\
&\quad \times \Delta_{H_2,G}^{-\frac{1}{2}}(h\exp(xX))\chi_f(h)dxd\nu(h) \\
&= \oint_{H_2/(H_1 \cap H_2)} \oint_{H_1/(H_1 \cap H_1')} \varphi(gh\exp(xX))\Delta_{H_1\cap H_2, H_2}^{-1}(\exp(xX))\Delta_{H_2,G}^{-\frac{1}{2}}(\exp(xX)) \\
&\quad \times \Delta_{H_2,G}^{-\frac{1}{2}}(h)\chi_f(h)dxd\nu(h) \\
&= \oint_{H_2/(H_1 \cap H_2)} [\oint_{H_1/(H_1 \cap H_1')} \varphi(gh\exp(xX))\Delta_{H_1,G}^{-\frac{1}{2}}(\exp(xX))dx] \\
&\quad \times \Delta_{H_2,G}^{-\frac{1}{2}}(h)\chi_f(h)d\nu(h) \\
&= I_{2,1} \circ I_{1,1'}\varphi(g).
\end{aligned}$$

Subcase 3.2 This case occurs when $(\mathfrak{h}_1 \cap \mathfrak{h}_2) \subset \mathfrak{g}^0$, which is equivalent to the fact $\mathfrak{a} \subset \mathfrak{h}_1 + \mathfrak{h}_2$. We first look at the following cases:

Subcase 3.2.1 If $\mathfrak{a} \subset \mathfrak{h}_1^0 + \mathfrak{h}_2^0$, then we have $Y = X_2 - X_1$ for some vectors $X_i \in \mathfrak{h}_i^0 \cap \ker(f)$, $i = 1, 2$. As in Lemma 2.6.3, we can choose X_1 and X_2 in $\mathfrak{g}_0 = \ker(\alpha)$. So, $[X_1, X_2] = [X_1, X_1 + Y] = 0$. We have that $\mathfrak{h}_1' \cap \mathfrak{h}_2 \simeq (\mathfrak{h}_1 \cap \mathfrak{h}_2) \oplus \mathbb{R}X_2$, $\mathfrak{h}_2' \cap \mathfrak{h}_1 \simeq (\mathfrak{h}_1 \cap \mathfrak{h}_2) \oplus \mathbb{R}X_1$ and $\mathfrak{h}_1' \cap \mathfrak{h}_2' \equiv (\mathfrak{h}_1 \cap \mathfrak{h}_2) \oplus \mathbb{R}X_1 \oplus \mathbb{R}X_2$. Furthermore

$$H_2' = H_2^0\exp(\mathbb{R}Y), \;\; H_1' \cap H_2' = (H_1 \cap H_2)\exp(\mathbb{R}X_2)\exp(\mathbb{R}Y)$$

and the space $H_2'/(H_1' \cap H_2')$ is isomorphic to $H_2^0/(H_1 \cap H_2)\exp\mathbb{R}X_2$. We now prove that:

$$\mathrm{Tr}(\mathrm{ad}_{\mathfrak{h}_1/(\mathfrak{h}_1\cap\mathfrak{h}_2)}(X_1)) = -\mathrm{Tr}(\mathrm{ad}_{\mathfrak{h}_2/(\mathfrak{h}_1\cap\mathfrak{h}_2)}(X_2)).$$

Indeed, using Theorem 2.2.2 and that $X_2 = X_1 + Y$, we get

$$\mathrm{Tr}(\mathrm{ad}_{\mathfrak{h}_1/(\mathfrak{h}_1\cap\mathfrak{h}_2')}(X_1)) + \mathrm{Tr}(\mathrm{ad}_{\mathfrak{h}_2'/(\mathfrak{h}_1\cap\mathfrak{h}_2')}(X_2)) = 0,$$

which leads to

$$\mathrm{Tr}(\mathrm{ad}_{\mathfrak{h}_1/(\mathfrak{h}_1\cap\mathfrak{h}_2)}(X_1)) + \mathrm{Tr}(\mathrm{ad}_{\mathfrak{h}_2'/(\mathfrak{h}_1\cap\mathfrak{h}_2')}(X_2)) = 0$$

as $\mathfrak{h}_1\cap\mathfrak{h}_2' = (\mathfrak{h}_1\cap\mathfrak{h}_2)\oplus\mathbb{R}X_1$. Using now that $\mathfrak{h}_2'/(\mathfrak{h}_1\cap\mathfrak{h}_2') = \mathfrak{h}_2'/((\mathfrak{h}_1\cap\mathfrak{h}_2)\oplus\mathbb{R}X_1)$ and that $[X_1, X_2] = 0$, we are done.

Knowing that $Y \in \mathfrak{h}_2'$ is a nilpotent element, we easily get

$$\mathrm{Tr}(\mathrm{ad}_{\mathfrak{g}/\mathfrak{h}_2'}(X_1)) = \mathrm{Tr}(\mathrm{ad}_{\mathfrak{g}/\mathfrak{h}_2'}(X_2)).$$

On the other hand, using $(\mathfrak{h}_1\cap\mathfrak{h}_2')\oplus\mathbb{R}Y = \mathfrak{h}_1'\cap\mathfrak{h}_2'$, $[X_1, Y] = 0$, and then that $\mathfrak{h}_2'/(\mathfrak{h}_2'\cap\mathfrak{h}_1') \simeq \mathfrak{h}_2^0/(\mathfrak{h}_1'\cap\mathfrak{h}_2)$, we obtain:

$$\mathrm{Tr}(\mathrm{ad}_{\mathfrak{h}_2'/(\mathfrak{h}_2'\cap\mathfrak{h}_1)}(X_1)) = \mathrm{Tr}(\mathrm{ad}_{\mathfrak{h}_2'/(\mathfrak{h}_2'\cap\mathfrak{h}_1')}(X_1)) = \mathrm{Tr}(\mathrm{ad}_{\mathfrak{h}_2'/(\mathfrak{h}_2'\cap\mathfrak{h}_1')}(X_1 + Y))$$
$$= \mathrm{Tr}(\mathrm{ad}_{\mathfrak{h}_2'/(\mathfrak{h}_2'\cap\mathfrak{h}_1')}(X_2)) = \mathrm{Tr}(\mathrm{ad}_{\mathfrak{h}_2^0/(\mathfrak{h}_1'\cap\mathfrak{h}_2)}(X_2)),$$

which implies

$$\Delta_{H_1\cap H_2', H_2'}(\exp(yX_1)) = \Delta_{H_1'\cap H_2, H_2^0}(\exp(yX_2))$$
$$=\Delta_{(H_1\cap H_2)\exp(\mathbb{R}X_2), H_2^0}(\exp(yX_2)).$$

It follows then for $\varphi \in \mathscr{K}(f, \mathfrak{h}_1, G)$, $g \in G$,:

$$\oint_{H_2/H_2^0} \Delta_{H_2,G}^{-\frac{1}{2}}(k)\chi_f(k) \oint_{H_2'/(H_2'\cap H_1')} \Delta_{H_2',G}^{-\frac{1}{2}}(h)\chi_f(h)$$
$$\times \int_{\mathbb{R}} \varphi(gkh\exp(yY))dy d\nu(h)d\nu(k)$$
$$= \oint_{H_2/H_2^0} \Delta_{H_2,G}^{-\frac{1}{2}}(k)\chi_f(k) \oint_{H_2'/(H_2'\cap H_1')} \Delta_{H_2',G}^{-\frac{1}{2}}(h)\chi_f(h)$$
$$\times \int_{\mathbb{R}} \varphi(gkh\exp(yX_2))\Delta_{H_1,G}^{-\frac{1}{2}}(\exp(yX_1))dy d\nu(h)d\nu(k)$$
$$= \oint_{H_2/H_2^0} \Delta_{H_2,G}^{-\frac{1}{2}}(k)\chi_f(k) \oint_{H_2^0/(H_2\cap H_1)\exp(\mathbb{R}X_2)} \Delta_{H_2',G}^{-\frac{1}{2}}(\exp(yX_1))\chi_f(h)$$
$$\times \Delta_{H_1\cap H_2', H_2'}^{-1}(\exp(yX_1))\Delta_{H_2^0, H_2}^{-1}(h)\Delta_{H_2,G}^{-\frac{1}{2}}(h)\varphi(gkh\exp(yX_2))dy d\nu(h)d\nu(k)$$

$$= \oint_{H_2/H_2^0} \chi_f(k)\Delta_{H_2,G}^{-\frac{1}{2}}(k) \oint_{H_2^0/(H_1\cap H_2)\exp(\mathbb{R}X_2)} \int_{\mathbb{R}} \chi_f(h)\Delta_{H_2',G}^{-\frac{1}{2}}(\exp(yX_2))$$

$$\times\Delta^{-1}_{(H_1\cap H_2)\exp(\mathbb{R}X_2),H_2^0}(\exp(yX_2))\Delta^{-1}_{H_2^0,H_2}(h)\Delta_{H_2,G}^{-\frac{1}{2}}(h)$$

$$\times\varphi(gkh\exp(yX_2))dyd\nu(h)d\nu(k)$$

$$= \oint_{H_2/H_2^0} \chi_f(k)\Delta_{H_2,G}^{-\frac{1}{2}}(k)$$

$$\times \oint_{H_2^0/(H_1\cap H_2)} \chi_f(h)\Delta^{-1}_{H_2^0,H_2}(h)\Delta_{H_2,G}^{-\frac{1}{2}}(h)\varphi(gkh)d\nu(h)d\nu(k)$$

$$= \oint_{H_2/(H_1\cap H_2)} \varphi(gh)\chi_f(h)\Delta_{H_2,G}^{-\frac{1}{2}}(h)d\nu(h).$$

In particular we get:

$$\oint_{H_2/H_2^0} \Delta_{H_2,G}^{-\frac{1}{2}}(k) \oint_{H_2'/(H_1'\cap H_2')} \Delta_{H_2',G}^{-\frac{1}{2}}(h) \int_{\mathbb{R}} |\varphi(gkh\exp(yY))|dyd\nu(h)d\nu(k)$$

$$= \oint_{H_2/(H_1\cap H_2)} |\varphi(gh)|\Delta_{H_2,G}^{-\frac{1}{2}}(h)d\nu(h) < \infty.$$

Hence for every $\varphi \in \mathscr{K}(f, \mathfrak{h}_1, G)$ and almost every $g \in G$:

$$\int_{\mathbb{R}} |\varphi(g\exp(yY))|dy < +\infty.$$

Arguing as in the proof of Proposition 2.8.1, we now see that every function contained in the subspace $I_{\mathfrak{h}_1'\mathfrak{h}_1}(\mathscr{K}(f, \mathfrak{h}_1, G))$ is continuous (and contained in $L^2(G/H_1', \chi_f)$, since $I_{\mathfrak{h}_1'\mathfrak{h}_1}$ is an isometry for the L^2-norms) and satisfies condition (2.8.2) for H_1' and H_2'. For $\varphi \in \mathscr{K}(f, \mathfrak{h}_1, G)$ and $g \in G$, let $I(g) := I_{2,2'}\circ I_{2',1'}\circ I_{1',1}\varphi(g)$. It turns out, using the previous computations and the induction hypothesis, that this integral converges, and furthermore:

$$I(g) = \oint_{H_2/H_2^0} \chi_f(k)\Delta_{H_2,G}^{-\frac{1}{2}}(k) \oint_{H_2'/(H_1'\cap H_2')} \chi_f(h)\Delta_{H_2',G}^{-\frac{1}{2}}(h)$$

$$\times \int_{\mathbb{R}} \varphi(gkh\exp(yY))dyd\nu(h)d\nu(k)$$

$$= I_{21}\varphi(g).$$

Hence $I_{2,1}$ is a multiple of an isometry.

Subcase 3.2.2 Assume now that $\mathfrak{a} \subset (\mathfrak{h}_1 + \mathfrak{h}_2) \setminus (\mathfrak{h}_1^0 + \mathfrak{h}_2^0)$, then we can find $X_i \in \mathfrak{h}_i \cap \ker(f), i = 1, 2$ and $a \in \mathbb{R}^*$, such that $Y = a(X_2 - X_1)$, $[X_1, Y] = [X_2, Y] = \delta Y + Z$ for some δ in $\mathbb{R}$, $\mathfrak{h}_1 = \mathfrak{h}_1^0 \oplus \mathbb{R}X_1$ and $\mathfrak{h}_2 = \mathfrak{h}_2^0 \oplus \mathbb{R}X_2$. Note that Lemmas 2.6.4, 2.6.5 are available.

In the case at hand $\mathfrak{h}_1 \cap \mathfrak{h}_2 = \mathfrak{h}_1' \cap \mathfrak{h}_2$ and $[X_2, \mathfrak{h}_1 \cap \mathfrak{h}_2] \subset [\mathfrak{h}_1, \mathfrak{h}_2]$. Let $\mathfrak{k} := \mathbb{R}X_2 \oplus (\mathfrak{h}_1 \cap \mathfrak{h}_2)$ and $K := \exp \mathfrak{k}$. This allows us to identify the space H_2/K with $H_2^0/(H_1 \cap H_2)$. Let $\varphi \in \mathscr{H}$, $g \in G$ and $I(g) = I_{2,1'} \circ I_{1',1}\varphi(g)$. Assume for moment that $I(g)$ converges. Using once more the transitivity of $\nu_{\cdot,\cdot}$ and Lemmas 2.6.4, 2.6.5, we get:

$$
\begin{aligned}
I(g) &= \oint_{H_2/H_1\cap H_2'} (I_{1',1}\varphi)(gh)\Delta_{H_2,G}^{-\frac{1}{2}}(h)\chi_f(h)d\nu(h) \\
&= \oint_{H_2/K}\oint_{K/H_1\cap H_2} (I_{1',1}\varphi)(ghk)\Delta_{H_2,G}^{-\frac{1}{2}}(hk)\Delta_{K,H_2}^{-1}(k)\chi_f(hk)d\nu(k)d\nu(h) \\
&= \oint_{H_2/K} \chi_f(h)\Delta_{H_2,G}^{-\frac{1}{2}}(h)d\nu(h) \\
&\times \int_{\mathbb{R}} (I_{1',1}\varphi)(gh\exp(xX_2))e^{-\frac{x}{2}\mathrm{Tr}\,\mathrm{ad}_{\mathfrak{g}/\mathfrak{h}_2}X_2}e^{-x\mathrm{Tr}\,\mathrm{ad}_{\mathfrak{h}_2/\mathfrak{k}}X_2}dx \\
&= \oint_{H_2/K} \chi_f(h)\Delta_{H_2,G}^{-\frac{1}{2}}(h)d\nu(h) \\
&\times \int_{\mathbb{R}}\int_{\mathbb{R}} \varphi(gh\exp(xX_2)\exp(yY))dy e^{-\frac{x}{2}\mathrm{Tr}\,\mathrm{ad}_{\mathfrak{g}/\mathfrak{h}_2}X_2}e^{-x\mathrm{Tr}\,\mathrm{ad}_{\mathfrak{h}_2/\mathfrak{k}}X_2}dx \\
&= \oint_{H_2/K} \chi_f(h)\Delta_{H_2,G}^{-\frac{1}{2}}(h)d\nu(h)\int_{\mathbb{R}} e^{-\frac{x}{2}\mathrm{Tr}\,\mathrm{ad}_{\mathfrak{g}/\mathfrak{h}_2}X_2}e^{-x\mathrm{Tr}\,\mathrm{ad}_{\mathfrak{h}_2/\mathfrak{k}}X_2}dx \\
&\times \int_{\mathbb{R}} \varphi(gh\exp((x+ay)X_2)\exp(-ayX_1)\exp(-\frac{1}{2}ay^2Z))dx \\
&= \oint_{H_2/K} \chi_f(h)\Delta_{H_2,G}^{-\frac{1}{2}}(h)d\nu(h)\int_{\mathbb{R}} e^{-\frac{x}{2}\mathrm{Tr}\,\mathrm{ad}_{\mathfrak{g}/\mathfrak{h}_2}X_2}e^{-x\mathrm{Tr}\,\mathrm{ad}_{\mathfrak{h}_2/\mathfrak{k}}X_2}dx \\
&\times \int_{\mathbb{R}} e^{\frac{i}{2}ay^2}e^{-\frac{1}{2}ay\mathrm{Tr}\,\mathrm{ad}_{\mathfrak{g}/\mathfrak{h}_1}X_1}\varphi(gh\exp((x+ay)X_2))dy \\
&= \frac{1}{|a|}\oint_{H_2/K} \chi_f(h)\Delta_{H_2,G}^{-\frac{1}{2}}(h)d\nu(h)\int_{\mathbb{R}} e^{-\frac{x}{2}\mathrm{Tr}\,\mathrm{ad}_{\mathfrak{g}/\mathfrak{h}_2}X_2}e^{-x\mathrm{Tr}\,\mathrm{ad}_{\mathfrak{h}_2/\mathfrak{k}}X_2}dx \\
&\times \int_{\mathbb{R}} e^{\frac{i}{2a}(y-x)^2}e^{-\frac{1}{2}(y-x)\mathrm{Tr}\,\mathrm{ad}_{\mathfrak{g}/\mathfrak{h}_1}X_1}\varphi(gh\exp(yX_2))dy \\
&= \frac{1}{|a|}\oint_{H_2/K} \chi_f(h)\Delta_{H_2,G}^{-\frac{1}{2}}(h)d\nu(h) \\
&\times \int_{\mathbb{R}} dx\int_{\mathbb{R}} e^{\frac{i}{2a}(y-x)^2}e^{-\frac{1}{2}y\mathrm{Tr}\,\mathrm{ad}_{\mathfrak{g}/\mathfrak{h}_1}X_1}\varphi(gh\exp(yX_2))dy
\end{aligned}
$$

$$= \sqrt{\frac{2\pi}{|a|}} e^{\frac{i\pi}{4}\operatorname{sgn}(a)} \oint_{H_2/K} \chi_f(h) \Delta_{H_2,G}^{-\frac{1}{2}}(h) d\nu(h)$$

$$\times \int_{\mathbb{R}} e^{-\frac{1}{2}x \operatorname{Tr} \operatorname{ad}_{\mathfrak{g}/\mathfrak{h}_2} X_2} e^{-x \operatorname{Tr} \operatorname{ad}_{\mathfrak{h}_2/\mathfrak{k}} X_2} \varphi(gh \exp(xX_2)) dx$$

$$= \sqrt{\frac{2\pi}{|a|}} e^{\frac{i\pi}{4}\operatorname{sgn}(a)} \oint_{H_2/H_1\cap H_2} \chi_f(h) \Delta_{H_2,G}^{-\frac{1}{2}}(h) \varphi(gh) d\nu(h)$$

$$= \sqrt{\frac{2\pi}{|a|}} e^{\frac{i\pi}{4}\operatorname{sgn}(a)} (I_{2,1}\varphi)(g).$$

Above we have used the integral formula

$$\int_{\mathbb{R}} \Big(\int_{\mathbb{R}} e^{\frac{i}{2a}(y-x)^2} h(y) dy \Big) dx = \sqrt{2\pi |a|} e^{\frac{i\pi}{4}\operatorname{sgn}(a)} \int_{\mathbb{R}} h(x) dx,$$

whose proof can be found in [98], for $h(y) = e^{-\frac{1}{2}y \operatorname{Tr} \operatorname{ad}_{\mathfrak{g}/\mathfrak{h}_1} X_1} \varphi(gh \exp(yX_2))$ in the Schwartz space $\mathscr{S}(\mathbb{R})$ for fixed g and h. Note also that the trace equality

$$\frac{1}{2}\operatorname{Tr} \operatorname{ad}_{\mathfrak{g}/\mathfrak{h}_2}(X_2) - \frac{1}{2}\operatorname{Tr} \operatorname{ad}_{\mathfrak{g}/\mathfrak{h}_1}(X_1) + \operatorname{Tr} \operatorname{ad}_{\mathfrak{h}_2/\mathfrak{k}}(X_2) = 0$$

has been used throughout. The latter's proof follows from the trace relation in Lemma 2.6.5 and the fact that

$$\operatorname{Tr} \operatorname{ad}_{\mathfrak{g}/\mathfrak{h}_i}(X_i) = \operatorname{Tr} \operatorname{ad}_{\mathfrak{g}/(\mathfrak{h}_1\cap\mathfrak{h}_2)}(X_i) - \operatorname{Tr} \operatorname{ad}_{\mathfrak{h}_i/(\mathfrak{h}_1\cap\mathfrak{h}_2)}(X_i), \; i = 1, 2.$$

We now prove that the integral $I_{2,1'}$ converges absolutely on $I_{1',1}(\mathscr{H})$. Indeed,

$$\oint_{H_2/H_1\cap H_2'} |I_{1',1}(\varphi)(gh)| \Delta_{H_2,G}^{-\frac{1}{2}}(h) d\nu(h)$$

$$= \frac{1}{|a|} \oint_{H_2/K} \Delta_{H_2,G}^{-\frac{1}{2}}(h) d\nu(h)$$

$$\times \int_{\mathbb{R}} \Big| \int_{\mathbb{R}} e^{\frac{i}{2a}(y-x)^2} e^{-\frac{1}{2}y \operatorname{Tr} \operatorname{ad}_{\mathfrak{g}/\mathfrak{h}_1} X_1} \varphi(gh \exp(yX_2)) dy \Big| dx$$

$$= \frac{1}{|a|} \oint_{H_2/K} \Delta_{H_2,G}^{-\frac{1}{2}}(h) d\nu(h)$$

$$\times \int_{\mathbb{R}} \Big| \int_{\mathbb{R}} e^{\frac{i}{2a}(y-x)^2} \varphi(gh \exp(\frac{y}{a}Y)) dy \Big| dx < \infty,$$

as $\mathscr{H}$ is $\mathscr{U}(\mathfrak{g})$-invariant. This finishes the proof in this case.

Regarding the cases where $\dim \mathfrak{a} = 3$, which can only occur when G is not completely solvable, they are treated in a similar way. It suffices to use the calculations of the previous section. This concludes the proof of Theorem 2.8.2. □

Corollary 2.8.3 *Under the assumptions of Theorem 2.8.2, we have*

$$\left(\mathcal{H}_{\pi_1}^{-\infty}\right)^{H_2,\chi_f\Delta_{H_2,G}^{\frac{1}{2}}} = \mathbb{C}a,$$

with

$$a(\phi) = \oint_{H_2/(H_1\cap H_2)} \overline{\phi(h)\chi_f(h)\Delta_{H_2,G}^{-\frac{1}{2}}}(h)d\nu(h)$$

for any $\phi \in \mathcal{H}_{\pi_1}^{\infty}$.

Finally, Theorem 2.5.3 provides us with the composition formula for this setting.

Theorem 2.8.4 *Let* $H_i = \exp \mathfrak{h}_i (1 \le i \le 3)$ *be three Pukanszky polarizations of* $\mathfrak{g}$ *at* $f \in \mathfrak{g}^*$ *such that the pairs* $(\mathfrak{h}_i, \mathfrak{h}_j)(1 \le i \ne j \le 3)$ *satisfy property (CP). If the operators* $I_{\mathfrak{h}_i\mathfrak{h}_j}$ *are isometric (which might require their normalization), then*

$$I_{\mathfrak{h}_1\mathfrak{h}_3} \circ I_{\mathfrak{h}_3\mathfrak{h}_2} \circ I_{\mathfrak{h}_2\mathfrak{h}_1} = e^{(i\pi/4)\tau(\mathfrak{h}_3,\mathfrak{h}_2,\mathfrak{h}_1)}\mathbb{I}.$$

Chapter 3
Intertwining Operators of Induced Representations and Restrictions of Representations of Exponential Solvable Lie Groups

3.1 Introduction

One of the most important problems in the disintegration theory of group representations is to write down explicit unitary operators that intertwine unitary representations and their decomposition into irreducibles. The answer to this question can help to solve several important problems, such as the solvability of invariant differential operators on homogeneous spaces, Frobenius reciprocity, the disintegration of generalized invariant-vectors etc. In this chapter we will describe instances of intertwining operators for induced representations, restrictions of representations and tensor products of unitary representations, in the setting of exponential solvable Lie groups.

3.2 Intertwining Operators of Induced Representations of Nilpotent Lie Groups

Let G be a connected and simply connected nilpotent Lie group with Lie algebra $\mathfrak{g}$. Consider, as in (1.6.1), the monomial representation $\tau = \mathrm{ind}_H^G \chi$ induced by the unitary character χ of an analytic subgroup H of G. Recall that $\mathfrak{h}$ is the Lie subalgebra corresponding to H, and χ is written in the form $\chi(\exp X) = e^{i\langle f,X\rangle}(X \in \mathfrak{h})$ with $f \in \mathfrak{g}^\star$. Let $\Gamma_f = f + \mathfrak{h}^\perp$ be as in Eq. (1.4.8). In parallel to decomposition formula (1.6.2), the disintegration of τ into irreducibles reads:

$$\tau \simeq \int_{\Gamma_f/H}^{\oplus} \pi_l dl, \tag{3.2.1}$$

A. Baklouti et al., *Representation Theory of Solvable Lie Groups and Related Topics*, Springer Monographs in Mathematics,
https://doi.org/10.1007/978-3-030-82044-2_3

where dl is some natural measure on the space Γ_f/H of H-orbits (cf. Theorem 1.4.2).

We construct in this section an intertwining operator of decomposition (3.2.1) for an arbitrary subgroup H. The idea is to make formula (3.2.1) explicit through the construction of certain affine subspaces

$$\mathscr{V}^{R,\mathfrak{B}} = \left\{\sum_{i=1}^{r} R_i(T) f_i;\ T \in \mathbb{R}^k\right\}$$

of Γ_f, and a measure $d\lambda = d\lambda^{R,\mathfrak{B}}$ on $\mathscr{V}^{R,\mathfrak{B}}$, so that

$$\tau \simeq \int_{\mathscr{V}^{R,\mathfrak{B}}}^{\oplus} \pi_\phi d\lambda^{R,\mathfrak{B}}(\phi). \tag{3.2.2}$$

Here R denotes a family $\{R_1, \ldots, R_r\}$ of affine functions defined on $\mathbb{R}^k$, for some non-negative integer $k \in \mathbb{N}$, and $\mathfrak{B}^\star = \{f_r, \ldots, f_1\}$ denotes the dual of a Malcev basis $\mathfrak{B}$ relative to $\mathfrak{h}$. Such a basis defines an invariant measure on G/H and so the norm on the space $\mathscr{H}_\tau$ of τ. We then determine a Zariski open set $\mathscr{V}_0$ of $\mathscr{V}^{R,\mathfrak{B}}$, a polarization $\mathfrak{b}(\phi)$ in ϕ for each $\phi \in \mathscr{V}_0$, a Malcev basis $\mathscr{X}(\phi)$ of $\mathfrak{b}(\phi)$, a Malcev basis $\mathscr{Y}(\phi)$ of $\mathfrak{b}(\phi)$ relative to $\mathfrak{h}\cap\mathfrak{b}(\phi)$ and a Malcev basis $\mathscr{U}$ of $\mathfrak{h}$ relative to $\mathfrak{h}\cap\mathfrak{b}(\phi)$. All these bases vary continuously in $\phi \in \mathscr{V}_0$, which allows to fix the invariant measure $d_{G,B(\phi)}$ on $G/B(\phi)$, where $B(\phi) = \exp(\mathfrak{b}(\phi))$ (and hence the norm on the space $\mathscr{H}_\phi$ of the irreducible representation $\pi_\phi = \mathrm{ind}_{B(\phi)}^G \chi_\phi$), $d_{B(\phi),B(\phi)\cap H}$ on $B(\phi)/B(\phi) \cap H$, and $d_{H,H\cap B(\phi)}$ on $H/B(\phi) \cap H$. This permits to define the disintegration space

$$\mathscr{H} = \int_{\mathscr{V}^{R,\mathfrak{B}}}^{\oplus} \mathscr{H}_\phi d\lambda(\phi)$$

of τ. We now associate, to any sufficiently regular function ξ of $\mathscr{H}_\tau$ and any $\phi \in \mathscr{V}_0$, the C^∞-vector of $\mathscr{H}_\phi$:

$$T_{B(\phi),H}\xi(g) = \int_{B(\phi)/B(\phi)\cap H} \xi(gb)\chi_\phi(b) d_{B(\phi),B(\phi)\cap H}(b),\ g \in G. \tag{3.2.3}$$

We will show that

$$\int_{\mathscr{V}_0} \|T_{B(\phi),H}\xi\|^2_{\mathscr{H}_\phi} d\phi = \|\xi\|^2_{\mathscr{H}_\tau}$$

therefore obtaining an isometric intertwining operator of (3.2.2). A (smooth) disintegration of $L^2(G)$ is also generated.

3.2.1 Generalities

Notation and Backgrounds

Let G be a connected and simply connected nilpotent Lie group with Lie algebra $\mathfrak{g}$. Let exp denote, as earlier, the exponential map, so that $G = \exp\mathfrak{g}$. Let V, W be real vector spaces of finite dimension with $W \subset V$. We denote by $V^\star$ the dual vector space of V and by $W^\perp$ the orthogonal to W in $V^\star$. If $u_1, \dots, u_p$, $(p \in \mathbb{N})$ indicate linearly independent vectors in V, we denote by $\mathbb{R}$-span$(u_1, \dots, u_p)$ the vector subspace of V they span, and we say the basis $\{u_1, \dots, u_p\}$ generates this space. Given $l \in \mathfrak{g}^\star$ and $X \in \mathfrak{g}$, we denote by $\langle l, X\rangle$ the image of X under l. Recall that the kernel of the bilinear form B_l, defined in Sect. 1.2.4 by $B_l(X, Y) = \langle l, [X, Y]\rangle$, is denoted by $\mathfrak{g}(l) = \{X \in \mathfrak{g}\,;\ B_l(X, Y) = 0 \text{ for all } Y \in \mathfrak{g}\}$. The largest ideal contained in $\mathfrak{g}(l)$ will be denoted by $\mathfrak{a}(l)$. Clearly

$$\mathfrak{a}(l) = \bigcap_{\phi\in G\cdot l} \mathfrak{g}(\phi). \tag{3.2.4}$$

Let $\mathfrak{h} \in S(l, \mathfrak{g})$ and let χ_l be the unitary character of the analytic subgroup $H = \exp\mathfrak{h}$ associated to l by

$$\chi_l(\exp X) = e^{-i\langle l,X\rangle}$$

for $X \in \mathfrak{h}$.

Let V be a real vector space of finite dimension and $\rho : G \longrightarrow \mathrm{End}(V)$ a unipotent action of G on V. We designate by

$$(0) = V_n \subset V_{n-1} \subset \dots \subset V_1 \subset V_0 = V$$

a Jordan–Hölder sequence for G and call $\mathscr{L} = \{v_1, \dots, v_n\}$ an associated Jordan–Hölder basis $(v_j \in V_{j-1}\backslash V_j)$. If $v \in V$, we write $\rho(x)v = x \cdot v$ for all $x \in G$ and

$$X \cdot v = \frac{d}{dt}\{\rho(\exp(tX)) \cdot v\}|_{t=0},\ (X \in \mathfrak{g}).$$

The set of indices $e^{\mathscr{L}}(v) = \{i_1 < \dots < i_d\}$ of an element v in V relative to $\mathscr{L}$ is the following subset of $\{1, \dots, n\}$:

$$i \in e^{\mathscr{L}}(v) \Leftrightarrow \exists X \in \mathfrak{g} : X \cdot v \in V_{i-1}\backslash V_i.$$

The cardinality d of $e^{\mathscr{L}}(v)$ is the dimension of the G-orbit Ω of v and does not vary with v in Ω.

Definition 3.2.1 Let W be a G-invariant subspace of V. Following the saturation property of orbits of Sect. 1.3.3, one says that an orbit Ω of G in V is saturated with respect to W if $\Omega + W = \Omega$.

Fix d in $\mathbb{N}$. For a subset e of $\{1, \ldots, n\}$ of cardinality d, we denote by $V_e \subset V$ the subset of all elements v such that $e^{\mathscr{L}}(v) = e$. Pukanszky has shown in [142] that there exists an ordering on subsets of indices of length d, and for each e there exists a polynomial P_e such that

$$V_e = \{v \in V;\ P_e(v) \neq 0,\ P_{e'}(v) = 0,\ \text{for all } e' > e\}.$$

The subsets $(V_d)_d$ of V which contain all the elements v such that G-orbits are d-dimensional form a disjoint union of V_e. There also exists a system of n real functions $\{p_1^{\mathscr{L}}, \ldots, p_n^{\mathscr{L}}\}$ on $\mathbb{R}^d \times V_e$ with the following properties:

(a) For $v \in V_e$ fixed, $p_i^{\mathscr{L}}(z, v)$ is a polynomial function in $z = (z_1, \ldots, z_d)$ and the coefficients are G-invariant functions on V_e, $i \in \{1, \ldots, n\}$;
(b) $P_i^{\mathscr{L}}(z, v) = z_k$ for $i = i_k \in e$, $v \in V_e$;
(c) If $1 \leq i \leq n$, $i_k \leq i < i_{k+1}$ then $p_i^{\mathscr{L}}(z, v)$ only depends on $z_1, \ldots, z_k$;
(d) For all $v \in V_e$ we have

$$\rho(G)v = \{p^{\mathscr{L}}(z, v) = \sum_{i=1}^{n} p_i^{\mathscr{L}}(z, v)v_i \,:\, z \in \mathbb{R}^d\}.$$

Take for example $V = \mathfrak{g}^\star$ and the coadjoint action. Let $\mathscr{D}^\star = \{l_1, \ldots, l_n\}$ be a Jordan–Hölder basis of $\mathfrak{g}^\star$ and let $\mathscr{D} = \{X_1, \ldots, X_n\}$ be the dual basis in $\mathfrak{g}$, ie

$$\langle l_i, X_j \rangle = \delta_{i,j};\ 1 \leq i, j \leq n.$$

Note that $\mathfrak{g}_j = \mathbb{R}\text{-span}(X_1, \ldots, X_j)$ for every $j \in \{1, \ldots, n\}$; the index set

$$e^{\mathscr{D}^\star}(l) = \{i_1 < \ldots < i_d\}$$

of an element l in $\mathfrak{g}^\star$ related to $\mathscr{D}^\star$ is also the following set of $\{1, \ldots, n\}$:

$$i \in e^{\mathscr{D}^\star}(l) \Leftrightarrow \mathfrak{g}(l) + \mathfrak{g}_{i-1} \subsetneq \mathfrak{g}(l) + \mathfrak{g}_i.$$

3.2.2 Disintegration of Monomial Representations

Choice of Measures

Let $B = \exp \mathfrak{b}$ be a connected closed subgroup of G and let $\{Y_1, Y_2, \ldots, Y_k\}$ be a Jordan–Hölder basis of $\mathfrak{b}$. Then the map

$$\psi : \mathbb{R}^k \longrightarrow B,\ (t_1, \ldots, t_k) \longrightarrow \exp t_1 Y_1 \cdots \exp t_k Y_k$$

is a diffeomorphism. We recall from Sect. 1.2.2 how to choose normalized measures db on B and G/B: for any C^∞-function with compact support η in B,

$$\int_B \eta(b)d(b) = \int_{\mathbb{R}^d} \eta(\exp(t_1Y_1)\dots\exp(t_kY_k))dt_1\dots dt_k.$$

This measure is left-invariant and therefore a Haar measure. On the other hand, if we choose a Malcev basis $\{X_1, \dots, X_d\}$ of $\mathfrak{g}$ relative to $\mathfrak{b}$, ie

$$\mathfrak{g} = \mathbb{R}X_1 \oplus \dots \oplus \mathbb{R}X_d \oplus \mathfrak{b} \text{ and } \sum_{k=j}^{d} \mathbb{R}X_k + \mathfrak{b}$$

is a subalgebra of $\mathfrak{g}$ for all j, then the measure $d_{G,B}$ on G/B is defined so that

$$\int_{G/B} \tilde{a}(g)d_{G,B}(g) = \int_{\mathbb{R}^d} \tilde{a}(\exp(t_1X_1)\cdots\exp(t_dX_d))dt_1\cdots dt_d$$

is left-invariant for any continuous function with compact support $\tilde{a}$ on G/B. By normalizing one of the vectors X_j, we have that

$$\int_G q(g)dg = \int_{G/B} (\int_B q(xb)db)d_{G,B}(x)$$

for any continuous function with compact support q on G. We always choose the invariant measures $d_{G,B}$ on the quotient spaces G/B in such a way that this identity holds.

Let $\mathfrak{b}_1, \mathfrak{b}_2$ be two polarizations at the point $\phi \in \mathfrak{g}^\star$, and B_1, B_2 the two associated subgroups. Notice

$$S(G/B_i, \phi) = \mathcal{H}^\infty_{\mathrm{ind}_{B_i}^G \chi_\phi}, \quad i \in \{1, 2\},$$

the space of C^∞-vectors of the representation spaces $\mathrm{ind}_{B_i}^G \chi_\phi$, $i \in \{1, 2\}$, which are in fact isomorphic to the Schwartz space of $\mathbb{R}^{\mathrm{codim}\,\mathfrak{b}_i}$. For the sake of simplicity, let $\pi_{\phi,B}$ denote the representation $\mathrm{ind}_B^G \chi_\phi$. If $d_{B_2,B_2\cap B_1}$ denotes the B_2-left-invariant measure on $B_2/B_2 \cap B_1$, for any function $\tilde{k}$ of $S(G/B_1, \phi)$ the integral

$$T_{B_2,B_1}\tilde{k}(g) = \int_{B_2/B_2\cap B_1} \tilde{k}(gb)\chi_\phi(b)d_{B_2,B_2\cap B_1}(b)$$

defined for every $g \in G$ is absolutely convergent and defines an isomorphism of $S(G/B_1, \phi)$ on $S(G/B_2, \phi)$ which extends by continuity into an intertwining operator between π_{ϕ,B_1} and π_{ϕ,B_2}. Furthermore, if the measures on the homogeneous spaces G/B_1 and G/B_2 are suitably normalized, T_{B_2,B_1} is an isometry.

Let $\mathcal{H}_\tau$ and $\mathcal{H}_\tau^\infty$ denote the space of τ and that of its C^∞ vectors respectively. Let $\{X_1, \ldots, X_r\}$ be a Malcev basis of $\mathfrak{g}$ relative to $\mathfrak{h}$. It is well known that the mapping

$$E : \mathbb{R}^r \mapsto G/H;\ E(t_1, \ldots, t_r) = \exp(t_1 X_1) \ldots \exp(t_r X_r) \cdot H$$

is a diffeomorphism of $\mathbb{R}^r$ to G/H. We denote by $S(G/H, f)$ the Schwartz space of the representation τ, ie the space of functions ξ which are C^∞ on G, verify the cocycle relation

$$\xi(gh) = \chi_f(h^{-1})\xi(g), (g \in G, h \in H) \tag{3.2.5}$$

and are such that $\xi \circ E$ is an ordinary Schwartz function on $\mathbb{R}^r$ for one, and hence for any, Malcev basis relative to $\mathfrak{h}$. The Fréchet space $S(G/H, f)$ is obviously isomorphic to the Fréchet space $S(G/H)$. In the same way, we define the spaces $C^\infty(G/H, f)$ of C^∞ functions and $C_c^\infty(G/H, f)$ of C^∞ functions with compact support modulo H satisfying (3.2.5). For simplicity, we will call by polarization both the subalgebra and the associated subgroup. Let $\phi \in \Gamma_f$ and suppose $B(\phi) = \exp(\mathfrak{b}(\phi))$ is a polarization of ϕ. The integral

$$T_{B(\phi),H}\xi(g) = \int_{B(\phi)/B(\phi)\cap H} \xi(gu)\chi_\phi(u) d_{B(\phi),B(\phi)\cap H}(u)$$

exists for any Schwartz function ξ in $S(G/H, f)$ and g in G and defines a C^∞ function on G which satisfies the required covariance relation for $B(\phi)$ and the character χ_ϕ. The function $T_{B(\phi),H}\xi$ is in fact an element of $\mathcal{H}_\phi^\infty$, that is to say a C^∞ vector of the induced representation $\pi_{\phi,B(\phi)}$. The following lemma is straightforward:

Lemma 3.2.2 *Let $\phi \in \Gamma_f$ and $B = \exp \mathfrak{b}$ be a polarization in ϕ. Then the map*

$$T_{B,H} : S(G/H, f) \to S(G/B, \phi),$$

$$T_{B,H}\xi(x) = \int_{B/B\cap H} f(xb)\chi_\phi(b) d_{B,B\cap H}(b)\ ;\ x \in G,$$

is linear, continuous and surjective.

A Canonical Disintegration Formula

Let $\mathscr{A}$ be a sequence of ideals of $\mathfrak{g}$ satisfying

$$(0) = \mathfrak{a}^{n+1} \subset \mathfrak{a}^n \subset \ldots \subset \mathfrak{a}^1 = \mathfrak{g}$$

and let $\mathscr{C} = \{Z_1, \ldots, Z_n\}$ be a Jordan–Hölder basis of $\mathfrak{g}$ extracted from $\mathscr{A}$, ie $Z_j \in \mathfrak{a}^j \backslash \mathfrak{a}^{j+1}$ and $\mathfrak{a}^j = \mathbb{R}\text{-span}(Z_j, \ldots, Z_n)$. We extract from $\mathscr{A}$ a Malcev sequence $\{\mathfrak{g}_j\}$ and a Malcev basis $\mathscr{B} = \{B_1, \ldots, B_r\}$ of $\mathfrak{g}$ relative to $\mathfrak{h}$ by the following procedure. Look at the subalgebras $\mathfrak{h}_j = \mathfrak{a}_j + \mathfrak{h}$, $j = 1, \ldots, n$, of $\mathfrak{g}$. Let $n \geq j_1 > \ldots > j_r \geq 1$ be the indices such that $\mathfrak{h}_j \neq \mathfrak{h}_{j+1}$. We set $B_i = Z_{ji}$, and we get $\mathscr{B} = \{B_1, \ldots, B_r\}$. We then take $\mathfrak{g}_k = \mathfrak{h} \oplus \mathbb{R}\text{-span}(B_1, \ldots, B_k)$, $k = 1, \ldots, r$ and we have

$$\mathfrak{h} = \mathfrak{g}_0 \subset \mathfrak{g}_1 \subset \ldots \subset \mathfrak{g}_r = \mathfrak{g}.$$

We will say that the Malcev basis $\mathscr{B}$ with respect to $\mathfrak{h}$ is associated with $\mathscr{A}$. We will also use $\mathscr{B}$ to determine a G-invariant measure on G/H and hence to fix the norm on $\mathscr{H}_\tau$. Taking the dual basis $\mathscr{B}^\star = \{f_1, \ldots, f_r\}$ of $\mathscr{B}$, we get a Jordan–Hölder basis of $\mathfrak{h}^\perp$ for the action of H.
Conversely, from a Jordan–Hölder basis $\mathscr{C} = \{Z_1, \ldots, Z_n\}$ of $\mathfrak{g}$, we can construct a flag of ideals $\mathscr{A} = (\mathfrak{a}_j)_{j=1}^{n+1}$ by setting $\mathfrak{a}_j = \mathbb{R}\text{-span}(Z_j, \ldots, Z_n)$, $j = 1, \ldots, n$ and $\mathfrak{a}^{n+1} = \{0\}$.

Let $G_j = \exp(\mathfrak{g}_j)$, $j = 1, \ldots, r$ and $\{B_0, \ldots, B_{-p}\}$ a basis of $\mathfrak{h}$. We consider $\phi_j = \phi_{|\mathfrak{g}_j}$, for any $j = 1, \ldots, r$ and any $\phi \in \Gamma_f$. Let

$$d_j(\phi) = \dim(G_j \cdot \phi_j) = \text{rk}((\langle \phi, [B_i, B_{i'}] \rangle)_{-p \leq i, i' \leq j})$$

and

$$d_j = \max_{\phi \in \Gamma_f} d_j(\phi), d_0 = 0.$$

Note that these dimensions depend only on the subalgebras $\mathfrak{g}_i$ which themselves depend only on the ideals of $\mathscr{A}$. The choice of the vectors $(B_i)_i$ is immaterial as regards the determination of these dimensions. We put

$$\begin{aligned} I^{H,\mathscr{C}} =& I^H = \{j \in \{1, \ldots, r\}; d_j > d_{j-1}\} \\ =& \{j \in \{1, \ldots, r\}\}; \text{ almost all } G_j\text{-orbits in } \Gamma_f^j = \Gamma_{f|\mathfrak{g}_j} \\ & \text{are saturated with respect to } \mathfrak{g}_{j-1}^\perp \subset \mathfrak{g}_j^\star\}. \end{aligned}$$

Let

$$\begin{aligned} L^{H,\mathscr{C}} =& L^H = \{1, \ldots, r\} \backslash I^H = \{j \in \{1, \ldots, r\}, d_j = d_{j-1}\} \\ =& \{j \in \{1, \ldots, r\}; \text{ almost any } G_j\text{-orbit in } \Gamma_f^j = \Gamma_{f|\mathfrak{g}_j} \\ & \text{is saturated with respect to } \mathfrak{g}_{j-1}^\perp\}. \end{aligned}$$

Here, saturation is understood as in Definition 3.2.1.

We now introduce the following definitions:

Definition 3.2.3 A linear form $\phi \in \Gamma_f$ is in general position if $d_j(\phi) = d_j$ for any $j \in \{1, \ldots, r\}$. The set of linear forms in general position is Zariski open in Γ_f.

Definition 3.2.4 We will disintegrate the representation τ over some affine subspaces $\mathscr{V}^{R,\mathscr{B}}$ of Γ_f described below. Let $L^H = \{i_1 < \cdots < i_k\}$ and let $R = \{R_1, \ldots, R_r\}$ be a family of affine functions defined on $\mathbb{R}^k$ as follows:

(i_1) For $j \in I^H$, $R_j(t_1, \ldots, t_k)$ depends only on $t_1, \ldots, t_{j'}$, where $i_{j'} \leq j < i_{j'+1}$.
(i_2) For all $i_j \in L^H$, $R_{i_j}(t_1, \ldots, t_k) = t_j$.

We then put

$$\mathscr{V}^{R,\mathscr{B}} = \{f + R(T);\ T = (t_1, \ldots, t_k) \in \mathbb{R}^k\} \subset \Gamma_f, \tag{3.2.6}$$

where $R(T) = \sum_{i=1}^{r} R_i(T) f_i$. We endow the space $\mathscr{V}^{R,\mathscr{B}}$ with the Lebesgue measure $d\lambda = d\lambda^{R,\mathscr{B}}$.

A subspace $\mathscr{V}$ of Γ_f will be called $\mathscr{C}$-admissible if there exist a Malcev basis $\mathscr{B}$ relative to $\mathfrak{h}$ associated to $\mathscr{A}$ (where $\mathscr{A}$ is the chain of ideals associated with $\mathscr{C}$) and a family of functions $R = \{R_1, \ldots, R_r\}$ verifying properties i_1 and i_2 and such that $\mathscr{V}$ is of the form $\mathscr{V}^{R,\mathscr{B}}$. We say that $\mathscr{V}$ is in general position if it contains elements of Γ_f in general position.

Proposition 3.2.5 *Let $\mathscr{A}$ be a chain of ideals of $\mathfrak{g}$ and $\mathscr{B}$ a Malcev basis relative to $\mathfrak{h}$ associated to $\mathscr{A}$. The affine space $\mathscr{V}^{R,\mathscr{B}}$ defined as in* (3.2.6) *is in general position in Γ_f. That is to say, for any $\phi \in \Gamma_f$ in general position as in Definition 3.2.3, there exists $g \in G$ such that* $\mathrm{Ad}^\star(g)\phi$ *is in general position in Γ_f and contained in $\mathscr{V}^{R,\mathscr{B}}$.*

Proof Let $\phi = \phi^0$ and $T^0 = 0$. Let us show by induction on $j = 0, \ldots, r$, that there exist $\phi^j \in \Gamma_f \cap G \cdot \phi$ in general position and a $T^j = (t_1^j, \ldots, t_k^j) \in \mathbb{R}^k$ such that, for any $i = 1, \ldots, j$,

$$\phi^j(B_i) = R_i(T^j) + f(B_i),$$

where $\mathscr{B} = \{B_1, \ldots, B_r\}$. Suppose ϕ^j and T^j are found and look at the index $j+1$. If $j+1 \in I^H$, then the orbit of $\phi^j_{|\mathfrak{g}_{j+1}}$ in $\mathfrak{g}^\star_{j+1}$ is saturated with respect to $\mathfrak{g}_j^\perp$, since ϕ^j is in general position. Then there exists $u \in G_j$ such that for $\phi^{j+1} = \mathrm{Ad}^\star(u)\phi^j$ we have

$$\phi^j_{|\mathfrak{g}_j} = \phi^{j+1}_{|\mathfrak{g}_j} \text{ and } \phi^{j+1}(B_{j+1}) = R_{j+1}(T^j) + f(B_{j+1}).$$

Then $\phi^{j+1} \in \Gamma_f$. Let us show that ϕ^{j+1} is in general position, ie

$$\dim\ G_p(\phi^{j+1})_{|\mathfrak{g}_p} = d_p,\ p = 1, \ldots, r.$$

If $p \leq j-1$, then $\phi^{j+1}_{|\mathfrak{g}_p} = \phi^j_{|\mathfrak{g}_p}$ and since ϕ^j is in general position, we get the result. If $p \geq j$, then as $u \in G_p$, $G_p(\phi^{j+1})_{|\mathfrak{g}_p} = G_p(\mathrm{Ad}^\star(u)\phi^j)_{|\mathfrak{g}_p} = G_p(\phi^j)_{|\mathfrak{g}_p}$ and therefore $\dim G_p(\phi^{j+1})_{|\mathfrak{g}_p} = d_p$. Set then $T^{j+1} = T^j$.
If $j+1 \in L^H$, then $j+1 = i_{j'}$ for some j' and $R_{j+1}(T^j) = t^j_{j'}$. We take $\phi^{j+1} = \phi^j$ and replace T^j with

$$T^{j+1} = (t^{j+1}_1 = t^j_1, \ldots, t^{j+1}_{j'-1} = t^j_{j'-1}, t^{j+1}_{j'} = \phi^j(B_{j+1}) + f(B_{j+1}), t^{j+1}_{j'+1} = t^j_{j'+1}, \ldots).$$

□

We will often need to change bases as in

$$Z'_j = Z_j + \sum_{k>j} c_{j,k} Z_k, \; j = 1, \ldots, n. \tag{3.2.7}$$

Then the Malcev basis $\{B_1, \ldots, B_r\}$ is changed to $\{B'_1, \ldots, B'_r\}$, where

$$B'_j = B_j + \sum_{i<j} a_{i,j} B_i + U_j, \; j = 1, \ldots, r,$$

for some $U_j \in \mathfrak{h}$. In the same way $\phi = \sum_{i=1}^r \phi_i f_i + f \in \Gamma_f$, in the new basis $\mathscr{B}'$, reads $\phi'_i = \phi_i + \sum_{i>j} a_{j,i}\phi_j$. Let $R = \{R_1, \ldots, R_r\}$ be a family of functions as in (3.2.4). For $i_j \in L^H$, $T \in \mathbb{R}^k$, let

$$t_j + \sum_{i<i_j}^{r} a_{i,i_j} R_i(T) = \langle R(T), B'_{i_j} \rangle = t'_j.$$

Thus $t'_j = t_j + A_j(t_1, \ldots, t_{j-1})$ where A_j is displayed and hence

$$t_j = t'_j + A'_j(t'_1, \ldots, t'_{j-1}),$$

where A'_j is affine. Let for $i_j \in L^H$, $j \in I^H$ and for $T' \in \mathbb{R}^k$

$$R'_{i_j}(T') = t'_j, \, R'_j(T') = \langle R(A'(T')), B'_j) \rangle.$$

Then $A'(T') = (t'_1 + A'_1, \ldots, t'_k + A'_k(t'_1, \ldots, t'_{k-1}))$, and we find a new family of affine functions $(R'_j)_j$ verifying i_1 and i_2, such that in the basis $\mathscr{B}'$ the spaces $\mathscr{V}^{R,\mathscr{B}}$ and $\mathscr{V}^{R',\mathscr{B}'}$ coincide, as well as their measures $d\lambda^{R,\mathscr{B}}$, $d\lambda^{R',\mathscr{B}'}$.

Theorem 3.2.6 *Let G be a connected and simply connected nilpotent Lie group with Lie algebra $\mathfrak{g}$, let $f \in \mathfrak{g}^\star$ and let $\mathfrak{h}$ be a subalgebra of $\mathfrak{g}$ subordinate to f, where $H = \exp \mathfrak{h}$ is the connected closed subgroup of G associated to $\mathfrak{h}$, τ the monomial* *representation induced from the unitary character* χ_f *of* H. *Let* $\mathscr{B} = \{B_1, \ldots, B_r\}$ *be a Malcev basis of* $\mathfrak{g}$ *relative to* $\mathfrak{h}$, $\mathscr{V}^{R,\mathscr{B}} \subset \Gamma_f$ *is the subspace described in* (3.2.6) *and* $d\lambda^{R,\mathscr{B}}$ *its measure. Then the disintegration of* τ *into irreducibles reads*

$$\tau \simeq \int_{\mathscr{V}^{R,\mathscr{B}}}^{\oplus} \pi_\phi d\lambda^{R,\mathscr{B}}(\phi).$$

Proof Note that the usual canonical decomposition occurs on the base space Γ_f/H. In the case of infinite multiplicities we have $\dim(\mathscr{V}^{R,\mathscr{B}}) < \dim(\Gamma_f/H)$, while the dimensions are equal in case of finite multiplicities. We use induction on the dimension of $\mathfrak{g}$. The subspace $\mathfrak{g}_0 = \mathbb{R}\text{-span}(\mathfrak{h}, B_1, \ldots, B_{r-1})$ is an ideal of codimension 1 of $\mathfrak{g}$ and $\mathscr{B}_0 = \{B_1, \ldots, B_{r-1}\}$ is a Malcev basis of $\mathfrak{g}_0$ relative to $\mathfrak{h}$. Let $R_0 = \{R_1, \ldots, R_{r-1}\}$ and $f_0 = f_{|\mathfrak{g}_0}$. Let $\mathscr{V}^{R_0,\mathscr{B}_0}$ denote the subset of Γ_{f_0} with respect to $\mathfrak{g}_0$, $\mathscr{B}_0$ and R_0. In fact, $\mathscr{V}^{R_0,\mathscr{B}_0}$ is the restriction of $\mathscr{V}^{R,\mathscr{B}}$ to $\mathfrak{g}_0$. Let $d\lambda^{R_0,\mathscr{B}_0}$ be the Lebesgue measure on $\mathscr{V}^{R_0,\mathscr{B}_0}$.

As $\mathrm{ind}_H^G \chi_f = \mathrm{ind}_{G_0}^G \mathrm{ind}_H^{G_0} \chi_f$ where $G_0 = \exp(\mathfrak{g}_0)$, using the induction hypothesis on G_0, one obtains

$$\tau \simeq \int_{\mathscr{V}^{R_0,\mathscr{B}_0}}^{\oplus} \mathrm{ind}_{G_0}^G \pi_{\phi_0} d\lambda^{R_0,\mathscr{B}_0}(\phi_0).$$

We shall therefore study the following two cases:

(1) $r \in I^H$: almost all G-orbits are saturated with respect to $\mathfrak{g}_0^\perp$. For almost all $\phi \in \mathscr{V}^{R,\mathscr{B}}$, by Proposition 3.2.5 for $\phi_0 = \phi_{|\mathfrak{g}_0}$ it follows $\pi_\phi = \mathrm{ind}_{G_0}^G \pi_{\phi_0}$ is irreducible. In addition, in view of the form of R_r, the map $\phi \mapsto \phi_0 = \phi_{|\mathfrak{g}_0}$ is an affine isomorphism that respects the measures of $d\lambda^{R,\mathscr{B}}$ and $d\lambda^{R_0,\mathscr{B}_0}$. Hence

$$\tau \simeq \int_{\mathscr{V}^{R_0,\mathscr{B}_0}}^{\oplus} \mathrm{ind}_{G_0}^G \pi_0 d\lambda^{R_0,\mathscr{B}_0}(\phi_0) \simeq \int_{\mathscr{V}^{R,\mathscr{B}}}^{\oplus} \pi_\phi d\lambda^{R,\mathscr{B}}(\phi).$$

(2) $r \in L^H$: the generic G-orbits are not saturated with respect to $\mathfrak{g}_0^\perp$. Let us choose for each $\phi_0 \in \Gamma_{f_0}$, the extension $\phi_0^1 \in \Gamma_f$ defined by $\phi_0^1(B_r) = 0$. We have then

$$\mathscr{V}^{R,\mathscr{B}} = \{\phi_0^1 + sB_r^\star,\ s \in \mathbb{R},\ \phi_0 \in \mathscr{V}^{R_0,\mathscr{B}_0}\} \simeq \mathscr{V}^{R_0,\mathscr{B}_0} \times \mathbb{R},$$

Then the measure $d\lambda^{R,\mathscr{B}}$ coincides with $d\lambda^{R_0,\mathscr{B}_0} \otimes ds$, where ds is the Lebesgue measure on $\mathbb{R}$. On the other hand we have, for ϕ_0 in general position in Γ_{f_0},

$$\tau \simeq \int_{\mathbb{R}}^{\oplus} \pi_{\phi_0+\alpha B_r^\star} d\alpha.$$

It follows that

$$\tau \simeq \int_{\mathscr{V}^{R_0,\mathscr{B}_0}}^{\oplus} \Big(\int_{\mathbb{R}}^{\oplus} \pi_{\phi_0+\alpha B_r^*} d\alpha \Big) d\lambda^{R_0,\mathscr{B}_0}(\phi_0) = \int_{\mathscr{V}^{R_0,\mathscr{B}_0}}^{\oplus} \pi_\phi d\lambda^{R,\mathscr{B}}(\phi),$$

which completes the proof. □

3.2.3 Construction of the Intertwining Operator

Construction of Polarizations and Malcev Bases

We now specify a flag $\mathscr{A}$ of ideals of $\mathfrak{g}$. The peculiarity comes from the fact that if $\mathscr{C} = \{Z_1, \ldots, Z_n\}$ is the Jordan–Hölder basis of $\mathfrak{g}$ extracted from $\mathscr{A}$, then $\mathscr{C}$ contains a Jordan–Hölder basis

$$\mathscr{D} = \{V_1 = Z_{l_1}, \ldots, V_{n-r} = Z_{l_{n-r}}\}$$

of $\mathfrak{h}$. We also extract from the latter the Malcev basis $\mathscr{B} = \{B_1, \ldots, B_r\}$ of $\mathfrak{g}$ relative to $\mathfrak{h}$ as above. The basis $\mathscr{C}$ gives us the index sets I^H and L^H and allows to choose a family $R = \{R_1, \ldots, R_r\}$ of real affine functions on $\mathbb{R}^k$ having the properties defined above for L^H. Moreover, by what we saw earlier the basis $\mathscr{B}$ gives us a G-invariant measure $d_{G,H}$ on G/H, which allows to fix the norm

$$\|\xi\|_{L^2(G/H,f)} = \Big(\int_{G/H} |\xi(g)|^2 d_{G,H}(g)\Big)^{\frac{1}{2}},$$

for $\xi \in \mathscr{H}_\tau = L^2(G/H, f)$. Let $\mathscr{V} = \mathscr{V}^{R,\mathscr{B}}$. We will now construct a Zariski open set $\mathscr{V}_0$ of $\mathscr{V}$ on which all of the following objects will be well defined. For $\phi \in \mathscr{V}_0$, we will construct a polarization $\mathfrak{b}(\phi)$ at ϕ, a Malcev basis $\mathscr{X}(\phi) = \{X_1(\phi), \ldots, X_l(\phi)\}$ of $\mathfrak{g}$ relative to $\mathfrak{b}(\phi)$, a Jordan–Hölder basis $\mathscr{D}(\phi) = \{V_1(\phi), \ldots, V_q(\phi)\}$ of $\mathfrak{h} \cap \mathfrak{b}(\phi)$, a Malcev basis $\mathscr{Y}(\phi) = \{Y_1(\phi), \ldots, Y_m(\phi)\}$ of $\mathfrak{b}(\phi)$ relative to $\mathfrak{h} \cap \mathfrak{b}(\phi)$ and finally a basis $\mathscr{U}(\phi) = \{U_1(\phi), \ldots, U_p(\phi)\}$ of $\mathfrak{h}$ relative to $\mathfrak{h} \cap \mathfrak{b}(\phi)$. Here the numbers l, m, p do not depend upon $\phi \in \mathscr{V}_0$. In addition, all vectors $X_j(\phi)$, $V_j(\phi)$, $Y_j(\phi)$ and $U_j(\phi)$ vary rationally and smoothly on $\phi \in \mathscr{V}_0$.
The vectors $(X_j(\phi))_j$ will define a G-invariant measure on $G/B(\phi)$, and hence the norm of the space π_ϕ. Likewise, the vectors $(Y_j(\phi))_j$ determine the $B(\phi)$-invariant measure on $B(\phi)/H \cap B(\phi)$, hence the infinitesimal intertwining operator $T_{B(\phi),H}$ and the vectors $(U_j(\phi))_j$ will determine the measure on $H/H \cap B(\phi)$. All objects mentioned above are going to be constructed inductively, step by step. Step $s = 0$ consists in defining the bases using certain subalgebras, set of indices and Zariski open sets of $\mathscr{V}$. We will also introduce the tools at this stage, while the objects will be given at the intermediate step $s \in \{0, \ldots, \dim(\mathfrak{g})\}$. In (3.2.3) we will explain in

detail the techniques for passing from s to $s+1$ and exhibit the newly constructed objects. At the very end we will show that the procedure actually stops, at some $s_0 \in \{0, \ldots, \dim(\mathfrak{g})\}$, and that the outcome bases are convenient. Note that our constructions depend only upon $\mathscr{C}$, f and $\mathfrak{h}$.

Step 0

Proceed by induction on $s \in \{0, \ldots, \dim(\mathfrak{g})\}$ starting with $s = 0$. Let

$$\mathscr{C}^0 = \mathscr{C},\ \mathscr{V}_{0,0} = \mathscr{V}$$

$$\mathscr{X}^0(\phi) = \mathscr{Y}^0(\phi) = \mathscr{U}^0(\phi) = \emptyset,\ \mathscr{D}^0(\phi) = \mathscr{D},$$

$$\mathfrak{k}^0(\phi) = (0),\ \mathfrak{h}^0(\phi) = \mathfrak{h},\ i_0 = 0,\ j_0 = n \text{ and } \mathfrak{g}^0(\phi) = \mathfrak{g},$$

$$I^0 = I^{H,\mathscr{C}},\ R^0 = R,\ r_0 = r,\ k_0 = k,\ n_0 = \dim(\mathfrak{g}).$$

Step s

Suppose we have found, for $s \in \{0, \ldots, n = \dim(\mathfrak{g})\}$, a Zariski open set $\mathscr{V}_{0,s}$ of $\mathscr{V}$, for $\phi \in \mathscr{V}_{0,s}$ a subalgebra $\mathfrak{g}^s(\phi)$ of $\mathfrak{g}$ dimension n_s satisfying

$$\mathrm{ad}^*(\mathfrak{g}^s(\phi)) \cdot \phi \supset (\mathfrak{g}^s(\phi))^{\perp},$$

indices $i_s \leq j_s$ in $\{0, \ldots, n_s\}$, and a Jordan–Hölder basis of $\mathfrak{g}^s(\phi)$ that we call $\mathscr{C}^s(\phi) = \{Z_1^s(\phi), \ldots, Z_{n_s}^s(\phi)\}$. Let $\phi_s = \phi_{|\mathfrak{g}^s(\phi)} \in \mathfrak{g}^s(\phi)^*$. The ideal

$$\sum_{j=j_s+1}^{n_s} \mathbb{R}Z_j^s(\phi)$$

of $\mathfrak{g}^s(\phi)$ is contained in the ideal $\mathfrak{a}(\phi_s)$ of (3.2.4). We also have a subalgebra $\mathfrak{h}^s(\phi)$ of $\mathfrak{g}^s(\phi)$ subordinated to ϕ and contained in $\mathfrak{g}^s(\phi)$. From a Jordan–Hölder basis

$$\mathscr{D}^s(\phi) = \{V_1^s(\phi), \ldots, V_{q_s}^s\}$$

of $\mathfrak{h} \cap \mathfrak{g}^s(\phi)$ contained in $\mathscr{C}^s(\phi)$, we extract a Malcev basis

$$\mathscr{X}^s(\phi) = \{X_1^s(\phi), \ldots, X_{l_s}^s(\phi)\}$$

of $\mathfrak{g}$ relative to $\mathfrak{g}^s(\phi)$, a Malcev basis

$$\mathscr{Y}^s(\phi) = \{Y_1^s(\phi), \ldots, Y_{m_s}^s\}$$

of $\mathfrak{h}^s(\phi)$ relative to $\mathfrak{h}^s(\phi)\cap\mathfrak{h}$ and contained in $\mathscr{C}^s(\phi)$, and finally a third Malcev basis $\mathscr{U}^s(\phi) = \{U_1^s(\phi), \ldots, U_{p_s}^s(\phi)\}$ of $\mathfrak{h}$ relative to $\mathfrak{h}\cap\mathfrak{g}^s(\phi)$ where q_s, l_s, m_s, p_s do not depend on ϕ. From an index set $I^s \subset \{1, \ldots, r_s\}$, where $r_s = \dim(\mathfrak{g}^s(\phi)/\mathfrak{h}^s(\phi))$ we obtain

$$I^{H^s(\phi),\mathscr{C}^s(\phi)} = I^s$$

for all $\phi \in \mathscr{V}_{0,s}$. Set $L^s = \{1, \ldots, r_s\}\backslash I^s$. Besides, we have for all $\phi \in \mathscr{V}_{0,s}$ a family

$$R^s(\phi) = \{R_1^s(\phi), \ldots, R_{r_s}^s(\phi)\}$$

of affine functions defined on $\mathbb{R}^{k_s}$ satisfying conditions i_1, i_2 of (3.2.4) with respect to I^s and L^s, and such that all coefficients of the functions $R_\alpha^s(\phi)$, $\alpha \in \{1, \ldots, r_s\}$, are rational maps whose denominators do not vanish on $\mathscr{V}_{0,s}$. We also assume that the vectors $Z_i^s(\phi)$, $X_i^s(\phi)$, $Y_i^s(\phi)$ and $U_i^s(\phi)$ obtained through the process are written as non-singular regular expressions in $\phi \in \mathscr{V}_{0,s}$, ie expressed by vector-valued rational maps $W'(\phi)$ defined over $\mathscr{V}$ whose denominators do not vanish on $\mathscr{V}_{0,s}$.

Step s+1

Suppose first that

$$\mathscr{V}'_{0,s} = \{\phi \in \mathscr{V}_{0,s};\ \langle\phi, [\mathfrak{g}^s(\phi), \mathfrak{g}^s(\phi)]\rangle \neq (0)\} = \emptyset.$$

The subalgebra $\mathfrak{g}^s(\phi)$ is then subordinate to ϕ for all $\phi \in \mathscr{V}_{0,s}$ and is therefore necessarily a polarization in ϕ as

$$(\mathfrak{g}(\phi))^\perp \subset ad^*(\mathfrak{g}^s(\phi)) \cdot \phi \subset (\mathfrak{g}^s(\phi))^\perp.$$

Let $\mathfrak{b}(\phi) = \mathfrak{g}^s(\phi)$. Then $\mathfrak{b}(\phi)$ is the Vergne polarization at ϕ with respect to the basis $\mathscr{C}$ (cf. [119]). Set $\mathscr{D}(\phi) = \mathscr{D}^s(\phi)$, $\mathscr{X}(\phi) = X^s(\phi)$, $\mathscr{U}(\phi) = \mathscr{U}^s(\phi)$, $\mathscr{Y}(\phi) = \mathscr{Y}^s(\phi)$ and $\mathscr{V}_0 = \mathscr{V}_{0,s}$. We can therefore assume $\mathscr{V}'_{0,s} \neq \emptyset$ in the rest of the construction.

Case 1 This first case occurs if

$$\mathfrak{k}^{s+1}(\phi) = \sum_{j>j_{s+1}} \mathbb{R}Z_j^s(\phi) \not\subset \mathfrak{h}^s(\phi)$$

for some $\phi \in \mathscr{V}_{0,s+1,1}$. Recall that $\mathscr{B}^* = \{f_1, \ldots, f_r\}$ is dual to the Malcev basis $\mathscr{B}$ relative to $\mathfrak{h}$. Let $u = h_{s+1}$ be the largest index such that $Z_{h_{s+1}}^s(\phi) \notin \mathfrak{h}^s(\phi)$. We consider the matrix $M(\phi) = (a_{i,j}(\phi))$ where

$$a_{i,j}(\phi) = \langle f_i, Y_j'^s(\phi)\rangle, i = 1, \ldots, r;\ j = 1, \ldots m_s$$

and the matrix $M_u(\phi)$ obtained by appending to $M(\phi)$ the column vector

$$a_{i,u}(\phi) = \langle f_i, Z_u'^s(\phi)\rangle, i = 1, \ldots, r.$$

Let $d(\phi) = \operatorname{rk} M(\phi)$ and $d_u(\phi) = \operatorname{rk} M_u(\phi)$ and let

$$\mathscr{V}_{0,s+1} = \{\phi \in \mathscr{V}_{0,s}; d(\phi) < d_u\} = \{\phi \in \mathscr{V}_{0,s}; Z^s_{h_{s+1}}(\phi) \notin \mathfrak{h}^s(\phi)\}.$$

Then it is clear that $\mathscr{V}_{0,s+1}$ is a Zariski open subset in $\mathscr{V}_{0,s}$. We put

$$\mathfrak{g}^{s+1}(\phi) = \mathfrak{g}^s(\phi), \mathscr{C}^{s+1} = \mathscr{C}^s(\phi), \mathfrak{h}^{s+1}(\phi) = \mathfrak{h}^s(\phi) + \mathbb{R}Z^s_{h_{s+1}}(\phi), \phi \in \mathscr{V}_{0,s+1}.$$

It is clear that $(\mathfrak{g}^{s+1}(\phi))^{\perp} \subset \operatorname{ad}^*(\mathfrak{g}^{s+1}(\phi)) \cdot \phi$. Since an element $X \in \mathfrak{g}$ belongs in $\mathfrak{h}$ if and only if $\langle f_i, X\rangle = 0$ for all i, and because the family $\{Y_j^s(\phi)\}_j$ forms a basis of $\mathfrak{h}^s(\phi)$ modulo $\mathfrak{h}^s(\phi) \cap \mathfrak{h}$, we have

$$\mathfrak{h}^{s+1}(\phi) = \mathbb{R}\text{-span}\{Z^s_{h_{s+1}}, Y_j^s(\phi), j = 1, \ldots, m_s\} + \mathfrak{h} \cap \mathfrak{h}^s(\phi).$$

Hence the family $\{Z^s_{h_{s+1}}(\phi)\} \cup \mathscr{Y}^s(\phi)$ gives us a Malcev basis $\mathscr{Y}^{s+1}(\phi)$ of $\mathfrak{h}^{s+1}(\phi)$ relative to $\mathfrak{h}^{s+1}(\phi) \cap \mathfrak{h}$. Let $k_{s+1} = k_s - 1$. The dimension r_{s+1} of $\mathfrak{g}^{s+1}(\phi)/\mathfrak{h}^{s+1}(\phi)$ is equal to $r_s - 1$ and for $\phi \in \mathscr{V}_{0,s+1}$, $T' \in \mathbb{R}^{k_s+1}$, let

$$\begin{aligned} R^{s+1}(\phi)(T') =& \{R_1^{s+1}(\phi)(T'), \ldots, R^{s+1}_{r_s+1}(\phi)(T')\} \\ =& \{R_2^s(\phi)(\phi(Z^s_{h_{s+1}}(\phi)), T'), \ldots, R^s_{r_s}(\phi)(\phi(Z^s_{h_{s+1}}(\phi)), T')\}. \end{aligned}$$

We see then that the family $R^{s+1}(\phi)$ fulfils properties i_1, i_2 of Definition 3.2.4 for $\mathfrak{h}^{s+1}(\phi)$ and $\mathscr{C}^{s+1}(\phi)$. We put $\mathfrak{D}^{s+1}(\phi) = \mathscr{D}^s(\phi) \cup \{Z^s_{h_{s+1}}(\phi)\}$. We leave all other objects unchanged, and this completes the first case.

We may henceforth assume $\mathfrak{h}^s(\phi) \supset \mathfrak{k}^{s+1}(\phi)$ for $\phi \in \mathscr{V}_{0,s}$. Let, for $\phi \in \mathscr{V}'_{0,s}$,

$$j_{s+1}(\phi) = \max\{j; Z_j^s(\phi) \notin \mathfrak{a}(\phi_s)\} = \max\{j; \langle \phi, [Z_j^s(\phi), \mathfrak{g}^s(\phi)]\rangle \neq 0\}$$

and let $j_{s+1} = \max_{\phi \in \mathscr{V}'_{0,s}} j_{s+1}(\phi)$. Let $Y_{s+1}(\phi) = Z^s_{j_{s+1}}(\phi), \phi \in \mathscr{V}'_{0,s}$. The set

$$\mathscr{V}_{0,s+1,1} = \{\phi \in \mathscr{V}'_{0,s}; j_{s+1}(\phi) = j_{s+1}\}$$

is also a Zariski open set, because

$$\mathscr{V}_{0,s+1,1} = \{\phi \in \mathscr{V}_{0,s}; \langle \phi, [\mathfrak{g}^s(\phi), Y'_{s+1}(\phi)]\rangle \neq 0\}.$$

Likewise, set

$$i_{s+1}(\phi) = \max\{i; \langle \phi, [Z_i^s(\phi), Y_{s+1}(\phi)]\rangle \neq 0\}, \phi \in \mathscr{V}_{0,s+1,1}$$

and

$$i_{s+1} = \max_{\phi \in \mathscr{V}_{0,s+1,1}} i_{s+1}(\phi).$$

Put

$$X_{s+1}(\phi) = Z^s_{i_{s+1}}(\phi), \phi \in \mathscr{V}_{0,s+1,1}$$

and

$$\begin{aligned}\mathscr{V}_{0,s+1,2} =&\{\phi \in \mathscr{V}_{0,s+1,1}; i_{s+1}(\phi) = i_{s+1}\}\\ =&\{\phi \in \mathscr{V}_{0,s+1,1}; \langle \phi, [X'_{s+1}(\phi), Y'_{s+1}(\phi)]\rangle \neq 0\},\end{aligned}$$

which is also a Zariski open set in $\mathscr{V}$. Let

$$\mathfrak{g}^{s+1,1}(\phi) = \{W \in \mathfrak{g}^s(\phi); \langle [W, Y_{s+1}(\phi)]\rangle = 0\}, \phi \in \mathscr{V}_{0,s+1,2}.$$

We thus obtain an ideal of codimension 1 in $\mathfrak{g}^s(\phi)$. Let $\phi_{s+1} = \phi_{|\mathfrak{g}^{s+1,1}}(\phi)$. To determine the other objects to order $s+1$, we shall distinguish two new cases:

Case 2 This case occurs when $\mathfrak{h}^s(\phi) \subset \mathfrak{g}^{s+1,1}(\phi)$ for $\phi \in \mathscr{V}_{0,s+1,2}$. Then we set

$$\mathfrak{h}^{s+1}(\phi) = \mathfrak{h}^s(\phi), \mathfrak{g}^{s+1}(\phi) = \mathfrak{g}_1^{s+1}(\phi).$$

Here again, we verify that $(\mathfrak{g}^{s+1}(\phi))^\perp \subset \mathrm{ad}^*(\mathfrak{g}^{s+1}(\phi)) \cdot \phi$ as $Y_{s+1}(\phi) \in \mathfrak{g}^{s+1}(\phi)$. We can also put

$$\mathscr{D}^{s+1}(\phi) = \mathscr{D}^s(\phi), \mathscr{Y}^{s+1}(\phi) = \mathscr{Y}^s(\phi), \mathscr{U}^{s+1}(\phi) = \mathscr{U}^s(\phi).$$

We now change $\mathscr{C}^s(\phi)$ by replacing the vector $Z^s_i(\phi)$ by

$$Z^s_i(\phi) - \frac{\langle \phi, [Z^s_i(\phi), Y_{s+1}(\phi)]\rangle}{\langle \phi, [X_{s+1}(\phi), Y_{s+1}(\phi)]\rangle} X_{s+1}(\phi), \quad i = 1, \ldots, i_{s+1} - 1, \quad \phi \in \mathscr{V}_{0,s+1,2}.$$

We can therefore assume that $Z^s_i(\phi) \in \mathfrak{g}^{s+1,1}(\phi)$ for all $i < i_{s+1}$. We get then for $\phi \in \mathscr{V}_{0,s+1,2}$ a new family of functions $R^s(\phi)$ using a similar process.

We replace the basis $\mathscr{C}^s(\phi)$ by the basis

$$\begin{aligned}{}_1\mathscr{C}^s(\phi) =&\{X_{s+1}(\phi), Z^s_1(\phi), \ldots, Z^s_{i_s-1}(\phi), Z^s_{i_s+1}, \ldots, Z^s_{n_s}(\phi)\}\\ =&\{{}_1Z^s_1(\phi), \ldots, {}_1Z^s_{n_s}\}.\end{aligned}$$

Let $\{B^s_j(\phi), j = 1, \ldots, r_s\}$ be the Malcev basis of $\mathfrak{g}^s(\phi)$ relative to $\mathfrak{h}^s(\phi)$ associated with $\mathscr{C}^s(\phi)$ and let α be such that $B^s_\alpha(\phi) = X_{s+1}(\phi)$.

Lemma 3.2.7 *For all $\phi \in \mathscr{V}_{0,s+1,2}$ and s, the subset I^s can be replaced by*

$$ {}_1I^s = I^{H^s(\phi),{}_1\mathscr{C}^s(\phi)} = \{i \in I^s, i < \alpha\} \cup \{i; \alpha \leq i < r, i+1 \in I^s\} \cup \{r\}. $$

Proof Let $\phi \in \mathscr{V}_{0,s+1,2}$. Recall that i is in I^s if and only if there exists $U = U(\phi) \in \mathfrak{g}_i^s(\phi) = \mathbb{R}\text{-span}\,\{B_i^s(\phi), \ldots, B_1^s(\phi), \mathfrak{h}^s(\phi)\}$ such that

$$ \langle U \cdot \phi, B_i^s(\phi)\rangle = 1; \quad \langle U \cdot \phi, B_j^s(\phi)\rangle = 0, j = i-1, \ldots, 1; \quad \langle U \cdot \phi, \mathfrak{h}^s(\phi)\rangle = (0). $$

Moreover, $i \notin I^s$ if and only if there exists $U \in \mathfrak{g}_i^s(\phi) \backslash \mathfrak{g}_{i+1}^s(\phi)$ such that U is an element of the stabilizer of $\phi_i = \phi_{|\mathfrak{g}_i^s(\phi)}$. Hence for $i < \alpha$ it is obvious that $i \in I^s$ if and only if $i \in {}_1I^s$.
Suppose now that $i \geq \alpha$ and that $i \in {}_1I^s$. There exists then a vector $U \in {}_1\mathfrak{g}_i^s(\phi) = \mathbb{R}\text{-span}\{B_{i+1}^s(\phi), \ldots, B_{\alpha+1}^s(\phi), B_{\alpha-1}^s(\phi), \ldots, B_1^s(\phi), \mathfrak{h}^s(\phi)\}$ such that

$$ \langle U \cdot \phi, B_{i+1}^s(\phi)\rangle = 1; \quad \langle U \cdot \phi, B_j^s(\phi)\rangle = 0, j = i, \ldots, 1, j \neq \alpha; \quad \langle U \cdot \phi, \mathfrak{h}^s(\phi)\rangle = (0). $$

Therefore for $t = \langle U \cdot \phi, X_{s+1}(\phi)\rangle$, the element

$$ V = U - \frac{t}{\langle \phi, [Y_{s+1}(\phi), X_{s+1}(\phi)]\rangle} Y_{s+1}(\phi) \in {}_1\mathfrak{g}_i^s(\phi) $$

satisfies

$$ \langle V \cdot \phi, B_{i+1}^s(\phi)\rangle = 1; \quad \langle U \cdot \phi, B_j^s(\phi)\rangle = 0, j = i, \ldots, 1; \quad \langle U \cdot \phi, \mathfrak{h}^s(\phi)\rangle = (0). $$

and so $i + 1 \in I^s$. On the other hand, if $i + 1 \in I^s$, there exists $U \in \mathfrak{g}_{i+1}^s(\phi)$ such that

$$ \langle U \cdot \phi, B_{i+1}^s(\phi)\rangle = 1; \quad \langle U \cdot \phi, B_j^s(\phi)\rangle = 0, j = i, \ldots, 1, j \neq \alpha; \quad \langle U \cdot \phi, \mathfrak{h}^s(\phi)\rangle = (0). $$

In particular $\langle U \cdot \phi, Y_{s+1}(\phi)\rangle = 0$, so $U \in \mathfrak{g}_{i+1}^s(\phi) \cap \mathfrak{g}^{s+1}(\phi) = {}_1\mathfrak{g}_{i+1}^s(\phi)$ and then $i \in {}_1I^s$. It is obvious that $r \in {}_1I^s$ and $\alpha \in I^s$. This finishes the proof. □

We get the Jordan–Hölder basis

$$ \mathscr{C}^{s+1}(\phi) = \{Z_1^{s+1}(\phi) = {}_1Z_2^s(\phi), \ldots, Z_{n_s-1}^{s+1}(\phi) = {}_1Z_{n_s}^s(\phi)\} $$

of $\mathfrak{g}^{s+1,1}(\phi)$. Put $k_{s+1} = k_s$. The dimension r_{s+1} of

$$ \mathfrak{g}^{s+1,1}(\phi)/\mathfrak{h}^{s+1}(\phi) $$

equals $r_s - 1$ and for $\phi \in \mathscr{V}_{0,s+1,2}$, $T' \in \mathbb{R}^{k_s+1}$, let

$$\begin{aligned} R^{s+1}(\phi)(T') =& \{R_1^{s+1}(\phi)(T'), \dots, R_{r_s+1}^{s+1}(\phi)(T')\} \\ =& \{R_1^s(\phi)(T'), \dots, R_{\alpha-1}^s(\phi)(T'), R_{\alpha+1}^s(\phi)(T'), \dots, R_{r_s}^s(\phi)(T')\}. \end{aligned}$$

We then see that the family $R^{s+1}(\phi)$ satisfies properties i_1, i_2 of Definition 3.2.4 for $\mathfrak{h}^{s+1}(\phi)$ and $\mathscr{C}^{s+1}(\phi)$. We can take for $\mathscr{X}^{s+1}(\phi)$ the set

$$\mathscr{X}^{s+1}(\phi) = \mathscr{X}^s(\phi) \cup \{X_{s+1}(\phi)\}$$

and $\mathscr{V}_{0,s+1} = \mathscr{V}_{0,s+1,2}$. This closes the second case.

Case 3 In this case, we assume that there exists at least one $\phi_0 \in \mathscr{V}_{0,s+1,2}$ such that

$$\mathfrak{h}^s(\phi_0) \not\subset \mathfrak{g}^{s+1,1}(\phi_0).$$

Let

$$\mathscr{V}'_{0,s+1,2} = \{\phi \in \mathscr{V}_{0,s+1,2};\ \langle \phi, [\mathfrak{h}^s(\phi), Y_{s+1}(\phi)] \rangle \neq 0\},$$

and for $\phi \in \mathscr{V}'_{0,s+1,2}$

$$i^{s+1}(\phi) = \max_i\{Z_i^s(\phi) \notin \mathfrak{g}^{s+1,1}(\phi),\ Z_i^s(\phi) \in \mathfrak{h}\}, \qquad i^{s+1} = \max_{\phi \in \mathscr{V}'_{0,s+1,2}} \{i^{s+1}(\phi)\}$$

and

$$\mathscr{V}_{0,s+1,3} = \{\phi \in \mathscr{V}'_{0,s+1,2};\ \langle \phi, [Z'^s_{i^s+1}(\phi), Y'_{s+1}(\phi)] \rangle \neq 0\}.$$

Then $\mathscr{V}_{0,s+1,3}$ is a non-empty Zariski open set of $\mathscr{V}'_{0,s+1,2}$. For all $\phi \in \mathscr{V}_{0,s+1,3}$, we put

$$\mathfrak{h}^{s+1}(\phi) = \mathfrak{h}^s(\phi) \cap \mathfrak{g}^{s+1,1}(\phi) + \mathbb{R}Y_{s+1}(\phi),\ \mathfrak{g}^{s+1}(\phi) = \mathfrak{g}^s(\phi).$$

It is not difficult to see that $\mathfrak{h}^{s+1}(\phi) \subset \mathfrak{g}^{s+1,1}(\phi)$ for ϕ and $Y_{s+1}(\phi) \notin \mathfrak{h}^s(\phi)$ as $\mathfrak{h}^s(\phi)$ is subordinate to ϕ. Similarly, $(\mathfrak{g}^{s+1}(\phi))^\perp \subset \mathrm{ad}^*(\mathfrak{g}^{s+1}(\phi)) \cdot \phi$. The Malcev basis $\mathscr{Y}^{s+1}(\phi)$ of $\mathfrak{h}^{s+1}(\phi)$ relative to $\mathfrak{h}^{s+1}(\phi) \cap \mathfrak{h}$ is therefore given by

$$\mathscr{Y}^{s+1}(\phi) = \mathscr{Y}^s(\phi) \cup \{Y_{s+1}(\phi)\}.$$

We also set $r_{s+1} = r_s$; $k_{s+1} = k_s$. We must distinguish two cases, according to whether $i^{s+1} = i_{s+1}$(case a) or $i^{s+1} < i_{s+1}$(case b). In the former case $X_{s+1}(\phi) \in \mathfrak{h}$, whilst in the latter $X_{s+1}(\phi) \notin \mathfrak{h}$ and $Z^s_{i^{s+1}}(\phi) = V^s_{k^{s+1}}(\phi) \in \mathfrak{h}^s(\phi)$ for some k^{s+1}.

Case (a) We first change $\mathscr{C}^s(\phi)$ by replacing the vector $Z_i^s(\phi)$ by

$$Z_i^s(\phi) - \frac{\langle \phi, [Z_i^s(\phi), Y_{s+1}(\phi)]\rangle}{\langle \phi, [X_{s+1}(\phi), Y_{s+1}(\phi)]\rangle} X_{s+1}(\phi), i = 1, \dots, i_{s+1} - 1, \phi \in \mathscr{V}_{0,s+1,3}.$$

We can therefore assume that $Z_i^s(\phi) \in \mathfrak{g}^{s+1,1}(\phi)$ for all $i < i_{s+1}$. We take as before $\{B_j^s(\phi), j = 1, \dots, r_s\}$ as Malcev basis of $\mathfrak{g}^s(\phi)$ relative to $\mathfrak{h}^s(\phi)$ associated with $\mathscr{C}^s(\phi)$ and ${}_1B_j^s(\phi), j = 1, \dots, r_s$ the Malcev basis of $\mathfrak{g}^s(\phi)$ relative to $\mathfrak{h}^{s+1}(\phi)$ associated with $\mathscr{C}^s(\phi)$. Let $\alpha \in \{1, \dots, r_s\}$ be such that $X_{s+1}(\phi) = {}_1B_\alpha^s(\phi)$ for ϕ. We can easily see that $B_1^s(\phi) = Y_{s+1}(\phi)$, ${}_1B_j^s(\phi) = B_{j+1}^s(\phi)$ for $j < \alpha$ and ${}_1B_j^s(\phi) = B_j^s(\phi)$ for $j > \alpha$.

Lemma 3.2.8 *For $\phi \in \mathscr{V}_{0,s+1,3}$ and s, the set I^s can be replaced by*

$$I^{H^{s+1}(\phi), \mathscr{C}^s(\phi)} =: I^{s+1} = \{i; i+1 \in I^s, 1 < i < \alpha\} \cup \{\alpha\} \cup \{i \in I^s, i > \alpha\}.$$

Proof Let $\phi \in \mathscr{V}_{0,s+1,3}$. For $i < \alpha$, it is obvious that $i \in I^s$ if and only if $i \in I^{s+1}$. We now take $i > \alpha$. If $i \in I^{s+1}$, then there exists $U \in \mathfrak{g}_i^s(\phi) = \mathbb{R}\text{-span}\{B_{i+1}^s(\phi), \dots, B_2^s(\phi), \mathfrak{h}^{s+1}(\phi)\}$ such that

$$\langle U\cdot\phi, B_{i+1}^s(\phi)\rangle = 0; \quad \langle U\cdot\phi, B_j^s(\phi)\rangle = 0, j = i, \dots, 2; \quad \langle U\cdot\phi, \mathfrak{h}^{s+1}(\phi)\rangle = (0).$$

Let $t = \langle \phi, [X_{s+1}(\phi), U]\rangle$ and $V = U + \frac{t}{\langle \phi, [Y_{s+1}(\phi), X_{s+1}(\phi)]\rangle} Y_{s+1}(\phi) \in \mathfrak{g}_{i+1}^s(\phi)$. So

$$\langle U\cdot\phi, B_{i+1}^s(\phi)\rangle = 0; \quad \langle U\cdot\phi, B_j^s(\phi)\rangle = 0, j = i, \dots, 1; \quad \langle U\cdot\phi, \mathfrak{h}^{s}(\phi)\rangle = (0),$$

and $i+1 \in I^s$. On the other hand, if $i+1 \in I^s$, then there exists $U \in \mathfrak{g}_{i+1}^s(\phi)$ such that

$$\langle U\cdot\phi, B_{i+1}^s(\phi)\rangle = 1; \quad \langle U\cdot\phi, B_j^s(\phi)\rangle = 0, j = i, \dots, 1; \quad \langle U\cdot\phi, \mathfrak{h}^{s}(\phi)\rangle = (0),$$

Then in particular $U \in \mathfrak{g}_i^{s+1}(\phi)$ and

$$\langle U\cdot\phi, B_{i+1}^s(\phi)\rangle = 1; \quad \langle U\cdot\phi, B_j^s(\phi)\rangle = 0, j = i, \dots, 2; \quad \langle U\cdot\phi, \mathfrak{h}^{s+1}(\phi)\rangle = (0),$$

and so $i \in I^s$. On the other hand, it is obvious that $1 \in I^s$ and $\alpha \in I^{s+1}$. This completes the proof. □

If we put

$$\begin{aligned} R^{s+1}(\phi) =& \{R_1^{s+1}(\phi), \dots, R_{r_s+1}^{s+1}(\phi)\} \\ =& \{R_2^s(\phi), \dots, R_\alpha^s(\phi), \phi(X_{s+1}(\phi)), R_{\alpha+1}^s(\phi), \dots, R_{r_s}(\phi)\}, \end{aligned}$$

then the new family $\{R^{s+1}(\phi), \phi \in \mathscr{V}_{0,s+1,3}\}$ verifies conditions i_1, i_2 of Definition 3.2.4 for I^{s+1} and $L^{s+1} = \{1, \ldots, r_{s+1}\} \backslash I^{s+1}$. We get the Jordan–Hölder basis

$$\mathscr{C}^{s+1}(\phi) = \mathscr{C}^s(\phi)$$

of $\mathfrak{g}^s(\phi)$ and the Jordan–Hölder basis $\mathscr{D}^{s+1}(\phi)$ of $\mathfrak{h} \cap \mathfrak{g}^{s+1,1}(\phi)$ as follows:

$$\mathscr{D}^{s+1}(\phi) = \{Z_j^{s+1}(\phi) \in \mathscr{C}^{s+1}(\phi) \text{ such that } Z_j^{s+1}(\phi) \in \mathfrak{h}\} = \mathscr{D}^s(\phi) \backslash \{X_{s+1}(\phi)\}.$$

We take

$$\mathscr{X}^{s+1}(\phi) = \mathscr{X}^s(\phi) \cup \{(\langle \phi, [X_{s+1}(\phi), Y_{s+1}(\phi)] \rangle)^{-1} X_{s+1}(\phi)\},$$
$$\mathscr{U}^{s+1}(\phi) = \mathscr{U}^s(\phi) \cup \{(\langle \phi, [X_{s+1}(\phi), Y_{s+1}(\phi)] \rangle)^{-1} X_{s+1}(\phi)\},$$

and $\mathscr{V}_{0,s+1} = \mathscr{V}_{0,s+1,3}$.

Case (b) We will prove that we can change the Jordan–Hölder basis $C^s(\phi)$ to a new basis ${}_1C^s(\phi)$, as we found in case (a). To do this, let us first put $V_{s+1}(\phi) = Z^s_{i^{s+1}}(\phi) \in \mathfrak{h}$ and look for $\alpha \in \{1, \ldots, r_s\}$ such that $X_{s+1}(\phi) = B^s_\alpha(\phi), \phi \in \mathscr{V}_{0,s+1,3}$. We replace $Z_i^s(\phi)$ by

$$Z_i^s(\phi) - \frac{\langle \phi, [Z_i^s(\phi), Y_{s+1}(\phi)] \rangle}{\langle \phi, [X_{s+1}(\phi), Y_{s+1}(\phi)] \rangle} X_{s+1}(\phi), i = i^{s+1} + 1, \ldots, i_{s+1} - 1.$$

Let us also determine the index $i_{\alpha+1} \in \{1, \ldots, i_{s+1} - 1\}$ such that $B^s_{\alpha+1}(\phi) = Z^s_{i_{\alpha+1}}(\phi)$. We will distinguish different cases:

Case b-1 We assume in this case that $i_{\alpha+1} < i^{s+1}$, which means that the vectors $Z_j^s(\phi), i_{\alpha+1} < j < i_{s+1}$ lie in $\mathfrak{h}^s(\phi)$. The basis

$${}_1C^s(\phi) = \{Z_1^s(\phi), \ldots, Z^s_{i^{s+1}-1}(\phi), \widetilde{X}_{s+1}(\phi), V_{s+1}(\phi), Z^s_{i^{s+1}+1}(\phi), \ldots, Z^s_{n_s}(\phi)\}$$

turns out to be a Jordan–Hölder basis of $\mathfrak{g}$, where

$$\widetilde{X}_{s+1}(\phi) = X_{s+1}(\phi) - \frac{\langle \phi, [X_{s+1}(\phi), Y_{s+1}(\phi)] \rangle}{\langle \phi, [V_{s+1}(\phi), Y_{s+1}(\phi)] \rangle} V_{s+1}(\phi).$$

We can replace $C^s(\phi)$ by ${}_1C^s(\phi)$ and we are back in case (a).

Case b-2 We assume in this case that $i_{\alpha+1} > i^{s+1}$. As we can swap the vectors $Z_j^s(\phi)$ of $\mathfrak{h}^s(\phi)$ for which $i^{s+1} < j < i_{s+1}$ with $Z_{i_{s+1}}(\phi)$, we can assume that $i_{\alpha+1} = i_{s+1} - 1$. In the following, we will see that we can change the vectors $Z_{i_{s+1}}$ and $Z_{i_{s+1}-1}$. This leads to changes at the level of $R_j^s(\phi), j \in \{\alpha, \alpha + 1\}$. For this, we will treat four subcases of b-2, denoted (b-2-1)–(b-2-4).

Case b-2-1 We assume in this case that $\alpha \in I^s$ and $\alpha + 1 \in I^s$, so

$$R^s_\alpha(\phi)(T) = R^s_\alpha(\phi)(t_1, \dots, t_j) \text{ and } R^s_{\alpha+1}(\phi)(T) = R^s_{\alpha+1}(\phi)(t_1, \dots, t_j)$$

for some j. So we see that we can swap $B^s_\alpha(\phi)$ and $B^s_{\alpha+1}(\phi)$. Moreover, it is not difficult to see that α and $\alpha + 1$ are also in I^s.

Case b-2-2 We assume in this case that $\alpha \in I^s$ and $\alpha + 1 \in L^s$, then

$$R^s_\alpha(\phi)(T) = R^s_\alpha(\phi)(t_1, \dots, t_j) \quad \text{and} \quad R^s_{\alpha+1}(\phi)(T) = t_{j+1},$$

for some j. So we see that we can exchange $B^s_\alpha(\phi)$ and $B^s_{\alpha+1}(\phi)$ which gives us $\alpha \in L^s$ and $\alpha + 1 \in I^s$. Thus $R^s_\alpha(\phi)$ will be transformed in $R^{s+1}_\alpha(\phi)$ and $R^{s+1}_\alpha(\phi)$ to $R^s_\alpha(\phi)$.

Case b-2-3 Suppose now that $\alpha \in L^s$ and $\alpha + 1 \in L^s$. So we can exchange $Z^s_{i_{s+1}}(\phi)$ with $Z^s_{i_{s+1}-1}(\phi)$, which will exchange $B^s_\alpha(\phi)$ with $B^s_{\alpha+1}(\phi)$ and $R^s_\alpha(\phi)$ with $R^s_{\alpha+1}(\phi)$ and conversely.

Case b-2-4 Assume now that $\alpha \in L^s$ and $\alpha + 1 \in I^s$. Write

$$R^s_{\alpha+1}(\phi)(T) = c^s_{\alpha+1,j}(\phi)t_j + (R^s_{\alpha+1})'(\phi)(t_1, \dots, t_{j-1}), \;\; R^s_\alpha(\phi)(T) = t_j$$

for some j and some rational function $c^s_{\alpha+1,j}(\phi)$ in ϕ. If $c^s_{\alpha+1,j}(\phi)$ is identically zero, we can exchange $X_{s+1}(\phi)$ with $Z^s_{i_{\alpha+1}}(\phi)$ and we continue the procedure. Otherwise, let

$$\mathscr{V}_{0,s+1,4} = \{\phi \in \mathscr{V}_{0,s+1,3};\; \text{the numerator of } c^s_{\alpha+1,j}(\phi) \neq 0\}.$$

We replace the vector $Z^s_{i_{\alpha+1}}(\phi)$ with the vector $X_{s+1}(\phi)$ and the vector $X_{s+1}(\phi)$ with the vector $Z^s_{i_{\alpha+1}}(\phi) - c^s_{\alpha+1,j}(\phi)X_{s+1}(\phi)$, which becomes the new vector $X_{s+1}(\phi)$. However, $R^s_\alpha(\phi)$ is changed to $R^s_{\alpha+1}(\phi) - c^s_{\alpha+1,j}(\phi)R^s_\alpha(\phi)$ which no longer depends on t_j and $R^s_{\alpha+1}(\phi)$ to $R^s_\alpha(\phi)$. So the new index set I^s contains α and $\alpha + 1 \in L^s$. We are then in case b-2-2. This ends the study of the cases. We finally take $\mathscr{V}_{0,s+1} = \mathscr{V}_{0,s+1,4}$.

Definition 3.2.9 Let $C(\mathscr{V}_0; L^2)_2$ denote the set of mappings

$$\psi : \mathscr{V}_0 \longrightarrow \bigcup_{\phi \in \mathscr{V}_0} L^2(G/B(\phi), \phi),$$

such that: the functions $(\phi, g) \longmapsto \psi(\phi, g) := \psi(\phi)(g)$ are continuous on $\mathscr{V}_0 \times G$, $\psi(\phi) \in L^2(G/B(\phi), \phi)$ for all $\phi \in \mathscr{V}_0$, and which verify

$$\|\psi\|_2^2 = \int_{\mathscr{V}^{R,\mathscr{B}}} \|\psi(\phi)\|^2_{L^2(G/B(\phi),\phi)} d\lambda^{R,\mathscr{B}}(\phi) < \infty.$$

Let

$$\mathscr{H}_{\tau_I} = \int_{\mathscr{V}^{R,\mathscr{B}}}^{\oplus} L^2(G/B(\phi), \phi) d\lambda^{R,\mathscr{B}}(\phi)$$

be the completion of $C(\mathscr{V}_0, L^2)_2$ for the norm $\| \ \|_2$.

Let also $C_c^\infty(\mathscr{V}_0; C_c)$ be the subspace of $C(\mathscr{V}_0, L^2)_2$ of all mappings $(\phi, g) \longmapsto \psi(\phi, g)$ that are in addition C^∞ on $\mathscr{V}_0 \times G$. On $\mathscr{V}_0 \times \mathbb{R}^l$ the map

$$(\phi, t_1, \ldots, t_l) \longmapsto \psi(\phi, \exp(t_1 X_1(\phi)) \ldots \exp(t_l X_l(\phi)))$$

is C^∞ and with compact support, where $\mathscr{X}(\phi) = \{X_1(\phi), \ldots, X_l(\phi)\}$ denotes the Malcev basis of Sect. 3.2.3. It is clear that $C_c^\infty(\mathscr{V}_0, C_c)$ is dense of $\mathscr{H}_{\tau_I}$.

Theorem 3.2.10 *Let G be a connected and simply connected nilpotent Lie group with Lie algebra $\mathfrak{g}$. Let $f \in \mathfrak{g}^\star$ and let $\mathfrak{h}$ be a subalgebra of $\mathfrak{g}$ subordinate to f, where $H = \exp(\mathfrak{h})$ is the connected closed subgroup of G associated to $\mathfrak{h}$. Let τ be the monomial representation of H induced by the unitary character χ_f. Call $\mathscr{B}$ the Malcev basis of $\mathfrak{g}$ relative to $\mathfrak{h}$, $\mathscr{V}^{R,\mathscr{B}} \subset \Gamma_f$ the affine subspace of (3.2.6) with measure $d\lambda^{R,\mathscr{B}}$, $\mathscr{V}_0$ the open Zariski set of $\mathscr{V}^{R,\mathscr{B}}$ and $\phi \longrightarrow B(\phi)$ the continuous choice of polarizations in $\phi \in \mathscr{V}_0$. Then*

$$\tau \simeq \tau_I = \int_{\mathscr{V}^{R,\mathscr{B}}}^{\oplus} \pi_{\phi, B(\phi)} d\lambda^{R,\mathscr{B}}(\phi). \tag{3.2.8}$$

The operator U defined for $\xi \in S(G/H, f)$ by:

$$U(\xi)(\phi) = T_{B(\phi),H}\xi$$

is an isometric linear operator with values in $\mathscr{H}_{\tau_1}$ which extends over $L^2(G/H, f)$ to an isometric intertwining operator of (3.2.8)*, provided the measures on $G/B(\phi)$ and on $B(\phi)/B(\phi) \cap H$ are suitably normalized.*

Proof It is clear that the operator U formally intertwines τ and its disintegration τ_I. Moreover, for every $\xi \in S(G/H, f)$, $\phi \in \mathscr{V}_0$ and $g \in G$, the function

$$U(\xi)(\phi)(g) = T_{B(\phi),H}\xi(g) =$$

$$\int_{\mathbb{R}^n} \xi(g \cdot \exp(t_1 Y_1(\phi)) \cdots \exp(t_m Y_m(\phi))) e^{-i(t_1\phi(Y_1(\phi))+\ldots+t_m\phi(Y_m(\phi)))} dt_1 \cdots dt_m$$

is obviously continuous on $\mathscr{V}_0 \times G$. On the other hand, for $\phi \in \mathscr{V}_0$, $\xi \in S(G/H, f)$, we know that $U(\xi)(\phi) \in S(G/B(\phi), \phi)$. If all the elements of Γ_f are characters of $\mathfrak{g}$, then $\mathscr{V}^{R,\mathscr{B}} = \Gamma_f$, $\mathfrak{b}(\phi) = \mathfrak{g}$ for all ϕ, $\mathfrak{h}$ contains $[\mathfrak{g}, \mathfrak{g}]$ and hence it is an ideal in $\mathfrak{g}$, and $\mathscr{Y} = \mathscr{B}$. Therefore U is simply a partial Fourier transform and everything is obvious in this case.

We now argue by induction on $\dim(G)+\dim(G/H)$ and we will prove that U is an isomorphism on $C_c(G/H, f)$, which will lead us to the fact that U takes its values in $C(\mathcal{V}_0; L^2)_2$.
We may assume that there exists j_1 in $\{1, \dots, n\}$ such that $Z_{j_1} = Y$ does not belong to $\mathfrak{a}(\phi)$ for ϕ in $\mathcal{V}_0$ and Z_k belongs to $\mathfrak{a}(\phi)$ for all $k > j_1$. We will study the following possible cases:

Case 1 There exists $h > j_1$, such that $Z_i \in \mathfrak{h}$ for all $i > h$ and $T = Z_h \notin \mathfrak{h}$. If $\mathfrak{h}^1 = \mathfrak{h} \oplus \mathbb{R}T$, from the construction of $\mathscr{B}$ we have $T = B_1$ and therefore $1 \in L^H$. Hence we have that $\Gamma_f = \bigcup_{t\in\mathbb{R}} \Gamma_{f_t}$ where $f_t = f + tT^\star$ and

$$\Gamma_{f_t} = f_t + \mathfrak{h}^{1\perp} = \{\phi \in \Gamma_f : \langle \phi, T\rangle = \langle f, T\rangle + t\}.$$

Let $\mathscr{B}' = \{B_2, \dots, B_r\}$ be the Malcev basis of $\mathfrak{g}$ relative to $\mathfrak{h}^1$ and for all $i \in \{2, \dots, r\}$

$${}_tR_i(t_2, \dots, t_k) = R_i(t, t_2, \dots, t_k).$$

We find that $\mathcal{V}^{{}_tR, \mathscr{B}'}$ is $\mathscr{C}$-admissible for almost all $t \in \mathbb{R}$, and given that $R_1(t, t_2, \dots, t_k) = t$, we have

$$\mathcal{V}^{R,\mathscr{B}} = \bigcup_{t\in\mathbb{R}} \mathcal{V}^{{}_tR,\mathscr{B}'} \tag{3.2.9}$$

and $d\lambda^{R,\mathscr{B}} = \int_{\mathbb{R}} d\lambda^{{}_tR,\mathscr{B}'} dt$. Furthermore, we prove that $\mathcal{V}_0 = \bigcup_{t\in\mathbb{R}_0} \mathcal{V}_0^t$ where $\mathbb{R}_0 = \{t \in \mathbb{R}, \mathcal{V}_0^t \neq \emptyset\}$ and $\mathcal{V}_0^t$ is Zariski open in $\mathcal{V}^{{}_tR,\mathscr{B}'}$. For this, it is sufficient to check that for $t \in \mathbb{R}_0$

$$\mathcal{V}_0^t = \mathcal{V}_0 \cap V^{{}_tR,\mathscr{B}'}.$$

In fact, let us recall that $\mathcal{V}_0 = \{\phi \in \mathcal{V}^{R,\mathscr{B}} : Q(\phi) \neq 0\}$ for a certain polynomial Q. Hence, as in Theorem 3.2.6, we can see that

$$\mathcal{V}_0^t = \{\phi_t \in \mathcal{V}^{{}_tR,\mathscr{B}'} : Q(\phi_t) \neq 0\}$$

with the same polynomial Q, taking into account the induction procedure from $s = 0$ to $s = 1$. Up to this point, note that the set

$$\{t \in \mathbb{R} : \mathcal{V}_0^t = \emptyset\} = \{t \in \mathbb{R}, T' \longmapsto Q(tB_1^* + \sum_{j=2}^{r} R_j(t, T')B_j^*) = 0\}$$

is finite and therefore negligible in $\mathbb{R}$. Hence for almost all $t \in \mathbb{R}$, $\mathcal{V}_0^t \neq \emptyset$.

We will check now that for $\phi \in \mathscr{V}_0$, all our constructions, namely the polarizations $\mathfrak{b}(\phi)$ and the bases $\mathscr{X}, \mathscr{U}$ made from $\mathfrak{h}$ or $\mathfrak{h}^1$, coincide. In fact, let us recall that stage $s = 0$ is defined by the data $(\mathfrak{g}, \mathfrak{h}, \phi)$, for the construction of $\mathfrak{b}(\phi)$ and the bases $\mathscr{X}$ and $\mathscr{U}$. The stage $s = 1$ is defined by the data $(\mathfrak{g}, \mathfrak{h}^1, \phi)$. We then proceed to the final stage $s = s_0$. It is clear that starting from $s = 0$ or $s = 1$ will lead to the same final result. Moreover, we have that $\mathscr{D}' = \mathscr{D} \cup \{T\}$, where $\mathscr{D}'$ is the Jordan–Hölder basis of $\mathfrak{h}^1$ which is contained in $\mathscr{C}$. As for the Malcev basis $\mathscr{Y}'$ of $b(\phi)$ relative to $\mathfrak{h}^1 \cap \mathfrak{b}(\phi)$, it is obtained as follows: the vector T is in $\mathscr{Y}$ and we have that $\mathscr{Y}' = \mathscr{Y}\backslash\{T\}$.
We apply the induction hypothesis to all pairs $(\mathfrak{h}^1, f_t)$; $t \in \mathbb{R}_0$. Note again for $\phi = \phi_t \in \mathscr{V}_0^t$, $t \in \mathbb{R}_0$ that

$$U(\xi)(\phi) = T_{B(\phi),H}\xi = T_{B(\phi_t),H^1}\xi^t = U_t(\xi^t)(\phi_t), \tag{3.2.10}$$

where

$$\xi^t(u) = \int_{\mathbb{R}} \xi(u \exp(\lambda T))e^{-it\lambda}d\lambda. \tag{3.2.11}$$

The induction hypothesis tells us that for every $t \in \mathbb{R}_0$,

$$U_t : S(G/H^1, f_t) \longrightarrow \int^{\oplus}_{\mathscr{V}_t{}^{R,\mathscr{B}'}} L^2(G/B(\phi_t), \phi_t)d\lambda^{t^{R,\mathscr{B}'}}(\phi_t) \tag{3.2.12}$$

is an isometry. For $\xi \in S(G/H, f)$, ξ^t defines one element of $S(G/H^1, f_t)$. It follows from Plancherel's theorem that

$$\begin{aligned}
\|\xi\|^2_{L^2(G/H,f)} &= \int_{\mathbb{R}} \|\xi^t\|^2_{L^2(G/H^1,f_t)}dt \qquad \text{(induction hypothesis)}\\
&= \int_{\mathbb{R}} \|U_t(\xi^t)\|_2^2 dt\\
&= \int_{\mathbb{R}}\int_{\mathscr{V}} \|T_{B(\phi_t),H^1}\xi\|^2_{L^2(G/B(\phi_t),\phi_t)} d\lambda^{t^{R,\mathscr{B}'}}(\phi_t)dt\\
&= \int_{\mathscr{V}^{R,\mathscr{B}}} \|T_{B(\phi),H}\xi\|^2_{L^2(G/B(\phi),\phi)} d\lambda^{R,\mathscr{B}}(\phi).
\end{aligned}$$

We can now assume that $Z_{j_1+1}, \ldots, Z_n \in \mathfrak{h}$. We apply the constructions made in Sect. 3.2.3, step s+1 with $s = 0$ for cases 2 and 3. For $\phi \in \Gamma_f$, we put

$$\mathfrak{g}_1^1(\phi) = \{T \in \mathfrak{g} \text{ such that } \langle \phi, [Y, T]\rangle = 0\}.$$

Then $\mathfrak{g}_1^1(\phi)$ is an ideal of codimension 1 in $\mathfrak{g}$, and is equal to $\mathfrak{g}_1^1(\psi)$ for all $\psi \in \Gamma_f$, denoted $\mathfrak{g}^1$ in the whole sequence. Note that if $Y = Y_1 \in \mathfrak{h}$ then $\mathfrak{h} \subset \mathfrak{g}^1$. We must therefore distinguish the two cases from this construction.

Case 2 $\mathfrak{h} \subset \mathfrak{g}^1$ and $Z_{j_1+1}, \ldots, Z_n \in \mathfrak{h}$.
Note that we can assume that there exists $i_1 \in \{1, \ldots, r\}$, such that $X = X_1 = Z_{i_1} \notin \mathfrak{g}^1$. We now change $\mathscr{C}$ by replacing the vector Z_i by

$$Z_i - \frac{\langle \phi, [Z_i, Y] \rangle}{\langle \phi, [X, Y] \rangle} X, \quad i = 1, \ldots, i_1 - 1, \quad \phi \in \mathscr{V}_0.$$

This changes the objects $\mathscr{B}$ and R, but these objects retain their properties of rationality and we can therefore assume that $Z_i \in \mathfrak{g}^1$ for all $i \neq i_1$. We replace the basis $\mathscr{C}$ by the following basis

$${}_1\mathscr{C} = \{X, Z_1, \ldots, Z_{i_1-1}, Z_{i_1+1}, \ldots, Z_n\} = \{{}_1Z_1, \ldots, {}_1Z_n\}.$$

We get the Jordan–Hölder basis

$$\mathscr{C}^1 = \{Z_1^1 = {}_1Z_2, \ldots, Z_{n-1}^1 = {}_1Z_n\}$$

of $\mathfrak{g}^1$. Let $\alpha \in \{1, \ldots, r\}$ be such that $X = B_\alpha$. According to our extraction procedure, we get the new Malcev basis

$${}_1\mathscr{B} = \{{}_1B_1 = B_1, \ldots, {}_1B_{\alpha-1} = B_{\alpha-1,1}B_\alpha = B_{\alpha+1}, \ldots, {}_1B_{r-1} = B_r = X\}$$

of $\mathfrak{g}$ relative to $\mathfrak{h}$. Then the Malcev basis $\mathscr{B}^1$ of $\mathfrak{g}^1$ relative to $\mathfrak{h}$ associated to $\mathscr{C}^1$ is given by $\mathscr{B}^1 = \{{}_1B_1, \ldots, {}_1B_{r-1}\}$. As in Lemma 3.2.7,

$${}_1I = I^{H,{}_1\mathscr{C}} = \{i \in I^{H,\mathscr{C}}, i < \alpha\} \cup \{i; \alpha \leq i < r, i+1 \in I^{H,\mathscr{C}}\} \cup \{r\}.$$

Similarly, we have ${}_1R = \{{}_1R_1, \ldots, {}_1R_r\}$ where ${}_1R_j = R_j$ for $j < \alpha$, ${}_1R_j = R_{j+1}$ for $r > j \geq \alpha$ and ${}_1R_r = R_\alpha$. Then $\mathscr{V}^{{}_1R,{}_1\mathscr{B}}$ is ${}_1\mathscr{C}$-admissible, according to the description of ${}_1I$, and in addition $\mathscr{V}^{R,\mathscr{B}} = \mathscr{V}^{{}_1R,{}_1\mathscr{B}}$ and the measures of $d\lambda^{R,\mathscr{B}}$ and $d\lambda^{{}_1R,{}_1\mathscr{B}}$ coincide. Let $R^1 = \{{}_1R_1, \ldots, {}_1R_{r-1}\}$. The space $\mathscr{V}^{R^1,\mathscr{B}^1} \subset (\mathfrak{g}^1)^\star$ is $\mathscr{C}^1$-admissible and more precisely it is the restriction of $\mathscr{V}^{R,\mathscr{B}}$ to $\mathfrak{g}^1$. In addition, the measure $d_{G,H}$ on G/H, defined from the Malcev basis $\mathscr{B}$, equals the measure coming from the Malcev basis $\{X\} \cup \mathscr{B}^1$. Every element g of G is written in a unique way in the form $g = \exp(tX) \cdot g'$, with $t \in \mathbb{R}$, and $g' \in G^1 = \exp(\mathfrak{g}^1)$. For a function ξ on G, we denote the function ξ_t on $\mathfrak{g}^1$ defined by $\xi_t(g') = \xi(\exp(tX)g')$ for all $g' \in \mathfrak{g}^1$. It is clear that for almost all $t \in \mathbb{R}$ the function $\xi_t \in L^2(G^1/H, f_1)$, $f_1 = f_{|\mathfrak{g}^1}$ when $\xi \in L^2(G/H, f)$.

By referring to the construction in Sect. 3.2.3, we note that

$$B(\phi) = B(\phi_1) \subset \mathfrak{g}^1$$

where $\phi_1 = \phi_{|\mathfrak{g}^1}$. Let $U_1 = T_{B(\phi_1),H}$, it is clear that

$$U(\xi)(\phi)(\exp(tX) \cdot g') = U_1(\xi_t)(\phi_1)(g'). \tag{3.2.13}$$

For $\eta \in S(G^1/H, f_1)$, the induction hypothesis says

$$\|\eta\|^2_{L^2(G^1/H,f_1)} = \int_{\mathscr{V}_{R^1,\mathscr{B}^1}} \|T_{B(\phi_1),H}\eta\|^2_{L^2(G^1/B(\phi_{\|}1),\phi_1)} d\lambda^{R^1,\mathscr{B}^1}(\phi_1).$$

So for $\xi \in S(G/H, f)$, one obtains

$$\begin{aligned}
\|U(\xi)\|_2^2 &= \int_{\mathscr{V}_{R,\mathscr{B}}} \|U(\xi)(\phi)\|^2_{L^2(G/B(\phi),\phi)} d\lambda^{R,\mathscr{B}}(\phi) \\
&= \int_{\mathscr{V}_{R,\mathscr{B}}} \|T_{B(\phi),H}(\xi)\|^2_{L^2(G/B(\phi),\phi)} d\lambda^{R,\mathscr{B}}(\phi) \\
&= \int_{\mathbb{R}} \int_{\mathscr{V}_{R^1,\mathscr{B}^1}} \|T_{B(\phi_1),H}(\xi_t)\|^2_{L^2(G^1/B(\phi_1),\phi_1)} d\lambda^{R^1,\mathscr{B}^1}(\phi_1) dt \\
&\quad \text{(apply the induction hypothesis on } G^1) \\
&= \int_{\mathbb{R}} \|\xi_t\|^2_{L^2(G^1/H,f_1)} dt = \|\xi\|^2_{L^2(G/H,f)}.
\end{aligned}$$

Case 3 $\mathfrak{h} \not\subset \mathfrak{g}^1$. In this case $Y \notin \mathfrak{h}$ and so $1 \in I = I^{H,\mathscr{C}}$. Let $\mathfrak{h}^1 = \mathfrak{h} \cap \mathfrak{g}^1 \oplus \mathbb{R}Y$ and $H^1 = \exp(\mathfrak{h}^1)$. We shall take up the two cases called (a) and (b) in Sect. 3.2.3.

Case a We first change in this case $\mathscr{C}$ by replacing the vector Z_i by

$$Z_i - \frac{\langle \phi, [Z_i, Y] \rangle}{\langle \phi, [X, Y] \rangle} X, \quad i = 1, \dots, i_1 - 1, \ \phi \in \mathscr{V}_0.$$

This changes $\mathscr{B}$ and R, but these objects retain their properties and we can therefore assume that $Z_i \in \mathfrak{g}^1$ for all $i \neq i_1$. Likewise, we obtain a Malcev basis $\mathscr{B}^1$ of $\mathfrak{g}$ relative to $\mathfrak{h}^1$ from $\mathscr{C}$ by our extraction process. We can easily see that $B_1 = Y \in \mathscr{B}$, that the first vector of $\mathscr{B}^1$ is the element B_2, that $X = B^1_\alpha$ for some α and that $B^1_j = B_{j+1}$ for all $j < \alpha$, $B^1_j = B_j$ for $j > \alpha$. We have shown that

$$I^1 = I^{H^1,\mathscr{C}} = \{i : i - 1 \in I = I^{H,\mathscr{C}}, 1 < i < \alpha\} \cup \{\alpha\} \cup \{i \in I^{H,\mathscr{C}}, i > \alpha\}$$

and we set

$$R^1 = \{R^1_1, \dots, R^1_r\} = \{R_2, \dots, R_\alpha, f(X), R_{\alpha+1}, \dots\}.$$

As $Y \notin \mathfrak{h}$, we can assume that f satisfies $f(Y) = R_1(0)$ and then $\mathscr{V}^{R,\mathscr{B}}$ is also an affine subspace of $\Gamma^1_f = f + (\mathfrak{h}^1)^\perp$. Then we see that $\mathscr{V}^{R^1,\mathscr{B}^1}$ is $\mathscr{C}$-admissible, $\mathscr{V}^{R,\mathscr{B}} = \mathscr{V}^{R^1,\mathscr{B}^1}$ and $d\lambda^{R,\mathscr{B}} = d\lambda^{R^1,\mathscr{B}^1}$.

Case b In this case we set $V = Z_{i^1} \in \mathfrak{h}$ and we replace Z_i by

$$Z_i - \frac{\langle \phi, [Z_i, Y] \rangle}{\langle \phi, [X, Y] \rangle} X, \quad i = i^1 + 1, \ldots, i_1 - 1.$$

We see that the family $R = \{R_1, \ldots, R_r\}$ satisfies Definition 3.2.4. We show that we can exchange the basis $\mathscr{C}$ with a new Jordan–Hölder basis ${}_1\mathscr{C}$ of $\mathfrak{g}$, which places us in case (a) and which gives us a Malcev basis ${}_1\mathscr{B}$ of $\mathfrak{g}$ relative to $\mathfrak{h}$ and a family of functions ${}_1R$ such that

$$\mathscr{V}^{R,\mathscr{B}} = \mathscr{V}^{{}_1R,{}_1\mathscr{B}}.$$

Furthermore, the induced representations $\mathrm{ind}_H^G \chi_f$ and $\mathrm{ind}_{H'}^G \chi_f$ are equivalent, and the operator $T_{H^1,H}$ defined for all $\xi \in\in S(G/H, f)$ and $g \in G$ by

$$T_{H^1,H}\xi(g) = \int_{H^1/H^1 \cap H} \xi(gh) d_{H^1,H^1\cap H}(h) = \int_{\mathbb{R}} \xi(g \exp(tY)) dt$$

is an intertwining operator. Indeed, it is an isometry, because we can easily check that the measure on G/H determined by $\mathscr{B}$ is the same as the one determined by $\mathscr{B}^1 = \{Y, {}_1B_2, \ldots, {}_1B_r\}$ and the measure on G/H^1 determined by ${}_1\mathscr{B}^1 = \{{}_1B_2, \ldots, {}_1B_r\} \cup \{X\}$. $T_{H^1,H}$ is then a partial Fourier transformation in the direction of X, as $\langle f, [X, Y] \rangle = 1$.

Let $U^1 = T_{B(\phi),H^1}$. We check that for $\phi \in \mathscr{V}_0$ and all $\xi \in S(G/H, f)$ that

$$U^1(T_{H^1,H}\xi) = U(\xi), \tag{3.2.14}$$

which follows from a direct integral calculation. We then obtain

$$\begin{aligned}
\|U(\xi)\|_2^2 &= \int_{\mathscr{V}^{R,\mathscr{B}}} \|U(\xi)(\phi)\|^2_{L^2(G/B(\phi),\phi)} d\lambda^{R,\mathscr{B}}(\phi) \\
&= \int_{\mathscr{V}^{R,\mathscr{B}}} \|T_{B(\phi),H}\xi\|^2_{L^2(G/B(\phi),\phi)} d\lambda^{R,\mathscr{B}}(\phi) \\
&= \int_{\mathscr{V}^{{}_1R,{}_1\mathscr{B}}} \|T_{B(\phi_1),H^1} \circ T_{H^1,H}\xi\|^2_{L^2(G/B(\phi_1),\phi_1)} d\lambda^{{}_1R,{}_1\mathscr{B}}(\phi_1) \\
&\quad \text{(apply Case 2)} \\
&= \|T_{H^1,H}\xi\|^2_{L^2(G/H^1,f)} = \|\xi\|^2_{L^2(G/H,f)}.
\end{aligned}$$

We have proved that U is an isometry on $S(G/H, f)$. Since this space is a dense subspace in $L^2(G/H, f)$, there exists a unique extension $\overline{U}$ of U (still denoted by U) to an isometry. The desired covariance relation will obviously be required on the whole space. This completes the proof of the theorem. □

We shall see later that we cannot consider arbitrary rational polarizations to get an intertwining operator.
The next step consists in proving that the operator U is invertible. For $\phi \in \mathscr{V}_0$, we define the operator $T_{H,B(\phi)}$ on $C_c^\infty(\mathscr{V}_0, C_c)$ by

$$T_{H,B(\phi)}K(g) = \int_{H/B(\phi)\cap H} K(\phi)(gb)\chi_\phi(b)d_{H,H\cap B(\phi)}(b),$$

$g \in G$, $K \in C_c^\infty(\mathscr{V}_0, C_c)$, or $\phi \longrightarrow B(\phi)$ denotes the rational choice of polarizations in $\phi \in \mathscr{V}_0$ and $d_{H,B(\phi)\cap H}$ is the measure on $H/B(\phi) \cap H$.

Theorem 3.2.11 *We keep the same hypothesis and notations as in Theorem 3.2.10. Then the intertwining operator U is invertible. Its inverse $V = U^{-1}$ is given for all $K \in C_c^\infty(\mathscr{V}_0; C_c)$ and $g \in G$ by*

$$V(K)(g) = \int_{\mathscr{V}^{R,\mathscr{B}}} T_{H,B(\phi)}K(\phi)(g)d\lambda^{R,\mathscr{B}}(\phi).$$

Proof It is clear that $V \circ \tau_I = \tau_I \circ V$ and that $V(K)$ fulfils the covariance relation (3.2.5) for G, H and f. We prove that for $K \in C_c^\infty(\mathscr{V}_0, C_c)$, $V(K)$ is an element of $L^2(G/H, f)$. We will in fact demonstrate that V is an isometry on $C_c^\infty(\mathscr{V}_0, C_c)$. We argue as before by induction on $\dim(\mathfrak{g}) + \dim(\mathfrak{g}/\mathfrak{h})$, and take up the three cases discussed in Theorem 3.2.10. Let us show that

$$\|V(K)\|^2_{L^2(G/H,f)} = \int_{\mathscr{V}^{R,\mathscr{B}}} \|K(\phi)\|^2_{L^2(G/B(\phi),\phi)} d\lambda^{R,\mathscr{B}}(\phi).$$

Case 1 We write for $s \in \mathbb{R}$ and $K \in C_c^\infty(\mathscr{V}_0, C_c)$, $K_s = K_{|V_0^s} = K_{|V_0^s}$. It is clear that $K_s \in C_c^\infty(\mathscr{V}_0^s; C_c)$. For $g \in G$

$$V_s(K_s)(g) = \int_{\mathscr{V}^{R_s,\mathscr{B}'}} T_{H',B(\phi_s)}K_s(\phi_s)(g)d\lambda^{R_s,\mathscr{B}'}(\phi_s) \tag{3.2.15}$$

The existence of integral (3.2.15) is obvious by the conditions on K_s. We remark that the mapping $(s, g) \longmapsto V_s(K_s)(g)$ is C^∞ with compact support in s and C^∞ in g, and that for all $g \in G$,

$$V(K)(g) = \int_{\mathbb{R}} V_s(K_s)(g)ds.$$

It follows that

$$a \longmapsto V(K)(g\exp(aT) = \int_{\mathbb{R}} V_s(K_s)(g\exp(aT))ds = \int_{\mathbb{R}} V_s(K_s)(g)e^{ias}ds$$

is a Schwartz function in a. By Fourier's inversion theorem we have

$$\begin{aligned}\int_{\mathbb{R}} V(K)(g\exp(aT))e^{-iat}da &= \int_{\mathbb{R}^2} V_s(K_s)(g)e^{-iat}e^{ias}dsda \\ = \int_{\mathbb{R}^2} V_s(K_s)(g)e^{iat}e^{-ias}dsda &= V_t(K_t)(g) = V(K)^t(g).\end{aligned} \tag{3.2.16}$$

for every $t \in \mathbb{R}$. Then:

$$\|V(K)\|^2_{L^2(G/H,f)}$$

$$= \int_{\mathbb{R}}\int_{G/H'} |V(K)(g_0 \cdot \exp(sT)|^2 d_{G,H'}(g_0)ds$$

$$= \int_{\mathbb{R}}\int_{G/H'} |\int_{\mathscr{V}^{R,\mathfrak{B}}} T_{H,B(\phi)}K(\phi)(g_0 \cdot \exp(sT))d\lambda^{R,\mathscr{B}}(\phi)|^2 d_{G,H'}(g_0)dg$$

$$= \int_{\mathbb{R}}\int_{G/H'} |\int_{\mathbb{R}}\int_{\mathscr{V}^{R_t,\mathfrak{B}'}} T_{H',B(\phi_t)}K_t(\phi_t)(g_0)e^{its}d\lambda^{R_t,\mathscr{B}'}(\phi_t)dt|^2 d_{G,H'}(g_0)ds$$

$$= \int_{\mathbb{R}}\int_{G/H'} |\int_{\mathbb{R}} V_t(K_t)(g_0)e^{-its}dt|^2 d_{G,H'}(g_0)ds$$

(we use the Plancherel formula)

$$= \int_{\mathbb{R}} \|V_s(K_s)\|^2_{L^2(G/H',f_s)}$$

(we apply the induction hypothesis)

$$= \int_{\mathbb{R}}\int_{V^{R_s,\mathscr{B}'}} \|K_s(\phi_s)\|^2_{L^2(G/B(\phi_s),\phi_s)} d\lambda^{R_s,\mathscr{B}'}(\phi_s)ds$$

$$= \int_{\mathscr{V}^{R,\mathscr{B}}} \|K(\phi)\|^2_{L^2(G/B(\phi),\phi)} d\lambda^{R,\mathscr{B}}(\phi) = \|K\|_2^2.$$

Case 2 For $K \in C_c^\infty(\mathscr{V}_0; C_c)$ and $t \in \mathbb{R}$, we note

$$K_t(\phi_1)(g_1) = K(\phi_1)(\exp(tX) \cdot g_1),\ \phi_1 \in \mathscr{V}_1 = \mathscr{V}_{0|\mathfrak{g}^1},\ g_1 \in G^1.$$

We have that $K_t \in C_c^\infty(\mathscr{V}_1; C_c)$ and

$$V(K)(g) = V(K)_t(g_1) = \int_{\mathscr{V}^{R^1,\mathscr{B}^1}} T_{H,B(\phi_1)}K_t(\phi_1)(g_1)d\lambda^{R^1,\mathscr{B}^1}(\phi_1) = V_1(K_t)(g_1).$$

On the other hand,

$$\|V(K)\|^2_{L^2(G/H,f)} = \int_{\mathbb{R}} \|V(K)_t\|^2_{L^2(G^1/H,f_1)} dt = \int_{\mathbb{R}} \|V_1(K_t)\|^2_{L^2(G^1/H,f_1)} dt$$

(we apply the induction hypothesis)

$$= \int_{\mathbb{R}} \int_{\mathscr{V}_{R^1,\mathscr{B}^1}} \|K_t(\phi_1)\|^2_{L^2(G^1/B(\phi_1),\phi_1)} d\lambda^{R^1,\mathscr{B}^1}(\phi_1) dt$$

$$= \int_{\mathscr{V}_{R,\mathscr{B}}} \|K(\phi)\|^2_{L^2(G/B(\phi),\phi)} d\lambda^{R,\mathscr{B}}(\phi).$$

Case 3 According to formula (3.2.14), we have

$$T_{H,H^1} \circ T_{H^1,B(\phi)} = T_{H,B(\phi)}, \ \ \phi \in \mathscr{V}_0,$$

where $V = T_{H,H^1} \circ V^1$ and V^1 denotes the operator corresponding to G, H^1 and f. So we see immediately that V is isometric on a dense subspace of $\mathscr{H}_{\tau_I}$, and there exists a unique isometric extension $\overline{V}$ of V (still denoted by V). The desired covariance relation is then ensured over the entire space and this completes the proof.

We now prove that $U \circ V = \mathbb{I}$, which implies that the operator U is surjective, so invertible, and its inverse is V. We prove this equality on $C_c^\infty(\mathscr{V}_0, C_c)$ and the result extends. We further argue by induction on $\dim(\mathfrak{g}/\mathfrak{h}) + \dim(\mathfrak{g})$.

Case 1 Let $\xi \in S(G/H, f)$, and as in (3.2.11),

$$\xi^t(u) = \int_{\mathbb{R}} \xi(u \exp(\lambda T)) e^{-it\lambda} d\lambda, \ t \in \mathbb{R}.$$

So it is clear that

$$\int_{\mathbb{R}} \xi^t(u) dt = \xi(u) \text{ and that } \int_{\mathbb{R}} \|\xi^t\|^2_{L^2(G/H',f_t)} dt = \|\xi\|^2_{L^2(G/H,f)}.$$

By extension, we can deduce that

$$L^2(G/H, f) \simeq \int_{\mathbb{R}}^{\oplus} L^2(G/H', f_t) dt.$$

Let Ψ be the operator field defined on $L^2(G/H, f)$ by

$$\Psi(\xi)(S) = \xi^s \in L^2(G/H', f_s), \ \ s \in \mathbb{R}.$$

Our operator U also decomposes as $U = \int_{\mathbb{R}}^{\oplus} U_t dt$. We then have

$$U = (\int_{\mathbb{R}}^{\oplus} U_t dt) \circ \Psi. \tag{3.2.17}$$

Now let $K \in C_c^{\infty}(\mathscr{V}_0, C_c)$ and $g \in G$, so for $\phi = \phi_t \in \mathscr{V}_0^t$ that

$$\begin{aligned} U \circ V(K)(\phi_t)(g) =& U_t(V(K)^t)(\phi_t)(g) \quad \text{(from (3.2.17))} \\ & \text{(we apply (3.2.16))} \\ =& U_t(V_t(K_t))(\phi_t)(g) \\ & \text{(we use the induction hypothesis)} \\ =& K_t(\phi_t)(g) = K(\phi)(g). \end{aligned}$$

Case 2 It is clear that $L^2(G/H, f) \simeq L^2(\mathbb{R}, L^2(G_1/H, f_1))$ and that

$$\int_{\mathscr{V}^{R,\mathscr{B}}} L^2(G/B(\phi), \phi) d\lambda^{R,\mathscr{B}}(\phi) = \int_{\mathscr{V}^{R^1,\mathscr{B}^1}} L^2(\mathbb{R}, L^2(G_1/B(\phi_1), \phi_1)) d\lambda^{R^1,\mathscr{B}^1}(\phi_1).$$

So, $U = \widetilde{U}_1 \circ W$ where $W : L^2(G/H, f) \longrightarrow L^2(\mathbb{R}, L^2(G_1/H, f_1))$ is the operator field defined by

$$W(\xi)(t)(g_1) = \xi(\exp(tX) \cdot g_1) = \xi_t(g_1), \; g_1 \in G$$

and $\widetilde{U}_1(\xi)(t)(g_1) = U_1(\xi_t)(g_1)$.

Let $\phi \in \mathscr{V}_0$ and $g = \exp(tX) \cdot g_1 \in G$:

$$\begin{aligned} U \circ V(K)(\phi)(g) =& U_1 \circ V(K)_t(\phi_1)(g_1) = U_1 \circ V_1(K_t)(\phi_1)(g_1) \\ & \text{(we apply the induction hypothesis)} \\ =& K_t(\phi_1)(g_1) = K(\phi_1)(\exp(tX) g_1) = K(\phi)(g). \end{aligned}$$

Case 3 We know that $V = T_{H,H^1} \circ V^1$. On the other hand, $U^1 = U \circ T_{H,H^1}$. So

$$U \circ V = U^1 \circ T_{H^1,H} \circ T_{H,H^1} \circ V^1 = \mathbb{I},$$

according to the choice of our normalization. This completes the proof of the theorem.

□

3.2.4 Examples

Example 3.2.12 We take the Heisenberg algebra $\mathfrak{g} = \mathbb{R}\text{-span}(X, Y, Z)$ where $[X, Y] = Z$, and denote by G the associated Lie group. Let $\mathfrak{h} = \mathbb{R}\text{-span}(X - Y)$, $H = \exp\mathfrak{h}$ and $f = 0$. We consider $\tau = \mathrm{ind}_H^G 1$ and take the family $\{X - Y, Y, Z\}$ as a Jordan–Hölder basis of $\mathfrak{g}$. We get the Malcev sequence

$$\mathfrak{h} = \mathbb{R}\text{-span}(X - Y) \subset \mathfrak{g}_1 = \mathbb{R}\text{-span}(X - Y, Z) \subset \mathfrak{g}.$$

It follows in this example that $V^{R,\mathscr{B}} = \mathscr{V} = \mathbb{R}Z^\star$, $\mathscr{V}_0 = \{\lambda Z^\star, \lambda \neq 0\}$. For all $\phi \in \mathscr{V}_0$, $\mathfrak{b}(\phi) = \mathbb{R}\text{-span}(Y, Z)$ is the rational polarization in ϕ. Let U be the intertwining operator of Theorem 3.2.10. We have for $\xi \in S(\mathbb{R}^2)$ and $g \in G$

$$U(\xi)(\lambda Z^\star)(g) = \int_{\mathbb{R}^2} \xi(g \exp(tY) \exp(sZ)) e^{-is\lambda} ds dt.$$

Let us show directly that this operator is isometric. Indeed,

$$\begin{aligned}
\|U(\xi)\|_2^2 &= \int_{\mathscr{V}} \|U(\xi)(\phi)\|^2_{L^2(G/B(\phi),\phi)} d\lambda(\phi) \\
&= \int_{\mathbb{R}^2} \left| \int_{\mathbb{R}^2} \xi(\exp(\frac{a}{\lambda}(X - Y)) \exp(tY) \exp(sZ)) e^{-is\lambda} ds dt \right|^2 da d\lambda.
\end{aligned}$$

Since $\exp(\frac{a}{\lambda}(X - Y)) \exp(tY) = \exp(tY) \exp(\frac{a}{\lambda}(X - Y)) \exp(\frac{a}{\lambda} tZ)$, we get

$$\begin{aligned}
\|U(\xi)\|_2^2 &= \int_{\mathbb{R}^2} \left| \int_{\mathbb{R}^2} \xi(\exp(tY) \exp((s + \frac{ta}{\lambda}) Z) e^{-is\lambda} ds dt \right|^2 da d\lambda \\
&= \int_{\mathbb{R}^2} \left| \int_{\mathbb{R}^2} \xi(\exp(tY) \exp(sZ)) e^{-is\lambda} e^{-it(-a)} ds dt \right|^2 da d\lambda \\
&= \int_{\mathbb{R}^2} |\widehat{\xi}(\exp(aY)) \exp(\lambda Z))|^2 da d\lambda
\end{aligned}$$

$\widehat{\xi}$ being the Fourier transform of ξ

$$= \int_{\mathbb{R}^2} |\xi(\exp(tY) \exp(sZ))|^2 dt ds = \|\xi\|^2_{L^2(G/H,f)},$$

according to Plancherel's formula. The inverse operator U^{-1} is given for every $K \in C_c^\infty(\mathscr{V}_0; C_c)$ and $g \in G$ by

$$U^{-1}(K)(g) = \int_{\mathbb{R}} \int_{\mathbb{R}} K(\lambda Z^\star)(g \cdot \exp(r \frac{(X - Y)}{\lambda})) dr d\lambda.$$

Example 3.2.13 Let $\mathfrak{g}$ be the algebra with generators X_1, X_2, X_3, X_4 such that

$$[X_1, X_2] = X_3 \text{ and } [X_1, X_3] = X_4,$$

and take the related Jordan–Hölder basis $\mathscr{C} = \{X_1, X_2, X_3, X_4\}$. Let $\mathfrak{b} = \mathbb{R}\text{-span}(X_2, X_3, X_4)$ and G the associated Lie group to $\mathfrak{g}$.

I. Let $\mathfrak{h} = \mathbb{R}\text{-span}(X_4)$ and $H = \exp(\mathfrak{h})$ and $f_a = aX_4^\star$ a linear form on $\mathfrak{h}$ with $a \neq 0$. We adop the following Malcev basis relative to $\mathfrak{h}$:

$$\begin{aligned}\mathfrak{h} = \mathbb{R}\text{-span}(X_4) \subset \mathfrak{g}_1 = \mathbb{R}\text{-span}(X_3, X_4) \subset \mathfrak{g}_2 = \mathbb{R}\text{-span}(X_2, X_3, X_4)\\ \subset \mathfrak{g}_3 = \mathbb{R}\text{-span}(X_1, X_2, X_3, X_4) = \mathfrak{g}.\end{aligned}$$

It is clear that almost all G-orbits are saturated with respect to $\mathfrak{g}_2^\perp$, while almost all H-orbits are not saturated. So,

$$\mathscr{V} = \mathscr{V}^{R,\mathscr{B}} = \{\phi \in f_a + \mathfrak{h}^\perp \; : \langle\phi, X_1\rangle = 0\} = aX_4^\star + \mathbb{R}\text{-span}(X_3^\star, X_2^\star) \simeq \mathbb{R}^2.$$

For $\phi \in \mathscr{V}$, the subalgebra $\mathfrak{b}$ is the Vergne polarization in ϕ for the basis $\mathscr{C}$. Our intertwining operator is given, for $\xi \in S(\mathbb{R}^3; f_a)$, by

$$U(\xi)(aX_4^\star + bX_3^\star + cX_2^\star)(g) = \int_{\mathbb{R}^2} \xi(g \exp tX_3 \exp sX_2)e^{-itb}e^{-isc}dtds, \;\; g \in G.$$

The inverse operator U^{-1} is given for all $K \in C_c^\infty(\mathscr{V}_0; C_c)$ and $g \in G$ by

$$U^{-1}(K)(g) = \int_{\mathbb{R}^2} K(aX_4^\star + tX_3^\star + sX_2^\star)(g)dtds.$$

II. Let $\mathfrak{h} = \mathbb{R}\text{-span}(X_3)$ and $H = \exp(\mathfrak{h})$ and $f_v = vX_3^\star$, $V \neq 0$. Our Jordan–Hölder basis $\mathscr{C}$ gives us the Malcev sequence relative to $\mathfrak{h}$

$$\begin{aligned}\mathfrak{h} = \mathbb{R}\text{-span}(X_3) \subset \mathfrak{g}_1 = \mathbb{R}\text{-span}(X_3, X_4) \subset \mathfrak{g}_2 = \mathbb{R}\text{-span}(X_3, X_4, X_2)\\ \subset \mathfrak{g}_3 = \mathbb{R}\text{-span}(X_3, X_4, X_2, X_1) = \mathfrak{g}.\end{aligned}$$

Almost all H-orbits in general position are saturated with respect to $\mathfrak{g}_2^\perp$, as are G-orbits. Therefore

$$\mathscr{V} = \mathscr{V}^{R,\mathscr{B}} = (vX_3^\star + \mathfrak{h}^\perp) \cap \{\phi; \langle\phi, X_1\rangle = 0\} = vX_3^\star + \mathbb{R}\text{-span}(X_4^\star, X_2^\star) \simeq \mathbb{R}^2.$$

Let $\phi_{c,d} = vX_3^\star + cX_4^\star + dX_2^\star \in \mathscr{V}$ and set $\mathscr{V}_0 = \{\phi_{c,d},\ cd \neq 0\}$. The subalgebra $\mathfrak{b}$ is the Vergne polarization in $\phi_{c,d} \in \mathscr{V}_0$ for the base $\mathscr{C}$. Our intertwining operator is therefore given for all $\xi \in S(\mathbb{R}^3, vX_3^\star)$ by

$$U(\xi)(\phi_{c,d})(g) = \int_{\mathbb{R}^2} \xi(g \exp tX_4\ \exp\ sX_2) e^{-itc} e^{-isd} dtds$$

for all $g \in G$. The inverse operator U^{-1} is given for all $K \in C_c^\infty(\mathscr{V}; C_c)$ and $g \in G$ by

$$U^{-1}(K)(g) = \int_{\mathbb{R}^2} K(\phi_{s,t})(g) dsdt.$$

III. We now take $\mathfrak{h} = \mathbb{R}\text{-span}(X_2)$, $H = \exp(\mathfrak{h})$ and $f = X_2^\star$. A Malcev sequence relative to $\mathfrak{h}$ is such that

$$\mathfrak{h} = \mathbb{R}\text{-span}(X_2) \subset \mathfrak{g}_1 = \mathbb{R}\text{-span}(X_2, X_4) \subset \mathfrak{g}_2 = \mathbb{R}\text{-span}(X_2, X_4, X_3) \subset \mathfrak{g}_3 = \mathfrak{g}.$$

Almost all H-orbits in general position are saturated with respect to $\mathfrak{g}_2^\perp$, as are G-orbits, so

$$\mathscr{V} = \mathscr{V}^{R,\mathscr{B}} = (f + \mathfrak{h}^\perp) \cap \{\phi;\ \langle \phi, X_1 \rangle = 0\} = X_2^\star + \mathbb{R}\text{-span}(X_4^\star, X_3^\star).$$

Let $\phi_{c,d} = X_2^\star + cX_3^\star + dX_4^\star \in \mathscr{V}_0$, the subalgebra $\mathfrak{b}$ is the Vergne polarization in $\phi_{c,d}$ for the basis $\mathscr{C}$. Our intertwining operator is therefore given for $\xi \in S(\mathbb{R}^3, f)$ by

$$U(\xi)(\phi_{c,d})(g) = \int_{\mathbb{R}^2} \xi(g \exp tX_3 \exp sX_4) e^{-itc} e^{-isd} dtds,\quad g \in G.$$

The inverse operator U^{-1} is given for all $K \in C_c^\infty(\mathscr{V}_0; C_c)$ and $g \in G$ by

$$U^{-1}(K)(g) = \int_{\mathbb{R}^2} K(X_2^\star + sX_3^\star + rX_4^\star)(g) dsdr.$$

IV. We finally treat the case where $\mathfrak{h} = \mathbb{R}\text{-span}(X_1)$, $f = 0$. Take the Malcev sequence

$$\mathfrak{g}_1 = \mathbb{R}\text{-span}(X_1, X_4) \subset \mathfrak{g}_2 = \mathbb{R}\text{-span}(X_1, X_4, X_3) \subset \mathfrak{g}_3 = \mathbb{R}\text{-span}(X_1, X_4, X_3, X_2) = \mathfrak{g}$$

relative to $\mathfrak{h}$. Here $\mathscr{V} = \mathscr{V}^{R,\mathscr{B}} = \mathbb{R}\text{-span}(X_4^\star, X_2^\star)$, and $\mathscr{V}_0 = \{cX_4^\star + dX_2^\star, cd \neq 0\}$. Again, the algebra $\mathfrak{g}_2$ is the Vergne polarization of any element of $\mathscr{V}_0$ for the base $\mathscr{C}$. Thus our intertwining operator is given for $\xi \in S(\mathbb{R}^3, 0)$ by

$$U(\xi)(cX_4^\star + dX_2^\star)(g) = \int_{\mathbb{R}^3} \xi(g \exp(tX_2) \exp(sX_3) \exp(rX_4)) e^{-i(cr+td)} drdsdt.$$

The inverse operator U^{-1} is given for all $K \in C_c^\infty(\mathscr{V}_0, C_c)$ and $g \in G$ by

$$V(K)(g) = \int_{\mathbb{R}^3} K(sX_4^\star + tX_2^\star)(g\exp(rX_1))drdsdt.$$

Rational Disintegration of $L^2(G)$

Let G be a connected and simply connected nilpotent Lie group with Lie algebra $\mathfrak{g}$. We know that $\mathrm{ind}_{\{e\}}^G 1$ is the left-regular representation λ_G on $L^2(G)$:

$$\lambda_G(g)\xi(x) = \xi(g^{-1}x),$$

for $g, x \in G$ and $\xi \in L^2(G)$. Here then $f = 0$ and $\Gamma_f = \mathfrak{g}^\star$. Let

$$(0) = \mathfrak{a}^{n+1} \subset \mathfrak{a} \subset \ldots \subset \mathfrak{a} = \mathfrak{g} \tag{3.2.18}$$

be a Jordan–Hölder sequence of $\mathfrak{g}$ and let $\mathscr{C} = \{Z_1, \ldots, Z_n\}$ an extracted Jordan–Hölder basis of $\mathfrak{g}$ in such a way that $Z_j \in \mathfrak{a}^j \backslash \mathfrak{a}^{j+1}$. We put $A^j = \exp(\mathfrak{a}^j)$, $j = 1, \ldots, n$. In this case

$$I^{\{e\}} = \{j \in \{1, \ldots, n\} : \text{almost all } A^{j-1} - \text{orbits are saturated with respect to } (\mathfrak{a}^j)^\perp\}$$

It follows that our disintegration space is described as

$$\mathscr{V} = \{\phi \in \mathfrak{g}^\star : \langle \phi, X_j \rangle = 0, \ j \in I^{\{e\}}\}.$$

Let $d\lambda$ denote the Lebesgue measure on $\mathscr{V}$. Let $\phi \in \mathscr{V}$, write $\phi_i = \phi_{|\mathfrak{a}^i}$ and call $B(\phi) = \exp(\mathfrak{b}(\phi))$ the Vergne polarization at ϕ. Moreover, we know from Pukanszky's condition that $Ad^\star(B(\phi))\phi = \phi + \mathfrak{b}(\phi)^\perp$, so it is clear that $Ad^\star(B(\phi))\phi \cap \mathscr{V} = \{\phi\}$. We get the following disintegration of $L^2(G)$:

$$(L^2(G), \mu_G) \simeq \int_{\mathscr{V}}^{\oplus} (L^2(G/B(\phi), \phi))d\lambda(\phi).$$

Example 3.2.14 We show here that the choice of polarization $B(\phi)$ for $\phi \in \mathscr{V}_0$ is relevant in the construction of the intertwining operator. In fact, let $\mathfrak{g}$ be the Lie algebra generated by the vectors X, P, Y, Q, Z with the following brackets:

$$[X, Y] = Z, [P, Q] = Z.$$

Let $\mathfrak{h} = \mathbb{R}\text{-span}(Z, P - X)$, $f = Z^\star$ and $\mathscr{C} = \{X, P - X, Y, Q, Z\}$ a Jordan–Hölder basis of $\mathfrak{g}$. Therefore $\mathscr{B} = \{Q, Y, X\}$ is the Malcev basis of $\mathfrak{g}$ relative to $\mathfrak{h}$.

The set L^H consists only of a single element, which is associated with Y. Take $\mathscr{V} = Z^\star + \mathbb{R}Y^\star$ and for all $s \in \mathbb{R}$, $\phi_s = Z^\star + sY^\star$. The subalgebra $\mathfrak{b} = \mathbb{R}\text{-span}(Z, X, P)$ is a polarization for all the $\phi_s \in \mathscr{V}$, but it is not a Vergne polarization at ϕ_s with respect to $\mathscr{C}$. Let $B = \exp(\mathfrak{b})$, then the infinitesimal intertwining operator U is defined for any $\xi \in S(G/H, f) = S(\mathbb{R}^3, Z^\star)$ and $g \in G$ by

$$U(\xi)(\phi_s)(g) = T_{B,H}\xi(g) = \int_{B/B\cap H} \xi(gb)\chi_{\phi_s}(b)d_{B,B\cap H}(b) = \int_{\mathbb{R}} \xi(g\exp(pP))dp.$$

Notice U does not depend on $s \in \mathbb{R}$, and therefore U cannot be an isometry.

3.3 The Case of Exponential Solvable Groups

We study in this section exponential solvable Lie groups $G = \exp\mathfrak{g}$ for which the inducing subgroup $H = \exp\mathfrak{h}$ is normal. We still consider a monomial representation $\tau = \mathrm{ind}_H^G \chi_f$, where χ_f denotes a character of H. Starting from a good sequence of subalgebras $\mathfrak{s} = (\mathfrak{a}_i)_{i=0}^n$ passing through $\mathfrak{h}$, we determine an affine subspace V of Γ_f and a measure $d\lambda$ on V such that

$$\tau \simeq \int_V^{\oplus} \pi_\phi d\lambda(\phi),$$

where π_ϕ are the irreducible representations associated to ϕ. We next construct an explicit unitary intertwining operator and we find its inverse. The construction of such an operator U goes through the following steps. We start from the good sequence $\mathfrak{s}$, we construct a coexponential basis B of $\mathfrak{h}$ in $\mathfrak{g}$, we obtain our disintegration space V with Lebesgue measure $d\lambda$. The basis B defines an invariant measure on G/H, hence the norm on the space $\mathscr{H}_\tau$ of τ. We build next a Zariski open set V_0 of V, and for each $\phi \in V_0$ the Vergne polarization $\mathfrak{b}(\phi)$ at ϕ relative to the good sequence $\mathfrak{s}$. These polarizations obviously contain $\mathfrak{h}$. We determine then for all $\phi \in V_0$ a coexponential basis $X(\phi)$ of $\mathfrak{b}(\phi)$ in $\mathfrak{g}$, which fixes a G-invariant positive form $\nu_{G,B(\phi)}$ on the space $K(G, B(\phi)) := \mathscr{E}(G, B(\phi))$ as in Sect. 1.2.2. The latter is the space of continuous numerical functions on G, with compact support modulo $B(\phi) = \exp(\mathfrak{b}(\phi))$ and satisfying:

$$F(gb) = \Delta_{B(\phi),G}(b)F(g) \quad (g \in G,\ b \in B(\phi)),$$

Then we have a norm on the space $\mathscr{H}_\phi$ of the irreducible representation $\pi_\phi = \mathrm{ind}_{B(\phi)}^G \chi_\phi$. We also construct a coexponential basis $Y(\phi)$ to $\mathfrak{h}$ in $\mathfrak{b}(\phi)$ then an invariant measure $d_{B(\phi),H}$ on $B(\phi)/H$. All these bases vary continuously with $\phi \in V_0$ and allow to define the set

$$\mathscr{H} = \int_V^{\oplus} \mathscr{H}_\phi d\lambda(\phi)$$

of the disintegration of τ. Now we associate, to each smooth function ξ on G with compact support modulo H satisfying the generalized covariance relation (1.2.2):

$$\xi(gh) = \chi_f(h^{-1})\Delta_{H,G}^{1/2}(h)\xi(g), \ \ (g \in G,\ h \in H) \tag{3.3.1}$$

and to each $\phi \in V_0$, the C^∞-vector

$$T_{\mathfrak{b}(\phi),\mathfrak{h}}\xi(g) = \int_{B(\phi)/H} \xi(gb)\chi_\phi(b)\Delta_{B(\phi),G}^{-1/2}(b)d_{B(\phi),H}(b), \ \ g \in G \tag{3.3.2}$$

of $\mathscr{H}_\phi$. We prove next that this operator is invertible. We will also examine some examples and describe a smooth disintegration of $L^2(G)$ for an exponential solvable Lie group G.

3.3.1 A Base Space of the Disintegration of Induced Representations

We fix in the whole section an exponential solvable Lie group $G = \exp \mathfrak{g}$ with Lie algebra $\mathfrak{g}$. Let f be an element of $\mathfrak{g}^\star$ and $H = \exp \mathfrak{h}$ a normal subgroup of G. Recall the monomial representation $\tau = \mathrm{ind}_H^G \chi_f$, which is realized by left translations in the Hilbert space $\mathscr{H}_\tau$ of continuous functions ξ on G that satisfy the covariance relation (3.3.1) for all g in G and h in H and are square-integrable on G/H for the G-invariant measure. A result on the disintegration of τ was obtained earlier (cf. Theorem 1.4.2). We first recall the precise disintegration formula:

Theorem 3.3.1 *Let $G = \exp \mathfrak{g}$ be an exponential solvable group and $H = \exp \mathfrak{h}$ a normal subgroup of G. Then*

$$\tau \simeq \int_{f+\mathfrak{h}^\perp/H}^{\oplus} \pi_l d\mu(l), \tag{3.3.3}$$

where μ is the image under the Kirillov–Bernat map of a finite positive measure on $\Gamma_f \subset \mathfrak{g}^\star$ equivalent to the Lebesgue measure. On the other hand, the multiplicities involved in this decomposition are identically 1 *or* $+\infty$, *depending on whether*

$$\dim(H \cdot l) = \dim(G \cdot l \cap \Gamma_f)$$

or not, for l generic in Γ_f. Equivalently, we might have

$$2\dim(H \cdot l) = \dim(G \cdot l)$$

or not. In either case the multiplicity of π_l in τ is the number of H-orbits in $G{\cdot}l \cap \Gamma_f$.

We now fix $\mathfrak{s} = (\mathfrak{a}_j)_{j=0}^n$, a good sequence of subalgebras of $\mathfrak{g}$ passing through the ideal $\mathfrak{h}$ (said to be adapted to $\mathfrak{h}$). We can extract a Malcev basis $\{Z_1, \ldots, Z_n\}$ of $\mathfrak{g}$ satisfying, for all $i \in \{1, \ldots, n\}$,

$$\mathfrak{a}_i = \mathbb{R}\text{-span}(Z_1, \ldots, Z_i)$$

and $B = \{X_1, \ldots, X_r\} = \{Z_{p+1}, \ldots, Z_n\}$, $(r = n - p)$, is a Malcev basis relative to $\mathfrak{h}$ (which is at the same time a coexponential basis of $\mathfrak{h}$ in $\mathfrak{g}$). As in Sect. 1.2.2, we use B to determinate a G-invariant measure on G/H, then to fix the norm on the space of the space τ. Recall that this G-invariant measure (that we will call $d_{G,H}$) will be defined as follows: for any continuous function with compact support $\tilde{a}$ on G/H, we have

$$\int_{G/H} \tilde{a}(g) d_{G,H}(g) = \int_{\mathbb{R}^r} \tilde{a}(\exp(t_r X_r) \ldots \exp(t_1 X_1)) dt_r \ldots dt_1. \tag{3.3.4}$$

We note $X = X_r$, $\mathfrak{g}_0 = \mathfrak{a}_{n-1}$ and $\mathfrak{s}_0 = (\mathfrak{a}_j)_{j=0}^{n-1}$ which is in fact a good sequence of subalgebras of $\mathfrak{g}_0$ adapted to $\mathfrak{h}$. We note likewise that the basis $B_0 = \{X_1, \ldots, X_{r-1}\}$ is coexponential to $\mathfrak{h}$ in $\mathfrak{g}_0$ and is defined as in (3.3.4). Moreover, there is a G_0-invariant measure $d_{G_0,H}$ on G_0/H, where $G_0 = \exp(\mathfrak{g}_0)$.

Call $B^\star = \{f_1, \ldots, f_r\} \subset \mathfrak{h}^\perp$ the dual basis and $A_j = \exp(\mathfrak{a}_j)$. Let $\phi_j = \phi_{|\mathfrak{a}_j}$, for $j = 0, \ldots, n$ and $\phi \in \Gamma_f$. Put

$$d_j(\phi) = \dim(A_j \cdot \phi_j) = \mathrm{rk}\left((< \phi, [Z_i, Z'_i] >)_{1 \le i, i' \le j}\right) \text{ and}$$

$$d_j = \max_{\phi \in \Gamma_f} d_j(\phi),\ d_0 = 0.$$

Let also

$$L^H = \{j \in \{1, \ldots, r\}, d_{j+p} = d_{j-1+p}\} \text{ and}$$

$$K^H = \{1, \ldots, r\} \setminus L^H = \{j \in \{1, \ldots, r\}, d_{j+p} > d_{j-1+p}\}.$$

It is clear that for all j in K^H, almost all A_{j+p}-orbits in $\Gamma_f^{j+p} = \Gamma_{f|\mathfrak{a}_j+p}$ are saturated with respect to $\mathfrak{a}_{j-1+p}$. We denote by Γ_f^0 the set of elements $\phi \in \Gamma_f$ satisfying

$$d_{j+p}(\phi) = d_{j+p}, \forall j \in \{1, \ldots, r\}.$$

Then Γ_f^0 is a Zariski open set in Γ_f. The elements of Γ_f^0 will be called in general position. It is clear that we can always take f in Γ_f^0 and then $V' = V \cap \Gamma_f^0$ is a non-empty Zariski open set in V where

$$V = \{f + \sum_{i \in L^H} t_i f_i,\ t_i \in \mathbb{R}\} \subset \Gamma_f. \tag{3.3.5}$$

Let $d\lambda$ be the Lebesgue measure on V.

3.3.2 *Construction of the Intertwining Operator*

Several Constructions

Let $\mathfrak{s} = (\mathfrak{a}_j)_{j=0}^n$ be a good sequence of subspaces of $\mathfrak{g}$ adapted to $\mathfrak{h}$, and extract a coexponential basis $B = \{X_1, \ldots, X_r\}$ to $\mathfrak{h}$ in $\mathfrak{g}$. Consider also the disintegration space V endowed with the Lebesgue measure $d\lambda$ as in Sect. 3.3.1 and formula (3.3.5). The basis B defines an invariant measure on G/H, which allows to fix the norm

$$\|\xi\|_{L^2(G/H,f)} = (\int_{G/H} |\xi(g)|^2 d_{G,H}(g))^{1/2}$$

on $\mathscr{H}_\tau = L^2(G/H, f)$ of τ. We now build a Zariski open set V_0 of V and for $\phi \in V_0$ the Vergne polarization $\mathfrak{b}(\phi)$ at ϕ relatively to the good sequence $\mathfrak{s}$ as in Theorem 1.2.4. Since this good sequence is adapted to $\mathfrak{h}$, we must have $\mathfrak{h} \subset \mathfrak{b}(\phi)$ for all $\phi \in V_0$. We next construct, for $\phi \in V_0$, a coexponential basis $X(\phi)$ to $\mathfrak{b}(\phi)$ in $\mathfrak{g}$ and a coexponential basis $Y(\phi)$ to $\mathfrak{h}$. All these bases vary continuously on V_0. For $\phi \in V'$ and $j = 1, \ldots, n$, we set

$$J_j(\phi) = \{k \in \{1, \ldots, j\} : \mathfrak{a}_j(\phi_j) + \mathfrak{a}_{k-1} \subsetneq \mathfrak{a}_j(\phi_j) + \mathfrak{a}_k\}.$$

The set of indices $J_j(\phi)$ is typically not constant for $\phi \in V'$, but its cardinality is constant and equal to d_j for all $j = 1, \ldots, n$. Note then

$$J_j(\phi) = \{i_1(\phi) < \cdots < i_{d_j}(\phi)\},$$

We endow the set $\{J_j(\phi), \phi \in V'\}$ with the lexicographic order defined by

$$\{i_1(\phi) < \cdots < i_{d_j}(\phi) < i_1(\phi') < \cdots < i_{d_j}(\phi')\},$$

if there exists $\sigma \in \{1, \ldots, d_j\}$ such that $i_1(\phi) = i_1(\phi'), \ldots, i_{\sigma-1}(\phi'), i_\sigma(\phi) < i_\sigma(\phi')$. Using this order, let

$$J_j = \min_{\phi \in V'} J_j(\phi) = \{i_1 < \cdots < i_{d_j}\}$$

$j = 1, \ldots, n$. Let V_0 be the set of elements of V' such that for all $j = 1, \ldots, n$, $J_j(\phi) = J_j$ and

$$Q_j(\phi) = \det\left((\langle\phi, [Z_i, Z_{i'}]\rangle)_{i,i' \in J_j}\right)$$

is a non-zero polynomial. It is clear that V_0 is a non-empty Zariski open set in V. For $\phi \in V_0$ and $k = 1, \ldots, n$, we set

$$M_k(\phi) = \left((\langle\phi, [Z_i, Z_{i'}]\rangle)_{1 \leq i,i' \leq k}\right), \; M_0(\phi) = 0,$$

and for all $j = 1, \ldots, n$,

$$I_j(\phi) = \{k \in \{1, \ldots, j\} \, : \, \mathfrak{a}_k(\phi_k) \not\subseteq \mathfrak{a}_{k-1}\}$$

$$\{k \in \{1, \ldots, j\} \, : \, \mathrm{rk}(M_k(\phi)) = \mathrm{rk}(M_{k-1}(\phi))\}.$$

Now the sets $I_j(\phi)$ do not depend upon ϕ, ie $I_j(\phi) = I_j$ for all $\phi \in V_0$. Let now $B_j(\phi) = Z_j$, for $j \leq p$ and

$$B_j(\phi) = Z_j + \sum_{k=1}^{j-1} a_k(\phi) Z_k, \text{ for } j > p \text{ and } j \in I_n \tag{3.3.6}$$

where a_k are the non-singular rational functions on V_0 solutions to

$$\langle\phi, [Z_j, Z_l]\rangle = \sum_{k=1}^{j-1} a_k(\phi) \langle\phi, [Z_l, Z_k]\rangle, \; l \in \{1, \ldots, j\}.$$

This system admits at least one solution since ϕ is in V_0. Then it is clear that the family $(B_j(\phi))_j$ forms a Jordan–Hölder basis of the polarization of $\mathfrak{b}(\phi)$.

Lemma 3.3.2 *For all $\phi \in V_0$, the vectors Z_k, $k \notin I_j$ for all $j = 1, \ldots, n$ constitute a coexponential basis at $\mathfrak{b}(\phi)$ as well.*

Proof We use induction on the dimension of G. We suppose first that the orbit of ϕ is saturated with respect to $\mathfrak{g}_0$, meaning $\mathfrak{g}(\phi) \subset \mathfrak{g}_0$ and then $\mathfrak{b}(\phi) \subset \mathfrak{g}_0$. In this case $n \notin I_n$ and then $I_n = I_{n-1}$. On the other hand,

$$\mathfrak{b}(\phi) = \mathfrak{b}(\phi_0), \tag{3.3.7}$$

where $\mathfrak{b}(\phi_0)$ is the Vergne polarization in $\phi_0 = \phi_{|\mathfrak{g}_0}$ with respect to the good sequence $\mathfrak{s}_0$. The induction hypothesis tells us that the vectors Z_k, $k \in I_j$ for all $j = 1, \ldots, n-1$ constitute also a coexponential basis at $\mathfrak{b}(\phi)$ in $\mathfrak{g}_0$, and since $\{Z_n\}$ is a coexponential basis of $\mathfrak{g}_0$ in $\mathfrak{g}$ we have the result.

Let us pass to the case where the orbit of ϕ is not saturated with respect to $\mathfrak{g}_0$, which means that $\mathfrak{g}(\phi) \nsubseteq \mathfrak{g}_0$ and then $\mathfrak{b}(\phi) \nsubseteq \mathfrak{g}_0$ and

$$\mathfrak{b}(\phi_0) = \mathfrak{b}(\phi) \cap \mathfrak{g}_0 \tag{3.3.8}$$

is the Vergne polarization in $\phi_0 = \phi_{|\mathfrak{g}_0}$ with respect to the good sequence $\mathfrak{s}_0$. In this case $n \in I_n$. The induction hypothesis tells us that the vectors Z_k, $k \in I_j$ $(j = 1, \ldots, n-1)$ constitute also a coexponential basis of $\mathfrak{b}(\phi_0)$ in $\mathfrak{g}_0$ which is a coexponential basis at $\mathfrak{b}(\phi)$ in $\mathfrak{g}$. □

Let $\{Z_{i_1}, \ldots, Z_{i_q}\}$, $(i_1 < \cdots < i_q)$ be the coexponential basis at $\mathfrak{b}(\phi)$ in $\mathfrak{g}$ determined in Lemma 3.3.6 and $B(\phi) = \exp(\mathfrak{b}(\phi))$. For $F \in K(G, B(\phi))$, the integral

$$\int_{\mathbb{R}^q} F(\exp(t_q Z_{i_q}) \cdots \exp(t_1 Z_{i_1}))dt_q \cdots dt_1. \tag{3.3.9}$$

defines a G-invariant positive linear form on $K(G, B(\phi))$, denoted by

$$\oint_{G/B(\phi)} F(g)d\nu_{G,B(\phi)}(g).$$

This form fixes the norm on the space $\mathscr{H}_\phi$ of the irreducible representation $\pi_\phi = \mathrm{ind}_{B(\phi)}^G \chi_\phi$. The vectors $\{B_j(\phi), j > p, j \in I_n\} = \{B_{j_1}(\phi), \ldots, B_{j_m}(\phi)\}$ constitute the coexponential basis $Y(\phi)$ for $\mathfrak{h}$ in $\mathfrak{b}(\phi)$ which defines, as in (3.3.4), a $B(\phi)$-invariant measure $d_{B(\phi),H}$ on the space $B(\phi)/H$. As $\mathfrak{b}(\phi) \subset \mathfrak{g}_0$, the vectors Z_k, $(k \notin I_j, j = 1, \ldots, n-1)$ constitute also a coexponential basis $X_0(\phi)$ to $\mathfrak{b}(\phi) = \mathfrak{b}(\phi_0)$ in virtue of (3.3.7), the polarization of ϕ_0 in $\mathfrak{g}_0$ and then a positive form $\nu_{G_0,B(\phi)}$ on $K(G_0, B(\phi_0))$ where $B(\phi_0) = \exp(\mathfrak{b}(\phi_0))$. When $\mathfrak{b}(\phi)$ in not contained in $\mathfrak{g}_0$, then $n \in I_n$, and as in (3.3.8) $\mathfrak{b}(\phi_0) = \mathfrak{b}(\phi) \cap \mathfrak{g}_0$ is the Vergne polarization in $\phi_0 = \phi_{|\mathfrak{g}_0}$ with respect to the good sequence $\mathfrak{s}_0$. We have then

$$B(\phi) = B(\phi_0) \cdot \exp(\mathbb{R}B_n(\phi)),\ \mathfrak{g} = \mathfrak{g}_0 \oplus \mathbb{R}B_n(\phi),$$

and then $G/B(\phi) \simeq G_0/B(\phi_0)$. We have then two positive forms $\nu_{G,B(\phi)}$ on $K(G, B(\phi))$ and $\nu_{G_0,B(\phi_0)}$ on $K(G_0, B(\phi_0))$. The family of vectors

$$\{B_j(\phi), j > p, j \in I_{n-1}\} = \{B_{j_1}(\phi), \ldots, B_{j_{m-1}}(\phi)\}$$

constitutes a coexponential basis $Y_0(\phi)$ of $\mathfrak{h}$ in $\mathfrak{b}(\phi_0)$. Likewise, this basis defines as in (3.3.4) a $B(\phi_0)$-invariant measure $d_{B(\phi_0),H}$ on the homogeneous space $B(\phi_0)/H$.

In view of the fact $\mathfrak{h} \subset \mathfrak{b}(\phi)$ for all $\phi \in V_0$, we have that

$$\Delta_{B(\phi),G}(h) = 1\ (h \in H).$$

However, if ξ is a vector of $C_c^\infty(G/H, f)$ (the space of C^∞ vectors with compact support modulo H) and g an element of G, the function Φ_g on $B(\phi)$ defined by:

$$\Phi_g(b) = \xi(gb)\chi_\phi(b)\Delta_{B(\phi),G}^{-1/2}(b)$$

satisfies the relation:

$$\Phi_g(bh) = \Phi_g(b) \ (h \in H,\ b \in B(\phi)).$$

So we are able to write the formal integral (3.3.2) where $d_{B(\phi),H}$ is the invariant measure on the homogeneous space $B(\phi)/H$ determined in Sect. 1.2.2. At least at the formal level, it is clear that $T_{\mathfrak{b}(\phi),\mathfrak{h}}\xi$ satisfies condition (3.3.1) for ϕ, B_ϕ and G required to belong to $\mathscr{H}(\phi, \mathfrak{b}(\phi), G)$ and that $T_{\mathfrak{b}(\phi),\mathfrak{h}}$ commutes with the action of G by left translations. Furthermore, integral (3.3.2) converges on $C_c^\infty(G/H, f)$. It is therefore no exaggeration to say that the only problem is to prove that (3.3.2) is in fact an isometric intertwining integral in τ and its decomposition into irreducibles provides an suitable measure normalization on homogeneous spaces. For $\xi \in C_c^\infty(G/H, f)$, we first prove that the element $T_{\mathfrak{b}(\phi),\mathfrak{h}}\xi$ is a C^∞ vector for the representation $\pi_\phi = \mathrm{ind}_{B(\phi)}^G \chi_\phi$.

Proposition 3.3.3 *Let $\phi \in V_0$ and suppose $B(\phi) = \exp(\mathfrak{b}(\phi))$ is the Vergne polarization at ϕ with respect to the good sequence $\mathfrak{s}$. Then the vector $T_{\mathfrak{b}(\phi),\mathfrak{h}}\xi$ is a C^∞-vector of the representation π_ϕ for all $\xi \in C_c^\infty(G/H, f)$.*

Proof For $\phi \in V_0$, the function

$$(t_1, \ldots, t_q) \mapsto \int_{B(\phi)/H} \xi(\exp(t_q Z_{i_q}) \cdots \exp(t_1 Z_{i_1}) \cdot b)\chi_\phi(b)\Delta_{B(\phi),G}^{-1/2}(b) d_{B(\phi),H}(b)$$

has compact support in $\mathbb{R}^q$. Hence $T_{\mathfrak{b}(\phi),\mathfrak{h}}\xi$ is C^∞ on G and its support is compact modulo $B(\phi)$. □

Similarly to Definition 3.2.9, let V be as in (3.3.5) and for all $\phi \in V_0$, $B(\phi)$ as in Sect. 3.3.2. We call $C(V_0, L^2)$ the set of maps

$$\psi : V_0 \to \bigcup_{\phi \in V_0} L^2(G/B(\phi), \phi)$$

such that $(\phi, g) \mapsto \psi(\phi)(g)$ are continuous on $V_0 \times G$, $\psi(\phi) \in L^2(G/B(\phi), \phi)$ and which satisfy

$$\|\psi\|_2^2 = \int_V \|\psi(\phi)\|_{L^2(G/B(\phi),\phi)}^2 d\lambda(\phi) < \infty. \tag{3.3.10}$$

Define

$$L^2(V, L^2) = \int_V^{\oplus} L^2(G/B(\phi), \phi) d\lambda(\phi)$$

as the completion of the space $C(V_0, L^2)$ with respect to the norm $\| \|_2$.
Let $\{Z_{i_1}, \dots, Z_{i_q}\}$ $(i_1 < \cdots < i_q)$ be the coexponential basis of $\mathfrak{b}(\phi)$ in $\mathfrak{g}$ for all $\phi \in V_0$ determined in Lemma 3.3.2. We define $C_c^\infty(V_0, C_c)$ as the subspace of $C(V_0, L^2)$ formed by elements ψ such that $(\phi, g) \mapsto \psi(\phi)(g)$ is C^∞ on $V_0 \times G$ and such that the map

$$(\phi, t_1, \dots, t_q) \mapsto \psi(\phi)(\exp(t_q Z_{i_q}) \cdots \exp(t_1 Z_{i_1}))$$

is C^∞ and compactly supported on $V_0 \times \mathbb{R}^q$. Relation (3.3.6) holds for all elements of $C_c^\infty(V_0, C_c)$. It is clear that the latter is dense in $L^2(V, L^2)$.
Now we are ready to prove our main result.

Theorem 3.3.4 *Let $G = \exp \mathfrak{g}$ be an exponential solvable Lie group. Let $H = \exp \mathfrak{h}$ be a normal subgroup of G, f an element of $\mathfrak{g}^\star$ such that $f([\mathfrak{h}, \mathfrak{h}]) = \{0\}$ and $\tau = \mathrm{ind}_H^G \chi_f$. Let $\mathfrak{s}$ be a good sequence of subalgebras of $\mathfrak{g}$ adapted to $\mathfrak{h}$, $V \subset \Gamma_f$ the affine subspace described in (3.3.5) and $d\lambda$ its measure. Let V_0 be the Zariski open set in V described in Sect. 3.3.2 and for $\phi \in V_0$, $B(\phi) = \exp(\mathfrak{b}(\phi))$ the Vergne polarization in ϕ with respect to $\mathfrak{s}$. Then we have*

$$\tau \simeq \int_V^{\oplus} \pi_{\phi, B(\phi)} d\lambda(\phi). \tag{3.3.11}$$

The operator U defined for all $\xi \in C_c^\infty(G/H, f)$ by:

$$U(\xi)(\phi) = T_{\mathfrak{b}(\phi), \mathfrak{h}} \xi$$

is isometric and valued in $C(V_0, L^2)$, and it extends by continuity over $\mathscr{H}(f, \mathfrak{h}, G)$ to an isometric intertwining operator of the form (3.3.11), when the measures on $G/B(\phi)$ and on $B(\phi)/H$ are suitably normalized as in Sect. 3.3.2.

Proof Note first that (3.3.11) is an immediate consequence of Theorem 3.3.1. On the other hand, it is clear that the operator U of the theorem formally intertwines (3.3.11). Let $\xi \in C_c^\infty(G/H, f)$ and $\phi \in V_0$. Let $Y(\phi) = \{B_{j_1}(\phi), \dots, B_{j_m}(\phi)\}$ be the coexponential basis of $\mathfrak{h}$ in $\mathfrak{b}(\phi)$ built above. For $g \in G$

$$U(\xi)(\phi)(g) = \int_{\mathbb{R}^m} \xi\Big(g \exp(t_m B_{j_m}(\phi)) \cdots \exp(t_1 B_{j_1}(\phi))\Big) e^{-i\phi(t_m B_{j_m}(\phi) + \cdots + t_1 B_{j_1}(\phi))}$$

$$\Delta_{B(\phi), G}^{-\frac{1}{2}}\Big(\exp(t_m B_{j_m}(\phi)) \cdots \exp(t_1 B_{j_1}(\phi))\Big) dt_m \cdots dt_1.$$

Then $(\phi, g) \mapsto U(\xi)(\phi)(g)$ is manifestly continuous on $V_0 \times G$. We will argue by induction on $\dim(G/H)$ and we will prove that U is an isometry, which leads to the fact that $U(\xi) \in C(V_0, L^2)$. Throughout the proof we write $V^0 = V_{0|\mathfrak{g}_0}$ and we will distinguish two cases: the first when for all $\phi \in V_0$ the stabilizer $\mathfrak{g}(\phi)$ is contained in $\mathfrak{g}_0$. In this case, for $\phi_0 = \phi_{|\mathfrak{g}_0}$, $\pi_\phi = \mathrm{ind}_{G_0}^G \pi_{\phi_0}$ is irreducible and then $r \in K^H$. Furthermore, the map $\phi \mapsto \phi_0 = \phi_{|\mathfrak{g}_0}$ is an isomorphism of affine spaces which respect the measures $d\lambda$ on V and $d\lambda^0$ on V^0. Likewise, the Vergne polarization $\mathfrak{b}(\phi)$ is still contained in $\mathfrak{g}_0$ for all $\phi \in V_0$ and it is equal to the Vergne polarization $\mathfrak{b}(\phi_0)$ at ϕ_0 in $\mathfrak{g}_0$ with respect to the good sequence of subalgebras $\mathfrak{s}_0$ which is adapted to $\mathfrak{h}$. Any element g in G is uniquely written as $g = \exp(tX) \cdot \mathfrak{g}_0$, with $t \in \mathbb{R}$ and $\mathfrak{g}_0 \in G_0$. For a function ξ on G, we call ξ_t the function on G_0 defined by $\xi_t(g_0) = \xi(\exp(tX)g_0)$ for all $g_0 \in G_0$. We set $f_0 = f_{|\mathfrak{g}_0}$. In view of

$$\Delta_{B(\phi_0),G}(b) = \Delta_{B(\phi_0),G_0}(b),\ b \in B(\phi_0), \tag{3.3.12}$$

we have for all $\xi \in C_c^\infty(G/H, f)$ that

$$U(\xi)(\phi)(g) = T_{\mathfrak{b}(\phi),\mathfrak{h}}\xi(g) = T_{\mathfrak{b}(\phi_0),\mathfrak{h}}\xi_t(g_0) = U_0(\xi_t)(\phi_0)(g_0). \tag{3.3.13}$$

Add to this that the measure $d_{G,H}$ on G/H, defined from the Malcev basis B, is equal to the measure defined from the Malcev basis $\{X\} \cup B_0$. By induction we then have, for all $\eta \in C_c^\infty(G_0/H, f_0)$,

$$\|\eta\|^2_{L^2(G_0/H,f_0)} = \int_{V_0} \|T_{\mathfrak{b}(\phi_0),\mathfrak{h}}\eta\|^2_{L^2(G_0/B(\phi_0),\phi_0)} d\lambda^0(\phi_0).$$

and then for all $\xi \in C_c^\infty(G/H, f)$

$$\begin{aligned}
\|U(\xi)\|_2^2 &= \int_V \|U(\xi)(\phi)\|^2_{L^2(G/B(\phi),\phi)} d\lambda(\phi) = \int_V \|T_{\mathfrak{b}(\phi),\mathfrak{h}}\xi\|^2_{L^2(G/B(\phi),\phi)} d\lambda(\phi)\\
&= \int_{\mathbb{R}}\int_{V^0} \|T_{\mathfrak{b}(\phi_0),\mathfrak{h}}\xi_t\|^2_{L^2(G_0/B(\phi_0),\phi_0)} d\lambda^0(\phi_0)dt\\
&\quad \text{(apply the induction hypothesis on } G_0)\\
&= \int_{\mathbb{R}} \|\xi_t\|^2_{L^2(G_0/H,f_0)} dt = \|\xi\|^2_{L^2(G/H,f)}.
\end{aligned}$$

We now pass to the second case, where $r \in L^H$, ie the G-orbits associated to the elements in general position are not saturated with respect to $\mathfrak{g}_0$. We have then

$$V_0 = \{\phi + sX^\star,\ s \in \mathbb{R},\ \phi \in V_0,\ \phi(X) = 0\} \simeq V^0 \times \mathbb{R}.$$

We remark in this case that for all $\phi \in V_0$, $\mathfrak{g}(\phi)$ is not contained in $\mathfrak{g}_0$, and so it is for the polarizations $\mathfrak{b}(\phi)$. As in (3.3.8),

$$\mathfrak{b}(\phi_0) = \mathfrak{b}(\phi) \cap \mathfrak{g}_0$$

is the Vergne polarization at $\phi_0 = \phi_{|\mathfrak{g}_0}$ in $\mathfrak{g}_0$ with respect to the good sequence $\mathfrak{s}_0$ and that $\mathfrak{b}(\phi) = \mathfrak{b}(\phi_0) \oplus \mathbb{R}B_n(\phi)$, where

$$B_n(\phi) = X + Z_0(\phi)$$

and $Z_0(\phi) \in \mathfrak{g}_0$ only depends on ϕ_0 by (4.1.1). For $\phi \in V_0$, we can write $\phi = \phi_0 + sX^\star$ where $\phi_0 + sX^\star$ is the linear form on $\mathfrak{g}$ whose restriction on $\mathfrak{g}_0$ takes value s at X.

Formula (3.3.12) is also true in this case and we have for $\xi \in C_c^\infty(G/H, f)$ that

$$\begin{aligned} T_{\mathfrak{b}(\phi),\mathfrak{h}}\xi(g) &= \int_{B(\phi)/H} \xi(gb)\chi_\phi(b)\Delta_{B(\phi),G}^{-1/2}(b)d_{B(\phi),H}(b) \\ &= \int_{\mathbb{R}}\int_{B(\phi_0)/H} \xi(g\exp(tB_n(\phi))b_0)\chi_{\phi_0}(b_0)e^{-its}\Delta_{B(\phi),G}^{-1/2}(\exp(tB_n(\phi))) \\ &\quad e^{-it\phi_0(Z_0(\phi))}\Delta_{B(\phi_0),G_0}^{-1/2}(b_0)d_{B(\phi_0),H}(b_0)dt \\ &= \int_{\mathbb{R}} T_{\mathfrak{b}(\phi_0),\mathfrak{h}}\xi(g\exp(tB_n(\phi)))e^{-its}\Delta_{B(\phi),G}^{-1/2}(\exp(tB_n(\phi)))e^{-it\phi_0}(Z_0(\phi))dt \end{aligned} \tag{3.3.14}$$

For all $\xi \in C_c^\infty(G/H, f)$ we have that

$$\|\xi\|^2_{L^2(G/H,f)} = \int_{\mathbb{R}} \|\xi_t\|^2_{L^2(G_0/H,f_0)}dt$$

(apply the induction hypothesis on G_0)

$$= \int_{\mathbb{R}}\int_{V^0}\oint_{G_0/B(\phi_0)} |T_{\mathfrak{b}(\phi_0),\mathfrak{h}}\xi_t(g_0)|^2 d\nu_{G_0,B(\phi_0)}(g_0)d\lambda^0(\phi_0)dt$$

$$= \int_{V^0}\int_{\mathbb{R}}\oint_{G_0/B(\phi_0)} |T_{\mathfrak{b}(\phi_0),\mathfrak{h}}\xi(\exp(tX)g_0)|^2 d\nu_{G_0,B(\phi_0)}(g_0)dtd\lambda^0(\phi_0)$$

(apply Lemma 2.2)

$$= \int_{V^0}\oint_{G/B(\phi_0)} |T_{\mathfrak{b}(\phi_0),\mathfrak{h}}\xi(g)|^2 d\nu_{G,B(\phi_0)}(g)d\lambda^0(\phi_0) \tag{3.3.15}$$

(apply again Lemma 2.2)

$$\begin{aligned} &= \int_{V^0}\oint_{G_0/B(\phi_0)}\int_{\mathbb{R}} |T_{\mathfrak{b}(\phi_0),\mathfrak{h}}\xi(g_0\exp(sB_n(\phi)))\Delta_{B(\phi),G}^{-1/2}(\exp(sB_n(\phi)))|^2 \\ &\quad ds d\nu_{G_0,B(\phi_0)}(g_0)d\lambda^0(\phi_0) \\ &= \int_{V^0}\oint_{G_0/B(\phi_0)}\int_{\mathbb{R}} |T_{\mathfrak{b}(\phi_0),\mathfrak{h}}\xi(g_0\exp(sB_n(\phi)))\Delta_{B(\phi),G}^{-1/2}(\exp(sB_n(\phi))) \\ &\quad e^{-is\phi_0(Z_0(\phi))}|^2 ds d\nu_{G_0,B(\phi_0)}(g_0)d\lambda^0(\phi_0) \end{aligned}$$

(apply the Plancherel formula)

$$= \int_{V^0} \oint_{G_0/B(\phi_0)} \int_{\mathbb{R}} | \int_{\mathbb{R}} T_{\mathfrak{b}(\phi_0),\mathfrak{h}}\xi(g_0 \exp(sB_n(\phi)))\Delta^{-1/2}_{B(\phi),G}(\exp(tB_n(\phi)))$$

$$e^{-its}e^{-it\phi_0(Z_0(\phi))}dt|^2 ds d\nu_{G_0,B(\phi_0)}(g_0)d\lambda^0(\phi_0)$$

$$= \int_{V^0} \int_{\mathbb{R}} \oint_{G/B(\phi_0+sX^\star)} |T_{\mathfrak{b}(\phi_0+sX^\star),\mathfrak{h}}\xi(g)|^2 d\nu_{G,B(\phi_0+sX^\star)}(g) ds d\lambda^0(\phi_0)$$

$$= \int_V \|T_{\mathfrak{b}(\phi),\mathfrak{h}}\xi\|^2_{L^2(G/B(\phi),\phi)} d\lambda(\phi) = \int_V \|U(\xi)(\phi)\|^2_{L^2(G/B(\phi),\phi)} d\lambda(\phi) = \|U(\xi)\|_2^2.$$

We note that the basis $X'(\phi) = X_0(\phi) \cup \{B_n(\phi)\}$ is coexponential for $\mathfrak{b}(\phi_0)$ in $\mathfrak{g}$, and it allows to build the positive form $\nu_{G,B(\phi_0)}$ with which the integral of (3.3.15) was calculated. Then U is an isometry on $C_c^\infty(G/H, f)$, which is in fact dense in $L^2(G/H, f)$. Then there exists a unique extension $\overline{U}$ of U to an isometry on the whole $\mathscr{H}(f, \mathfrak{h}, G)$. We still call this U. The desired covariance relation is required over the entire space, which completes the proof. □

3.3.3 The Inverse Operator

We now prove that formulas (3.3.13) and (3.3.14), established in the proof of the theorem on $C_c^\infty(G/H, f)$, actually hold on $L^2(G/H, f)$. We will resume the cases studied in the previous theorem.

In the first case, it is clear that $L^2(G/H, f) = L^2(G_0/H, f_0)$ and that

$$\int_V L^2(G/B(\phi), \phi)d\lambda(\phi) = \int_{V_0} L^2(\mathbb{R}, L^2(G_0/B(\phi_0), \phi_0))d\lambda^0(\phi_0).$$

Hence $U = \tilde{U}_0 \circ W$, where $W : L^2(G/H, f) \to L^2(\mathbb{R}, L^2(G_0/H, f))$ is the operator field defined by

$$W(\xi)(t)(g_0) = \xi(\exp(tX) \cdot g_0) = \xi_t(g_0), \; g_0 \in G_0$$

and $\tilde{U}_0(\xi)(t)(g_0) = U_0(\xi_t)(g_0)$.

We move to the second case, so let $\phi \in V_0$ and $\phi_0 = \phi_{|\mathfrak{g}_0}$. Then $\phi = \phi_s = \phi_0 + sX^\star$ for some $s \in \mathbb{R}$. For $\eta \in C_c^\infty(G/B(\phi_0), \phi_0)$, let η^s be the function defined on G by

$$\eta^s(g) = \int_{\mathbb{R}} \eta(g \exp(tB_n(\phi)))e^{-its}\Delta^{-1/2}_{B(\phi),G}(\exp(tB_n(\phi)))e^{-it\phi_0}(Z_0(\phi))dt, \; g \in G.$$

Then it is clear that the function η^s exists for all $s \in \mathbb{R}$, and it satisfies the required covariance relation (3.3.1) for G, $B(\phi_s)$ and ϕ_s. Since $G/B(\phi_s) \simeq G_0/B(\phi_0)$, the same arguments as in Proposition 3.3.3 lead to $\eta^s \in C_c^\infty(G/B(\phi_s)\phi_s)$ and then $\eta^s \in L^2(G/B(\phi_s)\phi_s)$. Observe that $s \mapsto \eta^s(g)$ is a Schwartz function for all $g \in G$ and then

$$\int_{\mathbb{R}} \eta^s(g) = \int_{\mathbb{R}^2} \eta(g \exp(tB_n(\phi)))e^{-its} \Delta_{B(\phi),G}^{-1/2}(\exp(tB_n(\phi)))e^{-it\phi_0(Z_0(\phi))} dtds = \eta(g).$$

On the other hand, we have that

$$\|\eta\|^2_{L^2(G/B(\phi_0),\phi_0)} = \oint_{G/B(\phi_0)} |\eta(g)|^2 d\nu_{G,B(\phi_0)}(g)$$

(apply Lemma 2.2)

$$= \oint_{G/B(\phi_s)} \int_{\mathbb{R}} |\eta(g \exp(tB_n(\phi)))|^2 \Delta^{-1}_{B(\phi_s),G}(\exp(tB_n(\phi)))dtd\nu_{G,B(\phi_s)}(g)$$

$$= \oint_{G_0/B(\phi_0)} \int_{\mathbb{R}} |\eta(g_0 \exp(tB_n(\phi)))\Delta^{-1/2}_{B(\phi),G}(\exp(tB_n(\phi)))e^{-it\phi_0(Z_0(\phi))}|^2 dtd\nu_{G_0,B(\phi_0)}(g_0)$$

(apply the Plancherel theorem)

$$= \oint_{G_0/B(\phi_0)} \int_{\mathbb{R}} \left| \int_{\mathbb{R}} \eta(g_0 \exp(tB_n(\phi)))e^{-its} \Delta^{-1/2}_{B(\phi),G}(\exp(tB_n(\phi)))e^{-it\phi_0(Z_0(\phi))} dt\right|^2 \times$$
$$\times\, dsd\nu_{G_0,B(\phi_0)}(g_0)$$

$$= \oint_{G_0/B(\phi_0)} \int_{\mathbb{R}} |\eta^s(g_0)|^2 dsd\nu_{G_0,B(\phi_0)}(g_0) = \int_{\mathbb{R}} \|\eta^s\|^2_{L^2(G/B(\phi_s),\phi_s)} ds.$$

By extension, we can deduce that

$$L^2(G/B(\phi_0), \phi_0) \simeq \int_{\mathbb{R}}^{\oplus} L^2(G/B(\phi_s), \phi_s)ds.$$

Let

$$\Psi : L^2(G/B(\phi_0), \phi_0) \to \int_{\mathbb{R}}^{\oplus} L^2(G/B(\phi_s), \phi_s)ds,\ \eta \mapsto \Psi(\eta)(s) = \eta^s.$$

Let also $\tilde{U}^0$ be the operator defined on $L^2(G/H, f)$ by

$$\tilde{U}^0(\xi)(\phi_0)(\exp(tX) \cdot g_0) = U_0(\xi_t)(\phi_0)(g_0).$$

It is not hard to verify that for almost all $\phi_0 \in V_0$, $\tilde{U}^0(\xi)(\phi_0)$ belongs to the space $L^2(G/B(\phi_0), \phi_0)$. We have then on $L^2(G/H, f)$ that

$$U = \Psi \circ \tilde{U}^0.$$

The next paragraph consists in proving that the operator U is invertible.

Theorem 3.3.5 *The intertwining operator U described in Theorem 3.3.4 is invertible. Its inverse $V = U^{-1}$ is given for all $K \in C_c^\infty(V_0, C_c)$ and $g \in G$ by:*

$$V(K)(g) = \int_V K(\phi)(g)d\lambda(\phi).$$

Proof It is clear that V formally intertwines formula (3.3.11) and that $V(K)$ has the covariance property (3.3.1) for G, H and f. We will prove then that for $K \in C_c^\infty(V_0, C_c)$, $V(K)$ is an element of $L^2(G/H, f)$. We prove in fact that V is an isometry on $C_c^\infty(V_0, C_c)$, and as in the proof of the previous theorem we can extend it in a unique way on $L^2(V, L^2)$ to an isometry $\bar{V}$, still denoted by V. We proceed as before by induction on $\dim(G/H)$ and we keep all notations of Theorem 3.3.4. Let us show that

$$\|V(K)\|_{L^2(G/H)} = \|K\|_2.$$

In the first case, when almost all orbits of elements in general position are saturated with respect to $\mathfrak{g}_0$, we compute for $g = \exp(tX)g_0$:

$$\begin{aligned} V(K)(g) &= \int_V K(\phi)(\exp(tX)_{g_0})d\lambda(\phi) = \int_{V^0} K_t(\phi_0)(g_0)d\lambda^0(\phi_0) \\ &= V_0(K_t)(g_0) = V(K)_t(g_0), \end{aligned} \tag{3.3.16}$$

where $K_t(\phi_0) = (K(\phi))_t$. Then

$$\|V(K)\|^2_{L^2(G/H,f)} = \int_{\mathbb{R}} \|V(K)_t\|^2_{L^2(G_0/H,f_0)}dt$$

(apply (3.3.16))

$$= \int_{\mathbb{R}} \|V_0(K)_t\|^2_{L^2(G_0/H,f_0)}dt$$

(apply the induction hypothesis)

$$= \int_{\mathbb{R}}\int_{V^0} \|K_t(\phi_0)\| = \|K\|_2^2.$$

We pass to the second case, and prove first this lemma.

Lemma 3.3.6 *Assume that we are in the second case. Let $K \in C_c^\infty(V_0, C_c)$. For $\phi_0 \in V_0$ and $r \in \mathbb{R}$, we call as before $\phi_0 + rX^\star$ the linear form on $\mathfrak{g}$ whose restriction at $\mathfrak{g}_0$ is ϕ_0 and whose value at X is r. The function K_1 defined on $V^0 \times G$ by*

$$K_1(\phi_0)(g) = \int_{\mathbb{R}} K(\phi_0 + rX^\star)(g)dr,\ \phi_0 \in V^0,\ g \in G$$

is C^∞, and its restriction to $V^0 \times G_0$ is in $C_c^\infty(V^0, C_c)$.

Proof It is clear that K_1 exists, since for $\phi_0 \in V^0 \subset \mathfrak{g}_0^\star$ fixed, the map $r \mapsto K(\phi_0 + rX^\star)(g)$ is continuous with compact support in $\mathbb{R}$. It is clear that the map $(\phi_0, g) \mapsto K_1(\phi_0)(g)$ is C^∞ on $V^0 \times G$. On the other hand and from what we saw above, the basis $\{Z_{i_1}, \ldots, Z_{i_q}\}$, $i_1 < \cdots < I_q$ is coexponential for $\mathfrak{b}(\phi)$ in $\mathfrak{g}$ for $\phi \in V_0$, and also for $\mathfrak{b}(\phi_0)$ in $\mathfrak{g}_0$ for all $\phi_0 \in V^0$. As

$$(\phi, t_1, \ldots, t_q) \mapsto K(\phi)(\exp(t_q Z_{i_q}) \cdots \exp(t_1 Z_{i_1}))$$

has compact support in $V_0 \times \mathbb{R}^q$, the map

$$(\phi_0, t_1, \ldots, t_q) \mapsto K_1(\phi_0)(\exp(t_q Z_{i_q}) \cdots \exp(t_1 Z_{i_1}))$$

turns out to have compact support on $V^0 \times \mathbb{R}^q$. The desired covariance relation is obtained from that of $K(\phi_0 + rX^\star)$ and from relation (3.3.12). □

We return to the proof of the theorem. Let $g = \exp(tX)g_0 \in G$, so

$$\begin{aligned} V(K)(g) &= \int_V K(\phi)(g)d\lambda(\phi) = \int_{V^0}\int_{\mathbb{R}} K(\phi_0 + sX^\star)(\exp(tX)g_0)dsd\lambda^0(\phi_0) \\ &= \int_{V^0} K_{t,1}(\phi_0)(g_0)d\lambda^0(\phi_0)V_0(K_{t,1})(g_0), \end{aligned} \tag{3.3.17}$$

where $K_{t,1}(\phi_0)(g_0) = \int_{\mathbb{R}} K_t(\phi_0 + rX^\star)(g_0)dr$. As the function $r \mapsto K_t(\phi_0 + rX^\star)(g)$ is C^∞ with compact support, we have from the Fourier inversion theorem that

$$\begin{aligned} K(\phi_0 + rX^\star)(g) &= \int_{\mathbb{R}^2} K(\phi_0 + xX^\star)(g)e^{ixt}e^{-irt}dtdx \\ &= \int_{\mathbb{R}^2} K(\phi_0 + xX^\star)(g\exp(tB_n(\phi)))\Delta_{B(\phi),G}^{-1/2}(\exp(tB_n(\phi))) \\ &\quad e^{-it\phi_0(Z_0(\phi))}e^{-irt}dtdx \\ &= \int_{\mathbb{R}} K_1(\phi_0)(g\exp(tB_n(\phi)))\Delta_{B(\phi),G}^{-1/2}(\exp(tB_n(\phi)))e^{-it\phi_0(Z_0(\phi))}e^{-irt}dt. \end{aligned} \tag{3.3.18}$$

Hence for $\phi_r = \phi_0 + rX^\star$ we get:

$$\int_{\mathbb{R}} |K(\phi_r)(g)|^2 dr = \int_{\mathbb{R}} |K_1(\phi_0)(g \exp(tB_n(\phi)))|^2 \Delta^{-1}_{B(\phi),G}(\exp(tB_n(\phi)))dt. \tag{3.3.19}$$

Then we have

$$\begin{aligned}
\|K\|_2^2 &= \int_V \|K(\phi)\|^2_{L^2(G/B(\phi),\phi)} d\lambda(\phi). \\
&= \int_{V^0}\int_{\mathbb{R}} \oint_{G/B(\phi_r)} |K(\phi_r)(u)|^2 d\nu_{G,B(\phi_r)}(u) dr d\lambda^0(\phi_0) \\
&= \int_{V^0}\int_{\mathbb{R}} \oint_{G/B(\phi_r)} |K(\phi_r)(u_0)|^2 d\nu_{G_0,B(\phi_0)}(u_0) dr d\lambda^0(\phi_0)
\end{aligned}$$

(apply(3.3.19))

$$= \int_{V^0} \oint_{G_0/B(\phi_0)} \int_{\mathbb{R}} |K_1(\phi_0)(g_0 \exp(tB_n(\phi)))|^2 \Delta^{-1}_{B(\phi),G}(\exp(tB_n(\phi)))dt$$

$$d\nu_{G_0,B(\phi_0)}(g_0) d\lambda^0(\phi_0)$$

(apply Lemma 2.2)

$$\begin{aligned}
&= \int_{V^0} \oint_{G/B(\phi_0)} |K_1(\phi_0)(g)|^2 d\nu_{G,B(\phi_0)}(g) d\lambda^0(\phi_0) \\
&= \int_{V^0}\int_{\mathbb{R}} \oint_{G_0/B(\phi_0)} |K_{t,1}(\phi_0)(g_0)|^2 d\nu_{G_0,B(\phi_0)}(g_0) dt d\lambda^0(\phi_0) \\
&= \int_{\mathbb{R}}\int_{V^0} \oint_{G_0/B(\phi_0)} |K_{t,1}(\phi_0)(g_0)|^2 d\nu_{G_0,B(\phi_0)}(g_0) d\lambda^0(\phi_0) dt
\end{aligned}$$

(apply the hypothesis induction and (3.3.17))

$$= \int_{\mathbb{R}} \|V_0(K_{t,1})\|^2_{L^2(G_0/H,f)} dt = \|V(K)\|^2_{L^2(G/H,f)}.$$

We prove in this last step that $U \circ V = \mathbb{I}$, which implies that the operator U is surjective, hence invertible, and that its inverse is V. We prove this equality on $C_c^\infty(V_0, C_c)$. For the first case, we get from (3.3.16) that:

$$U \circ V(K)(\phi)(g) = U_0 \circ V_0(K_t)(\phi_0)(g_0)$$

(apply the induction hypothesis)

$$= K_t(\phi_0)(g_0) = K(\phi)(\exp(tX)g_0) = K(\phi)(g).$$

We pass to the second case. We can write $\phi = \phi_r$ for some $r \in \mathbb{R}$. From (3.3.18) we have for $g = \exp(tX)g_0$ that

$$K(\phi_0 + rX^\star)(g) = \int_{\mathbb{R}} K_1(\phi_0)(g \exp(sB_n(\phi)))\Delta^{-1/2}_{B(\phi),G}(\exp(sB_n(\phi)))e^{-is\phi_0(Z_0(\phi))}e^{-irs}ds$$

(apply the induction hypothesis)

$$= \int_{\mathbb{R}} U_0 \circ V_0(K_{t,1})(\phi_0)(g_0 \exp(sB_n(\phi)))\Delta^{-1/2}_{B(\phi),G}(\exp(sB_n(\phi)))$$
$$e^{-is\phi_0(Z_0(\phi))}e^{-irs}ds$$

(apply (4.6.2) and (3.3.17))

$$= U \circ V(K)(\phi_0 + rX^\star)(g).$$

This concludes the proof of Theorem 3.3.5. □

3.3.4 *A Rational Disintegration of $L^2(G)$ for an Exponential Solvable Lie Group G*

Let G be an exponential solvable Lie group with Lie algebra $\mathfrak{g}$ and left-regular representation $\lambda_G = \mathrm{ind}_{\{e\}}^G 1$. Here $f = 0$ and $\Gamma_f = \mathfrak{g}^\star$. Take a good sequence of subalgebras of $\mathfrak{g}$

$$\mathfrak{a}_0 = \{0\} \subset \mathfrak{a}_1 \subset \mathfrak{a}_2 \subset \cdots \subset \mathfrak{a}_n = \mathfrak{g} \tag{3.3.20}$$

from which we extract a Malcev basis $\{X_1, \ldots, X_n\}$ of $\mathfrak{g}$, $X_i \in \mathfrak{a}_i \setminus \mathfrak{a}_{i-1}$. In this case and as in Sect. 3.3.1, $K^{\{e\}}$ is the set of all $j \in \{1, \ldots, n\}$ such that all A_j-orbits are saturated with respect to $\mathfrak{a}_{j-1}$, which implies $V = \{\phi \in \mathfrak{g}^\star : \langle \phi, X_j \rangle = 0,\ j \in K^{\{e\}}\}$. Let $\phi \in V$ and set $\phi_i = \phi_{|\mathfrak{a}_i}$. Let

$$\mathfrak{b}(\phi) = \sum_{i=1}^{n} \mathfrak{a}_i(\phi_i),$$

be the Vergne polarization at ϕ with respect to the Jordan–Hölder sequence (3.3.20) and $B(\phi)$ its associated Lie group. In addition, we have from the Pukanszky condition that

$$\mathrm{Ad}^\star(B(\phi))\phi = \phi + \mathfrak{b}(\phi)^\perp.$$

Let μ_G be the Haar measure on G. We have the following rational disintegration of $L^2(G)$:

$$\left(L^2(G), \mu_G\right) \simeq \int_V^{\oplus} \left(L^2(G/B(\phi)), \phi\right) d\lambda(\phi).$$

The isometry is given by:

$$U(\xi)(\phi)(g) = \int_{B(\phi)} \xi(gu)\chi_\phi(u)\Delta_{B(\phi),G}^{-\frac{1}{2}}(u) d_{B(\phi)}(u), \ g \in G$$

where $\xi \in C_c^\infty(G)$ is the set of C^∞ functions with compact support in G and $\phi \in V$, $d_{B(\phi)}$ is the Haar measure on $B(\phi)$.

3.3.5 *Examples*

Example 3.3.7 Let G be the Boidol group. The Lie algebra is generated by vectors A, X, Y, Z such that $[A, X] = X$, $[A, Y] = -Y$, $[X, Y] = Z$. Let $H = \exp(\mathfrak{h}) = \exp(\mathbb{R}Z)$ be the analytic subgroup of G. We put $f = \lambda Z^\star$, $\lambda \neq 0$. The good sequence of subalgebras passing through $\mathfrak{h}$ is

$$\mathfrak{h} = \mathbb{R}Z \subset \mathbb{R}\text{-span}(Z, Y) \subset \mathbb{R}\text{-span}(Z, Y, X) \subset \mathfrak{g}.$$

It is clear that $V \simeq f + \mathbb{R}Y^\star \oplus \mathbb{R}A^\star$. Let $V_0 = V$ and call $\phi_{\alpha,\beta}$ the element $\lambda Z^\star + \alpha Y^\star + \beta A^\star$. The Vergne polarization at an element $\phi_{\alpha,\beta}$ of V_0 is

$$B(\phi_{\alpha,\beta}) = \exp(\mathfrak{b}(\phi_{\alpha,\beta})) = \exp(\text{vect}(Z, Y, A + \frac{\alpha}{\beta}X)).$$

The expression of our operator is

$$U(\xi)(\phi_{\alpha,\beta})(g) = \int_{\mathbb{R}^2} \xi(g \cdot \exp(t(A + \frac{\alpha}{Y}X)) \exp(sY))e^{-it\beta}e^{-i\alpha s}e^{-\frac{t}{2}}dtds,$$

where ξ is C^∞ on G and has compact support modulo H, and $g \in G$. Let us check that U is an isometry:

$$\|U(\xi)\|_2^2 = \int_{\mathbb{R}^2}\int_{\mathbb{R}} |U(\xi)(\phi_{\alpha,\beta})(\exp(xX))|^2 dx d\alpha d\beta$$

$$= \int_{\mathbb{R}^3} |\int_{\mathbb{R}^2} \xi(\exp(xX) \cdot \exp(t(A + \frac{\alpha}{\lambda}X)) \exp(sY))e^{-it\beta}e^{-i\alpha s}e^{-\frac{t}{2}}dtds|^2 dx d\alpha d\beta$$

$$= \int_{\mathbb{R}^3} |\int_{\mathbb{R}} \xi(\exp(xX) \cdot \exp(\beta(A + \frac{\alpha}{\lambda}X)) \exp(sY))e^{-i\alpha s}ds|^2 e^{-\beta} dx d\alpha d\beta$$

$$= \int_{\mathbb{R}^3} | \int_{\mathbb{R}} \xi(\exp(xX) \cdot \exp(-\frac{\alpha}{\lambda}) \cdot \exp(\beta A) \cdot \exp(\frac{\alpha}{\lambda} X) \exp(sY)) e^{-i\alpha s} ds|^2 e^{-\beta} dx d\alpha d\beta$$

$$= \int_{\mathbb{R}^3} |\xi(\exp(\beta A) \cdot \exp(e^{-\beta} xX) \cdot \exp(\alpha Y))|^2 e^{-\beta} d\alpha dx d\beta.$$

$$= \int_{\mathbb{R}^3} |\xi(\exp(\alpha Y) \cdot \exp(xX) \exp(\beta A))|^2 d\alpha dx d\beta = \|\xi\|^2_{L^2(G/H,f)}.$$

The inverse operator is given for all $K \in C_c^\infty(V_0, C_c)$ and $g \in G$ by:

$$V(K)(g) = \int_{\mathbb{R}^2} K(\phi_{\alpha,\beta})(g) d\alpha d\beta.$$

Example 3.3.8 Let now $H = \exp(\mathfrak{h}) = \exp(\mathbb{R}Z \oplus \mathbb{R}X)$ be the analytic subgroup of G. We put $f = \lambda Z^\star$, $\lambda \neq 0$. The good sequence of subalgebras passing through $\mathfrak{h}$ is

$$\mathfrak{h} = \mathbb{R}\text{-span}(Z, X) \subset \mathbb{R}\text{-span}(Z, X, Y) \subset \mathfrak{g}.$$

Clearly

$$V \simeq f + \mathbb{R}A^\star.$$

Let $V_0 = V$ and call ϕ_α the element $\lambda Z^\star + \alpha A^\star$. The Vergne polarization at an element ϕ_α of V_0 is $B(\phi_\alpha) = \exp(\mathfrak{b}(\phi_\alpha)) = \exp(\mathbb{R}\text{-span}(Z, X, Y))$. The expression of our operator is

$$U(\xi)(\phi_\alpha)(g) = \int_{\mathbb{R}} \xi(g \cdot \exp(tA)) e^{-it\alpha} e^{\frac{t}{2}} dt,$$

where ξ is C^∞ on G with compact support modulo H and $g \in G$. The inverse operator is given for all $K \in C_c^\infty(V_0, C_c)$ and $g \in G$ by:

$$V(K)(g) = \int_{\mathbb{R}} K(\phi_\alpha)(g) d\alpha.$$

Example 3.3.9 Let now $H = \{e\}$ and $f = 0$. We consider the following good sequence of subalgebras adapted to $\mathfrak{h}$

$$\{0\} \subset \mathbb{R}Z \subset \mathbb{R}\text{-span}(Z, Y) \subset \mathbb{R}\text{-span}(Z, Y, X) \subset \mathfrak{g}.$$

It is clear that

$$V \simeq \mathbb{R}Z^\star \oplus \mathbb{R}Y^\star \oplus \mathbb{R}A^\star.$$

Write $\phi_{\lambda,\alpha,\beta}$ for $\lambda Z^\star + \alpha Y^\star + \beta A^\star$. Let $V_0 = \{\phi_{\lambda,\alpha,\beta} : \lambda^2 + \alpha^2 \neq 0\}$. The Vergne polarization at an element $\phi_{\lambda,\alpha,\beta}$ of V_0 is

$$B(\phi_{\lambda,\alpha,\beta}) = \exp(\mathfrak{b}(\phi_{\lambda,\alpha,\beta})) = \exp(\mathrm{vect}(Z, Y, \lambda A + \alpha X)).$$

Our operator reads, in this case:

$$U(\xi)(\phi_{\lambda,\alpha,\beta})(g) = \int_{\mathbb{R}^3} \xi(g\cdot\exp(t(\lambda A+\alpha X))\exp(sY)\exp(rZ))e^{-it\lambda\beta}e^{-i\alpha s}e^{-i\lambda r}e^{-\frac{t\lambda}{2}}dtdsdr,$$

where ξ is C^∞ on G with compact support in G and $g \in G$. The inverse operator is given for all $K \in C_c^\infty(V_0, C_c)$ and $g \in G$ by:

$$V(K)(g) = \int_{\mathbb{R}^3} K(\phi_{\lambda,\alpha,\beta})(g)d\lambda d\alpha d\beta.$$

Example 3.3.10 In the final example we do not assume $\mathfrak{h}$ is an ideal of $\mathfrak{g}$. Let $\mathfrak{g}$ be the "$ax + b$"-algebra, i.e $\mathfrak{g} = \mathbb{R}X \oplus \mathbb{R}Y$, $[X, Y] = Y$. We take $f = Y^\star$ and $\mathfrak{h} = \mathbb{R}X \in M(f, \mathfrak{g})$. The completely solvable group $G = \exp\mathfrak{g}$ has only two unitary irreducibles representations of infinite dimension $\pi_\pm$ associated with two open orbits $\pm G \cdot Y^\star$. The monomial representation $\tau = \mathrm{ind}_H^G \chi_f$, $H = \exp\mathfrak{h}$, is equivalent to the direct sum $\pi_+ \oplus \pi_-$. We realize $\pi_\pm$ as $\mathrm{ind}_B^G \chi_{\pm Y^\star}$, where $B = \exp(\mathbb{R}Y)$, and we identify $\mathscr{H}_{\pi_\pm}$ with $L^2(\mathbb{R})$ by

$$\xi \mapsto \tilde{\xi}(t) = \xi(\exp tX),\ (t \in \mathbb{R}).$$

By the way, we consider the operator U defined for such ξ and for $g \in G$ by

$$U(\xi)(Y^\star)(g) = \int_{\mathbb{R}} \xi(g \cdot \exp(tY))e^{-it}dt$$

and

$$U(\xi)(-Y^\star)(g) = \int_{\mathbb{R}} \xi(g \cdot \exp(tY))e^{it}dt.$$

We simply write $\xi(t)$ instead of $\xi(\exp(tY))$ and for $g = \exp(bX)$. These expressions become

$$U(\xi)(Y^\star)(\exp(bX)) = \int_{\mathbb{R}} \xi(t)e^{-ie^{-b}t}e^{-\frac{b}{2}}dt = e^{-\frac{b}{2}}\hat{\xi}(e^{-b})$$

and

$$U(\xi)(-Y^\star)(\exp(bX)) = \int_{\mathbb{R}} \xi(t)e^{ie^{-b}t}e^{-\frac{b}{2}}dt = e^{-\frac{b}{2}}\hat{\xi}(-e^{-b}).$$

We have then

$$\begin{aligned}\|U(\xi)\|_2^2 &= \int_{\mathbb{R}} |e^{-\frac{b}{2}}\hat{\xi}(e^{-b})|^2 db + \int_{\mathbb{R}} |e^{-\frac{b}{2}}\hat{\xi}(-e^{-b})|^2 db \\ &= \int_{\mathbb{R}} |e^{-b}\hat{\xi}(e^{-b})|^2 db + \int_{\mathbb{R}} |e^{-b}\hat{\xi}(-e^{-b})|^2 db \\ &= \int_0^{+\infty} |\hat{\xi}(b)|^2 db + \int_{+\infty}^{0} |\hat{\xi}(b)|^2 db \\ &= \int_{\mathbb{R}} |\xi(b)|^2 db = \|\xi\|^2_{L^2(G/H,f)}.\end{aligned}$$

We point out that this operator is indeed an intertwining operator. Now we might ask whether we can determine a subset V of $\mathfrak{g}^\star$ for which (3.3.2) is well defined and intertwines the equivalence (3.3.3).

3.4 Intertwining of Representations Induced from Maximal Subgroups of Exponential Solvable Lie Groups

Definition 3.4.1 Let G be a Lie group. A subgroup H of G is said to be a maximal subgroup if $H \neq G$ and for every subgroup K such that $H \subset K \subset G$, then either $K = H$ or $K = G$.

Remark 3.4.2 If G is a simply connected solvable Lie group and H is a maximal subgroup of G, then H has codimension one or two. In the latter case H cannot be a normal subgroup of G.

The following result describes the structure of maximal subalgebras of exponential solvable algebras, a proof of which can be found in [106].

Theorem 3.4.3 *Let $G = \exp \mathfrak{g}$ be an exponential solvable Lie group and $H = \exp \mathfrak{h}$ a non-normal maximal subgroup of G. Then*

1. *if $\mathfrak{h}$ is a hyperplane, there exist a codimension-one subalgebra $\mathfrak{g}_0$ of $\mathfrak{h}$ which is a codimension-two ideal in $\mathfrak{g}$, plus two elements $A \in \mathfrak{h} \setminus \mathfrak{g}_0$, $X \in \mathfrak{g} \setminus \mathfrak{h}$ such that*

$$[A, X] = X \bmod \mathfrak{g}_0.$$

2. *If $\mathfrak{h}$ has codimension two, there exists a codimension-one subalgebra $\mathfrak{g}_0$ of $\mathfrak{h}$ which is a codimension-three ideal in $\mathfrak{g}$, plus three nonzero vectors A, X, Y and a nonzero real number α such that*

$$\mathfrak{g} = \mathfrak{h} \oplus \mathbb{R}X \oplus \mathbb{R}Y, \quad \mathfrak{h} = \mathfrak{g}_0 \oplus \mathbb{R}A$$

$$[A, X] = X + \alpha Y \bmod \mathfrak{g}_0, \; [A, Y] = Y - \alpha X \bmod \mathfrak{g}_0,$$

and

$$[X, Y] = 0 \bmod \mathfrak{g}_0.$$

We now prove the following disintegration formula, which basically stems from Theorem 3.4.3.

Proposition 3.4.4 *Let G be an exponential solvable Lie group and H a maximal closed connected subgroup. Then there exists a measurable subset $\mathscr{O}(\tau) \subset \mathbb{R}^{\operatorname{codim} H}$ such that*

$$\tau \simeq \int_{\mathscr{O}(\tau)}^{\oplus} \pi_{\phi_t} d\lambda(t), \tag{3.4.1}$$

where $d\lambda$ is the Lebesgue measure on $\mathscr{O}(\tau)$. Except when H is normal in G, moreover, the set $\mathscr{O}(\tau)$ is compact.

Proof We keep the same notations of Theorem 3.4.3. Let $\theta = f_{|\mathfrak{g}_0}$ and

$$\mathfrak{g}_\theta = \{T \in \mathfrak{g} : \langle \theta, [T, \mathfrak{g}_0] \rangle = \{0\}\}.$$

So, obviously $\mathfrak{g}_\theta$ is a subalgebra containing $\mathfrak{h}$. As H is a maximal subgroup of G, then $\mathfrak{g}_\theta = \mathfrak{h}$ or $\mathfrak{g}_\theta = \mathfrak{g}$. Assume first of all that $\mathfrak{g}_\theta = \mathfrak{h}$. In this case τ is irreducible and $\mathscr{O}(\tau)$ consists only of a single point. Suppose next that $\mathfrak{g}_\theta = \mathfrak{g}$. If H is a normal subgroup of G then $\mathscr{O}(\tau) = \mathbb{R}$ (see [84]). Furthermore

$$\tau \simeq \int_{\mathbb{R}}^{\oplus} \pi_{\theta + t\delta} dt,$$

where $\delta \in \mathfrak{h}^\perp$, $\delta \neq 0$. If H is a non-normal subgroup and $\operatorname{codim} H = 1$, then according to [100] there exists $s_0 \in \mathbb{R}$ such that $\mathscr{O}(\tau) = \{s_1, s_2\}$ where $s_1, s_2 \in \mathbb{R}$ and $s_1 < s_0 < s_2$. Even more, we have the decomposition

$$\tau \simeq \pi_{\phi_{s_1}} \oplus \pi_{\phi_{s_2}},$$

where $\phi_{s_i} = f + s_i X^*$, $i = 0, 1, 2$. Note furthermore that

$$\dim G \cdot \phi_{s_1} = \dim G \cdot \phi_{s_2} > \dim G \cdot \phi_{s_0}. \tag{3.4.2}$$

Finally, if $\operatorname{codim} H = 2$, then according to [106] there exists a unique $P_0 = (s_0, t_0) \in \mathbb{R}^2$ such that $\mathscr{O}(\tau) = \mathscr{C} = \{(s, t) \in \mathbb{R}^2 : (s - s_0)^2 + (t - t_0)^2 = 1\}$ is the unit circle centred at P_0 and

$$\tau \simeq \int_{\mathscr{C}}^{\oplus} \pi_{\phi_{s,t}} ds dt, \tag{3.4.3}$$

where $\phi_{s,t} = f + sX^* + tY^*$. Note in addition that

$$\dim G \cdot \phi_{s,t} > \dim G \cdot \phi_{s_0,t_0}, (s,t) \neq (s_0, t_0). \tag{3.4.4}$$

This proves the proposition. □

Construction of an Intertwining Operator

In this section we construct an intertwining operator between the induced representation $\tau = \mathrm{ind}_H^G \chi_f$ and its decomposition into irreducibles explicitly. Let $\mathfrak{s} = (\mathfrak{a}_j)_{j=0}^n$ be a good sequence of subalgebras of $\mathfrak{g}$ passing through $\mathfrak{g}_0$, where $\mathfrak{g}_0$ is defined as in Theorem 3.4.3. With the notations above, we can choose $\mathfrak{s}$ as follows:

1. If $\mathfrak{h}$ is an ideal of $\mathfrak{g}$ we have $\operatorname{codim} \mathfrak{h} = 1$, then $\mathfrak{a}_{n-1} = \mathfrak{h} = \mathfrak{g}_0$.
2. If $\mathfrak{h}$ is not an ideal and $\operatorname{codim} \mathfrak{h} = 1$, then $\mathfrak{a}_{n-2} = \mathfrak{g}_0$ and $\mathfrak{a}_{n-1} = \mathfrak{g}_0 \oplus \mathbb{R}X$.
3. If $\operatorname{codim} \mathfrak{h} = 2$, then $\mathfrak{a}_{n-3} = \mathfrak{g}_0$, $\mathfrak{a}_{n-2} = \mathfrak{g}_0 \oplus \mathbb{R}X$ and $\mathfrak{a}_{n-1} = \mathfrak{g}_0 \oplus \mathbb{R}X \oplus \mathbb{R}Y$.

In the sequel, we shall identify $\mathscr{O}(\tau)$ with the set $\{\phi_t \in \Gamma_f : t \in \mathscr{O}(\tau)\}$. For l in $\mathscr{O}(\tau)$, let $\mathfrak{b}[l]$ be the Vergne polarization of l associated to $\mathfrak{s}$ and $B[l] = \exp \mathfrak{b}[l]$. We prove first the following

Lemma 3.4.5 *For any l in $\mathscr{O}(\tau)$, there exist a coexponential basis $\mathscr{Y}$ of $\mathfrak{b}[l] \cap \mathfrak{h}$ in $\mathfrak{b}[l]$, a coexponential basis $\mathscr{Z}$ of $\mathfrak{b}[l] \cap \mathfrak{h}$ in $\mathfrak{h}$ and a coexponential basis $\mathscr{X}$ of $\mathfrak{b}[l]$ in $\mathfrak{g}$ which do not depend on l.*

Proof As above, we distinguish two cases. We keep the same notations as in Proposition 3.4.4. If $\mathfrak{g}_\theta = \mathfrak{h}$, then τ is irreducible and $\mathscr{O}(\tau) = \{\phi\}$ where $\phi \in p^{-1}(\{f\})$. Hence $\dim \mathfrak{b}[\phi] = \dim \mathfrak{h}$. We are going to prove that in this situation $\mathfrak{b}[\phi] = \mathfrak{h}$, which implies

$$\mathscr{Y} = \mathscr{Z} = \emptyset.$$

Suppose for starters that H is a codimension-one subgroup of G. Then $\mathfrak{g} = \mathfrak{h} \oplus \mathbb{R}X$. If H is a normal subgroup of G, then as $\mathfrak{g}(\phi) \subset \mathfrak{g}_\theta = \mathfrak{h}$, we already get that $\mathfrak{b}[\phi] = \mathfrak{h}$. Now we suppose that H is a non-normal subgroup of G. It follows from the definition of the Vergne polarization $\mathfrak{b}[\phi]$ and for all $i = 0, \ldots, \operatorname{codim} \mathfrak{h}$, $\mathfrak{a}_{n-i}(\phi_{|\mathfrak{a}_{n-i}}) \subset \mathfrak{a}_{n-i} \cap \mathfrak{g}_\theta \subset \mathfrak{h}$, that $\mathfrak{b}[\phi] \subset \mathfrak{h}$, which implies $\mathfrak{b}[\phi] = \mathfrak{h}$. We conclude that if $\operatorname{codim} \mathfrak{h} = 1$, we have

$$\mathscr{X} = \{X\},$$

and if $\operatorname{codim} \mathfrak{h} = 2$, we have

$$\mathscr{X} = \{X, Y\}.$$

We now look at the case where $\mathfrak{g}_\theta = \mathfrak{g}$. We have $X \in \mathfrak{b}[l]$ and $\mathfrak{g}_0 \subset \mathfrak{g}(l)$, for all l in $\mathscr{O}(\tau)$. Suppose first that H is a codimension-one subgroup of G. If H is a normal subgroup of G, then $\mathfrak{g}(l) = \mathfrak{g}$ and then $\mathfrak{b}[l] = \mathfrak{g}$. Therefore,

$$\mathscr{Y} = \{X\}, \ \mathscr{Z} = \mathscr{X} = \emptyset.$$

Assume then that H is a non-normal subgroup of G, so $\mathfrak{g}(\phi_{s_1}) = \mathfrak{g}(\phi_{s_2}) = \mathfrak{g}_0$ from Eq. (3.4.2) and hence $\mathfrak{b}[\phi_{s_1}] = \mathfrak{b}[\phi_{s_2}] = \mathfrak{g}_0 \oplus \mathbb{R}X$. This implies that

$$\mathscr{Y} = \{X\}, \ \mathscr{Z} = \{A\} \text{ and } \mathscr{X} = \{A\}.$$

Secondly, we treat the case where $\mathfrak{h}$ has codimension two. Let (s, t) be an arbitrary element of $\mathscr{C}$, we consider the Vergne polarization $\mathfrak{b}[\phi_{s,t}]$ with respect to the good sequence $\mathfrak{s}$. Recall that in the case $\mathfrak{g}_0 \subset \mathfrak{g}(\phi_{s,t})$, either $\dim \mathfrak{g}(\phi_{s,t}) = \dim \mathfrak{g}_0 + 1$ or $\mathfrak{g}(\phi_{s,t}) = \mathfrak{g}$. Using (3.4.4) we get that $\dim G \cdot \phi_{s_0,t_0} = 0$ and $\dim G \cdot \phi_{s,t} = 2$. Hence $\dim \mathfrak{g}(\phi_{s,t}) = \dim \mathfrak{g}_0 + 1$ and then $\dim \mathfrak{b}[\phi_{s,t}] = \dim \mathfrak{g}_0 + 2$. Therefore there exists $(a, y) \in \mathbb{R}^2$ such that $\mathfrak{b}[\phi_{s,t}] = \mathfrak{g}_0 \oplus \mathbb{R}X \oplus \mathbb{R}(aA + yY)$. Since $[X, aA + yY] \in \mathfrak{b}[\phi_{s,t}]$, we get $a = 0$. Then $\mathfrak{b}[\phi_{s,t}] = \mathfrak{g}_0 \oplus \mathbb{R}X \oplus \mathbb{R}Y$ which is an ideal of $\mathfrak{g}$. We have then

$$\mathscr{Y} = \{X, Y\}, \ \mathscr{Z} = \{A\} \text{ and } \mathscr{X} = \{A\}.$$

This completes the proof. □

Corollary 3.4.6 *We keep the same hypotheses and notations. For each l in $\mathscr{O}(\tau)$ we have*

(1) $HB[l]$ is closed in G.
(2) $\Delta_{H\cap B[l],B[l]}\Delta_{H\cap B[l],H} \equiv 1$ on $H \cap B[l]$.

Proof Take l in $\mathscr{O}(\tau)$. From the proof of Lemma 3.4.5 we have $\mathfrak{h} + \mathfrak{b}[l] = \mathfrak{h}$ or $\mathfrak{g}$, which implies that $HB[l]$ is closed in G. On the other hand, we have that $\mathfrak{h}\cap\mathfrak{b}[l] = \mathfrak{h}$ or $\mathfrak{g}_0$ depending on whether $\mathfrak{b}[l] = \mathfrak{h}$ or not. Hence $H \cap B[l]$ is a normal subgroup of $B[l]$ and of H, and therefore $\Delta_{H\cap B[l],B[l]}\Delta_{H\cap B[l],H} \equiv 1$ on $H \cap B[l]$. □

Corollary 3.4.7 *For all l in $\mathscr{O}(\tau)$ and ξ in $C_c^\infty(G/H, f)$, the vector $T_l(\xi)$ defined by*

$$T_l(\xi)(g) = \int_{B[l]/B[l]\cap H} \xi(gb)\chi_l(b)\Delta_{B[l],G}^{-\frac{1}{2}}(b)d_{\mathscr{Y}}(b), \ g \in G$$

is an element of $C_c^\infty(G/B[l], l)$.

Proof It is clear that the last integral is convergent by Corollary 3.4.6. Furthermore, it is easy to check that the vector $T_l(\xi)$ has the required covariance relation (3.3.1) and that the support of $T_l(\xi)$ is compact modulo $B[l]$. □

Now we are ready to prove the following.

Theorem 3.4.8 *With the same notations above, the operator $\mathscr{U}$ defined on $C_c^\infty(G/H, f)$ by*

$$\mathscr{U}(\xi)(l)(g) = T_l\xi(g) = \int_{B[l]/B[l]\cap H} \xi(gb)\chi_l(b)\Delta_{B[l],G}^{-\frac{1}{2}}(b)\, d_{\mathscr{Y}}(b),\ g \in G$$

is an isometric linear operator which can be extended over $L^2(G/H, f)$ to an intertwining operator of τ. It decomposes as in (3.4.1) when the norms on the homogeneous spaces $G/B[l]$ and $B[l]/B[l] \cap H$ are normalized using the bases provided in Lemma 3.4.5.

Proof Observe first that when H is a normal subgroup of G, the theorem is proved in [4]. We assume then that H is not normal. We use induction on the dimension of G. We begin by considering the case where $\mathfrak{g}_\theta = \mathfrak{h}$, and recall that in this case τ is irreducible, $\mathscr{O}(\tau) = \{\phi\}$ and the Vergne polarization $\mathfrak{b}[\phi]$ is equal to $\mathfrak{h}$. Therefore, our operator $\mathscr{U}$ is the identity mapping of $L^2(G/H, f)$. Assume now that $\mathfrak{g}_\theta = \mathfrak{g}$. Suppose first that $\mathfrak{g}$ contains a non-trivial ideal $\mathfrak{d}$ contained in $\mathfrak{h}$ on which f vanishes. We write $\overline{\mathfrak{g}} = \mathfrak{g}/\mathfrak{d}$, $\overline{G} = \exp\overline{\mathfrak{g}}$ and $\overline{\mathfrak{h}} = \mathfrak{h}/\mathfrak{d}$. So it is clear that $\overline{H} = \exp\overline{\mathfrak{h}}$ remains a maximal, non-normal subgroup of $\overline{G}$. Let $\overline{\mathfrak{g}_0}$ be the codimension-one ideal of $\overline{\mathfrak{h}}$ defined as in Theorem 3.4.3 and $\overline{\mathfrak{s}}$ the good sequence of $\overline{\mathfrak{g}}$ passing through $\overline{\mathfrak{g}_0}$ constructed above. We consider now the canonical projections $q_1 : \mathfrak{g} \longrightarrow \overline{\mathfrak{g}}$ and $q_2 : G \longrightarrow \overline{G}$. Let $\overline{\tau} = \mathrm{Ind}_{\overline{H}}^{\overline{G}}\chi_{\overline{f}}$, where $\overline{f} \circ q_1 = f$, then $\overline{\tau} \circ q_2 = \tau$. It is also clear that $\mathscr{O}(\overline{\tau}) = \mathscr{O}(\tau)$, since $\mathfrak{d} \subset \mathfrak{b}[l] \cap \mathfrak{h}$ for all l in $\mathscr{O}(\tau)$, so $\overline{\mathscr{Y}} = \mathscr{Y}$ is a coexponential basis for $\mathfrak{b}[\overline{l}] \cap \overline{\mathfrak{h}}$ in $\mathfrak{b}[\overline{l}]$, $\overline{\mathscr{X}} = \mathscr{X}$ is a coexponential basis for $\mathfrak{b}[\overline{l}]$ in $\overline{\mathfrak{g}}$ and $\overline{\mathscr{Z}} = \mathscr{Z}$ is a coexponential basis for $\mathfrak{b}[\overline{l}] \cap \overline{\mathfrak{h}}$ in $\overline{\mathfrak{h}}$. So the induction hypothesis enables us to conclude.

So we assume that $\mathfrak{g}$ does not contain any non-trivial ideal contained in $\mathfrak{h}$ on which f vanishes. It follows that $\mathfrak{g}_0$ is Abelian as f vanishes on $[\mathfrak{g}_0, \mathfrak{g}_0]$. On the other hand, let

$$\mathfrak{k}_\theta = \mathrm{Ker}(\theta) = \{T \in \mathfrak{g}_0 : \theta(T) = 0\}.$$

Then $\mathfrak{k}_\theta$ is an ideal of $\mathfrak{g}$ contained in $\mathfrak{h}$ as $\langle\theta, [\mathfrak{g}, \mathfrak{g}_0]\rangle = \{0\}$. As f vanishes on $\mathfrak{k}_\theta$, it follows that $\mathfrak{k}_\theta = \{0\}$ according to our assumption. It follows that either $\mathfrak{g}_0 = \{0\}$ or $\mathfrak{g}_0$ is a one-dimensional ideal of $\mathfrak{g}$. In the last case, write $\mathfrak{g}_0 = \mathbb{R}Z$ for some Z in $\mathfrak{g}$ such that $f(Z) = 1$.

If $\operatorname{codim}\mathfrak{h} = 1$, then $\mathfrak{g} = \mathbb{R}\text{-span}(A, X) \oplus \mathfrak{g}_0$, $\mathfrak{h} = \mathbb{R}A \oplus \mathfrak{g}_0$ and $[A, X] = X$. In fact, if $\mathfrak{g}_0 = \mathbb{R}Z$ then by Theorem 3.4.3 there exist $a, b \in \mathbb{R}$ such that

$$[A, X] = X + aZ,\ \ [A, Z] = bZ.$$

Since $\mathfrak{h}$ is an isotropic subalgebra for B_f, we get $b = 0$. In addition, we can suppose that $a = 0$ (otherwise we substitute X with $X' = X + aZ$).

Moreover, we get from (3.4.2) that $s_0 = 0$ and for all pairs $(s_1, s_2) \in \mathbb{R}^2$ such that $s_1 < s_0 < s_2$ we have

$$\tau \simeq \pi_{\phi_{s_1}} \oplus \pi_{\phi_{s_2}},$$

where $\phi_{s_i} = f + s_i X^*,\ i = 1, 2$. As $\pi_{s_1} \simeq \pi_{-1}$ and $\pi_{s_2} \simeq \pi_{+1}$, we may choose $s_1 = -1$ and $s_2 = 1$. We denote $\phi_{s_1} = \phi^-$ and $\phi_{s_2} = \phi^+$. The Vergne polarization at ϕ^- and ϕ^+ associated to the good sequence $\mathfrak{s}$ is $\mathfrak{b}[\phi^-] = \mathfrak{b}[\phi^+] = \mathbb{R}X \oplus \mathfrak{g}_0$. The operator $\mathscr{U}$ is then given by

$$\begin{array}{rcl}\mathscr{U} : C_c^\infty(G/H, f) & \longrightarrow & C_c^\infty(G/B[\phi^-], \phi^-) \oplus C_c^\infty(G/B[\phi^+], \phi^+) \\ \xi & \longmapsto & \mathscr{U}^-(\xi) + \mathscr{U}^+(\xi),\end{array}$$

where for $g \in G$,

$$\mathscr{U}^-(\xi)(g) = \int_{\mathbb{R}} \xi(g \exp xX) e^{2i\pi x}\, dx$$

and

$$\mathscr{U}^+(\xi)(g) = \int_{\mathbb{R}} \xi(g \exp xX) e^{-2i\pi x}\, dx.$$

It is shown in ([4], Question 5.2.4) that

$$\|\mathscr{U}^-(\xi)\|^2_{L^2(G/B[\phi^-],\phi^-)} + \|\mathscr{U}^+(\xi)\|^2_{L^2(G/B[\phi^+],\phi^+)} = \|\xi\|^2_{L^2(G/H,f)},$$

which completes the proof in this case. Suppose now that $\operatorname{codim} \mathfrak{h} = 2$, then $\mathfrak{g} = \mathbb{R}\text{-span}(A, X, Y) \oplus \mathfrak{g}_0$. We can suppose in both cases that

$$[A, X] = X + \alpha Y,\ [A, Y] = Y - \alpha X \text{ and } [X, Y] = 0.$$

In fact, if $\dim \mathfrak{g}_0 = 1$, then there exist $a, b \in \mathbb{R}$ such that

$$[A, X] = X + \alpha Y + aZ, [A, Y] = Y - \alpha X + bZ$$

and all other brackets are in $\mathfrak{g}_0 = \mathbb{R}Z$. Since $\mathfrak{g}_\theta = \mathfrak{g}$, we have that $[Z, A] = [Z, X] = [Z, Y] = 0$, and by the circular relation of brackets, we obtain $[X, Y] = 0$. On the other hand, we can suppose that $a = b = 0$, otherwise we replace X by $\frac{X + a - \alpha b}{1+\alpha^2} Z$ and Y by $\frac{Y + b + \alpha a}{1+\alpha^2} Z$. From Proposition 3.4.4, we have

$$\tau \simeq \int_{\mathscr{C}}^{\oplus} \pi_{\phi_{s,t}} ds\, dt,$$

where $\mathscr{C}$ is the unit circle centred at $(s_0, t_0) \in \mathbb{R}^2$ satisfying inequality (3.4.4). In this way, since $\mathfrak{g}_0 \subset \mathfrak{g}(\phi_{s,t})$, there are two possibilities: either $\dim \mathfrak{g}(\phi_{s,t}) = \dim \mathfrak{g}_0 + 1$ or $\mathfrak{g}(\phi_{s,t}) = \mathfrak{g}$, $(s,t) \in \mathbb{R}^2$, which gives us $\dim G \cdot \phi_{s_0,t_0} = 0$ and $\dim G \cdot \phi_{s,t} = 2$, for all $(s,t) \in \mathscr{C}$. We get then $(s_0, t_0) = (0, 0)$. For $\omega \in [0, 2\pi[$, let $\phi_\omega = f + \cos\omega X^* + \sin\omega Y^*$, then $\mathfrak{b}[\phi_\omega] = \mathfrak{b} = \mathfrak{g}_0 \oplus \mathbb{R}X \oplus \mathbb{R}Y$ is the Vergne polarization of ϕ_ω associated to $\mathfrak{s}$ and $B = \exp \mathfrak{b}$. We consider the operator $\mathscr{U}$ defined on $C_c^\infty(G/H, f)$ by

$$\begin{aligned}\mathscr{U}(\xi)(\phi_\omega)(g) &= \int_{B/B\cap H} \xi(gb)\Delta_{B,G}^{-\frac{1}{2}}(b)\chi_{\phi_\omega}(b)d\dot{y}(b) \\ &= \int_{\mathbb{R}^2} \xi(g \exp xX \exp yY)e^{-2i\pi(x\cos w + y\sin w)}dxdy,\end{aligned}$$

where $\xi \in C_c^\infty(G/H, f)$ and $g \in G$. We have then

$$\begin{aligned}\|\mathscr{U}(\xi)\|_2^2 &= \int_0^{2\pi} \|\mathscr{U}(\xi)(\phi_\omega)\|^2_{L^2(G/B,\phi_\omega)}d\omega = \int_0^{2\pi}\int_{\mathbb{R}} |\mathscr{U}(\xi)(\phi_\omega)(\exp aA)|^2 da\, d\omega \\
&= \int_0^{2\pi}\int_{\mathbb{R}} \Big|\int_{\mathbb{R}^2} \xi(\exp aA \exp xX \exp yY)e^{-2i\pi(x\cos\omega + y\sin\omega)}dxdy\Big|^2 dad\omega \\
&= \int_0^{2\pi}\int_{\mathbb{R}} \Big|\int_{\mathbb{R}^2} \xi(\exp e^a(x\cos\alpha a - y\sin\alpha a)X \exp e^a(y\cos\alpha a + x\sin\alpha a)Y) \\
&\qquad e^a e^{2i\pi a\langle f,A\rangle} e^{-2i\pi(x\cos\omega + y\sin\omega)}dxdy\Big|^2 dad\omega \\
&= \int_0^{2\pi}\int_{\mathbb{R}} \Big|\int_{\mathbb{R}^2} \xi(\exp xX\ \exp yY)e^{-2i\pi e^{-a}(\cos\omega\ (x\cos\alpha a + y\sin\alpha a))} \\
&\qquad e^{-2i\pi e^{-a}(\sin\omega\ (y\cos\alpha a - x\sin\alpha a))}e^{-2a}dx\ dy\Big|^2 e^{2a} da\ d\omega \\
&= \int_0^{2\pi}\int_{\mathbb{R}} \Big|\int_{\mathbb{R}^2} \xi(\exp xX\ \exp yY)e^{-2i\pi e^{-a}x(\cos\omega\cos\alpha a - \sin\omega\sin\alpha a))} \\
&\qquad e^{-2i\pi e^{-a}y(\sin t\cos\alpha a + \cos t\sin\alpha a)}dxdy\Big|^2 e^{-2a} da\ d\omega \\
&= \int_0^{2\pi}\int_{\mathbb{R}} \Big|\int_{\mathbb{R}^2} \xi(\exp xX \exp yY)e^{-2i\pi e^{-a}(x\cos(\omega+\alpha a) + y\sin(\omega+\alpha a))}dxdy\Big|^2 e^{-2a} dad\omega \\
&= \int_{\mathbb{R}}\int_0^{2\pi} \Big|\int_{\mathbb{R}^2} \xi(\exp xX\ \exp yY)e^{-2i\pi e^{-a}(x\cos(\omega+\alpha a) + y\sin(\omega+\alpha a))}dx\ dy\Big|^2 e^{-2a} d\omega da \\
&= \int_{\mathbb{R}}\int_0^{2\pi} \Big|\int_{\mathbb{R}^2} \xi(\exp xX \exp yY)e^{-2i\pi e^{-a}(x\cos\omega + y\sin\omega)}dxdy\Big|^2 e^{-2a} d\omega da \\
&= \int_{\mathbb{R}}\int_0^{2\pi} |\hat{\xi}(\exp(e^{-a}\cos\omega)X \exp(e^{-a}\sin\omega)Y)|^2 e^{-2a} d\omega da \\
&= \int_{\mathbb{R}}\int_{\mathbb{R}} |\hat{\xi}(\exp rX \exp sY)|^2 drds = \|\hat{\xi}\|^2_{L^2(G/H,f)} = \|\xi\|^2_{L^2(G/H,f)}.\end{aligned}$$

Hence $\mathscr{U}$ is an intertwining operator of decomposition (3.4.3), which completes the proof. □

The Inverse Operator

In this section we show that the intertwining operator $\mathscr{U}$ has a dense range, and hence invertible. We also describe its inverse explicitly. We set $\mathscr{H} = \mathscr{U}(C_c^\infty(G/H, f))$.

Theorem 3.4.9 *The intertwining operator $\mathscr{U}$ given in Theorem 3.4.8 is invertible and has a dense range. Its inverse is given by*

$$\mathscr{V}(K)(g) = \int_{\mathscr{O}(\tau)} \int_{H/H\cap B[l]} K(l)(gh)\chi_l(h)\Delta_{H,G}^{-\frac{1}{2}}(h)d_{\mathscr{Y}}(h)d\lambda(l), \ K \in \mathscr{H}. \tag{3.4.5}$$

Proof We need to prove the theorem only in the case where H is not normal. Assume for a moment $\mathscr{H}$ is a dense subset, which obviously implies that $\mathscr{U}$ is invertible. As $\mathscr{O}(\tau)$ is compact, $(l, g) \longmapsto \mathscr{U}(\xi)(l)(g)$ is a continuous function and $g \longmapsto \mathscr{U}(\xi)(l)(g)$ has a compact support which does not depend on l for fixed $\xi \in C_c^\infty(G/H, f)$. The integral given by Eq. (3.4.5) is absolutely convergent and so $\mathscr{V}$ is well defined on $\mathscr{H}$. We shall prove that $\mathscr{V} \circ \mathscr{U} = \mathrm{Id}$ on $C_c^\infty(G/H, f)$, which merely entails that $\mathscr{V}$ is the inverse of $\mathscr{U}$. We use induction on the dimension of G and we treat the same cases as before. If $\mathfrak{g}$ contains a non-trivial ideal contained in $\mathfrak{h}$ on which f vanishes, then the induction hypothesis enables us to conclude. We suppose then that $\mathfrak{g}$ never contains such an ideal. In the case where $\operatorname{codim}\mathfrak{h} = 1$, we have for $\xi \in C_c^\infty(G/H, f)$ and $g \in G$ that

$$\begin{aligned}
\mathscr{V} \circ \mathscr{U}(\xi)(g) &= \int_{H/H\cap B[\phi^+]} \mathscr{U}(\xi)(\phi^+)(gh)\chi_{\phi^+}(h)\Delta_{H,G}^{-\frac{1}{2}}(h)d_{\mathscr{Y}}(h) + \\
&\qquad \int_{H/H\cap B[\phi^-]} \mathscr{U}(\xi)(\phi^-)(gh)\chi_{\phi^-}(h)\Delta_{H,G}^{-\frac{1}{2}}(h)d_{\mathscr{Y}}(h) \\
&= \int_{\mathbb{R}} (\mathscr{U}(\xi)(\phi^-) + \mathscr{U}(\xi)(\phi^+))(g \exp aA)\chi_f(\exp aA)\Delta_{H,G}^{-\frac{1}{2}}(\exp aA)da \\
&= \int_{\mathbb{R}}\int_{\mathbb{R}} \xi(g \exp aA \exp xX)(e^{2i\pi x} + e^{-2i\pi x})\, dx\chi_f(\exp aA)\Delta_{H,G}^{-\frac{1}{2}}(\exp aA)da \\
&= \int_{\mathbb{R}}\int_{\mathbb{R}} \xi(g \exp xe^{a}X \exp aA)(e^{2i\pi x} + e^{-2i\pi x})dx\chi_f(\exp aA) \\
&\qquad \Delta_{H,G}^{-\frac{1}{2}}(\exp aA)da \\
&= \int_{\mathbb{R}}\int_{\mathbb{R}} \xi(g \exp xX\,)(e^{2i\pi e^{-a}x} + e^{-2i\pi e^{-a}x})\, e^{-a}dxda \\
&= \int_{\mathbb{R}} (\hat{\xi}(g \exp e^{-a}X) + \hat{\xi}(g \exp -e^{-a}X))e^{-a}da
\end{aligned}$$

$$= \int_{\mathbb{R}^+} \hat{\xi}(g \exp aX)da + \int_{\mathbb{R}^-} \hat{\xi}(g \exp aX)da$$

$$= \int_{\mathbb{R}} \hat{\xi}(g \exp aX)da = \xi(g).$$

Assume now that $\mathfrak{h}$ has codimension two. Let $\xi \in C_c^\infty(G/H, f)$ and $g \in G$, we have

$$\mathscr{V} \circ \mathscr{U}(\xi)(g) = \int_{\mathscr{C}} \int_{H/H\cap B[l]} \mathscr{U}(\xi)(l)(gh)\chi_l(h)\Delta_{H,G}^{-\frac{1}{2}}(h)d_{\mathscr{X}}(h)d\lambda(l)$$

$$= \int_{\mathscr{C}} \int_{H/H\cap B[l]} \int_{B[l]/H\cap B[l]} \xi(ghb)\chi_l(b)\chi_f(h)\Delta_{H,G}^{-\frac{1}{2}}(h)d_{\mathscr{Y}}(b)d_{\mathscr{X}}(h)d\lambda(l)$$

$$= \int_{\mathscr{C}} \int_{H/G_0} \int_{B[l]/G_0} \xi(ghb)\chi_l(b)\chi_f(h)\Delta_{H,G}^{-\frac{1}{2}}(h)d_{\mathscr{Y}}(b)d_{\mathscr{X}}(h)d\lambda(l)$$

$$= \int_0^{2\pi} \int_{\mathbb{R}} \int_{\mathbb{R}^2} \xi(g \exp aA \exp xX \exp yY)e^{-2i\pi(x\cos\omega + y\sin\omega)}\chi_f(\exp aA)$$
$$\Delta_{H,G}^{-\frac{1}{2}}(\exp aA)dxdydad\omega$$

$$= \int_0^{2\pi} \int_{\mathbb{R}} \int_{\mathbb{R}^2} \xi(g \exp e^a(x\cos\alpha a - y\sin\alpha a)X \exp e^a(y\cos\alpha a + x\sin\alpha a)Y)$$
$$\Delta_{H,G}^{\frac{1}{2}}(\exp aA)e^{-2i\pi(x\cos\omega + y\sin\omega)}\Delta_{H,G}^{-\frac{1}{2}}(\exp aA)dxdydad\omega$$

$$= \int_0^{2\pi} \int_{\mathbb{R}} \int_{\mathbb{R}^2} \xi(g \exp e^a(x\cos\alpha a - y\sin\alpha a)X \exp e^a(y\cos\alpha a + x\sin\alpha a)Y)$$
$$e^{-2i\pi(x\cos\omega + y\sin\omega)}dxdydad\omega$$

$$= \int_0^{2\pi} \int_{\mathbb{R}} \int_{\mathbb{R}^2} \xi(g \exp xX \exp yY)e^{-2i\pi e^{-a}(\cos\omega(x\cos\alpha a - y\sin\alpha a) + \sin\omega\ (y\cos\alpha a + x\sin\alpha a))}$$
$$e^{-2a}dxdydad\omega$$

$$= \int_0^{2\pi} \int_{\mathbb{R}} \int_{\mathbb{R}^2} \xi(g \exp xX \exp yY)e^{-2i\pi e^{-a}(x\cos\omega + y\sin\omega)}e^{-2a}\ dxdydad\omega.$$

($u = e^{-a}\cos\omega$ and $v = e^{-a}\sin\omega$)

$$= \int_{\mathbb{R}} \int_{\mathbb{R}} \int_{\mathbb{R}^2} \xi(g \exp xX \exp yY)e^{-2i\pi(ux+vy)}dxdydudv$$

$$= \int_{\mathbb{R}^2} \hat{\xi}(g \exp uX \exp vY)dudv = \xi(g).$$

There remains to prove that $\mathscr{H}$ is a dense subspace. As previously, we do assume that $\mathfrak{h}$ contains no ideals of $\mathfrak{g}$ on which f vanishes. We treat first the case where $\operatorname{codim}\mathfrak{h} = 1$.

Let $\zeta \in (\mathrm{im}(\mathscr{U}))^{\perp}$, then for all $\xi \in C_c^{\infty}(G/H, f)$, we have

$$\begin{aligned}
\langle \zeta, \mathscr{U}(\xi)\rangle &= \int_{\mathbb{R}} \zeta(a)(\mathscr{U}^{-}(\xi) + \mathscr{U}^{+}(\xi))(\exp aA)da \\
&= \int_{\mathbb{R}}\int_{\mathbb{R}} \zeta(a)\xi(\exp aA \exp xX)(e^{-2i\pi x} + e^{2i\pi x})dxda \\
&= \int_{\mathbb{R}}\int_{\mathbb{R}} \zeta(a)\xi(\exp xe^{a}X\ \exp aA)(e^{-2i\pi x} + e^{2i\pi x})dxda \\
&= \int_{\mathbb{R}}\int_{\mathbb{R}} \zeta(a)\xi_X(xe^{a})e^{-2i\pi af(A)}e^{\frac{1}{2}a}(e^{-2i\pi x} + e^{2i\pi x})dxda \\
&= \int_{\mathbb{R}}\int_{\mathbb{R}} \zeta(a)\xi_X(x)e^{-2i\pi af(A)}e^{-\frac{1}{2}a}(e^{-2i\pi e^{-a}x} + e^{2i\pi e^{-a}x})dxda \\
&= \int_{\mathbb{R}} (\zeta(a)e^{-2i\pi af(A)}e^{-\frac{1}{2}a})\big(\hat{\xi}_X(e^{-a}) + \hat{\xi}_X(-e^{-a})\big)da \\
&= \int_{0}^{+\infty} \big(\frac{1}{\sqrt{a}}\zeta(-\operatorname{Log} a)e^{2i\pi f(A)\operatorname{Log} a}\big)\big(\hat{\xi}_X(a) + \hat{\xi}_X(-a)\big)da \\
&= \int_{\mathbb{R}} \big(\frac{H^{+}(a)}{\sqrt{|a|}}\zeta(-\operatorname{Log}|a|)e^{2i\pi f(A)\operatorname{Log}|a|}\big)\big(\hat{\xi}_X(a) + \hat{\xi}_X(-a)\big)da \\
&= \int_{\mathbb{R}} \tilde{\zeta}(a)\big(\hat{\xi}_X(a) + \hat{\xi}_X(-a)\big)da = 0,
\end{aligned}$$

where $\tilde{\zeta}(a) = \frac{H^{+}(a)}{\sqrt{|a|}}\zeta(-\operatorname{Log}|a|)e^{2i\pi f(A)\operatorname{Log}|a|}$, $\xi_X(t) = \xi(\exp tX)$, for all $t \in \mathbb{R}$ and H^{+} is the Heaviside function. For all odd functions $\varphi \in C_c^{\infty}(\mathbb{R})$, there exists a function $\eta \in C_c^{\infty}(\mathbb{R})$ such that $\varphi = \hat{\eta}$. Then we have

$$\int_{\mathbb{R}} \tilde{\zeta}(a)\varphi(a)da = \langle \tilde{\zeta}, \varphi\rangle = 0.$$

Let $\varphi \in C_c^{\infty}(\mathbb{R})$, we can write $\varphi = \varphi_{-} + \varphi_{+}$, where φ_{-} and φ_{+} are respectively even and odd. We then obtain

$$\langle \tilde{\zeta}, \varphi\rangle = 2\langle \tilde{\zeta}, \varphi_{+}\rangle = 0.$$

Now, let $\varphi \in L^2(\mathbb{R})$. Then there exists a sequence $(\varphi_n)_{n\in\mathbb{N}}$ in $C_c^{\infty}(\mathbb{R})$ which converges to φ in L^2-norm. Then

$$|\langle \tilde{\zeta}, \varphi\rangle| = |\langle \tilde{\zeta}, \varphi - \varphi_n\rangle + \langle \tilde{\zeta}, \varphi_n\rangle| \leq \| \tilde{\zeta} \|_{L^2(\mathbb{R})} \| \varphi - \varphi_n \|_{L^2(\mathbb{R})} .$$

Then $\langle \tilde{\zeta}, \varphi\rangle = 0$ for all $\varphi \in L^2(\mathbb{R})$. As $\tilde{\zeta} \in L^2(\mathbb{R})$, we get that $\tilde{\zeta} = 0$ and therefore $\zeta = 0$.

Assume now that $\mathfrak{h}$ has codimension two. Let $\zeta \in (\mathrm{im}(\mathscr{U}))^{\perp}$ and $\xi \in C_c^{\infty}(G/H, f)$, then

$$\langle \zeta, \mathscr{U}(\xi)\rangle = \int_0^{2\pi}\int_{\mathbb{R}}\int_{\mathbb{R}^2} \zeta(w,a)\xi(\exp aA \exp xX \exp yY)e^{-2i\pi(x\cos\omega + y\sin\omega)}dxdydad\omega$$

$$= \int_0^{2\pi}\int_{\mathbb{R}}\int_{\mathbb{R}^2} \zeta(w,a)\xi_{X,Y}(e^a(x\cos\alpha a - y\sin\alpha a), e^a(y\cos\alpha a + x\sin\alpha a))$$

$$e^a e^{-2i\pi a f(A)} e^{-2i\pi(x\cos\omega + y\sin\omega)}dxdydad\omega$$

$$= \int_0^{2\pi}\int_{\mathbb{R}}\int_{\mathbb{R}^2} \zeta(w,a)\xi_{X,Y}(x,y)e^{-2i\pi e^{-a}(x\cos(\omega+\alpha a) + y\sin(\omega+\alpha a))}$$

$$e^{-a}e^{-2i\pi a f(A)}dxdydad\omega$$

$$= \int_0^{2\pi}\int_{\mathbb{R}}\int_{\mathbb{R}^2} \zeta(w-\alpha a,a)\xi_{X,Y}(x,y)e^{-2i\pi e^{-a}(x\cos\omega + y\sin\omega)}$$

$$e^{-a}e^{-2i\pi a f(A)}dxdydad\omega$$

$$= \int_0^{2\pi}\int_{\mathbb{R}} (\zeta(w-\alpha a,a)e^{-a}e^{-2i\pi a f(A)})\hat{\xi}_{X,Y}(e^{-a}\cos\omega, e^{-a}\sin\omega)dad\omega$$

$$= \int_0^{2\pi}\int_0^{+\infty} \left(\frac{1}{r}\zeta(w+\alpha \operatorname{Log} r, -\operatorname{Log} r)e^{2i\pi f(A)\operatorname{Log} r}\right)\hat{\xi}_{X,Y}(r\cos\omega, r\sin\omega)rdrd\omega$$

$$=0,$$

where $\xi_{X,Y}(x,y) = \xi(\exp xX \exp yY)$. We note

$$\widetilde{\zeta}(r\cos\omega, r\sin\omega) = \frac{1}{r}\zeta(w+\alpha \operatorname{Log} r, -\operatorname{Log} r)e^{2i\pi f(A)\operatorname{Log} r}.$$

Then for all $\varphi \in C_c^{\infty}(\mathbb{R}^2)$, we have

$$\int_0^{2\pi}\int_0^{+\infty} \widetilde{\zeta}(r\cos\omega, r\sin\omega)\varphi(r\cos\omega, r\sin\omega)rdrd\omega = 0.$$

As in the previous case, the last equality can be extended for all $\varphi \in L^2(\mathbb{R}^2)$. On the other hand, it is not hard to see that $\widetilde{\zeta} \in L^2(\mathbb{R}^2)$, indeed:

$$\begin{aligned}\int_{\mathbb{R}^2} |\widetilde{\zeta}(u,v)|^2 dudv &= \int_0^{2\pi}\int_0^{+\infty} |\frac{1}{r}\zeta(w+\alpha \operatorname{Log} r, -\operatorname{Log} r)|^2 rdrd\omega \\ &= \int_0^{2\pi}\int_{\mathbb{R}} |\zeta(\omega+\alpha a, -a)|^2 dad\omega \\ &= \int_0^{2\pi}\int_{\mathbb{R}} |\zeta(\omega,a)|^2 dad\omega = \|\zeta\|_2^2 < \infty.\end{aligned}$$

It follows then that $\widetilde{\zeta} = 0$ and finally $\zeta = 0$, which wraps up the theorem. □

3.5 Intertwining Operators of the Restriction of Representations of Nilpotent Lie Groups

We return to the setting of a simply connected nilpotent Lie group $G = \exp \mathfrak{g}$ and a closed connected subgroup $H = \exp \mathfrak{h}$. Given an irreducible unitary representation π of G, we present an explicit disintegration of the restriction $\pi_{|H}$ of π to H, which is based on a precise description of the space of double cosets $H\backslash G/B$, where B is any closed connected subgroup of G, and the well-known smooth disintegration of monomial representations of nilpotent Lie groups. The aim is still to write down a smooth intertwining operator for the decomposition of $\pi_{|H}$ into irreducibles. As an application we produce a concrete disintegration of tensor products of irreducible representations of G and a criterium for the irreducibility of these representations.

One should first study the general problem of describing a concrete disintegration of the restriction of an irreducible unitary representation of G to a closed connected subgroup $H = \exp \mathfrak{h}$. Since by Kirillov's theory every $\pi \in \hat{G}$ is of the form $\pi = \pi_{l,\mathfrak{b}} = \mathrm{ind}_B^G \chi_l$, where $l \in \mathfrak{g}^*$ and $\mathfrak{b} \subset \mathfrak{g}$ is a polarization at l, it is known from Mackey [124] that the restriction of π to H disintegrates over the set of double cosets $H\backslash G/B$ and that the integrands are of the form $\mathrm{ind}_{B(x)}^H \chi_{l(x)}$, where $B(x) = H \cap \psi(x)B\psi(x)^{-1}, x \in H\backslash G/B$ and $l(x) = \mathrm{Ad}^*(\psi(x))l_{|\mathfrak{h}}$, $x \in H\backslash G/B$ and $\psi : H\backslash G/B \to G$ is a section for the double cosets. The idea is to describe an open dense subset of $H\backslash G/B$ and a section ψ which give us an explicit description of $\pi_{|H}$ in term of an integral over $H\backslash G/B$ of the representations $\mathrm{ind}_{B(x)}^H \chi_{l(x)}$ (see Proposition 3.5.27). The results concerning the explicit disintegration of monomial representations are used to obtain a concrete disintegration of the restriction. This is somehow needed to get an "abstract" disintegration of the restriction into irreducibles to connected closed subgroups of simply connected nilpotent Lie groups. 'Abstract' here means that the measure class in $\hat{G}$ for the disintegration of the restriction and the multiplicities of the irreducibles appearing in the disintegration are given. These constructions will be applied to the disintegration of the tensor product of two irreducible representations π and π'.

3.5.1 Double-Coset Space

Let $\mathfrak{g}$ be a nilpotent Lie algebra, $\mathfrak{b}$ any subalgebra, $B \subset G$ their simply connected Lie groups. Recall that the exponential mapping $\exp : \mathfrak{g} \longrightarrow G$ is a diffeomorphism. Given a sequence of ideals

$$\mathfrak{g}_{n+1} := \{0\} \subsetneq \ldots \subsetneq \mathfrak{g}_i \subsetneq \ldots \subsetneq \mathfrak{g}_1 = \mathfrak{g},\ \dim(\mathfrak{g}_i/\mathfrak{g}_{i+1}) = 1, \tag{3.5.1}$$

denote for every $i = 1, \ldots, n$, $G_i := \exp \mathfrak{g}_i$ and choose a vector $Z_i \in \mathfrak{g}_i \setminus \mathfrak{g}_{i+1}$, so that $\mathfrak{g}_i = \mathbb{R}\text{-span}(Z_i, \ldots, Z_n)$. One obtains in this way a Jordan-Hölder basis $\mathscr{Z} := (Z_1, \ldots, Z_n)$ of $\mathfrak{g}$. To simplify the notations, let

$$V_1 \cdot V_2 \cdots V_k := \exp(V_1) \cdot \exp(V_2) \cdots \exp(V_k) \in G,$$

for given vectors $V_1, \ldots, V_k \in \mathfrak{g}$. Denote as before dg the Haar measure on G. Using the basis $\mathscr{Z}$, one can express dg in the following way.

$$\int_G f(g)\, dg = \int_{\mathbb{R}^n} f\big(z_1 Z_1 \cdots z_n Z_n\big)\, dz, \ \ (f \in L^1(G)).$$

Since G is nilpotent, the quotient space G/B has a G-invariant measure which is unique up to a positive scalar multiple. This measure (denoted by $d\dot{g}$) is described in Chap. 1, Sect. 1.2.2. Let us recall such construction. Let

$$\mathscr{I}^{\mathfrak{g}/\mathfrak{b}} = \big\{k \in \{1, \ldots, n\},\ Z_k \notin \mathfrak{b} + \mathfrak{g}_{k+1}\big\} =: \{k_1 < \ldots < k_p\}. \tag{3.5.2}$$

One obtains the sequence of subalgebras

$$\mathfrak{b}^{p+1} := \mathfrak{b} \subsetneq \ldots \subsetneq \mathfrak{b}^j := \mathbb{R}Z_{k_j} \oplus \mathfrak{b}^{j-1} \subsetneq \ldots \subsetneq \mathfrak{b}^1 = \mathfrak{g} \tag{3.5.3}$$

and the Malcev basis $\mathscr{M} := (Z_{k_1}, \ldots, Z_{k_p})$ of $\mathfrak{g}$ relative to $\mathfrak{b}$. The invariant measure $d\dot{g}$ is then given for $\varphi \in \mathscr{C}_c(G/B)$ by:

$$\mu_{\mathscr{M}}(\varphi) = \mu_{\mathfrak{g}/\mathfrak{b}}(\varphi) = \int_{G/B} \varphi(\dot{g})\, d\dot{g} := \int_{\mathbb{R}^p} \varphi\big(\, w_1 Z_{k_1} \cdots w_p Z_{k_p} \cdot B\,\big)\, dw, \tag{3.5.4}$$

where $\mathscr{C}_c(G/B)$ denotes the space of complex-valued continuous functions with compact support on G/B. This is a consequence of the fact that the mapping

$$\begin{array}{rccc} \dot{E}_{\mathscr{M}}: & \mathbb{R}^p & \longrightarrow & G/B, \\ & w = (w_1, \ldots, w_p) & \longmapsto & w_1 Z_{k_1} \cdots w_p Z_{k_p} \cdot B =: \dot{E}_{\mathscr{M}}(w) \end{array} \tag{3.5.5}$$

is a diffeomorphism. If $\mathfrak{h}$ is another subalgebra of $\mathfrak{g}$, then denote by $\mathbb{I}^{\mathfrak{h}} \subset \{1, \ldots, n\}$ the index set

$$\mathbb{I}^{\mathfrak{h}} := \{i \in \{1, \ldots, n\};\ \mathfrak{h} + \mathfrak{g}_{i+1} = \mathfrak{h} + \mathfrak{g}_i\} = \{1, \ldots, n\} \setminus \mathbb{I}^{\mathfrak{g}/\mathfrak{h}}. \tag{3.5.6}$$

One can then assume that the vectors $Z_i,\ i \in \mathbb{I}^{\mathfrak{h}}$, lie in $\mathfrak{h}$ so that $\mathfrak{h} = \mathbb{R}\text{-span}\{Z_i, i \in \mathbb{I}^{\mathfrak{h}}\}$.

The Set of Double Cosets

Let $G = \exp \mathfrak{g}$ be a connected, simply connected, nilpotent Lie group, $\mathfrak{b}$ and $\mathfrak{h}$ two subalgebras of $\mathfrak{g}$, $B = \exp \mathfrak{b}$ and $H = \exp \mathfrak{h}$ their Lie subgroups. We examine the double quotient space $H\diagdown G\diagup B$, which will be used to describe in Sect. 3.5.2 an appropriate measure $d\gamma$ appearing in the disintegration of $\pi_{|H}$ of π to H, where π is an irreducible unitary representation of G which is induced from a character χ_l of B.

Description of $H\diagdown G\diagup B$

For $g \in G$, denote by $\tilde{g}$ its double coset $H \cdot g \cdot B = \{hgb, (h, b) \in H \times B\}$. The aim is to find an open dense subset of $H\diagdown G\diagup B$ which will support the measure $d\gamma(\tilde{g})$ and which is diffeomorphic to a Zariski open subset of $\mathbb{R}^d$ for some $d \in \mathbb{N}^*$. The following example illustrates this fact:

Example 3.5.1 Let $\mathfrak{g}$ be the 7-dimensional Lie algebra spanned by the Jordan-Hölder basis $\mathscr{Z} = (Z_1, \dots, Z_7)$, equipped with the brackets

$$[Z_1, Z_4] = Z_6, \qquad [Z_1, Z_5] = Z_7, \qquad [Z_2, Z_3] = Z_7.$$

Consider its Abelian subalgebras $\mathfrak{h} = \mathbb{R}\text{-span}(Z_4, Z_5, Z_7)$ and $\mathfrak{b} = \mathbb{R}\text{-span}(Z_3, Z_4, Z_7)$. Since many products commute, the element $g =: z_1Z_1 \cdots z_7Z_7 \in G$, $(z_1, \dots, z_7) \in \mathbb{R}^7$, can be described in the following way:

$$\begin{aligned} g &= (z_1Z_1 \cdot z_5Z_5) \cdot z_2Z_2 \cdot z_6Z_6 \cdot (z_3Z_3 \cdot z_4Z_4 \cdot z_7Z_7) \\ &= (z_1z_5Z_7 \cdot z_5Z_5 \cdot z_1Z_1) \cdot z_2Z_2 \cdot z_6Z_6 \cdot (z_3Z_3 \cdot z_4Z_4 \cdot z_7Z_7) \\ &\in H \cdot z_1Z_1 \cdot z_2Z_2 \cdot z_6Z_6 \cdot B. \end{aligned}$$

This implies that $\tilde{g} = H \cdot g \cdot B = H \cdot z_1Z_1 \cdot z_2Z_2 \cdot z_6Z_6 \cdot B$. On the other hand if $z_1 \neq 0$, the element $z_1Z_1 \cdot z_2Z_2 \cdot z_6Z_6$ can also be written as $z_1Z_1 \cdot z_2Z_2$ conjugated by $\exp(-\frac{z_6}{z_1})Z_4$, which is contained in $H \cap B$. Hence if $z_1 \neq 0$, then $\tilde{g} = H \cdot z_1Z_1 \cdot z_2Z_2 \cdot B$. As a conclusion,

$$\begin{aligned} H\diagdown G\diagup B &= \left\{H \cdot z_1Z_1 \cdot z_2Z_2 \cdot B, (z_1, z_2) \in \mathscr{V}\right\} \dot{\cup} \left\{H \cdot z_2Z_2 \cdot z_6Z_6 \cdot B, (z_2, z_6) \in \mathbb{R}^2\right\} \\ &= \qquad\qquad \mathbf{p}(G \setminus G_2) \qquad\qquad \dot{\cup} \qquad\qquad \mathbf{p}(G_2) \end{aligned}$$

where $\mathbf{p} : g \longmapsto \tilde{g}$, is the canonical projection of G on $H\diagdown G\diagup B$, and $\mathscr{V} := \mathbb{R}^\star \times \mathbb{R}$. Hence the space $H\diagdown G\diagup B$ is the disjoint union of two subsets, the first is the projection of a Zariski open subset of G and the second of a Zariski closed subset. The measure $d\gamma(\tilde{g})$ is shown to be supported on the first set.

The General Case We shall prove that there is a Zariski open $\mathscr{V}$ of $\mathbb{R}^d$ which is homeomorphic to a subset $\tilde{\mathscr{U}}$ of $H\backslash G/B$ containing the double cosets of almost all (with respect to Haar measure) elements of G.

Definition 3.5.2

1. A subset $\mathscr{W}$ of G:
   1. is symmetric if $\mathscr{W} = \mathscr{W}^{-1} := \{g^{-1}, g \in \mathscr{W}\}$,
   2. has property $(\mathscr{P})$ (with respect to H and B), if $H\mathscr{W}B \subseteq \mathscr{W}$.
2. It is easy to see that for any $i \in \{1, \ldots, n-1\}$, the subset

$$\{g \in G, Z_i \notin \mathfrak{h} + \mathrm{Ad}(g)\mathfrak{b} + \mathfrak{g}_{i+1}\}$$

   either contains a Zariski open subset or its complement in G does. One characterizes the indices of the subset $\mathscr{I}(\mathfrak{g}/\mathfrak{h}, \mathfrak{g}/\mathfrak{b})$ of $\{1, \ldots, n\}$ by

$$i \in \mathscr{I}(\mathfrak{g}/\mathfrak{h}, \mathfrak{g}/\mathfrak{b}) \iff \begin{array}{c}\text{the subset } \mathscr{W}_i := \{g \in G, Z_i \notin \mathfrak{h} + \mathrm{Ad}(g)\mathfrak{b} + \mathfrak{g}_{i+1}\} \\ = \{g \in G, Z_i \notin \mathrm{Ad}(g^{-1})\mathfrak{h} + \mathfrak{b} + \mathfrak{g}_{i+1}\} \\ \text{contains a non-empty Zariski open subset of } G.\end{array}$$

   Denote by $\mathscr{U}_i$ the maximal symmetric Zariski open subset contained in $\mathscr{W}_i$.
3. Let also for $i \in \{1, \ldots, n\} \setminus \mathscr{I}(\mathfrak{g}/\mathfrak{h}, \mathfrak{g}/\mathfrak{b}) =: \mathbb{I}(\mathfrak{h}, \mathfrak{b})$,

$$\mathscr{W}_i := \{g \in G, Z_i \in \mathfrak{h} + \mathrm{Ad}(g)\mathfrak{b} + \mathfrak{g}_{i+1}\}.$$

   For $i \in \mathbb{I}(\mathfrak{h}, \mathfrak{b})$, consider the maximal symmetric Zariski open subset $\mathscr{U}_i$ of $\mathscr{W}_i$, with the property that for every $g \in \mathscr{U}_i$, there exist rational mappings $G \ni g' \mapsto h(g') \in \mathfrak{h}, g' \to b(g') \in \mathfrak{b}, g \to b_{i+1}(g) \in \mathfrak{g}_{i+1}$, with no singularities in a Zariski neighbourhood $\mathscr{U}_i(g)$ of g such that $Z_i = h(g') + \mathrm{Ad}(g')b(g) + b_{i+1}(g')$ for all $g' \in \mathscr{U}_i(g)$.

The following is immediate:

Proposition 3.5.3

1. *The following two conditions are equivalent:*
   (i) $\mathscr{U}_i \subset \{g \in G, Z_i \in \mathfrak{h} + \mathrm{Ad}(g)\mathfrak{b} + \mathfrak{g}_{i+1}\}$
   (ii) $\mathscr{U}_i \subset \{g \in G, Z_i \in \mathfrak{b} + \mathrm{Ad}(g)\mathfrak{h} + \mathfrak{g}_{i+1}\}$.
2. *The index subset* $\mathbb{I}(\mathfrak{g}/\mathfrak{h}, \mathfrak{g}/\mathfrak{b})$ *is contained in* $\mathbb{I}^{\mathfrak{g}/\mathfrak{h}} \cap \mathbb{I}^{\mathfrak{g}/\mathfrak{b}}$.
3. *Every* $\mathscr{W}_i, i \in \{1, \ldots, n\}$, *has property* $(\mathscr{P})$. *Furthermore, the maximal Zariski open subsets* $\mathscr{U}_i, i \in \mathscr{I}(\mathfrak{g}/\mathfrak{h}, \mathfrak{g}/\mathfrak{b})$, *of* G *have property* $(\mathscr{P})$.

Proof

1. Suppose for instance that $\mathscr{U}_i \subset \{g \in G, Z_i \in \mathfrak{b} + \mathrm{Ad}(g)\mathfrak{h} + \mathfrak{g}_{i+1}\}$. Since $\mathscr{U}_i$ is symmetric, for $g \in \mathscr{U}_i$, one has that $Z_i = X + \mathrm{Ad}(g^{-1})Y + S_{i+1}$ for some

$X \in \mathfrak{b}, Y \in \mathfrak{h}$ and $S_{i+1} \in \mathfrak{g}_{i+1}$. One obtains that

$$\mathrm{Ad}(g)Z_i = \mathrm{Ad}(g)X + Y + \mathrm{Ad}(g)S_{i+1}$$

and $\mathrm{Ad}(g)Z_i = Z_i \bmod \mathfrak{g}_{i+1}$. By putting $T_{i+1} := (Z_i - \mathrm{Ad}(g)Z_i) + \mathrm{Ad}(g)S_{i+1} \in \mathfrak{g}_{i+1}$, one gets $Z_i = \mathrm{Ad}(g)X + Y + T_{i+1}$. Hence $Z_i \in \mathfrak{h} + \mathrm{Ad}(g)\ \mathfrak{b}\ + \mathfrak{g}_{i+1}$, which means that $\mathscr{U}_i \subset \{g \in G, Z_i \in \mathfrak{h} + \mathrm{Ad}(g)\ \mathfrak{b}\ + \mathfrak{g}_{i+1}\}$.

2. If $i \in \{1, \ldots, n\} \setminus \mathbb{I}^{\mathfrak{g}/\mathfrak{h}} \cap \mathbb{I}^{\mathfrak{g}/\mathfrak{b}} = \mathbb{I}^{\mathfrak{b}} \cup \mathbb{I}^{\mathfrak{h}}$, then $Z_i \in \mathfrak{b} + \mathfrak{g}_{i+1}$ or $Z_i \in \mathfrak{h} + \mathfrak{g}_{i+1}$. Therefore $\mathrm{Ad}(g)(Z_i) \in \mathfrak{b} + \mathfrak{g}_{i+1}$ for every $g \in G$ or $\mathrm{Ad}(g)(Z_i) \in \mathfrak{h} + \mathfrak{g}_{i+1}$ for every $g \in G$ and so $Z_i \in \mathrm{Ad}(g^{-1})\mathfrak{b} + \mathfrak{g}_{i+1} \subset \mathrm{Ad}(g^{-1})\mathfrak{b} + \mathfrak{h} + \mathfrak{g}_{i+1}$ for every $g \in G$ or $Z_i \in \mathrm{Ad}(g^{-1})\mathfrak{h} + \mathfrak{g}_{i+1} \subset \mathrm{Ad}(g^{-1})\mathfrak{h} + \mathfrak{b} + \mathfrak{g}_{i+1}$ for every $g \in G$. Hence $i \in \mathbb{I}(\mathfrak{h}, \mathfrak{b})$.
3. If $Z_i = \mathrm{Ad}(g)X + Y + T_{i+1} = X + \mathrm{Ad}(g^{-1})Y + S_{i+1}$ for some $X \in \mathfrak{b}, Y \in \mathfrak{h}$ and $T_{i+1}, S_{i+1} \in \mathfrak{g}_{i+1}$, then for every $b \in B, h \in H$, one has

$$Z_i = \mathrm{Ad}(gb)(\mathrm{Ad}(b^{-1})X) + Y + Z_{i+1}$$

and $Z_i = X + \mathrm{Ad}((hg)^{-1})(\mathrm{Ad}(h)Y) + S_{i+1}$. This shows that the subset of G consisting of elements g such that $Z_i \subset \mathrm{Ad}(g)\mathfrak{b} + \mathfrak{h} + \mathfrak{g}_{i+1}$ is H-invariant on the right and B- invariant on the left, and therefore also its complement is. □

Remark 3.5.4 Putting $\mathscr{U} := \bigcap_{1 \leq i \leq n} \mathscr{U}_i$, one obtains a "canonical" non-empty Zariski open of G having property ($\mathscr{P}$), which allows us to say that

$$\mathscr{I}(\mathfrak{g}/\mathfrak{h}, \mathfrak{g}/\mathfrak{b}) = \{1 \leq i \leq n,\ \forall g \in \mathscr{U},\ Z_i \notin \mathfrak{h} + \mathrm{Ad}(g)\mathfrak{b} + \mathfrak{g}_{i+1}\}. \tag{3.5.7}$$

Definition 3.5.5 Write $\mathscr{I}(\mathfrak{g}/\mathfrak{h}, \mathfrak{g}/\mathfrak{b}) = \{i_1 < \ldots < i_d\}$, and put

1.
$$\phi : \mathbb{R}^d \longrightarrow G, \quad (t_1, \ldots, t_d) \longmapsto t_1 Z_{i_1} \cdots t_d Z_{i_d}$$

2.
$$\mathbf{p} : G \longrightarrow H \diagdown G \diagup B, \quad g \longmapsto \tilde{g} = H \cdot g \cdot B$$

3.
$$\mathscr{V} = \phi^{-1}(\mathscr{U}) := \{t \in \mathbb{R}^d, \phi(t) \in \mathscr{U}\}, \ \tilde{\mathscr{U}} = \mathbf{p}(\mathscr{U}) := \{\tilde{g}, g \in \mathscr{U}\}$$

4.
$$\begin{array}{rccc} \tilde{\phi} := \mathbf{p} \circ \phi_{|\mathscr{V}} : & \mathscr{V} & \longrightarrow & \tilde{\mathscr{U}}, \\ & t = (t_1, \ldots, t_d) & \longmapsto & H \cdot \phi(t) \cdot B \end{array}$$

5. The space of double cosets $H\backslash G/B$ is endowed with the quotient topology, ie a subset $\tilde{U}$ in $H\backslash G/B$ is open if and only if the subset $U := \mathbf{p}^{-1}(\tilde{U})$ is open in G. The mapping $\mathbf{p}$ is then open, since for every open subset U of G, one has that $\mathbf{p}^{-1}(\mathbf{p}(U)) = HUB = \bigcup_{h\in H, b\in B} hUb$ is open in G.

Lemma 3.5.6 *In the particular case where $\mathfrak{b}$ is an ideal, one has that $\tilde{\phi}$ is a diffeomorphism between $\mathbb{R}^d$ and $H\backslash G/B$.*

Proof The closed subgroup $HB = BH$ has $\mathfrak{h} + \mathfrak{b}$ as a Lie algebra, and it satisfies $HB\backslash G = H\backslash G/B$. Indeed for every $g \in G$,

$$H \cdot g \cdot B = \{hgb, h \in H, b \in B\} = \{hb'g, h \in H, gbg^{-1} =: b' \in B\} = HB \cdot g. \tag{3.5.8}$$

Notice that $\mathscr{I}(\mathfrak{g}/\mathfrak{h}, \mathfrak{g}/\mathfrak{b}) = \mathscr{I}^{\mathfrak{g}/\mathfrak{h}+\mathfrak{b}} := \{Z_{i_1}, \ldots, Z_{i_d}\}$, since the condition $Z_i \notin \mathfrak{h} + \mathrm{Ad}\, g\ \mathfrak{b}\ + \mathfrak{g}_{i+1}$ is now equivalent to the condition $Z_i \notin \mathfrak{h} + \mathfrak{b} + \mathfrak{g}_{i+1}$. This tells us that $\mathscr{U}$ is all of G. Hence $(Z_{i_1}, \ldots, Z_{i_d})$ is a Malcev basis for $\mathfrak{g}$ relative to $\mathfrak{h} + \mathfrak{b}$. Finally, by Corwin and Greenleaf [38], one has diffeomorphism: $t = (t_1, \ldots, t_d) \longmapsto HB\cdot(-t_d Z_{i_d})\cdots(-t_1 Z_{i_1})$ between $\mathbb{R}^d$ and the set $HB\backslash G$ of the right cosets modulo HB. If this mapping is composed with the mapping $HB\backslash G \ni HB\cdot g^{-1} \longmapsto g\cdot HB \in G/HB$, which is also a diffeomorphism, then one obtains that the mapping $\mathbb{R}^d \ni t = (t_1, \ldots, t_d) \longmapsto t_1 Z_{i_1} \cdots t_d Z_{i_d} \cdot HB \ \in G/HB$, which is nothing else than $\tilde{\phi}$ by Definition 3.5.5 (4), is a diffeomorphism as well. □

Remark 3.5.7 In Theorem 3.5.10, the canonical mapping $\tilde{\phi} : \mathscr{V} \to \tilde{\mathscr{U}}$ is shown to be a homeomorphism. Consider first the index

$$l := \max\{i \in \mathscr{I}^{\mathfrak{g}/\mathfrak{b}}\}$$

and set

$$\mathfrak{a} := \mathfrak{g}_{l+1}, \quad \mathfrak{c} := \mathfrak{b} + \mathfrak{g}_l = \mathfrak{b} \oplus \mathbb{R}Z_l.$$

Since $Z_j \in \mathfrak{b}$ for every $j > l$, it follows that $\mathfrak{a}$ is an ideal of $\mathfrak{g}$ contained in $\mathfrak{b}$. Furthermore

$$\mathscr{I}^{\mathfrak{g}/\mathfrak{c}} = \mathscr{I}^{\mathfrak{g}/\mathfrak{b}} \setminus \{l\} = \{k_1 < \cdots < k_{p-1}\}.$$

Hence

$$\mathscr{I}^{\mathfrak{g}/\mathfrak{h}} \cap \mathscr{I}^{\mathfrak{g}/\mathfrak{c}} = \mathscr{I}^{\mathfrak{g}/\mathfrak{h}} \cap \mathscr{I}^{\mathfrak{g}/\mathfrak{b}} \setminus \{l\}.$$

The following two cases must be discussed in the proof of Theorem 3.5.10.

(a) $Z_l \notin \mathfrak{h} + \mathfrak{g}_{l+1}$, which means that $l \in \mathscr{I}^{\mathfrak{g}/\mathfrak{h}}$ and so

$$\mathscr{I}^{\mathfrak{g}/\mathfrak{h}} \cap \mathbb{I}^{\mathfrak{g}/\mathfrak{b}} = \mathscr{I}^{\mathfrak{g}/\mathfrak{h}} \cap \mathbb{I}^{\mathfrak{g}/\mathfrak{c}} \cup \{l\}.$$

(b) $Z_l \in \mathfrak{h} + \mathfrak{g}_{l+1}$, whence $Z_l \in \mathfrak{h}$ by Remark 3.5.4, (b). In this case one has $l \notin \mathscr{I}^{\mathfrak{g}/\mathfrak{h}}$ and so $\mathscr{I}^{\mathfrak{g}/\mathfrak{h}} \cap \mathscr{I}^{\mathfrak{g}/\mathfrak{b}} = \mathscr{I}^{\mathfrak{g}/\mathfrak{h}} \cap \mathscr{I}^{\mathfrak{g}/\mathfrak{c}}$.

Since $\mathfrak{g}_l \subset \mathfrak{g}_{i+1}$ for all $i < l$, it follows that

$$\mathfrak{h} + \mathrm{Ad}(g)\mathfrak{c} + \mathfrak{g}_{i+1} = \mathfrak{h} + \mathrm{Ad}(g)(\mathfrak{b} + \mathfrak{g}_l) + \mathfrak{g}_{i+1} = \mathfrak{h} + \mathrm{Ad}(g)\mathfrak{b} + \mathfrak{g}_{i+1},\ g \in G.$$

This tells us that $\mathscr{I}(\mathfrak{g}/\mathfrak{h}, \mathfrak{g}/\mathfrak{c}) = \mathscr{I}(\mathfrak{g}/\mathfrak{h}, \mathfrak{g}/\mathfrak{b}) \setminus \{l\}$. Hence in case (**b**), one has that

$$\mathscr{I}(\mathfrak{g}/\mathfrak{h}, \mathfrak{g}/\mathfrak{b}) = \mathscr{I}(\mathfrak{g}/\mathfrak{h}, \mathfrak{g}/\mathfrak{c}),$$

since $l \notin \mathscr{I}^{\mathfrak{g}/\mathfrak{h}} \supset \mathscr{I}(\mathfrak{g}/\mathfrak{h}, \mathfrak{g}/\mathfrak{b})$ by Proposition 3.5.3 (1). At first look the following two possibilities are at hand:

(a-1) $l \in \mathscr{I}(\mathfrak{g}/\mathfrak{h}, \mathfrak{g}/\mathfrak{b})$. This case is characterized by the existence of a non-empty Zariski open subset $\mathscr{U}_l$ of G, given by Proposition 3.5.3, such that for every $g \in \mathscr{U}_l$, one has $Z_l \notin \mathfrak{h} + \mathrm{Ad}(g)\mathfrak{b}$ (since $\mathfrak{g}_{l+1} \subseteq \mathrm{Ad}(g)\mathfrak{b}$ for $g \in G$).

(a-2) $l \notin \mathscr{I}(\mathfrak{g}/\mathfrak{h}, \mathfrak{g}/\mathfrak{b})$. Here one has $\mathscr{I}(\mathfrak{g}/\mathfrak{h}, \mathfrak{g}/\mathfrak{c}) = \mathscr{I}(\mathfrak{g}/\mathfrak{h}, \mathfrak{g}/\mathfrak{b})$ as in case b). Hence for every g in a non-empty Zariski open subset $\mathscr{U}_l$ of G, one has that $Z_l \in \mathfrak{h} + \mathrm{Ad}(g)\mathfrak{b}$.

Remark 3.5.8 For every $(w, \tilde{g}) \in G_l \times H\backslash G/B$, one has $H \cdot wgw^{-1} \cdot B = \tilde{g}$. This is an immediate consequence of the fact that $wgw^{-1} \equiv g\ mod[B]$, $(g \in G)$.

One has the following:

Lemma 3.5.9 *The mapping*

$$\tilde{\phi} : \mathscr{V} \ni (t_1, \ldots, t_d) \longmapsto H \cdot t_1 Z_{i_1} \ldots t_d Z_{i_d} \cdot B \in \tilde{\mathscr{U}}$$

is bijective.

Proof Proceed by induction on the codimension p of B. If $p = 1$, then the claim is trivial in view of Lemma 3.5.6, since $\mathfrak{b}$ is then an ideal of $\mathfrak{g}$. Let $l = \max \mathscr{I}^{\mathfrak{g}/\mathfrak{b}} = \{k_1 < \ldots < k_p = l\}$ and $\mathfrak{c} := \mathbb{R}Z_l + \mathfrak{b}$ as in Remark 3.5.7. Apply the induction hypothesis to $\mathfrak{g}$, $\mathfrak{h}$ and $\mathfrak{c}$, giving us the bijection $\tilde{\phi^{o}} : \mathscr{V}^{o} \ni s \longmapsto H \cdot \phi^{o}(s) \cdot C \in \tilde{\mathscr{U}^{o}}$. Let us introduce the projection

$$\begin{array}{rccc} \Lambda^{o}: & H\backslash G/B & \longrightarrow & H\backslash G/C \\ & H \cdot g \cdot B = \tilde{g} & \longmapsto & \tilde{\tilde{g}} = H \cdot g \cdot C \end{array} \tag{3.5.9}$$

Notice that the surjectivity of $\tilde{\phi}$ is equivalent to $H\phi(\mathscr{V})B = \mathscr{U}$. Since one has already the first inclusion "$\subseteq$" by property ($\mathscr{P}$), for the surjectivity it suffices to prove that one has the other inclusion: $\{h\phi(t)b, h \in H, t \in \mathscr{V}, b \in B\} \supseteq \mathscr{U}$. Discuss now the three cases of Remark 3.5.7.

Case (a-1) $l \in \mathscr{I}(\mathfrak{g}/\mathfrak{h}, \mathfrak{g}/\mathfrak{b}) = \{i_1, \ldots, i_d = l\}$. Let us recall that $\mathscr{I}(\mathfrak{g}/\mathfrak{h}, \mathfrak{g}/\mathfrak{c}) = \mathscr{I}(\mathfrak{g}/\mathfrak{h}, \mathfrak{g}/\mathfrak{b}) \setminus \{l\} = \{i_1, \ldots, i_{d-1}\}$. The map ϕ^o goes from $\mathbb{R}^{d-1}$ to G. One has also the inclusion $\bigcap_{i \in \mathscr{I}^{\mathfrak{h},\mathfrak{c}}} \mathscr{U}_i = \mathscr{U}^o \supseteq \mathscr{U} = \mathscr{U}^o \cap \mathscr{U}_l$. By the induction hypothesis,

$$\widetilde{\phi}^o : \mathscr{V}^o \ni s = (t_1, \ldots, t_{d-1}) \longmapsto H \cdot \phi^o(s) \cdot C \in \widetilde{\mathscr{U}}^o$$

is bijective. As defined in Definition 3.5.5, $\mathscr{V}^o$ is the non-empty Zariski open subset $\phi^{o-1}(\mathscr{U}^o)$ of $\mathbb{R}^{d-1}$ and $\widetilde{\mathscr{U}}^o$ is the open dense subset $\{\widetilde{g} = H \cdot g \cdot C, g \in \mathscr{U}^o\}$ of $H \backslash G / C$.

- Surjectivity of $\tilde{\boldsymbol{\phi}}$: let $u \in \mathscr{U} = \mathscr{U}^o \cap \mathscr{U}_l$. As $\mathscr{U}^o = H\phi^o(\mathscr{V}^o)C$, one can write $u = h\phi^o(s)c$ for certain $h \in H, s = (s_1, \ldots, s_{d-1}) \in \mathscr{V}^o$ and $c \in C$. One has $c = (\exp \gamma Z_l)b$ where $b \in B$ and $\gamma \in \mathbb{R}$. Denoting by $t = (s_1, \ldots, s_{d-1}, t_d := \gamma) \in \mathbb{R}^d$, one obtains that $u = h\phi(t)b$. But $t \in \mathscr{V}$ since $\phi(t) = h^{-1}ub^{-1}$ is in $\mathscr{U}$ because of $(\mathscr{P})$. Finally $u \in H\phi(\mathscr{V})B$.
- Injectivity of $\tilde{\boldsymbol{\phi}}$: At first, note that if $t = (t_1, \ldots, t_d) \in \mathbb{R}^d$ is in $\mathscr{V}$, then $s := (t_1, \ldots, t_{d-1}) \in \mathbb{R}^{d-1}$ must be in $\mathscr{V}^o$. Indeed, $\phi(t) \in \mathscr{U} = \mathscr{U}^o \cap \mathscr{U}_l \subset \mathscr{U}^o$ and so, by property $(\mathscr{P})$, the stability of $\mathscr{U}^o$ under right multiplication with elements of C gives us $\phi^o(s) = \phi(t)\exp(-t_d Z_l) \in \mathscr{U}^o$, hence $s \in \mathscr{V}^o$. Let now $t = (t_1, \ldots, t_d), t' = (t'_1, \ldots, t'_d) \in \mathscr{V}$ such that $H \cdot \phi(t) \cdot B = H \cdot \phi(t') \cdot B$. Then $H \cdot \phi(s) \cdot C = H \cdot \phi(s') \cdot C$, where $s = (t_1, \ldots, t_{d-1}), s' = (t'_1, \ldots, t'_{d-1}) \in \mathscr{V}^o$. Now since $\widetilde{\phi}^o$ is injective, $s = s'$. Finally, to show that $t_d = t'_d$, make the following calculation. Let $h = \exp Y \in H$ and $b = \exp X \in B$, such that $\phi(t) = h\phi(t')b$, and let $g = \phi^o(s) \in \mathscr{U}^o$. Then $g\exp(t_d Z_l)$ and $g\exp(t'_d Z_l) \in \mathscr{U}$. One then has, if $t'_d - t_d \neq 0$,

$$\begin{aligned}
g\exp(t_d Z_l) &= \exp(Y) g \exp(t'_d Z_l)\exp(X) \\
&\Leftrightarrow \\
g^{-1}\exp(-Y)g &= \exp(t'_d Z_l)\exp(X)\exp(-t_d Z_l) \\
&\Leftrightarrow \\
\exp(-\mathrm{Ad}(g^{-1})Y) &= \exp(X + (t'_d - t_d)Z_l) \bmod G_{l+1} \\
&\quad \text{(by the Campbell–Baker–Hausdorff formula)} \\
&\Rightarrow Z_l - \frac{1}{t'_d - t_d}\big(\mathrm{Ad}(g^{-1})(Y) + X\big) \in \mathfrak{g}_{l+1} \\
&\Rightarrow Z_l \in \mathfrak{b} + \mathrm{Ad}(g^{-1})(\mathfrak{b}) + \mathfrak{g}_{l+1}.
\end{aligned}$$

This last line contradicts the fact that $g \in \mathscr{U}^o$. Hence $t_d = t'_d$ and $\tilde{\phi}$ is injective.

Case (a-2) $\mathscr{I}^{\mathfrak{g}/\mathfrak{h}} \cap \mathbb{I}^{\mathfrak{g}/\mathfrak{b}} = I^{\mathfrak{g}/\mathfrak{h}} \cap \mathbb{I}^{\mathfrak{g}/\mathfrak{c}} \cup \{l\}$ but $l \notin \mathscr{I}(\mathfrak{g}/\mathfrak{h}, \mathfrak{g}/\mathfrak{b}) = \{i_1, \ldots, i_d\} = \mathscr{I}(\mathfrak{g}/\mathfrak{h}, \mathfrak{g}/\mathfrak{c})$. Here, one has $\phi^o = \phi : \mathbb{R}^d \longrightarrow G$, but the non-empty Zariski subsets of $\mathbb{R}^d$ defined by $\mathscr{V}^o = \phi^{-1}(\mathscr{U}^o)$ and $\mathscr{V} = \phi^{-1}(\mathscr{U})$ do not coincide in general. In fact, one knows that

$$\mathscr{U}^o = \bigcap_{i \in \mathscr{I}^{\mathfrak{h},\mathfrak{g}}} \mathscr{U}_i \supseteq \mathscr{U} = \mathscr{U}^o \cap \mathscr{U}_l,$$

where the Zariski open subset $\mathscr{U}_l$ of G satisfies $Z_l \in \mathfrak{h} + \mathrm{Ad}(g)\mathfrak{b}$ for any $g \in \mathscr{U}_l$. By the induction hypothesis, $\widetilde{\mathscr{U}}^o = \{\tilde{g} = H \cdot g \cdot C, g \in \mathscr{U}^o\}$ is in 1-1 correspondence with $\mathscr{V}^o$ because of $\widetilde{\phi}^o$.

- Surjectivity of $\tilde{\boldsymbol{\phi}}$: Let $u \in \mathscr{U} = \mathscr{U}^o \cap \mathscr{U}_l \subset \mathscr{U}^o$. Then u can be written as $h\phi^o(t)c = h\phi(t)c$, where $h \in H, c = \exp(\gamma Z_l)b \in C$ and $t \in \mathscr{V}^o$. Then also $v := \phi(t)\exp(\gamma Z_l) \in \mathscr{U}$, because of property $\mathscr{P}$. It will be shown that

$$v = h'\phi(t)b' \tag{3.5.10}$$

for some $h' \in H, b' \in B$. This then shows that $t \in \mathscr{V} = \phi^{-1}(\mathscr{U})$ and so one sees that $\tilde{\phi}$ is surjective. Now since $v \in \mathscr{U}_l$, one has $Z_l \in \mathrm{Ad}(v^{-1})\mathfrak{h} + \mathfrak{b}$. There exist, therefore, $Y =: \log h' \in \mathfrak{h}$ and $X =: \log x \in \mathfrak{b}$, such that

$$\begin{aligned}
\gamma Z_l &= \mathrm{Ad}(v^{-1})Y + X \qquad (3.5.11)\\
&\Leftrightarrow -X + \gamma Z_l = \mathrm{Ad}(v^{-1})Y\\
&\Rightarrow \exp(\gamma Z_l)x^{-1}q = v^{-1}h'v \text{ (for some } q \in G_{l+1} \subset B)\\
&\Rightarrow \exp(\gamma Z_l) = v^{-1}h'vb'' \text{ (for some } b'' \in B)\\
&\Rightarrow v = \phi(t)\exp(\gamma Z_l) = \phi(t)(\exp(-\gamma Z_l)\phi(t)^{-1}h'\phi(t)\exp(\gamma Z_l)b'')\\
&\Rightarrow v = \phi(t)(\phi(t)^{-1}h'\phi(t)b') = h'\phi(t)b' \text{ (for some } b' \in B).
\end{aligned}$$

- Injectivity of $\tilde{\boldsymbol{\phi}}$: This is an immediate consequence from the injectivity of $\widetilde{\phi}^o$, since $\mathscr{V} \subset \mathscr{V}^o$ and $\tilde{\phi} = \widetilde{\phi}^o_{|\mathscr{V}}$.

Case (b) $l \notin \mathscr{I}^{\mathfrak{g}/\mathfrak{h}}$. This means that $Z_l \in \mathfrak{h}$. Now one has $\mathscr{I}(\mathfrak{g}/\mathfrak{h}, \mathfrak{g}/\mathfrak{c}) = \mathscr{I}(\mathfrak{g}/\mathfrak{h}, \mathfrak{g}/\mathfrak{b}) = \{i_1, \ldots, i_d\}, \phi^o = \phi, \mathscr{U}^o = \mathscr{U} = \bigcap_{i \in \mathscr{I}} \mathscr{U}_i$ and $\mathscr{V}^o = \phi^{o-1}(\mathscr{U}^o) = \phi^{-1}(\mathscr{U}) = \mathscr{V}$. But of course the spaces $H\backslash G/B$ and $H\backslash G/C$ are different, and $\widetilde{\phi}^o$ and $\tilde{\phi}$ are related by the formula

$$\begin{aligned}
\widetilde{\phi}^o = \Lambda^o \circ \tilde{\phi} : \qquad &\mathscr{V} \longrightarrow \widetilde{\mathscr{U}}^o\\
t = (t_1, \ldots, t_d) &\longmapsto H \cdot \phi(t) \cdot C
\end{aligned}$$

- Surjectivity of $\tilde{\boldsymbol{\phi}}$: As one has already $\mathscr{U} = \mathscr{U}^o = H\phi^o(\mathscr{V}^o)C = H\phi(\mathscr{V})C$, it suffices to prove the inclusion $H\phi(\mathscr{V})C \subseteq H\phi(\mathscr{V})B$. Now if one takes $u = h\phi(t)c$, where the elements $h, t = (t_1, \ldots, t_d), c = \exp(\gamma Z_l)b, \gamma$ and b are respectively in $H, \mathscr{V}, C, \mathbb{R}$ and B. Hence

$$u = h\phi(t)c = h\phi(t)\exp(\gamma Z_l)b = \big(h\exp(\gamma Z_l)\big)\ \exp(-\gamma Z_l)\phi(t)\exp(\gamma Z_l)\ b$$

which gives

$$\begin{aligned} H\cdot u\cdot B &= H\cdot h\exp(\gamma Z_l)\exp(-\gamma Z_l)\phi(t)\exp(\gamma Z_l)b\cdot B \qquad (3.5.12)\\ &= H\cdot\exp(-\gamma Z_l)\phi(t)\exp(\gamma Z_l)\cdot B = H\phi(t)B, \end{aligned}$$

by Remark 3.5.8.

- Injectivity of $\tilde{\boldsymbol{\phi}}$: This is the same as in case a-2), in fact $\widetilde{\phi}^{o} = \Lambda^{o}\circ\tilde{\phi}$ is injective, and so is $\tilde{\phi}$.

□

Theorem 3.5.10 *The mapping* $\tilde{\phi} : \mathscr{V} \ni t = (t_1,\ldots,t_d) \longmapsto H\cdot t_1 Z_{i_1}\ldots t_d Z_{i_d}\cdot B \in \tilde{\mathscr{U}}$ *is a homeomorphism.*

Proof The mapping $\tilde{\phi} = \mathbf{p}\circ\phi_{|_{\mathscr{V}}}$ is evidently continuous since so are ϕ and $\mathbf{p}$. Hence it is enough to prove that the mapping

$$H\times\mathbb{R}^d\times B\to G, (h,t,b)\to h\phi(t)b$$

is open. Let us do this again by induction on the dimension of G/B, using the notations of the proof of Lemma 3.5.9. If the dimension of G/B =1, then B is normal and apply Lemma 3.5.6.

In **Case (a-1)**, it suffices to consider subsets $W_H\subset H$, which are open in H, $W = W_0\times]a,b[\subset\mathscr{V}$ which are open in $\mathbb{R}^{d-1}\times\mathbb{R}$ and subsets $W_B\subset B$, which are open in B, and then prove that $W_H\phi(W)W_B$ is open in G. But by induction hypothesis for the groups H and C, one has that

$$W_H\phi(W)W_B = W_B\phi^{o}(W_0)\exp(]a,b[Z_l)W_B$$

is open in G, since $\exp(]a,b[Z_l)W_B$ is open in B.

In case **Case (a-2)**, consider open subsets W_H of H, $W = W_0\times]a,b[$ open in $\mathscr{V}\subset\mathbb{R}^{d-1}\times\mathbb{R}$ and W_B open in B and prove that $W_H\phi(W)W_B$ is open in G. Let $b\in W_B, h\in W_H, t=(t_1,\cdots,t_d)\in W$. One has that $\phi = \phi^{o}_{|\mathscr{V}}$ and that W is open in $\mathscr{V}^{o}$. Since $l\in\mathscr{I}^{\mathfrak{g}/\mathfrak{h}}\cap I^{\mathfrak{g}/\mathfrak{b}}\setminus\mathscr{I}(\mathfrak{g}/\mathfrak{h},\mathfrak{g}/\mathfrak{b})$, there exist rational mappings

$$g\mapsto Y(g)\in\mathfrak{h}, g\mapsto X(g)\in\mathfrak{b},$$

defined on a Zariski open neighbourhood of $\phi(t)$, which can be assumed to contain W, such that for every g in this neighbourhood one has that

$$\begin{aligned} Z_l &= \mathrm{Ad}(g^{-1})Y(g) + X(g)\\ &\Rightarrow \exp(t_d Z_l - t_d X(g)) = g^{-1}\exp(t_d Y(g))g,\ t_d\in\mathbb{R}, \end{aligned}$$

$$\Rightarrow \exp(t_d Z_l)b(g, t_d)^{-1} = g^{-1}\exp(tY(g))g \text{ (where } (g, t_d) \mapsto b(g, t_d) \in B \text{ is rational)},$$

$$\Rightarrow \exp(t_d Z_l) = g^{-1}h(g, t_d)gb(g, t_d)(\text{where } h(t_d) = \exp(t_d Y(g) \in H).$$

Hence for $b' \in B, h' \in H, t' \in W, t_d \in \mathbb{R}$, one gets

$$h'\phi(t')\exp(t_d Z_l)b' = h'h(\phi(t'), t_d)\phi(t')b(g, t_d)b'. \tag{3.5.13}$$

Since the mapping

$$H \times B \times \mathbb{R} \times W \to H \times B \times \mathbb{R} \times W;$$

$$(h', b', t_d, t') \mapsto (h'h(\phi(t'), t_d), b(g, t_d)b', t_d, t),$$

is a diffeomorphism, one can find small neighbourhoods $W'_H \subset W_H$ of h, $W'_B \subset W_B$ of b and $W' \subset W$ of t, such that for every $t_d \in]-\varepsilon, \varepsilon[$ for some $\varepsilon > 0$, one has that

$$(h'h(\phi(t'), t_d), b(g, t_d)b') \subset W_H \times W_B$$

This implies by (3.5.13) that

$$h\phi(t)b \in W'_B\phi(W')\exp(]-\varepsilon, \varepsilon[Z_l)W'_B \subset W_H\phi(W)W_B.$$

By the induction hypothesis, the subset $W'_B\phi(W')\exp(]-\varepsilon, \varepsilon[Z_l)W'_B$ is an open neighbourhood of $h\phi(t)b$ and therefore the set $W_H\phi(W)W_H$ is open in G. In case b), one has that Z_l is an element of $\mathfrak{c}$. Consider again open subsets W_H of H, $W = W_0 \times]a, b[$ open in $\mathscr{V} \subset \mathbb{R}^d$ and W_B open in B and prove that $W_H\phi(W)W_B$ is open in G. Let $b \in W_B, h \in W_H, t = (t_1, \cdots, t_d) \in W$. For $h' \in H, b' \in B, s \in \mathbb{R}$ and $t \in W$, one has that

$$\begin{aligned} h'\phi(t)\exp(sZ_l)b' &= h'\exp(sZ_l)(\exp(-sZ_l)\phi(t)\exp(sZ_l))b' \\ &= h'\exp(sZ_l)\phi(t)b(s, t)b', \end{aligned} \tag{3.5.14}$$

where $b(s, t) \in B$ is polynomial in (s, t). This implies, as in case (a-2), that there exist small neighbourhoods $W'_H \subset W_H$ of h, $W'_B \subset W_B$ of b and $W' \subset W$ of t, such that for every $s \in]-\varepsilon, \varepsilon[$ (for some $\varepsilon > 0$), $h' \in W'_H, b' \in W'_B$, one has that

$$(h'h, b(s, t)b') \subset W_H \times W_B \tag{3.5.15}$$

Therefore, by (3.5.14) and (3.5.15),

$$h\phi(t)b \in W'_B\phi(W')\exp(]-\varepsilon, \varepsilon[Z_l)W'_B \subset W_H\phi(W)W_B$$

and we conclude as in case (a-2). □

Definition 3.5.11 Denote by

$$\psi : H\backslash G/B \to G,$$

the inverse of the mapping $\phi(\mathscr{V}) \to H\backslash G/B$. Then ψ is a continuous section of the open dense subset $\tilde{\mathscr{U}}$ of $H\backslash G/B$ into G.

Example 3.5.12 Returning to the previous example, one can see that $\mathscr{I}^{\mathfrak{g}/\mathfrak{h}} \cap \mathbb{I}^{\mathfrak{g}/\mathfrak{b}} = \{1, 2, 6\}$ and $\mathscr{I}(\mathfrak{g}/\mathfrak{h}, \mathfrak{g}/\mathfrak{b}) = \{1, 2\}$. In fact, if $g = z_1Z_1 \ldots z_7Z_7 \in G$, then $\mathrm{Ad}(g)\mathfrak{b} = \mathbb{R}\text{-span}(Z_3, Z_4 + z_1Z_6, Z_7)$, so in the Zariski open set $\mathscr{U}_6 := G \setminus G_2$ of G, one has $\forall g \in \mathscr{U}_6$, $Z_6 \in \mathfrak{h} + \mathrm{Ad}(g)\, \mathfrak{b} + \mathfrak{g}_7 = \mathbb{R}\text{-span}(Z_3, Z_4, Z_5, z_1Z_6, Z_7)$, but for $i = 1, 2$, it is clear that $Z_i \notin \mathfrak{h} + \mathrm{Ad}(g)\mathfrak{b} + \mathfrak{g}_{i+1}$, for any $g \in G =: \mathscr{U}_i$. In conclusion,

$$\mathscr{I}(\mathfrak{g}/\mathfrak{h}, \mathfrak{g}/\mathfrak{b}) = \big\{i \in \mathscr{I} = \{1, 2, 6\},\ Z_i \notin \mathfrak{h} + \mathrm{Ad}(g)\ \mathfrak{b} + \mathfrak{g}_{i+1}\ \ \forall g \in \mathscr{U} = G \setminus G_2\big\} = \big\{1, 2\big\}.$$

Now for $H\backslash G/B$, due to $\mathscr{V} = \phi^{-1}(\mathscr{U}) = \{t \in \mathbb{R}^2, t_1Z_1.t_2Z_2 \in G \setminus G_2\} = \mathbb{R}^\star \times \mathbb{R}$ and $\tilde{\mathscr{U}} = \mathbf{p}(\mathscr{U}) = \{H \cdot z_1Z_1 \cdots z_7Z_7 \cdot B, z = (z_1, \ldots, z_7) \in \mathbb{R}^\star \times \mathbb{R}^6\}$ where these maps are given by

$$\begin{array}{rccc} \phi : & \mathbb{R}^2 & \longrightarrow & G \\ & (t_1, t_2) & \longmapsto & t_1Z_1.t_2Z_2 \end{array} \quad \text{and} \quad \begin{array}{rcl} \mathbf{p} : G & \longrightarrow & H\backslash G/B, \\ g & \longmapsto & H \cdot g \cdot B \end{array}$$

one has that

$$\begin{array}{rcl} \tilde{\phi} : \mathbb{R}^\star \times \mathbb{R} & \longrightarrow & \tilde{\mathscr{U}} \\ (t_1, t_2) & \longmapsto & H \cdot t_1Z_1.t_2Z_2 \cdot B \end{array}$$

which is an homeomorphism by the previous theorem. In particular, this confirms the relation $\mathbf{p}(G \setminus G_2) = \tilde{\phi}(\mathbb{R}^\star \times \mathbb{R})$ of the first example's calculation.

3.5.2 A Measure on $H\backslash G/B$

The aim now is to disintegrate the restriction of the irreducible representation $\pi_{l,\mathfrak{b}}$. The first step is to realize the Hilbert space $L^2(G/B, \chi_l)$ as a direct integral. Therefore one needs a measure $d\gamma$ on the space $H\backslash G/B$ such that

$$L^2(G/B, \chi_l) \simeq \int_{H\backslash G/B}^{\oplus} L^2(H/B(x), \chi_{l(x)})d\gamma(x),$$

where the linear functional

$$l(x) = \mathrm{Ad}^*(\psi(x))l, x \in H\backslash G/B,$$

defines a character of the subalgebra $\mathfrak{b}(x) := \mathfrak{h} \cap \mathrm{Ad}(\psi(x))\mathfrak{b}$ and where $B(x) := H \cap \psi(x)B\psi(x)^{-1}$.

Below we recall the construction of measures on homogeneous spaces as explained in Chap. 1, Sect. 1.2.2. Some notations will differ.

Definition 3.5.13 Let $H = \exp \mathfrak{h}$ be a closed subgroup of G. Let $\mathscr{Y} = \{Y_1, \cdots, Y_m\}$ be a Jordan-Hölder basis of $\mathfrak{h}$. As was said in Sect. 3.5.1, the mapping

$$E_{\mathscr{Y}} : \mathbb{R}^m \to H, \quad (y_1, \cdots, y_m) \mapsto \exp(y_1 Y_1) \cdots \exp(y_m Y_m)$$

is a diffeomorphism and the measure $d_{\mathscr{Y}}$ defined on H by

$$\int_H f(h) d_{\mathscr{Y}}(h) := \int_{\mathbb{R}^m} f \circ E_{\mathscr{Y}}(y) dy, \quad f \in C_c(H),$$

is a Haar measure of H. Furthermore, one knows that for any subalgebra $\mathfrak{k} \subset \mathfrak{h}$, if

$$\mathbb{I}^{\mathfrak{h}/\mathfrak{k}} := \{i \in \{1, \cdots, m\}, Y_i \notin \mathfrak{k} + \mathfrak{h}_{i+1}\} = \{j_1 < \cdots < j_r\},$$

then the subset

$$\mathscr{N} := \{Y_j, j \in \mathbb{I}^{\mathfrak{h}/\mathfrak{k}}\} \tag{3.5.16}$$

is a Malcev basis of $\mathfrak{h}$ relative to $\mathfrak{k}$ and the mapping

$$E_{\mathscr{N}} : \mathbb{R}^r \to H, \quad (s_1, \cdots, s_r) \mapsto \exp(s_1 Y_{j_1}) \cdots \exp(s_r Y_{j_r}),$$

composed with the mapping : $H \ni h \mapsto \dot{h} = h \cdot K \in H/K$, gives the diffeomorphism

$$E^{\cdot}_{\mathscr{N}} : \mathbb{R}^r \to H/K, \quad (s_1, \cdots, s_r) \mapsto \exp(s_1 Y_{j_1}) \cdots \exp(s_r Y_{j_r}) \cdot K.$$

In particular, in analogy to (3.5.4), one obtains an H-invariant measure on H/K denoted $d_{\mathscr{N}}(\dot{h})$ or simply $d\dot{h}$, by putting

$$\int_{H/K} \varphi(\dot{h}) d_{\mathscr{N}}(\dot{h}) := \int_{\mathbb{R}^r} \varphi \circ E^{\cdot}_{\mathscr{N}}(s) ds, \quad \varphi \in C_c(H/K).$$

Proposition 3.5.14 *There exist a Zariski open subset $\mathscr{U}''$ of G with property $\mathscr{P}$, and an index set $\mathscr{J}^{\mathfrak{h}/\mathfrak{b}} \subset \mathbb{I}^{\mathfrak{h}}$, such that*

$$\mathscr{J}^{\mathfrak{h}/\mathfrak{b}} = \mathbb{I}^{\mathfrak{h}/\mathfrak{h} \cap \mathrm{Ad}(g)\mathfrak{b}} \quad \textit{for all } g \in \mathscr{U}''.$$

Proof For every $i \in \mathscr{I}^{\mathfrak{h}}$, the set

$$\mathscr{S}'_i = \{g \in G, Z_i \notin \mathfrak{h} \cap \mathrm{Ad}(g)\mathfrak{b} + \mathfrak{h} \cap \mathfrak{g}_{i+1}\}$$

has property $\mathscr{P}$ and is either contained in a Zariski closed subset of G or it contains a Zariski open subset $\mathscr{S}_i$ with property $\mathscr{P}$. The collection of all indices i verifying the second possibility gives us the subset $\mathscr{I}^{\mathfrak{h}/\mathfrak{b}}$ of $\mathscr{I}^{\mathfrak{h}}$ and $\mathscr{U}'' = \bigcap_{i \in \mathscr{I}^{\mathfrak{h}/\mathfrak{b}}} \mathscr{S}_i$.

□

Definition 3.5.15 Let $\{j_1 <, \cdots, < j_r\}$ be the elements of $\mathscr{I}^{\mathfrak{h}/\mathfrak{b}}$, then $\mathscr{N} := \{Y_{j_1}, \cdots, Y_{j_r}\}$ is a Malcev basis of $\mathfrak{h}$ relative to $\mathfrak{b}(g) := \mathfrak{h} \cap \mathrm{Ad}(g)\mathfrak{b}$ for all $g \in \mathscr{U}''$. Now put

$$\mathscr{U}' := \mathscr{U} \cap \mathscr{U}'' \subset G, \quad \mathscr{V}' := \phi^{-1}(\mathscr{U}') \subset \mathbb{R}^d, \quad \tilde{\mathscr{U}}' := \mathbf{p}(\mathscr{U}') \subset H\backslash G/B.$$

The subsets $\mathscr{U}'$ and $\mathscr{V}'$ respectively of G and $\mathbb{R}^d$ are Zariski open sets and $\tilde{\mathscr{U}}'$ is open and dense in $H\backslash G/B$.

Remark 3.5.16 Since the representations $\tau_{l,\mathfrak{b}}$ and $\tau_{\mathrm{Ad}^*(g)l,\mathrm{Ad}(g)\mathfrak{b}}$, $g \in G$, are equivalent, one can replace the subalgebra $\mathfrak{b}$ by a conjugated algebra $\mathrm{Ad}(g)\mathfrak{b}$, with $g \in \mathscr{U}'$. This allows to assume that e belongs to the Zariski open subset $\mathscr{U}'$. This means that 0 can be supposed to be an element of $\mathscr{V}'$.

Theorem 3.5.17 *The mapping*

$$\Phi : \mathbb{R}^r \times \mathscr{V}' \to G/B : (s, t) \mapsto E_{\mathscr{N}}(s)\phi(t) \cdot B$$

is injective and polynomial, its image is equal to $\mathscr{U}'$ mod B and it is regular on a Zariski open subset of $\mathbb{R}^r \times \mathscr{V}'$.

Proof Let $g \in \mathscr{U}'$. Then $g = h\phi(t)b$ with $t \in \mathscr{V}', h \in H$ and $b \in B$. One can write $h = E_{\mathscr{N}}(s)h_0$, where $h_0 = \phi(t)b_0\phi(t)^{-1}$ (with $b_0 \in B$), is an element of $H \cap \phi(t)B\phi(t)^{-1}$. Hence

$$\Phi(s,t) = E_{\mathscr{N}}(s)\phi(t) \cdot B = E_{\mathscr{N}}(s)\phi(t)b_0 \cdot B = E_{\mathscr{N}}(s)\phi(t)b_0\phi(t)^{-1}\phi(t) \cdot B = g \cdot B.$$

If $\Phi(s,t) = \Phi(s', t')$ for two pairs (s, t) and (s', t') in $\mathbb{R}^r \times \mathscr{V}'$, then $H \cdot \phi(t) \cdot B = H \cdot \phi(t') \cdot B$ and since $t \in \mathscr{V}$, it follows from the injectivity of $\tilde{\phi}$ that $t = t'$. Hence $E_{\mathscr{N}}(s) = E_{\mathscr{N}}(s')$ and the injectivity of the mapping $E_{\mathscr{N}}$ implies that $s = s'$. Now, take the Malcev basis $\mathscr{M} = \{Z_{k_1}, \cdots, Z_{k_p}\}$ of $\mathfrak{g}$ relative to $\mathfrak{b}$ defined in (3.5.3), so for

$$z = \exp(\sum_{j=1}^{n} z_j Z_j) \in G,$$

one has that

$$z \bmod B = E_{\mathscr{U}}(Q_1(z_1 \dots z_n) \dots Q_p(z_1 \dots z_n)) \bmod B,$$

where $Q_1, \cdots, Q_k : \mathbb{R}^n \to \mathbb{R}$ are real-valued polynomials. This tells us that Φ is a polynomial mapping, since the mapping $(s, t) \to E_{\mathscr{N}}(s)\phi(t)$ is clearly polynomial. The differential of the mapping Φ at zero is given by the family of vectors

$$\{Y_{j_1} \dots Y_{j_r}, Z_{i_1} \dots Z_{i_d}\} \bmod \mathfrak{b}.$$

These vectors form a basis of $\mathfrak{g}$ modulo $\mathfrak{b}$ (but not necessarily a Malcev basis). Hence Φ is regular at $(0, 0)$ and since Φ is polynomial, Φ is regular on a Zariski open subset of $\mathbb{R}^r \times \mathbb{R}^d$. □

Remark 3.5.18 As a consequence of the previous theorem, one has the following condition on cardinalities

$$\sharp \mathscr{I}^{\mathfrak{g}/\mathfrak{b}} = \sharp \mathscr{I}(\mathfrak{g}/\mathfrak{h}, \mathfrak{g}/\mathfrak{b}) + \sharp \mathscr{J}^{\mathfrak{h}/\mathfrak{b}},$$

albeit in general

$$\mathscr{I}^{\mathfrak{g}/\mathfrak{b}} \neq \mathscr{I}(\mathfrak{g}/\mathfrak{h}, \mathfrak{g}/\mathfrak{b}) \cup \mathscr{J}^{\mathfrak{h}/\mathfrak{b}}.$$

This is shown by the following example.

Example 3.5.19 Use again Example 3.5.1. Clearly one has that $\mathscr{I}^{\mathfrak{g}/\mathfrak{b}} = \{1, 2, 5, 6\}$ and as previously $\mathbb{I}(\mathfrak{g}/\mathfrak{h}, \mathfrak{g}/\mathfrak{b}) = \{1, 2\}$, so it easily follows that $\mathscr{J}^{\mathfrak{h}/\mathfrak{b}} = \{4, 5\}$.

Corollary 3.5.20 *Recall the definitions of the sets of indices:*

- $\mathscr{I}(\mathfrak{g}/\mathfrak{h}, \mathfrak{g}/\mathfrak{b}) = \{i_1, \dots, i_d\}$ *giving the directions* $(Z_{i_1}, \dots, Z_{i_d})$ *which characterize almost all elements of* $H \backslash G / B$ *by Theorem 3.5.10.*
- $\mathscr{J}^{\mathfrak{h}/\mathfrak{b}} = J = \{j_1, \cdots, j_r\}$ *giving the Malcev basis* $\mathscr{N} = (Y_{j_1}, \cdots, Y_{j_r})$ *of* $\mathfrak{h}$ *relative to* $\mathfrak{b}(g)$ *for every* $g \in \mathscr{U}''$ *defined in Definition 3.5.15.*
- $\mathscr{I}^{\mathfrak{g}/\mathfrak{b}} = \{k_1, \dots, k_p\}, (p = r + d)$, *giving the Malcev basis* $\mathscr{M} = (Z_{k_1}, \dots, Z_{k_p})$ *of* $\mathfrak{g}$ *relative to* $\mathfrak{b}$ *defined in* (3.5.3).

The mapping

$$\begin{array}{llcl} \Psi = \dot{E}_{\mathscr{M}}^{-1} \circ \Phi \ : & \mathbb{R}^r \times \mathscr{V}' & \longrightarrow & \mathbb{R}^p \\ & \big(s = (s_1, \dots, s_r), t = (t_1, \dots, t_d)\big) & \longmapsto & w = (w_1, \dots, w_p) \ : \\ E_{\mathscr{N}}(s)\phi(t) \cdot B = & s_1 Y_{j_1} \dots s_r Y_{j_r} . t_1 Z_{i_1} \dots t_d Z_{i_d} \cdot B & = & w_1 Z_{k_1} \dots w_p Z_{k_p} \cdot B = E_{\mathscr{M}}(w) \end{array}$$

is a polynomial diffeomorphism onto its image $\mathscr{W}$, *which is a Zariski open subset of* $\mathbb{R}^p$.

Proof This is a direct consequence of Theorem 3.5.17. □

Definition 3.5.21 Denote, for $t \in \mathcal{V}'$, $B(t) = H \cap \phi(t)B\phi(t)^{-1}$ the closed subgroup of H associated to the Lie algebra $\mathfrak{h} \cap \mathrm{Ad}(\phi(t))\mathfrak{b}$. Then one has a diffeomorphism E_t between H and $\mathbb{R}^r \times B(t)$ such that for $h \in H$, and $(s, b) \in \mathbb{R}^r \times B(t)$,

$$E_t(h) = (s, b) \iff h = E_{\mathcal{N}}(s)b.$$

Write also for $t \in \mathcal{V}'$ and $h \in H$, the element $s(h, t)$ of $\mathbb{R}^r$ verifying

$$h = E_{\mathcal{N}}(s(h, t)) \bmod B(t). \tag{3.5.17}$$

Fix now $t \in \mathcal{V}'$ and $h \in H$ in order to define the polynomial mapping $q_{h,t} : \mathbb{R}^r \to \mathbb{R}^r$ by

$$q_{h,t}(s) = \dot{E}_{\mathcal{N}}^{-1}(hE_{\mathcal{N}}(s)) = s(hE_{\mathcal{N}}(s), t), s \in \mathbb{R}^r. \tag{3.5.18}$$

It follows that

$$q_{hh',t} = q_{h,t} \circ q_{h',t}, \; h, h' \in H, t \in \mathcal{V}'.$$

Proposition 3.5.22 *The Jacobian $Jac(s, t)$ of the mapping Φ of Theorem 3.5.17 does not depend upon $s \in \mathbb{R}^r$.*

Proof Indeed, if $f \in C_c(\mathcal{U}'/B)$, $t \in \mathcal{U}'$, and $h' \in H$, using the left-invariant measure $d\dot{g}$ on G/B, one gets

$$\begin{aligned}
\int_{\mathbb{R}^d}\int_{\mathbb{R}^r} f(\Phi(s, t))|Jac(s, t)|dsdt &= \int_{G/B} f(g)d\dot{g} = \int_{G/B} f(h'g)d\dot{g} \\
&= \int_{\mathbb{R}^d}\int_{H/B(t)} f(h'h\phi(t))|Jac(s((h, t), t)|d_{\mathcal{N}}(h)dt \\
&= \int_{\mathbb{R}^d}\int_{H/B(t)} f(h\phi(t))|Jac(s(h'^{-1}h, t), t)|d_{\mathcal{N}}(h)dt \\
&= \int_{\mathbb{R}^d}\int_{\mathbb{R}^r} f(E_{\mathcal{N}}(s)\phi(t))|Jac(q_{h'^{-1},t}(s), t)|dsdt \\
&= \int_{\mathbb{R}^d}\int_{\mathbb{R}^r} f(\Phi(s, t))|Jac(q_{h'^{-1},t}(s), t)|dsdt.
\end{aligned}$$

This shows that $|Jac(q_{h'^{-1},t}(s), t)| = |Jac(s, t)|$ for every $h' \in H, s \in \mathbb{R}^r$ and $t \in \mathcal{V}'$. Therefore $Jac\ \Phi(s, t) = Jac\ \Phi(0, t)$ for every $s \in \mathbb{R}^r, t \in \mathcal{V}'$, since Φ is a real-valued polynomial function. □

Definition 3.5.23 Let

$$F(t) := Jac\ \Phi(0, t), t \in \mathscr{V}'.$$

Then $F : \mathscr{V}' \to \mathbb{R}$ is a function on $\mathscr{V}'$ which admits a polynomial extension on $\mathbb{R}^d$.

Definition 3.5.24 Define now our measure on the topological space $H \diagdown G \diagup B$. One has the Zariski dense open subset $\mathscr{U}$ of G, which has property $\mathscr{P}$ and therefore its complement $G \setminus C$ in G is Zariski closed and has property $\mathscr{P}$ too. The projection $\mathbf{p} : G \to H \diagdown G \diagup B$ maps $\mathscr{U}$ to the open dense subset $\tilde{\mathscr{U}}$ and $\mathscr{C}$ to the closed subset $\tilde{\mathscr{C}}$. Define a measure $d\gamma$, which is supported on $\tilde{\mathscr{U}}$, by the rule

$$\int_{H \diagdown G \diagup B} f(x) d\gamma(x) := \int_{\mathscr{V}} f(\phi(t))|F(t)|dt.$$

Proposition 3.5.25 *Let $f \in C_c(G/B)$. Then*

$$\int_{G/B} f(g)d\dot{g} = \int_{H \diagdown G \diagup B} \int_{H/B(x)} f(h\psi(x))d_{\mathscr{Y}}(h)d\gamma(x).$$

Proof Indeed, by Theorem 3.5.17, Proposition 3.5.22 and Definition 3.5.23, one has that

$$\begin{aligned}
\int_{G/B} f(g)d\dot{g} &= \int_{\mathbb{R}^d} \int_{\mathbb{R}^r} f(\Phi(s, t))|Jac\ \Phi(s, t)|dsdt \\
&= \int_{\mathbb{R}^d} \int_{\mathbb{R}^r} f(E_{\mathscr{N}}(s)\phi(t))|F(t)|dsdt \\
&= \int_{\mathbb{R}^d} \int_{H/B(t)} f(h\phi(t))|F(t)|d_{\mathscr{Y}}(h)dt \\
&= \int_{H \diagdown G \diagup B} \int_{H/B(x)} f(h\psi(x))d_{\mathscr{Y}}(h)d\gamma(x)
\end{aligned}$$

(by Definition 3.5.24).

□

Definition 3.5.26 Let $H = \exp \mathfrak{h}$ be a closed subgroup of the simply connected nilpotent Lie group $G = \exp \mathfrak{g}$. Let $l \in \mathfrak{g}^*$, let $B = \exp(\mathfrak{b})$ be a polarization at l and $\pi_l = \pi_{l,\mathfrak{b}} = \mathrm{ind}_B^G \chi_l$ the irreducible representation associated to $(l, \mathfrak{b})$. Let for $g \in \mathscr{U}'$,

$$l(g) = \mathrm{Ad}^*(g)l_{|\mathfrak{h}}, \quad B(g) = gBg^{-1} \cap H, \quad \tau_{l(g),\mathfrak{b}(g)} = \mathrm{ind}_{B(g)}^H \chi_{l(g)} =: \sigma(g).$$

Then one can show that the representations $\sigma(g')$ and $\sigma(g)$ of H are equivalent whenever g' and g are "equivalent", ie g and g' are in the same double coset. Hence define $\sigma(\tilde{g}) := [\sigma(g)]$ as the equivalence class of the unitary representation $\sigma(g)$ for $\tilde{g} \in \mathcal{U}'$.

Proposition 3.5.27 *Define, for $x \in \tilde{\mathcal{U}}' \subset H\backslash G/B$,*

$$l(x) := l(\psi(x)),\ B(x) := B(\psi(x)),$$

and the representation $\sigma(x) := \tau_{l(x),\mathfrak{b}(x)}$ whose Hilbert space is equal to

$$\mathcal{H}_{\sigma(x)} := L^2(H/B(x), \chi_{l(x)}).$$

Then

$$L^2(G/B, \chi_l) \simeq \int_{H\backslash G/B}^{\oplus} \mathcal{H}_{\sigma(x)} d\gamma(x),$$

and

$$\pi_{l,\mathfrak{b}|H} \simeq \int_{H\backslash G/B}^{\oplus} \sigma(x) d\gamma(x).$$

Proof Let L_0 be the space of vector fields $F = (f_x \in C_c(H/B(x))_{x\in\tilde{\mathcal{U}}'}$, such that the function $\mathbb{R}^r \times \tilde{\mathcal{U}}' \ni (s, x) \mapsto f_x(E_{\mathcal{N}}(s))$ is continuous with compact support. On L_0 one has the Hilbert-space norm

$$\|F\|_2^2 := \int_{H\backslash G/B} \int_{\mathbb{R}^r} |F_x(E_{\mathcal{N}}(s))|^2 ds d\gamma(x) = \int_{H\backslash G/B} \|F_x\|^2_{L^2(H/B(x),\chi_{l(x)})} d\gamma(x).$$

The direct integral $\int_{H\backslash G/B} L^2(H/B(x), \chi_{l(x)}) d\gamma(x)$ can then be defined as the completion of L_0 for the norm $\|.\|_2$ of L_0. Define now for $\xi \in C_c(G/B, \chi_l)$ the function

$$\mathbb{T}(\xi)(x)(h) := \xi(h\psi(x)), x \in \tilde{\mathcal{U}}', h \in H.$$

For $x \in \tilde{\mathcal{U}}'$, the function $\mathbb{T}(\xi)(x)$ is then an element of $C_c(H/B(t), \chi_{l(t)})$, since it satisfies the covariance relation

$$\mathbb{T}(\xi)(hb) = \xi(hh\phi(t)) = \xi(h\phi(t)\phi(t)^{-1}b\phi(t)) = \chi_{l(x)}\xi(h\phi(t)) = \chi_{l(x)}\mathbb{T}(\xi)(x)(h),$$

for $t = \psi(x), h \in H$, and $b \in B(x)$. It is obviously continuous with compact support modulo $B(x)$, for every x, since the mapping $\Phi : \mathbb{R}^r \times \mathcal{V}' \to \mathcal{U}'$ is a homeomorphism. One obtains in this way a vector field $(\mathbb{T}(\xi)(x) \in C_c(H/B(x), \chi_{l(x)}))_{x\in\tilde{\mathcal{U}}'}$, which is contained in L_0. The mapping $\mathbb{T} : C_c(G/B, \chi_l) \to L_0$ is isometric by Proposition 3.5.25, it is a linear isomorphism and it obviously intertwines the representation $\pi_{l|H}$ and the direct integral representation $\int_{H\diagdown G/B} \tau_{\chi_{l(x)},B(x)} d\gamma(x)$. □

3.5.3 A Concrete Intertwining Operator

We have seen in the previous section that the disintegration of the restriction of the irreducible representation $\pi_{l,\mathfrak{b}}$ to H reduces now to the disintegration of the monomial representations $\sigma(x) = \tau_{\chi_l(x),B(x)}$ of H into irreducibles. Let us use the techniques developed in Sect. 3.2. Let $\mathcal{N} = \{Y_{j_1}, \cdots, Y_{j_r}\}$ be the Malcev basis defined in (3.5.16) and let us use the following notations:

$$\mathfrak{b}(t) := \mathfrak{b}(\tilde{\phi}(t)) = \mathfrak{b}_{r+1} \subset \ldots \subset \mathfrak{b}_c(t) = \mathbb{R}\text{-span}(Y_{j_c}, \ldots, Y_{j_r}, \mathfrak{b}(t) >\subset \ldots \subset \mathfrak{b}_1(t) = \mathfrak{h}$$

$$B_c(t) := \exp \mathfrak{b}_c(t), \quad l(t) := \mathrm{Ad}^*(\phi(t))l_{|\mathfrak{h}}, \quad t \in \mathcal{V}', c = 1, \ldots, r.$$

One has for every $t \in \mathcal{V}'$ a Jordan-Hölder basis $p_{\mathfrak{a}}(G \cdot \Gamma_f)(t) = \{X_{r+1}(t), \ldots, X_m(t)\}$ of $\mathfrak{b}(t)$, such that the vectors $X_i(t)$, $i = r+1, \ldots, m$, vary rationally without singularities in $t \in \mathcal{V}'$. Let

$$\mathcal{Y}(t) = \{X_1(t) := Y_{j_1}, \ldots, X_r(t) := Y_{j_r}, X_{r+1}(t), \ldots, X_m(t)\}, \quad t \in \mathcal{V}',$$

be the corresponding Malcev basis of $\mathfrak{h}$ and let

$$\mathcal{Y}(t)^* = \{X_1(t)^*, \ldots, X_r(t)^*, X_{r+1}(t)^*, \ldots, X_m(t)^*\}, \quad t \in \mathcal{V}',$$

be its dual basis in $\mathfrak{h}^*$. For $t \in \mathcal{V}'$, let $\Gamma(t) = l(t) + \mathfrak{b}(t)^{\perp}$ and for any $c \in \{1, \ldots, r\}$ let

$$d_c(t) := \max(\dim \mathrm{Ad}^*(B_c(t))(f)); \ f \in \Gamma(t), t \in \mathcal{V}',$$

and let $d_{r+1}(t) = 0$. Let also

$$I(t) := \{c_1 < \ldots < c_{q(t)}\} := \{c \in \{1, \ldots, r\}, \quad d_c(t) = d_{c+1}(t)\} \tag{3.5.19}$$

and

$$\mathcal{R}(t) := \{l(t) + \sum_{c_i \in I(t)} v_i X_{c_i}(t)^*, \quad (v_1, \ldots, v_{q(t)}) \in \mathbb{R}^{q(t)}\}, t \in \mathcal{V}'. \tag{3.5.20}$$

One can again assume, if necessary by shrinking $\mathcal{V}'$, that there exists a fixed index set I such that for every $t \in \mathcal{V}'$ one has that

$$I(t) = I = c_1, \ldots, c_q \text{ and so } q(t) = q = |I|. \tag{3.5.21}$$

Use also the notation

$$l(v,t) := l(t) + \sum_{c\in I(t)} v_i Y_{j_c}(t)^* \in \Gamma(t), t \in \mathscr{V}', v \in \mathbb{R}^q. \tag{3.5.22}$$

It follows from Theorem 3.3.4 that there exists a Zariski open subset $\mathscr{S}$ of $\mathbb{R}^q \times \mathbb{R}^d$ such that, for every $(u,t) \in \mathscr{S}$, one has a Malcev basis

$$p_{\mathfrak{a}}(G\cdot\Gamma_f)(v,t) = \{X_1(v,t),\dots,X_d(v,t)\} \tag{3.5.23}$$

of $\mathfrak{h}$ relative to the Vergne polarization $\mathfrak{b}(v,t)$ associated to the Jordan-Hölder basis

$$\mathscr{Y} = \{Y_1,\dots,Y_m\} = \{Z_j, j \in \mathbb{I}^{\mathfrak{h}}\}$$

at $l(v,t) \subset \Gamma(t)$, and also a Malcev basis

$$\mathscr{Y}(v,t) = \{Y_1(v,t),\dots,Y_e(v,t)\} \tag{3.5.24}$$

of $\mathfrak{b}(v,t)$ relative to the subalgebra $\mathfrak{b}(v,t) \cap \mathfrak{b}(t)$ such that all vectors $X_j(v,t)$ and $Y_i(v,t)$ vary rationally without singularities in $(v,t) \in \mathscr{S}$, and such that the operator

$$\mathbb{T}(v,t) : \mathscr{S}(H/B(t),\chi_{l(t)}) \to \mathscr{S}(H/B(v,t),\chi_{l(v,t)})$$

defined by

$$\mathbb{T}(v,t)(\eta)(h) = \int_{B(v,t)/B(v,t)\cap B(t)} \eta(hb)\chi_{l(v,t)(b)} d_{\mathscr{Y}(v,t)}(b), \quad h \in H,$$

(where $\mathscr{Y}(v,t)$ is as in (3.5.24)) gives us an isometric intertwining operator

$$\mathbb{T}(t) : L^2(H/B(t),\chi_{l(t)}) \to \int_{\Gamma(t)} L^2(H/B(v,t),\chi_{l(v,t)}), \quad \mathbb{T}(t)(\eta) := \mathbb{T}(v,t)\eta,$$

(with the norm on $L^2(H/B(v,t)),\chi_{l(v,t)})$ via the measure $d_{p_{\mathfrak{a}}(G\cdot\Gamma_f)(v,t)}$ on $H/B(v,t))$. Combining these manipulations with Theorem 3.3.4 and Proposition 3.5.27, one obtains the following concrete decomposition of $\pi_{l|H}$:

Theorem 3.5.28 *Let $G = \exp\mathfrak{g}$ be a connected simply connected Lie group, $H = \exp\mathfrak{h}$ a closed connected subgroup of G. Let $\mathscr{Z}$ be a Jordan-Hölder basis of $\mathfrak{g}$, let $l \in \mathfrak{g}^*$ and let $\mathfrak{b}$ be the Vergne polarization at l with respect to $\mathscr{Z}$. Let for any $(v,t) \in \mathscr{R}$, for $\xi \in \mathscr{S}(G/B,\chi_l)$*

$$T_{v,t}\xi(h) = \int_{B(v,t)/B(v,t)\cap B(t)} \xi(hb\phi(t))\chi_{v,t}(b) d_{\mathscr{Y}(v,t)}(b),$$

where $\mathscr{Y}(v,t)$ *is as in* (3.5.24). *Then*

$$\int_{H\backslash G/B}\int_{H/B(v,\psi(x))}\|T_{v,t}\xi\|^2_{L^2(H/B(v,\psi(x)),\chi_{l(v,\psi(x))})}d\gamma(x)=\|\xi\|^2_{L^2(G/B,\chi_l)}$$

and the operator

$$\mathbb{T}:L^2(G/B,\chi_l)\to\int_{H\backslash G/B}L^2(H/B(\psi(x))),\chi_{l(v,\psi(x))}d\gamma(x)$$

defined by

$$\xi\mapsto\mathbb{T}_{v,\psi(x)}(\xi),\quad \xi\in\mathscr{S}(G/B,\chi_l),(x,\psi(x))\in\mathscr{R},$$

extends to a unitary intertwining operator.

3.6 Disintegrating Tensor Products of Irreducible Representations of Nilpotent Lie Groups

In this section we produce a concrete disintegration of the tensor product of two irreducible unitary representations of a simply connected, connected nilpotent Lie group $G = \exp\mathfrak{g}$ into irreducibles. "Concrete" means that one writes down an explicit integral of the spaces into irreducibles. Towards this, one uses the well-known fact that a tensor product $\pi\otimes\pi'$ of two irreducible representations π,π' of G corresponds to the restriction to the diagonal group D of the irreducible representation $\pi\times\pi'$ of $G\times G$. The disintegration is carried out over a subvariety of $\mathfrak{g}^*$, which is contained in the sum of the Kirillov orbits of π and π'. Using an example one shows that the disintegration formulas depend on the choice of the polarizations used to realize the representations π and π'. A necessary and sufficient condition for the irreducibility of $\pi\otimes\pi'$ will be derived.

3.6.1 A Concrete Intertwining Operator for Tensor Products of Unitary Representations

Let $G = \exp\mathfrak{g}$ be a simply connected, connected Lie group and let $\pi=\pi_l,\pi'=\pi_{l'}$ be two irreducible unitary representations of G. Take a Jordan-Hölder basis $\mathscr{Z}=\{Z_1,\cdots,Z_n\}$ of $\mathfrak{g}$ and obtain the corresponding Vergne polarizations $\mathfrak{b}$ at l and $\mathfrak{b}'$ at l'. Realize then π as $\pi_{l,\mathfrak{b}}$ resp. π' as $\pi_{l',\mathfrak{b}'}$.

Let $\mathbb{G} := G \times G$ and $\mathbb{B} := B \times B'$ be the direct products respectively of G with G and B with B'. Let also $\mathbf{\Pi} = \pi \times \pi'$ be the outer tensor product of π and π'. Realize Π as an induced representation. It acts as a left translation on the space

$$\mathscr{H}_{\Pi} := L^2(\mathbb{G}/\mathbb{B}, \chi_{(l,l')}) \simeq \mathscr{H}_{\pi}\hat{\otimes}\mathscr{H}_{\pi'},$$

where (l, l') is the linear functional defined on $\mathfrak{g} \times \mathfrak{g}$ by

$$(l, l')(Z, Z') := l(Z) + l'(Z'), Z, Z' \in \mathfrak{g},$$

and so $\mathfrak{b} \times \mathfrak{b}'$ is a polarization at (l, l'). Hence $\Pi = \pi_{(l,l'), \mathfrak{b}\times\mathfrak{b}'}$. Let $\mathscr{D} = \{(g, g); g \in G\}$ be the diagonal group of $\mathbb{G}$. This group is of course isomorphic to G and the tensor product representation $\pi \otimes \pi'$ is the representation of G acting on the space of Π through the rule

$$\pi \otimes \pi'(g) := \Pi(g, g), \ \ g \in G.$$

Then one should investigate the structure of the space of double cosets $\mathscr{D}\backslash\mathbb{G}/\mathbb{B}$. We first prove the following:

Proposition 3.6.1 *The mapping*

$$\mathbb{M} : B\backslash G/B' \to \mathscr{D}\backslash\mathbb{G}/\mathbb{B}, \ B \cdot g \cdot B' \mapsto \mathscr{D} \cdot (e, g) \cdot \mathbb{B} := \mathbb{M}(B \cdot g \cdot B'),$$

is well defined and is a bijection.

Proof Indeed, if $g \in G$ and $g' \in B \cdot g \cdot B'$, then $g' = bgb'$ for some $b \in B$ and $b' \in B'$ and so

$$\begin{aligned}\mathscr{D} \cdot (e, g') \cdot \mathbb{B} = \mathscr{D} \cdot (e, bgb') \cdot \mathbb{B} &= \{(ub_2, ubgb'b_2'), u \in G, b_2 \in B, b_2' \in B'\} \\ &= \{(ub^{-1}b_2, ugb_2'), u \in G, b_2 \in B, b_2' \in B'\} \\ &= \{(ub_2, ugb_2'), u \in G, b_2 \in B, b_2' \in B'\} \\ &= \mathscr{D} \cdot (e, g) \cdot \mathbb{B}.\end{aligned}$$

Hence our mapping is well defined. Since

$$\mathscr{D}\cdot(g, g')\cdot\mathbb{B} = \mathscr{D}\cdot(g, g)\cdot(e, g^{-1}g')\cdot\mathbb{B} = \mathscr{D}\cdot(e, g^{-1}g')\cdot\mathbb{B} = \mathbb{M}(B\cdot(g^{-1}g')\cdot B'),$$

it is also surjective. If $\mathbb{M}(B\cdot g\cdot B') = \mathbb{M}(B\cdot g'\cdot B')$ for two elements g, g' in G, then $(e, g') = (u, u)(e, g)(b, b')$ for some $u \in G$, $b \in B$ and $b' \in B'$. Hence $ub = e$ and $ugb' = g'$ which means that $g' = b^{-1}gb'$ is an element of $B \cdot g \cdot B'$. Therefore $\mathbb{M}$ is injective and so it is a bijection. □

According to the two sections above, we first need a Jordan-Hölder basis of $\mathfrak{g} \times \mathfrak{g}$ in order to perform our computations. Let

$$\mathbb{Z}_i := \begin{cases} (Z_j, Z_j) & \text{if } i = 2j - 1 \text{ is odd,} \\ (Z_j, -Z_j) & \text{if } i = 2j \text{ is even} \end{cases} \qquad (j \in \{1, \ldots, n\}).$$

The basis $\mathscr{A} = \{A_1, \ldots, A_n\}$ of the Lie algebra $\delta = \{(Z, Z);\ Z \in \mathfrak{g}\}$ of the diagonal group $\mathscr{D}$ is then given by the vectors

$$A_j := (Z_j, Z_j) = \mathbb{Z}_{2j-1}, \quad j = 1, \ldots, n.$$

Hence

$$\mathbb{I}^{\delta} = \{2j - 1,\ j \in \{1, \ldots, n\}\},\ \mathbb{I}^{\mathfrak{g}\times\mathfrak{g}/\delta} = \{2j,\ j \in \{1, \ldots, n\}\}. \tag{3.6.1}$$

Let us determine the index sets $\mathbb{I}(\mathfrak{g} \times \mathfrak{g}/\delta, \mathfrak{g} \times \mathfrak{g}/\mathfrak{b} \times \mathfrak{b}') \subset \mathbb{I}^{\mathfrak{g}\times\mathfrak{g}/\delta} \cap \mathbb{I}^{\mathfrak{g}\times\mathfrak{g}/\mathfrak{b}\times\mathfrak{b}'}$. By (3.6.1) one has that

$$\mathbb{I}(\mathfrak{g} \times \mathfrak{g}/\delta, \mathfrak{g} \times \mathfrak{g}/\mathfrak{b} \times \mathfrak{b}') \subset \{2j,\ j \in \{1, \ldots, n\}\}.$$

Proposition 3.6.2

$$\mathbb{I}(\mathfrak{g} \times \mathfrak{g}/\delta, \mathfrak{g} \times \mathfrak{g}/\mathfrak{b} \times \mathfrak{b}') = \{2j,\ j \in \mathbb{I}(\mathfrak{g}/\mathfrak{b}, \mathfrak{g}/\mathfrak{b}')\}. \tag{3.6.2}$$

Proof Let $\mathscr{U}$ be a Zariski open subset of G with property $\mathscr{P}$, such that for every $j \in \mathbb{I}(\mathfrak{g}/\mathfrak{b}, \mathfrak{g}/\mathfrak{b}')$,

$$Z_j \notin \mathfrak{b} + \mathrm{Ad}(g)\mathfrak{b}' + \mathfrak{g}_{j+1} \text{ for every } g \in \mathscr{U}.$$

Let

$$\mathbb{U} := \{(g, g') \in \mathbb{G},\ g^{-1}g' \in \mathscr{U}\}.$$

Then $\mathbb{U}$ is a Zariski open subset of $\mathbb{G}$. Let us show that for every $j \in \mathbb{I}(\mathfrak{g}/\mathfrak{b}, \mathfrak{g}/\mathfrak{b}')$, one has

$$\mathbb{Z}_{2j} = (Z_j, -Z_j) \notin \delta + \mathrm{Ad}(g, g')(\mathfrak{b} \times \mathfrak{b}') + (\mathfrak{g}\times\mathfrak{g}')_{2j+1}, \forall (g, g') \in \mathbb{U}. \tag{3.6.3}$$

Then necessarily the index $2j$ is in $\mathbb{I}(\mathfrak{g} \times \mathfrak{g}/\delta, \mathfrak{g} \times \mathfrak{g}/\mathfrak{b} \times \mathfrak{b}')$. Suppose that for some elements j and (g, g') respectively of $\mathbb{I}(\mathfrak{g}/\mathfrak{b}, \mathfrak{g}/\mathfrak{b}')$ and $\mathbb{U}$, relation (3.6.3) is false. Then one has that

$$\big(Z_j, -Z_j\big) = \big(Y, Y\big) + \big(\mathrm{Ad}(g)(X), \mathrm{Ad}(g')(X')\big) + \big(U, U'\big)$$

for certain vectors Y, X, X', U and U' respectively in $\mathfrak{g}$, $\mathfrak{b}$, $\mathfrak{b}'$ and $\mathfrak{g}_{j+1}$, since $(\mathfrak{g} \times \mathfrak{g}')_{2j+1} = \mathfrak{g}_{j+1} \times \mathfrak{g}_{j+1}$. Hence

$$\begin{cases} Z_j = Y + \mathrm{Ad}(g)(X) + U \\ -Z_j = Y + \mathrm{Ad}(g')(X') + U', \end{cases}$$

which gives us the identity

$$Z_j = \mathrm{Ad}(g)(\frac{X}{2}) + \mathrm{Ad}(g')(\frac{-X'}{2}) + \frac{U - U'}{2}.$$

This implies that

$$\begin{aligned} Z_j &\equiv \frac{X}{2} + \mathrm{Ad}(g^{-1}g')(\frac{-X'}{2}) &\quad \mod [\mathfrak{g}_{j+1}] \\ &\in \mathfrak{b} + \mathrm{Ad}(g^{-1}g')\,(\mathfrak{b}') &\quad \mod [\mathfrak{g}_{j+1}], \end{aligned}$$

where $g^{-1}g' \in \mathscr{U}$ (by definition of $\mathbb{U}$). This last relation contradicts the fact that $j \in \mathbb{I}(\mathfrak{g}/\mathfrak{b}, \mathfrak{g}/\mathfrak{b}')$. Let now $2j \in \mathbb{I}(\mathfrak{g} \times \mathfrak{g}/\delta, \mathfrak{g} \times \mathfrak{g}/\mathfrak{b} \times \mathfrak{b}')$. Then for every (g, g') in a Zariski open subset of $\mathbb{G}$, one has that

$$(Z_j, -Z_j) \notin \delta + \mathrm{Ad}(g)(\mathfrak{b}) \times \mathrm{Ad}(g')(\mathfrak{b}') + \mathfrak{g}_{j+1} \times \mathfrak{g}_{j+1}.$$

Let us show first that $j \in \mathbb{I}^{\mathfrak{g}/\mathfrak{b}} \cap \mathbb{I}^{\mathfrak{g}/\mathfrak{b}'}$. If not, assume that $j \notin \mathbb{I}^{\mathfrak{g}/\mathfrak{b}'}$. There exists $X' \in \mathfrak{b}'$, such that

$$Z_j = X' \bmod \mathfrak{g}_{j+1} = \mathrm{Ad}(g')X' \bmod \mathfrak{g}_{j+1}, \text{ for all } g' \in G.$$

Hence

$$\begin{aligned} \mathbb{Z}_{2j} = (Z_j, -Z_j) &= (Z_j, Z_j) + (0, \mathrm{Ad}(g')(-2X')) \bmod \mathfrak{g}_{j+1} \times \mathfrak{g}_{j+1} \\ &\in \quad \delta + \mathrm{Ad}(g, g')(\mathfrak{b} \times \mathfrak{b}') \qquad \bmod (\mathfrak{g} \times \mathfrak{g}')_{2j+1}, \end{aligned}$$

for any $(g, g') \in \mathbb{G}$. This relation contradicts the fact that $2j \in \mathbb{I}(\mathfrak{g} \times \mathfrak{g}/\delta, \mathfrak{g} \times \mathfrak{g}/\mathfrak{b} \times \mathfrak{b}')$. Suppose now that $j \notin \mathbb{I}(\mathfrak{g}/\mathfrak{b}, \mathfrak{g}/\mathfrak{b}')$. Then

$$Z_j \in \mathfrak{b} + \mathrm{Ad}(g)\mathfrak{b}' + \mathfrak{g}_{j+1},$$

for any $g \in \mathscr{U}$. Let (g, g') be one element of the Zariski open subset $\mathbb{U}$ of $\mathbb{G}$, which means that $g'' := g^{-1}g' \in \mathscr{U}$. There exist $X \in \mathfrak{b}$ and $X' \in \mathfrak{b}'$ such that

$$Z_j = X + \mathrm{Ad}(g'')(X') \bmod \mathfrak{g}_{j+1}.$$

Hence

$$\mathrm{Ad}(g)Z_j = \mathrm{Ad}(g)X + \mathrm{Ad}(g')X' \bmod \mathfrak{g}_{j+1}$$

and Z_j can then be written as

$$Z_j = \mathrm{Ad}(g)X + \mathrm{Ad}(g')X' \bmod \mathfrak{g}_{j+1}.$$

Let

$$Y = -\mathrm{Ad}(g)X + \mathrm{Ad}(g')X'.$$

It is easy to verify then that

$$\begin{aligned}\big(Z_j, -Z_j\big) &= \big(Y, Y\big) + \big(\mathrm{Ad}(g)(2X)\,,\ \mathrm{Ad}(g')(-2X')\,\big) \bmod \mathfrak{g}_{j+1} \times \mathfrak{g}_{j+1} \\ &\in \delta + \mathrm{Ad}(g, g')(\mathfrak{g} \times \mathfrak{g}') \bmod \mathfrak{g}_{j+1} \times \mathfrak{g}_{j+1}.\end{aligned}$$

This is impossible since $(g, g') \in \mathbb{U}$. Hence $j \in \mathbb{I}(\mathfrak{g}/\mathfrak{b}, \mathfrak{g}/\mathfrak{b}')$. □

The following is then immediate:

Corollary 3.6.3 *Let $\mathscr{U}$ be the Zariski open subset of G defined in Remark 3.5.4 for the subgroups B and B' and let $\mathbb{U} = \{(g, g') \in \mathbb{G};\ g^{-1}g' \in \mathscr{U}\}$. Let $\tilde{\mathscr{U}}$ be the projection of $\mathscr{U}$ into $B\backslash G/B'$ and $\tilde{\mathbb{U}}$ the projection of $\mathbb{U}$ onto $\mathscr{D}\backslash\mathbb{G}/\mathbb{B}$. Then the mapping $\mathbb{M}$ of Proposition 3.6.1 sends $\tilde{\mathscr{U}}$ onto $\tilde{\mathbb{U}}$ homeomorphically.*

Definition 3.6.4 Let now $\mathscr{U}$ be the Zariski open subset of G with property $\mathscr{P}$ such that $j \in \mathbb{I}(\mathfrak{g}/\mathfrak{b}, \mathfrak{g}/\mathfrak{b}')$ if and only if $Z_j \notin \mathfrak{b} + \mathrm{Ad}(g)\mathfrak{b}' + \mathfrak{g}_{j+1}$, for all $g \in \mathscr{U}$.
Let us write

$$\mathbb{I}(\mathfrak{g}/\mathfrak{b}, \mathfrak{g}/\mathfrak{b}') = \{i_1, \ldots, i_d\}$$

and

$$a(t) := \exp(t_1 Z_{i_1}) \cdots \exp(t_d Z_{i_d}),\ t \in \mathbb{R}^d.$$

Then the needed linear functionals $l(t)$ of $\mathfrak{g}$ are given for $t \in \mathbb{R}^d$ by

$$l(t) = \mathrm{Ad}^*(a(t))l + \mathrm{Ad}^*(a(-t))l' \in \mathrm{Ad}^*(G)l + \mathrm{Ad}^*(G)l',$$

and using the notation (Definition 3.5.5), one has

$$\phi(t) = \exp(t_1(Z_{i_1}, -Z_{i_1})) \cdots \exp(t_d(Z_{i_d}, -Z_{i_d})) = (a(t), a(-t)),$$

and

$$\langle l(t), (Z, Z)\rangle = \langle \mathrm{Ad}^*(\phi(t))(l, l'), Z, Z\rangle = \langle \mathrm{Ad}^*(a(t))l + \mathrm{Ad}^*(a(-t))l, Z\rangle,\ \ Z \in \mathfrak{g}, t \in \mathbb{R}^d.$$

Determine now the index set I defined as in Eq. (3.5.21) for the algebras $\mathfrak{b}$ and $\mathfrak{b}'$ and the elements $l(v, t)$ in $l(t) + \mathfrak{b}(t)^\perp$ defined in (3.5.22). Let us remark that

$$\mathfrak{b}(t) = \delta \cap \mathrm{Ad}(\phi(t))(\mathfrak{b} \times \mathfrak{b}') \simeq \mathrm{Ad}(a(t))b \cap \mathrm{Ad}(a(-t))b'. \tag{3.6.4}$$

Hence since $\mathrm{Ad}(a(t))\mathfrak{b}$ is a polarization at $\mathrm{Ad}^*(a(t))l$ and similarly for $\mathrm{Ad}^*(a(-t))\mathfrak{b}'$, one has that

$$\begin{aligned} l(t) + \mathfrak{b}(t)^\perp &= \mathrm{Ad}^*(a(t))l + \mathrm{Ad}^*(a(-t))l' + (\mathrm{Ad}(a(t))\mathfrak{b} \cap \mathrm{Ad}(a(-t))\mathfrak{b}')^\perp \\ &= \mathrm{Ad}^*(a(t))l + \mathrm{Ad}^*(a(-t))l' + \mathrm{Ad}(a(t))\mathfrak{b}^\perp + \mathrm{Ad}(a(-t))\mathfrak{b}'^\perp \\ &= \mathrm{Ad}^*(a(t)Ba(t)^{-1})l + \mathrm{Ad}^*(a(-t)B'a(-t)^{-1})l' \\ &\subset \mathrm{Ad}^*(G)l + \mathrm{Ad}^*(G)l'. \end{aligned}$$

Hence the affine subsets defined in (3.5.20)

$$\mathscr{R}(t) = \{l(t) + \sum_{i\in} v_i Y_{j_i}(t)^*,\ (v_1, \cdots, v_q) \in \mathbb{R}^q\}, t \in \mathscr{V}'.$$

are all contained in the sum of the orbits of l and of l'.

Definition 3.6.5 Let us also denote, for $t \in \mathscr{U}$,

$$\mathfrak{p}(t) = \mathrm{Ad}(a(t))\mathfrak{b}, \mathfrak{p}'(t) := \mathrm{Ad}(a(-t))\mathfrak{b}'.$$

Then $\mathfrak{p}(t)$ is a polarization at $\mathrm{Ad}^*(a(t))l$ and $\mathfrak{p}'(t)$ at $\mathrm{Ad}^*(a(-t))l'$. Let also

$$P(t) = \exp \mathfrak{p}(t),\ P'(t) = \exp \mathfrak{p}'(t),\ \ t \in \mathscr{U}.$$

On the Zariski open subset $\mathscr{R} \subset \mathbb{R}^q \times \mathbb{R}^d$, define the measure $d\nu$ by

$$d\nu = dp \otimes |F(t)|dt,$$

where F is the function defined as in (3.5.23) (for the subgroups B and B'). For $\theta = (v, t) \in \mathscr{R}$, let $t(\theta) = t$ and $v(\theta) = v$.

Rephrase now Theorem 3.5.28 for the special case of $\mathbf{\Pi}$ and $\mathscr{D}$.

Theorem 3.6.6 *Let* $G = \exp \mathfrak{g}$ *be a connected simply connected Lie group, let* $\mathscr{Z}$ *be a Jordan-Hölder basis of* $\mathfrak{g}$*, and let* $l, l' \in \mathfrak{g}^*$ *and let* $\mathfrak{b}$ *resp.* $\mathfrak{b}'$ *be the Vergne* *polarization at l, resp. at l', with respect to $\mathscr{Z}$. Let for any $\theta \in \mathscr{R} \subset \Omega_\pi + \Omega'_{\pi}$, for $\xi \in \mathscr{S}(\mathbb{G}/\mathbb{B}, \chi_{(l,l')})$*

$$T_\theta \xi(g) = \int_{B(\theta)/B(\theta)\cap P(t(\theta))\cap P'(t(\theta))} \xi(gba(t(\theta)), gba(-t(\theta)))\chi_\theta(b) d_{\mathscr{Y}(\theta)}(b),$$

where $\mathscr{Y}(\theta)$ is as in Eq. (3.5.24). Then

$$\int_{\mathscr{R}} \|T_\theta \xi\|^2_{L^2(G/B(\theta),\chi_{l(\theta)})} d\nu(\theta) = \|\xi\|^2_{L^2(\mathbb{G}/B,\chi_{(l,l')})}$$

and the operator

$$\mathbb{T} : L^2(G/B, \chi_l) \to \int_{\mathscr{R}} L^2(H/B(\theta), \chi_{l(\theta)}) d\nu(\theta)$$

defined by

$$\xi \mapsto \mathbb{T}_\theta(\xi), \quad \xi \in \mathscr{S}(G/B, \chi_l), \theta \in \mathscr{R},$$

extends to a unitary intertwining operator.

3.6.2 A Concrete Example

Through the following example one shows that the choice of a polarization for a given linear functional has a strong influence on the concrete form of the disintegration of the representation $\pi \otimes \pi'$. Let $\mathfrak{g} = \mathbb{R}\text{-span}(Z_1, \ldots, Z_7)$ be equipped with the non-trivial brackets

$$[Z_1, Z_4] = Z_7, \ [Z_2, Z_4] = Z_5, \ [Z_2, Z_6] = Z_7,$$

$$[Z_3, Z_4] = Z_6, \ [Z_3, Z_5] = Z_7.$$

Then $\mathscr{Z} = \{Z_1, \cdots, Z_7\}$ is a Jordan-Hölder basis of $\mathfrak{g}$ and one can easily verify that the coadjoint orbit Ω_l of any $l \in \mathfrak{g}^*$ with $l(Z_7) \neq 0$ is equal to $l + Z_7^\perp$. Let $\mathscr{Z}^* = \{Z_1^*, \cdots, Z_7^*\}$ be the dual basis of $\mathscr{Z}$. Choose now $l = Z_7^* + Z_6^* \in \mathfrak{g}^*$. Its stabilizer in $\mathfrak{g}$ is reduced to the center $\mathbb{R}Z_7$ of $\mathfrak{g}$. There are several possibilities for polarizations at l:

$$\mathfrak{b}_1 := \mathbb{R}\text{-span}(Z_7, Z_6, Z_5, Z_4)$$

$$\mathfrak{b}_2 := \mathbb{R}\text{-span}(Z_7, Z_3, Z_2, Z_1)$$

$$\mathfrak{b}_3 := \mathbb{R}\text{-span}(Z_7, Z_6, Z_4 - Z_5, Z_3)$$

$$\mathfrak{b}_4 := \mathbb{R}\text{-span}(Z_7, Z_6, Z_4, Z_1)$$

Then $\mathfrak{b}_1$ is an Abelian ideal of $\mathfrak{g}$, $\mathfrak{b}_2$ and $\mathfrak{b}_4$ are Abelian subalgebras and $\mathfrak{b}_3$ is non-Abelian. Let now

$$\pi_j := \mathrm{ind}_{B_j}^G \chi_l, \ j = 1, 2, 3, 4.$$

Disintegration of $\pi_1 \otimes \pi_2$

One has that $\mathfrak{b}_1$ is an ideal of $\mathfrak{g}$ and $\mathfrak{b}_1 + \mathfrak{b}_2 = \mathfrak{g}$. Hence $\mathbb{I}(\mathfrak{b}_1, \mathfrak{b}_2) = \emptyset$. Hence by Proposition 3.5.27, one has that

$$\pi_1 \otimes \pi_2 \simeq \mathrm{ind}_{B_1 \cap B_2}^G \chi_{2l}.$$

Determine the index set I as in Eq. (3.5.19). One has now that $2l + (\mathfrak{b}_1 \cap \mathfrak{b}_2)^\perp = 2l + Z_7^\perp$ and the family $\{Z_1, \cdots, Z_6\}$ is a Malcev basis of $\mathfrak{g}$ relative $\mathfrak{b}_1 \cap \mathfrak{b}_2$. Determine the indices set I. One has for a linear form $f = 2Z_7^* + \sum_{j=2}^{1} \beta Z_j^* \in 2l + \mathbb{Z}_7^\perp$

$$\begin{array}{lcl}
\mathfrak{g}_6(f_6) = & \mathfrak{g}_6 & \not\subset \mathfrak{g}_7 \Longrightarrow 6 \in I, \\
\mathfrak{g}_5(f_5) = & \mathfrak{g}_5 & \not\subset \mathfrak{g}_6 \Longrightarrow 5 \in I, \\
\mathfrak{g}_4(f_4) = & \mathfrak{g}_4 & \not\subset \mathfrak{g}_5 \Longrightarrow 4 \in I, \\
\mathfrak{g}_3(f_3) = & \mathbb{R}\text{-span}(Z_7, Z_6, Z_4 - \frac{\beta_6}{2} Z_5) & \subset \mathfrak{g}_4 \Longrightarrow 3 \notin I, \\
\mathfrak{g}_2(f_2) = & \mathbb{R}\text{-span}(Z_7, Z_4 - \frac{\beta_6}{2} Z_5 + \frac{\beta_6}{2} Z_6) & \subset \mathfrak{g}_3 \Longrightarrow 2 \notin I, \\
\mathfrak{g}(f) \ = & \mathfrak{g}_7 & \subset \mathfrak{g}_2 \Longrightarrow 1 \notin I.
\end{array}$$

where $f_j := f_{|\mathfrak{g}_j}$. This implies that $I = \{k_1 = 4, k_2 = 5, k_3 = 6\}$ and the disintegration space of $\pi_1 \otimes \pi_2$ is given by $\mathscr{R} = 2Z_7^* + \mathbb{R}Z_6^* + \mathbb{R}Z_5^* + \mathbb{R}Z_4^*$. Take for every $\theta \in \mathscr{R}$ the Vergne polarization $\mathfrak{b} = \mathfrak{b}_1$ associated to Jordan-Hölder basis $\mathscr{Z}$. The mapping $T_\theta : L^2(\mathbb{G}/B_1 \times B_2, \chi_{l,l})$ of Theorem 3.6.6 is then given by:

$$\mathbb{T}_\theta(\xi)(g) = \int_{\mathbb{R}^3} \xi(g, g(u_4 Z_4)(u_5 Z_5)(u_6 Z_6)) e^{-2\pi i (v_4 u_4 + v_5 u_5 + (v_6+1) u_6 + 2Z_7^*)} du_4 du_5 du_6, \ g \in G.$$

and

$$\mathscr{H}_{\pi_1} \hat{\otimes} \mathscr{H}_{\pi_2} \simeq \int_{\mathbb{R}^3} L^2(G/B, \chi_{v_4 Z_4^* + v_5 Z_5^* + (v_6+1) Z_6^* + 2Z_7^*}) du_4 du_5 du_6.$$

Disintegration of $\pi_3 \otimes \pi_4$

Here

$$\mathscr{I}^{\mathfrak{g}/\mathfrak{b}_3} = \{1, 2, 5\}, \mathscr{I}^{\mathfrak{g}/\mathfrak{b}_4} = \{2, 3, 5\}$$

and so $\mathbb{I}^{\mathfrak{g}/\mathfrak{b}_3} \cap \mathbb{I}^{\mathfrak{g}/\mathfrak{b}_4} = \{2, 5\}$. Since for every $g = \prod_{j=1}^{7}(g_j Z_j),\ g_2 \neq 0$,

$$Z_5 = Z_5 - Z_4 + \mathrm{Ad}((g_1 Z_1)(g_2 Z_2)(g_3 Z_3)(g_4 Z_4))(\frac{1}{g_2}(-Z_4)) \mathrm{mod}\, \mathfrak{g}_6 \subset \mathfrak{b}_3 + \mathrm{Ad}(g)(\mathfrak{b}_4) \mathrm{mod}\, \mathfrak{g}_6,$$

it follows that $5 \notin \mathbb{I}(\mathfrak{b}_3, \mathfrak{b}_4)$. Hence since $[\mathfrak{g}, \mathfrak{g}] = \mathbb{R}\text{-span}(Z_5, Z_6, Z_7)$, one has that $\mathbb{I}(\mathfrak{b}_3, \mathfrak{b}_4) = \{2\}$. Compute the function F of (3.5.23). Let $\Phi : \mathbb{R}^2 \times \mathbb{R} \to G/B_4$ the mapping such that

$$\Phi(s_1, s_2, t) = (s_1 Z_1)(s_2(Z_4 - Z_5))(t Z_2) \cdot B_4 = (s_1 Z_1)(t Z_2)((s_2 t + 1) Z_5) \cdot B_4.$$

Therefore $\Phi(s_1, s_2, t) \equiv (s_1, t, 1 + s_2 t)$ and so $F(t) = t$. One has that

$$\mathrm{Ad}^*(t Z_2) l = l - t Z_6^* \Rightarrow l(t) = \mathrm{Ad}^*(t Z_2) l + \mathrm{Ad}^*(-t Z_2) l = 2l,\ t \in \mathbb{R}.$$

The Lie algebra $\mathfrak{b}(t)$ of definition (3.6.4) is given for $t \in \mathbb{R}$ by:

$$\mathfrak{b}(t) = \mathrm{Ad}(t Z_2)\mathfrak{b}_3 \cap \mathrm{Ad}(-t Z_2)\mathfrak{b}_4 = \mathbb{R}\text{-span}(Z_6, Z_7).$$

Hence the disintegration space this time is $\mathscr{R} = \mathbb{R}^2 \times \mathbb{R}$ and the mapping $\mathbb{T}_\theta,\ \theta \in \mathscr{R}$ is defined by

$$T_{(v_1, v_2), t}\xi(g) = \int_{\mathbb{R}^2} \xi(g(u_4 Z_4)(u_5 Z_5)(t Z_2), g(u_4 Z_4)(u_5 Z_5)(-t Z_2)) e^{-2\pi i (u_4 v_4 + u_5 v_5)} du_4 du_5$$

and

$$\mathscr{H}_{\pi_3} \hat{\otimes} \mathscr{H}_{\pi_4} \simeq \int_{\mathbb{R}} \int_{\mathbb{R}^2} L^2(G/B, \chi_{v_4 Z_4^* + v_5 Z_5^* + 2 Z_7^*}) |t| du_4 du_5 dt.$$

3.6.3 *Criteria for Irreducibility of Tensor Products*

This section is devoted to a necessary and sufficient condition for the irreducibility of the tensor representation $\pi \otimes \pi'$. We prove the following:

Theorem 3.6.7 *Let $G = \exp \mathfrak{g}$ be a simply connected nilpotent Lie group, let $\pi = \pi_l$ and $\pi' = \pi_{l'}$ be two irreducible unitary representations of G. The following conditions are equivalent*

1. $\pi \otimes \pi' \in \hat{G}$
2. *For every polarization $\mathfrak{b}$ at l and for every polarization $\mathfrak{b}'$ at l', one has that*

$$\begin{cases} \mathfrak{b} + \mathfrak{b}' = \mathfrak{g}, \\ \text{and } \mathfrak{b} \cap \mathfrak{b}' \text{ is a polarization at } l + l'. \end{cases}$$

3. *There exist some polarizations* $\mathfrak{b}$ *at* l *and* $\mathfrak{b}'$ *at* l', *such that*

$$\begin{cases} \mathfrak{b} + \mathfrak{b}' = \mathfrak{g}, \\ \textit{and } \mathfrak{b} \cap \mathfrak{b}' \textit{ is a polarization at } l + l'. \end{cases}$$

4.

$$\begin{cases} \mathrm{Ad}^*(G)l + \mathrm{Ad}^*(G)l' = \mathrm{Ad}^*(G)(l + l'), \\ \textit{and } \mathfrak{g}(l) + \mathfrak{g}(l') = \mathfrak{g}. \end{cases}$$

Proof (1) $\Longrightarrow$ (2) By Theorem 3.5.28, the space of double cosets $\mathscr{D}\diagdown\mathbb{G}\diagup\mathbb{B} \simeq B\diagdown G\diagup B'$ must reduce to one point, which implies by Lemma 3.5.9 that $\mathbb{I}(\mathfrak{g}/\mathfrak{b}, \mathfrak{g}/\mathfrak{b}') = \emptyset$ and $G = BB'$. In particular this tells us that $L^2(\mathbb{G}/\mathbb{B}, \chi_{(l,l')})$ is isometrically isomorphic to $L^2(G/B \cap B', \chi_{l+l'})$ and so $\pi \otimes \pi' \simeq \mathrm{ind}_{B\cap B'}^G \chi_{l+l'}$. Hence the representation $\mathrm{ind}_{B\cap B'}^G \chi_{l+l'}$ is irreducible, which forces $\mathfrak{b} \cap \mathfrak{b}'$ to be a polarization at $l + l'$ by Kirillov's theorem.

(2) $\Longrightarrow$ (3) is trivial.

(3) $\Longrightarrow$ (4) Let us first show that $\mathrm{Ad}^*(G)l + \mathrm{Ad}^*(G)l' = \mathrm{Ad}^*(G)(l + l')$. Do this by induction on $\dim \mathfrak{g}$. If this dimension is 1, the claim is evident since one works then with characters. If l' is a character, then one has that $\mathrm{Ad}^*(G)l + \mathrm{Ad}^*(G)l' = \mathrm{Ad}^*(G)l + l' = \mathrm{Ad}^*(G)(l + l')$. One can thus assume that $\mathfrak{b}'$ is a proper subalgebra of $\mathfrak{g}$. Let $\mathfrak{h}$ be again an ideal of codimension 1 in $\mathfrak{g}$ containing $\mathfrak{b}'$. Then $\mathfrak{b} \not\subset \mathfrak{h}$ by the condition $\mathfrak{b} + \mathfrak{b}' = \mathfrak{g}$. Hence $\mathfrak{b} \cap \mathfrak{h}$ is a polarization at $l_{|\mathfrak{h}}$ and $\mathfrak{b}'$ is a polarization at $l'_{|\mathfrak{h}}$ and also $\mathfrak{b} \cap \mathfrak{h} + \mathfrak{b}'$ is a polarization at $l_{|h} + l'_{|\mathfrak{h}}$. It follows from the induction hypothesis that

$$\mathrm{Ad}^*(H)l_{|\mathfrak{h}} + \mathrm{Ad}^*(H)l'_{|\mathfrak{h}} = \mathrm{Ad}^*(H)(l_{|\mathfrak{h}} + l'_{|\mathfrak{h}}).$$

Therefore

$$\mathrm{Ad}^*(H)l + \mathrm{Ad}^*(H)l' + \mathfrak{h}^{\perp} = \mathrm{Ad}^*(H)(l + l') + \mathfrak{h}^{\perp} \subset \mathrm{Ad}(G)(l + l'),$$

since the G-orbit $\mathrm{Ad}(G)(l + l')$ is saturated with respect to $\mathfrak{h}$. But since $\mathfrak{b}' \subset \mathfrak{h}$, one also has that $\mathrm{Ad}^*(H)l' + \mathfrak{h}^{\perp} = \mathrm{Ad}^*(G)l'$. Finally

$$\mathrm{Ad}^*(G)l + \mathrm{Ad}^*(G)l' \subset \mathrm{Ad}(G)(l + l').$$

Since $l + l' \in \mathrm{Ad}^*(G)l + \mathrm{Ad}^*(G)l'$ it follows immediately that

$$\mathrm{Ad}(G)(l + l') \subset \mathrm{Ad}^*(G)l + \mathrm{Ad}^*(G)l'$$

Let us show now that $\mathfrak{g}(l) + \mathfrak{g}(l') = \mathfrak{g}$. One has for any polarization $\mathfrak{b}$ at l that

$$\dim \mathfrak{b} = \frac{\dim \mathfrak{g} + \dim \mathfrak{g}(l)}{2}.$$

Hence since $\mathfrak{b} + \mathfrak{b}' = \mathfrak{g}$ and $\mathfrak{b} \cap \mathfrak{b}'$ is a polarization at $l + l'$ and since

$$\dim \mathfrak{b} + \dim \mathfrak{b}' = \dim \mathfrak{g} + \dim \mathfrak{b} \cap \mathfrak{b}',$$

one has that

$$\frac{2 \dim \mathfrak{g} + \dim \mathfrak{g}(l) + \dim \mathfrak{g}(l')}{2} = \dim \mathfrak{g} + \frac{\dim \mathfrak{g} + \dim \mathfrak{g}(l + l')}{2}.$$

Therefore, since $\mathfrak{g}(l) \cap \mathfrak{g}(l') \subset \mathfrak{g}(l + l')$,

$$\begin{aligned} \dim \mathfrak{g}(l) + \mathfrak{g}(l') + \dim \mathfrak{g}(l + l') &\geq \dim \mathfrak{g}(l) + \mathfrak{g}(l') + \dim \mathfrak{g}(l) \cap \mathfrak{g}(l') \\ &= \dim \mathfrak{g}(l) + \dim \mathfrak{g}(l') \\ &= \dim \mathfrak{g}(l + l') + \dim \mathfrak{g}, \end{aligned}$$

and so $\dim \mathfrak{g}(l) + \mathfrak{g}(l') \geq \dim \mathfrak{g}$ which implies that $\mathfrak{g} = \mathfrak{g}(l) + \mathfrak{g}(l')$.

(4) $\Longrightarrow$ (1) Let $\mathfrak{b}$ (resp. $\mathfrak{b}'$) be a polarization at l (resp. at l'). Then $\mathfrak{g}(l) \subset \mathfrak{b}$, $\mathfrak{g}(l') \subset \mathfrak{b}'$ and hence $\mathfrak{b} + \mathfrak{b}' = \mathfrak{g}$. Therefore BB' contains an open subset of G, whence it is itself open. Since the product of two closed connected subgroups of G is always closed, it follows that BB' is open and closed which implies that $BB' = G$. Therefore $\pi_{l,\mathfrak{b}} \otimes \pi_{l',\mathfrak{b}'} \simeq \mathrm{ind}_{B \cap B'}^{G} \chi_{l+l'}$. If $\mathfrak{b} \cap \mathfrak{b}'$ is shown to be a polarization at $l + l'$, and $\pi \otimes \pi'$ is irreducible. The condition $\mathrm{Ad}^*(G)l + \mathrm{Ad}^*(G)l' = \mathrm{Ad}^*(G)(l + l')$ implies that one has an injective mapping

$$E : G/G(l) \times G/G(l') \ni (g \cdot G(l), g' \cdot G(l')) \mapsto \mathrm{Ad}^*(g)l + \mathrm{Ad}^*(g')l' \in \mathrm{Ad}^*(G)(l + l').$$

Hence

$$\dim \mathfrak{g}/\mathfrak{g}(l) + \dim \mathfrak{g}/\mathfrak{g}(l') \leq \dim \mathfrak{g}/\mathfrak{g}(l + l'),$$

and from that

$$\begin{aligned} 2 \dim \mathfrak{g} - \dim \mathfrak{g}(l) - \dim \mathfrak{g}(l) &\leq \dim \mathfrak{g} - \dim \mathfrak{g}(l + l') \\ &\Rightarrow \\ \dim \mathfrak{g}(l) + \mathfrak{g}(l') - \dim \mathfrak{g}(l) - \dim \mathfrak{g}(l) &\leq - \dim \mathfrak{g} - \dim \mathfrak{g}(l + l') \\ &\Rightarrow \\ \dim \mathfrak{g}(l + l') &\leq \dim \mathfrak{g}(l) \cap \mathfrak{g}(l'). \end{aligned}$$

Therefore $\mathfrak{g}(l) \cap \mathfrak{g}(l') = \mathfrak{g}(l + l')$, since $\mathfrak{g}(l) \cap \mathfrak{g}(l') \subset \mathfrak{g}(l + l')$. This tells us then that

$$\begin{aligned} \dim \mathfrak{b} \cap \mathfrak{b}' &= \dim \mathfrak{b} + \dim \mathfrak{b}' - \dim \mathfrak{b} + \mathfrak{b}' \\ &= \dim \mathfrak{b} + \dim \mathfrak{b}' - \dim \mathfrak{g} \end{aligned}$$

$$= \frac{2\dim \mathfrak{g} + \dim \mathfrak{g}(l) + \dim \mathfrak{g}(l')}{2} - \dim \mathfrak{g}$$
$$= \frac{\dim \mathfrak{g}(l) + \dim \mathfrak{g}(l')}{2}$$
$$= \frac{\dim \mathfrak{g} + \dim \mathfrak{g}(l) \cap \dim \mathfrak{g}(l')}{2}$$
$$= \frac{\dim \mathfrak{g} + \dim \mathfrak{g}(l + l')}{2}.$$

Hence $\mathfrak{b} \cap \mathfrak{b}'$ is a polarization at $l + l'$. □

Example 3.6.8 Let $\mathfrak{g} = \mathbb{R}\text{-span}(Z_1, \ldots, Z_6)$ be the Lie algebra equipped with brackets

$$[Z_2, Z_3] = Z_5 \quad [Z_1, Z_4] = Z_6.$$

Let $l = Z_5^\star, l' = Z_6^\star \in \mathfrak{g}^\star$. Take the polarizations

$$\mathfrak{b} := \mathbb{R}\text{-span}(Z_6, Z_5, Z_4, Z_3, Z_1)$$

at l and

$$\mathfrak{b}' := \mathbb{R}\text{-span}(Z_6, Z_5, Z_4, Z_3, Z_2)$$

at l'. Then $\mathfrak{b} + \mathfrak{b}' = \mathfrak{g}$ and

$$\mathfrak{b} \cap \mathfrak{b}' = \mathbb{R}\text{-span}(Z_6, Z_5, Z_4, Z_3),$$

which is a polarization at $l + l'$. Hence $\pi_l \otimes \pi_{l'} \in \hat{G}$.

3.7 Intertwining of Quasi-Regular Representations of Nilmanifolds

Let G be a locally compact unimodular topological group and let Γ be a co-compact discrete subgroup of G. The space G/Γ of left cosets has a unique invariant Borel probability measure ν. Let R_Γ be the unitary representation of G by left translations on the corresponding Hilbert space $L^2(G/\Gamma) = L^2(G/\Gamma, \nu)$. Thus,

$$\big(R_\Gamma(g)\phi\big)(x) = \phi(g^{-1}x)$$

for all $\phi \in L^2(G/\Gamma)$ and $g, x \in G$. The representation R_Γ is called *the quasi-regular representation* of G induced from Γ. According to Gornet [83, page 125], one says that two co-compact, discrete subgroups Γ_1 and Γ_2 of a Lie group G are *representation-equivalent* (cf. Definition 3.7.20), if the quasi-regular representations R_{Γ_1} and R_{Γ_2} are unitarily equivalent. A natural problem is to obtain conditions for Γ_1 and Γ_2 to be representation-equivalent.

In the case $G = \mathbb{R}^n$, there are no representation-equivalent subgroups. In [83, Theorem 1], Gornet gave a necessary and sufficient condition for a pair of co-compact discrete subgroups of a two-step nilpotent Lie group to be representation-equivalent. More precisely, given a co-compact discrete subgroup Γ of G, she defines a group of automorphisms of G up to Γ-equivalence (cf. Definition 3.7.21). A discrete co-compact subgroup of G is representation-equivalent to Γ if and only if it differs from Γ by a Γ-equivalence.

The basic problem studied in this chapter is the construction of an explicit intertwining operator between the unitary representation R_Γ and its decomposition into irreducibles, in the case of nilpotent Lie groups of step-length less than or equal to three, and then also between the representations R_{Γ_1} and R_{Γ_2} when Γ_1 and Γ_2 are representation-equivalent subgroups for two-step nilpotent Lie groups. This reduces to consider a unitary operator $\mathbb{T} : L^2(G/\Gamma_1) \longrightarrow L^2(G/\Gamma_2)$ such that the following diagram is commutative

$$\begin{array}{ccc} L^2(G/\Gamma_1) & \xrightarrow{\quad R_{\Gamma_1}(g) \quad} & L^2(G/\Gamma_1) \\ {\scriptstyle \mathbb{T}}\big\downarrow & & {\scriptstyle \mathbb{T}}\big\downarrow \\ L^2(G/\Gamma_2) & \xrightarrow{\quad R_{\Gamma_2}(g) \quad} & L^2(G/\Gamma_2) \end{array} \tag{3.7.1}$$

for every $g \in G$.

3.7.1 *Rational Structures and Uniform Subgroups*

In this subsection we present some results on discrete uniform subgroups of a nilpotent Lie group. Let G be a nilpotent, connected and simply connected real Lie group and let $\mathfrak{g}$ be its Lie algebra. We say that $\mathfrak{g}$ (or G) has a *rational structure* if there is a Lie algebra $\mathfrak{g}_\mathbb{Q}$ over $\mathbb{Q}$ such that $\mathfrak{g} \cong \mathfrak{g}_\mathbb{Q} \otimes \mathbb{R}$. It is clear that $\mathfrak{g}$ has a rational structure if and only if $\mathfrak{g}$ has an $\mathbb{R}$-basis $\{X_1, \ldots, X_n\}$ with rational structure constants.

Let $\mathfrak{g}$ have a fixed rational structure given by $\mathfrak{g}_\mathbb{Q}$ and let $\mathfrak{h}$ be an $\mathbb{R}$-subspace of $\mathfrak{g}$. Define $\mathfrak{h}_\mathbb{Q} = \mathfrak{h} \cap \mathfrak{g}_\mathbb{Q}$. We say that $\mathfrak{h}$ is *rational* if $\mathfrak{h} = \mathbb{R}\text{-span}\left\{\mathfrak{h}_\mathbb{Q}\right\}$. A connected, closed subgroup H of G is *rational* if its Lie algebra $\mathfrak{h}$ is rational. The elements of $\mathfrak{g}_\mathbb{Q}$ (or $G_\mathbb{Q} = \exp \mathfrak{g}_\mathbb{Q}$) are called *rational elements* (or *rational points*) of $\mathfrak{g}$ (or G).

A discrete subgroup Γ is called *uniform* in G if the quotient space G/Γ is compact. The homogeneous space G/Γ is called a *compact nilmanifold*. A criterion of Malcev states that G admits a uniform subgroup Γ if and only if $\mathfrak{g}$ admits a rational structure. Furthermore, if G has a uniform subgroup Γ, then $\mathfrak{g}$ (hence G) has a rational structure such that $\mathfrak{g}_{\mathbb{Q}} = \mathbb{Q}\text{-span}\{\log(\Gamma)\}$. Conversely, if $\mathfrak{g}$ has a rational structure given by some $\mathbb{Q}$-algebra $\mathfrak{g}_{\mathbb{Q}} \subset \mathfrak{g}$, then G has a uniform subgroup Γ such that $\log(\Gamma) \subset \mathfrak{g}_{\mathbb{Q}}$ (see [39] and [126]). If G is endowed with the rational structure induced by a uniform subgroup Γ and if H is a Lie subgroup of G, then H is rational if and only if $H \cap \Gamma$ is a uniform subgroup of H. Note that the notion of rationality depends on Γ. A real linear functional $f \in \mathfrak{g}^*$ is called rational ($f \in \mathfrak{g}^*_{\mathbb{Q}}$, $\mathfrak{g}_{\mathbb{Q}} = \mathbb{Q}\text{-span}(\log(\Gamma))$) if $\langle f, \mathfrak{g}_{\mathbb{Q}}\rangle \subset \mathbb{Q}$, or equivalently $\langle f, \log(\Gamma)\rangle \subset \mathbb{Q}$.

Let Γ be a uniform subgroup of G. A strong Malcev basis $\{X_1, \ldots, X_n\}$ for $\mathfrak{g}$ is said to be *strongly based on* Γ if

$$\Gamma = \exp \mathbb{Z}X_1 \cdots \exp \mathbb{Z}X_n. \tag{3.7.2}$$

Such a basis always exists (cf. [39, 127]).

Finally, a proof of the next result can be found in Proposition 5.3.2 of [39].

Proposition 3.7.1 *Let Γ be a uniform subgroup in a nilpotent Lie group G, and let $H_1 \subsetneq H_2 \subsetneq \cdots \subsetneq H_k = G$ be rational normal subgroups of G. Let $\mathfrak{h}_1, \ldots, \mathfrak{h}_{k-1}, \mathfrak{h}_k = \mathfrak{g}$ be the corresponding Lie algebras. Then there exists a strong Malcev basis $\{X_1, \ldots, X_n\}$ for $\mathfrak{g}$ that is strongly based on Γ and that passes through $\mathfrak{h}_1, \ldots, \mathfrak{h}_{k-1}$.*

Let $\mathscr{B} = (X_1, \ldots, X_n)$ be a strong Malcev basis for $\mathfrak{g}$. For $i = 1, \ldots, n$, put

$$\mathfrak{a}_i = \mathbb{R}\text{-span}\{X_1, \ldots, X_i\}.$$

For $l \in \mathfrak{g}^*$ given, let

$$\mathfrak{m}(l) = \sum_{i=1}^{n} \mathfrak{a}_i(l_i), \tag{3.7.3}$$

be the Vergne polarization at l associated to $\mathscr{B}$ (cf. Theorem 1.2.4), where $\mathfrak{a}_i(l_i)$ denotes the stabilizer of $l_i = l|_{\mathfrak{a}_i}$ inside the subalgebra $\mathfrak{a}_i$. The following proposition (cf. [39, Proposition 5.2.6] and [35, Theorem 3.4]) will be used in the sequel.

Proposition 3.7.2 *Let $G = \exp \mathfrak{g}$ be a connected, simply connected nilpotent Lie group and Γ a discrete uniform subgroup of G. Let $\mathscr{B}$ be a strong Malcev basis of $\mathfrak{g}$ strongly based on Γ. If $l \in \mathfrak{g}^*_{\mathbb{Q}}$, then the Vergne polarization at l associated to $\mathscr{B}$ is rational.*

Before stating the next result, some notations and definitions are needed. Recall first the concept of a fundamental domain.

Definition 3.7.3 Let G be a topological group and let H be a subgroup of G. A fundamental domain for H is a Borel subset Ω of G such that

$$G = \bigsqcup_{h \in H} \Omega h$$

is the disjoint union of Borel subsets Ωh, $h \in H$.

It is clear that a Borel subset Ω of G is a fundamental domain for H in G if and only if the natural map $\Omega \longrightarrow G/H$ defined by $g \longmapsto gH$ is bijective.

Fundamental Domains for Uniform Subgroups

Let Γ be a discrete uniform subgroup of a connected, simply connected nilpotent Lie group G, and let $\mathscr{B} = (X_1, \ldots, X_n)$ be a strong Malcev basis for the Lie algebra $\mathfrak{g}$ of G strongly based on Γ. Define the mapping $\mathcal{E}_{\mathscr{B}} : \mathbb{R}^n \longrightarrow G$ by

$$\mathcal{E}_{\mathscr{B}}(T) = \exp t_n X_n \cdots \exp t_1 X_1 \tag{3.7.4}$$

where $T = (t_1, \ldots, t_n) \in \mathbb{R}^n$. It is well known that $\mathcal{E}_{\mathscr{B}}$ is a diffeomorphism ([39]). Let

$$\mathbb{I} = [0, 1) = \{t \in \mathbb{R} : 0 \le t < 1\} \tag{3.7.5}$$

and let

$$\Omega = \mathcal{E}_{\mathscr{B}}(\mathbb{I}^n) \tag{3.7.6}$$

Then Ω is a fundamental domain for Γ in G ([35, Lemma 3.6]), and the mapping $\mathcal{E}_{\mathscr{B}}$ maps the Lebesgue measure dt on $\mathbb{I}^n$ to the G-invariant probability measure ν on G/Γ, that is, for φ in $\mathscr{C}(G/\Gamma)$, one has

$$\int_{G/\Gamma} \varphi(\dot{g}) d\nu(\dot{g}) = \int_{\mathbb{I}^n} \varphi(\mathcal{E}_{\mathscr{B}}(t)) dt. \tag{3.7.7}$$

Furthermore, for $\phi \in \mathscr{C}_c(G)$ one has

$$\int_G \phi(g) d\nu(g) = \sum_{s \in \mathbb{Z}^n} \int_{\mathbb{I}^n} \phi\Big(\mathcal{E}_{\mathscr{B}}(t)\mathcal{E}_{\mathscr{B}}(s)\Big) dt. \tag{3.7.8}$$

Proposition 3.7.4 *Let $G = \exp \mathfrak{g}$ be a connected, simply connected nilpotent Lie group and let Γ be a discrete uniform subgroup of G. Let $\mathfrak{m}$ be a rational ideal of $\mathfrak{g}$ of dimension k. Let $M = \exp \mathfrak{m}$ and let $(X_1, \ldots, X_n)$ be a strong Malcev basis of $\mathfrak{g}$*

strongly based on Γ *passing through* $\mathfrak{m}$. *For every* $\xi \in \mathscr{C}_c(G/M)$, *one has*

$$\int_{G/M} \xi(g)dg = \sum_{s\in\mathbb{Z}^{n-k}} \int_{\mathbb{I}^{n-k}} \xi\Big(\exp x_{n-k}\overline{X}_n \cdots \exp x_1\overline{X}_{k+1}$$

$$\exp s_{n-k}\overline{X}_n \cdots \exp s_1\overline{X}_{k+1}\Big)dx_n \cdots dx_{k+1},$$

where $\overline{X}_i$ *is the image of* X_i *under the canonical projection* $p : \mathfrak{g} \longrightarrow \mathfrak{g}/\mathfrak{m}$.

Proof Since M is rational, it follows from Lemma 5.1.4 of [39] that $P(\Gamma)$ is a uniform subgroup of $P(G)$, where $P : G \longrightarrow G/M$ is the canonical projection. On the other hand, $(\overline{X}_{k+1}, \ldots, \overline{X}_n)$ is a strong Malcev basis of $\mathfrak{g}/\mathfrak{m}$ strongly based on $P(\Gamma)$. □

3.7.2 *Intertwining Operators*

In the sequel, if $F = \mathbb{R}\text{-span}\{e_1, \ldots, e_m\}$ denotes a subspace of $E = \mathbb{R}\text{-span}\{e_1, \ldots, e_n\}$, then any element $l \in F^*$ can be extended to an element $\bar{l}$ of E^* by trivial extension, ie, $\langle \bar{l}, e_i\rangle = 0$ for every $i = m+1, \ldots, n$.

Let $G = \exp \mathfrak{g}$ be a k-step, connected and simply connected nilpotent Lie group such that $k \leq 3$, which contains a uniform subgroup Γ. Let $\mathcal{D}(\mathfrak{g}) = [\mathfrak{g}, \mathfrak{g}]$. Let $\mathfrak{z}$ or $\mathfrak{z}(\mathfrak{g})$ denote the center of $\mathfrak{g}$. Since $\mathcal{D}(\mathfrak{g}) \cap \mathfrak{z}(\mathfrak{g})$ and $\mathcal{D}(\mathfrak{g})$ are rational, by Proposition 3.7.1 there exists a strong Malcev basis $\mathscr{B} = (X_1, \ldots, X_n)$ of $\mathfrak{g}$ strongly based on Γ passing through $\mathcal{D}(\mathfrak{g}) \cap \mathfrak{z}(\mathfrak{g})$ and $\mathcal{D}(\mathfrak{g})$. One then has $\mathcal{D}(\mathfrak{g}) \cap \mathfrak{z}(\mathfrak{g}) = \mathbb{R}\text{-span}\{X_1, \ldots, X_\alpha\}$ for some $\alpha \in \{1, \ldots, n\}$. Let

$$\mathscr{V}^* = \mathbb{R}X_1^* + \ldots + \mathbb{R}X_\alpha^* \subset \mathfrak{g}^*$$

and

$$\mathscr{V}_\Gamma^* = \mathbb{Z}X_1^* + \ldots + \mathbb{Z}X_\alpha^*,$$

where $(X_1^*, \ldots, X_n^*)$ is the dual basis of $\mathscr{B}$. If $\mathfrak{a}$ is a subspace of $\mathfrak{g}$, let $\mathfrak{a}_\Gamma^*$ denote the set

$$\mathfrak{a}_\Gamma^* = \left\{l \in \mathfrak{a}^* : \langle l, \mathfrak{a} \cap \log(\Gamma)\rangle \subset \mathbb{Z}\right\}.$$

For $l \in \mathfrak{g}^*$, let $g_l = \dim \mathcal{D}(\mathfrak{g}) \cap \mathfrak{g}(l)$. Let also for $f \in \mathscr{V}$

$$\delta(f) = \{U \in \delta; \langle f, [U, \mathfrak{g}]\rangle = \{0\}\}.$$

The following lemma is immediate.

Lemma 3.7.5 *Let $f \in \mathcal{V}_\Gamma^*$, $l \in \mathfrak{g}^*$, such that $l_{|\delta\cap\mathfrak{z}(\mathfrak{g})} = f_{|\delta\cap\mathfrak{z}(\mathfrak{g})}$. Then one has that*

$$\delta(f) = \mathfrak{D}(\mathfrak{g}) \cap \mathfrak{g}(l).$$

In particular, one has $g_f = g_l$.

Proof Indeed, since $[\delta, \mathfrak{g}] \subset \mathfrak{z}(\mathfrak{g})$, $\mathfrak{g}$ being of step ≤ 3, one has that:

$$\begin{aligned}\mathfrak{D}(\mathfrak{g}) \cap \mathfrak{g}(l) &= \{U \in \delta, \langle l, [U, \mathfrak{g}]\rangle = \{0\}\} \\ &= \{U \in \delta, \langle l_{|\mathfrak{z}(\mathfrak{g})}, [U, \mathfrak{g}]\rangle = \{0\}\} \\ &= \{U \in \delta, \langle f_{|\mathfrak{z}(\mathfrak{g})}, [U, \mathfrak{g}]\rangle = \{0\}\} \\ &= \delta(f).\end{aligned}$$

□

For $f \in \mathcal{V}_\Gamma^*$, let

$$\mathcal{V}_f = \{l \in \delta(f)_\Gamma^* : l|_{\mathfrak{D}(\mathfrak{g})\cap\mathfrak{z}(\mathfrak{g})} = f\}. \tag{3.7.9}$$

Lemma 3.7.6 *For $f \in \mathcal{V}_\Gamma^*$ and $l_0 \in \mathcal{V}_f$, let $\mathfrak{m}(l_0)$ be the Vergne polarization at l_0 associated to the basis $\mathscr{B}$. Then $\mathfrak{m}(l_0)$ is a rational ideal in $\mathfrak{g}$ which contains $\mathfrak{D}(\mathfrak{g})$.*

Proof Since the basis $\mathscr{B}$ passes through δ, there exists an index $i \in \{1, \ldots, n-2\}$ such that $\mathfrak{a}_i = \mathfrak{D}(\mathfrak{g})$. By definition of the Vergne polarization associated to $\mathscr{B}$, $\mathfrak{m}(l_0)$ contains $\mathfrak{a}_i(l_0|_{\mathfrak{a}_i})$. On the other hand, since the group G is at most 3-step, the ideal $\delta = \mathfrak{a}_i$ is Abelian and so $\mathfrak{a}_i(l_0|_{\mathfrak{a}_i}) = \mathfrak{a}_i = \delta$. Consequently, $\mathfrak{m}(l_0) \supset \delta$ is an ideal of $\mathfrak{g}$. Finally, the rationality of $\mathfrak{m}(l_0)$ follows from Proposition 3.7.2. □

For $l \in \mathfrak{g}^*$, let for $f \in \mathcal{V}_\Gamma$ and $l_0 \in \mathcal{W}_f$,

$$\mathfrak{a}_\Gamma^*(f, l_0) = \{l \in \mathfrak{m}(l_0)_\Gamma^* : l|_{\delta(f)} = l_0\}. \tag{3.7.10}$$

Lemma 3.7.7 *With the above notation, the sets $\mathfrak{m}(l_0)_\Gamma^*$ and $\mathfrak{a}_\Gamma^*(f, l_0)$ are* $\mathrm{Ad}^*(\Gamma)$-*invariant.*

Proof Let $\exp\gamma \in \Gamma$ and $l \in \mathfrak{m}(l_0)_\Gamma^*$. Let $X \in \mathfrak{m}(l_0) \cap \log(\Gamma)$. By definition of the coadjoint representation one has

$$\langle \exp -\gamma \cdot l, X\rangle = \langle l, e^{\mathrm{ad}\gamma}(X)\rangle.$$

On the other hand, since $M(l_0)$ is normal in G, one has that

$$\exp e^{\mathrm{ad}\gamma}(X) = \exp\gamma \exp X \exp -\gamma \in \Gamma \cap M(l_0).$$

It follows that $e^{\mathrm{ad}\gamma}(X) \in \log(\Gamma) \cap \mathfrak{m}(l_0)$ and therefore $\langle l, e^{\mathrm{ad}\gamma}(X)\rangle \in \mathbb{Z}$. Consequently, $\exp -\gamma \cdot l \in \mathfrak{m}(l_0)^*_\Gamma$. Now let $l \in \mathfrak{a}^*_\Gamma(f, l_0)$. Then

$$\begin{aligned}\left(\exp -\gamma \cdot l\right)\Big|_{\mathcal{D}(\mathfrak{g})\cap\mathfrak{g}(f)} &= \left(\exp -\gamma \cdot l\right)\Big|_{\mathcal{D}(\mathfrak{g})\cap\mathfrak{g}(l)} \qquad \text{(by Lemma 3.7.5)}\\ &= l|_{\mathcal{D}(\mathfrak{g})\cap\mathfrak{g}(l)}\\ &= l_0|_{\mathcal{D}(\mathfrak{g})\cap\mathfrak{g}(l)}\\ &= l_0|_{\mathcal{D}(\mathfrak{g})\cap\mathfrak{g}(f)}.\end{aligned}$$

Consequently, $\exp -\gamma \cdot l \in \mathfrak{a}^*_\Gamma(f, l_0)$. □

Lemma 3.7.8 *Let $f \in \mathcal{V}^*, l, l' \in \mathfrak{g}^*$ such that $l_{|\delta(f)} = l'_{|\delta(f)}$. Then the Vergne polarization $\mathfrak{m}(l')$ is also a polarization at l.*

Proof Let us show that $\mathfrak{m}(l')$ is subordinated to l. Let $X, Y \in \mathfrak{m}(l')$. Then for $Z \in \mathfrak{g}$, one has that

$$\begin{aligned}\langle l', [Z, [X, Y]]\rangle &= \langle l', [X, [Z, Y]] + [Y, [X, Z]]\rangle\\ &\qquad \text{(by the Jacobi identity)}\\ &= 0, \ \text{since } \mathfrak{m}(l') \text{ is an ideal of } \mathfrak{g}.\end{aligned}$$

This implies that $[X, Y] \in \mathcal{D}(\mathfrak{g}) \cap \mathfrak{g}(l')$. On the other hand, by Lemma 3.7.5, one obtains then that $[X, Y] \in \mathcal{D}(\mathfrak{g}) \cap \mathfrak{g}(l)$. As $l|_{\mathcal{D}(\mathfrak{g})\cap\mathfrak{g}(l)} = l'|_{\mathcal{D}(\mathfrak{g})\cap\mathfrak{g}(l')}$, it follows that

$$\langle l, [X, Y]\rangle = \langle l', [X, Y]\rangle = 0.$$

Hence $\mathfrak{m}(l')$ is subordinated to l and therefore $d_{l'} \leq d_l$. Similarly, one proves that $d_l \leq d_{l'}$. Consequently, $\mathfrak{m}(l')$ is a polarization at l. □

Lemma 3.7.8 shows that for two elements $l, l' \in \mathfrak{g}^*$ such that $l_{|\mathcal{D}(\mathfrak{g})\cap\mathfrak{z}(\mathfrak{g})} = l'_{|\mathcal{D}(\mathfrak{g})\cap\mathfrak{z}(\mathfrak{g})} =: f$ and $l_{|\delta(f)} = l'_{|\delta(f)}$ the Vergne polarization $\mathfrak{m}(l')$ is also a polarization at l. Therefore for any $l_0 \in \mathcal{W}_f$, choose any extension $\overline{l_0}$ in $\mathfrak{g}^*$ and the Vergne polarization $\mathfrak{m}(\overline{l_0})$ gives us then a polarization at every $l \in \mathfrak{a}^*_\Gamma(f, l_0)$. Since this polarization depends only on l_0, write $\mathfrak{m}(l_0) := \mathfrak{m}(\overline{l_0})$. The set $\mathfrak{a}^*_\Gamma(f, l_0)$ is $\mathrm{Ad}^*\Gamma$-invariant (see Lemma 3.7.7). Let $\Sigma(f, l_0)$ be a cross-section for the Γ-orbits in $\mathfrak{a}^*_\Gamma(f, l_0)$.

One can now formulate the following:

Theorem 3.7.9 *Let G be a connected, simply connected nilpotent Lie group of step-length ≤ 3. With the above notations, one has*

$$\tau = ind_\Gamma^G 1 \simeq \rho = \bigoplus_{f\in\mathcal{V}^*_\Gamma} \bigoplus_{l_0\in\mathcal{V}_f} \bigoplus_{l\in\Sigma(f,l_0)} ind_{M(l_0)}^G \chi_l. \tag{3.7.11}$$

Furthermore, the operator U defined for all $\xi \in \mathscr{C}(G/\Gamma)$ and $g \in G$ by

$$U(\xi)(f, l_0, l)(g) = \int_{M(l_0)/M(l_0)\cap\Gamma} \xi(gm)\chi_l(m)dm \tag{3.7.12}$$

is an isometric linear operator with values in the Hilbert space $\mathscr{H}_\rho$ of ρ, and it can be extended over $\mathscr{H}_\tau = L^2(G/\Gamma)$ to an operator intertwining τ and ρ.

Proof First, it is clear that U intertwines τ and ρ, that is, for every $g \in G$ the following diagram is commutative:

$$\begin{array}{ccc} \mathscr{H}_\tau & \xrightarrow{\quad \tau(g) \quad} & \mathscr{H}_\tau \\ {\scriptstyle U}\big\downarrow & & {\scriptstyle U}\big\downarrow \\ \mathscr{H}_\rho & \xrightarrow{\quad \rho(g) \quad} & \mathscr{H}_\rho \end{array}$$

Next, one establishes that U is well defined and isometric. Let $\xi \in \mathscr{C}(G/\Gamma)$ be given. Then

$$\begin{aligned} \|U(\xi)\|^2_{\mathscr{H}_\rho} &= \sum_{f\in\mathscr{V}_\Gamma^*}\sum_{l_0\in\mathcal{V}_f}\sum_{l\in\Sigma(f,l_0)} \|U(\xi)(f, l_0, l)\|^2_{L^2(G/M(l_0),\chi_l)} \\ &= \sum_{f\in\mathscr{V}_\Gamma^*}\sum_{l_0\in\mathcal{V}_f}\sum_{l\in\Sigma(f,l_0)} \int_{G/M(l_0)} \left|U(\xi)(f, l_0, l)(g)\right|^2 dg \\ &= \sum_{f\in\mathscr{V}_\Gamma^*}\sum_{l_0\in\mathcal{V}_f}\sum_{l\in\Sigma(f,l_0)} \int_{G/M(l_0)} \left|\int_{M(l_0)/M(l_0)\cap\Gamma} \xi(gm)\chi_l(m)dm\right|^2 dg. \end{aligned}$$

Fix $f \in \mathscr{V}_\Gamma^*$ and l_0 in $\mathcal{V}_f$. Let $\mathfrak{m}(l_0)$ be the Vergne polarization at l_0 associated to $\mathscr{B}$. As before, since $\mathcal{D}(\mathfrak{g}) \cap \mathfrak{z}(\mathfrak{g})$, $\delta(f)$ and $\mathfrak{m}(l_0)$ are rational, there exists a strong Malcev basis $\mathscr{B}_{l_0} = (e_1(l_0), \ldots, e_n(l_0))$ of $\mathfrak{g}$ strongly based on Γ passing through $\mathcal{D}(\mathfrak{g}) \cap \mathfrak{z}(\mathfrak{g})$, $\delta(f)$ and $\mathfrak{m}(l_0)$, such that

$$\begin{aligned} e_i(l_0) &= X_i \qquad (1 \le i \le \alpha) \\ \mathcal{D}(\mathfrak{g}) \cap \mathfrak{g}(l_0) &= \mathbb{R}\text{-span}\left\{e_1(f), \ldots, e_{g_f}(f)\right\} \\ \mathfrak{m}(l_0) &= \mathbb{R}\text{-span}\left\{e_1(l_0), \ldots, e_{d_{l_0}}(l_0)\right\}, \end{aligned}$$

where $d_{l_0} = \dim \mathfrak{m}(l_0)$. Write

$$\begin{aligned} \delta_f &= g_f - \alpha = \dim \delta(f)/\delta \cap \mathfrak{z}, \\ \beta_{l_0} &= d_{l_0} - \alpha - \delta_f = \dim \mathfrak{m}(l_0)/\delta(f), \\ r_{l_0} &= n - d_{l_0} = \dim \mathfrak{g}/\mathfrak{m}(l_0). \end{aligned}$$

For $z = (z_1, \ldots, z_\alpha) \in \mathbb{R}^\alpha$, $y = (y_1, \ldots, y_{\delta_f}) \in \mathbb{R}^{\delta_f}$, $x = (x_1, \ldots, x_{\beta_{l_0}}) \in \mathbb{R}^{\beta_{l_0}}$ and $t = (t_1, \ldots, t_{r_{l_0}}) \in \mathbb{R}^{r_{l_0}}$, define:

$$\mathrm{j}_{\mathscr{B}}(t, x, y, z) = \mathcal{E}_{\mathscr{B}}(T) \quad \text{and} \quad \mathrm{j}_{\mathscr{B}_{l_0}}(t, x, y, z) = \mathcal{E}_{\mathscr{B}_{l_0}}(T) \tag{3.7.13}$$

where $T = (t_{r_{l_0}}, \ldots, t_1, x_{\beta_{l_0}}, \ldots, x_1, y_{\delta_f}, \ldots, y_1, z_\alpha, \ldots, z_1) \in \mathbb{R}^n$.

Using Proposition 3.7.4, one obtains

$$\|U(\xi)\|^2_{\mathscr{H}_\rho}$$

$$= \sum_{f\in\mathscr{V}_\Gamma^*} \sum_{l_0\in\mathcal{V}_f} \sum_{l\in\Sigma(f,l_0)} \sum_{s\in\mathbb{Z}^{r_{l_0}}} \int_{\mathbb{I}^{r_{l_0}}} \left| \int_{M(l_0)/M(l_0)\cap\Gamma} \xi(\mathrm{j}_{\mathscr{B}_{l_0}}(t,0,0,0)\mathrm{j}_{\mathscr{B}_{l_0}}(s,0,0,0)m) \chi_l(m)dm \right|^2 dt$$

$$= \sum_{f\in\mathscr{V}_\Gamma^*} \sum_{l_0\in\mathcal{V}_f} \sum_{l\in\Sigma(f,l_0)} \sum_{s\in\mathbb{Z}^{r_{l_0}}} \int_{\mathbb{I}^{r_{l_0}}} \left| \int_{M(l_0)/M(l_0)\cap\Gamma} \xi(\mathrm{j}_{\mathscr{B}_{l_0}}(t,0,0,0)\mathrm{j}_{\mathscr{B}_{l_0}}(s,0,0,0)m\, \mathrm{j}_{\mathscr{B}_{l_0}}(s,0,0,0)^{-1}) \chi_l(m)dm \right|^2 dt$$

(since $\xi(g\gamma) = \xi(g)$, $\forall\, g \in G, \gamma \in \Gamma$)

$$= \sum_{f\in\mathscr{V}_\Gamma^*} \sum_{l_0\in\mathcal{V}_f} \sum_{l\in\Sigma(f,l_0)} \sum_{s\in\mathbb{Z}^{r_{l_0}}} \int_{\mathbb{I}^{r_{l_0}}} \left| \int_{M(l_0)/M(l_0)\cap\Gamma} \xi(\mathrm{j}_{\mathscr{B}_{l_0}}(t,0,0,0)m) \chi_l(\mathrm{j}_{\mathscr{B}_{l_0}}(s,0,0,0)^{-1}m\mathrm{j}_{\mathscr{B}_{l_0}}(s,0,0,0))dm \right|^2 dt$$

(by making the change of variable $m \longmapsto \mathrm{j}_{\mathscr{B}_{l_0}}(s,0,0,0)m\mathrm{j}_{\mathscr{B}_{l_0}}(s,0,0,0)^{-1}$)

$$= \sum_{f\in\mathscr{V}_\Gamma^*} \sum_{l_0\in\mathcal{V}_f} \sum_{l\in\Sigma(f,l_0)} \sum_{s\in\mathbb{Z}^{r_{l_0}}} \int_{\mathbb{I}^{r_{l_0}}} \left| \int_{M(l_0)/M(l_0)\cap\Gamma} \xi(\mathrm{j}_{\mathscr{B}_{l_0}}(t,0,0,0)m) \chi_{\mathrm{Ad}^*(\mathrm{j}_{\mathscr{B}_{l_0}}(s,0,0,0))l}(m)dm \right|^2 dt$$

Since the mapping $\Sigma(f, l_0) \times \mathbb{Z}^{r_{l_0}} \longrightarrow \mathfrak{a}_\Gamma^*(f, l_0)$; $(l, s) \longmapsto \mathrm{Ad}^*(\mathrm{j}_{\mathscr{B}_{l_0}}(s, 0, 0, 0))l$ is bijective, one has that

$$\|U(\xi)\|^2_{\mathscr{H}_\rho}$$

$$= \sum_{f\in\mathscr{V}_\Gamma^*} \sum_{l_0\in\mathcal{V}_f} \sum_{l\in\mathfrak{a}_\Gamma^*(f,l_0)} \int_{\mathbb{I}^{r_{l_0}}} \left| \int_{M(l_0)/M(l_0)\cap\Gamma} \xi(\mathrm{j}_{\mathscr{B}_{l_0}}(t,0,0,0)m) \chi_l(m)dm \right|^2 dt$$

$$= \sum_{f\in\mathcal{V}_\Gamma^*}\sum_{l_0\in\mathcal{V}_f}\sum_{l\in\mathfrak{a}_\Gamma^*(f,l_0)}\int_{\mathbb{I}^{r_{l_0}}}\left|\int_{\mathbb{I}^{d_{l_0}}}\xi(\mathrm{j}_{\mathscr{B}_{l_0}}(t,x,y,z))\chi_l(\mathrm{j}_{\mathscr{B}_{l_0}}(0,x,y,z))dxdydz\right|^2 dt$$

$$= \sum_{f\in\mathcal{V}_\Gamma^*}\sum_{l_0\in\mathcal{V}_f}\int_{\mathbb{I}^{r_{l_0}}}\sum_{l\in\mathfrak{a}_\Gamma^*(f,l_0)}\left|\int_{\mathbb{I}^{d_{l_0}}}\xi(\mathrm{j}_{\mathscr{B}_{l_0}}(t,x,y,z))\chi_l(\mathrm{j}_{\mathscr{B}_{l_0}}(0,x,y,z))dxdydz\right|^2 dt$$

$$= \sum_{f\in\mathcal{V}_\Gamma^*}\sum_{l_0\in\mathcal{V}_f}\int_{\mathbb{I}^{r_{l_0}}}\int_{\mathbb{I}^{d_{l_0}-g_f}}\left|\int_{\mathbb{I}^{g_f}}\xi(\mathrm{j}_{\mathscr{B}_{l_0}}(t,x,y,z))\chi_{l_0}(\mathrm{j}_{\mathscr{B}_{l_0}}(0,0,y,z))dydz\right|^2 dtdx$$

(by the Parseval identity with respect to the variable x)

$$= \sum_{f\in\mathcal{V}_\Gamma^*}\sum_{l_0\in\mathcal{V}_f}\int_{\mathbb{I}^{n-g_f}}\left|\int_{\mathbb{I}^{g_f}}\xi(\mathrm{j}_{\mathscr{B}_{l_0}}(t,x,y,z))\chi_{l_0}(\mathrm{j}_{\mathscr{B}_{l_0}}(0,0,y,z))dydz\right|^2 dtdx$$

$$= \sum_{f\in\mathcal{V}_\Gamma^*}\sum_{l_0\in\mathcal{V}_f}\int_{\mathbb{I}^{n-g_f}}\left|\int_{\mathbb{I}^{g_f}}\xi(\mathrm{j}_{\mathscr{B}_{l_0}}(t,x,y,z))\chi_{l_0}(\mathrm{j}_{\mathscr{B}_f}(0,0,y,z))dydz\right|^2 dtdx$$

since $\mathrm{j}_{\mathscr{B}_{l_0}}(0,0,y,z) = \mathrm{j}_{\mathscr{B}_f}(0,0,y,z)$.

Let now $D(f) = \exp\delta(f)$. The function

$$g \mapsto \tilde{\xi}(g) := \left|\int_{\mathbb{I}^{g_f}}\xi(g\mathrm{j}_{\mathscr{B}_f}(0,0,y,z))\chi_{l_0}(\mathrm{j}_{\mathscr{B}_f}(0,0,y,z))dydz\right|$$

belongs to $\mathscr{C}_c(G/\Gamma D(f))$. As $D(f) = \mathfrak{D}(\mathfrak{g})\,\mathfrak{g}\cap G(f)$ is a rational normal subgroup of G, it follows that $\Gamma D(f)/D(f)$ is a discrete uniform subgroup of $G/D(f)$ (see [39, Lemma 5.1.4]). On the other hand,

$$\prod_{g_f<i\le n}(\exp\mathbb{I}e_i(l_0))/D(f) \text{ resp. } \prod_{g_f<i\le n}(\exp\mathbb{I}e_i(f))/D(f)$$

are fundamental domains for $\Gamma D(f)/D(f)$ in $G/D(f)$. Therefore

$$\int_{\mathbb{I}^{n-g_f}}\int_{\mathbb{I}^{g_f}}\phi(\mathrm{j}_{\mathscr{B}_{l_0}}(t,x,0,0))dtdx = \int_{\mathbb{I}^{n-g_f}}\int_{\mathbb{I}^{g_f}}\phi(\mathrm{j}_{\mathscr{B}_f}(t,x,0,0)dtdx$$

for every continuous function ϕ on $G/\Gamma D(f)$. It follows that

$$\|U(\xi)\|^2_{\mathscr{H}_\rho} = \sum_{f\in\mathcal{V}_\Gamma^*}\int_{\mathbb{I}^{n-g_f}}\sum_{l_0\in\mathcal{V}_f}\left|\int_{\mathbb{I}^{g_f}}\xi(\mathrm{j}_{\mathscr{B}_f}(t,x,y,z))\chi_{l_0}(\mathrm{j}_{\mathscr{B}_{l_0}}(0,0,y,z))dydz\right|^2 dtdx$$

At this point, an explicit description of the set $\mathcal{V}_f$ is needed.

Lemma 3.7.10 *One has*

$$\mathcal{V}_f = f + \mathbb{Z}\text{-span}\left\{e_i^*(f) : \alpha < i \le g_f\right\}$$

where $(e_1^*(f), \dots, e_n^*(f))$ *is the dual basis of* $(e_1(f), \dots, e_n(f))$.

Proof One can assume that $\alpha \neq g_f$. Let us denote

$$\mathscr{F}^* = f + \mathbb{Z}\text{-span}\left\{e_i^*(f) : \alpha < i \leq g_f\right\}.$$

Let $l \in \mathcal{V}_f$. Note that for every $i = \alpha + 1, \dots, g_f$, the vector $e_i(f)$ belongs to $\mathfrak{D}(\mathfrak{g}) \cap \mathfrak{g}(f) \cap \mathrm{Log}(\Gamma)$, hence $\langle l, e_i(f)\rangle \in \mathbb{Z}$. Therefore

$$l = f + \sum_{i=\alpha+1}^{g_f} \langle l, e_i(f)\rangle e_i^*(f) \in \mathscr{F}^*.$$

Conversely, let $l = f + \sum_{i=\alpha+1}^{g_f} \alpha_i e_i^*(f) \in \mathscr{F}^*$. One has for $D = \exp \delta$ that

$$\mathfrak{D}(\mathfrak{g})\, g \cap G(f) \cap \Gamma = \exp \mathbb{Z} e_1(f) \cdots \exp \mathbb{Z} e_{g_f}(f).$$

Since $\mathfrak{D}(\mathfrak{g})\, g \cap G(f) \cap \Gamma$ is Abelian, then necessarily

$$\mathfrak{D}(\mathfrak{g}) \cap \mathfrak{g}(f) \cap \log(\Gamma) = \mathbb{Z}\text{-span}\,\{e_1(f), \dots, e_{g_f}(f)\}$$

and therefore $\langle l, \mathfrak{D}(\mathfrak{g}) \cap \mathfrak{g}(f) \cap \mathrm{Log}(\Gamma)\rangle \subset \mathbb{Z}$. Thus $l \in \mathcal{V}_f$. □

Let us now finish the proof of the theorem. Using Parseval's identity in the variable y, one obtains

$$\begin{aligned}
\|U(\xi)\|^2_{\mathscr{H}_\rho} &= \sum_{f \in \mathscr{V}_\Gamma^*} \int_{\mathbb{I}^{n-g_f}} \int_{\mathbb{I}^{\delta_f}} \left| \int_{\mathbb{I}^{g_f-\delta_f}} \xi(\mathrm{j}_{\mathscr{B}_f}(t,x,y,z))\chi_f(\mathrm{j}_{\mathscr{B}_f}(0,0,0,z))dz \right|^2 dt\, dx\, dy \\
&= \int_{\mathbb{I}^{n-\alpha}} \sum_{f \in \mathscr{V}_\Gamma^*} \left| \int_{\mathbb{I}^{\alpha}} \xi(\mathrm{j}_{\mathscr{B}}(t,x,y,z))\chi_f(\mathrm{j}_{\mathscr{B}}(0,0,0,z))dz \right|^2 dt\, dx\, dy
\end{aligned}$$

(since $g_f - \delta_f = \alpha$)

$$= \int_{\mathbb{I}^n} \left|\xi(\mathrm{j}_{\mathscr{B}}(t,x,y,z))\right|^2 dt\, dx\, dy\, dz$$

(By Parseval's identity in z)

$$= \|\xi\|^2_{\mathscr{H}_\tau}$$

□

The operator U being now isometric, it can be extended to an isometry from $L^2(G/\Gamma)$ to $\mathscr{H}_\rho$. Next, let us show that U is an invertible operator. Let

$$\mathscr{H}_\rho^c = \bigoplus_{f \in \mathscr{V}_\Gamma^*} \bigoplus_{l_0 \in \mathcal{V}_f} \bigoplus_{l \in \Sigma(f,l_0)} \mathscr{C}_c(G/M(l_0), \chi_l)$$

and let

$$\mathscr{H}_\rho^0 = \{K \in \mathscr{H}_\rho^c : K(f, l_0, l) \text{ is } 0, \text{ except for finitely many } (f, l_0, l)\}.$$

The space $\mathscr{H}_\rho^0$ is dense in $\mathscr{H}_\rho$. Define the operator V on the dense subspace $\mathscr{H}_\rho^0$ by

$$V(K)(g) = \sum_{f \in \mathscr{V}_\Gamma^*} \sum_{l_0 \in \mathcal{V}_f} \sum_{l \in \Sigma(f,l_0)} \sum_{\gamma \in \Gamma/\Gamma \cap M(l_0)} K(f, l_0, l)(g\gamma) \qquad (g \in G). \tag{3.7.14}$$

Let

$$S_g = \{\gamma \in \Gamma : g\gamma \in \text{supp}(K(f, l_0, l))\}.$$

Then there is an integer $n_{K(f,l_0,l)}$, independent of $g \in G$, such that S_g is the union of at most $n_{K(f,l_0,l)}$ cosets of $\Gamma \cap M(l_0)$ ([36, Lemma 3.2]). This observation shows that the sum over $\Gamma/\Gamma \cap M(l_0)$ defining $V(K)$ in (3.7.14) has at most $n_{K(f,l_0,l)}$ non-zero entries for each $g \in G$. Consequently, (3.7.14) is a sum of finitely many terms, and hence is well defined for any $K \in \mathscr{H}_\rho^0$.

Proposition 3.7.11 *The operator V is the inverse of U.*

Proof First, observe that $V(K)$ satisfies the required covariance relation in $L^2(G/\Gamma)$. In fact, let $g \in G$ and $\gamma_0 \in \Gamma$, so

$$\begin{aligned} V(K)(g\gamma_0) &= \sum_{f \in \mathscr{V}_\Gamma^*} \sum_{l_0 \in \mathcal{V}_f} \sum_{l \in \Sigma(f,l_0)} \sum_{\gamma \in \Gamma/\Gamma \cap M(l_0)} K(f, l_0, l)(g\gamma_0\gamma) \\ &\text{(use the change of variable } \gamma \longmapsto \gamma_0^{-1}\gamma) \\ &= \sum_{f \in \mathscr{V}_\Gamma^*} \sum_{l_0 \in \mathcal{V}_f} \sum_{l \in \Sigma(f,l_0)} \sum_{\gamma \in \Gamma/\Gamma \cap M(l_0)} K(f, l_0, l)(g\gamma) \\ &= V(K)(g). \end{aligned}$$

Now, compute $U \circ V(K)$ for $K \in \mathscr{H}_\rho^0$. For this let $K \in \mathscr{H}_\rho^0$ and $(f, l_0, l) \in \mathscr{V}_\Gamma^* \times \mathcal{V}_f \times \Sigma(f, l_0)$. Then

$$\begin{aligned} &U \circ V(K)(f, l_0, l)(e) = U(V(K))(f, l_0, l)(e) \\ &= \int_{M(l_0)/M(l_0)\cap\Gamma} V(K)(m)\chi_l(m)dm \\ &= \int_{M(l_0)/M(l_0)\cap\Gamma} \Big(\sum_{f' \in \mathscr{V}_\Gamma^*} \sum_{l_0' \in \mathcal{V}_{f'}} \sum_{l' \in \Sigma(f',l_0')} \sum_{\overline{\gamma} \in \Gamma/\Gamma \cap M(l_0')} K(f', l_0', l')(m\gamma)\Big)\chi_l(m)dm \end{aligned}$$

$$= \int_{M(l_0)/M(l_0)\cap\Gamma} \Big(\sum_{f'\in\mathcal{V}_\Gamma^*} \sum_{l_0'\in\mathcal{V}_{f'}} \sum_{l'\in\Sigma(f',l_0')} \sum_{\overline{\gamma}\in\Gamma/\Gamma\cap M(l_0')} K(f', l_0', l')(\gamma\gamma^{-1}m\gamma)\Big)\chi_l(m)dm$$

$$= \int_{M(l_0)/M(l_0)\cap\Gamma} \Big(\sum_{f'\in\mathcal{V}_\Gamma^*} \sum_{l_0'\in\mathcal{V}_{f'}} \sum_{l'\in\Sigma(f',l_0')} \sum_{\overline{\gamma}\in\Gamma/\Gamma\cap M(l_0')} K(f', l_0', l')(\gamma)\chi_{l'}^{-1}(\gamma^{-1}m\gamma)\Big)\chi_l(m)dm$$

$$= \int_{M(l_0)/M(l_0)\cap\Gamma} \Big(\sum_{f'\in\mathcal{V}_\Gamma^*} \sum_{l_0'\in\mathcal{V}_{f'}} \sum_{l'\in\Sigma(f',l_0')} \sum_{\overline{\gamma}\in\Gamma/\Gamma\cap M(l_0')} K(f', l_0', l')(\gamma)\chi_{\mathrm{Ad}^*\gamma l'}^{-1}(m)\Big)\chi_l(m)dm$$

$$= \sum_{f'\in\mathcal{V}_\Gamma^*} \sum_{l_0'\in\mathcal{V}_{f'}} \sum_{l'\in\Sigma(f',l_0')} \sum_{\overline{\gamma}\in\Gamma/\Gamma\cap M(l_0')} K(f', l_0', l')(\gamma)\Big(\int_{M(l_0)/M(l_0)\cap\Gamma} \chi_{\mathrm{Ad}^*\gamma l'}^{-1}(m)\chi_l(m)dm\Big)$$

Swapping integration and summation is justified by the fact that the sum only contains a finite number of terms because $K \in \mathscr{H}_\rho^0$. On the other hand, one has the identity:

$$\int_{M(l_0)/M(l_0)\cap\Gamma} \chi_{\mathrm{Ad}^*\gamma l'}^{-1}(m)\chi_l(m)dm = \begin{cases} 1 \text{ if } \mathrm{Ad}^*\gamma l' = l \\ 0 \text{ otherwise.} \end{cases}$$

The condition $\mathrm{Ad}^*\gamma l' = l$ implies that $l_0' = l_0$ and in particular $f' = f$. It follows that $l' = l$ and hence $\overline{\gamma} = \overline{e}$. Finally, one obtains

$$U \circ V(K)(f, l_0, l)(e) = K(f, l_0, l)(e).$$

Then for every K in $\mathscr{H}_\rho^0$, one has $U \circ V(K) = K$. Next, for $\xi \in L^2(G/\Gamma)$ such that $U(\xi) \in \mathscr{H}_\rho^0$ one has

$$V \circ U(\xi)(e)$$

$$= \sum_{f\in\mathcal{V}_\Gamma^*} \sum_{l_0\in\mathcal{W}_f^*} \sum_{l\in\Sigma(f,l_0)} \sum_{\overline{\gamma}\in\Gamma/\Gamma\cap M(l_0)} U(\xi)(f, l_0, l)(\gamma)$$

$$= \sum_{f\in\mathcal{V}_\Gamma^*} \sum_{l_0\in\mathcal{W}_f^*} \sum_{l\in\Sigma(f,l_0)} \sum_{\overline{\gamma}\in\Gamma/\Gamma\cap M(l_0)} \int_{M(l_0)/M(l_0)\cap\Gamma} \xi(\gamma m)\chi_l(m)dm \tag{3.7.15}$$

$$= \sum_{f\in\mathcal{V}_\Gamma^*} \sum_{l_0\in\mathcal{W}_f^*} \sum_{l\in\Sigma(f,l_0)} \sum_{\overline{\gamma}\in\Gamma/\Gamma\cap M(l_0)} \int_{M(l_0)/M(l_0)\cap\Gamma} \xi(m)\chi_{\mathrm{Ad}^*\gamma l}(m)dm$$

$$= \sum_{f\in\mathcal{V}_\Gamma^*} \sum_{l_0\in\mathcal{W}_f^*} \sum_{l\in\mathfrak{a}_\Gamma^*(f,l_0)} \int_{M(l_0)/M(l_0)\cap\Gamma} \xi(m)\chi_l(m)dm$$

$$= \sum_{f\in\mathcal{V}_\Gamma^*} \sum_{l_0\in\mathcal{W}_f^*} \sum_{l\in\mathfrak{a}_\Gamma^*(f,l_0)} \int_{\mathbb{T}^{d_{l_0}}} \xi(\mathrm{j}_{\mathscr{B}_{l_0}}(0, x, y, z))\chi_l(\mathrm{j}_{\mathscr{B}_{l_0}}(0, x, y, z))dx\, dy\, dz$$

$$= \sum_{f \in \mathscr{V}_\Gamma^*} \sum_{l_0 \in \mathscr{W}_f^*} \int_{\mathbb{I}^{g_f}} \xi(\mathrm{j}_{\mathscr{B}_{l_0}}(0,0,y,z))\chi_{l_0}(\mathrm{j}_{\mathscr{B}_{l_0}}(0,0,y,z))dy\,dz$$

(by the Fourier inversion formula for the variable x)

$$= \sum_{f \in \mathscr{V}_\Gamma^*} \sum_{l_0 \in \mathscr{W}_f^*} \int_{\mathbb{I}^{g_f}} \xi(\mathrm{j}_{\mathscr{B}_f}(0,0,y,z))\chi_{l_0}(\mathrm{j}_{\mathscr{B}_f}(0,0,y,z))dy\,dz$$

(since $\mathrm{j}_{\mathscr{B}_{l_0}}(0,0,y,z) = \mathrm{j}_{\mathscr{B}_f}(0,0,y,z)$)

$$= \sum_{f \in \mathscr{V}_\Gamma^*} \int_{\mathbb{I}^{\alpha}} \xi(\mathrm{j}_{\mathscr{B}_f}(0,0,0,z))\chi_{l_0}(\mathrm{j}_{\mathscr{B}_f}(0,0,0,z))dy\,dz$$

(by the Fourier inversion formula for the variable y)

$$= \sum_{f \in \mathscr{V}_\Gamma^*} \int_{\mathbb{I}^{\alpha}} \xi(\mathrm{j}_{\mathscr{B}}(0,0,0,z))\chi_f(\mathrm{j}_{\mathscr{B}}(0,0,0,z))dz$$

(since $\mathrm{j}_{\mathscr{B}_f}(0,0,0,z) = \mathrm{j}_{\mathscr{B}}(0,0,0,z)$)

$$= \xi(e),$$

by the Fourier inversion formula for the variable z. Consequently, the isometric linear operator U has a dense range and is therefore unitary. □

Remark 3.7.12 Keep the notations of Theorem 3.7.9.

1. When $k = 1$ (ie $\mathfrak{g}$ is Abelian), one has $\mathscr{V}_\Gamma^* = \mathcal{V}_0 = \{0\}$ and $\mathfrak{a}_\Gamma^*(0,0) = \Sigma(0,0) = \mathfrak{g}_\Gamma^* = \mathbb{Z}\text{-span}\{X_1^*, \ldots, X_n^*\}$. Consequently

$$\tau = \bigoplus_{l \in \mathfrak{g}_\Gamma^*} \chi_l.$$

2. When $k = 2$, for $f \in \mathscr{V}_\Gamma^* = \mathbb{Z}\text{-span}\{X_1^*, \ldots, X_\alpha^*\}$ one has

$$\mathcal{D}(\mathfrak{g}) \cap \mathfrak{z}(\mathfrak{g}) = \mathcal{D}(\mathfrak{g}) \cap \mathfrak{g}(f) = \mathcal{D}(\mathfrak{g}).$$

It follows that $\mathcal{V}_f = \{f\}$ and $\mathfrak{a}_\Gamma^*(f,f) = \{l \in \mathfrak{m}(f)_\Gamma^* : l|_{\mathcal{D}(\mathfrak{g})} = f\}$. Consequently

$$\tau = \bigoplus_{f \in \mathscr{V}_\Gamma^*} \bigoplus_{l \in \Sigma(f,f)} \mathrm{ind}_{M(f)}^G \chi_l.$$

An immediate consequence of Theorem 3.7.9 is the following result due to C. C. Moore [130].

Corollary 3.7.13 *If $f \in \mathfrak{g}^*$ and $\pi_f \in \mathcal{S}(G : \Gamma)$, then $\mathcal{O}_{\pi_f} \cap \mathfrak{g}^*_{\mathbb{Q}}$ is non-empty.*

3.7.3 On the Multiplicity Formula

The following proposition is an immediate consequence of Theorem 3.7.9. Let $\#A$ denote the cardinality of a set A and let $\mathcal{O}_\pi \in \mathfrak{g}^*/G$ be the coadjoint orbit of a representation $\pi \in \hat{G}$.

Proposition 3.7.14 *Keeping the notation of Theorem 3.7.9,*

(1) *the support of the representation $\mathrm{ind}_\Gamma^G 1$ in $\widehat{G}$ is the set*

$$\begin{aligned}\mathcal{S}(G:\Gamma) = \{\pi \in \widehat{G};\ & \mathcal{O}_{\pi|\mathfrak{s}\cap\mathfrak{z}(\mathfrak{g})} = \{f\} \subset \mathcal{V}_\Gamma,\\ & \mathcal{O}_{\pi|\mathfrak{s}(f)} = \{l_0\} \subset \mathcal{W}_f^*,\\ & \mathcal{O}_{\pi|\mathfrak{m}(l_0)} \cap \mathfrak{m}(l_0)^*_\Gamma \neq \emptyset\}.\end{aligned}$$

(2) *Let $l \in \Sigma(f, l_0)$. The multiplicity mult(π_l) of the representation π_l is given by*

$$mult(\pi_l) = \#\big(\Sigma(f, l_0) \cap \mathcal{O}_{\pi_l}\big) = \#\big(\mathfrak{a}^*_\Gamma(f, l_0) \cap \mathcal{O}_{\pi_l}/\Gamma\big). \tag{3.7.16}$$

Proof It follows from Theorem 3.7.9 that the support of the representation ρ is the set $\{[\pi_l];\ l \in \mathfrak{g}^*, f = l_{|\mathfrak{s}\cap\mathfrak{z}} \in \mathcal{V}_\Gamma, l_0 = l_{|\mathfrak{s}(f)} \in \mathcal{W}_f, l_{|\mathfrak{m}(l_0)} \in \mathfrak{m}(l_0)^*_\Gamma\}$. This proves the first point. If $l \in \Sigma(f, l_0)$ resp. $l' \in \Sigma(f', l'_0)$ are in the same G-orbit Ω, then $l_{|\mathfrak{s}\cap\mathfrak{z}} = l'_{|\mathfrak{s}\cap\mathfrak{z}} =: f$ and $l_{|\mathfrak{s}(f)} = l'_{|\mathfrak{s}(f)} = l_0$ and so the multiplicity mult(π_l) is equal to $\#(\Sigma(f, l_0) \cap \mathcal{O}_{\pi_l})$. □

Applying Proposition 3.7.14 to a two-step nilpotent Lie group one can derive the following result due to H. Pesce (see [135, Proposition A. 9]).

Corollary 3.7.15 *Let notations be as in Theorem 3.7.9. Assume that $k = 2$. For $f \in \mathcal{V}^*_\Gamma$, let $\mathcal{B}_f$ be a strong Malcev basis of $\mathfrak{g}$ strongly based on Γ passing through $\mathfrak{g}(f)$ and $\mathfrak{m}(f)$. Let us denote by $(Y_1(f), \ldots, Y_r(f))$ a basis of $\mathfrak{g}$ modulo $\mathfrak{m}(f)$ and by $(e_1(f), \ldots, e_r(f))$ the basis of $\mathfrak{m}(f)$ modulo $\mathfrak{g}(f)$ extracted from $\mathcal{B}_f$. Then the multiplicity function is given for l in $\Sigma(f, f)$ by*

$$mult(\pi_l) = \begin{cases} 1, & \text{if } f = 0;\\ \big|\det(\mathcal{A}_l)\big|, & \text{if } f \neq 0,\end{cases} \tag{3.7.17}$$

where

$$\mathcal{A}_l = \mathrm{Mat}\big(\langle l, [Y_j(f), e_i(f)]\rangle;\ 1 \leq i, j \leq r\big).$$

Proof Let $l \in \Sigma(f, f)$. By Proposition 3.7.14, one has

$$\mathrm{mult}(\pi_l) = \#\big(\mathcal{O}_{\pi_l} \cap \Sigma(f, f)\big).$$

If $f = 0$ then $\pi_l = \chi_l$ and hence $\mathrm{mult}(\pi_l) = 1$. Now, assume that $f \neq 0$. It is not hard to verify that

$$\Gamma \cdot l - l = \mathbb{Z}\text{-span}\left\{\sum_{i=1}^{r} \langle l, [Y_j(f), e_i(f)]\rangle e_i^*(f),\ j = 1, \dots, r\right\}.$$

Denote by $\mathcal{Q}$ the lattice $\Gamma \cdot l - l$. Moreover the vectors

$$Y_j' = \sum_{i=1}^{r} \langle l, [Y_j(f), e_i(f)]\rangle e_i^*(f), \qquad (1 \leq j \leq r)$$

are linearly independent. In fact, suppose that one has

$$\sum_{j=1}^{r} \lambda_j Y_j' = 0 \tag{3.7.18}$$

for some coefficients $\lambda_j \in \mathbb{R}$. Then one has to show that all λ_j vanish. Formula (3.7.18) implies that

$$\left\langle l, \left[\sum_{j=1}^{r} \lambda_j Y_j(f), e_i(f)\right]\right\rangle = 0, \qquad (\forall\, 1 \leq i \leq r) \tag{3.7.19}$$

One deduces from this that $\sum_{i=1}^{r} \lambda_j Y_j(f) \in \mathfrak{m}(l) = \mathfrak{m}(f)$. Since $(Y_1(f), \dots, Y_r(f))$ is a basis of $\mathfrak{g}$ modulo $\mathfrak{m}(f)$, one concludes that $\sum_{i=1}^{r} \lambda_j Y_j(f) = 0$. This gives in turn that $\lambda_j = 0$ for all $j = 1, \dots, r$. It follows from the previous reasoning that $\mathcal{Q}$ is a full sub-lattice of

$$\mathcal{R} = \mathbb{Z}\text{-span}\left\{e_1^*(f), \dots, e_r^*(f)\right\}.$$

Then there exist linearly independent vectors $\varepsilon_1^*, \dots, \varepsilon_r^*$ of $\mathfrak{g}^*$ and $d_1, \dots, d_r \in \mathbb{N}^*$ such that $\mathcal{R} = \mathbb{Z}\text{-span}\left\{\varepsilon_1^*, \dots, \varepsilon_r^*\right\}$ and $\mathcal{Q} = \mathbb{Z}\text{-span}\left\{d_1\varepsilon_1^*, \dots, d_r\varepsilon_r^*\right\}$. It follows that for each $i = 1, \dots, r$, one has $\langle l, \varepsilon_i\rangle \in \mathbb{Z}$ and

$$\Gamma \cdot l = l|_{\mathfrak{g}(f)} + \bigoplus_{i=1}^{r} \big(\langle l, \varepsilon_i\rangle + d_i\mathbb{Z}\big)\varepsilon_i^*.$$

One concludes that $m(\pi_l) = d_1 \cdots d_r$. On the other hand, there exists a matrix $T \in GL(n, \mathbb{Z})$ such that

$$[d_1\varepsilon_1^*, \ldots, d_r\varepsilon_r^*] = [Y_1', \ldots, Y_r']T,$$

where $[d_1\varepsilon_1^*, \ldots, d_r\varepsilon_r^*]$ (resp. $[Y_1', \ldots, Y_r']$) denotes the matrix with column vectors $d_1\varepsilon_1^*, \ldots, d_r\varepsilon_r^*$ (resp. $Y_1', \ldots, Y_r'$) expressed in the basis $e_1^*(f), \ldots, e_r^*(f)$. Therefore

$$[\varepsilon_1^*, \ldots, \varepsilon_r^*] \begin{pmatrix} d_1 & & 0 \\ & \ddots & \\ 0 & & d_r \end{pmatrix} = [Y_1', \ldots, Y_r']T.$$

In particular, one has

$$(d_1 \cdots d_r) \det([\varepsilon_1^*, \ldots, \varepsilon_r^*]) = \det([Y_1', \ldots, Y_r']) \det(T).$$

Since $\det([\varepsilon_1^*, \ldots, \varepsilon_r^*]) = \pm 1$ and $\det(T) = \pm 1$, then one has

$$d_1 \cdots d_r = \left|\det([Y_1', \ldots, Y_r'])\right| = \left|\det(\langle l, [Y_j(f), e_i(f)]\rangle;\ 1 \leq i, j \leq r)\right|.$$

This finishes the proof of the corollary. □

A discrete uniform subgroup Γ of G is said to be a lattice subgroup if $\log(\Gamma)$ is an additive subgroup of $\mathfrak{g}$. We have the following:

Corollary 3.7.16 ([130]) *Let Γ be a lattice subgroup of a k-step, connected and simply connected nilpotent Lie group G with $k \leq 3$. Then one has*

$$mult(\pi) \leq \#\left(\{\mathcal{O}_\pi \cap \mathfrak{g}_\Gamma^*/\Gamma\}\right),$$

for every $\pi \in \mathcal{S}(G : \Gamma)$, where $\mathfrak{g}_\Gamma^* = \{l \in \mathfrak{g}^* : \langle l, \log(\Gamma)\rangle \subset \mathbb{Z}\}$.

Proof Let $\pi \in \mathcal{S}(G : \Gamma)$. Let $f \in \mathcal{V}_\Gamma^*$, $l_0 \in \mathcal{V}_f$ and $l \in \mathfrak{a}_\Gamma^*(f, l_0)$ such that $\pi = \pi_l$. Let $(\varepsilon_1, \ldots, \varepsilon_n)$ be a strong Malcev basis of $\mathfrak{g}$ strongly based on Γ and passing through $\mathfrak{m}(l_0)$. By Lemma 3.9 of [35] one has

$$\mathfrak{m}(l_0)_\Gamma^* = \mathbb{Z}\varepsilon_1^* + \ldots + \mathbb{Z}\varepsilon_{d_{l_0}}^*$$

and

$$\mathfrak{g}_\Gamma^* = \mathbb{Z}\varepsilon_1^* + \ldots + \mathbb{Z}\varepsilon_n^*.$$

It is clear that $\overline{\mathfrak{m}(l_0)^*_\Gamma} \subset \mathfrak{g}^*_\Gamma$. On the other hand, the mapping

$$\mathcal{O}_\pi \cap \mathfrak{m}(l_0)^*_\Gamma/\Gamma \longrightarrow \mathcal{O}_\pi \cap \mathfrak{g}^*_\Gamma/\Gamma, \quad \Gamma \cdot \varphi \longmapsto \overline{\Gamma \cdot \varphi}$$

is injective. The proof is complete. □

3.7.4 Primary Projections

For any $\pi \in \hat{G}$ appearing in the decomposition of $L^2(G/\Gamma)$, the Hilbert space $L^2(G/\Gamma)$ can be decomposed into a direct sum $\mathscr{K}_\pi \oplus \mathscr{L}_\pi$ in such a way that $\mathscr{K}_\pi$ and $\mathscr{L}_\pi$ are both invariant under R_Γ, the restriction of R_Γ to $\mathscr{K}_\pi$ is equivalent to a finite multiple of π and π is not a sub-representation of the restriction of R_Γ to $\mathscr{L}_\pi$. The space $\mathscr{K}_\pi$ is called the *π-primary subspace of* $L^2(G/\Gamma)$. The projection $P_\pi : L^2(G/\Gamma) \longrightarrow \mathscr{K}_\pi$ is called the *π-primary projection* on $L^2(G/\Gamma)$ (cf. [31, 42, 43, 132] and [146]).

Proposition 3.7.17 *If π occurs in the decomposition of τ, the π-primary projection is given for each $\xi \in \mathscr{C}(G/\Gamma)$ by*

$$P_\pi(\xi)(g) = \sum_{l \in \Sigma(f,l_0) \cap \mathcal{O}_{\pi_\phi}} \sum_{\overline{\gamma} \in \Gamma/\Gamma \cap M(l_0)} \int_{M(l_0)/M(l_0)\cap\Gamma} \xi(g\gamma m)\chi_l(m)dm$$

where $\mathcal{O}_{\pi|\mathfrak{s}\cap\mathfrak{z}(\mathfrak{g})} = \{f\}$ *and* $\mathcal{O}_{\pi|\mathfrak{s}(f)} = \{l_0\}$.

Proof This is an immediate consequence of Proposition 3.7.14 and the fact that the π-primary subspace of $L^2(G/\Gamma)$ is the image under V of the subspace $\sum_{l \in \Sigma(f,l_0) \cap \mathcal{O}_\pi} \mathscr{H}_{\pi_l}$. □

Example 3.7.18 Let us discuss an example to illustrate Theorem 3.7.9. Let $(n, p) \in \mathbb{N}^* \times \mathbb{N}$. Let us denote by $\mathfrak{g}_{n,p}$ the $(2n + p + 1)$-dimensional Lie algebra spanned over the reals by the elements

$$X'_1, \ldots, X'_n, Y'_1, \ldots, Y'_n, Z', W'_1, \ldots, W'_p,$$

with the following commutation relations

$$[X'_i, Y'_i] = Z', \quad 1 \leq i \leq n, \tag{3.7.20}$$

and all other brackets being zero or obtained by skew-symmetry. It is clear that $\mathfrak{g}_{n,p}$ is a two-step nilpotent Lie algebra with $\mathfrak{z}(\mathfrak{g}_{n,p}) = \mathbb{R}\text{-span}\left\{Z', W'_1, \ldots, W'_p\right\}$ and $[\mathfrak{g}_{n,p}, \mathfrak{g}_{n,p}] = \mathbb{R}\text{-span}\left\{Z'\right\}$. Note that $\mathfrak{g}_{n,p} = \mathfrak{h}_n \oplus \mathbb{R}^p$ where $\mathfrak{h}_n$ is the $(2n + 1)$-

dimensional Heisenberg algebra. Let $G_{n,p}$ be the Lie group associated to $\mathfrak{g}_{n,p}$. One should classify first the uniform subgroups of $G_{n,p}$. Let

$$\mathcal{D}_n = \{r = (r_1, \dots, r_n) \in (\mathbb{N}^*)^n : r_j \text{ divides } r_{j+1}, 1 \leq j \leq n-1\}.$$

For $r = (r_1, \dots, r_n) \in \mathcal{D}_n$, define

$$\Gamma_r = \prod_{i=1}^{p} \exp \mathbb{Z}W'_i \, \exp \mathbb{Z}Z' \prod_{i=1}^{n} \exp \mathbb{Z}Y'_i \prod_{i=1}^{n} \exp r_i \mathbb{Z}X'_i. \tag{3.7.21}$$

The following theorem was proved by H. Pesce ([135, Proposition B. 1]) from earlier result of C. S. Gordon and E. N. Wilson ([82, Theorem 2.4]) (see also [77]).

Theorem 3.7.19 *With the notations used above,*

1. *let $r \in \mathcal{D}_n$. Then Γ_r is a discrete uniform subgroup of $G_{n,p}$.*
2. *If Γ is a discrete uniform subgroup of $G_{n,p}$, then there exist $r \in \mathcal{D}_n$ and $\Phi \in Aut(G_{n,p})$ such that $\Phi(\Gamma) = \Gamma_r$.*
3. *For r and s in $\mathcal{D}_n$, Γ_r and Γ_s are isomorphic groups if and only if $r = s$.*

Let $r = (r_1, \dots, r_n) \in \mathcal{D}_n$ and Γ_r the associated discrete uniform subgroup of $G_{n,p}$ and let $X_1 = r_1 X'_1, \dots, X_n = r_n X'_n, Y_1 = Y'_1, \dots, Y_n = Y'_n, Z = Z', W_1 = W'_1, \dots, W_p = W'_p$. It is clear that $(Z, W_1, \dots, W_p, Y_1, \dots, Y_n, X_1, \dots, X_n)$ is a strong Malcev basis of $\mathfrak{g}_{n,p}$ strongly based on Γ_r passing through $\mathfrak{D}(\mathfrak{g})(\mathfrak{g}_{n,p})$. In this case, one has $\mathcal{V}^* = \mathbb{R}Z^*$, $\mathcal{V}^*_{\Gamma_r} = \mathbb{Z}Z^*$ and for $f = cZ^* \in \mathcal{V}^*_{\Gamma_r}$, one has $\mathcal{V}_f = \{f\}$ and

$$\mathfrak{a}^*_{\Gamma_r}(f, f) = \begin{cases} f + \mathbb{Z}\text{-span}\left\{Y_1^*, \dots, Y_n^*, W_1^*, \dots, W_p^*\right\}, & \text{if } f \neq 0 \\ \mathbb{Z}\text{-span}\left\{Y_1^*, \dots, Y_n^*, X_1^*, \dots, X_n^*, W_1^*, \dots, W_p^*\right\}, & \text{if } f = 0 \end{cases}$$

Let

$$\gamma = \prod_{i=1}^{p} \exp t_i W_i \, \exp zZ \prod_{i=1}^{n} \exp y_i Y_i \prod_{i=1}^{n} \exp x_i X_i \in \Gamma_r$$

where $x = (x_1, \dots, x_n), y = (y_1, \dots, y_n) \in \mathbb{Z}^n, (t_1, \dots, t_p) \in \mathbb{Z}^p, z \in \mathbb{Z}$ and $l = l_1 Y_1^* + \dots l_n Y_n^* + cZ^* + w_1 W_1^* + \dots + w_p W_p^* \in \mathfrak{a}^*_{\Gamma_r}(f, f)$. One has

$$\mathrm{Ad}^* \gamma^{-1} l = (l_1 + c r_1 x_1) Y_1^* + \dots + (l_n + c r_n x_n) Y_n^* + cZ^* + w_1 W_1^* + \dots + w_p W_p^*$$

and therefore

$$\mathrm{Ad}^*(\Gamma) l = (l_1 + c r_1 \mathbb{Z}) Y_1^* + \dots + (l_n + c r_n \mathbb{Z}) Y_n^* + cZ^* + w_1 W_1^* + \dots + w_p W_p^*.$$

Let

$$\Sigma_1 = \mathbb{Z}\text{-span}\left\{Y_1^*, \ldots, Y_n^*, X_1^*, \ldots, X_n^*, W_1^*, \ldots, W_p^*\right\}$$

and

$$\Sigma_2 = \{l = cZ^* + y_1 Y_1^* + \ldots + y_n Y_n^* + w_1 W_1^* + \ldots + w_p W_p^* :\\ c \in \mathbb{Z}^*, w_j \in \mathbb{Z}, \forall j = 1, \ldots, p,\ y_i \in \mathbb{Z}, 0 \le y_i < r_i|c|, \forall\, i = 1, \ldots, n\}.$$

Then

$$\Sigma(f, f) = \Sigma_1 \cup \Sigma_2$$

is a cross-section of Γ-orbits in $\mathfrak{a}_{\Gamma_r}^*(f, f)$. By Theorem 3.7.9 one has

$$\tau \simeq \rho = \bigoplus_{l \in \Sigma_1} \chi_l \oplus \bigoplus_{l \in \Sigma_2} \text{ind}_M^G \chi_l, \tag{3.7.22}$$

where M is the connected normal subgroup of G with Lie algebra

$$\mathfrak{m} = \mathbb{R}\text{-span}\{Z, W_1, \ldots, W_p, Y_1, \ldots, Y_n\}.$$

Finally, one has that two representations π_l and $\pi_{l'}$ $(l, l' \in \mathfrak{g}_{n,p}^*)$ are equivalent in the group $G_{n,p}$ if and only if $l|_{\mathfrak{z}(\mathfrak{g}_{n,p})} = l'|_{\mathfrak{z}(\mathfrak{g}_{n,p})}$. The multiplicity formula for $G_{n,p}$ is then:

$$\text{mult}(\pi_l) = \begin{cases} |l(Z)|^n r_1 \cdots r_n, & \text{if } l \in \Sigma_2 \\ \\ 1, & \text{if } l \in \Sigma_1. \end{cases}$$

3.7.5 *Characterization of Two-Step Nilmanifolds with Equivalent Quasi-Regular Representations*

Let $\mathfrak{g}$ still denote a two-step nilpotent, finite-dimensional, real Lie algebra with center $\mathfrak{z}(\mathfrak{g})$. Recall that a Lie algebra $\mathfrak{g}$ is two-step nilpotent if $\mathfrak{g}$ is non-Abelian and $[X, Y] \in \mathfrak{z}(\mathfrak{g})$ for all $X, Y \in \mathfrak{g}$. Let G denote the unique simply connected Lie group with Lie algebra $\mathfrak{g}$. The Campbell-Baker-Hausdorff theorem gives the product rule

$$\exp X \exp Y = \exp X + Y + \frac{1}{2}[X, Y]$$

for all $X, Y \in \mathfrak{g}$. Conjugation in G is therefore given by

$$\exp X \exp Y (\exp X)^{-1} = \exp Y + [X, Y].$$

Let $Z(G)$ denote the center of G. We now introduce the following two definitions (cf. [83]).

Definition 3.7.20 We say that two co-compact, discrete subgroups Γ_1 and Γ_2 of a Lie group G are representation-equivalent if the quasi-regular representations R_{Γ_1} and R_{Γ_2} are unitarily equivalent.

Definition 3.7.21 (Γ-Equivalence Automorphism) Let G be a two-step nilpotent Lie group. Let Φ be an automorphism of G and let Γ be a uniform subgroup of G. Φ is called a Γ-equivalence if for all γ in Γ there exists a_γ in G and z_γ in $\Gamma \cap [G, G]$ such that $\Phi(\gamma) = a_\gamma \gamma a_\gamma^{-1} z_\gamma$.

Under the isomorphism $\Phi \longmapsto \Phi_*$ between automorphisms of a simply connected Lie group and automorphisms of its Lie algebra, a Lie group automorphism Φ of G is a Γ-equivalence if and only if

$$\Phi_*(X) \in X + [X, \mathfrak{g}] + \log(\Gamma) \cap [\mathfrak{g}, \mathfrak{g}] \tag{3.7.23}$$

for all X in $\log(\Gamma)$.

Let Φ be a Γ-equivalence of G. From the Definition 3.7.21, the following relations hold

$$\Phi_*([X, Y]) = [\Phi_*(X), Y] = [X, \Phi_*(Y)] = [X, Y] \tag{3.7.24}$$

for all $X, Y \in \mathfrak{g}$, and

$$\Gamma \cap Z(G) = \Phi(\Gamma) \cap Z(G). \tag{3.7.25}$$

The next theorem, due to Gornet ([83, Theorem 1]), gives a necessary and sufficient condition for a pair of uniform subgroups in a two-step nilpotent Lie group to be representation-equivalent.

Theorem 3.7.22 *Let G be a two-step nilpotent Lie group. Let Γ_1 and Γ_2 be uniform subgroups of G. The quasi-regular representations R_{Γ_1} and R_{Γ_2} are unitarily equivalent if and only if there exists Φ, a Γ_1-equivalence of G, such that $\Phi(\Gamma_1) = \Gamma_2$.*

3.7.6 Decomposition of the Quasi-Regular Representation R_Γ

Let $G = \exp \mathfrak{g}$ be a two-step nilpotent Lie group which contains a discrete uniform subgroup Γ. Since $\mathfrak{z}(\mathfrak{g})$ is rational (see [39, Theorem 5.2.3]) then by

Proposition 3.7.1 there exists a strong Malcev basis $\mathscr{B}_\Gamma = \{X_1, \dots, X_n\}$ of $\mathfrak{g}$ strongly based on Γ, passing through $\mathfrak{z}(\mathfrak{g})$. Put

$$\mathfrak{z}(\mathfrak{g}) = \mathbb{R}\text{-span}(X_1, \dots, X_\alpha)$$

and

$$\mathfrak{z}(\mathfrak{g})^* = \mathbb{R}\text{-span}(X_1^*, \dots, X_\alpha^*)$$

where $\{X_1^*, \dots, X_n^*\}$ is the dual basis of $\mathscr{B}_\Gamma$. Let

$$\mathcal{V}_\Gamma = \mathbb{Z}\text{-span}\left\{X_1^*, \dots, X_\alpha^*\right\} \tag{3.7.26}$$

be the dual lattice of the lattice $\mathfrak{z}(\mathfrak{g}) \cap \log(\Gamma)$.

As regards the rest of the section, we make the following observation. For $f \in \mathcal{V}_\Gamma$, let $\mathfrak{m}[f]$ be the Vergne polarization at f associated to $\mathscr{B}_\Gamma$. Since G is two-step nilpotent, any subgroup containing the center contains $[G, G]$ and hence is normal. In particular, $M[f] = \exp \mathfrak{m}[f]$ is normal in G. As before, since $\mathfrak{z}(\mathfrak{g})$ and $\mathfrak{m}[f]$ are rational, then there exists a Jordan–Hölder basis $\mathscr{B}_f = \{e_1(f), \dots, e_n(f)\}$ of $\mathfrak{g}$ strongly based on Γ passing through $\mathfrak{z}(\mathfrak{g})$ and $\mathfrak{m}[f]$, such that

$$\begin{aligned} e_i(f) &= X_i && (1 \le i \le \alpha) \\ \mathfrak{z}(\mathfrak{g}) &= \mathbb{R}\text{-span}\,\{e_1(f), \dots, e_\alpha(f)\} && (\alpha = \dim \mathfrak{z}(\mathfrak{g})) \\ \mathfrak{m}[f] &= \mathbb{R}\text{-span}\left\{e_1(f), \dots, e_{d(f)}(f)\right\} && (d(f) = \dim \mathfrak{m}[f]). \end{aligned}$$

Put

$$\mathfrak{m}[f]_\Gamma^* = \left\{l \in \mathfrak{m}[f]^* : \langle l, \mathfrak{m}[f] \cap \log \Gamma\rangle \subset \mathbb{Z}\right\}, \tag{3.7.27}$$

where $\left\{e_1^*(f), \dots, e_n^*(f)\right\}$ is the dual basis of $\mathscr{B}_f$, and

$$\mathcal{A}(\mathrm{f}, \Gamma) = \left\{l \in \mathfrak{m}[f]_\Gamma^* : l|_{\mathfrak{z}(\mathfrak{g})} = f|_{\mathfrak{z}(\mathfrak{g})}\right\}. \tag{3.7.28}$$

The following is a reformulation of Lemmas 4.4 and 4.5 of [79] in the setting of two-step nilpotent Lie groups.

Lemma 3.7.23 *Let the notation be as above. Then one has*

(a) Γ *acts on* $\mathfrak{m}[f]_\Gamma^*$ *via the coadjoint representation and* $\mathcal{A}(\mathrm{f}, \Gamma)$ *is* Γ*-invariant.*
(b) $\mathfrak{m}[f]$ *is the Vergne polarization associated to* $\mathscr{B}_\Gamma$ *at each* l *in* $\mathcal{A}(\mathrm{f}, \Gamma)$.

Proof

(a) Let $\exp\gamma \in \Gamma$ and $l \in \mathfrak{m}[f]^*_\Gamma$. Let $i = 1, \dots, d(f)$. By definition of the coadjoint representation one has

$$\langle \exp -\gamma \cdot l, e_i(f)\rangle = \langle l, e^{\mathrm{ad}\gamma}(e_i(f))\rangle.$$

On the other hand, since $M[f]$ is normal in G, then

$$\exp e^{\mathrm{ad}\gamma}(e_i(f)) = \exp\gamma \exp e_i(f) \exp -\gamma \in \Gamma \cap M[f].$$

It follows that $e^{\mathrm{ad}\gamma}(e_i(f)) \in \log(\Gamma) \cap \mathfrak{m}[f]$ and therefore $\langle l, e^{\mathrm{ad}\gamma}(e_i(f))\rangle \in \mathbb{Z}$. Hence $\exp -\gamma \cdot l \in \mathfrak{m}[f]^*_\Gamma$. The second statement is obvious.

(b) Since $\mathfrak{g}$ is two-step nilpotent, the center $\mathfrak{z}(\mathfrak{g})$ contains $[\mathfrak{g}, \mathfrak{g}]$ and hence

$$\langle l, X\rangle = \langle f, X\rangle,$$

for every $X \in [\mathfrak{g}, \mathfrak{g}]$ and $l \in \mathcal{A}(\mathrm{f}, \Gamma)$. The desired result follows. □

Next, let

$$\mathscr{H} = \bigoplus_{f\in\mathcal{V}_\Gamma} \bigoplus_{l\in\mathcal{E}(f,\Gamma)} L^2(G/M[f], \chi_l)$$

and

$$\mathscr{H}_0 = \Big\{F = \sum_{f\in\mathcal{V}_\Gamma} \sum_{l\in\mathcal{E}(f,\Gamma)} F(f,l) \in \mathscr{H} :$$
$$F(f,l) = 0 \text{ except for finitely many } (f,l)\Big\}.$$

Let $\mathscr{U}_\Gamma$ and $\mathscr{V}_\Gamma$ be the intertwining operators described in Theorem 3.7.9. Then the following consequences are immediate:

Corollary 3.7.24 *Let the notation be as in Theorem 3.7.9. Then the following hold:*

1. *One has the decomposition*

$$R_\Gamma \simeq \bigoplus_{f\in\mathcal{V}_\Gamma} \bigoplus_{l\in\mathcal{E}(f,\Gamma)} \pi_l \tag{3.7.29}$$

where π_l is the irreducible unitary representation associated to l by the Kirillov mapping Kir_G.

2. *The multiplicity function is given by the formula*

$$m(\pi) = \#[\mathcal{E}(f,\Gamma) \cap \mathcal{O}_\pi]$$

where $\{f\} = p_{\mathfrak{g},\mathfrak{z}(\mathfrak{g})}(\mathcal{O}_\pi)$ and $p_{\mathfrak{g},\mathfrak{z}(\mathfrak{g})} : \mathfrak{g}^* \longrightarrow \mathfrak{z}(\mathfrak{g})^*$ is the canonical projection.

Proof This follows from the fact that the operator $\mathscr{U}_\Gamma$ defined in the above theorem is a unitary intertwining operator between R_Γ and the representation $\bigoplus_{f\in\mathcal{V}_\Gamma}\bigoplus_{l\in\mathcal{E}(f,\Gamma)}\pi_l$. □

3.7.7 Intertwining Operators

Throughout this section G denotes a two-step nilpotent Lie group with Lie algebra $\mathfrak{g}$. Let Γ_1 and Γ_2 two representation-equivalent discrete uniform subgroups of G. Let Φ be a Γ_1-equivalence automorphism of G such that $\Gamma_2 = \Phi(\Gamma_1)$ (Theorem 3.7.22). Let

$$\mathscr{B}_{\Gamma_1} = \{X_1, \ldots, X_n\}$$

denote a strong-Malcev basis of $\mathfrak{g}$ strongly based on Γ_1 and

$$\mathscr{B}_{\Gamma_2} = \{\Phi_*(X_1), \ldots, \Phi_*(X_n)\}.$$

It is obvious that $\mathscr{B}_{\Gamma_2}$ is a strong-Malcev basis of $\mathfrak{g}$ strongly based on Γ_2.

To obtain a unitary intertwining operator between R_{Γ_1} and R_{Γ_2} there remains to construct a unitary operator $\mathbb{T}_{\Gamma_1,\Gamma_2}$ between the spaces

$$\mathscr{H}_{\Gamma_1} \stackrel{\text{def}}{=} \bigoplus_{f\in\mathcal{V}_{\Gamma_1}} \bigoplus_{l\in\mathcal{E}(f,\Gamma_1)} L^2(G/M[f], \chi_l)$$

and

$$\mathscr{H}_{\Gamma_2} \stackrel{\text{def}}{=} \bigoplus_{f\in\mathcal{V}_{\Gamma_2}} \bigoplus_{l\in\mathcal{E}(f,\Gamma_2)} L^2(G/M[f], \chi_l).$$

The relative diagram is the following:

$$\begin{array}{ccc} L^2(G/\Gamma_1) & \longrightarrow & L^2(G/\Gamma_2) \\ \mathscr{U}_{\Gamma_1}\big\downarrow & & \mathscr{V}_{\Gamma_2}\big\uparrow \\ \mathscr{H}_{\Gamma_1} & \xrightarrow{\mathbb{T}_{\Gamma_1,\Gamma_2}} & \mathscr{H}_{\Gamma_2} \end{array}$$

In order to do this, the relations between the spaces $\mathcal{V}_{\Gamma_1}$, $\mathcal{A}(f, \Gamma_1)$, $\mathcal{E}(f, \Gamma_1)$ and $\mathcal{V}_{\Gamma_2}$, $\mathcal{A}(f, \Gamma_2)$, $\mathcal{E}(f, \Gamma_2)$ should be described.

Let G be a two-step nilpotent Lie group with Lie algebra $\mathfrak{g}$ and let Γ_1 be a uniform subgroup of G. Let Φ be a Γ_1-equivalence automorphism of G. Let $\Gamma_2 = \Phi(\Gamma_1)$. By virtue of (3.7.25), one has that $\mathcal{V}_{\Gamma_1} = \mathcal{V}_{\Gamma_2}$. In the sequel, the notation

$$\mathcal{V} = \mathcal{V}_{\Gamma_1} = \mathcal{V}_{\Gamma_2}$$

will often be employed.

Lemma 3.7.25 *Let G be a two-step nilpotent Lie group with Lie algebra $\mathfrak{g}$ and let Γ_1 be a uniform subgroup of G. Let Φ be a Γ_1-equivalence automorphism of G. Let $\Gamma_2 = \Phi(\Gamma_1)$. Let $f \in \mathcal{V}$ be arbitrary.*

(i) If $\mathfrak{m}[f]$ is the Vergne polarization at f associated to $\mathscr{B}_{\Gamma_1}$, then $\mathfrak{m}[f]$ is also the Vergne polarization at f associated to $\mathscr{B}_{\Gamma_2}$.

(ii) The mapping

$$q : \mathfrak{z}(\mathfrak{g})^* \longrightarrow \mathfrak{z}(\mathfrak{g})^*,\ l \longmapsto l \circ \left(\Phi_*^{-1}|_{\mathfrak{z}(\mathfrak{g})}\right)$$

is bijective and keeps the subset $\mathcal{V}$ stable.

(iii) $\mathfrak{m}[f] = \mathfrak{m}[f \circ \left(\Phi_*^{-1}|_{\mathfrak{z}(\mathfrak{g})}\right)]$.

(iv) The mapping

$$p : \mathfrak{m}[f]^*_{\Gamma_1} \longrightarrow \mathfrak{m}[f]^*_{\Gamma_2},\ l \longmapsto l \circ \left(\Phi_*^{-1}|_{\mathfrak{m}[f]}\right)$$

is bijective and $p(\mathcal{A}(f, \Gamma_1)) = \mathcal{A}(f \circ \left(\Phi_*^{-1}|_{\mathfrak{z}(\mathfrak{g})}\right), \Gamma_2)$. Moreover, $p(\mathcal{E}(f, \Gamma_1))$ is a cross-section of Γ_2-orbits in $\mathcal{A}(f \circ \left(\Phi_*^{-1}|_{\mathfrak{z}(\mathfrak{g})}\right), \Gamma_2)$, which will be denoted by $\mathcal{E}(f \circ \left(\Phi_*^{-1}|_{\mathfrak{z}(\mathfrak{g})}\right), \Gamma_2)$.

Proof

(i) Let $\mathscr{B}_{\Gamma_1} = \{X_1, \ldots, X_n\}$ and let $\mathfrak{a}_i = \mathbb{R}\text{-span}\,\{X_1, \ldots, X_i\}$, $(1 \le i \le n)$. Since Φ is Γ_1-equivalence automorphism then

$$\Phi_*(X_j) \equiv X_j \ \mathrm{mod}(\mathfrak{z}(\mathfrak{g})) \qquad (1 \le j \le n)$$

(cf. Formula (3.7.23)). It follows that $\mathfrak{a}_i = \mathbb{R}\text{-span}\,\{\Phi_*(X_1), \ldots, \Phi_*(X_i)\}$ for each $i = \dim(\mathfrak{z}(\mathfrak{g})) + 1, \ldots, n$ and hence $\mathfrak{m}[f]$ is also the Vergne polarization at f associated to $\mathscr{B}_{\Gamma_2}$. This shows (i).

(ii) It is clear that the mapping $\mathfrak{z}(\mathfrak{g})^* \longrightarrow \mathfrak{z}(\mathfrak{g})^*,\ l \longmapsto l \circ \left(\Phi_*|_{\mathfrak{z}(\mathfrak{g})}\right)$ is the inverse of q. On the other hand, let $l \in \mathcal{V}$ and $X \in \mathfrak{z}(\mathfrak{g}) \cap \log(\Gamma_1)$. By (3.7.25) one has $\Phi_*^{-1}(X) \in \mathfrak{z}(\mathfrak{g}) \cap \log(\Gamma_1)$ and therefore

$$\langle q(l), X\rangle = \langle l, \Phi_*^{-1}(X)\rangle \in \mathbb{Z}.$$

Then $q(l)$ belongs to $\mathcal{V}$. Deduce that $q(\mathcal{V}) \subset \mathcal{V}$ and therefore $q(\mathcal{V}) = \mathcal{V}$.

(iii) This follows from the definition of Vergne polarization and relation (3.7.24).

(iv) Begin by observing that $p(\mathfrak{m}[f]^*_{\Gamma_1}) \subset \mathfrak{m}[f]^*_{\Gamma_2}$ and the inverse of p is the mapping

$$s : \mathfrak{m}[f]^*_{\Gamma_2} \longrightarrow \mathfrak{m}[f]^*_{\Gamma_1},\ l \longmapsto l \circ \left(\Phi_*|_{\mathfrak{m}[f]}\right).$$

On the other hand, let $l \in \mathcal{A}(f, \Gamma_1)$. Since $\mathfrak{m}[f] = \mathfrak{m}[f \circ \left(\Phi_*^{-1}|_{\mathfrak{z}(\mathfrak{g})}\right)]$ then $p(l) = l \circ \left(\Phi_*^{-1}|_{\mathfrak{m}[f]}\right) \in \mathfrak{m}[f \circ \left(\Phi_*^{-1}|_{\mathfrak{z}(\mathfrak{g})}\right)]^*_{\Gamma_2}$. Let $X \in \mathfrak{z}(\mathfrak{g})$, compute

$$\begin{aligned}
\langle p(l), X\rangle &= \langle l \circ \Phi_*^{-1}, X\rangle \\
&= \langle l, \Phi_*^{-1}(X)\rangle \\
&= \langle f, \Phi_*^{-1}(X)\rangle \qquad\qquad \text{(since } \mathfrak{z}(\mathfrak{g}) \text{ is stable under } \Phi_*^{-1}) \\
&= \langle f \circ \Phi_*^{-1}, X\rangle.
\end{aligned}$$

Then $p(l)|_{\mathfrak{z}(\mathfrak{g})} = f \circ \Phi_*^{-1}|_{\mathfrak{z}(\mathfrak{g})}$ and hence $p(l) \in \mathcal{A}(f \circ \left(\Phi_*^{-1}|_{\mathfrak{z}(\mathfrak{g})}\right), \Gamma_2)$. It follows that $p(\mathcal{A}(f, \Gamma_1)) \subset \mathcal{A}(f \circ \left(\Phi_*^{-1}|_{\mathfrak{z}(\mathfrak{g})}\right), \Gamma_2)$. Similarly, one proves $s(\mathcal{A}(f \circ \left(\Phi_*^{-1}|_{\mathfrak{z}(\mathfrak{g})}\right), \Gamma_2)) \subset \mathcal{A}(f, \Gamma_1)$. From this, deduce that $\mathcal{A}(f \circ \left(\Phi_*^{-1}|_{\mathfrak{z}(\mathfrak{g})}\right), \Gamma_2) \subset p(\mathcal{A}(f, \Gamma_1))$. The two inclusions imply equality, $p(\mathcal{A}(f, \Gamma_1)) = \mathcal{A}(f \circ \left(\Phi_*^{-1}|_{\mathfrak{z}(\mathfrak{g})}\right), \Gamma_2)$. Next we prove that $p(\mathcal{E}(f, \Gamma_1))$ is a cross-section of Γ_2-orbits in $\mathcal{A}(f \circ \left(\Phi_*^{-1}|_{\mathfrak{z}(\mathfrak{g})}\right), \Gamma_2)$. One has

$$\mathcal{A}(f, \Gamma_1) = \bigsqcup_{l \in \mathcal{E}(f, \Gamma_1)} \Gamma_1 \cdot l,$$

where the symbol $\sqcup$ denotes disjoint union. As p is bijective, then

$$p(\mathcal{A}(f, \Gamma_1)) = \bigsqcup_{l \in \mathcal{E}(f, \Gamma_1)} p(\Gamma_1 \cdot l).$$

On the other hand, the mapping p is a Γ_1-morphism. In fact, since the elements of Γ_1 differ from elements of Γ_2 by central elements, then $\Gamma_2 \cdot l = \Gamma_1 \cdot l$ for every $l \in \mathfrak{m}[f]^*_{\Gamma_2}$ and hence $\mathfrak{m}[f]^*_{\Gamma_2}$ is a Γ_1-space. Furthermore, let $l \in \mathfrak{m}[f]^*_{\Gamma_1}$ and $\gamma = \exp -X \in \Gamma_1$. Compute

$$\begin{aligned}
\langle \gamma \cdot p(l), Y\rangle &= \langle \gamma \cdot (l \circ \Phi_*^{-1}), Y\rangle \\
&= \langle l \circ \Phi_*^{-1}, e^{\operatorname{ad} X} Y\rangle \\
&= \langle l \circ \Phi_*^{-1}, Y + [X, Y]\rangle \\
&= \langle l, \Phi_*^{-1}(Y) + \Phi_*^{-1}([X, Y])\rangle \\
&= \langle l, \Phi_*^{-1}(Y) + [X, \Phi_*^{-1}(Y)]\rangle \qquad\qquad \text{(by (3.7.24))}
\end{aligned}$$

$$
\begin{aligned}
&= \langle \gamma \cdot l, \Phi_*^{-1}(Y) \rangle \\
&= \langle (\gamma \cdot l) \circ \Phi_*^{-1}, Y \rangle \\
&= \langle p(\gamma \cdot l), Y \rangle,
\end{aligned}
$$

for every $Y \in \mathfrak{m}[f]$. Then $p(\gamma \cdot l) = \gamma \cdot p(l)$. It follows that $p(\Gamma_1 \cdot l) = \Gamma_1 \cdot p(l)$. Therefore

$$
\mathcal{A}(f \circ (\Phi_*^{-1}|_{\mathfrak{z}(\mathfrak{g})}), \Gamma_2) = \bigsqcup_{l \in \mathcal{E}(f, \Gamma_1)} \Gamma_1 \cdot p(l) = \bigsqcup_{l \in \mathcal{E}(f, \Gamma_1)} \Gamma_2 \cdot p(l),
$$

as required.

□

Definition 3.7.26 Let Φ be a Γ-equivalence automorphism of a two-step nilpotent Lie group $G = \exp \mathfrak{g}$ and let $\mathscr{B} = \{X_1, \ldots, X_n\}$ be a strong Malcev basis of $\mathfrak{g}$ strongly based on Γ. Consider the Lie algebra automorphism

$$
[\Phi_*]^{\mathscr{B}} : \mathfrak{g} \longrightarrow \mathfrak{g}
$$

defined as

$$
[\Phi_*]^{\mathscr{B}}(X_i) = X_i + [a_i, X_i],
$$

if

$$
\Phi_*(X_i) = X_i + [a_i, X_i] + z_i
$$

for all $i = 1, \ldots, n$. The Lie group automorphism $[\Phi]^{\mathscr{B}}$ associated to $[\Phi_*]^{\mathscr{B}}$ is called a *pseudo-Γ-almost inner automorphism associated to Φ relative to the basis $\mathscr{B}$.*

The notion of almost inner automorphisms of nilpotent Lie groups was introduced in [81] and [80].

Lemma 3.7.27 *Keep the notations of Definition 3.7.26. One has*

$$
[\Phi_*]^{\mathscr{B}}([X, Y]) = [X, Y],
$$

for every $X, Y \in \mathfrak{g}$.

Proof The proof is straightforward. □

To describe the operator $\mathbb{T}_{\Gamma_1, \Gamma_2}$, the following lemma is needed.

Lemma 3.7.28 *Let G be a two-step nilpotent Lie group with Lie algebra $\mathfrak{g}$ and let Γ_1 be a uniform subgroup of G. Let Φ be a Γ_1-equivalence automorphism of G. Let $f \in \mathcal{V}$ and let $\mathscr{B}_f$ be strong Malcev basis of $\mathfrak{g}$ strongly based on Γ_1 and* *passing through* $\mathfrak{z}(\mathfrak{g})$, $\mathfrak{g}(f)$ *and* $\mathfrak{m}[f]$. *Let* $l \in \mathcal{A}(f, \Gamma_1)$. *The linear functionals* l *and* $l \circ ([\Phi_*]^{\mathscr{B}_f}|_{\mathfrak{m}[f]})$ *(as elements of* $\mathfrak{g}^*$*) lie in the same coadjoint* G*-orbit.*

Proof One has

$$\langle l \circ ([\Phi_*]^{\mathscr{B}_f}|_{\mathfrak{m}[f]}), [X, Y]\rangle = \langle l \circ [\Phi_*]^{\mathscr{B}_f}, [X, Y]\rangle = \langle l, [X, Y]\rangle$$

for all $X, Y \in \mathfrak{g}$. This implies that

$$\mathfrak{g}(l \circ ([\Phi_*]^{\mathscr{B}_f}|_{\mathfrak{m}[f]})) = \mathfrak{g}(l). \tag{3.7.30}$$

Moreover, for every $X \in \mathfrak{g}(l) \cap \mathscr{B}_f$ one has

$$\begin{aligned}\langle l \circ ([\Phi_*]^{\mathscr{B}_f}|_{\mathfrak{m}[f]}), X\rangle &= \langle l, ([\Phi_*]^{\mathscr{B}_f}|_{\mathfrak{m}[f]})(X)\rangle \\ &= \langle l, [\Phi_*]^{\mathscr{B}_f}(X)\rangle \\ &= \langle l, X + [a_X, X]\rangle \qquad (a_X \in \mathfrak{g}) \\ &= \langle l, X\rangle,\end{aligned}$$

finally

$$l \circ ([\Phi_*]^{\mathscr{B}_f}|_{\mathfrak{m}[f]})|_{\mathfrak{g}(l)} = l|_{\mathfrak{g}(l)}. \tag{3.7.31}$$

As G is two-step, conditions (3.7.30) and (3.7.31) imply that

$$G \cdot \left(l \circ ([\Phi_*]^{\mathscr{B}_f}|_{\mathfrak{m}[f]})\right) = G \cdot l.$$

This completes the proof. □

Note that, in view of the above result, if $f \in \mathcal{V}$ and $l \in \mathcal{E}(f, \Gamma_1)$ then there exists $a_f \in G$ (depending only on f because $l|_{\mathfrak{z}(\mathfrak{g})} = f|_{\mathfrak{z}(\mathfrak{g})}$ and G is two-step) such that

$$l \circ ([\Phi_*]^{\mathscr{B}_f}|_{\mathfrak{m}[f]}) = \mathrm{Ad}^*(a_f^{-1})l. \tag{3.7.32}$$

For simplicity we set, in the rest of this section,

$$\begin{aligned} f \circ \Phi_* &= f \circ (\Phi_*|_{\mathfrak{z}(\mathfrak{g})}), \\ f \circ \Phi_*^{-1} &= f \circ (\Phi_*^{-1}|_{\mathfrak{z}(\mathfrak{g})}) && (\text{if } f \in \mathcal{V}) \\ l \circ \Phi_* &= l \circ (\Phi_*|_{\mathfrak{m}[f]}) && (\text{if } l \in \mathcal{E}(\cdot, \Gamma_2)) \\ l \circ \Phi_*^{-1} &= f \circ (\Phi_*^{-1}|_{\mathfrak{m}[f]}) && (\text{if } l \in \mathcal{E}(\cdot, \Gamma_1)) \end{aligned}$$

Proposition 3.7.29 *Let G be a two-step nilpotent Lie group with Lie algebra $\mathfrak{g}$ and let Γ_1 be a uniform subgroup of G. Let Φ be a Γ_1-equivalence automorphism of G. Let $\Gamma_2 = \Phi(\Gamma_1)$. The operator*

$$\mathbb{T}_{\Gamma_1,\Gamma_2} : \bigoplus_{f\in\mathcal{V}} \bigoplus_{l\in\mathcal{E}(f,\Gamma_1)} L^2(G/M[f], \chi_l) \longrightarrow \bigoplus_{f\in\mathcal{V}} \bigoplus_{l\in\mathcal{E}(f\circ\Phi_*^{-1},\Gamma_2)} L^2(G/M[f], \chi_l)$$

defined on $F = \sum_{f\in\mathcal{V}} \sum_{l\in\mathcal{E}(f,\Gamma_1)} F(f, l)$ *by*

$$\mathbb{T}_{\Gamma_1,\Gamma_2}(F)(g) = \sum_{f\in\mathcal{V}} \sum_{l\in\mathcal{E}(f\circ\Phi_*^{-1},\Gamma_2)} \chi_{f\circ\theta_f}(g) F(f \circ \Phi_*, l \circ \Phi_*)(ga_f)$$

where $\theta_f = \Phi_* - [\Phi_*]^{\mathscr{B}_f}$, *is a unitary operator.*

Proof Let $F = \sum_{f\in\mathcal{V}} \sum_{l\in\mathcal{E}(f,\Gamma_1)} F(f, l) \in \bigoplus_{f\in\mathcal{V}} \bigoplus_{l\in\mathcal{E}(f,\Gamma_1)} L^2(G/M[f], \chi_l)$ and let $f \in \mathcal{V}, l \in \mathcal{E}(f \circ \Phi_*^{-1}, \Gamma_2)$. The mapping

$$g \longmapsto \chi_{f\circ\theta_f}(g) F(f \circ \Phi_*, l \circ \Phi_*)(ga_f)$$

has the desired covariance (3.2.5) in $L^2(G/M[f], \chi_l)$. In fact, let $g \in G$ and $m \in M[f]$. Compute

$$\begin{aligned}
&\chi_{f\circ\theta_f}(gm) F(f \circ \Phi_*, l \circ \Phi_*)(gma_f)\\
&=\chi_{f\circ\theta_f}(gm) F(f \circ \Phi_*, l \circ \Phi_*)(ga_f a_f^{-1} m a_f)\\
&=\chi_{f\circ\theta_f}(gm) \chi_{l\circ\Phi_*}^{-1}(a_f^{-1} m a_f) F(f \circ \Phi_*, l \circ \Phi_*)(ga_f)\\
&=\chi_{f\circ\theta_f}(gm) \chi_{\mathrm{Ad}^* a_f(l\circ\Phi_*)}^{-1}(m) F(f \circ \Phi_*, l \circ \Phi_*)(ga_f)\\
&=\chi_{f\circ\theta_f}(g) \chi_{f\circ\theta}(m) \chi_{\mathrm{Ad}^* a_f(l\circ\Phi_*)}^{-1}(m) F(f \circ \Phi_*, l \circ \Phi_*)(ga_f)\\
&\quad (\chi_{f\circ\theta_f} \text{ is a unitary character of } G \text{ since } \langle f \circ \theta_f, [\mathfrak{g}, \mathfrak{g}]\rangle = \{0\})\\
&=\chi_{f\circ\theta_f}(g) \chi_{(\mathrm{Ad}^* a_f(l\circ\Phi_*) - f\circ\theta_f)}^{-1}(m) F(f \circ \Phi_*, l \circ \Phi_*)(ga_f)\\
&=\chi_{f\circ\theta_f}(g) \chi_{\mathrm{Ad}^* a_f(l\circ\Phi_* - f\circ\theta_f)}^{-1}(m) F(f \circ \Phi_*, l \circ \Phi_*)(ga_f)\\
&\quad (\text{since } \mathrm{Ad}^* a_f(f \circ \theta_f) = f \circ \theta_f)\\
&=\chi_{f\circ\theta_f}(g) \chi_{\mathrm{Ad}^* a_f(l\circ([\Phi_*]^{\mathscr{B}_f}{}_{|\mathfrak{m}[f]}))}^{-1}(m) F(f \circ \Phi_*, l \circ \Phi_*)(ga_f)\\
&=\chi_{f\circ\theta_f}(g) \chi_l^{-1}(m) F(f \circ \Phi_*, l \circ \Phi_*)(ga_f)
\end{aligned}$$

(by (3.7.32))

$$=\chi_l^{-1}(m)\Big(\chi_{f\circ\theta_f}(g)F(f\circ\Phi_*,l\circ\Phi_*)(ga_f)\Big).$$

Next, compute

$$\begin{aligned}
\|\mathbb{T}_{\Gamma_1,\Gamma_2}(F)\|^2 &= \sum_{f\in\mathcal{V}}\sum_{l\in\mathcal{E}(f\circ\Phi_*^{-1},\Gamma_2)}\int_{G/M[f]}\left|\chi_{f\circ\theta_f}(g)F(f\circ\Phi_*,l\circ\Phi_*)(ga_f)\right|^2 dg\\
&= \sum_{f\in\mathcal{V}}\sum_{l\in\mathcal{E}(f\circ\Phi_*^{-1},\Gamma_2)}\int_{G/M[f]}\left|F(f\circ\Phi_*,l\circ\Phi_*)(g)\right|^2 dg\\
&= \sum_{f\in\mathcal{V}}\sum_{l\in\mathcal{E}(f,\Gamma_1)}\int_{G/M[f]}\left|F(f,l)(g)\right|^2 dg \qquad \text{(by Lemma 3.7.25)}\\
&= \|F\|^2.
\end{aligned}$$

Finally, one proves that the operator $\mathbb{R}_{\Gamma_1,\Gamma_2}$ defined by

$$\mathbb{R}_{\Gamma_1,\Gamma_2}(K)(g)=\sum_{f\in\mathcal{V}}\sum_{l\in\mathcal{E}(f\circ\Phi_*,\Gamma_1)}\chi^{-1}_{(f\circ\theta_f)}(ga_f^{-1})K(f\circ\Phi_*^{-1},l\circ\Phi_*^{-1})(ga_f^{-1})$$

is the inverse of $\mathbb{T}_{\Gamma_1,\Gamma_2}$ where $K=\sum_{f\in\mathcal{V}}\sum_{l\in\mathcal{E}(f,\Gamma_2)}K(f,l)$. For this, there remains to show that $\mathbb{T}_{\Gamma_1,\Gamma_2}\circ\mathbb{R}_{\Gamma_1,\Gamma_2}=\mathbf{id}$. Similarly to what was done above, one can show that if $f\in\mathcal{V}$ and $l\in\mathcal{E}(f,\Gamma_1)$, the function

$$g\longmapsto\chi^{-1}_{(f\circ\theta_f)}(ga_f^{-1})K(f\circ\Phi_*^{-1},l\circ\Phi_*^{-1})(ga_f^{-1})$$

belongs to $L^2(G/M[f],\chi_l)$. Now, calculate

$$\begin{aligned}
\mathbb{T}_{\Gamma_1,\Gamma_2}\circ\mathbb{R}_{\Gamma_1,\Gamma_2}(K)(g) &= \sum_{f\in\mathcal{V}}\sum_{l\in\mathcal{E}(f,\Gamma_2)}\chi_{(f\circ\Phi_*^{-1}\circ\theta_f)}(g)\chi^{-1}_{(f\circ\theta_f)}(g)K(f,l)(g).\\
&= \sum_{f\in\mathcal{V}}\sum_{l\in\mathcal{E}(f,\Gamma_2)}\chi_{(f\circ\theta_f)}(g)\chi^{-1}_{(f\circ\theta_f)}(g)K(f,l)(g)\\
&\qquad\qquad (\text{ since }\Phi_*^{-1}\circ\theta_f=\theta_f)\\
&= \sum_{f\in\mathcal{V}}\sum_{l\in\mathcal{E}(f,\Gamma_2)}K(f,l)(g)\\
&= K(g).
\end{aligned}$$

This completes the proof of the proposition. □

By Theorem 3.7.9 and Proposition 3.7.29, one easily derives the key fact of our discussion.

Theorem 3.7.30 *Let G be a two-step nilpotent Lie group with Lie algebra $\mathfrak{g}$ and let Γ_1 be a uniform subgroup of G. Let Φ be a Γ_1-equivalence automorphism of G. Let $\Gamma_2 = \Phi(\Gamma_1)$. The operator*

$$\mathbb{T} = \mathscr{V}_{\Gamma_2} \circ \mathbb{T}_{\Gamma_1,\Gamma_2} \circ \mathscr{U}_{\Gamma_1}$$

is an intertwining operator between R_{Γ_1} and R_{Γ_2}.

Proof It is not hard to show that the unitary operator $\mathbb{T}$ satisfies (3.7.1). □

As an immediate consequence of Theorem 3.7.30, one obtains the following classical result.

Corollary 3.7.31 *If $\Phi = I_a : G \longrightarrow G, x \longmapsto axa^{-1}$ is the inner automorphism of G defined by $a \in G$, then the operator $\mathbb{T}$ is given by*

$$\mathbb{T}(\xi)(g) = \xi(ga)$$

for every $\xi \in L^2(G/\Gamma_1)$ and $g \in G$.

Proof Keep the notation of Proposition 3.7.30 and assume that $\Phi = I_a$. In this case, one has $\theta_f = 0$ for any $f \in \mathcal{V}$. On the other hand, for any $f \in \mathcal{V}$ and $l \in \mathcal{E}(f, \Gamma_1)$, one has $a_f = a$ (see (3.7.32)). Then one obtains

$$\begin{aligned}
\mathbb{T}(\xi)(g) &= \sum_{f\in\mathcal{V}} \sum_{l\in\mathcal{E}(f\circ\Phi_*^{-1},\Gamma_2)} \sum_{\overline{\gamma}\in\Gamma_2/M[f]\cap\Gamma_2} \int_{M[f]/M[f]\cap\Gamma_1} \xi(g\gamma am)\chi_{l\circ\Phi_*}(m)dm \\
&= \sum_{f\in\mathcal{V}} \sum_{l\in\mathcal{E}(f,\Gamma_1)} \sum_{\overline{\gamma}\in\Gamma_2/M[f]\cap\Gamma_2} \int_{M[f]/M[f]\cap\Gamma_1} \xi(g\gamma am)\chi_l(m)dm \\
&\qquad \text{(by Lemma)} \\
&= \sum_{f\in\mathcal{V}} \sum_{l\in\mathcal{E}(f,\Gamma_1)} \sum_{\overline{\gamma}\in\Gamma_2/M[f]\cap\Gamma_2} \int_{M[f]/M[f]\cap\Gamma_1} \xi(gaa^{-1}\gamma am)\chi_l(m)dm \\
&= \sum_{f\in\mathcal{V}} \sum_{l\in\mathcal{E}(f,\Gamma_1)} \sum_{\overline{\gamma}\in\Gamma_1/M[f]\cap\Gamma_1} \int_{M[f]/M[f]\cap\Gamma_1} \xi(ga\gamma m)\chi_l(m)dm \\
&\qquad \text{(changing variables } \gamma \longmapsto a^{-1}\gamma a) \\
&= \mathscr{V}_{\Gamma_1} \circ \mathscr{U}_{\Gamma_1}(\xi)(ga) \\
&= \xi(ga). \qquad \text{(by Theorem)}
\end{aligned}$$

This completes the proof. □

Remark 3.7.32 Corollary 3.7.31 holds for any induced representation $\mathrm{ind}_{H_1}^{G}\sigma_1$, where $H_2 = aH_1a^{-1}$, $\sigma_2(h) = \sigma_1(a^{-1}ha)$.

Example 3.7.33 We consider a six-dimensional two-step nilpotent Lie group containing a pair of representation-equivalent discrete uniform subgroups that are not conjugate. The example is the smallest possible because pairs of representation-equivalent discrete uniform subgroups of two-step nilpotent Lie groups of dimension up to five are necessarily conjugate. Consider the Lie algebra $\mathfrak{g} = \mathbb{R}\text{-span}(Z_1, Z_2, U_1, U_2, U_3, U_4)$ with Lie brackets

$$[U_1, U_3] = 6Z_1 \qquad [U_1, U_4] = -2(Z_1 + 2Z_2)$$

$$[U_2, U_3] = -4Z_1 \qquad [U_2, U_4] = 8(Z_1 + 2Z_2)$$

and all other brackets of basis elements zero. Let G be the simply connected Lie group with Lie algebra $\mathfrak{g}$. Note that $Z(G) = [G, G] = \exp \mathbb{R}Z_1 \exp \mathbb{R}Z_2$. Let Γ_1 be the discrete uniform subgroup of G generated by

$$\mathscr{B}_{\Gamma_1} = \{\exp Z_1, \exp Z_2, \exp U_1, \exp U_2, \exp U_3, \exp U_4\}.$$

Let Γ_2 be the discrete uniform subgroup of G generated by

$$\mathscr{B}_{\Gamma_2} = \left\{\exp Z_1, \exp Z_2, \exp U_1 + \tfrac{1}{4}Z_2, \exp U_2 - \tfrac{1}{4}Z_1, \exp U_3, \exp U_4\right\}.$$

Let Φ_* be the Lie algebra automorphism of $\mathfrak{g}$ defined by mapping:

$$\begin{aligned} U_1 &\longmapsto U_1 + \tfrac{1}{4}Z_2 \\ U_2 &\longmapsto U_1 - \tfrac{1}{4}Z_1 \\ U_3 &\longmapsto U_3 \\ U_4 &\longmapsto U_4 \\ Z_2 &\longmapsto Z_2 \\ Z_1 &\longmapsto Z_1. \end{aligned}$$

Let Φ be the automorphism of G associated to Φ_*. Φ is a Γ_1-equivalence automorphism of G such that $\Phi(\Gamma_1) = \Gamma_2$ (see Claim 1 of [83]). Then for every element γ of Γ_1, there exists $a_\gamma \in G$ such that

$$\Phi(\gamma) \equiv a_\gamma \gamma a_\gamma^{-1} \mathrm{mod}(\Gamma_1 \cap [G, G]). \tag{3.7.33}$$

An element $a_\gamma = \exp t_1 U_1 \exp t_2 U_2 \exp t_3 U_3 \exp t_4 U_4$ satisfies (3.7.33) if and only if $t_1, \ldots, t_4 \in \mathbb{R}$ satisfy the following two conditions

$$-\tfrac{\gamma_1}{4} - 4\gamma_4 t_1 + 16\gamma_4 t_2 + 4(\gamma_1 - 4\gamma_2)t_4 \in \mathbb{Z}$$
$$\tfrac{\gamma_2}{4} + (6\gamma_3 - 2\gamma_4)t_1 - 4(\gamma_3 - 2\gamma_4)t_2 + (4\gamma_2 - 6\gamma_1)t_3 + (2\gamma_1 - 8\gamma_2)t_4 \in \mathbb{Z}.$$

where $\gamma \equiv \exp \gamma_1 U_1 \exp \gamma_2 U_2 \exp \gamma_3 U_3 \exp \gamma_4 U_4 \mathrm{mod}(Z(G))$. On the other hand, one has

$$\mathscr{V} = \mathscr{V}_{\Gamma_1} = \mathscr{V}_{\Gamma_2} = \mathbb{Z}\text{-span}\,\{Z_1^*, Z_2^*\}.$$

For $f = \alpha Z_1^* + \beta Z_2^*$, let $\mathfrak{m}[f]$ be the Vergne polarization at f associated to $\mathscr{B}_{\Gamma_1}$. One has

$$\mathfrak{m}[f] = \begin{cases} \mathfrak{g} & \text{if } \alpha = \beta = 0 \\ \mathbb{R}\text{-span}(Z_1, Z_2, U_1, U_2, U_3) & \text{if } \alpha = 0, \beta \neq 0 \\ \mathbb{R}\text{-span}(Z_1, Z_2, U_1, U_2) & \text{if } \alpha \neq 0, \beta = 0 \\ \mathbb{R}\text{-span}(Z_1, Z_2, U_1, U_2) & \text{if } \alpha \neq 0, \beta \neq 0, \alpha \neq -2\beta \\ \mathbb{R}\text{-span}(Z_1, Z_2, U_1, U_2, U_4) & \text{if } \alpha \neq 0, \beta \neq 0, \alpha = -2\beta. \end{cases}$$

For every $\xi \in \mathscr{C}(G/\Gamma_1)$ and $g \in G$, as $\Phi_*|_{\mathfrak{z}(\mathfrak{g})} = \mathrm{id}_{\mathfrak{z}(\mathfrak{g})}$, one has

$$\mathbb{T}(\xi)(g) = \sum_{f \in \mathcal{V}} \sum_{l \in \mathcal{E}(f,\Gamma_2)} \sum_{\overline{\gamma} \in \Gamma_2/M[f]\cap\Gamma_2} \int_{M[f]/M[f]\cap\Gamma_1} \xi(g\gamma a_f m)\chi_{l\circ\Phi_*}(m)dm.$$

Chapter 4
Variants of Plancherel Formulas for Monomial Representations of Exponential Solvable Lie Groups

4.1 Layout of the Problems

Let π be a unitary representation of a connected Lie group G acting on a Hilbert space $\mathscr{H}_\pi$. Denote by $\mathscr{H}_\pi^\infty$ the Fréchet space of smooth vectors for π, and by $\mathscr{H}_\pi^{-\infty}$ the space of continuous, anti-linear functionals on $\mathscr{H}_\pi^\infty$. Let α be any positive distribution on G of finite order. The Bonnet-Plancherel formula (cf. [28]) tells us that for any test function φ

$$\alpha(\varphi) = \int_{\hat{G}} \mathrm{tr}(\pi(\varphi)U_\pi)d\nu(\pi), \tag{4.1.1}$$

where, for ν-almost everywhere, $U_\pi : \mathscr{H}_\pi^\infty \to \mathscr{H}_\pi^{-\infty} (\pi \in \hat{G})$ is a certain uniquely determined nuclear operator (cf. [4, Theorem 4-1]).

In the case of exponential Lie groups, determining the Bonnet operators U_π is difficult. One of the reasons is that there exists no easy way to determine explicitly the C^∞-vectors of a representation. The first precise formulas in this case were given by Currey in [46], where the author studied monomial representations induced from a "Levi" factor. Chabot-Chevalarias [33] determined these operators for an exponential solvable group, a normal subgroup H and a character χ_f such that $\tau = \tau_f = \mathrm{ind}_H^G \chi_f$ is of finite multiplicity. As we will explain later, there is a way to generalize this result to any character of a connected, closed and normal subgroup H of G.

The general distribution-theoretical Plancherel formula is due to Penney (cf. [136]). It is associated to a disintegration of τ and it is of the form

$$\langle \tau(\omega)\alpha_\tau, \alpha_\tau \rangle = \int_{\mathscr{V}} \langle \pi_\phi(\omega)\beta_\phi, \beta_\phi \rangle d\phi, \ \omega \in C_c(G), \tag{4.1.2}$$

A. Baklouti et al., *Representation Theory of Solvable Lie Groups and Related Topics*, Springer Monographs in Mathematics,
https://doi.org/10.1007/978-3-030-82044-2_4

where α_τ is the canonical cyclic generalized vector for τ, and β_ϕ is an (suitably H-covariant) generalized vector for π_ϕ. In general the determination of such distributions is problematic. In case G is nilpotent, (4.1.2) was obtained by Fujiwara in a different form (see [62]) when the multiplicities occurring in decomposition (4.1.1) are finite. Groundbreaking progress on extending the results of [62] to other classes of homogeneous spaces has been made by Fujiwara-Yamagami [71] and Lipsman [101, 103, 105]. Beyond the nilpotent case, however, the technical difficulties involved in (4.1.2) are considerable.

In [55] Duflo and Rais computed the Plancherel formula for $L^2(G)$ of an exponential solvable group. The formulas of Penney and Bonnet-Plancherel have been described for nilpotent and exponential groups in [62, 64, 65, 67].

The aim of this chapter is to compute explicitly the Bonnet-Plancherel formula associated to a monomial representation of a nilpotent Lie group, and more generally for an exponential solvable Lie group when the inducing subgroup is normal. We will also prove a variant of the Penney-Plancherel formula associated to monomial representations of nilpotent Lie groups.

4.2 The Penney-Plancherel Formula for Nilpotent Lie Groups

Let us recall first the Penney formula for monomial representations of a nilpotent Lie group G. Let G be a nilpotent, connected and simply connected Lie group with Lie algebra $\mathfrak{g}$. Recall that the canonical central disintegration of τ is written as follows: recall $\Gamma_f = f + \mathfrak{h}^\perp$ in $\mathfrak{g}^*$. Take a finite measure $\widetilde{\mu}$ on Γ_f equivalent to the Lebesgue measure and consider $\widetilde{\mu}$ as a measure on $\mathfrak{g}^*$. Let μ be the image of $\widetilde{\mu}$ by the Kirillov map $\mathfrak{g}^* \to \widehat{G}$. For $\rho \in \widehat{G}$, let $\Omega(\rho)$ be the coadjoint orbit associated to ρ and $m(\rho)$ the number of H-orbits contained in $\Gamma_f \cap \Omega(\rho)$. Theorem 1.6.2 asserts that

$$\tau \simeq \int_{\widehat{G}}^{\oplus} m(\rho)\rho d\mu(\rho). \tag{4.2.1}$$

For a closed subgroup N of G and its character χ, one puts

$$\left(\mathscr{H}_\gamma^{-\infty}\right)^{N,\chi} = \{a \in \mathscr{H}_\gamma^{-\infty} : \gamma(x)a = \chi(x)a, \ \forall\, x \in N\}.$$

The Dirac distribution δ_τ given by $\delta_\tau(\varphi) = \overline{\varphi(e)}$ belongs to $\left(\mathscr{H}_\tau^{-\infty}\right)^{H,\chi_f}$ (e is the neutral element of G). And here is the abstract Plancherel formula for τ due to Penney (cf. [136]).

Theorem 4.2.1 *We use the same set-up and notation. Assume that τ has finite multiplicities. Following the disintegration (4.2.1) of τ, the generalized vector δ_τ disintegrates as*

$$\delta_\tau \simeq \int_{\widehat{G}}^{\oplus} \sum_{j=1}^{m(\rho)} a_\rho^j d\mu(\rho)$$

for some elements a_ρ^j of $\left(\mathscr{H}_\rho^{-\infty}\right)^{H,\chi_f}$.

In order to write the generalized vectors a_ρ^j intervening in the abstract Plancherel formula for τ explicitly, let $\rho \in \widehat{G}$ be in the spectrum of μ and call $\Omega(\rho)$ the coadjoint orbit associated to ρ. Then $\Gamma_f \cap \Omega(\rho)$ consists of $m(\rho)$ connected components C_j $(1 \leq j \leq m(\rho))$, each of which is an H-orbit. We fix $\ell \in \Omega(\rho)$ where we take a polarization $\mathfrak{b}[\ell]$ of $\mathfrak{g}$ and realize ρ as $\rho = \mathrm{ind}_{B[\ell]}^G \chi_\ell$ with $B[\ell] = \exp \mathfrak{b}[\ell]$. Then let us choose $g_j \in G$ $(1 \leq j \leq m(\rho))$ so that $g_j \cdot \ell \in C_j$, and define $a_\rho^j : \mathscr{H}_\rho^\infty \to \mathbb{C}$ by

$$\langle a_\rho^j, \psi \rangle = \int_{H/(H \cap g_j B[\ell] g_j^{-1})} \overline{\psi(hg_j)\chi_f(h)} d\dot{h} \tag{4.2.2}$$

for $\psi \in \mathscr{H}_\rho^\infty$. Next is a version of Frobenius reciprocity (cf. [69]).

Theorem 4.2.2 *In the same set-up and notation as in Theorem 4.2.1,*

$$m(\rho) = dim\left(\mathscr{H}_\rho^{-\infty}\right)^{H,\chi_f}$$

almost everywhere for μ. In addition, if τ has finite multiplicities, the generalized vectors $\{a_\rho^j;\ 1 \leq j \leq m(\rho)\}$ defined by formula (4.2.2) form a basis of $\left(\mathscr{H}_\rho^{-\infty}\right)^{H,\chi_f}$ almost everywhere for μ.

4.2.1 Tempered Distributions of Positive Type

Let $\mathscr{S}(\mathfrak{g})$ denote the Schwartz space of G, i.e., the space of complex-valued functions φ on G such that $f \circ \exp$ is an ordinary Schwartz function on the vector space $\mathfrak{g}$. Let $\mathscr{B} = \{X_1, \ldots, X_r\}$ be a Malcev-basis relative to $\mathfrak{h}$, then $\mathfrak{g} = \sum_{1 \leq i \leq r}^{\oplus} \mathbb{R}X_i \oplus \mathfrak{h}$ and for any $j = 1, \ldots, r$, the subspace $\mathfrak{g}_j = \mathrm{span}\{X_j, \ldots, X_r, \mathfrak{h}\}$ is a subalgebra. The mapping $E_{\mathscr{B}} : \mathbb{R}^r \to G/H : E_{\mathscr{B}}(t_1, \ldots, t_r) = \exp(t_1X_1) \cdots \exp(t_rX_r)H$ is then a diffeomorphism. As explained earlier, one obtains a G-invariant measure $d\dot{g} = d_{\mathscr{B}}\dot{g}$ on the quotient space G/H by setting

$$\int_{G/H} \xi(g) d_{\mathscr{B}}\dot{g} = \int_{\mathbb{R}^r} \xi(E_{\mathscr{B}}(T))dT,\ \xi \in C_c(G/H),$$

where $C_c(G/H)$ denotes the space of the continuous functions with compact support on G/H. Let $\mathscr{S}(G/H, f)$ be the space of C^∞-functions ξ on G such that $\xi(gh) = \chi_f(h^{-1})\xi(g)$ for all $g \in G$, $h \in H$ and such that the function $T \mapsto \xi(E_{\mathscr{B}}(T))$ is a Schwartz function on $\mathbb{R}^r$. Pick a Haar measure dh of H and let for $\varphi \in \mathscr{S}(G)$

$$P_{H,f}(\varphi)(g) = P(\varphi)(g) = \int_H \varphi(gh)\chi_f(h)dh, \ g \in G.$$

It is easy to see that $P(\varphi)$ is in $\mathscr{S}(G/H, f)$ and that the mapping P is linear, surjective and continuous, if we provide our spaces with the standard Fréchet topologies. Let $S_{H,f}$ be the tempered distribution on G defined by

$$\langle S_{H,f}, \varphi\rangle = P_{H,f}(\varphi)(e) = \int_H \varphi(h)\chi_f(h)dh, \ \varphi \in \mathscr{S}(G). \tag{4.2.3}$$

We observe that the distribution $S_{H,f}$ is $\overline{\chi_f}$-H invariant, i.e., for any $h \in H$ we have $\lambda_h(S_{H,f}) = \overline{\chi_f(h)}S_{H,f}$, where λ_h denotes the left translation by h. Indeed, for $\varphi \in \mathscr{S}(G)$

$$\begin{aligned}\langle\lambda_h(S_{H,f}), \varphi\rangle &= \langle S_{H,f}, \lambda_{h^{-1}}\varphi\rangle \\ &= \int_H \varphi(hh')\chi_f(h')dh' = \overline{\chi_f(h)}\int_H \varphi(h')\chi_f(h')dh' = \overline{\chi_f(h)}\langle S_{H,f}, \varphi\rangle.\end{aligned}$$

Let now $H = \exp(\mathfrak{h})$ and $K = \exp(\mathfrak{k})$ be two closed connected subgroups of G such that $\mathfrak{h}$ and $\mathfrak{k}$ are subordinated to some $f \in \mathfrak{g}^*$. We can construct a $\overline{\chi_f}$-invariant distribution $S^H_{K,f}$ on $\mathscr{S}(G/H, f)$ in the following way. Pick a K-invariant measure $d\dot{k}$ on $K/K \cap H$ and let

$$\langle S^H_{K,f}, \xi\rangle = \int_{K/H\cap K} \xi(k)\chi_f(k)d\dot{k}.$$

It follows as above that for all $k \in K$

$$\langle S^H_{K,f}, \lambda_{k^{-1}}\xi\rangle = \overline{\chi_f(k)}\langle S^H_{K,f}, \xi\rangle, \ \xi \in \mathscr{S}(G/H, f), \ k \in K.$$

Let $\phi \in \mathfrak{g}^*$ and let $\mathfrak{b}$ be a polarization at ϕ. Let $B = \exp(\mathfrak{b})$ and let χ_ϕ be the character of B associated to ϕ. The representation $\pi_\phi = \mathrm{ind}_B^G \chi_\phi$ is irreducible and the space $\mathscr{S}(G/H, \phi)$ is the space of the C^∞-vectors of π_ϕ. It was shown in [Chap. 3, Theorem 3.2.6] that there exists a certain affine subspace $\mathscr{V}$ of $\Gamma_f = f + \mathfrak{h}^\perp \subset \mathfrak{g}^*$ such that

$$\tau \simeq \int_{\mathscr{V}}^{\oplus} \pi_\phi d\phi = \tau', \tag{4.2.4}$$

where $d\lambda$ is the Lebesgue measure on $\mathscr{V}$ and π_ϕ is the irreducible representation associated to ϕ ($\phi \in \mathscr{V}$).

The first aim is to disintegrate the distribution $S_{H,f}$ into an integral $\int_{\mathscr{V}} S_{B(\phi),\phi} d\phi$ of pure distributions $S_{B(\phi),\phi}$ of positive type associated with the irreducible representations π_ϕ, where $\mathscr{V}$ is a certain affine subspace of $\mathfrak{g}^*$. In other words, we are going to prove (4.1.2) without taking into account the multiplicities occurring in (4.1.1).

4.2.2 Well-Adapted Bases

As in Chap. 3, we know how the representation $\tau = \mathrm{ind}_H^G \chi_f$ can be smoothly disintegrated into irreducibles. There exists a Zariski-open subset $\mathscr{V}_0$ of $\mathscr{V}$ with the following properties. For every $\phi \in \mathscr{V}_0$, there exists a polarization $B(\phi) = \exp(\mathfrak{b}(\phi))$ at ϕ, a Malcev basis

$$\mathscr{X} = \{X_1(\phi), \ldots, X_l(\phi)\}$$

of $\mathfrak{g}$ relative to $\mathfrak{b}(\phi)$, a Malcev basis

$$\mathscr{Y} = \{Y_1(\phi), \ldots, Y_m(\phi)\}$$

of $\mathfrak{b}(\phi)$ relative to $\mathfrak{h} \cap \mathfrak{b}(\phi)$ and a Malcev basis

$$\mathscr{U} = \{U_1(\phi), \ldots, U_p(\phi)\}$$

of $\mathfrak{h}$ relative to $\mathfrak{h} \cap \mathfrak{b}(\phi)$, such that the mappings

$$\phi \mapsto X_j(\phi); \phi \mapsto Y_j(\phi); \phi \mapsto U_j(\phi)$$

are rational and continuous on $\mathscr{V}_0$ for all j. The projections

$$T_\phi : \mathscr{S}(G/H, f) \to \mathscr{S}(G/B(\phi), \phi)(\phi \in \mathscr{V}_0)$$

given by

$$T_\phi(\xi)(g) = \int_{B(\phi)/H\cap B(\phi)} \xi(gb)\chi_\phi(b) d_{\mathscr{Y}(\phi)}\dot{b}, \xi \in \mathscr{S}(G/H, f), g \in G,$$

allow us to define an operator

$$U : S(G/H, f) \to \int_{\mathscr{V}_0}^{\oplus} \mathscr{H}_\phi d\phi = \mathscr{H}_{\tau'},$$

(where $\mathscr{H}_\phi = L^2(G/B(\phi), \phi)$ denotes the Hilbert space of the irreducible representation π_ϕ) by setting

$$U(\xi)(\phi) = T_\phi(\xi) \in \mathscr{H}_\phi, \phi \in \mathscr{V}_0, \xi \in \mathscr{S}(G/H, f).$$

This mapping U is in fact an isometry for the L^2-norms and extends to a unitary operator from $\mathscr{H}_\tau = L^2(G/H, f)$ onto $\mathscr{H}_{\tau'}$ (cf. [Chap. 3, Theorem 3.2.10]). Let dh be a Haar measure of H. We choose now for any $\phi \in \mathscr{V}_0$ a Malcev basis

$$\mathscr{Z}(\phi) = \{Z_1(\phi), \ldots, Z_r(\phi)\}$$

of $\mathfrak{b}(\phi) \cap \mathfrak{h}$ such that for the Malcev basis $\mathscr{B}(\phi) = \mathscr{U} \cup \mathscr{Z}(\phi)$ of $\mathfrak{h}$ the measure $d_{\mathscr{B}(\phi)}$ is just the given measure dh. Let also $\mathscr{Y}'(\phi) = \mathscr{Z} \cup \mathscr{Y}(\phi)$ be the Malcev basis of $\mathfrak{b}(\phi)$.

We shall use this isometry to prove the Penney-Plancherel theorem.

Theorem 4.2.3 *Let G be a connected and simply connected nilpotent Lie group, H a connected Lie subgroup, and $\chi = \chi_f$ a unitary character on H associated with some $f \in \mathfrak{g}^*$ such that $f_{|\mathfrak{h}}$ is a Lie homomorphism. Let $\mathscr{V}$ (respectively $\mathscr{V}_0$) be the affine subspace of Γ_f (resp. the open dense subset of $\mathscr{V}$ as in Sect. 4.2.2 for $\phi \in \mathscr{V}_0$). With the measure normalizations given by these bases, one has for any $\varphi \in \mathscr{S}(G)$*

$$\langle S_{H,f}, \varphi\rangle = \int_{\mathscr{V}_0} \langle S_\phi, \varphi\rangle d\phi,$$

where S_ϕ denotes the tempered distribution on $S(G)$ defined by

$$\begin{aligned}\langle S_\phi, \varphi\rangle &= \int_{H/B(\phi)\cap H} T_\phi(P_{H,f}(\phi))(h)\chi_f(h)d_{\mathscr{U}(\phi)}\dot{h} \\ &= \langle S_{H,f}^{B(\phi)}, T_\phi(P_{H,f}(\varphi))\rangle, \varphi \in \mathscr{S}(G), \phi \in \mathscr{V}_0.\end{aligned}$$

Proof Let $\phi, \psi \in \mathscr{S}(G)$. We will show that

$$\langle S_{H,f}, \varphi^* * \psi\rangle = \int_{\mathscr{V}_0} \langle S_\phi, \varphi^* * \psi\rangle d\phi. \tag{4.2.5}$$

Since the factorization theorem of Dixmier-Malliavin says that every Schwartz function ρ is of the form $\rho = \varphi^* * \psi$ for some elements φ, ψ in $\mathscr{S}(G)$ (see [51]), the theorem follows from (4.2.5). A standard computation tells us that

$$\begin{aligned}\langle S_{H,f}, \varphi^* * \psi\rangle &= \int_{G/H} (\int_H \overline{\varphi(gh')\chi_f(h')dh'})(\int_H \psi(gh)\chi_f(h)dh)d\dot{g} \\ &= \langle P_{H,f}(\psi), P_{H,f}(\varphi)\rangle_{L^2(G/H,f)},\end{aligned}$$

where $d\dot{g}$ is the G-invariant measure on G/H chosen so that $dg = d\dot{g}dh$.

Let now $\xi = P_{H,f}(\varphi), \eta = P_{H,f}(\psi) \in \mathcal{S}(G/H, f)$. The fact that the map U is an isometry tells us that

$$\int_{\mathcal{V}_0} \langle T_\phi(\eta), T_\phi(\xi)\rangle_{\mathcal{H}_\phi} d\phi = \langle U(\eta), U(\xi)\rangle_{\mathcal{H}_{\tau'}} = \langle \eta, \xi\rangle_{L^2(G/H,f)}.$$

Hence in order to prove the theorem, it suffices to show that for every $\phi \in \mathcal{V}_0$ we have that

$$\langle T_\phi(\eta), T_\phi(\xi)\rangle = \langle S_\phi, \varphi^* * \psi.\rangle \tag{4.2.6}$$

We write $P = P_{H,f}$, $B = B(\phi)$, $\mathcal{X} = \mathcal{X}(\phi)$, $\mathcal{Y} = \mathcal{Y}(\phi)$, $\mathcal{U} = \mathcal{U}(\phi)$, $\mathcal{B} = \mathcal{B}(\phi)$, $\mathcal{Y}' = \mathcal{Y}'(\phi)$, $\mathcal{Z} = \mathcal{Z}(\phi)$. We see that

$$\begin{aligned}
\langle T_\phi(\eta), T_\phi(\xi)\rangle_{\mathcal{H}_\phi} &= \int_{G/B} [(\int_{B/B\cap H} \eta(gb)\chi_\phi(b)d_{\mathcal{Y}}\dot{b})\overline{(\int_{B/B\cap H} \xi(gb)\chi_\phi(b)d_{\mathcal{Y}}\dot{b})}]d_{\mathcal{X}}\dot{g} \\
&= \int_{G/B} [(\int_{B/B\cap H} (\int_H \psi(gbh)\chi_f(h)d_{\mathcal{B}}h)\chi_\phi(b)d_{\mathcal{Y}_{\dot{b}}}) \\
&\quad \times \overline{(\int_{B/B\cap H} (\int_H \varphi(gbh')\chi_f(h')d_{\mathcal{B}}h')\chi_\phi(b)d_{\mathcal{Y}}\dot{b})}]d_{\mathcal{X}}\dot{g}.
\end{aligned} \tag{4.2.7}$$

On the other hand,

$$\begin{aligned}
T_\phi(P(\varphi^* * \psi)(h)) &= \int_{B/B\cap H} (\int_H (\varphi^* * \psi)(hbh')\chi_f(h')d_{\mathcal{B}}h'\chi_\phi(b))d_{\mathcal{Y}}\dot{b} \\
&= \int_{B/B\cap H} (\int_H \int_G \varphi^*(g)\psi(g^{-1}hbh')dg\chi_f(h')d_{\mathcal{B}}h'\chi_\phi(b))d_{\mathcal{Y}}\dot{b} \\
&= \int_{B/B\cap H} (\int_H \int_G \overline{\varphi(gh^{-1})}\psi(gbh')dg\chi_f(h')d_{\mathcal{B}}h'\chi_\phi(b))d_{\mathcal{Y}}\dot{b} \\
&= \int_{B/B\cap H} (\int_G \overline{\varphi(gh^{-1})}\eta(gb)dg\chi_\phi(b))d_{\mathcal{Y}}\dot{b} \\
&= \int_G T_\phi(\eta)(g)\overline{\varphi(gh^{-1})}dg \\
&= \int_{G/B}\int_B \overline{\varphi(gbh^{-1})}T_\phi(\eta)(gb)d_{\mathcal{Y}'}bd_{\mathcal{X}}\dot{g} \\
&= \int_{G/B}\int_B \overline{\varphi(gbh^{-1})}T_\phi(\eta)(g)\chi_\phi(b^{-1})d_{\mathcal{Y}'}bd_{\mathcal{X}}\dot{g}.
\end{aligned}$$

It follows that

$$\begin{aligned}\langle S_\varphi, \varphi^* * \psi\rangle &= \int_{H/B\cap H} T_\phi(P(\varphi^* * \psi))(h)\chi_f(h)d_{\mathcal{U}}\dot{h}\\ &= \int_{H/B\cap H}[\int_{G/B}(\int_B \overline{\varphi(gbh^{-1})T_\eta(g)\chi_\phi(b^{-1})}d_{\mathcal{Y}'}b)d_{\mathcal{X}}\dot{g}]\chi_f(h)d_{\mathcal{U}}\dot{h}\\ &= \int_{H/B\cap H}[\int_{G/B}(\int_B \overline{\varphi(gbh^{-1})\chi_\phi(b)}T_\phi(\eta)(g)d_{\mathcal{Y}'}b)d_{\mathcal{X}}\dot{g}]\chi_f(h)d_{\mathcal{U}}\dot{h}.\end{aligned}$$

The operator $\pi_\phi(\varphi^*)$ is Hilbert-Schmidt, its kernel is the function

$$I(g, g') = \int_B \overline{\varphi(gbg^{-1})\chi_\phi(b)}d_{\mathcal{Y}'}b,$$

and the function $(g, h) \mapsto I(g, h)$ is in $\overline{\mathcal{S}(G/B, \phi)} \otimes \mathcal{S}(H/B\cap H, f)$. Hence using Fubini, we can deduce that

$$\langle S_\phi, \varphi^* * \psi\rangle = \int_{G/B}\int_{H/B\cap H}\int_B \overline{\varphi(gbh^{-1})}T_\phi(\eta)(g)\overline{\chi_\phi(b)}d_{\mathcal{Y}'}b\chi_f(h)d_{\mathcal{U}}\dot{h}d_{\mathcal{X}}\dot{g}. \tag{4.2.8}$$

Now for any $q \in C_c(G)$ we have that

$$\int_{H/B\cap H}\int_B q(b'h^{-1})\chi_f(h)\overline{\chi_\phi(b')})d_{\mathcal{B}}hd_{\mathcal{Y}}\dot{g}' = \int_B\int_{H/H\cap B} q(b'h^{-1})\chi_\phi(h)\overline{\chi_\phi(b')}d_{\mathcal{U}}\dot{h}d_{\mathcal{Y}'b'}. \tag{4.2.9}$$

Indeed,

$$\begin{aligned}\int_{H/B\cap H}\int_H q(b'h^{-1})\chi_\phi(h)\overline{\chi_\phi(b')})d_{\mathcal{B}}hd_{\mathcal{Y}}\dot{b}' &= \int_{\mathbb{R}^m}\int_{\mathbb{R}^{r+q}} q(E_{\mathcal{Y}}(T)(E_{\mathcal{B}}(S))^{-1})\\ &\quad\times \overline{\chi_\phi(E_{\mathcal{Y}}(T))}\chi_\phi(E_{\mathcal{B}}(S))dSdT\\ &= \int_{\mathbb{R}^m}\int_{\mathbb{R}^p}\int_{\mathbb{R}^r} q(E_{\mathcal{Y}}(T))(E_{\mathcal{U}}(S)E_{\mathcal{X}}(R))^{-1}\overline{\chi_\phi(E_{\mathcal{Y}}(T))}\chi_\phi(E_{\mathcal{U}}(S)E_{\mathcal{X}}(R))dRdSdT\\ &= \int_{\mathbb{R}^m}\int_{\mathbb{R}^r}\int_{\mathbb{R}^p} q(E_{\mathcal{Y}}(T))(E_{\mathcal{X}}(R)^{-1}(E_{\mathcal{U}}(S))^{-1})\overline{\chi_\phi(E_{\mathcal{Y}}(T))}\chi_\phi(E_{\mathcal{U}}(S)E_{\mathcal{X}}(R))dSdRdT\\ &= \int_{\mathbb{R}^{m+r}}\int_{\mathbb{R}^p} q(E_{\mathcal{Y}'}(T))(E_{\mathcal{U}}(S)^{-1})\overline{\chi_\phi(E_{\mathcal{Y}'}(T))}\chi_\phi(E_{\mathcal{U}}(S))dSdT\\ &= \int_{H/H\cap B}\int_B q(b'h^{-1})\chi_\phi(h)\overline{\chi_\phi(b')}d_{\mathcal{U}}\dot{h}d_{\mathcal{Y}'}b'.\end{aligned}$$

Hence by (4.2.8) and (4.2.9), we have that

$$\begin{aligned}\langle S_\phi, \varphi^* * \psi\rangle &= \int_{G/B}\Big(\int_{H/B\cap H}\int_B \overline{\varphi(gbh^{-1})}T_\phi(\eta)(g)\overline{\chi_\phi(b)}d_{\mathscr{Y}'}b\chi_f(h)d_{\mathscr{U}}\dot{h}\Big)d_{\mathscr{X}}\dot{g}\\ &= \int_{G/B} T_\phi(\eta)(g)\Big(\int_{B/B\cap H}\int_H \overline{\varphi(gbh^{-1})}\chi_f(h)d_{\mathscr{B}}h\overline{\chi_\phi(b)}d_{\mathscr{Y}}\dot{b}\Big)d_{\mathscr{X}}\dot{g}\\ &= \int_{G/B} T_\phi(\eta)(g)\overline{T_\phi(\xi)(g)}d_{\mathscr{X}(\phi)}\dot{g}\\ &= \langle T_\phi(\eta), T_\phi(\xi)\rangle_{\mathscr{H}_\phi}.\end{aligned}$$

□

Corollary 4.2.4 *We keep the same hypotheses and notations as above. Let $\phi \in \mathscr{V}_0$ and $\psi \in \mathscr{S}(G/B(\phi), \phi)$. Let*

$$\beta_\phi(\psi) = \overline{\langle S_{H,f}^{B(\phi)}, \psi\rangle} = \int_{H/B(\phi)\cap H} \overline{\psi(h)\chi_f(h)}d_{\mathscr{U}}\dot{h}. \tag{4.2.10}$$

Then for all $\omega \in \mathscr{D}(G)$

$$\langle S_{H,f}, \omega\rangle = \langle \tau(\omega)\alpha_\tau, \alpha_\tau\rangle = \int_{\mathscr{V}} \langle \pi_\phi(\omega)\beta_\phi, \beta_\phi\rangle d\phi,$$

where α_τ is the canonical cyclic generalized vector for τ, i.e., $\alpha_r(\xi) = \overline{\xi(e)}$, $\xi \in \mathscr{S}(G/H, f)$.

Proof It is not difficult to see that $\langle S_{H,f}, \omega\rangle = \langle \tau(\omega)\alpha_\tau, \alpha_\tau\rangle$ (see [62, 71]). On the other hand the following computation from [71, page 177] tells us that for $\phi \in \mathscr{V}_0$ we have

$$\langle \pi_\phi(\omega)\beta_\phi, \beta_\phi\rangle = \int_{H/B(\phi)\cap H} T_\phi(P_{H,f}(\omega))(h)\chi_f(h)d_{\mathscr{U}(\phi)}\dot{h} = \langle S_{H,f}^{B(\phi)}, T_\phi(P_{H,f}(\omega))\rangle$$

for all $\omega \in \mathscr{D}(G)$. and then Theorem 4.2.3 allows to conclude. □

4.3 The Plancherel-Bonnet Formula for Normal Inducing Subgroups of Exponential Solvable Lie Groups

In this section we take, as above, a closed connected normal subgroup $H = \exp \mathfrak{h}$ of an exponential solvable group $G = \exp \mathfrak{g}$, a unitary character $\chi = \chi_f : \chi(h) = e^{-2\pi i\langle f, \log(h)\rangle}$ of H (where $f \in \mathfrak{g}^*$ is such that $\langle f, [\mathfrak{h}, \mathfrak{h}]\rangle = (0)$) and we consider, as in (4.2.3), the finite-order positive distribution

$$\langle S_{H,f}, \varphi \rangle = \int_H \varphi(h)\chi_f(h)dh.$$

To describe the measure ν in Formula (4.1.1), we use the orbit picture of the dual space $\widehat{G}$ of G. The coadjoint representation Ad^* of G on $\mathfrak{g}^*$ is defined by

$$\langle s \cdot \phi, x \rangle := \langle Ad^*(s)\phi, x \rangle := \langle \phi, Ad(s^{-1})x \rangle, \; s \in G, \; \phi \in \mathfrak{g}^*, \; x \in \mathfrak{g},$$

and $\mathfrak{g}^*/G$ denotes the space of coadjoint orbits. Let us recall that every irreducible representation π of G is equivalent to a representation $\pi_{\phi,B(\phi)}$ induced by a character χ_ϕ of a polarization satisfying Pukanszky's condition, and the equivalence class $[\pi_{\phi,B(\phi)}]$ of $\pi_{\phi,B(\phi)}$ depends only on the coadjoint orbit of ϕ.

Let H^f be the stabilizer in G of the restriction $f_{|\mathfrak{h}}$ of f to $\mathfrak{h}$. There exists a locally closed cross-section Σ of the H^f-orbits in Γ_f, and it turns out that the measure ν in Bonnet's formula is supported on the subset $Ad^*(H^f)\Sigma$ of $\mathfrak{g}^*$. We choose for every $\sigma \in \Sigma$ a Vergne polarization $C(\sigma)$ and we take the point evaluations $\delta_{s,\sigma}$, $s \in H^f$, on the space of the smooth vectors $\mathscr{H}_\sigma^\infty$ of π_σ. We show that the operators $U_{\pi_\sigma} = U_\sigma$ in Bonnet's formula are integrals of the rank-one operators $P_{s,\sigma} = \Delta_{G/A}(s)P_{\delta_{s,\sigma},\delta_{s,\sigma}}$ (where $\mathfrak{a} = \mathfrak{h} \cap \ker(f)$, $A = \exp \mathfrak{a}$ and the function $\Delta_{G/A}$ will be defined in Sect. 4.3.5):

$$U_\sigma = \oint_{H^f/C(\sigma)} P_{s,\sigma} ds.$$

We then describe the distribution $S_{H,f}$ in a different way and show that there exists a certain affine subspace V of Γ_f such that

$$S_{H,f} = \int_V S_{B(\phi),\phi} d\phi.$$

This allows to recover a disintegration formula of the representation

$$\tau_f =\simeq \int_V^\oplus \pi_\phi d\phi,$$

which was obtained in [Chap. 3, Theorem 3.2.6] by other means.

4.3.1 G-equivariant Projections

Let us begin by introducing some basic notions and formulas. Let $G = \exp \mathfrak{g}$ be an exponential solvable Lie group with Lie algebra $\mathfrak{g}$. Let dg be a left Haar measure of G and Δ_G the modular function of G. The Fourier transform $\hat{\psi} : \mathfrak{g}^* \to \mathbb{C}$ of an L^1-function $\psi : \mathfrak{g} \to \mathbb{C}$ is defined by:

$$\hat{\psi}(\phi) = \int_{\mathfrak{g}} \psi(x) e^{-2\pi i \langle \phi, x \rangle} dx.$$

For $s \in G$ we then have

$$\widehat{\psi \circ \mathrm{Ad}(s)} = \Delta_G(s) \hat{\psi} \circ \mathrm{Ad}^*(s) \tag{4.3.1}$$

Let K be a closed connected subgroup of G. For two functions $\varphi, \psi \in C_c(G)$, let

$$\varphi * \psi(x) = \int_G \varphi(y)\psi(y^{-1}x)dx, \ \varphi^*(x) = \Delta_G(x)^{-1}\overline{\varphi(x^{-1})}, \ x \in G.$$

Let now $K = \exp \mathfrak{k}$ be a closed subgroup of G and $\phi \in \mathfrak{g}^*$, such that $\langle \phi, [\mathfrak{k}, \mathfrak{k}] \rangle = (0)$. Let $\chi_\phi = \chi = e^{-2\pi i \phi \circ \log}$ be the corresponding unitary character of K and $\tau_\phi = \tau_\chi$ the monomial representation of G induced from χ. Define the projection T_ϕ of τ_ϕ from $C_c(G)$ to the Hilbert space $\mathcal{H}_\phi = \mathcal{H}_{\tau_\phi}$—actually to $\mathcal{E}(G, K, \chi_\phi)$—by

$$T_\phi(\varphi)(x) = \int_K \varphi(xk)\chi_\phi(k)\Delta_{K,G}^{-1/2}(k)dk, \ \varphi \in C_c(G), \ x \in G. \tag{4.3.2}$$

This projection is G-equivariant, and we have:

$$\tau_\phi(\psi)T_\phi(\varphi) = T_\phi(\psi * \varphi), \ \psi, \varphi \in C_c(G) \tag{4.3.3}$$

For $\varphi \in \mathcal{C}_c(G)$ the operator $\tau_\phi(\varphi) = \int_G \varphi(x)\tau_\phi(x)dx$ has a continuous kernel $K_{\tau_\phi}(\varphi)$ which is given by:

$$K_{\tau_\phi(\varphi)}(x, y) = \Delta_G(y^{-1}) \int_K \varphi(xby^{-1})\chi_\phi(b)\Delta_{K,G}^{-1/2}(b)db, \ x, y \in G. \tag{4.3.4}$$

We can also obtain the representation τ_ϕ through the positive distribution $S_{K,\phi} = S_\phi$, where

$$\langle S_\phi, \varphi \rangle = \int_K \varphi(k)\Delta_{G,K}^{-1/2}(k)\chi_\phi(k)dk = K_{\tau_\phi(\varphi)}(e, e), \ \varphi \in \mathcal{C}_c(G). \tag{4.3.5}$$

The Hilbert space $\mathcal{H}_\phi$ can be realized as the completion of $C_c(G)$ for the semi-norm

$$\|\varphi\|^2 = \langle S_\phi, \varphi^* * \varphi \rangle, \ \varphi \in C_c(G). \tag{4.3.6}$$

Weyl's composition formula tells us that

$$\int_G \psi(x)dx = \oint_{G/K}\int_K \psi(xk)\Delta_{G,K}(k)dkdx,\ \psi \in C_c(G).$$

Hence for $\varphi, \psi \in C_c(G)$, it follows that

$$\langle S_\phi, \psi^* * \varphi\rangle = \int_G T_\phi(\varphi)(x)\overline{\psi(x)}dx = \oint_{G/K} T_\phi(\varphi)(x)\overline{T_\phi(\psi)(x)}dx = \langle T_\phi(\varphi), T_\phi(\psi)\rangle_{\mathcal{H}_\phi}. \tag{4.3.7}$$

If K is a polarization at ϕ which has the Pukanszky property, then τ_ϕ is irreducible. We shall write in this case π_ϕ instead of τ_ϕ.

4.3.2 Sobolev Spaces

Let G be a connected Lie group. Denote by $\mathcal{D}(G)$ the space of test functions, that is complex-valued C^∞-functions with compact support on G. Let $Z_1, \ldots, Z_n \in \mathfrak{g}$ be a basis of $\mathfrak{g}$, and put

$$L := \sum_{j=1}^n Z_j^2 \in U(\mathfrak{g}),$$

where $U(\mathfrak{g})$ is the enveloping algebra of $\mathfrak{g}$. Since $(1-L)^N$ is hypo-elliptic for every $N \in \mathbb{N}^*$, there exists a local fundamental solution $E_N \in \mathcal{D}'(U)$ of $(1-L)^N$ on a neighbourhood U of $e \in G$, i.e.,

$$(1-L)^N E_N = \delta_e \text{ on } U.$$

We can take

$$E_N = \frac{1}{(N-1)!}\int_0^\infty t^{N-1}e^{-t}p_t dt \in L^1(G),$$

where $(p_t)_{t\in\mathbb{R}_+^*} \subset L^1(G)$ denotes the heat kernel associated to L. Since $(1-L)^N$ is hypo-elliptic, we have that E_N is C^∞ on $G \setminus \{e\}$, and if N is big enough, the function $E_N \in L^1(G)$ is even continuous in e. We shall fix this integer N in the sequel.

Choose $\beta \in \mathcal{D}(U)$, $\beta \equiv 1$ near e, and put $F := \beta E_N$ considered as a distribution on G. Then F is a continuous function with compact support and

$$(1-L)^N F = \delta_e - R,$$

where $R \in \mathscr{D}(G)$. This implies

$$F^* * (1 - L)^N = \delta_e - R^*, \tag{4.3.8}$$

and

$$\pi(F^*) \circ \pi((1 - L)^N) = \mathbb{I}_{\mathscr{H}_\pi} - \pi(R^*), \tag{4.3.9}$$

for any unitary representation $(\pi, \mathscr{H}_\pi)$ of G.

For $M \in \mathbb{N}^*$, the operator $A_\pi^M = A^M = (\mathbb{I}_{\mathscr{H}_\pi} - d\pi(L))^M$ is positive and invertible. We define the $2M$-th Sobolev space $\mathscr{H}_\pi^M$ of π as the image of the bounded operator $A^{-M} = \pi(E_M)$, and we equip it with the Sobolev norm $\|\xi\|_M = \|A^M \xi\|_{\mathscr{H}_\pi}$. Hence $\mathscr{H}_\pi^M$ is a Hilbert space and $A^M : \mathscr{H}_\pi^M \to \mathscr{H}_\pi$ is a unitary operator. In particular, if we choose an orthonormal basis $\mathscr{B} = (\xi_i)_{i \in \mathbb{N}}$ of $\mathscr{H}_\pi^M$, then $A^M \mathscr{B} = (A^M \xi_i)_{i \in \mathbb{N}}$ is an orthonormal basis of $\mathscr{H}_\pi$ and for $\varphi, \psi \in \mathscr{H}_\pi^M$ we have that

$$\langle \varphi, \psi \rangle_M = \langle A^M \varphi, A^M \psi \rangle_{\mathscr{H}_\pi}.$$

Furthermore an operator u on $\mathscr{H}_\pi^M$ is trace-class if and only if $v = A^M \circ u \circ A^{-M}$ is trace-class on $\mathscr{H}_\pi$, and we have that

$$\operatorname{tr}(u) = \operatorname{tr}(v), \tag{4.3.10}$$

where $\operatorname{tr}(u) = \sum_{i \in \mathbb{N}} \langle u\xi_i, \xi_i \rangle_M$ denotes the ordinary trace of u.

We denote the anti-dual space of $\mathscr{H}_\pi^M$ by $\mathscr{H}_\pi^{-M}$ and we write $\langle \nu, \xi \rangle = \overline{\langle \xi, \nu \rangle}$, $\nu \in \mathscr{H}_\pi^{-M}$, $\xi \in \mathscr{H}_\pi^M$, to express the duality between $\mathscr{H}_\pi^M$ and $\mathscr{H}_\pi^{-M}$.

A bounded operator u acts on $\mathscr{H}_\pi^{-M}$ in the following way:

$$\langle u(\nu), \xi \rangle = \langle \nu, u^* \xi \rangle, \ \nu \in \mathscr{H}_\pi^{-M}, \ \xi \in \mathscr{H}_\pi^M.$$

Since for $\xi \in \mathscr{H}_\pi^\infty$ and $\varphi \in C_c^\infty(G)$ we have that

$$\varphi \langle \pi(\varphi)\nu, A^{M'} \xi \rangle \leq \|\nu\|\text{op}\|\pi(\varphi^*) A^{M'} \xi\|_M \leq \|\nu\|\text{op}\|(1-L)^M * \varphi^* * (1-L)^{M'}\|_1 \|\xi\|_{\mathscr{H}_\pi}$$

for any $M' \in \mathbb{N}$, it follows that $\pi(\varphi)\nu \in \mathscr{H}_\pi^\infty$.

Let u be now a bounded linear operator of rank one

$$u = P_{\mu,\nu} : \ P_{\mu,\nu}(\xi) = \langle \xi, \nu \rangle \mu, \ \xi \in \mathscr{H}_\pi^M, \tag{4.3.11}$$

from $\mathscr{H}_\pi^M$ to $\mathscr{H}_\pi^{-M}$ (for some $\mu, \nu \in \mathscr{H}_\pi^{-M}$). Then $v_L = \pi(\varphi) \circ u = P_{\pi(\varphi)\mu,\nu}$, $\varphi \in \mathscr{D}(G)$, is a bounded operator from $\mathscr{H}_\pi^L$ to $\mathscr{H}_\pi^L$ for any $L \geq M$, $L \in \mathbb{N}$, and it is easy to check that

$$\operatorname{tr}(v_L) = \langle \pi(\varphi)\mu, \nu \rangle = \operatorname{tr}(v_M) \tag{4.3.12}$$

is independent of L.

For a bounded linear operator $v : \mathcal{H}_\pi^M \to \mathcal{H}_\pi^{-M}$ we have that $A^{-2M} \circ v : \mathcal{H}_\pi^M \to \mathcal{H}_\pi^M$ and we say that v is trace-class if $A^{-2M} \circ v$ is so. In particular for $\varphi \in \mathcal{D}(G)$, it follows that

$$\pi(\varphi) \circ v = \pi(\varphi * (1 - L)^{2M}) \circ A^{-2M} \circ v$$

and so

$$|\mathrm{tr}(\pi(\varphi) \circ v)| \leq \|\varphi * (1 - L)^{2M}\|_1 \|A^{-2M} \circ v\|_1,$$

where $\|w\|_1$ denotes the trace of the absolute value of an operator w.

4.3.3 Sobolev Spaces and Monomial Representations

Let α_ϕ be the generalized cyclic vector for τ. This anti-distribution on the Hilbert space $\mathcal{H}_\phi$ of τ_ϕ is defined by

$$\langle \alpha_\phi, \xi \rangle = \overline{\xi(e)}, \ \xi \in \mathcal{H}_\phi^\infty.$$

More generally, for $s \in G$, let

$$\langle \mathfrak{d}_{s,\phi}, \xi \rangle = \overline{\xi(s)}, \ \xi \in \mathcal{H}_\phi^\infty. \tag{4.3.13}$$

We shall see now that the $\mathfrak{d}_{s,\phi}$ are in $\mathcal{H}_\phi^{-N}$. For $\theta \in \mathcal{H}_\tau$, $s \in G$ and a continuous function with compact support φ on G, we have:

$$\tau(\varphi)(\theta)(s) = \oint_{G/K} K_{\tau(\varphi)}(s, x)\theta(x)dx.$$

As in (4.3.4), $K_{\tau(\varphi)}$ denotes the kernel function of the operator $\tau(\varphi)$, i.e.,

$$K_{\tau(\varphi)}(x, s) = \int_K \Delta_G(s^{-1})\varphi(xks^{-1})\Delta_{K,G}^{-1/2}(k)\chi_\phi(k)dk, \ x, s \in G \tag{4.3.14}$$

Let, for $s \in G$,

$$\varphi_{s,\phi}(x) = K_{\tau(\varphi)}(x, s) = \Delta_G(s^{-1}) \int_K \varphi(xks^{-1})\Delta_{K,G}^{-1/2}(k)\chi(k)dk, \ x \in G. \tag{4.3.15}$$

Then $\varphi_{s,\phi}$ is in $\mathscr{E}(G, H, \chi_\phi)$ and hence it is an element of the Hilbert space $\mathscr{H}_\phi$. Since $\tau(\varphi)^* = \tau(\varphi^*)$, we see that for any $\theta \in \mathscr{H}_\phi$:

$$\langle \tau(\varphi)\delta_{s,\phi}, \theta\rangle = \overline{\tau(\varphi^*)(\theta)(s)} = \overline{\oint_{G/K} \overline{K_{\tau(\varphi)}(x, s)}\theta(x)dx} = \langle \varphi_{s,\phi}, \theta\rangle.$$

In particular this tells us that

$$\tau(\varphi)\mathfrak{d}_{s,\phi} = \varphi_{s,\phi}, \ \tau(\varphi)\alpha_\tau = \varphi_{e,\phi}. \tag{4.3.16}$$

If we write, as in (4.3.8), $(1-L)^N * F = \delta_e - R$, where $F \in C_c(G)$ and $R \in C_c^\infty(G)$, then we see that

$$A_\phi^N \circ \tau_\phi(F) + \tau_\phi(R) = \mathbb{I}_{\mathscr{H}_\phi}$$

and so

$$\begin{aligned}\mathfrak{d}_{s,\phi} &= A_\phi^N \circ \tau_\phi(F)\mathfrak{d}_{s,\phi} + \tau_\phi(R)\mathfrak{d}_{s,\phi}\\ &= A_\phi^N F_{s,\phi} + R_{s,\phi}\\ &= A_\phi^N (F_{s,\phi} + A_\phi^{-N} R_{s,\phi}).\end{aligned}$$

Let us put

$$\xi_{s,\phi} = F_{s,\phi} + A_\phi^{-N} R_{s,\phi} \in \mathscr{H}_\phi.$$

Then

$$\mathfrak{d}_{s,\phi} = A_\phi^N \xi_{s,\phi}, \ s \in G, \tag{4.3.17}$$

and the distributions $\mathfrak{d}_{s,\phi}, \alpha_\tau$ are contained in the dual space $\mathscr{H}_\phi^{-N}$ of the Sobolev space $\mathscr{H}_\phi^N$. We shall see next that if we take for K a Vergne polarization $B(\phi)$ at ϕ, then the integral $A_\phi^{-N} \circ U_\phi \circ A_\phi^{-N} = \oint_{H^f/B(\phi)} A_\phi^{-N} \circ P_{s,\phi} \circ A_\phi^{-N} ds$ converges in the trace norm $\| \|_1$. Hence by (4.3.12), for $\varphi \in \mathscr{D}(G)$,

$$\operatorname{tr}(\pi_\phi(\varphi) \circ U_\phi) = \operatorname{tr}(\pi_\phi(\varphi) \circ \oint_{H^f/B(\phi)} P_{s,\phi} ds) = \oint_{H^f/B(\phi)} \operatorname{tr}(\pi_\phi(\varphi) \circ P_{s,\phi}) ds \tag{4.3.18}$$

does not depend on the particular Sobolev space $\mathscr{H}_\pi^M$, $M \geq N$.

4.3.4 Polarizations

Let $H = \exp \mathfrak{h}$ be a normal subgroup of the exponential solvable Lie group $G = \exp \mathfrak{g}$. Let $\chi = \chi_f$ be a unitary character of H associated with some $f \in \mathfrak{g}^*$ such that $\langle f, [\mathfrak{h}, \mathfrak{h}]\rangle = 0$. Let $S = (\mathfrak{a}_i)_{i=1}^n$ be a good decreasing sequence for $\mathfrak{g}$ adapted to $\mathfrak{h}$ (i.e., $\mathfrak{a}_p = \mathfrak{h}$ for some $1 \leq p \leq n$). From this good sequence we can extract a Malcev basis $\mathscr{Z} = \{Z_1, ..., Z_n\}$ of $\mathfrak{g}$ and we denote by $\mathscr{X} := \{X_k, ..., X_1\}$ the coexponential basis of $\mathfrak{h}$ in $\mathfrak{g}$, as above). For $j = 1, \ldots, n$ and $\phi \in \Gamma_f$, we put $\phi_j = \phi_{|\mathfrak{a}_j}$. Let

$$d_j(\phi) = \dim \mathfrak{a}_j \cdot \phi_j = \dim A_j \cdot \phi_j \text{ (where } A_j = \exp \mathfrak{a}_j).$$

Let $\mathfrak{b}(\phi) = \sum_{j=1}^n \mathfrak{a}_j(\phi_j)$ be the Vergne polarization associated to $\mathscr{Z}$ and $B(\phi) = \exp(\mathfrak{b}(\phi))$, where $\mathfrak{a}(\phi_{|\mathfrak{a}})$ denotes the stabilizer of $\phi_{|\mathfrak{a}}$ inside the subalgebra $\mathfrak{a}$. The index set

$$L^H(\phi) = \{j \in \{n, ..., 2\},\ d_j(\phi) = d_{j-1}(\phi)\}$$

tells us how to find a basis of $\mathfrak{b}(\phi)$. Indeed, for every $j \in L^H(\phi)$, we find an element

$$Y_j(\phi) = Z_j + \sum_{i=j-1}^{1} a_i(\phi, j) Z_i \in \mathfrak{a}_j(\phi_j)$$

(since $d_j(\phi) = d_{j-1}(\phi)$ means that $\mathfrak{a}_j(\phi_j) \not\subset \mathfrak{a}_{j-1}$) and so $\mathscr{Y}(\phi) = \{Y_j(\phi),\ j \in L^H(\phi)\}$ is a basis of $\mathfrak{b}(\phi)$. Since these vectors $Y_j(\phi)$ are solutions to linear systems whose coefficients are of the form $\langle \phi, [Z_k, Z_l]\rangle$, it is easy to see that there exists a finite partition $\Gamma_f = \cup_{r=0}^l \Gamma_f^r$ into locally closed subsets such that on each subset the dimension of $\mathfrak{b}(\phi)$ is constant and the maps $\phi \mapsto Y_j(\phi)$ are all continuous. In particular we can assume that the first set Γ_f^0 is Zariski open in Γ_f. Since Vergne polarizations have the Pukanszky property, we have that

$$B(\phi) \cdot \phi = \phi + \mathfrak{b}(\phi)^\perp,\ \phi \in \Gamma_f.$$

We can write the Lebesgue measure on the affine subspace $\phi + \mathfrak{b}(\phi)^\perp$ in the following way:

Proposition 4.3.1 (cf. [24, 55]) *For every $\phi \in \mathfrak{g}^*$ and for any polarization $B = \exp \mathfrak{b}$ at ϕ which satisfies Pukanszky's condition, we have*

$$\int_{\phi+\mathfrak{b}^\perp} \psi(\phi + l) dl = \oint_{B/G(\phi)} \Delta_{G,B(\phi)}(b)\psi(b \cdot \phi) db,\ \psi \in C_c(\phi + \mathfrak{b}^\perp).$$

Let now

$$d_j = \max_{\phi \in \Gamma_f}, \ d_j(\phi), \ j = 1, \dots, n.$$

Define the index set

$$L^H = \{j \in \{n, ..., 1\}, d_j = d_{j-1}\}$$

and let

$$V = \{f + \sum_{i \in L^H, i > p} t_i Z_i^*, t_i \in \mathbb{R}\}, \tag{4.3.19}$$

which is an affine subspace of Γ_f. The subset

$$\Gamma_f^1 = \{\phi \in \Gamma_f; \ d_j(\phi) = d_j \ \forall j \in L^H\}$$

is Zariski open in Γ_f.

4.3.5 Decomposition of Measures

Let $\mathfrak{h}^f = \{X \in \mathfrak{g} : \ f([X, \mathfrak{h}]) = (0)\}$ be the stabilizer in $\mathfrak{g}$ of $f|_{\mathfrak{h}}$. Then $H^f = \exp \mathfrak{h}^f = \{g \in G; Ad^*(g) f|_{\mathfrak{h}} = f|_{\mathfrak{h}}\}$. We put $\mathfrak{a} = \ker(f) \cap \mathfrak{h}$, and $A = \exp \mathfrak{a}$. If f is zero on $\mathfrak{h}$ then $A = H$; if it is not zero, then H/A is central in H^f/A and of dimension 1. We define the function $i_A : H^f \to \mathbb{R}_+^*$ by the relation

$$\int_A F(sxs^{-1}) dx = i_A(s) \int_A F(x) dx \quad (\forall s \in H^f), \tag{4.3.20}$$

where F is in $C_c(A)$. This function i_A is then a continuous homomorphism of H^f and $i_A(\exp(x)) = e^{-\mathrm{tr}(ad_{\mathfrak{a}}(x))}$, $x \in \mathfrak{a}$. For $s \in H^f$, we have

$$\int_{\Gamma_f} \beta(s \cdot \phi) d\phi = \Delta_{G/H}(s^{-1}) \int_{f + \mathfrak{h}^\perp} \beta(\phi) d\phi. \tag{4.3.21}$$

Since H/A is central in H^f/A it follows that

$$\Delta_{G/H}(s) = \Delta_G(s) i_A(s^{-1}) =: \Delta_{G/A}(s), \ s \in H^f.$$

Let $\mathfrak{b}$ be a polarization at $\phi \in \mathfrak{g}^*$. The skew-symmetric form

$$B_\phi(u + \mathfrak{b}, v + \mathfrak{g}(\phi)) = \langle \phi, [u, v] \rangle, \ u \in \mathfrak{g}, \ v \in \mathfrak{b},$$

is a pairing between $\mathfrak{g}/\mathfrak{b}$ and $\mathfrak{b}/\mathfrak{g}(\phi)$, and so

$$\mathrm{tr}(ad_{\mathfrak{g}/\mathfrak{b}})(s) = -\mathrm{tr}(ad_{\mathfrak{b}/\mathfrak{g}(\phi)}(s)),\ s \in \mathfrak{g}(\phi).$$

Hence

$$\Delta_G(s) = \Delta_{G(\phi)}(s),\ s \in G(\phi).$$

Since the linear functional $s \mapsto \mathrm{tr}(ad_{\mathfrak{h}^f/\mathfrak{a}}(s))$ on $\mathfrak{h}^f$ is a linear combination of the real parts of the roots of $\mathfrak{h}$, we can find a linear extension t of this functional to $\mathfrak{g}$ which is a character. In particular the kernel $\mathfrak{t}$ of t is an ideal of codimension 1 or 0 of $\mathfrak{g}$ (which contains $\mathfrak{h}$ since it is central in $\mathfrak{h}^f$ modulo $\mathfrak{a}$).

Now, let $(\mathfrak{a}_i)$ be a good sequence for $\mathfrak{g}$ passing through $\mathfrak{h}$ and $\mathfrak{t}$, and let $\mathfrak{b}_i = \mathfrak{a}_i \cap \mathfrak{h}^f$, $i = n, \ldots, 1$. It is easy to check that the two Vergne polarizations $\mathfrak{b}(\phi)$, $\mathfrak{b}(\phi)'$, respectively associated to $(\phi,\ (\mathfrak{a}_i))$ and $(\phi_{|\mathfrak{h}},\ (\mathfrak{b}_i)_{i=1}^n)$ actually coincide, since both must be contained in $\mathfrak{h}^f$. It is well known that for any solvable Lie algebra $\mathfrak{k}$ and every $\phi \in \mathfrak{k}^*$ whose coadjoint orbit has maximal dimension the stabilizer $\mathfrak{k}(\phi)$ is nilpotent (see for instance [55]). We choose now $\mathfrak{k} = \mathfrak{h}^f/\mathfrak{a}$ and we let

$$m_{\mathfrak{k}} = \min_{\tilde{\phi} \in \mathfrak{k}^*} \dim \mathfrak{k}(\tilde{\phi}).$$

For any $\phi \in \Gamma_f$, there exists a unique $\tilde{\phi} \in \mathfrak{k}^*$ such that $\phi_{|\mathfrak{h}^f} = \tilde{\phi} \circ p_{\mathfrak{a}}$, where $p_{\mathfrak{a}} : \mathfrak{h}^f \to \mathfrak{k}$ denotes the canonical projection. Since $\mathfrak{h}/\mathfrak{a}$ is central in $\mathfrak{k}$, the subset $\{\tilde{\phi} \in \tilde{f} + (\mathfrak{h}/\mathfrak{a})^{\perp},\ \dim \mathfrak{k}(\tilde{\phi}) = d_{\mathfrak{k}}\}$ of $\tilde{f} + (\mathfrak{h}/\mathfrak{a})^{\perp}$ is Zariski open. Hence the subset of Γ_f of all ϕ such that $\mathfrak{g}(\phi)$ is nilpotent is Zariski dense. Therefore, since $\mathfrak{a} \subset \mathfrak{g}(\phi)$ and since $\mathfrak{g}(\phi)/\mathfrak{a}$ is contained in $\mathfrak{k}(\tilde{\phi})$, we see that $\mathrm{tr}(\mathrm{ad}_{\mathfrak{h}^f/\mathfrak{a}}(x)) = 0$ and $\mathrm{tr}(\mathrm{ad}_{\mathfrak{g}/\mathfrak{a}}(x)) = \mathrm{tr}(\mathrm{ad}_{\mathfrak{g}(f)/\mathfrak{a}}(x)) = 0$ for almost all $\phi \in \Gamma_f$, for all $x \in \mathfrak{g}(\phi)$. But this implies that

$$\Delta_{G/A}(s) = 1,\ s \in G(\phi),\ \mathfrak{b}(\phi) \subset \mathfrak{t} \cap \mathfrak{h}^f\ \Rightarrow B(\phi) \subset \ker(\Delta_{H^f/A}), \tag{4.3.22}$$

which means that $\Delta_{G/A}(b)\Delta_{H^f,B(\phi)}(b) = \Delta_{G,B(\phi)}(b),\ b \in B(\phi)$, for all those ϕ.

We can now define a measurable function $m : f + \mathfrak{h}^{\perp} \to \mathbb{R}_+^*$, such that $m(Ad^*(s)\phi) = \Delta_{G/A}(s^{-1})m(\phi)$ for almost all ϕ and all $s \in H^f$. Indeed, by Currey [45], we have in $f + \mathfrak{h}^{\perp}$ a locally closed cross-section Σ of the H^f-orbits in $f + \mathfrak{h}^{\perp}$. So, for $\sigma \in \Sigma \cap \Gamma_f^0$ and $s \in H^f$, we can define

$$m(s \cdot \sigma) = \Delta_{G/A}(s^{-1}). \tag{4.3.23}$$

Then for $\phi = s \cdot \sigma \in \Gamma_f^0$ and for $t \in H^f$, we have

$$m(t \cdot \phi) = m(ts \cdot \sigma) = \Delta_{G/A}((ts)^{-1}) = \Delta_{G/A}(t^{-1})m(\phi).$$

Since, for almost all $\phi \in \Gamma_f$, the stabilizer $\mathfrak{s}_\phi$ of $\phi_{|\mathfrak{h}^f}$ in $\mathfrak{h}^f$ is nilpotent modulo $\mathfrak{a}$ and since $\mathfrak{a} \subset \mathfrak{g}(\phi)$ is contained in this stabilizer, it follows for $s \in \mathfrak{g}(\phi)$ that

$$\mathrm{tr}(ad_{\mathfrak{h}^f}(s)) = \mathrm{tr}(ad_{\mathfrak{s}_\phi}(s)) = \mathrm{tr}(ad_{\mathfrak{s}_\phi/\mathfrak{g}(\phi)}(s)) + \mathrm{tr}(ad_{\mathfrak{g}(\phi)}(s)) = \mathrm{tr}(ad_{\mathfrak{g}(\phi)}(s)).$$

This implies that

$$\Delta_{H^f,G(\phi)}(s) = 1, \ s \in G(\phi). \tag{4.3.24}$$

In particular the homogeneous space $H^f/G(\phi)$ and the orbit space $H^f \cdot \phi \subset \mathfrak{g}^*$ admit unique H^f-invariant Borel measures.

We can now disintegrate the Lebesgue measure on Γ_f:

Proposition 4.3.2 (cf. [121, 122]) *There exists a Borel measure ν on Σ such that*

$$\int_{\Gamma_f} m(l)dl = \int_\Sigma \int_{H^f\cdot\sigma} d\mu_\sigma(l)d\nu(\sigma),$$

where μ_σ is the invariant measure on the H^f-orbit $H^f \cdot \sigma$, for any $\sigma \in \Sigma$.

In particular, it follows from this decomposition of the Lebesgue measure dl on Γ_f, that for almost all $\sigma \in \Sigma$,

$$\int_{H^f/G(\sigma)} \Delta_{G/A}(s)\frac{1}{(1+\|s\{\sigma,\cdots,\|\})^q}ds < \infty,$$

where $q = \dim \mathfrak{h}^\perp + 1$ and where $\| \ \|$ denotes any norm on $\mathfrak{g}^*$.

For the Bonnet formula, we need another set of polarizations, which we shall denote by $\mathfrak{c}(\phi)$, $\phi \in \Gamma_f$. We shall put $C(\phi) = \exp(\mathfrak{c}(\phi))$. Let for $\phi = s \cdot \sigma$, $\sigma \in \Sigma$, $s \in H^f$,

$$\mathfrak{c}(\phi) = Ad(s)\mathfrak{b}(\sigma).$$

These polarizations have the agreeable property that

$$Ad(s)\mathfrak{c}(\phi) = \mathfrak{c}(s\cdot\phi), \ s \in H^f, \ \phi \in \Gamma_f.$$

We provide the groups $C(s\cdot\sigma) = sC(\sigma)s^{-1}$ with the following Haar measures:

$$\int_{C(s\cdot\sigma)} \varphi(c)dc := \int_{C(\sigma)} \varphi(scs^{-1})dc, \ \varphi \in C_c(C(s\cdot\sigma)).$$

We shall need in the sequel a special C^∞-function j on $\mathfrak{g}$, which has the following property. For every $\phi \in \Gamma_f$ in general position, we have that

$$j_{|\mathfrak{c}(\phi)}(u) = \Delta_{G,C(\phi)}(\exp(u))^{1/2} \det \frac{1 - e^{-\mathrm{ad}_{\mathfrak{c}(\phi)}(u)}}{\mathrm{ad}_{\mathfrak{c}(\phi)}(u)}, \quad u \in \mathfrak{c}(\phi). \tag{4.3.25}$$

Such a function can easily be constructed by using the roots of $\mathfrak{g}$, (cf. [55] for details).

Let us recall that for a Lie algebra $\mathfrak{c}$, the function

$$C = \exp(\mathfrak{c}) : u \mapsto \det \frac{1 - e^{-\mathrm{ad}_{\mathfrak{c}}(u)}}{\mathrm{ad}_{\mathfrak{c}}(u)},$$

is the Jacobian of the exponential mapping $\exp : \mathfrak{c} \to C$.

4.3.6 *The Bonnet-Plancherel Formula*

The aim of this section is to describe explicitly the Bonnet-Plancherel Formula associated to the disintegration formula (2.6.5). Let G, H, f, $S_{H,\chi_f}, \ldots$ be as in the introduction. We recall that the distribution S_{H,χ_f} is positive. Indeed, by (4.3.7), for $\varphi \in \mathscr{D}(G)$ we have:

$$\langle S_{H,\chi_f}, \varphi^* * \varphi \rangle = \langle T_f(\varphi), T_f(\varphi) \rangle \geq 0.$$

By a theorem of P. Bonnet [28], there exist positive nuclear operators $U_\pi : \mathscr{H}_\pi^\infty \to \mathscr{H}_\pi^{-\infty}$, such that

$$\int_H \varphi(h)\chi_f(h)dh = \int_{\hat{G}} \mathrm{tr}(\pi(\varphi)U_\pi)d\mu(\pi), \quad \varphi \in \mathscr{D}(G).$$

We shall show that the operators U_π are integral operators of rank one. The first step is a disintegration of the distribution $S_{H,f}$.

Theorem 4.3.3 *For $\varphi \in \mathscr{D}(G)$, we have*

$$\langle S_{H,f}, \varphi \rangle = \int_\Sigma \oint_{H^f/C(\sigma)} i_A(s)^{-1} \langle S_{C(s\cdot\sigma),s\cdot\sigma}, \varphi \rangle ds d\nu(\sigma).$$

Proof Let $\varphi \in \mathscr{D}(G/A)$. Put $\psi = j\varphi \circ \exp \in \mathscr{D}(\mathfrak{g})$ (j as in (4.3.25)). We apply Poisson's formula to ψ:

$$\int_{\mathfrak{h}} \psi(h)e^{-2\pi i\langle f,h\rangle}dh = \int_{f+\mathfrak{h}^{\perp}} \hat{\psi}(\phi)d\phi = \int_{\Sigma}\int_{H^f/G(\sigma)} m^{-1}(s\cdot\sigma)\hat{\psi}(s\cdot\sigma)dsd\nu(\sigma)$$

(by Proposition 4.3.2)

$$= \int_{\Sigma}\oint_{H^f/C(\sigma)} \Delta_{G/A}(s)\oint_{C(\sigma)/G(\sigma)} \Delta_{G/A}(b)\Delta_{H^f,C(\sigma)}(b)\hat{\psi}(sb\cdot\sigma)dbdsd\nu(\sigma)$$

$$= \int_{\Sigma}\oint_{H^f/C(\sigma)} \Delta_{G/A}(s)\oint_{C(\sigma)/G(\sigma)} \Delta_{G,C(\sigma)}(b)\hat{\psi}(sb\cdot\sigma)dbdsd\nu(\sigma)$$

(by (4.3.22))

$$= \int_{\Sigma}\oint_{H^f/C(\sigma)} i_A(s)^{-1}\oint_{C(\sigma)/G(\sigma)} \Delta_{G,C(\sigma)}(b)\widehat{\psi\circ \mathrm{Ad}(s)}(b\cdot\sigma)dbdsd\nu(\sigma)$$

(by (4.3.1))

$$= \int_{\Sigma}\oint_{H^f/C(\sigma)} i_A(s)^{-1}\int_{\sigma+\mathfrak{c}(\sigma)^{\perp}} \widehat{\psi\circ \mathrm{Ad}(s)}(\sigma+l)dldsd\nu(\sigma)$$

$$= \int_{\Sigma}\oint_{H^f/C(\sigma)} i_A(s)^{-1}(\widehat{\psi\circ \mathrm{Ad}(s)}_{|\mathfrak{c}(\sigma)})(\sigma_{|\mathfrak{c}(\sigma)})dsd\nu(\sigma)$$

$$= \int_{\Sigma}\oint_{H^f/C(\sigma)} i_A(s)^{-1}\int_{\mathfrak{c}(\sigma)} \psi\circ \mathrm{Ad}(s)(c)e^{-2\pi i\langle\sigma,c\rangle}dcdsd\nu(\sigma)$$

$$= \int_{\Sigma}\oint_{H^f/C(\sigma)} i_A(s)^{-1}\int_{C(\sigma)} \varphi(scs^{-1})\Delta^{1/2}_{G,C(\sigma)}(c)\chi_\sigma(c)dcdsd\nu(\sigma)$$

$$= \int_{\Sigma}\oint_{H^f/C(\sigma)} i_A(s)^{-1}\langle S_{C(s\cdot\sigma),s\cdot\sigma},\varphi\rangle dsd\nu(\sigma). \qquad \text{(by (4.3.5))}$$

Since $\mathfrak{h}$ is an ideal in $\mathfrak{g}$, it follows that

$$j_{|\mathfrak{h}} = \det\frac{1-e^{\mathrm{ad}_{\mathfrak{h}}}}{\mathrm{ad}_{\mathfrak{h}}}.$$

Hence

$$\int_{\mathfrak{h}} \psi(h)e^{-2\pi i\langle f,h\rangle}dh = \int_H \varphi(h)\chi_f(h)dh.$$

Finally

$$\langle S_{H,f},\varphi\rangle = \int_{\Sigma}\oint_{H^f/C(\sigma)} i_A(s)^{-1}\langle S_{C(s\cdot\sigma),s\cdot\sigma},\varphi\rangle dsd\nu(\sigma).$$

□

We can now use this identity to establish the Bonnet-Plancherel formula.

Before that we need to clarify certain details regarding the measures $\oint_{G/sCs^{-1}} dx$. Let G be any locally compact group and let C be a closed subgroup of G. If we fix Haar measures dx and dc on G and C respectively, then there exists a unique translation-invariant "measure" $\oint_{G/C} dx$ on G/C such that:

$$\int_G \varphi(x)dx = \oint_{G/C} \int_C \varphi(xc)\Delta_{G,C}(c)dcdx, \ \varphi \in C_c(G). \tag{4.3.26}$$

Let now $s \in G$. We have taken on sCs^{-1} the Haar measure

$$\int_{sCs^{-1}} \varphi(u)du := \int_C \varphi(scs^{-1})dc, \ \varphi \in C_c(C).$$

For $\psi \in \mathscr{E}(G, sCs^{-1})$, the function $x \mapsto \psi(xs^{-1}), \ x \in G$ is in $\mathscr{E}(G, C)$ and so

$$\psi \mapsto \oint_{G/C} \psi(xs^{-1})dx, \ \psi \in \mathscr{E}(G, sCs^{-1}),$$

defines an invariant positive linear functional on $\mathscr{E}(G, sCs^{-1})$. Furthermore the relation

$$\begin{aligned}
\int_G \varphi(x)dx &= \Delta_G(s)^{-1} \int_G \varphi(xs^{-1})dx \\
&= \Delta_G(s)^{-1} \oint_{G/C} \int_C \varphi((xs^{-1})scs^{-1})\Delta_{G,C}(c)dcdx \\
&= \Delta_G(s)^{-1} \oint_{G/C} \int_{sCs^{-1}} \varphi((xs^{-1})u)\Delta_{G,sCs^{-1}}(u)dudx
\end{aligned}$$

tells us that

$$\oint_{G/sCs^{-1}} \psi(x)dx = \Delta_G(s)^{-1} \oint_{G/C} \psi(xs^{-1})dx, \ \psi \in \mathscr{E}(G, sCs^{-1}). \tag{4.3.27}$$

Proposition 4.3.4 *We have for $\varphi, \psi \in C_c(G)$, that*

$$\begin{aligned}
\langle T_f(\varphi), T_f(\psi)\rangle &= \int_\Sigma \oint_{H^f/C(\sigma)} i_A(s)^{-1} \langle T_{s\cdot\sigma}(\varphi), T_{s\cdot\sigma}(\psi)\rangle_{\mathscr{H}_{s\cdot\sigma}} ds d\nu(\sigma) \\
&= \int_\Sigma \oint_{H^f/C(\sigma)} \Delta_{G/A}(s) \langle \varphi_{s,\sigma}, \psi_{s,\sigma}\rangle_{\mathscr{H}_\sigma} ds d\nu(\sigma).
\end{aligned}$$

Proof Let us recall that (see (4.3.15))

$$\begin{aligned}
\varphi_{s,\phi}(u) = K_{\pi_\phi(\varphi)}(u, s) &= \Delta_G(s)^{-1} \int_{C(\phi)} \varphi(ucs^{-1})\Delta^{1/2}_{G,C(\phi)}(c)\chi_\phi(c)dc \\
&= \Delta_G(s)^{-1} T_{s\cdot\phi}\varphi(us^{-1}), \ u \in G.
\end{aligned}$$

In particular, it follows for $u = s$, that

$$\varphi_{s,\phi}(s) = \Delta_G(s)^{-1}\langle S_{s\cdot\sigma}, \varphi\rangle, \ \varphi \in C_c(G). \tag{4.3.28}$$

Hence:

$$\begin{aligned}\int_{G/H} T_f(\varphi)(x)\overline{T_f(\psi)(x)}dx &= \langle S_{f,H}, \psi^* * \varphi\rangle && \text{(by (4.3.7))}\\ &= \int_\Sigma \oint_{H^f/C(\sigma)} i_A(s)^{-1}\langle S_{C(s\cdot\sigma),s\cdot\sigma}, \psi^* * \varphi\rangle ds d\nu(\sigma) && \text{(by Theorem 4.3.3)}\\ &= \int_\Sigma \oint_{H^f/C(\sigma)} i_A(s)^{-1}\langle T_{s\cdot\sigma}(\varphi), T_{s\cdot\sigma}(\psi)\rangle_{\mathcal{H}_{s\cdot\sigma}} ds d\nu(\sigma) && \text{(by (4.3.7))}\end{aligned}$$

Now:

$$\begin{aligned}\langle T_{s\cdot\sigma}(\varphi), T_{s\cdot\sigma}(\psi)\rangle_{\mathcal{H}_{s\cdot\sigma}} &= \oint_{G/C(s\cdot\sigma)} T_{s\cdot\sigma}(\varphi)(u)\overline{T_{s\cdot\sigma}(\psi)(u)}du\\ &= \Delta_G(s)^{-1}\oint_{G/C(\sigma)} T_{s\cdot\sigma}(\varphi)(us^{-1})\overline{T_{s\cdot\sigma}(\psi)(us^{-1})}du && \text{(by (4.3.27))}\\ &= \Delta_G(s)^{-1}\oint_{G/C(\sigma)} \Delta_G(s)^2\varphi_{s,\sigma}(u)\overline{\psi_{s,\sigma}(u)}du && \text{(by (4.3.28))}\\ &= \Delta_G(s)\langle\varphi_{s,\sigma}, \psi_{s,\sigma}\rangle_{\mathcal{H}_\sigma}.\end{aligned}$$

Therefore

$$\oint_{H^f/C(\sigma)} i_A(s)^{-1}\langle T_{s\cdot\sigma}(\varphi), T_{s\cdot\sigma}(\psi)\rangle_{\mathcal{H}_{s\cdot\sigma}} ds = \oint_{H^f/C(\sigma)} \Delta_{G/A}(s)\langle\varphi_{s,\sigma}, \psi_{s,\sigma}\rangle_{\mathcal{H}_\sigma} ds$$

and so

$$\begin{aligned}\langle T_f(\varphi), T_f(\psi)\rangle &= \langle S_{H,f}, \psi^* * \varphi\rangle\\ &= \int_\Sigma \oint_{H^f/C(\sigma)} i_A(s)^{-1}\langle T_{s\cdot\sigma}(\varphi), T_{s\cdot\sigma}(\psi)\rangle_{\mathcal{H}_{s\cdot\sigma}} ds d\nu(\sigma)\\ &= \int_\Sigma \oint_{H^f/C(\sigma)} \Delta_{G/A}(s)\langle\varphi_{s,\sigma}, \psi_{s,\sigma}\rangle_{\mathcal{H}_\sigma} ds d\nu(\sigma).\end{aligned}$$

□

As in (4.3.13) we define, for $s \in H^f$ and for $\phi \in \Gamma_f$, the element $\mathfrak{d}_{s,\phi}$ of the Sobolev space $\mathcal{H}_\phi^N = L^2(G/C(\phi), \phi)^N$ by

$$\langle \mathfrak{d}_{s,\phi}, \xi \rangle = \overline{\xi(s)},\ \xi \in L^2(G/C(\phi), \phi)^N.$$

This gives us the rank-one operators

$$P_{s,\phi} : \mathcal{H}_\phi^N \to \mathcal{H}_\phi^{-N};\ P_{s,\phi}(\xi) = \Delta_{G/A}(s)\langle \xi, \mathfrak{d}_{s,\phi}\rangle \mathfrak{d}_{s,\phi},\ \xi \in \mathcal{H}_\phi^N. \tag{4.3.29}$$

In particular, as in (4.3.16), $\pi_\phi(\varphi) \circ P_{s,\phi}$ is the rank-one operator $\Delta_{G/A} P_{\varphi_{s,\phi}, \mathfrak{d}_{s,\phi}} : \mathcal{H}_\phi^N \to \mathcal{H}_\phi^N$, where

$$P_{\varphi_{s,\phi}, \mathfrak{d}_{s,\phi}}(\eta) = \Delta_{G/A}(s)\eta(s)\varphi_{s,\phi},\ \eta \in \mathcal{H}_\phi^N. \tag{4.3.30}$$

Here $\varphi_{s,\phi}$ is as in (4.3.15). The trace of $P_{s,\phi}$ is the number

$$\begin{aligned} \mathrm{tr}(P_{s,\phi}) &= \Delta_{G/A}(s)\langle \varphi_{s,\phi}, \mathfrak{d}_{s,\phi}\rangle \\ &= \Delta_{G/A}(s)\varphi_{s,\phi}(s) \\ &= i_A(s)^{-1}\langle S_{C(s\cdot\phi), s\cdot\phi}, \varphi\rangle. \qquad \text{(by (4.3.14))} \end{aligned}$$

Hence

$$\mathrm{tr}(\pi_\phi(\varphi) \circ P_{s,\phi}) = i_A(s)^{-1}\langle S_{C(s\cdot\phi), s\cdot\phi}, \varphi\rangle. \tag{4.3.31}$$

For $\sigma \in \Sigma$, we define the operator $U_\sigma : \mathcal{H}_\sigma^N \to \mathcal{H}_\sigma^{-N}$ as the integral of these rank-one operators:

$$U_\sigma = \oint_{H^f/C(\sigma)} P_{s,\sigma} ds.$$

Proposition 4.3.5 *For almost all $\sigma \in \Sigma$, the operators $A_\sigma^{-N} \circ U_\sigma \circ A_\sigma^{-N} : \mathcal{H}_\sigma \to \mathcal{H}_\sigma$ (where $A_\sigma = d\pi_\sigma(1 - L)$) are trace-class and*

$$\int_\Sigma tr(A_\sigma^{-N} \circ U_\sigma \circ A_\sigma^{-N}) d\nu(\sigma) < \infty.$$

Proof Indeed, we have by (4.3.17) that

$$A_\sigma^{-N} \circ P_{s,\sigma} \circ A_\sigma^{-N} = \Delta_{G/A}(s) P_{\xi_{s,\sigma}, \xi_{s,\sigma}} : \mathcal{H}_\sigma \to \mathcal{H}_\sigma$$

is a bounded positive operator of rank one and

$$\mathrm{tr}(P_{s,\sigma}) = \mathrm{tr}(A_\sigma^{-N} \circ P_{s,\sigma} \circ A_\sigma^{-N}) = \Delta_{G/A}(s)\mathrm{tr}(P_{\xi_{\sigma,s}\xi_{\sigma,s}}) = \Delta_{G/A}(s)\|\xi_{\sigma,s}\|_{\mathcal{H}_\sigma}^2.$$

Let us recall that $\xi_{s,\sigma} = F_{s,\sigma} + A_\sigma^{-N} R_{s,\sigma}$ and so by (4.3.3), we have that

$$\begin{aligned}\|T_f(\alpha)\|^2_{\mathcal{H}_f} &= \int_\Sigma \oint_{H^f/C(\sigma)} \Delta_{G/A}(s)\|\alpha_{s,\sigma}\|^2_{\mathcal{H}_\sigma} ds d\nu(\sigma) \\ &= \int_\Sigma \oint_{H^f/C(\sigma)} i_A(s)^{-1}\|T_{s\cdot\sigma}\alpha\|^2_{\mathcal{H}_{s\cdot\sigma}} ds d\nu(\sigma) < \infty\end{aligned}$$

for $\alpha = F$ or R. Hence for almost all $\sigma \in \Sigma$, we have that

$$\oint_{H^f/C(\sigma)} \Delta_{G/A}(s)\|F_{s,\sigma}\|^2_{\mathcal{H}_\sigma} ds < \infty$$

and

$$\oint_{H^f/C(\sigma)} \Delta_{G/A}(s)\|A_\sigma^{-N} R_{s,\sigma}\|^2_{\mathcal{H}_\sigma} ds \le \|A_\sigma^{-N}\|^2_{\text{op}} \oint_{H^f/C(\sigma)} \Delta_{G/A}(s)\|R_{s,\sigma}\|^2_{\mathcal{H}_\sigma} ds < \infty$$

and finally

$$\begin{aligned}\text{tr}(A_\sigma^{-N} \circ U_\sigma \circ A_\sigma^{-N}) &= \oint_{H^f/C(\sigma)} \Delta_{G/A}(s)\text{tr}(P_{\xi_{s,\sigma},\xi_{s,\sigma}}) ds \\ &= \oint_{H^f/C(\sigma)} \Delta_{G/A}(s)\|\xi_{s,\sigma}\|^2_{\mathcal{H}_\sigma} ds < \infty.\end{aligned}$$

Hence also

$$\begin{aligned}\int_\Sigma \text{tr}(A_\sigma^{-N} \circ U_\sigma \circ A_\sigma^{-N}) d\nu(\sigma) &= \int_\Sigma \oint_{H^f/C(\sigma)} \Delta_{G/A}(s)\|\xi_{s,\sigma}\|^2 ds d\nu(\sigma) \\ &= \|T_f(F + A_f^{-N} R)\|^2_{\mathcal{H}_f} < \infty,\end{aligned}$$

where $A_f = d\tau_f(1 - L)$. □

Theorem 4.3.6 *Let $G = \exp \mathfrak{g}$ be an exponential solvable Lie group and $H = \exp \mathfrak{h}$ a closed connected normal subgroup of G. Let f be a linear functional of $\mathfrak{g}$ which is subordinated to $\mathfrak{h}$, $\chi_f = \chi$ the corresponding unitary character of H and let H^f be the stabilizer in G of $f_{|\mathfrak{h}}$. Let $\Sigma \subset \mathfrak{g}^*$ be a locally closed cross-section of the H^f-orbits in Γ_f. Then there exist $N \in \mathbb{N}$ and a Borel measure ν on Σ such that for almost all $\sigma \in \Sigma$ the positive bounded linear operator*

$$A_\sigma^{-N} \circ U_\sigma \circ A_\sigma^{-N} = \oint_{H^f/C(\sigma)} A_\sigma^{-N} \circ P_{\sigma,s} \circ A_\sigma^{-N} ds : \mathcal{H}_\sigma \to \mathcal{H}_\sigma$$

is trace-class and such that

$$\int_{\Sigma} tr(A_\sigma^{-N} \circ U_\sigma \circ A_\sigma^{-N}) d\nu(\sigma) < \infty.$$

Furthermore, for every $\varphi \in \mathscr{D}(G)$ *we have that* $\pi_\sigma(\varphi) \circ U_\sigma : \mathscr{H}_\sigma^N \to \mathscr{H}_\sigma^\infty$ *is trace-class for almost all* $\sigma \in \Sigma$ *and*

$$\langle S_{H,f}, \varphi \rangle = \int_{\Sigma} tr(\pi_\sigma(\varphi) \circ U_\sigma) d\nu(\sigma).$$

Proof We have seen in (4.3.5) that $U_\sigma = \oint_{H^f/C(\sigma)} P_{s,\sigma} ds$ is trace-class for almost all $\sigma \in \Sigma$. Hence

$$\begin{aligned}
A_\sigma^N \circ \pi_\sigma(\varphi) \circ U_\sigma \circ A_\sigma^{-N} &= (A_\sigma^N \circ \pi_\sigma(\varphi) \circ A_\sigma^N) \circ (A_\sigma^{-N} \circ U_\sigma \circ A_\sigma^{-N}) \\
&= (A_\sigma^N \circ \pi_\sigma(\varphi) \circ A_\sigma^N) \circ (A_\sigma^{-N} \circ U_\sigma \circ A_\sigma^{-N}) \\
&= (\pi_\sigma((1-L)^N * \varphi * (1-L)^N) \circ (A_\sigma^{-N} \circ U_\sigma \circ A_\sigma^{-N}) \\
&= \oint_{H^f/C(\sigma)} (\pi_\sigma((1-L)^N * \varphi * (1-L)^N) \circ (A_\sigma^{-N} \circ P_{s,\sigma} \circ A_\sigma^{-N}) ds \\
&= \oint_{H^f/C(\sigma)} A_\sigma^N \circ \pi_\sigma(\varphi) \circ P_{s,\sigma} \circ A_\sigma^{-N} ds
\end{aligned}$$

is a trace-class operator on $\mathscr{H}_\sigma$ for all those σ. From that,

$$\begin{aligned}
\mathrm{tr}(\pi_\sigma(\varphi) \circ U_\sigma) &= \mathrm{tr}(A_\sigma^N \circ \pi_\sigma(\varphi) \circ U_\sigma \circ A_\sigma^{-N}) \\
&= \oint_{H^f/C(\sigma)} \mathrm{tr}(A_\sigma^N \circ \pi_\sigma(\varphi) \circ P_{s,\sigma} \circ A_\sigma^{-N}) ds \\
&= \oint_{H^f/C(\sigma)} \mathrm{tr}(\pi_\sigma(\varphi) \circ P_{s,\sigma}) ds \\
&= \oint_{H^f/C(\sigma)} i_A(s)^{-1} \langle S_{C(s\cdot\sigma), s\cdot\sigma}, \varphi \rangle ds. \qquad \text{(by (4.3.28))}
\end{aligned}$$

Therefore

$$\mathrm{tr}(\pi_\sigma(\varphi) \circ U_\sigma) = \oint_{H^f/C(\sigma)} i_A(s)^{-1} \langle S_{C(s\cdot\sigma), s\cdot\sigma}, \varphi \rangle ds. \tag{4.3.32}$$

Hence for any $\varphi \in \mathscr{D}(G)$,

$$\langle S_{H,f}, \varphi\rangle = \int_{\Sigma} \oint_{H^f/C(\sigma)} i_A(s)^{-1}\langle S_{C(s\cdot\sigma),s\cdot\sigma}, \varphi\rangle ds d\nu(\sigma) \quad \text{(by Theorem 4.3.3)}$$

$$= \int_{\Sigma} \text{tr}(\pi_\sigma(\varphi) \circ U_\sigma) d\nu(\sigma). \quad \text{(by (4.3.32))}$$

□

Corollary 4.3.7 *Let G be an exponential solvable Lie group. Let $\varphi \in \mathscr{D}(G)$. We have that*

$$\varphi(e) = \int_{\mathfrak{g}^*/G} tr(\pi_\phi(\varphi) \circ U_\phi) d\nu(\phi).$$

4.3.7 A Variant of Penney's Plancherel Formula

We shall obtain the disintegration of τ_f described in [9] as a consequence of the following expression of $S_{H,f}$ as integrals of the distributions $S_{B(\phi),f}$ over the affine subspace $V \subset \Gamma_f$ defined in (4.3.19). First we need another disintegration formula for the Lebesgue measure on Γ_f.

Proposition 4.3.8 *Let $C_c(\Gamma_f)$ denote the space of the continuous functions with compact support on Γ_f. Let φ in $C_c(\Gamma_f)$. We have*

$$\int_{\Gamma_f} \varphi(l) dl = \int_V \int_{\mathfrak{b}(\phi)^\perp} \varphi(\phi + w) dw d\phi,$$

where $\mathfrak{b}(\phi)$ is the Vergne polarization for ϕ ($\phi \in V$).

Proof Let $\mathscr{Z} = \{Z_i; i = 1, \ldots, n\}$ be a basis of $\mathfrak{g}$ extracted from a good sequence adapted to $\mathfrak{h}$ as in 1.5. Call $\mathscr{Z}^* = \{Z_1^*, \ldots, Z_n^*\}$ the dual basis in $\mathfrak{g}^*$. We have

$$\int_{\Gamma_f} \varphi(l) dl = \int_V \int_W \varphi(\phi + w) dw d\phi$$

where as before $V = f \oplus_{i \in L_0^H} \mathbb{R}Z_i^*$, $W = \oplus_{i \notin L^H} \mathbb{R}Z_i^*$, and where we have set $L_0^H = L^H \setminus \{p, \ldots, 1\}$. The subset $V_1 = V \cap \Gamma_f^1$ is Zariski dense in V, and for $\phi \in V_1$ the Vergne polarization $\mathfrak{b}(\phi)$ at ϕ is given by

$$\mathfrak{b}(\phi) = \oplus_{i \in L^H} \mathbb{R}Y_i(\phi),$$

where $Y_i(\phi) = Z_i + \sum_{j<i} a_{j,i}(\phi) Z_j$. It is clear that the family

$$\mathscr{B} = \{Z_i, i \notin L^H\} \cup \{Y_i(\phi), i \in L^H\} := \{Y_i(\phi), i = 1, \ldots, n\}$$

is another basis of $\mathfrak{g}$. We denote by $\mathscr{B}^* = \{Y_i^*(\phi); i = 1, \ldots, n\}$ its dual basis in $\mathfrak{g}^*$. Hence

$$\mathfrak{b}(\phi)^\perp = \text{span}\{Y_i^*(\phi); i \notin L^H\}.$$

For an element $\phi = \sum_{i \in L_0^H} \phi_i Z_i^*$, let $\phi^i = \sum_{j \in L_0^H, j<i} \phi_j Z_j^*$. With these notations we remark that for $k \in L^H$, the vector Z_k can be written as

$$Z_k = Y_k(\phi) + \sum_{l=1}^{k-1} \alpha_{l,k}(\phi^{k-1}) Y_l(\phi)$$

for some scalars $\alpha_{l,k}(\phi^{k-1})$, $l = p+1, \ldots, n$ and that $Z_k = Y_k(\phi)$ for $p \leq k \leq l$. Hence if for $i = 1, \ldots, n$, we write

$$Z_i^* = \sum_{j=1}^{n} u_{j,i}(\phi) Y_j^*(\phi),$$

so we have

$$\begin{aligned}
\eth_{i,k} = \langle Z_i^*, Z_k \rangle &= \sum_{j=1}^{n} u_{j,i}(\phi) \langle Y_j(\phi)^*, Z_k \rangle \\
&= \sum_{j=1}^{n} u_{j,i}(\phi) \sum_{l=1}^{k-1} \alpha_{l,k}(\phi) \langle Y_j(\phi)^*, Y_l(\phi) \rangle + \sum_{j=1}^{n} u_{j,i}(\phi) \langle Y_j(\phi)^*, Y_k(\phi) \rangle \\
&= \sum_{l=1}^{k-1} u_{l,i}(\phi) \alpha_{l,k}(\phi) + u_{k,i}(\phi).
\end{aligned}$$

We deduce from this relation, first, that $u_{1,i}(\phi) = 0$, and then, inductively, that $u_{i,k}(\phi) = 0$ for every $1 \leq k \leq i-1$, that $u_{i,i}(\phi) = 1$ and that $u_{i,k}(\phi) = u_{i,k}(\phi^{k-1})$ for $k > i$. Hence for every $i = 1, \ldots, n$,

$$Z_i^* = Y_i^*(\phi) + \sum_{i+1}^{j=n} u_{j,i}(\phi^{i-1}) Y_j^*(\phi).$$

Hence we see that the subset $\mathscr{C}^* = \{C_1(\phi), \ldots, C_n(\phi)\}$ of $\mathfrak{g}^*$, where

$$C_i^*(\phi) = \begin{cases} Z_i^*, & \text{if } i \in L^H \\ Y_i^*(\phi), & \text{if } i \notin L^H \end{cases}$$

is also a basis of $\mathfrak{g}^*$ and that for $i \notin L^H$

$$Z_i^* = \sum_{j>i} v_{j,i}(\phi^{j-1}) C_j^*(\phi) + Y_i^*(\phi).$$

Hence for $\psi \in C_c(\Gamma_f)$,

$$\int_{\Gamma_f} \psi(\phi) d\phi = \int \psi \left(\sum_{i \in L_0^H} \phi_i Z_i^* + \sum_{i \notin L^H} t_i Z_i^* \right) \prod_{i \in L_0^H} d\phi_i \prod_{i \notin L^H} dt_i$$

$$= \int \psi \left(\sum_{i \in L_0^H} (\phi_i + \sum_{k<i} t_k v_{i,k}(\phi^{i-1})) Z_i^* + \sum_{i \notin L^H} (t_i + \sum_{k<i} t_k v_{i,k}(\phi^{i-1})) Z_i^* \right) \prod_{i \in L_0^H} d\phi_i \prod_{i \notin L^H} dt_i$$

$$= \int_V \int_{\mathfrak{b}(\phi)^\perp} \psi(\phi + w) dw d\phi.$$

□

Theorem 4.3.9 *Let G be an exponential solvable Lie group, H a normal closed connected subgroup, and $\chi = \chi_f$ a unitary character on H associated with some $f \in \mathfrak{g}^*$ such that $\langle f, [\mathfrak{h}, \mathfrak{h}] \rangle = (0)$. Let V be the affine subspace of Γ_f as in (4.3.19). Then one has for any $\varphi \in C_c^\infty(G)$:*

$$\langle S_{H,\chi_f}, \varphi \rangle = \int_V \langle S_\phi, \varphi \rangle d\phi,$$

where $S_\phi = S_{B(\phi),\chi_\phi}$ is defined on $\mathscr{D}(G)$ as in (4.3.5).

Proof This is similar to the proof of Theorem 4.3.3. Let $\varphi \in \mathscr{D}(G)$. Put $\psi = j\phi \circ \exp \in \mathscr{D}(\mathfrak{g})$, where j is as in (4.3.25). We apply Poisson's formula to ψ:

$$\begin{aligned} \int_{\mathfrak{h}} \psi(h) e^{-2\pi i \langle f,h \rangle} dh &= \int_{\Gamma_f} \hat{\psi}(\phi) d\phi \\ &= \int_V \int_{\mathfrak{b}(\phi)^\perp} \hat{\psi}(\phi + l) dl d\phi && \text{(by (4.3.8))} \\ &= \int_V \widehat{\psi_{|\mathfrak{b}(\phi)}}(\phi_{|\mathfrak{b}(\phi)}) d\phi \end{aligned}$$

$$= \int_V \int_{\mathfrak{b}(\phi)} \psi(b) e^{-2\pi i \langle \phi, b \rangle} db d\phi$$

$$= \int_V \int_{B(\phi)} \varphi(b) \Delta_{G,B(\phi)}^{1/2}(b) \chi_\phi(b) db d\phi$$

$$= \int_V \langle S_{B(\phi),\phi}, \varphi \rangle d\phi.$$

□

Theorem 3.3.4 of Chap. 3, due to Baklouti [4] is now an immediate consequence of Theorem 4.3.9. It states that the operator

$$U : \mathscr{E}(G, H, \chi_f) \to \int_V^{\oplus} \mathscr{H}_\phi d\lambda(\phi) = \mathscr{H}_{\tau'}$$

(where $\mathscr{H}_\phi = L^2(G/B(\phi), \phi)$ denotes the Hilbert space of the irreducible representation π_ϕ) defined for $\phi \in V$, $\psi \in \mathscr{E}(G, H, \chi_f)$ and $g \in G$ by:

$$U(\psi)(\phi)(g) = T_{B(\phi),\phi}(\psi)(g) = \int_{B(\phi)/H} \psi(gb) \chi_\phi(b) \Delta_{B(\phi),G}^{-1/2}(b) d_{Y(\phi)} b$$

is an isometry for the L^2-norms and extends to a unitary intertwining operator from $\mathscr{H}_\tau$ onto $\mathscr{H}_{\tau'}$. Indeed, it suffices to observe that the canonical projection:

$$T_f : C_c(G) \to \mathscr{E}(G, H, \chi_f); T_f(\varphi)(x) = \int_H \phi(xh) \chi_f(h) dh, \ x \in G,$$

is surjective. Hence for $\xi \in \mathscr{E}(G, H, \chi_f)$, there exists $\varphi \in C_c(G)$ such that $T_f(\varphi) = \xi$. Furthermore, we have that

$$T_\phi(\varphi)(x) = \int_{B(\phi)} \varphi(xb) \Delta_{G,B(\phi)}^{1/2}(b) \chi_\phi(b) db$$

$$= \int_{B(\phi)/H} \int_H \varphi(xbh) \chi_f(h) dh \Delta_{G,B(\phi)}^{1/2}(b) \chi_\phi(b) db$$

$$= T_{B(\phi),\phi}(\xi)(x),$$

and so

$$\|U(\xi)\|^2 = \int_V \|T_{B(\phi),\phi} \xi\|_{\mathscr{H}_\phi}^2 d\phi$$

$$= \int_V \langle S_{B(\phi),\phi}, \varphi * \varphi^* \rangle d\phi \qquad \text{(by (4.3.7))}$$

$$= \langle S_{H,f}, \phi * \phi^* \rangle = \|\xi\|^2.$$

Another consequence of Theorem 4.3.9 is the following:

Corollary 4.3.10 *We have that for all* $\varphi \in C_c^\infty(G)$

$$\langle \tau(\varphi)\alpha_\tau, \alpha_\tau\rangle = \int_V \langle \pi_\phi(\varphi)\alpha_\phi, \alpha_\phi\rangle d\lambda(\phi),$$

where α_ϕ *is the canonical cyclic generalized vector for* π_ϕ.

We close the section with a property of the affine subspace V.

Proposition 4.3.11 *For all* ϕ *in* Γ_f^1, *there exists a unique* b *in* $B(\phi)$ *modulo* $G(\phi)$ *such that* $b \cdot \phi \in V$. *In particular* $\Gamma_f^1 = \dot{\bigcup}_{\phi \in V} B(\phi) \cdot \phi$.

Proof We first prove the existence. We use induction on dim G. If $\dim G = 1$, there is nothing to prove. Let $\mathfrak{g}_{n-1} = \mathfrak{g}_0$. Let us write for an element $\phi \in \mathfrak{g}^*$, $\phi = \phi_0 + \alpha Z_n^*$, where $\phi_0 = \sum_{j=1}^{n-1} \phi(Z_j)Z_j^*$ and $\alpha = \phi(Z_n)$ and let $V_0 = \{\phi_0;\ \phi \in V\}$. Then we can apply the induction hypothesis to $V_{|\mathfrak{g}_0}$ and to $f_{|\mathfrak{g}_0}$. Let $\phi \in \Gamma_\phi^1$. We distinguish between the two cases:

Case 1: $n \in L^H$ We have that $\mathfrak{b}(\phi) \cap \mathfrak{g}_0 = \mathfrak{b}(\phi_{|\mathfrak{g}_0})$ for all $\phi \in \Gamma_f^1$ and furthermore $V = \mathbb{R}Z_n^* + V_0$. The induction hypothesis gives us $b_0 \in B(\phi_{|\mathfrak{g}_0})$ such that $b_0 \cdot \phi_{|\mathfrak{g}_0} \in V_{|\mathfrak{g}_0}$. Then $b_0 \cdot \phi = b_0 \cdot (\phi_0 + \alpha Z_n^*) = b_0 \cdot \phi_0 + \alpha b_0 \cdot Z_n^* = b_0 \cdot \phi_0 + (\alpha + \beta)Z_n^* \in V$ for some $\beta \in \mathbb{R}$.

Case 2: $n \notin L^H$ We have that $\mathfrak{g}(\phi) \subset \mathfrak{g}_0$ and $\mathfrak{b}(\phi) = \mathfrak{b}(\phi_{|\mathfrak{g}_0})$. By the induction hypothesis, there exists $b_0 \in B(\phi_{|\mathfrak{g}_0})$, such that $b_0 \cdot \phi_{|\mathfrak{g}_0} \in V_{|\mathfrak{g}_0}$. Since $\mathfrak{g}(b_0 \cdot \phi) = \mathrm{Ad}(b_0)\mathfrak{g}(\phi) \subset \mathfrak{g}_0$, there exists an element $y \in \mathfrak{g}_0$ such that $b_0 \cdot \phi([y, \mathfrak{g}_0]) = 0$ and $b_0 \cdot \phi([y, Z_n]) = 1$. It is clear that $y \in B(b_0 \cdot \phi_{|\mathfrak{g}_0}) = B(\phi_{|\mathfrak{g}_0}) = B(\phi)$. Hence we have $\exp \beta y \cdot b_0 \cdot \phi|\mathfrak{g}_0 = b_0 \cdot \phi_0$ and $\exp \beta y \cdot b_0 \cdot \phi(Z_n) = \alpha - \beta$ for some $\beta \in \mathbb{R}$. Then $\exp \alpha y \cdot b_0 \cdot \phi = b_0 \cdot \phi_0 \in V_0 = V$.

As for uniqueness, let $\phi \in \Gamma_f$. Assume that there exist $b, b' \in B(\phi)$ such that $b \cdot \phi = f + \sum_{i \in L^H} t_i Z_i^* \in V$ and $b' \cdot \phi = f + \sum_{i \in L^H} s_i Z_i^* \in V$. We have $b \cdot \phi = bb'^{-1}b'\phi = b'\phi + w$, where $w \in \mathfrak{b}(\phi)^\perp$. Then $b \cdot \phi - b' \cdot \phi = \sum_{i \in L^H}(t_i - s_i)Z_i^* \in \mathfrak{b}(\phi)^\perp$. We get therefore $t_i - s_i = 0$ for all $i \in L^H$, hence $b'b^{-1} \in G(\phi)$. □

4.4 The Penney-Plancherel Formula for Finite-Multiplicity Restrictions of Nilpotent Lie Groups

The general context in the present section is that of connected simply connected nilpotent Lie groups. We study the *Plancherel formula* and the *Penney-Plancherel formula for finite-multiplicity restrictions* to an analytic subgroup with finite multiplicities. Basically, the main motivation comes from the strong similarity between the disintegration formulas of induced and restricted representations. As suggested by Frobenius reciprocity, there is a sort of duality between the induction and the restriction of unitary representations in term of branching laws. As usual, the orbit method will be very useful in our study.

4.4.1 On Restrictions of Unitary Representations

Let $G = \exp \mathfrak{g}$ be a nilpotent Lie group with Lie algebra $\mathfrak{g}$. Let $\widehat{G}$ be the unitary dual of G and $\pi \in \widehat{G}$ associated to a coadjoint orbit $\Omega \subset \mathfrak{g}^*$. Let $K = \exp \mathfrak{k}$ be an analytic subgroup of G and $\pi|_K$ the restriction of π to K. Now we consider the restriction $\pi|_K$ and proceed in analogy to the case of induced representations. Let $p : \mathfrak{g}^* \to \mathfrak{k}^*$ be the restriction map, $\widetilde{\nu}$ the finite measure on Ω equivalent to the G-invariant measure, and ν the image of $p_*(\widetilde{\nu})$, $\widetilde{\nu}$ viewed as a measure on $\mathfrak{g}^*$ under the Kirillov map $\mathfrak{k}^* \to \widehat{K}$. For $\sigma \in \widehat{K}$, denote by $\omega(\sigma)$ the coadjoint orbit of K associated to σ and by $m(\sigma)$ the number of K-orbits contained in $\Omega \cap p^{-1}(\omega(\sigma))$. As described earlier, in Sect. 1.2.7, the canonical central disintegration of $\pi|_K$ is written as follows:

Theorem 4.4.1

$$\pi|_K \simeq \int_{\widehat{K}}^{\oplus} m(\sigma)\sigma d\nu(\sigma). \tag{4.4.1}$$

Here, once again, the multiplicities $m(\sigma)$ are either uniformly bounded or uniformly equal to infinity. In agreement with these possibilities we say that $\pi|_K$ is with finite multiplicities or infinite multiplicities. Then it is well known that $\pi|_K$ is with finite multiplicities if and only if

$$\dim \Omega = 2 \dim K \cdot \ell - \dim K \cdot p(\ell) \tag{4.4.2}$$

generically on Ω. Assume now that $\pi|_K$ is with finite multiplicities. We fix a flag of ideals

$$\{0\} = \mathfrak{g}_0 \subset \mathfrak{g}_1 \subset \cdots \subset \mathfrak{g}_{n-1} \subset \mathfrak{g}_n = \mathfrak{g}, \tag{4.4.3}$$

verifying $\dim \mathfrak{g}_j = j$ $(0 \le j \le n)$ of $\mathfrak{g}$. Let $\mathscr{I} = \{i_1 < \cdots < i_d\}$ $(d = \dim \mathfrak{k})$ be the set of indices $1 \le i \le n$ such that $\mathfrak{k} \cap \mathfrak{g}_i \neq \mathfrak{k} \cap \mathfrak{g}_{i-1}$. Considering $\mathfrak{k}_s = \mathfrak{k} \cap \mathfrak{g}_{i_s}$ $(1 \le s \le d)$, we form a flag of ideals

$$\{0\} = \mathfrak{k}_0 \subset \mathfrak{k}_1 \subset \cdots \subset \mathfrak{k}_{d-1} \subset \mathfrak{k}_d = \mathfrak{k} \tag{4.4.4}$$

of $\mathfrak{k}$. For ν-almost all $\sigma \in \widehat{K}$, the closed set $\Omega \cap p^{-1}(\omega(\sigma))$ consists of $m(\sigma)$ connected components $C_1, \ldots, C_{m(\sigma)}$, each of which is a K-orbit. In addition, write $\ell|_{\mathfrak{k}} = p(\ell)$ for $\ell \in \Omega \cap p^{-1}(\omega(\sigma))$, so each intersection $\widetilde{C}_j = C_j \cap p^{-1}(\ell|_{\mathfrak{k}})$ $(1 \le$

$j \leq m(\sigma)$) is a $K(\ell|_{\mathfrak{k}})$-orbit. Let us fix ℓ arbitrarily in Ω and take, using the flag (4.4.3), a Vergne polarization $\mathfrak{b} = \mathfrak{b}[\ell]$ of $\mathfrak{g}$ at ℓ to realize π as $\pi = \mathrm{ind}_B^G \chi_\ell$ with $B = \exp \mathfrak{b}$. Then we take $\ell^\sigma \in \Omega \cap p^{-1}(\omega(\sigma))$ for ν-almost all $\sigma \in \widehat{K}$ and let

$$\Omega \cap p^{-1}(\ell^\sigma|_{\mathfrak{k}}) = \widetilde{C}_1^\sigma \amalg \cdots \amalg \widetilde{C}_{m(\sigma)}^\sigma$$

as above. For $1 \leq j \leq m(\sigma)$, we choose $g_j^\sigma \in G$ so to have $\ell_j^\sigma = g_j^\sigma \cdot \ell \in \widetilde{C}_j^\sigma$. Now, by means of the Vergne polarization $\mathfrak{b}[\ell_\sigma]$ of $\mathfrak{k}$ at ℓ_σ built from the flag (4.4.4), where $\ell_\sigma = \ell^\sigma|_{\mathfrak{k}} \in \mathfrak{k}^*$, we set $B[\ell_\sigma] = \exp \mathfrak{b}[\ell_\sigma]$ and we define the following vectors a_σ^j $(1 \leq j \leq m(\sigma))$:

$$\langle a_\sigma^j, \psi \rangle = \int_{B[\ell_\sigma]/(B[\ell_\sigma] \cap g_j^\sigma B(g_j^\sigma)^{-1})} \overline{\psi(b g_j^\sigma) \chi_{\ell_j^\sigma}(b)} d\dot{b} \tag{4.4.5}$$

for $\psi \in \mathscr{H}_\pi^\infty$. Once again we are in presence of a Frobenius reciprocity of sorts [16].

Theorem 4.4.2 *In the disintegration formula (4.4.1) of $\pi|_K$, the multiplicities $m(\sigma)$ are ν-almost everywhere equal to the dimension of $\left(\mathscr{H}_\pi^{-\infty}\right)^{B[\ell_\sigma],\chi_{\ell_\sigma}}$. In addition, if $\pi|_K$ is of finite multiplicities, then ν-almost everywhere the generalized vectors $\{a_\sigma^j; 1 \leq j \leq m(\sigma)\}$ form a basis of $\left(\mathscr{H}_\pi^{-\infty}\right)^{B[\ell_\sigma],\chi_{\ell_\sigma}}$.*

4.4.2 The Plancherel and Penney-Plancherel Formulas

Keep all hypotheses and notations. For ν-almost all $\sigma \in \widehat{K}$ and all $1 \leq j \leq m(\sigma)$, we choose a coexponential basis $\{X_1, \ldots, X_r\}$ $(r = \frac{1}{2}\dim \Omega)$ of $G(\ell_j^\sigma)$ in $g_j^\sigma B(g_j^\sigma)^{-1}$. By a partial section of the homogeneous space $g_j^\sigma B(g_j^\sigma)^{-1}/G(\ell_j^\sigma)$ we mean a set $S_j(\sigma) := \exp(\mathbb{R}X_{i_1}) \cdots \exp(\mathbb{R}X_{i_w})$, where $\{X_{i_1}, \ldots, X_{i_w}\}$ is a subfamily of $\{X_1, \ldots, X_r\}$.

We propose to show the Plancherel formula for $\pi|_K$ in the following form:

Theorem 4.4.3 *For ν-almost all $\sigma \in \widehat{K}$ and all $1 \leq j \leq m(\sigma)$, there exists a partial section $S_j(\sigma)$ so that the following holds. We use a measure on $\widehat{K}$, still called ν, equivalent to the measure of the disintegration formula (4.4.1) of $\pi|_K$. Then for all $\psi \in \mathscr{H}_\pi^\infty$ we have*

$$\psi(e) = \int_{\widehat{K}} \Big(\sum_{j=1}^{m(\sigma)} \int_{S_j(\sigma)} \langle \psi, a_\sigma^j(k) \rangle d\dot{k} \Big) d\nu(\sigma)$$

with generalized vectors $a_\sigma^j(k)$ defined by formula (4.4.5) and properly normalized and modified by

$$\langle \psi, a_\sigma^j(k)\rangle = \int_{B[\ell_\sigma]/(B[\ell_\sigma]\cap kg_j^\sigma B(kg_j^\sigma)^{-1})} \psi(bkg_j^\sigma)\chi_{g_j^\sigma\cdot\ell}(b)d\dot{b}\ (k \in S_j(\sigma)). \tag{4.4.6}$$

Once we have related the generalized vectors a_σ^j to the intertwining operators between $\pi|_K$ and σ in $\widehat{K}$ occurring in the disintegration of $\pi|_K$, it is natural to go back and ask whether we could have done the same for induced representations. This might be explored in the near future in view of applications aimed at the construction of an explicit intertwining operator between $\pi|_K$ and its disintegration. Our second result concerns the Penney-Plancherel formula for the restriction $\pi|_K$, and is expressed as follows.

Theorem 4.4.4 *For ν-almost all $\sigma \in \widehat{K}$ and all $1 \le j \le m(\sigma)$, we can choose a partial section $S_j(\sigma)$ and, for all $k \in S_j(\sigma)$, a projector $R_\sigma^j(k)$ of $\mathscr{H}_\pi$ onto an isotypic σ-component, in a way such that*

$$\langle(\pi|_K)(\theta)(\psi), \delta_\pi\rangle_{\mathscr{H}_\pi} = \int_{\widehat{K}}\sum_{j=1}^{m(\sigma)}\Big(\int_{S_j(\sigma)}\langle\sigma(\theta)\cdot R_\sigma^j(k)(\psi), \delta_\sigma\rangle_{\mathscr{H}_\sigma}d\dot{k}\Big)d\nu(\sigma),$$

for $\theta \in \mathscr{D}(K)$ (the space of infinitely differentiable functions with compact supports) and $\psi \in \mathscr{H}_\pi^\infty$. Above, δ_π designates the Dirac operator on $\mathscr{H}_\pi^\infty$.

Returning to our goal and before proceeding to the proofs, let us consider a few examples.

4.4.3 Examples

Let us see what happens in the case of threadlike Lie algebras. First, let $\mathfrak{g} = \langle X, V, Y, Z\rangle_{\mathbb{R}} : [X, V] = Y, [X, Y] = Z$. Take $f = Z^* \in \mathfrak{g}^*$ and let $\Omega = G\cdot f$ be the coadjoint orbit of G passing through f. Let $\mathfrak{b} = \langle V, Y, Z\rangle_{\mathbb{R}}$ and $B = \exp\mathfrak{b}$. Then $\mathfrak{b}$ is a polarization of $\mathfrak{g}$ at f. We define as usual a unitary character χ_f of B by $\chi_f(\exp x) = e^{if(x)}$ $(x \in \mathfrak{b})$. Then we induce a monomial representation $\pi = \mathrm{ind}_B^G\chi_f$ of G which is irreducible. The Hilbert space $\mathscr{H}_\pi$ of π is identified with $L^2(\mathbb{R})$ by the map $R : \mathscr{H}_\pi \ni \psi \mapsto \Psi \in L^2(\mathbb{R}),\ \Psi(t) = \psi(\exp(tX))\ (\forall\, t \in \mathbb{R})$. Call $\mathscr{D}(G)$ the space of infinitely differentiable functions with compact support on G. A direct computation gives:

$$\Omega = \{\mathbb{R}X^* + vV^* + yY^* + Z^*; 2v = y^2, y \in \mathbb{R}\}.$$

Example 4.4.5 First, let $\mathfrak{k} = \mathbb{R}Y$. Then $\pi|_K$ is without multiplicities by the disintegration formula (4.4.1). Indeed, for $\psi \in \mathscr{H}_\pi$ we have

$$((\pi|_K)(\exp(sY))\psi)(\exp(tX)) = \psi(\exp(-sY)\exp(tX))$$
$$=\psi(\exp(tX)\exp(-sY)\exp(stZ)) = e^{-ist}\psi(\exp(tX)).$$

That is, for $\Psi \in L^2(\mathbb{R})$,

$$((\pi|_K)(\exp(sY))\Psi)(t) = e^{-ist}\Psi(t) \ (s,\ t \in \mathbb{R}).$$

Moreover,

$$\pi|_K \simeq \int_{\mathbb{R}}^{\oplus} \chi_t dt$$

with unitary characters χ_t of K defined by $\chi_t(\exp sY) = e^{-ist}$ $(s,\ t \in \mathbb{R})$. We interpret these simple facts as follows. For $t \in \mathbb{R}$ arbitrary, let $\ell_t = \exp(tX)\cdot f \in \Omega$. Then

$$(\ell_t|_{\mathfrak{k}})(Y) = \exp(tX)\cdot f(Y) = f(\exp(-tX)(Y)) = -t.$$

Define a generalized vector a_t of π by the formula

$$\langle a_t, \psi\rangle = \int_{K/K\cap B} \overline{\psi(k\exp(tX))\chi_{\ell_t}(k)d\dot{k}} = \overline{\psi(\exp(tX))}$$

for $\psi \in \mathscr{H}_\pi^\infty$. We know [16] that

$$\left(\mathscr{H}_\pi^{-\infty}\right)^{K,\chi_{\ell_t}} = \mathbb{C}a_t$$

for almost all t. It seems that these a_t $(t \in \mathbb{R})$ are not sufficient to recover the value $\psi(e)$. In order to integrate on the fiber of the projection $p : \mathfrak{g}^* \to \mathfrak{k}^*$, put

$$\langle\psi, a_t(s)\rangle = \psi(\exp(sY)\exp(tX)) \ (s \in \mathbb{R}).$$

Hence for $\psi \in \mathscr{H}_\pi^\infty$,

$$\int_{\widehat{K}}\int_{\mathbb{R}} \langle\psi, a_t(s)\rangle ds dt$$
$$= \int_{\mathbb{R}}\int_{\mathbb{R}} \psi(\exp(sY)\exp(tX))dsdt = \int_{\mathbb{R}^2} \psi(\exp(tX)\exp(sY - stZ))dsdt$$
$$= \int_{\mathbb{R}^2} e^{ist}\psi(\exp(tX))dsdt = 2\pi\psi(e).$$

On the other hand, the projector operator $R_t(\exp(sY))$ on $\mathscr{H}_\pi$ is nothing but the multiplication on the complex line by the number $\psi(\exp(sY)\exp(tX))$. The Penney-Plancherel formula now reads as follows. If $\theta \in \mathscr{D}(K)$,

$$\begin{aligned}\langle(\pi|_K)(\theta)(\psi), \delta_\pi\rangle =&(\pi|_K)(\theta)(\psi)(e)\\ =&\frac{1}{2\pi}\int_{\widehat{K}}\int_{S(\sigma)}\langle(\pi|_K)(\theta)(\psi), a_\sigma(k)\rangle d\dot{k}d\nu(\sigma)\\ =&\frac{1}{2\pi}\int_{\mathbb{R}}\int_{\mathbb{R}}\psi(\exp(sY)\exp(tX))dtds\int_{\mathbb{R}}\theta(r)dr\\ =&\,\psi(e)\int_{\mathbb{R}}\theta(r)dr.\end{aligned}$$

Example 4.4.6 Secondly, let $\mathfrak{k} = \mathbb{R}X$. For $s \in \mathbb{R}$ arbitrary, write χ_s for the character of K given by $\chi_s(\exp(tX)) = e^{ist}$ $(s,\ t \in \mathbb{R})$. This time the disintegration formula shows

$$\pi|_K \simeq \int_{\mathbb{R}}^{\oplus}\chi_s ds.$$

In fact, $\pi|_K$ is nothing else than the regular representation of $\mathbb{R}$. Interpret this fact as follows. For all $s \in \mathbb{R}$, put $\ell_s = \exp(sY)\cdot f \in \Omega$. Then $\ell_s(X) = f(\exp(-sY)X) = s$. Consequently, if we define a generalized vector a_s of π by the formula

$$\begin{aligned}\langle a_s, \psi\rangle =&\int_{K/K\cap B}\overline{\psi(k\exp(sY))\chi_{\ell_s}(k)}d\dot{k}\\ =&\int_{\mathbb{R}}\overline{\psi(\exp(tX))e^{ist}}dt\end{aligned}$$

for $\psi \in \mathscr{H}_\pi^\infty$ as we did before, we know (cf. [16]) that

$$\left(\mathscr{H}_\pi^{-\infty}\right)^{K,\chi_{\ell_s}} = \mathbb{C}a_s$$

almost everywhere for ds. Here, take $S(\sigma) = \{e\}$ and define on the fiber the generalized vectors $a_s(e)$ $(s \in \mathbb{R})$ by

$$\langle\psi, a_s(e)\rangle = \overline{\langle a_s, \psi\rangle} = \widehat{\psi}(\exp(sX)),$$

where $\widehat{\psi}(\exp(sX))$ designates the classic Fourier transform of the function Ψ. Altogether, we see that

$$2\pi\psi(e) = \int_{\mathbb{R}}\Big(\int_{\mathbb{R}}\psi(\exp(tX))e^{ist}dt\Big)ds = \int_{\widehat{K}}\langle\psi, a_s(e)\rangle ds.$$

Hence the projector operator $R_s(e)$ on $\mathscr{H}_\pi$ is the multiplication on $\mathbb{C}$ by $\widehat{\psi}(\exp(sX))$. This is the Penney-Plancherel formula, which may be rewritten as follows. For $\theta \in \mathscr{D}(K)$,

$$\begin{aligned}\langle(\pi|_K)(\theta)(\psi), \delta_\pi\rangle =&(\pi|_K)(\theta)(\psi)(e)\\ =&\int_{\mathbb{R}} \widehat{\theta}(\exp(sX))\widehat{\psi}(\exp(sX))ds.\end{aligned}$$

Example 4.4.7 Finally, let $\mathfrak{k} = \mathbb{R}V$. By calling $p : \mathfrak{g}^* \to \mathfrak{k}^*$ the restriction map, we observe that

$$p(\Omega) = \{\ell \in \mathfrak{k}^*; \ell(V) \geq 0\}$$

and that, for $\ell \in \mathfrak{k}^*$ satisfying $\ell(V) > 0$, $p^{-1}(\ell) \cap \Omega$ consists of two straight lines each of which is a K-orbit. By the disintegration formula, we see that

$$\pi|_K \simeq \int_{\mathbb{R}_+}^{\oplus} 2\chi_s ds$$

with unitary characters χ_s of K defined by $\chi_s(\exp(vV)) = e^{isv}$ $(s,\ v \in \mathbb{R})$. For $s > 0$, we set $g_s^{\pm} = \exp(\pm\sqrt{2s}X)$ and consider

$$a_s^{\pm} : \psi \mapsto \pm\frac{1}{\sqrt{2s}}\overline{\psi(g_s^{\pm})}$$

for $\psi \in \mathscr{H}_\pi^\infty$. Then we know that

$$\left(\mathscr{H}_\pi^{-\infty}\right)^{K,\chi_s} = \mathbb{C}a_s^+ \oplus \mathbb{C}a_s^-$$

almost everywhere for ds. Finally, to integrate $a_s^{\pm}$ on the fiber, put

$$\begin{aligned}\langle\psi, a_s^{\pm}(y)\rangle =& \frac{1}{\sqrt{2s}}\psi(\exp(yY)\exp(\pm\sqrt{2s}X))\\ =&\frac{1}{\sqrt{2s}}\psi(\exp(\pm\sqrt{2s}X)\exp(yY \mp y\sqrt{2s}Z)\\ =&\frac{1}{\sqrt{2s}}e^{\pm iy\sqrt{2s}}\psi(\exp(\pm\sqrt{2s}X))\ (y \in \mathbb{R}).\end{aligned}$$

Then

$$\int_{\mathbb{R}} \Big(\int_{\mathbb{R}_+} (\langle \psi, a_s^+(y) \rangle + \langle \psi, a_s^-(y) \rangle) ds \Big) dy$$
$$= \int_{\mathbb{R}} \Big(\int_{\mathbb{R}_+} \frac{1}{\sqrt{2s}} (e^{i\sqrt{2s}y} \psi(\exp(\sqrt{2s}X)) + e^{-i\sqrt{2s}y} \psi(\exp(-\sqrt{2s}X))) ds \Big) dy$$
$$= \int_{\mathbb{R}^2} e^{ity} \psi(\exp(tX)) dt dy = 2\pi \psi(e).$$

In this example, we have two projection operators on the representation space π:

$$R_s^{\pm}(y)(\psi) = \frac{1}{\sqrt{2s}} e^{\pm iy\sqrt{2s}} \psi(\exp(\pm\sqrt{2s}X)),$$

and as above:

$$\begin{aligned} \langle (\pi|_K)(\theta)(\psi), \delta_\pi \rangle =& (\pi|_K)(\theta)(\psi)(e) \\ =& \psi(e) \int_{\mathbb{R}} \theta(r) dr. \end{aligned}$$

Remark 4.4.8 Let now $\mathfrak{g} = \langle X_1, X_2, \ldots, X_n \rangle_{\mathbb{R}} : [X_1, X_j] = X_{j+1}$ $(2 \le j \le n-1)$. Take $f = X_n^*$ in $\mathfrak{g}^*$. Then the coadjoint orbit Ω of $G = \exp \mathfrak{g}$ passing through f is calculated as

$$\Omega = G{\cdot}f = \Big\{ \mathbb{R}X_1^* + \frac{t^{n-2}}{(n-2)!} X_2^* + \frac{t^{n-3}}{(n-3)!} X_3^* + \cdots + tX_{n-1}^* + X_n^*;\, t \in \mathbb{R} \Big\}.$$

Let $\mathfrak{b} = \sum_{i=2}^n \mathbb{R}X_i$, $B = \exp \mathfrak{b}$ and $\pi = \mathrm{ind}BG\chi_f$. By taking $\mathfrak{k} = \mathbb{R}X_j$ $(1 \le j \le n-1)$, we can repeat the same arguments as before. Let

$$\pi|_K \simeq \int_{\widehat{K}}^{\oplus} m(\sigma) \sigma d\nu(\sigma)$$

be the central canonical disintegration of the restriction $\pi|_K$. For this threadlike Lie algebra it is easy to choose $\mathfrak{k} \subset \mathfrak{b}$ so that the multiplicities are not uniform, even though it is well known that their parity is always the same.

4.4.4 *Proof of the Main Results*

Proof of Theorem 4.4.3

Proceed by induction on the integer $\dim G/K + \dim G$. If this number is one or two, our assertion is clear. First of all we can assume that $\mathfrak{k}$ is contains the center $\mathfrak{z}$ of $\mathfrak{g}$ by applying the induction hypothesis. Indeed, one sets $\mathfrak{l} = \mathfrak{k} + \mathfrak{z}$ and $L = \exp(\mathfrak{l})$. Then $\pi|_K$ and $\pi|_L$ have the same multiplicities, as the representations involved in their spectrum only differ by tensoring with a unitary character of $Z(G) = \exp(\mathfrak{z})$. The rest of the formula is clear. If $\mathfrak{a} = \mathfrak{z} \cap \ker(\ell)$ is non-trivial, we can pass to the quotient G/A, $A = \exp\mathfrak{a}$, to which the induction hypothesis applies. Hence we go back to the case where $\mathfrak{z} = \mathbb{R}Z$ with $\ell(Z) = 1$. In other words, $\mathfrak{g}_1 = \mathfrak{z}$ in flag (4.4.3). For the same flag (4.4.3) let $\mathfrak{g}_2 = \mathbb{R}Y \oplus \mathbb{R}Z$ with $Y \in \mathfrak{g}_2 \setminus \mathfrak{z}$ satisfying $\ell(Y) = 0$. Call $\widetilde{\mathfrak{g}}$ the centralizer of Y in $\mathfrak{g}$, which is an ideal of codimension 1 in $\mathfrak{g}$. Write $\mathfrak{g} = \mathbb{R}X + \widetilde{\mathfrak{g}}$ with X verifying $[X, Y] = Z$ and remark that $\mathfrak{b}$ is contained in $\widetilde{\mathfrak{g}}$. We set $\widetilde{\ell} = \ell|_{\widetilde{\mathfrak{g}}} \in \widetilde{\mathfrak{g}}^*$ and $\widetilde{G} = \exp\widetilde{\mathfrak{g}}$. For all $t \in \mathbb{R}$, let $\widetilde{\ell}_t = \exp(tX) \cdot \widetilde{\ell}$ and call $\widetilde{\pi}_t$ the unitary irreducible representation of $\widetilde{G}$ associated to $\widetilde{\ell}_t$. That is to say,

$$\widetilde{\pi}_t(\widetilde{g}) = \exp(tX) \cdot \widetilde{\pi}(\widetilde{g}) = \widetilde{\pi}(\exp(-tX)\widetilde{g}\exp(tX)) \; (\widetilde{g} \in \widetilde{G}),$$

where $\widetilde{\pi} = \widetilde{\pi}_0$. Let us put $B_t = \exp(tX)B\exp(-tX)$ $(t \in \mathbb{R})$, which is also a Vergne polarization of $\widetilde{\mathfrak{g}}$ at $\widetilde{\ell}_t$. Then it is well known that

$$\pi|_{\widetilde{G}} \simeq \int_{\mathbb{R}}^{\oplus} \widetilde{\pi}_t dt.$$

First, assume that $\mathfrak{k} \subset \widetilde{\mathfrak{g}}$. We indicate by

$$\widetilde{\pi}_t|_K \simeq \int_{\widehat{K}}^{\oplus} m_t(\sigma)\sigma d\widetilde{\nu}_t(\sigma) \; (t \in \mathbb{R})$$

the disintegration of $\widetilde{\pi}_t$ restricted to K, so

$$\nu = \int_{\mathbb{R}}^{\oplus} \widetilde{\nu}_t dt.$$

Let $\widetilde{\nu} = \widetilde{\nu}_0$. Consider the case where $Y \in \mathfrak{k}$. As Y belongs to the center of $\mathfrak{k}$, it follows that $m_t(\sigma) = m(\sigma)$ for ν-almost all $\sigma \in \widehat{K}$ and there exists a unique $t \in \mathbb{R}$ such that $m_t(\sigma) = m(\sigma)$, while $m_t(\sigma) = 0$ for all other $t \in \mathbb{R}$. In this situation, in fact:

$$\pi|_K = (\pi|_{\widetilde{G}})|_K \simeq \int_{\mathbb{R}}^{\oplus} \widetilde{\pi}_t|_K dt \simeq \int_{\mathbb{R}}^{\oplus} \int_{\widehat{K}}^{\oplus} m_t(\sigma)\sigma d\widetilde{\nu}_t(\sigma) dt \simeq \int_{\widehat{K}}^{\oplus} m(\sigma)\sigma d\nu(\sigma).$$

Let $\widetilde{S}^t_j(\sigma)$ be the section given by the induction hypothesis for the triple $(\widetilde{G}, \widetilde{\pi}_t, K)$ for $t \in \mathbb{R}$. As $\sigma = \sigma_t$ for some $t \in \mathbb{R}$, put $S_j(\sigma) = \exp(\mathbb{R}Y)\widetilde{S}^t_j(\sigma)$. We obtain then:

$$\int_{\widehat{K}} (\sum_{j=1}^{m(\sigma)} \int_{S_j(\sigma)} \langle \psi, a^j_\sigma(k)\rangle d\dot{k}) d\nu(\sigma)$$

$$= \int_{\mathbb{R}} dt \int_{\widehat{K}} (\sum_{j=1}^{m(\sigma)} \int_{\mathbb{R}} \int_{\widetilde{S}^t_j(\sigma)} \langle \psi, a^j_\sigma(\exp(yY)k)\rangle dyd\dot{k}) d\widetilde{\nu}_t(\sigma)$$

$$= \int_{\mathbb{R}} dt \int_{\widehat{K}} (\sum_{j=1}^{m(\sigma)} \int_{\mathbb{R}} \int_{\widetilde{S}^t_j(\sigma)} \langle \psi(\cdot \exp(tX)), \widetilde{a}^j_\sigma(k) e^{ity}\rangle dyd\dot{k}) d\widetilde{\nu}_t(\sigma)$$

(we apply the induction hypothesis to $(\widetilde{G}, \widetilde{\pi}_t, K)$)

$$= \int_{\mathbb{R}} \int_{\mathbb{R}} \psi(\exp(tX)) e^{ity} dydt$$

$$= \psi(e),$$

from the Fourier inversion formula for $\mathbb{R}$.

Now let $Y \notin \mathfrak{k}$. Set $\mathfrak{l} = \mathfrak{k} + \mathbb{R}Y$ and $L = \exp \mathfrak{l}$. L is an analytic subgroup of G and the restriction $\pi|_L$ is with finite multiplicities by formula (4.4.2) and the fact that the orbit Ω is saturated with respect to $\tilde{\mathfrak{g}}$. Let

$$\pi|_L \simeq \int_{\widehat{L}}^{\oplus} m'(\sigma)\sigma d\widehat{\nu}(\sigma)$$

be the disintegration of $\pi|_L$. Let $q : \mathfrak{l}^* \to \mathfrak{k}^*$ be the restriction map and $\kappa : \widehat{L} \to \widehat{K}$. Then ν is equivalent to the image $\kappa_*(\widehat{\nu})$, namely

$$\widehat{\nu} \simeq \int_{\widehat{K}}^{\oplus} \nu_\sigma d\nu(\sigma).$$

Here, for ν-almost everywhere, the intersection of the fiber $\kappa^{-1}(\sigma)$ with the spectrum of $\pi|_L$ consists of a finite set and the measure ν_σ is the counting measure of these points. Combining all these observations with our assertion for $\pi|_L$ (true by induction), we obtain the claim. Secondly, suppose that $\mathfrak{k} \not\subset \widetilde{\mathfrak{g}}$. Take X in $\mathfrak{k}$. Let $\widetilde{\mathfrak{k}} = \mathfrak{k} \cap \widetilde{\mathfrak{g}}$ and $\widetilde{K} = \exp \widetilde{\mathfrak{k}}$. By the Mackey subgroup theorem, we have

$$\pi|_K = \mathrm{ind}_{\widetilde{K}}^K (\widetilde{\pi}|_{\widetilde{K}}) \tag{4.4.7}$$

and then $\widetilde{\pi}|_{\widetilde{K}}$ turns out to have finite multiplicities. Assume first that the orbits $\omega(\sigma)$ are ν-almost everywhere non-saturated with respect to $\widetilde{\mathfrak{k}}$. If so, in the claimed formula we can ensure that $X \in \mathfrak{k}(\ell_\sigma)$ when we fix $\ell_\sigma|_{\widetilde{\mathfrak{k}}}$. Moreover, the elements g_j appearing there can be written as $g_j = \exp(yY)\widetilde{g}_j$, where y is in $\mathbb{R}$ and $\widetilde{g}_j \in \widetilde{G}$ does not depend on y. These $\exp(yY)$ are in the variable of ψ, and come out of ψ with the character $\chi_\ell(\exp(yY)) = e^{iy\ell(Y)} = 1$. On the other hand, there is the integral on

$$B[\ell_\sigma]/(B[\ell_\sigma] \cap kg_j^\sigma B(kg_j^\sigma)^{-1}) \cong \mathbb{R} \times B[\widetilde{\ell}_\sigma]/(B[\widetilde{\ell}_\sigma] \cap kg_j^\sigma B(kg_j^\sigma)^{-1}),$$

where $B[\widetilde{\ell}_\sigma] = B[\ell_\sigma] \cap \widetilde{\mathfrak{k}}$ is a polarization of $\widetilde{\mathfrak{k}}$ at $\widetilde{\ell}_\sigma = \ell_\sigma|_{\widetilde{\mathfrak{k}}}$. This integral concerns the variable $b = \exp(tX)\widetilde{b}$ for $t \in \mathbb{R}$, $\widetilde{b} \in B[\widetilde{\ell}_\sigma]/(B[\widetilde{\ell}_\sigma] \cap kg_j^\sigma B(kg_j^\sigma)^{-1})$ and the integral ψ is multiplied by the character

$$\chi_{\ell_j^\sigma}(\exp(tX)\widetilde{b}) = \chi_{\widetilde{g}_j \cdot \ell}(\widetilde{b})e^{ity}.$$

In the end, the theorem stems once again from the Fourier inversion formula for $\mathbb{R}$ and from the induction hypothesis. Finally, assume that the orbits $\omega(\sigma)$ are ν-almost everywhere saturated with respect to $\widetilde{\mathfrak{k}}$. Then in view of formula (4.4.2), the restriction $\pi|_{\widetilde{K}}$ emphasizes the finite multiplicities and all elements ℓ_j^σ are chosen so that $\ell_j^\sigma(X) = \ell_j^\sigma(Y) = 0$. In this situation, we have:

$$\pi|_{\widetilde{K}} = (\pi|_K)|_{\widetilde{K}} = \int_{\hat{K}}^{\oplus} m(\sigma)\sigma|_{\widetilde{K}} d\nu(\sigma)$$

$$\simeq \int_{\hat{K}}^{\oplus} \int_{\mathbb{R}}^{\oplus} m(\sigma)\sigma_t d\nu(\sigma) dt,$$

where $\sigma_t = \exp(tX) \cdot \sigma_0 \in \widehat{\widetilde{K}}$. On the other hand, all the $\widetilde{g}_j$ are taken in $\widetilde{G}$ and that allows us to bring everything back to the triple $(\widetilde{G}, \widetilde{\pi}, \widetilde{K})$: here the induction hypothesis applies, as $\widetilde{\pi}|_{\widetilde{K}}$ is with finite multiplicities. In fact, let

$$\widetilde{\pi}|_{\widetilde{K}} = \int_{\widehat{\widetilde{K}}}^{\oplus} m(\sigma)\sigma d\widetilde{\nu}(\sigma).$$

By formula (4.4.7), we have:

$$\pi|_K = \int_{\widehat{\widetilde{K}}}^{\oplus} m(\sigma) \mathrm{ind}_{\widetilde{K}}^K \sigma d\widetilde{\nu}(\sigma)$$

with all mutually inequivalent irreducible representations $\mathrm{ind}\widetilde{K}K\sigma$. Let us use twiddle symbols for all objects, and let $\psi \in \mathscr{H}_\pi^\infty$. For $\widetilde{\psi} := \psi_{|\widetilde{G}}$, we have:

$$\psi(e) = \widetilde{\psi}(e) = \int_{\widehat{\widetilde{K}}} \Big(\sum_{j=1}^{m(\sigma)} \int_{S_j(\sigma)} \langle \widetilde{\psi}, \widetilde{a}_\sigma^j(k)\rangle d\dot{k}\Big) d\widetilde{\nu}(\sigma)$$

$$= \int_{\widehat{\widetilde{K}}} \Big(\sum_{j=1}^{m(\sigma)} \int_{S_j(\sigma)} \langle \psi, a_\sigma^j(k)\rangle d\dot{k}\Big) d\widetilde{\nu}(\sigma)$$

$$= \int_{\widehat{K}} \Big(\sum_{j=1}^{m(\sigma)} \int_{S_j(\sigma)} \langle \psi, a_\sigma^j(k)\rangle d\dot{k}\Big) d\nu(\sigma),$$

all expected. This completes the proof of this theorem. □

Proof of Theorem 4.4.4

For $\theta \in \mathscr{D}(K)$ and $\psi \in \mathscr{H}_\pi^\infty$, we will calculate $\langle(\pi|_K)(\theta)(\psi), a_\sigma^j\rangle_{\mathscr{H}_\pi}$ as follows:

$$\langle(\pi|_K)(\theta)(\psi), a_\sigma^j\rangle_{\mathscr{H}_\pi} = \int_{B[\ell_\sigma]/(B[\ell_\sigma]\cap g_j^\sigma B(g_j^\sigma)^{-1})} ((\pi|_K)(\theta)\psi)(bg_j^\sigma)\chi_{\ell_j^\sigma}(b)d\dot{b}$$

$$= \int_{B[\ell_\sigma]/(B[\ell_\sigma]\cap g_j^\sigma B(g_j^\sigma)^{-1})} \int_K \theta(x)\psi(x^{-1}bg_j^\sigma)dx\,\chi_{\ell_j^\sigma}(b)d\dot{b}$$

$$= \int_K \int_{B[\ell_\sigma]/(B[\ell_\sigma]\cap g_j^\sigma B(g_j^\sigma)^{-1})} \theta(x^{-1})\psi(xbg_j^\sigma)\chi_{\ell_j^\sigma}(b)d\dot{b}dx$$

$$= \int_{K/B[\ell_\sigma]} \int_{B[\ell_\sigma]} \overline{\theta^*(\dot{x}y)}$$

$$\times \int_{B[\ell_\sigma]/(B[\ell_\sigma]\cap g_j^\sigma B(g_j^\sigma)^{-1})} \psi(\dot{x}ybg_j^\sigma)\chi_{\ell_j^\sigma}(b)d\dot{b}dyd\dot{x}$$

$$= \int_{K/B[\ell_\sigma]} \int_{B[\ell_\sigma]} \overline{\theta^*(\dot{x}y)\chi_{\ell_\sigma}(y)}dy$$

$$\times \int_{B[\ell_\sigma]/(B[\ell_\sigma]\cap g_j^\sigma B(g_j^\sigma)^{-1})} \psi(\dot{x}bg_j^\sigma)\chi_{\ell_j^\sigma}(b)d\dot{b}d\dot{x}$$

$$= \langle R_\sigma^j(\psi), \sigma(\theta^*)\delta_\sigma\rangle_{\mathscr{H}_\sigma}$$

$$= \langle \sigma(\theta)R_\sigma^j(\psi), \delta_\sigma\rangle_{\mathscr{H}_\sigma},$$

where

$$R^j_\sigma(\psi)(x) := \int_{B[\ell_\sigma]/(B[\ell_\sigma]\cap g^\sigma_j B(g^\sigma_j)^{-1})} \psi(\dot{x}bg^\sigma_j)\chi_{\ell^\sigma_j}(b)d\dot{b}, \ (x \in K)$$

designates the projector of $\mathscr{H}_\pi$ on a σ-component build using the semi-invariant generalized vector a^j_σ. In particular

$$\langle(\pi|_K)(\theta)(\psi), a^j_\sigma(k)\rangle_{\mathscr{H}_\pi} = \langle\sigma(\theta)R^j_\sigma(k)(\psi), \delta_\sigma\rangle_{\mathscr{H}_\sigma} \ (k \in S_j(\sigma)),$$

where $R^j_\sigma(k)(\psi)(x) := \langle\pi(x^{-1})\psi, a^j_\sigma(k)\rangle$ $(x \in K)$ is as in formula (4.4.6). Hence

$$\begin{aligned}\langle(\pi|_K)(\theta)(\psi), \mathfrak{d}_\pi\rangle &= (\pi|_K)(\theta)(\psi)(e)\\ &= \int_{\widehat{K}}\Big(\sum_{j=1}^{m(\sigma)}\int_{S_j(\sigma)}\langle(\pi|_K)(\theta)(\psi), a^j_\sigma(k)\rangle d\dot{k}\Big)d\nu(\sigma)\\ &= \int_{\widehat{K}}\Big(\sum_{j=1}^{m(\sigma)}\int_{S_j(\sigma)}\langle\sigma(\theta)R^j_\sigma(k)(\psi), \delta_\sigma\rangle_{\mathscr{H}_\sigma} d\dot{k}\Big)d\nu(\sigma),\end{aligned}$$

using Theorem 4.4.3, and this ends the proof. □

4.4.5 The Case of Normal Subgroups

We finally look at the situation where the subgroup K is normal in G. It follows that $\pi|_K$ disintegrates without multiplicities once we take in account our assumption. Let

$$\pi|_K \simeq \int_{\widehat{K}}^{\oplus} \sigma d\nu(\sigma),$$

be as in (4.4.1). The flag of ideals (4.4.3) can be chosen to pass through $\mathfrak{k}$, and then all polarizations $\mathfrak{b}[\ell_{|\mathfrak{k}}]$ can be chosen in $\mathfrak{b}[\ell]$ for all $\ell \in \mathfrak{O}$. Then by the finite-multiplicity condition, $\mathfrak{b}[\ell]$ coincides with $\mathfrak{b}[\ell_{|\mathfrak{k}}] + \mathfrak{g}(\ell)$. Hence the generalized vectors a_σ are just Dirac operators. Let

$$\langle a_\sigma, \psi\rangle = \overline{\psi(e)}$$

for $\psi \in \mathscr{H}_\pi^{+\infty}$. For $k \in S(\sigma)$, $\langle\psi, a_\sigma(k)\rangle = \mathfrak{d}_k(\psi)$, where $\mathfrak{d}_k$ designates the Dirac operator at k. Hence our Plancherel formula reads

$$\psi(e) = \int_{\widehat{K}}\int_{S(\sigma)} \psi(k)d\dot{k}d\nu(\sigma).$$

On the other hand,

$$\langle(\pi|_K)(\theta)(\psi), \mathfrak{d}_\pi\rangle_{\mathscr{H}_\pi} = \int_{\widehat{K}}\int_{S(\sigma)} \langle\sigma(\theta)R_\sigma(k)(\psi), \delta_\sigma\rangle_{\mathscr{H}_\sigma} d\dot{k}d\nu(\sigma),$$

where $R_\sigma(k)(\psi) = \langle\psi, a_\sigma(k)\rangle$.

4.4.6 An Intertwining Operator

In this section we will write an explicit intertwining operator between $\pi|_K$ and its decomposition (4.4.1). Specifically, we shall prove:

Theorem 4.4.9 *Assume that the restriction $\pi|_K$ is with finite multiplicities. Then*

$$(R\psi)(k) = \int_{\widehat{K}} \sum_{j=1}^{m(\sigma)} \langle\pi(k^{-1})\psi, a_\sigma^j\rangle d\nu(\sigma), \tag{4.4.8}$$

for all $\psi \in \mathscr{H}_\pi^\infty$ and $k \in K$, defines intertwining operator between $\pi|_K$ and its canonical central disintegration (4.4.1).

Remark first that the operator R commutes with the actions of K. Indeed, for all x and $k \in K$,

$$\begin{aligned}\Big(\Big(\int_{\widehat{K}}^{\oplus} m(\sigma)\sigma(x)d\nu(\sigma)\Big)(R\psi)\Big)(k) =&(R\psi)(x^{-1}k)\\ =&\int_{\widehat{K}} \sum_{j=1}^{m(\sigma)} \langle\pi((x^{-1}k)^{-1})\psi, a_\sigma^j\rangle d\nu(\sigma)\\ =&\int_{\widehat{K}} \sum_{j=1}^{m(\sigma)} \langle\pi(k^{-1})\pi(x)\psi, a_\sigma^j\rangle d\nu(\sigma)\\ =&(R(\pi|_K(x)\psi))(k).\end{aligned}$$

As for the norms, we will follow the same steps and keep the same notations used in Theorem 4.4.3, proceeding by induction on $\dim G + \dim G/K$. If this number is one or two, our assertion is clear. As before, we can assume that $\mathfrak{k}$ contains the center $\mathfrak{z}$ of $\mathfrak{g}$ and that $\mathfrak{a} = \mathfrak{z} \cap \ker(\ell)$ is trivial. Hence we fall back in the case $\mathfrak{z} = \mathbb{R}Z$ with $\ell(Z) = 1$. First, suppose that $\mathfrak{k} \subset \widetilde{\mathfrak{g}}$. When $Y \in \mathfrak{k}$, we see that $g_j^\sigma = \exp(-\ell_\sigma(Y)X)\widetilde{g}_j^\sigma$, $\widetilde{g}_j^\sigma$ designate the elements of $\widetilde{G}$. Let $\psi \in \mathscr{H}_\pi^\infty$. For all $t \in \mathbb{R}$, put

$$\psi_t(\widetilde{g}) = \psi(\exp(tX)\widetilde{g}) \ (\widetilde{g} \in \widetilde{G}).$$

Then $\psi_t \in \mathscr{H}_{\widetilde{\pi}}^{\infty}$ and, for all $k \in K, t \in \mathbb{R}, \widetilde{g} \in \widetilde{G}$,

$$((\pi|_K)(k)\psi)(\exp(tX)\widetilde{g}) = \psi(k^{-1}\exp(tX)\widetilde{g})$$
$$=\psi(\exp(tX)\exp(-tX)k^{-1}\exp(tX)\widetilde{g}) = (\widetilde{\pi}(\exp(-tX)k\exp(tX))\psi_t(\widetilde{g}).$$

By setting $K_t = \exp(-tX)K\exp(tX)$ $(t \in \mathbb{R})$ and by transferring the structure, we see that the disintegration

$$\pi|_K \simeq \int_{\mathbb{R}}^{\oplus} \widetilde{\pi}_t|_K dt$$

becomes

$$\pi|_K \simeq \int_{\mathbb{R}}^{\oplus} \widetilde{\pi}|_{K_t} dt.$$

If so,

$$\|\psi\|_{\mathscr{H}_\pi}^2 = \int_{\mathbb{R}} \|\psi_t\|_{\mathscr{H}_{\widetilde{\pi}}}^2 dt = \int_{\mathbb{R}} \|(R_t)(\psi_t)\|^2 dt = \|R(\psi)\|^2 \tag{4.4.9}$$

by the induction hypothesis. Now let $Y \notin \mathfrak{k}$. Put $\mathfrak{l} = \mathfrak{k}+\mathbb{R}Y$ and $L = \exp \mathfrak{l}$, as before. L is an analytic subgroup of G and the restriction $\pi|_L$ is with finite multiplicities. Recall the disintegration of

$$\pi|_L \simeq \int_{\widehat{L}}^{\oplus} m'(\sigma)\sigma d\widehat{\nu}(\sigma),$$

the restriction map $q : \mathfrak{l}^* \to \mathfrak{k}^*$ and $\kappa : \widehat{L} \to \widehat{K}$. Then ν is equivalent to the image $\kappa_*(\widehat{\nu})$:

$$\widehat{\nu} \simeq \int_{\widehat{K}}^{\oplus} \nu_\sigma d\nu(\sigma).$$

This corresponds to the decomposition of the measure dt in formula (4.4.9) using the projection κ:

$$dt = \int_{\widehat{K}}^{\oplus} \mu_\sigma d\nu(\sigma)$$

with the counting measure on the intersection of the fiber $\kappa^{-1}(\sigma)$ with the spectrum of $\pi|_L$. By combining all these observations with the assertion for $\pi|_L$, which holds by induction, we obtain the theorem.

Second, suppose that $\mathfrak{k} \not\subset \widetilde{\mathfrak{g}}$. Let $\widetilde{\mathfrak{k}} = \mathfrak{k} \cap \widetilde{\mathfrak{g}}$ and $\widetilde{K} = \exp \widetilde{\mathfrak{k}}$. We fix $X \in \mathfrak{k}$ such that $\mathfrak{k} = \mathbb{R}X + \widetilde{\mathfrak{k}}$. Assume first that all coadjoint orbits $\omega(\sigma)$ of K associated to σ are ν-almost everywhere non-saturated with respect to $\widetilde{\mathfrak{k}}$. If this holds, we can ensure that there is $X(\widetilde{\sigma}) \in \mathfrak{k}(\ell_\sigma)$ verifying $\mathfrak{k} = \mathbb{R}X(\widetilde{\sigma}) + \widetilde{\mathfrak{k}}$ provided we fix $\widetilde{\ell_\sigma} = \ell_\sigma|_{\widetilde{\mathfrak{k}}}$. Let $\psi \in \mathscr{H}_\pi^\infty$. Put

$$\widetilde{\psi}_y(\widetilde{g}) = \int_{\mathbb{R}} \psi(\exp(tX(\widetilde{\sigma}))\widetilde{g})e^{ity}dt \ (y \in \mathbb{R}, \widetilde{g} \in \widetilde{G})$$

and

$$\psi_y(\widetilde{g}) = \int_{\mathbb{R}} \psi(\exp(tX)\widetilde{g})e^{ity}dt \ (y \in \mathbb{R}, \widetilde{g} \in \widetilde{G}).$$

Then $\widetilde{\psi}_y$, ψ_y belong to $\mathscr{H}_{\widetilde{\sigma}}^\infty$. Let

$$\Psi_\sigma^j(x) = \langle \pi(x^{-1})\psi, a_\sigma^j \rangle \ (x \in K)$$

$$\Psi_{\widetilde{\sigma}}^y(x) = \langle \pi(x^{-1})(\widetilde{\psi}_y), a_{\widetilde{\sigma}}^j \rangle \ (x \in \widetilde{K})$$

and

$$\Psi_\sigma^y(x) = \langle \pi(x^{-1})\psi_y, a_{\widetilde{\sigma}}^j \rangle \ (x \in \widetilde{K}).$$

Then

$$\|\Psi_\sigma^j\|^2_{\mathscr{H}_\sigma} = \|\Psi_{\widetilde{\sigma}}^y\|^2_{\mathscr{H}_{\widetilde{\sigma}}} = \|\Psi_\sigma^y\|^2_{\mathscr{H}_{\widetilde{\sigma}}}.$$

Now applying the induction hypothesis to the triple $(\widetilde{G}, \widetilde{K}, \widetilde{\pi})$ and recalling the Fourier transform for $\mathbb{R}$ is an isometry, gives the claim. Finally, suppose that the orbits $\omega(\sigma)$ are ν-almost everywhere saturated with respect to $\widetilde{\mathfrak{k}}$. Then the restriction $\pi|_{\widetilde{K}}$ is of finite multiplicities and all elements ℓ_j^σ are such that $\ell_j^\sigma(X) = \ell_j^\sigma(Y) = 0$. This allows use induction on the triple $(\widetilde{G}, \widetilde{K}, \widetilde{\pi})$. In fact, let $\psi \in \mathscr{H}_\pi^\infty$. For $t \in \mathbb{R}$, we define as before an element ψ_t of $\mathscr{H}_{\widetilde{\pi}}^\infty$ by

$$\psi_t(\widetilde{g}) = \psi(\exp(tX)\widetilde{g}) \ (\widetilde{g} \in \widetilde{G}).$$

There only remains to apply induction to ψ_t and integrate the formula thus obtained in $t \in \mathbb{R}$.

Chapter 5
Polynomial Conjectures

5.1 Introduction

The aim of this chapter is to present two conjectures which are interesting and still open (cf. [10, 13, 41, 66]). Here we briefly explain them. Let $G = \exp \mathfrak{g}$ be a connected and simply connected nilpotent Lie group with Lie algebra $\mathfrak{g}$. Let $f \in \mathfrak{g}^*$, $\mathfrak{h} \in S(f, \mathfrak{g})$ and $H = \exp \mathfrak{h}$. As usual, we define the unitary character χ_f of H by $\chi_f(\exp X) = e^{if(X)}$, $X \in \mathfrak{h}$ and construct the monomial representation $\tau = \mathrm{ind}_H^G \chi_f$ of G. Now, we consider the algebra $D_\tau(G/H)$ of the G-invariant differential operators acting on the line bundle associated to the data (G, H, χ_f) over the base space G/H. Assume that τ is of finite multiplicities, namely that τ has finite multiplicities in its canonical central decomposition. Then the polynomial conjecture says that the algebra $D_\tau(G/H)$ is isomorphic to the algebra of H-invariant polynomial functions on the affine space $f + \mathfrak{h}^\perp$ of $\mathfrak{g}^*$.

We also consider the corresponding statement regarding the restriction. Let $G = \exp \mathfrak{g}$ be as above, π an irreducible unitary representation of G and $K = \exp \mathfrak{k}$ an analytic subgroup of G. Let $\mathscr{U}(\mathfrak{g})$ be the enveloping algebra of $\mathfrak{g}_\mathbb{C}$. We denote by $(\mathscr{U}(\mathfrak{g})/\ker(\pi))^\mathfrak{k}$ the subalgebra of $\mathfrak{k}$-invariant elements of the quotient algebra $\mathscr{U}(\mathfrak{g})/\ker(\pi)$ and by $D_\pi(G)^K$ its image under the homomorphism π. Assume that the restriction $\pi|_K$ of π to K is of finite multiplicities. The polynomial conjecture says that the algebra $D_\pi(G)^K$ is isomorphic to the algebra of K-invariant polynomial functions on the coadjoint G-orbit associated to π.

Corresponding circumstances can be found in [70].

A. Baklouti et al., *Representation Theory of Solvable Lie Groups and Related Topics*, Springer Monographs in Mathematics,
https://doi.org/10.1007/978-3-030-82044-2_5

5.2 The Case of Induced Representations

Let $G = \exp \mathfrak{g}$ be a connected and simply connected nilpotent Lie group with Lie algebra $\mathfrak{g}$. Given a closed connected subgroup $H = \exp \mathfrak{h}$ of G and its unitary character χ, we consider the induced representation $\tau = \mathrm{ind}_H^G \chi$ of G. Recall the disintegration formula (1.6.2):

$$\tau \simeq \int_{\widehat{G}}^{\oplus} m(\pi) d\mu(\pi), \tag{5.2.1}$$

i.e. the canonical central decomposition of τ. As in Chap. 1, the multiplicities $m(\pi)$ are either uniformly bounded, or $m(\pi) \equiv \infty$ almost everywhere with respect to the measure μ. Depending on these two situations, we say that τ has finite or infinite multiplicities. We shall now define an algebra of invariant differential operators. We denote by $C^\infty(G, \tau)$ the space of C^∞-functions ψ on G which satisfy the covariance relation

$$\psi(gh) = \chi^{-1}(h)\psi(g) \ (g \in G, h \in H).$$

Let Diff(G, τ) be the algebra of C^∞-differential operators which fix the space $C^\infty(G, \tau)$. The object of our study is $D_\tau(G/H)$, which is constituted by the restrictions $D|C^\infty(G, \tau)$ of $D \in$ Diff(G, τ) that commute with the left translations of G in $C^\infty(G, \tau)$. We know [74] that the algebra $D_\tau(G/H)$ is commutative if and only if τ has finite multiplicities. The character χ is written as

$$\chi(\exp X) = e^{if(X)} \ (X \in \mathfrak{h}, \ i = \sqrt{-1})$$

using a certain linear form $f \in \mathfrak{g}^*$ satisfying $f([\mathfrak{h}, \mathfrak{h}]) = \{0\}$. The affine space $\Gamma_\tau = f + \mathfrak{h}^\perp$ in $\mathfrak{g}^*$ was introduced in (1.4.8). The following conjecture of Corwin and Greenleaf [40, 41] is well known:

Polynomial Conjecture Assume that τ has finite multiplicities. Is the algebra $D_\tau(G/H)$ isomorphic to the algebra $\mathbb{C}[\Gamma_\tau]^H$ of H-invariant polynomial functions on Γ_τ?

We first describe $D_\tau(G/H)$ in terms of the enveloping algebra of $\mathfrak{g}_\mathbb{C}$. For $A \in \mathscr{U}(\mathfrak{g})$, we denote by $R(A)$ its action from the right. Let $\mathscr{U}(\mathfrak{g}, \tau)$ be the set of $A \in \mathscr{U}(\mathfrak{g})$ such that $R(A)$ fixes $C^\infty(G, \tau)$. Let $\mathfrak{a}_\tau$ be the vector subspace of $\mathscr{U}(\mathfrak{g})$ spanned by $Y + if(Y)(Y \in \mathfrak{h})$ and $\mathscr{U}(\mathfrak{g})\mathfrak{a}_\tau$ the left ideal of $\mathscr{U}(\mathfrak{g})$ generated by $\mathfrak{a}_\tau$. Then

$$\mathscr{U}(\mathfrak{g}, \tau) = \{A \in \mathscr{U}(\mathfrak{g}); [Y, A] \in \mathscr{U}(\mathfrak{g})\mathfrak{a}_\tau, \forall Y \in \mathfrak{h}\}$$

and the mapping $A \mapsto R(A)|C^\infty(G, \tau)$ gives us a homomorphism of $\mathscr{U}(\mathfrak{g}, \tau)$ onto $D_\tau(G/H)$ with kernel $\mathscr{U}(\mathfrak{g})\mathfrak{a}_\tau$. To sum up,

$$D_\tau(G/H) \cong \mathscr{U}(\mathfrak{g}, \tau)/\mathscr{U}(\mathfrak{g})\mathfrak{a}_\tau.$$

Assume hereafter that τ has finite multiplicities and recall briefly Penney's Plancherel formula [136] for τ. For a unitary representation ρ of G, we denote by $\mathscr{H}_\rho$, $\mathscr{H}_\rho^\infty$ and $\mathscr{H}_\rho^{-\infty}$ the Hilbert space of ρ, the space of C^∞-vectors of ρ and the anti-dual space of $\mathscr{H}_\rho^\infty$, whose elements are sometimes called generalized vectors of ρ. For $a \in \mathscr{H}_\rho^{\pm\infty}$ and $b \in \mathscr{H}_\rho^{\mp\infty}$, we denote by $\langle a, b\rangle$ the image of b under a. Hence $\langle a, b\rangle = \overline{\langle b, a\rangle}$. Let $\mathscr{D}(G)$ be the space of C^∞-functions on G with compact support. For $\phi \in \mathscr{D}(G)$, we construct an element ϕ^χ of $\mathscr{H}_\tau^\infty$:

$$\phi^\chi(g) = \int_H \phi(gh)\chi(h)dh \ (g \in G)$$

for a fixed Haar measure dh on H. Then Penney's Plancherel formula for τ reads

$$\phi^\chi(g) = \int_{\widehat{G}} \sum_{k=1}^{m(\pi)} \langle \pi(\phi_g)a_\pi^k, a_\pi^k\rangle d\mu(\pi) \ (\forall g \in G), \tag{5.2.2}$$

where $\phi_g(x) = \phi(gx)$ for any $x \in G$. Here the tempered distributions a_π^k belong to

$$(\mathscr{H}_\pi^{-\infty})^{H,\chi} = \{a \in \mathscr{H}_\pi^{-\infty};\, \pi(h)a = \chi(h)a,\, h \in H\}.$$

In particular, if we take $g = e$, the unit element of G, and $\phi = \varphi^* * \varphi$, $\varphi \in \mathscr{D}(G)$, formula (5.2.2) gives

$$\int_{G/H} |\varphi^\chi(x)|^2 d\dot{x} = \int_{\widehat{G}} \sum_{k=1}^{m(\pi)} \|\pi(\varphi)a_\pi^k\|^2 d\mu(\pi).$$

This means that the map

$$\mathscr{H}_\tau^\infty \ni \varphi^\chi \mapsto \int_{\widehat{G}} \sum_{k=1}^{m(\pi)} \pi(\varphi)a_\pi^k d\mu(\pi) \ (\varphi \in \mathscr{D}(G))$$

gives an intertwining operator between equivalence (5.2.1).

We can make a_π^k explicit as follows [62]. Almost everywhere in Γ_τ with respect to the measure μ, the coadjoint orbit $\Omega(\pi)$ of G corresponding to π meets $m(\pi)$ H-orbits, say $\omega_\pi^1, \omega_\pi^2, \ldots, \omega_\pi^{m(\pi)}$. We take $\ell \in \omega_\pi^k$ and a polarization $\mathfrak{b}$ at ℓ in order to realize $\pi = \mathrm{ind}_B^G \chi_\ell$, where $B = \exp \mathfrak{b}$ and $\chi_\ell(\exp X) = e^{i\ell(X)} (X \in \mathfrak{b})$. Now, there exists an invariant measure $d\dot{h}$ on the homogeneous space $H/(H\cap B)$ such that

$$\langle a_\pi^k, \psi\rangle = \int_{H/(H\cap B)} \overline{\psi(h)\chi(h)} d\dot{h} \tag{5.2.3}$$

for any $\psi \in \mathscr{H}_\pi^\infty$. Applying $R(X), X \in \mathscr{U}(\mathfrak{g}, \tau)$, to formula (5.2.2), an easy calculation leads us to

$$(R(X)\phi^X)(g) = \int_{\widehat{G}} \sum_{k=1}^{m(\pi)} \langle \pi(\phi_g)a_\pi^k, \pi(\overline{X})a_\pi^k \rangle d\mu(\pi) \ (g \in G). \tag{5.2.4}$$

Here we realize $\pi = \pi_\ell$ at $\ell \in \Gamma_\tau$ and define a tempered distribution a_ℓ by the same formula as (5.2.3). We expect, for any $X \in \mathscr{U}(\mathfrak{g}, \tau)$, that $\pi(\overline{X})a_\ell$ is a multiple of a_ℓ almost everywhere in Γ_τ. In fact, we have:

Theorem 5.2.1 ([67]) *Assume that τ has finite multiplicities. Let $X \in \mathscr{U}(\mathfrak{g}, \tau)$. Then $\pi(\overline{X})a_\ell$ is a multiple of a_ℓ at μ-almost every point $\ell \in \Gamma_\tau$. Namely, these a_ℓ are H-semi-invariant tempered eigendistributions for all operators $\pi(\overline{X})$, $X \in \mathscr{U}(\mathfrak{g}, \tau)$. Therefore there is a character $\lambda_\ell : \mathscr{U}(\mathfrak{g}, \tau) \to \mathbb{C}$ such that*

$$\overline{X}a_\ell = \pi(\overline{X})a_\ell = \overline{\lambda_\ell(X)}a_\ell$$

for any $X \in \mathscr{U}(\mathfrak{g}, \tau)$.

Proof We set $P_X(\ell) = \lambda_\ell(X)$ for $X \in \mathscr{U}(\mathfrak{g}, \tau)$. Using an intertwining operator, we easily see that the function P_X is H-invariant. In fact, τ has finite multiplicities if and only if $\mathfrak{h} + \mathfrak{g}(\ell)$ is a Lagrangian subspace for the bilinear form B_ℓ almost everywhere in Γ_τ. And under the latter condition, the function $P_X(\ell)$ does not depend on the choice of the polarization which realizes π at $\ell \in \Omega(\pi)$ (cf. [62]). Note first that $\pi(\bar{A})a_\ell = 0$ for $A \in \mathscr{U}(\mathfrak{g})\mathfrak{a}_\tau$. Moreover, we can suppose that $\mathfrak{h}$ contains the center $\mathfrak{z}$ of $\mathfrak{g}$. We proceed by induction on dim G. Passing to the quotient by $\exp(\mathfrak{z} \cap \ker(f))$ if necessary, it suffices to treat the case where $\mathfrak{z} \cap \ker(f) = \{0\}$. Let $\mathfrak{z} = \mathbb{R}Z$, $f(Z) = 1$. It is well known [93] that there are two elements X, Y in $\mathfrak{g} \cap \ker(f)$ such that $[X, Y] = Z$ and $\mathfrak{g} = \mathbb{R}X + \mathfrak{g}_0$, where $\mathfrak{g}_0$ is the centralizer of Y in $\mathfrak{g}$. Furthermore we can take $\mathfrak{b}$ in $\mathfrak{g}_0$. Put $G_0 = \exp(\mathfrak{g}_0)$, $\ell_0 = \ell|_{\mathfrak{g}_0} \in \mathfrak{g}_0^*$ and $\pi_0 = \mathrm{ind}_B^{G_0} \chi_{\ell_0}$.

We distinguish two cases. First, let $\mathfrak{h} \subset \mathfrak{g}_0$. Theorem 5.4 (b) in [40] says that $\mathscr{U}(\mathfrak{g}, \tau) = \mathscr{U}(\mathfrak{g}_0, \tau_0) + \mathscr{U}(\mathfrak{g})\mathfrak{a}_\tau$ with $\tau_0 = \mathrm{ind}_H^{G_0} \chi$. Hence it suffices to apply the induction hypothesis to G_0, because a_ℓ is regarded as a_{ℓ_0} in $(\mathscr{H}_{\pi_0}^{-\infty})^{H,\chi}$.

Second, let $\mathfrak{h} \not\subset \mathfrak{g}_0$. We take X in $\mathfrak{h} \cap \ker(f)$ and put $\mathfrak{h}_0 = \mathfrak{h} \cap \mathfrak{g}_0$, $H_0 = \exp(\mathfrak{h}_0)$, $\tau_0 = \mathrm{ind}_{H_0}^{G_0} \chi_{f_0}$ and $\Gamma_{\tau_0} = f_0 + (\mathfrak{h}_0)^\perp \subset (\mathfrak{g}_0)^*$. Then $\mathfrak{h}_0 + \mathfrak{g}_0(\ell_0)$ is a Lagrangian subspace almost everywhere in Γ_{τ_0}. Therefore, τ_0 has finite multiplicity. Each element $A \in \mathscr{U}(\mathfrak{g})$ is written as $A = \sum_{j=0}^d A_j X^j$ for some $A_j \in \mathscr{U}(\mathfrak{g}_0)$, $0 \le j \le d$. So, $A \equiv A_0$ modulo $\mathscr{U}(\mathfrak{g})\mathfrak{a}_\tau$ and $\pi(\overline{A})a_\ell = \pi(\overline{A_0})a_\ell$. Moreover, if $A \in \mathscr{U}(\mathfrak{g}, \tau)$, $[W, A_0] \in \mathscr{U}(\mathfrak{g}_0)\mathfrak{a}_{\tau_0}$ for any $W \in \mathfrak{h}$. Recall that $\mathfrak{b} \subset \mathfrak{g}_0$. This means that $\mathfrak{b}$ is a polarization at both $\ell \in \mathfrak{g}^*$ and $\ell_0 \in (\mathfrak{g}_0)^*$. Let us compute $\langle \pi(\overline{A_0})a_\ell, \psi \rangle$ for $\psi \in \mathscr{H}_\pi^\infty$ having the form $\psi(\exp(xX)g_0) = \zeta(x)\xi(g_0)(x \in \mathbb{R}, g_0 \in G_0)$ with $\zeta \in \mathscr{S}(\mathbb{R})$, $\xi \in \mathscr{H}_{\pi_0}^\infty$. For $V \in \mathscr{U}(\mathfrak{g})$, we denote by $g{\cdot}V$ the image of V under the adjoint action of $g \in G$ and by ${}^tV$ the image of V under the principal anti-automorphism of $\mathcal{U}(\mathfrak{g})$. Put $h(x) = \exp(xX) \in H (\forall x \in \mathbb{R})$. We then verify

$$\begin{aligned}
\langle \pi(\overline{A_0}) a_\ell, \psi \rangle &= \langle a_\ell, \pi({}^t A_0)\psi \rangle \\
&= \int_{\mathbb{R}} \overline{\zeta(x)} \left(\int_{H_0/(H_0 \cap B)} \overline{(\pi_0({}^t(h(-x)\cdot A_0))\xi)(h_0)\chi(h_0) d\dot{h}_0} \right) dx \\
&= \int_{\mathbb{R}} \overline{\zeta(x)} \langle \pi_0(\overline{h(-x)\cdot A_0}) a_{\ell_0}, \xi \rangle dx \\
&= \Big(\int_{\mathbb{R}} \overline{\zeta(x)} dx \Big) \langle \pi_0(\overline{A_0}) a_{\ell_0}, \xi \rangle,
\end{aligned}$$

where $a_{\ell_0} \in (\mathcal{H}_{\pi_0}^{-\infty})^{H_0, \chi}$ is given by

$$\langle a_{\ell_0}, \xi \rangle = \int_{H_0/(H_0 \cap B)} \overline{\xi(h_0)\chi(h_0)} d\dot{h}_0$$

if $d\dot{h} = dx d\dot{h}_0$ for $h = \exp(xX)h_0 \in H$. By induction hypothesis, $\pi_0(\overline{A_0}) a_{\ell_0} = \overline{\lambda_{\ell_0}(A_0)} a_{\ell_0}$ and hence $\langle \pi(\overline{A}) a_\ell, \psi \rangle = \overline{\lambda_{\ell_0}(A_0)} \langle a_\ell, \psi \rangle$ for $\psi = \zeta \otimes \xi$. Since $\mathcal{S}(\mathbb{R}) \widehat{\otimes} \mathcal{H}_{\pi_0}^\infty \cong \mathcal{H}_\pi^\infty$, it follows that $\pi(\overline{A}) a_\ell = \overline{\lambda_{\ell_0}(A_0)} a_\ell$ for $A \in \mathcal{U}(\mathfrak{g}, \tau)$. □

5.2.1 Towards the Conjecture

It descends from the Plancherel formula for τ that the homomorphism

$$D_\tau(G/H) \cong \mathcal{U}(\mathfrak{g}, \tau)/\mathcal{U}(\mathfrak{g})\mathfrak{a}_\tau \ni [X] \mapsto P_X,$$

where $[X]$ denotes the canonical image of X, is injective. To establish the conjecture, we have to address following two questions.

1. Does the function P_X extend to a polynomial function on Γ_τ?
2. If the answer to (1) is yes, is the homomorphism

$$\mathcal{U}(\mathfrak{g}, \tau)/\mathcal{U}(\mathfrak{g})\mathfrak{a}_\tau \ni [X] \mapsto P_X \in \mathbb{C}[\Gamma_\tau]^H$$

surjective?

We denote by $S(\mathfrak{g})$ the symmetric algebra of $\mathfrak{g}_\mathbb{C}$ and regard its elements as polynomial functions on $i\mathfrak{g}^*, i = \sqrt{-1}$. Then $\mathbb{C}[\Gamma_\tau]^H$ is identified with $(S(\mathfrak{g})/S(\mathfrak{g})\overline{\mathfrak{a}_\tau})^H$, the set of all H-invariant elements under the adjoint action. Therefore we have $(S(\mathfrak{g})/S(\mathfrak{g})\overline{\mathfrak{a}_\tau})^H$ on one side and $\mathcal{U}(\mathfrak{g}, \tau)/\mathcal{U}(\mathfrak{g})\mathfrak{a}_\tau = (\mathcal{U}(\mathfrak{g})/\mathcal{U}(\mathfrak{g})\mathfrak{a}_\tau)^H$ on the other, and the above two questions are actually asking about an isomorphism

$$(\mathscr{U}(\mathfrak{g})/\mathscr{U}(\mathfrak{g})\mathfrak{a}_\tau)^H \ni [X] \mapsto P_X \in (S(\mathfrak{g})/S(\mathfrak{g})\overline{\mathfrak{a}_\tau})^H.$$

So, our fundamental question is:

Question 1 If we are given $X \in \mathscr{U}(\mathfrak{g}, \tau)$, how do we compute the function $P_X(\ell)$?

Since G is nilpotent, the symmetrization map $\beta : S(\mathfrak{g}) \to \mathscr{U}(\mathfrak{g})$ will be useful. In fact, this is the case for symmetric spaces [21, 22]. If $X \in \mathscr{U}(\mathfrak{g}, \tau)$ is of degree less than or equal to 1 in $\mathscr{U}(\mathfrak{g})$, then $X = \lambda W + \alpha$ with $W \in \mathfrak{g}$ and $\lambda, \alpha \in \mathbb{C}$ such that $[\mathfrak{h}, W] \subset \mathfrak{h} \cap \ker(f)$. Hence $W \in \mathfrak{h} + \mathfrak{g}(\ell)$ almost everywhere in Γ_τ, because $\mathfrak{h} + \mathfrak{g}(\ell)$ is Lagrangian. We compute:

$$\begin{aligned}\langle \overline{P_X(\ell)} a_\ell, \phi\rangle &= \langle \overline{X} a_\ell, \phi\rangle = \langle \overline{\alpha} a_\ell, \phi\rangle - \langle \overline{\lambda} a_\ell, W\phi\rangle \\ &= \langle \overline{\alpha} a_\ell, \phi\rangle - \overline{\lambda} \int_{H/(H\cap B)} \overline{(L(W)\phi)(h)\chi_f(h)}dh,\end{aligned}$$

where L represents the left action. Put $\mathfrak{h}' = \mathfrak{h} + \mathbb{R}W$ and $H' = \exp(\mathfrak{h}')$. Then the last integral is identified with

$$\begin{aligned}\int_{H'/(H'\cap B)} \overline{(L(W)\phi)(h')\chi_\ell(h')}dh' &= \overline{i\ell(W)} \int_{H'/(H'\cap B)} \overline{\phi(h')\chi_\ell(h')}dh' \\ &= \overline{i\ell(W)} \int_{H/(H\cap B)} \overline{\phi(h)\chi_\ell(h)}dh.\end{aligned}$$

Namely,

$$\overline{P_X(\ell)} a_\ell = \overline{(\alpha - \lambda i\ell(W))} a_\ell = \beta^{-1}(\overline{X})(i\ell) a_\ell,$$

that is to say, $P_X(\ell) = \overline{\beta^{-1}(\overline{X})(i\ell)}$. When $X \in \mathscr{U}(\mathfrak{g})$ belongs to the center $\mathscr{Z}(\mathfrak{g})$ of $\mathscr{U}(\mathfrak{g})$, the result of Kirillov says that we have the same conclusion.

Example 5.2.2 Let $\mathfrak{g} = \langle X, U, V, E, Z\rangle_\mathbb{R}$ with non-zero brackets:

$$[U, V] = E, [X, U] = V, [X, V] = Z.$$

Hence the center $\mathfrak{z}$ of $\mathfrak{g}$ is equal to $\mathbb{R}Z \oplus \mathbb{R}E$. Let $G = \exp \mathfrak{g}$, $f = E^*$, $\mathfrak{h} = \mathbb{R}X + \mathbb{R}E$, $H = \exp \mathfrak{h}$ and $\tau = \mathrm{ind}_H^G \chi_f$. If we take $A = 2UZ - V^2$ and $W = A - 2EX$, we get $[\mathfrak{h}, A] = \{0\}$ and $W \in \mathscr{Z}(\mathfrak{g})$. For $\ell \in \Gamma_\tau = f + \mathfrak{h}^\perp$, we have

$$\mathfrak{g}(\ell) = \mathfrak{z} + \mathbb{R}(X - \ell(Z)U + \ell(V)V),$$

and this means that τ has finite multiplicities. We see that A and W are two representatives of the same element in $D_\tau(G/H) \cong \mathscr{U}(\mathfrak{g}, \tau)/\mathscr{U}(\mathfrak{g})\mathfrak{a}_\tau$ and

$$Aa_\ell = Wa_\ell = \beta^{-1}(W)(i\ell)a_\ell = \beta^{-1}(A)(i\ell)a_\ell = (\ell(V)^2 - 2\ell(Z)\ell(U))a_\ell.$$

Consequently,

$$A^3a_\ell = W^3a_\ell = (\ell(V)^2 - 2\ell(Z)\ell(U))^3a_\ell = \beta^{-1}(W^3)(i\ell)a_\ell$$

since $W^3 \in \mathscr{Z}(\mathfrak{g})$. On the other hand, a direct calculation in $S(\mathfrak{g})$ shows

$$\beta^{-1}(A^3) = (2UZ - V^2)^3 - 2Z^2E^2.$$

Hence

$$A^3a_\ell \neq \beta^{-1}(A^3)(i\ell)a_\ell.$$

This example forces us, if we use the symmetrization map β in full, to carefully choose a representative in $\mathscr{U}(\mathfrak{g}, \tau)$.

For a moment we focus on the rational generators of $D_\tau(G/H)$ introduced in [40]. Let us consider a unipotent representation of G on a real finite-dimensional vector space V. Let $v \in V$ be a G-invariant vector. For a fixed vector $x \in V$, let $L_x = x + \mathbb{R}v$ be the line in V of direction v passing through x. Then we have two possibilities: either $L_x \cap G{\cdot}x = L_x$ or $L_x \cap G{\cdot}x = \{x\}$ (cf. [142]). In the first case we say that the G-orbit $G{\cdot}x$ is saturated in the direction $\mathbb{R}v$, in the second case that it is non-saturated in the direction $\mathbb{R}v$. We shall apply this fact to the coadjoint representation of G (or a subgroup of G). Here, the invariant vector is a linear form which vanishes on an ideal $\mathfrak{g}'$ of codimension 1 in $\mathfrak{g}$. In this situation we also say that the orbit in question is either saturated or non-saturated with respect to $\mathfrak{g}'$.

We recall here e-central elements defined by Corwin and Greenleaf [40]. Let $\{X_1, \dots, X_n\}$ be a strong Malcev basis of $\mathfrak{g}$, $\{X_1^*, \dots, X_n^*\}$ the dual basis of $\mathfrak{g}^*$ and $(\ell_1, \dots, \ell_n)$ the dual coordinates $\ell_j = \ell(X_j)$ of an element $\ell \in \mathfrak{g}^*$. Then for $1 \le j \le n$, $\mathfrak{g}_j = \langle X_1, \dots, X_j\rangle_{\mathbb{R}} = \sum_{i=1}^j \mathbb{R}X_i$ is an ideal of $\mathfrak{g}$, $(\mathfrak{g}_j)^\perp = \langle X_{j+1}^*, \dots, X_n^*\rangle_{\mathbb{R}} \subset \mathfrak{g}^*$ and $\mathfrak{g}_j^* \cong \mathfrak{g}^*/(\mathfrak{g}_j)^\perp$. Let $p_j : \mathfrak{g}^* \to \mathfrak{g}_j^*$ be the restriction mapping, which intertwines the coadjoint actions of G on $\mathfrak{g}^*$ and on $(\mathfrak{g}_j)^*$. For $\ell \in \mathfrak{g}^*$, put $e_j(\ell) = \dim(G{\cdot}p_j(\ell))$, $e(\ell) = (e_1(\ell), \dots, e_n(\ell))$ and define the set of dimension indices $\mathscr{E} = \{e(\ell); \ell \in \mathfrak{g}^*\}$. For $e \in \mathscr{E}$, define the G-invariant e-layer $U_e = \{\ell \in \mathfrak{g}^*; e(\ell) = e\}$ and, setting $e_0 = 0$, define

$$S(e) = \{1 \le j \le n; e_j = e_{j-1} + 1\},\ T(e) = \{1 \le j \le n; e_j = e_{j-1}\}.$$

Let $e \in \mathscr{E}$. We say that $A \in \mathscr{U}(\mathfrak{g})$ is e-central if $\pi_\ell(A)$ is scalar for all $\ell \in U_e$. Then there is a Zariski open set $\mathscr{O} \subset \mathfrak{g}^*$ such that $\mathscr{O} \cap U_e$ is non-empty and G-invariant, and for each $j \in T(e)$ there exists $A_j \in \mathscr{U}(\mathfrak{g}_j)$ with the following properties:

1. Each A_j is e-central on $\mathcal{O} \cap U_e$, i.e. $\pi_\ell(A_j)$ is scalar for $\ell \in \mathcal{O} \cap U_e$; furthermore $A_j = P_j X_j + Q_j$ is such that
   (a) P_j is a polynomial in the A_k such that $k \in T(e)$ and $k < j$. In particular, $P_j \in \mathscr{U}(\mathfrak{g}_{j-1})$,
   (b) $Q_j \in \mathscr{U}(\mathfrak{g}_{j-1})$. In particular, $P_1, Q_1 \in \mathbb{C}$.
2. $\pi_\ell(P_j) \neq 0$ for all $\ell \in \mathcal{O} \cap U_e$.
3. $\pi_\ell(A_j) = \phi_j(\ell)I$ for $\ell \in \mathcal{O} \cap U_e$, where $\phi_j(\ell) = \widetilde{p}_j(\ell')\ell_j + \widetilde{q}_j(\ell')$. Here $\widetilde{p}, \widetilde{q}$ are rational non-singular functions on $\mathcal{O} \cap U_e$ and depend only on $\ell' = (\ell_1, \ldots, \ell_{j-1})$.
4. $\widetilde{p}_j(\ell')$ is G-invariant and is never zero on $\mathcal{O} \cap U_e$.

Having taken the Zariski open set $\mathcal{O} \cap U_e$ out of U_e, we may repeat the whole process to get the same result on the remaining part. Thus we can refine the layering, by keeping the same orbit parametrization within each sub-layer of U_e, and treat each piece as a layer in its own right on which the above result remains valid.

Let $\mathfrak{g}_0$ be an ideal of codimension 1 in $\mathfrak{g}$ containing $\mathfrak{h}$. Assume that almost all H-orbits in Γ_τ are non-saturated in the direction $(\mathfrak{g}_0)^\perp$. Let $\mathfrak{g} = \mathbb{R}X + \mathfrak{g}_0$, and consider a layer U_e in $\mathfrak{g}^*$ such that $\Gamma_\tau \cap U_e$ is a non-empty Zariski open set of Γ_τ. Then there exists [40] $A = XU + V \in \mathscr{U}(\mathfrak{g}, \tau)$ with $V \in \mathscr{U}(\mathfrak{g}_0)$, $U \in \mathscr{U}(\mathfrak{g}_0, \tau_0)$, where $\tau_0 = \mathrm{ind}_H^{G_0} \chi_f$, such that ${}^tA$ is e-central. Here, ${}^tA$ denotes as above the image of A by the principal anti-automorphism of $\mathscr{U}(\mathfrak{g})$, and e-central means that $\pi({}^tA)$ is a scalar operator for $\pi \in \widehat{G}$ associated to $\ell \in U_e$. If we introduce a flag of subalgebras:

$$\mathfrak{h} = \mathfrak{k}_0 \subset \mathfrak{k}_1 \subset \mathfrak{k}_2 \subset \cdots \subset \mathfrak{k}_{p-1} \subset \mathfrak{k}_p = \mathfrak{g},$$

where $\dim(\mathfrak{k}_j/\mathfrak{k}_{j-1}) = 1, 1 \leq j \leq p = \dim(\mathfrak{g}/\mathfrak{h})$, we construct an element A_j at each step $\mathfrak{k}_j$ of the non-saturated situation. These elements A_j form a system of rational generators of $\mathscr{U}(\mathfrak{g}, \tau)$. Now, let $A \in \mathscr{U}(\mathfrak{g}, \tau)$ be e-central. Let $W \in \mathscr{U}(\mathfrak{g})$ and $\phi \in \mathscr{D}(G)$. We write simply ϕ^f instead of ϕ^{χ_f}. From Plancherel's formula (5.2.4):

$$(R([W, A])\phi^f)(g) = \int_{\widehat{G}} \sum_{k=1}^{m(\pi)} \langle \pi(\phi_g)a_\pi^k, \overline{[W, A]}a_\pi^k \rangle d\mu(\pi)$$

$$= \int_{\widehat{G}} \sum_{k=1}^{m(\pi)} (\langle \pi(\phi_g)a_\pi^k, \overline{WA}a_\pi^k \rangle - \langle {}^tA\pi(\phi_g)a_\pi^k, \overline{W}a_\pi^k \rangle) d\mu(\pi)$$

$$= \int_{\widehat{G}} \sum_{k=1}^{m(\pi)} (P_A(\ell) - P_A(\ell)) \langle \pi(\phi_g)a_\pi^k, \overline{W}a_\pi^k \rangle d\mu(\pi) = 0 \ (g \in G).$$

This means $[W, A] \in \mathcal{U}(\mathfrak{g})\mathfrak{a}_\tau$. In this way, we are led to consider

$$\mathcal{J} = \{A \in \mathcal{U}(\mathfrak{g}, \tau);\ [V, A] = VA - AV \in \mathcal{U}(\mathfrak{g})\mathfrak{a}_\tau, \forall V \in \mathcal{U}(\mathfrak{g})\}$$

Then for $A \in \mathcal{J}$, where X_j are elements of $\mathfrak{g}$,

$$[A, X_1X_2\cdots X_m] = \sum_{i=1}^{m} X_1\cdots X_{i-1}[A, X_i]X_{i+1}\cdots X_m$$

$$\equiv \sum_{i=1}^{m} X_1\cdots X_{i-1}\mathrm{ad}([A, X_i])(X_{i+1}\cdots X_m)$$

$$= \sum_{i<j} X_1\cdots X_{i-1}\hat{X}_i X_{i+1}\cdots X_{j-1}[[A, X_i], X_j]X_{j+1}\cdots X_m$$

modulo $\mathcal{U}(\mathfrak{g})\mathfrak{a}_\tau$, where $\hat{X}_i$ means the term is omitted. Hence

$$[[\cdots[[A, X_1], X_2]\cdots], X_m] \in \mathcal{U}(\mathfrak{g})\mathfrak{a}_\tau.$$

In other words,

$$\mathcal{J} = \{A \in \mathcal{U}(\mathfrak{g}, \tau);\ (\mathrm{ad}(X_1))\cdots(\mathrm{ad}(X_m))(A) \in \mathcal{U}(\mathfrak{g})\mathfrak{a}_\tau, X_j \in \mathfrak{g}, m \in \mathbb{N}\}.$$

Specializing Question 1, we pose the following:

Question 2 For any $A \in \mathcal{I} = \mathcal{J} \cap \mathcal{U}(\mathfrak{g})\mathfrak{a}_\tau$, is $\beta^{-1}(\overline{A})(i\ell)$ equal to 0 in Γ_τ?

Question 3 For any $X \in \mathcal{J}$, does $P_X(\ell) = \overline{\beta^{-1}(\overline{X})(i\ell)}$?

If we leave the symmetrization map β and try to tackle the polynomial conjecture by induction on $\dim G + \dim(G/H)$, the essential case is that where $\dim \mathfrak{z} = 2$ and $\dim(\mathfrak{h} \cap \mathfrak{z}) = 1$ so that $f(\mathfrak{h} \cap \mathfrak{z}) \neq \{0\}$. Here, $\mathfrak{z}$ denotes as before the center of $\mathfrak{g}$. Moreover, there is no ideal of codimension 1 for which the G-orbits are saturated μ-almost everywhere. In fact, if there exists such an ideal containing $\mathfrak{h}$, Theorem 5.4, (b) in [40] shows that there is nothing left to do. If $\mathfrak{p} = \mathfrak{h} \cap \mathfrak{z} \cap \ker(f) \neq \{0\}$, we can go down to the quotient $\mathfrak{g}/\mathfrak{p}$ and apply induction. On the other hand, if $\dim(\mathfrak{z}/(\mathfrak{h}\cap\mathfrak{z})) \geq 2$, let Z_1, Z_2 be two linearly independent central elements modulo $\mathfrak{h} \cap \mathfrak{z}$. Put $\mathfrak{h}_j = \mathfrak{h} + \mathbb{R}Z_j (1 \leq j \leq 2)$. If we apply the induction hypothesis to $\mathfrak{h}_1$, the denominator of the rational function $P_X(\ell)$, $X \in \mathcal{U}(\mathfrak{g}, \tau)$, contains only one variable $\ell(Z_1)$. The same thing holds for $\mathfrak{h}_2$, and consequently $P_X(\ell)$ should be a polynomial function.

Concerning the surjectivity of the mapping which sends $X \in \mathcal{U}(\mathfrak{g}, \tau)$ to $P_X(\ell) \in \mathbb{C}[\Gamma_\tau]^H \cong (S(\mathfrak{g})/S(\mathfrak{g})\overline{\mathfrak{a}_\tau})^H$, let us consider for the moment the case where $\mathfrak{z} \not\subset \mathfrak{h}$. Let $0 \neq Z \in \mathfrak{z}\backslash(\mathfrak{h} \cap \mathfrak{z})$. Put $\mathfrak{h}' = \mathfrak{h} + \mathbb{R}Z$, $H' = \exp(\mathfrak{h}')$ and $\tau_\alpha = \mathrm{ind}_{H'}^G \chi_\ell$ with $\alpha = \ell(Z)$. Given a polynomial function $F(\ell) \in \mathbb{C}[\Gamma_\tau]^H$, there are $W, A \in$

$\mathscr{U}(\mathfrak{g}, \tau)$ satisfying $F(\ell) = P_W(\ell)/P_A(\ell)$, because the image of the mapping in question rationally generates the field $\mathbb{C}(\Gamma_\tau)^H$ of H-invariant rational functions on Γ_τ [40]. The induction hypothesis applied to τ_α says that there is $Y_\alpha \in \mathscr{U}(\mathfrak{g}, \tau_\alpha)$ satisfying $F(\ell)|_{\Gamma_{\tau_\alpha}} = P_{Y_\alpha}(\ell)$. Taking in $\mathfrak{g}$ a complementary basis $\{X_j\}_{j=1}^p$, $p = \dim(\mathfrak{g}/\mathfrak{h}')$, to $\mathfrak{h}'$, and agreeing to write $X^I = X_1^{i_1} X_2^{i_2} \cdots X_p^{i_p}$ for the multi-index $I = (i_1, i_2, \ldots, i_p)$, we write modulo $\mathscr{U}(\mathfrak{g})\mathfrak{a}_\tau$

$$W = \sum_{I,r} c_{I,r} X^I Z^r \equiv \sum_{I,r} c_{I,r} X^I (-i\ell(Z))^r + V(Z + i\ell(Z)) (c_{I,r} \in \mathbb{C}, V \in \mathscr{U}(\mathfrak{g}))$$

and

$$A \equiv \sum_{J,q} a_{J,q} X^J Z^q \ (a_{J,q} \in \mathbb{C}).$$

Thus,

$$\sum_{I,r} c_{I,r} X^I (-i\ell(Z))^r \equiv (\sum_{J,q} a_{J,q} X^J (-i\ell(Z))^q) Y_\alpha + V_\alpha (Z + i\ell(Z)),$$

$$W \equiv (\sum_{J,q} X^J (-i\ell(Z))^q) Y_\alpha + (V_\alpha + V)(Z + i\ell(Z))$$

for a certain $V_\alpha \in \mathscr{U}(\mathfrak{g})$. Namely,

$$W \equiv (\sum_{J,q} a_{J,q} X^J (-i\alpha)^q) Y_\alpha + (V_\alpha + V)(Z + i\alpha),$$

where Y_α does not contain Z. From these observations, we deduce that Y_α is rational in α. The substitution $Z = -i\alpha$ gives

$$W_{Z=-i\alpha} \equiv (\sum_{J,q} a_{J,q} X^J (-i\alpha)^q) Y_\alpha.$$

Hence we see that Y_α is indeed rational in α only if A possesses a polynomial factor in Z only. If Y_α is polynomial in α, we make the substitution $\alpha \to iZ$ to get an element Y of $\mathscr{U}(\mathfrak{g})$ such that $W \equiv AY$. This implies that $P_Y = F$ and $Y \in \mathscr{U}(\mathfrak{g}, \tau)$, which means F belongs to the image of the map.

When $\dim(\mathfrak{h}/(\mathfrak{h} \cap \mathfrak{z})) \geq 2$, let $\mathfrak{z} = (\mathfrak{h} \cap \mathfrak{z}) + \sum_j \mathbb{R} Z_j$ for a complementary basis $\{Z_j\}$ in $\mathfrak{z}$ to $\mathfrak{h} \cap \mathfrak{z}$. Take $X \in (S(\mathfrak{g})/S(\mathfrak{g})\overline{\mathfrak{a}_\tau})^H$. From what we have seen so far and by the induction hypothesis, there exist two polynomials $F_1(Z_1)$, $F_2(Z_2)$ such that $P_{F_1(Z_1)} X = P_{W_1}$, $P_{F_2(Z_2)} X = P_{W_2}$ for certain $W_1, W_2 \in \mathscr{U}(\mathfrak{g}, \tau)$. Then $P_{F_1(Z_1)} P_{F_2(Z_2)} X \in \mathbb{C}[\Gamma_\tau]^H$ is associated to $F_1(Z_1) W_2$ and also to $F_2(Z_2) W_1$. So, $F_1(Z_1) W_2 \equiv F_2(Z_2) W_1$ modulo $\mathscr{U}(\mathfrak{g})\mathfrak{a}_\tau$. If we express W_1, W_2 by means of a complementary basis to $\mathfrak{h}$ in $\mathfrak{g}$, W_2 is divisible by $F_2(Z_2)$ and W_1 is divisible by $F_1(Z_1)$. This means that there is $V \in \mathscr{U}(\mathfrak{g}, \tau)$ such that $W_2 = F_2(Z_2)V$, $W_1 = F_1(Z_1)V$. Finally, $X = P_V$.

Now suppose that there is an ideal $\mathfrak{g}_0$ of codimension 1 in $\mathfrak{g}$ not containing $\mathfrak{h}$ and such that the G-orbits are saturated in the direction $(\mathfrak{g}_0)^\perp$ almost everywhere in Γ_τ. Let $T \in \mathfrak{h}$ such that $\mathfrak{g} = \mathbb{R}T + \mathfrak{g}_0$. We know that each generic $\ell \in \Gamma_\tau$ possesses a polarization $\mathfrak{b}$ contained in $\mathfrak{g}_0$. Put $\mathfrak{h}_0 = \mathfrak{h} \cap \mathfrak{g}_0$, $H_0 = \exp(\mathfrak{h}_0)$, $G_0 = \exp(\mathfrak{g}_0)$, $\ell_0 = \ell|_{\mathfrak{g}_0} \in (\mathfrak{g}_0)^*$ and $\pi_0 = \mathrm{ind}_B^{G_0} \chi_{\ell_0}$. It follows that $\tau_0 = \mathrm{ind}_{H_0}^{G_0} \chi_{f_0}$, $f_0 = f|_{\mathfrak{g}_0}$ has finite multiplicities. We can replicate the proof of the previous theorem in order to show $P_X(\ell) \in \mathbb{C}[\Gamma_\tau]^H$ for any $X \in \mathscr{U}(\mathfrak{g}, \tau)$. Conversely, let $F(\ell) \in \mathbb{C}[\Gamma_\tau]^H$. In this case, $F(\ell)$ is identified with an H-invariant polynomial function $F(\ell_0)$ on $\Gamma_{\tau_0} = f_0 + (\mathfrak{h}_0)^\perp \subset (\mathfrak{g}_0)^*$. By induction hypothesis, there is $X \in \mathscr{U}(\mathfrak{g}_0, \tau_0)$ such that $P_X(\ell_0) = F(\ell_0)$ for generic $\ell_0 \in \Gamma_{\tau_0}$. For $t \in \mathbb{R}$, put $h(t) = \exp(tT)$ and $X(t) = \mathrm{Ad}(h(t))X \in \mathscr{U}(\mathfrak{g}_0, \tau_0)$. Because of formula (5.2.3), which gives us the distribution $a_{\ell_0} \in (\mathscr{H}_{\pi_0}^{-\infty})^{H_0, \chi_{f_0}}$, we have

$$\pi_0(\overline{X(t)})a_{\ell_0} = \pi_0(\overline{X})a_{h(-t)\cdot\ell_0} = \overline{F(h(-t)\cdot\ell_0)}a_{\ell_0} = \overline{F(\ell)}a_{\ell_0}.$$

The Plancherel formula for τ guarantees that the mapping

$$\mathbb{R} \ni t \mapsto [X(t)] \in \mathscr{U}(\mathfrak{g}_0, \tau_0)/\mathscr{U}(\mathfrak{g}_0)\mathfrak{a}_{\tau_0}$$

is constant. Hence differentiating in t, we find that $X \in \mathscr{U}(\mathfrak{g}, \tau)$. Next, applying X to a_ℓ we can pass the first coordinate with respect to T in the integral (5.2.3) of a_ℓ, giving us $P_X(\ell) = F(\ell)$. We prove in this way the surjectivity of the mapping in question.

From what we have observed, if we employ this reasoning by induction, the essential case is $\dim \mathfrak{z} = 2$, $\dim(\mathfrak{h} \cap \mathfrak{z}) = 1$, $f(\mathfrak{h} \cap \mathfrak{z}) \neq \{0\}$. Let $\mathfrak{z} = \mathbb{R}Z + (\mathfrak{h} \cap \mathfrak{z})$. We can suppose by induction that, for any $W \in \mathscr{U}(\mathfrak{g}, \tau)$, there is a polynomial $F(Z) = F_W(Z)$ of Z such that $F(Z)W$ corresponds to a polynomial function in $\mathbb{C}[\Gamma_\tau]^H$. Besides, let $\mathfrak{g}_0$ be an ideal of codimension 1 in $\mathfrak{g}$ and $X \in \mathfrak{g}$ such that $\mathfrak{h} \subset \mathfrak{g}_0$, $\mathfrak{g} = \mathbb{R}X + \mathfrak{g}_0$. In the case where the G-orbits are non-saturated in the direction $(\mathfrak{g}_0)^\perp$, we have in $\mathscr{U}(\mathfrak{g}, \tau)$ an e-central element $A = XU + V$, as mentioned before. Induction applied to $U \in \mathscr{U}(\mathfrak{g}_0, \tau_0)$, where $\tau_0 = \mathrm{ind}_H^{G_0} \chi$, says U corresponds to a polynomial function in $\mathbb{C}[\Gamma_{\tau_0}]^H$. So, let $P_U(\ell) = (I(Z)J)(i\ell)$ for a certain polynomial $I(Z)$ of Z and a certain $J \in S(\mathfrak{g})$ whose restriction to Γ_τ admits no polynomial of Z as factor. Since $P_A(\ell)$ is a polynomial function (cf. [73]), using induction on the degree of W in X, we can choose for F a power of I. That is to say, there is a non-negative integer m such that $I(Z)^m W$ corresponds to a polynomial function in $\mathbb{C}[\Gamma_\tau]^H$.

How can we eliminate the factor $I(Z)^m$? We examine an example where dim $\mathfrak{h} = 1$.

5.2.2 Special Cases

In what follows we go to establish the conjecture when $\dim(H\cdot\ell) \leq 1$ for generic $\ell \in \Gamma_\tau$. Let us treat first a special case.

Claim If $\dim \mathfrak{h} = 1$, the conjecture holds.

Proof We first show that $P_X(\ell)$ is a polynomial function for every $X \in \mathscr{U}(\mathfrak{g}, \tau)$. We denote by Q the set of all $\ell \in \Gamma_\tau$ such that $\mathfrak{h} + \mathfrak{g}(\ell)$ is a Lagrangian subspace for the bilinear form B_ℓ, and that $\dim G\cdot\ell$ is maximal. We see that Q is a non-empty Zariski open set of Γ_τ and, by the proof of the previous theorem, that all functions P_X, $X \in \mathscr{U}(\mathfrak{g}, \tau)$, are well defined on Q. Let us begin by choosing a polarization $\mathfrak{b}[\ell]$ at each point $\ell \in \Gamma_\tau$. Then $\dim(\mathfrak{g}/\mathfrak{b}[\ell]) \leq 1$ and $[\mathfrak{g}, \mathfrak{g}] \subset \mathfrak{b}[\ell]$. Evidently, it suffices to assume that $\mathfrak{g} = \sum_{\ell\in Q} \mathfrak{b}[\ell]$ and $\dim(\mathfrak{g}/\mathfrak{b}[\ell]) = 1$ for every $\ell \in Q$. Set $\mathfrak{g}_0 = \cap_{\ell\in Q}\mathfrak{b}[\ell]$ and $d = \dim(\mathfrak{g}/\mathfrak{g}_0)$. We have $[\mathfrak{g}, \mathfrak{g}] \subset \mathfrak{g}_0$. If $d = 1$, then $\mathfrak{g}_0$ is a common polarization for all $\ell \in Q$, which gives the result as we shall see later. Suppose now $d \geq 2$.

Assume first $\mathfrak{h} \subset \mathfrak{g}_0$. It follows that $[\mathfrak{g}_0, \mathfrak{g}_0] \subset \cap_{\ell\in Q} \ker(\ell) \subset \mathfrak{h} = \mathbb{R}Y$. In fact, it suffices to suppose that $\mathfrak{h}$ is not central in $\mathfrak{g}$, what implies $[\mathfrak{g}_0, \mathfrak{g}_0] = \{0\}$, because $[\mathfrak{g}_0, \mathfrak{g}_0]$ is an ideal of $\mathfrak{g}$. Let $\{X_1, \ldots, X_d\}$ be elements of $\mathfrak{g}$, linearly independent modulo $\mathfrak{g}_0$. It follows that $\mathfrak{g} = (\sum_{j=1}^d \mathbb{R}X_j) + \mathfrak{g}_0$ and $W_j = [X_j, Y](1 \leq j \leq d)$ are linearly independent. In fact, if $W \in \mathfrak{g}$ verifies $[W, Y] = 0$, W belongs to $\mathfrak{b}[\ell]$ for every $\ell \in Q$ and hence to $\mathfrak{g}_0$. Let $\mathfrak{n} = (\sum_{j=2}^d \mathbb{R}X_j) + \mathfrak{g}_0$.

We see immediately that

$$X_j W_k - X_k W_j \in \mathscr{U}(\mathfrak{g})$$

are H-invariant. On the other hand, the Penney distribution may be written as

$$a_\ell : \psi \mapsto \overline{\psi(e)}$$

for $\psi \in \mathscr{H}_\pi^\infty$, where $\pi = \pi_\ell = \mathrm{ind}_B^G \chi_\ell$ with $B = B[\ell] = \exp(\mathfrak{b}[\ell])$ (and e is the neutral element of G). Now, it is straightforward that

$$Ua_\ell = i\ell(U)a_\ell, \ \forall U \in \mathfrak{b}[\ell].$$

As $\ell(W_k)X_j - \ell(W_j)X_k \in \mathfrak{b}[\ell]$ for all $1 \leq j, k \leq d$, it follows that

$$\begin{aligned}(X_j W_k - X_k W_j)a_\ell &= i(\ell(W_k)X_j - \ell(W_j)X_k)a_\ell \\ &= -(\ell(W_k)\ell(X_j) - \ell(W_j)\ell(X_k))a_\ell = \beta^{-1}(X_j W_k - X_k W_j)(i\ell)a_\ell.\end{aligned}$$

Now assume $\mathfrak{h} \not\subset \mathfrak{g}_0$. Remember that $\mathfrak{g}_0$ is commutative. As $d \geq 2$, we have $\mathfrak{g} \neq \mathfrak{h} + \mathfrak{g}_0$. Take an ideal $\mathfrak{n}$ of codimension 1 containing $\mathfrak{h} + \mathfrak{g}_0$. Let $\mathfrak{g} = \mathbb{R}e_1 + \mathfrak{n}$ and $\mathfrak{h} = \mathbb{R}Y$. Define inductively $e_{j+1} = [Y, e_j]$. Let k be the first positive integer such that $e_{k+1} = 0$. We can choose e_1 in such a manner that $k \geq 2$, if $\mathfrak{g}$ is non-commutative. If $k = 2$ for any e_1, $L = \operatorname{ad} Y$ verifies $L^2 = 0$. If furthermore $L = 0$ on $\mathfrak{g}_0$, which contains $[\mathfrak{g}, \mathfrak{g}]$, then $\mathfrak{p} = \mathfrak{h} + \mathfrak{g}_0$ becomes an Abelian ideal of $\mathfrak{g}$. If necessary we swap $\mathfrak{b}[\ell]$ with another term that contains $\mathfrak{p}$, so we are led back to the previous case. Hence suppose that $L \neq 0$ on $\mathfrak{g}_0$. Let $\widetilde{e}_3 \in \mathfrak{g}_0$ be such that $\widetilde{e}_4 = [Y, \widetilde{e}_3] \neq 0$. Then we see that $X = e_1\widetilde{e}_4 - e_2\widetilde{e}_3 \in \mathscr{U}(\mathfrak{g})$ is H-invariant. Let now $k \geq 3$. We define $X \in \mathscr{U}(\mathfrak{g})$ by

$$\begin{aligned} X = {} & 2e_1e_k - 2e_2e_{k-1} + 2e_3e_{k-2} - 2e_4e_{k-3} + \cdots \\ & \pm 2e_{m-1}e_{m+3} \mp 2e_me_{m+2} \pm e_{m+1}^2 \end{aligned}$$

if $k = 2m + 1$ is odd, and

$$\begin{aligned} X = {} & e_1e_k^{k-2} - e_2e_{k-1}e_k^{k-3} + e_3e_{k-1}^2e_k^{k-4}/2! - e_4e_{k-1}^3e_k^{k-5}/3! \\ & + e_5e_{k-1}^4e_k^{k-6}/4! - \cdots - e_{k-2}e_{k-1}^{k-3}e_k/(k-3)! + e_{k-1}^{k-1}/(k-3)!(k-1) \end{aligned}$$

if $k = 2m$ is even. Here,

$$\langle a_\ell, \psi \rangle = \int_{\mathbb{R}} \overline{\psi(\exp(tY))e^{itf(Y)}}dt,$$

and if R denotes the right action,

$$\langle Xa_\ell, \psi \rangle = \int_{\mathbb{R}} \overline{(R(\overline{X})\psi)(\exp(tY))e^{itf(Y)}}dt.$$

On the other hand, for almost all $\ell \in \Gamma_\tau$ there is $\alpha(\ell) \in \mathbb{R}$ such that $e_1 - \alpha(\ell)Y$ belongs to $\mathfrak{g}(\ell)$ and we have

$$\begin{aligned} \int_{\mathbb{R}} (R(e_1)\psi)(\exp(tY))e^{itf(Y)}dt = {} & \int_{\mathbb{R}} (R(e_1 - \alpha(\ell)Y)\psi)(\exp(tY))e^{itf(Y)}dt \\ & + \alpha(\ell) \int_{\mathbb{R}} (R(Y)\psi)(\exp(tY))e^{itf(Y)}dt \\ = {} & (-i\ell(e_1 - \alpha(\ell)Y) - i\alpha(\ell)f(Y)) \int_{\mathbb{R}} \psi(\exp(tY))e^{itf(Y)}dt \\ = {} & -i\ell(e_1) \int_{\mathbb{R}} \psi(\exp(tY))e^{itf(Y)}dt = \langle -i\ell(e_1)a_\ell, \psi \rangle. \end{aligned}$$

Therefore, we get

$$Xa_\ell = \beta^{-1}(X)(i\ell)a_\ell \text{ ie } P_X(\ell) = \overline{\beta^{-1}(\overline{X})(i\ell)}.$$

Now for any $W \in \mathscr{U}(\mathfrak{g},\tau)$, we write $W = \sum_{j=0}^{r} X_1^j V_j$ if $\mathfrak{h} \subset \mathfrak{g}_0$ or $W = \sum_{j=0}^{r} e_1^j V_j$ if $\mathfrak{h} \not\subset \mathfrak{g}_0$ for some $V_j \in \mathscr{U}(\mathfrak{n})(0 \le j \le r)$. We claim $P_W(\ell)$ is a polynomial function. This assertion is proved by induction on the degree r of W in X_1 or e_1. If $r = 0$, it suffices to apply the induction hypothesis to $\mathfrak{n}$. Since $V_r \in \mathscr{U}(\mathfrak{g},\tau)$, we see that $P_{X^r V_r}(\ell) = P_{X^r}(\ell)P_{V_r}(\ell)$, where $X = X_1 W_d - X_d W_1$ if $\mathfrak{h} \subset \mathfrak{g}_0$, is polynomial by induction. As $WW_d^r - X^r V_r$, $W\widetilde{e}_4^r - X^r V_r$, or $We_k^p - X^r V_r$ for some non-negative integer p, are of degree less than r in X_1 or e_1, $(W_d(-i\ell))^r P_W(\ell)$, $(\widetilde{e}_4(-i\ell))^r P_W(\ell)$ or $(e_k(-i\ell))^p P_W(\ell)$ are polynomial. In the end, $P_W(\ell)$ is a rational function whose denominator is a factor of $(W_d(-i\ell))^r$, $(\widetilde{e}_4(-i\ell))^r$ or $(e_k(-i\ell))^p$.

Now we use the Plancherel formula again. The canonical central decomposition of τ is equivalent to the following [99]:

$$\tau \simeq \int_{\Gamma_\tau/H} \pi_\ell d\widetilde{\mu}(\dot{\ell}),$$

where the measure $\widetilde{\mu}$ is the push-forward of the Lebesgue measure on Γ_τ. Corresponding to this decomposition, the Plancherel formula is written as:

$$\phi^\chi(e) = \int_{\Gamma_\tau/H} \langle \pi_\ell(\phi)a_\omega, a_\omega\rangle d\widetilde{\mu}(\dot{\ell}) \ (\forall \phi \in \mathscr{D}(G)),$$

where ω denotes the H-orbit passing through $\ell \in Q \subset \Gamma_\tau$ and a_ω is the corresponding Penney distribution. It follows that, for any $X \in \mathscr{U}(\mathfrak{g},\tau)$,

$$\int_{G/H} (R(X)\phi^\chi)(g)\overline{\phi^\chi(g)}d\dot{g} = R(X)(\phi^* * \phi)^\chi(e) = \int_{\Gamma_\tau/H} P_X(\ell)\|\pi_\ell(\phi)a_\omega\|^2 d\widetilde{\mu}(\dot{\ell}),$$

where $\phi^*(g) = \overline{\phi(g^{-1})}(g \in G)$. Substituting $X = W^m (m \in \mathbb{N})$, we obtain:

$$\int_{G/H} (R(W^m)\phi^\chi)(g)\overline{\phi^\chi(g)}d\dot{g} = \int_{\Gamma_\tau/H} P_{W^m}(\ell)\|\pi_\ell(\phi)a_\omega\|^2 d\widetilde{\mu}(\ell)$$

$$= \int_{\Gamma_\tau/H} (F_1(\ell)/F_2(\ell))^m \|\pi_\ell(\phi)a_\omega\|^2 d\widetilde{\mu}(\dot{\ell}), \qquad (5.2.5)$$

where $F_1(\ell)$ is a polynomial function and $F_2(\ell)$ is $(W_d(-i\ell))^{r'}$, $(\widetilde{e}_4(-i\ell))^{r'}$ or $(e_k(-i\ell))^{p'}$ for some non-negative integer $r' \le r$ or $p' \le p$. The last integral would diverge, for m large enough, if there was a denominator. This gives a contradiction. We shall see the details later.

Now, we examine the surjectivity of the mapping $\mathscr{U}(\mathfrak{g},\tau) \ni X \mapsto P_X(\ell) \in \mathbb{C}[\Gamma_\tau]^H$. We fix a flag of ideals:

$$\{0\} = \mathfrak{g}_0 \subset \mathfrak{g}_1 \subset \cdots \subset \mathfrak{g}_{n-1} \subset \mathfrak{g}_n = \mathfrak{g}, \dim(\mathfrak{g}_j) = j(0 \le j \le n).$$

Let j_0 be the index such that $\mathfrak{h} \not\subset \mathfrak{g}_{j_0-1}$ but $\mathfrak{h} \subset \mathfrak{g}_{j_0}$. Take non-zero elements $X_j \in \mathfrak{g}_j \backslash \mathfrak{g}_{j-1} (j \neq j_0)$ and set $\mathfrak{m} = \sum_{j \neq j_0} \mathbb{R}X_j$. Then $\mathfrak{m}$ is a complement subspace to $\mathfrak{h}$ in $\mathfrak{g}$, and invariant by the action of ad$\mathfrak{h}$. Namely, the pair $(\mathfrak{g}, \mathfrak{h})$ is reductive. So, each X in $(S(\mathfrak{g})/S(\mathfrak{g})\overline{\mathfrak{a}_\tau})^H$ or $\mathscr{U}(\mathfrak{g}, \tau)/\mathscr{U}(\mathfrak{g})\mathfrak{a}_\tau = (\mathscr{U}(\mathfrak{g})/\mathscr{U}(\mathfrak{g})\mathfrak{a}_\tau)^H$ admits an H-invariant representative in $S(\mathfrak{m})$ or $\beta(S(\mathfrak{m}))$.

Among the multi-indices $I = (i_n, \dots, i_{j_0+1}, i_{j_0-1}, \dots, i_1)$ of length $n-1$, we introduce the lexicographic order:

$$I < J = (j_n, \dots, j_{j_0+1}, j_{j_0-1}, \dots, j_1) \Leftrightarrow i_n = j_n, \dots, i_{k+1} = j_{k+1}, i_k < j_k.$$

We say that a monomial

$$cX_n^{i_n} \cdots X_{j_0+1}^{i_{j_0+1}} X_{j_0-1}^{i_{j_0-1}} \cdots X_1^{i_1} \ (c \in \mathbb{C})$$

is of index $I = (i_n, \dots, i_{j_0+1}, i_{j_0-1}, \dots, i_1)$, and we agree that the index of an element

$$X = \sum_I c_I X^I = \sum_I c_I X_n^{i_n} \cdots X_{j_0+1}^{i_{j_0+1}} X_{j_0-1}^{i_{j_0-1}} \cdots X_1^{i_1} \ (c_I \in \mathbb{C})$$

in $S(\mathfrak{m})$ indicates the largest index of the monomials appearing in X. We get the surjectivity by induction on the index. Let $X = \sum_I c_I X^I \in S(\mathfrak{m})^H$ be of index I_0. Take $X' \in S(\mathfrak{m})$ so that we have $X(\ell) = \overline{X'(i\ell)}$ for $\ell \in \mathfrak{g}^*$. Then X' and hence $\beta(\overline{X'}) \in \mathscr{U}(\mathfrak{g})$ is H-invariant and gives us a polynomial function $P_{\beta(\overline{X'})}$ on Γ_τ which is H-invariant, namely $P_{\beta(\overline{X'})} \in (S(\mathfrak{m})/S(\mathfrak{m})\overline{\mathfrak{a}_\tau})^H$. If we take in $S(\mathfrak{m})^H$ a representative $\widetilde{P}_{\beta(\overline{X'})}$ of $P_{\beta(\overline{X'})}$, we verify that the index of $X - \widetilde{P}_{\beta(\overline{X'})}$ is smaller than that of X. From the induction hypothesis, there is $W \in \mathscr{U}(\mathfrak{g}, \tau)$ such that

$$P_W(\ell) = (X - \widetilde{P}_{\beta(\overline{X'})})(\ell) = (X - P_{\beta(\overline{X'})})(\ell) \ (\ell \in \Gamma_\tau).$$

Finally, $X = P_{W+\beta(\overline{X'})}$. □

Next, we slightly generalize this result. We first consider the case where $\dim(\mathfrak{h}/(\mathfrak{h} \cap \mathfrak{z})) \leq 1$. It is immediate that we can suppose $\dim(\mathfrak{h}/(\mathfrak{h} \cap \mathfrak{z})) = 1$. Let $\mathfrak{h} = \mathbb{R}Y + (\mathfrak{h} \cap \mathfrak{z})$ and $\mathfrak{h} \cap \mathfrak{z} = \oplus_{k=1}^s \mathbb{R}Z_k$. Focus on the element Y. As in the case where $\dim \mathfrak{h} = 1$, $P_V(\ell)(\ell \in Q)$ extends to a polynomial function on Γ_τ for any $V \in \mathscr{U}(\mathfrak{g}, \tau)$.

In order to show the surjectivity of the map sending $V \in \mathscr{U}(\mathfrak{g}, \tau)$ to the function $P_V(\ell) \in \mathbb{C}[\Gamma_\tau]^H$, we take a flag of ideals:

$$\mathfrak{h} \cap \mathfrak{z} = \mathfrak{k}_0 \subset \mathfrak{k}_1 \subset \cdots \subset \mathfrak{k}_{m-1} \subset \mathfrak{k}_m = \mathfrak{g}, \dim(\mathfrak{k}_j/\mathfrak{k}_{j-1}) = 1 (1 \leq j \leq m).$$

Let j_0 be the first index j such that $Y \in \mathfrak{k}_j$. For every $j \neq j_0$, we extract an $X_j \in \mathfrak{k}_j \setminus \mathfrak{k}_{j-1}$. Given $F(\ell) \in \mathbb{C}[\Gamma_\tau]^H \cong (S(\mathfrak{g})/S(\mathfrak{g})\overline{\mathfrak{a}_\tau})^H$, we choose a representative

$W \in S(\mathfrak{g})$ verifying $[W, Y] \in S(\mathfrak{g})\overline{\mathfrak{a}_\tau}$ and express

$$W = \sum_I c_I X^I = \sum_I c_I X_m^{i_m} \cdots X_{j_0+1}^{i_{j_0+1}} X_{j_0-1}^{i_{j_0-1}} \cdots X_1^{i_1} \ (c_I \in \mathbb{C}).$$

Then $[Y, W] \in S(\mathfrak{g})\mathfrak{p}$, where $\mathfrak{p} = \sum_{k=1}^{s} \mathbb{C}(Z_k - if(Z_k))$. Hence $\overline{\beta(W)} \in \mathscr{U}(\mathfrak{g}, \tau)$, which allows us to reason as in the case where $\dim \mathfrak{h} = 1$.

Let us take a step forward. Assume that the dimension of the generic G-orbits passing Γ_τ is equal to 2. Let $\mathfrak{b}[\ell]$ be a polarization at generic $\ell \in \Gamma_\tau$. Put as before $\mathfrak{g}_0 = \cap_{\ell \in Q} \mathfrak{b}[\ell]$. Then just as before $\dim(\mathfrak{g}/\mathfrak{b}[\ell]) = 1$ and we can suppose that $\mathfrak{g}_0$ is an Abelian ideal of $\mathfrak{g}$. We remarked earlier if we set out to prove the conjecture, the previous reasoning leads to the case where $\dim \mathfrak{z} = 2$, $\dim(\mathfrak{h} \cap \mathfrak{z}) = 1$ and f does not vanish on $\mathfrak{h} \cap \mathfrak{z}$. This reduction is also possible in the present framework. Let $\mathfrak{h} \cap \mathfrak{z} = \mathbb{R}E$, $f(E) = 1$ and $\mathfrak{z} = \mathbb{R}E \oplus \mathbb{R}Z$. Take $Y \in \mathfrak{g}\backslash\mathfrak{z}$ such that $[Y, \mathfrak{g}] \subset \mathfrak{z}$. If $\dim([\mathfrak{g}, Y]) = 1$, we can argue as in the case where $\dim \mathfrak{z} = 1$. Assume $\dim([\mathfrak{g}, Y]) = 2$, that is to say $\mathfrak{z} = [\mathfrak{g}, Y]$. Fix two elements X_1, X_2 in $\mathfrak{g}$ verifying $[X_1, Y] = E$, $[X_2, Y] = Z$. We see that $\mathfrak{q} = \mathbb{R}Y + \mathfrak{z}$ is an Abelian ideal of $\mathfrak{g}$, and we can suppose that $\mathfrak{b}[\ell]$ is contained in the orthogonal $\mathfrak{q}^\ell$ of $\mathfrak{q}$ with respect to the bilinear form B_ℓ. This implies $\mathfrak{b}[\ell] = \mathfrak{q}^\ell = \{X \in \mathfrak{g}; \ell([X, \mathfrak{q}]) = \{0\}\}$ for $\ell \in Q$ and consequently $\mathfrak{g}_0$ coincides with the centralizer of $\mathfrak{q}$ in $\mathfrak{g}$.

We are in one of the following two situations: (i) $\mathfrak{h} \subset \mathfrak{g}_0$, (ii) $\mathfrak{h} \not\subset \mathfrak{b}[\ell]$ for almost all $\ell \in \Gamma_\tau$. In case (i), we have $Y \in \mathfrak{h} + \mathfrak{g}(\ell)$ for generic $\ell \in \Gamma_\tau$. If we substitute $Z \mapsto -i\ell(Z)$ in each $U \in \mathscr{U}(\mathfrak{g}, \tau)$, then U belongs to $\mathscr{U}(\mathfrak{q}^\ell)$, since the orbit $G{\cdot}\ell$ is saturated in the direction $(\mathfrak{q}^\ell)^\perp$. So, U is written as

$$U = \sum_{k=0}^{m} c_k (iZX_1 - X_2)^k U_k + A = \sum_{k=0}^{m} c_k (\ell(Z)X_1 - X_2)^k U_k + A + V(Z + i\ell(Z))$$

for certain $c_k \in \mathbb{C}$, $U_k \in \mathscr{U}(\mathfrak{g}_0) (0 \le k \le m)$, $A \in \mathscr{U}(\mathfrak{g})\mathfrak{a}_\tau$ and $V \in \mathscr{U}(\mathfrak{g})$.

On the other hand, the Penney distribution a_ℓ for $\pi = \pi_\ell = \mathrm{ind}_{B[\ell]}^G \chi_\ell$, $B[\ell] = \exp(\mathfrak{b}[\ell])$, is obtained by the formula $\langle a_\ell, \psi \rangle = \overline{\psi(e)}$ for arbitrary $\psi \in \mathscr{H}_\pi^\infty$. These observations allow us to calculate

$$\langle \overline{U} a_\ell, \psi \rangle = \beta^{-1}(\overline{U})(i\ell)\langle a_\ell, \psi \rangle, \text{ ie } \overline{U} a_\ell = \beta^{-1}(\overline{U})(i\ell) a_\ell,$$

meaning $P_U(\ell) = \overline{\beta^{-1}(\overline{U})(i\ell)}$ is a polynomial function on Q.

We remain in the case where $\mathfrak{h} \subset \mathfrak{g}_0$. We have $\mathscr{U}(\mathfrak{g}_0) \subset \mathscr{U}(\mathfrak{g}, \tau)$ and, if we set $V = X_2 - iZX_1 \in \mathscr{U}(\mathfrak{g})$, we verify that $\overline{V} a_\ell = \overline{V}(i\ell) a_\ell$ for $\ell \in Q$. This implies $V \in \mathscr{U}(\mathfrak{g}, \tau)$ and consequently $V \in S(\mathfrak{g})$ gives us an H-invariant polynomial function on Γ_τ belonging to the image of the mapping

$$\mathscr{U}(\mathfrak{g}, \tau) \ni X \mapsto P_X(\ell) \in \mathbb{C}[\Gamma_\tau]^H \cong (S(\mathfrak{g})/S(\mathfrak{g})\overline{\mathfrak{a}_\tau})^H.$$

To show the surjectivity it now suffices to use induction on the degree in X_2 of the elements in $(S(\mathfrak{g})/S(\mathfrak{g})\overline{\mathfrak{a}_\tau})^H$.

We next consider case (ii). Assume first $\dim(\mathfrak{h}/(\mathfrak{h}\cap\mathfrak{g}_0)) = 2$. On one side, each coset of $\mathscr{U}(\mathfrak{g},\tau)/\mathscr{U}(\mathfrak{g},\tau)\mathfrak{a}_\tau$ admits a representative $W \in \mathfrak{g}_0$. On the other side, there is $X \in \mathfrak{h}$ such that $\mathfrak{g} = \mathbb{R}X + \mathfrak{b}[\ell]$ with the polarization $\mathfrak{b}[\ell] = \mathfrak{q}^\ell$ at $\ell \in Q$. We therefore have

$$\langle a_\ell, \psi\rangle = \int_{\mathbb{R}} \overline{\psi(\exp(tX))e^{itf(X)}}dt$$

for all $\psi \in \mathscr{H}_\pi^\infty$. It follows that

$$\begin{aligned}\langle \overline{W}a_\ell, \psi\rangle &= \int_{\mathbb{R}} \overline{(R(W)\psi)(\exp(tX))e^{itf(X)}}dt \\ &= \overline{W}(i\ell)\langle a_\ell, \psi\rangle = \beta^{-1}(\overline{W})(i\ell)\langle a_\ell, \psi\rangle.\end{aligned}$$

Hence $P_W(\ell) = \overline{\beta^{-1}(\overline{W})(i\ell)}$ extends to a polynomial function on Γ_τ. We go proceed similarly for the surjectivity of

$$(\mathscr{U}(\mathfrak{g})/\mathscr{U}(\mathfrak{g})\mathfrak{a}_\tau)^H = \mathscr{U}(\mathfrak{g},\tau)/\mathscr{U}(\mathfrak{g})\mathfrak{a}_\tau \ni [X] \mapsto P_X(\ell) \in \mathbb{C}[\Gamma_\tau]^H.$$

Every $F(\ell) \in \mathbb{C}[\Gamma_\tau]^H$ admits a representative $V \in S(\mathfrak{g}_0)$ such that $F(\ell) = \overline{V(i\ell)}$ and $[\mathfrak{h}, V] \subset S(\mathfrak{g}_0)\overline{\mathfrak{a}_\tau}$. Since $\mathfrak{g}_0$ is Abelian, $\beta(\overline{V}) \in \mathscr{U}(\mathfrak{g},\tau)$ and

$$\langle \beta(V)a_\ell, \psi\rangle = V(i\ell)\langle a_\ell, \psi\rangle,\ \ (\ell \in Q)$$

for any $\psi \in \mathscr{H}_\pi^\infty$. That is to say, $P_{\overline{V}}(\ell) = \overline{V(i\ell)} = F(\ell)$ on Γ_τ.

Finally, let $\dim(\mathfrak{h}/(\mathfrak{h}\cap\mathfrak{g}_0)) = 1$. Take an element $X \in \mathfrak{g}$ so that

$$\mathfrak{g} = \mathbb{R}X + \mathfrak{h} + \mathfrak{g}_0.$$

Then each coset of $\mathscr{U}(\mathfrak{g},\tau)/\mathscr{U}(\mathfrak{g})\mathfrak{a}_\tau$ or $(S(\mathfrak{g})/S(\mathfrak{g})\overline{\mathfrak{a}_\tau})^H$ admits a representative W of the form

$$W = \sum_{j=0}^{m} X^j U_j,\ \ U_j \in \mathscr{U}(\mathfrak{g}_0) \cong S(\mathfrak{g}_0)$$

such that $U_m \notin \mathscr{U}(\mathfrak{g})\mathfrak{a}_\tau$ or $U_m \notin S(\mathfrak{g})\overline{\mathfrak{a}_\tau}$. If $m \geq 1$ and if $\mathfrak{h}\cap\mathfrak{g}_0 \neq \{0\}$, we calculate for $V \in \mathfrak{h}\cap\mathfrak{g}_0$:

$$[V, W] = mX^{m-1}U_m[V, X] + A$$

for a certain A in $\mathscr{U}(\mathfrak{g})$ or $S(\mathfrak{g})$, whose degree in X is at most $m-2$. In case $m=1$ we agree that $A=0$. In order that $[V, W]$ belongs to $\mathscr{U}(\mathfrak{g})\mathfrak{a}_\tau$, it is necessary that $[V, X]$ falls in $\mathfrak{p}=\mathfrak{h}\cap\mathfrak{g}_0\cap\ker(f)$. This forces $\mathfrak{p}$ to be an ideal of $\mathfrak{g}$. We hence pass to the quotient $\mathfrak{g}/\mathfrak{p}$ if $\mathfrak{p}\neq\{0\}$, and find ourselves in the case where $\dim(\mathfrak{h}/(\mathfrak{h}\cap\mathfrak{z}))\leq 1$ if $\mathfrak{p}=\{0\}$.

To finish the proof, we examine integral (5.2.5) in more detail. A simple calculation shows (cf. [62])

$$(\pi_\ell(\phi)a_\omega)(g)=\int_{B[\ell]/(B[\ell]\cap H)}\phi^\chi(gb)\chi_\ell(b)d\dot{b}\ (\phi\in\mathscr{D}(G), g\in G)$$

and hence

$$\|\pi_\ell(\phi)a_\omega\|^2_{\mathscr{H}_{\pi_\ell}}=\int_{G/B[\ell]}\Big|\int_{B[\ell]/(B[\ell]\cap H)}\phi^\chi(gb)\chi_\ell(b)d\dot{b}\Big|d\dot{g}.$$

From the previous observations we can suppose, in formula (5.2.5), that $W_d, \widetilde{e}_4$ or e_k belong to $\mathfrak{h}+\mathfrak{z}$ but not to $\mathfrak{h}$. Then, replacing $Z\in\mathfrak{z}\backslash(\mathfrak{h}\cap\mathfrak{z})$ if necessary, we may assume $F_2(\ell)=\ell(Z)^r(\ell\in Q)$ for a certain non-negative integer r. Namely,

$$\int_{G/H}(R(W^m)\phi^\chi)(g)\overline{\phi^\chi(g)}d\dot{g}=\int_{\Gamma_\tau/H}\Big(\frac{F(\ell)}{\ell(Z)^r}\Big)^m\|\pi_\ell(\phi)a_\omega\|^2d\widetilde{\mu}(\dot{\ell})(m\in\mathbb{N}) \tag{5.2.6}$$

with $F(\ell)$ polynomial. In order to determine the measure $\widetilde{\mu}$, let us calculate both sides of the Plancherel formula:

$$\int_{G/H}|\phi^\chi(g)|^2d\dot{g}=\int_{\Gamma_\tau/H}\|\pi_\ell(\phi)a_\omega\|^2d\widetilde{\mu}(\dot{\ell}) \tag{5.2.7}$$

for test functions $\phi\in\mathscr{D}(G)$. Let $\mathfrak{v}$ be a complement in $\mathfrak{g}_0$ to $\mathfrak{g}_0\cap(\mathfrak{h}+\mathfrak{z})$, and $\{X, X'\}$ a complementary basis to $\mathfrak{g}_0$ in $\mathfrak{g}$. Let us introduce on G a coordinate system.

Consider first the case (i), i.e. $\mathfrak{h}\subset\mathfrak{g}_0$. Here, we choose for instance $X'=X_1$ and $X=X_2$. Then every element $g\in G$ is uniquely written as

$$g=\exp(x_1X_1)\exp(x_2X_2)\exp V\exp(wZ)h, x_1, x_2, w\in\mathbb{R}, V\in\mathfrak{v}, h\in H.$$

By means of this expression we identify G with $\mathbb{R}^2\times\mathfrak{v}\times\mathbb{R}\times H$, and consider $\phi\in\mathscr{D}(G)$ of the form

$$\phi(g)=\phi_1(x_1)\phi_2(x_2)\zeta(V)\kappa(h)\phi_3(w)$$

with $\phi_j \in \mathscr{D}(\mathbb{R})$, $\zeta \in \mathscr{D}(\mathfrak{v})$ and $\kappa \in \mathscr{D}(H)$. Then

$$\int_{G/H} |\phi^\chi(g)|^2 d\dot{g} = \Big(\prod_{j=1}^{3} \|\phi_j\|^2_{L^2(\mathbb{R})}\Big)\|\zeta\|^2_{L^2(\mathfrak{v})}\Big|\int_H \kappa(h)\chi(h)dh\Big|^2$$
$$= |\kappa^\chi(e)|^2\Big(\prod_{j=1}^{3} \|\phi_j\|^2_{L^2(\mathbb{R})}\Big)\|\zeta\|^2_{L^2(\mathfrak{v})}.$$

On the other hand,

$$\|\pi_\ell(\phi)a_\omega\|^2_{\mathscr{H}_{\pi_\ell}} = \int_{G/B[\ell]}\Big|\int_{B[\ell]/(B[\ell]\cap H)} \phi^\chi(gb)\chi_\ell(b)d\dot{b}\Big|^2 d\dot{g}$$
$$= \int_{\mathbb{R}}\Big|\int_{\mathbb{R}\times(G_0/(G_0\cap H))} \phi^\chi(\exp(x_1X_1)\exp(x_2(\ell(Z)X_1 - X_2))g_0)$$
$$\times e^{ix_2(\ell(Z)\ell(X_1)-\ell(X_2))}\chi_\ell(g_0)dx_2d\dot{g}_0\Big|^2 dx_1$$
$$= \int_{\mathbb{R}}\Big|\int_{\mathbb{R}} dx_2 \int_{G_0/(G_0\cap H)} \phi^\chi(\exp((x_1 + x_2\ell(Z))X_1)\exp(-x_2X_2)b(x_2,\alpha)g_0)$$
$$\times e^{ix_2(\ell(Z)\xi_1-\xi_2)}\chi_\ell(g_0)d\dot{g}_0\Big|^2 dx_1$$

with $\xi_j = \ell(X_j)(1 \leq j \leq 2)$ and a certain element $b = b(x_2,\alpha)$, $\alpha = \ell(Z)$, of G_0. Finally, the last integral is equal to

$$\int_{\mathbb{R}}\Big|\int_{\mathbb{R}} \overline{\chi_\ell(b)}dx_2 \int_{G_0/(G_0\cap H)} \phi^\chi(\exp((x_1 + x_2\ell(Z))X_1)\exp(-x_2X_2)g_0)$$
$$\times e^{ix_2(\ell(Z)\xi_1-\xi_2)}\chi_\ell(g_0)d\dot{g}_0\Big|^2 dx_1$$
$$= \int_{\mathbb{R}}\Big|\int_{\mathbb{R}} \overline{\chi_\ell(b)}dx_2 \int_{\mathfrak{v}} \phi^\chi(\exp((x_1 + x_2\ell(Z))X_1)\exp(-x_2X_2)\exp(V))$$
$$\times e^{ix_2(\ell(Z)\xi_1-\xi_2)}e^{i\ell(V)}dV\Big|^2 dx_1$$
$$= |\kappa^\chi(e)|^2|\widehat{\zeta}(\ell|_{\mathfrak{v}})|^2|\widehat{\phi_3}(\ell(Z))|^2\int_{\mathbb{R}} |\mathscr{F}_2(\Phi)(x_1,\ell(Z)\xi_1 - \xi_2)|^2 dx_1,$$

where the hat indicates the Fourier transform and $\mathscr{F}_2(\Phi)$ denotes the partial Fourier transform relative to the second variable of the function

$$\Phi(x_1,x_2) = \phi_1(x_1 + x_2\ell(Z))\phi_2(-x_2)\overline{\chi_\ell(b)}.$$

Eventually, we obtain that the product of Lebesgue measures $d\alpha d(\ell|_{\mathfrak{v}})d\xi_2$ is the measure $\widetilde{\mu}$ on Γ_τ/H realized as $\{\ell \in \Gamma_\tau;\, \xi_1 = \text{constant}\}$, for instance. Indeed,

$$\begin{aligned}
&\int_{\Gamma_\tau/H} \|\pi_\ell(\phi)a_\omega\|^2 d\alpha d(\ell|_{\mathfrak{v}})d\xi_2 \\
=&|\kappa^\chi(e)|^2\|\zeta\|^2 \int_{\mathbb{R}} |\widehat{\phi_3}(\alpha)|^2 d\alpha \int_{\mathbb{R}} dx_1 \int_{\mathbb{R}} |\mathscr{F}_2(\Phi)(x_1, \alpha\xi_1 - \xi_2)|^2 d\xi_2 \\
=&|\kappa^\chi(e)|^2\|\zeta\|^2 \int_{\mathbb{R}} |\widehat{\phi_3}|^2 d\alpha \int_{\mathbb{R}} dx_1 \int_{\mathbb{R}} |\Phi(x_1, x_2)|^2 dx_2 \\
=&|\kappa^\chi(e)|^2\|\zeta\|^2 \Big(\prod_{j=1}^{3} \|\phi_j\|^2\Big),
\end{aligned}$$

i.e. this measure gives us the Plancherel formula (5.2.7).

Let us examine case (ii), so assume $\mathfrak{h} \not\subset \mathfrak{g}_0$. If $\dim(\mathfrak{h}/(\mathfrak{h} \cap \mathfrak{g}_0)) = 2$, we choose $X' = X_1, X = X_2$ in $\mathfrak{h}$. Every $g \in G$ is uniquely written as $g = \exp V \exp(wZ)h$ with $V \in \mathfrak{v}, w \in \mathbb{R}$ and $h \in H$. In this way we identify G with $\mathfrak{v} \times \mathbb{R} \times H$. Consider $\phi \in \mathscr{D}(G)$ such that $\phi(g) = \zeta(V)\theta(w)\kappa(h)$ with $\zeta \in \mathscr{D}(\mathfrak{v}), \theta \in \mathscr{D}(\mathbb{R})$ and $\kappa \in \mathscr{D}(H)$. Evidently,

$$\int_{G/H} |\phi^\chi|^2 d\dot{g} = \|\zeta\|^2\|\theta\|^2|\kappa^\chi(e)|^2.$$

Concerning the right-hand side of formula (5.2.7),

$$\begin{aligned}
\|\pi_\ell(\phi)a_\omega\|^2_{\mathscr{H}_{\pi_\ell}} &= \int_{G/B[\ell]} \Big| \int_{B[\ell]/(B[\ell]\cap H)} \phi^\chi(gb)\chi_\ell(b)d\dot{b}\Big|^2 d\dot{g} \\
&= \int_{\mathbb{R}} dx \Big| \int_{B[\ell]/(B[\ell]\cap H)} \phi^\chi(\exp(xX_1)b)\chi_\ell(b)d\dot{b}\Big|^2 \\
&= \int_{\mathbb{R}} dx \Big| \int_{G_0/(G_0\cap H)} \phi^\chi(\exp(xX_1)g_0)\chi_\ell(g_0)d\dot{g_0}\Big|^2 \\
&= \int_{\mathbb{R}} \Big| \int_{G_0/(G_0\cap H)} \phi^\chi(g_0)\chi_{h(x)\cdot\ell}(g_0)d\dot{g_0}\Big|^2,
\end{aligned}$$

with $h(x) = \exp(xX_1) \in H$. We disintegrate the Lebesgue measure $d\ell$ on Γ_τ in agreement with the action of H. Let $\Gamma_\tau^\alpha = \{\ell \in \Gamma_\tau;\, \ell(Z) = \alpha\}$. We know [30] that there is a positive measure ν_α on Γ_τ^α/H such that

$$d\ell = \int_{\Gamma_\tau^\alpha/H} d\nu_\alpha(\dot{\ell}) \int_{\mathbb{R}} h(x)\cdot\ell dx$$

on Γ_τ^α. Then $\widetilde{\mu} = \int_{\mathbb{R}} \nu_\alpha d\alpha$ suits our purpose. Indeed, taking into account the formula written above

$$\begin{aligned}
&\int_{\Gamma_\tau/H} \|\pi_\ell(\phi)a_\omega\|^2 d\widetilde{\mu}(\dot{\ell}) \\
=&\int_{\mathbb{R}} d\alpha \int_{\Gamma_\tau^\alpha/H} d\nu_\alpha \int_{\mathbb{R}} dx \Big| \int_{G_0/(G_0\cap H)} \phi^X(g_0)\chi_{h(x)\cdot\ell}(g_0)d\dot{g}_0\Big|^2 \\
=&\int_{\mathbb{R}} d\alpha \int_{\Gamma_\tau^\alpha} d\ell \Big| \int_{\mathfrak{v}\times\mathbb{R}} \phi^X(\exp V \exp(wZ))e^{i\ell(V)}e^{i\alpha w} dV\Big|^2 \\
=&|\kappa^X(e)|^2 \int_{\mathbb{R}} \widehat{\theta}(\alpha)d\alpha \int_{\mathfrak{v}^*} d\ell \Big| \int_{\mathfrak{v}} \zeta(V)e^{i\ell(V)}dV\Big|^2 \\
=&|\kappa^X(e)|^2\|\theta\|^2\|\zeta\|^2.
\end{aligned}$$

Assume finally $\dim(\mathfrak{h}/(\mathfrak{h}\cap\mathfrak{g}_0)) = 1$. Modifying X by a multiple of E if necessary, take $X \in \mathfrak{h}\cap\ker(f)$. Take $X = X_2, X' = X_1$ if $[X,\mathfrak{q}] \subset \mathbb{R}Z$, and $X = X_1 + \lambda X_2$, $X' = X_2$ otherwise. Put $h(t) = \exp(tX) \in H$ for all $t \in \mathbb{R}$. Every $g \in G$ is uniquely written as

$$g = \exp(x'X')\exp V \exp(wZ)h,\ \ x', w \in \mathbb{R}, V \in \mathfrak{v}, h \in H.$$

Considering $\phi \in \mathscr{D}(G)$ of the form

$$\phi(g) = \gamma(x')\zeta(V)\theta(w)\kappa(h),\ \ \gamma,\theta \in \mathscr{D}(\mathbb{R}), \zeta \in \mathscr{D}(\mathfrak{v}), \kappa \in \mathscr{D}(H),$$

we repeat the same argument. First of all,

$$\int_{G/H} |\phi^X(g)|^2 d\dot{g} = |\kappa^X(e)|^2\|\gamma\|^2\|\zeta\|^2\|\theta\|^2.$$

Assume first $X = X_2, X' = X_1$. Let us calculate the right-hand side of Plancherel's formula (5.2.7):

$$\begin{aligned}
\|\pi_\ell(\phi)a_\omega\|^2_{\mathscr{H}_{\pi_\ell}} =& \int_{G/B[\ell]} \Big| \int_{B[\ell]/(B[\ell]\cap H)} \phi^X(gb)\chi_\ell(b)d\dot{b}\Big|^2 d\dot{g} \\
=\int_{\mathbb{R}} \Big| \int_{\mathbb{R}} dx_2 & \int_{G_0/(G_0\cap H)} \phi^X(\exp(x_1X_1)\exp(x_2(\ell(Z)X_1 - X_2))g_0) \\
&\times e^{ix_2\ell(Z)\ell(X_1)}\chi_\ell(g_0)dx_2 d\dot{g}_0\Big|^2 dx_1 \\
=\int_{\mathbb{R}} dx_1 \Big| \int_{\mathbb{R}} dx_2 & \int_{G_0/(G_0\cap H)} \phi^X(\exp((x_1 + x_2\ell(Z))X_1)\exp(-x_2X_2)b(x_2,\alpha)g_0) \\
&\times e^{ix_2\ell(Z)\xi_1}\chi_\ell(g_0)d\dot{g}_0\Big|^2,
\end{aligned}$$

where, as before, $\xi_1 = \ell(X_1)$ and $b = b(x_2, \alpha)$ denotes a certain element of G_0. The last integral becomes

$$\int_{\mathbb{R}} dx_1 \Big| \int_{\mathbb{R}} \overline{\chi_\ell(b)} dx_2 \int_{G_0/(G_0 \cap H)} \phi^\chi(\exp((x_1 + x_2\ell(Z))X_1)\exp(-x_2X_2)g_0)$$
$$\times e^{ix_2\ell(Z)\xi_1} \chi_\ell(g_0) d\dot{g}_0 \Big|^2$$
$$= \int_{\mathbb{R}} dx_1 \Big| \int_{\mathbb{R}} \overline{\chi_\ell(b)} dx_2 \int_{G_0/(G_0 \cap H)} \phi^\chi(\exp((x_1 + x_2\ell(Z))X_1)g_0)$$
$$\times e^{ix_2\ell(Z)\xi_1} \chi_{h(-x_2)\cdot\ell}(g_0) d\dot{g}_0 \Big|^2.$$

We denote by f_0 the restriction of f to $\mathfrak{g}_0$, by Λ the hyperplane $\{\ell \in \mathfrak{g}^*; \ell(Y) = 0\}$ of $\mathfrak{g}^*$ and by Ξ the affine subspace $f_0 + (\mathfrak{h} \cap \mathfrak{g}_0)^\perp$ of $\mathfrak{g}_0^*$. Take $\Xi_\alpha = \{\ell \in \Xi; \ell(Z) = \alpha\}$ and disintegrate the Lebesgue measure $d\ell$ on Ξ_α by the action of H:

$$d\ell = \int_{\Xi_\alpha/H} d\nu_\alpha(\dot{\ell}) \int_{\mathbb{R}} h(x)\cdot\dot{\ell} dx.$$

It is easy to see that the Borel space Γ_τ/H is identified with $\Gamma_\tau \cap \Lambda$ and then with $\mathbb{R} \times (\Xi/H)$ by the mapping

$$\Gamma_\tau \cap \Lambda \ni \ell \mapsto (\ell(X_1), H\cdot(\ell|_{\mathfrak{g}_0})) \in \mathbb{R} \times (\Xi/H).$$

After these remarks the measure $\widetilde{\mu}$ is $d\xi_1 \times \int_{\mathbb{R}} |\alpha| \nu_\alpha d\alpha$. Indeed,

$$\int_{\mathbb{R}} |\alpha| d\alpha \int_{\Xi_\alpha/H} d\nu_\alpha(\dot{\ell}) \int_{\mathbb{R}} \|\pi_\ell(\phi)a_\omega\|^2 d\xi_1$$
$$= \int_{\mathbb{R}} d\alpha \int_{\Xi_\alpha/H} d\nu_\alpha(\dot{\ell}) \int_{\mathbb{R}^2} dx_1 dx_2$$
$$\times \Big| \int_{G_0/(G_0 \cap H)} \phi^\chi(\exp((x_1 + \alpha x_2)X_1)g_0) \chi_{h(-x_2)\cdot\ell}(g_0) d\dot{g}_0 \Big|^2$$
$$= \int_{\mathbb{R}^2} dx_1 d\alpha \int_{\Xi_\alpha/H} d\nu_\alpha(\dot{\ell}) \int_{\mathbb{R}} dx_2$$
$$\times \Big| \int_{G_0/(G_0 \cap H)} \phi^\chi(\exp(x_1X_1)g_0) \chi_{h(x_2)\cdot\ell}(g_0) d\dot{g}_0 \Big|^2$$
$$= \int_{\mathbb{R}^2} \int_{\Xi_\alpha} \Big| \int_{G_0/(G_0 \cap H)} \phi^\chi(\exp(x_1X_1)g_0) \chi_\ell(g_0) d\dot{g}_0 \Big|^2 d\ell$$
$$= \int_{\mathbb{R}^2} dx_1 d\alpha \int_{\mathfrak{v}^*} \Big| \int_{\mathfrak{v} \times \mathbb{R}} \phi^\chi(\exp(x_1X_1)\exp V \exp(wZ)) e^{i\ell(V)} e^{i\alpha w} dV dw \Big|^2 d\ell$$

$$
\begin{aligned}
&=|\kappa^X(e)|^2 \int_{\mathbb{R}} |\gamma(x_1)|^2 dx_1 \int_{\mathbb{R}} |\widehat{\theta}(\alpha)|^2 d\alpha \int_{\mathfrak{v}^*} |\widehat{\zeta}(\ell)|^2 d\ell \\
&=|\kappa^X(e)|^2 \|\gamma\|^2 \|\zeta\|^2 \|\theta\|^2.
\end{aligned}
$$

Assume next $X = X_1 + \lambda X_2 (\lambda \in \mathbb{R})$, $X' = X_2$, and go about as in the previous case:

$$
\begin{aligned}
\|\pi_\ell(\phi)a_\omega\|^2 &= \int_{G/B[\ell]} \Big| \int_{B[\ell]/(B[\ell]\cap H)} \phi^X(gb)\chi_\ell(b)d\dot{b}\Big|^2 d\dot{g} \\
&= \int_{\mathbb{R}} dx_1 \Big| \int_{\mathbb{R}\times(G_0/(G_0\cap H))} \phi^X(\exp(x_1X_1)\exp(x_2(\alpha X_1 - X_2))g_0) \\
&\quad \times e^{ix_2(\alpha\ell(X_1)-\ell(X_2))}\chi_\ell(g_0)dx_2 d\dot{g}_0\Big|^2 \\
&= \int_{\mathbb{R}} dx_1 \Big| \int_{\mathbb{R}\times(G_0/(G_0\cap H))} \Phi(x_1, x_2, g_0)e^{-ix_2\xi_2(1+\lambda\alpha)}\chi_\ell(g_0)dx_2 d\dot{g}_0\Big|^2,
\end{aligned}
$$

where

$$
\Phi(x_1, x_2, g_0) = \phi^X(\exp((-x_2(1+\lambda\alpha) - \lambda x_1)X_2)\exp((x_1 + x_2\alpha)X)bg_0).
$$

Here as before $\xi_2 = \ell(X_2), \alpha = \ell(Z)$ and $b = b(x_1, x_2, \alpha)$ denotes a certain element of G_0. The last term gives us:

$$
\begin{aligned}
&\int_{\mathbb{R}} dx_1 \Big| \int_{\mathbb{R}} \overline{\chi_\ell(b)}dx_2 \int_{G_0/(G_0\cap H)} \phi^X(\exp((-x_2(1+\lambda\alpha) - \lambda x_1)X_2)g_0) \\
&\times e^{-ix_2\xi_2(1+\lambda\alpha)}\chi_{h(x_1+\alpha x_2)\cdot\ell}(g_0)d\dot{g}_0\Big|^2 \\
&= \frac{1}{|1+\lambda\alpha|}\int_{\mathbb{R}} dx_1 \Big| \int_{\mathbb{R}} \overline{\chi_\ell(b)}dx_2 \int_{G_0/(G_0\cap H)} \phi^X(\exp((x_2 - \lambda x_1)X_2)g_0) \\
&\times e^{ix_2\xi_2}\chi_{h(x_1-\alpha x_2/(1+\lambda\alpha))\cdot\ell}(g_0)d\dot{g}_0\Big|^2.
\end{aligned}
$$

This time $\widetilde{\mu} = d\xi_2 \times \int_{\mathbb{R}} \nu_\alpha d\alpha$ is suitable, for

$$
\begin{aligned}
&\int_{\mathbb{R}^2} d\xi_2 d\alpha \int_{\Xi_\alpha/H} \|\pi_\ell(\phi)a_\omega\|^2 d\nu_\alpha(\dot{\ell}) \\
&= \frac{1}{|1+\lambda\alpha|}\int_{\mathbb{R}^2} dx_1 d\alpha \int_{\Xi_\alpha/H} d\nu_\alpha(\dot{\ell}) \int_{\mathbb{R}} d\xi_2 \\
&\times \Big| \int_{\mathbb{R}} \overline{\chi_\ell(b)}dx_2 \int_{G_0/(G_0\cap H)} \Psi(x_1, x_2, \alpha, \xi_2, g_0)d\dot{g}_0\Big|^2,
\end{aligned}
$$

where

$$\Psi(x_1, x_2, \alpha, \xi_2, g_0) = \phi^\chi(\exp((x_2 - \lambda x_1)X_2)g_0)e^{ix_2\xi_2}\chi_{h(x_1-\alpha x_2/(1+\lambda\alpha))\cdot\ell}(g_0).$$

The last integral is equal to

$$\frac{1}{|1+\lambda\alpha|}\int_{\mathbb{R}^2} dx_1 d\alpha \int_{\Xi_\alpha/H} d\nu_\alpha(\dot{\ell}) \int_{\mathbb{R}} dx_2$$
$$\times \Big|\int_{G_0/(G_0\cap H)} \phi^\chi(\exp((x_2 - \lambda x_1)X_2)g_0)\chi_{h(x_1-\alpha x_2/(1+\lambda\alpha))\cdot\ell}(g_0)d\dot{g_0}\Big|^2$$
$$= \int_{\mathbb{R}^2} dx_1 d\alpha \int_{\Xi_\alpha/H} d\nu_\alpha(\dot{\ell}) \int_{\mathbb{R}} dx_2$$
$$\times \Big|\int_{G_0/(G_0\cap H)} \phi^\chi(\exp(x_2X_2)g_0)\chi_{h(x_1)\cdot\ell}(g_0)d\dot{g_0}\Big|^2$$
$$= \int_{\mathbb{R}^2} dx_2 d\alpha \int_{\Xi_\alpha} d\ell \Big|\int_{G_0/(G_0\cap H)} \phi^\chi(\exp(x_2X_2)g_0)\chi_\ell(g_0)d\dot{g_0}\Big|^2$$
$$= \int_{\mathbb{R}^2} dx_2 d\alpha \int_{\mathfrak{v}^*} d\ell \Big|\int_{\mathfrak{v}} \phi^\chi(\exp(x_2X_2)\exp V)e^{i\ell(V)}e^{i\alpha w}dVdw\Big|^2$$
$$= |\kappa^\chi(e)|^2 \int_{\mathbb{R}} |\gamma(x_2)|^2 dx_2 \int_{\mathfrak{v}^*} |\widehat{\zeta}(\ell)|^2 d\ell \int_{\mathbb{R}} |\widehat{\theta}(\alpha)|^2 d\alpha$$
$$= |\kappa^\chi(e)|^2 \|\gamma\|^2 \|\zeta\|^2 \|\theta\|^2.$$

If $F(\ell)$ is not divisible by α^r in formula (5.2.6), we arrive at a contradiction, as mentioned earlier. In fact, in all cases examined until now, the normalization of the measures we have employed means, roughly speaking, that the Plancherel measure $\widetilde{\mu}$ is rational in $\alpha = \ell(Z)$. Assume $r > 0$ in formula (5.2.6), and that the polynomial $F(\ell)$ is not divisible by $\ell(Z)$. Choosing appropriately the test functions, especially ϕ_3 or θ, whose Fourier transform does not vanish at $\alpha = 0$, we deduce that the right-hand side of (5.2.6) may diverge, although the left-hand side is convergent.

In fact, to show divergence on the right in (5.2.6), we evaluate $\|\pi_\ell(\phi)a_\omega\|^2_{\mathscr{H}_{\pi_\ell}}$ when $\alpha = \ell(Z)$ tends to 0. Examining the expression of this quantity, the crucial case is that where $\dim(\mathfrak{h}/(\mathfrak{h}\cap\mathfrak{g}_0)) = 1$ and $[\mathfrak{h}, \mathfrak{q}] \subset \mathbb{R}Z$. Keeping the notations, we have

$$\liminf_{\alpha\to 0}\|\pi_\ell(\phi)a_\omega\|^2_{\mathscr{H}_{\pi_\ell}} \geq \int_C \liminf_{\alpha\to 0}\Big|\int_{\mathbb{R}} \overline{\chi_\ell(b)}dx_2$$
$$\times \int_{G_0/(G_0\cap H)} \phi^\chi(\exp((x_1+\alpha x_2)X_1)g_0)e^{i\alpha x_2\xi_1}\chi_{h(-x_2)\cdot\ell}(g_0)d\dot{g_0}\Big|^2 dx_1$$

for any compact set $C \subset \mathbb{R}$. For $\alpha \neq 0$, we can identify the quotient Ξ_α/H also with the section $\widetilde{\Xi}_\alpha = \{\ell \in \Xi_\alpha; \ell(Y) = 1\}$. In the last integral, if we choose $\gamma \in \mathscr{D}(\mathbb{R})$ appropriately, the matter boils down to the variable x_2 ensuring that αx_2 remains in a compact set. In what follows, according to the choice of C and γ, we take into account the fact that this non-empty compact set can be made as small as we want in $\mathbb{R}_+$ (or $\mathbb{R}_-$ if need be). Furthermore, if $[\mathfrak{h}, \mathfrak{g}_0] \not\subset (\mathfrak{h} \cap \ker(f)) + \mathbb{R}Z$, the H-orbit $H\cdot(\ell|_{\mathfrak{g}_0}) \subset \mathfrak{g}_0^*$ is not bounded for generic $\ell \in \widetilde{\Xi}_\alpha$. As the Plancherel formula holds also for Schwartz functions satisfying the covariance condition under the right action of H, we can choose a Schwartz function $\zeta \in \mathscr{S}(\mathfrak{v})$ such that the Fourier transform

$$\widehat{\zeta}(h(-x_2)\cdot\ell) = \int_{\mathfrak{v}} \zeta(V)e^{i(h(-x_2)\cdot\ell)(V)}dV$$

vanishes for all x_2 outside a compact set that does not depend on α. Then projecting Ξ_α onto Ξ_0 along $\mathfrak{v}^\perp$, we have

$$\begin{aligned}&\liminf_{\alpha\to 0}\|\pi_\ell(\phi)a_\omega\|^2_{\mathscr{H}_{\pi_\ell}}\\ \geq&|\kappa^X(e)|^2|\widehat{\theta}(0)|^2 \int_C |\gamma(x_1)|^2dx_1\Big|\int_{\mathbb{R}} dx_2 \int_{\mathfrak{v}} \zeta(V)e^{i(h(-x_2)\cdot\ell_0)(V)}dV\Big|^2,\end{aligned}$$

where $\ell_0 \in \Xi_0$ designates the projection's image of $\ell \in \Xi_\alpha$. We can arrange things so that the right-hand side of this last integral is positive for our function $\zeta \in \mathscr{S}(\mathfrak{v})$, which gives the desired divergence.

Assume finally $[\mathfrak{h}, \mathfrak{g}_0] \subset (\mathfrak{h} \cap \ker(f)) + \mathbb{R}Z$. Let $\mathfrak{d} = \mathfrak{h} \cap \mathfrak{g}_0 \cap \ker(f)$. Then $\mathfrak{j} = \mathfrak{d} + \mathbb{R}Z$ is an ideal of $\mathfrak{g}$. In fact, there is an $X_\ell \in \mathfrak{g}_0$ such that $X(\ell) = \alpha X_1 - X_2 + X_\ell \in \mathfrak{g}(\ell)$ with $\alpha = \ell(Z)$. Then $[X(\ell), \mathfrak{j}] \equiv \alpha[Z_1, \mathfrak{j}]$ modulo $\mathfrak{d}$. Consequently, $[X_1, \mathfrak{j}] \subset \ker(\ell)$ and hence $[X_1, \mathfrak{j}] \subset \mathfrak{d}$ for generic $\ell \in \Gamma_\tau$. If $\mathfrak{j} \neq \mathbb{R}Z$, there exists $\widetilde{Y} \in \mathfrak{h}$ such that $[\mathfrak{h}, \widetilde{Y}] = \mathbb{R}Z$ since $\mathfrak{j} \cap \mathfrak{z} = \mathbb{R}Z$. In this way, we find ourselves in the situation where almost all G-orbits through Γ_τ are saturated in the direction $\mathfrak{k}^\perp$, where $\mathfrak{k}$ denotes the centralizer of $\widetilde{Y}$ in $\mathfrak{g}$. Hence we can pass to the normal subgroup $K = \exp \mathfrak{k}$ to prove the conjecture.

Let now $\mathfrak{j} = \mathbb{R}Z$, which amounts to saying that $\mathfrak{h} = \mathbb{R}X_2 + \mathbb{R}E$. In particular, $\mathfrak{d} = \{0\}$. Then it follows that $\mathfrak{g}_0 = \mathfrak{q}$. If this is not the case, we write $\mathfrak{g}_0 = \mathfrak{m} + \mathfrak{q}$ for a certain subspace $\mathfrak{m}$ satisfying $[X_2, \mathfrak{m}] = \{0\}$. Hence $\alpha[X_1, \mathfrak{m}] = [X(\ell), \mathfrak{m}] \subset \ker(\ell)$ for generic $\ell \in \Gamma_\tau$, i.e. $[X_1, \mathfrak{m}] \subset \mathfrak{d} = \{0\}$. As the center $\mathfrak{z}$ is contained in $\mathfrak{q}$, $\mathfrak{m}$ turns out to be trivial.

There only remains to examine the example $\mathfrak{g} = \mathbb{R}\text{-span}(X_1, X_2, Y, E, Z)$, $\mathfrak{h} = \mathbb{R}\text{-span}(X_2, E)$ with $f(X_2) = 0$, $f(E) = 1$. The non-zero bracket relations are

$$[X_1, Y] = E, [X_2, Y] = Z, [X_1, X_2] \in \mathfrak{g}_0$$

where $\mathfrak{g}_0 = \mathbb{R} - \text{span}(Y, E, Z)$. Assume $[X_1, X_2] \neq 0$, otherwise everything is clear. Let $X = ZX_1 + [X_1, X_2]Y$ if $[X_1, X_2] \in \mathfrak{z}$, or taking $Y \in \mathfrak{q}\backslash\mathfrak{z}$ if necessary,

$X = 2ZX_1 + \delta Y^2$ if $[X_1, X_2] = \delta Y \notin \mathfrak{z}$. We have already seen, when $\dim \mathfrak{h} = 1$, that $X \in \mathscr{U}(\mathfrak{g})^H$ and that $P_X(\ell)$ is a polynomial function. Let us consider a representative element

$$W = P_m(Z)X_1^m + \sum_{k=0}^{m-1} P_k(Y, Z)X_1^k$$

in $\mathscr{U}(\mathfrak{g}, \tau)$, where P_m is a polynomial of one variable and $P_k (0 \leq k \leq m-1)$ are polynomials of two variables. Then $[X_2, W] \in \mathscr{U}(\mathfrak{g})(E + if(E))$ and hence $[X_2, \overline{\beta^{-1}(W)}] \in S(\mathfrak{g})(E - if(E))$. In $S(\mathfrak{g})$ we write

$$\overline{\beta^{-1}(W)} = P_m(Z)X_1^m + \sum_{k=0}^{m-1} Q_k(Y, Z)X_1^k$$

modulo $S(\mathfrak{g})(E - if(E))$, using certain polynomials $Q_k (0 \leq k \leq m-1)$ of two variables. Calculating the condition $[X_2, \overline{\beta^{-1}(W)}] \in S(\mathfrak{g})(E - if(E))$, we easily verify that $P_m(Z)$ is divisible by Z^m. It follows that there is a polynomial $\widetilde{P}_m$ of one variable such that $W - \widetilde{P}_m(Z)X^m \in \mathscr{U}(\mathfrak{g}, \tau)$ is of degree at most $m-1$ in X_1. By induction on this degree, $P_W(\ell)$ extends to an element of $\mathbb{C}[\Gamma_\tau]^H$.

In this way, we arrive at the following:

Theorem 5.2.3 *Assume that $\tau = ind_H^G \chi$ has finite multiplicities. If* $\dim(G{\cdot}\ell) \leq 2$ *for generic $\ell \in \Gamma_\tau$, then the polynomial conjecture holds.*

Let us justify the fact we used in the proof of the theorem. We will keep our line of thought, but let us mention there is another method involving intertwining operators [10].

Theorem 5.2.4 *The polynomial conjecture holds if there exists a common polarization* $\mathfrak{b}$ *for μ-almost all $\ell \in \Gamma_\tau$.*

Proof Take an ideal $\mathfrak{n}$ of codimension 1 containing $\mathfrak{b}$. Let $N = \exp \mathfrak{n}$, $B = \exp \mathfrak{b}$, $\mathfrak{h}_0 = \mathfrak{h} \cap \mathfrak{n}$, $H_0 = \exp(\mathfrak{h}_0)$ and $\tau_0 = \mathrm{ind}_{H_0}^N \chi$. If $\mathfrak{h} \subset \mathfrak{n}$, everything goes down to N to which we apply the induction hypothesis. Suppose that $\mathfrak{h} \not\subset \mathfrak{n}$. For any $X \in D_\tau(G/H) \cong \mathscr{U}(\mathfrak{g}, \tau)/\mathscr{U}(\mathfrak{g})\mathfrak{a}_\tau$, we choose its representative in $\mathscr{U}(\mathfrak{n})$ and identify $\Gamma_\tau = f + \mathfrak{h}^\perp$ with $\Gamma_{\tau_0} = f|_\mathfrak{n} + (\mathfrak{h}_0)^\perp \subset \mathfrak{n}^*$. Then as in the proof of Theorem 1 in [67], we see that $P_X(\ell)$ gives us a polynomial function on Γ_τ.

Conversely, start from a polynomial function $F(\ell) \in \mathbb{C}[\Gamma_\tau]^H$. Because of the H_0-invariance of $F(\ell)$, the induction hypothesis says that there exists $X \in \mathscr{U}(\mathfrak{n}, \tau_0)$ satisfying $P_X(\ell) = F(\ell)$ for generic $\ell = \ell_0 \in \Gamma_{\tau_0}$, i.e. $\overline{X a_{\ell_0}} = \overline{F(\ell) a_{\ell_0}}$, where a_{ℓ_0} denotes the Penney distribution for (H_0, π_{ℓ_0}) with $\pi_{\ell_0} = \mathrm{ind}_B^N \chi_\ell$. In these set-up we may use the earlier argument. Definition (5.2.3) of a_{ℓ_0} gives

$$\langle \overline{X a_{\ell_0}}, \psi \rangle = \int_{H_0/(H_0 \cap B)} \overline{(L({}^tX)\psi)(h)\chi(h)} d\dot{h}$$

for all $\psi \in \mathscr{H}^{\infty}_{\pi_{\ell_0}}$. Let $\mathfrak{h} = \mathfrak{h}_0 + \mathbb{R}Y$ for a certain $Y \notin \mathfrak{n}$. Set $h_s = \exp(sY)(\forall s \in \mathbb{R})$ and $\psi_s(n) = \psi(h_s^{-1} n h_s)(\forall n \in N)$. As F is H-invariant,

$$\begin{aligned}
\langle \overline{F(h_s{\cdot}\ell)} a_{h_s\cdot\ell_0}, \psi_s\rangle &= \overline{F(\ell)}\langle a_{h_s\cdot\ell_0}, \psi_s\rangle \\
&= \overline{F(\ell)} \int_{H_0/(H_0\cap h_s B h_s^{-1})} \overline{\psi_s(h)\chi(h)} d\dot{h} = \overline{F(\ell)} \int_{H_0/(H_0\cap B)} \overline{\psi(h)\chi(h)} d\dot{h} \\
&= \overline{F(\ell)}\langle a_{\ell_0}, \psi\rangle = \langle \overline{X} a_{\ell_0}, \psi\rangle.
\end{aligned}$$

On the other hand,

$$\begin{aligned}
\langle \overline{F(h_s{\cdot}\ell)} a_{h_s\cdot\ell_0}, \psi_s\rangle &= \langle \overline{X} a_{h_s\cdot\ell_0}, \psi_s\rangle \\
&= \int_{H_0/(H_0\cap h_s B h_s^{-1})} \overline{(L({}^t X)\psi_s)(h)\chi(h)} d\dot{h} \\
&= \int_{H_0/(H_0\cap B)} \overline{(L({}^t(h_{-s}{\cdot}X))\psi)(h)\chi(h)} d\dot{h} = \langle \overline{(h_{-s}{\cdot}X)} a_{\ell_0}, \psi\rangle.
\end{aligned}$$

Hence

$$\overline{(h_{-s}{\cdot}X)} a_{\ell_0} = \overline{X} a_{\ell_0} \ (\forall s \in \mathbb{R})$$

for almost all $\ell \in Q$. This means, differentiating, that $[Y, X] \in \mathscr{U}(\mathfrak{n})\mathfrak{a}_\tau$. In other words $X \in \mathscr{U}(\mathfrak{g}, \tau)$. □

Corollary 5.2.5 *If $\mathfrak{h}$ is an ideal of $\mathfrak{g}$, the polynomial conjecture holds.*

Proof It is well known that if τ has finite multiplicities, then the stabilizer of $f|_{\mathfrak{h}} \in \mathfrak{h}^*$ in $\mathfrak{g}$ is a common polarization for almost all $\ell \in \Gamma_\tau$. □

Corollary 5.2.6 *If* $\dim \mathfrak{h} = \frac{1}{2}(\dim \mathfrak{g} + \dim \mathfrak{g}(\ell)) - 1$ *for generic* $\ell \in \Gamma_\tau$, *the polynomial conjecture holds.*

Proof The condition implies that $\dim \mathfrak{h} = \dim \mathfrak{b}[\ell] - 1$ for generic $\ell \in \Gamma_\tau$. Hence if the center $\mathfrak{z}$ of $\mathfrak{g}$ is not contained in $\mathfrak{h}$, $\mathfrak{b} = \mathfrak{h} + \mathfrak{z}$ is a common polarization for generic $\ell \in \Gamma_\tau$. If $\mathfrak{z} \subset \mathfrak{h}$ and $\dim \mathfrak{z} \geq 2$, we can pass to the quotient $\mathfrak{g}/(\mathfrak{z} \cap \ker(f))$ to which we apply the induction hypothesis.

Let $\mathfrak{z} \subset \mathfrak{h}$ and $\dim \mathfrak{z} = 1$. If $f|_{\mathfrak{z}} = 0$, we go once again to the quotient $\mathfrak{g}/\mathfrak{z}$. Let $f|_{\mathfrak{z}} \neq 0$. We take $Y \in \mathfrak{g}\backslash\mathfrak{z}$ such that $[Y, \mathfrak{g}] \subset \mathfrak{z}$ and introduce the centralizer $\mathfrak{g}_0$ of Y in $\mathfrak{g}$. If $\mathfrak{h} \subset \mathfrak{g}_0$, it suffices to apply the induction hypothesis to $\mathfrak{g}_0$.

Finally, we consider the case where $\mathfrak{h} \not\subset \mathfrak{g}_0$. To obtain the desired result, we modify $\mathfrak{h}$ to $\mathfrak{h}' = \mathfrak{h}_0 + \mathbb{R}Y$ with $\mathfrak{h}_0 = \mathfrak{h} \cap \mathfrak{g}_0$. Let $Z \in \mathfrak{z}$, $X \in \mathfrak{h}$ be such that $\mathfrak{h} = \mathbb{R}X + \mathfrak{h}_0$, $f(Z) = 1$ and $[X, Y] = Z$. We fix $\ell_0 \in Q$ and let $f' \in \mathfrak{g}^*$ satisfy $f'|_{\mathfrak{h}_0} = f|_{\mathfrak{h}_0}$ and $f'(Y) = \ell_0(Y)$. Replacing Y by $Y - \ell_0(Y)Z$ if necessary, we can assume $\ell_0(Y) = 0$.

We then see that $\tau' = \operatorname{ind}_{H'}^{G} \chi_{f'}$, where $H' = \exp(\mathfrak{h}')$, has finite multiplicities. Given an element of $D_\tau(G/H)$, we choose its representative $W \in \mathscr{U}(\mathfrak{g}, \tau)$ in $\mathscr{U}(\mathfrak{g}_0)$. Then $W \in \mathscr{U}(\mathfrak{g}, \tau')$ and the function $P'_W(\ell)$ built out of the pair $(f', \mathfrak{h}')$ is an H'-invariant polynomial function on $\Gamma_{\tau'}$. Combining this with the H-invariance of $P_W(\ell)$, it follows that $P_W(\ell) = P'_W(\exp(-\ell(Y)X)\cdot\ell)$ for $\ell \in Q$. As a consequence, $P_W(\ell)$ is a polynomial function on Γ_τ.

Now we start from $F(\ell) \in \mathbb{C}[\Gamma_\tau]^H$. For $\ell \in Q$, we choose a polarization $\mathfrak{b}[\ell]$ contained in $\mathfrak{g}_0$. Let $H_0 = \exp(\mathfrak{h}_0)$, $G_0 = \exp(\mathfrak{g}_0)$, $B[\ell] = \exp(\mathfrak{b}[\ell])$ and $\pi_{\ell_0} = \operatorname{ind}_{B[\ell]}^{G_0} \chi_\ell$. We denote by Φ the affine subspace $f|_{\mathfrak{g}_0} + (\mathfrak{h}_0)^\perp$ in $(\mathfrak{g}_0)^*$. The surjectivity of the mapping for the pair $(f', \mathfrak{h}')$ and the H-invariance of $F(\ell)$ show that there exists $W \in \mathscr{U}(\mathfrak{g}_0, \tau')$ such that $F(\ell) = P'_W(\exp(-\ell(Y)X)\cdot\ell)(\ell \in Q)$. But then definition (5.2.3) of the Penney distribution a_{ℓ_0} for the pair (H_0, π_{ℓ_0}) gives us $\overline{W} a_{\ell_0} = \overline{F(\ell)} a_{\ell_0}$ for generic $\ell \in \Phi$. Now we can go over the proof of the theorem to show that $W \in \mathscr{U}(\mathfrak{g}, \tau)$ and $P_W(\ell) = F(\ell)$. □

Corollary 5.2.7 *If* $\dim G \leq 6$, *the polynomial conjecture holds.*

Proof Assume for example $\dim G = 6$. Let $d = \dim \mathfrak{h}$. If $d = 6$, there is nothing to do. If $d = 1$, the result follows from Theorem 5.2.3, likewise for $d = 2$ because, as we remarked in the course of the reduction, we are led to the case where $\dim(\mathfrak{h} \cap \mathfrak{z}) = 1$, $\mathfrak{z}$ being the center of $\mathfrak{g}$. If $d = 5$, it suffices to apply Lemma 5.2.5. If $d = 4$, we apply again Theorem 5.2.3, supposing that $\tau = \operatorname{ind}_H^G \chi$ is not irreducible. There remains to examine the case $d = 3$. But if $\dim(G\cdot\ell) = 4$ for generic $\ell \in \Gamma_\tau$, the result follows from Corollary 5.2.6. In all other cases, $\dim(G\cdot\ell) \leq 2$ for generic $\ell \in \Gamma_\tau$ and the result descends once again from Theorem 5.2.3. □

Before finishing this section, we give a simple proof of a theorem due to Corwin and Greenleaf [41], itself a special case of our last theorem.

Theorem 5.2.8 *The polynomial conjecture holds, provided there exists* μ*-almost everywhere in* Γ_τ *a common polarization* $\mathfrak{b}$ *which is normalized by* $\mathfrak{h}$.

Proof First of all, let us remark that $\mathfrak{c} = \mathfrak{h} + \mathfrak{b}$ is a subalgebra of $\mathfrak{g}$, as $\mathfrak{h}$ normalizes $\mathfrak{b}$. Let $C = \exp(\mathfrak{c})$. Then $\tau = \operatorname{ind}_C^G \tau_0$ where $\tau_0 = \operatorname{ind}_H^C \chi_f$ and so by Corwin and Greenleaf [40] the algebra $D_\tau(G/H)$ is isomorphic to the algebra $D_{\tau_0}(C/H)$. On the other hand, by the finite-multiplicity condition, we know that $\operatorname{ad}^*(\mathfrak{h})(f) \supset \mathfrak{c}^\perp$ and so $f + \mathfrak{c}^\perp$ is contained in the H-orbit of f. Hence the restriction map defines an H-equivariant isomorphism between the algebra of H-invariant polynomial functions defined on Γ_τ and the algebra of H-invariant polynomial functions defined on $f|_\mathfrak{c} + \mathfrak{h}^{\perp,\mathfrak{c}} \subset \mathfrak{c}^*$. Hence we can suppose that $G = C$. In particular $\mathfrak{b}$ is now an ideal of $\mathfrak{g}$ and $\mathfrak{g} = \mathfrak{h} + \mathfrak{b}$.

The Fourier transform, here denoted by U, maps the space $L^2(G/H, \chi_f)$ onto the Hilbert space $L^2(\Gamma_\tau)$. The transformation U is defined for $\xi \in \mathscr{S}(G/H, \chi_f)$ by

$$U(\xi)(\phi) = \int_{B/B\cap H} \xi(b)\chi_\phi(b)d\dot{b}, \ \phi \in \Gamma_\tau.$$

Let us take a Malcev basis $\mathscr{X} = \{X_1, \ldots, X_r\}$ of $\mathfrak{g}$ relative to $\mathfrak{h}$. Since $\mathfrak{g} = \mathfrak{h} + \mathfrak{b}$, we can assume that $\mathscr{X} \subset \mathfrak{b}$. But then for any $\phi \in \Gamma_\tau$, the set $\mathscr{X}$ is also a Malcev basis of $\mathfrak{b}$ relative to $\mathfrak{h} \cap \mathfrak{b} = \mathfrak{h} \cap \mathfrak{g}(\phi)$. Then we can write U in the following form:

$$U(\xi)(\phi) = \int_{\mathbb{R}^r} \xi(\exp(t_1 X_1) \cdots \exp(t_r X_r)) e^{-i(\sum_{k=1}^r t_k \phi(X_k))} dt_1 \cdots dt_r.$$

Hence in these coordinates U is just the ordinary Fourier transform on $\mathbb{R}^r$. Using U we can transfer the representation τ of G from $L^2(G/H, \chi_f)$ to $L^2(\Gamma_\tau)$, and we get a representation of G on $L^2(\Gamma_\tau)$. In particular,

$$\rho(h)\eta(\phi) = \chi_f(h)\eta(\mathrm{Ad}^*(h^{-1})\phi), \rho(b)\eta(\phi) = \chi_\phi(b)\eta(\phi)$$

for $b \in B, h \in H, \eta \in L^2(\Gamma_\tau)$.

Let now D be an element of $D_\tau(G/H)$. Then D commutes with $\tau(b)$ for all $b \in B$. Furthermore, D is represented by an element of the universal enveloping algebra $\mathscr{U}(\mathfrak{g})$, hence, it can be written as a differential operator on $\mathscr{S}(G/H, \chi_f)$ with polynomial coefficients. Let $D' = U \circ D \circ U^{-1}$ be the corresponding operator acting on the Schwartz space $\mathscr{S}(\Gamma_\tau)$ on Γ_τ. Then since U is the ordinary Fourier transform, D' too is a differential operator with polynomial coefficients, and it commutes with the multiplication by $e^{i(\cdot)(X)}$, $X \in \mathfrak{b}$. Therefore D' itself is a multiplication operator by a polynomial function P_D. As D commutes with the action of H, the function P_D must be H-invariant. Then P_D is an H-invariant polynomial function on Γ_τ. On the other hand, if P is an H-invariant polynomial function on Γ_τ, then the multiplication by P defines an operator D' on $\mathscr{S}(\Gamma_\tau)$ which commutes with the action of G. Hence $D = U^{-1} \circ D' \circ U$ is an element of $D_\tau(G/H)$. Consequently the algebra $D_\tau(G/H)$ is isomorphic to the algebra of H-invariant polynomial functions defined on Γ_τ. □

5.3 The Case of Restricted Representations

Let $G = \exp \mathfrak{g}$ be a connected and simply connected nilpotent Lie group with Lie algebra $\mathfrak{g}$. Let π be an irreducible unitary representation of G. We consider the restriction $\pi|_K$ of π to a closed connected subgroup $K = \exp \mathfrak{k}$ of G. Let

$$\pi|_K \simeq \int_{\widehat{K}}^{\oplus} m(\sigma)\sigma d\mu(\sigma) \tag{5.3.1}$$

be the canonical central decomposition of $\pi|_K$ as in Chap. 4. It is well known that the multiplicities $m(\sigma)$ are either uniformly bounded or $m(\sigma) \equiv \infty$ almost everywhere with respect to the measure μ. According to these possibilities, we say that $\pi|_K$ has finite or infinite multiplicities. Let us define an algebra of invariant differential operators. Let $\ker(\pi)$ be the primitive ideal of $\mathscr{U}(\mathfrak{g})$ associated to π and

$$\mathscr{U}_\pi(\mathfrak{g})^{\mathfrak{k}} = \{A \in \mathscr{U}(\mathfrak{g}); [\mathfrak{k}, A] \subset \ker(\pi)\}.$$

Evidently, $\ker(\pi)$ is a two-sided ideal of $\mathscr{U}_\pi(\mathfrak{g})^{\mathfrak{k}}$. Our focus is the algebra $D_\pi(G)^K$, which is the image of $\mathscr{U}_\pi(\mathfrak{g})^K$ under the homomorphism π. That is to say,

$$D_\pi(G)^K \cong \mathscr{U}_\pi(\mathfrak{g})^{\mathfrak{k}}/\ker(\pi) \cong (\mathscr{U}(\mathfrak{g})/\ker(\pi))^K,$$

where the last expression represents the algebra of K-invariant elements of $\mathscr{U}(\mathfrak{g})/\ker(\pi)$. The algebra $D_\pi(G)^K$ is regarded as the algebra of differential operators which leave $\mathscr{H}_\pi^\infty$ invariant and commute with the action of K on this space. It is known [7] that the algebra $D_\pi(G)^K$ is commutative if and only if $\pi|_K$ has finite multiplicities. We denote by $\Omega = \Omega(\pi)$ the coadjoint orbit of G corresponding to $\pi \in \widehat{G}$. As in the previous case we ask the following question.

Polynomial Conjecture (Cf. [18]) Assume that $\pi|_K$ has finite multiplicities. Is the algebra $D_\pi(G)^K$ isomorphic to the algebra $\mathbb{C}[\Omega(\pi)]^K$ of K-invariant polynomial functions on $\Omega(\pi)$?

Let $\mathfrak{g}'$ be an ideal of codimension 1 in $\mathfrak{g}$, $G' = \exp(\mathfrak{g}')$ and $\gamma : \mathfrak{g}^* \to (\mathfrak{g}')^*$ the restriction mapping. If the orbit Ω is saturated with respect to $\mathfrak{g}'$, then $\gamma(\Omega)$ is the union of a one-parameter family $\{\omega_t\}, t \in \mathbb{R}$, of G'-orbits, and there exists an element $Y_\ell \in \mathfrak{g}'$, which depends smoothly on ℓ, such that $\exp(\mathbb{R}Y_\ell) \cdot \ell = \mathrm{Ad}^*(\exp(\mathbb{R}Y_\ell))(\ell) = \ell + (\mathfrak{g}')^\perp$ for all $\ell \in \Omega$. Fix a vector $X \in \mathfrak{g}\backslash\mathfrak{g}'$ and define the mapping $\iota : (\mathfrak{g}')^* \to \mathfrak{g}^*$ by

$$\iota(\ell')(X) = 0,\ \iota(\ell')|_{\mathfrak{g}'} = \ell'\ (\ell' \in (\mathfrak{g}')^*).$$

The mapping $\psi : \mathbb{R} \times \mathbb{R} \times \omega_0 \to \Omega$, $\omega_0 = G' \cdot \ell' \subset (\mathfrak{g}')^*$, defined by

$$\psi(s, t, \ell') = \mathrm{Ad}^*(\exp(tX)) \circ \mathrm{Ad}^*(\exp(sY_{\iota(\ell')}))\iota(\ell')$$

is then a diffeomorphism and the invariant measure $d\mu_\pi$ on Ω can be decomposed as

$$\int_\Omega \varphi(\ell) d\mu_\pi(\ell) = \int_{\mathbb{R}}\int_{\mathbb{R}}\int_{\omega_0} (\varphi \circ \psi)(s, t, \ell') d\mu_{\pi'}(\ell') dt ds,\ \varphi \in C_c(\Omega). \quad (5.3.2)$$

Here π' denotes the irreducible unitary representation of G' corresponding to ω_0, while $d\mu_{\pi'}$ is the invariant measure on ω_0 and $C_c(\Omega)$ is the space of continuous functions with compact support on Ω. According to this decomposition of Ω, the representation $\pi|_{G'}$ disintegrates into an direct integral

$$\pi|_{G'} \simeq \int_{\mathbb{R}}^{\oplus} \pi'_t dt$$

of irreducible unitary representations π'_t of G' corresponding to the orbit $\omega_t = \exp(tX) \cdot \omega_0 (t \in \mathbb{R})$.

5.3.1 Frobenius Vectors

Lemma 5.3.1 *The representation $\pi|_K$ has finite multiplicities if and only if for μ_π-almost every $\ell \in \Omega$ the subspace $\mathfrak{b}[\ell|_{\mathfrak{k}}]+\mathfrak{g}(\ell)$ is Lagrangian for the skew-symmetric bilinear form B_ℓ, where $\mathfrak{b}[\ell|_{\mathfrak{k}}]$ denotes any polarization of $\mathfrak{k}$ at $\ell|_{\mathfrak{k}}$.*

Proof Let us proceed by induction on $\dim \mathfrak{g}$. Let $\mathfrak{g}'$ be an ideal of codimension 1 in $\mathfrak{g}$ containing $\mathfrak{k}$, $G' = \exp(\mathfrak{g}')$ and $\gamma : \mathfrak{g}^* \to (\mathfrak{g}')^*$ the restriction mapping. We write $\ell' = \gamma(\ell) \in (\mathfrak{g}')^*$. If the orbit Ω is non-saturated with respect to $\mathfrak{g}'$, then $\gamma(\Omega) = G' \cdot \ell'$ and $\pi' = \pi|_{G'}$ is irreducible. Since $\dim(\mathfrak{g}(\ell)) = \dim(\mathfrak{g}'(\ell')) + 1$ and since a subspace $\mathfrak{p}'$ of $\mathfrak{g}'$ is Lagrangian for $B_{\ell'}$ if and only if $\mathfrak{p}' + \mathfrak{g}(\ell)$ is Lagrangian for B_ℓ, we see that the induction hypothesis applied to $(\mathfrak{g}', \ell')$ gives us the desired result.

If the orbit Ω is saturated with respect to $\mathfrak{g}'$, then $\gamma(\Omega)$ is the union of a one-parameter family $\{\omega_t\}, t \in \mathbb{R}$, of G'-orbits. We write π'_t for the irreducible unitary representation of G' corresponding to the orbit ω_t. It follows from [7] that $\pi|_K$ has finite multiplicities if and only if $\pi'_t|_K$ has finite multiplicities for almost all $t \in \mathbb{R}$ and the orbit $K \cdot \ell$ is saturated with respect to $\mathfrak{g}'$ for μ_π-almost all $\ell \in \Omega$. By the induction hypothesis we know that for every $t \in \mathbb{R}$ such that $\pi'_t|_K$ has finite multiplicities, the subspace $\mathfrak{b}[\ell'_t|_{\mathfrak{k}}] + \mathfrak{g}'(\ell'_t)$ is Lagrangian for $B_{\ell'_t}$ for $\mu_{\pi'_t}$-almost all $\ell'_t \in \omega_t$. We conclude from this and from description (5.3.2) of Ω that, if $\pi|_K$ has finite multiplicities, then μ_π-almost everywhere $\mathfrak{b}[\ell'|_{\mathfrak{k}}] + \mathfrak{g}'(\ell')$ is Lagrangian with respect to $B_{\ell'}$. Furthermore, since Ω is saturated with respect to $\mathfrak{g}'$, for every $\ell \in \Omega$ and $\ell' = \gamma(\ell) \in (\mathfrak{g}')^*$, we have $\mathfrak{g}'(\ell') = \mathfrak{g}(\ell) + \mathbb{R}Y_\ell$, where Y_ℓ has been introduced above. Hence for every $\ell \in \Omega$ such that $K \cdot \ell \supset \ell + (\mathfrak{g}')^\perp$, we can find a Z_ℓ in $\mathfrak{k}(\ell|_{\mathfrak{k}}) \subset \mathfrak{b}[\ell|_{\mathfrak{k}}]$ for which $0 \neq Z_\ell \cdot \ell = \mathrm{ad}^*(Z_\ell)(\ell) \in (\mathfrak{g}')^\perp$. It follows from these considerations that the finite multiplicities of $\pi|_K$ imply that

$$\mathfrak{b}[\ell|_{\mathfrak{k}}] + \mathfrak{g}(\ell) = \mathfrak{b}[\ell'|_{\mathfrak{k}}] + \mathfrak{g}'(\ell')$$

is Lagrangian for B_ℓ for μ_π-almost all $\ell \in \Omega$.

If on the other hand $\mathfrak{b}[\ell|_{\mathfrak{k}}] + \mathfrak{g}(\ell)$ is Lagrangian μ_π-almost everywhere, then we can choose a Z_ℓ in $\mathfrak{k}$ such that $\ell([Z_\ell, \mathfrak{g}']) = \{0\}$ and $\ell([X, Z_\ell]) = 1$ and so $K \cdot \ell$ is saturated with respect to $\mathfrak{g}'$ for all these ℓ. Hence we also have that $\mathfrak{b}[\ell'|_{\mathfrak{k}}]+\mathfrak{g}'(\ell') = \mathfrak{b}[\ell|_{\mathfrak{k}}] + \mathfrak{g}(\ell)$ is Lagrangian for $B_{\ell'}$ for those ℓ's. The induction hypothesis and the aforementioned structure of Ω tell us that, for almost all $t \in \mathbb{R}$, $\pi'_t|_K$ has finite multiplicities. Hence $\pi|_K$ has finite multiplicities too. □

Remark 5.3.2 Frobenius' reciprocity between the induced representation and the restricted representation seems to suggest that, generically, $\pi|_K$ has finite multiplicities if and only if $\mathrm{ind}_K^G\sigma$ has finite multiplicities for μ-almost all $\sigma \in \widehat{K}$. The last statement implies that $\mathfrak{b}[\ell|_{\mathfrak{k}}] + \mathfrak{g}(\ell)$ is generically a Lagrangian subspace for B_ℓ.

Let

$$\mathscr{S} : \{0\} = \mathfrak{g}_0 \subset \mathfrak{g}_1 \subset \cdots \subset \mathfrak{g}_{n-1} \subset \mathfrak{g}_n = \mathfrak{g}$$

be a flag of ideals of $\mathfrak{g}$ such that $\dim(\mathfrak{g}_k) = k$ for $1 \le k \le n$. Choosing for every $1 \le j \le n$ a vector Z_j in $\mathfrak{g}_j \setminus \mathfrak{g}_{j-1}$, we obtain a Jordan-Hölder basis $\mathscr{Z} = \{Z_1, \ldots, Z_n\}$ of $\mathfrak{g}$. Let

$$\mathscr{I}^{\mathfrak{k}} = \mathscr{I} = \{i_1 < \cdots < i_d\},\ d = \dim \mathfrak{k},$$

be the set of indices $1 \le i \le n$ such that $\mathfrak{k} \cap \mathfrak{g}_i \neq \mathfrak{k} \cap \mathfrak{g}_{i-1}$. Let us put

$$\mathscr{J}^{\mathfrak{g}/\mathfrak{k}} = \mathscr{J} = \{j_1 < \cdots, < j_q\} = \{1, \ldots, n\} \setminus \mathscr{I}$$

with $q = \dim(\mathfrak{g}/\mathfrak{k})$.

We obtain an increasing sequence of subalgebras $\mathfrak{l}_r$, $0 \le r \le q$, of $\mathfrak{g}$ such that $\dim(\mathfrak{l}_r/\mathfrak{l}_{r-1}) = 1$, $\mathfrak{l}_q = \mathfrak{g}$ by setting

$$\mathfrak{l}_0 = \mathfrak{k},\ \mathfrak{l}_r = \mathfrak{k} + \mathfrak{g}_{j_r},\ 1 \le r \le q. \tag{5.3.3}$$

Furthermore, considering $\mathfrak{k}_s = \mathfrak{k} \cap \mathfrak{g}_{i_s}$, we produce a flag of ideals of $\mathfrak{k}$:

$$\{0\} = \mathfrak{k}_0 \subset \mathfrak{k}_1 \subset \cdots \subset \mathfrak{k}_{d-1} \subset \mathfrak{k}_d = \mathfrak{k},\ \dim(\mathfrak{k}_s) = s\ (0 \le s \le d). \tag{5.3.4}$$

We now fix a vector Y_s of $\mathfrak{k}_s \setminus \mathfrak{k}_{s-1}$ for $1 \le s \le d$ and obtain a Jordan-Hölder basis $\{Y_1, \ldots, Y_d\}$ of $\mathfrak{k}$. In the same way, extracting an element Y_{d+r} of $\mathfrak{l}_r \setminus \mathfrak{l}_{r-1}$ for $1 \le r \le q$, we form a Malcev basis $\{Y_{d+1}, \ldots, Y_n\}$ of $\mathfrak{g}$ relative to $\mathfrak{k}$.

Let $\ell \in \Omega = \Omega(\pi)$. Taking a real polarization $\mathfrak{b}[\ell]$ at ℓ, we realize the representation π as $\pi = \mathrm{ind}_{B[\ell]}^G \chi_\ell$, where $B[\ell] = \exp(\mathfrak{b}[\ell])$ and $\chi_\ell(\exp X) = e^{i\ell(X)}$, $X \in \mathfrak{b}[\ell]$. We can use the flag (5.3.4) to construct the Vergne polarization [150] $\mathfrak{b}[\ell|_{\mathfrak{k}}]$ at $\ell|_{\mathfrak{k}} \in \mathfrak{k}^*$. Let $B[\ell|_{\mathfrak{k}}] = \exp(\mathfrak{b}[\ell|_{\mathfrak{k}}])$. It is easy to see that there exists a Malcev basis of $\mathfrak{b}[\ell|_{\mathfrak{k}}]$ relative to $\mathfrak{b}[\ell|_{\mathfrak{k}}] \cap \mathfrak{b}[\ell]$ which is contained in a Malcev basis of $\mathfrak{g}$ relative to $\mathfrak{b}[\ell]$. From this it follows directly that the formula

$$\langle a_\ell, \varphi \rangle = \int_{B[\ell|_{\mathfrak{k}}]/(B[\ell|_{\mathfrak{k}}] \cap B[\ell])} \overline{\varphi(b)\chi_\ell(b)}\, d\dot{b}\ (\varphi \in \mathscr{H}_\pi^\infty), \tag{5.3.5}$$

where $d\dot{b}$ denotes an invariant measure on the homogeneous space $B[\ell|_{\mathfrak{k}}]/(B[\ell|_{\mathfrak{k}}] \cap B[\ell])$, defines a generalized semi-invariant vector $a_\ell \in (\mathscr{H}_\pi^{-\infty})^{B[\ell|_{\mathfrak{k}}], \chi_\ell}$. Because of the following theorem, a_ℓ is sometimes called Frobenius vector. Suppose that $\pi|_K$

has finite multiplicities. Then according to Lemma 5.3.1, we know that $\mathfrak{b}[\ell|_{\mathfrak{k}}]+\mathfrak{g}(\ell)$ is μ_π-almost everywhere a Lagrangian subspace for the bilinear form B_ℓ. For such an ℓ, we know that the choice of the polarization does not matter for the definition of the distribution a_ℓ (cf. [62]). Indeed, if $\mathfrak{b}'[\ell]$ is another polarization at ℓ and $B'[\ell] = \exp(\mathfrak{b}'[\ell])$, we can form the distribution a'_ℓ as in (5.3.5) and if T denotes an intertwining operator for these two realizations of π, then

$$a'_\ell \circ T = c a_\ell \tag{5.3.6}$$

for some complex number c.

Theorem 5.3.3 *Suppose that $\pi|_K$ has finite multiplicities. Then μ_π-almost everywhere on Ω, the distribution a_ℓ is an eigenvector for every element of $D_\pi(G)^K$. In other words, for $W \in \mathscr{U}_\pi(\mathfrak{g})^{\mathfrak{k}}$, we have*

$$\pi(W)a_\ell = P_W(\ell)a_\ell$$

for certain complex scalars $P_W(\ell)$ for μ_π-almost all $\ell \in \Omega$. Furthermore the function $P_W(\ell)$ of ℓ is K-invariant.

Proof If $G = K$, the algebra $D_\pi(G)^K$ reduces to $\mathbb{C}$ by Schur's lemma, and there is nothing to prove. Let now $q > 0$ in the sequence (5.3.3) and let us proceed by induction on $\dim G$. Let $G' = \exp(\mathfrak{l}_{q-1})$ and $\ell' = \ell|_{\mathfrak{l}_{q-1}}$. Suppose first that Ω is saturated with respect to $\mathfrak{l}_{q-1}$. According to (5.3.6), we can choose the polarization $\mathfrak{b}[\ell]$ as we want, so we can assume that $\mathfrak{b}[\ell] \subset \mathfrak{l}_{q-1}$. Let $\pi' = \mathrm{ind}_{B[\ell]}^{G'} \chi_\ell$ with $B[\ell] = \exp(\mathfrak{b}[\ell])$. Then a_ℓ can be identified with $a_{\ell'}$. We know from [7] that $\mathscr{U}_\pi(\mathfrak{g})^{\mathfrak{k}} \subset \mathscr{U}(\mathfrak{l}_{q-1}) + \ker(\pi)$. Hence we can apply the induction hypothesis to $(\pi', a_{\ell'})$.

Suppose now that Ω is non-saturated with respect to $\mathfrak{l}_{q-1}$. Then

$$\pi' = \pi|_{G'} = \mathrm{ind}_{B[\ell']}^{G'} \chi_{\ell'},$$

where $B[\ell'] = \exp(\mathfrak{b}[\ell] \cap \mathfrak{l}_{q-1})$, is irreducible and a result of Pedersen [134] says that there exists an element $A \in \ker(\pi)$ of the form $A = Y_n + V$ with $V \in \mathscr{U}(\mathfrak{l}_{q-1})$. Let $W = \sum_{j=0}^m Y_n^j V_j$, with $V_j \in \mathscr{U}(\mathfrak{l}_{q-1}), 0 \le j \le m$, be an element of $\mathscr{U}_\pi(\mathfrak{g})^{\mathfrak{k}}$. Replacing in the expression of W the vector Y_n by the element A, we see that $W \in \mathscr{U}_{\pi'}(\mathfrak{l}_{q-1})^{\mathfrak{k}} + \ker(\pi)$. On the other hand, since $\mathfrak{k} \subset \mathfrak{l}_{q-1}$, we can identify a_ℓ with $a_{\ell'} \in \mathscr{H}_{\pi'}^{-\infty}$. These considerations allow us to pass to (G', π'), where the induction hypothesis applies. The function $P_W(\ell)$ is easily checked to be K-invariant. □

Remark 5.3.4 We can see that the intertwining operator R defined by the integral formula (4.4.9) diagonalizes the action of the elements of the algebra $D_\pi(G)^K$. Let $X \in \mathscr{U}_\pi(\mathfrak{g})^{\mathfrak{k}}$ and $\psi \in \mathscr{H}_\pi^\infty$. By using the intertwining operator R introduced in

Theorem 4.4.9, we have for all $k \in K$

$$(R(\pi(X)\psi))(k) = \int_{\widehat{K}} \sum_{j=1}^{m(\sigma)} \langle \pi(k^{-1})(\pi(X)\psi), a_\sigma^j \rangle d\nu(\sigma)$$

$$= \int_{\widehat{K}} \sum_{j=1}^{m(\sigma)} \langle \pi(k^{-1})\psi, \pi(X^*)a_\sigma^j \rangle d\nu(\sigma)$$

since $\pi(X)$ commutes with the action of K. But from [13], the generalized vectors a_σ^j are the eigenvectors of the elements of $\mathscr{U}_\pi(\mathfrak{g})^{\mathfrak{k}}$, with eigenvalues that extends to K-invariant rational functions on Ω. Thus one proves the commutativity of the algebra $\mathscr{U}_\pi(\mathfrak{g})^{\mathfrak{k}}$, which was already obtained in Theorem 5.3.3.

5.3.2 The Function P_W on $\Omega(\pi)$

We suppose again that $\pi|_K$ has finite multiplicities. Putting $\mathfrak{l}_j = \mathfrak{k}_{d+j}$ for $1 \leq j \leq q$, we have a sequence of subalgebras:

$$\{0\} = \mathfrak{k}_0 \subset \mathfrak{k}_1 \subset \cdots \subset \mathfrak{k}_d = \mathfrak{k} \subset \cdots \subset \mathfrak{k}_{d+1} \subset \cdots \subset \mathfrak{k}_{n-1} \subset \mathfrak{k}_n = \mathfrak{g}, \tag{5.3.7}$$

where $\dim(\mathfrak{k}_r/\mathfrak{k}_{r-1}) = 1 (1 \leq r \leq n)$, and a Malcev basis $\{Y_1, \ldots, Y_n\}$ of $\mathfrak{g}$. Let $K_j = \exp(\mathfrak{k}_j)$ for $0 \leq j \leq n$. Let $W \in \mathscr{U}_\pi(\mathfrak{g})^{\mathfrak{k}}$ and let $P_W(\ell)$ be the function defined μ_π-almost everywhere on $\Omega = \Omega(\pi)$ by Theorem 5.3.3. We shall show that this function is rational on Ω. First we need

Lemma 5.3.5 *Let $W \in \mathscr{U}_\pi(\mathfrak{g})^{\mathfrak{k}}$. Then $P_W \equiv 0$ if and only if $W \in \ker(\pi)$.*

Proof Assume first $P_W \equiv 0$. Let us proceed by induction on $\dim G$. Put $\mathfrak{g}' = \mathfrak{l}_{q-1}$, and $G' = \exp(\mathfrak{g}')$. Let $\theta_{G'} : (\mathfrak{g}')^* \to \widehat{G'}$ be the Kirillov mapping. We denote by $\gamma : \mathfrak{g}^* \to (\mathfrak{g}')^*$ the restriction mapping. The proof of Theorem 5.3.3 shows that W reduces modulo $\ker(\pi)$ to an element $W' \in \mathscr{U}(\mathfrak{g}')$ which is contained in $\mathscr{U}_{\pi'}(\mathfrak{g}')^{\mathfrak{k}}$ for $(\theta_{G'} \circ \gamma)_*(\mu_\pi)$-almost all $\pi' \in \widehat{G'}$ and such that $P_{W'}(\ell') = 0$ for $\gamma_*(\mu_\pi)$-almost all $\ell' \in \gamma(\Omega)$. The induction hypothesis tells us that $\pi'(W') = 0$ for $(\theta_{G'} \circ \gamma)_*(\mu_\pi)$-almost all $\pi' \in \widehat{G'}$. This implies that $\pi(W') = \pi(W) = 0$.

Assume now that $W \in \ker(\pi)$. Then of course ${}^t\overline{W} \in \ker(\pi)$ too, and for every $\xi \in \mathscr{H}_\pi^\infty$ we have

$$P_W(\ell)\langle a_\ell, \xi \rangle = \langle \pi(W)a_\ell, \xi \rangle = \langle a_\ell, \pi(\overline{{}^tW})\xi \rangle = 0.$$

Hence $P_W(\ell) \equiv 0$. □

Considering the K-action on Ω, let us define two sets of indices S_K and T_K contained in $\{1, \ldots, n\}$. S_K is the set of 'jump' indices and T_K is that of 'non-jump' indices. Explicitly, $j \in S_K$ if and only if there exists a non-empty Zariski open subset $\mathscr{S}_j \subset \Omega$ such that

$$\dim(K\cdot(\ell_{\mathfrak{k}_j})) = \dim(K\cdot(\ell|_{\mathfrak{k}_{j-1}})) + 1 \ (\forall\, \ell \in \mathscr{S}_j),$$

while $j \in T_K$ if and only if there exists a non-empty Zariski open subset $\mathscr{T}_j \subset \Omega$ such that

$$\dim(K\cdot(\ell|_{\mathfrak{k}_j})) = \dim(K\cdot(\ell|_{\mathfrak{k}_{j-1}})) \ (\forall\, \ell \in \mathscr{T}_j)$$

Putting

$$\mathscr{U}_\pi(\mathfrak{k}_j)^{\mathfrak{k}} = \mathscr{U}_\pi(\mathfrak{g})^{\mathfrak{k}} \cap \mathscr{U}(\mathfrak{k}_j),\ 1 \le j \le n,$$

we have from [7] that for $j \in S_K$

$$\mathscr{U}_\pi(\mathfrak{k}_j)^{\mathfrak{k}} = \mathscr{U}_\pi(\mathfrak{k}_{j-1})^{\mathfrak{k}} + \mathscr{U}(\mathfrak{k}_j)(\mathscr{U}(\mathfrak{k}_{j-1}) \cap \ker(\pi)),$$

and for $j \in T_K$

$$\mathscr{U}_\pi(\mathfrak{k}_j)^{\mathfrak{k}} \neq \mathscr{U}_\pi(\mathfrak{k}_{j-1})^{\mathfrak{k}} + \mathscr{U}(\mathfrak{k}_j)(\mathscr{U}(\mathfrak{k}_{j-1}) \cap \ker(\pi)).$$

Let $j \in T_K$ and $d+1 \le j \le n$, and assume that $\pi|_K$ has finite multiplicities. Then for μ_π-almost all $\ell \in \Omega$ the subalgebra $\mathfrak{k}_j(\ell|_{\mathfrak{k}_j})$ is not contained in $\mathfrak{k}_{j-1}$ and so there exists a Corwin-Greenleaf e-central element $W_j \in \mathscr{U}_\pi(\mathfrak{k}_j)^{\mathfrak{k}_j}$ having the form

$$W_j = a_j Y_j + b_j,\ a_j, b_j \in \mathscr{U}(\mathfrak{k}_{j-1}),\ a_j \in \mathscr{U}_\pi(\mathfrak{k}_{j-1})^{\mathfrak{k}_{j-1}},\ a_j \notin \ker(\pi).$$

Moreover, concerning these Corwin-Greenleaf e-central elements W_j, $j \in T_K$, we have $P_{W_j}(\ell) = \varphi_j(\ell)\ell_j + \psi_j(\ell)$, where $\ell_i = \ell(Y_i)$, $1 \le i \le n$, $\ell \in \Omega$ and φ_j, ψ_j are rational functions in $\ell_1, \dots, \ell_{j-1}$.

Theorem 5.3.6 *Suppose that $\pi|_K$ has finite multiplicities. Let $W \in \mathscr{U}_\pi(\mathfrak{g})^{\mathfrak{k}}$. Then the function P_W, defined μ_π-almost everywhere on Ω by Theorem 5.3.3, extends to a K-invariant rational function on Ω. Furthermore, the homomorphism $\mathscr{U}_\pi(\mathfrak{g})^{\mathfrak{k}} \ni W \mapsto P_W$ defines an embedding of $D_\pi(G)^K$ in the field $\mathbb{C}(\Omega)^K$ of K-invariant rational functions on Ω.*

Proof The definition of the function P_W implies $P_{W\cdot W'} = P_W P_{W'}$ for any $W, W' \in \mathscr{U}_\pi(\mathfrak{g})^{\mathfrak{k}}$. Let $1 \le m \le n$ be the smallest index in sequence (5.3.7) such that $W \in \mathscr{U}(\mathfrak{k}_m) + \ker(\pi)$. Let us proceed by induction on m to show that P_W is rational on Ω. We first remark that in order to compute the eigenvalue $P_W(\ell)$ we can consider $W \in \mathscr{U}(\mathfrak{k}_m)$ and replace $a_\ell \in \mathscr{H}_\pi^{-\infty}$ by $a_{\ell|_{\mathfrak{k}_m}} \in \mathscr{H}_\sigma^{-\infty}$, $\sigma = \theta_{K_m}(\ell|_{\mathfrak{k}_m}) \in \widehat{K_m}$, repeating step by step the observation made in the proof of Theorem 5.3.3 to descend from $\mathfrak{g}$ to $\mathfrak{l}_{q-1}$.

If $m = 1$, then W is in the center of $\mathscr{U}(\mathfrak{k})$ modulo $\ker(\pi)$ and $\theta_K(f)(W) = \widehat{W}(f)I$ for every $f \in \mathfrak{k}^*$, where the function $\mathfrak{k}^* \ni f \mapsto \widehat{W}(f)$ is a K-invariant polynomial function [39]. If $p : \mathfrak{g}^* \to \mathfrak{k}^*$ stands for the restriction mapping, then P_W is the restriction to $p(\Omega)$ of the polynomial function $\widehat{W}$.

Assuming now $m > 1$, let us write $W = \sum_{k=0}^{r} Y_m^k w_k$ with $r > 0$ and $w_k \in \mathscr{U}(\mathfrak{k}_{m-1})$ for $0 \le k \le r$ satisfying $w_r \notin \ker(\pi)$. We shall use another induction on the degree r of W in Y_m. If $w_r \notin \mathscr{U}_\pi(\mathfrak{g})^{\mathfrak{k}}$, we can find elements $X_1, \dots, X_q \in \mathfrak{k}$ such that $[X_1, [\cdots, [X_q, w_r]\cdots]] \notin \ker(\pi)$ but $[X, [X_1, [\cdots, [X_q, w_r]\cdots]]] \in \ker(\pi)$ for all $X \in \mathfrak{k}$. Namely, $u_r = [X_1, [\cdots, [X_q, w_r]\cdots]] \notin \ker(\pi)$ is contained in $\mathscr{U}_\pi(\mathfrak{g})^{\mathfrak{k}} \cap \mathscr{U}(\mathfrak{k}_{m-1})$. The element $V = [X_1, [\cdots, [X_q, W]\cdots]]$ of $\mathscr{U}(\mathfrak{g})$ belongs to $\ker(\pi)$ and has the form $V = Y_m^r u_r + V'$ for a certain $V' \in \sum_{k=0}^{r-1} Y_m^k \mathscr{U}(\mathfrak{k}_{m-1})$. We apply the induction hypothesis to $\widetilde{W} = u_r W - V w_r \in \mathscr{U}_\pi(\mathfrak{g})^{\mathfrak{k}}$ which is of degree smaller than r in Y_m. By induction on m and r, and because of the multiplicative property of P_W, both the function

$$P_{\widetilde{W}} = P_{u_r} P_W - P_V P_{w_r} = P_{u_r} P_W$$

and also, since $P_{u_r} \not\equiv 0$ by Lemma 5.3.5,

$$P_W = \frac{P_{\widetilde{W}}}{P_{u_r}},$$

admit extensions to rational functions on Ω. Thus, we can assume now that $w_r \in \mathscr{U}_\pi(\mathfrak{g})^{\mathfrak{k}}$. Since $W \in \mathscr{U}(\mathfrak{k}_m) \setminus \mathscr{U}(\mathfrak{k}_{m-1})$, we know from [7] that $m \in T_K$ and so, generically for $\ell \in \Omega$, $\dim(K{\cdot}\ell|_{\mathfrak{k}_m}) = \dim(K{\cdot}\ell|_{\mathfrak{k}_{m-1}})$ in the case where $m > d = \dim \mathfrak{k}$. Anyhow we have a Corwin-Greenleaf e-central element $W_m = a_m Y_m + b_m$ with $a_m, b_m \in \mathscr{U}(\mathfrak{k}_{m-1})$, $a_m \notin \ker(\pi)$ and $a_m \in \mathscr{U}_\pi(\mathfrak{k}_{m-1})^{\mathfrak{k}}$. Hence the element

$$\widetilde{W} = a_m^r W - (W_m)^r w_r$$

of $\mathscr{U}_\pi(\mathfrak{k}_m)^{\mathfrak{k}}$ is of degree smaller than r in Y_m modulo $\ker(\pi)$ and so we can apply the induction hypothesis to it. Therefore $P_{\widetilde{W}} = (P_{a_m})^r P_W - (P_{W_m})^r P_{w_r}$, P_{a_m}, P_{w_r} and

$$P_W = \frac{P_{\widetilde{W}} + (P_{W_m})^r P_{w_r}}{(P_{a_m})^r}$$

are rational functions on Ω as well. □

We have seen that for the Corwin-Greenleaf e-central elements W_j, $j \in T_K$, the functions P_{W_j} have the form $P_{W_j}(\ell) = \varphi_j(\ell)\ell_j + \psi_j(\ell)$, where $\ell_i = \ell(Y_i)$, $1 \le i \le n$, $\ell \in \Omega = \Omega(\pi)$ and φ_j, ψ_j are rational functions in $\ell_1, \dots, \ell_{j-1}$. Hence we obtain as a corollary of Theorem 5.3.6 the following result (see Theorem 5.4 in [40] for the case of monomial representations).

Proposition 5.3.7 *Suppose that $\pi|_K$ has finite multiplicities. Let $A \in \mathscr{U}_\pi(\mathfrak{k}_m)^{\mathfrak{k}}$. Then there exist two polynomials β_A, γ_A in the variables W_j, $j \in T_K$, $j \leq m$, such that $\beta_A A \equiv \gamma_A$ modulo $\ker(\pi)$. In particular, the functions $\{P_{W_j};\ j \in T_K\}$ are rational generators of the field $\mathbb{C}(\Omega(\pi))^K$.*

Remark 5.3.8 Recalling the polynomial conjecture for monomial representations, it is natural to ask whether the functions P_W are polynomial functions or not.

Polynomial Conjecture We recall the question for clarity. Assume that $\pi|_K$ has finite multiplicities. Is the algebra $D_\pi(G)^K$ isomorphic to the algebra $\mathbb{C}[\Omega(\pi)]^K$ of K-invariant polynomial functions on $\Omega(\pi)$?

Proposition 5.3.9 *Assume that* $\dim \Omega \leq 2$. *If $\pi|_K$ has finite multiplicities, the algebra $D_\pi(G)^K$ is isomorphic to $\mathbb{C}[\Omega]^K$.*

Proof As usual we use induction on $\dim G$. It suffices to examine the case where $\mathfrak{k} \neq \mathfrak{g}$ and $\dim \Omega = 2$. Let $\ell \in \Omega$. If ℓ vanishes on a non-trivial ideal $\mathfrak{a}$ of $\mathfrak{g}$, we can descend to the quotient $\mathfrak{g}/\mathfrak{a}$ and apply the induction hypothesis. Suppose then that ℓ does not vanish on any non-trivial ideal of $\mathfrak{g}$, and take the sequence of subalgebras (5.3.7). As argued earlier, if Ω is non-saturated with respect to $\mathfrak{k}_{n-1}$, we can immediately descend to the subgroup $K_{n-1} = \exp(\mathfrak{k}_{n-1})$ to which applies the induction hypothesis.

Now assume that Ω is saturated with respect to $\mathfrak{k}_{n-1}$. For almost all $\ell \in \Omega$, $\mathfrak{k}_{n-1}$ is a polarization. This together with our assumption implies that $\mathfrak{k}_{n-1}$ is Abelian. The Frobenius vector turns out to be the Dirac measure at the identity of G, and $\mathscr{U}_\pi(\mathfrak{g})^{\mathfrak{k}}$ is, modulo $\ker(\pi)$, contained in $\mathscr{U}(\mathfrak{k}_{n-1})$, which is identified with the symmetric algebra $S(\mathfrak{k}_{n-1})$ of $(\mathfrak{k}_{n-1})_{\mathbb{C}}$. We obtain $P_W(\ell) = \widehat{W}(\ell)(\ell \in \Omega)$ for any $W \in \mathscr{U}_\pi(\mathfrak{g})^{\mathfrak{k}}$, hence the result. □

Remark 5.3.10 Let us stick to the second situation of the above proposition, i.e. Ω saturated with respect to $\mathfrak{k}_{n-1}$, and assume that $\mathfrak{g}$ has one-dimensional center $\mathfrak{z} = \mathbb{R}Z$ on which Ω is not trivial. Take a new Jordan-Hölder sequence $\{\mathfrak{g}'_j\}_{j=1}^n$ such that $\mathfrak{g}'_{n-1} = \mathfrak{k}_{n-1}$. Then we have $\mathfrak{g}'_1 = \mathfrak{z}$ and $\mathfrak{g}'_2 = \mathfrak{z} + \mathbb{R}Y$ with $\ell([Y_n, Y]) \neq 0$ for any $\ell \in \Omega$. By the finite multiplicity condition there exists for almost all $\ell \in \Omega$ a vector $X(\ell) \in \mathfrak{g}(\ell)$ satisfying $Y + X(\ell) \in \mathfrak{k}$. It follows from this that $D_\pi(G)^K \cong \mathbb{C}[Y] \cong \mathbb{C}[P_Y]$, where P_Y is a polynomial function defined by $P_Y(\ell) = \ell(Y)$. In the above proposition, we are essentially reducing to the case where $\mathfrak{k}$ is contained in a codimension-one polarization at a generic $\ell \in \Omega$. In fact, we get the result in the following slightly more general setting. Assume that π is realized by a normal polarization which contains $\mathfrak{k}$. Then the same result holds.

Proposition 5.3.11 *Assume that the orbit $\Omega = \Omega(\pi)$ is flat, i.e. $\Omega = \ell + (\mathfrak{g}(\ell))^\perp$ for any $\ell \in \Omega$. Equivalently, $\mathfrak{g}(\ell)$ is an ideal of $\mathfrak{g}$. Suppose that $\pi|_K$ has finite multiplicities. Then the polynomial conjecture holds.*

Proof If we consider the subalgebra $\mathfrak{h} = \mathfrak{k} + \mathfrak{g}(\ell)$, it is straightforward to show that any $W \in \mathscr{U}_\pi(\mathfrak{g})^{\mathfrak{k}}$ belongs, modulo $\ker(\pi)$, to $\mathscr{U}(\mathfrak{h})$. Besides, $X - i\ell(X) \in \ker(\pi)$ for every $X \in \mathfrak{g}(\ell)$. So, W reduces to an element of $\mathscr{U}(\mathfrak{k})$ and turns out to be a

Corwin- Greenleaf e-central element. It follows that $P_W(\ell)$ is a polynomial function [73].

□

Example 5.3.12 Let $\mathfrak{g} = \langle X_1, X_2, \ldots, X_{n-1}, X_n\rangle_{\mathbb{R}}$ with non-zero brackets $[X_n, X_j] = X_{j-1}$ for $2 \leq j \leq n-1$ (threadlike algebra). Let $\pi \in \widehat{G}$. We have $\dim(\Omega(\pi)) \leq 2$. Let $\mathfrak{k}$ be a non-central subalgebra of $\mathfrak{g}$, namely $\mathfrak{k} \not\subset \mathfrak{z} = \mathbb{R}X_1$. If $\pi(X_1) \neq 0$, then $\pi|_K$ has finite multiplicities and $D_\pi(G)^K \cong \mathbb{C}[\Omega(\pi)]^K$. More precisely, if $\mathfrak{k} \subset \mathfrak{g}_{n-1} = \sum_{j=1}^{n-1} \mathbb{R}X_j$, then

$$D_\pi(G)^K \cong \mathbb{C}[\Omega(\pi)]^K \cong \mathbb{C}[P_{X_2}].$$

If $\mathfrak{k} \not\subset \mathfrak{g}_{n-1}$, let us take X_n in $\mathfrak{k}$. Two cases can occur. When all K-orbits in $p(\Omega(\pi)) \subset \mathfrak{k}^*$ are points, $\mathfrak{k}$ must be Abelian and

$$D_\pi(G)^K \cong \mathbb{C}[\Omega(\pi)]^K \cong \mathbb{C}[P_{X_n}].$$

When $\dim(K\cdot p(\ell)) = 2$ for generic $\ell \in \Omega(\pi)$, we have $\dim(K\cdot\ell) = \dim(\Omega(\pi)) = 2$ and

$$D_\pi(G)^K \cong \mathbb{C}[\Omega(\pi)]^K \cong \mathbb{C}.$$

In the latter situation $\pi|_K$ turns out to be irreducible.

We have an affirmative answer in yet another case: when $\mathfrak{k}$ is an ideal of $\mathfrak{g}$.

Theorem 5.3.13 *Assume that $\pi|_K$ has finite multiplicities. If $\mathfrak{k}$ is an ideal of $\mathfrak{g}$, then the polynomial conjecture holds.*

Proof We can assume that our flag of ideals $(\mathfrak{g}_j)_{j=0}^n$ passes through $\mathfrak{k}$, i.e. $\mathfrak{g}_d = \mathfrak{k}$. In particular, the index set $\mathscr{I}$ is now equal to $\{1, \ldots, d\}$ and $\mathfrak{k}_j = \mathfrak{g}_j$ for $1 \leq j \leq n$. Fix a base point $\ell_0 \in \Omega = \Omega(\pi)$ and let $\mathfrak{b} = \mathfrak{b}[\ell_0]$, $B = \exp\mathfrak{b}$, $\pi = \mathrm{ind}_B^G \chi_{\ell_0}$. Then for every point $\ell = g\cdot\ell_0 \in \Omega$, the Vergne polarization $\mathfrak{b}[\ell]$ contains the Vergne polarization $\mathfrak{b}[\ell|_{\mathfrak{k}}]$ and so the distribution a_ℓ, passed to $\mathscr{H}_\pi^{-\infty}$, becomes the evaluation at the point g:

$$\langle a_\ell, \varphi\rangle = \int_{B[\ell|_{\mathfrak{k}}]/(B[\ell|_{\mathfrak{k}}]\cap B[\ell])} \overline{\varphi(bg)\chi_\ell(b)}d\dot{b} = \overline{\varphi(g)}\ (\varphi \in \mathscr{H}_\pi^\infty).$$

Furthermore, by the finite-multiplicity condition, we know that $\mathfrak{g}(\ell) + \mathfrak{b}[\ell|_{\mathfrak{k}}]$ is Lagrangian with respect to B_ℓ for μ_π-almost all $\ell \in \Omega$. Since these spaces are now conjugate, $\mathfrak{k}$ being an ideal, it follows that $\mathfrak{b}[\ell] = \mathfrak{g}(\ell) + \mathfrak{b}[\ell|_{\mathfrak{k}}]$ for all $\ell \in \Omega$. Therefore $B[\ell] = B[\ell|_{\mathfrak{k}}]G(\ell)$ since $[\mathfrak{b}[\ell|_{\mathfrak{k}}], \mathfrak{g}(\ell)] \subset \mathfrak{b}[\ell|_{\mathfrak{k}}]$. Choosing a Malcev basis $\mathscr{Q} = \{Q_1, \ldots, Q_q\}$ of $\mathfrak{g}$ relative to $\mathfrak{b}$ which is extracted from our Jordan-Hölder basis $\mathscr{X}$ of $\mathfrak{g}$, we obtain a polynomial diffeomorphism

$$E_{\mathscr{Q}} : \mathbb{R}^q \to G/B;\ E_{\mathscr{Q}}(u_1, \ldots, u_q) = \prod_{i=q}^{1} \exp(u_i Q_i)$$

and an identification of $\mathscr{H}_\pi$ with $L^2(\mathbb{R}^q)$. We also know by Kirillov's theorem that $\mathscr{H}_\pi^\infty$ is isomorphic to the Schwartz space $\mathscr{S}(\mathbb{R}^q)$. In particular, for any $W \in \mathscr{U}(\mathfrak{g})$, $\pi(\overline{{}^t W})$ becomes a partial differential operator with polynomial coefficients:

$$(\pi(\overline{{}^t W})\varphi)\circ E_{\mathscr{Q}} = \sum_{\alpha\in\mathbb{N}^q} P_\alpha \partial^\alpha(\varphi\circ E_{\mathscr{Q}}). \tag{5.3.8}$$

Take now $W \in \mathscr{U}_\pi(\mathfrak{g})^{\mathfrak{k}} \setminus \ker(\pi)$. We know from Theorem 5.3.3 that

$$P_W(g\cdot\ell_0)\overline{\varphi(g)} = \langle \pi(W)a_\ell, \varphi\rangle = \langle a_\ell, \pi(\overline{{}^t W})\varphi\rangle = \overline{\pi(\overline{{}^t W})\varphi(g)}.$$

Hence by (5.3.8)

$$P_W(E_{\mathscr{Q}}(u)\cdot\ell_0)\overline{\varphi(E_{\mathscr{Q}}(u))} = \sum_{\alpha\in\mathbb{N}^q} \overline{P_\alpha(u)}\partial^\alpha\overline{(\varphi\circ E_{\mathscr{Q}})(u)},\ u \in \mathbb{R}^q.$$

Now there exists $\gamma \in \mathbb{N}^q$ such that $P_\gamma \neq 0$ since $\overline{{}^t W} \notin \ker(\pi)$. We can assume that the length $|\gamma|$ of γ is maximal. Suppose that $|\gamma| > 0$. Take $u \in \mathbb{R}^q$ such that $P_\gamma(u) \neq 0$. We choose $\psi \in \mathscr{S}(G/B)$ for which $\partial^\gamma(\psi\circ E_{\mathscr{Q}})(u) = 1$ but $\partial^\beta(\psi\circ E_{\mathscr{Q}})(u) = 0$ for every $\beta \in \mathbb{N}^q$, $\beta \neq \gamma$ and $|\beta| \leq |\gamma|$. Then

$$0 = P_W(E_{\mathscr{Q}}(u)\cdot\ell_0)\overline{\psi(E_{\mathscr{Q}}(u))\varphi(E_{\mathscr{Q}}(u))} = \sum_{|\alpha|\leq|\gamma|} \overline{P_\alpha(u)}\partial^\alpha\overline{((\psi\varphi)\circ E_{\mathscr{Q}})(u)}$$

$$= \overline{P_\gamma(u)\varphi(E_{\mathscr{Q}}(u))},\ \varphi \in \mathscr{H}_\pi^\infty.$$

This contradiction tells us that $|\gamma| = 0$. Hence $P_W(E_{\mathscr{Q}}(u)\cdot\ell_0) = \overline{P_0(u)}, u \in \mathbb{R}^q$ and thus P_W is a polynomial function on Ω.

Take now a polynomial function $P : \Omega \to \mathbb{C}$ on Ω which is K-invariant, i.e. $P(k\cdot\ell) = P(\ell), k \in K, \ell \in \Omega$. Define the polynomial function $\widetilde{P}$ on G by the formula

$$\widetilde{P}(g) = P(g\cdot\ell_0),\ g \in G.$$

Then $\widetilde{P}$ is K-invariant. Since $B = B[\ell_0|_{\mathfrak{k}}]G(\ell_0) \subset KG(\ell_0)$, it follows that $\widetilde{P}(gb) = \widetilde{P}(g), g \in G, b \in B$, and $\widetilde{P}$ is in fact a polynomial function on G/B. The operator $M : \mathscr{H}_\pi^\infty \to \mathscr{H}_\pi^\infty$ defined by means of the multiplication by $\widetilde{P}$ is, by Kirillov's theorem, contained in $\pi(\mathscr{U}(\mathfrak{g}))$. Hence there exists $W \in \mathscr{U}(\mathfrak{g})$ such that $\pi(W)\varphi = \widetilde{P}\varphi, \varphi \in \mathscr{H}_\pi^\infty$. The fact that $\widetilde{P}$ is K-invariant tells us that $W \in \mathscr{U}_\pi(\mathfrak{g})^{\mathfrak{k}}$. It is obvious now that $P = P_W$. This shows that the mapping

$$\mathscr{U}_\pi(\mathfrak{g})^{\mathfrak{k}} \ni W \mapsto P_W \in \mathbb{C}[\Omega]^K$$

is a surjective homomorphism whose kernel is $\ker(\pi)$ by Lemma 5.3.5. $\square$

Corollary 5.3.14 *Assume that* $\mathfrak{g}$ *is two-step nilpotent. Then* $\pi|_K$ *has finite multiplicities if and only if* $D_\pi(G)^K \cong \mathbb{C}[\Omega(\pi)]^K$.

Proof In fact, adding the center of $\mathfrak{g}$ to $\mathfrak{k}$, we find ourselves in the case where $\mathfrak{k}$ is an ideal. □

5.3.3 *Further Study of the Conjecture*

When $G = K$, our question is trivial. To study it we proceed by induction on $\dim G$. We start from the flag $\mathscr{S}$ of ideals $\{\mathfrak{g}_j\}_{j=0}^n$ in $\mathfrak{g}$ such that $\mathfrak{g}_{n-1}$ contains $\mathfrak{k}$. Parametrizing the orbit $\Omega = \Omega(\pi)$, we define the jump index set S_Ω and the non-jump index set T_Ω. Let $\{X_1, \dots, X_n\}$ be a Malcev basis adapted to the flag $\mathscr{S}$, $\ell_j = \ell(X_j)(1 \leq j \leq n)$ for $\ell \in \Omega$, $S_\Omega = \{s_1 < \cdots < s_r\}$, $r = \dim \Omega$ and $x_k = \ell_{s_k}$ for $1 \leq k \leq r$. Describe as before the orbit Ω by the polynomial relations

$$\ell_j = F_j(x_1, \dots, x_k),\ s_k < j < s_{k+1},$$

where $x = (x_1, \dots, x_r)$ runs over $\mathbb{R}^r$. In these circumstances rational functions on Ω are nothing but rational functions in the variables $(x_1, \dots, x_r)$. For $1 \leq k \leq r$, let $I^{(k)}$ be the set of K-invariant polynomial functions on Ω which depend only on the variables $\{x_i;\ 1 \leq i \leq k\}$. The arguments developed in [142, pp.60-61] tell us that every $R \in \mathbb{C}(\Omega)^K$ verifying

$$\frac{\partial R}{\partial x_k} \neq 0,\ \frac{\partial R}{\partial x_i} = 0\ (k+1 \leq i \leq r)$$

is written in the form P/Q with P, Q belonging to $I^{(k)}$. Now the existence of such an element R means that $I^{(k-1)}$ is strictly contained in $I^{(k)}$. So, let $Q = \sum_{i=0}^m Q_i x_k^i (m > 0)$ be an element of $I^{(k)} \setminus I^{(k-1)}$, where $Q_i (0 \leq i \leq m)$ are polynomials of $(x_1, \dots, x_{k-1})$ verifying $Q_m \neq 0$. Then Q_m and $m Q_m x_k + Q_{m-1}$ are K-invariant polynomials.

Put $\mathfrak{g}' = \mathfrak{g}_{n-1}$ and $G' = \exp(\mathfrak{g}')$ in the flag $\mathscr{S}$. Assume that the orbit Ω is non-saturated with respect to $\mathfrak{g}'$, namely that $n \in T_\Omega$. Then the restriction map $p : \mathfrak{g}^* \to (\mathfrak{g}')^*$ gives us a K-equivariant homeomorphism between Ω and $\omega = p(\Omega)$ which is a G'-orbit. Hence $\mathbb{C}[\Omega]^K \cong \mathbb{C}[\omega]^K$. On the other hand, $\pi' = \pi|_{G'}$ is irreducible and there exists in $\ker(\pi)$ an element of the form $W = X_n + A$ with $A \in \mathscr{U}(\mathfrak{g}')$, which serves to identify $D_\pi(G)^K$ with $D_{\pi'}(G')^K$. Since ω is the coadjoint orbit of G' corresponding to π', the induction hypothesis ensures the desired result.

Assume now that Ω is saturated with respect to $\mathfrak{g}'$, namely $n \in S_\Omega$. Then there exists one, and only one, index $j \in S_\Omega, 2 \leq j \leq n-1$, having the property that we find in $\mathscr{U}_\pi(\mathfrak{g}_j)^{\mathfrak{g}'}$ an element $W = aX_j + b$ with $a, b \in \mathscr{U}(\mathfrak{g}_{j-1})$ verifying $\pi(a) \neq 0$. From this $a \in \mathscr{U}_\pi(\mathfrak{g}_{j-1})^{\mathfrak{g}'}$, otherwise there would be in $\ker(\pi)$ an element of the form $[X, W] = [X, a]X_j + b'$ with $b' \in \mathscr{U}(\mathfrak{g}_{j-1})$, $X \in \mathfrak{g}'$ verifying

$[X, a] \notin \ker(\pi)$, against Pedersen's result [134] because of $j \in S_\Omega$. Taking into account the fact that

$$\mathscr{U}_\pi(\mathfrak{g}_{j-1})^{\mathfrak{g}'} = \mathscr{U}_\pi(\mathfrak{g}_{j-1})^{\mathfrak{g}} \equiv \mathbb{C}$$

modulo $\ker(\pi)$, we can assume that a is a constant modulo $\ker(\pi)$ and hence $a = 1$.

Likewise, if $j = s_i (1 \leq i \leq r-1)$, there exists a G'-invariant polynomial function

$$y = x_i + \varphi(x_1, \ldots, x_{i-1})$$

on Ω which separates the G'-orbits contained in $p(\Omega)$. This means that $p(\Omega) = \sqcup_{y \in \mathbb{R}} \omega_y$, the disjoint union of G'-orbits ω_y. Accordingly, $\pi|_{G'} \simeq \int_{\mathbb{R}}^{\oplus} \pi_y dy$ with $\pi_y = \theta_{G'}(\omega_y)$ for all $y \in \mathbb{R}$.

Let now $W \in \mathscr{U}_\pi(\mathfrak{g})^{\mathfrak{k}}$. Since $\pi|_K$ has finite multiplicities, W belongs modulo $\ker(\pi)$ to $\mathscr{U}(\mathfrak{g}')$ (cf. [7]) and $P_W(\ell), \ell \in \Omega$, is a rational function of $x_k (1 \leq k \leq r-1)$ and hence of y and x_k $(1 \leq k \leq r-1, k \neq i)$. By the induction hypothesis applied to G', the restriction of $P_W(\ell)$ to ω_y is a polynomial of $x_k (k \neq i)$ for almost all $y \in \mathbb{R}$. It follows that $P_W(\ell)$ is a rational function whose denominator is a polynomial of y. Indeed, for a multi-index $I = (\alpha_1, \ldots, \alpha_{i-1}, \alpha_{i+1}, \ldots, \alpha_{r-1})$ of non-negative integers, we write

$$x^I = x_1^{\alpha_1} \cdots x_{i-1}^{\alpha_{i-1}} x_{i+1}^{\alpha_{i+1}} \cdots x_{r-1}^{\alpha_{r-1}}.$$

From what we have seen, then,

$$P_W(\ell) = \frac{\sum_I p_I(y) x^I}{\sum_J q_J(y) x^J} = \sum_\Gamma r_\Gamma(y) x^\Gamma \tag{5.3.9}$$

for generic $\ell \in \Omega$, where $p_I(y), q_J(y)$ are polynomials of y and $r_\Gamma(y)$ are certain functions of y. It suffices to confirm that the $r_\Gamma(y)$ are rational functions of y. To that end we introduce a lexicographic order on the set of $(r-2)$-indices: for $I = (\alpha_1, \ldots, \alpha_{r-2})$ and $J = (\beta_1, \ldots, \beta_{r-2})$, we define

$$I < J \Leftrightarrow 1 \leq \exists k \leq r-2; \alpha_{r-2} = \beta_{r-2}, \ldots, \alpha_{k+1} = \beta_{k+1}, \alpha_k < \beta_k.$$

Let us employ induction on this ordering of the right-hand side of (5.3.9). When the right side does not depend on x, it is obvious that $r_{(0,\ldots,0)}(y)$ is rational in y. Let I_0, J_0 and Γ_0 respectively be the dominant multi-indices of I, J and Γ appearing in (5.3.9). From

$$\sum_I p_I(y) x^I = \Big(\sum_J q_J(y) x^J\Big)\Big(\sum_\Gamma r_\Gamma(y) x^\Gamma\Big),$$

we obtain $p_{I_0}(y) = q_{J_0}(y) r_{\Gamma_0}(y)$, which means that $r_{\Gamma_0}(y)$ is rational in y. Passing the term $r_{\Gamma_0}(y) x^{\Gamma_0}$ from right to left in (5.3.9), we get a new relation, in the right side of which the dominant multi-index's order is less than Γ_0. We can then apply to that the induction hypothesis and get the desired result.

In the same way we assess the surjectivity of the homomorphism

$$\Theta : \mathscr{U}_\pi(\mathfrak{g})^{\mathfrak{k}} \ni W \mapsto P_W(\ell) \in \mathbb{C}[\Omega]^K,$$

supposing for the moment $P_W(\ell)$ are polynomial functions for all $W \in \mathscr{U}_\pi(\mathfrak{g})^{\mathfrak{k}}$. Let E be the vector subspace of $\mathscr{U}(\mathfrak{g})$ spanned by the elements of the form $X^A = X_{s_1}^{\alpha_1} \cdots X_{s_r}^{\alpha_r}$, with $A = (\alpha_1, \dots, \alpha_r)$ running over all r-indices of non-negative integers. We know [134] that $\pi|_E$ is faithful. Note that every element of $D_\pi(G)^K \cong \mathscr{U}_\pi(\mathfrak{g})^{\mathfrak{k}}/\ker(\pi)$ admits a unique representative in E. We keep the notations introduced above, and let

$$\zeta(x_1, \dots, x_{i-1}, y', x_{i+1}, \dots, x_{r-1})$$

be a K-invariant polynomial function on Ω. Here, Theorem 5.3.13 says that

$$y' = x_i + \varphi(x_1, \dots, x_{i-1}) = P_{Y'}(\ell)$$

for a certain

$$Y' = -\sqrt{-1} X_{s_i} + \sum c_{k_1,\dots k_{i-1}} X_{s_1}^{k_1} \cdots X_{s_{i-1}}^{k_{i-1}} \ (c_{k_1,\dots k_{i-1}} \in \mathbb{C})$$

belonging to $\mathscr{U}_\pi(\mathfrak{g})^{\mathfrak{g}'}$ with $c_{k_1,\dots k_{i-1}} \in \mathbb{C}$. For multi-indices $(\gamma_1, \dots, \gamma_{i-1}, \gamma_{i+1}, \dots, \gamma_{r-1})$ of $r-2$ made of non-negative integers, we put

$$\widehat{X}^I = X_{s_1}^{\gamma_1} \cdots X_{s_{i-1}}^{\gamma_{i-1}} X_{s_{i+1}}^{\gamma_{i+1}} \cdots X_{s_{r-1}}^{\gamma_{r-1}}$$

and consider the vector subspace E' of $\mathscr{U}(\mathfrak{g}')$ spanned by these $\widehat{X}^I$. The induction hypothesis applied to G' says that for almost all $y' \in \mathbb{R}$ there exists an element

$$W(y') = \sum_I r_I(y') \widehat{X}^I \in \mathscr{U}_{\pi_{y'}}(\mathfrak{g}')^{\mathfrak{k}}$$

such that

$$P_{W(y')}(\ell|_{\omega_{y'}}) = \zeta(x_1, \dots, x_{i-1}, y', x_{i+1}, \dots, x_{r-1})$$

for generic $\ell \in \Omega$. Here, $r_I(y')$ are certain functions of y'. On the other hand, Proposition 5.3.7 says that there exist

$$U = \sum_J p_J(Y') \widehat{X}^J, \ V = \sum_\Gamma q_\Gamma(Y') \widehat{X}^\Gamma$$

in $\mathcal{U}_\pi(\mathfrak{g})^{\mathfrak{k}}$ satisfying

$$P_U(\ell)\zeta(x_1,\ldots,x_{i-1},y',x_{i+1},\ldots,x_{r-1}) = P_V(\ell)$$

for generic $\ell \in \Omega$. Here, $p_J(Y'), q_\Gamma(Y')$ are certain polynomials of Y'. For almost all $y' \in \mathbb{R}$, because $\pi_{y'}$ is faithful on E' the E'-component of

$$\Big(\sum_J p_J(y')\widehat{X}^J\Big)\Big(\sum_I r_I(y')\widehat{X}^I\Big)$$

is equal to $\sum_\Gamma q_\Gamma(y')\widehat{X}^\Gamma$. It follows just as before that $r_I(y')$ are rational in y'. Then there exists a certain polynomial $s(y')$ of y' so that we have $s(y')W(y') = \sum_I t_I(y')\widehat{X}^I$, with polynomials $t_I(y')$ of y'. This means that $Q = \sum_I t_I(Y')\widehat{X}^I$ belongs to $\mathcal{U}_\pi(\mathfrak{g})^{\mathfrak{k}}$ and

$$P_Q(\ell) = s(y')\zeta(x_1,\ldots,x_{i-1},y',x_{i+1},\ldots,x_{r-1}).$$

To sum up, we have:

Proposition 5.3.15 *For any $\zeta(x) \in \mathbb{C}[\Omega]^K$ there exists a polynomial $s(y')$ of y' such that the product $s(y')\zeta(x)$ lies in the image of the homomorphism in question.*

Assume for the moment that $\mathfrak{k}$ is contained in an ideal of codimension 2. We can suppose that the orbit $\Omega(\pi)$ is two times saturated in these steps. Consequently, the denominator of the rational function $P_W(\ell)$ is a polynomial in

$$y = x_i + \varphi(x_1,\ldots,x_{i-1})$$

on one side and in

$$z = x_j + \psi(x_1,\ldots,x_{j-1})$$

on the other, with $i \neq j$. Hence, the function $P_W(\ell)$ is polynomial.

We keep the notations and always assume that $\pi|_K$ has finite multiplicities. Recall the subalgebra sequence (5.3.7), which is a Jordan-Hölder sequence for the action of K. We consider the image $p(\Omega)$ of $\Omega = \Omega(\pi)$ by the projection $p : \mathfrak{g}^* \to \mathfrak{k}^*$ and the layer of K-orbits which encounters $p(\Omega)$ in a non-empty Zariski open set of $p(\Omega)$. Using the notations introduced just before Theorem 5.3.6, let

$$T_K \cap \{1,\ldots,d\} = \{k_1 < \cdots < k_u\}.$$

Theorem 5.3.16 *Assume that $\pi|_K$ has finite multiplicities. Then any $W \in \mathcal{U}_\pi(\mathfrak{g})^{\mathfrak{k}}$ is algebraic modulo* $\ker(\pi)$ *over $\{W_{k_1},\ldots,W_{k_u}\}$.*

Proof In the sequence (5.3.7) of subalgebras, W is supposed to be contained modulo ker(π) in $\mathscr{U}(\mathfrak{k}_p), d < p$ but not in $\mathscr{U}(\mathfrak{k}_{p-1})$. We assume also that the theorem holds for the elements of $\mathscr{U}(\mathfrak{k}_{p-1})$. Since $\pi|_K$ has finite multiplicities, $p \in T_K$ and there exists [7] in ker(π) an element $\kappa = UX_p^m + V$, where $U \in \mathscr{U}(\mathfrak{k}_{p-1}), \pi(U) \neq 0, m \geq 1$ and V is an element of $\mathscr{U}(\mathfrak{k}_p)$ whose degree in X_p is smaller than or equal to $m - 1$. Choose κ in such a way that $m \geq 1$ is the smallest possible. If necessary, we apply repeatedly the adjoint action of $\mathfrak{k}$ on κ so that we can suppose $U \in \mathscr{U}_\pi(\mathfrak{g})^\mathfrak{k}$. Let us use the Corwin-Greenleaf e-central element $W_p = aX_p + b, a \in \mathscr{U}_\pi(\mathfrak{k}_{p-1})^\mathfrak{k}, b \in \mathscr{U}(\mathfrak{k}_{p-1})$ introduced before. From Proposition 5.3.7 it suffices to show that the theorem is true for $W = W_p$. Clearly,

$$UW_p^m - a^m\kappa = \lambda X_p^u + V',$$

where $u \leq m - 1, \lambda \in \mathscr{U}(\mathfrak{k}_{p-1})$ and V' is an element of $\mathscr{U}(\mathfrak{k}_p)$ with degree in X_p smaller than or equal to $m - 2$, belongs to $\mathscr{U}_\pi(\mathfrak{k}_p)^\mathfrak{k}$. The minimality of m implies λ belongs to $\mathscr{U}_\pi(\mathfrak{k}_{p-1})^\mathfrak{k}$. Next,

$$a^u(UW_p^m - a^m\kappa) - \lambda W_p^u$$

belongs to $\mathscr{U}_\pi(\mathfrak{k}_p)^\mathfrak{k}$ and its degree in X_p is smaller than or equal to $m-2$. Repeating this process, we arrive at the desired result. □

For every $j \in T_K$ relative to sequence (5.3.7), there exists in $\mathbb{C}[\Omega]^K$ a function

$$f_j(\ell) = g_j(\ell_1, \ldots, \ell_{j-1})\ell_j + h_j(\ell_1, \ldots, \ell_{j-1}),$$

where $\ell_i = \ell(X_i)(1 \leq i \leq j)$ and g_j, h_j are two polynomials verifying $0 \neq g_j \in \mathbb{C}[\Omega]^K$. A similar reasoning to the above proof leads to:

Theorem 5.3.17 *Assume that $\pi|_K$ has finite multiplicities. Then every $F(\ell) \in \mathbb{C}[\Omega]^K$ is algebraic over $\{f_{k_1}(\ell), \ldots, f_{k_u}(\ell)\}$.*

Besides, we also know the following. Let $W \in \mathscr{U}_\pi(\mathfrak{k}_v)^\mathfrak{k}$. If $v \leq d$, W belongs to $\mathscr{U}_\pi(\mathfrak{k})^\mathfrak{k}$ and the operator $\sigma(W)$ is a scalar for almost all $\sigma \in \widehat{K}$ with respect to the measure μ used in the irreducible decomposition of $\pi|_K$. Then we can apply Theorem 2.1.1 in [73] to get:

Proposition 5.3.18 *For any $W \in \mathscr{U}_\pi(\mathfrak{k})^\mathfrak{k}$, the function $P_W(\ell)$ is polynomial.*

Henceforth we will suppose that there exists a common polarization $\mathfrak{b}$ for almost all $\ell \in \Omega(\pi)$ with respect to the invariant measure. In particular, this is the case if $\mathfrak{b}$ is either Abelian or normal in $\mathfrak{g}$. In fact, under this hypothesis $\mathfrak{b}$ becomes essentially Abelian. Indeed, the intersection $\mathfrak{a}$ of every ker(ℓ) as ℓ varies in $\Omega(\pi)$ is G-invariant and contains $[\mathfrak{b}, \mathfrak{b}]$. If we try to show the polynomial conjecture by induction on $\dim G + \dim G/K$, we can suppose that $\mathfrak{k}$ contains $\mathfrak{a}$ which is contained in ker(π). In this way, we can pass to the quotient $\mathfrak{g}/\mathfrak{a}$ and use induction. Hence we can suppose $\mathfrak{a}$, and consequently $[\mathfrak{b}, \mathfrak{b}]$, are trivial. We shall consider the case where $\dim \mathfrak{g}/\mathfrak{b} \geq$

2, otherwise $\dim \Omega(\pi) \leq 2$ and the polynomial conjecture holds [13]. Now, our objective is:

Theorem 5.3.19 *Let G be a connected and simply connected nilpotent Lie group with Lie algebra $\mathfrak{g}$ and π an irreducible unitary representation of G. We denote by Ω the coadjoint orbit of G associated to π. Let K be an analytic subgroup of G such that the restriction $\pi|_K$ of π to K has finite multiplicities. Suppose that Ω admits a polarization $\mathfrak{b}$ which is an ideal of $\mathfrak{g}$. Then the polynomial conjecture holds.*

Proof The proof will be divided in several steps. We first show the following proposition.

Proposition 5.3.20 *For any $W \in \mathscr{U}_\pi(\mathfrak{g})^{\mathfrak{k}}$, the function $\Omega \ni \ell \mapsto P_W(\ell)$ is polynomial.*

Proof As usual we employ induction on $\dim G + \dim G/K$. We assume in particular that the polynomial conjecture holds for proper analytic subgroups $G'' = \exp(\mathfrak{g}'')$ of G which properly contain K and such that $\mathfrak{b} \cap \mathfrak{g}''$ is a polarization at $\ell'' = \ell|_{\mathfrak{g}''}$. Assuming $G \neq K$, we consider the sequence of subalgebras

$$\mathfrak{k}_d = \mathfrak{k} \subset \mathfrak{k}_{d+1} \subset \cdots \subset \mathfrak{k}_n = \mathfrak{g},\ \dim(\mathfrak{k}_j) = \dim(\mathfrak{k}_{j-1}) + 1\ (d+1 \leq j \leq n)$$

and the restriction mappings $p_j : \mathfrak{g}^* \to (\mathfrak{k}_j)^*$ for $d \leq j \leq n-1$, where $p_d = p$. Let as before $\mathfrak{k}_j = \mathbb{R}X_j + \mathfrak{k}_{j-1}$ for $1 \leq j \leq n$ and put $\mathfrak{g}' = \mathfrak{k}_{n-1}$, $p' = p_{n-1}$ and $G' = \exp(\mathfrak{g}')$.

Suppose first that Ω is not saturated with respect to $\mathfrak{g}'$. Let $\mathfrak{g} = \mathbb{R}X' + \mathfrak{g}'$. Then there exists in $\ker(\pi)$ an element $\eta = X' + \zeta$ with $\zeta \in \mathscr{U}(\mathfrak{g}')$. One can use η to eliminate from W the part outside $\mathscr{U}(\mathfrak{g}')$. This means that we can suppose W in $\mathscr{U}(\mathfrak{g}')$. Since the restriction π' of π to G' is irreducible and $\Omega' = p'(\Omega)$ is the coadjoint orbit of G' associated to π', our goal is a direct consequence of the induction hypothesis.

Suppose next that Ω is saturated with respect to $\mathfrak{g}'$. Let $\mathfrak{g} \neq \mathfrak{k} + \mathfrak{b}$. Then we are able to suppose that $\mathfrak{g}'$ contains the subalgebra $\mathfrak{k}+\mathfrak{b}$. As $\pi|_K$ has finite multiplicities, W belongs modulo $\ker(\pi)$ to $\mathscr{U}(\mathfrak{g}')$. Then the induction hypothesis says that the denominator of the rational function $P_W(\ell)$ is a polynomial of the e-central element $P_\alpha(\ell)$, $\alpha = X_{i_0} + c$, for a certain index $2 \leq i_0 \leq n-1$ and $c \in \mathscr{U}(\mathfrak{g}_{i_0-1})$ for G' which separates the G'-orbits with one-parameter in $\Omega' = p'(\Omega)$. Considering W multiplied by a certain polynomial of α if necessary, it suffices to study the case where the denominator of $P_W(\ell)$ is a polynomial of $P_\alpha(\ell)$ of degree 1.

Now we prove a lemma that plays a key role toward the proposition.

Lemma 5.3.21 *Let $H = \exp \mathfrak{h}$, $N = \exp \mathfrak{n}$ be subgroups of G such that $\mathfrak{n}$ contains $\mathfrak{h}_{j_0} = \mathfrak{h} + \mathfrak{g}_{j_r}$ and ρ an irreducible representation of N whose coadjoint orbit Ω_ρ is contained in the projection of Ω on $\mathfrak{n}^*$. Suppose that $\rho|_H$ has finite multiplicities, $\alpha \in \mathscr{U}(\mathfrak{n})$, and $\rho(\alpha)$ is not a constant. Suppose further that the coadjoint orbit Ω_ρ of N admits a polarization $\mathfrak{p}$ which is an ideal of $\mathfrak{n}$. Then if the algebra $\mathscr{U}_\rho(\mathfrak{n})^{\mathfrak{h}}$*

really extends in the passage from $\mathfrak{h}_{j_0-1}$ *to* $\mathfrak{h}_{j_0}$*, i.e. if*

$$\mathscr{U}_\rho(\mathfrak{h}_{j_0})^{\mathfrak{h}} \not\subset \mathscr{U}_\rho(\mathfrak{h}_{j_0-1})^{\mathfrak{h}} + ker(\rho),$$

then there exists an $\mathfrak{h}_{j_0}$*-invariant element*

$$\nu = aX_{j_0} + b,\ a, b \in \mathscr{U}(\mathfrak{h}_{j_0-1}).$$

Put equivalently, ν *is a Corwin-Greenleaf e-central element for* $p_{j_0}(\Omega)$ *such that* a *is not divisible by* α *modulo* $\ker(\rho)$.

Proof Let us proceed by induction on $\dim N$. Let $\mathfrak{n}' = \mathfrak{n} \cap \mathfrak{g}'$. Because $\rho(\alpha)$ is not a constant, $\mathfrak{n}'$ is an ideal of codimension 1 in $\mathfrak{n}$. It suffices to treat the case where $\dim(\mathfrak{n}/\mathfrak{p}) \geq 2$. Indeed, if $\dim(\mathfrak{n}/\mathfrak{p}) = 1$, then modulo $\ker(\rho)$ we can suppose that $\mathfrak{p}$ is Abelian. In this case, we are essentially in the threadlike case, namely when the center of $\mathfrak{n}$ is one-dimensional. Furthermore, if $\mathscr{U}_\rho(\mathfrak{n})^{\mathfrak{h}} \neq \mathbb{C}$ modulo $\ker(\rho)$ then $\mathfrak{h} \subset \mathfrak{p}$, and at the level j_0 there is a new entry in $\ker(\rho)$, except when j_0 corresponds to the first jump index for ρ. This says that either this situation does not occur, or $\nu = X_{j_0}$ suits us.

Take an ideal $\widetilde{\mathfrak{n}}$ of codimension 1 in $\mathfrak{n}$ different from $\mathfrak{n}'$ and containing $\mathfrak{p}$. Let $\widetilde{N} = \exp\widetilde{\mathfrak{n}}$, $\mathfrak{n} = \mathbb{R}X + \widetilde{\mathfrak{n}}$ and call $\widetilde{p} : \mathfrak{n}^* \to (\widetilde{\mathfrak{n}})^*$ the restriction mapping. To simplify the notation, we agree that $\widetilde{p}(\Omega_\rho)$ is the disjoint union of a one-parameter family $\{\omega_\lambda\}$, in $\lambda \in \mathbb{R}$, of coadjoint orbits of $\widetilde{N}$. Let ρ_λ be the irreducible unitary representation of $\widetilde{N}$ associated to ω_λ. Taking the intersection of the Jordan-Hölder sequence $\{\mathfrak{g}_j\}_{j=1}^n$ of $\mathfrak{g}$ with $\mathfrak{n}$ and $\mathfrak{h}$, we consider the following Jordan-Hölder sequences: $\{\mathfrak{n}_j\}_{j=1}^{n'}$ for $\mathfrak{n}$, where $n' = \dim\mathfrak{n}$, and $\{\mathfrak{h}_j\}_{j=1}^{d'}$ for $\mathfrak{h}$, where $d' = \dim\mathfrak{h}$. Put as before

$$J = \{j_1 < \cdots < j_{n'-d'}\} = \{1 \leq j \leq n' : \dim(\mathfrak{n}_j \cap \mathfrak{h}) = \dim(\mathfrak{n}_{j-1} \cap \mathfrak{h})\}$$

and $\mathfrak{h}_k = \mathfrak{h} + \mathfrak{n}_{j_{k-d'}}$ for $d' + 1 \leq k \leq n'$. Write again $\mathfrak{h}_k = \mathbb{R}X_k + \mathfrak{h}_{k-1}$ and $H_k = \exp(\mathfrak{h}_k)$. Put $\widetilde{\mathfrak{h}} = \mathfrak{h} \cap \widetilde{\mathfrak{n}}$, $\widetilde{H} = \exp\widetilde{\mathfrak{h}}$, $\widetilde{\mathfrak{h}}_{j_0} = \widetilde{\mathfrak{h}} + \mathfrak{g}_{j_r}$ and $\widetilde{H}_{j_0} = \exp(\widetilde{\mathfrak{h}}_{j_0})$. Since $\rho_\lambda|_{\widetilde{H}_{j_0}}$ has finite multiplicities, there exists by induction hypothesis an element

$$\widetilde{\nu}_\lambda = a(\lambda)X_{j_0} + b(\lambda)$$

which is $\widetilde{\mathfrak{h}}_{j_0}$-invariant for the projection image of ω_λ on $(\widetilde{\mathfrak{h}}_{j_0})^*$. Here $a(\lambda), b(\lambda)$ are two polynomial functions of λ with values in $\mathscr{U}(\mathfrak{h}_{j_0-1} \cap \widetilde{\mathfrak{n}})$ because $\omega_\lambda = \exp(\lambda X) \cdot \omega_0$, and $a(\lambda)$ is not divisible by α. For the image $\widetilde{p}(\Omega_\rho)$ of Ω_ρ on $(\widetilde{\mathfrak{n}})^*$ there exists an e-central element

$$\widetilde{\kappa} = X_{k_0} + u\ (k_0 \leq n - 1)$$

for a certain $u \in \mathcal{U}(\mathfrak{h}_{k_0-1} \cap \widetilde{\mathfrak{n}})$ such that $\rho(\widetilde{\kappa})$ is not a constant. Choose such $\widetilde{\kappa}$ so to make k_0 as small as possible. We may suppose, by modifying $\widetilde{\kappa}$ by a scalar, that

$$\rho_\lambda(\widetilde{\kappa}) = \lambda,$$

so the element

$$\widetilde{\nu} = a(\widetilde{\kappa})X_{j_0} + b(\widetilde{\kappa})$$

verifies

$$[\widetilde{\mathfrak{h}}_{j_0}, \widetilde{\nu}] \subset \ker(\rho)$$

and $a(\widetilde{\kappa})$ is not divisible by α modulo $\ker(\rho)$.

5.3.4 *Case 1.* $\mathfrak{h} \subset \widetilde{\mathfrak{n}}$

We reduce to the case where $k_0 < j_0$. For if $k_0 > j_0$, suppose $\mathfrak{h}$ is contained in $\widetilde{\mathfrak{n}}$, so $[\mathfrak{h}, \widetilde{\kappa}] \subset \ker(\rho)$ and, when we write $\widetilde{\nu}$ as a polynomial of $\widetilde{\kappa}$, all the coefficients commute with $\mathfrak{h}$ modulo $\ker(\rho)$ and there is at least one which suits us. Moreover, if $k_0 = j_0$, the element $\widetilde{\kappa}$ itself qualifies as ν. That is how $\widetilde{\nu}$ supplies us with the required element.

5.3.5 *Case 2.* $\mathfrak{h} \not\subset \widetilde{\mathfrak{n}}$

Suppose now that $\mathfrak{h}$ is not contained in $\widetilde{\mathfrak{n}}$ and $\widetilde{\nu}$ is not $\mathfrak{h}$-invariant modulo $\ker(\rho)$. We expand on the argument used in [7]. Let $\mathfrak{h} = \mathbb{R}X + \widetilde{\mathfrak{h}}$. As $[X, \widetilde{\kappa}]$ is e-central for $\widetilde{p}(\Omega_\rho)$, it is constant modulo $\ker(\rho)$. Then $[X, [X, \widetilde{\kappa}]]$ belongs to $\ker(\rho)$. For a sufficiently large integer $v \in \mathbb{N}$, we consider

$$\psi = \widetilde{\nu} + F(\widetilde{\kappa}),$$

where $F(t)$ is a polynomial of degree $2v$ in one variable t. For $k \in \mathbb{N}$, put

$$\psi_0 = \psi, \ \psi_k = (\mathrm{ad}X)^k(\psi).$$

Since $(\mathrm{ad}X)^k(F(\widetilde{\kappa}))$ are e-central for $\widetilde{p}(\Omega_\rho)$, if they lie outside $\ker(\rho)$, they are not divisible by α. Otherwise, taking X' in $\widetilde{\mathfrak{n}}$, $[X', (\mathrm{ad}X)^k(F(\widetilde{\kappa}))]$ does not belong to $\ker(\rho)$. Now, as the integer v is sufficiently large, we see that

$$\psi_{2v} \notin \ker(\rho), \ \psi_{2v+1} \in \ker(\rho).$$

Everything being as in [7], let us construct an element of $\mathscr{U}_\rho(\mathfrak{n}_{j_0})^{\mathfrak{h}}$ by the formula

$$\begin{aligned}\nu =&(\psi_0\psi_{2v} + \psi_{2v}\psi_0) - (\psi_1\psi_{2v-1} + \psi_{2v-1}\psi_1) + \cdots \\ &+ (-1)^{v-2}(\psi_{v-2}\psi_{v+2} + \psi_{v+2}\psi_{v-2}) \\ &+ (-1)^{v-1}(\psi_{v-1}\psi_{v+1} + \psi_{v+1}\psi_{v-1}) + (-1)^v\psi_v^2.\end{aligned}$$

Here, we remark that $k_0 < j_0$. Because, $[\widetilde{\kappa}, \widetilde{\mathfrak{h}}_{j_0}] \subset \ker(\rho)$. If $k_0 \geq j_0$, there would exist a new element of $\mathscr{U}_\rho(\mathfrak{n})^{\mathfrak{h}}$ appearing at the level of $\mathfrak{h}_{k_0}$. Since $\rho|_H$ has finite multiplicities, this means that there exists at almost all $\ell \in \Omega_\rho$ an element in $\mathfrak{h}_{k_0}$ outside $\mathfrak{h}_{k_0-1}$ which is written as $a(\ell)+b(\ell)$ with $a(\ell) \in \mathfrak{h}(\ell|_{\mathfrak{h}})$ and $b(\ell) \in \mathfrak{g}(\ell)$. But this is impossible, because $\mathfrak{h} \not\subset \widetilde{\mathfrak{n}}$. Recall that v is large. This guarantees that ν is of degree 1 in X_{j_0}. Finally, choosing suitably the polynomial F, the coefficient of X_{j_0} in ν is not divisible by α modulo $\ker(\rho)$. Indeed, let

$$F(t) = \lambda_0 + \lambda_1 t + \cdots + \lambda_{2v-1}t^{2v-1} + t^{2v},\ \lambda_j \in \mathbb{C}\ (0 \leq j \leq 2v-1).$$

Suppose that $(\mathrm{ad}X)^k(a(\widetilde{\kappa}))$ $(0 \leq k \leq m)$ are not divisible modulo $\ker(\rho)$ by α, whereas $(\mathrm{ad}X)^{m+1}(a(\widetilde{\kappa}))$, hence $(\mathrm{ad}X)^k(a(\widetilde{\kappa})), k \geq m+1$, are divisible by it. Concerning the coefficient of X_{j_0} in ν, the coefficient λ_{2v-m} of F comes only from the terms

$$\sum_{k\geq m}(-1)^k(\psi_k\psi_{2v-k} + \psi_{2v-k}\psi_k) \equiv 2\sum_{k\geq m}(-1)^k\psi_{2v-k}\psi_k\ (\mathrm{mod}\ \ker(\rho))$$

and λ_{2v-m} appears in the term $\psi_{2v-m}\psi_m$ as

$$\lambda_{2v-m}(\mathrm{ad}X)^{2v-m}(\widetilde{\kappa}^{2v-m})\psi_m.$$

If so, in order for the coefficient of X_{j_0} in ν to always be divisible modulo $\ker(\rho)$ by α for any value of λ_{2v-m}, necessarily the coefficient of X_{j_0} in $(\mathrm{ad}X)^{2v-m}(\widetilde{\kappa}^{2v-m})\psi_m$, and hence in ψ_m, must be divisible by α modulo $\ker(\rho)$. This contradicts the choice of index m. In this way we find the required element ν, and this finishes the proof of the lemma. □

We now continue the proof of the proposition. Let j_0 be the smallest index j such that W belongs modulo $\ker(\pi)$ to $\mathscr{U}(\mathfrak{k}_j)$. When $j_0 \leq d$, W belongs to $\mathscr{U}(\mathfrak{k})^{\mathfrak{k}}$, namely an e-central element for $p(\Omega) \subset \mathfrak{k}^*$ and consequently the function $P_W(\ell)$ is polynomial by Proposition 5.3.18.

Suppose $j_0 \geq d+1$ and write again

$$W = v_sX_{j_0}^s + v_{s-1}X_{j_0}^{s-1} + \cdots + v_1X_{j_0} + v_0,\ v_i \in \mathscr{U}(\mathfrak{k}_{j_0-1})\ (0 \leq i \leq s)$$

with $v_s \notin \ker(\pi)$. Remark that v_s is $\mathfrak{k}$-invariant modulo $\ker(\pi)$. In fact, let us forget for a while the existence of a normal polarization and consider the general case. If

Ω is not saturated with respect to $\mathfrak{g}'$, the restriction π' of π to G' is irreducible and associated to the coadjoint orbit $\Omega' = p'(\Omega)$ of G'. Since $\pi'|_K$ has finite multiplicities and W is in $\mathscr{U}_{\pi'}(\mathfrak{g}')^{\mathfrak{k}}$, v_s is $\mathfrak{k}$-invariant by induction hypothesis. If Ω is saturated with respect to $\mathfrak{g}'$, let $\pi|_{G'} \simeq \int_{\mathbb{R}}^{\oplus} \pi_t' dt$ be the usual irreducible decomposition. As $\pi|_K$ has finite multiplicities, the same is true for $\pi_t'|_K$ and W belongs to $\mathscr{U}(\mathfrak{g}')$ hence to $\mathscr{U}_{\pi_t'}(\mathfrak{g}')^{\mathfrak{k}}$ for generic $t \in \mathbb{R}$. Then the induction hypothesis says that $[\mathfrak{k}, v_s] \subset \ker(\pi_t')$ for almost all $t \in \mathbb{R}$, hence giving $[\mathfrak{k}, v_s] \subset \ker(\pi)$.

We will show by induction on s that $P_W(\ell)$ is a polynomial function on Ω. If there exists in $\mathscr{U}(\mathfrak{k}_{j_0})^{\mathfrak{k}}$ a $\mathfrak{k}$-invariant element $\nu = aX_{j_0} + b$ which corresponds to a polynomial function, where a, b belong to $\mathscr{U}(\mathfrak{k}_{j_0-1})$ such that $a \notin \ker(\pi)$, it follows that $a, a^s W - v_s \nu^s$ belong to $\mathscr{U}(\mathfrak{k}_{j_0})^{\mathfrak{k}}$. As the degree of $a^s W - v_s \nu^s$ in X_{j_0} is smaller than or equal to $s-1$, this implies that W multiplied by a certain power of a corresponds to a polynomial function. Therefore if $P_{W_1}(\ell)$ is the denominator of $P_W(\ell)$, W_1 appears as a factor γ of a power of a. That is to say, α is an irreducible factor of γ. In order to show that $P_W(\ell)$ is a polynomial function, it suffices to show the existence of such ν with a not divisible by α and such that the function $P_\nu(\ell)$ is polynomial.

Assume that $\mathfrak{k}_{j_0} = \mathfrak{k} + \mathfrak{g}_{j_r}$. If $r = 1$, there exists in $\mathscr{U}_\pi(\mathfrak{g})^{\mathfrak{k}}$ an element $A = uX_{d+1} + v$ with $u, v \in \mathscr{U}(\mathfrak{k})$. This means that $[X_{d+1}, W] \in \ker(\pi)$ and hence $[\mathfrak{k}_{d+1}, W] \subset \ker(\pi)$, which means by induction that $P_W(\ell)$ is a polynomial function, i.e. α is trivial. Therefore $2 \le r \le q$.

When $\mathfrak{g} = \mathfrak{k} + \mathfrak{b}$, we are led as usual to the case where $\mathfrak{g}_1 = \mathbb{R}Z$ is the center of $\mathfrak{g}$, $\mathfrak{g}_2 = \mathbb{R}Y + \mathfrak{g}_1$ and $\mathfrak{g} = \mathbb{R}X + \mathfrak{g}'$, for the centralizer $\mathfrak{g}'$ of Y in $\mathfrak{g}$. Here, $X \in \mathfrak{k}$ satisfies $[X, Y] = Z$. Moreover, it is possible to take $\mathfrak{b}$ in $\mathfrak{g}'$. We write $\mathfrak{k} = \mathbb{R}X + \mathfrak{k}'$ with $\mathfrak{k}' = \mathfrak{k} \cap \mathfrak{g}'$. Let also $G' = \exp(\mathfrak{g}')$, $K' = \exp(\mathfrak{k}')$ and $\ell' = p'(\ell)$ for $\ell \in \Omega$. Recall the usual decomposition $\pi|_{G'} \simeq \int_{\mathbb{R}}^{\oplus} \pi_t' dt$ into irreducibles. Then the restriction $\pi_t'|_{K'}$ has finite multiplicities for almost all $t \in \mathbb{R}$. First assume that $W \in \mathscr{U}(\mathfrak{g}')$. From the induction hypothesis, $\pi_t'(W)a_{\ell'} = P_W'(\ell')a_{\ell'}$ for almost all $t \in \mathbb{R}$ for some functions $P_W'(\ell')$ such that $P_W'(\ell')|_{\Omega_t}$ is polynomial, where Ω_t denotes the orbit associated to π_t'. The K-invariance of W implies $P_W(\ell) = P_W'(\ell')$ for generic $\ell \in \Omega$. As the denominator of $P_W(\ell)$ is a polynomial function in the variable $\ell(Y)$, it is indeed a polynomial function by K-invariance.

Let $W \notin \mathscr{U}(\mathfrak{g}')$. This means that the generic K-orbits in $p(\Omega)$ are not saturated with respect to $\mathfrak{k}'$. Hence there exists a Corwin-Greenleaf e-central element $\gamma = aX + b \in \mathscr{U}_\pi(\mathfrak{k})^{\mathfrak{k}}, a, b \in \mathscr{U}(\mathfrak{k}')$, where $a \notin \ker(\pi)$ such that $[\mathfrak{k}, a] \subset \ker(\pi)$. Making use of γ to reduce the degree of W relative to X, we see by induction hypothesis that $a^m W$ corresponds to a polynomial function for a certain positive integer m. Hence α comes this time from an irreducible factor of a. If so, we can use the argument of Case 2 in Lemma 5.3.21, with Y playing the role of $\widetilde{\kappa}$. At the level of the disintegration $\pi|_{G'} \simeq \int_{\mathbb{R}}^{\oplus} \pi_t' dt$, we take the reference point π_0' so that $\pi_0'|_{K'}$ has finite multiplicities. From the induction hypothesis applied to the triple (G', K', π_0'), there exists an ν' in $\mathscr{U}_{\pi_0'}(\mathfrak{k}_{j_0})^{\mathfrak{k}'}$ of degree 1 in X_{j_0} whose coefficient is not divisible by α. Then it suffices to extend ν' to a $\mathfrak{k}$-invariant element. Namely, substituting $-Y$ to t in the expression of $\exp(tX) \cdot \nu'$, we obtain the desired element ν.

Taking $\mathfrak{n} = \mathfrak{g}, \rho = \pi, \mathfrak{h} = \mathfrak{k}$, we can apply Lemma 5.3.21 to conclude that the function $P_W(\ell)$ is in fact polynomial. After constructing ν, in fact, the function $P_\nu(\ell)$ is polynomial on the orbit Ω, and Proposition 5.3.20 follows. □

In order to prove the theorem, we examine the surjectivity of the homomorphism

$$\Theta : \mathscr{U}_\pi(\mathfrak{g})^{\mathfrak{k}} \ni W \mapsto P_W(\ell) \in \mathbb{C}[\Omega]^K.$$

Take an arbitrary $q(\ell)$ in $\mathbb{C}[\Omega]^K$. Just as before, we may look at the last possibility only:

$$\mathfrak{g} = \mathbb{R}X + \widetilde{\mathfrak{g}},\ X \in \mathfrak{k},$$

with an ideal $\widetilde{\mathfrak{g}}$ of codimension 1 for which the orbit Ω is saturated. Let $\widetilde{\kappa} \in \mathscr{U}(\widetilde{\mathfrak{g}})$ be the e-central element for $\widetilde{p}(\Omega)$ used above. Then there exists a polynomial $F(y')$ of

$$y' = P_{\widetilde{\kappa}}(\ell)\ (\ell \in \Omega)$$

such that $F(y')q(\ell)$ is in the image of the homomorphism Θ by means of Proposition 5.3.15. Suppose that the function $q(\ell)$ depends only on

$$x_1 = \ell(X_1), \ldots, x_{j_0} = \ell(X_{j_0})$$

and $\ell|_{\mathfrak{k}}$. Write

$$q(\ell) = r(\ell)x_{j_0}^{v} + s(\ell)$$

for a certain integer $v \in \mathbb{N}$, where $r(\ell)$ does not depend on x_{j_0} and $s(\ell)$ is of degree smaller than or equal to $v-1$ in x_{j_0}. Let α be a polynomial of $\widetilde{\kappa}$ such that $P_\alpha(\ell)$ is an irreducible factor of $F(y')$.

In this situation the algebra $\mathscr{U}_\pi(\mathfrak{g})^{\mathfrak{k}}$ extends when we pass through $\mathfrak{k}_{j_0-1}$ to $\mathfrak{k}_{j_0}$ and as we proved above, there exists in $\mathscr{U}_\pi(\mathfrak{k}_{j_0})^{\mathfrak{k}}$ an element

$$\sigma = aX_{j_0} + b\ (a, b \in \mathscr{U}(\mathfrak{k}_{j_0-1})),$$

where a is not divisible by α modulo $\ker(\pi)$. Then

$$(P_a(\ell))^v q(\ell) - r(\ell)P_{\sigma^v}(\ell)$$

belongs to the image of Θ by induction hypothesis on the degree v. This says that $(P_a(\ell))^v q(\ell)$ belongs to the image of Θ, because $r(\ell)$ does by induction hypothesis. Let

$$P_{W_1}(\ell) = F(y')q(\ell),\ P_{W_2}(\ell) = (P_a(\ell))^v q(\ell)$$

for certain $W_1, W_2 \in \mathscr{U}_\pi(\mathfrak{g})^{\mathfrak{k}}$. Hence

$$F(y')(P_a(\ell))^v q(\ell) = (P_a(\ell))^v P_{W_1}(\ell) = F(y')P_{W_2}(\ell).$$

This implies

$$P_{a^v W_1}(\ell) = P_{F(\widetilde{\kappa})W_2}(\ell)$$

and then

$$a^v W_1 \equiv F(\widetilde{\kappa})W_2$$

modulo $\ker(\pi)$. Since a is not divisible modulo $\ker(\pi)$ by α, W_1 must be divisible by α modulo $\ker(\pi)$, and we can eliminate α from $F(\widetilde{\kappa})$. In this way we conclude that $q(\ell)$ belongs to the image of homomorphism Θ and this achieves the proof of the theorem. □

The case where $\mathfrak{g}$ is two-step nilpotent was treated in Corollary 5.3.14. More generally,

Corollary 5.3.22 *Suppose that $\mathfrak{g}$ is three-step nilpotent. Then the polynomial conjecture holds for $G = \exp \mathfrak{g}$.*

Proof In this case $[\mathfrak{g}, \mathfrak{g}]$ is an Abelian ideal of $\mathfrak{g}$. This allows to choose a normal polarization at any linear form on $\mathfrak{g}$. □

Here is another proof of this result.

Theorem 5.3.23 *Assume that $\mathfrak{g}$ is three-step nilpotent. Then the polynomial conjecture holds.*

Proof We keep the notations and denote by $\mathfrak{z}$ the center of $\mathfrak{g}$. Going down to $\mathfrak{g}/(\mathfrak{z} \cap \ker(\pi))$ if necessary, we can suppose that $\mathfrak{z} = \mathbb{R}Z$, $f(Z) = 1 (\forall f \in \Omega)$, $[X, Y] = Z$, $\mathfrak{g} = \mathbb{R}X + \mathfrak{g}_0$, with $\mathfrak{g}_0$ being the centralizer of Y in $\mathfrak{g}$. Put $\mathfrak{m} = \{T \in \mathfrak{g}; [\mathfrak{g}, T] \subset \mathfrak{z}\}$. By our assumption $[\mathfrak{g}, \mathfrak{g}] \subset \mathfrak{m}$. Let $\mathfrak{k}$ be a subalgebra of $\mathfrak{g}$ such that $\pi|_K$, $K = \exp \mathfrak{k}$, has finite multiplicities. Let us see that the function $P_W(\ell)$ is polynomial for any $W \in \mathscr{U}_\pi(\mathfrak{g})^{\mathfrak{k}}$. To this aim, we can also assume that $\mathfrak{k}$ contains $\mathfrak{z}$. When $\dim((\mathfrak{k} \cap \mathfrak{m})/\mathfrak{z}) \geq 1$, we take Y in $\mathfrak{k} \cap \mathfrak{m}$ and consequently it suffices to suppose that $\mathfrak{k}$ is contained in $\mathfrak{g}_0$.

Suppose first that $\dim((\mathfrak{k} \cap \mathfrak{m})/\mathfrak{z}) \geq 2$ and take in $\mathfrak{k} \cap \mathfrak{m}$ an element Y' not belonging to $\mathbb{R}Y + \mathfrak{z}$ and $X' \in \mathfrak{g}$ satisfying $[X', Y'] = Z$. If $[\mathfrak{g}_0, Y'] = \{0\}$, $Y' - f([X, Y'])Y$ is central in $\mathfrak{g}$, which is absurd. Hence $[\mathfrak{g}_0, Y'] \neq \{0\}$. This being the case, we take X' in $\mathfrak{g}_0$ and denote by $\mathfrak{g}_0'$ the centralizer of Y' in $\mathfrak{g}$. Now $\mathfrak{g}_0, \mathfrak{g}_0'$ are two ideals of codimension 1 in $\mathfrak{g}$ verifying

$$\mathfrak{g} = \mathbb{R}X + \mathfrak{g}_0 = \mathbb{R}X' + \mathfrak{g}_0'.$$

As the generic K-orbits in $\Omega = \Omega(\pi)$ are saturated with respect to $\mathfrak{g}_0$ and $\mathfrak{g}_0'$, our algebra $\mathscr{U}_\pi(\mathfrak{g})^{\mathfrak{k}}$ is contained modulo $\ker(\pi)$ in $\mathscr{U}(\mathfrak{g}_0 \cap \mathfrak{g}_0')$. Since $\mathfrak{g}_0 \cap \mathfrak{g}_0'$ is an ideal of codimension 2, the result follows.

Suppose now that $\dim((\mathfrak{k} \cap \mathfrak{m})/\mathfrak{z}) = 1$ but $\dim(\mathfrak{m}/\mathfrak{z}) \geq 2$. Choose Y' in $\mathfrak{m} \backslash (\mathfrak{m} \cap \mathfrak{k})$ and define its centralizer as before. When $\mathfrak{k}$ is contained in $\mathfrak{g}_0'$, we can reason just as before. Let $\mathfrak{k} \not\subset \mathfrak{g}_0'$ and take in $\mathfrak{k}$ the element X' introduced above. Since $[\mathfrak{k}, \mathfrak{k}]$ is contained in $\mathbb{R}Y + \mathfrak{z}$, $\mathfrak{k}$ being two-step, we can choose X' central in $\mathfrak{k}$ if $\dim([\mathfrak{k}, \mathfrak{k}]) \leq 1$. It follows that $X' \in \mathscr{U}_\pi(\mathfrak{g})^{\mathfrak{k}}$ and we can easily pass to $\mathfrak{g}_0 \cap \mathfrak{g}_0'$.

Suppose henceforth $\dim([\mathfrak{k}, \mathfrak{k}]) = 2$, namely $[\mathfrak{k}, \mathfrak{k}] = \mathbb{R}Y + \mathfrak{z}$. Since we can suppose that there is in $\mathfrak{g}$ no ideal of codimension 2 containing $\mathfrak{k}$, we have $X \notin \mathfrak{k}$ and $\mathfrak{g} = \mathbb{R}X + \mathfrak{k} + [\mathfrak{g}, \mathfrak{g}]$. Thus, $\mathfrak{g}_0 = \mathfrak{k} + [\mathfrak{g}, \mathfrak{g}]$. For any $V \in \mathfrak{k}$, either $[\mathfrak{k}, V] = \{0\}$ or $[\mathfrak{k}, V] = \mathbb{R}Y + \mathfrak{z}$. Otherwise, the generic K-orbits in Ω are saturated with respect to $\widetilde{\mathfrak{k}} = \mathfrak{k} \cap \widetilde{\mathfrak{g}}$, where $\widetilde{\mathfrak{g}}$ is a certain ideal of codimension 1 in $\mathfrak{g}$. Let $\mathfrak{k} = \mathbb{R}\widetilde{X} + \widetilde{\mathfrak{k}}$, $\widetilde{K} = \exp(\widetilde{\mathfrak{k}})$ and $\widetilde{G} = \exp(\widetilde{\mathfrak{g}})$. If Ω is saturated with respect to $\widetilde{\mathfrak{g}}$, then $W \in \mathscr{U}(\widetilde{\mathfrak{g}})$ and $\pi|_{\widetilde{G}} \simeq \int_{\mathbb{R}}^{\oplus} \rho_t dt$ with one-parameter family $\{\rho_t\}$ of irreducible unitary representations of $\widetilde{G}$. Because $\rho_t|_{\widetilde{K}}$ generically has finite multiplicities, the induction hypothesis applied to $(\widetilde{G}, \widetilde{K}, \rho_t, W)$ says that the denominator of $P_W(\ell)$ is a polynomial of a variable other than $\ell(Y)$. In this way, the rational function $P_W(\ell)$ turns out to be polynomial. If Ω is non-saturated with respect to $\widetilde{\mathfrak{g}}$, there exists in $\ker(\pi)$ an element $w = \widetilde{X} + \widetilde{w}$ with $\widetilde{w} \in \mathscr{U}(\widetilde{\mathfrak{g}})$ and we can modify W if necessary so that $W \in \mathscr{U}(\widetilde{\mathfrak{g}})$. Then as $(\pi|_{\widetilde{G}})|_{\widetilde{K}}$ has finite multiplicities, we can apply the induction hypothesis to $(\widetilde{G}, \widetilde{K}, \pi|_{\widetilde{G}}, W)$. Now, if X' is central in $\mathfrak{k}$, we can pass to $\mathfrak{g}_0 \cap \mathfrak{g}_0'$ as before. Therefore, we can suppose that $[\mathfrak{k}, X'] = \mathbb{R}Y + \mathfrak{z}$. Then for generic $\ell \in \Omega$, the stabilizer of $\ell|_{\mathfrak{k}}$ in $\mathfrak{k}$ is contained in $\mathfrak{k}' = \mathfrak{k} \cap \mathfrak{g}_0'$. This means that the generic K-orbits in Ω are saturated with respect to $\mathfrak{k}'$ and we can repeat the arguments developed above.

Let now $\dim \mathfrak{m} > 2$ and $[\mathfrak{k}, \mathfrak{k}] \subset \mathfrak{z}$. Since $\mathfrak{k}$ is a direct product of a Heisenberg algebra by an Abelian ideal, we can take X (resp. X') in the center of $\mathfrak{k}$ when the generic K-orbits in Ω are non-saturated with respect to $\mathfrak{g}_0$ (resp. $\mathfrak{g}_0'$), which allows to descend to $\mathfrak{g}_0 \cap \mathfrak{g}_0'$.

Suppose finally that $\mathfrak{m} = \mathbb{R}Y + \mathfrak{z}$. If $Y \notin \mathfrak{k}$, the requirement $[\mathfrak{k}, \mathfrak{k}] \subset \mathfrak{z}$ permits us to repeat the same argument as before. Let $Y \in \mathfrak{k}$. Replacing $\mathfrak{k}$ by $\mathfrak{k} \cap \mathfrak{g}_0$ if necessary, we can suppose $X \notin \mathfrak{k}$ and $\mathfrak{g} = \mathbb{R}X + \mathfrak{k} + [\mathfrak{g}, \mathfrak{g}]$. Hence $\mathfrak{k}$ coincides with the ideal $\mathfrak{g}_0$ and the result is known (Theorem 5.3.13).

Lastly, we can verify in a similar way that our mapping $W \mapsto P_W(\ell)$ is surjective on the algebra $\mathbb{C}[\Omega]^K$. □

We treat in what follows the case where $\mathfrak{k}$ is an Abelian subalgebra. First we have:

Lemma 5.3.24 *Assume that $K = \exp \mathfrak{k}$ is an Abelian subgroup of G. Let $\pi \in \widehat{G}$ be realized as $\pi = \mathrm{ind}_{B[\ell]}^G \chi_\ell$ by means of a Vergne polarization $\mathfrak{b}[\ell]$, $B[\ell] = \exp(\mathfrak{b}[\ell])$, constructed at $\ell \in \Omega(\pi)$ from the flag $\mathscr{S}$. We define $a_\ell \in \mathscr{H}_\pi^{-\infty}$ by the formula*

$$\langle a_\ell, \varphi \rangle = \int_{K/(K \cap B[\ell])} \overline{\varphi(k)\chi_\ell(k)} d\dot{k}, \ \varphi \in \mathscr{H}_\pi^\infty.$$

Let $W \in \mathcal{U}_\pi(\mathfrak{g})^{\mathfrak{k}}$ *such that*

$$W{\cdot}a_\ell = P_W(\ell)a_\ell$$

on a non-empty Zariski open set of $\Omega(\pi)$ *for a certain function* P_W*. Then* $P_W(\ell)$ *is a rational function.*

Proof The polarizations $\mathfrak{b}[\ell]$ and the subalgebras $\mathfrak{k} \cap \mathfrak{b}[\ell]$ vary rationally with $\ell \in \Omega(\pi)$. We can find a non-empty Zariski open set $\mathcal{U}$ of $\Omega(\pi)$ and a family of elements $\mathfrak{B} = \{X_1, \cdots, X_r\}$ of $\mathfrak{g}$ which constitutes a Malcev basis of $\mathfrak{g}$ modulo the polarization $\mathfrak{b}[\ell]$ at ℓ and such that $\mathfrak{B} \cap \mathfrak{k}$ is a Malcev basis of $\mathfrak{k}$ modulo $\mathfrak{k} \cap \mathfrak{b}[\ell]$ for $\ell \in \mathcal{U}$. Consider the applications $E_{\mathfrak{B}} : \mathbb{R}^r \to G$ defined by

$$E_{\mathfrak{B}}(t_1, \cdots, t_r) = \exp(t_1 X_1) \cdots \exp(t_r X_r)$$

and $E_{\mathfrak{B}\cap K} : \mathbb{R}^m \to K$ defined by

$$E_{\mathfrak{B}\cap K}(t) = \exp(t_1 X_{i_1}) \cdots \exp(t_m X_{i_m}),$$

where $\{X_{i_1}, \cdots, X_{i_m}\} := \mathfrak{k} \cap \mathfrak{B}$. For any $g \in G$ we can write

$$g = E_{\mathfrak{B}}(t(g, \ell)) \cdot b(g, \ell)$$

with $t(g, \ell) \in \mathbb{R}^r$ and $b(g, \ell) \in B[\ell]$. The maps $(g, \ell) \mapsto t(g, \ell)$ and $(g, \ell) \mapsto b(g, \ell)$ are then polynomial in g and rational in $\ell \in \Omega(\pi)$ and the mapping $G \to \mathbb{R}^r \times B[\ell] : g \mapsto (t(g, \ell), b(g, \ell))$ is polynomial in g and rational in ℓ.

Let us fix a Schwartz function $\psi : \mathbb{R}^r \to \mathbb{C}$. Then for any $\ell \in \mathcal{U}$ the function defined on G by

$$\psi_\ell(E_{\mathfrak{B}}(t) \cdot b_\ell) = \psi(t)\chi_\ell(b_\ell^{-1}),\ t \in \mathbb{R}^r, b_\ell \in B[\ell],$$

is a C^∞-vector of the representation $\pi_\ell = \mathrm{ind}_{B[\ell]}^G \chi_\ell$. If $V \in \mathcal{U}(\mathfrak{g})$, then there exist for $\alpha, \beta \in \mathbb{N}^r$, some rational functions $p^V_{\alpha,\beta}$ in $\ell \in \Omega(\pi)$ such that

$$d\pi_\ell(V)(\psi_\ell)(E_{\mathfrak{B}}(t)) = \sum_{\alpha,\beta} p^V_{\alpha,\beta}(\ell) t^\alpha \partial^\beta \psi(t),\ \ell \in \mathcal{U}, t \in \mathbb{R}^r, \psi \in \mathcal{S}(\mathbb{R}^r).$$

In this way, for $W \in \mathcal{U}(\mathfrak{g})$ and $\psi_\ell \in \mathcal{H}^\infty_{\pi_\ell}$ we see that

$$\langle W \cdot a_\ell, \psi_\ell \rangle = \sum_{\alpha,\beta} \overline{p^{W*}_{\alpha,\beta}(\ell)} \int_{\mathbb{R}^m} t^\alpha \overline{(\partial^\beta \psi_\ell(E_{\mathfrak{B}\cap K}(t)))\chi_\ell(E_{\mathfrak{B}\cap K}(t))} dt.$$

Hence for $W \in \mathscr{U}_\pi(\mathfrak{g})^{\mathfrak{k}}$ of our lemma and for $\ell \in \Omega(\pi)$, we have that

$$\begin{aligned}\langle W \cdot a_\ell, \psi_\ell\rangle &= P_W(\ell) \int_{\mathbb{R}^m} \overline{\psi_\ell(E_{\mathfrak{B}\cap K}(t))\chi_\ell(E_{\mathfrak{B}\cap K}(t))dt} \\ &= \sum_{\alpha,\beta} \overline{p^{W^*}_{\alpha,\beta}(\ell)} \int_{\mathbb{R}^m} t^\alpha \overline{(\partial^\beta \psi_\ell(E_{\mathfrak{B}\cap K}(t)))\chi_\ell(E_{\mathfrak{B}\cap K}(t))}dt,\end{aligned}$$

where W^* denotes the image of W by the anti-automorphism of $\mathscr{U}(\mathfrak{g})$. This tells us that the function P_W is rational. □

We next prove the:

Lemma 5.3.25 *Assume that $\mathfrak{k}$ is Abelian and let $W \in \mathscr{U}_\pi(\mathfrak{g})^{\mathfrak{k}}$. Then $W \in \ker(\pi)$ if and only if $W\cdot a_\ell = \pi(W)a_\ell = 0$ for generic $\ell \in \Omega(\pi)$.*

Proof When $W \in \ker(\pi)$, it is evident that $W\cdot a_\ell = 0$ because $a_\ell \in \mathscr{H}_\pi^{-\infty}$. Suppose now that $W\cdot a_\ell = 0$ for generic $\ell \in \Omega = \Omega(\pi)$. Let $\mathfrak{z}$ be the center of $\mathfrak{g}$. Dividing if necessary by $\mathfrak{z}\cap\ker(\pi)$, we are led as usual to the case where $\dim\mathfrak{z} = 1$ and $\mathfrak{z}$ is not contained in $\ker(\pi)$. Thus, in the Jordan-Hölder sequence $\mathscr{S}$, $\mathfrak{g}_1 = \mathfrak{z}$ and $\mathfrak{g}_2 = \mathbb{R}Y + \mathfrak{z}$. Put $\mathfrak{g} = \mathbb{R}X + \mathfrak{g}'$, where $\mathfrak{g}'$ denotes the centralizer of Y in $\mathfrak{g}$, $G' = \exp(\mathfrak{g}')$, $\pi' = \mathrm{ind}_{B[\ell]}^{G'}\chi_\ell$ and denote by ℓ' the restriction of $\ell \in \Omega$ to $\mathfrak{g}'$. We make use of vectors $\varphi \in \mathscr{H}_\pi^\infty$ having the form

$$\varphi(g) = \phi(t)\psi(g'),\ g = \exp(tX)g' \in G,\ t \in \mathbb{R},\ g' \in G'$$

with $\phi \in \mathscr{S}(\mathbb{R})$, $\psi \in \mathscr{H}_{\pi'}^\infty$. If $\mathfrak{k} \subset \mathfrak{g}'$, write

$$W = \sum_{j=0}^m X^j w_j,\ w_j \in \mathscr{U}(\mathfrak{g}')\ (0 \le j \le m).$$

We have

$$\langle \pi(W)a_\ell, \varphi\rangle = \sum_{j=0}^m (-1)^j \frac{d^j\overline{\phi}}{dt^j}(0)\langle \pi'(w_j)a_{\ell'}, \psi\rangle. \tag{5.3.10}$$

Hence $\pi(W)a_\ell = 0$ entails $\pi'(w_j)a_{\ell'} = 0$ for $0 \le j \le m$. Remark that the $\mathfrak{k}$-invariance of W forces the same property on w_m. Then the induction hypothesis means $w_m \in \ker(\pi')$. As this happens at almost every point $\ell \in \Omega$, we conclude that $X^m w_m \in \ker(\pi)$. Repeating this argument, we see that W belongs to $\ker(\pi)$.

If $\mathfrak{k} \not\subset \mathfrak{g}'$, we choose X in $\mathfrak{k}$ so that $X\cdot a_\ell = i\ell(X)a_\ell$. Let $\mathfrak{k} = \mathbb{R}X + \mathfrak{k}'$, $\mathfrak{k}' = \mathfrak{k}\cap\mathfrak{g}'$ and $K' = \exp(\mathfrak{k}')$. Let us write

$$W = \sum_{j=0}^m w_j X^j,\ w_j \in \mathscr{U}(\mathfrak{g}')\ (0 \le j \le m).$$

Since W is $\mathfrak{k}$-invariant, all $w_j, 0 \leq j \leq m$, are $\mathfrak{k}$-invariant too. Then we have

$$\langle \pi(W)a_\ell, \varphi\rangle = \sum_{j=0}^{m}(i\ell(X))^j\Big(\int_{\mathbb{R}} \overline{\phi(t)e^{it\ell(X)}}dt\Big)\langle \pi'(w_j)a_{\ell'}, \psi\rangle,$$

where

$$\langle a_{\ell'}, \psi\rangle = \int_{K'/(K'\cap B[\ell])} \overline{\psi(k')\chi_\ell(k')}d\dot{k}'.$$

Hence $\pi(W)a_\ell = 0$ entails $\pi'(w_j)a_{\ell'} = 0$ for $0 \leq j \leq m$. Then the induction hypothesis says that $w_j, 0 \leq j \leq m$, belong to $\ker(\pi')$ at almost every point $\ell \in \Omega$ and finally W belongs to $\ker(\pi)$. □

We now prove the following result.

Theorem 5.3.26 *Let $G = \exp \mathfrak{g}$ be a nilpotent Lie group and $K = \exp \mathfrak{k}$ an Abelian subgroup of G. Let $\pi \in \widehat{G}$ be realized as $\pi = ind_{B[\ell]}^G \chi_\ell$ by means of the Vergne polarization $\mathfrak{b}[\ell]$, $B[\ell] = \exp(\mathfrak{b}[\ell])$, constructed from the flag $\mathscr{S}$ at the point $\ell \in \Omega = \Omega(\pi)$. Let us define $a_\ell \in (\mathscr{H}_\pi^{-\infty})^{K,\chi_\ell}$ by the formula*

$$\langle a_\ell, \varphi\rangle = \int_{K/(K\cap B[\ell])} \overline{\varphi(k)\chi_\ell(k)}d\dot{k},\ \varphi \in \mathscr{H}_\pi^\infty.$$

Let $W \in \mathscr{U}_\pi(\mathfrak{g})^{\mathfrak{k}}$ satisfying

$$W{\cdot}a_\ell = P_W(\ell)a_\ell$$

almost everywhere on Ω with a rational function $P_W(\ell)$. Then $P_W(\ell)$ is a polynomial function. In particular, the polynomial conjecture holds when $\mathfrak{k}$ is Abelian.

Proof Let us proceed as usual by induction on $n = \dim G$. When $n = 1$ or $G = K$, there is nothing to do. Suppose $n > 1$ and $G \neq K$. We may moreover reduce to the case where the center $\mathfrak{z}$ of $\mathfrak{g}$ is one-dimensional and is not contained in $\ker(\pi)$. Thus, $\mathfrak{g}_1 = \mathfrak{z}$ in the flag $\mathscr{S}$. We put there $\mathfrak{g}_2 = \mathbb{R}Y + \mathfrak{z}$ and let $\mathfrak{g}'$ be the centralizer of Y in $\mathfrak{g}$. Let $G' = \exp(\mathfrak{g}')$ and $p' : \mathfrak{g}^* \to (\mathfrak{g}')^*$ be the restriction mapping. Suppose first that $\mathfrak{k} \subset \mathfrak{g}'$. Because $W{\cdot}a_\ell = P_W(\ell)a_\ell$ almost everywhere on Ω, W belongs modulo $\ker(\pi)$ to $\mathscr{U}(\mathfrak{g}')$. Then following the beginning of Sect. 5.3.3, the induction hypothesis says that the denominator of the rational function $P_W(\ell)$ is a polynomial of the Corwin-Greenleaf e-central element $\ell(Y)$ for G' which separates the one-parameter family of G'-orbits in $\Omega' = p'(\Omega)$.

Suppose next that $\mathfrak{k} \not\subset \mathfrak{g}'$. Let $\mathfrak{k}' = \mathfrak{k} \cap \mathfrak{g}'$, $K' = \exp(\mathfrak{k}')$ and $\mathfrak{k} = \mathbb{R}X + \mathfrak{k}'$. Then for $\varphi \in \mathscr{H}_\pi^\infty$ having the form

$$\varphi(\exp(tX)g') = \phi(t)\psi(g')\ (t \in \mathbb{R}, g' \in G')$$

with $\phi \in \mathscr{S}(\mathbb{R})$ and $\psi \in \mathscr{H}_{\pi'}^{\infty}$, where $\pi' = \mathrm{ind}_{B[\ell]}^{G'}\chi_\ell$, we have

$$\langle a_\ell, \varphi\rangle = \int_{\mathbb{R}} \overline{\phi(t)e^{it\ell(X)}}dt \int_{K'/(K'\cap B[\ell])} \overline{\psi(g')\chi_\ell(g')}d\dot{g}'.$$

Now, if we write

$$W = \sum_{j=0}^{r'} W_j X^j,\ \ W_j \in \mathscr{U}(\mathfrak{g}')(0 \le j \le r'),$$

we see immediately that $[\mathfrak{k}, W_j] \subset \ker(\pi)(0 \le j \le r')$, because $\mathfrak{k}$ is Abelian. Hence

$$P_W(\ell) = \sum_{j=0}^{r'} (i\ell(X))^j P_{W_j}(\ell).$$

From the $\mathfrak{k}$-invariance, W_j acts on the functions $\psi \in \mathscr{H}_{\pi'}^{\infty}$ and the induction hypothesis says that the denominator of $P_W(\ell)$ is a polynomial function in the variable $\ell(Y)$.

Now we choose another ideal $\widetilde{\mathfrak{g}}$ of codimension 1 in $\mathfrak{g}$ and put $\widetilde{G} = \exp(\widetilde{\mathfrak{g}})$. Suppose first that Ω is not saturated with respect to $\widetilde{\mathfrak{g}}$. Let $\mathfrak{g} = \mathbb{R}\widetilde{X} + \widetilde{\mathfrak{g}}$. Then there exists in $\ker(\pi)$ an element $\eta = \widetilde{X} + \zeta$ with $\zeta \in \mathscr{U}(\widetilde{\mathfrak{g}})$. Using this η, we can eliminate from W the part outside $\mathscr{U}(\widetilde{\mathfrak{g}})$. That is to say, we can suppose that W belongs to $\mathscr{U}(\widetilde{\mathfrak{g}})$. Since the restriction $\widetilde{\pi}$ of π to $\widetilde{G} = \exp(\widetilde{\mathfrak{g}})$ is irreducible, the claim follows directly from the induction hypothesis if $\mathfrak{k} \subset \widetilde{\mathfrak{g}}$. Suppose that $\mathfrak{k} \not\subset \widetilde{\mathfrak{g}}$. Let $\widetilde{\mathfrak{k}} = \mathfrak{k} \cap \widetilde{\mathfrak{g}}$, $\mathfrak{k} = \mathbb{R}\widetilde{X} + \widetilde{\mathfrak{k}}$ and $\widetilde{K} = \exp(\widetilde{\mathfrak{k}})$. Let $\widetilde{\ell} = \ell|_{\widetilde{\mathfrak{g}}}$, $\mathfrak{b}[\widetilde{\ell}] = \mathfrak{b}[\ell] \cap \widetilde{\mathfrak{g}}$, $B[\widetilde{\ell}] = \exp(\mathfrak{b}[\widetilde{\ell}])$ for $\ell \in \Omega$ and $\widetilde{\Omega} = \Omega|_{\widetilde{\mathfrak{g}}}$. We remark that $\mathfrak{b}[\widetilde{\ell}]$ is the Vergne polarization at $\widetilde{\ell}$ constructed by means of the flag $\mathscr{S} \cap \widetilde{\mathfrak{g}}$. We identify $\mathscr{H}_{\pi}^{-\infty}$ with $\mathscr{H}_{\widetilde{\pi}}^{-\infty}$. Taking a Malcev basis in $\mathfrak{g}$ relative to $\mathfrak{b}[\ell]$ which contains a Malcev basis in $\mathfrak{k}$ relative to $\mathfrak{k} \cap \mathfrak{b}[\ell]$, we identify the space $\mathscr{H}_\pi$ of π with $\mathbb{R}^m$, $m = \dim(\mathfrak{g}/\mathfrak{b}[\ell])$. Since

$$K/(K \cap B[\ell]) \simeq KB[\ell]/B[\ell] = \exp(\mathbb{R}\widetilde{X})\widetilde{K}B[\ell]/B[\ell],$$

there are two possibilities: either

$$K/(K \cap B[\ell]) \simeq \widetilde{K}/(\widetilde{K} \cap B[\widetilde{\ell}])$$

or

$$K/(K \cap B[\ell]) \simeq \exp(\mathbb{R}\widetilde{X})\widetilde{K}B[\ell]/B[\ell] \simeq \exp(\mathbb{R}\widetilde{X}) \times \widetilde{K}/(\widetilde{K} \cap B[\ell])$$

for almost all $\ell \in \Omega$. In the former case, the generalized vector a_ℓ of π is identified with the generalized vector $a_{\widetilde{\ell}}$ of $\widetilde{\pi}$. In the latter case, for $\varphi \in \mathscr{H}_\pi^{\infty}$ verifying

$$\varphi(\exp(t\widetilde{X})\widetilde{g}) = \phi(t)\psi(\widetilde{g}),\ \ t \in \mathbb{R}, \widetilde{g} \in \widetilde{G},$$

with $\phi \in C_c(\mathbb{R})$, $\psi \in \mathscr{H}_{\widetilde{\pi}}^{\infty}$, we have by the $\mathfrak{k}$-invariance of W that

$$\begin{aligned}\langle W\cdot a_\ell, \varphi\rangle &= \Big(\int_{\mathbb{R}} \overline{\phi(t)e^{it\ell(\widetilde{X})}dt}\Big)\langle W\cdot a_{\widetilde{\ell}}, \psi\rangle \\ &= P_W(\widetilde{\ell})\Big(\int_{\mathbb{R}} \overline{\phi(t)e^{it\ell(\widetilde{X})}dt}\Big)\langle a_{\widetilde{\ell}}, \psi\rangle = P_W(\widetilde{\ell})\langle a_\ell, \varphi\rangle.\end{aligned}$$

In any way, the induction hypothesis tells us that $P_W(\ell) = P_W(\widetilde{\ell})$ is a polynomial function.

Suppose finally that Ω is saturated with respect to $\widetilde{\mathfrak{g}}$. We can assume that $\mathfrak{b}[\ell]$ is in $\widetilde{\mathfrak{g}}$. Put $\widetilde{\pi} = \mathrm{ind}_{B[\ell]}^{\widetilde{G}}\chi_\ell$. We first consider the case where $\mathfrak{k} \subset \widetilde{\mathfrak{g}}$. Because $W\cdot a_\ell = P_W(\ell)a_\ell$ for generic $\ell \in \Omega$, it follows that $W \in \mathscr{U}(\widetilde{\mathfrak{g}})$ modulo $\ker(\pi)$. Indeed, write $W = \sum_{j=0}^{m} X^j w_j$, $w_j \in \mathscr{U}(\widetilde{\mathfrak{g}})(0 \le j \le m)$. Looking at formula (5.3.10) and choosing ϕ such that

$$\frac{d^j\bar{\phi}}{dt^j}(0) = 0\ (0 \le j \le m-1),\ \frac{d^m\bar{\phi}}{dt^m}(0) = 1,$$

we arrive at $w_m \in \ker(\pi)$ by Lemma 5.3.25. From the disintegration

$$\pi|_{\widetilde{G}} \simeq \int_{\mathbb{R}}^{\oplus} \widetilde{\pi}_t dt,\ \widetilde{\pi}_t \in \widehat{\widetilde{G}}, t \in \mathbb{R},$$

where $\widetilde{\pi}_0 = \widetilde{\pi}, \widetilde{\pi}_t = \exp(t\widetilde{X})\cdot\widetilde{\pi}$, we deduce by induction hypothesis that the denominator of $P_W(\ell)$ is a polynomial of $\ell(\widetilde{\kappa})$, $\widetilde{\kappa}$ being an e-central element on the projection of Ω in $(\widetilde{\mathfrak{g}})^*$. But this means that $P_W(\ell)$ is a polynomial function, since $\widetilde{\kappa}$ is not a polynomial in Y. We next consider the case where $\mathfrak{k} \not\subset \widetilde{\mathfrak{g}}$ and write $\mathfrak{k} = \mathbb{R}\widetilde{X} + \widetilde{\mathfrak{k}}$ with $\widetilde{\mathfrak{k}} = \mathfrak{k} \cap \widetilde{\mathfrak{g}}$. Put $\widetilde{K} = \exp(\widetilde{\mathfrak{k}})$ as before. For $\varphi \in \mathscr{H}_\pi^\infty$ having the form

$$\varphi(\exp(t\widetilde{X})\widetilde{g}) = \phi(t)\psi(\widetilde{g}),\ t \in \mathbb{R}, \widetilde{g} \in \widetilde{G},$$

with $\phi \in \mathscr{S}(\mathbb{R})$, $\psi \in \mathscr{H}_{\widetilde{\pi}}^\infty$, we have

$$\langle a_\ell, \varphi\rangle = \int_{\mathbb{R}} \overline{\phi(t)e^{it\ell(\widetilde{X})}dt} \int_{\widetilde{K}/(\widetilde{K}\cap B[\ell])} \overline{\psi(\widetilde{g})\chi_\ell(\widetilde{g})d\dot{\widetilde{g}}}.$$

Therefore, if $W \in \mathscr{U}(\widetilde{\mathfrak{g}})$, we conclude as before that $P_W(\ell)$ is a polynomial function. If $W \notin \mathscr{U}(\widetilde{\mathfrak{g}})$, we write

$$W = \sum_{j=0}^{v} W_j\widetilde{X}^j,\ W_j \in \mathscr{U}(\widetilde{\mathfrak{g}}), 0 \le j \le v.$$

Since $\mathfrak{k}$ is Abelian, we see for every $Y' \in \mathfrak{k}$ that

$$[Y', W] = \sum_{j=0}^{v} [Y', W_j]\widetilde{X}^j,$$

from which immediately $W_j, 0 \leq j \leq v$, are all $\mathfrak{k}$-invariant. Then the above argument proves that the denominators of $P_{W_j}(\ell)$ are polynomial functions in $\ell(\widetilde{\kappa})$, and consequently $P_{W_j}(\ell)$ are polynomial functions. Hence

$$P_W(\ell) = \sum_{j=0}^{v} (i\ell(\widetilde{X}))^j P_{W_j}(\ell)$$

is a polynomial function.

Now, let us return to the theorem's set-up. When $\pi|_K$ has finite multiplicities, the mapping $\mathscr{U}_\pi(\mathfrak{g})^{\mathfrak{k}} \ni W \mapsto P_W$ gives, by passage to the quotient, an injective homomorphism of algebras from $D_\pi(G)^K$ into the algebra $\mathbb{C}[\Omega(\pi)]^K$ of K-invariant polynomial functions on the orbit $\Omega(\pi)$ (cf. [16]). There remains to prove the surjectivity of the homomorphism

$$\Theta : \mathscr{U}_\pi(\mathfrak{g})^{\mathfrak{k}} \ni W \mapsto P_W(\ell) \in \mathbb{C}[\Omega]^K,$$

We first prove the following result:

Lemma 5.3.27 *Assume that $K = \exp \mathfrak{k}$ is Abelian. Let $q(x) \in \mathbb{C}[\Omega]^K$. Suppose that there exists $u(x) \in \mathbb{C}[\Omega]$ belonging to the image of Θ such that $u(x)q(x)$ belongs to the image of Θ. Then $q(x)$ itself belongs to the image of Θ.*

Proof We keep the notations of the previous lemma. When $\mathfrak{k} \subset \mathfrak{g}'$, our condition requires that $q(x), u(x)$ do not contain the variable $x_r = \ell(X)$. Let

$$\pi|_{G'} \simeq \int_{\mathbb{R}}^{\oplus} \pi'_t dt, \ \Omega|_{\mathfrak{g}'} = \sqcup_{t\in\mathbb{R}} \omega_t, \ t = \ell(Y),$$

be the usual decompositions. For almost all $t \in \mathbb{R}$, there exists by induction hypothesis $W_t \in \mathscr{U}_{\pi'_t}(\mathfrak{g}')^{\mathfrak{k}}$ satisfying $P_{W_t}(\ell') = q(x)|_{x_1=t}$, where $x = (x_1, \ldots)$. Actually, x_1 corresponds to the variable associated to the vector Y. As the Vergne polarization depends rationally on ℓ, W_t depends rationally on t. In fact, we take a reference point $\ell_0 \in \Omega$ such that $\ell_0(Y) = 0$ and consider the section

$$\Omega/G' \simeq \mathbb{R} \ni t \mapsto \ell_t = \exp(-tX) \cdot \ell_0 \in \Omega.$$

Let $\{Y_1(t), \ldots, Y_{m_1}(t)\}$, $m_1 = \dim \mathfrak{b}[\ell_t]$, be a basis of $\mathfrak{b}[\ell_t]$ and $\{X_1, \ldots, X_{m_2}\}$, $m_2 = \dim \mathfrak{k}/(\mathfrak{k} \cap \mathfrak{b}[\ell_t])$, a Malcev basis of $\mathfrak{k}$ relative to $\mathfrak{k} \cap \mathfrak{b}[\ell_t]$ for generic $t \in \mathbb{R}$. For $I = (i_1, \ldots, i_{m_2})$ and $J = (j_1, \ldots, j_{m_1})$, set

$$X^I = X_1^{i_1} \cdots X_{m_2}^{i_{m_2}}, \ Y^J = (Y_1(t))^{j_1} \cdots (Y_{m_1}(t))^{j_{m_1}}.$$

Then W_t is written as $W_t = \sum_{I,J} c_{I,J}(t) X^I Y^J$ in terms of certain rational functions $c_{I,J}(t)$ of $t \in \mathbb{R}$. Multiplying W_t by its denominator and substituting t with $-iY$, we see that there exists a polynomial $s(x_1)$ of $x_1 = \ell(Y)$ so that the product $s(x_1)q(x)$ belongs to the image of Θ.

When $\mathfrak{k} \not\subset \mathfrak{g}'$, we write

$$q(x) = \sum_{j=0}^{m} q_j(x_1, \ldots, x_{r-1}) x_r^j, \; u(x) = \sum_{j=0}^{m'} u_j(x_1, \ldots, x_{r-1}) x_r^j$$

with certain polynomials $q_j (0 \le j \le m)$, $u_j (0 \le j \le m')$ of $(x_1, \ldots, x_{r-1})$ which are all $\mathfrak{k}$-invariant. Hence it follows that the $u_j (0 \le j \le m')$ belong to the image of Θ. As

$$u(x)q(x) = \sum_{j=0}^{m+m'} p_j(x_1, \ldots, x_{r-1}) x_r^j, \; p_j = \sum_{0 \le k \le j} u_k q_{j-k},$$

belongs to the image of Θ, $p_{m+m'} = u_{m'} q_m$ does too. For generic $t \in \mathbb{R}$, there exists by induction hypothesis $w_m(t) \in \mathscr{U}_{\pi_t'}(\mathfrak{g}')^{\mathfrak{k}'}$ verifying $P_{w_m(t)}(\ell') = q_m(\ell')$ for almost all $\ell' \in \omega_t$. Since $q(x)$ is K-invariant,

$$P_{W_{\exp(tX)\cdot y'}}(\exp(tX)\cdot\ell) = q(\exp(tX)\cdot x) = q(x)$$

for all $t \in \mathbb{R}$. At the same time the definition of the function $P_{W'}(\ell')$, $W' \in \mathscr{U}_{\pi_t'}(\mathfrak{g}')^{\mathfrak{k}'}$, gives

$$P_{\exp(sX)\cdot w_m(t)}(\exp(sX)\cdot\ell') = P_{w_m(t)}(\ell'), \; \forall\, s \in \mathbb{R}.$$

Hence $w_m(\exp(sX)\cdot\ell') = \exp(sX)\cdot(w_m(t))$ and

$$\begin{aligned}
&\langle P_{w_m(\exp(sX)\cdot\ell')}(\exp(sX)\cdot\ell')a_{\exp(sX)\cdot\ell'}, \varphi\rangle = \langle(\exp(sX)\cdot(w_m(\ell'))a_{\exp(sX)\cdot\ell'}, \varphi\rangle \\
&= \int_{K'/(K'\cap B[\exp(sX)\cdot\ell'])} \overline{\exp(sX)\cdot {}^t(w_m(\ell'))\varphi(k')\chi_{\exp(sX)\cdot\ell'}(k')}d\dot{k}' \\
&= \int_{K'/(K'\cap B[\ell'])} \overline{\exp(sX)\cdot {}^t(\overline{w_m(\ell')})\varphi(k')\chi_{\ell'}(k')}d\dot{k}' \\
&= \langle(\exp(sX)\cdot(w_m(\ell')))a_{\ell'}, \varphi\rangle.
\end{aligned}$$

Namely,

$$\begin{aligned}
&\langle P_{w_m(\exp(sX)\cdot\ell')}(\exp(sX)\cdot\ell')a_{\ell'}, \varphi\rangle = \langle(\exp(sX)\cdot(w_m(\ell')))a_{\ell'}, \varphi\rangle \\
&= \langle P_{\exp(sX)\cdot(w_m(\ell'))}(\ell')a_{\ell'}, \varphi\rangle
\end{aligned}$$

for all $\varphi \in \mathscr{H}_{\pi'}^{\infty}$. In this way,

$$P_{\exp(sX)\cdot(w_m(\ell'))}(\ell') = P_{w_m(\exp(sX)\cdot\ell')}(\exp(sX)\cdot\ell') = q(x).$$

Differentiating this equality in s, we find by the previous lemma $[X, w_m(t)] \in \ker(\pi_t')$ for almost all $t = \ell(Y)$, that is to say $w_m(t) \in \mathscr{U}_\pi(\mathfrak{g})^{\mathfrak{k}}$. Moreover, we find as before that $w_m(t)$ depends rationally on $t = \ell(Y) \in \mathbb{R}$. Summing up, there exists a polynomial $s_m(x_1)$ of $x_1 = \ell(Y)$ so that the product $s_m(x_1)q_m(x_1, \ldots, x_{r-1})$ belongs to the image of Θ. Next, $p_{m+m'-1} = u_{m'-1}q_m + u_{m'}q_{m-1}$ belongs to the image of Θ, as does $s_m(x_1)u_{m'}q_{m-1}$. We can repeat the argument to show that there exists a polynomial $s_{m-1}(x_1)$ of x_1 so that $s_{m-1}q_{m-1}$ is in the image of Θ. We are able to continue the process to conclude that there exists a polynomial $s(x_1)$ of $x_1 = \ell(Y)$ so that the product $s(x_1)q(x)$ belongs to the image of Θ.

Now, we take another ideal $\widetilde{\mathfrak{g}} \neq \mathfrak{g}'$ of codimension 1 in $\mathfrak{g}$. Let $\widetilde{G} = \exp(\widetilde{\mathfrak{g}})$. For $\ell \in \mathfrak{g}^*$, put $\widetilde{\ell} = \ell|_{\widetilde{\mathfrak{g}}} \in (\widetilde{\mathfrak{g}})^*$. Suppose first that Ω is not saturated with respect to $\widetilde{\mathfrak{g}}$. Let $\widetilde{\Omega}$ be the projection image of Ω in $(\widetilde{\mathfrak{g}})^*$, which is the $\widetilde{G}$-orbit corresponding to the irreducible representation $\widetilde{\pi} = \pi|_{\widetilde{G}}$ of $\widetilde{G}$. For $\ell \in \Omega$, put $\widetilde{\ell} = \ell|_{\widetilde{\mathfrak{g}}} \in \widetilde{\Omega}$. If $\mathfrak{k} \subset \widetilde{\mathfrak{g}}$, it suffices to apply the induction hypothesis to the triple $(\widetilde{G}, \widetilde{\pi}, K)$.

If $\mathfrak{k} \not\subset \widetilde{\mathfrak{g}}$, let $\widetilde{\mathfrak{k}} = \mathfrak{k} \cap \widetilde{\mathfrak{g}}$. The induction hypothesis says that there exists $W \in \mathscr{U}_{\widetilde{\pi}}(\widetilde{\mathfrak{g}})^{\widetilde{\mathfrak{k}}}$ verifying $P_W(\widetilde{\ell}) = q(x)$ almost everywhere on $\widetilde{\Omega}$. Since $q(x)$ is K-invariant, so is the case for W, namely $W \in \mathscr{U}_\pi(\mathfrak{g})^{\mathfrak{k}}$. Suppose next that Ω is saturated with respect to $\widetilde{\mathfrak{g}}$. We can suppose that the Vergne polarization $\mathfrak{b}[\ell]$ is contained in $\widetilde{\mathfrak{g}}$ and $q(x)$ is rewritten [10] as

$$q(x) = q(x_1, \ldots, x_{i-1}, y', x_{i+1}, \ldots, x_{r-1})$$

with

$$y' = x_i + \varphi(x_1, \ldots, x_{i-1}) = P_{Y'}(\ell)$$

for a certain

$$Y' = -\sqrt{-1}X_{s_i} + \sum c_{k_1,\ldots,k_{i-1}} X_{s_1}^{k_1} \cdots X_{s_{i-1}}^{k_{i-1}}$$

belonging to $\mathscr{U}_\pi(\mathfrak{g})^{\widetilde{\mathfrak{g}}}$ with $c_{k_1,\ldots,k_{i-1}} \in \mathbb{C}$. Then there exists a polynomial $r(y')$ of y' so that $r(y')q(x)$ belongs to the image of Θ. Summing up, there exist two elements W_1, W_2 in $\mathscr{U}_\pi(\mathfrak{g})^{\mathfrak{k}} \cap E$ such that

$$P_{W_1}(\ell) = s(x_1)q(x), \;\; P_{W_2}(\ell) = r(y')q(x)$$

for generic $\ell \in \Omega$. Taking into account the previous lemma,

$$r(Y')W_1 \equiv s(-iY)W_2 \tag{5.3.11}$$

modulo $\ker(\pi)$. As $\pi|_E$ is faithful and $r(Y')$ is not divisible by $s(-iY)$, W_1 must be divisible by $s(-iY)$ because Y is central in $\mathfrak{g}'$. To see this, it suffices to invoke induction on the degree d_1 of W_1. If $d_1 = 1$, there is nothing to prove. If $d_1 > 1$, we expand identity (5.3.11) according to a vector $X_j \not\equiv Y$ modulo $\ker(\pi)$ and use the induction hypothesis on the factors that appear, which are of smaller degrees.

Write now $W_1 = s(-iY)W_0$ and $s(-iY) = \sum_{m\geq 0} a_m Y^m$. We easily show that $W_0 \in \mathscr{U}_\pi(\mathfrak{g})^{\mathfrak{k}}$. Indeed, the result is immediate if $\mathfrak{k} \subset \mathfrak{g}'$. Otherwise we can suppose $X \in \mathfrak{k}$, and hence

$$\ker(\pi) \ni \sum_{m\geq 1} m a_m Y^{m-1} Z W_0 + \sum_{m\geq 0} a_m Y^m [X, W_0].$$

Since $\pi(Y)$ acts as a multiplication operator, the result follows. These observations imply that $q(x)$ belongs to the image of Θ, which ends the lemma's proof. □

Assume now that $\pi|_K$ has finite multiplicities and we can show the surjectivity of Θ. Consider an arbitrary element $q(x)$ in $\mathbb{C}[\Omega]^K$. This time instead of $\widetilde{\mathfrak{g}}$ we use the $\mathfrak{k}_{n-1}$ appearing in sequence (5.3.7). It is an ideal of codimension 1 in $\mathfrak{g}$ containing $\mathfrak{k}$. Then by Proposition 5.3.15 there exists a polynomial $s(y')$ of $y' = P_{Y'}(\ell)$ such that $s(y')q(x)$ belongs to the image of Θ. Therefore, by the above lemma $q(x)$ belongs to the image of Θ, and the proof finishes. □

Remark 5.3.28 Assume that $\mathscr{I} \subset S_\Omega$ regarding the flag $\mathscr{S}$. Then the projection of $\Omega(\pi)$ on $\mathfrak{k}^*$ coincides with the whole $\mathfrak{k}^*$, and our algebra $\mathscr{U}_\pi(\mathfrak{g})^{\mathfrak{k}}$ is isomorphic to the center of $\mathscr{U}(\mathfrak{k})$. It follows that the algebra $D_\pi(G)^K$ is isomorphic to the algebra $\mathbb{C}[\Omega(\pi)]^K$.

5.3.6 Examples

Here we give some examples. Let us begin with some elements in $\ker(\pi)$.

Example 5.3.29 Let $\mathfrak{g} = \mathbb{R}\text{-span}(X, P_1, P_2, Q, Y, Z)$ with non-zero brackets

$$[X, P_1] = P_2, [X, Y] = Z, [P_1, Q] = Y, [P_2, Q] = Z.$$

Let $\pi \in \widehat{G}$ such that $Z^* \in \Omega(\pi)$ and $\mathfrak{k} = \mathbb{R}\text{-span}(P_1, Y, Z)$. Then a direct computation shows that an element $\ell \in \Omega(\pi)$ is written as

$$\ell = Z^* + x_1 Y^* + x_2 Q^* + x_3 (P_2)^* + x_1 x_3 (P_1)^* + x_4 X^*,\ x_j \in \mathbb{R}(1 \leq j \leq 4).$$

From this, $P_1 - x_1 P_2 - x_3 Y \in \mathfrak{g}(\ell)$. On the other hand, we have

$$\langle a_\ell, \varphi\rangle = \int_{\mathbb{R}} \overline{\varphi(\exp(tP_1))} e^{itx_1x_3} dt$$

for $\varphi \in \mathscr{H}_\pi^\infty$. So,

$$\langle P_2 a_\ell, \varphi\rangle = \frac{d}{ds}\int_{\mathbb{R}} \overline{\varphi(\exp(sP_2)\exp(tP_1))}e^{itx_1x_3}dt\Big|_{s=0}$$

$$= \frac{d}{ds}\int_{\mathbb{R}} \overline{\varphi\left(\exp\left(\left(t+\frac{s}{x_1}\right)P_1\right)\exp\left(sP_2 - \frac{s}{x_1}P_1\right)\right)}e^{itx_1x_3}dt\Big|_{s=0}$$

$$= \frac{d}{ds}e^{-is(x_3+1)}\int_{\mathbb{R}} \overline{\varphi\left(\exp\left(\left(t+\frac{s}{x_1}\right)P_1\right)\right)}e^{itx_1x_3}dt\Big|_{s=0}$$

$$= \frac{d}{ds}\int_{\mathbb{R}} \overline{\varphi\left(\exp\left(\left(t+\frac{s}{x_1}\right)P_1\right)\right)}\Big|_{s=0}$$

$$= \frac{d}{ds}\int_{\mathbb{R}} \overline{\varphi(\exp(tP_1))}e^{i\left(t-\frac{s}{x_1}\right)x_1x_3}dt\Big|_{s=0}$$

$$= \left(\frac{d}{ds}e^{-isx_3}\right)\Big|_{s=0}\int_{\mathbb{R}} \overline{\varphi(\exp(tP_1))}e^{itx_1}dt = -ix_3\langle a_\ell, \varphi\rangle.$$

for $\ell \in \Omega(\pi)$ satisfying $x_1 = \ell(Y) \neq 0$. Put $\Phi(x,t) = \varphi(\exp(xX)\exp(tP_2))$ for $(x,t) \in \mathbb{R}^2$ and compute:

$$(P_1\Phi)(x,t) = \frac{d}{ds}\varphi(\exp(-sP_1)\exp(xX)\exp(tP_2))|_{s=0}$$

$$= \frac{d}{ds}\varphi(\exp(xX)\exp(-sP_1 + sxP_2)\exp(tP_2))|_{s=0}$$

$$= \frac{d}{ds}\varphi(\exp(xX)\exp((t+sx)P_2)\exp(-sP_1 + sx_1P_2$$

$$+ sx_3Y)\exp(-sx_1P_2 - sx_3Y))|_{s=0}$$

$$= \frac{d}{ds}\varphi(\exp(xX)\exp((t+sx-sx_1)P_2))|_{s=0} = (x-x_1)\frac{\partial\Phi}{\partial t}(x,t),$$

$$(P_2\Phi)(x,t) = \frac{d}{ds}\varphi(\exp(-sP_2)\exp(xX)\exp(tP_2))|_{s=0} = -\frac{\partial\Phi}{\partial t}(x,t),$$

$$(Y\Phi)(x,t) = \frac{d}{ds}\varphi(\exp(-sY)\exp(xX)\exp(tP_2))|_{s=0}$$

$$= \frac{d}{ds}\varphi(\exp(xX)\exp(-sY + xsZ)\exp(tP_2))|_{s=0} = i(x_1 - x)\Phi(x,t).$$

Taking these observations into account, $P_1 + iYP_2$ belongs to $\ker(\pi)$.

Example 5.3.30 Let $\mathfrak{g} = \mathbb{R}\text{-span}(X, P_1, P_2, P_3, Q_1, Q_2, Y, Z)$ with non-zero brackets

$$[X, P_1] = P_2, [X, P_2] = P_3, [P_1, Q_1] = Q_2, [X, Q_2] = [P_2, Q_1] = Y, [X, Y] = [P_3, Q_1] = Z.$$

Let $\pi \in \widehat{G}$ such that $Z^* \in \Omega(\pi)$ and $\mathfrak{k} = \mathbb{R}\text{-span}(P_1, Y, Z)$. Then a direct computation shows that an element $\ell \in \Omega(\pi)$ is written as

$$\ell = Z^* + x_1 Y^* + \frac{x_1^2}{2} Q_2^* + x_2 Q_1^* + x_3 P_3^* \\ + x_1 x_3 P_2^* + \frac{x_1^2 x_3}{2} P_1^* + x_4 X^*, \; x_j \in \mathbb{R}\ (1 \leq j \leq 4).$$

From this, $P_1 - \frac{x_1^2}{2} P_3 - x_1 x_3 Y$, $P_2 - x_1 P_3 - x_3 Y$ belong to $\mathfrak{g}(\ell)$. On the other hand,we have

$$\langle a_\ell, \varphi \rangle = \int_{\mathbb{R}} \overline{\varphi(\exp(tP_1))} e^{\frac{i x_1^2 x_3}{2} t} dt$$

for $\varphi \in \mathscr{H}_\pi^\infty$. So,

$$\begin{aligned}
\langle P_3 a_\ell, \varphi \rangle &= \frac{d}{ds} \int_{\mathbb{R}} \overline{\varphi(\exp(sP_3)\exp(tP_1))} e^{\frac{i x_1^2 x_3}{2} t} dt \Big|_{s=0} \\
&= \frac{d}{ds} \int_{\mathbb{R}} \overline{\varphi\left(\exp(tP_1)\exp\left(\frac{2s}{x_1^2} P_1\right)\exp\left(sP_3 - \frac{2s}{x_1^2} P_1\right)\right)} e^{\frac{i x_1^2 x_3}{2} t} dt \Big|_{s=0} \\
&= \frac{d}{ds} \int_{\mathbb{R}} \overline{\varphi\left(\exp\left(\left(t + \frac{2s}{x_1^2}\right) P_1\right)\right)} e^{\frac{i x_1^2 x_3}{2} t} e^{2isx_3} dt \Big|_{s=0} \\
&= \frac{d}{ds} \int_{\mathbb{R}} \overline{\varphi(\exp(tP_1))} e^{\frac{i x_1^2 x_3}{2}\left(t - \frac{2s}{x_1^2}\right)} e^{2isx_3} dt \Big|_{s=0} \\
&= \left(\frac{d}{ds} e^{isx_3}\right)\Big|_{s=0} \int_{\mathbb{R}} \overline{\varphi(\exp(tP_1))} e^{\frac{i x_1^2 x_3}{2} t} dt = i x_3 \langle a_\ell, \varphi \rangle.
\end{aligned}$$

for $\ell \in \Omega(\pi)$ satisfying $x_1 = \ell(Y) \neq 0$. Put $\Phi(x, t) = \varphi(\exp(xX)\exp(tP_3))$ for $(x, t) \in \mathbb{R}^2$ and compute:

$$\begin{aligned}
(P_1 \Phi)(x, t) &= \frac{d}{ds} \varphi(\exp(-sP_1)\exp(xX)\exp(tP_3)) \Big|_{s=0} \\
&= \frac{d}{ds} \varphi\left(\exp(xX)\exp\left(-sP_1 + sxP_2 - \frac{sx^2}{2} P_3\right)\exp(tP_3)\right)\Big|_{s=0} \\
&= \frac{d}{ds} \varphi\left(\exp(xX)\exp\left(\left(t - \frac{sx^2}{2}\right) P_3\right)\exp(sxP_2 - sP_1)\right)\Big|_{s=0}
\end{aligned}$$

$$= \frac{d}{ds}\varphi\left(\exp(xX)\exp\left(\left(\left(t - \frac{sx^2}{2} + s\left(xx_1 - \frac{x_1^2}{2}\right)\right)P_3\right) \times \right.\right.$$

$$\left.\left.\times \exp\left(sxP_2 - sP_1 - s\left(xx_1 - \frac{x_1^2}{2}\right)P_3\right)\right)\right|_{s=0}$$

$$= \frac{d}{ds}\varphi\left(\exp(xX)\exp\left(\left(t - \frac{sx^2}{2} + s\left(xx_1 - \frac{x_1^2}{2}\right)\right)P_3\right)\right)\Big|_{s=0}$$

$$= \left(xx_1 - \frac{x_1^2}{2} - \frac{x^2}{2}\right)\frac{\partial\Phi}{\partial t},$$

$$(P_3\Phi)(x,t) = \frac{d}{ds}\varphi(\exp(-sP_3)\exp(xX)\exp(tP_3))\Big|_{s=0} = -\frac{\partial\Phi}{\partial t},$$

$$(Y\Phi)(x,t) = \frac{d}{ds}\varphi(\exp(-sY)\exp(xX)\exp(tP_3))\Big|_{s=0}$$

$$= \frac{d}{ds}\varphi(\exp(xX)\exp(tP_3)\exp(-sY + sxZ))\Big|_{s=0}$$

$$= \left(\frac{d}{ds}e^{-i(-sx_1+sx)}\right)\Big|_{s=0}\Phi(x,t) = i(x_1 - x)\Phi(x,t).$$

Taking these observations into account, $P_1 - \frac{Y^2}{2}P_3$ belongs to $\ker(\pi)$. Finally,

$$(P_2\Phi)(x,t) = \frac{d}{ds}\varphi(\exp(-sP_2)\exp(xX)\exp(tP_3))\Big|_{s=0}$$

$$= \frac{d}{ds}\varphi(\exp(xX)\exp(-sP_2 + sxP_3)\exp(tP_3))\Big|_{s=0}$$

$$= \frac{d}{ds}\varphi(\exp(xX)\exp((t + sx + -sx_1)P_3)\exp(-sP_2 + sx_1P_3))\Big|_{s=0}$$

$$= \frac{d}{ds}\Phi(x, t + s(x - x_1))\Big|_{s=0} = (x - x_1)\frac{\partial\Phi}{\partial t}.$$

Accordingly, $P_2 + iYP_3$ belongs to $\ker(\pi)$.

Now we give some illustrative examples for the polynomial conjecture.

Example 5.3.31 Let $\mathfrak{g} = \mathbb{R}\text{-span}(X, T, U, V, Y, Z)$ with non-zero brackets $[X, Y] = Z, [X, T] = U, [T, V] = Y, [U, V] = Z$. Let $G = \exp\mathfrak{g}$, $f = Z^* \in \mathfrak{g}^*$, $\Omega = G{\cdot}f$, $\mathfrak{k} = \mathbb{R}\text{-span}(T, Y, Z)$ and $K = \exp\mathfrak{k}$. Then $\mathfrak{k}$ is Abelian and $\mathfrak{g}(f) = \mathbb{R}T \oplus \mathbb{R}Z$. It follows that $\dim\Omega = 4$ and $\mathfrak{b} = \mathbb{R}V + \mathfrak{k}$ is a polarization of $\mathfrak{g}$ at f. We construct the irreducible unitary representation π of G associated to Ω as $\pi = \mathrm{ind}_B^G\chi_f$ where $B = \exp\mathfrak{b}$. Simple calculations show

$$\Omega = \{\ell = \alpha X^* + \beta\lambda T^* + \beta U^* + \gamma V^* + \lambda Y^* + Z^*; \alpha, \beta, \gamma, \lambda \in \mathbb{R}\}.$$

For $\ell \in \Omega$, we have

$$\mathfrak{g}(\ell) = \mathbb{R}(T - \ell(Y)U - \ell(U)Y) + \mathbb{R}Z.$$

Moreover,

$$\mathfrak{b}[\ell] = \mathbb{R} - \text{span}(T - \ell(Y)U, V, Y, Z)$$

is a polarization of $\mathfrak{g}$ at ℓ and $\mathfrak{b}[\ell|_{\mathfrak{k}}] = \mathfrak{k}$ is a polarization of $\mathfrak{k}$ at the restriction $\ell|_{\mathfrak{k}} \in \mathfrak{k}^*$. Let $p : \mathfrak{g}^* \to \mathfrak{k}^*$ be the restriction mapping. We immediately see that

$$p(\Omega) = \mathbb{R}T^* + \mathbb{R}Y^* + Z^*$$

except for the subset $(\mathbb{R} \setminus \{0\})T^* + Z^*$. Furthermore, for $\delta = aT^* + bY^* + Z^* \in p(\Omega)$ $(a \in \mathbb{R}, b \in \mathbb{R} \setminus \{0\})$, $p^{-1}(\delta)$ is a K-orbit. In this way, the restriction $\pi|_K$ is multiplicity-free. In more details, for $\delta = aT^* + bY^* + Z^* \in \mathfrak{k}^*$ $(a, b \in \mathbb{R})$, let χ_δ be the unitary character of K defined by

$$\chi_\delta(\exp(tT + yY + zZ)) = e^{i(at+by+z)}.$$

Then

$$\pi|_K \simeq \int_{\mathbb{R}^2}^{\oplus} \chi_\delta da db.$$

Define the Penney distribution a_ℓ at $\ell \in \Omega$ verifying $\ell(Y) \neq 0$ by the formula

$$a_\ell(\varphi) = \int_{\mathbb{R}} \overline{\varphi(\exp(tT))e^{i\ell(T)t}} dt$$

for $\varphi \in \mathscr{H}_\pi^\infty$.

Now we consider a Jordan-Hölder sequence

$$\mathfrak{g}_1 = \mathbb{R}Z \subset \mathfrak{g}_2 = \mathbb{R}Z + \mathbb{R}Y \subset \mathfrak{g}_3 = \mathbb{R} - \text{span}(V, Y, Z)$$
$$\subset \mathfrak{g}_4 = \mathbb{R} - \text{span}(U, V, Y, Z) \subset \mathfrak{g}_5 = \mathbb{R} - \text{span}(Z, Y, V, U, T) \subset \mathfrak{g}_6 = \mathfrak{g}.$$

of $\mathfrak{g}$ and the associated sequence of subalgebras

$$\mathfrak{k}_1 = \mathfrak{g}_1 \subset \mathfrak{k}_2 = \mathfrak{g}_2 \subset \mathfrak{k}_3 = \mathfrak{k} \subset \mathfrak{k}_4 = \mathfrak{k} + \mathbb{R}V \subset \mathfrak{k}_5 = \mathfrak{k}_4 + \mathbb{R}[X, T] \subset \mathfrak{k}_6 = \mathfrak{g}.$$

When we pass from $\mathfrak{k}_4$ to $\mathfrak{k}_5$, the element

$$\sigma = YU - ZT$$

appears in the center of the enveloping algebra $\mathscr{U}(\mathfrak{g})$ and satisfies the relation $\pi(\sigma) = 0$. Moreover, U is $\mathfrak{k}$-invariant and

$$\pi(Z)\cdot a_\ell = i a_\ell, \pi(Y)\cdot a_\ell = i\ell(Y) a_\ell, \pi(T)\cdot a_\ell = i\ell(T) a_\ell$$

for all $\ell \in \Omega$ verifying $\ell(Y) \neq 0$. Therefore $\pi(U)\cdot a_\ell = i\ell(U) a_\ell$, and the algebra $D_\pi(G)^K$ is isomorphic to the algebra $\mathbb{C}[\Omega]^K$.

Example 5.3.32 Let $\mathfrak{g} = \mathbb{R}\text{-span}(X, T, U, V, Y, Z)$ with non-zero brackets

$$[X, Y] = Z, [X, T] = U, [U, T] = Z, [T, V] = Y, [U, V] = Z.$$

Let $G = \exp \mathfrak{g}$, $f = Z^* \in \mathfrak{g}^*$, $\Omega = G\cdot f$, $\mathfrak{k} = \mathbb{R}\text{-span}(T, Y, Z)$ and $K = \exp \mathfrak{k}$. Then $\mathfrak{k}$ is Abelian and $\mathfrak{g}(f) = \mathbb{R}(T - V) \oplus \mathbb{R}Z$. It follows that $\dim \Omega = 4$ and $\mathfrak{b} = \mathbb{R}V + \mathfrak{k}$ is a polarization of $\mathfrak{g}$ at f. We construct the irreducible unitary representation π of G associated to Ω as $\pi = \mathrm{ind}_B^G \chi_f$ where $B = \exp \mathfrak{b}$. Simple calculations show

$$\Omega = \{\ell = \alpha X^* + (\beta\lambda + \gamma)T^* + \beta U^* + \gamma V^* + \lambda Y^* + Z^*; \alpha, \beta, \gamma, \lambda \in \mathbb{R}\}.$$

Let $p : \mathfrak{g}^* \to \mathfrak{k}^*$ be the restriction mapping. We immediately see that

$$p(\Omega) = \mathbb{R}T^* + \mathbb{R}Y^* + Z^*.$$

Furthermore, for $\delta = aT^* + bY^* + Z^* \in p(\Omega)$ $(a, b \in \mathbb{R})$, $p^{-1}(\delta)$ is a K-orbit. In this way the restriction $\pi|_K$ is multiplicity-free.

Now we consider a Jordan-Hölder sequence

$$\begin{aligned}\mathfrak{g}_1 &= \mathbb{R}Z \subset \mathfrak{g}_2 = \mathbb{R}Z + \mathbb{R}Y \subset \mathfrak{g}_3 = \mathbb{R} - \text{span}(V, Y, Z) \\ &\subset \mathfrak{g}_4 = \mathbb{R} - \text{span}(U, V, Y, Z) \subset \mathfrak{g}_5 = \mathbb{R} - \text{span}(Z, Y, V, U, T) \subset \mathfrak{g}_6 = \mathfrak{g}\end{aligned}$$

of $\mathfrak{g}$ and the associated sequence of subalgebras

$$\mathfrak{k}_1 = \mathfrak{g}_1 \subset \mathfrak{k}_2 = \mathfrak{g}_2 \subset \mathfrak{k}_3 = \mathfrak{k} \subset \mathfrak{k}_4 = \mathfrak{k} + \mathbb{R}V \subset \mathfrak{k}_5 = \mathfrak{k}_4 + \mathbb{R}U \subset \mathfrak{k}_6 = \mathfrak{g}.$$

From $\mathfrak{k}_4$ to $\mathfrak{k}_5$, the element

$$\sigma = YU - ZT + VZ$$

appears in the center of the enveloping algebra $\mathscr{U}(\mathfrak{g})$. The Hilbert space $\mathscr{H}_\pi$ of π is identified with $L^2(\mathbb{R}^2)$ by the mapping $\mathscr{H}_\pi \ni \varphi \mapsto \Phi \in L^2(\mathbb{R}^2)$ given by

$$\Phi(x, s) = \varphi(\exp(xX)\exp(sU)) \ (x, s \in \mathbb{R}).$$

In this situation, simple calculations show

$$\pi(Z) = i, (\pi(Y)\Phi)(x, s) = -ix\Phi(x, s),$$

$$(\pi(V)\Phi)(x, s) = -is\Phi(x, s), (\pi(T)\Phi)(x, s) = \big(- is + x\frac{\partial}{\partial s}\big)\Phi(x, s)$$

$$(\pi(U)\Phi)(x, s) = -\frac{\partial \Phi}{\partial s}(x, s).$$

Hence we confirm $\pi(\sigma) = 0$ and the $\mathfrak{k}$-invariant element $w = YU + VZ$ is nothing but the element ZT modulo $\ker(\pi)$.

Example 5.3.33 Let finally $\mathfrak{g} = \mathbb{R}\text{-span}(x, t, s, r, u, v, y, z)$ have non-zero brackets

$$[x, y] = z, [x, u] = v, [x, v] = y, [t, u] = v$$
$$[x, t] = s, [x, s] = r, [s, u] = y, [r, u] = z.$$

Then

$$[\mathfrak{g}, \mathfrak{g}] = \mathbb{R} - \text{span}(s, r, v, y, z)$$

is Abelian. Let

$$G = \exp \mathfrak{g}, f = Z^* \in \mathfrak{g}^*, \Omega = G{\cdot}f,$$
$$\mathfrak{k} = \mathbb{R} - \text{span}(t, u, v, y, z)$$

and $K = \exp \mathfrak{k}$. We have

$$\mathfrak{g}(f) = \mathbb{R} - \text{span}(t, s, v, z),$$

$\dim \Omega = 4$ and $\mathfrak{b} = \mathbb{R}t + [\mathfrak{g}, \mathfrak{g}]$ is a polarization of $\mathfrak{g}$ at f. We realize the irreducible unitary representation π of G associated to Ω as $\pi = \text{ind}_B^G \chi_f$ where $B = \exp \mathfrak{b}$. Next, take a Jordan-Hölder sequence

$$\mathfrak{g}_1 = \mathbb{R}z \subset \mathfrak{g}_2 = \mathbb{R}z + \mathbb{R}y \subset \mathfrak{g}_3 = \mathbb{R} - \text{span}(z, y, v)$$
$$\subset \mathfrak{g}_4 = \mathbb{R} - \text{span}(z, y, v, r) \subset \mathfrak{g}_5 = [\mathfrak{g}, \mathfrak{g}]$$
$$\subset \mathfrak{g}_6 = [\mathfrak{g}, \mathfrak{g}] + \mathbb{R}u \subset \mathfrak{g}_7 = [\mathfrak{g}, \mathfrak{g}] + \mathbb{R}u + \mathbb{R}t \subset \mathfrak{g}_8 = \mathfrak{g}$$

of $\mathfrak{g}$ and the corresponding sequence of subalgebras

$$\mathfrak{k}_1 = \mathfrak{g}_1 \subset \mathfrak{k}_2 = \mathfrak{g}_2 \subset \mathfrak{k}_3 = \mathfrak{g}_3 \subset \mathfrak{k}_4 = \mathfrak{k}_3 + \mathbb{R}u \subset \mathfrak{k}_5 = \mathfrak{k}$$
$$\subset \mathfrak{k}_6 = \mathfrak{k} + \mathbb{R}r \subset \mathfrak{k}_7 = \mathfrak{k} + \mathbb{R}r + \mathbb{R}s \subset \mathfrak{k}_8 = \mathfrak{g}.$$

For

$$\ell = at^* + bu^* + cv^* + dy^* + z^* \in \mathfrak{k}^* \ (a, b, c, d \in \mathbb{R})$$

satisfying $c \neq 0$, $\mathfrak{b}[\delta] = \mathbb{R}\text{-span}(u, v, y, z)$ is the Vergne polarization relative to the Jordan-Hölder sequence $\{\mathfrak{k}_j\}_{j=1}^5$ of $\mathfrak{k}$.

Now the Hilbert space $\mathscr{H}_\pi$ of π is identified with $L^2(\mathbb{R}^2)$ by the mapping $\mathscr{H}_\pi \ni \varphi \mapsto \Phi \in L^2(\mathbb{R}^2)$ given by

$$\Phi(a, b) = \varphi(\exp(ax)\exp(bu)) \ (a, b \in \mathbb{R}).$$

So, simple calculations show

$$\pi(z) = i, \ (\pi(y)\Phi)(a, b) = -ia\Phi(a, b),$$

$$(\pi(v)\Phi)(a, b) = \frac{ia^2}{2}\Phi(a, b), \ (\pi(r)\Phi)(a, b) = ib\Phi(a, b)$$

$$(\pi(s)\Phi)(a, b) = -iab\Phi(a, b), \ (\pi(u)\Phi)(a, b) = -(\frac{\partial}{\partial b} + \frac{ia^3}{6})\Phi(a, b),$$

$$(\pi(t)\Phi)(a, b) = \frac{ia^2b}{2}\Phi(a, b), \ (\pi(x)\Phi)(a, b) = -\frac{\partial\Phi}{\partial a}(a, b).$$

From this, $\ker(\pi)$ is spanned by

$$z - i, 2zv - y^2, s + iyr, t + ivr.$$

On the other hand, our algebra $\mathscr{U}_\pi(\mathfrak{g})^{\mathfrak{k}}$ is generated modulo $\ker(\pi)$ by $\{yt - vs, v, y, z\}$. But

$$yt - vs \equiv yt - \frac{y^2}{2i}s \equiv \frac{y}{2i}(2it - ys)$$
$$\equiv \frac{y}{2i}(2it + iy^2r) \equiv \frac{y}{2}(2t + y^2r) \equiv y(t + ivr) \equiv 0$$

modulo $\ker(\pi)$. Further computations show

$$\Omega = \{\ell = ax^* + \frac{cd^2}{2}t^* + bu^* + cds^* + cr^*$$
$$+ \frac{d^2}{2}v^* + dy^* + z^* \ (a, b, c, d \in \mathbb{R})\}.$$

Here also the restriction $\pi|_K$ turns out to be multiplicity-free. For $\ell \in \Omega$ verifying $\ell(v) \neq 0$, the Penney distribution a_ℓ is defined by the formula

$$a_\ell(\varphi) = \int_{\mathbb{R}} \overline{\varphi(\exp(\lambda u))\chi_\ell(\exp(\lambda u))} d\lambda \ (\varphi \in \mathscr{H}_\pi^\infty).$$

At last,

$$\pi(z)a_\ell = ia_\ell, \pi(y)a_\ell = i\ell(y)a_\ell, \pi(v)a_\ell = i\ell(v)a_\ell.$$

Hence the algebra $D_\pi(G)^K$ is isomorphic to $\mathbb{C}[\Omega]^K$.

Chapter 6
Holomorphically Induced Representations of Solvable Lie Groups

6.1 Introduction

Once we leave the exponential picture, we know the fundamental results by Auslander-Kostant [3] and Pukanszky [144]. Here we study some intertwining operators and real polarizations. First, let $N = \exp \mathfrak{n}$ be a connected and simply connected nilpotent Lie group with Lie algebra $\mathfrak{n}$ and $\mathfrak{h}$ a complex positive polarization of $\mathfrak{n}$ at $f \in \mathfrak{n}^*$. We will construct the holomorphically induced representation $\rho(f, \mathfrak{h}, N)$ of N and determine the space of C^∞-vectors of $\rho(f, \mathfrak{h}, G)$. Let $\mathfrak{h}'$ be an another complex positive polarization of $\mathfrak{n}$ at f. We shall describe an explicit intertwining operator between $\rho(f, \mathfrak{h}, N)$ and $\rho(f, \mathfrak{h}', N)$, which satisfies the composition formula for three polarizations in terms of Maslov index. Next, let G be a connected solvable Lie group with Lie algebra $\mathfrak{g}$ and let $\mathfrak{n}$ be a nilpotent ideal of $\mathfrak{g}$ containing $[\mathfrak{g}, \mathfrak{g}]$. Let $\mathfrak{h}_j$ be complex positive polarizations of $\mathfrak{g}$ at $f \in \mathfrak{g}^*$ which are strongly admissible for $\mathfrak{n}$ and $\rho_j = \rho(f, \mathfrak{h}_j, G)(j = 1, 2)$. We shall describe an explicit intertwining operator between ρ_1 and ρ_2 satisfying the composition formula for three polarizations.

We will finally study the holomorphically induced representation $\rho(f, \mathfrak{h}, G)$ for a connected solvable Lie groups G when $\mathfrak{h}$ is a real polarization of $\mathfrak{g}$ at $f \in \mathfrak{g}^*$. We will discuss irreducibility issues, equivalence and related intertwining operators.

6.2 Intertwining Operators

In this section, we follow [72] to study some intertwining operators and apply them to calculate the Mackey obstruction.

A. Baklouti et al., *Representation Theory of Solvable Lie Groups and Related Topics*, Springer Monographs in Mathematics,
https://doi.org/10.1007/978-3-030-82044-2_6

6.2.1 Nilpotent Lie Groups and Maslov Index

Let N denote a connected and simply connected nilpotent Lie group with Lie algebra $\mathfrak{n}$. The exponential mapping identifies N with its Lie algebra $\mathfrak{n}$ and we can talk about polynomial mappings on N or with values in N. We denote by $\mathfrak{z}$ the center of $\mathfrak{n}$. Let $f \in \mathfrak{n}^*$ and let B_f be the alternating bilinear form on $\mathfrak{n}$ or $\mathfrak{n}_{\mathbb{C}}$ defined by $B_f(X, Y) = f([X, Y])$, $X, Y \in \mathfrak{n}_{\mathbb{C}}$. For $\mathfrak{k} \subset \mathfrak{n}$, we denote by $\mathfrak{k}^f$ the orthogonal of $\mathfrak{k}$ in $\mathfrak{n}$ with respect to B_f.

Let $\mathfrak{n}(f)$ be the radical of B_f and χ the function on N defined by $\chi(\exp X) = e^{if(X)}$ for all $X \in \mathfrak{n}$. The letter $\mathfrak{h}$ denotes an element of the set $P^+ = P^+(f, \mathfrak{n})$ of positive polarizations of $\mathfrak{n}$ at f, namely:

(i) $\mathfrak{h}$ is a maximally isotropic subspace of $\mathfrak{n}_{\mathbb{C}}$ for B_f;
(ii) the Hermitian form $(x, y) \mapsto iB_f(x, \overline{y})$ is positive on $\mathfrak{h}$;
(iii) $\mathfrak{h}$ is a subalgebra of $\mathfrak{n}_{\mathbb{C}}$.

We say that $\mathfrak{h}$ is real if $\mathfrak{h} = \overline{\mathfrak{h}}$ and denote by $M = M(f, \mathfrak{n})$ the set of real polarizations of $\mathfrak{n}$ at f. The previous conditions imply that $\mathfrak{h} + \overline{\mathfrak{h}}$ is a subalgebra of $\mathfrak{n}_{\mathbb{C}}$. In what follows, the letters $\mathfrak{e}$, $\mathfrak{d}$ and $\mathfrak{b}$ designate the subalgebras of $\mathfrak{n}$ such that $\mathfrak{e}_{\mathbb{C}} = \mathfrak{h} + \overline{\mathfrak{h}}$, $\mathfrak{d}_{\mathbb{C}} = \mathfrak{h} \cap \overline{\mathfrak{h}}$ and $\mathfrak{b} = \mathfrak{d} \cap \ker(f)$. We have $\mathfrak{n}(f) \subset \mathfrak{d} \subset \mathfrak{e}$ and $\mathfrak{d}^f = \mathfrak{e}$. Hence B_f induces a non-degenerate alternating bilinear form B on $\mathfrak{e}/\mathfrak{d}$.

Let us recall some basic facts. There exists a canonical complex structure J on $\mathfrak{e}/\mathfrak{d}$. Indeed, we have

$$(\mathfrak{e}/\mathfrak{d})_{\mathbb{C}} \cong \mathfrak{h}/\mathfrak{d}_{\mathbb{C}} \oplus \overline{\mathfrak{h}}/\mathfrak{d}_{\mathbb{C}}.$$

We consider the operator J on $\mathfrak{e}/\mathfrak{d}$ such that $\mathfrak{h}/\mathfrak{d}_{\mathbb{C}}$ (resp. $\overline{\mathfrak{h}}/\mathfrak{d}_{\mathbb{C}}$) is the eigenspace of J corresponding to the eigenvalue $-i$ (resp. $+i$). Then J defines a real endomorphism of $\mathfrak{e}/\mathfrak{d}$ such that $J^2 = -1$.

Lemma 6.2.1 *B is invariant under J.*

Proof It suffices to verify the assertion on $(\mathfrak{e}/\mathfrak{d})_{\mathbb{C}}$. As $\mathfrak{h}/\mathfrak{d}_{\mathbb{C}}$ and $\overline{\mathfrak{h}}/\mathfrak{d}_{\mathbb{C}}$ are totally isotropic for B, we have

$$\begin{aligned} B(X' + \overline{X}, Y' + \overline{Y}) &= B(X', \overline{Y}) + B(\overline{X}, Y') \\ &= B(J(X' + \overline{X}), J(Y' + \overline{Y})) \end{aligned}$$

for $X, X', Y, Y' \in \mathfrak{h}$. □

We define the Hermitian form H_f on $\mathfrak{e}_{\mathbb{C}}$ by

$$H_f(x, y) = 2iB(x, \overline{y}) = 2if([x, \overline{y}]).$$

Thus a polarization $\mathfrak{h}$ at f is positive if the restriction of H_f to $\mathfrak{h}$ is positive. Assume that $\mathfrak{h}$ is a positive polarization at f, then

$$\mathfrak{d}_{\mathbb{C}} = \{x \in \mathfrak{h};\ f([x, \overline{x}]) = 0\}.$$

If $\mathfrak{h}$ is a polarization at f, H_f induces a non-degenerate Hermitian form H on $\mathfrak{h}/\mathfrak{d}_{\mathbb{C}}$. On the other hand, every element x of $\mathfrak{e}/\mathfrak{d}$ is uniquely written in the form $x = p(x) + \overline{p(x)}$ with $p(x) \in \mathfrak{h}/\mathfrak{d}_{\mathbb{C}}$ and the mapping p is an isomorphism of $\mathfrak{e}/\mathfrak{d}$ with complex structure J and $\mathfrak{h}/\mathfrak{d}_{\mathbb{C}}$ with complex structure given by multiplication by $-i$. In fact $p(x) = \frac{1}{2}(x + iJx)$, because $x + iJx$ has eigenvalue $-i$ and hence belongs to $\mathfrak{h}/\mathfrak{d}_{\mathbb{C}}$ and $x = p(x) + \overline{p(x)}$. Using p we transport H from $\mathfrak{h}/\mathfrak{d}_{\mathbb{C}}$ and define a Hermitian form on $\mathfrak{e}/\mathfrak{d}$ which we denote again by H. For $x, y \in \mathfrak{e}/\mathfrak{d}$, we have

$$H(x, y) = \frac{1}{4}H(x + iJx, y + iJy) = \frac{i}{2}B(x + iJx, y + iJy).$$

Since B is J-invariant,

$$H(x, y) = B(x, Jy) + iB(x, y).$$

We denote $S(x, y)$ the real part of the Hermitian form $H(x, y)$, i.e., $S(x, y) = B(x, Jy)$. Clearly, S is a non-degenerate symmetric form on $\mathfrak{e}/\mathfrak{d}$ and $\mathfrak{h}$ is a positive polarization at f if and only if S is positive.

From now on $\mathfrak{h}$ will denote a positive polarization at f. For $x \in \mathfrak{d}$, we consider the adjoint action $A = \operatorname{ad} x$ on $\mathfrak{e}$. This action keeps $\mathfrak{h}/\mathfrak{d}_{\mathbb{C}}, \overline{\mathfrak{h}}/\mathfrak{d}_{\mathbb{C}}$ unchanged and hence commutes with J. For $y, z \in \mathfrak{e}$,

$$0 = B(x, [y, z]) = B([x, y], z) + B(y, [x, z]).$$

Therefore, $S(Ay, z) = -S(y, Az)$ and A is skew-symmetric. Since A is nilpotent, this means that $\mathfrak{d}$ and $\mathfrak{b}$ are ideals of $\mathfrak{e}$ and $\mathfrak{e}/\mathfrak{b}$ is a Heisenberg algebra with center $\mathfrak{d}/\mathfrak{b}$. The analytic subgroups of N corresponding to $\mathfrak{b}, \mathfrak{d}, \mathfrak{e}$ are denoted respectively by B, D, E.

For $X \in \mathfrak{n}$ and $\phi \in C^{\infty}(N)$, put

$$(R(X)\phi)(n) = \frac{d}{dt}\phi(n \exp(tX)) \big|_{t=0} \ (n \in N)$$

and for $X = X_1 + iX_2 \in \mathfrak{n}_{\mathbb{C}}$, $R(X) = R(X_1) + iR(X_2)$. We consider the space of functions $\phi \in C^{\infty}(N)$ satisfying

(i) $R(X)\phi = -if(X)\phi, \forall X \in \mathfrak{h}$,
(ii) $\|\phi\|^2 = \int_{N/D} |\phi|^2 dm_{N/D} < \infty$,

where $dm_{N/D}$ is an N-invariant measure on the homogeneous space N/D. We denote by $\mathscr{H} = \mathscr{H}(f, \mathfrak{h}, N)$ the completion of this space in the norm $\| \cdot \|$. From the theory of holomorphically induced representations we obtain in this fashion a non-trivial Hilbert space, whose Hermitian product is denoted by $(\cdot|\cdot) = (\cdot|\cdot)_{\mathscr{H}}$, and an irreducible unitary representation $\rho = \rho^N = \rho(f, \mathfrak{h}, N)$ of N on $\mathscr{H}$ by left translations. We denote by $\mathscr{H}^\infty$ the Fréchet space of C^∞-vectors of ρ, and still by ρ the representation on $\mathscr{H}^\infty$ of the enveloping algebra $\mathscr{U}(\mathfrak{n})$ of $\mathfrak{n}_{\mathbb{C}}$. When $\mathfrak{h}$ varies in P^+, the various representations obtained in this manner are all mutually equivalent. In what follows, we indicate the $\mathfrak{h}$-associated objects $\mathfrak{b}, \mathfrak{d}, \mathfrak{e}, B, D, E, \rho, \mathscr{H}, \mathscr{H}^\infty, (\cdot|\cdot)$, etc. as $\rho_1 = \rho(f, \mathfrak{h}_1, N)$ for example. We shall construct canonical intertwining operators between the various representations.

Let C be an analytic subgroup of N with Lie algebra $\mathfrak{c}$ which contains $\mathfrak{e}$. Then $\mathfrak{h} \in P^+(f, \mathfrak{c})$. We write $\rho^C = \rho(f, \mathfrak{h}, C)$ and $\mathscr{H}^C = \mathscr{H}(f, \mathfrak{h}, C)$. If $\phi \in \mathscr{H}$, we consider the vector field $n \mapsto \phi_n$ defined almost everywhere on N with values in $\mathscr{H}^C$ such that $\phi_n(k) = \phi(nk) (k \in C)$. This vector field is square integrable on N/C. If $\phi \in \mathscr{H}^\infty$, the mapping $n \mapsto \phi_n$ with values in $(\mathscr{H}^C)^\infty$ is differentiable on N.

Let K be an analytic subgroup of N with Lie algebra $\mathfrak{k}$. There exists a complementary basis (CB) of $\mathfrak{k}$ in $\mathfrak{n}$, by which we mean that a CB is an ordered family $(L_i)_{1 \le i \le p}$, $p = \dim(\mathfrak{n}/\mathfrak{k})$, such that the mapping

$$(t_p, \dots, t_1; k) \mapsto \exp(t_p L_p) \cdots \exp(t_1 L_1) k$$

is a polynomial diffeomorphism, namely polynomial with polynomial inverse, of $\mathbb{R}^p \times K$ onto N. In order to simplify the notations, we write this correspondence as $(t, k) \mapsto \exp(tL)k$ $(t \in \mathbb{R}^p, k \in K)$. Then the mapping $t \mapsto \exp(tL)K$ establishes a diffeomorphism F of $\mathbb{R}^p$ onto N/K, by which the Lebesgue measure on $\mathbb{R}^p$ corresponds to the N-invariant measure $dm_{N/K}$ on N/K. So, if $\phi \in L^1(N/K)$,

$$\int_{N/K} \phi(n) dm_{N/K}(n) = \int_{\mathbb{R}^p} \phi(\exp(tL)) dt.$$

Remark 6.2.2 If K, H are two analytic subgroup of N such that $K \subset H$ and if $\phi \in L^1(N/K)$, we have

$$\int_{N/K} \phi(n) dm_{N/K}(n) = \int_{N/H} \Big(\int_{H/K} \phi(nh) dm_{H/K}(h) \Big) dm_{N/H}(h).$$

Remark 6.2.3 If $(L_i')_{1 \le i \le p}$ is another CB of $\mathfrak{k}$ in $\mathfrak{n}$ associated to the diffeomorphism $F' \colon \mathbb{R}^p \to N/K$, then $F^{-1} \circ F'$ is a polynomial diffeomorphism of $\mathbb{R}^p$. In particular, it preserves the Schwartz space $\mathscr{S}(\mathbb{R}^p)$, it leaves invariant the space of differential operators with polynomial coefficients, and its Jacobian is a non-zero constant.

Now, we study the structure of $\mathscr{H}^\infty$. When $\mathfrak{h}$ is given, we choose a CB

$$(P_1, Q_1, P_2, Q_2, \ldots, P_s, Q_s)$$

of $\mathfrak{d}$ in $\mathfrak{e}$ such that, if $X \mapsto \dot{X}$ designates the canonical projection of $\mathfrak{e}$ to $\mathfrak{e}/\mathfrak{b}$ and $\dot{Z}$ the generator of the center of $\mathfrak{e}/\mathfrak{b}$ satisfying $f(Z) = 1$, we have

$$[\dot{P}_i, \dot{Q}_j] = \delta_{ij}\dot{Z}, [\dot{P}_i, \dot{P}_j] = [\dot{Q}_i, \dot{Q}_j] = 0 \ (1 \le i, j \le s).$$

Next we choose a CB $(L_1, \ldots, L_r)$ of $\mathfrak{e}$ in $\mathfrak{n}$. For $x_j, y_j \in \mathbb{R}$ $(1 \le j \le s)$, we put $z_j = x_j + iy_j$. In what follows, we use the standard notation for multi-indices:

$$z = (z_1, \ldots, z_s), t = (t_1, \ldots, t_s), \alpha = (\alpha_1, \ldots, \alpha_s),$$

$$z^\alpha = z_1^{\alpha_1} \cdots z_s^{\alpha_s}, |z|^2 = \sum_{i=1}^{s} |z_i|^2, e^{xP+yQ} = e^{x_sP_s+y_sQ_s} \cdots e^{x_1P_1+y_1Q_1}.$$

Then every element n of N is uniquely written as

$$n = e^{tL}e^{xP+yQ}d \ (t \in \mathbb{R}^r, x, y \in \mathbb{R}^s, d \in D).$$

In these conditions let $\mathscr{O}$ be the space of differential operators on N spanned by finite $\mathbb{C}$-linear combinations of

$$z^\alpha \frac{\partial}{\partial z^\beta} t^\gamma \frac{\partial}{\partial t^\delta} \ (\alpha, \beta \in \mathbb{N}^s, \gamma, \delta \in \mathbb{N}^r),$$

where for example

$$\begin{aligned}&\Big(\frac{\partial}{\partial z^\beta}\phi\Big)(e^{tL}e^{xP+yQ}d) \\ &= \frac{\partial^{\beta_1}}{\partial z_1^{\beta_1}} \cdots \frac{\partial^{\beta_s}}{\partial z_s^{\beta_s}} \phi(e^{tL}e^{x_sP_s+y_sQ_s} \cdots e^{x_1P_1+y_1Q_1}d)\end{aligned}$$

with $\frac{\partial}{\partial z_j} = \frac{1}{2}\big(\frac{\partial}{\partial x_j} - i\frac{\partial}{\partial y_j}\big)(1 \le j \le s)$.

Let $\varphi_0 \in C^\infty(N)$ be defined by

$$\varphi_0(e^{tL}e^{xP+yQ}d) = \chi^{-1}(d)e^{-\frac{|z|^2}{4}}.$$

If $\phi \in \mathscr{H}^\infty$, $\varphi_0^{-1}\phi$ is invariant under the right translation by d.

Theorem 6.2.4 *The space $\mathcal{H}^\infty$ consists of all functions $\phi \in C^\infty(N)$ such that*

(i) $\phi(nd) = \chi^{-1}(d)\phi(n)$ $(n \in N, d \in D)$,
(ii) $\varphi_0^{-1}\phi$ *is holomorphic in* z.
(iii) $\forall P \in \mathcal{O}, \varphi_0 P(\varphi_0^{-1}\phi) \in \mathcal{H}$.

For every $U \in \mathcal{U}(\mathfrak{n})$, there exists an element $P(U) \in \mathcal{O}$ such that

$$\rho(U)\phi = \varphi_0 P(U)(\varphi_0^{-1}\phi)\ \forall \phi \in \mathcal{H}^\infty.$$

This mapping $\mathcal{U}(\mathfrak{n}) \to \mathcal{O}$ is surjective.

Proof

(a) If $\mathfrak{h}$ is real, $s = 0$ and $\varphi_0(n) = \chi^{-1}(d)$. Then the theorem is nothing but the classical result of Kirillov [93].
(b) If $\mathfrak{e} = \mathfrak{n}$, $r = 0$. As the restriction of ρ to the normal subgroup B is trivial, we reduce to the case where N is a Heisenberg group by taking the quotient by B. Then it is well known that the elements of $\mathcal{H}$ are represented by functions ϕ on N such that $\phi(ne^{uZ}) = \phi(n)e^{-iu}$, $\varphi_0^{-1}\phi$ is holomorphic in z and $\phi \in L^2(N/D)$. Moreover

$$\rho(P_j + iQ_j)\phi = z_j\phi,\ \rho(P_j - iQ_j) = -2\varphi_0\frac{\partial}{\partial z_j}(\varphi_0^{-1}\phi)\ (1 \le j \le s).$$

From these observations we deduce the theorem in this case.
(c) Now $\mathfrak{h}$ is arbitrary. We put $\mathfrak{h} = \mathfrak{h}_2$ and take an element $\mathfrak{h}_1 \in M(f, \mathfrak{e})$, for example

$$\mathfrak{h}_1 = \oplus_{j=1}^s \mathbb{C}Q_j \oplus \mathfrak{d}_\mathbb{C}.$$

We make use of the notations introduced before and consider an intertwining operator T^E between the representations $\rho_i^E = \rho(f, \mathfrak{h}_i, E)(i = 1, 2)$ such that $T^E\rho_1^E = \rho_2^E T^E$. Then T^E induces an operator T which intertwines ρ_1 and ρ_2. For any $\phi \in \mathcal{H}_1^\infty$ we have $T\phi \in \mathcal{H}_2^\infty$ with $(T\phi)_n = T^E\phi_n(\forall n \in N)$.

We know from (a) that the image of $\mathcal{U}(\mathfrak{n})$ under ρ_1 is generated by the differential operators

$$x^\alpha\frac{\partial}{\partial x^\beta},\ t^\gamma\frac{\partial}{\partial t^\delta}\ (\alpha, \beta \in \mathbb{N}^s; \gamma, \delta \in \mathbb{N}^r).$$

Let us clarify what the image under T of these two types of operators is.
First of all we have

$$T\big(t^\gamma\frac{\partial}{\partial t^\delta}\phi\big) = t^\gamma\frac{\partial}{\partial t^\delta}(T\phi)$$

if $\phi \in \mathcal{H}_1^\infty$. Let us verify this in a particular case of the operator $\frac{\partial}{\partial t}$. Remark first that from the properties of C^∞-vectors of the induced representations,

$$\frac{\partial}{\partial t}(\phi_{\exp(tL)}) = (\frac{\partial}{\partial t}\phi)_{\exp(tL)},$$

which means that the derivative in t of the vector field $t \mapsto \phi_{\exp(tL)}$ with values in $\mathcal{H}_1^E$ coincides with the pointwise derivative in t of the functions $\phi_{\exp(tL)}$ defined on E. It follows that

$$\begin{aligned}(T\frac{\partial}{\partial t}\phi)_{\exp(tL)} &= T^E(\frac{\partial}{\partial t}\phi)_{\exp(tL)} = T^E\frac{\partial}{\partial t}(\phi_{\exp(tL)}) = \frac{\partial}{\partial t}T^E\phi_{\exp(tL)} \\ &= \frac{\partial}{\partial t}(T\phi)_{\exp(tL)} = (\frac{\partial}{\partial t}T\phi)_{\exp(tL)}\end{aligned}$$

as expected.

Now we denote by $\mathcal{O}_1$ (resp. $\mathcal{O}_2$) the set of differential operators on N generated by those of the form $x^\alpha \frac{\partial}{\partial x^\beta}$ (resp. $z^\alpha \frac{\partial}{\partial z^\beta}$). If $P_1 \in \mathcal{O}_1$, there exist $W \in \mathcal{U}(\mathfrak{e})$ and $P_2 \in \mathcal{O}_2$ such that

$$\begin{aligned}&(TP_1\phi)_{\exp(tL)} = T^E(P_1\phi)_{\exp(tL)} = T^E\rho_1^E(W)\phi_{\exp(tL)} \\ =&\rho_2^E(W)(T\phi)_{\exp(tL)} = \varphi_0 P_2\varphi_0^{-1}(T\phi)_{\exp(tL)} = (\varphi_0 P_2\varphi_0^{-1}T\phi)_{\exp(tL)}\end{aligned}$$

for $\forall\phi \in \mathcal{H}_1^\infty$, $\forall t \in \mathbb{R}^r$. Therefore to every $P_1 \in \mathcal{O}_1$ we can associate an element $P_2 \in \mathcal{O}_2$ such that $TP_1 = \varphi_0 P_2\varphi_0^{-1}T$. Doing the same for T^{-1} we see that this correspondence is in fact a bijection. From these observations, we get the theorem. □

Corollary 6.2.5 *Let $(X_i)_{1\le i\le m}$ $(m = r+2s)$ be any CB of $\mathfrak{d}$ in $\mathfrak{n}$. To every function ϕ on N we associate the function $R\phi$ on $\mathbb{R}^m$ given by $R\phi(x) = \phi(e^{xX})$. Then R establishes an isomorphism of topological vector spaces between $\mathcal{H}^\infty$ and a closed vector subspace of $\mathcal{S}(\mathbb{R}^m)$.*

Proof It suffices to prove the corollary for a CB of the type used in the above theorem. The topology of $\mathcal{S}(\mathbb{R}^m)$ is given by the semi-norms $\|P\varphi\|_{L^2(\mathbb{R}^m)}$ where P runs over the set of differential operators on $\mathbb{R}^m$ with polynomial coefficients. Well, the result comes from the following inequalities and the fact that the space $\mathcal{H}^\infty$ is complete. For $\phi \in \mathcal{H}^\infty$, $x, y \in \mathbb{R}^s$ and $z \in \mathbb{C}^s$,

$$\begin{aligned}&|x\phi| \le |z\phi|,\ |\frac{\partial}{\partial x}\phi| \le \frac{1}{2}|\varphi_0\frac{\partial}{\partial z}(\varphi_0^{-1}\phi)| + \frac{1}{4}|z\phi|, \\ &|z\phi| \le |x\phi| + |y\phi|,\ |\varphi_0\frac{\partial}{\partial z}(\varphi_0^{-1}\phi)| \le |\frac{\partial}{\partial x}\phi| + |\frac{\partial}{\partial y}\phi| + |x\phi| + |y\phi|.\end{aligned}$$

□

We would like to construct a canonical intertwining operator. We use the following general result due to Poulsen [140].

Theorem 6.2.6 *If ρ_1, ρ_2 are equivalent irreducible unitary representations of an arbitrary Lie group N acting on Hilbert spaces $(\mathcal{H}_1, (\cdot|\cdot)_1)$, $(\mathcal{H}_2, (\cdot|\cdot)_2)$, and if $\langle\cdot|\cdot\rangle$ is an N-invariant, separately continuous sesquilinear form on $\mathcal{H}_1^\infty \times \mathcal{H}_2^\infty$, then there exists a unitary intertwining operator T_{21} between two representations and a positive scalar α such that*

$$\alpha(T_{21}\phi_1|\phi_2)_2 = \langle\phi_1|\phi_2\rangle,\ \forall\phi_1 \in \mathcal{H}_1^\infty, \forall\phi_2 \in \mathcal{H}_2^\infty.$$

Returning to our study, we have the following result.

Theorem 6.2.7 *For any $\mathfrak{h}_1, \mathfrak{h}_2 \in P^+$, the integral*

$$\langle\phi_1|\phi_2\rangle = \int_{N/(D_1\cap D_2)} \phi_1\overline{\phi_2}dm,\ \phi_1 \in \mathcal{H}_1^\infty, \phi_2 \in \mathcal{H}_2^\infty$$

is well defined and gives a sesquilinear, separately continuous form on $\mathcal{H}_1^\infty \times \mathcal{H}_2^\infty$ (which is manifestly N-invariant).

Proof We will not give a complete proof. We shall content ourselves with pointing out the role of Corollary 6.2.5 and just state a series of lemmas which eventually lead up to the result. Refer to [97] for details. □

Lemma 6.2.8 *There is a CB $(X_i)_{1\le i\le m+p}$ of $\mathfrak{d}_1 \cap \mathfrak{d}_2$ in $\mathfrak{n}$, the first p elements $(X_i)_{1\le i\le p}$ of which form a CB of $\mathfrak{d}_1 \cap \mathfrak{d}_2$ in $\mathfrak{d}_2$.*

Lemma 6.2.9 *With the notations of the theorem, the integral $\int_{D_1/(D_1\cap D_2)} |\phi|dm$ converges for any $\phi \in \mathcal{H}_2^\infty$, and on the latter space it defines a continuous semi-norm.*

Lemma 6.2.10 *Let p be a continuous semi-norm on $\mathcal{H}^\infty$. There exist a finite family $(p_j)_{j\in J}$ of continuous semi-norms and a finite family $(Q_j)_{j\in J}$ of polynomial functions on N such that*

$$p(\rho(n)\phi) \le \sum_{j\in J} |Q_j(n)|p_j(\phi),\ \forall\phi \in \mathcal{H}^\infty, \forall n \in N.$$

Remark 6.2.11 From Lemma 6.2.9 and Remark 6.2.2,

$$\langle\phi_1|\phi_2\rangle = \int_{N/D_1} \phi_1(n)\Big(\int_{D_1/(D_1\cap D_2)} \overline{\phi_2(nh)\chi(h)}dm(h)\Big)dm(n).$$

Using the last two theorems and choosing appropriately the measure $dm_{N/(D_1 \cap D_2)}$, we have:

Theorem 6.2.12 *There exists a unique unitary intertwining operator* T_{21} *from* $(\mathscr{H}_1, \rho_1)$ *to* $(\mathscr{H}_2, \rho_2)$ *such that*

$$(T_{21}\phi_1|\phi_2) = \int_{N/(D_1 \cap D_2)} \phi_1 \overline{\phi_2} dm_{N/(D_1 \cap D_2)}.$$

Let us describe a composition formula for our intertwining operators. For three positive polarizations $\mathfrak{h}_j \in P^+ (1 \leq j \leq 3)$, we denote by

$$c_{\mathfrak{n}}(\mathfrak{h}_1, \mathfrak{h}_2, \mathfrak{h}_3) = c(\mathfrak{h}_1, \mathfrak{h}_2, \mathfrak{h}_3) = c(1, 2, 3)$$

the complex number of modulus 1 satisfying $T_{12}T_{23} = c(1, 2, 3)T_{13}$.

Proposition 6.2.13

(i) $(T_{12})^{-1} = T_{21}$.
(ii) $c(1, 2, 3) = c(2, 3, 1) = \overline{c(3, 2, 1)}$.
(iii) *c verifies the cocycle identity:*

$$c(1, 2, 3) = c(1, 2, 4)c(2, 3, 4)c(3, 1, 4).$$

Proof Computing $T_{12}T_{23}T_{34}$ in two ways, we verify $c(1, 2, 3)c(1, 3, 4) = c(1, 2, 4)c(2, 3, 4)$. All other relations come from the fact that the adjoint, hence the inverse of T_{12}, is equal to T_{21} because

$$\langle \phi_1 | \phi_2 \rangle = (T_{21}\phi_1|\phi_2)_2 = (\phi_1|T_{12}\phi_2)_1.$$

□

Let (V, B) be a real symplectic vector space of dimension $2m$. We associate to it the Heisenberg Lie algebra $\mathfrak{n}_m = V \oplus \mathbb{R}Z$, where Z is any vector in $\mathbb{R}$, such that

$$[X + tZ, Y + uZ] = B(X, Y)Z \ (X, Y \in V, t, u \in \mathbb{R}).$$

We denote by $\perp$ the orthogonality relation on (V, B). The positive polarizations of $\mathfrak{n}_m$ are the vector subspaces of $(\mathfrak{n}_m)_{\mathbb{C}}$ which can be written in the form $\lambda + \mathbb{C}Z$, where λ is an element of the set $\overline{\mathfrak{d}}$ of positive Lagrangian subspaces of $(\mathfrak{n}_m)_{\mathbb{C}}$. The notion of Maslov index, usually defined on triples of real Lagrangian subspaces, generalizes [125] to a real-valued map $\tau = \tau_{V_m}$ on $\overline{\mathfrak{d}} \times \overline{\mathfrak{d}} \times \overline{\mathfrak{d}}$. We have

$$c_{\mathfrak{n}_m}(\lambda_1 \oplus \mathbb{C}Z, \lambda_2 \oplus \mathbb{C}Z, \lambda_3 \oplus \mathbb{C}Z) = e^{i\frac{\pi}{4}\tau_{V_m}(\lambda_1, \lambda_2, \lambda_3)}.$$

If $(\mathfrak{n}, B_f)$ is a real vector space equipped with an alternating bilinear form (not necessarily non-degenerate) and if $\mathfrak{h}_j (1 \leq j \leq 3)$ are positive maximally totally isotropic subspaces of $(\mathfrak{n}, B_f)$ (namely if they satisfy conditions (i), (ii) of positive polarizations), we define

$$\tau_{\mathfrak{n}}(\mathfrak{h}_1, \mathfrak{h}_2, \mathfrak{h}_3) = \tau_{\mathfrak{n}/\ker(B_f)}(\mathfrak{h}_1/\ker(B_f), \mathfrak{h}_2/\ker(B_f), \mathfrak{h}_3/\ker(B_f)).$$

Proposition 6.2.14 *If the positive polarizations $\mathfrak{h}_j (1 \leq j \leq 3)$ are contained in a subalgebra $\mathfrak{c}$ of $\mathfrak{n}$, then*

$$c_{\mathfrak{n}}(\mathfrak{h}_1, \mathfrak{h}_2, \mathfrak{h}_3) = c_{\mathfrak{c}}(\mathfrak{h}_1, \mathfrak{h}_2, \mathfrak{h}_3).$$

Proof We use the notations introduced before. Let T_{ji}^C indicate the canonical intertwining operator from ρ_i^C to ρ_j^C. We verify

$$T_{ji}^C(\phi_n) = (T_{ji}\phi)_n, \ \forall n \in N.$$

Indeed,

$$\int_{N/(D_1 \cap D_2)} \phi_i \overline{\phi_j} dm = \int_{N/C} (T_{ji}^C((\phi_i)_n) | (\phi_j)_n) dm(n), \phi_i \in \mathscr{H}_i^\infty, \phi_j \in \mathscr{H}_j^\infty,$$

and

$$\int_{N/D_j} (T_{ji}\phi_i)\overline{\phi_j} dm = \int_{N/C} ((T_{ji}\phi_i)_n | (\phi_j)_n) dm(n).$$

The result follows from this. □

Besides,

Proposition 6.2.15 *If $\mathfrak{a}$ is an ideal of $\mathfrak{n}$ contained in $\mathfrak{n}(f) \cap \ker(f)$,*

$$c_{\mathfrak{n}}(\mathfrak{h}_1, \mathfrak{h}_2, \mathfrak{h}_3) = c_{\mathfrak{n}/\mathfrak{a}}(\mathfrak{h}_1/\mathfrak{a}, \mathfrak{h}_2/\mathfrak{a}, \mathfrak{h}_3/\mathfrak{a}).$$

These two propositions applied to the Heisenberg algebra give:

Proposition 6.2.16 *Let $\mathfrak{h}_j (1 \leq j \leq 3)$ be three positive maximally isotropic subspaces of $(\mathfrak{n}, B_f)$ and $\mathfrak{a}$ a vector subspace of $\mathfrak{n}$ contained in $\mathfrak{h}_1 \cap \mathfrak{h}_2 \cap \mathfrak{h}_3$. Then*

$$e^{i\frac{\pi}{4}\tau_{\mathfrak{n}}(\mathfrak{h}_1,\mathfrak{h}_2,\mathfrak{h}_3)} = e^{i\frac{\pi}{4}\tau_{\mathfrak{a}^f}(\mathfrak{h}_1,\mathfrak{h}_2,\mathfrak{h}_3)}.$$

Theorem 6.2.17 *Let $\mathfrak{a}$ be a subalgebra of $\mathfrak{n}$ such that $\mathfrak{a}^f$ is a subalgebra containing $\mathfrak{a}$ as ideal and let $\mathfrak{h} \in P^+$. Then*

$$\mathfrak{h}^{\mathfrak{a}} = \mathfrak{h} \cap (\mathfrak{a}^f)_{\mathbb{C}} + \mathfrak{a}_{\mathbb{C}}$$

is an element of P^+. *Now, let* $\mathfrak{h}_j \in P^+ (1 \le j \le 3)$ *be such that* $\mathfrak{a} \subset \mathfrak{h}_1 \cap \mathfrak{h}_2$. *Then*

$$c(\mathfrak{h}_3, (\mathfrak{h}_3)^{\mathfrak{a}}, \mathfrak{h}_1) = 1,$$
$$c(\mathfrak{h}_1, \mathfrak{h}_2, \mathfrak{h}_3) = c(\mathfrak{h}_1, \mathfrak{h}_2, (\mathfrak{h}_3)^{\mathfrak{a}}).$$

Applying this result to the case of Heisenberg algebra, we obtain:

Corollary 6.2.18 *Let* η *be an isotropic subspace of* (V, B) *and* $\lambda \in \bar{\mathfrak{d}}$. *Then*

$$\lambda^{\eta} = \lambda \cap (\eta^{\perp})_{\mathbb{C}} + \eta_{\mathbb{C}}$$

is an element of $\bar{\mathfrak{d}}$. *If* $\lambda_j \in \bar{\mathfrak{d}} (1 \le j \le 3)$ *are such that* $\eta \subset \lambda_1 \cap \lambda_2$, *then*

$$e^{i\frac{\pi}{4}\tau(\lambda_1,\lambda_2,\lambda_3)} = e^{i\frac{\pi}{4}\tau(\lambda_1,\lambda_2,(\lambda_3)^{\eta})}.$$

Proof It is clear that $\mathfrak{h}^{\mathfrak{a}} \in P^+$. The second formula is easily deduced from the first one by the cocycle formula applied to $(\mathfrak{h}_1, \mathfrak{h}_2, \mathfrak{h}_3, (\mathfrak{h}_3)^{\mathfrak{a}})$. So, we prove the first one making use of the following fact.

Lemma 6.2.19 *Let* A *be the closed connected subgroup associated to* $\mathfrak{a}$. *Then*

$$\big(T_{(\mathfrak{h}_3)^{\mathfrak{a}}\mathfrak{h}_3}\phi\big)(n) = \int_{A/(D_3\cap A)} \phi(nh)\chi(h)dm(h), \ \forall \phi \in \mathcal{H}_3^{\infty}.$$

Proof We denote by $\phi'(n)$ the right-hand side of the formula to be proved. To simplify the notations, put $\mathfrak{h}_4 = (\mathfrak{h}_3)^{\mathfrak{a}}$. We shall verify:

(a) $(T_{43}\phi|\psi)_4 = \int_{N/D_4} \phi' \overline{\psi} dm, \ \forall \psi \in \mathcal{H}_4^{\infty}$.
(b) $R(X)\phi' = -if(X)\phi', \ \forall X \in \mathfrak{h}_4$.

If we admit (a) and (b), in order to obtain the lemma it suffices to prove that a function ϕ' on N that verifies (b) of $\mathcal{H}_4$ and which defines an element of the dual of $\mathcal{H}_4$ by the formula $\int_{N/D_4} \phi' \overline{\psi} dm$, and hence an element of $\mathcal{H}_4$, is equal to this last element as function on N. This result is easily verified when $\mathfrak{e}_4 = \mathfrak{n}$, using the properties of holomorphic functions. In general, we can use the properties of square-integrable vector fields on N/E_4 with values in $\mathcal{H}^{E_4}$.

Now we prove (a). Remark 6.2.2 gives us:

$$(T_{43}\phi|\psi)_4 = \int_{N/D_4} \overline{\psi}(n)\Big(\int_{D_4/(D_4\cap D_3)} \phi(nh)\chi(h)dm(h)\Big)dm(n)$$

for any $\phi \in \mathcal{H}_3^{\infty}$ and $\psi \in \mathcal{H}_4^{\infty}$.

We have $\mathfrak{h}_4 = \mathfrak{h}_3 \cap (\mathfrak{a}^f)_{\mathbb{C}} + \mathfrak{a}_{\mathbb{C}}$ and $\mathfrak{e}_4 = \mathfrak{e}_3 \cap \mathfrak{a}^f + \mathfrak{a}$ by projection on $\mathfrak{n}$. Then by orthogonality $\mathfrak{d}_4 = \mathfrak{d}_3 \cap \mathfrak{a}^f + \mathfrak{a}$. Consequently,

$$\mathfrak{d}_4/(\mathfrak{d}_4 \cap \mathfrak{d}_3) = (\mathfrak{d}_3 \cap \mathfrak{a}^f + \mathfrak{a})/(\mathfrak{d}_3 \cap \mathfrak{a}^f) \simeq \mathfrak{a}/(\mathfrak{d}_3 \cap \mathfrak{a}).$$

We can construct a CB of $\mathfrak{d}_4 \cap \mathfrak{d}_3$ in $\mathfrak{d}_4$ which is at the same time a CB of $\mathfrak{d}_3 \cap \mathfrak{a}$ in $\mathfrak{a}$. In this way,

$$\int_{D_4/(D_4 \cap D_3)} \cdot\, dm = \int_{A/(D_3 \cap A)} \cdot\, dm,$$

and we get (a).

It suffices now to prove (b) for $X \in \mathfrak{a}$ and $X \in \mathfrak{h}_3 \cap (\mathfrak{a}^f)_{\mathbb{C}}$. For $X \in \mathfrak{a}$, the claim comes from $\phi'(n \exp X) = e^{-if(X)}\phi'(n)$. If $X \in \mathfrak{e}_3 \cap \mathfrak{a}^f$, we have $R(X)\phi' = (R(X)\phi)'$. Indeed, in this case

$$\int_{A/(D_3 \cap A)} \phi(n \exp(tX)h)\chi(h)dm(h) = \int_{A/(D_3 \cap A)} \phi(nh \exp(tX))\chi(h)dm(h).$$

In order to see this, it suffices to make the change of variable $h' = \mathrm{Ad}(\exp(tX))(h)$ and confirm $\chi(h) = \chi(h')$ because of $\mathfrak{a} \subset \mathfrak{d}_3 = (\mathfrak{e}_3)^f$. Further, if we identify $A/(D_3 \cap A)$ to $\mathbb{R}^p$ by means of a CB, the Jacobian of the transformation $h \mapsto h'$ is equal to 1 by the fact that $\mathfrak{a}$ is an ideal in $\mathfrak{a}^f$. Hence if $X \in \mathfrak{h}_3 \cap (\mathfrak{a}^f)_{\mathbb{C}}$, we have $R(X)\phi' = (R(X)\phi)' = -if(X)\phi'$. □

With the lemma proved, we tackle the theorem by showing that

$$T_{\mathfrak{h}_1\mathfrak{h}_3^{\mathfrak{a}}} T_{\mathfrak{h}_3^{\mathfrak{a}}\mathfrak{h}_3} = T_{\mathfrak{h}_1\mathfrak{h}_3}.$$

Remark 6.2.2 gives us

$$(T_{\mathfrak{h}_1\mathfrak{h}_3}\phi|\psi) = \int_{N/(D_1 \cap D_3^{\mathfrak{a}})} \overline{\psi}(n)\Big(\int_{(D_1 \cap D_3^{\mathfrak{a}})/(D_1 \cap D_3)} \phi(nh)\chi(h)dm(h)\Big)dm(n).$$

Paying attention to

$$(\mathfrak{d}_1 \cap \mathfrak{d}_3^{\mathfrak{a}})/(\mathfrak{d}_1 \cap \mathfrak{d}_3) \simeq \mathfrak{a}/(\mathfrak{d}_3 \cap \mathfrak{a}),$$

we verify

$$\int_{(D_1 \cap D_3^{\mathfrak{a}})/(D_1 \cap D_3)} \cdot dm = \int_{A/(D_3 \cap A)} \cdot dm$$

and the last lemma gives the desired formula. □

We now proceed to the main theorem of this subsection. If $\mathfrak{h} \in P^+$, $\mathfrak{h}/\mathfrak{n}(f)$ is a positive Lagrangian subspace of the symplectic space $\mathfrak{n}/\mathfrak{n}(f)$. We have the following result (aside, this is *the* definition when $\mathfrak{n}$ is Heisenberg).

Theorem 6.2.20 *If* $\mathfrak{h}_j \in P^+ (1 \leq j \leq 3)$, *we have*

$$c(\mathfrak{h}_1, \mathfrak{h}_2, \mathfrak{h}_3) = e^{i\frac{\pi}{4}\tau(\mathfrak{h}_1, \mathfrak{h}_2, \mathfrak{h}_3)}.$$

Proof We proceed by induction on $\dim \mathfrak{n}$. The assertion is trivial if $\mathfrak{n}$ is commutative, hence if $\dim \mathfrak{n} = 1$. Assume $\dim \mathfrak{n} = p$ and that the theorem holds up to dimension $p - 1$. If there exists a subalgebra $\mathfrak{a}$ of $\mathfrak{n}$ contained in $\mathfrak{h}_1 \cap \mathfrak{h}_2$ such that $\mathfrak{a}^f$ is a proper subalgebra of $\mathfrak{n}$ containing $\mathfrak{a}$ as an ideal, the desired formula holds. In fact, using successively Theorem 6.2.17, Proposition 6.2.14, induction on p and Proposition 6.2.16, we have

$$\begin{aligned} c_{\mathfrak{n}}(\mathfrak{h}_1, \mathfrak{h}_2, \mathfrak{h}_3) &= c_{\mathfrak{n}}(\mathfrak{h}_1, \mathfrak{h}_2, (\mathfrak{h}_3)^{\mathfrak{a}}) = c_{\mathfrak{a}^f}(\mathfrak{h}_1, \mathfrak{h}_2, \mathfrak{h}_3) \\ &= e^{i\frac{\pi}{4}\tau_{\mathfrak{a}^f}(\mathfrak{h}_1,\mathfrak{h}_2,\mathfrak{h}_3)} = e^{i\frac{\pi}{4}\tau_{\mathfrak{n}}(\mathfrak{h}_1,\mathfrak{h}_2,\mathfrak{h}_3)}. \end{aligned}$$

When $\mathfrak{h}_j (1 \leq j \leq 3)$ are arbitrary, we can suppose $\mathfrak{n}$ non-commutative. Take $Y \in \mathfrak{n}$ such that $[Y, \mathfrak{n}] = \mathfrak{z}$ and put $\mathfrak{h}' = \mathfrak{h}^{\mathfrak{a}}$ with $\mathfrak{a} = \mathbb{R}Y$. Iterated applications of the cocycle relation give

$$\begin{aligned} c(\mathfrak{h}_1, \mathfrak{h}_2, \mathfrak{h}_3) =& c(\mathfrak{h}'_1, \mathfrak{h}_1, \mathfrak{h}_2)c(\mathfrak{h}'_2, \mathfrak{h}'_1\mathfrak{h}_2)c(\mathfrak{h}'_2, \mathfrak{h}_2, \mathfrak{h}_3) \\ &\times c(\mathfrak{h}'_2, \mathfrak{h}_3, \mathfrak{h}'_1)c(\mathfrak{h}_1, \mathfrak{h}'_1, \mathfrak{h}_3). \end{aligned}$$

Since $e^{i\frac{\pi}{4}\tau(\cdot,\cdot,\cdot)}$ also verifies the cocycle relation, it suffices to check the theorem for the terms of the right-hand side. Up to a permutation of indices, they are of two forms $c(\mathfrak{h}'_1, \mathfrak{h}'_2, \mathfrak{h}_3)$ and $c(\mathfrak{h}_1, \mathfrak{h}'_1, \mathfrak{h}_3)$.

For the first type $c(\mathfrak{h}'_1, \mathfrak{h}'_2, \mathfrak{h}_3)$, put $\mathfrak{a} = \mathbb{R}Y$. Then $\mathfrak{a} \subset \mathfrak{h}'_1 \cap \mathfrak{h}'_2$ and $\mathfrak{a}^f \neq \mathfrak{n}$. Consequently, $c = e^{i\frac{\pi}{4}\tau}$ by the remark at the beginning.

For the second type $c(\mathfrak{h}_1, \mathfrak{h}'_1, \mathfrak{h}_3)$, put $\mathfrak{a} = \mathfrak{h}_1 \cap \mathfrak{h}'_1 \cap \mathfrak{n} = \mathfrak{d}_1 \cap (\mathbb{R}Y)^f$. Two things can happen. Either $\mathfrak{a}^f = \mathfrak{e}_1 + \mathbb{R}Y$ is strictly contained in $\mathfrak{n}$, hence necessarily $c = e^{i\frac{\pi}{4}\tau}$, or $\mathfrak{n} = \mathfrak{e}_1 + \mathbb{R}Y$. In this latter case, there are two further possibilities. Either $\mathfrak{e}_1 = \mathfrak{n}$, hence $\mathfrak{e}_1/\mathfrak{b}_1$ is a Heisenberg and $c = e^{i\frac{\pi}{4}\tau}$ by Proposition 6.2.16 and Theorem 6.2.17, or $\mathfrak{n} = \mathfrak{e}_1 \oplus \mathbb{R}Y$. In the last eventuality, since $f|\mathfrak{n}(f)$ is not identically zero, the ideal $\mathfrak{b}_1 \cap (\mathbb{R}Y)^f$ is of codimension 1 in $\mathfrak{b}_1$ and $\mathfrak{n}/(\mathfrak{b}_1 \cap (\mathbb{R}Y)^f)$ is Heisenberg. We then deduce the result from Proposition 6.2.16 and Theorem 6.2.17. □

6.2.2 *Study of Connected Solvable Lie Groups*

In this subsection, we fix the following notations. Let G be a real connected solvable Lie group with Lie algebra $\mathfrak{g}$, $\mathfrak{n}$ a nilpotent ideal of $\mathfrak{g}$ containing $[\mathfrak{g}, \mathfrak{g}]$, N the analytic subgroup of G with Lie algebra $\mathfrak{n}$ and f a given linear form on $\mathfrak{g}$. We put $f' = f|_{\mathfrak{n}}$ and denote by $G(f)$ (resp. $G(f')$) the stabilizer of f (resp. f') in G. We denote by $\overline{G(f)}$ the reduced stabilizer of f defined as follows. Let χ_f be a unitary character of the identity component $G(f)^\circ$ of $G(f)$ with differential if and Z being the center of $G(f)/\ker(\chi_f)$. $\overline{G(f)}$ is the pre-image of Z in $G(f)$ and χ_f can be extended (in various ways) to a unitary character of $\overline{G(f)}$. Let $G(f)^1$ be a subgroup of $G(f)$

containing $\overline{G(f)}$ such that χ_f can be extended to a unitary character of $G(f)^1$. We shall denote by η_f such an extension (fixed in what follows). In general, K° denotes the identity component of a topological group K and $G(f')^1$ the subgroup $G(f)^1G(f')^\circ$. Let $\mathfrak{h}$ (or $\mathfrak{h}_i$ when i runs over a set of indices) be a polarization of $\mathfrak{g}$ at f, namely a totally isotropic subalgebra of $\mathfrak{g}_\mathbb{C}$ for the form $B_f(X,Y) = f([X,Y])$, of maximal dimension among all isotropic vector subspaces, and such that $\mathfrak{h}+\overline{\mathfrak{h}}$ is a subalgebra of $\mathfrak{g}_\mathbb{C}$. We consider only positive polarizations $\mathfrak{h}$, meaning $if([X,\overline{X}]) \geq 0$ if $X \in \mathfrak{h}$, which are strongly admissible for $\mathfrak{n}$. That is to say, they are stable under $G(f)$ and $\mathfrak{h} \cap \mathfrak{n}_\mathbb{C}$ is a polarization of $\mathfrak{n}$ at f' stable under $G(f')$. Auslander-Kostant [3] showed the existence of such polarizations and Vergne [150] constructed them.

If $\mathfrak{h}_i$ is a polarization, we define $\mathfrak{d}_i = \mathfrak{h}_i \cap \mathfrak{g}$, $\mathfrak{e}_i = (\mathfrak{h}_i + \overline{\mathfrak{h}_i}) \cap \mathfrak{g}$, the connected subgroups D_i°, E_i° with Lie algebras $\mathfrak{d}_i$, $\mathfrak{e}_i$ respectively and $D_i = G(f)^1 D_i^0$, $E_i = G(f)^1 E_i^\circ$.

We put $\mathfrak{n}^f = \{X \in \mathfrak{g}; B_f(X,\mathfrak{n}) = \{0\}\}$. This is nothing but the Lie algebra of $G(f')$. In general, a dash will indicate objects associated with $\mathfrak{n}$ and two dashes objects associated with $\mathfrak{n}^f$. In this fashion, we put

$$\mathfrak{h}_i' = \mathfrak{h}_i \cap \mathfrak{n}_\mathbb{C}, \mathfrak{h}_i'' = \mathfrak{h}_i \cap (\mathfrak{n}^f)_\mathbb{C}, \mathfrak{e}_i' = \mathfrak{e}_i \cap \mathfrak{n},$$

$$\mathfrak{e}_i'' = \mathfrak{e}_i \cap \mathfrak{n}^f, \mathfrak{d}_i' = \mathfrak{d}_i \cap \mathfrak{n}, \mathfrak{d}_i'' = \mathfrak{d}_i \cap \mathfrak{n}^f.$$

We denote by E_i', $(E_i'')^\circ$, D_i' and $(D_i'')^\circ$ the connected subgroups with Lie algebras $\mathfrak{e}_i'$, $\mathfrak{e}_i''$, $\mathfrak{d}_i'$ and $\mathfrak{d}_i''$ respectively. Finally, let $E_i'' = G(f)^1(E_i'')^\circ$, $D_i'' = G(f)^1(D_i'')^\circ$ and $f'' = f|_{\mathfrak{n}^f}$.

Let us recall some facts [24]. The admissibility of $\mathfrak{h}$ for $\mathfrak{n}$ gives $\mathfrak{h} = \mathfrak{h}' + \mathfrak{h}''$, where $\mathfrak{h}'$, $\mathfrak{h}''$ are polarizations of $\mathfrak{n}$, $\mathfrak{n}^f$ at f', f'' respectively. Besides, $\mathfrak{n} \cap \mathfrak{n}^f$ is a subalgebra of $\mathfrak{g}$ contained in $\mathfrak{h}' \cap \mathfrak{h}''$. Remark that $\mathfrak{h}$ is a polarization of $\mathfrak{n} + \mathfrak{n}^f$ at f restricted to this subalgebra. A polarization $\mathfrak{h}$ is said to be real if $\mathfrak{d} = \mathfrak{e}$.

Finally, let S be a group and j a homomorphism of S to the automorphism group of G. We suppose $\mathfrak{n}$, f, η_f invariant under the action of S. For the applications, S will be locally compact and separable, and j will be continuous.

Left-invariant Haar measures on a (Lie) group H are denoted by dm_H and the modular function by Δ_H:

$$\int_H F(xyx^{-1})dm_H(y) = \Delta_H(x) \int_H F(y)dm_H(y).$$

For a closed subgroup K of H, we put

$$\Delta_{K,H}(u) = \Delta_K(u)\Delta_H^{-1}(u), \ u \in K.$$

Let $C_0(H,K)$ be the set of complex-valued continuous functions F on H with compact support modulo K and satisfying

$$F(xu) = \Delta_{K,H}(u)F(x), \ x \in H, u \in K.$$

There exists a unique (up to a scalar multiple) positive H-invariant linear form $dm_{H,K}$ on $C_0(H, K)$ and we write its action as an integral:

$$dm_{H,K}(F) = \int_{H/K} F(x)dm_{H,K}(x).$$

To the data $(\mathfrak{g}, G, f, \eta_f, \mathfrak{h}_i)$ we shall associate the holomorphically induced representation $\rho_i^G = \rho(f, \eta_f, \mathfrak{h}_i, G)$. Here is its definition. We consider the space $\mathscr{C}_i^G = \mathscr{C}_0(f, \eta_f, \mathfrak{h}_i, G)$ of complex-valued continuous functions F on G with compact support modulo D_i and satisfying:

$$F(xd) = \Delta_{D_i,G}^{1/2}(d)\eta_f^{-1}(d)F(x),\ x \in G, d \in D_i,$$

$$(R(Y)F)(x) = \big(- if(Y) + \frac{1}{2}\mathrm{tr}_{\mathfrak{g}_\mathbb{C}/\mathfrak{e}_\mathbb{C}}\mathrm{ad}Y\big)F(x),\ Y \in \mathfrak{h}_i.$$

The space $\mathscr{C}_i^G$ is equipped with the Hermitian product

$$(F_1, F_2) = \int_{G/D_i} F_1(x)\overline{F_2(x)}dm_{G,D_i}(x)$$

and we denote by $\mathscr{H}_i^G = \mathscr{H}(f, \eta_f, \mathfrak{h}_i, G)$ the associated Hilbert space. Then ρ_i^G acts by left translations on $\mathscr{H}_i^G$:

$$(\rho_i^G(g)F)(x) = F(g^{-1}x),\ g \in G, x \in G.$$

Theorem 6.2.21 *Let $\mathfrak{h}_i (1 \le i \le 3)$ be three positive polarizations of $\mathfrak{g}$ at f, strongly admissible for $\mathfrak{n}$, and η_f a unitary character of $G(f)^1$ with differential if. There exist canonical unitary operators $T_{j,i}^G (1 \le i, j \le 3)$ from $\mathscr{H}_i^G$ to $\mathscr{H}_j^G$ intertwining ρ_i^G and ρ_j^G. The operators $T_{j,i}^G$ satisfy the relation (I), and are characterized by relation (II), or (III), when $\mathfrak{h}_i$, $\mathfrak{h}_j$ are real. The functions F_i describe dense subspaces of $\mathscr{H}_i^G$ (guaranteeing the convergence of the integrals), where c, c' are positive real constants depending on the choice of measures.*

$$(I)\ T_{1,2}^G \circ T_{2,3}^G \circ T_{3,1}^G = e^{i\frac{\pi}{4}\tau(\mathfrak{h}_1,\mathfrak{h}_2,\mathfrak{h}_3)}Id$$

$$(II)\ \langle T_{i,j}^G F_j, F_i\rangle = c\int_{G/(D_i\cap D_j)} F_j(g)\overline{F_i(g)}dm_{G,D_i\cap D_j}(g)$$

$$(III)\ (T_{i,j}^G F_j)(g) = c'\int_{D_i/(D_i\cap D_j)} F_j(gh)\eta_f(h)\Delta_{D_i,G}^{-1/2}(h)dm_{D_i,D_i\cap D_j}(h)$$

Remark 6.2.22 The technique of this proof may give us similar results when the positive polarizations are admissible for a nilpotent ideal $\mathfrak{n}$ of $\mathfrak{g}$ such that $\mathfrak{g}/\mathfrak{n}$ is nilpotent.

Proof The proof follows a classical pattern [3]. The group N is normal in G and of type I. Let $\rho' \in \widehat{N}$ be the equivalence class of $\rho(f, \mathfrak{h}', N)$ and $G(\rho')$ the stabilizer of ρ' in G. We study the situation at the level of $G(\rho')$ and then pass to G by means of induction. Following Pukanszky [144], we rather look at $G(f')^1 N$ which is a subgroup of G with the same Lie algebra as $G(\rho')$. The Lie algebra of $G(f')^1$ is $\mathfrak{n}^f$, and in fact what is relevant is only a certain nilpotent quotient $\mathfrak{n}^f/\mathfrak{a}$ of $\mathfrak{n}^f$. So we reduce to the case of nilpotent groups and the study of the previous subsection.

We begin by supposing that G is connected and simply connected. Let $G^1(f'')$ be the stabilizer of $f'' = f|_{\mathfrak{n}^f}$ in $G(f')^1$. The character η_f of $G(f)^1$ is uniquely extended to a character of $G^1(f'')$, denoted again by η_f, with differential if. In fact, $G^1(f'') = G(f)^1 N(f')$ and $N(f')$ is connected and simply connected. We put

$$\rho_i^{G(f')^1} = \rho(f'', \eta_f, \mathfrak{h}_i'', G(f')^1).$$

Let $G(f')^1 \ltimes N$ be the semi-direct product of $G(f')^1$ by N with multiplication

$$(g_1, n_1)(g_2, n_2) = (g_1 g_2, g_2^{-1} n_1 g_2 n_2).$$

Lemma 6.2.23 *Both the stabilizer $\mathfrak{n}^f(f'')$ of f'' in $\mathfrak{n}^f$ and $\mathfrak{a} = \mathfrak{n}^f(f'') \cap ker(f'')$ are ideals of $\mathfrak{n}^f$.*

Proof Remark that $\mathfrak{n}^f$ is a subalgebra and $[\mathfrak{n}^f, \mathfrak{n}^f] \subset \mathfrak{n} \cap \mathfrak{n}^f$. But $\mathfrak{n} \cap \mathfrak{n}^f$ is contained in $\mathfrak{n}^f(f'')$ since $\mathfrak{n}$ is orthogonal to $\mathfrak{n}^f$. Hence $[\mathfrak{n}^f, \mathfrak{n}^f(f'')] \subset \mathfrak{n}^f(f'')$, which proves the first assertion. Now, let $X \in \mathfrak{a}$ and $Y \in \mathfrak{n}^f$. Clearly $[X, Y] \in \mathfrak{n} \cap \mathfrak{n}^f \cap \ker(f'')$, which gives the second assertion. □

One consequence of the lemma is that $\mathfrak{n}^f/\mathfrak{a}$ is a nilpotent Lie algebra of step-length at most 2. Let A be the analytic subgroup of G with Lie algebra $\mathfrak{a}$. It is a connected and simply connected closed subgroup of G.

Lemma 6.2.24 *The subgroup A is normal in $G(f')^1$.*

Proof It is clear that the identity component of $G(f')^1 = G(f)^1 G(f')^\circ$ normalizes A. On the other hand, for $g \in G(f)^1$ (or $G(f)$ in general), $X \in \mathfrak{a}$ and $Y \in \mathfrak{n}^f$, the equalities

$$\langle f'', \mathrm{Ad}(g)X\rangle = \langle f, \mathrm{Ad}(g)X\rangle = \langle \mathrm{Ad}(g)^* f, X\rangle = 0$$

as well as

$$\langle f'', [\mathrm{Ad}(g)X, Y]\rangle = \langle f, \mathrm{Ad}(g)[X, \mathrm{Ad}(g^{-1})Y]\rangle = \langle f, [X, \mathrm{Ad}(g^{-1})Y]\rangle = 0$$

show that $G(f)^1$ normalizes A. □

Let $M = G(f')^1/A$, the Lie algebra of which is nilpotent. The identity component M° of M is simply connected. Let π be the canonical projection of $G(f')^1$ onto M. f'' and η_f pass to the quotient and the representations

$\rho_i^{G(f')^1}$, $\rho(f'', \eta_f, \pi(\mathfrak{h}_i''), M)$ are naturally deduced from each other. These two will be denoted by $\rho_i^{G(f')^1}$ and act on the space $\mathcal{H}_i^{G(f')^1}$. Let $\mathcal{H}_i^{M^\circ}$ be the space of $\rho_i^\circ = \rho(f, \pi(\mathfrak{h}_i''), M^\circ)$. The elements of $\mathcal{H}_i^{M^\circ}$ are the restrictions to M° of the functions of $\mathcal{H}_i^M = \mathcal{H}_i^{G(f')^1}$.

Lemma 6.2.25 *Let $r : \mathcal{H}_i^M \to \mathcal{H}_i^{M^\circ}$ be the mapping which associates to an element of $\mathcal{H}_i^M$ its restriction to M°, and $e : \mathcal{H}_i^{M^\circ} \to \mathcal{H}_i^M$ the extension mapping which associates to F° the function F given by*

$$F(mg) = \Delta_{D_i'',M}^{1/2}(g)\eta_f^{-1}(g)F^\circ(m),\ m \in M^\circ, g \in G(f)^1.$$

The mappings r, e are inverse isometries.

This lemma shows in particular the fact that the information on the representation $\rho(f'', \eta_f, \mathfrak{h}_i'', M)$ is subsumed in a comprehensive manner by the data $\rho(f'', \mathfrak{h}_i'', M^\circ)$ and η_f. This fact is related to the existence of the isomorphism

$$G(f')^\circ G(f)^1/D_i'' \simeq G(f')^\circ/(D_i'')^\circ.$$

Remark also the isomorphism

$$G(f')^\circ G(f)^1/(D_i'' \cap D_j'') \simeq G(f')^\circ/((D_i'')^\circ \cap (D_j'')^\circ).$$

There exist a positive linear form $dm_{ij} = dm_{G(f')^1, D_i \cap D_j}$ and an invariant measure $dm_{ij}^\circ = dm_{G(f')^\circ,(D_j'')^\circ}$ such that, for every pair $(F_i, F_j) \in \mathcal{H}_i^{G(f')^1} \times \mathcal{H}_j^{G(f')^1}$, we have

$$\int_{G(f')^1/(D_i''\cap D_j'')} F_i(g)\overline{F_j(g)}dm_{ij}(g) = \int_{G(f')^\circ/((D_i'')^\circ\cap(D_j'')^\circ)} F_i(g)\overline{F_j(g)}dm_{ij}^\circ(g).$$

The results of the previous subsection applied to the connected nilpotent group M°, and equality $T_{ij}^{G(f')^1} = e \circ T_{ij}^{G(f')^\circ} \circ r$, give:

Proposition 6.2.26 *If the positive form dm_{ij} is suitably normalized, the integral*

$$\langle T_{ji}^{G(f')^1} F_i, F_j\rangle = \int_{G(f')^1/(D_i''\cap D_j'')} F_i(g)\overline{F_j(g)}dm_{ij}(g)$$

converges for any pair (F_i, F_j) of differentiable vectors in $\mathcal{H}_i^{G(f')^1} \times \mathcal{H}_j^{G(f')^1}$, and uniquely determines an intertwining operator between $\rho_i^{G(f')^1}$ and $\rho_j^{G(f')^1}$.

Furthermore, we have the relation

$$T_{ij}^{G(f')^1} \circ T_{jk}^{G(f')^1} \circ T_{ki}^{G(f')^1} = e^{i\frac{\pi}{4}\tau(\mathfrak{h}_i'',\mathfrak{h}_j'',\mathfrak{h}_k'')} Id.$$

Lemma 6.2.27 *Let $\mathfrak{h}_i (1 \leq i \leq 3)$ be three positive polarizations at $f \in \mathfrak{g}^*$ admissible for $\mathfrak{n}$. Put $\rho_i^{G(f')} = \rho(f, \eta_f, \mathfrak{h}_i'', G(f'))$. The statement of the last proposition remains valid if we replace $G(f')^1$ by $G(f')$.*

This is an immediate consequence of the induction in stages [70].

Remark 6.2.28 When F_i is a differentiable vector in $\mathscr{H}_i^{G(f')}$, we have

$$(T_{ji}^{G(f')} F_i)(g) = \int_{D_j''/(D_i''\cap D_j'')} F_i(gd)\eta_f(d)dm_{D_j''/(D_i''\cap D_j'')}(d),$$

where $dm_{D_j''/(D_i''\cap D_j'')}$ is a suitably normalized, positive invariant measure on $D_j''/(D_i'' \cap D_j'')$.

Let $\Delta^{1/2}(g)$ be the norm in $\mathscr{H}_i^N = \mathscr{H}(f', \mathfrak{h}_i', N)$ of the application $F \mapsto F\circ g$, where $F\circ g(n) = F(gng^{-1}), n \in N$. For $F \in \mathscr{H}_i^N$ and $(g, n) \in G(f') \ltimes N$, put

$$(\overline{\rho}_i^N(g, n)F)(x) = \Delta_{D_i',N}^{-1/2}(g)F(n^{-1}g^{-1}xg), \ x \in N.$$

This formula defines a unitary representation of $G(f') \ltimes N$ which extends ρ_i^N. Besides, by the Auslander-Kostant theorem, which will be proved in the next subsection, we have

$$\overline{\rho}_j^N = T_{ji}^N \circ \overline{\rho}_i^N \circ T_{ij}^N. \tag{6.2.1}$$

Let us extend the representation $\rho_i^{G(f')}$ trivially on N. Thus we have at our disposal two representations $\rho_i^{G(f')}, \overline{\rho}_i^N$ of $G(f') \ltimes N$. Consider $\widehat{\rho_i} = \rho_i^{G(f')} \otimes \overline{\rho}_i^{G(f')}$ which acts on the space $\mathscr{H}_i^{G(f')} \otimes \mathscr{H}_i^N$. This representation is trivial at the elements $(n^{-1}, n), n \in N(f')$, of the kernel of the canonical projection $(g, n) \mapsto gn$ of $G(f') \ltimes N$ onto $G(\rho') = G(f')N$. On the other hand, if we put

$$\widehat{T}_{ji}^{G(\rho')} = T_{ji}^{G(f')} \otimes T_{ji}^N,$$

(6.2.1) implies

$$\widehat{\rho_j} = \widehat{T}_{ji}^{G(\rho')} \circ \widehat{\rho_i} \circ \widehat{T}_{ij}^{G(\rho')}.$$

Lemma 6.2.29 *Let $\mathfrak{h}_i (1 \leq i \leq 3)$ be positive polarizations at $f \in \mathfrak{g}^*$ admissible for $\mathfrak{n}$ and stable by $G(f')$. Then the operators $\widehat{T}_{ji}^{G(\rho')}$ intertwine $\widehat{\rho'}_j$ and $\widehat{\rho'}_i$ and satisfy the relations*

$$\widehat{T}_{12}^{G(\rho')} \circ \widehat{T}_{23}^{G(\rho')} \circ \widehat{T}_{31}^{G(\rho')} = e^{i\frac{\pi}{4}\tau(\mathfrak{h}_1,\mathfrak{h}_2,\mathfrak{h}_3)} Id.$$

$$\langle \widehat{T}_{21}^{G(\rho')} F_1, F_2 \rangle = c \int_X F_1(a,n)\overline{F_2(a,n)}dm(a)dm(n)$$

for a certain $c > 0$, $X = (G(f')/(D_1'' \cap D_2'')) \times N/(D_1' \cap D_2')$, where $F_i (i = 1, 2)$ are in certain dense subspaces of $\mathscr{H}_i^{G(f')} \otimes \mathscr{H}_i^N$.

These properties are immediate by construction, provided

$$e^{i\frac{\pi}{4}\tau(\mathfrak{h}_1,\mathfrak{h}_2,\mathfrak{h}_3)} = e^{i\frac{\pi}{4}\tau(\mathfrak{h}_1'',\mathfrak{h}_2'',\mathfrak{h}_3'')} e^{i\frac{\pi}{4}\tau(\mathfrak{h}_1',\mathfrak{h}_2',\mathfrak{h}_3')}.$$

This follows from the fact that $(\mathfrak{n} \cap \mathfrak{n}^f) \times (\mathfrak{n} \cap \mathfrak{n}^f)$ is in the kernel of the form $B_{f''} \otimes B_{f'}$ on $\mathfrak{n}^f \times \mathfrak{n}$ and from the properties of the Maslov index.

Formulas (I), (II) of Theorem 6.2.21 descend from the previous lemma. Indeed, it is known [3] that the representations $\overline{\rho}_i' = \rho_i^{G(f')} \otimes \rho_i^M$ and $\rho(f, \eta_f, \mathfrak{h}_i, G(\rho'))$ are intertwined by the operator

$$\Lambda_i : \mathscr{H}_i^{G(f')} \otimes \mathscr{H}_i^N \to \mathscr{H}_i^{G(\rho')}$$

given by

$$\Lambda_i(F''\otimes F')(a,b) = J_{N/D_i'}^{1/2}(a)F''(a)F'(aba^{-1}), \ (a,b) \in G(f') \times N,$$

where $J_{N/D_i'}^{1/2}(a)$ is the square root of the modulus of the action of a on N/D_i' (in fact, $a \in G(f')$ stabilizes D_i'). A formal calculation shows:

$$\begin{aligned}
&\langle \widehat{T}_{21}^{G(\rho')} F_1''\otimes F_1', F_2''\otimes F_2' \rangle \\
=&\Big(\int_{G(f')/(D_1''\cap D_2'')} F_1''(a)\overline{F_2''(a)}dm(a)\Big)\Big(\int_{N/(D_1'\cap D_2')} F_1'(b)\overline{F_2'(b)}dm(b)\Big) \\
=&\int_{G(f')/(D_1''\cap D_2'')} \Big(\int_{N/(D_1'\cap D_2')} F_1''(a)\overline{F_2''(a)} J_{N/(D_1'\cap D_2')}(a) \\
&\times F_1'(aba^{-1})\overline{F_2'(aba^{-1})}dm(b)\Big)dm(a) \\
=&\int_{G(\rho')/(D_1\cap D_2)} \Lambda_1(F_1''\otimes F_1')(n)\overline{\Lambda_2(F_2''\otimes F_2')(n)}dm(n),
\end{aligned}$$

since $J_{N/(D'_1 \cap D'_2)} = J^{1/2}_{N/D'_1} J^{1/2}_{N/D'_2}$ from Lemma 6.2.34 in the next subsection. The integrals converge absolutely if F''_i, F'_i are differentiable vectors.

Now, if we define $T^{G(\rho')}_{ji} = \Lambda_j \circ \widehat{T}^{G(\rho')}_{ji} \circ \Lambda_i^{-1}$, the above equalities give us the theorem's (I), (II) by induction from $G(\rho')$ to G. Finally, (III) follows from (II) by Fubini's theorem.

Evidently all of these extend to the case where G is connected but not necessarily simply connected. We must then suppose the existence of the character η_f of $G(f)^1$, namely that f is integral. □

Let S be a group acting on G by automorphisms and leaving $\mathfrak{n}$, f, η_f stable. Let $\mathfrak{h}$ be a positive polarization at f strongly admissible for $\mathfrak{n}$. If F is a function on G, we define for $s \in S$ the operator $D(s)$ by

$$(D(s)F)(g) = (F \circ s^{-1})(g) = F(s^{-1}(g)), \ g \in G.$$

The image of $\mathscr{H}(f, \eta_f, \mathfrak{h}, G)$ under $D(s)$ is $\mathscr{H}(f, \eta_f, s(\mathfrak{h}), G)$ and we denote by $\|s\|_{\mathfrak{h}}$ the norm of this operator. We put

$$R_{\mathfrak{h}}(s) = \|s\|_{\mathfrak{h}}^{-1} T_{\mathfrak{h}, s(\mathfrak{h})} \circ D(s).$$

$R_{\mathfrak{h}}(s)$ acts in $\mathscr{H}(f, \eta_f, \mathfrak{h}, G)$ and verifies

$$\rho \circ s^{-1}(g) = R_{\mathfrak{h}}(s)^{-1} \circ \rho(g) \circ R_{\mathfrak{h}}(s) \ (g \in G)$$

with $\rho = \rho(f, \eta_f, \mathfrak{h}, G)$. We define in this way a projective representation $R_{\mathfrak{h}}$ of S on the space of ρ.

Theorem 6.2.30 *The cocycle of the projective representation $R_{\mathfrak{h}}$ of S is related to the Maslov index by*

$$R_{\mathfrak{h}}(s_1) \circ R_{\mathfrak{h}}(s_2) = e^{i\frac{\pi}{4}\tau(\mathfrak{h}, s_1(\mathfrak{h}), s_1 s_2(\mathfrak{h}))} R_{\mathfrak{h}}(s_1 s_2).$$

Corollary 6.2.31 *Assume that G is a closed normal subgroup of a separable locally compact group K and that S is the stabilizer in K of f, η_f, $G(f)^1$. Then the Maslov index is the obstruction to extend $\rho(f, \eta_f, \mathfrak{h}, G)$ to a representation of SG.*

6.2.3 Explicit Expression of Intertwining Operators

According to whether the groups of concern are nilpotent or solvable, we adopt the hypotheses and notations of previous subsections.

Let $\mathfrak{h}_i (i = 1, 2)$ be two positive polarizations of $\mathfrak{n}$ at f. We have defined the unitary intertwining operator $T_{21} = c I_{21}$, c being a positive constant, between ρ_1^N

and ρ_2^N by

$$(I_{21}F_1, F_2)_{\mathscr{H}_2} = \int_{N/(D_1\cap D_2)} F_1\overline{F_2}dm, \ (F_1, F_2) \in \mathscr{H}_1 \times \mathscr{H}_2.$$

Theorem 6.2.32 *There exists a unique function $\widetilde{K}_2$ on E_2 satisfying for any $X \in \mathfrak{h}$ the conditions*

$$R(\overline{X})\widetilde{K}_2 = if(\overline{X})\widetilde{K}_2, \ L(X)\widetilde{K}_2 = -if(X)\widetilde{K}_2$$

and $\widetilde{K}_2(1_N) = 1$, where L is the left action and 1_N the unit element of N. The operator T_{21} is of the form $c'\widetilde{T}_{21}$, c' being a positive constant, where $\widetilde{T}_{21}$ is given for $F_1 \in \mathscr{H}_1^\infty$, $g \in N$ by the convergent integral

$$(\widetilde{T}_{21}F_1)(g) = \int_{E_2/(D_1\cap D_2)} F_1(gh)\widetilde{K}_2(h)dm(h). \tag{6.2.2}$$

Proof First we construct the function $\widetilde{K}_2$. Since $\mathfrak{e}_2/\mathfrak{b}_2$ is Heisenberg, it suffices to consider an invariant function under translations by B_2. Define

$$\widetilde{K}_2(\exp(xP + yQ)\exp(wZ)) = e^{-\frac{x^2+y^2}{4}}e^{iw},$$

where (P_i, Q_i, Z) corresponds to a canonical basis of $\mathfrak{e}_2/\mathfrak{d}_2$. This is but one convenient function of many. The techniques developed in the first subsection, especially Lemmas 6.2.8–6.2.10 guarantee the convergence of (6.2.2). Furthermore, $\widetilde{T}_{21}F_1$ satisfies the covariance relations by $\mathfrak{h}_2$ of the elements of $\mathscr{H}_2$. The mapping $F_1 \mapsto \widetilde{T}_{21}F_1$ commutes with the action of N, sends $\mathscr{H}_1^\infty$ to the dual of $\mathscr{H}_2^\infty$ continuously and, by Poulsen [140], it intertwines ρ_1^N and ρ_2^N. We shall see afterwards that $\widetilde{K}_2$ is related to the Bergman kernel of $\mathscr{H}_2$, and that $\widetilde{T}_{21}$ is not a trivial operator.

There remains to show that the proportionality factor c' between T_{21} and $\widetilde{T}_{21}$ is strictly positive. Let $(H_k)_{k\in\mathbb{N}}$ be an orthonormal basis of $\mathscr{H}_2$. Then

$$I_{21}F_1 = \sum_{k\in\mathbb{N}}(I_{21}F_1, H_k)H_k, \ F_1 \in \mathscr{H}_1^\infty.$$

Put

$$K^n(g_1, g_2) = \sum_{k\leq n} H_k(g_1)\overline{H_k(g_2)}, \ (g_1, g_2) \in N \times N$$

and

$$(I^n F_1)(g_1) = \int_{N/(D_1\cap D_2)} F_1(g_2)K^n(g_1, g_2)dm(g_2).$$

On $\mathscr{H}_2$, we have $I_{21} = \lim_{n\to\infty} I^n$ pointwise. Suppose now $\mathfrak{h}_2$ totally complex, namely $\mathfrak{e}_2 = \mathfrak{n}$, and let $2n+1$ be the dimension of $\mathfrak{e}_2/\mathfrak{b}_2$. Write $z = x + iy \in \mathbb{C}^n$. Then

$$\widetilde{K}_2(\exp(xP + yQ)\exp(wZ)) = e^{-\frac{|z|^2}{4}} e^{iw}.$$

For $g = \exp(xP + yQ)$ and $g' = \exp(x'P + y'Q)$, we have

$$\begin{aligned}\widetilde{K}_2(g'^{-1}g) &= \exp\Big(\frac{i}{2}\sum_j (x_j y'_j - x'_j y_j) - \frac{1}{4}\sum_j ((x_j - x'_j)^2 + (y_j - y'_j)^2)\Big)\\ &= \overline{\widetilde{K}_2}\widetilde{K}_2(g)\exp\Big(\frac{1}{2}\sum_j (x_j x'_j + y_j y'_j + i(x_j y'_j - x'_j y_j)\Big).\end{aligned}$$

That is to say,

$$\widetilde{K}_2(g'^{-1}g) = \overline{\widetilde{K}_2(g')}e^{\frac{1}{2}\sum_j z'_j z_j}. \tag{6.2.3}$$

On the other hand, the functions $(H_m)_{m\in\mathbb{N}}$ defined by

$$H_m(g) = \frac{1}{(2\pi)^{n/2}}\frac{1}{(m_1!\cdots m_n!)^{1/2}}\frac{1}{2^{|m|/2}}\overline{z^m\widetilde{K}_2(g)}$$

form an orthonormal basis of $\mathscr{H}_2$. At the same time the sequence of kernels $K^n(g, g')$ defining I^n converges to a kernel

$$\begin{aligned}K(g', g) &= \sum_{m\in\mathbb{N}^n} H_m(g')\overline{H_m(g)}\\ &= \frac{1}{(2\pi)^n}\Big(\sum_{m\in\mathbb{N}^n}\frac{1}{2^{|m|}m_1!\cdots m_n!}\overline{z'^m \bar{z}^m}\Big)\overline{\widetilde{K}_2(g')}\widetilde{K}_2(g)\\ &= \frac{1}{(2\pi)^n}e^{\frac{1}{2}\sum_j \overline{z'_j \bar{z}_j}}\overline{\widetilde{K}_2(g')}\widetilde{K}_2(g) = \frac{1}{(2\pi)^n}\widetilde{K}_2(g'^{-1}g)\end{aligned}$$

by (6.2.3). This shows $\widetilde{T}_{21} = (2\pi)^n I_{21}$. Remark that $K(\cdot, g) \in \mathscr{H}_2^\infty$ for every fixed $g \in N$,

$$K(g', g) = K(g^{-1}g', 1_N) = K(1_N, g'^{-1}g)$$

for $(g, g') \in N \times N$, and

$$(F_2, \overline{\widetilde{K}_2(g'^{-1}\cdot)})_{\mathcal{H}_2} = (2\pi)^n (F_2, K(\cdot, g'))_{\mathcal{H}_2} = (2\pi)^n F_2(g'), \; g' \in N,$$

for any $F_2 \in \mathcal{H}_2$. In this way, we recover the reproducing kernel of $\mathcal{H}_2$.

The theorem is hence established for totally complex $\mathfrak{h}_2$, and we now forget this hypothesis. From what we have seen,

$$(\widetilde{T}_{21} F_1, F_2)_{\mathcal{H}_2} = \int_{N/D_2} \overline{F_2(g)} \Big(\int_{E_2/(D_1 \cap D_2)} F_1(gh) \widetilde{K}_2(h) dm(h) \Big) dm(g)$$

$$= \int_{N/E_2} \Big(\int_{E_2/D_2} \Big(\int_{E_2/(D_1 \cap D_2)} \overline{F_2(gk)} F_1(gh) \widetilde{K}_2(k^{-1}h) dm(h) \Big) dm(k) \Big) dm(g) \tag{6.2.4}$$

for $F_i \in \mathcal{H}_i^\infty (i = 1, 2)$. It follows from the proof of Theorem 6.2.7 in the first subsection that the function $(h, k) \mapsto \overline{F_2(gk)} F_1(gh) \widetilde{K}_2(k^{-1}h)$ is integrable over $(E_2/D_2) \times (E_2/(D_1 \cap D_2))$. If we exchange the integrals, taking into account (6.2.4) and the fact that $\mathfrak{h}_2$ is totally complex in $\mathfrak{e}_2$, we have

$$\begin{aligned}(\widetilde{T}_{21} F_1, F_2)_{\mathcal{H}_2} &= \int_{N/E_2} \Big(\int_{E_2/(D_1 \cap D_2)} c F_1(gh) (K(\cdot, h), F_2(g\cdot))_{\mathcal{H}_2} dm(h) \Big) dm(g) \\ &= c \int_{N/E_2} \Big(\int_{E_2/(D_1 \cap D_2)} F_1(gh) \overline{F_2(gh)} dm(h) \Big) dm(g) \\ &= c (I_{21} F_1, F_2)_{\mathcal{H}_2}\end{aligned}$$

with $c > 0$. □

We prove here a theorem of Auslander and Kostant. Consider as in the second subsection the semi-direct product $G(f') \ltimes N$ and its representations $\overline{\rho}'_i$ in $\mathcal{H}_i^N$ defined by

$$(\overline{\rho}'_i(a, b) F)(n) = \Delta_{D'_i, N}^{-1/2}(a) F(b^{-1} a^{-1} n a), \; (a, b) \in G(f') \ltimes N, \; F \in \mathcal{H}_i^N.$$

Theorem 6.2.33 *The representations $\overline{\rho}'_i (i = 1, 2)$, which extend to $G(f') \ltimes N$ the representations $\rho'_i = \rho(f', \mathfrak{h}'_i, N)$ associated to two positive polarizations $\mathfrak{h}'_i$ stable under $G(f')$, are intertwined by T_{ij}^N:*

$$\overline{\rho}'_1(g) = T_{12}^N \circ \overline{\rho}'_2(g) \circ T_{21}^N, \; g = (a, b) \in G(f') \ltimes N.$$

Proof The kernel $\widetilde{K}_2^N$ given by Theorem 6.2.32 is invariant under the action of $G(f')$ by inner automorphisms. Indeed, $\mathfrak{h}'_2$ is invariant under $G(f')$ and the given information of $\mathfrak{h}'_2$ determines $\widetilde{K}_2^N$. Then we have, for $g = (a, b) \in G(f') \ltimes N$ and a differentiable vector F in $\mathscr{H}_1^N$,

$$(\overline{\rho}_2'(g) \circ \widetilde{T}_{21}^N F)(n) = \Delta_{D_2',N}^{-1/2}(a) \int_{E_2'/(D_1' \cap D_2')} F(b^{-1}a^{-1}nak)\widetilde{K}_2^N(k)dm(k).$$

On the other hand, the change of variable $h \mapsto a^{-1}ha$ gives

$$\begin{aligned}(T_{21}^N \circ \rho_1'(g)F)(n) &= \int_{E_2'/(D_1' \cap D_2')} \Delta_{D_1',N}^{-1/2}(a)F(b^{-1}a^{-1}nah)\widetilde{K}_2^N(h)dm(h) \\ &= \Delta_{D_1',N}^{-1/2}(a)\Delta_{D_1' \cap D_2', E_2'}(a) \int_{E_2'/(D_1' \cap D_2'')} \\ &\quad F(b^{-1}a^{-1}nah)\widetilde{K}_2^N(h)dm(h).\end{aligned}$$

So, the theorem will be proved if we show

$$\Delta_{D_1',N}^{-1/2}(a)\Delta_{D_1' \cap D_2', E_2'}(a) = \Delta_{D_2',N}^{-1/2}(a), \ a \in G(f'). \tag{6.2.5}$$

This is a consequence of the following lemma.

Lemma 6.2.34 *Let Q be a Lie group with Lie algebra $\mathfrak{q}$, $f \in \mathfrak{q}^*$, s an automorphism of $\mathfrak{q}$ which leaves the form B_f invariant and $\mathfrak{h}_i (i = 1, 2)$ two positive polarizations of $\mathfrak{q}$ at f stable under s. Then the subspaces $\mathfrak{d}_i$, $\mathfrak{e}_i$ are stable under s, and*

(1) $(\det_{(\mathfrak{e}_1+\mathfrak{e}_2)/\mathfrak{e}_j} s)(\det_{\mathfrak{d}_j/(\mathfrak{d}_1 \cap \mathfrak{d}_2)} s) = 1 \ (j = 1, 2)$,
(2) $|\det_{(\mathfrak{e}_1+\mathfrak{e}_2)/\mathfrak{e}_1} s||\det_{(\mathfrak{e}_1+\mathfrak{e}_2)/\mathfrak{e}_2} s| = 1$,
(3) $|\det_{\mathfrak{d}_1/(\mathfrak{d}_1 \cap \mathfrak{d}_2)} s||\det_{\mathfrak{d}_2/(\mathfrak{d}_1 \cap \mathfrak{d}_2)} s| = 1$.

Proof The form B_f is a pairing of $(\mathfrak{e}_1 + \mathfrak{e}_2)/\mathfrak{e}_j$ and $\mathfrak{d}_j/(\mathfrak{d}_1 \cap \mathfrak{d}_2) (j = 1, 2)$. Hence (1) follows from the equality

$$B_f(sX, Y) = B_f(X, s^{-1}Y), \ X \in (\mathfrak{e}_1 + \mathfrak{e}_2)/\mathfrak{e}_j, Y \in \mathfrak{d}_j/(\mathfrak{d}_1 \cap \mathfrak{d}_2),$$

coming from the invariance of B_f under s.

Recall that the positive polarization $\mathfrak{h}_i$ is associated to an inner product S_i on $\mathfrak{e}_i/\mathfrak{d}_i$ and s acts as an orthogonal transformation on this space. Hence if $\mathfrak{a}_i$ is a subspace invariant by s and containing $\mathfrak{d}_i$, we have $|\det_{\mathfrak{a}_i/\mathfrak{d}_i} s| = 1$. In particular,

$$|\det_{(\mathfrak{d}_1 \cap \mathfrak{e}_2)/(\mathfrak{d}_1 \cap \mathfrak{d}_2)} s| = |\det_{((\mathfrak{d}_1 \cap \mathfrak{e}_2)+\mathfrak{d}_2)/\mathfrak{d}_2} s| = 1.$$

By duality,

$$|\det_{(\mathfrak{e}_1+\mathfrak{e}_2)/(\mathfrak{e}_1+\mathfrak{d}_2)} s| = |\det_{(\mathfrak{e}_1+\mathfrak{e}_2)/(\mathfrak{d}_1+\mathfrak{e}_2)} s| = 1.$$

Now, $(\mathfrak{e}_1+\mathfrak{d}_2)/\mathfrak{e}_1 \simeq \mathfrak{d}_2/(\mathfrak{e}_1\cap\mathfrak{d}_2)$ is paired with $(\mathfrak{d}_1+\mathfrak{e}_2)/\mathfrak{e}_2$, and the action of s on the first space is obtained by transposing the action of s^{-1} on the second. Hence

$$\begin{aligned}|\det_{(\mathfrak{e}_1+\mathfrak{e}_2)/\mathfrak{e}_1} s| &= |\det_{(\mathfrak{e}_1+\mathfrak{d}_2)/\mathfrak{e}_1} s| \\ &= |\det_{(\mathfrak{d}_1+\mathfrak{e}_2)/\mathfrak{e}_2} s^{-1}| = |\det_{(\mathfrak{e}_1+\mathfrak{e}_2)/\mathfrak{e}_2} s|^{-1},\end{aligned}$$

and (2) follows. Equality (3) is clear by duality. □

Relation (6.2.5) descends from this lemma applied to $Q = N, s = a \in G(f')$. Indeed,

$$\begin{aligned}\Delta^{-1/2}_{D_1',N}(a)\Delta_{D_1'\cap D_2',E_2'}(a) &= \Delta^{-1/2}_{D_1'\cap D_2',N}(a)\Delta^{1/2}_{D_1'\cap D_2',D_1'}(a)\Delta_{D_1'\cap D_2',D_2'}(a) \\ &= \Delta^{-1/2}_{D_1'\cap D_2',N}(a)\Delta^{1/2}_{D_1'\cap D_2',D_2'}(a) = \Delta^{-1/2}_{D_2',N}(a).\end{aligned}$$

Let us proceed to the solvable case. We retake the notations of the second subsection and assume $G(f)^1 = G(f)$ for simplicity. The polarizations $\mathfrak{h}_i$ are strongly admissible for $\mathfrak{n}$. Lemma 6.2.24 permits us to consider the quotient $M = G(f')/A$ which is nilpotent, and in fact only the connected component M° of M intervenes. Hence we have the (unique) kernel $K_2^{M^\circ}$ given by Theorem 6.2.32. Put

$$K_2^{G(f')}(mg) = K_2^{M^\circ}(m)\eta_f(g),\ \ g \in G(f), m \in G(f')^\circ.$$

If F_1'' is a differentiable vector of $\mathscr{H}_1^{G(f')}$,

$$(T_{21}^{G(f')}F_1'')(g) = \int_{E_2''/(D_1''\cap D_2'')} F_1''(gk)K_2^{G(f')}(k)dm(k),\ \ g \in G(f').$$

We also have at disposal the kernel K_2^N, for which

$$(T_{21}^N F_1')(n) = \int_{E_2'/(D_1'\cap D_2')} F_1'(nh)K_2^N(h)dm(h),\ \ n \in N,$$

for any differentiable vector F_1' in $\mathscr{H}_1^N$. We have already remarked that $\widetilde{K}_2$, and hence K_2^N too, are invariant by $G(f')$. We then define K_2^G by

$$K_2^G(h) = K_2^{G(f')}(h')K_2^N(h'')\Delta^{-1/2}_{E_2,G}(h),\ \ h = h'h'', (h', h'') \in E_2' \times E_2''.$$

Remark that $K_2^G(h)$ does not depend on the expression $h = h'h''$.

Theorem 6.2.35 *Let $\mathfrak{h}_i\,(i=1,2)$ be two polarizations at f strongly admissible for $\mathfrak{n}$. There exist a dense subspace $\mathscr{C}_1$ of $\mathscr{H}_1^G=\mathscr{H}(f,\eta_f,\mathfrak{h}_1,G)$ and a positive form $dm_{E_2,D_1\cap D_2}$ on $C_0(E_2,D_1\cap D_2)$ such that:*

(1) The integral

$$(\overline{T}_{21}F_1)(g)=\int_{E_2/(D_1\cap D_2)}F_1(gh)K_2^G(h)dm_{E_2,D_1\cap D_2}(h)$$

converges for any F_1 in $\mathscr{C}_1$.

(2) The operator $\overline{T}_{21}$ extends continuously to a bijective isometry from $\mathscr{H}_1^G$ to $\mathscr{H}_2^G$, which coincides with the intertwining operator T_{21} defined in Theorem 6.2.21.

Proof We verify

$$F(ghd)K_2^G(hd)=F(gh)K_2^d(h)\Delta_{D_1\cap D_2,E_2}(d),\ d\in D_1\cap D_2,\ (g,h)\in G\times E_2,$$

for $F\in\mathscr{H}_1^G$, and we wish to interpret the integral of statement (1). Consider the group $G(\rho')=G(f')N$. The representation $\rho_i^{G(f')}\otimes\overline{\rho}_i'$ of $G(f')\ltimes N$ defines by passing to the quotient a representation of $G(\rho')$, denoted by $\widehat{\rho_i}$ in the second subsection. It turns out (cf. [3, 24]) that $\widehat{\rho_i}$ is equivalent to $\rho(f,\eta_f,\mathfrak{h}_i,G(\rho'))$, the intertwining operator Λ_1 being given by

$$\Lambda_1(F'\otimes F'')(g)=\Delta_{D_i',N}^{1/2}(a)F'(a)F''(aba^{-1})$$

with $g=ab\in G(\rho')$, $(a,b)\in G(f')\ltimes N$, $F'\otimes F''\in\mathscr{H}_i^{G(f')}\otimes\mathscr{H}_i^N$. On the other hand, we have defined $\widehat{T}_{ji}^{G(\rho')}=T_{ji}^{G(f')}\otimes T_{ji}^N$ as an operator intertwining $\widehat{\rho_i}$ and $\widehat{\rho}_j$. Hence we have a diagram

$$\begin{array}{ccc}\mathscr{H}_1^{G(f')}\otimes\mathscr{H}_1^N & \xrightarrow{\Lambda_1} & \mathscr{H}_1^{G(\rho')}\\ \widehat{T}_{21}^{G(\rho')}\downarrow & & \downarrow T_{21}^{G(\rho')}\\ \mathscr{H}_2^{G(f')}\otimes\mathscr{H}_2^N & \xrightarrow{\Lambda_2} & \mathscr{H}_2^{G(\rho')}\end{array}$$

and we want to obtain the operator $T_{21}^{G(\rho')}$ which makes this commutative. For $F=\Lambda_1(F''\otimes F')$ with differentiable vectors F'', F' in $\mathscr{H}_1^{G(f')}$, $\mathscr{H}_1^N$ respectively,

$$(T_{21}^{G(\rho')}\circ\Lambda_1)(F''\otimes F')(gn)=\Lambda_2\circ\widehat{T}_{21}^{G(\rho')}(F''\otimes F')(g,n),\ \ (g,n)\in G(f')\ltimes N,$$
$$=\Delta_{D_2',N}^{1/2}(g)(T_{21}^{G(f')}F'')(g)(T_{21}^NF')(gng^{-1})$$

$$
\begin{aligned}
=&\Delta_{D_2',N}^{1/2}(g)\int_{E_2''/(D_2''\cap D_1'')} F''(gh)K_2^{G(f')}(h)dm(h)\\
&\times\int_{E_2'/(D_2'\cap D_1')} F'(gng^{-1}k)K_2^N(k)dm(k)\\
=&\Delta_{D_2',N}^{1/2}(g)\int_{E_2''/(D_2''\cap D_1'')} F''(gh)K_2^{G(f')}(h)\Delta_{D_2'\cap D_1',E_2'}(gh)dm(h)\\
&\times\int_{E_2'/(D_2'\cap D_1')} F'(gnhkh^{-1}g^{-1})K_2^N(k)dm(k).
\end{aligned}
$$

But from Lemma 6.2.34 and equality (6.2.5),

$$\Delta_{D_2',N}^{1/2}(g)\Delta_{D_2'\cap D_1',E_2'}(gh)=\Delta_{D_2',N}^{-1/2}(h)\Delta_{D_1',N}^{1/2}(gh).$$

On the other hand, $\Delta_{D_2',N}^{-1/2}(h)=\Delta_{E_2',G(\rho')}^{-1/2}(h)$ and the last integral is written as

$$\int_{E_2/(D_1\cap D_2)} \Lambda_1(F''\otimes F')(t)dm(t).$$

The absolute convergence of the integral is clear. The theorem on the group $G(\rho')$ (in general non-connected) follows from this last expression. We obtain the desired result on G by induction from $G(\rho')$ to G. □

6.2.4 Examples

Example 6.2.36 Suppose that $\mathfrak{h}_2$ is totally complex. Then $G=E_2=G(f')N$ and $K_2^G(g)=K_2^{G(f')}(a)K_2^N(n)$ if $g=an,(a,n)\in G(f')\ltimes N$. We can use the coordinate expressions of the kernels $K_2^{G(f')}$, K_2^N to obtain the expression of K_2^G that defines a reproducing kernel of the space $\mathscr{H}_2^G$.

Example 6.2.37 Let $\mathfrak{g}=\mathbb{R}\text{-span}(X_1,X_2,P,Q,Z)$ with non-zero bracket relations

$$[P,Q]=Z,[X_1,X_2]=Z,[X_i,P]=-P,[X_i,Q]=Q,\ i=1,2.$$

$\mathfrak{g}$ is completely solvable of dimension 5. Let $f=Z^*\in\mathfrak{g}^*$ with the dual basis. Put

$$\mathfrak{h}_i=\mathbb{R}X_i+\mathbb{R}P+\mathbb{R}Z\ (i=1,2),\ \mathfrak{h}_3=\mathbb{R}X_2+\mathbb{R}Q+\mathbb{R}Z.$$

Let G be the connected and simply connected Lie group with Lie algebra $\mathfrak{g}$. We have $G(f)=\exp(\mathbb{R}Z)$ and $G(f')=\exp(\mathbb{R}X_1+\mathbb{R}X_2+\mathbb{R}Z)$. Then $\mathfrak{h}_i(1\le i\le 3)$

are real polarizations of $\mathfrak{g}$ at f strongly admissible for $\mathfrak{n} = \langle P, Q, Z\rangle_{\mathbb{R}}$. Let $F_i \in \mathscr{H}_i^G$ $(i = 1, 2)$. The function F_i is determined by its values on the elements of G having the form $\exp(uQ)\exp(x_jX_j)$, $(j = 1, 2,\ j \neq i)$. Besides,

$$\|F_i\|^2_{\mathscr{H}_i} = \int_{G/H_i} |F_i(g)|^2 dm_{G,H_i}(g) = \int_{\mathbb{R}^2} |F_i(\exp(uQ)\exp(x_iX_i))|^2 e^{-x_j} dx_j du.$$

For well-chosen $F_1 \in \mathscr{H}_1^G$,

$$(T_{21}F_1)(\exp(uQ)\exp(sX_1)) = \int_{\mathbb{R}} F_1(\exp(uQ)\exp(xX_2))e^{-x/2}e^{-isx}dx.$$

Concerning the expression of $T_{31}F_1$, it is given by

$$\begin{aligned}(T_{31}F_1)(\exp(aX_1)\exp(bP)) &= e^{-a/2}\int_{\mathbb{R}^2} F_1(\exp(sX_2)\exp(tQ))e^{s/2}e^{isa-itbe^{s-a}}dsdt \\ &= e^{a/2}\int_{\mathbb{R}^2} F_1(\exp(sX_2)\exp(te^{a-s}Q))e^{-s/2}e^{i(as-tb)}dsdt.\end{aligned}$$

6.3 Real Polarizations

Following [61] we study in this section real polarizations for solvable Lie groups and the associated holomorphically induced representations. We assume all Lie groups and Hilbert spaces are separable. Let V be a finite-dimensional vector space and $W \subset U$ subspaces of V. When $x \in \mathrm{End}(V)$ leaves U, W stable, $x_{U/W}$ denotes the induced element of $\mathrm{End}(U/W)$. Let G be a Lie group with Lie algebra $\mathfrak{g}$, and π a representation of G on V. For $g \in G$ (resp. $X \in \mathfrak{g}$), we shall write $\pi_{U/W}(g)$ (resp. $\pi_{U/W}(X)$) instead of $\pi(g)_{U/W}$ (resp. $\pi(X)_{U/W}$). We denote by $M(f, \mathfrak{g})$ (resp. $M(f, \mathfrak{g}_{\mathbb{C}})$) the set of subalgebras (resp. complex subalgebras) of $\mathfrak{g}$ (resp. $\mathfrak{g}_{\mathbb{C}}$) which are maximally totally isotropic subspaces for the form B_f, and by $P^+(f, \mathfrak{g})$ the set of positive polarizations at $f \in \mathfrak{g}^*$. Let H be a closed subgroup of G with Lie algebra $\mathfrak{h}$. On the space of numerical functions F on G satisfying

$$F(gh) = \Delta_{H,G}(h)F(g) \ (g \in G, h \in H),$$

there exists a unique (up to scalar multiples) positive G-invariant linear form denoted by

$$F \mapsto \nu_{G,H}(F) = \oint_{G/H} F(g)d\nu_{G,H}(g),$$

which we define even when F takes values in a Banach space. If there exists an invariant measure on the homogeneous space G/H, we shall often write it

$d\dot{g}$. Recall how we constructed the induced representations. Let π be a unitary representation of H on a Hilbert space $\mathscr{H}$. The induced representation $\operatorname{ind}_H^G \pi$ of G is obtained by left translations in the space of $\mathscr{H}$-valued measurable functions ϕ on G satisfying

$$\phi(gh) = \Delta_{H,G}^{1/2}(h)\pi(h)^{-1}\phi(g) \ (g \in G, h \in H)$$

and $\nu_{G,H}(\|\phi\|^2) < \infty$.

Using the notations introduced in the previous section, fix $\mathfrak{h} \in P^+(f, \mathfrak{g})$ invariant by $G(f)$. We say that $\mathfrak{h}$ satisfies the Pukanszky condition if $D \cdot f = f + \mathfrak{e}^\perp$. Suppose that there exists a unitary representation η_f of $G(f)$ on a Hilbert space $\mathscr{H}$ with differential $(if|_{\mathfrak{g}(f)})\mathrm{id}_{\mathscr{H}}$. If $\mathfrak{h}$ satisfies the Pukanszky condition, η_f uniquely extends to a unitary representation χ_f of D with differential $(if|_{\mathfrak{d}})\mathrm{id}_{\mathscr{H}}$. We define the holomorphically induced representation $\rho(f, \eta_f, \mathfrak{h}, G)$ as sub-representation of $\operatorname{ind}_D^G \chi_f$. It is obtained on the subspace $\mathscr{H}(f, \eta_f, \mathfrak{h}, G)$, completion of the space of C^∞-functions ϕ belonging to the space of $\operatorname{ind}_D^G \chi_f$ and verifying the relation

$$R(X)\phi = \big(-if(X) + \frac{1}{2}\mathrm{tr}\ \mathrm{ad}_{\mathfrak{g}_\mathbb{C}/\mathfrak{e}_\mathbb{C}} X\big), \ \forall X \in \mathfrak{h}.$$

When η_f is a character, and if it is unique, we write $\rho(f, \mathfrak{h}, G)$ instead of $\rho(f, \eta_f, \mathfrak{h}, G)$.

Let $\mathfrak{h} \in M(f, \mathfrak{g})$ be stable under $G(f)$. We say that $\mathfrak{h}$ satisfies the Pukanszky condition if $\mathfrak{h}_\mathbb{C} \in P^+(f, \mathfrak{g})$ satisfies it. We rewrite this condition in another form which we shall use later, namely: $\mathfrak{h}$ satisfies the Pukanszky condition if and only if $\mathfrak{h} \in M(f + \ell, \mathfrak{g})$ for any $\ell \in \mathfrak{h}^\perp$ (cf. [24]). In this situation, putting $H = G(f)H^\circ$, the holomorphically induced representation $\rho(f, \eta_f, \mathfrak{h}, G)$ is nothing but $T(f, \eta_f, \mathfrak{h}, G) = \operatorname{ind}_H^G \chi_f$.

6.3.1 *Preliminaries*

Let G be a connected solvable Lie group with Lie algebra $\mathfrak{g}$. Let $f \in \mathfrak{g}^*$ and η_f a unitary representation of $G(f)$ with differential $(if|_{\mathfrak{g}(f)})\mathrm{id}_{\mathscr{H}}$, where $\mathscr{H}$ denotes the Hilbert space of η_f. Now we consider $\mathfrak{h} \in P^+(f, \mathfrak{g})$ stable under $G(f)$ and satisfying the Pukanszky condition. If $s \in \mathrm{Aut}(G)$ stabilizes $f, \mathfrak{h}$ and the class of η_f, we put, for $\phi \in \mathscr{H}_\mathfrak{h} = \mathscr{H}(f, \eta_f, \mathfrak{h}, G)$ and $g \in G$,

$$(\nu_\mathfrak{h}(s)\phi)(g) = \theta_\mathfrak{h}(s)\phi(s^{-1}(g))$$

and

$$(\widetilde{\nu}_\mathfrak{h}(s)(\phi))(g) = R(s)((\nu_\mathfrak{h}(s)\phi)(g)).$$

Above $R(s)$ is an isometry chosen once for all in $\mathscr{H}$, satisfying the relation

$$\eta_f(s(x)) = R(s)\eta_f(x)R(s)^{-1} \ (x \in G(f)),$$

and equal to $\eta_f(s)$ for $s \in G(f)$. On the other hand $\theta_{\mathfrak{h}}(s) \in \mathbb{R}^+$ is fixed so that $\widetilde{\nu}_{\mathfrak{h}}(s)$ is a unitary operator in $\mathscr{H}_{\mathfrak{h}}$, i.e.,

$$\theta_{\mathfrak{h}}(s)^2 d\nu_{G,D}(s(g)) = d\nu_{G,D}(g)$$

or again

$$\theta_{\mathfrak{h}}(s) = \sqrt{\frac{d\nu_{G,D}(s^{-1}(g))}{d\nu_{G,D}(g)}}.$$

Remark that $\widetilde{\nu}_{\mathfrak{h}}(x) = \rho(f, \eta_f, \mathfrak{h}, G)(x)$ for $x \in G(f)$.

We suppose that $\mathfrak{h}$ is admissible for a nilpotent ideal $\mathfrak{n}$ containing $[\mathfrak{g}, \mathfrak{g}]$, that is to say, $\mathfrak{h}_0 = \mathfrak{h} \cap \mathfrak{n}_{\mathbb{C}} \in P^+(f_0, \mathfrak{n})$ with $f_0 = f|_{\mathfrak{n}} \in \mathfrak{n}^*$, and develop some arguments borrowed from [3] (see also [53]). We set $\mathfrak{g}_1 = \mathfrak{g}(f_0) = \mathfrak{n}^f$, $f_1 = f|_{\mathfrak{g}_1} \in (\mathfrak{g}_1)^*$ and $\mathfrak{h}_1 = \mathfrak{h} \cap (\mathfrak{g}_1)_{\mathbb{C}}$. Then we know $\mathfrak{h} = \mathfrak{h}_1 + \mathfrak{h}_0$ and $\mathfrak{h}_1 \in P^+(f_1, \mathfrak{g}_1)$. Let $G_1 = G(f_0)$ and call $N = \exp \mathfrak{n}$ the analytic subgroup corresponding to $\mathfrak{n}$. Since $N(f_0)$ is connected, η_{f_0} is well determined as character. Besides, we see $G_1(f_1) = G(f)N(f_0)$. Consequently, there exists only one unitary representation η_{f_1} of $G_1(f_1)$ which extends η_f and verifies $d\eta_{f_1} = if_1|_{\mathfrak{g}_1(f_1)}$.

Since there is $\mathfrak{h}_2 \in P^+(f_0, \mathfrak{n})$ stable under G_1, we construct a unitary representation $\nu_{\mathfrak{h}_0}$ of G_1 in $\mathscr{H}_0 = \mathscr{H}(f_0, \mathfrak{h}_0, N)$ by the formula

$$\nu_{\mathfrak{h}_0}(g) = T_{\mathfrak{h}_2\mathfrak{h}_0}^{-1} \nu_{\mathfrak{h}_2}(g) T_{\mathfrak{h}_2\mathfrak{h}_0} \ (g \in G_1),$$

where $T_{\mathfrak{h}_2\mathfrak{h}_0}$ is an intertwining operator from $\rho_0 = \rho(f_0, \mathfrak{h}_0, N)$ to $\rho_2 = \rho(f_0, \mathfrak{h}_2, N)$. Putting $\sigma(g, n) = \nu_{\mathfrak{h}_0}(g)\rho_0(n)$ for $g \in G_1, n \in N$, we define a unitary representation σ of the semi-direct product $G_1 \ltimes N$ on $\mathscr{H}_0$ which extends ρ_0. On the other hand, we trivially extend $\rho_1 = \rho(f_1, \eta_{f_1}, \mathfrak{h}_1, G_1)$ to a representation, denoted again by ρ_1, of $G_1 \ltimes N$ on $\mathscr{H}_1 = \mathscr{H}(f_1, \eta_{f_1}, \mathfrak{h}_1, G_1)$. We then see that the representation $\rho_1 \otimes \sigma$ comes from a representation, denoted again by $\rho_1 \otimes \sigma$, of the closed subgroup $K = G_1N$ of G.

From $\mathfrak{h} \in P^+(f, \mathfrak{g})$ (resp. $\mathfrak{h}_0 \in P^+(f_0, \mathfrak{n})$; $\mathfrak{h}_1 \in P^+(f_1, \mathfrak{g}_1)$), we introduce objects $\mathfrak{d}$, $\mathfrak{e}$, D°, E°, D, E (resp. $\mathfrak{d}_0$, $\mathfrak{e}_0$, $D_0 = D_0^\circ$, $E_0 = E_0^\circ$; $\mathfrak{d}_1$, $\mathfrak{e}_1$, D_1°, E_1°, D_1, E_1). For instance, $E_1 = G_1(f_1)E_1^\circ$, where E_1° designates the analytic subgroup of G associated to $\mathfrak{e}_1 = (\mathfrak{h}_1 + \overline{\mathfrak{h}_1}) \cap \mathfrak{g}_1$, etc. We say that $\mathfrak{h}$ is semi-strongly admissible for $\mathfrak{n}$ if $\mathfrak{h}_0 \in P^+(f_0, \mathfrak{n})$ is stable under E_1°, which we will assume from now on.

Proposition 6.3.1 *The representation $\rho = \rho(f, \eta_f, \mathfrak{h}, G)$ is equivalent to $ind_K^G(\rho_1 \otimes \sigma)$ irrespective of the choice of $\mathfrak{h}$, and its commutator is isomorphic to that of η_f.*

If η_f is a character, ρ is also equivalent to the representation $T(f, \eta_f)$ constructed by Auslander and Kostant [3].

Proof In fact, at the level of the closed subgroup K, an isomorphism $\Lambda_{\mathfrak{h}}$ intertwining $\rho_1 \otimes \sigma$ with ρ is given on a certain dense subspace of $\mathscr{H}_1 \otimes \mathscr{H}_0$ by

$$\Lambda_{\mathfrak{h}}(\psi \otimes \phi) = \psi(g)(\nu_{\mathfrak{h}_0}(g^{-1})\phi)(n) \; (g \in G_1, n \in N),$$

where $\psi \in \mathscr{H}_1, \phi \in \mathscr{H}_0$ (see [3, 53]). Put $\mathfrak{q} = \mathfrak{n}(f_0) \cap \ker(f)$ and let Q be the corresponding analytic subgroup. Then Q is a normal subgroup of G_1°. As $\rho(f_1, \eta_1|_{G_1^\circ(f_1)}, \mathfrak{h}_1, G_1^\circ)$ is trivial on $\overline{Q} \subset G_1(f_1)^\circ$ and $G_1^\circ/\overline{Q}$ is a connected nilpotent group, we see that f_1 vanishes on the Lie algebra $\overline{\mathfrak{q}}$ of $\overline{Q}$. We consider the irreducible representation $\rho(f_1)$ of G_1° associated to f_1 and, putting

$$\sigma(f_1)(x, g) = \nu_{\mathfrak{h}_1}(x)\rho(f_1)(g) \; (x \in G(f), g \in G_1^\circ),$$

we have an extension $\sigma(f_1)$ of $\rho(f_1)$ to the semi-direct product $G(f) \ltimes G_1^\circ$. Extending η_{f_1} trivially to a representation of $G(f) \ltimes G_1^\circ$, we form $\eta_{f_1} \otimes \sigma(f_1)$. Passing to the quotient, this gives a representation of $G(f)G_1^\circ$. Finally, $\rho(f_1, \eta_{f_1}, \mathfrak{h}_1, G(f)G_1^\circ)$ is equivalent to $\eta_{f_1} \otimes \sigma(f_1)$ under

$$U(v \otimes \phi)(gh) = \phi(g)\eta_f(h)^{-1}v$$

for $v \in \mathscr{H}, \phi \in \mathscr{H}_1, g \in G_1^\circ, h \in G(f)$.

Lastly, since $G(f)G_1^\circ = G_1(f_1)G_1^\circ$ is nothing but the stabilizer of $\rho(f_1)$ in G_1, Mackey's theory proves the commutator statement. □

Take now another $\mathfrak{h}' \in P^+(f, \mathfrak{g})$, stable under $G(f)$ and semi-strongly admissible for $\mathfrak{n}$. We know that ρ (resp. ρ_0, ρ_1) is equivalent to $\rho' = \rho(f, \eta_f, \mathfrak{h}', G)$ (resp. $\rho_0' = \rho(f_0, \mathfrak{h}_0', N), \rho_1' = \rho(f_1, \mathfrak{h}_1', G_1)$ with $\mathfrak{h}_0' = \mathfrak{h}' \cap \mathfrak{n}_{\mathbb{C}}, \mathfrak{h}_1' = \mathfrak{h}' \cap (\mathfrak{g}_1)_{\mathbb{C}}$). Let T_0 (resp. T_1) be an isomorphism intertwining ρ_0 and ρ_0' (resp. ρ_1 and ρ_1'). It is easy to see that the isomorphism $T = \Lambda_{\mathfrak{h}'}(T_1 \otimes T_0)\Lambda_{\mathfrak{h}}^{-1}$ intertwines $\rho_K = \rho(f, \eta_f, \mathfrak{h}, K)$ and $\rho_K' = \rho(f, \eta_f, \mathfrak{h}', K)$. Let us introduce $\mathfrak{d}', \mathfrak{d}_0', \mathfrak{d}_1'$ etc.

Take the intertwining operator T_1' (defined in the previous section) between $\rho(f_1)$ in $\mathscr{H}(f_1, \mathfrak{h}_1, G_1^\circ)$ and the one in $\mathscr{H}(f_1, \mathfrak{h}_1', G_1^\circ)$. T_1' is defined using the kernel $K_{\mathfrak{h}_1'}$ given by Theorem 6.2.32 applied to the connected nilpotent group $G_1^\circ/\overline{Q}$. Then if we remain on the subgroup $G_1' = G(f)G_1^\circ$, T_1 is obtained by

$$(T_1\phi)(g) = (U(I \otimes T_1')U^{-1}\phi)(g) = \int_{(E_1')^\circ/(D_1^\circ \cap (D_1')^\circ)} \phi(gh)K_{\mathfrak{h}_1'}(h)d\dot{h} \; (g \in G_1')$$

on a certain dense subspace of $\mathscr{H}(f_1, \eta_{f_1}, \mathfrak{h}_1, G_1')$.

We set, for $h_0 \in (E_1')^\circ$ and $h_1 \in G(f)$,

$$\widetilde{K}_{\mathfrak{h}_1'}(h_0h_1) = K_{\mathfrak{h}_1'}(h_0)\eta_f(h_1),$$

which is well defined on $E_1' = (E_1')^\circ G(f)$. For fixed $\phi \in \mathscr{H}_1$ and $g \in G_1$, we look at the mapping

$$\Phi(h) = \widetilde{K}_{\mathfrak{h}_1'}(h)\phi(gh)$$

on E_1'. It is direct to check that Φ is right invariant by $D_1 \cap D_1'$. Since $E_1'/(D_1 \cap D_1')$ is isomorphic to $(E_1')^\circ/(D_1^\circ \cap (D_1')^\circ)$, we find in this way a formula

$$(T_1\phi)(g) = \int_{E_1'/(D_1\cap D_1')} \widetilde{K}_{\mathfrak{h}_1'}(h)\phi(gh)d\dot{h} \ (g \in G_1),$$

which is valid, for example, for ϕ belonging to $\mathscr{H}_1^\infty$.

Writing $h \in E'$ in the form $h = gn$ with $g \in E_1'$ and $n \in E_0'$, we introduce a kernel $\widetilde{K}_{\mathfrak{h}'}$ on E' by

$$\widetilde{K}_{\mathfrak{h}'}(h) = \widetilde{K}_{\mathfrak{h}'}(gn) = \Delta_{E',G}^{-1/2}(g)\widetilde{K}_{\mathfrak{h}_1'}(g)K_{\mathfrak{h}_0'}(n),$$

which is well defined.

Proposition 6.3.2 *Formula*

$$(T_{\mathfrak{h}'\mathfrak{h}}\alpha)(g) = \oint_{E'/(D\cap D')} \widetilde{K}_{\mathfrak{h}'}(h)\alpha(gh)d\nu_{E',D\cap D'}(h) \ (g \in G),$$

for α in some dense subspace of $\mathscr{H}_{\mathfrak{h}}$, defines an isomorphism intertwining ρ and ρ'.

Proof It suffices to show the formula at the level of K. We denote, for example, by $\mathscr{H}_0^\infty$ the space of C^∞-vectors of ρ_0. Let $\alpha \in \mathscr{H}_{\mathfrak{h}}$ be such that $\Lambda_{\mathfrak{h}}^{-1}\alpha = \psi \otimes \phi$ with $\phi \in \mathscr{H}_0^\infty$, $\psi \in \mathscr{H}_1^\infty$. Let us calculate $T\alpha = \Lambda_{\mathfrak{h}'}(T_1 \otimes T_0)\Lambda_{\mathfrak{h}}^{-1}\alpha$. For $g \in G_1$ and $n \in N$,

$$\begin{aligned}(T\alpha)(gn) =&(T_1\psi)(g)(\nu_{\mathfrak{h}_0'}(g^{-1})T_0\phi)(n) = (T_1\psi)(g)(T_0\nu_{\mathfrak{h}_0}(g^{-1})\phi)(n)\\ =&\int_{E_1'/(D_1\cap D_1')} \widetilde{K}_{\mathfrak{h}_1'}(h_1)\psi(gh_1)d\dot{h}_1\\ &\times \int_{E_1'/(D_0\cap D_0')} (\nu_{\mathfrak{h}_0}(g^{-1})\phi)(nh_0)K_{\mathfrak{h}_0'}(h_0)d\dot{h}_0.\end{aligned}$$

On the other hand, the right-hand side of the formula to be shown is well defined, and

$$
\begin{aligned}
(T_{\mathfrak{h}'\mathfrak{h}}\alpha)(gn) &= \oint_{E'/(D\cap D')} \widetilde{K}_{\mathfrak{h}'}(h)\alpha(gnh)d\nu_{E',D\cap D'}(h) \\
&= \int_{E'_1/(D_1\cap D'_1)} \widetilde{K}_{\mathfrak{h}'_1}(h_1)\psi(gh_1)\Delta_{E'_0,N}^{-1/2}(h_1)d\dot{h}_1 \\
&\times \int_{E'_0/(D_0\cap D'_0)} (\nu_{\mathfrak{h}_0}(h_1^{-1}g^{-1})\phi)(h_1^{-1}nh_1h_0)K_{\mathfrak{h}'_0}(h_0)d\dot{h}_0 \\
&= \int_{E'_1/(D_1\cap D'_1)} \widetilde{K}_{\mathfrak{h}'_1}(h_1)\psi(gh_1)\Delta_{E'_0,N}^{-1/2}(h_1) \\
&\times (\nu_{\mathfrak{h}'_0}(h_1^{-1})[T_{\mathfrak{h}'_0\mathfrak{h}_0}\nu_{\mathfrak{h}_0}(g^{-1})\phi])(h_1^{-1}nh_1)d\dot{h}_1 \\
&= \int_{E'_1/(D_1\cap D'_1)} \widetilde{K}_{\mathfrak{h}'_1}(h_1)\psi(gh_1)d\dot{h}_1(T_{\mathfrak{h}'_0\mathfrak{h}_0}\nu_{\mathfrak{h}_0}(g^{-1})\phi)(n).
\end{aligned}
$$

from which the desired result follows. □

Let $s \in \mathrm{Aut}(G)$ stabilize f, $\mathfrak{h}$, $\mathfrak{h}'$, $\mathfrak{n}$ and the class of η_f. We have a fundamental lemma, shown in [3] for the nilpotent case.

Lemma 6.3.3 $T_{\mathfrak{h}'\mathfrak{h}}\widetilde{\nu}_{\mathfrak{h}}(s) = \widetilde{\nu}_{\mathfrak{h}'}(s)T_{\mathfrak{h}'\mathfrak{h}}$.

Proof Take $\alpha \in \mathscr{H}_{\mathfrak{h}}$ as in the last proposition. Taking into account the invariance of $K_{\mathfrak{h}'_0}$ and $K_{\mathfrak{h}'_1}$ by s, we then have, for $g \in G$,

$$
\begin{aligned}
(T_{\mathfrak{h}'\mathfrak{h}}\widetilde{\nu}_{\mathfrak{h}}(s)\alpha(g)) &= \oint_{E'/(D\cap D')} \widetilde{K}_{\mathfrak{h}'}(h)(\widetilde{\nu}_{\mathfrak{h}}(s)\alpha)(gh)d\nu_{E',D\cap D'}(h) \\
=&\theta_{\mathfrak{h}}(s)\oint_{E'/(D\cap D')} \widetilde{K}_{\mathfrak{h}'}(h)R(s)(\alpha(s^{-1}(gh)))d\nu_{E',D\cap D'}(h) \\
=&c(s)\theta_{\mathfrak{h}}(s)\oint_{E'/(D\cap D')} R(s)\widetilde{K}_{\mathfrak{h}'}(h)\alpha(s^{-1}(g)h)d\nu_{E',D\cap D'}(h) \ (c(s) \in \mathbb{R}^+).
\end{aligned}
$$

Hence

$$
T_{\mathfrak{h}'\mathfrak{h}}\widetilde{\nu}_{\mathfrak{h}}(s) = c(s)\theta_{\mathfrak{h}}(s)\theta_{\mathfrak{h}'}(s)^{-1}\widetilde{\nu}_{\mathfrak{h}'}(s)T_{\mathfrak{h}'\mathfrak{h}}.
$$

By comparing the norms, we obtain the result. □

In what follows, everything happens on K. By means of the intertwining operator $\Lambda_{\mathfrak{h}}$, we transform $\widetilde{\nu}_{\mathfrak{h}}(s)$ into an operator on $\mathscr{H}_1 \otimes \mathscr{H}_0$. Let $\phi \in \mathscr{H}_0^\infty$ and $\psi \in \mathscr{H}_1^\infty$. We have, for $g \in G_1$ and $n \in N$,

$$\begin{aligned}\widetilde{\nu}_{\mathfrak{h}}(s)\Lambda_{\mathfrak{h}}(\psi \otimes \phi)(gn) &= \theta_{\mathfrak{h}}(s)R(s)(\Lambda_{\mathfrak{h}}(\psi \otimes \phi)(s^{-1}(gn)))\\ &= \theta_{\mathfrak{h}}(s)R(s)(\psi(s^{-1}(g)))(\nu_{\mathfrak{h}_0}(s^{-1}(g^{-1}))\phi)(s^{-1}(n))\\ &= \theta_{\mathfrak{h}}(s)\theta_{\mathfrak{h}_1}(s)^{-1}(\widetilde{\nu}_{\mathfrak{h}_1}(s)\psi)(g)(\nu_{\mathfrak{h}_0}(s^{-1}(g^{-1}))\phi)(s^{-1}(n)).\end{aligned}$$

We introduce $\mathfrak{h}_2 \in P^+(f_0, \mathfrak{n})$ stable under G_1. The precise forms of intertwining operators obtained in the previous section enable the following calculations. Up to a positive real constant,

$$\begin{aligned}&(\nu_{\mathfrak{h}_0}(s^{-1}(g^{-1}))\phi)(s^{-1}(n)) = (T_{\mathfrak{h}_0\mathfrak{h}_2}\nu_{\mathfrak{h}_2}(s^{-1}(g^{-1}))T_{\mathfrak{h}_2\mathfrak{h}_0}\phi)(s^{-1}(n))\\ &= \int_{E_0/(D_2\cap D_0)} (\nu_{\mathfrak{h}_2}(s^{-1}(g^{-1}))T_{\mathfrak{h}_2\mathfrak{h}_0}\phi)(s^{-1}(n)n_0)K_{\mathfrak{h}_0}(n_0)d\dot{n}_0\\ &= \int_{E_0/(D_2\cap D_0)} (T_{\mathfrak{h}_2\mathfrak{h}_0}\phi)(s^{-1}(gn)n_0s^{-1}(g^{-1}))K_{\mathfrak{h}_0}(n_0)d\dot{n}_0\\ &= \int_{E_0/(D_2\cap D_0)} K_{\mathfrak{h}_0}(n_0)d\dot{n}_0 \int_{E_2/(D_0\cap D_2)} \phi(s^{-1}(gn)n_0s^{-1}(g^{-1})n_1)K_{\mathfrak{h}_2}(n_1)d\dot{n}_1\\ &= \int_{E_0/(D_2\cap D_0)} [T_{s(\mathfrak{h}_2)\mathfrak{h}_0}(\nu_{\mathfrak{h}_0}(s)\phi)](gns(n_0)g^{-1})K_{\mathfrak{h}_0}(n_0)d\dot{n}_0\\ &= \int_{E_0/(sD_2\cap D_0)} [T_{s(\mathfrak{h}_2)\mathfrak{h}_0}(\nu_{\mathfrak{h}_0}(s)\phi)](gnn_0g^{-1})K_{\mathfrak{h}_0}(n_0)d\dot{n}_0\\ &= \int_{gE_0/(sD_2\cap gD_0)} [T_{s(\mathfrak{h}_2)\mathfrak{h}_0}(\nu_{\mathfrak{h}_0}(s)\phi)](gnn_0g^{-1})K_{g(\mathfrak{h}_0)}(n_0)d\dot{n}_0\\ &=[T_{g(\mathfrak{h}_0)s(\mathfrak{h}_2)}\cdot T_{s(\mathfrak{h}_2)\mathfrak{h}_0}(\nu_{\mathfrak{h}_0}(s)\phi)](gng^{-1})\\ &=c(g\mathfrak{h}_0, s\mathfrak{h}_2, \mathfrak{h}_0)(T_{g(\mathfrak{h}_0)\mathfrak{h}_0}\nu_{\mathfrak{h}_0}(s)\phi)(gng^{-1})\end{aligned}$$

in terms of the Maslov index $c(g\mathfrak{h}_0, s\mathfrak{h}_2, \mathfrak{h}_0) \in \mathbb{T}$.

Likewise, we have

$$\Lambda_{\mathfrak{h}}(\widetilde{\nu}_{\mathfrak{h}_1}(s)\psi \otimes \nu_{\mathfrak{h}_0}(s)\phi)(gn) = (\widetilde{\nu}_{\mathfrak{h}_1}(s)\psi)(g)(\nu_{\mathfrak{h}_0}(g^{-1})\nu_{\mathfrak{h}_0}(s)\phi)(n)$$

and

$$\begin{aligned}&(\nu_{\mathfrak{h}_0}(g^{-1})\nu_{\mathfrak{h}_0}(s)\phi)(n) = [T_{\mathfrak{h}_0\mathfrak{h}_2}\nu_{\mathfrak{h}_2}(g^{-1})T_{\mathfrak{h}_2\mathfrak{h}_0}\nu_{\mathfrak{h}_0}(s)\phi](n)\\ &= \int_{E_0/(D_2\cap D_0)} (\nu_{\mathfrak{h}_2}(g^{-1})T_{\mathfrak{h}_2\mathfrak{h}_0}\nu_{\mathfrak{h}_0}(s)\phi)(nn_0)K_{\mathfrak{h}_0}(n_0)d\dot{n}_0\end{aligned}$$

$$= \int_{E_0/(D_2 \cap D_0)} (T_{\mathfrak{h}_2\mathfrak{h}_0} \nu_{\mathfrak{h}_0}(s)\phi)(gnn_0g^{-1}) K_{\mathfrak{h}_0}(n_0) d\dot{n}_0$$

$$= \int_{gE_0/(D_2 \cap gD_0)} (T_{\mathfrak{h}_2\mathfrak{h}_0} \nu_{\mathfrak{h}_0}(s)\phi)(gng^{-1}n_0) K_{g\mathfrak{h}_0}(n_0) d\dot{n}_0$$

$$= [T_{g(\mathfrak{h}_0)\mathfrak{h}_2} \cdot T_{\mathfrak{h}_2\mathfrak{h}_0} \nu_{\mathfrak{h}_0}(s)\phi](gng^{-1})$$

$$= c(g\mathfrak{h}_0, \mathfrak{h}_2, \mathfrak{h}_0)(T_{g(\mathfrak{h}_0)\mathfrak{h}_0} \nu_{\mathfrak{h}_0}(s)\phi)(gng^{-1})$$

for the Maslov index $c(g\mathfrak{h}_0, \mathfrak{h}_2, \mathfrak{h}_0) \in \mathbb{T}$. From what we have seen, we obtain the relation

$$\begin{aligned} & c(g\mathfrak{h}_0, \mathfrak{h}_1, \mathfrak{h}_0)\widetilde{\nu}_{\mathfrak{h}}(s)\Lambda_{\mathfrak{h}}(\psi \otimes \phi)(gn) \\ =& c(g\mathfrak{h}_0, s\mathfrak{h}_2, \mathfrak{h}_0)\Lambda_{\mathfrak{h}}(\widetilde{\nu}_{\mathfrak{h}_1}(s)\psi \otimes \nu_{\mathfrak{h}_0}(s)\phi)(gn). \end{aligned}$$

We establish in this way the following:

Lemma 6.3.4

$$\begin{aligned} \widetilde{\nu}_{\mathfrak{h}}(s)\Lambda_{\mathfrak{h}}(\psi \otimes \phi) &= c_{\mathfrak{h}}(\cdot, s)\Lambda_{\mathfrak{h}}(\widetilde{\nu}_{\mathfrak{h}_1}(s)\psi \otimes \nu_{\mathfrak{h}_0}(s)\phi) \\ &= \Lambda_{\mathfrak{h}}(c_{\mathfrak{h}}(\cdot, s)\widetilde{\nu}_{\mathfrak{h}_1}(s)\psi \otimes \nu_{\mathfrak{h}_0}(s)\phi) \end{aligned}$$

for the well-defined map function $c_{\mathfrak{h}}(\cdot, s)$

$$K \ni gn \mapsto c(g\mathfrak{h}_0, s\mathfrak{h}_2, \mathfrak{h}_0)c(g\mathfrak{h}_0, \mathfrak{h}_2, \mathfrak{h}_0)^{-1} \in \mathbb{T},$$

$g \in G_1$ and $n \in N$.

In fact, the function $c_{\mathfrak{h}}(\cdot, s)$ vanishes on E_1N.

From now on we do not suppose that $\mathfrak{h}$ is s-invariant. According to Lemma 6.3.3, the unitary operator $\widetilde{\nu}_{\mathfrak{h}}(s) = T_{\mathfrak{h}'\mathfrak{h}}^{-1}\widetilde{\nu}_{\mathfrak{h}'}(s)T_{\mathfrak{h}'\mathfrak{h}}$ on $\mathscr{H}_{\mathfrak{h}}$ does not depend on $\mathfrak{h}' \in P^+(f, \mathfrak{g})$ stable under s and semi-strongly admissible for $\mathfrak{n}$. To make the next diagram commutative, we transform $\widetilde{\nu}_{\mathfrak{h}}(s)$ into an operator $\widehat{\nu}_{\mathfrak{h}}(s)$ in $\mathscr{H}_1 \otimes \mathscr{H}_0$:

$$\begin{aligned} \widehat{\nu}_{\mathfrak{h}}(s) &= (T_1^{-1} \otimes T_0^{-1})(c_{\mathfrak{h}'}(\cdot, s)\widetilde{\nu}_{\mathfrak{h}'_1}(s) \otimes \nu_{\mathfrak{h}'_0}(s))(T_1 \otimes T_0) \\ &= T_1^{-1}(c_{\mathfrak{h}'}(\cdot, s)\widetilde{\nu}_{\mathfrak{h}'_1}(s))T_1 \otimes T_0^{-1}\nu_{\mathfrak{h}'_0}(s)T_0 \\ &= \tau_{\mathfrak{h}}(s) \otimes \nu_{\mathfrak{h}_0}(s) \end{aligned}$$

with $\tau_{\mathfrak{h}}(s) = T_1^{-1}(c_{\mathfrak{h}'}(\cdot, s)\widetilde{\nu}_{\mathfrak{h}'_1}(s))T_1$.

$$
\begin{array}{ccccccccc}
 & \longrightarrow & & & \xrightarrow{\widehat{\nu}_{\mathfrak{h}}(s)} & & & \longrightarrow & \\
\uparrow & & & & & & & & \downarrow \\
\mathscr{H}_1\otimes\mathscr{H}_0 & \xrightarrow{T_1\otimes T_0} & \mathscr{H}_1'\otimes\mathscr{H}_0' & & \xrightarrow{c_{\mathfrak{h}'}(\cdot,s)\widetilde{\nu}_{\mathfrak{h}_1'}(s)\otimes\nu_{\mathfrak{h}_0'}(s)} & & \mathscr{H}_1'\otimes\mathscr{H}_0' & \xrightarrow{T_1^{-1}\otimes T_0^{-1}} & \mathscr{H}_1\otimes\mathscr{H}_0 \\
\downarrow{\scriptstyle\Lambda_{\mathfrak{h}}} & & \downarrow{\scriptstyle\Lambda_{\mathfrak{h}'}} & & & & {\scriptstyle\Lambda_{\mathfrak{h}'}}\downarrow & & {\scriptstyle\Lambda_{\mathfrak{h}}}\downarrow \\
\mathscr{H} & \xrightarrow{T_{\mathfrak{h}'\mathfrak{h}}} & \mathscr{H}' & & \xrightarrow{\widetilde{\nu}_{\mathfrak{h}'}(s)} & & \mathscr{H}' & \xrightarrow{T_{\mathfrak{h}'\mathfrak{h}}^{-1}} & \mathscr{H} \\
\downarrow & & & & & & & & \uparrow \\
 & \longrightarrow & & & \xrightarrow{\widetilde{\nu}_{\mathfrak{h}}(s)} & & & \longrightarrow &
\end{array}
$$

6.3.2 *Irreducibility and Equivalence*

Let G be a connected solvable Lie group with Lie algebra $\mathfrak{g}$ and $f \in \mathfrak{g}^*$. We suppose that there exists a unitary representation η_f of $G(f)$ on a Hilbert space $\mathscr{H}$ whose restriction to $G(f)^\circ$ is the character with differential $if|_{\mathfrak{g}(f)}$. Let $\mathfrak{h} \in M(f, \mathfrak{g})$ be stable under $G(f)$ and satisfy the Pukanszky condition. We construct

$$T(f, \eta_f, \mathfrak{h}, G) = \mathrm{ind}_H^G \chi_f$$

defined above and propose to prove the following theorems.

Theorem 6.3.5 *The representation $T(f, \eta_f, \mathfrak{h}, G)$ is independent of the choice of $\mathfrak{h}$ and its commutator is isomorphic to that of η_f. In fact, it is equivalent to the representation $\rho(f, \eta_f, \mathfrak{h}'', G)$ constructed by means of $\mathfrak{h}'' \in P^+(f, \mathfrak{g})$ stable under $G(f)$ and strongly admissible for a nilpotent ideal of $\mathfrak{g}$ containing $[\mathfrak{g}, \mathfrak{g}]$, and hence to $T(f, \eta_f)$ when η_f is a character.*

Theorem 6.3.6 *Let $s \in Aut(G)$ stabilize f and the class of η_f. Let $\mathfrak{h}_1, \mathfrak{h}_2 \in M(f, \mathfrak{g})$ be stable under $G(f)$, invariant by s and satisfy the Pukanszky condition. There exists an isomorphism T_{21} intertwining $T(f, \eta_f, \mathfrak{h}_1, G)$ with $T(f, \eta_f, \mathfrak{h}_2, G)$ and possessing the property $T_{21}\widetilde{\nu}_{\mathfrak{h}_1}(s) = \widetilde{\nu}_{\mathfrak{h}_2}(s)T_{21}$. If $\mathfrak{h}_3 \in P^+(f, \mathfrak{g})$ is $G(f)$-stable, invariant by s and semi-strongly admissible for a nilpotent ideal of $\mathfrak{g}$, which is stable under s and contains $[\mathfrak{g}, \mathfrak{g}]$, there exists an isomorphism T_{31} intertwining $T(f, \eta_f, \mathfrak{h}_1, G)$ with $\rho(f, \eta_f, \mathfrak{h}_3, G)$ and such that $T_{31}\widetilde{\nu}_{\mathfrak{h}_1}(s) = \widetilde{\nu}_{\mathfrak{h}_3}(s)T_{31}$.*

We intend to prove these theorems simultaneously, by induction on $\dim G$. Note that it suffices to show them for connected and simply connected G, which we assume in the sequel. We prepare the ground with a lemma concerning a particular case.

Lemma 6.3.7 *Let $\mathfrak{n}$ be an ideal of $\mathfrak{g}$ which is a Heisenberg algebra with center $\mathfrak{z}$ such that $f|_{\mathfrak{z}} \neq 0$ and $\mathfrak{z}$ is contained in the center of $\mathfrak{g}$. This means $\mathfrak{g} = \mathfrak{n}^f + \mathfrak{n}$ and $\mathfrak{n}^f \cap \mathfrak{n} = \mathfrak{z}$. Let $\mathfrak{a}$ be an Abelian subalgebra of $\mathfrak{n}^f$ verifying either $\mathfrak{a} \subset \mathfrak{g}(f)$ or $[\mathfrak{a}, \mathfrak{n}^f] \subset \mathfrak{z}$. Put $\mathfrak{m} = \mathfrak{a} + \mathfrak{n}$ and suppose $\mathfrak{m} \supset [\mathfrak{g}, \mathfrak{g}]$, $[\mathfrak{a}, \mathfrak{n}] = \{0\}$ and that $ad_{\mathfrak{n}} X$ is semi-simple for $X \in \mathfrak{n}^f$. Let $s \in Aut(G)$ stabilize f, $\mathfrak{n}^f$ and $\mathfrak{m}$.*

(i) There is a polarization $\mathfrak{h}' \in P^+(f, \mathfrak{g})$ stable under $G(f)$, invariant by s and semi-strongly admissible for $\mathfrak{n}$.
(ii) $T(f, \eta_f, \mathfrak{h}, G)$ is equivalent to $\rho(f, \eta_f, \mathfrak{h}', G)$, and consequently to $T(f, \eta_f)$ if η_f is a character.
(iii) There exists an isomorphism J intertwining $T(f, \eta_f, \mathfrak{h}, G)$ and $\rho(f, \eta_f, \mathfrak{h}', G)$ and verifying $J\widetilde{\nu}_{\mathfrak{h}}(s) = \widetilde{\nu}_{\mathfrak{h}'}(s)J$.

Proof Let us proceed by induction on $\dim G$. Let $f_0 = f|_{\mathfrak{m}} \in \mathfrak{m}^*$. Then $\mathfrak{m}(f_0) = \mathfrak{a} + \mathfrak{z}$ is an ideal of $\mathfrak{g}$, commutative and s-invariant. If $\mathfrak{a} \not\subset \mathfrak{g}(f)$, we invoke Proposition 6.3.9 (which will be proved later) to pass from $\mathfrak{h}$ to $\mathfrak{h} \cap \mathfrak{a}^f + \mathfrak{a}$ contained in the subalgebra $\mathfrak{a}^f \neq \mathfrak{g}$, to which induction applies. We simply point out that $G(f)$ is contained in the analytic subgroup associated to $\mathfrak{a}^f$.

Suppose that $\mathfrak{a} \subset \mathfrak{g}(f)$. Replacing $\mathfrak{a}$ by $\mathfrak{a} + \mathfrak{z}$ and passing if necessary to the quotient by $\mathfrak{a} \cap \ker(f)$, we can restrict ourselves to the case where $\mathfrak{m} = \mathfrak{n}$. If $\mathfrak{g} \neq \mathfrak{h} + \mathfrak{n}$, we only have to invoke the induction hypothesis at the level of $\mathfrak{g}_0 = \mathfrak{h} + \mathfrak{n}$. Suppose from now on that $\mathfrak{g} = \mathfrak{h} + \mathfrak{n}$. We keep the notation $f_0 = f|_{\mathfrak{n}} \in \mathfrak{n}^*$. If $\mathfrak{b} = \mathfrak{h} \cap \mathfrak{n} \supsetneq \mathfrak{z}$, $\mathfrak{b}$ is an Abelian ideal of $\mathfrak{g}$ and $\mathfrak{b}^{f_0} = \mathfrak{b}^f \cap \mathfrak{n} = \mathfrak{b} + \mathfrak{n}_1$ with a Heisenberg algebra $\mathfrak{n}_1$. It suffices to apply the induction hypothesis to $\mathfrak{g}_1 = \mathfrak{h} + \mathfrak{b}^{f_0}$, because $\dim(\mathfrak{g}_1) < \dim \mathfrak{g}$ and $\mathfrak{g}_1$ is of the required form.

There remains to see the case where $\mathfrak{z} = \mathfrak{h} \cap \mathfrak{n}$. Then $\mathfrak{h}$ is Abelian. Put $\mathfrak{n}_0 = \{Y \in \mathfrak{n}; [X, Y] = 0, \forall X \in \mathfrak{n}^f\}$. It turns out that the ideal $\mathfrak{n}_0$ of $\mathfrak{g}$ is Heisenberg. We show $\mathfrak{n}_0 = \mathfrak{n}$, which implies that $\mathfrak{g}$ is nilpotent and the lemma follows immediately. This is the same as $(\mathfrak{n}_0)^{f_0} = \mathfrak{z}$. If not, $W = (\mathfrak{n}_0)^{f_0} \cap \ker(f_0) \neq \{0\}$ and $\mathrm{ad}_{W_{\mathbb{C}}} \mathfrak{n}^f$ would be diagonalizable with all non-zero weights. Let V be a minimal subspace among the invariant subspaces by $\mathscr{S} = \mathrm{ad}_W \mathfrak{n}^f$, and let U be a $\mathscr{S}$-stable complementary subspace of V in W.

If $[V, V] = \{0\}$, the radical V' of $B_f|_U$ is $\mathscr{S}$-stable and B_f pairs V, V'. As $\mathfrak{q}$ is the $\mathscr{S}$-stable complementary subspace $U \cap V^{f_0}$ to V' in U, we write

$$W = V \oplus V' \oplus \mathfrak{q} = \mathfrak{p} + \mathfrak{q}$$

with $\mathfrak{p} = V \oplus V'$. Let $\mathrm{pr} : \mathfrak{g} \to \mathfrak{p}$ be the projection of $\mathfrak{g}$ onto $\mathfrak{p}$ parallel to $\mathfrak{n}^f \oplus \mathfrak{q}$. As $\mathfrak{z} = \mathfrak{h} \cap \mathfrak{n}$,

$$\dim \mathrm{pr}(\mathfrak{h}) = \dim \mathfrak{p} = 2 \dim V.$$

Well, with $\mathfrak{h}_0 = \{X \in \mathfrak{h}; [X, Y] \in \mathfrak{z}, \forall Y \in \mathfrak{p}\}$, we have $1 \leq \dim(\mathfrak{h}/\mathfrak{h}_0) \leq \dim V$. It follows that there are two elements $X_1, X_2 \in \mathfrak{h}$ such that $X_1 \notin \mathfrak{h}_0, X_2 \in \mathfrak{h}_0$ verifying $\mathrm{pr}\, X_2 \neq 0$. Consequently, $[X_1, X_2] \neq 0$ which is absurd.

If $[V, V] \neq \{0\}$, take $V^f \cap W$ as U. Let $\mathfrak{h}_0 = \{X \in \mathfrak{h};\ [X, Y] \in \mathfrak{z}, \forall Y \in V\}$ and $\mathrm{pr} : \mathfrak{g} \to V$ the projection of $\mathfrak{g}$ onto V parallel to $\mathfrak{n}^f \oplus U$. The weights corresponding to V being purely imaginary, we see $\dim(\mathfrak{h}/\mathfrak{h}_0) = 1$. On the other hand, $\dim \mathrm{pr}(\mathfrak{h}) = \dim V = 2$. This allows us to reason as before to find a contradiction. □

Now we prove our two theorems. Assume that they hold for groups of dimension less than $\dim G$. Suppose that there exists an s-invariant ideal $\mathfrak{a} \neq \{0\}$ on which f vanishes. Denoting by A the corresponding analytic subgroup of G, we can pass to the quotient G/A to which applies the induction hypothesis. Now we place ourselves in the case where there is no non-trivial ideal on which f vanishes, because if there were an ideal $\mathfrak{a} \neq \{0\}$ contained in $\ker(f)$, then $\sum_{k=-\infty}^{+\infty} s^k\mathfrak{a}$ would be an s-invariant ideal on which f vanishes. In particular, if we denote by $\mathfrak{z}$ the center of $\mathfrak{g}$, $\dim \mathfrak{z} \leq 1$.

The socle $\mathfrak{s}$ of $\mathfrak{g}$ is by definition the sum of all minimal ideals of $\mathfrak{g}$. Then $\mathfrak{s}$ is a characteristic Abelian ideal of $\mathfrak{g}$. Suppose first $\mathfrak{s} \neq \mathfrak{z}$. If $\mathfrak{a}$ is a minimal ideal of $\mathfrak{g}$, $\mathfrak{a}$ is commutative and contained in $\mathfrak{h}$. Indeed, if $\mathrm{ad}_{\mathfrak{a}}\mathfrak{h}$ is trivial, $\mathfrak{h} + \mathfrak{a}$ is manifestly subordinate to f, namely totally isotropic for B_f, and so $\mathfrak{h} \supset \mathfrak{a}$. If $\mathfrak{h} \cap \mathfrak{a} = \{0\}$, let $\ell \in f + \mathfrak{h}^{\perp}$ verify $\ell|_{\mathfrak{a}} = 0$. It is evident that $\mathfrak{h} + \mathfrak{a}$ is subordinate to ℓ and hence $\mathfrak{h} \supset \mathfrak{a}$ from the Pukanszky condition. If $\dim \mathfrak{a} = 2$, $\dim(\mathfrak{h} \cap \mathfrak{a}) = 1$ and if $\mathrm{ad}_{\mathfrak{a}}\mathfrak{h}$ is not trivial, then $\mathrm{ad}_{\mathfrak{a}}\mathfrak{h}$ is reducible and diagonalizable. Let $\{A_1, A_2\}$ be a basis of $\mathfrak{a}$ such that $\mathfrak{h} \cap \mathfrak{a} = \mathbb{R}A_1$, $[\mathfrak{h}, A_2] \subset \mathbb{R}A_2$. Then let $\ell \in f + \mathfrak{h}^{\perp}$ verify $\ell(A_2) = 0$. A_2 is orthogonal to $\mathfrak{h}$ with respect to B_{ℓ}, and belongs to $\mathfrak{h}$ by the Pukanszky condition. In all cases we arrive at a contradiction. Consequently, $\mathfrak{h}$ is contained in $\mathfrak{a}^f$ and then in $\mathfrak{s}^f$. Furthermore, in our situation, $\mathfrak{a}^f$ coincides with the centralizer of $\mathfrak{a}$ in $\mathfrak{g}$ and, consequently, $\mathfrak{s}^f$ coincides with the centralizer of $\mathfrak{s}$ in $\mathfrak{g}$.

Let S^f be the analytic subgroup associated to $\mathfrak{s}^f$. Consider the closed subgroup $G' = G(f)S^f$ of G and $f' = f|_{\mathfrak{s}^f} \in (\mathfrak{s}^f)^*$. It is easy to observe that $S^f(f') = S(G(f) \cap S^f)$ with the analytic subgroup S corresponding to $\mathfrak{s}$ and

$$H' = H^\circ S^f(f') = H^\circ(G(f) \cap S^f) = H \cap S^f.$$

By induction hypothesis, the representation $\pi = T(f', \eta_{f'}, \mathfrak{h}, S^f)$, where $\eta_{f'} = \chi_f|_{S^f(f')}$, is independent of the choice of $\mathfrak{h}$. Pushing a little further, π is equivalent to $\rho(f', \eta_{f'}, \mathfrak{h}'', S^f)$ with $\mathfrak{h}'' \in P^+(f', S^f)$ satisfying the required conditions in the statement of Theorem 6.3.5. Well, there exists such a polarization $\mathfrak{h}'' \in P^+(f, \mathfrak{g})$ contained in $(\mathfrak{s}^f)_{\mathbb{C}}$.

Applying Theorem 6.3.6 to S^f, the representation $\mathrm{ind}_H^{G'}\chi_f$, which is just the extension of π to G' given by $G' \ni gs \mapsto \widetilde{\nu_{\mathfrak{h}}}(g)\pi(s)$ with $g \in G(f)$ and $s \in S^f$, is equivalent to $\rho(f', \eta_f, \mathfrak{h}'', G')$. Inducing the two representations, we find Theorem 6.3.5. In order to show Theorem 6.3.6, it suffices to insert two intermediate stages at S^f and then G'.

We now engage in the case where $\mathfrak{s} = \mathfrak{z}$. Here $\dim \mathfrak{z} = 1$ and $f|_{\mathfrak{z}} \neq 0$. Suppose that there are commutative, non-central and s-invariant ideals of $\mathfrak{g}$ in $[\mathfrak{g}, \mathfrak{g}]$. Among them we take a minimal $\mathfrak{a}$. If $\mathfrak{h}$ contains $\mathfrak{a}$, introducing the connected subgroup A^f associated to $\mathfrak{a}^f$ and the closed subgroup $G' = G(f)A^f$ of G, the above argument applies.

Suppose from now on $\mathfrak{h} \not\supset \mathfrak{a}$, and put

$$\mathfrak{h}_0 = \mathfrak{h} \cap \mathfrak{a}^f, \mathfrak{h}' = \mathfrak{h}_0 + \mathfrak{a}, \mathfrak{k} = \mathfrak{h} + \mathfrak{h}' = \mathfrak{h} + \mathfrak{a}.$$

We easily see that $\mathfrak{h}' \in M(f, \mathfrak{g})$ is stable under $G(f)$ and satisfies the Pukanszky condition. Let $H' = G(f)(H')^\circ$ with the analytic subgroup $(H')^\circ$ of G corresponding to $\mathfrak{h}'$. Likewise, let $K = G(f)K^\circ$, where K° denotes the analytic subgroup of G associated to $\mathfrak{k}$. We aim to make sense of the formal operator

$$(T_{\mathfrak{h}'\mathfrak{h}}\phi)(g) = \oint_{H'/(H\cap H')} \Delta_{H',G}^{-1/2}(h)\chi'_f(h)(\phi(gh))d\nu_{H',H\cap H'}(h)$$

for $g \in G$ and $\phi \in \mathscr{H}_{\mathfrak{h}} = \mathscr{H}(f, \eta_f, \mathfrak{h}_{\mathbb{C}}, G)$, where χ'_f designates the unique representation of H' extending η_f and having the differential $if|_{\mathfrak{h}'}$. The operator $T_{\mathfrak{h}'\mathfrak{h}}\phi$ manifestly intertwines $T(f, \eta_f, \mathfrak{h}, G)$ with $T(f, \eta_f, \mathfrak{h}', G)$.

The commutative subalgebra $\mathfrak{b} = \mathfrak{h} \cap \mathfrak{a} \cap \ker(f)$ is stable under the closure $\overline{K}$ of K. As we may work at the level of $\overline{K}$, this allows us, by passing to the quotient if necessary, to suppose $\mathfrak{b}$ is trivial. That is to say, $\mathfrak{z} = \mathfrak{h} \cap \mathfrak{a}$.

Lemma 6.3.8 *We can choose a complement* $\mathfrak{p}$ *to* $\mathfrak{h}_0$ *in* $\mathfrak{h}$ *in such a way that* $[\mathfrak{p}, \mathfrak{a}] \subset \mathfrak{z}$.

Proof Let $\mathfrak{z} = \mathbb{R}Z$ with $f(Z) = 1$. The space $\mathfrak{a}/\mathfrak{z}$ is written as direct sum of $\mathfrak{g}$-stable subspaces of the same dimension, 1 or 2. Moreover, $\mathfrak{a}/\mathfrak{z} = \sum_{j=1}^n \oplus\mathfrak{a}_j$ with $\mathfrak{a}_j$ stable under $\mathfrak{g}$ and $\dim(\mathfrak{a}_j) = 1$ for all j or $\dim(\mathfrak{a}_j) = 2$ for all j.

The first case is $\dim(\mathfrak{a}_j) = 1$ for all j. There exist in $\mathfrak{a}$ elements $A_1, A_2, \ldots, A_n$ such that $\mathfrak{a} = \sum_{j=1}^n \oplus\mathbb{R}A_j \oplus \mathfrak{z}$ and that, for each j, $[X, A_j] = \alpha_j(X)A_j + \beta_j(X)Z$ $(X \in \mathfrak{g})$ with $\alpha_j, \beta_j \in \mathfrak{g}^*$. We will show, by contradiction, that $\beta'_j = \beta_j|_{\mathfrak{h}} \in \mathfrak{h}^*$, $1 \le j \le n$, are linearly independent in $\mathfrak{h}^*$ modulo $\{\alpha'_j = \alpha_j|_{\mathfrak{h}} \in \mathfrak{h}^*; 1 \le j \le n\}$. Indeed, take a linear relation

$$\sum_{j=1}^n (a_j\alpha'_j + b_j\beta'_j) = 0 \tag{6.3.1}$$

such that the b_j are not all zero. By Pukanszky's condition, there is an index j_0 verifying $a_{j_0} \neq 0$ and $b_{j_0} = 0$. Indeed, for any $X \in \mathfrak{h}$ and $\ell \in f + \mathfrak{h}^\perp$,

$$B_\ell(X, \sum_{j=1}^n c_jA_j) = \sum_{j=1}^n c_j(\alpha'_j(X)\lambda_j + \beta'_j(X))\ (c_j \in \mathbb{R})$$

with $\lambda_j = \ell(A_j)$. Hence if $a_j \neq 0$ implies $b_j \neq 0$ for any j, we shall find, choosing c_j and λ_j appropriately, an element $\sum_{j=1}^n c_jA_j \notin \mathfrak{z}$ belonging to $\mathfrak{h}$, which is not possible.

Among the linear relations of α'_j, β'_j, take (6.3.1), so that the cardinality of $J = \{j; 1 \le j \le n, b_j \ne 0\}$ is positive and minimal. Remark that the $\beta'_j (j \in J)$ are linearly independent modulo $\{\alpha'_j; j \in J\}$. Let $j_0 \in J$ be arbitrary. If $\alpha'_{j_0} = 0$, then $\alpha'_j = 0$ for all j and the lemma is evident. If not, there exist in $\mathfrak{h}$ two elements

$$X \in \cap_{j \in J} \ker(\beta_j), \ Y \in \ker(\alpha_{j_0}) \cap (\cap_{j \in (J \setminus \{j_0\})} \ker(\beta_j))$$

verifying $\alpha_{j_0}(X) = 1$ and $\beta_{j_0}(Y) = 1$. We then have

$$[X, Y] \in \mathfrak{h} \cap (\cap_{j=1}^n \ker(\alpha_j)) \cap (\cap_{j \in (J \setminus \{j_0\})} \ker(\beta_j)),$$

and relation (6.3.1) reduces to $b_{j_0}\beta_{j_0}([X, Y]) = 0$, which is absurd because of $\beta_{j_0}([X, Y]) \ne 0$.

The second case is $\dim(\mathfrak{a}_j) = 2$ for all j. The argument in this case is completely similar. There are elements $A_j, B_j (1 \le j \le n)$ in $\mathfrak{a}$ such that

$$\mathfrak{a} = \sum_{j=1}^{n} \oplus(\mathbb{R}A_j \oplus \mathbb{R}B_j) \oplus \mathfrak{z}$$

and, for each j,

$$[X, A_j] = \alpha_j(X)A_j - \beta_j(X)B_j + \lambda_j(X)Z,$$
$$[X, B_j] = \beta_j(X)A_j + \alpha_j(X)B_j + \mu_j(X)Z \ (X \in \mathfrak{g})$$

with $\alpha_j, \beta_j, \lambda_j, \mu_j \in \mathfrak{g}^*$. Following the same reasoning as above, we see that

$$\lambda'_j = \lambda_j|_{\mathfrak{h}}, \mu'_j = \mu_j|_{\mathfrak{h}} \in \mathfrak{h}^*(1 \le j \le n)$$

are linearly independent modulo

$$\{\alpha'_j = \alpha_j|_{\mathfrak{h}}, \beta'_j = \beta_j|_{\mathfrak{h}} \in \mathfrak{h}^*; 1 \le j \le n\}.$$

Consider a non-trivial linear relation

$$\sum_{j=1}^{n}(a_j\alpha'_j + b_j\beta'_j + c_j\lambda'_j + d_j\mu'_j) = 0 \tag{6.3.2}$$

The Pukanszky condition implies there exists an index j_0 verifying $a_{j_0}^2 + b_{j_0}^2 \neq 0$ and $c_{j_0} = d_{j_0} = 0$. Indeed, for any $X \in \mathfrak{h}$ and $\ell \in f + \mathfrak{h}^\perp$, we calculate

$$B_\ell(X, \sum_{j=1}^{n}(e_j A_j + f_j B_j))$$

$$= \sum_{j=1}^{n}\{(e_j\xi_j + f_j\zeta_j)\alpha'_j(X) + (f_j\xi_j - e_j\zeta_j)\beta'_j(X) + e_j\lambda'_j(X) + f_j\mu'_j(X)\}$$

with $\xi_j = \ell(A_j)$ and $\zeta_j = \ell(B_j)$. Hence if $c_j = d_j = 0$ implied $a_j = b_j = 0$ for any j, we could choose $\xi_j, \zeta_j (1 \le j \le n)$ satisfying the relations

$$c_j\xi_j + d_j\zeta_j = a_j, \; d_j\xi_j - c_j\zeta_j = b_j.$$

Consequently, $\sum_{j=1}^{n}(c_j A_j + d_j B_j) \in \mathfrak{h}$, which is absurd.

Among the linear relations of $\alpha'_j, \beta'_j, \lambda'_j, \mu'_j$, take now (6.3.2), so that the cardinality of

$$J = \{j; 1 \le j \le n, c_j^2 + d_j^2 \neq 0\}$$

is positive and minimal. Then it follows from the above that $\lambda'_j, \mu'_j (j \in J)$ are linearly independent modulo $\{\alpha'_j, \beta'_j; j \in J\}$. Let $j_0 \in J$ arbitrary. If $\alpha'_{j_0} = \beta'_{j_0} = 0$, the same holds for all α'_j, β'_j and evidently $[\mathfrak{h}, \mathfrak{a}] \subset \mathfrak{z}$. If not, we find elements X, Y_1, Y_2 in $\mathfrak{h}$ possessing the properties

$$X \in \cap_{j\in J}(\ker(\lambda_j) \cap \ker(\mu_j)), \; \alpha_{j_0}(X)^2 + \beta_{j_0}(X)^2 \neq 0,$$

$$Y_1 \in \ker(\alpha_{j_0}) \cap \ker(\beta_{j_0}) \cap (\cap_{j\in(J\setminus\{j_0\})}\ker(\lambda_j)) \cap (\cap_{j\in J}\ker(\mu_j)), \; \lambda_{j_0}(Y_1) = 1,$$

$$Y_2 \in \ker(\alpha_{j_0}) \cap \ker(\beta_{j_0}) \cap (\cap_{j\in J}\ker(\lambda_j)) \cap (\cap_{j\in(J\setminus\{j_0\})}\ker(\mu_j)), \; \mu_{j_0}(Y_2) = 1.$$

Then we obtain the equalities:

$$\lambda_{j_0}([X, Y_1]) = -\alpha_{j_0}(X), \; \mu_{j_0}([X, Y_1]) = -\beta_{j_0}(X)$$
$$\lambda_{j_0}([X, Y_2]) = \beta_{j_0}(X), \; \mu_{j_0}([X, Y_2]) = -\alpha_{j_0}(X).$$

Relation (6.3.2) then gives us

$$c_{j_0}\alpha_{j_0}(X) + d_{j_0}\beta_{j_0}(X) = 0, \; c_{j_0}\beta_{j_0}(X) - d_{j_0}\alpha_{j_0}(X) = 0,$$

which lead to the contradiction $c_{j_0} = d_{j_0} = 0$. The proof is complete. □

As regards this lemma, $K\cdot f = \exp \mathfrak{a}\cdot(f + \mathfrak{h}^{\perp})$ is closed in $\mathfrak{g}^*$, and the subgroup K is closed as well. Indeed, let

$$\exp(X_n)\cdot(f + \ell_n)\ (X_n \in \mathfrak{a}, \ell_n \in \mathfrak{h}^{\perp}, n = 1, 2, \dots)$$

tend to $\ell \in \mathfrak{g}^*$. Evaluating this limit on $\mathfrak{p}$, the sequence $\{X_n\}_{n=1}^{\infty}$ is convergent modulo $\mathfrak{z}$.

Lemma 6.3.8 allows to replicate the reasoning of [151] in order to realize the formal intertwining operator $T_{\mathfrak{h}'\mathfrak{h}}$ as an ordinary Fourier transform, which proves Theorem 6.3.5. As for Theorem 6.3.6, it suffices to remark that if $\mathfrak{h}$ is stable under s, the operator $T_{\mathfrak{h}'\mathfrak{h}}$ commutes with the actions of s. This means $T_{\mathfrak{h}'\mathfrak{h}}\widetilde{\nu}_{\mathfrak{h}}(s) = \widetilde{\nu}_{\mathfrak{h}'}(s)T_{\mathfrak{h}'\mathfrak{h}}$.

From what we said, we even obtain the following:

Proposition 6.3.9 *We keep the notations. Let $s \in Aut(G)$ leave f, η_f and $\mathfrak{h}$ stable. Let $\mathfrak{a}$ be an ideal of $\mathfrak{g}$, commutative and s-invariant. We set $\mathfrak{h}' = \mathfrak{h} \cap \mathfrak{a}^f + \mathfrak{a} \in M(f, \mathfrak{g})$. Then $\mathfrak{h}'$ satisfies the Pukanszky condition, $T(f, \eta_f, \mathfrak{h}, G)$ is equivalent to $T(f, \eta_f, \mathfrak{h}', G)$, and there exists an isomorphism $T_{\mathfrak{h}'\mathfrak{h}}$ intertwining these two representations and verifying $T_{\mathfrak{h}'\mathfrak{h}}\widetilde{\nu}_{\mathfrak{h}}(s) = \widetilde{\nu}_{\mathfrak{h}'}(s)T_{\mathfrak{h}'\mathfrak{h}}$.*

There remains to examine the case where there is no commutative, non-central and s-invariant ideal of $\mathfrak{g}$ in $[\mathfrak{g}, \mathfrak{g}]$. Then we know that the derived algebra $\mathfrak{n} = [\mathfrak{g}, \mathfrak{g}]$ is Heisenberg with center $\mathfrak{z}$, hence $\mathfrak{g} = \mathfrak{n}^f + \mathfrak{n}$ and $\mathfrak{n}^f \cap \mathfrak{n} = \mathfrak{z}$. Let $\mathfrak{a}$ be minimal among the non-central ideals of $\mathfrak{g}$ contained in $\mathfrak{n}$. Under our hypotheses, $\mathfrak{a}$ contains $\mathfrak{z}$ and $\widetilde{\mathfrak{a}} = \sum_{k=-\infty}^{+\infty} s^k(\mathfrak{a})$ is an s-stable ideal such that the adjoint action of $\mathfrak{g}$ on $\widetilde{\mathfrak{a}}/\mathfrak{z}$ is semi-simple. It follows that $\widetilde{\mathfrak{a}}$ is Heisenberg, for otherwise the center of $\widetilde{\mathfrak{a}}$ would be an Abelian, s-invariant ideal of $\mathfrak{g}$ different from $\mathfrak{z}$, which cannot be. Put $f_0 = f|_{\mathfrak{n}} \in \mathfrak{n}^*$. The algebra $(\widetilde{\mathfrak{a}})^{f_0} = (\widetilde{\mathfrak{a}})^f \cap \mathfrak{n}$, which is Heisenberg, is an s-stable ideal.

Repeating the above argument, we finally see that the adjoint action of $\mathfrak{n}^f$ in $\mathfrak{n}$ is semi-simple. Then a nilpotent ideal of $\mathfrak{g}$ which contains $\mathfrak{n}$ has the form $\mathfrak{m}$ of Lemma 6.3.7, which gives us the desired results. The proofs of Theorems 6.3.5, 6.3.6 are now complete. □

We no longer suppose that G is connected.

Corollary 6.3.10 *Let G be a solvable Lie group with Lie algebra $\mathfrak{g}$. With the previous data and notations, the representation $T(f, \eta_f, \mathfrak{h}, G)$ is independent of the choice of $\mathfrak{h}$. Let $\mathfrak{h}_1, \mathfrak{h}_2$ be two such real polarizations at $f \in \mathfrak{g}^*$. Let $s \in Aut(G)$ stabilize $f, \mathfrak{h}_1, \mathfrak{h}_2$ and the class of η_f. There exists in this situation an isomorphism T_{21} intertwining $T(f, \eta_f, \mathfrak{h}_1, G)$ with $T(f, \eta_f, \mathfrak{h}_2, G)$ and transforming $\widetilde{\nu}_{\mathfrak{h}_1}(s)$ to $\widetilde{\nu}_{\mathfrak{h}_2}(s)$, whence $T_{21}\widetilde{\nu}_{\mathfrak{h}_1}(s) = \widetilde{\nu}_{\mathfrak{h}_2}(s)T_{21}$.*

Proof Put $\eta_f^\circ = \eta_f|_{G^\circ(f)}$ and $G' = G(f)G^\circ$. For $s \in G(f)$ and $g \in G^\circ$, we have

$$T(f, \eta_f, \mathfrak{h}, G')(sg) = \widetilde{\nu}_{\mathfrak{h}}(s)T(f, \eta_f^\circ, \mathfrak{h}, G^\circ)(g).$$

Hence taking Theorems 6.3.5, 6.3.6 into account, $T(f, \eta_f, \mathfrak{h}, G) = \mathrm{ind}_{G'}^G T(f, \eta_f, \mathfrak{h}, G')$ is independent of the choice of $\mathfrak{h}$.

The second statement is easily derived from Theorem 6.3.6 and from an argument we have used implicitly many times. In fact, let T_{21}° be an isomorphism intertwining $\pi_1^\circ = T(f, \eta_f^\circ, \mathfrak{h}_1, G^\circ)$ with $\pi_2^\circ = T(f, \eta_f^\circ, \mathfrak{h}_2, G^\circ)$ and verifying $T_{21}^\circ \widetilde{\nu}_{\mathfrak{h}_1}(s) = \widetilde{\nu}_{\mathfrak{h}_2}(s)T_{21}^\circ$. Such T_{21}° exists by Theorem 6.3.6. Since $\sigma_i = T(f, \eta_f, \mathfrak{h}', G')$ $(i = 1, 2)$ is nothing but the extension of π_i° by $\eta_f = \widetilde{\nu}_{\mathfrak{h}_i}|_{G(f)}$, T_{21}° intertwines σ_1 with σ_2 and satisfies the same relationship $T_{21}^\circ \widetilde{\nu}_{\mathfrak{h}_1}(s) = \widetilde{\nu}_{\mathfrak{h}_2}(s)T_{21}^\circ$.

For $i = 1$ or 2, we identify $\pi_i = T(f, \eta_f, \mathfrak{h}_i, G)$ with $\mathrm{ind}_{G'}^G \sigma_i$ under the mapping S_i, which to $\phi \in \mathscr{H}(f, \eta_f, \mathfrak{h}_i, G)$ associates Φ belonging to the space $\mathscr{H}_i$ of the $\mathrm{ind}_{G'}^G \sigma_i$, and given by $\Phi(g)(g_0) = \phi(gg_0)$ for $g \in G, g_0 \in G'$. We now provide $\overline{T}_{21} : \mathscr{H}_1 \to \mathscr{H}_2$ defined by $(\overline{T}_{21}\Phi)(g) = T_{21}^\circ(\Phi(g))$ for $\Phi \in \mathscr{H}_1, g \in G$. Transporting $\overline{T}_{21}$ by means of S_i, we manifestly obtain an isomorphism $T_{21} = S_2^{-1}\overline{T}_{21}S_1$ intertwining π_1 with π_2.

On the other hand, we calculate $\overline{\nu}_{\mathfrak{h}_i}(s) = S_i \widetilde{\nu}_{\mathfrak{h}_i}(s) S_i^{-1}$; for $g \in G$ and $\Phi \in \mathscr{H}_i$, $(\overline{\nu}_{\mathfrak{h}_i}(s)\Phi)(g) = \widetilde{\nu}_{\mathfrak{h}_i}(s)(\Phi(s^{-1}(g)))$. Then

$$\begin{aligned}(\overline{\nu}_{\mathfrak{h}_2}(s)\overline{T}_{21}\Phi)(g) &= \widetilde{\nu}_{\mathfrak{h}_2}(s)((\overline{T}_{21}\Phi)(s^{-1}(g))) = \widetilde{\nu}_{\mathfrak{h}_2}(s)T_{21}^\circ(\Phi(s^{-1}(g)))\\ &= T_{21}^\circ\widetilde{\nu}_{\mathfrak{h}_1}(s)(\Phi(s^{-1}(g))) = T_{21}^\circ((\overline{\nu}_{\mathfrak{h}_1}(s)\Phi)(g)) = (\overline{T}_{21}\overline{\nu}_{\mathfrak{h}_1}(s)\Phi)(g),\end{aligned}$$

from which $T_{21}\widetilde{\nu}_{\mathfrak{h}_1}(s) = \widetilde{\nu}_{\mathfrak{h}_2}(s)T_{21}$ follows. □

Lemma 6.3.3 and Theorem 6.3.6 allow us to establish results similar to Corollary 6.3.10. Their proof is identical to that of Corollary 6.3.10.

Corollary 6.3.11 *Let $\mathfrak{n}$ be a nilpotent ideal invariant by $G(f)$ and containing $[\mathfrak{g}, \mathfrak{g}]$. Then the representation $\rho(f, \eta_f, \mathfrak{h}', G)$ is independent of the choice of $\mathfrak{h}' \in P^+(f, \mathfrak{g})$ stable under $G(f)$ and semi-strongly admissible for $\mathfrak{n}$. We consider two such polarizations $\mathfrak{h}_1', \mathfrak{h}_2' \in P^+(f, \mathfrak{g})$. Let $s \in Aut(G)$ stabilize $f, \mathfrak{h}_1', \mathfrak{h}_2', \mathfrak{n}$ and the class of η_f. Then there exists an isomorphism T_{21} intertwining $\rho(f, \eta_f, \mathfrak{h}_1', G)$ with $\rho(f, \eta_f, \mathfrak{h}_2', G)$ and verifying the relation $T_{21}\widetilde{\nu}_{\mathfrak{h}_1}(s) = \widetilde{\nu}_{\mathfrak{h}_2}(s)T_{21}$.*

Corollary 6.3.12 *Take $\mathfrak{h} \in M(f, \mathfrak{g})$ (resp. $\mathfrak{h}' \in P^+(f, \mathfrak{g})$) as in Corollary 6.3.10 (resp. Corollary 6.3.11). Then $\pi = T(f, \eta_f, \mathfrak{h}, G)$ is equivalent to $\rho = \rho(f, \eta_f, \mathfrak{h}', G)$. Let $s \in Aut(G)$ be as in Corollaries 6.3.10 and 6.3.11. Then there exists an isomorphism T intertwining π with ρ and verifying $T\widetilde{\nu}_{\mathfrak{h}}(s) = \widetilde{\nu}_{\mathfrak{h}'}(s)T$.*

Let us examine more in detail the non-connected case. Our aim is to prove:

Theorem 6.3.13 *Let G be a solvable Lie group with Lie algebra $\mathfrak{g}$ and $f \in \mathfrak{g}^*$. Suppose that there exist a unitary representation η_f of $G(f)$ on a Hilbert space with differential $d\eta_f = if|_{\mathfrak{g}(f)}$, and a real polarization $\mathfrak{h} \in M(f, \mathfrak{g})$ stable under $G(f)$ and satisfying the Pukanszky condition. Then the representation $T(f, \eta_f, \mathfrak{h}, G) = ind_H^G \chi_f$ is independent of the choice of such $\mathfrak{h}$ and its commutator is isomorphic to that of η_f. In particular, $T(f, \eta_f, \mathfrak{h}, G)$ is irreducible when η_f is a character.*

Proof According to what we have seen until now, there remains to establish the statement on the commutator of $T(f, \eta_f, \mathfrak{h}, G)$, which will be written more simply $T(f, \eta_f)$ or $T(f, \eta_f, G)$. We employ once again induction on $\dim G$. We first look at the case where $\dim \mathfrak{z} > 1$, $\mathfrak{z}$ denoting the center of $\mathfrak{g}$. Put $f' = f|_{\mathfrak{z}} \in \mathfrak{z}^*$ and $\mathfrak{a} = \mathfrak{z} \cap \ker(f')$. The analytic subgroup A associated to $\mathfrak{a}$ is closed in G. Indeed, denote by $\bar{\mathfrak{a}}$ the Lie algebra of the closure $\overline{A}$ of A. The relations $\overline{A} \subset H$ and $\chi_f|_A = 1$ lead to $\chi_f|_{\overline{A}} = 1$, from which $\bar{\mathfrak{a}} \subset \ker(f')$ and $\overline{A} = A$. Visibly, A is normal in $G' = G(f')$. The induction hypothesis applied to G'/A says that the commutator of $T(f, \eta_f, G')$ is isomorphic to that of η_f. The restriction of $T(f, \eta_f, G')$ to the connected component Z° of the center Z of G is a multiple of the character of Z° having differential if', of which G' is the stabilizer in G. Hence according to the theory of Mackey [123], the commutator of

$$T(f, \eta_f) = \mathrm{ind}_{G'}^G T(f, \eta_f, G')$$

is isomorphic to that of $T(f, \eta_f, G')$.

Assume henceforth $\dim \mathfrak{z} \leq 1$. In a similar way, we are led to the case where there is no G-stable commutative ideal in $\mathfrak{g}$ on which f vanishes. For if there existed such an ideal, let $\mathfrak{a}$ be a maximal one. Then the analytic subgroup A of G corresponding to $\mathfrak{a}$ would be closed and we could pass to the quotient G/A, to which the induction hypothesis would apply.

Suppose that there exist in $\mathfrak{g}$ non-central G-stable Abelian ideals. We choose among them a maximal one, say $\mathfrak{a}$. Let $\mathfrak{h}' = \mathfrak{h} \cap \mathfrak{a}^f + \mathfrak{a} \in M(f, \mathfrak{a})$. According to Corollary 6.3.10, $T(f, \eta_f) = T(f, \eta_f, \mathfrak{h}', G)$. Put $f' = f|_{\mathfrak{a}^f} \in (\mathfrak{a}^f)^*$ and $f_0 = f|_{\mathfrak{a}} \in \mathfrak{a}^*$. The closed subgroup $G' = G(f_0)$ has $\mathfrak{a}^f$ as Lie algebra (note $\mathfrak{a}^f$ is not equal to $\mathfrak{g}$), so the commutator of $T(f', \eta_{f'}, \mathfrak{h}', G')$ is isomorphic to the commutator of $\eta_{f'}$ by induction hypothesis. Here, $\eta_{f'}$ is given as follows. Let A (resp. $(H')^\circ$) be the analytic subgroup corresponding to $\mathfrak{a}$ (resp. $\mathfrak{h}'$). Then

$$G'(f') = G(f)A \subset H' = G(f)(H')^\circ.$$

Denote by $\chi_{f'}$ the representation of H' extending η_f and with differential $if|_{\mathfrak{h}'}$. The representation $\eta_{f'}$ is nothing but $\chi_{f'}$ restricted to $G'(f')$. Furthermore, A is closed in G, G' is the stabilizer of the character σ of A with differential $d\sigma = if|_{\mathfrak{a}}$, and the restriction of $T(f', \eta_{f'}, \mathfrak{h}', G')$ to A is a multiple of σ. So, the theory of Mackey

[123] warrants that

$$T(f,\eta_f)=\mathrm{ind}_{G'}^{G}T(f',\eta_{f'},\mathfrak{h}',G')$$

has commutator isomorphic to the commutator of $T(f',\eta_{f'},\mathfrak{h}',G')$ and hence to the commutator of η_f.

There remains to examine the case where $\dim\mathfrak{z}=1$, $f|_{\mathfrak{z}}\neq 0$ and there is no non-central G-stable Abelian ideal. Then the derived algebra $\mathfrak{n}=[\mathfrak{g},\mathfrak{g}]$ is Heisenberg, implying $\mathfrak{g}=\mathfrak{n}^f+\mathfrak{n}$ and $\mathfrak{n}^f\cap\mathfrak{n}=\mathfrak{z}$. Let $f_0=f|_{\mathfrak{n}}\in\mathfrak{n}^*$ and N the analytic subgroup corresponding to $\mathfrak{n}$. Evidently, $G=G(f_0)N$. Let $\mathfrak{a}$ be a minimal ideal among the non-central ideals of $\mathfrak{g}$ contained in $\mathfrak{n}$. If $\mathfrak{a}$ is not G-stable, there is $s_1\in G(f_0)$ such that $s_1\mathfrak{a}\cap\mathfrak{a}=\mathfrak{z}$. Put $\mathfrak{a}_1=\mathfrak{a}+s_1\mathfrak{a}$. If $\mathfrak{a}_1$ is still not G-stable, take $s_2\in G(f_0)$ such that $s_2\mathfrak{a}\cap\mathfrak{a}_1=\mathfrak{z}$ and put

$$\mathfrak{a}_2=\mathfrak{a}_1+s_2\mathfrak{a}=\mathfrak{a}+s_1\mathfrak{a}+s_2\mathfrak{a}.$$

Repeating this process, we reach a G-stable ideal

$$\widetilde{\mathfrak{a}}=\mathfrak{a}_k=\mathfrak{a}+s_1\mathfrak{a}+\cdots+s_k\mathfrak{a}$$

with $s_j\in G(f_0)$ verifying $s_j\mathfrak{a}\cap\mathfrak{a}_{j-1}=\mathfrak{z}$ for $1\leq j\leq k$, where $\mathfrak{a}_0=\mathfrak{a}$. It is clear that $\widetilde{\mathfrak{a}}$ is Heisenberg, otherwise the center of $\widetilde{\mathfrak{a}}$ would be a non-central G-stable Abelian ideal of $\mathfrak{g}$. We have

$$\mathfrak{n}=(\widetilde{\mathfrak{a}})^{f_0}+\widetilde{\mathfrak{a}},\ (\widetilde{\mathfrak{a}})^{f_0}\cap\widetilde{\mathfrak{a}}=\mathfrak{z},$$

and $(\widetilde{\mathfrak{a}})^{f_0}$ is Heisenberg, stable under G. Using for $(\widetilde{\mathfrak{a}})^{f_0}$ the above reasoning, we find that $\mathrm{ad}_{\mathfrak{n}}X$ is semi-simple for $X\in\mathfrak{n}^f$. Hence we are in the situation described in Lemma 6.3.7.

To continue the proof, we need the results obtained in the precedent subsection. We retain the notations used there. We introduce $\mathfrak{h}'\in P^+(f,\mathfrak{g})$ stable under $G^\circ(f)$, admissible for $\mathfrak{n}$ and such that $\mathfrak{h}'_0=\mathfrak{h}'\cap\mathfrak{n}_{\mathbb{C}}$ is stable under $G^\circ(f_0)$. Put $\mathfrak{g}_1=\mathfrak{g}(f_0)=\mathfrak{n}^f$, $f_1=f|_{\mathfrak{g}_1}\in\mathfrak{g}_1{}^*$, $G_1=G(f_0)$ and $\mathfrak{h}'_1=\mathfrak{h}'\cap(\mathfrak{g}_1)_{\mathbb{C}}\in P^+(f_1,\mathfrak{g}_1)$. Let $\eta_f{}^\circ=\eta_f|_{G^\circ(f)}$. By Theorem 6.3.5, $\widehat{\rho}=T(f,\eta_f{}^\circ,\mathfrak{h},G^\circ)$ is equivalent to $\widehat{\rho'}=\rho(f,\eta_f{}^\circ,\mathfrak{h}',G^\circ)$. We remark $G^\circ=G^\circ(f_0)N=(G_1)^\circ N$ and put $K=G(f)G^\circ(f_0)$, $M=G(f)G^\circ=KN$. Then $G^\circ(f_0)=(G_1)^\circ$, written as G_1°, and $K=G(f)G_1^\circ=M(f_0)$. As G_1° is connected and nilpotent, $G^\circ(f)=G_1^\circ(f_1)$ is connected. It follows that $\eta_f{}^\circ$ comes from the character $\widehat{\eta_f}$. Let $\rho=T(f,\mathfrak{h},G^\circ)$, $\rho'=\rho(f,\mathfrak{h}',G^\circ)$ and $\mathbf{1}_{\mathscr{H}}$ the trivial representation on the space $\mathscr{H}$ of η_f. We have $\widehat{\rho}=\mathbf{1}_{\mathscr{H}}\otimes\rho$ and $\widehat{\rho'}=\mathbf{1}_{\mathscr{H}}\otimes\rho'$. Denote by U an isometry which intertwines ρ' and ρ. For $s\in G(f)$ and $g\in G^\circ$,

$$T(f,\eta_f,\mathfrak{h},M)(sg)=\widetilde{\nu}_{\mathfrak{h}}(s)\widehat{\rho}(g)=\eta_f(s)\otimes\nu_{\mathfrak{h}}(s)\rho(g)$$

with $\nu_{\mathfrak{h}}$ constructed in $\mathscr{H}(f, \mathfrak{h}, G^\circ)$. This says that, putting $\nu_{\mathfrak{h}'}(s) = U^{-1}\nu_{\mathfrak{h}}(s)U$, we extend $\widehat{\rho'}$ to a representation π of M given by

$$\pi(sg) = \eta_f(s) \otimes \nu_{\mathfrak{h}'}(s)\rho'(g).$$

Hence it is important to show that the commutators of $\widetilde{\pi} = \mathrm{ind}_M^G \pi$ and η_f are isomorphic. The space of $\widetilde{\pi}$ is denoted by $\widetilde{\mathscr{H}}$.

For each $s \in G(f)$, according to Lemma 6.3.7, there exists $\mathfrak{h}'' \in P^+(f, \mathfrak{g})$ invariant by s and semi-strongly admissible for $\mathfrak{n}$. We know that $\nu_{\mathfrak{h}'}(s)$ is also obtained by the formula

$$\nu_{\mathfrak{h}'}(s) = T_{\mathfrak{h}''\mathfrak{h}'}^{-1}\nu_{\mathfrak{h}''}(s)T_{\mathfrak{h}''\mathfrak{h}'}$$

with $\nu_{\mathfrak{h}''}$ constructed in $\mathscr{H}(f, \mathfrak{h}'', G^\circ)$. Denote by σ the cocycle of the projective representation $\nu_{\mathfrak{h}_0'}$ of G_1 constructed in $\mathscr{H}_0' = \mathscr{H}(f_0, \mathfrak{h}_0', N)$: for any $g_1, g_2 \in G_1$,

$$\nu_{\mathfrak{h}_0'}(g_1g_2) = \sigma(g_1, g_2)\nu_{\mathfrak{h}_0'}(g_1)\nu_{\mathfrak{h}_0'}(g_2).$$

From what we saw at the end of the previous subsection, π is realized on the space $\mathscr{H} \otimes \mathscr{H}_1' \otimes \mathscr{H}_0'$ with $\mathscr{H}_1' = \mathscr{H}(f_1, \mathfrak{h}_1', G_1^\circ)$ by the formula

$$\begin{aligned}\pi(m) &= \eta_f(k_1) \otimes \tau_{\mathfrak{h}'}(k_1)\rho_1'(k_2) \otimes \nu_{\mathfrak{h}_0'}(k_1)\nu_{\mathfrak{h}_0'}(k_2)\rho_0'(n)\\ &= \sigma(k_1, k_2)^{-1}\eta_f(k_1) \otimes \tau_{\mathfrak{h}'}(k_1)\rho_1'(k_2) \otimes \nu_{\mathfrak{h}_0'}(k)\rho_0'(n)\end{aligned}$$

for $m = kn = k_1k_2n \in M$ with $k = k_1k_2 \in K, k_1 \in G(f), k_2 \in G_1^\circ$ and $n \in N$, where

$$\tau_{\mathfrak{h}'}(k_1) = T_1^{-1}(c_{\mathfrak{h}'}(\cdot, k_1)\nu_{\mathfrak{h}_1''}(k_1))T_1$$

with $\nu_{\mathfrak{h}_1''}$ constructed in $\mathscr{H}_1'' = \mathscr{H}(f_1, \mathfrak{h}_1'', G_1^\circ)$.

Since π is a true representation, formula

$$\kappa(k) = \sigma(k_1, k_2)^{-1}\eta_f(k_1) \otimes \tau_{\mathfrak{h}'}(k_1)\rho_1'(k_2)$$

for $k = k_1k_2$ with $k_1 \in G(f), k_2 \in G_1^\circ$, gives us a projective representation κ of K with cocycle σ^{-1}, and the restriction of κ to G_1° is a multiple of ρ_1'. We consider the projective representation $\rho_2 = \mathrm{ind}_K^{G_1}\kappa$ of G_1 with cocycle σ^{-1}. Its space is denoted by $\mathscr{H}_2$. When we regard the action of G_1 in the dual of G_1°, which is connected nilpotent hence of type I, and K is nothing but the stabilizer of ρ_1'. According to Mackey [123], we assume that the commutator of ρ_2 is isomorphic to the commutator of κ and consequently to that of η_f. We extend ρ_2 and σ trivially to two corresponding objects, called by the same symbols, of the semi-direct product $G_1 \ltimes N$.

As $\nu_{\mathfrak{h}'_0}$ was constructed so that

$$\nu_{\mathfrak{h}'_0}(g)\rho'_0(n)\nu_{\mathfrak{h}'_0}(g^{-1}) = \rho'_0(gng^{-1}) \ (g \in G_1, n \in N),$$

the map $\rho_3 : G_1 \ltimes N \ni (g, n) \mapsto \nu_{\mathfrak{h}'_0}(g)\rho'_0(n)$ gives a projective representation with cocycle σ. Finally, $\rho_2 \otimes \rho_3$ is a true representation providing, by passing to the quotient, a representation again called $\rho_2 \otimes \rho_3$, of G. Its commutator is isomorphic to that of η_f. There remains only to see that $\tilde{\pi}$ is equivalent to $\rho_2 \otimes \rho_3$, which is almost evident. Indeed, this is realized by means of the isometry Γ from $\mathscr{H}_2 \otimes \mathscr{H}'_0$ to $\widetilde{\mathscr{H}}$ defined by

$$\Gamma(\psi \otimes \phi)(gn) = \psi(g) \otimes \rho'_0(n^{-1})(\nu_{\mathfrak{h}'_0}(g^{-1})\phi) \ (g \in G_1, n \in N)$$

for $\psi \in \mathscr{H}_2, \phi \in \mathscr{H}'_0$. This ends the proof of the theorem. □

In what follows, η_f is assumed to be a character, namely of dimension 1. For $g \in G$, we denote by $g \cdot \eta_f$ the character of $G(g \cdot f)$ deduced from η_f by g.

Proposition 6.3.14 *For $j = 1, 2$, let $f_j \in \mathfrak{g}^*$ and η_{f_j} a character of $G(f_j)$ with differential $d\eta_{f_j} = i f_j|_{\mathfrak{g}(f_j)}$. Assume that there exist $\mathfrak{h}_j \in M(f_j, \mathfrak{g})$ stable under $G(f_j)$ and satisfying the Pukanszky condition. Then $T(f_1, \eta_{f_1})$ is equivalent to $T(f_2, \eta_{f_2})$ if and only if there exists $g \in G$ such that $f_2 = g \cdot f_1, \eta_{f_2} = g \cdot \eta_{f_1}$.*

Proof The result is well known in the connected case (cf. [3]). Besides, it is clear that $T(f_1, \eta_{f_1})$ is equivalent to $T(g \cdot f_1, g \cdot \eta_{f_1})$. To prove the converse, we use the idea of Theorem 6.3.13. Suppose that there exists a G-stable commutative ideal $\mathfrak{a}$ such that the corresponding analytic subgroup A is closed. According to the proof of Theorem 6.3.13 and Mackey's theory, $T(f_j, \eta_{f_j})$ restricted to A is supported on the orbit $G \cdot f'_j \subset \mathfrak{a}^*$ with $f'_j = f_j|_{\mathfrak{a}}$. Hence $G \cdot f'_1 = G \cdot f'_2$, which permits us to assume $f'_1 = f'_2$. Denote by A' the closed subgroup $G(f'_1) = G(f'_2)$ and by $\mathfrak{a}'$ its Lie algebra $\mathfrak{a}^{f_1} = \mathfrak{a}^{f_2}$. According to [123], the equivalence

$$T(f_1, \eta_{f_1}) = \mathrm{ind}_{A'}^G T(f_1, \eta_{f_1}, A') \simeq T(f_2, \eta_{f_2}) = \mathrm{ind}_{A'}^G T(f_2, \eta_{f_2}, A')$$

implies the equivalence $T(f_1, \eta_{f_1}, A') \simeq T(f_2, \eta_{f_2}, A')$. If $\dim(A') < \dim G$, the induction hypothesis ensures the existence of $g \in A'$ such that $f_2 = g \cdot f_1$ on $\mathfrak{a}'$ and $\eta_{f_2} = g \cdot \eta_{f_1}$. This means $g \cdot f_1 \in f_2 + (\mathfrak{a}')^{\perp} = A \cdot f_2$. Since $a \cdot \eta_{f_2} = \eta_{f_2}$ for any $a \in A$, we find the desired result.

According to what we have seen until now, it suffices to examine the last step of the proof of Theorem 6.3.13. The theory of Mackey requires

$$T(f_1, \eta_{f_1}, G(f)G^{\circ}) \simeq T(f_2, \eta_{f_2}, G(f)G^{\circ})$$

and consequently, restricting this equivalence to G°, we immediately see

$$T(f_1, \eta^\circ_{f_1}, G^\circ) \simeq T(f_2, \eta^\circ_{f_2}, G^\circ)$$

with $\eta^\circ_{f_j} = \eta_{f_j}|_{G^\circ(f_j)} (j = 1, 2)$. Then from [3], there exists $g \in G^\circ$ such that $f_2 = g \cdot f_1$. Hence we can suppose $f_1 = f_2$. What must be proved is $\eta_f = \eta'_f$ if $T(f, \eta_f) \simeq T(f, \eta'_f)$.

In fact, put $\sigma = T(f, \eta_f, G(f)G^\circ)$ and $\sigma' = T(f, \eta'_f, G(f)G^\circ)$. In our case, $\sigma|_{G^\circ} = \sigma'|_{G^\circ} = T(f, G^\circ)$ because G° is connected. The equivalence relation $U : \sigma \simeq \sigma'$ shows, by restriction to G°, that U commutes with $T(f, G^\circ)$. Since $T(f, G^\circ)$ is irreducible, it follows that U is a scalar. We conclude $\eta_f = \eta'_f$. □

The existence of a character η_f and a real polarization $\mathfrak{h}$ at $f \in \mathfrak{g}^*$ possessing the required properties depends only on the orbit $G \cdot f$. Let us denote by X the set of such orbits. For each $\Omega \in X$, we choose a representative $f_\Omega \in \Omega$ and denote by $\widehat{G(f_\Omega)}^\circ$ the set of the characters of $G(f_\Omega)$ with differential $i f_\Omega|_{\mathfrak{g}(f_\Omega)}$. We see in this way that the mapping

$$\widehat{G(f_\Omega)}^\circ \ni \eta_\Omega \mapsto T(f_\Omega, \eta_\Omega)$$

provides an injective mapping from

$$\widetilde{X} = \sqcup_{\Omega \in X} \widehat{G(f_\Omega)}^\circ$$

into the dual $\widehat{G}$ of G.

Remark 6.3.15 As the next example shows, the Pukanszky condition is not necessary for the irreducibility of the associated representation.

Example 6.3.16 The complex $ax + b$ group: let $\mathfrak{g}$ be the complex $ax + b$ algebra regarded as real Lie algebra. That is to say, $\mathfrak{g}$ is the solvable Lie algebra of dimension 4 with basis (e_1, e_2, e_3, e_4) and non-zero brackets

$$[e_1, e_3] = e_3, [e_1, e_4] = e_4, [e_2, e_3] = e_4, [e_2, e_4] = -e_3.$$

We denote by G the connected and simply connected solvable Lie group with Lie algebra $\mathfrak{g}$.

Put $f = e_4^* \in \mathfrak{g}^*, \mathfrak{h} = \mathbb{R}e_1 \oplus \mathbb{R}e_2$ and $\mathfrak{n} = [\mathfrak{g}, \mathfrak{g}] = \mathbb{R}e_3 \oplus \mathbb{R}e_4$. Let $H = \exp \mathfrak{h}$ and $N^\circ = \exp \mathfrak{n}$ be the corresponding analytic subgroups. We then have $G = HN^\circ = N^\circ H$. It is evident that $G(f) = \exp(2\pi \mathbb{Z} e_2)$, $\mathfrak{h} \in M(f, \mathfrak{g})$ and that $\mathfrak{h}$ does not satisfy the Pukanszky condition. More precisely, $H \cdot f$ is nothing but $f + \mathfrak{h}^\perp$ minus the origin. We start from the trivial character η_0 of $G(f)$ and extend it to the unique character $\chi_{f,\mathfrak{h}}$ (resp. $\chi_{f,\mathfrak{n}}$) of H (resp. $N = G(f)N^\circ$) with differential if.

It is well known that

$$T(f, \eta_0, \mathfrak{n}, G) = \mathrm{ind}_N^G \chi_{f,\mathfrak{n}}$$

is irreducible. In order to show the irreducibility of

$$T(f, \eta_0, \mathfrak{h}, G) = \mathrm{ind}_H^G \chi_{f,\mathfrak{h}},$$

we interpret the operator U formally defined by

$$(U\phi)(g) = \int_{N^\circ} \phi(gn)\chi_{f,\mathfrak{n}}(n)dn \ (g \in G).$$

where ϕ belongs in the space $\mathscr{H}_\mathfrak{h}$ of $T(f, \eta_0, \mathfrak{h}, G)$. This U manifestly intertwines the two representations.

In the usual way, the space $\mathscr{H}_\mathfrak{n}$ of $T(f, \eta_0, \mathfrak{n}, G)$ is identified with the space $L^2(\mathbb{R} \times [0, 2\pi); dt_1dt_2)$ and $\mathscr{H}_\mathfrak{h}$ to $L^2(\mathbb{R}^2; dt_2dt_4)$ under the applications $R_\mathfrak{n}$ and $R_\mathfrak{h}$ given by $(R_\mathfrak{n}\psi)(t_1, t_2) = \psi(\exp(t_1e_1+t_2e_2))$ and $(R_\mathfrak{h}\phi)(t_3, t_4) = \phi(\exp(t_3e_3+t_4e_4))$. Under these identifications, the operator U does make sense. Indeed, for $\phi \in \mathscr{H}_\mathfrak{h}$,

$$\begin{aligned}(U\phi)(t_1, t_2) &= e^{-t_1} \int_{\mathbb{R}^2} \phi(t_3, t_4) \exp(ie^{-t_1}(-t_3 \sin t_2 + t_4 \cos t_2))dt_3dt_4 \\ &= e^{-t_1}(\mathscr{F}\phi)(-e^{-t_1} \sin t_2, e^{-t_1} \cos t_2),\end{aligned}$$

where $\mathscr{F}$ denotes the ordinary Fourier transform. We deduce from this that

$$\begin{aligned}\|U\phi\|^2 &= \int_{\mathbb{R}\times[0,2\pi)} e^{-2t_1}|(\mathscr{F}\phi)(-e^{-t_1} \sin t_2, e^{-t_1} \cos t_2)|^2dt_1dt_2 \\ &= (2\pi)^2 \int_{\mathbb{R}^2} |\phi(t_3, t_4)|^2dt_3dt_4 = (2\pi)^2\|\phi\|^2,\end{aligned}$$

from which the desired result.

In fact, $\chi_{f,\mathfrak{h}}$ is trivial and the result says that the quasi-regular representation associated to the homogeneous space G/H is irreducible.

Chapter 7
Monomial Representations of Discrete Type of Exponential Solvable Lie Groups

7.1 Introduction

When investigating a concrete problem in the representation theory of solvable Lie groups, it is easy to realize that we have enough tools only for the nilpotent case. Here we shall consider Penney's Plancherel formula for exponential solvable Lie groups. Let G be an exponential solvable Lie group, H an analytic subgroup of G and χ a unitary character of H. We will study the monomial representation $\tau = \mathrm{ind}_H^G \chi$ when τ has either finite multiplicity or infinite multiplicity of discrete type. In particular, we are interested in Penney's Plancherel formula for τ and Duflo's commutativity question [54] for the algebra $D_\tau(G/H)$ of G-invariant differential operators on the line bundle associated to the data (G, H, χ). The orbit method is crucial for our study, and our general reference is [70].

7.2 Preliminaries

Let $G = \exp \mathfrak{g}$ be an exponential solvable Lie group with Lie algebra $\mathfrak{g}$, $H = \exp \mathfrak{h}$ a closed connected subgroup of G and $\chi = \chi_f, f \in \mathfrak{g}^*$ a unitary character of H, where $\chi_f(\exp X) = e^{if(X)}$ for $\forall X \in \mathfrak{h}$. Our objective is the monomial representation $\tau = \mathrm{ind}_H^G \chi$ of G, which is irreducible if and only if the subalgebra $\mathfrak{h}$ is a polarization of $\mathfrak{g}$ at f satisfying the Pukanszky condition. We denote by $K(G, H)$ the space of functions $\varphi : G \to \mathbb{C}$ with compact support modulo H verifying the covariance relation $\varphi(gh) = \Delta_{H,G}(h)\varphi(g)(g \in G, h \in H)$. We also denote by $K(G, H, f)$ the space of continuous functions $\varphi : G \to \mathbb{C}$ with compact support modulo H verifying the covariance relation

$$\varphi(gh) = \overline{\chi_f(h)}\Delta_{H,G}^{1/2}(h)\varphi(g) \ (g \in G, h \in H).$$

A. Baklouti et al., *Representation Theory of Solvable Lie Groups and Related Topics*, Springer Monographs in Mathematics,
https://doi.org/10.1007/978-3-030-82044-2_7

We begin by recalling the canonical central decomposition of τ. Let $\mathfrak{h}^\perp$ be the annihilator of $\mathfrak{h}$ in $\mathfrak{g}^*$, $\widetilde{\mu}$ a finite positive measure on the affine space $\Gamma = f + \mathfrak{h}^\perp$ equivalent to the Lebesgue measure on Γ. We regard $\widetilde{\mu}$ as a measure on $\mathfrak{g}^*$ and take its image μ by the Kirillov-Bernat map $\theta : \mathfrak{g}^* \to \widehat{G}$, $\widehat{G}$ being the unitary dual of G. The orbit method shows that $\widehat{G}$ is realized as the space $\mathfrak{g}^*/G$ of the coadjoint orbits of G by means of the induced bijection $\bar{\theta} = \bar{\theta}_G : \mathfrak{g}^*/G \to \widehat{G}$. In what follows, we often identify an equivalence class of unitary representations with its representatives, and denote the equivalence relation by $\simeq$ or even by an equality sign. For $\pi \in \widehat{G}$, we denote by $\Omega(\pi)$ the coadjoint orbit of G associated to π. Then

$$\tau \simeq \int_{\widehat{G}}^{\oplus} m(\pi)\pi d\mu(\pi). \tag{7.2.1}$$

Here the multiplicity $m(\pi)$ is given by the number of connected components of $\Gamma \cap \Omega(\pi)$ if each component is a manifold of dimension $\frac{1}{2}\dim\Omega(\pi)$. When this condition is not satisfied, $m(\pi)$ is equal to $+\infty$. In all cases, $m(\pi)$ is obtained as the number of H-orbits contained in $\Gamma \cap \Omega(\pi)$. We have ignored, up to now, how to characterize μ-almost everywhere the finiteness of these multiplicities, except for some particular cases [14]. When G is nilpotent, it is well known that the multiplicity function $\widehat{G} \ni \pi \to m(\pi)$ is either bounded or uniformly equal to $+\infty$.

Now we briefly recall the first objective in this chapter, i.e. Penney's Plancherel formula. Suppose for a moment that the monomial representation $\tau = \mathrm{ind}_H^G \chi_f$ has finite multiplicities. For a unitary representation ρ of G, we call $\mathscr{H}_\rho$ its Hilbert space, $\mathscr{H}_\rho^\infty$ the space of C^∞-vectors of ρ equipped with the usual topology and $\mathscr{H}_\rho^{-\infty}$ the anti-dual space of $\mathscr{H}_\rho^\infty$. For $a \in \mathscr{H}_\rho^{\pm}$ and $b \in \mathscr{H}_\rho^{\mp}$, we denote by $\langle a, b\rangle$ the value of a at b, so that $\langle b, a\rangle = \overline{\langle a, b\rangle}$. Designating by e the unit element of G, we see that the Dirac measure

$$\delta_\tau : \mathscr{H}_\tau^\infty \ni \psi \mapsto \overline{\psi(e)} \in \mathbb{C}$$

gives an element of $\left(\mathscr{H}_\tau^{-\infty}\right)^{H,\chi_f \Delta_{H,G}^{1/2}}$. Then according to the disintegration of τ, δ_τ disintegrates as

$$\delta_\tau \simeq \int_{\widehat{G}}^{\oplus} \Big(\sum_{k=1}^{m(\pi)} a_\pi^k\Big) d\mu(\pi) \tag{7.2.2}$$

with $a_\pi^k \in \left(\mathscr{H}_\pi^{-\infty}\right)^{H,\chi_f \Delta_{H,G}^{-1/2}}$. This means, for all $\varphi \in C_c^\infty(G)$,

$$\varphi_H^f(e) = \int_{\widehat{G}} \sum_{k=1}^{m(\pi)} \langle \pi(\varphi)a_\pi^k, a_\pi^k\rangle d\mu(\pi) \tag{7.2.3}$$

with

$$\varphi_H^f(g) = \int_H \varphi(gh)\chi_f(h)\Delta_{H,G}^{-1/2}(h)dh \ (g \in G)$$

The chapter's second objective is the algebra $D_\tau(G/H)$ of G-invariant differential operators on the line bundle over the base space G/H associated with the data (G, H, χ). Let $\mathfrak{a}_\tau$ be the subspace

$$\{Y + if(Y) - \frac{1}{2}\mathrm{trad}_{\mathfrak{g}/\mathfrak{h}}Y;\ Y \in \mathfrak{h}_\mathbb{C}\}$$

of $\mathscr{U}(\mathfrak{g})$, and $\mathscr{U}(\mathfrak{g})\mathfrak{a}_\tau$ the left ideal of $\mathscr{U}(\mathfrak{g})$ generated by $\mathfrak{a}_\tau$. Now we set

$$\mathscr{U}(\mathfrak{g}, \tau) = \{W \in \mathscr{U}(\mathfrak{g});\ [\mathfrak{h}, W] \subset \mathscr{U}(\mathfrak{g})\mathfrak{a}_\tau\}.$$

Then we know [40] that, under the right action of the elements of $\mathscr{U}(\mathfrak{g})$, the algebra $D_\tau(G/H)$ is isomorphic to $\mathscr{U}(\mathfrak{g}, \tau)/\mathscr{U}(\mathfrak{g})\mathfrak{a}_\tau$. The following problem was proved in the nilpotent case. For $\ell \in \mathfrak{g}^*$, let B_ℓ be the bilinear form on $\mathfrak{g}$ given by $B_\ell(X, Y) = \ell([X, Y])$ and $\mathfrak{g}(\ell)$ the radical of B_ℓ.

Duflo's Problem [54] The algebra $D_\tau(G/H)$ is commutative if and only if the subspace $\mathfrak{h} + \mathfrak{g}(\ell)$ is Lagrangian with respect to B_ℓ almost everywhere on Γ.

7.3 Monomial Representations of Discrete Type

Definition 7.3.1 We say that τ is of discrete type (or has multiplicities of discrete type) if each connected component of $\Gamma \cap \Omega(\pi)$ is an H-orbit μ-almost everywhere.

Abuse the language, we will often say that the multiplicities $m(\pi)$ are of discrete type. When G is completely solvable, a monomial representation is of discrete type if and only if it has finite multiplicities [15]. We also know [63] that τ is of discrete type if and only if $\mathfrak{h} + \mathfrak{g}(\ell)$ is $\widetilde{\mu}$-almost everywhere a Lagrangian subspace for B_ℓ. Irrespective of the fact that the monomial representation $\mathrm{ind}_H^G \chi_f$ is of discrete type, we have the following lemma.

Lemma 7.3.2 *Let $\mathfrak{g}$ be a real Lie algebra, $\mathfrak{h}$ a subalgebra of $\mathfrak{g}$ and $\mathfrak{n}$ an ideal of $\mathfrak{g}$. Let $f \in \mathfrak{g}^*$ such that $\langle f, [\mathfrak{h}, \mathfrak{h}]\rangle = \{0\}$. We put*

$$m_1 = \inf_{\Lambda \in \mathfrak{h}^\perp} \dim(\mathfrak{g}(f + \Lambda))$$

$$m_2 = \inf_{\Lambda \in \mathfrak{h}^\perp} \dim(\mathfrak{g}((f + \Lambda)|_\mathfrak{n}))$$

$$m_3 = \inf_{\Lambda \in \mathfrak{h}^\perp} \dim(\mathfrak{n} \cap \mathfrak{g}(f + \Lambda)).$$

Then the sets $\mathscr{U}_i \subset f + \mathfrak{h}^\perp (i = 1, 2, 3)$ *made of elements* $\ell \in \Gamma$ *such that* $\dim(\mathfrak{g}(\ell)) = m_1$, *resp.* $\dim(\mathfrak{g}(\ell|_\mathfrak{n})) = m_2$, *resp.* $\dim(\mathfrak{g}(\ell) \cap \mathfrak{n}) = m_3$ *are Zariski open sets. For every* $\ell \in \mathscr{U}_1$, *we have* $[\mathfrak{g}(\ell), \mathfrak{g}(\ell)] \subset \mathfrak{h} \cap \mathfrak{g}(\ell)$, *and for every* $\ell \in \mathscr{U}_2 \cap \mathscr{U}_3$ *we have* $[\mathfrak{g}(\ell|_\mathfrak{n}), \mathfrak{g}(\ell) \cap \mathfrak{n}] \subset \mathfrak{h} \cap \mathfrak{g}(\ell) \cap \mathfrak{n}$.

Proof One part of this lemma is already proved in [24], Chapter II. We can also apply the proof of proposition 1.11.7 of [50]. Here we will give another simple proof whose idea was suggested by Duflo. Let $d = \dim \mathfrak{g} - m_1$. For $\ell \in \mathscr{U}_1$, we take a basis $\mathscr{Z} = \{Z_1, \dots, Z_n\}$ of $\mathfrak{g}$ such that the matrix

$$M(\ell) = \begin{pmatrix} 0 & \langle \ell, [Z_1, Z_2] \rangle \dots & & \dots & \langle \ell, [Z_1, Z_d] \rangle \\ \langle \ell, [Z_2, Z_1] \rangle & 0 & \dots & \dots & \langle \ell, [Z_2, Z_d] \rangle \\ \vdots & \ddots & \ddots & \ddots & \vdots \\ \langle \ell, [Z_{d-1}, Z_1] \rangle & \dots & \dots & 0 & \langle \ell, [Z_{d-1}, Z_d] \rangle \\ \langle \ell, [Z_d, Z_1] \rangle & \dots & \dots \langle \ell, [Z_d, Z_{d-1}] \rangle & & 0 \end{pmatrix}$$

is of rank d, namely invertible. Then there is a neighbourhood $\mathscr{V}$ of 0 in $\mathfrak{h}^\perp$ such that the corresponding matrix $M(\ell + \Lambda)$ is also invertible for every $\Lambda \in \mathscr{V}$. For $r = d + 1, \dots, n$, let

$$\xi_r(\Lambda) = (\xi_r^1(\Lambda), \dots, \xi_r^d(\Lambda)) \in \mathbb{R}^d,$$

for which

$$\xi_r(\Lambda) M(\ell + \Lambda) = (-a_{1,r}(\Lambda), \dots, -a_{d,r}(\Lambda))$$

with $a_{k,r}(\Lambda) = \langle \ell + \Lambda, [Z_r, Z_k] \rangle$. Then the vectors

$$X_r(\Lambda) = Z_r + \sum_{j=1}^{d} \xi_r^j(\Lambda) Z_j, \ r = d + 1, \dots, n$$

form a basis of $\mathfrak{g}(\ell + \Lambda)$, and they depend rationally on Λ. So, for $\ell \in \mathscr{U}_1$ and two arbitrary elements x, y of $\mathfrak{g}(\ell)$, there exist two differentiable curves $x(t), y(t) \in \mathfrak{g}(\ell + t\Lambda)$, on a neighbourhood of $t = 0$, that satisfy $x(0) = x, y(0) = y$. Differentiating $\langle \ell + t\Lambda, [x(t), y(t)] \rangle = 0$ at $t = 0$, we obtain

$$\langle \Lambda, [x, y] \rangle + \langle \ell, [x'(0), y] \rangle + \langle \ell, [x, y'(0)] \rangle = 0,$$

from which $\langle \Lambda, [x, y] \rangle = 0$. This means that $[x, y] \in \mathfrak{h}$. Likewise, for $\ell \in \mathscr{U}_2 \cap \mathscr{U}_3$ and $x \in \mathfrak{g}(\ell|_\mathfrak{n}), y \in \mathfrak{g}(\ell) \cap \mathfrak{n}$, there exist two differentiable curves $x(t) \in \mathfrak{g}((\ell + t\Lambda)|_\mathfrak{n}), y(t) \in \mathfrak{g}(\ell + t\Lambda) \cap \mathfrak{n}$, in a neighbourhood of $t = 0$, satisfying

$x(0) = x, y(0) = y$. Then for t small enough, we have $\langle \ell + t\Lambda, [x(t), y(t)] \rangle = 0$. Differentiating this relation at $t = 0$, we have

$$\langle \Lambda, [x, y] \rangle + \langle \ell, [x'(0), y] \rangle + \langle \ell, [x, y'(0)] \rangle = 0,$$

from which we get $\langle \Lambda, [x, y] \rangle = 0$. This means that $[x, y] \in \mathfrak{h}$. □

7.3.1 Generic and Strongly Generic Elements

Let

$$\mathscr{S} : \{0\} = \mathfrak{g}_0 \subset \mathfrak{g}_1 \subset \cdots \subset \mathfrak{g}_{n-1} \subset \mathfrak{g}_n = \mathfrak{g}, \ \dim \mathfrak{g}_j = j \ (0 \le j \le n),$$

be a good sequence of subalgebras of $\mathfrak{g}$. That is to say, if $\mathfrak{g}_j$ is not an ideal of $\mathfrak{g}$, $\mathfrak{g}_{j\pm1}$ are ideals of $\mathfrak{g}$ and the adjoint action of $\mathfrak{g}$ on $\mathfrak{g}_{j+1}/\mathfrak{g}_{j-1}$ is irreducible. If $\mathfrak{g}_j$ and $\mathfrak{g}_{j-1}$ are ideals of $\mathfrak{g}$, we have a root $\alpha_j : \mathfrak{g} \to \mathbb{R}$ of $\mathfrak{g}$:

$$\mathrm{ad}(X)(X_j) - \alpha_j(X)X_j \in \mathfrak{g}_{j-1}, \ X_j \in \mathfrak{g}_j, \ X \in \mathfrak{g}.$$

If $\mathfrak{g}_j$ is not an ideal of $\mathfrak{g}$, we take a subspace $\mathfrak{v}_j$ of dimension 2 in $\mathfrak{g}_{j+1}$ such that $\mathfrak{g}_{j+1} = \mathfrak{v}_j + \mathfrak{g}_{j-1}$. Then there is a homomorphism α_j of $\mathfrak{g}$ into the endomorphism algebra of $\mathfrak{v}_j$ such that

$$\mathrm{ad}(X)v - \alpha_j(X)v \in \mathfrak{g}_{j-1}, \ X \in \mathfrak{g}, \ v \in \mathfrak{v}_j.$$

For $\ell \in \mathfrak{g}^*$, let

$$\mathfrak{g}_j(\ell_j) = \{X \in \mathfrak{g}_j; \langle \ell, [X, \mathfrak{g}_j] = \{0\}\}$$

with $\ell_j = \ell|_{\mathfrak{g}_j} (1 \le j \le n)$. The Vergne polarization $\mathfrak{b} = \mathfrak{b}[\ell] = \mathfrak{b}_{\mathscr{S}}[\ell]$ (with respect to $\mathscr{S}$) at ℓ was introduced in Theorem 1.2.4.

Let us return to our monomial representation τ of discrete type. Let $\pi \in \widehat{G}$ be such that $\Gamma \cap \Omega(\pi)$ decomposes into finite connected components as

$$\Gamma \cap \Omega(\pi) = \coprod_{k=1}^{m(\pi)} C_k,$$

where C_k is an H-orbit of dimension $(1/2) \dim \Omega(\pi)$. We proceed as follows. We fix $\ell \in \Omega(\pi)$ and take there a Vergne polarization $\mathfrak{b} = \mathfrak{b}[\ell]$ to realize $\pi = \mathrm{ind}_B^G \chi_\ell$ with $B = \exp \mathfrak{b}$. Let us choose $g_k \in G$ $(1 \le k \le m(\pi))$ satisfying $g_k \cdot \ell \in C_k$. A candidate of our generalized vector a_π^k would be

$$\langle a_\pi^k, \varphi\rangle = \oint_{H/(H\cap g_k B g_k^{-1})} \overline{\varphi(hg_k)\chi_f(h)\Delta_{H,G}^{-1/2}(h)d\nu_k(h)}, \ \varphi \in \mathscr{H}_\pi^\infty. \tag{7.3.1}$$

The first problem regarding this integral concerns the relationship of the traces. For simplicity, take $g_k = e$. Our problem is the convergence of the integral

$$\langle a_\pi, \varphi\rangle = \oint_{H/(H\cap B)} \overline{\varphi(h)\chi_f(h)\Delta_{H,G}^{-1/2}(h)d\nu(h)}. \tag{7.3.2}$$

We expect that the following trace relation holds for $X \in \mathfrak{b}[\ell] \cap \mathfrak{h}$:

$$\operatorname{Tr} \operatorname{ad}_{\mathfrak{b}[\ell]/(\mathfrak{b}[\ell]\cap\mathfrak{h})} X + \operatorname{Tr} \operatorname{ad}_{\mathfrak{h}/(\mathfrak{b}[\ell]\cap\mathfrak{h})} X = 0. \tag{7.3.3}$$

The next example shows that this is not always the case even if $\mathfrak{h} + \mathfrak{g}(\ell)$ is Lagrangian.

Example 7.3.3 Let $\mathfrak{g} = \mathbb{R}\text{-span}(X, Y_1, Y_2, Z, C)$; $[X, Y_1] = Y_2, [X, Y_2] = Z, [Y_1, C] = C$. Take $\mathfrak{h} = \mathbb{R}\text{-span}(Y_1, Y_2, Z)$ and $\ell = Z^*$. Then $\mathfrak{g}(\ell) = \mathbb{R}\text{-span}(Y_1, Z, C)$ and for a choice of $\mathscr{S}$, $\mathfrak{b}[\ell] = \mathbb{R}\text{-span}(Y_1, Y_2, Z, C)$. Clearly, $\mathfrak{h} + \mathfrak{g}(\ell)$ is Lagrangian for the bilinear form B_ℓ but $\operatorname{Tr} \operatorname{ad}_{\mathfrak{b}[\ell]/(\mathfrak{b}[\ell]\cap\mathfrak{h})} Y_1 + \operatorname{Tr} \operatorname{ad}_{\mathfrak{h}/(\mathfrak{b}[\ell]\cap\mathfrak{h})} Y_1 = 1$.

So, we introduce the following definition.

Definition 7.3.4 Let $\mathfrak{g}$ be an exponential solvable Lie algebra, $\mathfrak{h}$ a subalgebra of $\mathfrak{g}$ and $\mathfrak{n}$ a nilpotent ideal of $\mathfrak{g}$ containing $[\mathfrak{g}, \mathfrak{g}]$. Let $f \in \mathfrak{g}^*$ such that $\langle f, [\mathfrak{h}, \mathfrak{h}]\rangle = \{0\}$ and

$$\mathscr{S} : \{0\} \subset \mathfrak{g}_1 \subset \cdots \subset \mathfrak{g}_n = \mathfrak{g}$$

a good sequence of subalgebras containing $\mathfrak{n}$. For $\ell \in \Gamma$, we denote by $\mathfrak{b}[\ell]$ the Vergne polarization at ℓ associated to $\mathscr{S}$. Then ℓ is said to be generic with respect to $\mathfrak{n}$ and $\mathscr{S}$, or merely generic, if $\ell \in \mathscr{U}_1 \cap \mathscr{U}_2 \cap \mathscr{U}_3$ and if

$$\dim(\mathfrak{b}[\ell] \cap \mathfrak{h}) = \inf_{\Lambda\in\mathfrak{h}^\perp} \dim(\mathfrak{b}[f + \Lambda] \cap \mathfrak{h}).$$

Let Δ be the set of roots of $\mathfrak{g}$ that vanish on every $\mathfrak{g}(\ell|_\mathfrak{n})$ with generic $\ell \in \Gamma$. Let $\mathfrak{g}_0 = \cap_{\alpha\in\Delta} \ker(\alpha)$. We say that $\ell \in \Gamma$ is strongly generic if ℓ is generic and if $\alpha(\mathfrak{g}(\ell|_\mathfrak{n})) \neq \{0\}$ for every root α not contained in Δ.

Remark 7.3.5 From Lemma 7.3.2, for every generic linear form $\ell \in \Gamma$, we have

$$[\mathfrak{g}(\ell|_\mathfrak{n}), \mathfrak{g}(\ell) \cap \mathfrak{n}] \subset \mathfrak{h} \text{ and } [\mathfrak{g}(\ell), \mathfrak{g}(\ell)] \subset \mathfrak{h}.$$

Moreover, the set of the strongly generic elements of Γ contains a non-empty Zariski open set of Γ. Indeed, $\mathscr{U}_1 \cap \mathscr{U}_2 \cap \mathscr{U}_3$ is a Zariski open set of Γ and $\mathfrak{g}(\ell|_{\mathfrak{n}})$ admits a basis which varies rationally in ℓ on a Zariski open set of Γ. On the other hand, the set of linear forms $\ell \in \Gamma$ verifying

$$\dim(\mathfrak{b}[\ell] \cap \mathfrak{h}) = \inf_{\Lambda \in \mathfrak{h}^\perp} \dim(\mathfrak{b}[f + \Lambda] \cap \mathfrak{h})$$

is the set of linear forms which do not vanish on a finite family of polynomials on Γ.

Proposition 7.3.6 *With the previous notations, let $\mathfrak{h}_0 = \mathfrak{h} \cap \mathfrak{g}_0$ and $\Gamma_0 = f|_{\mathfrak{g}_0} + (\mathfrak{h}_0)^\perp \subset (\mathfrak{g}_0)^*$. Let $\mathscr{C}$ be a composition sequence passing through $\mathfrak{n}$, which does not change any Vergne polarization $\mathfrak{b}[\ell]$ constructed from the initial composition sequence, for $\ell \in \Gamma$ generic. Then for every strongly generic $\ell \in \Gamma$, the restriction ℓ_0 to $\mathfrak{g}_0$ is generic (with respect to $\mathfrak{h}_0, \mathscr{C} \cap \mathfrak{g}_0$ and $\mathfrak{n}$).*

Proof In fact, for every strongly generic $\ell \in \Gamma$, we have:

- $\mathfrak{g}(\ell|_{\mathfrak{n}}) = \mathfrak{g}_0(\ell_0|_{\mathfrak{n}})$,
- $\dim(\mathfrak{g}_0(\ell_0)/\mathfrak{g}(\ell)) = \dim(\mathfrak{g}/\mathfrak{g}_0)$,
- $\dim((\mathfrak{g}_0(\ell_0) \cap \mathfrak{n})/(\mathfrak{g}(\ell) \cap \mathfrak{n})) = \dim(\mathfrak{g}/\mathfrak{g}_0)$,
- $\mathfrak{h} \cap \mathfrak{b}[\ell] = \mathfrak{h}_0 \cap \mathfrak{b}[\ell]$.

□

The following proposition intends to prove the trace relation (7.3.3) for generic elements.

Proposition 7.3.7 *We keep the notations of the previous definition. Let $\ell \in \Gamma$ be generic with respect to $\mathfrak{n}$ and $\mathscr{S}$. Then (7.3.3) holds on $\mathfrak{b}[\ell] \cap \mathfrak{h}$.*

Proof We proceed by induction on $p = \dim G + \dim G/H$. Remark first that our polarization $\mathfrak{b} = \mathfrak{b}[\ell]$ is contained in $\mathfrak{g}(\ell|_{\mathfrak{n}}) + \mathfrak{n}$, since $\mathfrak{b} \cap \mathfrak{n}$ is a polarization at $\ell|_{\mathfrak{n}}$. Let

$$\mathscr{C} : \{0\} = \mathfrak{g}^0 \subset \mathfrak{g}^1 \subset \cdots \subset \mathfrak{g}^d = \mathfrak{g}$$

be the composition sequence of $\mathfrak{g}$ determined by $\mathscr{S}$ such that $\mathfrak{g}^k = \mathfrak{n}$. For $p \in \{1, 2\}$ or more generally if $\mathfrak{g}$ is Abelian, our assertion is clear. Let $j > 0$ be the smallest index such that $\mathfrak{g}^j \not\subset \mathfrak{g}(\ell)$. If $j > k$, then $\mathfrak{n} \subset \mathfrak{g}(\ell)$ and $\mathfrak{b}[\ell]$ is an ideal of $\mathfrak{g}$. Replacing $\mathfrak{h}$ by $\mathfrak{h}' = \mathfrak{h} + \mathfrak{n}$, we have $[\mathfrak{b}[\ell], \mathfrak{n}] \subset \mathfrak{h}$ by Lemma 7.3.2 because ℓ is generic and $\mathfrak{b}[\ell] \subset \mathfrak{g}(\ell|_{\mathfrak{n}})$. Hence

$$\mathrm{Tr}\, \mathrm{ad}_{\mathfrak{b}[\ell]/(\mathfrak{b}[\ell] \cap \mathfrak{h})} X = \mathrm{Tr}\, \mathrm{ad}_{\mathfrak{h}/(\mathfrak{b}[\ell] \cap \mathfrak{h})} X = 0$$

for every $X \in \mathfrak{b}[\ell] \cap \mathfrak{h}$.

So, we can suppose $j \leq k$, namely $\mathfrak{g}^j \subset \mathfrak{n}$. If $j > 1$ and if ℓ vanishes on a non-trivial ideal $\mathfrak{a}$ of $\mathfrak{g}$ contained in $\mathfrak{g}^{j-1}$, then $\mathfrak{a} \subset \mathfrak{g}(\ell)$. If $\mathfrak{a}$ is contained in $\mathfrak{h}$ and so in $\mathfrak{b} \cap \mathfrak{h}$, everything works by induction. If not, we can replace $\mathfrak{h}$ by $\mathfrak{h}' = \mathfrak{h} + \mathfrak{a}$ and apply the induction hypothesis. In fact, we have $\mathfrak{b}[\ell] \cap \mathfrak{h}' = \mathfrak{b}[\ell] \cap \mathfrak{h} + \mathfrak{a}$, from which

$$\mathfrak{h}'/(\mathfrak{b}[\ell] \cap \mathfrak{h}') \simeq \mathfrak{h}/(\mathfrak{b}[\ell] \cap \mathfrak{h})$$

and

$$\mathfrak{b}[\ell]/(\mathfrak{b}[\ell] \cap \mathfrak{h}) \simeq \mathfrak{b}[\ell]/(\mathfrak{b}[\ell] \cap \mathfrak{h}') + \mathfrak{a}/(\mathfrak{a} \cap \mathfrak{h}).$$

Using the induction hypothesis, we have, for every $X \in \mathfrak{b}[\ell] \cap \mathfrak{h}$,

$$\mathrm{Tr}\, \mathrm{ad}_{\mathfrak{b}[\ell]/(\mathfrak{b}[\ell] \cap \mathfrak{h}')} X + \mathrm{Tr}\, \mathrm{ad}_{\mathfrak{h}/(\mathfrak{b}[\ell] \cap \mathfrak{h})} X = 0,$$

as X is a sum of a nilpotent vector and an element of $\mathfrak{g}(\ell|_{\mathfrak{n}})$ and $\mathrm{Tr}\, \mathrm{ad}_{\mathfrak{a}/(\mathfrak{a} \cap \mathfrak{h})} X = 0$ because of $[\mathfrak{g}(\ell|_{\mathfrak{n}}), \mathfrak{a}] \subset \mathfrak{b} \cap \mathfrak{h}$.

Taking everything into account, we can suppose that either $j = 1$ or $j = 2$ where $\mathfrak{g}^1$ is central. We are now in the situation where $\ell(\mathfrak{g}^1) \neq \{0\}$ and the root α^1 of $\mathfrak{g}$ on $\mathfrak{g}^1$ is non-zero. Let $\mathfrak{g}^{(1)} = \{X \in \mathfrak{g};\, [X, \mathfrak{g}^1] = \{0\}\}$. Then $\mathfrak{g}^{(1)}$ is an ideal of $\mathfrak{g}$ of codimension 1 containing $\mathfrak{g}(\ell|_{\mathfrak{n}}) + \mathfrak{n}$. Let $\mathscr{S}^{(1)}$ be the good sequence determined by the sequence $\mathfrak{g}_j \cap \mathfrak{g}^{(1)}$. Let $\mathfrak{h}^{(1)} = \mathfrak{h} \cap \mathfrak{g}^{(1)} + \mathfrak{g}^1$ and $\ell^{(1)} = \ell|_{\mathfrak{g}^{(1)}}$. Then $\ell^{(1)}$ is generic with respect to $\mathfrak{h}^{(1)}$ and $\mathscr{S}^{(1)}$. Since $\mathfrak{b} \cap \mathfrak{h} \subset \mathfrak{g}^{(1)}$, we can apply the induction hypothesis to $\mathfrak{h}^{(1)}$, $\mathscr{S}^{(1)}$, $\ell^{(1)}$ and $\mathfrak{n}$.

There remains for us to study the case where $\mathfrak{g}^1 = \mathbb{R}Z$, Z being central, $Z \in \mathfrak{h}$ and $\ell(Z) = 1$. Let $\mathfrak{d}$ be the minimal non-central ideal $\mathfrak{g}^2$. If $\mathfrak{d} = \mathbb{R}Z \oplus \mathfrak{y}$ for an ideal $\mathfrak{y}$ of dimension 1 or 2, then we proceed as before, according to whether $\ell(\mathfrak{y}) = \{0\}$ or not. Finally, if none of the previous cases occurs, the subalgebra

$$\mathfrak{g}^{(1)} = \mathfrak{d}^{\ell} = \{X \in \mathfrak{g};\, B_\ell(X, Y) = 0,\ Y \in \mathfrak{d}\}$$

is proper and contains $\mathfrak{b}$. Let $\mathscr{S}^{(1)}$ be the good sequence of $\mathfrak{g}^{(1)}$ obtained from the sequence $\mathscr{S}$. The subalgebra $\mathfrak{b}$ is a polarization at $\ell^{(1)} = \ell|_{\mathfrak{g}^{(1)}}$. Let $\mathfrak{n}^{(1)} = \mathfrak{n} \cap \mathfrak{g}^{(1)}$. If $\mathfrak{h} \not\subset \mathfrak{g}^{(1)}$, set $\mathfrak{h}^{(1)} = \mathfrak{h} \cap \mathfrak{g}^{(1)} + \mathfrak{d}$. Then we easily verify that $\ell^{(1)}$ is generic with respect to $\mathfrak{h}^{(1)}$, $\mathfrak{n}^{(1)}$, $\mathscr{S}^{(1)}$. The induction hypothesis applied to $\mathfrak{g}^{(1)}$ ensures the trace relation

$$\mathrm{Tr}\, \mathrm{ad}_{\mathfrak{b}[\ell]/(\mathfrak{b}[\ell] \cap \mathfrak{h}^{(1)})} X + \mathrm{Tr}\, \mathrm{ad}_{\mathfrak{h}^{(1)}/(\mathfrak{b}[\ell] \cap \mathfrak{h}^{(1)})} X = 0$$

for every $X \in \mathfrak{b}[\ell] \cap \mathfrak{h}$, which is nothing but the desired relation. Indeed, $\mathfrak{b}[\ell] \cap \mathfrak{h}^{(1)} = \mathfrak{b}[\ell] \cap \mathfrak{h} + \mathfrak{d}$ and we have

$$\mathrm{Tr}\, \mathrm{ad}_{\mathfrak{d}/(\mathfrak{d} \cap \mathfrak{g}^1)} X + \mathrm{Tr}\, \mathrm{ad}_{\mathfrak{h}/(\mathfrak{h} \cap \mathfrak{g}^{(1)})} X = 0$$

for any $X \in \mathfrak{b}[\ell] \cap \mathfrak{h}$.

Now suppose that $\mathfrak{h} \subset \mathfrak{g}^{(1)}$. Let again $\mathfrak{h}^{(1)} = \mathfrak{h} + \mathfrak{d}$ and $\ell^{(1)} = \ell|_{\mathfrak{g}^{(1)}}$. If $\mathfrak{d} \subset \mathfrak{h}$, $\ell^{(1)}$ is generic and we can apply the induction hypothesis. If the root α^2 of $\mathfrak{g}$ on $\mathfrak{g}^2/\mathfrak{g}^1$ vanishes on $\mathfrak{h}$, then, considering the subalgebra $\mathfrak{h}^{(1)}$, $\ell^{(1)}$ is again generic with respect to $\mathfrak{n}^{(1)}$ and $\mathscr{S}^{(1)}$. Hence it suffices to apply the induction hypothesis, because

$$\mathrm{Tr}\ \mathrm{ad}_{\mathfrak{b}[\ell]/(\mathfrak{b}[\ell]\cap\mathfrak{h})}X = \mathrm{Tr}\ \mathrm{ad}_{\mathfrak{b}[\ell]/(\mathfrak{b}[\ell]\cap\mathfrak{h}^{(1)})}X,$$
$$\mathrm{Tr}\ \mathrm{ad}_{\mathfrak{h}^{(1)}/(\mathfrak{b}[\ell]\cap\mathfrak{h}^{(1)})}X = \mathrm{Tr}\ \mathrm{ad}_{\mathfrak{h}/(\mathfrak{b}[\ell]\cap\mathfrak{h})}X.$$

If the root α^2 does not vanish on $\mathfrak{h}$ and if $\mathfrak{d} \not\subset \mathfrak{h}$, then $\mathfrak{d} \cap \mathfrak{h} = \mathbb{R}Z$. Take $\Lambda \in \mathfrak{h}^{\perp}$ such that $\Lambda|_{\mathfrak{d}} \neq 0$. Since ℓ is generic, we have

$$\dim(\mathfrak{b}[\ell]) = \dim(\mathfrak{b}[\ell + t\Lambda]),\ \dim(\mathfrak{b}[\ell] \cap \mathfrak{h}) = \dim(\mathfrak{b}[\ell + t\Lambda] \cap \mathfrak{h})$$

for t small enough. If there is $U \in \mathfrak{b}[\ell]\cap\mathfrak{h}$ such that $\alpha^2(U) \neq 0$, this means that there exists $U_t \in \mathfrak{b}[\ell + t\Lambda] \cap \mathfrak{h}$ such that $\alpha^2(U_t) \neq 0$ for $t \neq 0$ small enough. But then $U_t \notin \mathfrak{g}^{(1)}$, which contradicts the fact that $\mathfrak{h} \subset \mathfrak{g}^{(1)}$. Therefore, $\alpha^2(\mathfrak{b}[\ell] \cap \mathfrak{h}) = \{0\}$ and we finally obtain the relation

$$\mathrm{Tr}\ \mathrm{ad}_{\mathfrak{b}[\ell]/(\mathfrak{b}[\ell]\cap\mathfrak{h})}X + \mathrm{Tr}\ \mathrm{ad}_{\mathfrak{h}/(\mathfrak{b}[\ell]\cap\mathfrak{h})}X = 0$$

for $X \in \mathfrak{b}[\ell] \cap \mathfrak{h}$. □

Proposition 7.3.8 *For $\ell \in \Gamma$, put $\mathfrak{b} = \mathfrak{b}[\ell]$ and $B = \exp\mathfrak{b}$. Let $g_k \in G$ be such that $\ell_k = Ad^*(g_k)\ell \in \Gamma$, $\mathfrak{b}_k = Ad(g_k)\mathfrak{b}$ and $B_k = g_k B g_k^{-1}$. Let us consider for $\varphi \in K(G, B, \ell)$ the function*

$$\Phi(h) = \overline{\varphi(hg_k)\chi_f(h)}\Delta_{H,G}^{-1/2}(h),\ \ h \in H.$$

If the trace relation (7.3.3) holds for ℓ_k, $\mathfrak{h}$, then we have for $x \in H \cap B_k$ and $h \in H$,

$$\Phi(hx) = \Delta_{H\cap B_k,H}(x)\Phi(h).$$

Proof The subalgebra $\mathfrak{b}_k$ is a Vergne polarization at ℓ_k associated with the good sequence $\mathrm{Ad}(g_k)(\mathscr{S})$. Then the trace relation (7.3.3) says that

$$\Delta_{B_k\cap H,H}(x) = \Delta_{B_k\cap H,B_k}^{-1}(x), \tag{7.3.4}$$

namely,

$$\Delta_{B_k\cap H}(x)\Delta_H^{-1/2}(x) = \Delta_{B_k}^{1/2}(x),\ x \in B_k \cap H.$$

Hence

$$\begin{aligned}
\overline{\Phi(hx)} &= \varphi(hxg_k)\chi_f(hx)\Delta_{H,G}^{-1/2}(hx) \\
&= \chi_\ell(g_k^{-1}xg_k)^{-1}\chi_f(x)\Delta_{B,G}^{1/2}(g_k^{-1}xg_k)\Delta_{H,G}^{-1/2}(hx)\chi_f(h)\varphi(hg_k) \\
&= \Delta_{B_k}^{1/2}(x)\Delta_H(x)^{-1/2}\Delta_{H,G}^{-1/2}(h)\chi_f(h)\varphi(hg_k) \\
&= \Delta_{B_k\cap H,H}(x)\Delta_{H,G}^{-1/2}(h)\chi_f(h)\varphi(hg_k) \text{ (using(7.3.4))} \\
&= \Delta_{B_k\cap H,H}(x)\overline{\Phi(h)}, \ x \in B_k \cap H \ h \in H.
\end{aligned}$$

□

This proves that (7.3.2) has, at least formally, a meaning. We still need to prove the convergence of the integral. If the simple product HB_k was a closed subset of G, as the homogeneous space $H/(H \cap B_k)$ is homeomorphic to HB_k/B_k, the integral (7.3.2) would be convergent for any continuous $\varphi \in \mathscr{H}_\pi^\infty$ with compact support modulo B. Unfortunately, the simple product in question is not always closed, as the following example shows.

Example 7.3.9 Let $\mathfrak{g} = \mathbb{R}\text{-span}(X, T, A, Z)$; $[X, T] = -T$, $[X, A] = A$, $[T, A] = Z$. Take $\mathfrak{h} = \mathbb{R}X$ and $f \in \mathfrak{g}^*$ such that $f|_\mathfrak{h} = 0$. Let $\ell = \alpha X^* + \beta T^* + \gamma A^* + \lambda Z^*$ ($\alpha, \beta, \gamma, \lambda \in \mathbb{R}$) be an arbitrary element of $\mathfrak{g}^*$. By simple calculations, we see that

$$G{\cdot}\ell = \left\{\left(\alpha - \frac{\beta\gamma}{\lambda} + \frac{uw}{\lambda}\right)X^* + uT^* + wA^* + \lambda Z^*;\ u, w \in \mathbb{R}\right\}$$

when $\lambda \neq 0$. Thus,

$$G{\cdot}\ell \cap \mathfrak{h}^\perp = \{uT^* + wA^* + \lambda Z^*;\ uw = \zeta\}$$

with $\zeta = \beta\gamma - \lambda\alpha$. Take again $\ell = uT^* + wA^* + \lambda Z^*$ ($\lambda \neq 0$) in $\mathfrak{h}^\perp$. If $uw \neq 0$, $\mathfrak{h} + \mathfrak{g}(\ell)$ is a Lagrangian subspace for the bilinear form B_ℓ and each connected component of $G{\cdot}\ell \cap \mathfrak{h}^\perp$ is an H-orbit, where $H = \exp\mathfrak{h}$. Then $\tau = \mathrm{ind}_H^G\chi_f$ has uniformly finite multiplicities equal to 2. Now, let us consider the Jordan-Hölder sequence of ideals of $\mathfrak{g}$,

$$\mathscr{S} : \{0\} \subset \mathbb{R}Z \subset \mathbb{R}Z + \mathbb{R}A \subset \mathbb{R}Z + \mathbb{R}A + \mathbb{R}T \subset \mathfrak{g}.$$

Then at the point ℓ, the associated Vergne polarization is

$$\mathfrak{b}_\mathscr{S}[\ell] = \mathfrak{b}[\ell] = \mathbb{R}(\lambda X - wT) + \mathbb{R}A + \mathbb{R}Z.$$

Putting

$$\pi_\ell = \pi_{\zeta,\lambda} = \mathrm{ind}_{B[\ell]}^G \chi_\ell \in \widehat{G}\ (\zeta = uw),$$

we have

$$\tau \simeq \int_{\mathbb{R}^2}^{\oplus} 2\pi_{\zeta,\lambda} d\zeta d\lambda.$$

Hence τ has finite multiplicities, but the simple product $HB[\ell]$, $B[\ell] = \exp(\mathfrak{b}[\ell])$, is a proper open subset of G.

Definition 7.3.10 Let G be an exponential solvable Lie group with Lie algebra $\mathfrak{g}$, $H = \exp \mathfrak{h}$ a closed connected subgroup of G and assume $\ell \in \mathfrak{g}^*$ vanishes on $[\mathfrak{h}, \mathfrak{h}]$. Let $\mathfrak{b}$ be a subalgebra of $\mathfrak{g}$ subordinate to ℓ and satisfying

$$\mathrm{Tr}\,\mathrm{ad}_{\mathfrak{b}/(\mathfrak{b}\cap\mathfrak{h})} X + \mathrm{Tr}\,\mathrm{ad}_{\mathfrak{h}/(\mathfrak{b}\cap\mathfrak{h})} X = 0,\ X \in \mathfrak{b} \cap \mathfrak{h}.$$

Let $B = \exp \mathfrak{b}$. We can define the linear form $a_{\ell,\mathfrak{b},\mathfrak{h}}$ on the vector space

$$D(B, H, \ell) := \{\varphi : G \to \mathbb{C};\ \text{continuous},\ \varphi(gc) = \Delta_{H,B\cap H}^{1/2}(c)\overline{\chi_\ell(c)}\varphi(g),\ c \in B \cap H,\ g \in G,$$

$$\|\varphi\|_1 = \oint_{H/(H\cap B)} |\varphi(h)|\Delta_{H,G}^{-1/2}(h) d\nu(h) < \infty\}$$

by the (convergent) integral

$$\langle a_{\ell,\mathfrak{b},\mathfrak{h}}, \varphi\rangle = \oint_{H/(H\cap B)} \overline{\varphi(h)\chi_\ell(h)}\Delta_{H,G}^{-1/2}(h) d\nu(h)$$

(see Proposition 7.3.8).

Our objective is to prove the convergence of (7.3.2), which will allow to describe the Plancherel formula. More precisely, we prove the following theorem, since the explicit determination of $\mathcal{H}_\pi^\infty$ is unknown.

Theorem 7.3.11 *Let G be an exponential solvable Lie group with Lie algebra $\mathfrak{g}$, $H = \exp \mathfrak{h}$ a closed connected subgroup of G, $f \in \mathfrak{g}^*$ vanishing on $[\mathfrak{h}, \mathfrak{h}]$ and $\tau = \mathrm{ind}_H^G \chi_f$ a monomial representation of discrete type. Let $\mathscr{C}$ be a composition sequence of $\mathfrak{g}$ passing through a nilpotent ideal $\mathfrak{n}$ which contains $[\mathfrak{g}, \mathfrak{g}]$ and $\mathscr{S}$ a good sequence of subalgebras of $\mathfrak{g}$ constructed from $\mathscr{C}$. For any strongly generic $\ell \in \Gamma$, the integral*

$$\langle a_{\ell,\mathfrak{b},\mathfrak{h}}, \varphi\rangle = \langle a_\pi, \varphi\rangle = \oint_{H/(H\cap B)} \overline{\varphi(h)\chi_f(h)}\Delta_{H,G}^{-1/2}(h) d\nu(h) \qquad (7.3.5)$$

is well defined and is convergent for any $\varphi \in K(G, B, \ell)$. *Namely,* $K(G, B, \ell) \subset D(B, H, \ell)$. *Here,* $\mathfrak{b} = \mathfrak{b}[\ell]$ *denotes the Vergne polarization at* ℓ *with respect to* $\mathscr{S}$, $B = \exp \mathfrak{b}$ *and* $\pi = ind_B^G \chi_\ell \in \widehat{G}$.

Corollary 7.3.12 *For any strongly generic* $\ell \in \Gamma$ *and any* $\varphi \in K(G, B, \ell)$, *the function defined by*

$$g \mapsto \varphi_{H,B}^f(g) = \oint_{H/(H\cap B)} \overline{\varphi(gh)\chi_f(h)}\Delta_{H,G}^{-1/2}(h)d\nu(h)$$

is continuous on G.

Proof In fact, since the mapping $C_c(G) \to K(G, B, \ell); \xi \mapsto \xi_B^\ell$ defined by

$$\xi_B^\ell(g) = \int_B \xi(gb)\chi_\ell(b)\Delta_{B,G}^{-1/2}(b)db,\ g \in G$$

is surjective, there is a function $\varphi_0 \in C_c(G)$ such that $\varphi = (\varphi_0)_B^\ell$. Then for any compact set K of G, we can take a function $\psi_0 \geq 0$ in $C_c(G)$ such that $|\varphi_0(kg)| \leq \psi_0(g)$ for all $g \in G, k \in K$. Then

$$\begin{aligned} |\varphi(kg)| &\leq \int_B |\varphi_0(kgb)|\Delta_{B,G}^{-1/2}(b)db \\ &\leq \int_B \psi_0(gb)\Delta_{B,G}^{-1/2}(b)db =: \psi(g). \end{aligned}$$

As $|\varphi|^2, \psi^2$ belong to $K(G, B)$, Theorem 7.3.11 gives

$$\oint_{H/(H\cap B)} |\varphi(kgh)|\Delta_{H,G}^{-1/2}(h)d\nu(h) \leq \oint_{H/(H\cap B)} \psi(gh)\Delta_{H,G}^{-1/2}(h)d\nu(h) < \infty$$

for all $k \in K, g \in G$. An application of Lebesgue's dominated convergence theorem guarantees the continuity of $\varphi_{H,B}^f$ on G. □

7.3.2 *A Basis for* $\mathfrak{h}/(\mathfrak{h} \cap \mathfrak{b})$

1. On Relative Indices Let $\mathfrak{g}$ be an exponential solvable Lie algebra and $\mathfrak{n}$ a nilpotent ideal of $\mathfrak{g}$ which contains $[\mathfrak{g}, \mathfrak{g}]$. Let us take a composition sequence

$$\mathscr{C} : \{0\} = \mathfrak{g}^0 \subset \mathfrak{g}^1 \subset \cdots \subset \mathfrak{g}^{d-1} \subset \mathfrak{g}^d = \mathfrak{g},\ 1 \leq \dim(\mathfrak{g}^j/\mathfrak{g}^{j-1}) \leq 2\ (1 \leq j \leq d),$$

where $\mathfrak{g}^k = \mathfrak{n}$ for a certain k. For every $j \in \{1, \ldots, d\}$, we determine a subspace $\mathfrak{w}^j$ of $\mathfrak{g}^j$ so that $\mathfrak{g}^j = \mathfrak{w}^j \oplus \mathfrak{g}^{j-1}$. We denote by $\alpha^j : \mathfrak{g} \to \mathrm{End}(\mathfrak{w}^j)$ the irreducible representation of $\mathfrak{g}$ in $\mathfrak{w}^j$ given by the adjoint action on $\mathfrak{g}^j/\mathfrak{g}^{j-1}$. By abuse of notation, we shall write $\dim(\alpha^j)$, which is actually 1 or 2, to designate the dimension of the space of the root α^j in question. Let

$$D = \{j \in \{1, \cdots, d\}; \mathfrak{g}^j/\mathfrak{g}^{j-1} \text{ is 2 dimensional}\}$$

and $r : \{0, \cdots, d\} \to \{0, \cdots, n\}$ the mapping defined by $r(0) = 0$ and inductively on $j = 1, \cdots, d$ by

$$r(j) = \begin{cases} r(j-1) + 1 & \text{if } j \notin D, \\ r(j-1) + 2 & \text{otherwise.} \end{cases}$$

This gives us a good sequence of subalgebras $(\mathfrak{g}_r)_{r=1}^n$ in the following way. Each $r \in \{1, \ldots, n\}$ is written as $r = r(j)$ or $r = r(j) - 1$ for a certain $j \in \{1, \ldots, d\}$. For every $j \in \{1, \ldots, d\} \setminus D$, take a vector $Z^j \in \mathfrak{g}^j \setminus \mathfrak{g}^{j-1}$ and put

$$\mathfrak{w}^j = \mathbb{R}Z^j \text{ and } Z_{r(j)} = Z^j.$$

For $j \in D$, we have a linear form $a_j : \mathfrak{g} \to \mathbb{R}$, a real number $\omega \neq 0$ and two vectors $X^j, Y^j \in \mathfrak{g}^j \setminus \mathfrak{g}^{j-1}$ such that, for $T \in \mathfrak{g}$,

$$\alpha^j(T)(X^j) \equiv a_j(T)(X^j - \omega Y^j), \ \alpha^j(T)(Y^j) \equiv a_j(T)(Y^j + \omega X^j)$$

modulo $\mathfrak{g}^{j-1}$. Put

$$\mathfrak{w}^j = \mathbb{R}X^j \oplus \mathbb{R}Y^j \text{ and } Z_{r(j)-1} = X^j, \ Z_{r(j)} = Y^j \in \mathfrak{g}^j \setminus \mathfrak{g}^{j-1}.$$

Then the vectors

$$\{Z_r, r = r(j), j \in \{1, \cdots, d\} \setminus D \text{ and } r = r(j), r(j) - 1, \text{ if } j \in D\}$$

give us a basis $\mathscr{B}$ of $\mathfrak{g}$ and also a good sequence of subalgebras $\mathscr{S} = (\mathfrak{g}_r)_{r=1}^n$. Denote by $\mathfrak{b}$ the Vergne polarization at ℓ with respect to $\mathscr{S}$. For a subspace $\mathfrak{v}$ of $\mathfrak{g}$, let

$$I_{\mathscr{C}}^{\mathfrak{v}} = \{i \in \{1, \cdots, d\}; \mathfrak{v}^i = \mathfrak{v} \cap \mathfrak{g}^i \neq \mathfrak{v}^{i-1}\}.$$

Likewise, let

$$I_{\mathscr{S}}^{\mathfrak{v}} = \{r \in \{1, \cdots, n\}; \mathfrak{v}_r = \mathfrak{v} \cap \mathfrak{g}_r \neq \mathfrak{v}_{r-1}\}.$$

For a linear form ℓ on $\mathfrak{g}$ and $X \in \mathfrak{g}$, we write $\ell(X)$ or often $\langle \ell, X \rangle$ to designate the value of ℓ at X.

2. Construction of an Algorithm of Coexponential Bases Let $\mathfrak{g}$ be an exponential solvable Lie algebra, $\mathfrak{h}$ a subalgebra of $\mathfrak{g}$ subordinate to $f \in \mathfrak{g}^*$ such that $\mathrm{ind}_H^G \chi_f$ is of discrete type and let $\ell \in \Gamma$ be strongly generic. In this part, we suppose that no root of $\mathfrak{g}$ vanishes on $\mathfrak{g}(\ell|_\mathfrak{n})$. We intend to construct an algorithm which determines the Vergne polarization $\mathfrak{b} = \mathfrak{b}_\mathscr{S}[\ell]$ at ℓ associated with a good sequence of subalgebras $\mathscr{S}$ related to the composition sequence $\mathscr{C}$, a coexponential basis of $\mathfrak{h}$ modulo $\mathfrak{b} \cap \mathfrak{h}$ and a coexponential basis of $\mathfrak{g}$ modulo $\mathfrak{b}$. This construction is mainly built on the sequence $\mathscr{C}$ and the ensuing root system. We now construct at each step of the induction an intermediate subalgebra $\mathfrak{g}^{(i)}$ which converges at the end of the algorithm to the polarization $\mathfrak{b}$, a subalgebra $\mathfrak{h}^{(i)}$ taking the position of $\mathfrak{h}$ at each step, and the sets of indices J, L which indicate the positions of our desired coexponential basis. For more details on these different constructions, we advise the readers to consult [96]. Start from our composition sequence $\mathscr{C} = (\mathfrak{g}^j)_{j=1}^d$ such that $\mathfrak{g}^k = \mathfrak{n}$. Let $\ell \in \mathfrak{g}^*$ and $\mathscr{B} = \{Z_1, \dots, Z_n\}$ a Jordan-Hölder basis of $\mathfrak{g}$ extracted from the sequence $\mathscr{S}$, namely $Z_j \in \mathfrak{g}_j \setminus \mathfrak{g}_{j-1}$. For $j \in I_\mathscr{S}^{\mathfrak{h}\cap\mathfrak{b}}$ (or $j \in I_\mathscr{S}^{\mathfrak{g}(\ell)}$), the vector Z_j is taken in $\mathfrak{h} \cap \mathfrak{b}$ (or $\mathfrak{g}(\ell)$ respectively).

2.1. Initial Step
Set $\mathfrak{g}^{(0)} = \mathfrak{g}$, $\mathfrak{h}^{(0)} = \mathfrak{h}$, $\mathscr{C}^{(0)} = \mathscr{C}$, $\ell^{(0)} = \ell$ and $L = \{1, \dots, n\}$.

2.2. First Step of the Algorithm

1. Let us consider the first index j_1 such that $\mathfrak{a} = \mathfrak{g}^{j_1} \not\subset \mathfrak{g}(\ell)$. If this index does not exist, then $\mathfrak{g} = \mathfrak{g}(\ell)$ and the situation is trivial. If $j_1 > k$, then $\mathfrak{n} \subset \mathfrak{g}(\ell)$. The polarization $\mathfrak{b}$ at ℓ and $\mathfrak{g}(\ell)$ are ideals of $\mathfrak{g}$ and $\mathfrak{g}/\mathfrak{g}(\ell)$ is Abelian. This situation is also easy to treat.

   If $j_1 = 1$, then $\ell(\mathfrak{g}^{j_1}) \neq \{0\}$ and $\mathfrak{a} = \mathfrak{g}^{j_1}$, whose dimension is 1 or 2, is not central. Then $\mathfrak{g}^{(1)} = \mathfrak{a}^\ell$ is the kernel of the root α^1, which is an ideal of codimension 1 in $\mathfrak{g}$ containing $\mathfrak{b} + [\mathfrak{g}, \mathfrak{g}]$. Since ℓ is strongly generic, this case does not occur.
2. Now, let $k \geq j_1 > 1$, then necessarily $\ell(\mathfrak{g}^{j_1}) \neq \{0\}$. Suppose first that $\ell(\mathfrak{g}^{j_1 - 1}) = \{0\}$. If the dimension of α^{j_1} is 1, take $Y_1 \in \mathfrak{g}^{j_1}$ such that $\ell(Y_1) = 1$. Then $Y_1 \in \mathfrak{g}_{r(j_1)}(\ell_{r(j_1)}) \subset \mathfrak{b}$. Furthermore, $[X, Y_1] = \alpha^{j_1}(X)Y_1$ modulo $\mathfrak{g}^{j_1 - 1}$ and $\ell([X, Y_1]) = \alpha^{j_1}(X)$ for all $X \in \mathfrak{g}$. Hence $\mathfrak{b} \subset \ker(\alpha^{j_1})$. So neither this case occurs. The same happens if the dimension of α^{j_1} is 2.
3. In this way, we can suppose that $\ell(\mathfrak{g}^{j_1 - 1}) \neq \{0\}$. We find $C_1 \in \mathfrak{g}^{j_1 - 1}$ with $\ell(C_1) = 1$ and $Y_1 \in \mathfrak{g}^{j_1} \setminus \mathfrak{g}^{j_1 - 1}$ with $\ell(Y_1) = 0$. We also find a non-zero linear form $\beta_1 : \mathfrak{g} \to \mathbb{R}$ such that

$$[A, Y_1] = \alpha^{j_1}(A)Y_1 + \beta_1(A)C_1 \bmod \mathfrak{g}^{j_1 - 1} \cap \ker(\ell),\ A \in \mathfrak{g}.$$

   Then

$$\beta_1(A) = \langle \ell, [A, Y_1] \rangle,\ A \in \mathfrak{g}.$$

If $\ker(\beta_1)$ is an ideal of $\mathfrak{g}$, then $\mathfrak{g}^{(1)} = \mathfrak{a}^\ell$ is an ideal of codimension 1 in $\mathfrak{g}$ containing $\mathfrak{b}[\ell] + [\mathfrak{g}, \mathfrak{g}]$. We set $\mathfrak{h}^{(1)} = \mathfrak{h} + \mathfrak{a}$ if $\mathfrak{h} \subset \mathfrak{a}^\ell$ and $\mathfrak{h}^{(1)} = \mathfrak{h} \cap \mathfrak{a}^\ell + \mathfrak{a}$ otherwise. In this last case, take a vector $U_1 \in \mathfrak{h} \setminus \mathfrak{a}^\ell$. In order to construct a basis of $\mathfrak{g}/\mathfrak{b}[\ell]$, take as first vector $V_1 = U_1$ if $\mathfrak{h} \not\subset \mathfrak{a}^\ell$ and otherwise an arbitrary vector V_1 of $\mathfrak{g} \setminus \mathfrak{a}^\ell$. The starting composition sequence $\mathscr{C}^0$ of $\mathfrak{g}^{(0)}$ is replaced by the new sequence $\mathscr{C}^1$ of $\mathfrak{g}^{(1)}$. Let $\ell^{(1)} = \ell|_{\mathfrak{g}^{(1)}}$ and $\mathfrak{n}^{(1)} = \mathfrak{n} \cap \mathfrak{g}^{(1)}$. We remark that $\alpha^{j_1} = 0$ in this case, because otherwise β_1 does not vanish on $[\mathfrak{g}, \mathfrak{g}]$.

4. Now, if $\ker(\beta_1)$ is not an ideal, we set $\mathfrak{g}^{(1)} = \ker(\beta_1)$, $\mathfrak{n}^{(1)} = \mathfrak{n} \cap \mathfrak{g}^{(1)}$ and take the smallest index r_1 such that $\mathfrak{g}^{r_1} \not\subset \ker(\beta_1)$ and $V_1 \in \mathfrak{g}^{r_1}$ such that $\beta_1(V_1) = 1$. If $\mathfrak{h} \not\subset \mathfrak{g}^{(1)}$, we also choose the smallest index s_1 such that $\mathfrak{h} \cap \mathfrak{g}^{s_1} \not\subset \ker(\beta_1)$ and $U_1 \in \mathfrak{h} \cap \mathfrak{g}^{s_1}$ with $\beta_1(U_1) = 1$. If $\mathfrak{h} \cap \ker(\beta_1) \not\subset \ker(\alpha^{j_1})$, we can find U_1 in $\mathfrak{n}$. Set $\mathfrak{h}^{(1)} = \mathfrak{h} + \mathfrak{g}^{j_1}$ if $\mathfrak{h} \subset \ker(\beta_1)$ and $\mathfrak{h}^{(1)} = \mathfrak{h} \cap \mathfrak{g}^{(1)} + \mathfrak{g}^{j_1}$ otherwise.
5. Suppose now that $\dim(\alpha^{j_1}) = 2$. Then $\mathfrak{g}^{j_1} \subset \mathfrak{b}$ and we have two vectors $Y_1, Y_1' \in (\mathfrak{g}^{j_1} \cap \ker(\ell)) \setminus \mathfrak{g}^{j_1 - 1}$, $C_1 \in \mathfrak{g}^{j_1-1}$ verifying $\ell(C_1) = 1$ and two linear forms β_1, β_1' such that, for $A \in \mathfrak{g}$,

$$[A, Y_1] = \alpha^{j_1}(A)(Y_1 - \omega Y_1') + \beta_1(A) C_1 \mod \mathfrak{g}^{j_1 - 1} \cap \ker(\ell),$$

$$[A, Y_1'] = \alpha^{j_1}(A)(Y_1' + \omega Y_1) + \beta_1'(A) C_1 \mod \mathfrak{g}^{j_1 - 1} \cap \ker(\ell).$$

Moreover, the linear forms β_1, β_1' are linearly independent of α^{j_1}. Indeed, if β_1, β_1' are linearly dependent, there exists a constant $\lambda \in \mathbb{R}$ such that $Y_1 - \lambda Y_1' \in \mathfrak{g}(\ell)$. Taking $A \in \mathfrak{g}(\ell|_{\mathfrak{n}}) \setminus \ker(\alpha^{j_1})$, we see that Y_1, Y_1' belong to $\mathfrak{g}(\ell)$ since ℓ is strongly generic. This is impossible. Set $\mathfrak{g}^{(1)} = \ker(\beta_1) \cap \ker(\beta_1')$, which is of codimension 2, and $\mathfrak{n}^{(1)} = \mathfrak{n} \cap \mathfrak{g}^{(1)}$.
6. Let r_1 be the minimal index such that $\mathfrak{g}^{r_1} \not\subset \mathfrak{g}^{(1)}$. Then it is easy to see $\dim(\mathfrak{g}^{r_1}/\mathfrak{g}^{r_1-1}) = 2$ and that there exist vectors $V_1, V_1' \in \mathfrak{g}^{r_1}$ such that

$$\mathfrak{g}^{r_1} = \mathfrak{g}^{r_1 - 1} + \mathbb{R}V_1 + \mathbb{R}V_1'$$

and

$$\beta_1(V_1) = 1, \beta_1(V_1') = 0, \ \beta_1'(V_1) = 0, \beta_1'(V_1') = 1.$$

Similarly, we set $\mathfrak{h}^{(1)} = \mathfrak{h} + \mathfrak{g}^{j_1}$ if $\mathfrak{h} \subset \mathfrak{g}^{(1)}$ and $\mathfrak{h}^{(1)} = \mathfrak{h} \cap \mathfrak{g}^{(1)} + \mathfrak{g}^{j_1}$ otherwise. If $\mathfrak{h} \not\subset \mathfrak{g}^{(1)}$, let s_1 be the smallest index such that $\mathfrak{h} \cap \mathfrak{g}^{s_1} \not\subset \mathfrak{g}^{(1)}$. If further $\mathfrak{h}^{(1)} \not\subset \ker(\alpha^{j_1})$, we have two vectors $U_1, U_1' \in \mathfrak{g}^{s_1}$ such that

$$\mathfrak{h} \cap \mathfrak{g}^{s_1} = \mathfrak{h} \cap \mathfrak{g}^{s_1 - 1} + \mathbb{R}U_1 + \mathbb{R}U_1'$$

and

$$\beta_1(U_1) = \beta_1'(U_1') = 1, \beta_1(U_1') = \beta_1'(U_1) = 0, \tag{7.3.6}$$

modifying U_1 by a multiple of U'_1 if necessary. Moreover, we can choose U_1, U'_1 in $\mathfrak{n}$. If $\mathfrak{h}$ is not contained in $\mathfrak{g}^{(1)}$ and if α^{j_1} vanishes on $\mathfrak{h}^{(1)}$, the dimension of $\mathfrak{h}/(\mathfrak{h} \cap \mathfrak{g}^{(1)})$ is 1 or 2. We find again the minimal index s_1 and a vector U_1, or the complementary minimal index s'_1 and two vectors $U_1 \in \mathfrak{g}^{s_1}$ and $U'_1 \in \mathfrak{g}^{s'_1}$ with relations (7.3.6). But we can no longer say that U_1 or U_1, U'_1 are in $\mathfrak{n}$. At this step, set

$$L_1 = \{1, \ldots, n\} \setminus \{r(j_1)\}, \text{ resp. } L_1 = \{1, \ldots, n\} \setminus \{r(j_1), r(j_1) - 1\}.$$

2.3. Second Step of the Algorithm
Take as before $\ell^{(1)} = \ell|_{\mathfrak{g}^{(1)}}$ and replace in the previous reasoning $\mathfrak{g}$ by $\mathfrak{g}^{(1)}$, $\mathfrak{n}$ by $\mathfrak{n}_1$. In cases (4) and (6) we make complementary changes. First we repeat the construction and consider as before the smallest index j_2 such that $\mathfrak{g}^{(1)} \cap \mathfrak{g}^{j_2} \not\subset \mathfrak{g}^{(1)}(\ell^{(1)})$. Suppose first that the root α^{j_2} is of dimension 1. We then are in case 4). Take

$$Y_2 \in (\mathfrak{g}^{j_2} \cap \mathfrak{g}^{(1)} \cap \ker(\ell)) \setminus \mathfrak{g}^{j_2-1}, \; C_2 \in \mathfrak{g}^{j_2-1} \cap \mathfrak{g}^{(1)}$$

verifying $\langle \ell, C_2 \rangle = 1$. Then for $A \in \mathfrak{g}$,

$$[A, Y_2] = \alpha^{j_2}(A) Y_2 + \beta_2(A) C_2 \bmod \mathfrak{g}^{j_2-1} \cap \ker(\ell),$$

where β_2 is the linear form given by $\beta_2(A) = \langle \ell, [A, Y_2] \rangle$. Being free to replace C_2 by $C_2 + \zeta Y_1$ (resp. $C_2 + \zeta Y_1 + \zeta' Y'_1$) for some suitably chosen $\zeta \in \mathbb{R}$ (resp. $\zeta, \zeta' \in \mathbb{R}$), we can suppose

$$\langle \ell, [U_1, C_2] \rangle = 0 \text{ (resp. } \langle \ell, [U_1, C_2] \rangle = 0, \; \langle \ell, [U'_1, C_2] \rangle = 0).$$

Likewise, being free to modify Y_2 by a multiple of Y_1 (resp. a linear combination of Y_1, Y'_1), we can assume

$$\langle \ell, [U_1, Y_2] \rangle = 0, \text{ resp. } \langle \ell, [U'_1, Y_2] \rangle = 0.$$

Set $\mathfrak{g}^{(2)} = \mathfrak{g}^{(1)} \cap \ker(\beta_2)$ and $\mathfrak{h}^{(2)} = \mathfrak{h}^{(1)} \cap \ker(\beta_2) + \mathfrak{g}^{(1)} \cap \mathfrak{g}^{j_2}$ if $\mathfrak{h}^{(1)} \not\subset \mathfrak{g}^{(2)}$, otherwise $\mathfrak{h}^{(2)} = \mathfrak{h}^{(1)} + (\mathfrak{g}^{(1)} \cap \mathfrak{g}^{j_2})$. We obtain the maximal index r_2 such that $\mathfrak{g}^{r_2-1} \cap \mathfrak{g}^{(1)} \subset \mathfrak{g}^{(2)}$ and a vector $V_2 \in \mathfrak{g}^{r_2} \cap \mathfrak{g}^{(1)}$ satisfying $\langle \ell, [V_2, Y_2] \rangle = 1$. In the same way, if $\mathfrak{h}^{(1)} \not\subset \mathfrak{g}^{(2)}$, we have an index s_2 and an element $U_2 \in \mathfrak{h} \cap \mathfrak{g}^{s_2} \cap \mathfrak{g}^{(1)}$ such that $\mathfrak{h}^{(1)} \cap \mathfrak{g}^{s_2-1} \subset \mathfrak{g}^{(2)}$ and $\langle \ell, [U_2, Y_2] \rangle = 1$. Finally, put $L_2 = L_1 \setminus \{r(j_2)\}$.

In the case where the dimension of α^{j_2} is 2, we shall have the indices r_2, s_2 and occasionally s'_2. We also have vectors V_2, V'_2, U_2 and occasionally U'_2 etc. If further $\mathfrak{h}^{(2)} \not\subset \ker(\alpha^{j_2})$, we can suppose that U_2 (resp. U_2 and U'_2) belongs to $\mathfrak{n}$. Finally, set $L_2 = L_1 \setminus \{r(j_2), r(j_2) - 1\}$.

2.4. Intermediate Step of the Algorithm
We continue this process until we arrive at the number l for which $\mathfrak{g}^{(l)} = \mathfrak{b}$.

Let $\ell^{(i)} = \ell|_{\mathfrak{g}^{(i)}}$. For $1 \leq i \leq l$, we find the smallest index j_i such that $\mathfrak{g}^{j_i} \not\subset \mathfrak{g}^{(i-1)}(\ell^{(i-1)})$, the root α^{j_i}, the vectors C_i, Y_i (resp. C_i, Y_i, Y_i') in $\mathfrak{g}^{j_i} \cap \mathfrak{g}^{(i-1)}$ such that $\ell(C_i) = 1$ and $\ell(Y_i) = 0$ (resp. $\ell(Y_i) = \ell(Y_i') = 0$), and the linear form β_i (resp. β_i, β_i') such that, for $A \in \mathfrak{g}$,

$$[A, Y_i] \equiv \alpha^{j_i}(A)Y_i + \beta_i(A)C_i \bmod \mathfrak{g}^{j_i-1} \cap \ker(\ell),$$

respectively

$$[A, Y_i] \equiv \alpha^{j_i}(A)(Y_i - \omega Y_i') + \beta_i(A)C_i$$

$$[A, Y_i'] \equiv \alpha^{j_i}(A)(Y_i' + \omega Y_i) + \beta_i'(A)C_i$$

modulo $\mathfrak{g}^{j_i-1} \cap \ker(\ell)$. Besides, we get the subalgebra $\mathfrak{g}^{(i)} = \mathfrak{g}^{(i-1)} \cap \ker(\beta_i)$ (resp. $\mathfrak{g}^{(i)} = \mathfrak{g}^{(i-1)} \cap \ker(\beta_i) \cap \ker(\beta_i')$, the smallest index r_i such that $\mathfrak{g}^{r_i} \cap \mathfrak{g}^{(i-1)} \not\subset \mathfrak{g}^{(i)}$ and the vector V_i (resp. V_i, V_i') in $\mathfrak{g}^{r_i} \cap \mathfrak{g}^{(i-1)}$ such that

$$\beta_i(V_i) = 1 \text{ (resp. } \beta_i(V_i) = \beta_i'(V_i') = 1, \beta_i'(V_i) = \beta_i(V_i') = 0).$$

Set $\mathfrak{h}^{(i)} = \mathfrak{h}^{(i-1)} \cap \mathfrak{g}^{(i)} + \mathfrak{g}^{j_i} \cap \mathfrak{g}^{(i-1)}$ if $\mathfrak{h}^{(i-1)} \not\subset \mathfrak{g}^{(i)}$. Otherwise, $\mathfrak{h}^{(i)} = \mathfrak{h}^{(i-1)} + \mathfrak{g}^{j_i} \cap \mathfrak{g}^{(i-1)}$. Then just as before, if $\mathfrak{h}^{(i-1)} \not\subset \mathfrak{g}^{(i)}$, we find the minimal index s_i (resp. s_i, s_i') and U_i (resp. U_i, U_i') in $\mathfrak{h}^{(i-1)}$ such that $\mathfrak{g}^{s_i} \cap \mathfrak{h}^{(i-1)} \not\subset \mathfrak{h}^{(i)}$ and

$$\beta_i(U_i) = 1 \text{ (resp. } \beta_i(U_i) = \beta_i'(U_i') = 1, \beta_i'(U_i) = \beta_i(U_i') = 0).$$

We have also the relations:

$$\langle \ell, [U_i, Y_j] \rangle = \delta_{i,j} \ (1 \leq i, j \leq l), \ \mathfrak{h}^{(i-1)} \neq \mathfrak{h}^{(i)}$$
$$\langle \ell, [U_i, C_j] \rangle = 0, \forall j \geq i. \tag{7.3.7}$$

Then

$$\langle \ell, [\mathfrak{h}, C_j] \rangle = \{0\} \tag{7.3.8}$$

for all j. Set finally $L_i = L_{i-1} \setminus \{r(j_i)\}$ or $L_i = L_{i-1} \setminus \{r(j_i), r(j_i) - 1\}$ according to the situation. We can impose that U_i (resp. U_i, U_i') is contained in $\mathfrak{n}$ if $\alpha^{j_i}(\mathfrak{h}^{(i)}) \neq \{0\}$.

2.5. Output of the Algorithm
Here we summarise the outcome of the previous algorithm, presented in the form of lemmas. Suppose that $\mathfrak{g} \neq \mathfrak{g}(\ell)$. We begin with:

Lemma 7.3.13 *There is a sequence of subalgebras*

$$\mathfrak{g} = \mathfrak{g}^{(0)} \supsetneq \mathfrak{g}^{(1)} \supsetneq \cdots \supsetneq \mathfrak{g}^{(s)} = \mathfrak{b}$$

and a sequence of integers $2 \leq j_1 \leq \cdots \leq j_s \leq d$ *such that, putting* $\ell^{(i)} = \ell|_{\mathfrak{g}^{(i)}} (1 \leq i \leq s-1)$, *we have* $\mathfrak{g}^{j_{i+1}-1} \cap \mathfrak{g}^{(i)} \subset \mathfrak{g}^{(i)}(\ell^{(i)})$, $\mathfrak{g}^{j_{i+1}} \cap \mathfrak{g}^{(i)} \not\subset \mathfrak{g}^{(i)}(\ell^{(i)})$ *and* $\mathfrak{g}^{(i+1)} = \mathfrak{g}^{(i)}(\ell^{(i)}|_{\mathfrak{g}^{j_{i+1}}\cap\mathfrak{g}^{(i)}})$.

We define the subalgebra $\mathfrak{h}^{(i)}$ of $\mathfrak{g}^{(i)}$ inductively on $0 \leq i \leq s$ by setting $\mathfrak{h}^{(0)} = \mathfrak{h}$ and $\mathfrak{h}^{(i)} = \mathfrak{h}^{(i-1)} \cap \mathfrak{g}^{(i)} + \mathfrak{g}^{j_i} \cap \mathfrak{g}^{(i-1)}$ for $i > 0$. For $0 \leq i \leq s$, we denote by $\mathscr{S}^{(i)}$ the trace of $\mathscr{S}$ on $\mathfrak{g}^{(i)}$ and put $\mathfrak{n}^{(i)} = \mathfrak{n} \cap \mathfrak{g}^{(i)}$, $G^{(i)} = \exp(\mathfrak{g}^{(i)})$, $H^{(i)} = \exp(\mathfrak{h}^{(i)})$, $f^{(i)} = f|_{\mathfrak{g}^{(i)}}$ and $\Gamma^{(i)} = f^{(i)} + (\mathfrak{h}^{(i-1)})^{\perp}$.

Lemma 7.3.14

(i) $f|_{[\mathfrak{h}^{(i)},\mathfrak{h}^{(i)}]} = \{0\}$, *so we can consider the character* $\chi^{(i)}$ *of* $H^{(i)}$ *with differential* $if|_{\mathfrak{h}^{(i)}}$ *and the monomial representation* $\tau^{(i)} = ind_{H^{(i)}}^{G^{(i)}} \chi^{(i)}$.
(ii) $\tau^{(i)}$ *is of discrete type.*
(iii) $\ell^{(i)}$ *is a strongly generic element of* $\Gamma^{(i)}$ *with respect to* $\mathfrak{h}^{(i)}$, $\mathscr{S}^{(i)}$ *and* $\mathfrak{n}^{(i)}$.
(iv) $\ell|_{\mathfrak{g}^{j_{i+1}-1}} \neq 0$.
(v) $\mathfrak{g}^{j_i} \subset \mathfrak{b} \subset \mathfrak{g}^{(i)}$.
(vi) B_ℓ *induces a* $\mathfrak{g}^{(i+1)}$*–invariant pairing between* $\mathfrak{g}^{(i)}/\mathfrak{g}^{(i+1)}$ *and* $\mathfrak{g}^{j_{i+1}}/\mathfrak{g}^{j_{i+1}-1}$. *In particular, the vector space* $\mathfrak{g}^{(i)}/\mathfrak{g}^{(i+1)}$ *is of dimension* 1 *or* 2.
(vii) $\mathfrak{h}^{(i)} = (\mathfrak{h} \cap \mathfrak{g}^{(i+1)}) + \mathfrak{g}^{j_1} + (\mathfrak{g}^{j_2} \cap \mathfrak{g}^{(1)}) + \cdots + (\mathfrak{g}^{j_i} \cap \mathfrak{g}^{(i-1)})$. *In particular, we have* $\mathfrak{h}^{(i)}/(\mathfrak{h}^{(i)} \cap \mathfrak{g}^{(i+1)}) = (\mathfrak{h} \cap \mathfrak{g}^{(i)})/(\mathfrak{h} \cap \mathfrak{g}^{(i+1)})$.
(viii) No root of $\mathfrak{g}$ *vanishes on* $\mathfrak{g}^{(i)}(\ell^{(i)}|_{\mathfrak{n}^{(i)}})$.

We set $I_k = \{1 \leq i \leq s; \dim(\mathfrak{g}^{j_i}/\mathfrak{g}^{j_i-1}) = k\}$, $J_k = \{1 \leq i \leq s; \dim(\mathfrak{h} \cap \mathfrak{g}^{(i-1)})/(\mathfrak{h} \cap \mathfrak{g}^{(i)}) = k\}$ for $k = 1, 2$, $I = I_1 \sqcup I_2 = \{1, \ldots, s\}$ and $J = J_1 \sqcup J_2 \subset I$. For $i \in I$ (resp. $i \in J$), we denote by r_i (resp. s_i) the smallest index such that $\mathfrak{g}^{r_i} \cap \mathfrak{g}^{(i-1)} \not\subset \mathfrak{g}^{(i)}$ (resp. $\mathfrak{g}^{s_i} \cap \mathfrak{h}^{(i-1)} \not\subset \mathfrak{g}^{(i)}$) and if $i \in J_2$ verifies $\dim(\mathfrak{g}^{s_i} \cap \mathfrak{h}^{(i-1)})/(\mathfrak{g}^{(i)} \cap \mathfrak{g}^{s_i} \cap \mathfrak{h}^{(i-1)}) = 1$, we denote by s_i' the smallest integer such that $\mathfrak{g}^{s_i} \cap \mathfrak{h}^{(i-1)} \subsetneq \mathfrak{g}^{s_i'} \cap \mathfrak{h}^{(i-1)}$.

Lemma 7.3.15 *There is* $C \in \mathfrak{g}^{j_1-1}$ *verifying* $\langle \ell, C \rangle = 1$, *and for* $i \in I_1$ *(resp.* $i \in I_2$*) there is* $Y_i \in (\mathfrak{g}^{j_i} \cap \mathfrak{g}^{(i-1)}) \setminus \mathfrak{g}^{(i)}$ *(resp.* $Y_i, Y_i' \in (\mathfrak{g}^{j_i} \cap \mathfrak{g}^{(i-1)}) \setminus \mathfrak{g}^{(i)}$ *and* $w_i \in \mathbb{R} \setminus \{0\}$*),* $\beta_i \in \mathfrak{g}^*$ *(resp.* $\beta_i, \beta_i' \in \mathfrak{g}^*$*),* $V_i \in (\mathfrak{g}^{r_i} \cap \mathfrak{g}^{(i-1)}) \setminus \mathfrak{g}^{(i)}$ *(resp.* $V_i, V_i' \in (\mathfrak{g}^{r_i} \cap \mathfrak{g}^{(i-1)}) \setminus \mathfrak{g}^{(i)}$*) such that, on the one hand, if* $i \in I_1$,

$$[X, Y_i] \equiv \alpha^{j_i}(X) Y_i + \beta_i(X) C, \ X \in \mathfrak{g},$$

modulo $\mathfrak{g}^{j_i-1} \cap \ker(\ell)$, $\ell(Y_i) = 0$, $\beta_i(V_i) = 1$, *and if* $i \in I_2$,

$$[X, Y_i] = \alpha^{j_i}(X)(Y_i + w_i Y_i') + \beta_i(X) C,$$

$$[X, Y_i'] = \alpha^{j_i}(X)(Y_i' - w_i Y_i) + \beta_i'(X) C, \ X \in \mathfrak{g},$$

modulo $\mathfrak{g}^{j_i-1} \cap \ker(\ell)$, $\ell(Y_i) = \ell(Y_i') = 0$, $\beta_i(V_i) = \beta_i'(V_i') = 1$, $\beta_i(V_i') = \beta_i'(V_i) = 0$. *On the other hand, if* $i \in J_1$, *there is* $U_i \in (\mathfrak{h} \cap \mathfrak{g}^{(i-1)} \cap \mathfrak{g}^{s_i}) \setminus \mathfrak{g}^{(i)}$ *verifying* $\beta_i(U_i) = 1$. *If* $i \in J_2$ *and* $\dim((\mathfrak{g}^{s_i} \cap \mathfrak{h}^{(i-1)})/(\mathfrak{g}^{(i)} \cap \mathfrak{g}^{s_i} \cap \mathfrak{h}^{(i-1)})) = 2$, *there are* $U_i, U_i' \in (\mathfrak{h} \cap \mathfrak{g}^{(i-1)} \cap \mathfrak{g}^{s_i}) \setminus \mathfrak{g}^{(i)}$ *such that* $\beta_i(U_i) = \beta_i'(U_i') = 1$ *and* $\beta_i(U_i') = \beta_i'(U_i) = 0$. *If* $i \in J_2$ *and* $\dim((\mathfrak{g}^{s_i} \cap \mathfrak{h}^{(i-1)})/(\mathfrak{g}^{(i)} \cap \mathfrak{g}^{s_i} \cap \mathfrak{h}^{(i-1)})) = 1$, *there are* $U_i \in (\mathfrak{h} \cap \mathfrak{g}^{(i-1)} \cap \mathfrak{g}^{s_i}) \setminus \mathfrak{g}^{(i)}$ *and* $U_i' \in (\mathfrak{h} \cap \mathfrak{g}^{(i-1)} \cap \mathfrak{g}^{s_i'}) \setminus (\mathfrak{h} \cap \mathfrak{g}^{s_i})$ *satisfying* $\beta_i(U_i) = \beta_i'(U_i') = 1$ *and* $\beta_i(U_i') = \beta_i'(U_i) = 0$. *Moreover, if* $i \in J_1$, $\langle \ell, [U_i, C] \rangle = 0$, *and if* $i \in J_2$, $\langle \ell, [U_i, C] \rangle = \langle \ell, [U_i', C] \rangle = 0$. *We have*

$$\langle \ell, [U_j, Y_i] \rangle = \langle \ell, [U_j', Y_i] \rangle = \langle \ell, [U_j, Y_i'] \rangle = \langle \ell, [U_j', Y_i'] \rangle = 0,$$

when the expressions are defined.

Let $i \in J$ *satisfy* $\alpha^{j_1}(\mathfrak{h}^{(i)}) \neq \{0\}$. *Then we have*

$$\dim((\mathfrak{g}^{s_i} \cap \mathfrak{h}^{(i-1)})/(\mathfrak{g}^{(i)} \cap \mathfrak{g}^{s_i} \cap \mathfrak{h}^{(i-1)})) = \dim((\mathfrak{h} \cap \mathfrak{g}^{(i-1)})/(\mathfrak{h} \cap \mathfrak{g}^{(i)}))$$

and we can choose $U_i \in \mathfrak{n}$ *(resp.* $U_i, U_i' \in \mathfrak{n}$*) if* $i \in J_1$ *(resp.* $i \in J_2$*). For* $i \in I_1$ *(resp.* $i \in I_2$*) satisfying* $\alpha^{j_i} \neq 0$, *the linear forms* α^{j_i} *and* β_i *(resp.* β_i, β_i'*) are linearly independent.*

Definition 7.3.16 Let $J = \{1 \leq i \leq s;\, \mathfrak{h}^{(i-1)} \neq \mathfrak{h}^{(i)}\}$. Remark that, for $i \in J$, the dimension of $\mathfrak{h}^{(i)}/(\mathfrak{h}^{(i-1)} \cap \mathfrak{g}^{(i)})$ is 1 or 2. Let $B = \exp \mathfrak{b}$ be the Vergne polarization at ℓ. For $s \in B$, we set

$$\delta_\ell(s) = \delta_{\ell,\mathscr{S}}(s) = \prod_{i \in J} e^{-\epsilon_i \mathrm{Tr}(\alpha^{j_i}(\log s))},$$

where $\epsilon_i = 1$ if $\dim(\alpha^{j_i}) = 1$ or $\dim(\alpha^{j_i}) = 2$ and $\dim(\mathfrak{h}^{(i)}/(\mathfrak{h}^{(i-1)} \cap \mathfrak{g}^{(i)}) = 2$. Otherwise, $\epsilon_i = 1/2$.

Corollary 7.3.17 *The process described by the above algorithm gives us the following families of coexponential bases:* $\{V_i, i \in I_1\} \sqcup \{V_i, V_i', i \in I_2\}$ *is a coexponential basis of* $\mathfrak{g}/\mathfrak{b}$,

$\{U_i, i \in J_1\} \sqcup \{U_i, U_i', i \in J_2\}$ *is a coexponential basis of* $\mathfrak{h}/(\mathfrak{b} \cap \mathfrak{h})$,

$\{Y_i, i \in I_1\} \sqcup \{Y_i, Y_i', i \in I_2\}$ *is a coexponential basis of* $\mathfrak{b}/\mathfrak{g}(\ell)$,

$\{Y_i, i \in I_1\} \sqcup \{Y_i, Y_i', i \in I_2\} \sqcup \{C\}$ *is a coexponential basis of* $\mathfrak{b}/(\mathfrak{g}(\ell) \cap \ker(\ell))$.

Corollary 7.3.18 *We can describe the quotient space* $H/(H \cap B)$ *and the integral*

$$\oint_{H/(H \cap B)} \varphi(h) d\nu_{H,H\cap B}(h)$$

for a continuous function φ *on* H *with compact support modulo* $H \cap B$ *in the following way:*

$$H/H \cap B = \prod_{j \in J} \exp(\mathbb{R}U_j) \;\; (or \prod_{j \in J} \exp(\mathbb{R}U_j) \exp(\mathbb{R}U_j'))$$

and

$$\oint_{H/(H\cap B)} \varphi(h) d\nu_{H,H\cap B}(h) = \int_{\mathbb{R}^{\#J}} \varphi(\prod_{i\in J} \exp(u_i U_i)) \prod_i du_i,$$

$$\left(or \oint_{H/H\cap B} \varphi(h) d\nu_{H,H\cap B}(h) = \int_{\mathbb{R}^{\#J}} \varphi(\prod_{i\in J} \exp(u_i U_i) \exp(u_i' U_i')) \prod_i du_i du_i'\right)$$

with

$$\#J = \sum_{i\in J} \dim(\mathfrak{h}^{(i)}/(\mathfrak{h}^{(i-1)} \cap \mathfrak{g}^{(i)})).$$

2. A Trace Formula Starting from the Root System We now prove a relative trace formula, which will turn out to be very useful in the sequel.

Proposition 7.3.19 *Let $\mathscr{C}$ be a composition sequence of $\mathfrak{g}$ and $\mathfrak{h}$ a subalgebra of $\mathfrak{g}$. Take $\ell \in \Gamma$ generic and such that the subspace $\widehat{\mathfrak{h}} = \mathfrak{h} + \mathfrak{g}(\ell)$ is Lagrangian for the alternating bilinear form B_ℓ. We denote by $\mathfrak{b} = \mathfrak{b}[\ell]$ the Vergne polarization at ℓ determined by $\mathscr{C}$. Let $U = T + V$ be in $\mathfrak{h}$ with $T \in \mathfrak{b}$ and $V \in \mathfrak{n}$. Assume further that*

$$\beta_i(U) = 0, \ resp.\ \beta_i(U) = \beta_i'(U) = 0, \ i \in J, \ i \geq 2$$

using previous notation. Then

$$Tr\ ad_{\mathfrak{g}/\mathfrak{h}}(U) - Tr\ ad_{\mathfrak{g}/\mathfrak{b}}(T) = 2\sum_{i\in J} Tr\ \alpha^{j_i}(T).$$

Proof We can assume that no root of $\mathfrak{g}$ vanishes on $\mathfrak{g}(\ell|_{\mathfrak{n}})$. We use the composition sequence $\mathscr{C}$ and the decomposition $\mathfrak{g}^j = \mathfrak{w}^j + \mathfrak{g}^{j-1}$, $1 \leq j \leq d$. We choose $\mathfrak{w}^j$ in $\mathfrak{g}(\ell)$ if $j \in I_{\mathscr{C}}^{\mathfrak{g}(\ell)}$, and in $\mathfrak{b}$ if $j \in I_{\mathscr{C}}^{\mathfrak{b}}$. If $\mathfrak{h} \cap \mathfrak{b} \cap \mathfrak{g}^j \not\subset \mathfrak{g}^{j-1}$, we determine a subspace $\mathfrak{v}^j$ in $\mathfrak{h} \cap \mathfrak{b} \cap \mathfrak{g}^j$ such that

$$\mathfrak{h} \cap \mathfrak{b} \cap \mathfrak{g}^j = \mathfrak{v}^j + (\mathfrak{h} \cap \mathfrak{b} \cap \mathfrak{g}^{j-1}).$$

For $i \in J$, we take a subspace $\mathfrak{u}_i \subset \mathfrak{h} \cap \mathfrak{g}^{s_i}$ such that $\mathfrak{h} \cap \mathfrak{g}^{s_i} = \mathfrak{u}_i + (\mathfrak{h} \cap \mathfrak{g}^{s_i - 1})$ (possibly with s_i' and $\mathfrak{u}_i'$ such that $\mathfrak{h} \cap \mathfrak{g}^{s_i'} = \mathfrak{u}_i' + (\mathfrak{h} \cap \mathfrak{g}^{s_i'-1})$). Then for $U_i \in \mathfrak{u}_i$, $i \in J$ and for $Z_i \in \mathfrak{v}^i$, $i \in I_{\mathscr{C}}^{\mathfrak{h}\cap\mathfrak{b}}$, we have

$$\mathrm{ad}(U)U_i = \sum_{i'\in J} \mu_{i',i}(U_i)(\in \mathfrak{u}_{i'}) + \sum_{i'\in I_{\mathscr{C}}^{\mathfrak{b}\cap\mathfrak{h}}} \nu_{i',i}(U_i)(\in \mathfrak{v}^{i'}),$$

$$\mathrm{ad}(U)Z_i = \sum_{i'\in J} \sigma_{i',i}(Z_i)(\in \mathfrak{u}_{i'}) + \sum_{i'\in I_{\mathscr{C}}^{\mathfrak{b}\cap\mathfrak{h}}} \tau_{i',i}(Z_i)(\in \mathfrak{v}^{i'}).$$

Let us determinate the vectors $\mu_{i,i}(U_i)$, $i \in J$. Relations (7.3.7), where we chose C_i so that $[U_i, Y_i] = \alpha^{j_i}(U_i)Y_i + \beta_i(U_i)C_i$, and the properties of U tell us that, for $1 < i \in J$, we have

$$\langle \ell, [[U, U_i], Y_i]\rangle = \langle \ell, [U, [U_i, Y_i]]\rangle - \langle \ell, [U_i, [U, Y_i]]\rangle$$
$$= \langle \ell, [U, \alpha^{j_i}(U_i)Y_i + \beta_i(U_i)C_i]\rangle - \langle \ell, [U_i, \alpha^{j_i}(U)Y_i + \beta_i(U)C_i] \bmod \ker(\ell|_{\mathfrak{g}^{j_i-1}})\rangle. \tag{7.3.9}$$

Now using (7.3.7) and (7.3.8),

$$\langle \ell, [[U, U_i], Y_i]\rangle = \alpha^{j_i}(U_i)\beta_i(U) - \alpha^{j_i}(U)\langle \ell, [U_i, Y_i] \bmod \ker(\ell|_{\mathfrak{g}^{j_i-1}})\rangle = -\alpha^{j_i}(U).$$

If $\dim \alpha^{j_i} = 2$, we have in the same way

$$\begin{aligned}\langle \ell, [[U, U_i], Y_i]\rangle &= \langle \ell, [U, [U_i, Y_i]]\rangle - \langle \ell, [U_i, [U, Y_i]]\rangle \\ &= \langle \ell, [U, \alpha^{j_i}(U_i)(Y_i - \omega Y_i') + \beta_i(U_i)C_i]\rangle \\ &\quad -\langle \ell, [U_i, \alpha^{j_i}(U)(Y_i - \omega Y_i') + \beta_i(U)C_i] \bmod \ker(\ell|_{\mathfrak{g}^{j_i-1}})\rangle \\ &= -\alpha^{j_i}(U),\end{aligned}$$

and eventually $\langle \ell, [[U, U_i'], Y_i']\rangle = -\alpha^{j_i}(U)$. In this way,

$$\mu_{i,i}(U_i) = -\alpha^{j_i}(U)U_i. \tag{7.3.10}$$

The indices $s_i, s_i' (i \in J)$ being minimal, we can assume that the vectors in $\mathfrak{u}^i$ have no component in the subspaces $\mathfrak{v}^j$, $j \in I_{\mathscr{C}}^{\mathfrak{h}\cap\mathfrak{b}}$. Hence we see that

$$\tau_{j,j}(Z_j) = \alpha^j(T)(Z_j),\ j \in I_{\mathscr{C}}^{\mathfrak{h}\cap\mathfrak{b}}. \tag{7.3.11}$$

Formulas (7.3.10) and (7.3.11) permit us to write the trace formula in the following way.

$$\begin{aligned}&\operatorname{Tr} \operatorname{ad}_{\mathfrak{g}/\mathfrak{h}}(U) - \operatorname{Tr} \operatorname{ad}_{\mathfrak{g}/\mathfrak{b}}(T) = \operatorname{Tr} \operatorname{ad}_{\mathfrak{b}}(T) - \operatorname{Tr} \operatorname{ad}_{\mathfrak{h}}(U) \\ &= \sum_{i\in\{1,\cdots,l\}} \operatorname{Tr} \alpha^{j_i}(T) + \sum_{j\in I_{\mathscr{C}}^{\mathfrak{g}(\ell)}} \operatorname{Tr} \alpha^j(T) + \sum_{i\in J} \operatorname{Tr} \alpha^{j_i}(T) - \sum_{i\in I_{\mathscr{C}}^{\mathfrak{h}\cap\mathfrak{b}}} \operatorname{Tr} \alpha^i(T).\end{aligned}$$

This gives us:

$$\begin{aligned}\operatorname{Tr}\operatorname{ad}_{\mathfrak{b}}(T) - \operatorname{Tr}\operatorname{ad}_{\mathfrak{h}}(U) &= 2\sum_{i\in J}\operatorname{Tr}\alpha^{j_i}(T) + \sum_{i\notin J, i\in\{1,\ldots,l\}}\operatorname{Tr}\alpha^{j_i}(T)\\ &\quad + \sum_{j\in I_{\mathscr{C}}^{\mathfrak{g}(\ell)}}\operatorname{Tr}\alpha^{j}(T) - \sum_{j\in I_{\mathscr{C}}^{\mathfrak{h}\cap\mathfrak{b}}}\operatorname{Tr}\alpha^{j}(T). \qquad (7.3.12)\end{aligned}$$

As $\langle \ell, [U_i, Y_i']\rangle = \delta_{i,i'}$ for all $i \in J$, we have $I_{\mathscr{C}}^{\widehat{\mathfrak{h}}\cap\mathfrak{b}} \cap J = \emptyset$. On the other hand, as $\widehat{\mathfrak{h}}$ is Lagrangian, we have the relations $\dim\widehat{\mathfrak{h}} = \dim\mathfrak{b}$ and

$$\sum_{i\in J}\dim(\mathfrak{h}^{(i)}/(\mathfrak{h}^{(i-1)}\cap\mathfrak{g}^{(i)})) = \dim(\mathfrak{h}/(\mathfrak{h}\cap\mathfrak{b})) = \dim(\widehat{\mathfrak{h}}/(\widehat{\mathfrak{h}}\cap\mathfrak{b})) = \dim(\mathfrak{b}/(\widehat{\mathfrak{h}}\cap\mathfrak{b})).$$

These equalities say that

$$\sum_{j\in I_{\mathscr{C}}^{(\widehat{\mathfrak{h}}\cap\mathfrak{b})/\mathfrak{g}(\ell)}}\operatorname{Tr}\alpha^{j}(T) = \sum_{i\notin J, i\in\{1,\ldots,l\}}\operatorname{Tr}\alpha^{j_i}(T).$$

Moreover, by Lemma 7.3.2,

$$\begin{aligned}\sum_{j\in I_{\mathscr{C}}^{\mathfrak{h}\cap\mathfrak{b}}}\operatorname{Tr}\alpha^{j}(T) &= \sum_{j\in I_{\mathscr{C}}^{\widehat{\mathfrak{h}}\cap\mathfrak{b}}}\operatorname{Tr}\alpha^{j}(T) - \sum_{j\in I_{\mathscr{C}}^{\mathfrak{g}(\ell)/(\mathfrak{h}\cap\mathfrak{g}(\ell))}}\operatorname{Tr}\alpha^{j}(T)\\ &= \sum_{j\in I_{\mathscr{C}}^{\widehat{\mathfrak{h}}\cap\mathfrak{b}}}\operatorname{Tr}\alpha^{j}(T)\\ &= \sum_{j\in I_{\mathscr{C}}^{\widehat{\mathfrak{h}}\cap\mathfrak{b}/\mathfrak{g}(\ell)}}\operatorname{Tr}\alpha^{j}(T) + \sum_{j\in I_{\mathscr{C}}^{\mathfrak{g}(\ell)}}\operatorname{Tr}\alpha^{j}(T).\end{aligned}$$

Finally,

$$\sum_{i\notin J, i\in\{1,\ldots,l\}}\operatorname{Tr}\alpha^{j_i}(T) + \sum_{j\in I_{\mathscr{C}}^{\mathfrak{g}(\ell)}}\operatorname{Tr}\alpha^{j}(T) = \sum_{j\in I_{\mathscr{C}}^{\mathfrak{h}\cap\mathfrak{b}}}\operatorname{Tr}\alpha^{j}(T)$$

and so we obtain the result from (7.3.12). □

7.4 A Convergence Proof

We keep our notations and prove in this section Theorem 7.3.11. The proof will be carried out in different steps each involving induction on $p = \dim G + \dim G/H$. If G is Abelian or more generally nilpotent, the theorem is evident. If the ideal $\mathfrak{g}_0$ of Definition 7.3.4 is different from $\mathfrak{g}$, then $G_0 = \exp(\mathfrak{g}_0)$ is a normal subgroup of G containing the polarization $B = \exp(\mathfrak{b}[\ell])$ and the normal subgroup $N = \exp\mathfrak{n}$. Besides, the modular functions Δ_G and Δ_{G_0} coincide on G_0. Hence if $H \subset G_0$, we have the identity

$$\oint_{H/(H\cap B)} |\varphi(h)|\Delta_{H,G}^{-1/2}(h)d\nu(h) = \oint_{H/(H\cap B)} |\varphi(h)|\Delta_{H,G_0}^{-1/2}(h)d\nu(h).$$

Applying the induction hypothesis to (G_0, H), we obtain the result through Proposition 7.3.6. If $\mathfrak{h} \not\subset \mathfrak{g}_0$, we take a subspace $\mathfrak{u}$ of $\mathfrak{h}$ such that $\mathfrak{h} = \mathfrak{u} \oplus (\mathfrak{h} \cap \mathfrak{g}_0)$ and write as before $\mathfrak{h}_0 = \mathfrak{h} \cap \mathfrak{g}_0$, $H_0 = \exp(\mathfrak{h}_0)$. Then

$$\begin{aligned}
&\oint_{H/(H\cap B)} |\varphi(h)|\Delta_{H,G}^{-1/2}(h)d\nu(h)\\
&= \int_{\mathfrak{u}} \oint_{H_0/(H_0\cap B)} |\varphi((\exp U)h)|\Delta_{H,G}^{-1/2}(\exp U)\Delta_{H_0,G_0}^{-1/2}(h)d\nu(h)dU\\
&\leq \int_{\mathfrak{u}} \Delta_{H,G}^{-1/2}(\exp U) \oint_{H_0/(H_0\cap B)} |\varphi((\exp U)h)|\Delta_{H_0,G_0}^{-1/2}(h)d\nu(h)dU.
\end{aligned}$$

This again permits us to apply the induction hypothesis, because the linear form $\ell_0 = \ell|_{\mathfrak{g}_0}$ is strongly generic and the function

$$U \mapsto \Delta_{H,G}^{-1/2}(\exp U) \oint_{H_0/(H_0\cap B)} |\varphi((\exp U)h)|\Delta_{H_0,G_0}^{-1/2}(h)d\nu(h)$$

is compactly supported and continuous by Lebesgue's dominated convergence theorem. Indeed, we can make use of an another continuous function with compact support modulo B having the necessary covariance relation, whose variables are separated in $\exp U$ and $h \in H_0$, and the modulus dominates $|\varphi|$.

Hence we can suppose that the roots of $\mathfrak{g}$ do not vanish on $\mathfrak{g}(\ell|_{\mathfrak{n}})$. Recall the composition sequence $\mathscr{C} = (\mathfrak{g}^j)_{1\leq j\leq d}$ of $\mathfrak{g}$ which passes through $\mathfrak{n} = \log N$ containing $[\mathfrak{g}, \mathfrak{g}]$. We use the previous constructions and notations. Let us prove the following stronger assertion: for any generic $\ell \in \Gamma$ and any $\varphi \in K(G, B, \ell)$, we have

$$\sup_{\substack{s\in B[\ell]\\ m\in N}} \delta_{\ell,\mathscr{S}}(s) \oint_{H/(H\cap B)} |\varphi(mshs^{-1})|\Delta_{H,G}^{-1/2}(h)d\nu(h) < \infty. \tag{7.4.1}$$

Everything is clear if $p = 1$ or if $N \subset G(\ell)$, namely $j_1 > k$. In this case, the function $\delta_{\ell,\mathscr{S}}$ is the constant 1 and $[G, G] \subset N \subset G(\ell) \subset B$. Hence $H \cap B$ is normal in H, HB is a closed subgroup of G and

$$\begin{aligned}&\sup_{\substack{s\in B[\ell]\\ m\in N}} \delta_{\ell,\mathscr{S}}(s) \oint_{H/(H\cap B)} |\varphi(mshs^{-1})|\Delta_{H,G}^{-1/2}(h)d\nu(h)\\ &= \int_{H/(H\cap B)} |\varphi(h)|\Delta_{H,G}^{-1/2}(h)d(h) < \infty.\end{aligned} \tag{7.4.2}$$

So, we can suppose $j_1 \leq k$.

a. We Can Suppose $\mathfrak{g}^{j_1-1} \subset \mathfrak{h}$ Indeed, if this is not the case, we take $\mathfrak{h}_1 = \mathfrak{h} + \mathfrak{g}^{j_1-1}$ which is also subordinate to ℓ and put $H_1 = \exp(\mathfrak{h}_1)$. The trace formula (7.3.3) holds also for $(\mathfrak{b}, \mathfrak{h}_1)$, because $[\mathfrak{g}(\ell|_\mathfrak{n}), \mathfrak{g}^{j_1-1}] \subset \mathfrak{h}$ as ℓ is generic and $\mathfrak{g}^{j_1-1} \subset \mathfrak{n}$. On the other hand, $\Delta_{H,G}(h) = \Delta_{H_1,G}(h)\chi(h)(h \in H)$, where χ is a real character of H verifying

$$\chi(\exp U) = e^{\mathrm{Tr}\,\mathrm{ad}_{\mathfrak{a}/(\mathfrak{a}\cap\mathfrak{h})}(U)}, \ U \in \mathfrak{h},$$

with $\mathfrak{a} = \mathfrak{g}^{j_1-1}$ and $\chi|_B = 1$ (as ℓ is generic). Then the function $\chi^{-1/2}\varphi$ belongs to the representation space of π_ℓ. Besides,

$$\begin{aligned}&\oint_{H/(H\cap B)} |\varphi(mshs^{-1})|\Delta_{H,G}^{-1/2}(h)d\nu(h)\\ &= \oint_{H_1/(H_1\cap B)} |\chi^{-1/2}(mshs^{-1})\varphi(mshs^{-1})|\Delta_{H_1,G}^{-1/2}(h)d\nu(h),\end{aligned}$$

which essentially gives us the desired property.

In cases (3) and (5), the subalgebra $\mathfrak{g}^{(1)}$ is an ideal of $\mathfrak{g}$ containing $[\mathfrak{g}, \mathfrak{g}] + \mathfrak{b}[\ell]$. Therefore, $G^{(1)} = \exp(\mathfrak{g}^{(1)})$ is a normal subgroup of G containing the polarization $B = \exp(\mathfrak{b}[\ell])$ and the normal subgroup $N^{(1)} = \exp(\mathfrak{n}^{(1)})$. Besides, the modular functions Δ_G and $\Delta_{G^{(1)}}$ coincide on $G^{(1)}$. In the cases where $H \subset G^{(1)}$, we hence have the identity

$$\begin{aligned}S :&= \sup_{\substack{s\in B[\ell]\\ m\in N}} \delta_{\ell,\mathscr{S}}(s) \oint_{H/(H\cap B)} |\varphi(mshs^{-1})|\Delta_{H,G}^{-1/2}(h)d\nu(h)\\ &= \sup_{\substack{s\in B[\ell]\\ m\in N}} \delta_{\ell,\mathscr{S}^{(1)}}(s) \oint_{H^{(1)}/(H^{(1)}\cap B)} |\varphi(mshs^{-1})|\Delta_{H,G^{(1)}}^{-1/2}(h)d\nu(h).\end{aligned}$$

Applying the induction hypothesis to $(G^{(1)}, H = H^{(1)})$, we get the result. If $\mathfrak{h} \not\subset \mathfrak{g}^{(1)}$, we take a subspace $\mathfrak{u}$ of dimension 1 or 2 in $\mathfrak{h}$ such that $\mathfrak{h} = \mathfrak{u} \oplus (\mathfrak{h} \cap \mathfrak{g}^{(1)})$ and we write as before $\mathfrak{h}^{(1)} = (\mathfrak{h} \cap \mathfrak{g}^{(1)}) + \mathfrak{g}^{j_1}$ and $\mathfrak{n}^{(1)} = \mathfrak{n} \cap \mathfrak{g}^{(1)}$. Here we treat only the case where $\mathfrak{u}$ is of dimension 2 and $\mathfrak{u} = \mathbb{R}A + \mathbb{R}X$ with $X \in \mathfrak{n}$. We then have

$$S = \sup_{\substack{s\in B[\ell]\\ m^{(1)}\in N^{(1)},x\in\mathbb{R}}} \delta_{\ell,\mathscr{S}}(s) \int_{\mathbb{R}^2} \oint_{H^{(1)}/(H^{(1)}\cap B)} \Delta_{H,G}^{-1/2}(\exp(aA))$$

$$|\varphi(\exp(xX)m^{(1)}s\exp(x'X)\exp(aA)h^{(1)}s^{-1})|\Delta_{H^{(1)},G^{(1)}}^{-1/2}(h^{(1)})d\nu(h^{(1)})dadx'$$

$$\le \sup_{x\in\mathbb{R}} \int_{\mathbb{R}^2} \Delta_{H,G}^{-1/2}(\exp(aA))\Big(\sup_{\substack{s\in B[\ell]\\ m^{(1)}\in N^{(1)}}} \delta_{\ell,\mathscr{S}^{(1)}}(s) \oint_{H^{(1)}/(H^{(1)}\cap B)}$$

$$|\varphi(\exp(xX)m^{(1)}s\exp(x'X)\exp(aA)h^{(1)}s^{-1})|\Delta_{H^{(1)},G^{(1)}}^{-1/2}(h^{(1)})d\nu(h^{(1)})\Big)dadx'.$$

For $m^{(1)} \in N^{(1)}, x', a \in \mathbb{R}, s \in B[\ell], h^{(1)} \in H^{(1)}$, we write

$$m^{(1)}s\exp(x'X)\exp(aA)h^{(1)}s^{-1} = \exp(x'X)\exp(aA)b(x', a, s, m^{(1)})m^{(1)}sh^{(1)}s^{-1}$$

with $b(x', a, s, m^{(1)}) \in [G, G] \subset N^{(1)}$. Since G is the topological product of $\mathbb{R}^2$ with $G^{(1)}$, we can take a function $\varphi_1 \in C_c(G/G^{(1)})$ and a function $\varphi_2 \in K(G^{(1)}, B, \ell)$ so that

$$|\varphi(exp(xX)\exp(aA)g^{(1)})| \le |\varphi_1(\exp(xX)\exp(aA))||\varphi_2(g^{(1)})|,\ g^{(1)} \in G^{(1)},\ x, a \in \mathbb{R}.$$

This again allows us to apply the induction hypothesis to $G^{(1)}, B, H^{(1)}$ and we have

$$S \le \sup_{x\in\mathbb{R}} \int_{\mathbb{R}^2} \Delta_{H,G}^{-1/2}(\exp(aA))\Big(\sup_{\substack{s\in B[\ell]\\ m^{(1)}\in N^{(1)}}} \delta_{\ell,\mathscr{S}^{(1)}}(s) \oint_{H^{(1)}/(H^{(1)}\cap B)} |\varphi(\exp(xX)\exp(x'X)$$

$$\times \exp(aA)b(x', a, s, m^{(1)})m^{(1)}sh^{(1)}s^{-1})|\Delta_{H^{(1)},G^{(1)}}^{-1/2}(h^{(1)})d\nu(h^{(1)})\Big)dadx'$$

$$\le \Big(\sup_{x\in\mathbb{R}} \int_{\mathbb{R}^2} \Delta_{H,G}^{-1/2}(\exp(aA))|\varphi_1(\exp((x+x')X)\exp(aA))|dx'da\Big)$$

$$\times \Big(\sup_{\substack{s\in B[\ell]\\ m^{(1)}\in N^{(1)},x\in\mathbb{R}}} \delta_{\ell,\mathscr{S}^{(1)}}(s) \oint_{H^{(1)}/(H^{(1)}\cap B)} |\varphi_2(b(x', a, s, m^{(1)})m^{(1)}sh^{(1)}s^{-1})|$$

$$\times \Delta_{H^{(1)},G^{(1)}}^{-1/2}(h^{(1)})d\nu(h^{(1)})\Big)$$

$$\le \Big(\sup_{x\in\mathbb{R}} \int_{\mathbb{R}^2} \Delta_{H,G}^{-1/2}(\exp(aA))|\varphi_1(\exp((x+x')X)\exp(aA))|dx'da\Big)$$

$$\times \Big(\sup_{\substack{s\in B[\ell]\\ m^{(1)}\in N^{(1)},x\in\mathbb{R}}} \delta_{\ell,\mathscr{S}^{(1)}}(s) \oint_{H^{(1)}/(H^{(1)}\cap B)} |\varphi_2(m^{(1)}sh^{(1)}s^{-1})|\Delta_{H^{(1)},G^{(1)}}^{-1/2}(h^{(1)})d\nu(h^{(1)})\Big)$$

$$< \infty.$$

There remains for us to study cases (4) and (6). We consider the first index j_1 such that $\mathfrak{g}^{j_1} \not\subset \mathfrak{g}(\ell)$. We already know $j_1 > 1$. We begin with:

b. Case Where $\dim(\alpha^{j_1}) = 1$ Take $Y_1 \in \mathfrak{g}^{j_1} \setminus \mathfrak{g}^{j_1-1}$. Let $\mathfrak{g}_0^{j_1-1} = \mathfrak{g}^{j_1-1} \cap \ker(\ell)$ and $C_1 \in \mathfrak{g}^{j_1-1} \setminus \mathfrak{g}_0^{j_1-1}$ if $\ell|_{\mathfrak{g}^{j_1-1}} \neq 0$. As $j_1 > 1$ we can always suppose $\ell(Y_1) = 0$. We then have a linear form $\beta_1 : \mathfrak{g} \to \mathbb{R}$ such that

$$[A, Y_1] \equiv \alpha^{j_1}(A)Y_1 + \beta_1(A)C_1 \bmod \mathfrak{g}_0^{j_1-1}, A \in \mathfrak{g}.$$

Let $V_1 \in \mathfrak{g}$ verify $\beta_1(V_1) = 1$. With the notations of Sect. 7.3.2.2, put $\mathfrak{g}^{(1)} = \ker(\beta_1)$. We have the relation

$$[T, V_1] = -\alpha^{j_1}(T)V_1 \bmod \mathfrak{g}^{(1)}.$$

Starting from the sequence $(\mathfrak{g}^j \cap \mathfrak{g}^{(1)})_j$, we construct a composition sequence $\mathscr{C}^{(1)}$ and a good sequence of subalgebras $\mathscr{S}^{(1)}$ of $\mathfrak{g}^{(1)}$. Let ℓ be generic, then as we saw in the proof of Proposition 7.3.7, $\ell^{(1)}$ is generic in $(\mathfrak{g}^{(1)})^*$ and

$$\mathfrak{b}_{\mathscr{S}^{(1)}}[\ell^{(1)}] = \mathfrak{b}_{\mathscr{S}}[\ell] \subset \mathfrak{g}^{(1)}.$$

b.1. Case Where $\mathfrak{h} \subset \mathfrak{g}^{(1)}$ Set $\mathfrak{h}^{(1)} = \mathfrak{h} + \mathbb{R}Y_1$. Let $\mathfrak{g}_0 = \ker(\alpha^{j_1}) \cap \mathfrak{g}^{(1)}$, which is an ideal of codimension 1 in $\mathfrak{g}^{(1)}$, $\mathfrak{h}_0 = \mathfrak{h} \cap \mathfrak{g}_0$ and $\mathfrak{h}_0^{(1)} = \mathfrak{h}_0 + \mathbb{R}Y_1$. Clearly, $\mathfrak{h}_0$ is an ideal of $\mathfrak{h}$ and of $\mathfrak{h}_0^{(1)}$. On the other hand, as the root α^{j_1} vanishes on $\mathfrak{h}_0$, we have

$$\Delta_{H_0,G}^{-1/2}(h_0) = e^{1/2\alpha^{j_1}(\log(h_0))}\Delta_{H_0,G^{(1)}}^{-\frac{1}{2}}(h_0) = \Delta_{H_0,G^{(1)}}^{-\frac{1}{2}}(h_0) = \Delta_{G^{(1)},G}^{-\frac{1}{2}}(h_0)\Delta_{H_0,G^{(1)}}^{-1/2}(h_0)$$

for $h_0 \in H_0 = \exp(\mathfrak{h}_0)$. For $\varphi \in K(G, B, \ell)$, the function $(\Delta_{G^{(1)},G}^{-1/2}\varphi)|_{G^{(1)}}$ is an element of $K(G^{(1)}, B, \ell^{(1)})$. As $\mathfrak{n} \not\subset \mathfrak{g}^{(1)}$, we write $\mathfrak{n} = \mathbb{R}X + \mathfrak{n}^{(1)}$, $N^{(1)} = \exp(\mathfrak{n}^{(1)})$, where $\mathfrak{n}^{(1)} = \mathfrak{n} \cap \mathfrak{g}^{(1)}$ and $X \in \mathfrak{g} \setminus \mathfrak{g}^{(1)}$. Then $G = \exp(\mathbb{R}X)G^{(1)}$ and we can write every $m \in N$ as $m = \exp(xX)m'$ with $x \in \mathbb{R}$ and $m' \in N^{(1)}$. Let $H_0^{(1)} = \exp(\mathfrak{h}_0^{(1)})$. If $B[\ell] \cap H \not\subset G_0 = \exp(\mathfrak{g}_0)$, then $\mathfrak{h}_0^{(1)} + \mathfrak{g}^{(1)}(\ell^{(1)})$ is Lagrangian and we have

$$\begin{aligned}
&\oint_{H/H\cap B} |\varphi(mshs^{-1})|\Delta_{H,G}^{-1/2}(h)d\nu(h)\\
&= \oint_{H_0/H_0\cap B} |\Delta_{G^{(1)},G}^{-1/2}\varphi(\exp(xX)m'sh_0s^{-1})|\Delta_{H_0,G^{(1)}}^{-1/2}(h_0)d\nu(h_0)\\
&= \oint_{H_0^{(1)}/H_0^{(1)}\cap B} |\Delta_{G^{(1)},G}^{-1/2}\varphi(\exp(xX)m'sh_0s^{-1})|\Delta_{H_0^{(1)},G^{(1)}}^{-1/2}(h_0)d\nu(h_0).
\end{aligned}$$

This allows us to conclude by use of the induction hypothesis. Indeed, we take a function $\psi \in K(G, B, \ell)$ whose modulus dominates the modulus of φ and whose variables are separated in $x \in \mathbb{R}$ and $g' \in G^{(1)}$. In this way, we see that the root associated to $U \in \mathfrak{h} \setminus \ker(\alpha^{j_1})$ has no influence on the expression of δ_ℓ. So, we can apply the induction hypothesis to $G^{(1)}, H_0^{(1)}, \ell^{(1)}$ for ψ to get also the result for φ. Let now $B \cap H \subset G_0$ and $H \not\subset G_0$. Then it is clear that $\Delta_{H,G} = \Delta_{H^{(1)},G^{(1)}}\Delta_{G^{(1)},G}$ and $\Delta_G, \Delta_{G^{(1)}}$ coincide on $H^{(1)} \cap B$. By these facts, the function $h' \mapsto |\varphi(h')|\Delta^{-1/2}_{H^{(1)},G^{(1)}}(h')$ is in $K(H^{(1)}, H^{(1)} \cap B)$ and satisfies the necessary trace property by Proposition 7.3.8. Hence our integral converges by arguments similar to those of the previous cases. The case where $H \subset G_0$ is easier and is similarly treated.

Now we suppose $\mathfrak{h} \not\subset \mathfrak{g}^{(1)}$. Put $\mathfrak{h}^{(1)} = \mathfrak{h} \cap \mathfrak{g}^{(1)} + \mathfrak{g}^{j_1}$. Using the notations of Sect. 7.3.2.2, we choose an element $U = U_1 \in \mathfrak{h}$ verifying $\beta_1(U) = 1$ and $\beta_i(U) = 0$ (resp. $\beta_i(U) = \beta_i'(U) = 0$) for any $i > 1, i \in J$.

b.2. Case Where $\mathfrak{h} \not\subset \mathfrak{g}^{(1)}$ and α^{j_1} Vanishes on $\mathfrak{h}^{(1)}$ We have a vector $U = U_1 \in \mathfrak{h} \setminus \mathfrak{h}^{(1)}$ which is written, if U is not in $\mathfrak{n}$, as $U = T + V$ with $T \in \mathfrak{b}[\ell]$ and $V \in [\mathfrak{g}, \mathfrak{g}]$ such that $[T, V] \equiv -\alpha^{j_1}(T)V$ modulo $\mathfrak{g}^{(1)}$. We therefore obtain a decomposition $\mathfrak{g} = \mathbb{R}V \oplus \mathfrak{g}^{(1)}$ and $G = \exp(\mathbb{R}V)G^{(1)}$. In the same way, we have

$$\exp(uU) = \exp\Big(\frac{e^{-u\alpha^{j_1}(T)} - 1}{-\alpha^{j_1}(T)} V\Big) n(u) \exp(uT), \ u \in \mathbb{R},$$

where $n(u)$ is analytic on u and it belongs to $G^{(1)} \cap [G, G]$, and

$$\frac{e^{-\alpha^{j_1}(uT)} - 1}{-\alpha^{j_1}(T)}$$

must be simply replaced by u if $\alpha^{j_1}(T) = 0$. From Proposition 7.3.19 we have

$$\Delta^{1/2}_{B,G}(\exp(uT))\Delta^{-1/2}_{H,G}(\exp(uU)) = \delta_\ell(\exp(uU)).$$

On the other hand, we have

$$\Delta_H(h) = \Delta_{H^{(1)}}(h), \ \Delta_G(h) = \Delta_{G^{(1)}}(h)$$

for $h \in H \cap G^{(1)}$. The transitivity rule of $\nu_{\cdot,\cdot}$ gives us

$$\begin{aligned}
&\oint_{H/H\cap B} \varphi(h)\Delta^{-1/2}_{H,G}(h)d\nu(h)\\
&= \int_{\mathbb{R}} \Delta^{-1/2}_{H,G}(\exp(uU)) \oint_{H^{(1)}/(H^{(1)}\cap B)} \varphi(\exp(uU)h')\Delta^{-1/2}_{H,G}(h')d\nu(h')du\\
&= \int_{\mathbb{R}} \Delta^{-1/2}_{H,G}(\exp(uU)) \oint_{H^{(1)}/(H^{(1)}\cap B)} \varphi(\exp(uU)h')\Delta^{-1/2}_{H^{(1)},G^{(1)}}(h')d\nu(h')du.
\end{aligned}$$

for continuous functions φ on H with compact support modulo $B \cap H$. Hence we obtain, for $\varphi \in K(G, B, \ell)$,

$$K(s, m) = \delta_\ell(s) \oint_{H/(H\cap B)} |\varphi(mshs^{-1})| \Delta_{H,G}^{-1/2}(h) d\nu(h)$$

$$= \delta_\ell(s) \int_{\mathbb{R}} du \oint_{H^{(1)}/(H^{(1)}\cap B)} |\varphi(ms\exp(uU)h's^{-1})| \Delta_{H,G}^{-1/2}(\exp(uU)) \Delta_{H^{(1)},G^{(1)}}^{-1/2}(h') d\nu(h')$$

$$= e^{-\alpha^{j_1}(\log s)} \delta_{\ell^{(1)}}(s) \int_{\mathbb{R}} du \oint_{H^{(1)}/(H^{(1)}\cap B)}$$

$$\times \left|\varphi\Big(m\exp\big(\frac{e^{-\alpha^{j_1}(uT)}-1}{-\alpha^{j_1}(T)} e^{-\alpha^{j_1}(\log s)} V\big) n(s,u)\exp(uT)sh's^{-1}\exp(-uT)\Big)\right|$$

$$\times \Delta_{H,G}^{-1/2}(\exp(uU)) \Delta_{B,G}^{1/2}(\exp(uT)) \Delta_{H^{(1)},G^{(1)}}^{-1/2}(h') d\nu(h')$$

with $n(s, u) \in N \cap G^{(1)}$. As an immediate consequence of these relations, we obtain

$$K(s, m) = \int_{\mathbb{R}} du\, \delta_{\ell^{(1)}}(s) \oint_{H^{(1)}/(H^{(1)}\cap B)}$$

$$\times \left|\varphi\Big(m\exp\big(\frac{e^{-\alpha^{j_1}(uT)}-1}{-\alpha^{j_1}(T)} e^{-\alpha^{j_1}(\log s)} V\big) n(s,u)\exp(uT)sh's^{-1}\exp(-uT)\Big)\right|$$

$$\times e^{-\alpha^{j_1}(uT)} e^{-\alpha^{j_1}(\log s)} \delta_{\ell^{(1)}}(\exp(uT)) \Delta_{H^{(1)},G^{(1)}}^{-1/2}(h') d\nu(h').$$

As we can suppose that $\alpha^{j_1}(T) \neq 0$, we can choose T such that $\alpha^{j_1}(T) = -1$. In such a case, we obtain, for $m = \exp(xV)m'$ with $x \in \mathbb{R}$ and $m' \in N^{(1)}$,

$$K(s, m) = \int_J dv \oint_{H^{(1)}/(H^{(1)}\cap B)} \delta_{\ell^{(1)}}(s)$$

$$\times \left|\varphi\Big(\exp((x+v)V) n(s, f(v), m)\exp(f(v)T)sh's^{-1}\exp(-f(v)T)\Big)\right|$$

$$\times \delta_{\ell^{(1)}}(\exp(f(v)T)) \Delta_{H^{(1)},G^{(1)}}^{-1/2}(h') d\nu(h'),$$

where J is an interval of $\mathbb{R}$, $n(s, f(v), m) \in N^{(1)}$ and $f(v) = \log(1 + ve^{\alpha^{j_1}(\log s)})$ is well defined on J. In this way,

$$\sup_{\substack{s\in B(\ell)\\ m\in N}} K(s, m) \leq \sup_{\substack{s\in B(\ell)\\ r\in\mathbb{R},\, m'\in N^{(1)}}} \int_{\mathbb{R}} dv\, \delta_{\ell^{(1)}}(s\exp(rT)) \oint_{H^{(1)}/(H^{(1)}\cap B)}$$

$$\times \left|(\Delta_{G^{(1)},G}^{-1/2}\varphi)\big(\exp(vV) n(s, r, m')\exp(rT)sh's^{-1}\exp(-rT)\big)\right|$$

$$\times \Delta_{H^{(1)},G^{(1)}}^{-1/2}(h') d\nu(h') < \infty,$$

applying the induction hypothesis to $\mathfrak{g}^{(1)}$, $\mathfrak{h}^{(1)}$ and $\ell^{(1)}$.

b.3. Case Where α^{j_1} Is Non-zero on $\mathfrak{h}^{(1)}$ In this case we can suppose $U \in \mathfrak{n}$. Write $G = \exp(\mathbb{R}U)G^{(1)}$. Keeping the same notations, we have $\Delta_H(h') = e^{2\alpha^{j_1}(\log h')}\Delta_{H^{(1)}}(h')$ and $\Delta_G(h') = e^{\alpha^{j_1}(\log h')}\Delta_{G^{(1)}}(h')$ for any $h' \in H^{(1)}$ and hence

$$\Delta_{H,G}(h') = e^{\alpha^{j_1}(\log h')}\Delta_{H^{(1)},G^{(1)}}(h').$$

For any continuous function φ on H with compact support modulo $B \cap H$, the transitivity formula of $\nu_{\cdot,\cdot}$ gives us

$$\begin{aligned}
&\oint_{H/H\cap B} \varphi(h)\Delta_{H,G}^{-1/2}(h)d\nu(h)\\
&= \int_{\mathbb{R}}\oint_{H^{(1)}/(H^{(1)}\cap B)} e^{\alpha^{j_1}(\log h')}\varphi(\exp(uU)h')\Delta_{H,G}^{-1/2}(h')d\nu(h')du\\
&= \int_{\mathbb{R}}\oint_{H^{(1)}/(H^{(1)}\cap B)} e^{\frac{1}{2}\alpha^{j_1}(\log h')}\varphi(\exp(uU)h')\Delta_{H^{(1)},G^{(1)}}^{-1/2}(h')d\nu(h')du\\
&= \int_{\mathbb{R}}\oint_{H^{(1)}/(H^{(1)}\cap B)} (\Delta_{G^{(1)},G}^{-1/2}\varphi)(\exp(uU)h')\Delta_{H^{(1)},G^{(1)}}^{-1/2}(h')d\nu(h')du.
\end{aligned}$$

So, for $\varphi \in K(G, B, \ell)$ we obtain with $m = \exp(xU)m'$, $x \in \mathbb{R}$, $m' \in N^{(1)}$:

$$\begin{aligned}
&\delta_\ell(s)\oint_{H/(H\cap B)} |\varphi(mshs^{-1})|\Delta_{H,G}^{-1/2}(h)d\nu(h)\\
&= \int_{\mathbb{R}} due^{-\alpha^{j_1}(\log s)}\delta_{\ell^{(1)}}(s)\oint_{H^{(1)}/(H^{(1)}\cap B)}\\
&\times |(\Delta_{G^{(1)},G}^{-1/2}\varphi)(\exp((x+ue^{-\alpha^{j_1}(\log s)})U)m'(s,u,m)sh's^{-1})|\Delta_{H^{(1)},G^{(1)}}^{-1/2}(h')d\nu(h')\\
&= \int_{\mathbb{R}} du\ \delta_{\ell^{(1)}}(s)\oint_{H^{(1)}/((H^{(1)}\cap B))}\\
&\times |(\Delta_{G^{(1)},G}^{-1/2}\varphi)(\exp((x+u)U)m'(s,u,m)sh's^{-1})|\Delta_{H^{(1)},G^{(1)}}^{-1/2}(h')d\nu(h'),
\end{aligned}$$

with $m'(s,u,m) \in N^{(1)}$. Therefore, we get the result by applying the induction hypothesis to $\mathfrak{g}^{(1)}$, $\mathfrak{h}^{(1)}$ and $\ell^{(1)}$.

c. Case Where $\dim(\alpha^{j_1}) = 2$ Exactly as before, we have two vectors $Y_1, Y_1' \in (\mathfrak{g}^{j_1} \cap \ker(\ell)) \setminus \mathfrak{g}^{j_1-1}$, a vector $C_1 \in \mathfrak{g}^{j_1-1}$ verifying $\ell(C_1) = 1$ and two linear forms β_1, β_1' such that

$$\begin{aligned}
[A, Y_1] &\equiv \alpha^{j_1}(A)(Y_1 - \omega Y_1') + \beta_1(A)C_1\\
[A, Y_1'] &\equiv \alpha^{j_1}(A)(Y_1' + \omega Y_1) + \beta_1'(A)C_1
\end{aligned}$$

modulo $\mathfrak{g}^{j_1-1} \cap \ker(\ell)$ for $A \in \mathfrak{g}$. Moreover, the linear forms β_1, β_1' are linearly independent of α^{j_1}. The situation where $\mathfrak{h} \subset \mathfrak{g}^{(1)} = \ker(\beta_1) \cap \ker(\beta_1')$ is treated similarly to case b.1. Hence suppose that $\mathfrak{h} \not\subset \mathfrak{g}^{(1)}$. If the root α^{j_1} is non-zero on $\mathfrak{h}^{(1)}$, then $\dim(\mathfrak{h}/(\mathfrak{h} \cap \mathfrak{g}^{(1)})) = 2$ and we can choose two linearly independent vectors U_1, U_1' in $\mathfrak{h} \cap \mathfrak{n}$ outside $\mathfrak{g}^{(1)}$. Then $\Delta_{H,G}(h') = e^{2\alpha^{j_1}(\log h')}\Delta_{H^{(1)},G^{(1)}}(h')$ and $\Delta_{G^{(1)},G}(h') = e^{-2\alpha^{j_1}(\log h')}$ for any $h' \in H^{(1)}$. On the other hand, we have

$$\delta_\ell(s) = \delta_{\ell^{(1)}}(s)e^{-2\alpha^{j_1}(\log s)}$$

for $s \in B[\ell]$. This is clearly sufficient to compute the terms coming out from the integral, in which we consider the two variables coming from the vectors U_1, U_1'. If the root α^{j_1} vanishes on $\mathfrak{h}^{(1)}$, then $\dim(\mathfrak{h}/(\mathfrak{h} \cap \mathfrak{g}^{(1)}))$ could be 1 or 2. In the case where this dimension is 2,

$$\delta_\ell(s) = \delta_{\ell^{(1)}}(s)e^{-2\alpha^{j_1}(\log s)}$$

for all $s \in B[\ell]$. This compensates the change of variables at the level of the expression

$$\exp\Big(\frac{e^{-u\alpha^{j_1}(T_1)}-1}{-\alpha^{j_1}(T_1)}e^{-\alpha^{j_1}(\log s)}V_1\Big)\exp\Big(\frac{e^{-u\alpha^{j_1}(T_2)}-1}{-\alpha^{j_1}(T_2)}e^{-\alpha^{j_1}(\log s)}V_2\Big),$$

where $U_i = T_i + V_i, T_i \in \mathfrak{b}, V_i \in [\mathfrak{g}, \mathfrak{g}]$ for $i = 1, 2$. When, instead, $\dim(\mathfrak{h}/(\mathfrak{h} \cap \mathfrak{g}^{(1)})) = 1$, we have

$$\delta_\ell(s) = \delta_{\ell^{(1)}}(s)e^{-\alpha^{j_1}(\log s)}$$

by Definition 7.3.16. Therefore, we see that everything goes just as in case b.2.

7.5 The Concrete Plancherel Formula

Following [62, 63, 71], we would like to describe explicitly the Plancherel formula for $\tau = \mathrm{ind}_H^G \chi_f$, always supposed to be of discrete type. We set out to prove:

Theorem 7.5.1 *We keep the same hypotheses and notations. Fixing from now on a good sequence of subalgebras, we can choose the measure μ used in the disintegration of τ, assumed of discrete type, in such a manner that the sum $\sum_{k=1}^{m(\pi)} \langle \pi(\varphi)a_\pi^k, a_\pi^k\rangle$ is well defined μ-almost everywhere. Moreover,*

$$\varphi_H^f(e) = \int_{\widehat{G}} \sum_{k=1}^{m(\pi)} \langle \pi(\varphi)a_\pi^k, a_\pi^k\rangle d\mu(\pi)$$

for all $\varphi \in C_c^\infty(G)$.

The proof of this theorem will be divided in several independent steps.

1. Computing the Matrix Coefficients Let $\ell \in \Gamma$ be strongly generic. For $\psi \in K(G, B, \ell)$, the function $\psi^f_{H,B}$ is continuous by Corollary 7.3.12 and we have $\varphi^f_H \in K(G, H, f)$ for any $\varphi \in C^\infty_c(G)$. This means

$$\oint_{G/H} |\varphi^f_H(g)| d\nu_{G,H}(g) \oint_{H/(H\cap B_k)} |\psi(ghg_k)| \Delta^{-1/2}_{H,G}(h) d\nu_{H,H\cap B_k}(h) < +\infty.$$

Using Fubini's theorem, we get

$$\begin{aligned}
\langle \pi(\varphi)a^k_\pi, \psi\rangle &= \int_G \varphi(g)\langle a^k_\pi, \pi(g^{-1})\psi\rangle dg \\
&= \oint_{G/H} d\nu_{G,H}(g) \int_H \varphi(gh')\Delta^{-1}_{H,G}(h')dh' \\
&\times \oint_{H/(H\cap B_k)} \overline{\psi(gh'hg_k)\chi_f(h)}\Delta^{-1/2}_{H,G}(h) d\nu_{H,H\cap B_k}(h) \\
&= \oint_{G/H} \varphi^f_H(g) d\nu_{G,H}(g) \oint_{H/(H\cap B_k)} \overline{\psi(ghg_k)\chi_f(h)}\Delta^{-1/2}_{H,G}(h) d\nu_{H,H\cap B_k}(h) \\
&= \oint_{G/(H\cap B_k)} \varphi^f_H(g)\overline{\psi(gg_k)} d\nu_{G,H\cap B_k}(g) \\
&= \oint_{G/B} d\nu_{G,B}(g) \oint_{B/(B\cap g_k^{-1}Hg_k)} \varphi^f_H(gbg_k^{-1})\overline{\psi(gb)}\Delta^{-1}_{B,G}(b) d\nu_{B,B\cap g_k^{-1}Hg_k}(b) \\
&= \langle \oint_{B/(B\cap g_k^{-1}Hg_k)} \varphi^f_H(gbg_k^{-1})\chi_\ell(b)\Delta^{-1/2}_{B,G}(b) d\nu_{B,B\cap g_k^{-1}Hg_k}(b), \psi\rangle.
\end{aligned}$$

Since $\psi \in \mathscr{H}^\infty_\pi$ has a compact support modulo B, these calculations hold. Indeed, using Theorem 7.3.11,

$$\begin{aligned}
&\oint_{G/(H\cap B_k)} |\varphi^f_H(g)\psi(gg_k)| d\nu_{G,H\cap B_k}(g) \\
&= \oint_{G/H} d\nu_{G,H}(g) \oint_{H/(H\cap B_k)} |\varphi^f_H(gh)\psi(ghg_k)| \Delta^{-1}_{H,G}(h) d\nu_{H,H\cap B_k}(h) \\
&= \oint_{G/H} |\varphi^f_H(g)| d\nu_{G,H}(g) \oint_{H/(H\cap B_k)} |\psi(ghg_k)| \Delta^{-1/2}_{H,G}(h) d\nu_{H,H\cap B_k}(h) < +\infty.
\end{aligned}$$

Therefore,

$$(\pi(\varphi)a^k_\pi)(g) = \oint_{B/(B\cap g_k^{-1}Hg_k)} \varphi^f_H(gbg_k^{-1})\chi_\ell(b)\Delta^{-1/2}_{B,G}(b) d\nu_{B,B\cap g_k^{-1}Hg_k}(b).$$

As an immediate consequence of these observations, we can deduce that $\langle \pi(\varphi)a_\pi^k, a_\pi^k\rangle$ is formally written as follows:

$$\begin{aligned}
\langle \pi(\varphi)a_\pi^k, a_\pi^k\rangle &= \oint_{H/(H\cap B_k)} \pi(\varphi)a_\pi^k(h)\chi_f(h)\Delta_{H,G}^{-1/2}(h)d\nu_{H,H\cap B_k}(h) \\
&= \oint_{H/(H\cap B_k)} \chi_f(h)\Delta_{H,G}^{-1/2}(h)d\nu_{H,H\cap B_k}(h) \\
&\times \oint_{B/(B\cap g_k^{-1}Hg_k)} \varphi_H^f(hg_kbg_k^{-1})\chi_\ell(b)\Delta_{B,G}^{-1/2}(b)d\nu_{B,B\cap g_k^{-1}Hg_k}(b) \\
&= \oint_{H/(H\cap B_k)} \chi_f(h)\Delta_{H,G}^{-1/2}(h)d\nu_{H,H\cap B_k}(h) \\
&\times \oint_{B_k/(B_k\cap H)} \varphi_H^f(hb)\chi_{g_k\cdot\ell}(b)\Delta_{B_k,G}^{-1/2}(b)d\nu_{B_k,B_k\cap H}(b).
\end{aligned}$$

In the sequel we intend to give a meaning to these matrix coefficients by using the induction and intertwining operators. We shall tackle every step of the induction case by case. Besides, as in [51], every $\varphi \in C_c^\infty(G)$ is written as a finite linear combination of functions of type $\psi^* * \psi$ for some $\psi \in C_c^\infty(G)$ and

$$\langle \pi(\psi^* * \psi)a_\pi^k, a_\pi^k\rangle = \|\pi(\psi)a_\pi^k\|^2 \geq 0.$$

This allows us to establish equalities between certain integrals in what follows.

2. An Intertwining Formula Here we suppose that there exists a vector subspace $\mathfrak{g}'$ of codimension 1 in $\mathfrak{g}$ which contains $\mathfrak{n}+\mathfrak{g}(\ell)+\mathfrak{h}$ for all ℓ in a non-empty Zariski open set $\mathscr{V}$ of Γ, where as before $\mathfrak{n}$ is a nilpotent ideal of $\mathfrak{g}$ containing $[\mathfrak{g}, \mathfrak{g}]$. Let $k_0 \leq n$ be the smallest index k verifying $\mathfrak{g}_k \not\subset \mathfrak{g}'$. Truncating the sequence $\mathscr{S}$ at $\mathfrak{g}'$ we have the good sequence:

$$\mathscr{S}' : \mathfrak{g}_1 \subset \cdots \subset \mathfrak{g}_{k_0-1} \subset \mathfrak{g}'_{k_0} \subset \cdots \subset \mathfrak{g}'_{n-1} = \mathfrak{g}' \subset \mathfrak{g},$$

where $\mathfrak{g}'_k = \mathfrak{g}_{k+1} \cap \mathfrak{g}'$ for $k_0 \leq k \leq n-1$. Let $\mathfrak{b}' = \mathfrak{b}'[\ell] = \mathfrak{b}_{\mathscr{S}'}[\ell_0]$ for $\ell \in \mathscr{V}$ where $\ell_0 = \ell|_{\mathfrak{g}'} \in (\mathfrak{g}')^*$ and $B' = \exp(\mathfrak{b}')$. We put $G' = \exp(\mathfrak{g}')$.

Suppose first that $\mathfrak{b}[\ell] \subset \mathfrak{g}'$ generically in Γ. In this case we have $\mathfrak{b}_{\mathscr{S}}[\ell] = \mathfrak{b}_{\mathscr{S}'}[\ell_0]$ for almost every $\ell \in \mathscr{V}$, $\tau' = \mathrm{ind}_H^{G'}\chi_{\ell_0}$ has multiplicities of discrete type, G' is a normal subgroup of G and the modular functions Δ_G, $\Delta_{G'}$ coincide on G'. The Penney distributions a_π for π and $a_{\pi'}$ for $\pi' = \mathrm{ind}_B^{G'}\chi_\ell$ defined by (7.3.5) are the same.

Now consider the case where $\mathfrak{b}[\ell] \not\subset \mathfrak{g}'$ generically in Γ. Writing $\mathfrak{g}'(\ell_0) = \mathbb{R}X(\ell) + \mathfrak{g}(\ell)$, we see that

$$\mathfrak{b}'[\ell] = \mathbb{R}X(\ell) + (\mathfrak{b}[\ell] \cap \mathfrak{g}')$$

and hence

$$\mathfrak{b}'[\ell] = \mathbb{R}X(\ell) + (\mathfrak{b}[\ell] \cap \mathfrak{b}'[\ell]).$$

Since $\mathfrak{h} + \mathfrak{g}(\ell)$ is a Lagrangian subspace, the element $X(\ell)$ belongs to $\mathfrak{h} + \mathfrak{g}(\ell)$. So we write $X(\ell) = T(\ell) + V(\ell)$ with $T(\ell) \in \mathfrak{h}$ and $V(\ell) \in \mathfrak{g}(\ell)$. Remark that $T(\ell) \notin \mathfrak{n}$ because $\mathfrak{b}[\ell] \not\subset \mathfrak{g}'$. Fixing ℓ for the moment, write simply X, T, V instead of $X(\ell), T(\ell), V(\ell)$. Let $\pi = \mathrm{ind}_B^G \chi_\ell$ and $\pi' = \mathrm{ind}_{B'}^G \chi_\ell$. Then as $\mathfrak{b}'[\ell] = \mathbb{R}T + (\mathfrak{b}[\ell] \cap \mathfrak{b}'[\ell])$, the intertwining operator R from π to π' is given by

$$(R\varphi)(g) = \int_{\mathbb{R}} \varphi(g \exp(tT)) e^{it\ell(T)} e^{-\frac{t}{2}\mathrm{Tr}\,\mathrm{ad}_{\mathfrak{g}/\mathfrak{b}'[\ell]}T} dt$$

for $\varphi \in \mathcal{H}_\pi$ continuous with compact support modulo B. The following lemma allows us to conclude the proof of convergence in this context, because R is an intertwining isomorphism.

Lemma 7.5.2 *Let a_π (resp. $a_{\pi'}$) be the Penney distribution for π (resp. π'), as defined in (7.3.5). Then $a_\pi = a_{\pi'} \circ R$.*

Proof Since $B_\ell(T, [\mathfrak{g}, \mathfrak{g}]) = \{0\}$ and $[T, \mathfrak{g}] \subset \mathfrak{g}'$, we have

$$\begin{aligned}
\mathrm{Tr}\,\mathrm{ad}_{\mathfrak{g}/\mathfrak{b}'[\ell]}T &= -\mathrm{Tr}\,\mathrm{ad}_{\mathfrak{b}'[\ell]/\mathfrak{g}'(\ell_0)}T \\
&= -\mathrm{Tr}\,\mathrm{ad}_{\mathfrak{b}'[\ell]/(\mathfrak{b}'[\ell]\cap(\mathfrak{h}+\mathfrak{g}'(\ell_0)))}T - \mathrm{Tr}\,\mathrm{ad}_{(\mathfrak{b}'[\ell]\cap(\mathfrak{h}+\mathfrak{g}'(\ell_0)))/\mathfrak{g}'(\ell_0)}T \\
&= \mathrm{Tr}\,\mathrm{ad}_{\mathfrak{h}/(\mathfrak{h}\cap\mathfrak{b}'[\ell])}T - \mathrm{Tr}\,\mathrm{ad}_{(\mathfrak{h}\cap\mathfrak{b}'[\ell])/(\mathfrak{h}\cap\mathfrak{g}'(\ell_0))}T.
\end{aligned}$$

Therefore by the transitivity property of $\nu_{\cdot,\cdot}$,

$$\begin{aligned}
&\oint_{H/(H\cap B')} \left| \left(\int_{\mathbb{R}} \overline{\varphi(h\exp(tT))} e^{-it\ell(T)} e^{-\frac{t}{2}\mathrm{Tr}\,\mathrm{ad}_{\mathfrak{g}/\mathfrak{b}'[\ell]}T} dt \right) \right| \overline{\chi_\ell(h)\Delta_{H,G}^{-1/2}}(h) d\nu(h) \\
&\le \oint_{H/(H\cap B')} \int_{\mathbb{R}} |\varphi(h\exp(tT))| e^{-\frac{t}{2}\mathrm{Tr}\,\mathrm{ad}_{\mathfrak{g}/\mathfrak{b}'[\ell]}T} dt\, \Delta_{H,G}^{-1/2}(h) d\nu(h) \\
&= \int_{\mathbb{R}} \oint_{H/(H\cap B')} |\varphi(\exp(tT)h)| \Delta_{H,G}^{-1/2}(h) d\nu(h) e^{\frac{t}{2}\left(\mathrm{Tr}\,\mathrm{ad}_{\mathfrak{h}/(\mathfrak{h}\cap\mathfrak{b}'[\ell])}T + \mathrm{Tr}\,\mathrm{ad}_{(\mathfrak{h}\cap\mathfrak{b}'[\ell])/(\mathfrak{h}\cap\mathfrak{g}'(\ell_0))}T\right)} dt \\
&= \int_{\mathbb{R}} \oint_{H/(H\cap B')} |\varphi(\exp(tT)h)| \Delta_{H,G}^{-1/2}(h) d\nu(h) e^{\frac{t}{2}\mathrm{Tr}\,\mathrm{ad}_{\mathfrak{h}/(\mathfrak{h}\cap\mathfrak{g}'(\ell_0))}T} dt \\
&= \int_{\mathbb{R}} \oint_{H/(H\cap B')} |\varphi(\exp(tT)h)| \Delta_{H,G}^{-1/2}(h) d\nu(h) e^{-\frac{t}{2}\mathrm{Tr}\,\mathrm{ad}_{\mathfrak{g}/\mathfrak{h}}T} dt \\
&= \oint_{H/(H\cap B)} |\varphi(h)| \Delta_{H,G}^{-1/2}(h) d\nu(h) \\
&< \infty.
\end{aligned}$$

Hence the function $R(\varphi)$ is in the domain of the distribution $a_{\pi'}$ and

$$(a_{\pi'} \circ R)(\varphi) = \oint_{H/(H\cap B')} \overline{\int_{\mathbb{R}} \varphi(h\exp(tT)) e^{-it\ell(T)} e^{-\frac{t}{2}\mathrm{Tr}\,\mathrm{ad}_{\mathfrak{g}/\mathfrak{b}'[\ell]}T} dt \chi_\ell(h)} \Delta_{H,G}^{-1/2}(h) d\nu(h).$$

This means that

$$\begin{aligned}(a_{\pi'} \circ R)(\varphi) &= \oint_{H/(H\cap B')} \int_{\mathbb{R}} \overline{\varphi(h\exp(tT))} e^{-it\ell(T)} e^{-\frac{t}{2}\left(\mathrm{Tr}\,\mathrm{ad}_{\mathfrak{h}/(\mathfrak{h}\cap\mathfrak{b}'[\ell])}T\right)} \\ &\times e^{\frac{t}{2}\mathrm{Tr}\,\mathrm{ad}_{(\mathfrak{h}\cap\mathfrak{b}'[\ell])/(\mathfrak{h}\cap\mathfrak{g}'(\ell_0)))}T} dt \overline{\chi_\ell(h)} \Delta_{H,G}^{-1/2}(h) d\nu(h) \\ &= \int_{\mathbb{R}} \oint_{H/(H\cap B')} \overline{\varphi(\exp(tT)h)\chi_\ell(h)} \Delta_{H,G}^{-1/2}(h) d\nu(h) \\ &\times e^{-it\ell(T)} e^{\frac{t}{2}\left(\mathrm{Tr}\,\mathrm{ad}_{\mathfrak{h}/(\mathfrak{h}\cap\mathfrak{b}'[\ell])}T + \mathrm{Tr}\,\mathrm{ad}_{(\mathfrak{h}\cap\mathfrak{b}'[\ell])/(\mathfrak{h}\cap\mathfrak{g}'(\ell_0))}T\right)} dt \\ &= \int_{\mathbb{R}} \oint_{H/(H\cap B')} \overline{\varphi(\exp(tT)h)\chi_\ell(h)} \Delta_{H,G}^{-1/2}(h) d\nu(h) \\ &\times e^{-it\ell(T)} e^{\frac{t}{2}\mathrm{Tr}\,\mathrm{ad}_{\mathfrak{h}/(\mathfrak{h}\cap\mathfrak{g}'(\ell_0))}T} dt \\ &= \int_{\mathbb{R}} \oint_{H/(H\cap B')} \overline{\varphi(\exp(tT)h)\chi_\ell(h)} \Delta_{H,G}^{-1/2}(h) d\nu(h) \\ &\times e^{-it\ell(T)} e^{-\frac{t}{2}\mathrm{Tr}\,\mathrm{ad}_{\mathfrak{g}/\mathfrak{h}}T} dt \\ &= \oint_{H/(H\cap B)} \overline{\varphi(h)\chi_\ell(h)} \Delta_{H,G}^{-1/2}(h) d\nu(h) = a_\pi(\varphi).\end{aligned}$$

□

3. Proof of Theorem 7.5.1 We proceed by induction on $m = \dim G + \dim G/H$ and follow closely the proof of the disintegration of τ given in [70]. If $m = 1$, everything is evident and we suppose that $m > 1$. Since $\mathfrak{g}$ is solvable, it contains at least an ideal $\mathfrak{g}_0$ containing $[\mathfrak{g}, \mathfrak{g}]$ and of codimension 1 in $\mathfrak{g}$. Put $G_0 = \exp(\mathfrak{g}_0)$. We shall treat separately the following cases.

3.1. Case Where $\mathfrak{g}_0$ Is an Ideal Containing $\mathfrak{h}$ for Which the Generic G-Orbits are Saturated In this case we have almost everywhere saturated orbits with respect to $\mathfrak{g}_0$. Put $f_0 = p(f) = f|_{\mathfrak{g}_0} \in \mathfrak{g}_0^*$ and

$$\Gamma_0 = f_0 + \mathfrak{h}^{\perp,\mathfrak{g}_0} = \{\ell_0 \in \mathfrak{g}_0^*; \ell_0|_{\mathfrak{h}} = f_0|_{\mathfrak{h}}\}.$$

Since $\mathfrak{h} + \mathfrak{g}(\ell)$ is almost everywhere Lagrangian, the representation $\tau_0 = \mathrm{ind}_H^{G_0}\chi_{f_0}$ is of discrete type. Let

$$\tau_0 \simeq \int_{\widehat{G_0}}^{\oplus} m_0(\pi_0)\pi_0 d\mu_0(\pi_0)$$

be the canonical central decomposition of τ_0. As the subgroup G_0 is normal, G acts on $\widehat{G_0}$ in the usual way and, up to a μ-negligible set, $\widehat{G}$ is identified with $\widehat{G_0}/G$. Besides, the disintegration of μ_0 relative to μ, which is the image of μ_0 by the projection $q : \widehat{G_0} \to \widehat{G_0}/G$, leads to

$$\tau \simeq \mathrm{ind}_{G_0}^G \tau_0 \simeq \int_{\widehat{G_0}}^{\oplus} m_0(\pi_0) \left(\mathrm{ind}_{G_0}^G \pi_0\right) d\mu_0(\pi_0)$$

$$= \int_{\widehat{G}}^{\oplus} d\mu(\pi) \int_{\hat{G_0}}^{\oplus} m_0(\pi_0) \left(\mathrm{ind}_{G_0}^G \pi_0\right) d\mu_0^{\pi}(\pi_0),$$

where the fiber measure μ_0^{π} passes to $q^{-1}(\pi)$. From what we have seen until now, the representation

$$\int_{\widehat{G_0}}^{\oplus} m_0(\pi_0) \left(\mathrm{ind}_{G_0}^G \pi_0\right) d\mu_0^{\pi}(\pi_0)$$

is factorial, namely a multiple of an irreducible representation, and

$$m(\pi)\pi \simeq \int_{\widehat{G_0}}^{\oplus} m_0(\pi_0) \left(\mathrm{ind}_{G_0}^G \pi_0\right) d\mu_0^{\pi}(\pi_0)$$

for μ-almost all π. It follows that $\tau = \mathrm{ind}_H^G \chi_f$ is of discrete type if and only if $\tau_0 = \mathrm{ind}_H^{G_0} \chi_f$ is supported on discrete points such that $m_0(\pi_0)$ are of discrete type. These are hence given by the number of connected components of $\Gamma_0 \cap \omega(\pi_0)$, where $\omega(\pi_0)$ denotes the G_0-orbit corresponding π_0. From these observations we deduce that $q^{-1}(\pi)$ is a discrete set and

$$m(\pi) = \sum_{\pi_0 \in q^{-1}(\pi)} m_0(\pi_0).$$

Then for each connected component C of $\Gamma \cap \Omega(\pi)$, $p(C)$ coincides with a connected component of $\Gamma_0 \cap \omega(\pi_0)$ for a certain $\pi_0 \in q^{-1}(\pi)$. As this component is a manifold of dimension $(1/2)\dim(\omega(\pi_0))$, C is a manifold of the desired dimension.

Let us return to the proof of the Plancherel formula (7.2.3). Going back to Sect. 7.5.2, we take the Vergne polarization $\mathfrak{b} \subset \mathfrak{g}_0$ realizing $\pi = \pi_\ell \in \widehat{G}$ ($\ell \in \Gamma$). Then taking into account the expression of the coefficients $\langle \pi(\varphi)a_\pi^k, a_\pi^k \rangle$ of Sect. 7.5.1, we see that these coefficients exist by applying directly the induction hypothesis at the level of G_0 for the coefficients $\langle \pi_0(\varphi)a_{\pi_0}^k, a_{\pi_0}^k \rangle$. Indeed,

$$\langle \pi(\varphi)a_\pi^k, a_\pi^k \rangle = \oint_{H/(H\cap B_k)} \chi_f(h)\Delta_{H,G}^{-1/2}(h) d\nu_{H,H\cap B_k}(h)$$

$$\times \oint_{B_k/(B_k\cap H)} \varphi_H^f(hb)\chi_{g_k\cdot\ell}(b)\Delta_{B_k,G}^{-1/2}(b) d\nu_{B_k,B_k\cap H}(b)$$

$$= \oint_{H/(H\cap B_k)} \chi_f(h)\Delta_{H,G_0}^{-1/2}(h)d\nu_{H,H\cap B_k}(h)$$

$$\times \oint_{B_k/(B_k\cap H)} \widetilde{\varphi}_H^f(hb)\chi_{g_k\cdot\ell}(b)\Delta_{B_k,G_0}^{-1/2}(b)d\nu_{B_k,B_k\cap H}(b),$$

which converges by the induction hypothesis, where $\widetilde{\varphi} = \varphi|_{G_0}$. Then from what we have observed, we are able to interpret the Plancherel formula for τ_0 (valid by induction and described for a test function φ by our claim (7.2.3)). More precisely, the induction hypothesis ensures us that the sum $\sum_{k=1}^{m_0(\pi_0)}\langle \pi_0(\varphi)a_{\pi_0}^k, a_{\pi_0}^k\rangle$ makes sense μ_0-almost everywhere, and the Plancherel formula

$$\varphi_H^f(e) = \int_{\widehat{G_0}} \sum_{k=1}^{m_0(\pi_0)} \langle \pi_0(\varphi)a_{\pi_0}^k, a_{\pi_0}^k\rangle d\mu_0(\pi_0)$$

holds. As μ_0 is described by the integral decomposition $\int_{\widehat{G}}^{\oplus} \mu_0^\pi d\mu(\pi)$, we obtain

$$\varphi_H^f(e) = \int_{\widehat{G}} \sum_{k=1}^{m(\pi)} \langle \pi(\varphi)a_\pi^k, a_\pi^k\rangle d\mu(\pi).$$

This allows us to conclude in this case.

3.2. Case Where $\mathfrak{h}+[\mathfrak{g},\mathfrak{g}] \neq \mathfrak{g}$ Let $\mathfrak{g}_0$ be an ideal of codimension 1 in $\mathfrak{g}$ containing $\mathfrak{h}+[\mathfrak{g},\mathfrak{g}]$ and $G_0 = \exp(\mathfrak{g}_0)$. If $\mathfrak{g}_0$ is an ideal of saturation, we are in the previous case 3.1 which has been settled. So we tackle the case where almost all orbits are non-saturated with respect to $\mathfrak{g}_0$. For almost all $\pi \in \widehat{G}$, the restriction π_0 of π to G_0 is irreducible, $m(\pi) = m(\pi_0)$ and the Borel space $\widehat{G}$ is identified with $\widehat{G_0} \times \mathbb{R}$ up to a negligible set. The induction hypothesis ensures the existence of a measure μ_0 which establishes the concrete Plancherel formula for τ_0. Then under our identification, the measure $\mu = \mu_0 \times d\lambda$, where $d\lambda$ is a Lebesgue measure on $\mathbb{R}$, will suit us. In fact, we can take g_k in G_0 and, at every point $\widetilde{\ell} \in \Gamma$ such that

$$p(\widetilde{\ell}) = \ell_0 = g_k \cdot \ell|_{\mathfrak{g}_0} \in \mathfrak{g}_0^*,$$

take an element $X = X(\ell_0)$, depending on ℓ_0, in $\mathfrak{g}$ so that $\mathfrak{g}(\widetilde{\ell}) = \mathbb{R}X \oplus \mathfrak{g}_0(\ell_0)$ and the Borel mapping $\pi \mapsto (\pi_0, \widetilde{\ell}(X))$ gives the identification between $\widehat{G}$ and $\widehat{G_0} \times \mathbb{R}$. Besides, it is possible to suppose that $B_k = \exp(\mathbb{R}X)g_kB_0g_k^{-1}$ with $B_0 = \exp(\mathfrak{b}_0)$, $\mathfrak{b}_0$ being the Vergne polarization of $\mathfrak{g}_0$ at $\ell|_{\mathfrak{g}_0} \in \mathfrak{g}_0^*$. In this way, the homogeneous space $B_k/(B_k \cap H)$ is isomorphic to the product of $\mathbb{R}$ with the homogeneous space $B_0^k/(B_0^k \cap H)$, where $B_0^k = g_kB_0g_k^{-1}$. Writing $b \in B_k$ as $b = \exp(xX)b_0$ with

$x \in \mathbb{R}$ and $b_0 \in B_0$, we have

$$
\begin{aligned}
\langle \pi(\varphi)a_\pi^k, a_\pi^k\rangle = & \oint_{H/(H\cap B_k)} \chi_f(h)\Delta_{H,G}^{-1/2}(h)d\nu_{H,H\cap B_k}(h) \\
& \times \oint_{B_k/(B_k\cap H)} \varphi_H^f(hb)\chi_{g_k\cdot\ell}(b)\Delta_{B_k,G}^{-1/2}(b)d\nu_{B_k,B_k\cap H}(b) \\
= & \oint_{H/(H\cap B_k^0)} \chi_f(h)\Delta_{H,G_0}^{-1/2}(h)d\nu_{H,H\cap B_k^0}(h) \\
& \times \oint_{B_k^0/(B_k^0\cap H)} \int_{\mathbb{R}} \Delta_{B_k^0,G_0}^{-1/2}(\exp(xX))\varphi_H^f(h\exp(xX)b_0)\chi_{g_k\cdot\ell}(b_0) \\
& \times \chi_\ell(\exp(xX))\Delta_{B_k^0,G_0}^{-1/2}(b_0)dxd\nu_{B_k^0,B_k^0\cap H}(b_0) \\
= & \oint_{B_k^0/(B_k^0\cap H)} \widetilde{\varphi}_H^f(hb_0)\chi_{g_k\cdot\ell}(b_0)\Delta_{B_k^0,G_0}^{-1/2}(b_0)d\nu_{B_k^0,B_k^0\cap H}(b_0),
\end{aligned}
$$

which converges by the induction hypothesis, where

$$\widetilde{\varphi}(g_1) = \int_{\mathbb{R}} \Delta_{B_k^0,G_0}^{-1/2}(\exp(xX))\varphi(g_1\exp(xX))\chi_\ell(\exp(xX))dx.$$

This allows us to conclude that the coefficients $\langle \pi(\varphi)a_\pi^k, a_\pi^k\rangle$ from Sect. 7.5.1 do exist. Now, putting $\lambda = \widetilde{\ell}(X)$, we can get formula (7.2.3) with the help of Fourier's inversion formula for $\mathbb{R}$ applied to dx and $d\lambda$, in order to achieve the concrete Plancherel formula for τ.

3.3. Case Where $\mathfrak{h} + [\mathfrak{g}, \mathfrak{g}] = \mathfrak{g}$ In this situation, it turns out that, μ-almost everywhere, the multiplicities of τ are finite and the sum $\sum_{k=1}^{m(\pi)} \langle \pi(\varphi)a_\pi^k, a_\pi^k\rangle$ is defined. There is only the Plancherel formula left to see. Remark that it suffices to show formula (7.2.3) for $\mathfrak{h}$ containing the center $\mathfrak{z}$ of $\mathfrak{g}$. Otherwise, let Z be a central element which is not contained in $\mathfrak{h}$. Put $\mathfrak{h}' = \mathfrak{h} + \mathbb{R}Z$ and $H^{(1)} = \exp(\mathfrak{h}')$. For arbitrary $\lambda \in \mathbb{R}$, let $f_\lambda \in \Gamma$ such that $f_\lambda(Z) = \lambda$ and

$$\Gamma_\lambda = \{\tilde{\ell} \in \Gamma; \tilde{\ell}(Z) = \lambda\}.$$

Hence we see here that the $\langle \pi(\varphi)a_\pi^k, a_\pi^k\rangle$ from Sect. 7.5.1 do exist. Assume the concrete Plancherel formula for $\tau' = \mathrm{ind}_{H^{(1)}}^G \chi_{f_\lambda}$ on Γ_λ with a measure μ_λ. We integrate this formula in the measure $d\lambda$ in order to pass from $\varphi_{H^{(1)}}^{f_\lambda}$ to φ_H^f, applying Fourier's inversion formula for $\mathbb{R}$. Next, we rewrite the right-hand side, passing from $B_k/(B_k \cap H^{(1)})$ to $B_k/(B_k \cap H)$ and from Γ_λ to Γ by the inversion formula once again.

In other respects, if $\mathfrak{h}$ contains an ideal $\mathfrak{a}$ of $\mathfrak{g}$ on which f vanishes, it suffices to apply the induction hypothesis to the quotient group $G/\exp\mathfrak{a}$. All these facts will be assumed in the sequel. It follows that $\dim \mathfrak{z} \leq 1$. Let $\mathfrak{a}$ be a minimal non-central ideal. We find that $\mathfrak{a}$ is Abelian and $\dim \mathfrak{a} \leq 3$. The adjoint action of $\mathfrak{g}$ on $\mathfrak{a}/(\mathfrak{a} \cap \mathfrak{z})$ gives us a root, or its real part $\lambda \in \mathfrak{g}^*$, and there always exists $X \in \mathfrak{h}$ verifying $\lambda(X) = 1$. The kernel of λ, denoted by $\mathfrak{g}_0$, is an ideal of $\mathfrak{g}$ and we have $\mathfrak{g} = \mathbb{R}X \oplus \mathfrak{g}_0$. Now we put $A = \exp \mathfrak{a}$, $G_0 = \exp(\mathfrak{g}_0)$ and go to examine different cases.

(I) Suppose that $\mathfrak{a} \cap \mathfrak{z} = \{0\}$. We find $\mathfrak{h} \cap \mathfrak{a} = \{0\}$. Let $\mathfrak{h}' = \mathfrak{h} \cap \mathfrak{g}_0 + \mathfrak{a}$ and $H^{(1)} = \exp(\mathfrak{h}')$. Reasoning at the level of the subgroup $K = \exp \mathfrak{k}$, $\mathfrak{k} = \mathfrak{h} + \mathfrak{a}$, we see immediately that the situation is entirely similar to the low-dimensional examples of [70].

If $\dim \mathfrak{a} = 1$, let $\mathfrak{a} = \mathbb{R}Y$ and $f_\pm \in \Gamma$ be such that $f_\pm(Y) = \pm 1$. It is evident that $f_\pm([\mathfrak{h}', \mathfrak{h}']) = \{0\}$. Let $\tau_\pm$ be the monomial representations of G induced from the unitary characters $\chi_\pm$ of $H^{(1)}$ coming from $f_\pm$. As $\mathfrak{g} \neq \mathfrak{h}' + [\mathfrak{g}, \mathfrak{g}]$, the result concerning the disintegration is obtained [70] from the relation

$$\tau \simeq \tau_+ \oplus \tau_- \simeq \int_{\widehat{G}}^{\oplus} m_+(\pi)\pi d\nu_+(\pi) \oplus \int_{\widehat{G}}^{\oplus} m_-(\pi)\pi d\nu_-(\pi)$$

with well-understood notations. But we know that intertwining operators $R_\pm$ from $\tau_\pm$ to τ are given by the formulas

$$(R_\pm \psi)(g) = \oint_{H/H_0} \psi(gh)\chi_\pm(h)\Delta_{H,G}^{-1/2}(h) d\nu_{H,H_0}(h) \ (g \in G)$$

with $H_0 = \exp(\mathfrak{h}_0)$ where $\mathfrak{h}_0 = \mathfrak{h} \cap \mathfrak{g}_0$. We denote by $a_\pi^{k,\pm}$ the Penney distributions associated to $\tau_\pm$. Then an application of the transitivity of $\nu_{\cdot,\cdot}$ and a simple verification show $a_\pi^k = R_\pm \circ a_\pi^{k,\pm}$ and $\pi(\varphi)a_\pi^k = R_\pm(\pi(\varphi)a_\pi^{k,\pm})$ for $\varphi \in C_c^\infty(G)$. In this case, the coefficients $\langle \pi(\varphi)a_\pi^k, a_\pi^k \rangle$ are understood as

$$\langle \pi(\varphi)a_\pi^k, a_\pi^k \rangle = \langle R_\pm(\pi(\varphi)a_\pi^{k,\pm}), R_\pm \circ a_\pi^{k,\pm} \rangle = \langle \pi(\varphi)a_\pi^{k,\pm}, a_\pi^{k,\pm} \rangle.$$

Put $a_\pm = \delta_\tau \circ R_\pm$, namely

$$a_\pm : \mathscr{H}_{\tau_\pm}^\infty \ni \psi \mapsto \int_{H/H_0} \overline{\psi(h)\chi_{f_\pm}(h)} \Delta_{H,G}^{-1/2}(h) d\dot{h}.$$

Appropriately normalized, $a_\pm$ satisfy

$$\langle \tau(\varphi)\delta_\tau, \delta_\tau \rangle = \langle \tau_+(\varphi)a_+, a_+ \rangle + \langle \tau_-(\varphi)a_-, a_- \rangle$$

for any $\varphi \in C_c^\infty(G)$. Then formula (7.2.3) is obtained as in [71].

If $\dim \mathfrak{a} = 2$, let $Y_1, Y_2 \in \mathfrak{a}$ be such that the three elements X, Y_1, Y_2 verify

$$[X, Y_1] = Y_1 - \alpha Y_2, \ [X, Y_2] = Y_2 + \alpha Y_1 \ (0 \neq \alpha \in \mathbb{R}).$$

Take $\widehat{\theta} \in \Gamma$ such that

$$\widehat{\theta}|_{\mathfrak{a}} = (\cos\theta)Y_1^* + (\sin\theta)Y_2^*.$$

Trivially, $\widehat{\theta}([\mathfrak{h}', \mathfrak{h}']) = \{0\}$ and

$$\tau \simeq \int_{[0,2\pi]}^{\oplus} \mathrm{ind}_{H^{(1)}}^G \chi_{\widehat{\theta}} d\theta \simeq \int_{[0,2\pi]}^{\oplus} d\theta \int_{\widehat{G}}^{\oplus} m_\theta(\pi)\pi d\nu_\theta(\pi). \tag{7.5.1}$$

Put $\Gamma'_\theta = \widehat{\theta} + \mathfrak{h}'^{\perp}$ and $\Gamma^0_\theta = \widehat{\theta} + \mathfrak{k}^{\perp}$. Take now a coadjoint orbit Ω of G which does not vanish on $\mathfrak{a}$. We then have $\dim(\mathfrak{g}(\ell) \cap \mathfrak{a}) = 1$ for $\ell \in \Omega$. For such an Ω, there exists a unique $\theta \in [0, 2\pi)$ such that

$$\Gamma \cap \Omega = \exp(\mathbb{R}X) \cdot (\Gamma^0_\theta \cap \Omega),\ \Gamma'_\theta \cap \Omega = A\cdot(\Gamma^0_\theta \cap \Omega),\ A = \exp \mathfrak{a}.$$

Taking these observations into account, the above formula (7.5.1) is interpreted as the desired disintegration. According to this disintegration, we form an intertwining operator R_θ from $\tau_\theta = \mathrm{ind}_{H^{(1)}}^G \chi_{\widehat{\theta}}$ to τ by the formula

$$(R_\theta\psi)(g) = \int_{H/H_0} \psi(gh)\chi_{\widehat{\theta}}(h)\Delta_{H,G}^{-1/2}(h)d\dot{h} \ \ (g \in G).$$

We next put $a_\theta = \delta_\tau \circ R_\theta$, namely

$$a_\theta : \mathscr{H}_{\tau_\theta}^{\infty} \ni \psi \mapsto \int_{H/H_0} \overline{\psi(h)\chi_f(h)\Delta_{H,G}^{-1/2}}(h)d\dot{h}.$$

Reasoning at the level of the subgroup K, we directly verify as in the case of the $ax + b$ group that a_θ belongs to the space $(\mathscr{H}_{\tau_\theta}^{-\infty})^{H,\chi_f\Delta_{H,G}^{1/2}}$. Example 2 in [63] then shows

$$\varphi_H^f(e) = \langle \tau(\varphi)\delta_\tau, \delta_\tau\rangle = (2\pi)^{-2} \int_0^{2\pi} \langle \tau_\theta(\varphi)a_\theta, a_\theta\rangle d\theta \tag{7.5.2}$$

for any $\varphi \in C_c^\infty(G)$. As in the previous case, we show the existence of the coefficients $\langle \pi(\varphi)a_\pi^k, a_\pi^k\rangle$ through the intertwining operator.

Let us continue the proof of formula (7.2.3). A direct computation gives us, for $g \in G$,

$$(\tau_\theta(\varphi)a_\theta)(g) = \int_{\mathbb{R}}\int_{H^{(1)}} \varphi(gh\exp(t'X))\chi_f(h)\Delta_{H,G}^{-1/2}(h)e^{it'f(X)}e^{-\frac{t'}{2}(\mathrm{Tr}\,\mathrm{ad}_{\mathfrak{g}}X + \mathrm{Tr}\,\mathrm{ad}_{\mathfrak{h}}X)}dhdt'.$$

Under the unique expression of $h \in H$ in the form $h = h_0\exp(sX)$ with $h_0 \in H_0$ and $s \in \mathbb{R}$, we have $dh = e^{-s\mathrm{Tr}\,\mathrm{ad}_{\mathfrak{h}}X}dh_0ds$. Let $\varphi \in C_c^\infty(G)$. For

$t \in \mathbb{R}$ arbitrarily fixed, we form $\Phi_t \in C_c^\infty(G_0)$ by

$$\Phi_t(g_0) = \int_{\mathbb{R}} \varphi(\exp(tX)g_0 \exp(t'X))e^{it'f(X)}e^{-\frac{t'}{2}(\mathrm{Tr}\,\mathrm{ad}_{\mathfrak{g}} X + \mathrm{Tr}\,\mathrm{ad}_{\mathfrak{h}} X)}dt'$$

and put

$$(\Phi_t)^{\widehat{\theta}}_{H^{(1)}}(g_0) = \int_{H^{(1)}} \Phi_t(g_0h')\chi_{\widehat{\theta}}(h')\Delta^{-1/2}_{H^{(1)},G_0}(h')dh' \quad (g_0 \in G_0).$$

From the induction hypothesis, we see

$$\begin{aligned}(\Phi_t)^{\widehat{\theta}}_{H^{(1)}}(e) = \int_{\widehat{G}} d\mu_\theta(\pi)\\ \sum_{k=1}^{m_\theta(\pi)} \oint_{H^{(1)}/(H^{(1)}\cap B_k)} \chi_{\widehat{\theta}}(h)\Delta^{-1/2}_{H^{(1)},G_0}(h)P^k_{\widehat{\theta}}(t,h)d\nu_{B_k,H^{(1)}\cap B_k}(h)\end{aligned}$$

with

$$\begin{aligned}P^k_{\widehat{\theta}}(t,h) &= \oint_{B_k/(H^{(1)}\cap B_k)} (\Phi_t)^{\widehat{\theta}}_{H^{(1)}}(hb)\chi_{g_k\cdot\ell}(b)\Delta^{-1/2}_{B_k,G_0}(b)d\nu_{B_k,H^{(1)}\cap B_k}(b)\\ &= \oint_{B_k/(H^{(1)}\cap B_k)} \chi_{g_k\cdot\ell}(b)\Delta^{-1/2}_{B_k,G_0}(b)d\nu_{B_k,H^{(1)}\cap B_k}(b)\\ &\times \int_{H^{(1)}} \Phi_t(hbh')\chi_{\widehat{\theta}}(h')\Delta^{-1/2}_{H^{(1)},G_0}(h')dh'\\ &= \oint_{B_k/(H^{(1)}\cap B_k)} \chi_{g_k\cdot\ell}(b)\Delta^{-1/2}_{B_k,G_0}(b)d\nu_{B_k,H^{(1)}\cap B_k}(b)\\ &\times \int_{H^{(1)}} \chi_{\widehat{\theta}}(h')\Delta^{-1/2}_{H^{(1)},G_0}(h') \int_{\mathbb{R}} \varphi(\exp(tX)hbh'\exp(t'X))\\ &\times e^{it'f(X)}e^{-\frac{t'}{2}(\mathrm{Tr}\,\mathrm{ad}_{\mathfrak{g}} X + \mathrm{Tr}\,\mathrm{ad}_{\mathfrak{h}} X)}dt'.\end{aligned}$$

Taking the transitivity of $\nu_{\cdot,\cdot}$ into account, we deduce from (7.5.2) that

$$\begin{aligned}\varphi^f_H(e) = \langle\tau(\varphi)\delta_\tau,\delta_\tau\rangle = (2\pi)^{-2}\int_0^{2\pi} d\theta \int_{\widehat{G}} d\mu_\theta(\pi)\\ \times \sum_{k=1}^{m(\pi)} \oint_{H/(H\cap B_k)} \chi_{\widehat{\theta}}(h)\Delta^{-1/2}_{H,G}(h)d\nu_{H,H\cap B_k}(h)\\ \times \oint_{B_k/(H\cap B_k)} \chi_{g_k\cdot\ell}(b)\Delta^{-1/2}_{B_k,G}(b)d\nu_{B_k,H\cap B_k}(b)\int_{H_0} dh_0\end{aligned}$$

$$\times \int_{\mathbb{R}} \varphi(hbh_0 \exp(t'X))\chi_f(h_0 \exp(t'X))\Delta_{H,G}^{-1/2}(h_0 \exp(t'X))e^{-t' \mathrm{Tr}\, \mathrm{ad}_{\mathfrak{h}} X} dt'$$

$$= \int_{\widehat{G}} d\mu(\pi) \sum_{k=1}^{m(\pi)} \oint_{H/(H\cap B_k)} \chi_f(h)\Delta_{H,G}^{-1/2}(h) d\nu_{H,H\cap B_k}(h)$$

$$\times \oint_{B_k/(H\cap B_k)} \chi_{g_k\cdot \ell}(b)\Delta_{B_k,G}^{-1/2}(b) d\nu_{B_k,H\cap B_k}(b)$$

$$\times \int_H \varphi(hbh')\chi_f(h')\Delta_{H,G}^{-1/2}(h')dh' = \int_{\widehat{G}} \sum_{k=1}^{m(\pi)} \langle \pi(\varphi)a_\pi^k, a_\pi^k\rangle d\mu(\pi).$$

(II) Suppose that $\mathfrak{a}\cap\mathfrak{z} = \mathfrak{z} \neq \{0\}$. Then $\mathfrak{h}\cap\mathfrak{a} = \mathfrak{a}$ or $\mathfrak{h}\cap\mathfrak{a} = \mathfrak{z}$. Suppose first $\mathfrak{a} \subset \mathfrak{h}$. We introduce the proper subalgebra $\mathfrak{g}_1 = \mathfrak{a}^f$, orthogonal to $\mathfrak{a}$ with respect to B_f, which contains $\mathfrak{h}$. From the induction hypothesis applied to the subgroup $G_1 = \exp(\mathfrak{g}_1)$, it follows that

$$\tau_1 = \mathrm{ind}_H^{G_1} \chi_f \simeq \int_{\widehat{G_1}}^{\oplus} m_1(\rho)\rho d\nu_1(\rho).$$

Then

$$\tau \simeq \mathrm{ind}_{G_1}^G \tau_1 \simeq \int_{\widehat{G_1}}^{\oplus} m_1(\rho)(\mathrm{ind}_{G_1}^G \rho) d\nu_1(\rho). \tag{7.5.3}$$

Here $\rho|_A$ is a multiple of $\chi_f|_A$, so these $\mathrm{ind}_{G_1}^G \rho$ are all irreducible and inequivalent to each other. On the other hand, $A\cdot\ell = \ell + (\mathfrak{a}^\ell)^\perp$ for any $\ell \in \mathfrak{g}^*$. Now, formula (7.5.3) gives us the disintegration. Our formula (7.2.3) holds for τ_1 by the induction hypothesis applied to the test function $\varphi_1(g_1) = \varphi(g_1)\Delta_{G_1,G}^{-1/2}(g_1)$ $(g_1 \in G_1)$. The existence of the coefficients $\langle \pi(\varphi)a_\pi^k, a_\pi^k\rangle$ is assured by the induction hypothesis, exactly as in case 3.1.

We now consider the case where $\mathfrak{h}\cap\mathfrak{a} = \mathfrak{z}$. Suppose first that $\mathfrak{h}_0 = \mathfrak{h}\cap\mathfrak{g}_0 \subset \mathfrak{g}_1$. As $[\mathfrak{h}_0, \mathfrak{a}] = \{0\}$, the subalgebra $\mathfrak{h}'$ is subordinate to $f_\pm$ or $\widehat{\theta}$, which depends on dim $\mathfrak{a}$. The above arguments, followed by the last term rearrangement under the coadjoint action of G, lead us to the desired formula (7.2.3).

Let us examine the last case where $\mathfrak{h}_0 \not\subset \mathfrak{g}_1$. Let again $\mathfrak{h}_0 = \mathfrak{h}\cap\mathfrak{g}_1$, $H_0 = \exp(\mathfrak{h}_0)$, $\mathfrak{h}' = \mathfrak{h}_0 + \mathfrak{a}$ and $H^{(1)} = \exp(\mathfrak{h}')$. Evidently, $f([\mathfrak{h}', \mathfrak{h}']) = \{0\}$. We know that an intertwining operator R from $\tau' = \mathrm{ind}_{H^{(1)}}^G \chi_f$ to τ is obtained by the formula

$$(R\psi)(g) = \oint_{H/H_0} \psi(gh)\chi_f(h)\Delta_{H,G}^{-1/2}(h) d\nu_{H,H_0}(h) \ (g \in G).$$

Also here we use the intertwining operator to ensure the existence of the coefficients $\langle \pi(\varphi)a_\pi^k, a_\pi^k\rangle$. Putting $a' = \delta_\tau \circ R$ and reasoning at the level of the subgroup K, we see that

$$a' : \mathscr{H}_{\tau'}^{\infty} \ni \psi \mapsto \oint_{H/H_0} \overline{\psi(h)\chi_f(h)\Delta_{H,G}^{-1/2}(h)} d\nu_{H,H_0}(h)$$

gives us an element of the space $(\mathscr{H}_{\tau'}^{-\infty})^{H,\chi_f \Delta_{H,G}^{-1/2}}$. Lemma 2 in [71] says that up to a normalization

$$\varphi_H^f(e) = \langle \tau(\varphi)\delta_\tau, \delta_\tau\rangle = \langle \tau'(\varphi)a', a'\rangle$$

for all $\varphi \in C_c^\infty(G)$. Finally, formula (7.2.3) is proved as in [71]. This finishes the proof of Theorem 7.5.1.

Example 7.5.3 Let us recall the data of Example 7.3.9. We denote by $\mathscr{H}_\ell$ the Hilbert space of π_ℓ. Since $G = \exp(\mathbb{R}X)B[\ell]$, $\mathscr{H}_\ell$ is identified with $L^2(\mathbb{R})$ by the mapping $\Phi(t) = \phi(\exp(tT))(t \in \mathbb{R})$. If we define

$$\langle a_\ell, \Phi\rangle = \int_H \overline{\phi(h)\chi_f(h)\Delta_{H,G}^{-1/2}(h)} dh = \int_{\mathbb{R}} \overline{\phi(\exp(xX))} dx$$

for $\Phi \in L^2(\mathbb{R})$, the relation

$$\exp(xX) = \exp\left(\frac{w}{\lambda}(1-e^{-x})T\right)\exp\left(\frac{x}{\lambda}(\lambda X - wT)\right)$$

gives

$$\begin{aligned}\langle a_\ell, \Phi\rangle &= \int_{\mathbb{R}} \overline{\phi\left(\exp\left(\frac{w}{\lambda}(1-e^{-x})T\right)\exp\left(\frac{x}{\lambda}(\lambda X - wT)\right)\right)} dx \\ &= \int_{\mathbb{R}} \overline{\Phi\left(\frac{w}{\lambda}(1-e^{-x})\right)e^{-i\frac{x\zeta}{\lambda}} e^{-\frac{x}{2}}} dx \\ &= \int_0^\infty \overline{\Phi\left(\frac{w}{\lambda}(1-y)\right)e^{-i\frac{\zeta \log y}{\lambda}}} \frac{dy}{\sqrt{y}}.\end{aligned}$$

On the other hand, simple computations show

$$(\lambda X - wT)\cdot\Phi = \lambda t\frac{d\Phi}{dt} + \left(\frac{\lambda}{2} - i\zeta\right)\Phi$$

$$T\cdot\Phi = -\frac{d\Phi}{dt},\ A\cdot\Phi = i(w-\lambda t)\Phi,\ Z\cdot\Phi = i\lambda\Phi.$$

Hence the space of C^∞-vectors of π_ℓ is identified with the Schwartz space $\mathscr{S}(\mathbb{R})$ of fast decaying C^∞-functions. Since a_ℓ is a tempered distribution, we conclude that

$$a_\ell \in \left(\mathscr{H}_\ell^{-\infty}\right)^{H,\chi_f \Delta_{H,G}^{1/2}}.$$

Now, for $\varphi \in C_c^\infty(G)$, put

$$\varphi_H^f(g) = \int_H \varphi(gh)\chi_f(h)\Delta_{H,G}^{-1/2}(h)dh = \int_{\mathbb{R}} \varphi(\exp(xX))dx.$$

Here $G \cdot \ell \cap \mathfrak{h}^\perp$ has two connected components of $\ell_i (i = 1, 2)$, where

$$\ell_1 = \zeta T^* + A^* + \lambda Z^* \text{ et } \ell_2 = -\zeta T^* - A^* + \lambda Z^*.$$

Then we find the formula:

$$\langle \pi_{\ell_1}(\varphi)a_{\ell_1}, a_{\ell_1}\rangle = \int_{\mathbb{R}} dx \int_{B[\ell_1]} \varphi_H^f(\exp(xX)b)\chi_{\ell_1}(b)\Delta_{B[\ell_1],G}^{-1/2}(b)db$$

$$= \int_{\mathbb{R}} dx \int_{\mathbb{R}^3} \varphi_H^f(\exp(xX)\exp(t(\lambda X - T))\exp(aA)\exp(zZ))$$
$$\times\, e^{-it\zeta} e^{ia} e^{i\lambda z} e^{\frac{\lambda t}{2}} dtdadz$$
$$= \int_{\mathbb{R}} dx \int_{\mathbb{R}^3} \varphi_H^f\left(\exp\left(\frac{e^{-x}(e^{-t\lambda} - 1)}{\lambda}T\right)\exp\left(ae^x e^{t\lambda}A\right)\exp(zZ)\right)$$
$$\times\, e^{-it\zeta} e^{ia} e^{i\lambda z} e^{\frac{\lambda t}{2}} dtdadz$$
$$= \int_{\mathbb{R}} dx \int_{\mathbb{R}^3} \varphi_H^f\left(\exp\left(\frac{e^{-x}(e^{-t\lambda} - 1)}{\lambda}T\right)\exp(bA)\exp(zZ)\right)$$
$$\times\, e^{-it\zeta} e^{ibe^{-x}e^{-t\lambda}} e^{i\lambda z} e^{-x} e^{-\frac{\lambda t}{2}} dtdbdz.$$

In the same way,

$$\langle \pi_{\ell_2}(\varphi)a_{\ell_2}, a_{\ell_2}\rangle$$
$$= \int_{\mathbb{R}} dx \int_{\mathbb{R}^3} \varphi_H^f\left(\exp\left(\frac{e^{-x}(1 - e^{-t\lambda})}{\lambda}T\right)\exp(bA)\exp(zZ)\right)$$
$$\times\, e^{it\zeta} e^{-ibe^{-x}e^{-t\lambda}} e^{i\lambda z} e^{-x} e^{-\frac{\lambda t}{2}} dtdbdz.$$

Finally, since the measures are normalized appropriately, we arrive at the Plancherel formula for τ:

$$\int_{\mathbb{R}^2} (\langle \pi_{\ell_1}(\varphi)a_{\ell_1}, a_{\ell_1}\rangle + \langle \pi_{\ell_2}(\varphi)a_{\ell_2}, a_{\ell_2}\rangle) d\zeta d\lambda$$

$$= \int_{\mathbb{R}^4} \varphi_H^f(\exp(bA)\exp(zZ)) \left(e^{ibe^{-x}} + e^{-ibe^{-x}}\right) e^{i\lambda z} e^{-x} dbdxdzd\lambda$$

$$= \int_{\mathbb{R}^2} \varphi_H^f(\exp(bA)) \left(e^{ibe^{-x}} + e^{-ibe^{-x}}\right) e^{-x} dbdx$$

$$= \int_{\mathbb{R}^2} \varphi_H^f(\exp(bA)) e^{iby} dbdy = \varphi_H^f(e),$$

using the ordinary Fourier inversion formula.

7.6 Invariant Differential Operators

We retain the notations introduced thus far and suppose that our monomial representation $\tau = \mathrm{ind}_H^G \chi_f$ has multiplicities of discrete type. By means of the disintegration formula (7.2.2) of δ_τ, it is immediate to see, as in the case of finite multiplicities (cf. [105]), that the mapping

$$\tau(\varphi)\delta_\tau \mapsto \int_{\widehat{G}}^{\oplus} \left(\sum_{k=1}^{m(\pi)} \pi(\varphi) a_\pi^k \right) d\mu(\pi) \ (\varphi \in C_c^\infty(G))$$

gives us an intertwining operator between τ and its disintegration (7.2.1). We are interested in studying Duflo's problem. Let $W \in \mathscr{U}(\mathfrak{g}, \tau)$ act on $\tau(\varphi)\delta_\tau \in \mathscr{H}_\tau^{\infty}$ on the right, and transfer this action to the disintegration using the intertwining operator. This action is decomposable into factors. This means that, for μ-almost all $\pi \in \widehat{G}$, there exists a complex matrix $A(\pi) = (a_{ij})$ of order $m(\pi)$ verifying

$$R(W)(\pi(\varphi)a_\pi^k) = \sum_{j=1}^{m(\pi)} a_{kj}\pi(\varphi)a_\pi^j \tag{7.6.1}$$

for all $\varphi \in C_c(G)$, where $R(W)$ denotes the right action of W. Indeed, according to disintegration (7.2.1), the operator $U = R(W)$ splits into an integral $U = \int_{\widehat{G}}^{\oplus} U_\pi d\mu(\pi)$, which grants the existence of a continuous operator $U_\pi^{k,l}(k, l \in \mathbb{N})$ on the space $\mathscr{H}_\pi^{k,\infty}$. Actually it suffices to consider the composition of U_π with the partial l-projection of the space $\sum_{p=1}^{m(\pi)} \mathscr{H}_\pi^{p,\infty}$ on $\mathscr{H}_\pi^{l,\infty}$. Here the spaces $\mathscr{H}_\pi^{p,\infty}(1 \le p \le m(\pi))$ denote the spaces of C^∞ vectors corresponding to the isotypic components of the decomposition of τ. On the other hand, it is easy to see that the operators $U_\pi^{k,l}(k, l \in \mathbb{N})$ commute with the action of G, which allows us to confirm that these operators are scalars by corollary 3.5 in [140].

At this stage, the double cosets $Bg_iH = B[\ell]g_iH$ $(1 \leq i \leq m(\pi))$ have, μ-almost everywhere in Γ, the same dimension. In fact, it suffices to consider the H-invariant Zariski open set $\{\ell \in \Gamma; \dim(\mathfrak{h} + \mathfrak{b}[\ell])$ is maximal$\}$.

As each term of Eq. (7.6.1) is a C^∞-function, we can evaluate each term at the point $e \in G$. Let $j \neq k$. Take a test function φ that vanishes on the union of the double cosets $\coprod_{i \neq j} Bg_i^{-1}H$, together with its derivatives up to a certain determined order. Then $a_{ij} = 0$ except when $i = j$. Indeed, as each double coset in the union is locally closed, $Bg_j^{-1}H$ can intersect the closure $\overline{Bg_i^{-1}H}$ only on its frontier for $i \neq j$. Hence

$$\pi(\overline{W})a_\pi^k = \overline{P_W(g_k \cdot \ell)}a_\pi^k$$

for an H-invariant function P_W defined almost everywhere on Γ. The mapping

$$\mathscr{U}(\mathfrak{g}, \tau) \ni W \mapsto P_W(\ell)$$

induces a homomorphism from the algebra $D_\tau(G/H)$ into the algebra of H-invariant functions on Γ. Besides, the Plancherel formula for τ implies that this homomorphism is injective. Summarizing, we have one implication of Duflo's problem.

Theorem 7.6.1 *If $\tau = ind_H^G \chi_f$ has multiplicities of discrete type, the algebra $D_\tau(G/H)$ is commutative.*

Corollary 7.6.2 *Suppose that $\tau = ind_H^G \chi_f$ has multiplicities of discrete type. Let $\mathfrak{g}_0$ be an ideal of codimension 1 in $\mathfrak{g}$ such that the G-orbits are μ-almost everywhere saturated in the direction $\mathfrak{g}_0^\perp$. Then the algebra $\mathscr{U}(\mathfrak{g}, \tau)$ is contained in $\mathscr{U}(\mathfrak{g}_0)$ modulo the kernel $\mathscr{U}(\mathfrak{g})\mathfrak{a}_\tau$.*

Proof Let $\mathfrak{g} = \mathbb{R}X + \mathfrak{g}_0$. If $\mathscr{U}(\mathfrak{g}, \tau)$ is not contained in $\mathscr{U}(\mathfrak{g}_0)$, there exists an element W having the form $W = Xa + b$ with $a, b \in \mathscr{U}(\mathfrak{g}_0)$ and a belonging to $\mathscr{U}(\mathfrak{g}, \tau) \setminus \mathscr{U}(\mathfrak{g})\mathfrak{a}_\tau$. In fact, if $W = X^m a_m + X^{m-1}a_{m-1} + \cdots + a_0 \in \mathscr{U}(\mathfrak{g}, \tau)$ with $a_i \in \mathscr{U}(\mathfrak{g}_0)(0 \leq i \leq m)$ and $a_m \notin \mathscr{U}(\mathfrak{g})\mathfrak{a}_\tau$, then it is clear that $mXa_m + a_{m-1}$ has the same property. Write $W = Xa + b$.

Generally, we take ℓ in the Zariski open set of Γ given by Theorem 7.3.11 and let $\pi = \theta_G(\ell) \in \widehat{G}$. We consider the associated Penney distribution a_ℓ. We realize π as $\pi = \mathrm{ind}_B^G \chi_\ell$, $B = \exp \mathfrak{b}$, by means of a Vergne polarization at ℓ, and we calculate the differential operator W,

$$\langle \pi(\overline{W})a_\ell, \psi \rangle = \oint_{H/(H\cap B)} \left(\sum_J \lambda_J(h) \frac{\partial^J \bar{\psi}}{\partial x^J}(h) \right) \Delta_{H,G}^{-1/2}(h) \overline{\chi_f(h)} d\nu(h)$$

for some multi-indices J and functions $\lambda_J(h)$. Since

$$\sum_J \lambda_J(h)\frac{\partial^J}{\partial x^J} = c\frac{\partial}{\partial x_1} + A,$$

where c is a non-zero complex number, x_1 is the first coordinate and A is a partial differential operator in $(x_2, \ldots, x_p)$, $p = \dim(G/B)$, it is evident that $\pi(\overline{W})a_\ell$ is not a multiple of a_ℓ, which is absurd. Remark here that $\pi(\overline{W})$ is a differential operator on the representation space, which we express by means of the coordinates $(x_1, \ldots, x_p)$. □

Example 7.6.3 Let us come back to Example 7.3.9 with all the notations used there. Let $W \in \mathscr{U}(\mathfrak{g})$. As $X \in \mathfrak{a}_\tau$, W takes the form

$$W = \sum_m \left(\sum_{p,q} c^m_{p,q} T^p A^q\right) Z^m \ (c^m_{p,q} \in \mathbb{C}).$$

Then W belongs to $\mathscr{U}(\mathfrak{g}, \tau)$ if and only if $c^m_{p,q} = 0$ if $p \neq q$. Namely, W is a polynomial in Z and TA. Let $\ell \in \Gamma$ such that a_ℓ is well defined. Let us calculate $(TA)\cdot a_\ell$. For $\phi \in \mathscr{H}^\infty_{\pi_\ell}$,

$$\begin{aligned}
\langle (TA)\cdot a_\ell, \phi\rangle = \langle a_\ell, (AT)\cdot\phi\rangle &= \int_{\mathbb{R}} \overline{(AT)\cdot\phi(\exp(xX))}dx \\
&= -\int_{\mathbb{R}} \overline{\frac{d}{dt}(T\cdot\phi)(\exp(tA)\exp(xX))\Big|_{t=0}} dx \\
&= -\int_{\mathbb{R}} \overline{\frac{d}{dt}(T\cdot\phi)(\exp(xX)\exp(te^{-x}A))\Big|_{t=0}} dx \\
&= -\int_{\mathbb{R}} \frac{d}{dt}\left(\overline{(T\cdot\phi)(\exp(xX))}e^{ite^{-x}\ell(A)}\right)\Big|_{t=0} dx \\
&= -i\ell(A)\int_{\mathbb{R}} e^{-x}\overline{(T\cdot\phi)(\exp(xX))}dx \\
&= -i\ell(A)\int_{\mathbb{R}} e^{-x}\overline{\frac{d}{dt}\phi(\exp(-tT)\exp(xX))\Big|_{t=0}} dx.
\end{aligned}$$

Supposing $\ell(A) \neq 0$, we put $\alpha = \frac{\ell(Z)}{\ell(A)}$. Then $\alpha X - T \in \mathfrak{b}[\ell]$. Here, a simple computation shows

$$\exp(-tT)\exp(xX) = \exp(yX)\exp s(\alpha X - T)$$

with $e^{\alpha s} = 1 + \alpha t e^x,\ y + \alpha s = x$ for $t \in \mathbb{R}$ in a neighbourhood of 0. Therefore, $\alpha s = \log(1 + \alpha t e^x)$. So we have

$$\langle (TA)\cdot a_\ell, \phi\rangle = -i\ell(A)\int_{\mathbb{R}} e^{-x}\, \overline{\frac{d}{dt}\phi(\exp(yX)\exp s(\alpha X - T))\Big|_{t=0}}\, dx$$

$$= -i\ell(A)\int_{\mathbb{R}} e^{-x}\,\frac{d}{dt}\left(\overline{\phi(\exp(yX))}e^{-\frac{\alpha s}{2}}e^{-is\ell(T)}\right)\Big|_{t=0} dx$$

$$= -i\ell(A)\int_{\mathbb{R}} e^{-x}\,\frac{d}{dt}\left(\overline{\phi(\exp(yX))}(1+\alpha t e^x)^{-1/2-i\frac{\ell(T)}{\alpha}}\right)\Big|_{t=0} dx$$

$$= -i\ell(A)\int_{\mathbb{R}} e^{-x}\left\{\overline{\phi(\exp(xX))}\left(-1/2 - i\frac{\ell(T)}{\alpha}\right)\alpha e^x\right.$$

$$\left. + \alpha e^x\,\overline{\frac{d\phi}{dx}(\exp(xX))}\right\} dx$$

$$= \left(\frac{i}{2}\ell(Z) - \ell(T)\ell(A)\right)\int_{\mathbb{R}}\overline{\phi(\exp(xX))}dx - i\ell(Z)\int_{\mathbb{R}}\overline{\frac{d\phi}{dx}(\exp(xX))dx}.$$

Since

$$\lim_{x\to\pm\infty}\phi(\exp(xX)) = \lim_{x\to\pm\infty}\phi\left(\exp\left(\left(\frac{1-e^{-x}}{\alpha}\right)T\right)\right)e^{-\frac{x}{2}}e^{i\frac{x}{\alpha}\ell(T)} = 0,$$

we finally get

$$\langle (TA)\cdot a_\ell, \phi\rangle = \left(\frac{i}{2}\ell(Z) - \ell(T)\ell(A)\right)\langle a_\ell, \phi\rangle.$$

Namely,

$$(TA)\cdot a_\ell = \left(\frac{i}{2}\ell(Z) - \ell(T)\ell(A)\right)a_\ell.$$

Now, it follows that our mapping

$$\mathscr{U}(\mathfrak{g}, \tau) \ni W \mapsto P_W(\ell)$$

induces an isomorphism from the algebra $D_\tau(G/H)$ onto the algebra $\mathbb{C}[\Gamma]^H$ of H-invariant polynomial functions on Γ.

Example 7.6.4 Take now $\mathfrak{h} = \mathbb{R}T,\ H = \exp\mathfrak{h}$ and $f \in \mathfrak{g}^*$ such that $f|_{\mathfrak{h}} = 0$ in the same example. Put $\ell = \alpha X^* + \beta Y^* + \lambda Z^* \in \mathfrak{h}^\perp$ $(\alpha, \beta, \lambda \in \mathbb{R})$ and compute the orbit $G{\cdot}\ell$. Writing $g = \exp(xX)\exp(tT)\exp(aA) \in G$ with $x, t, a \in \mathbb{R}$, we have

$$g^{-1}{\cdot}\ell(X) = \ell(g{\cdot}X) = \alpha - ae^x\beta - at\lambda,$$
$$g^{-1}{\cdot}\ell(T) = \ell(g{\cdot}T) = -a\lambda,$$
$$g^{-1}{\cdot}\ell(A) = \ell(g{\cdot}A) = \beta e^x + t\lambda.$$

Hence

$$G{\cdot}\ell = x^*X^* + t^*T^* + a^*A^* + \lambda Z^*,\ x^* - \alpha = \frac{t^*a^*}{\lambda}$$

for $\lambda \neq 0$. So,

$$G{\cdot}\ell \cap \mathfrak{h}^\perp = \alpha X^* + \mathbb{R}A^* + \lambda Z^* = H{\cdot}\ell.$$

It follows that $\tau = \mathrm{ind}_H^G \chi_f$ is without multiplicity. On the other hand the algebra $\mathscr{U}(\mathfrak{g}, \tau)$ is, modulo $\mathscr{U}(\mathfrak{g})\mathfrak{a}_\tau$, the polynomial algebra in two variables X and Z. For $\phi \in \mathscr{H}_{\pi_\ell}^\infty$, put

$$\Phi(t) = \phi(\exp(tT))\ (t \in \mathbb{R})$$

and consider the Penney distribution

$$a_\ell(\phi) = \int_{\mathbb{R}} \overline{\Phi(t)}dt.$$

Then

$$(X{\cdot}a_\ell)(\phi) = -\int_{\mathbb{R}} \overline{X{\cdot}\phi(\exp(tT))}dt = \int_{\mathbb{R}} \overline{\frac{d}{dt}\phi(\exp(xX)\exp(tT))\Big|_{x=0}}dt$$
$$= \int_{\mathbb{R}} \overline{\frac{d}{dt}\phi\left(\exp\left(te^{-x} + \frac{\beta}{\lambda}(1-e^{-x})\right)T\exp\left(x\left(X - \frac{\beta}{\lambda}T\right)\right)\right)\Big|_{x=0}}dt$$
$$= \int_{\mathbb{R}} \overline{\frac{d}{dt}\phi\left(\exp\left(te^{-x} + \frac{\beta}{\lambda}(1-e^{-x})\right)T\right)e^{-\frac{x}{2}}e^{i\alpha x}\Big|_{x=0}}dt$$
$$= \int_{\mathbb{R}} \left((-1/2 + i\alpha)\,\overline{\Phi(t)} - \left(t - \frac{\beta}{\lambda}\right)\overline{\frac{d\Phi}{dt}(t)}\right)dt$$
$$= (-1/2 + i\alpha)\int_{\mathbb{R}} \overline{\Phi(t)}dt - \left\{\left[\left(t - \frac{\beta}{\lambda}\right)\overline{\Phi(t)}\right]_{-\infty}^{\infty} - \int_{\mathbb{R}} \overline{\Phi(t)}dt\right\}$$
$$= (1/2 + i\alpha)\int_{\mathbb{R}} \overline{\Phi(t)}dt = (1/2 + i\alpha)\,a_\ell(\phi).$$

Consequently,

$$X \cdot a_\ell = (1/2 + i\alpha)\, a_\ell.$$

In the end, $D_\tau(G/H)$ is once again isomorphic to the algebra $\mathbb{C}[\Gamma]^H$ of H-invariant polynomial functions on Γ.

Example 7.6.5 Let G be an exponential solvable Lie group. Consider the case where H is normal in G. Then it is well known that τ is either multiplicity-free or it has uniformly infinite multiplicities almost everywhere on Γ. Then it is shown in [6] that the algebras $D_\tau(G/H)$ and $\mathbb{C}[\Gamma]^H$ are isomorphic when τ is multiplicity-free.

Far from the nilpotent context, the following example shows that the Duflo problem can have a negative solution in the exponential framework.

Example 7.6.6 When $G = \exp \mathfrak{g}$ is no longer nilpotent, the algebra $D_\tau(G/H)$ can be very small. Let

$$\mathfrak{g} = \mathbb{R} - \operatorname{span}(T, X, Y, Z) : [T, X] = X, [T, Y] = Y, [T, Z] = 2Z, [X, Y] = Z.$$

Therefore,

$$G = \left\{ \begin{pmatrix} a^2 & x & z \\ 0 & a & y \\ 0 & 0 & 1 \end{pmatrix} ; a > 0,\ x, y, z \in \mathbb{R} \right\}$$

is completely solvable. Take $f = 0 \in \mathfrak{g}^*$ and $\mathfrak{h} = \mathbb{R}T$. Then it follows immediately that $\mathfrak{g}(\ell) = \{0\}$ for $\ell \in \mathfrak{g}^*$ verifying $\ell(Z) \neq 0$. Consequently $\tau = \operatorname{ind}_H^G 1$ has infinite multiplicities of continuous type. However, $D_\tau(G/H)$ is trivial. Indeed, each monomial $X^p Y^q Z^r$ $(p, q, r \in \mathbb{N})$ is an eigenvector for ad T with eigenvalue $p + q + 2r$, which immediately proves our claim. Hence Theorem 7.6.1 is the best possible generalization to exponential Lie groups of the Corwin-Greenleaf result for nilpotent Lie groups.

7.7 Polarizations

As an example we treat a particular case of monomial representations of discrete type, where $\mathfrak{h}$ is a polarization at $f \in \mathfrak{g}^*$ which does not satisfy the Pukanszky condition. Recall the results of [151]. We designate by Γ/G the set of coadjoint orbits of G whose intersection with the affine subspace Γ are non-empty open sets in Γ. It turns out that Γ/G is a finite set. For $\Omega \in \Gamma/G$, consider the connected components of $\Gamma \cap \Omega$: each is an H-orbit and their number $m(\Omega)$ is finite. The result of [151] says that $\tau = \mathrm{ind}_H^G \chi_f$ splits as a direct sum

$$\tau \simeq \sum_{\Omega \in \Gamma/G}^{\oplus} m(\Omega)\pi(\Omega),$$

where we wrote $\pi(\Omega)$ in place of $\bar{\theta}(\Omega) \in \widehat{G}$.

For an intertwining operator $R : \mathscr{H}_{\pi(\Omega)} \to \mathscr{H}_\tau$, we define $a_R \in \mathscr{H}_{\pi(\Omega)}^{-\infty}$ by

$$a_R(\psi) = \langle \delta_\tau, R(\psi) \rangle = \overline{R(\psi)(e)} \; (\psi \in \mathscr{H}_{\pi(\Omega)}^{\infty}).$$

A simple calculation shows that

$$a_R \in \left(\mathscr{H}_{\pi(\Omega)}^{-\infty}\right)^{H, \chi_f \Delta_{H,G}^{1/2}}$$

and the mapping $R \mapsto a_R$ is anti-linear and injective. Hence

$$\dim \left(\mathscr{H}_{\pi(\Omega)}^{-\infty}\right)^{H, \chi_f \Delta_{H,G}^{1/2}} \geq m(\Omega)$$

(for more general result, see [136]).

Now let $a \in \left(\mathscr{H}_{\pi(\Omega)}^{-\infty}\right)^{H, \chi_f \Delta_{H,G}^{1/2}}$. For any $\varphi \in C_c^\infty(G)$, a direct calculation shows

$$\pi(\Omega)(\varphi)a = \pi(\Omega)(\varphi_H^f)a.$$

Therefore the mapping

$$\tau(\varphi)(\delta_\tau) = \varphi_H^f \mapsto \pi(\Omega)(\varphi_H^f)a = \pi(\Omega)(\varphi)a$$

gives us an intertwining operator from the space

$$\mathscr{H}_\tau^\infty = \sum_{\Omega \in \Gamma/G}^{\oplus} m(\Omega)\mathscr{H}_{\pi(\Omega)}^\infty$$

into the space $\mathscr{H}_{\pi(\Omega)}^\infty$.

Let

$$\delta_\tau \simeq \sum_{\Omega \in \Gamma/G}^{\oplus} \left(\sum_{k=1}^{m(\pi)} a_{\pi(\Omega)}^k \right)$$

be the concrete Plancherel formula described by the Penney distributions a_π^k (cf. [71]). Denote by $\mathcal{I}(\mathcal{H}_\tau^\infty, \mathcal{H}_{\pi(\Omega)}^\infty)$ the space of intertwining operators. For $\Omega \in \Gamma/G$, let us consider the projections

$$p_\Omega^j : \mathcal{H}_\tau^\infty \to \mathcal{H}_{\pi(\Omega)}^\infty \ (1 \leq j \leq m(\Omega))$$

defined by

$$p_\Omega^j(\tau(\varphi)\delta_\tau) = \pi(\Omega)(\varphi)a_{\pi(\Omega)}^j \ (\varphi \in C_c^\infty(G)).$$

Then we know [140] that

$$\mathcal{I}(\mathcal{H}_\tau^\infty, \mathcal{H}_{\pi(\Omega)}^\infty) = \sum_{j=1}^{m(\Omega)} \mathbb{C}p_\Omega^j.$$

In this way, we arrive to a variant of Frobenius reciprocity, which was studied in [71] under an additional condition.

Theorem 7.7.1 *Assume that $\mathfrak{h}$ is a polarization at $f \in \mathfrak{g}^*$. Then for $\Omega \in \Gamma/G$, we obtain*

$$m(\Omega) = \dim \left(\mathcal{H}_{\pi(\Omega)}^{-\infty}\right)^{H,\chi_f \Delta_{H,G}^{1/2}}.$$

On the other hand, for $\Omega \notin \Gamma/G$,

$$\left(\mathcal{H}_{\pi(\Omega)}^{-\infty}\right)^{H,\chi_f \Delta_{H,G}^{1/2}} = \{0\}.$$

Remark 7.7.2 When $\mathfrak{h}$ is a polarization at $f \in \mathfrak{g}^*$, it is easy to check (cf. [71]) that the algebra $D_\tau(G/H)$ is trivial, namely $D_\tau(G/H) = \mathbb{C}1$.

Chapter 8
Bounded Irreducible Representations

8.1 Introduction

Let $G = \exp(\mathfrak{g})$ be an exponential solvable Lie group. In this chapter we characterize bounded, topologically irreducible Banach-space representations of G using triples $(\Omega, \tau, \| \|)$, where $\Omega \subset \mathfrak{g}^*$ is a coadjoint orbit of G, τ is a topologically irreducible representation of the algebra $L^1(\mathbb{R}^n, \omega)$ for a certain $n \in \mathbb{N}^*$ and a weight ω on $\mathbb{R}^n$, and $\| \|$ is a so-called extension norm. In particular we determine the simple modules of the group algebra $L^1(G)$. If the group G is not symmetric, then a new type of bounded irreducible Banach-space representations of G crop up. These are related to the invariant-subspace problem and are fundamentally different from monomial representations.

In the first two Sects. 8.2 and 8.3 we provide some background information on simple, and more generally irreducible, representations of Banach algebras A with two-sided, bounded approximate units. We look at the particular situation where the irreducible module (T, X) of the Banach algebra A admits a multiplier p such that $T(p) \neq 0$. If the module T is simple, it is completely determined by the simple module (S, Y) of the subalgebra $A_p = p \cdot A \cdot p$ of A, where S is the restriction of $T_{|A_p}$ to $Y := T(p)X$. We choose an element $y \neq 0$ in Y and define the left ideal A_y of A by

$$A_y := \{a \in A | S(p \cdot (ba))y = 0, \forall b \in A\}.$$

Then A_y is a maximal modular left ideal of A, and the left module A/A_y of A is equivalent to the module (X, T).

One way to obtain simple modules is to use the finite-rank submodule of an irreducible Banach module:

A. Baklouti et al., *Representation Theory of Solvable Lie Groups and Related Topics*, Springer Monographs in Mathematics,
https://doi.org/10.1007/978-3-030-82044-2_8

Let A be a Banach algebra and let (T, X) be a Banach module. We denote by I_T^{fin} the ideal of A defined by

$$I_T^{fin} := \{a \in A;\ \text{the operator } T(a) \in B(X) \text{ has finite rank}\}.$$

Let also X^{fin} be the (A)-invariant subspace of X given by

$$X^{fin} := \operatorname{span}\{(T(a)(X));\ a \in I_T^{fin}\}.$$

The following method produces simple modules.

Let (T, X) be an irreducible Banach module of A. Suppose that the finite-rank subspace X^{fin} is different from $\{0\}$. Then X^{fin} is the unique simple submodule of (T, X). We show in Sect. 8.15 that every simple module of the algebra $L^1(G)$ is, for exponential groups, the finite-rank submodule of $L^{\overline{p}}(G/P, \ell)$, for a Pukansky polarization P at some $\ell \in \mathfrak{g}^*$ and some $\overline{p} \in \mathbb{R}^{d_\ell}_{\geq 1}$, where $d_\ell = \dim \mathfrak{g}(\ell)/\mathfrak{g}(\ell \cap [\mathfrak{g}, \mathfrak{g}])$ with $\sum_{i=1}^{d_\ell} \frac{1}{p_i} = 1$ (see Sect. 8.11.3 and Theorem 8.15.4 for details).

If (T, X) is an irreducible Banach module, we can proceed similarly in the case where $T(p)X \not\equiv 0$ and $(T|_{pAp}, T(p)X)$ is an irreducible (pAp)-module.

Every irreducible A-module (T, X) with $T(p)(X) \not\equiv 0$ can then be identified with an A-module of the form $\overline{Ap/M}^{\|\!|\cdot|\!\|_X}$, where

$$M = M_w = \big\{x \in Ap;\quad T(pAx)w = \{0\}\big\} \quad (0 \neq w \in T(p)X \text{ fixed})$$

is a left ideal A contained in Ap and where the norm $|\!|\!|\cdot|\!|\!|_X$ satisfies the condition

$$|\!|\!|\cdot|\!|\!|_X \leq C_T \|\cdot\|_{Ap/M}$$

with

$$\|\dot{a}\|_{Ap/M} = \inf_{m \in M} \|a + m\|_A \leq \|a\|_A \quad \forall a \in Ap.$$

In order to analyze now the irreducible Banach $L^1(G)$-modules (T, X), we have to develop an analogue of Mackey's little group method. Let $N = \exp \mathfrak{n}$ be a nilpotent subgroup of G containing $[G, G]$. An essential step in the analysis is to determine the kernel $\ker_{L^1(N)}(T)$ in $L^1(N)$ of the representation $T_{|N}$. It is easy to see that $\ker_{L^1(N)}(T)$ is a closed G-prime ideal in $L^1(N)$. Using functional calculus (8.5) and some abstract arguments, we show in Sect. 8.9 that the kernel of T in the Schwartz algebra $S(N)$ of N is the kernel of a G-orbit in $\hat{N}$. This means that there exists an element $\ell \in \mathfrak{g}^*$ such that

$$\ker_{S(N)}(T) = \bigcap_{g \in G} \ker_{S(N)}(\pi_{g \cdot \ell_{|\mathfrak{n}}}).$$

Here π_q denotes the irreducible unitary representation of N associated to the coadjoint orbit of $q \in \mathfrak{n}^*$.

The main difficulty is then to prove that $\ker_{S(N)}(T)$ is dense in $\ker_{L^1(N)}(T)$. For that we need Sects. 8.4.1, 8.7, and 8.8 and Theorem 8.10.9. The proof of this statement will be finalized in Sect. 8.10.3.

Let now $\sigma \subset \mathfrak{g}(\ell)$ be a subspace of $\mathfrak{g}(\ell)$ such that $\sigma \oplus \mathfrak{g}(\ell) \cap \mathfrak{n} = \mathfrak{g}(\ell)$. We assume now that $\sigma \neq \{0\}$, the case $\sigma = \{0\}$ being easier to handle. Let $\mathfrak{n} \subset \mathfrak{h} \subset \mathfrak{g}$ be an ideal of $\mathfrak{g}$ such that $\mathfrak{g} = \sigma \oplus \mathfrak{h}$. Let $H = \exp \mathfrak{h}$. Then H is a closed, connected, invariant subgroup of G containing $[G, G]$. We let $\gamma = \gamma_\ell$ be the irreducible unitary representation of H attached to the coadjoint orbit of $\ell_{|\mathfrak{h}}$. It is shown in Theorem 8.11.6 that $\ker_{L^1(H)}(\gamma) = \ker_{L^1(H)}(T_{|H})$. This then implies that $\ker_{L^1(G)}(T)$ contains the kernel in $L^1(G)$ of the unitary representation $\pi^\gamma = \mathrm{ind}_H^G \gamma$. It suffices then to study the irreducible Banach space representations of the algebra $A = L^1(G)/\ker_{L^1(G)}(\pi^\gamma)$.

We take now a smooth element $\lambda \in \mathcal{H}_\gamma$ such that there exists an $f_\lambda = f_\lambda^* \in L^1(H)$, so that $\gamma(f_\lambda)$ is the one-dimensional projection onto the one-dimensional subspace $\mathbb{C}\lambda$ of the Hilbert space $\mathcal{H}_\gamma$ of γ. Then f_λ defines by convolution an idempotent multiplier $p = p_\lambda$ of the algebra A. The retract theorem (Sect. 8.10.2) tells us that there exists for every $s \in G$ an element $v(s) \in L^1(H)$ such that $\gamma(v(s))$ is the rank-one operator $P_{\pi_\ell(s)\lambda,\lambda}$ and the mapping $s \to v(s)$ is continuous. We define the function ω on $G(\ell)/G(\ell) \cap N$ by $\omega(s) := \|v(s)/\ker_{L^1(H)}(\gamma)\|$. It turns out that ω is then a continuous weight on $G(\ell)/G(\ell) \cap \mathbb{N}$ (see Proposition 8.12.2) and that the algebra $p \cdot A \cdot p$ is isometrically isomorphic to the weighted convolution algebra $L^1(\mathbb{R}^n, \omega)$, where we identify the group $S := G(\ell)/G(\ell \cap N)$ with $S \simeq \mathbb{R}^n$, the number n being the dimension of S (see Theorem 8.12.3). We then estimate the growth of the weight ω, which depends on the roots of S acting on $\mathfrak{n}$ (see Proposition 8.13.3). This growth determines the characters of the algebra $L^1(S, \omega)$ and so also the simple modules of A. There are always uncountably many non-equivalent irreducible bounded representations of G attached to an $\ell \in \mathfrak{g}^*$ whenever $\langle \ell, [\mathfrak{g}, \mathfrak{g}] \rangle \neq \{0\}$.

8.2 Simple Modules of Banach Algebras

8.2.1 Elementary Definitions

1. Let $(A, +, \cdot, \|\cdot\|_A)$ be a complex Banach algebra. We always assume that A has no global zero divisors. This means that

$$\text{if } a \cdot b = 0 \ \text{ or if } b \cdot a = 0 \text{ for all } a \in A$$
$$\Rightarrow b = 0.$$

Let X be a complex vector space. We denote by $L(X)$ the algebra of linear endomorphisms of X.

We say that X is an *A-module* or a *left* A-module, if there exists a non-trivial algebra homomorphism $T : A \to L(X)$.

We say that X is a *right A-module*, if there exists a non-trivial linear mapping $T : A \to L(X)$, such that

$$T(a \cdot b) = T(b) \circ T(a), a, b \in A.$$

In both cases we write

$$a \cdot x := T(a)x, a \in A, x \in X.$$

Example 8.2.1 Let $\varphi \in A^*$. We put for $a \in A$

$$\langle \varphi \cdot a, b \rangle := \langle \varphi, ba \rangle, b \in A$$

and

$$\langle a \cdot \varphi, b \rangle := \langle \varphi, ab \rangle, b \in A.$$

Then

$$T(a)\varphi \ := \ \varphi \cdot a, \ a \in A,$$

resp.

$$T(a)\varphi \ := \ \varphi \cdot a, \ a \in A,$$

defines a right-module structure (resp. a left-module structure) on the algebraic dual A^* of A.

Remark 8.2.2 Let A be an involutive algebra and let (X, T) be a right A-module. Consider the conjugate $\overline{X}$: as Abelian groups $\overline{X} = X$, but now the multiplication by scalars $\mathbb{C} \times \overline{X} \to \overline{X}$ reads

$$\lambda \cdot x := \overline{\lambda} x, x \in X, \lambda \in \mathbb{C}.$$

Then $\overline{X}$ admits a left A-module structure $(\overline{T}, \overline{X})$ by putting

$$\overline{T}(a) := T(a^*), a \in A.$$

In this way, every right module (X^r, T^r) of A can be obtained from its left-module version $(X^l, T^l) := (\overline{X}, \overline{T})$.

2. If X is equipped with a norm $\| \|_X$, we say that the A-module (T, X) is *bounded* if there exists a constant $C > 0$ such that

$$\|a \cdot x\|_X \leq C\|a\|_A\|x\|_X, a \in A, x \in X.$$

   Let $B(X)$ be the subalgebra of $L(X)$ consisting of all bounded linear operators on the normed vector space $(X, \| \|_X)$. We see that the A-module (T, X) is bounded if and only if the homomorphism T maps A to $B(X)$ and T itself is bounded as a linear operator from $(A, \| \|_A)$ to $(B(X), \| \|_{\text{op}})$.
3. We say that the A-module (T, X) is a *Banach* module if X is a Banach space (with norm $\| \|_X$) and the homomorphism $T : A \to B(X)$ is bounded.
4. Let A be an algebra and I a left ideal in A. The algebra A acts on the vector space A/I by left multiplication:

$$l(a)(b + I) := a \cdot b + I, a, b \in A.$$

   We obtain in this way an A-module structure on $X = A/I$. If A is a Banach algebra and the left ideal I is closed in A, then $X = A/I$ is a Banach module for the left multiplication l.
5. We say that two A-modules (resp. two Banach modules) (T, X), (T', X') are *equivalent*, if there exists a linear bijection (resp. a bounded linear bijection) $L : X \to X'$ such that

$$L(a \cdot x) = a \cdot L(x), x \in X, a \in A.$$

6. We say that the A-module (T, X) is *simple* if $X \neq \{0\}$ and the only A-invariant subspaces of X are the two trivial ones, namely X and $\{0\}$.
7. We say that the Banach A-module (T, X) is *irreducible*, or *topologically simple*, if the only A-invariant *closed* subspaces of X are the trivial ones X and $\{0\}$.

   Definition 8.2.3 Let $Simple(A)$ be the family of all equivalence classes of simple A-modules.

   Let also $\widehat{A}^{\text{top}}$ denote the family of equivalence classes of irreducible Banach-space modules.

   We say that an ideal I of A is *primitive* if there exists a simple module (T, X) of A such that

$$I = \ker(T).$$

8. Let (T, X) be an A-module. Let $0 \neq x \in X$. The annihilator A_x of x in A is the left ideal

$$A_x := \{a \in A; a \cdot x = 0\}.$$

9. Two simple A-modules (T, X) and (T', X') are equivalent if and only if there exist $x \in X, x' \in X'$ such that the annihilator ideals A_x and $A_{x'}$ coincide. Indeed, if $A_x = A_{x'}$ then the modules A/A_x and $A_{x'}$ are the same, (T, X) is equivalent to A/A_x and (T', X') is equivalent to $A/A_{x'}$. If the modules are equivalent, take an intertwining operator $\Psi : X \to X'$. Fix $x \in X$ and let $x' := \Psi(x)$. Then

$$\begin{aligned} A_{x'} &= \{a \in A;\, a \cdot x' = 0\} \\ &= \{a \in A;\, a \cdot \Psi(x) = 0\} \\ &= \{a \in A;\, \Psi(a \cdot x) = 0\} \\ &= \{a \in A;\, a \cdot x = 0\} \\ &= A_x. \end{aligned}$$

10. A left ideal I of the algebra A is called *modular*, if there exists $u \in A$ such that

$$a \cdot u - a \in I, a \in A.$$

The element $u \in A$ is called a *(right) modular unit* (for I). We remark that if u is a modular unit of the left ideal I, then for every $i \in I$ the element $u + i$ is also a modular unit for I.

Proposition 8.2.4

1. *Let (T, X) be an A-module. Let $x \neq 0$ be a cyclic vector of X, which means that the subspace $A \cdot x$ is equal to X itself. Then the A-modules (T, X) and A/A_x are equivalent.*
2. *Let (T, X) be an A-module. Then T is simple if and only if every $x \neq 0$ in X is a cyclic vector.*
3. *If (T, X) is a simple A-module, then for every $x \neq 0$ in X the annihilator A_x of x is a maximal modular left ideal of A.*
4. *If I is a maximal modular left ideal of A, then the left A-module A/I is simple.*

Proof

1. Let $L_x : A/A_x \to X$ be the mapping defined by

$$L_x(a + A_x) = a \cdot x.$$

Then L_x is obviously linear, injective, and for every a, b in A, we have that

$$\begin{aligned} L_x((a \cdot (b + A_x)) &= L_x((a \cdot b) + A_x) \\ &= (a \cdot b) \cdot x \end{aligned}$$

$$= a \cdot (b \cdot x)$$
$$= a \cdot L_x(b + A_x).$$

Hence the linear mapping L_x intertwines the modules (T, X) and A/A_x. Since x is a cyclic vector of X, the mapping L_x is also surjective.

2. Suppose that the module (T, X) is simple. Then for $x \neq 0$ in X, the subspace $A \cdot x$ is A-invariant. If $A \cdot x = \{0\}$, the A-invariant subspace $Z_X := \{y \in X; A \cdot y = \{0\}\}$ of X contains x and so $Z_X = X$, since (T, X) is simple. But then T is trivial, which is excluded by definition. Hence $A \cdot x \neq \{0\}$ and then $A \cdot x = X$ by the simplicity of T.

   If every non-trivial $x \in X$ is cyclic, then every non-trivial A-invariant subspace of X is equal to X, which means that (T, X) is simple.
3. If (T, X) is simple, then for every $x \neq 0$ in X we have that $A \cdot x = X$. Hence there exists $u \in A$ such that $u \cdot x = x$. But then for every $a \in A$ we have that

$$\begin{aligned}(a - a \cdot u) \cdot x &= a \cdot x - (a \cdot u) \cdot x \\ &= a \cdot x - a \cdot (u \cdot x) \\ &= a \cdot x - a \cdot x \\ &= 0.\end{aligned}$$

   Hence u is a modular unit for the left ideal A_x. Suppose that A_x is not maximal, i.e., there exists a proper left ideal J of A containing A_x. Then the subspace $J \cdot x$ of X is different from $\{0\}$ and A-invariant. Hence $J \cdot x = X = A \cdot x$. But then for every $a \in A$ there exists $j \in J$, such that

$$a \cdot x = j \cdot x \Rightarrow a - j \in A_x \Rightarrow A \subset J.$$

   This contradiction shows that A_x is a maximal left ideal in A.
4. Suppose that I is a maximal modular ideal of A (with modular unit u for I). Let $Y \subset A/I$ be a proper A-invariant subspace of $X := A/I$. Let $P_I : A \to A/I$ be the canonical projection and let $J := P_I^{-1}(Y)$. Then J is a left ideal of A (containing I), since Y is A-invariant. Furthermore $J \neq A$, since Y is proper and $P_I(J) = Y \neq X = P_I(A)$. Hence $J = I$ by the maximality of I and so $Y = \{0\}$ in A/I. Therefore the left module A/I is simple. □

Theorem 8.2.5 *Let A be a Banach algebra and let I be a proper maximal modular left ideal of A. Then I is closed in A.*

Proof If the closure $\overline{I}$ of I is a proper subspace of A, then $\overline{I} = I$, since I is a maximal left ideal. So suppose that $\overline{I} = A$. Let v be a modular unit for I. Since $v \in \overline{I}$, there exists $i \in I$ such that $\|v - i\|_A < \frac{1}{2}$. Let $u := v - i$. Then u is also a modular unit for I and the series $\sum_{j=1}^{\infty} u^j$ converges in A. Let $x \in A$ and

$y := x + \sum_{j=1}^{\infty} xu^j$. Then $y - yu \in I$, since u is a modular unit. But

$$\begin{aligned} y - yu &= x + \sum_{j=1}^{\infty} xu^j - xu - \sum_{j=2}^{\infty} xu^j \\ &= x. \end{aligned}$$

Hence $x \in I$ and we see that $A = I$, which contradicts the fact that I is proper. □

Corollary 8.2.6 *Let (T, X) be a simple A-module. Then there exists a norm $\| \|_X$ on X turning $(T, X, \| \|_X)$ in a Banach A-module.*

Proof Fix an $x \neq 0$ in X. The bijective linear mapping $L_x : A/A_x \to X$ is an intertwining operator and A/A_x is Banach space, since A_x is closed (being a maximal modular left ideal of A). Let $\| \|_{A/A_x}$ be the canonical norm of A/A_x. Then the function

$$\|x\|_X := \|L_x^{-1}(x)\|_{A_x}$$

is a norm on X. Furthermore X is automatically a Banach space for this norm and X becomes a (simple) A-Banach module. □

Remark 8.2.7 Let (T, X) be a simple module of the Banach algebra A. Schur's Lemma (see [29]) says that any linear operator $D : X \to X$ which commutes with $T(a), a \in A$, is a multiple of the identity $\mathbb{I}_X$.

Example 8.2.8 Let $G = \exp \mathfrak{g}$ be a connected Lie group with Lie algebra $\mathfrak{g}$ and let $\mathcal{U}(\mathfrak{g})$ be the enveloping algebra of $\mathfrak{g}$. Let $A := L^1(G)$, $A^\infty := L^{1,\infty}(G)$ be the subalgebra of A consisting of smooth $L^1(G)$ functions:

$$A^\infty = \{F : G \to \mathbb{C};\ F \text{ smooth}, U * F \in L^1(G),\ U \in \mathcal{U}(\mathfrak{g})\}.$$

Lemma 8.2.9 *Suppose that (T, X) is a simple $L^1(G)$-module. Let D be central in $\mathcal{U}(\mathfrak{g})$. Then $dT(D)$ is a scalar multiple of the identity operator of X^∞.*

Proof Let $F \in C_c^\infty(G)$. Then for $x \in X^\infty$ and $U \in \mathfrak{g}$ we have:

$$\begin{aligned} dT(U) \circ T(F)(x) &= \frac{d}{dt}(T(\exp(tU))(T(F)x))_{t=0} \\ &= \frac{d}{dt}(\int_G (F(\exp(tU)^{-1}g)T(g))(x)dg)_{t=0} \\ &= \frac{d}{dt}(\int_G F(g)T(g)T(g^{-1}\exp(tU)g)(x)dg)_{t=0} \\ &= \int_G F(g)T(g)dT(\mathrm{Ad}(g^{-1})U)(x)dg. \end{aligned}$$

This shows that for any $E \in \mathscr{U}(\mathfrak{g})$ we have

$$dT(D) \circ T(F)(x) = \int_G F(g)T(g)dT(\mathrm{Ad}(g^{-1})E)(x)dg.$$

For our central element D we have $\mathrm{Ad}(g^{-1})D = D$, so

$$dT(D)(T(F)x) = T(F)(dT(D)x).$$

We can approximate any $F \in A^\infty$ by a sequence $(F_k)_{k\in\mathbb{N}} \subset C_c^\infty(G)$ by multiplying F with smooth compactly supported functions, which are identically 1 on an increasing sequence of compact subsets of G. Then

$$\lim_{k\to\infty} \|D * F - D * F_k\|_1 = 0$$

and so

$$\begin{aligned} dT(D)(T(F)x) &= T(D * F)(x) \\ &= \lim_{k\to\infty} T(D * F_k)(x) \\ &= \lim_{k\to\infty} dT(D)(T(F_k)x) \\ &= \lim_{k\to\infty} T(F_k)(dT(D)x) \\ &= T(F)(dT(D)(x)). \end{aligned}$$

Take now $F \in L^1(G)$ and $F' \in C_c^\infty(G)$, so $F' * F \in A^\infty$. If now

$$T(F)x = 0,$$

then

$$\begin{aligned} 0 &= dT(D)(T(F')(T(F))(x)) \\ &= dT(D)(T(F' * F)x) \\ &= T(F' * F)(dT(D)x) \\ &= T(F')(T(F)(dT(D)x)). \end{aligned}$$

Hence

$$T(F)(dT(D)x) = 0.$$

We see that the annihilator A_x of x in A is contained in $A_{dT(D)x}$. Since A_x is a maximal left ideal, we have

$$A_x = A_{dT(D)x}.$$

Hence for any $F \in A^\infty$

$$\begin{aligned} \|dT(D)(T(F)x)\| &= \|T(F)(dT(D)(x))\| \\ &\leq \|F \bmod A_x\|_{A/A_x} \|dT(D)x\| \\ &= \|T(F)x\|_X \|dT(D)x\|. \end{aligned}$$

This inequality says that $dT(D)$ is a bounded operator on X^∞, which can then be extended to the whole Banach space X. Since $dT(D)$ commutes with $T(F)$, $F \in A^\infty$, and hence with every $F \in A$, we can apply Schur's lemma. Therefore $dT(D)$ is a scalar multiple of the identity on X and on X^∞. □

Lemma 8.2.10 *Let M be a maximal modular left ideal of the Banach algebra A. Let $0 \neq \varphi \in M^\perp \subset A^*$. Then*

$$M = \{a \in A; \varphi \cdot a = 0\}.$$

Proof Indeed, for any $m \in M$

$$\varphi \cdot m = 0$$

since for any $b \in A$ the element bm of A is contained in M and so $\langle \varphi, ba \rangle = 0$. Hence M is contained in the annihilator $A_\varphi := \{c \in A; \varphi \cdot c = 0\}$. This annihilator is a left ideal in A. The maximality of M tells us that $M = A_\varphi$. □

8.2.2 *The Spectrum in Banach Algebras*

Definition 8.2.11 Let A be a Banach algebra without unit. We denote by $\tilde{A}$ the vector space

$$\tilde{A} := \mathbb{C} \oplus A$$

and we equip it with the multiplication

$$(s, a) \cdot (t, b) := (s + t, s \cdot b + t \cdot a + a \cdot b), s, t \in \mathbb{C}, a, b \in A.$$

The function

$$\|(s,a)\|_{\tilde{A}} := |s| + \|a\|_A, \quad s \in \mathbb{C}, a \in A,$$

is then a Banach-algebra norm on $\tilde{A}$. The Banach algebra $\tilde{A}$ is unital, with unit element $\mathbb{I}_{\tilde{A}} = (1, 0)$. We identify A with the subalgebra $\{(0, a); a \in A\}$ of $\tilde{A}$.

Definition 8.2.12

1. Let A be a Banach algebra with unit $\mathbb{I}_A$. Let $a \in A$. The spectrum $\sigma_A(a) = \sigma(a) \subset \mathbb{C}$ of a is by definition given by

$$\sigma_A(a) := \{\lambda \in \mathbb{C}; \lambda\mathbb{I}_A - a \text{ is not invertible in } A\}.$$

If a is invertible the spectrum $\sigma_A(a)$ coincides with the subset

$$\{\lambda \in \mathbb{C}^*; \text{ there is no } b \in A \text{ such that } \frac{a}{\lambda} + \lambda b - a \cdot b = 0 \text{ and } a \cdot b = b \cdot a\}.$$

If a is not invertible, then

$$\sigma_A(a) := \{0\} \cup \{\lambda \in \mathbb{C}^*; \text{ there is no } b \in A \text{ such that } \frac{a}{\lambda} + \lambda b - a \cdot b = 0 \text{ and } a \cdot b = b \cdot a\}.$$

2. Let A be a Banach algebra without unit. Let $a \in A$. The spectrum $\sigma_A(a) = \sigma(a) \subset \mathbb{C}$ of a is by definition given by

$$\sigma_A(a) := \sigma_{\tilde{A}}(0, a).$$

3. For a unital Banach algebra A and for $a \in A$, we denote by

$$\rho_A(a) = \rho(a) := \{\lambda \in \mathbb{C}; \lambda\mathbb{I}_A - a \text{ exists }\} = \mathbb{C} \setminus \sigma_A(a),$$

the *resolvent set* of a.

Theorem 8.2.13 *Let A be a Banach algebra. Then the spectrum $\sigma_A(a)$ is a compact non-empty subset of $\mathbb{C}$ for any $a \in A$.*

Proof We can assume that A is unital. Let $a \in A$. Let $\lambda \in \mathbb{C}$ be such that $|\lambda| > \|a\|_A$. Then the series $\sum_{j=1}^{\infty} \frac{a^j}{\lambda^{j+1}}$ converges in A, since A is a Banach space. Let

$$b := -\sum_{j=1}^{\infty} \frac{a^j}{\lambda^{j+1}}.$$

We see that $\frac{a}{\lambda} + \lambda b = a \cdot b$ and $a \cdot b = b \cdot a$. Hence $\lambda\mathbb{I}_A - a$ is invertible. So the spectrum of a is contained in the disk $D_1 := \{z \in \mathbb{C}; |z| \leq 1\}$.

Let us show that the resolvent set $\rho(a)$ of a, which is the complement of $\sigma(a)$ in $\mathbb{C}$, is open.

Let $a \in A$ be an invertible element. Let $r := \|a^{-1}\|_A$. Then for every $b \in A$ with $\|b\|_A < \frac{1}{r}$ we have that $\|a^{-1}b\|_A < 1$ and so

$$c := \sum_{j=0}^{\infty}(-(a^{-1}b))^j \text{ converges in } A$$

and

$$c = (\mathbb{I}_A + a^{-1} \cdot b)^{-1}.$$

Therefore

$$(a+b)^{-1} = (a \cdot (\mathbb{I}_A + a^{-1} \cdot b))^{-1} = (\mathbb{I}_A + a^{-1} \cdot b)^{-1} \cdot a^{-1} = c \cdot a^{-1} \tag{8.2.1}$$

is well defined in A. Hence the set

$$A^{reg} := \{a \in A; a^{-1} \text{ is well defined}\}$$

is open in A. Let now for $a \in A$ and $\lambda \in \rho(a)$

$$r(\lambda, a) := (\lambda\mathbb{I}_A - a)^{-1}. \tag{8.2.2}$$

Let $\mu \in \mathbb{C}$ be such that $|\mu - \lambda|\|r(\lambda, a)\|_A < \frac{1}{2}$. We have that

$$\begin{aligned}(\mu\mathbb{I}_A - a)r(\lambda, a) &= (\mu - \lambda)r(\lambda, a) + (\lambda\mathbb{I}_A - a)r(\lambda, a)\\ &= (\mu - \lambda)r(\lambda, a) + \mathbb{I}_A.\end{aligned}$$

Hence $\mu \in \rho(a)$ by (8.2.1) and so $\rho(a)$ is open in $\mathbb{C}$. Furthermore

$$\begin{aligned}r(\mu, a) - r(\lambda, a) &= (\mu\mathbb{I}_A - a)^{-1} - (\lambda\mathbb{I}_A - a)^{-1}\\ &= (\lambda - \mu)(\mu\mathbb{I}_A - a)^{-1}(\lambda\mathbb{I}_A - a)^{-1}\\ &= (\lambda - \mu)(\mu\mathbb{I}_A - a)^{-1}r(\lambda, a)\\ &= (\lambda - \mu)\Big((\mu - \lambda)\mathbb{I}_A + (\lambda\mathbb{I}_A - a)\Big)^{-1} r(\lambda, a)\\ &= (\lambda - \mu)\Big((\mu - \lambda)r(\lambda, a) + \mathbb{I}_A\Big)^{-1} r(\lambda, a)^2.\end{aligned}$$

Hence

$$\|r(\mu, a) - r(\lambda, a)\|_A \leq |\mu - \lambda| \|\Big(\mu - \lambda)r(\lambda, a) + \mathbb{I}_A\Big)^{-1}\|_A \|r(\lambda, a)^2\|_A$$
$$\leq 2|\mu - \lambda| \|r(\lambda, a)^2\|_A.$$

This shows that the mapping $\mu \to r(\mu, a)$ is continuous on the open subset $\rho(a)$ of $\mathbb{C}$ and even differentiable, since

$$\lim_{\mu \to \lambda} \frac{r(\mu, a) - r(\lambda, a)}{\mu - \lambda} = \lim_{\mu \to \lambda} \frac{(\lambda - \mu)\Big((\mu - \lambda)r(\lambda, a) - \mathbb{I}_A\Big)^{-1} r(\lambda, a)^2}{\mu - \lambda}$$
$$= - \lim_{\mu \to \lambda} \Big((\mu - \lambda)r(\lambda, a) + \mathbb{I}_A\Big)^{-1} r(\lambda, a)^2$$
$$= r(\lambda, a)^2.$$

This means that the mapping

$$r_a : \rho(a) \to A; r_a(\lambda) := r(\lambda, a)$$

is analytic.

If now for some $a \in A$ the spectrum $\sigma(a)$ is empty, then the function r_a is entire. Since

$$\lim_{\lambda \to \infty} r_a(\lambda) = \lim_{k \to \infty} \frac{1}{\lambda}(\mathbb{I}_A - \frac{a}{\lambda})^{-1} = 0$$

it follows that the entire mapping r_a is uniformly bounded. Hence by Liouville's theorem it is constant, and so $r_a(\mu) = \lim_{\lambda \to \infty} r_a(\lambda) = 0$ for every $\mu \in \mathbb{C}$. In particular $a^{-1} = r_a(0) = 0$. This contradiction tells us that $\sigma(a)$ cannot be empty.

□

8.2.3 Simple Modules and the Spectrum

Definition 8.2.14 Let $(A, +, \cdot)$ be an algebra. We denote by A^{op} the algebra $(A, +, \cdot_{op})$; where

$$a \cdot_{op} b := b \cdot a, a, b \in A.$$

Then every left (right) ideal in A is a right (left) ideal in A^{op}. Furthermore for any $a \in A$ and $\lambda \in \mathbb{C}$ we have that

$$\lambda \in \sigma_A(a) \Leftrightarrow \lambda \in \sigma_{A^{op}}(a).$$

Theorem 8.2.15 *Let A be a Banach algebra and $a \in A$. Take any $0 \neq \lambda \in \mathbb{C}$. Then $\lambda \in \sigma(a)$ if and only there exists a simple module (T, X) of A and an element $x \neq 0$ in X such that*

$$a \cdot x = \lambda x.$$

Proof We can assume that A is unital. Let $a \in A, a \neq 0, 0 \neq \lambda \in \sigma(a)$ and let

$$I_\lambda^l := \{b - b \cdot \frac{a}{\lambda}; b \in B\}; \ I_\lambda^r := \{b - \frac{a}{\lambda} \cdot b; b \in B\}.$$

Suppose that $I_\lambda^l = A = I_\lambda^r$. Then there exists $b, c \in A$ such that

$$\lambda b - \lambda b \cdot \frac{a}{\lambda} = -\frac{a}{\lambda} = \lambda c - \frac{a}{\lambda} \cdot \lambda c$$
$$\Leftrightarrow$$
$$(\frac{1}{\lambda}\mathbb{I}_A - b)(\lambda \mathbb{I}_A - a) = \mathbb{I}_A = (\lambda \mathbb{I}_A - a)(\frac{1}{\lambda}\mathbb{I}_A - c)$$

But then $\frac{a}{\lambda}$ has a right and left inverse in A and therefore a is invertible. This contradiction tells us that either

$$I_\lambda^l \text{ or } I_\lambda^r \neq A.$$

Suppose that $I_\lambda^l \neq A$. Since I_λ^l is a modular left ideal and proper, it is contained in a maximal left ideal I with modular right unit $\frac{a}{\lambda}$. This gives us the simple module A/I and the modular unit $\frac{a}{\lambda}$ is not contained in I. Let $x := a + I$. then

$$\begin{aligned} a \cdot x &= \lambda(a \cdot \frac{a}{\lambda} + I) \\ &= \lambda((a \cdot \frac{a}{\lambda} - a) + a + I) \\ &= \lambda(a + I) \\ &= \lambda x. \end{aligned}$$

Similarly, if $I_\lambda^r \neq A$, then we find a modular maximal right ideal I of A containing I_λ^r. Then A/I defines a simple right A-module.

The "sufficient part" of the theorem is trivial. □

8.2.4 Construction of Simple Modules

Definition 8.2.16 Let A be a Banach algebra and let (T, X) be a Banach module. We denote by I_T^{fin} the ideal of A defined by

$$I_T^{fin} := \{a \in A;\ \text{the operator } T(a) \in B(X) \text{ has finite rank}\}.$$

Let also X^{fin} be the (A)-invariant subspace of X given by

$$X^{fin} := \{\text{span}(T(a)(X));\, a \in I_T^{fin}\}.$$

We have the following method to obtain simple modules.

Theorem 8.2.17 *Let A be a Banach algebra and let (T, X) be an irreducible Banach module. Suppose that the finite-rank subspace X^{fin} is different from $\{0\}$. Then X^{fin} is the unique simple submodule of (T, X).*

Proof Since T is irreducible, for any $x^{fin} \neq 0$ in X^{fin}, the A-invariant subspace $A \cdot x^{fin}$ is dense in X. Let $y^{fin} \in X^{fin}$. Then

$$y^{fin} = \sum_{j=1}^{m} T(a_j)x_j$$

for some $a_j \in I_T^{fin}$, $m \in \mathbb{N}$, $x_j \in X$, $j = 1, \cdots, m$. Since the subspace $T(a_j)X$ of X is finite-dimensional, we have then that

$$T(a_j)X = T(a_j)(A \cdot x^{fin}),\ j = 1, \cdots, m.$$

Hence for any j, there exists $b_j \in A$ such that

$$T(a_j)x_j = T(a_j)(T(b_j)x^{fin}).$$

Therefore

$$y^{fin} = \sum_{j=1}^{m} T(a_j \cdot b_j) \cdot x^{fin} \in A \cdot x^{fin}.$$

Hence

$$X^{fin} \subset A \cdot x^{fin} \subset X^{fin}$$
$$\Rightarrow X^{fin} = A \cdot x^{fin} \text{ for every } x^{fin} \neq 0 \text{ in } X^{fin}.$$

This shows that X^{fin} is a simple submodule of (T, X). Let now $Y \subset X$ be any proper A-invariant subspace of X. Suppose that $I_T^{fin} \cdot Y = \{0\}$. Then the closed A-invariant subspace

$$Z_{fin} := \{z \in X; I_T^{fin} \cdot z = 0\}$$

is different from $\{0\}$. Hence the irreducibility of T tells us that $Z_{fin} = X$. This contradiction tells that the A-invariant subspace-3

$$I_T^{fin} \cdot Y$$

of $Y \cap X^{fin}$ is different from $\{0\}$ and so, if $(T_{|Y}, Y)$ is a simple A-module,

$$Y = Y \cap X^{fin} = X^{fin}.$$

□

Example 8.2.18 The Heisenberg group.
Let H_n be the $(2n+1)$-dimensional Heisenberg Lie group, i.e., the manifold $\mathbb{R}^n \times \mathbb{R}^n \times \mathbb{R}$ with smooth multiplication

$$(x, y, t)(x', y', t') = (x + x', y + y', t + t' + \frac{1}{2}(x \cdot y' - x' \cdot y)),$$

where $x \cdot y = x_1 y_1 + \cdots + x_n y_n$ denotes the Euclidean dot product on $\mathbb{R}^n$. The center of H_n is the subgroup $\mathscr{Z} := \{0_n\} \times \{0_n\} \times \mathbb{R}$ and the commutator subgroup $[H_n, H_n]$ of H_n is given by $[H_n, H_n] = \mathscr{Z}$. The Lie algebra $\mathfrak{g}$ of H_n has basis

$$\mathscr{B} := \{X_j, Y_j, j = 1 \cdots, n, Z = (0_n, 0_n, 1)\},$$

where $X_j = (e_j, 0_n, 0)$, $Y_j = (0_n, e_j, 0)$, $j = 1, \cdots, n$ and e_j is the j-th canonical basis vector of $\mathbb{R}^n$, with non-trivial brackets

$$[X_i, Y_j] = \delta_{i,j} Z.$$

For every $\lambda \in \mathbb{R}^*$ there exists an irreducible unitary representation π_λ of H_n on the Hilbert space $\mathscr{H}_\lambda = L^2(\mathbb{R}^n)$, given by formula (see [70])

$$\pi_\lambda(x, y, t)\xi(s) := e^{-2\pi i \lambda t - 2\pi i \frac{\lambda}{2} x \cdot y + 2\pi i \lambda s \cdot y} \xi(s - x), \ s \in \mathbb{R}^n, \xi \in L^2(\mathbb{R}^n), (x, y, t) \in H_n.$$

Hence for $F \in L^1(G)$ and $\xi \in L^2(\mathbb{R}^n)$ we get

$$\begin{aligned}\pi_\lambda(F)\xi(s) &= \int_{H_n} F(h)\pi_\lambda(h)\xi(s)dh \\ &= \int_{\mathbb{R}^n \times \mathbb{R}^n} \hat{F}^3(x, y, \lambda) e^{-2\pi i \frac{\lambda}{2} x \cdot y + 2\pi i \lambda s \cdot y} \xi(s - x) dy dx\end{aligned}$$

$$= \int_{\mathbb{R}^n \times \mathbb{R}^n} \hat{F}^3(s-x,y,\lambda) e^{-2\pi i \frac{\lambda}{2} x\cdot y + -2\pi i \frac{\lambda}{2}(s-x)\cdot y + 2\pi i \lambda s \cdot y} \xi(x) dy dx$$

$$= \int_{\mathbb{R}^n \times \mathbb{R}^n} \hat{F}^3(s-x,y,\lambda) e^{\frac{2\pi i}{2} \lambda s \cdot y} \xi(x) dy dx$$

$$= \int_{\mathbb{R}^n} \hat{F}^3(s-x, -\frac{\lambda}{2}(s+x), \lambda) dx, \, s \in \mathbb{R}^n.$$

It is easy to see that π_λ is equivalent to π_ν if and only if $\lambda = \nu$.

Let $\lambda \in \mathbb{R}^*$. Take any $\xi, \eta \in L^2(\mathbb{R}^n)$ and let $P_{\xi,\eta}$ be the rank-one operator

$$P_{\xi,\eta}(\varphi) := \langle \varphi, \eta \rangle \cdot \xi, \;\; \varphi \in L^2(\mathbb{R}^n).$$

We determine now a necessary and sufficient condition on ξ, η for the existence of an $f \in L^1(H_n)$ such that

$$\pi_\lambda(f) = P_{\xi,\eta}.$$

The kernel $\ker_\lambda$ of π_λ in $L^1(H_n)$ is given by

$$\ker_\lambda = \{F \in L^1(G_n); \int_{\mathbb{R}} F(a,b,t) e^{i\lambda t} dt = 0, \;\; (a,b,u) \in H_n\}.$$

Then the Banach algebra $L^1(H_n)/\ker(\lambda)$ is isometrically isomorphic to the Banach algebra $L^1(H_n)_\lambda$, where

$$L^1(H_n)_\lambda := \{f : H_n \to \mathbb{C}; \, f \text{ measurable}, \, f(g \cdot (0,0,t)) = e^{i\lambda t} f(g), \, g \in H_n, t \in \mathbb{R},$$
$$\|f\|_1 := \int_{\mathbb{R}^{2n}} |f(g,0)| dg < \infty\}.$$

If now $\pi_\lambda(f) = P_{\xi,\eta}$ for some $f \in L^1(H_n)_\lambda$, then

$$\hat{f}^2(s-t, -\frac{\lambda}{2}(s+t)) = \xi(s)\overline{\eta(t)}, \, s,t \in \mathbb{R}^{2n}.$$

Here

$$\hat{f}^2(s,t) := \int_{\mathbb{R}^n} f(s,u) e^{-2i\pi t \cdot u} du, \, s,t \in \mathbb{R}^n.$$

This condition is equivalent to

$$\hat{f}^2(u,v) = \xi(\frac{u}{2} - \frac{v}{\lambda}) \overline{\eta(-\frac{v}{\lambda} - \frac{u}{2})}.$$

Hence

$$f(u,y) = \int_{\mathbb{R}^n} e^{i2\pi v\cdot y}\xi(\frac{u}{2}-\frac{v}{\lambda})\overline{\eta(-\frac{v}{\lambda}-\frac{u}{2})}dv \quad (8.2.3)$$

$$\overset{(v=\frac{\lambda u}{2}-\lambda v')}{=} |\lambda|^n e^{2\pi i\frac{\lambda u}{2}\cdot y}\int_{\mathbb{R}^n} e^{i2\pi\lambda y\cdot v'}\xi(v')\overline{\eta(v'-u)}dv'.$$

Therefore we see that for $\xi, \eta \in L^2(\mathbb{R}^n)$ there exists a (unique) $p_{\xi,\eta} \in L^1(H_n)_\lambda$ such that

$$\pi_\lambda(f_{\xi,\eta}) = P_{\xi,\eta}$$

if and only if

$$\int_{\mathbb{R}^{2n}} |(\chi_y\cdot\xi) * \eta^*(u)|dudy < \infty.$$

In particular, if ξ and η are Schwartz functions then the function $f_{\xi,\eta}$ is contained in the Schwartz space of $\mathbb{R}^n \times \mathbb{R}^n$.

8.2.5 Simple *A*-modules and Simple *pAp*-modules

Definition 8.2.19 Let A be a Banach algebra with bounded two-sided approximate units. Let m be a bounded linear operator of the Banach space A. We say that m is a *multiplier* of A if there exists a bounded linear operator $r : A \to A$ such that

$$r(a)\cdot b = a\cdot m(b),\ a, b \in A.$$

Remark 8.2.20

1. Let m be a multiplier of the Banach algebra A. Then the mapping $r : A \to A$ given in the definition of a multiplier is unique. Indeed if we have a second such mapping r', then

$$\begin{aligned} r(a)\cdot b &= a\cdot m(b) = r'(a)\cdot b \\ &\Rightarrow (r(a)-r'(a))\cdot b = 0,\ a, b\in A \\ &\Rightarrow r(a) = r'(a),\ a\in A. \end{aligned}$$

   We can therefore write

$$r = r(m)$$

   for any multiplier m.

2. Let m be a multiplier of A. Then

$$m(a \cdot b) = m(a) \cdot b, a, b \in A.$$

Indeed, for $a, b, c \in A$ we have that

$$\begin{aligned} a \cdot m(b \cdot c) &= r(a) \cdot (b \cdot c) \\ &= (r(a) \cdot b) \cdot c \\ &= (a \cdot m(b)) \cdot c \\ &= a \cdot (m(b) \cdot c) \\ &\Rightarrow m(b \cdot c) = m(b) \cdot c, \ b, c \in A. \end{aligned}$$

Similarly we can show that

$$r(b \cdot c) = b \cdot r(c), b, c \in A.$$

3. Suppose that A has a bounded two-sided approximate unit $(e_k)_{k \in K}$. Let m be a multiplier of A. Then

$$\begin{aligned} r(a) &= \lim_{k \to \infty} a \cdot m(e_k), \\ m(a) &= \lim_{k \to \infty} m(e_k) \cdot a, a \in A. \end{aligned}$$

Indeed, we have that

$$\begin{aligned} r(a) &= \lim_{k \to \infty} r(a) \cdot e_k \\ &= \lim_{k \to \infty} a \cdot m(e_k), \\ m(a) &= \lim_{k \to \infty} m(e_k \cdot a) \\ &= \lim_{k \to \infty} m(e_k) \cdot a. \end{aligned}$$

Lemma 8.2.21 *Let I be a closed left ideal of A. Then for every multiplier m of A we have that*

$$m(I) \subset I.$$

Proof Indeed if $i \in I$ then by the above:

$$\begin{aligned} m(i) &= \lim_{k \to \infty} m(e_k i) \\ &= \lim_{k \to \infty} m(e_k) \cdot i \in I. \end{aligned}$$

□

4.

Definition 8.2.22 Let

$$M(A) := \{m : A \to A; m \text{ is a multiplier of } A\}.$$

It follows from Remark 8.2.20 1. that $M(A)$ is a Banach subspace of the algebra $\mathscr{B}(A)$ of bounded linear operators on A, and that the mapping $r : M(A) \to \mathscr{B}(A)$ is linear.

Proposition 8.2.23 *The subspace $M(A)$ of $\mathscr{B}(X)$ is a subalgebra of the algebra of bounded linear operators $\mathscr{B}(A)$ of A, and $M(A)$ "contains" A.*

Proof Let $m, m' \in M(A)$ and $a, b \in A$. Then

$$\begin{aligned} a \cdot m \circ m'(b) &= r(a) \cdot (m'(b)) \\ &= r'(r(a)) \cdot b. \end{aligned}$$

Hence $m \circ m'$ is also a multiplier and

$$r(m \circ m') = r(m') \circ r(m).$$

Let $a \in A$. We define the element $m(a)$ of $\mathscr{B}(A)$ by

$$m(a)(b) := a \cdot b, b \in B.$$

Then for $c \in A$

$$\begin{aligned} b \cdot m(a)(c) &= b \cdot (a \cdot c) \\ &= (b \cdot a) \cdot b. \end{aligned}$$

Hence

$$r(m(a)) = \text{ right multiplication by } a.$$

□

5. Let $m \in M(A)$. Then

$$m \circ r(m) = r(m) \circ m.$$

Indeed for any $a, b \in A$ we have that

$$\begin{aligned} m(r(a)) \cdot b &= m(r(a) \cdot b) \\ &= m(a \cdot m(b)) \end{aligned}$$

$$= m(a) \cdot m(b)$$
$$= r(m(a)) \cdot b.$$

6. If the algebra A has bounded approximate units, then we shall use the following notation:

$$ma = m \cdot a := m(a)$$
$$am = a \cdot m := r(m)(a), m \in M(A), a \in A.$$

For an A-module (T, X) we also write as before:

$$a \cdot x := T(a)x, a \in A, x \in X.$$

7. Let G be a locally compact group and let $A = L^1(G)$. Let $\mathcal{M}(G) = C_\infty(G)^*$ be the measure algebra of G. Then

$$M(L^1(G)) = \mathcal{M}(G).$$

Indeed, if $\mu \in \mathcal{M}(G)$ and $\varphi \in C_c(G)$ then by the definition of the convolution product of $\mathcal{M}(G)$ with $C_\infty(G)$ we have that

$$\mu * \varphi(s) = \langle \mu, \lambda(s)\check{\varphi} \rangle, \ \ s \in G,$$

where as usual the function $\check{\varphi}$ is defined by $\check{\varphi}(s) = \varphi(s^{-1}), s \in G$. Hence $\mu * \varphi$ is a continuous bounded function of G.

For every $\psi \in C_c(G)$ we then have that

$$\langle \psi, \mu * \varphi \rangle = \int_G \langle \mu, \lambda(t)\check{\varphi} \rangle \psi(t)dt$$
$$= \langle \mu, \int_G \psi(t)\lambda(t)\check{\varphi} \rangle$$
$$= \langle \mu, \psi * \check{\varphi} \rangle.$$

Therefore

$$|\langle \psi, \mu * \varphi \rangle| \leq \|\mu\| \|\psi\|_\infty \|\varphi\|_1.$$

This shows that $\mu * \varphi \in L^1(G)$ and $\|\mu * \varphi\|_1 \leq \|\mu\| \|\varphi\|_1$. Finally, since $C_c(G)$ is dense in $L^1(G)$, we have that $\mu * L^1(G) \subset L^1(G)$ and that the linear mapping

$L^1(G) \to L^1(G);\ f \to \mu * f$ is bounded by $\|\mu\|$. We also define $f * \mu$ by

$$\begin{aligned} f * \mu(t) &:= \int_G \rho(t)(\Delta_G \check{} f) d\mu \\ &= \langle \mu, \rho(t)(\Delta_G f)\check{}\,\rangle. \end{aligned}$$

for $f \in C_c(G)$ (where $\rho(t)\varphi(x) = \varphi(xt)$) and then we define by the usual limit procedure $\mu * f \in L^1(G)$ for $f \in L^1(G)$.

It follows then that

$$(f * \mu) * g = f * (\mu * g), \ \ f, g \in L^1(G).$$

Let now $m \in M(L^1(G))$. First we observe that for $f \in L^1(G), t \in G$:

$$\begin{aligned} m(\rho(t)f) &= \lim_{k\to\infty} m(f * \rho(t)e_k) \\ &= \lim_{k\to\infty} m(f) * \rho(t)e_k \\ &= \rho(t)m(f). \end{aligned}$$

Since $L^1(G)$ admits bounded two-sided approximate units, we find for every $\psi \in C_\infty(G)$ another $\psi' \in C_\infty(G)$ and an $f \in L^1(G)$, such that $\psi = f * \psi'$. More precisely, for every $\varphi \in C_c(G)$ and $\varepsilon > 0$ there exist a function $f_\varepsilon \in L^1(G)$ with compact support and norm ≤ 2 and $\varphi_\varepsilon \in C_c(G)$ with $\|\varphi - \varphi_\varepsilon\|_\infty \leq \varepsilon$ such that $\varphi = f_\varepsilon * \varphi_\varepsilon$ (see [29, Section 11, Theorem 10]). Hence the $L^1(G)$-function $m(\varphi) = m(f_\varepsilon * \varphi_\varepsilon) = m(f_\varepsilon) * \varphi_\varepsilon$ is continuous. We can then define the linear functional μ_m on $C_c(G)$ by

$$\langle \mu_m, \check{\varphi} \rangle := m(\varphi)(e).$$

For any $\varphi \in C_c(G)$ we then have that

$$\begin{aligned} \|\mu_m * \varphi\|_\infty &= \sup_{\|f\|_1 \leq 1} \left| \int_G f(t)\mu_m * \varphi(t) dt \right| \\ &= \sup_{\|f\|_1 \leq 1} \left| \int_G f(t) \langle \mu_m, \lambda(t)\check{\varphi} \rangle dt \right| \\ &= \sup_{\|f\|_1 \leq 1} |\langle \mu_m, f * \check{\varphi} \rangle| \\ &= \sup_{\|f\|_1 \leq 1} |m(f * \check{\varphi})(e)| \\ &\leq \sup_{\|f\|_1 \leq 1} \|m(f)\|_1 \|\varphi\|_\infty \\ &\leq \|m\| \|\varphi\|_\infty. \end{aligned}$$

In particular

$$\begin{aligned}|\langle \mu_m, \varphi\rangle| &= |\langle \mu_m * \check{\varphi}(e)| \\ &\leq \|\mu_m * \check{\varphi}\|_\infty \\ &\leq \|m\| \|\varphi\|_\infty.\end{aligned}$$

This shows that $\|\mu_m\| \leq \|m\|$ and μ_m is a measure on G. Furthermore for $t \in G$

$$\begin{aligned}\mu_m * \varphi(t) &= \langle \mu_m, \lambda(t)\check{\varphi}\rangle \\ &= \langle \mu_m, (\rho(t)\varphi)\check{}\rangle \\ &= m((\rho(t)\varphi))(e) \\ &= \rho(t)m(\varphi) \\ &= m(\varphi)(t).\end{aligned}$$

Hence

$$\mu_m * \varphi = m(\varphi), \varphi \in L^1(G).$$

Definition 8.2.24 Let (T, X) be a simple module of the Banach algebra A.
Let $m \in M(A)$. We define the linear operator $T(m) \in L(X)$ by

$$T(m)(T(a)x) := T(m(a)x), a \in A, x \in X.$$

This definition is justified, since if

$$T(a)x = T(b)y,$$

then $y = T(c)x$ for some $c \in A$, because (T, X) is a simple module. Hence $T(a - bc)x = 0$. By Lemma 8.2.21 we then have that

$$\begin{aligned}0 = T(m(a - bc))x &= 0 \\ &\Rightarrow \\ T(m(a))x &= T(m(bc))x \\ &= T(m(b)c)x \\ &= T(m(b))y.\end{aligned}$$

Furthermore, if we realize the simple module (T, X) using the left multiplication on A/I for some modular maximal ideal I with modular unit u, then $m(I) \subset I$ arises as above, and so

$$T(m)T(a)(u + I) = m(a) + I.$$

Hence

$$\begin{aligned}\|T(m)(a+I)\| &= \|m(a)+I\| \\ &= \inf_{i\in I}\|m(a)+i\| \\ &\le \inf_{i\in I}\|(m(a)+m(i))\| \\ &\le \inf_{i\in I}\|m\|\,\|a+i\| \\ &= \|m\|\,\|a+I\|.\end{aligned}$$

Therefore $T(m)$ is a bounded linear operator on the space $X = A/I$. We observe the following relations

$$\begin{aligned}T(m)\circ T(a) &= T(m(a)) \\ T(r(a)) &= T(a)\circ T(m), a\in A, m\in M(A).\end{aligned} \tag{8.2.4}$$

The first relation is evident. For the second one, choose for $x \in X$ an element $a_x \in A$ such that $T(a_x)(x) = x$. Then by definition of $T(m)$:

$$\begin{aligned}T(a)\circ T(m)(x) &= T(a)\circ T(m)\circ T(a_x)(x) \\ &= T(a\cdot m(a_x)(x) \\ &= T(r(a)\cdot a_x)(x) \\ &= T(r(a))(x).\end{aligned}$$

Definition 8.2.25 Let A be a Banach algebra with bounded approximate units and let (T, X) be an A-module. Let us write for simplicity

$$a\cdot x := T(a)x, a\in A, x\in X.$$

Proposition 8.2.26 *Let A be a Banach algebra and take $p \in M(A)$ such that $p^2 = p$. Let (T, X) be a simple A-module such that $X_p := T(p)X \neq \{0\}$.*

(a) Then $(T|_{pAp}, X_p)$ is a simple (pAp)-module.
(b) Take $0 \neq y \in X_p$. The ideal $A_y \subset A$ is then given by

$$A_y = \{a\in A|\ T(pAap)y = \{0\}\}.$$

(c) Let (T, X) and (T', X') be two simple A-modules. Let us assume that there is a (pAp)-linear isomorphism

$$\Phi : X_p \longrightarrow X'_p,$$

i.e., such that

$$\Phi(T(pap)y) = T'(pap)\Phi(y),\ y \in X_p.$$

Then there is a unique extension of Φ to an A-linear isomorphism between X and X'.

Proof

(a) Let $x \in X_p, x \neq 0$ and let $y \in X_p$. Since X is simple A-module, there exists $a \in A$ such that $a \cdot x = y$. Then $y = p \cdot y = p \cdot (a \cdot x) = (p \cdot a \cdot p) \cdot x$. Therefore $(T|_{pAp}, X_p)$ is a simple $p \cdot A \cdot p$ module.

(b) If $a \in A_y$ then $a \cdot p \cdot y = a \cdot y = 0$. Therefore

$$p \cdot A \cdot a \cdot p \cdot y = p \cdot A \cdot 0 = \{0\}.$$

If now $p \cdot A \cdot a \cdot p \cdot y = \{0\}$ but $a \cdot y \neq 0$, then there exists $b \in A$ such that $b \cdot a \cdot y = y$ and then

$$p \cdot b \cdot a \cdot p \cdot y = y \neq 0.$$

This contradiction shows that $a \in A_y$.

(c) We have now for $y' := \Phi(y)$ that

$$\begin{aligned} A_y &= \{a \in A;\ p \cdot Aa \cdot p \cdot y = \{0\}\} \\ &= \{a \in A;\ \Phi(p \cdot Aa \cdot p \cdot y) = \{0\}\} \\ &= \{a \in A;\ p \cdot Aa \cdot p \cdot y' = \{0\}\} \\ &= A_{y'}. \end{aligned}$$

Therefore the A-modules (T, X) and (T', X') are equivalent and the isomorphism $\tilde{\Phi}$ between them is necessarily unique, since $\tilde{\Phi}_{|X_p} = \Phi$. □

Let us now start with a simple (pAp)-module (S, Y) and let us show that it is equivalent to a (pAp)-module of the form $(T|_{X_p}, X_p)$, where (T, X) is a simple A-module. For any left ideal I of A let $(L, A/I)$ be the left-multiplication module of A defined on A/I.

Proposition 8.2.27 *Assume that (S, Y) is a simple (pAp)-module.*

(a) Then A/A_y is a simple A-module, where

$$A_y = \{a \in A \mid S(pAap)y = 0\}.$$

(b) The simple (pAp)-module (S, X_p) is equivalent to $(L|_{(p \cdot A/A_y)}, p \cdot A/A_y)$.

(c) If (T, X) is a simple A-module such that $(T|_{pAp}, X_p)$ is equivalent to (S, X_p), then (T, X) is equivalent to the simple A-module $(L, A/A_y)$.

Proof

(a) Let us take $a_1, a_2 \in A$ such that $a_1 \notin A_y$. Then there is $a_0 \in A$ such that $S(pa_0a_1p)y \neq 0$. As (S, Y) is a simple (pAp)-module, there exists $b_0 \in pAp$ such that $S(b_0(pa_0a_1p))y = y$. Then

$$S(pa(a_2 - (a_2b_0pa_0)a_1)p)y = 0, \quad \forall a \in A$$

and

$$a_2 = (a_2b_0pa_0)a_1 \bmod A_y.$$

This shows that the vector $a_1 + A_y$ is cyclic. Hence the module $(L, A/A_y)$ is simple.

(b) The map

$$p \cdot A/A_y \longrightarrow Y; \; p \cdot (a \bmod A_y) \longmapsto S(pap)y$$

is a (pAp)-linear isomorphism between $(L|_{p\cdot A/A_y}, p \cdot A/A_y)$ and (S, Y).

(c) follows from assertion (b) and Proposition 8.2.26, (c). □

Case of Boidol's Group

Let $\mathfrak{b}$ be the real four-dimensional Lie algebra

$$\mathfrak{b} := \{T, X, Y, Z\}$$

with non-trivial Lie brackets

$$[T, X] = -X, \; [T, Y] = Y, [X, Y] = Z.$$

The connected, simply connected group B with Lie algebra $\mathfrak{b}$ is called Boidol group. It may be realized on $\mathbb{R}^4$ with multiplication

$$(t, x, y, z) \cdot (t', x', y', z) = (t + t', e^{t'}x + x', e^{-t'}y + y', z + z' + \frac{1}{2}(e^{t'}xy' - e^{-t'}x'y)).$$

The inverse of (t, x, y, z) is given by

$$(t, x, y, z)^{-1} = (-t, -e^{-t}x, -e^{t}y, -z).$$

We see that B is an extension of the three-dimensional Heisenberg group H by $\mathbb{R}$.

Define the element D in the enveloping algebra of $\mathfrak{b}$ by

$$D := TZ - XY.$$

Then D is central. Indeed

$$\begin{aligned}[D, X] &= [T, X]Z - X[Y, X] \\ &= -XZ - X(-Z) = 0,\end{aligned}$$

$$\begin{aligned}[D, Y] &= [T, Y]Z - [X, Y]Y \\ &= YZ - ZY = 0,\end{aligned}$$

$$\begin{aligned}[D, T] &= 0 - [X, T]Y - X[Y, T] \\ &= -XY - X(-Y) = 0.\end{aligned}$$

Let

$$H := \{(0, x, y, z); x, y, z \in \mathbb{R}\}, \ \mathscr{Z} := \{(0, 0, 0, z); z \in \mathbb{R}\}$$

Then H is the Heisenberg group realized as normal closed subgroup of B and $\mathscr{Z}$ is the center of B.

The Boidol Group's L^p-representations Let for $(\rho, \lambda) \in \mathbb{R} \times \mathbb{R}^*$, $p \in [1, +\infty[$

$$\pi^p_{\rho,\lambda} := \operatorname{ind}_P^G \chi_{(\rho,\lambda),p}$$

on the space $L^p(B/P, \chi_{\rho,\lambda})$, where $P = \exp \mathfrak{p}$, $\mathfrak{p} := \operatorname{span}\{T, Y, Z\}$ and where

$$\chi_{(\rho,\lambda)}(\exp tT + yY + zZ) := e^{-i\rho t - \lambda z}, t, y, z \in \mathbb{R}.$$

Let $(t, x, y, z) \in B$, $\xi \in L^p(\mathbb{R}) = L^p(B/P, \chi_{\rho,\lambda})$, $u \in \mathbb{R}$. Then

$$\pi^p_{\rho,\lambda}(t, x, y, z)\xi(u) = e^{t/p} e^{-i\rho t} e^{-i\lambda z} e^{-i\lambda xy/2} e^{i\lambda e^t yu} \xi(e^t u - x).$$

Hence for $F \in L^1(B)$, $\xi \in L^p(\mathbb{R})$, $u \in \mathbb{R}$, we have that

$$\begin{aligned}\pi^p_{\rho,\lambda}(F)\xi(u) &= \int_G F(g)\pi_{\rho,\lambda}(g)\xi(u)dg \\ &= \int_{\mathbb{R}} \Big(\int_{\mathbb{R}} e^{t/p} e^{-i\rho t} \hat{F}^{3,4}(t, e^t u - x, -\frac{\lambda}{2}(x + e^t u), \lambda)dt \Big)\xi(x)dx.\end{aligned}$$

Therefore $\pi^p_{\rho,\lambda}(F)$, $F \in L^1(B)$, is a kernel operator with kernel function

$$F^p_{\rho,\lambda}(u,x) = \int_{\mathbb{R}} e^{t/p} e^{-i\rho t} \hat{F}^{3,4}(t, e^t u - x, -\frac{\lambda}{2}(x + e^t u), \lambda) dt, \, u, x \in \mathbb{R}.$$

For $p = +\infty$ let us consider the space $C_0(B/P, \chi_{\rho,\lambda})$ and we obtain the irreducible representation $\pi^\infty_{\rho,\lambda}$ of the group B. Then

$$\pi^\infty_{\rho,\lambda}(F)\xi(u) = \int_{\mathbb{R}} \int_H (\pi^\infty_{\rho,\lambda}(\exp tT)\pi_\lambda(F(t))\xi)(u) dt.$$

Hence $\pi^\infty_{\rho,\lambda}(F)$, $F \in L^1(B)$, is a kernel operator with kernel function

$$F_{\rho,\lambda}(u,x) = \int_{\mathbb{R}} e^{-i\rho t} \hat{F}^{3,4}(t, e^t u - x, -\frac{\lambda}{2}(x + e^t u), \lambda) dt, \, u, x \in \mathbb{R}.$$

Let us compute for $(\rho, \lambda) \in \mathbb{R} \times \mathbb{R}^*$ and $p \in [1, \infty]$ the operator $d\pi^p_{\rho,\lambda}(D)$, which acts on the smooth vectors ξ of the representation space. We have

$$\begin{aligned} d\pi^p_{\rho,\lambda}(T)\xi(u) &= \frac{d}{dt}(\pi^p_{\rho,\lambda}(\exp tT))_{t=0}\xi(u) \\ &= ((1/p - i\rho)\xi(u) + u\xi'(u) \\ d\pi^p_{\rho,\lambda}(X)\xi(u) &= -\xi'(u) \\ d\pi^p_{\rho,\lambda}(Y)\xi(u) &= i\lambda e^t u\xi(u) \\ d\pi^p_{\rho,\lambda}(Z)\xi(u) &= -i\lambda\xi(u), \, u \in \mathbb{R}. \end{aligned}$$

Hence

$$\begin{aligned} d\pi^p_{\rho,\lambda}(D)\xi(u) &= -i\lambda((1/p - i\rho)\xi(u)) - i\lambda u\xi'(u)) + (i\lambda u\xi)'(u) \\ &= (-i\lambda/p) - \rho\lambda)\xi(u) - i\lambda u\xi'(u) + i\lambda\xi(u) + i\lambda u\xi'(u) \\ &= (-i\lambda(1/p - 1) + \rho\lambda)\xi(u). \end{aligned}$$

Therefore

$$d\pi^p_{\rho,\lambda}(D) = (i\lambda(1 - 1/p) + \rho\lambda)\mathbb{I}.$$

We see from this relation that the simple $L^1(B)$ modules $(\pi^{p,fin}_{\rho,\lambda}, (L^p(\mathbb{R}))^{fin})$, $p \in [1, \infty]$, are all inequivalent (see also Lemma 8.2.9).

Some Elements of $Prim(B)$ Let (T, X) be a simple $L^1(B)$ module. Then by Schur's lemma, there exists a unitary character $\chi_\lambda(\exp zZ) = e^{-i\lambda z}$, $z \in \mathbb{R}$, of the center of B such that

$$T(\exp zZ) = \chi_\lambda(\exp zZ)\mathbb{I}_X, \; z \in \mathbb{R}.$$

Let for $\lambda \neq 0$

$$\pi^\lambda := \mathrm{ind}_H^B \pi_\lambda$$

be the induced representation π_λ of $\hat{H}$ from the Heisenberg subgroup H to the whole group B. This representation acts on $L^2(\mathbb{R}^2) = L^2(\mathbb{R}, L^2(\mathbb{R}))$. Then for $\psi \in L^2(\mathbb{R}, L^2(\mathbb{R}))$, $s \in \mathbb{R}$:

$$\begin{aligned}\pi^\lambda(t,h)\varphi(s) &= \varphi(h^{-1}\exp(-tT)\exp sT)\\ &= \pi_\lambda(\exp(t-s)Th\exp(s-t)T)(\varphi(s-t))\\ &= a_{t-s} \circ \pi_\lambda(h) \circ a_{s-t}(\varphi(s-t)).\end{aligned}$$

Here a_t, $t \in \mathbb{R}$, is the unitary operator

$$a_t(\xi)(u) := e^{t/2}\xi(e^t u), \; u, t \in \mathbb{R}, \xi \in L^2(\mathbb{R}). \tag{8.2.5}$$

So for $F \in L^1(B)$

$$\pi^\lambda(F)\varphi(s) = \int_{\mathbb{R}} (a_{-t} \circ (\pi_\lambda(F(s-t)) \circ a_t)(\varphi(t))dt,$$

where $F(t)(h) := f(\exp(tT)h)$, $h \in H, t \in \mathbb{R}$.

The kernel $\ker_{L^1(B)}\pi^\lambda$ is the ideal

$$\ker_{L^1(B)}\pi^\lambda = \{F \in L^1(B); \int_{\mathbb{R}} F(g\exp zZ)\chi_\lambda(\exp zZ)dz, \; g \in G\}$$

For any $s \in \mathbb{R}$, it follows that

$$\begin{aligned}\ker_{L^1(B)}\pi^\lambda &= \{F \in L^1(B); \pi_\lambda(F(t)) = 0, t \in \mathbb{R}\}\\ &= \{F \in L^1(B); \int_{\mathbb{R}} F(t)(h\exp(zZ))\chi_\lambda(\exp zZ)dz = 0, t \in \mathbb{R}, h \in H\}\\ &= \{F \in L^1(B); \int_{\mathbb{R}} F(g\exp(zZ))\chi_\lambda(\exp zZ)dz = 0, g \in B\}.\end{aligned}$$

Let

$$\begin{aligned}L^1(B)_\lambda := \{F : B \to \mathbb{C};\ \text{measurable}\ , F(g\exp zZ) = \chi_\lambda(\exp -zZ), z \in \mathbb{R}, g \in G\\ \|F\|_1 := \int_{G/\mathscr{Z}} |F(g)|d\dot{g} < \infty\}.\end{aligned}$$

Let $P_\lambda : L^1(B) \to L^1(B)_\lambda$ be defined by

$$P_\lambda(F)(g) := \int_{\mathbb{R}} F(gz)\chi_\lambda(z)dz, \, g \in G.$$

Then P_λ is a bounded surjective homomorphism of the Banach algebra $L^1(B)$ onto the Banach algebra $L^1(B)_\lambda$. The kernel of P_λ is given by

$$\begin{aligned}\ker_{L^1(B)}(P_\lambda) &= \{F \in L^1(B); \int_{\mathbb{R}} F(g \exp zZ)e^{-i\lambda z}dz = 0, \, g \in G\}\\ &= \ker_{L^1(B)}(\pi^\lambda).\end{aligned}$$

Hence $L^1(B)_\lambda$ is isomorphic to the Banach algebra $L^1(B)/\ker_{L^1(B)}\pi^\lambda$.

Let now $\xi, \eta \in \mathcal{S}(\mathbb{R})$ and define the element $p_{\xi,\eta} \in L^1(H)_\lambda$ as in (8.2.3) through the relation

$$\pi_\lambda(p_{\xi,\eta}) = P_{\xi,\eta}.$$

Then we know that ${}_{\xi,\eta}$ is a Schwartz function. Since

$$P_{\xi,\xi} \circ P_{\eta,\eta} = \langle \eta, \xi \rangle P_{\xi,\eta},$$

we have that

$$p_{\xi,\xi} * p_{\eta,\eta} = \langle \eta, \xi \rangle p_{\xi,\eta}. \tag{8.2.6}$$

In particular, if $\|\xi\| = 1$ it follows that $p_\xi := p_{\xi,\xi}$ is a self-adjoint idempotent element in the algebra $L^1(H)_\lambda$, and it defines by convolution a self-adjoint idempotent multiplier of the algebra $L^1(B)_\lambda$. Let

$$L^1(B)_\xi := p_\xi * L^1(B)_\lambda * p_\xi.$$

Then $L^1(B)_\xi$ is a closed self-adjoint subalgebra of $L^1(B)_\lambda$. We have, for $\varphi \in C_c(\mathbb{R}, L^2(\mathbb{R}))$ and $F \in L^1(B)_\lambda$ with compact support in the first variable, that

$$\begin{aligned}\pi^\lambda(p_\xi * F * p_\xi)\varphi(s) &= a_{-s} \circ P_\xi \circ a_s(\int_{\mathbb{R}} (a_{-t} \circ \pi_\lambda(F(s-t)) \circ a_t \circ a_{-t} \circ P_\xi \circ a_t)(\varphi(t))dt)\\ &= a_{-s} \circ P_\xi((\int_{\mathbb{R}} (a_{s-t} \circ \pi_\lambda((F(s-t)) \circ P_\xi \circ a_t)(\varphi(t))dt)\\ &= a_{-s}(\xi) \int_{\mathbb{R}} \langle a_t(\varphi(t)), \xi \rangle \langle (a_{s-t}(\pi_\lambda((F(s-t))\xi)), \xi \rangle)dt.\end{aligned}$$

Let us write for $F = p_\xi * F * p_\xi \in L^1(B)_\xi$:

$$q(t) = \langle a_t(\pi_\lambda(F(t))(\xi)), \xi\rangle, t \in \mathbb{R}.$$

Then

$$|q(t)| \leq \|F(t)\|_1, t \in \mathbb{R},$$

and

$$\pi^\lambda(F) = \pi^\lambda(p_\xi * F * p_\xi)\varphi(s) = a_{-s}(\xi) \int_{\mathbb{R}} q(t)\langle a_t(\varphi(t)), \xi\rangle dt \quad (8.2.7)$$

Let also

$$F_q(t) := q(t) p_{a_{-t}(\xi),\xi}, t \in \mathbb{R}.$$

Then

$$\begin{aligned} \pi^\lambda(F_q)\varphi(s) &= \int_{\mathbb{R}} (a_{-t} \circ (\pi_\lambda(F_q(s-t)) \circ a_t)(\varphi(t))dt \\ &= \int_{\mathbb{R}} q(t)(a_{-t} \circ (P_{a_{t-s}(\xi),\xi}) \circ a_t)(\varphi(t))dt \\ &= \int_{\mathbb{R}} q(t)((P_{a_{-s}(\xi),\xi}) \circ a_t)(\varphi(t))dt \\ &= a_{-s}(\xi) \int_{\mathbb{R}} q(t)\langle a_t(\varphi(t)), \xi\rangle dt. \end{aligned} \quad (8.2.8)$$

Therefore, comparing (8.2.7) with (8.2.8) we see that

$$\pi^\lambda(F) = \pi^\lambda(F_q).$$

Since the representation π^λ is injective on $L^1(B)_\lambda$ it follows that

$$F = F_q.$$

Hence

$$\|F\|_1 = \int_{\mathbb{R}} |q(t)| \|p_{a_{-t}\xi,\xi}\|_1 dt.$$

Let us estimate the $L^1(H)_\lambda$-norm of the functions $f(t) := p_{a_t\xi,\xi}, t \in \mathbb{R}$, for some $\xi \in \mathscr{S}(\mathbb{R})$ of L^2-norm equal to one. We have, by (8.2.6),

$$p_{a_t\xi,\xi} = \langle \xi, a_t\xi\rangle^{-1} p_{a_t\xi} * p_\xi, t \in \mathbb{R}.$$

Let now $i_t, t \in \mathbb{R}$, be the automorphism of the Heisenberg group H defined by

$$i_t(x, y, z) := (e^{-t}x, e^t y, z), (x, y, z) \in H_n.$$

For $f \in L^1(H)$ let

$$i_t(f)(h) := f(i_t^{-1}(h)), h \in H, t \in \mathbb{R}.$$

Then i_t is a unimodular isomorphism of the algebras $L^1(H)$ and $L^1(H)_\lambda$. Furthermore we check that

$$\begin{aligned}\pi_\lambda(i_t(x, y, z)) &= a_t \circ \pi_\lambda(x, y, z) \circ a_{-t}, , (x, y, z) \in H\\ &\Rightarrow\\ \pi_\lambda(i_t(f)) &= a_{-t} \circ \pi_\lambda(f) \circ a_t, f \in L^1(H), t \in \mathbb{R}.\end{aligned}$$

In particular

$$p_{a_t\xi,\xi} = \langle a_t(\xi), \xi\rangle^{-1} i_{-t}(p_\xi) * p_\xi, t \in \mathbb{R}.$$

Therefore

$$\|p_{a_t\xi,\xi}\|_1 \leq |\langle a_t(\xi), \xi\rangle^{-1}| \|p_\xi\|_1^2, t \in \mathbb{R}.$$

If we choose for ξ an even non-negative continuous function which decreases on $\mathbb{R}_+$, then for $t \geq 0$:

$$\begin{aligned}|\langle a_t(\xi), \xi\rangle| &= e^{\frac{t}{2}} \int_{\mathbb{R}} \xi(e^t x)\xi(x)dx\\ &\geq e^{\frac{t}{2}} | \int_{\mathbb{R}} \xi(e^t x)\xi(e^t x)dx\\ &= e^{\frac{-t}{2}} \|\xi\|_2^2, t \in \mathbb{R}.\end{aligned}$$

For $t \leq 0$ we obtain then

$$\begin{aligned}|\langle a_t(\xi), \xi\rangle| &= |\langle \xi, a_{-t}(\xi)\rangle|\\ &= |\langle a_{-t}(\xi), \xi\rangle|\\ &\geq e^{\frac{t}{2}} \|\xi\|_2^2, t \in \mathbb{R}.\end{aligned}$$

This shows that

$$\|p_{a_{-t}\xi,\xi}\|_1 \le C_\xi e^{\frac{|t|}{2}}, t \in \mathbb{R},$$

for some constant $C_\xi > 0$.

Let us write

$$\omega_\xi(t) = \omega(t) := \|p_{a_{-t}\xi,\xi}\|_1, t \in \mathbb{R}, \tag{8.2.9}$$

The representations $\pi^p_{\rho,\lambda}$, $p \in [1, \infty]$, are bounded on $L^1(G)_\lambda$ by the L^1-norm. For $F_q \in L^1(B)_\xi$ we then have, by (8.2.5),

$$\begin{aligned}\pi^p_{\rho,\lambda}(F_q) &= \int_{\mathbb{R}} q(t)e^{-i\rho t}e^{(1/p-1/2)t}a_t \circ P_{a_{-t}\xi,\xi}dt\\ &= (\int_{\mathbb{R}} q(t)e^{-i\rho t}e^{(1/p-1/2)t}dt)P_{\xi,\xi}\end{aligned}$$

for unit $\xi \in \mathscr{S}(\mathbb{R})$. Hence for every $p \in [1, \infty]$, $\rho \in \mathbb{R}$, and $q \in C_c(\mathbb{R})$ we get:

$$\begin{aligned}|\int_{\mathbb{R}} q(t)e^{-i\rho t}e^{(1/p-1/2)t}dt| &= |\langle\pi^p_{\rho,\lambda}(F_q)\xi, \xi\rangle|\\ &\le \|\xi\|_p\|\xi\|_{\frac{p}{p-1}}\|F_q\|_1\\ &= \|\xi\|_p\|\xi\|_{\frac{p}{p-1}}\int_{\mathbb{R}}|q(t)|\omega(t)dt.\end{aligned}$$

Hence we obtain the estimate

$$e^{(1/p-1/2)t} \le \|\xi\|_p\|\xi\|_{\frac{p}{p-1}}\omega(t), t \in \mathbb{R}, p \in [1, \infty].$$

Finally we see that the function ω is equivalent to the function

$$t \mapsto e^{\frac{|t|}{2}}, t \in \mathbb{R}.$$

Let now $F_q(t) := q(t)p_{a_{-t}(\xi),\xi} t \in \mathbb{R}$, and similarly $F_{q'}(t) := q'(t)p_{a_{-t}(\xi),\xi}, t \in \mathbb{R}$, where $q, q' \in L^1(\mathbb{R}, e^{|t|/2})$. Then

$$\begin{aligned}F_q * F_{q'}(s)(h) &= \int_{\mathbb{R}}\int_H F(u,k)F'((u,k)^{-1}(s,h))dkdu\\ &= \int_{\mathbb{R}}\int_H F_q(u,k)F_{q'}(s-u,((u-s)\cdot k^{-1})h)dkdu\\ &= \int_{\mathbb{R}}\int_H F_q(u,(s-u)\cdot k)F'_{q'}(s-u,k^{-1})h)dkdu\end{aligned}$$

$$= \int_{\mathbb{R}} F_q^{(s-u)}(u) *_H F_{q'}(s-u)(h)du$$
$$= \int_{\mathbb{R}} q(u)q'(s-u)p_{a_{u-s}(a_{-u}(\xi),a_{u-s}(\xi))} * p_{a_{u-s}\xi,\xi}(h)du$$
$$= \int_{\mathbb{R}} p_{a_{-s}(\xi),a_{u-s}(\xi))} p_{a_{u-s}\xi,\xi} du$$
$$= q * q'(s) p_{a(-s)\xi,\xi}(h)$$
$$= F_{q*q'}(s)(h), s \in \mathbb{R},$$

where $F^s(u, h) := F(u, s \cdot h), h \in H, s \in \mathbb{R}$.

Let $\xi \in \mathscr{S}(\mathbb{R})$ and let $t \in \mathbb{R}$. Let

$$\xi_t := a_t(\xi) = t \cdot \xi, t \in \mathbb{R}.$$

We have seen above that ω is a continuous function on $\mathbb{R}$ such that $\omega(t) \geq 1, t \in \mathbb{R}$.

Furthermore, for $s, t \in \mathbb{R}$ we have that

$$\begin{aligned} P_{a_{-s-t}\xi,\xi} &= a_{-t} \circ P_{a_{-s}\xi,\xi} \circ a_t \circ a_{-t} \circ P_{\xi,\xi} \\ &= a_{-t} \circ P_{a_{-s}\xi,\xi} \circ a_t \circ P_{a_{-t}\xi,\xi} \\ &= a_{-t} \circ (P_{a_{-s}\xi,\xi} \circ a_t \circ P_{a_{-t}\xi,\xi} \circ a_{-t}) \circ a_t. \end{aligned}$$

Hence

$$p_{a_{-s-t}\xi,\xi} = i_t(p_{a_{-s}\xi,\xi} * i_{-t}(p_{a_{-t}\xi,\xi})), t \in \mathbb{R},$$

and so

$$\begin{aligned} \omega(s+t) &= \|p_{a_{-t-s}\xi,\xi}\|_1 \\ &= \|i_t(p_{a_{-s}\xi,\xi} * i_{-t}(p_{a_{-t}\xi,\xi}))\|_1 \\ &\leq \|p_{a_{-s}\xi,\xi}\|_1 \|i_{-t}(p_{a_{-t}\xi,\xi})\|_1 \\ &= \omega(s)\omega(t). \end{aligned}$$

Furthermore

$$\begin{aligned} \omega(t) &= \|P_{a_{-t}\xi,\xi}\|_{\text{op}} \\ &= \|P^*_{a_{-t}\xi,\xi}\|_{\text{op}} \\ &= \|P_{\xi,a_{-t}\xi}\|_{\text{op}} \\ &= \|P_{a_t\xi,\xi}\|_{\text{op}} \\ &= \omega(-t), t \in \mathbb{R}. \end{aligned}$$

Therefore ω is a symmetric weight on $(\mathbb{R}, +)$, and the mapping

$$F_q \to q$$

is an algebra isomorphism between $L^1(G)_\xi$ and $L^1(\mathbb{R}, e^{|t|/2})$.

Now every character χ of the algebra $L^1(\mathbb{R}, \omega)$ is of the form:

$$\langle \chi, F \rangle = \int_{\mathbb{R}} e^{-\alpha t} q(t) dt, q \in L^1(\mathbb{R}, \omega),$$

where $\alpha = a + i\rho \in \mathbb{C}$ with $|a| \leq \frac{1}{2}$. On the other hand there exists exactly one $p \in [1, \infty]$ such that $a = \frac{1}{p} - \frac{1}{2}$.

As a conclusion, using Propositions 8.2.26 and 8.2.27, we can say that the simple modules of the Banach algebra $L^1(G)$ which are not trivial on the center of the group B are given by the finite-rank submodules of the Banach-space representations $\pi^p_{\rho,\lambda}$, for $(\rho, \lambda) \in \mathbb{R} \times \mathbb{R}^*$ and $p \in [1, \infty]$.

8.3 Irreducible Banach-Space Representations and Projections

8.3.1 Submodules of an Irreducible Module

In the following, A denotes a normed algebra and $(a_j)_j$ a two-sided bounded approximate unit in A. We write

$$\|a_j\|_A \leq C$$

where C is a positive constant. Let p be a multiplier of A as in Sect. 8.2.19. Hence due to the existence of the bounded approximate unit, we have that

$$p(ab) = (pa)b, (ab)p = a(bp); \quad \forall a, b \in A.$$

Let us assume furthermore that p is idempotent, i.e.,

$$p^2 = p, \quad pAp \neq 0.$$

Hence $(pa_jp)_j$ is a bounded approximate unit of the algebra pAp.

Let (T, X) be a bounded non-degenerate A-module, i.e., such that $T(A)X$ is dense in X. In particular there exists a constant $C_T > 0$ such that

$$\|T(a)w\|_X \leq C_T \cdot \|a\|_A \cdot \|w\|_X \quad \forall a \in A, \forall w \in X.$$

Lemma 8.3.1 *Let (T, X) be a non-degenerate bounded A-module X on a Banach space. Then p acts in a natural way on X. Furthermore the operator $T(p)$ is a bounded projection and the subspace $T(p)X$ is closed in X. We have that*

$$T(p)T(a) = T(pa),\ T(a)T(p) = T(ap), \quad \forall a \in A.$$

Proof We define the action of p on $T(A)X$ by

$$T(p)(T(a)u) = T(pa)u \quad \forall a \in A, \forall u \in X.$$

Since we have a bounded approximate unit, this action is well defined, bounded and can be extended to an action of p on the whole space X, since $T(A)X$ is dense in X by hypothesis. To show that $T(p)X$ is closed, suppose that

$$T(p)(u_j) \longrightarrow u$$

in X. Then

$$T(p)(u_j) = T(p)(T(p)(u_j)) \longrightarrow T(p)u$$

and $u = T(p)u \in T(p)X$. □

Proposition 8.3.2 *Let (T, X) be an irreducible A-module. If $T(p)X \not\equiv 0$, then*

$$(T|_{pAp}, T(p)X)$$

is an irreducible (pAp)-module.

Proof As (T, X) is irreducible, (T, X) is not degenerate. Take $T(p)u \neq 0$ in $T(p)X$. Then $T(A)(T(p)u) = T(Ap)(T(p)u)$ is dense in X by the irreducibility of T, and $T(pAp)(T(p)u) = T(p)(T(A)(T(p)u))$ is dense in $T(p)X$. □

Proposition 8.3.3 *Every irreducible A-module (T, X) with $T(p)(X) \not\equiv 0$ can be identified with an A-module of the form $\overline{Ap/M}^{\|\!|\cdot|\!\|_X}$, where*

$$M = M_w = \{x \in Ap; \quad T(pAx)w = \{0\}\} \quad (0 \neq w \in T(p)X \textit{ fixed})$$

is a left ideal A contained in Ap and where the new norm $|\!|\!| \cdot |\!|\!|_X$ of X satisfies the condition

$$|\!|\!| \cdot |\!|\!|_X \le C_T \| \cdot \|_{Ap/M}$$

with

$$\|\dot{a}\|_{Ap/M} = \inf_{m \in M} \|a + m\|_A \le \|a\|_A \quad \forall a \in Ap.$$

Proof Let us fix $w = T(p)w \in T(p)X$ such that $\|w\|_X = 1$. Put

$$M = \{a \in Ap\ ;\ T(a)w = 0\}.$$

Hence M is a closed left ideal in A contained in Ap. The mapping

$$\varphi : Ap \longrightarrow T(Ap)wa \longmapsto T(a)w$$

is a linear surjection with kernel M. Hence

$$\Phi : Ap/M \longrightarrow T(Ap)w\dot{a} \longmapsto T(a)w$$

is a linear bijection. Define $\|\!|\cdot|\!\|_X$ on Ap/M by

$$\|\!|\dot{a}|\!\|_X = \|T(a)w\|_X \quad \forall a \in A.$$

As $T(Ap)w$ is dense in the irreducible module X, Φ extends to a linear isometry between $X^{\|\!|} = \overline{Ap/M}^{\|\!|\cdot|\!\|_X}$ and $(X, \|\cdot\|_X)$. Furthermore, there exists a constant C_T such that

$$\|\!|\dot{a}|\!\|_X = \|T(a+m)w\|_X \leq C_T \cdot \|a+m\|_A \cdot \|w\|_X = C_T \cdot \|a+m\|_A$$

for every $m \in M$. Hence

$$\|\!|\dot{a}|\!\|_X \leq C_T \cdot \inf_{m \in M} \|a+m\|_A = C_T \cdot \|\dot{a}\|_{Ap/M} \leq C_T \cdot \|a\|_A.$$

Finally, the action of A on Ap/M is given by $L(a)\dot{x} = a \cdot \dot{x} = (ax)^{\dot{}}$ and can be extended to a continuous action of A on $\overline{Ap/M}^{\|\!|\cdot|\!\|_{\mathcal{U}}}$. The mapping Φ is then an isomorphism of A-modules. Let now

$$M_0 = \big\{x \in Ap\ ;\ T(pAx)w = \{0\}\big\}.$$

We remark that $M \subset M_0$ and that M_0/M is a A-invariant subspace of Ap/M. Hence $(M_0/M)^- = \{0\}$ in $\overline{Ap/M}^{\|\!|\cdot|\!\|_X}$, whence $M_0 = M$. □

Remark 8.3.4 In the previous proposition, the left ideal M is completely determined by the (pAp)-module $T(p)X$. In order to determine every irreducible module of A, for which $T(p)X$ is equivalent to the given (pAp)-module (S, Y), it suffices, so to speak, to find all norms $\|\!|\cdot|\!\|$ on Ap/M whose restriction to pAp/M is equivalent to Y and such that $\overline{(Ap/M)}^{\|\!|\cdot|\!\|}$ is an irreducible A-module.

8.3.2 Minimal Norm and Extension Norms

Definition 8.3.5 Let (S, Y) be an irreducible (pAp)-module and let

$$M_w = M = \{x \in Ap\ ;\ S(pAxp)w = \{0\}\}$$

where $Y \ni w \neq 0$ has been arbitrarily fixed. Then M is a closed left ideal in A contained in Ap and

$$\|\dot{x}\|_{\min} = \sup_{\|a\|_{\mathbb{I}_{\mathscr{A}}} \leq 1} \|S(pax)w\|$$

defines a norm on Ap/M.

One easily checks the following properties of $\|\cdot\|_{\min}$

$$\begin{aligned} \|a\dot{x}\|_{\min} &\leq \|a\|_{\mathbb{I}_{\mathscr{A}}} \cdot \|\dot{x}\|_{\min}, \\ \|p\dot{x}\|_{\min} &\leq K\|\dot{x}\|_{\min}, \\ \|\dot{x}\|_{\min} &\leq C'\|x\|_{\mathbb{I}_{\mathscr{A}}}, \\ \frac{1}{C''}\|S(px)w\|_{\mathbb{I}_{\mathscr{W}}} \leq \|\dot{px}\|_{\min} &\leq C''\|S(px)w\|, \\ & a \in \mathscr{A}, \dot{x} = x + M \in Ap/M, \end{aligned}$$

for some positive constants K, C' and C''.

Definition 8.3.6 We call *extension norm* every norm $|||\cdot|||$ on Ap/M which satisfies the relations

$$|||a \cdot \dot{x}||| \leq C_{|||}\|a\|_A|||\dot{x}|||,$$

for any $\dot{x} = x + M \in Ap/M$, $a \in A$ and some positive constant $C_{|||}$, and

$$|||\cdot|||_{|pAp/M} \simeq \|S(\cdot)w\|_Y.$$

Example 8.3.7 The norm $\|\cdot\|_{\min}$ is evidently an extension norm according to (8.3.1).

Proposition 8.3.8 *Let $|||\cdot|||$ be an extension norm on Ap/M. Then there exists $C' = C'_{|||}$ such that*

$$\|\dot{x}\|_{min} \leq C'|||\dot{x}|||, \quad |||p\dot{x}||| \leq C'|||\dot{x}|||,$$

$$|||\dot{x}||| \leq C'\|\dot{x}\|_{Ap/M} \leq C'\|x\|_{\mathscr{A}} \quad \forall \dot{x} = x + M \in Ap/M.$$

Proof According to our hypothesis on $|||\cdot|||$, there exists a positive constant D such that

$$\frac{1}{D}|||b+M||| \leq \|S(b)w\| \leq D|||b+M|||, \quad \forall b \in pAp.$$

Hence for every $\dot{x} = x + M \in Ap/M$, for all $m \in M$,

$$\|\dot{x}\|_{\min} = \sup_{\|a\|_A \leq 1} \|S(pa(x+m))w\|_Y$$

$$\leq D \sup_{\|a\|_A \leq 1} |||pa(x+M)||| \leq DC_{|||} K |||x+M|||.$$

On the other hand, according to the factorization theorem in [29] we have that $Ap = Ap \cdot pAp$ and there exists for every $\epsilon > 0$, for all $x \in Ap$, a $x_1 \in Ap$ such that $\|x_1 - x\|_A \leq \epsilon$ and a $b \in pAp$ such that $\|b\|_A \leq CK^2$ and such that $x = x_1 \cdot b$. Hence

$$|||\dot{x}||| = |||x_1 \cdot (b+M)||| \leq C_{|||} \|x_1\|_{\mathscr{A}} \cdot |||b+M|||$$

$$\leq DC_{|||} \|x_1\|_A \|S(b)w\|_Y \leq CK^2 DC_{|||} C_S \|w\|_Y (\|x\|_A + \epsilon).$$

Finally, when ϵ tends to 0, by choosing $C' = C'_{|||}$ sufficiently large we find

$$|||\dot{x}||| \leq C'\|x\|_A,$$

$$\|\dot{x}\|_{\min} \leq C'|||\dot{x}|||$$

and

$$|||p\dot{x}||| = \lim_j |||pa_j\dot{x}||| \leq C_{|||} \sup_j \|pa_j\|_{\mathscr{A}} |||\dot{x}||| \leq C_{|||} KC|||\dot{x}||| \leq C'|||\dot{x}|||.$$

□

Definition 8.3.9 Let $|||\cdot|||$ be an extension norm on Ap/M. We denote by $X^{|||}$ the completion Ap/M for the norm $|||\cdot|||$. We write $(L, X^{|||})$ for the A-module obtained by continuously extending from Ap/M to $\overline{(Ap/M)}^{|||\cdot|||}$ both the left multiplication map and the action of p.

8.3.3 Topologically Simple Norms

Proposition 8.3.10 *Let $|||\cdot|||$ be an extension norm on Ap/M. The module $(L, X^{|||})$ is cyclic and the (pAp)-module $(L, L(p)X^{|||}) = (L, \overline{(pAp/M)}^{|||})$ is isomorphic to the module (S, Y). Furthermore one can choose the cyclic vector of $(L, X^{|||})$ to live in $L(p)X^{|||}$.*

Proof The mapping

$$\mathscr{F} : pAp/M \to Y, \quad \mathscr{F}(b + M) = S(b)w,$$

is injective, since

$$Y_0 = \{b \in pAp\ ;\ S(pApb)w = \{0\}\} = \{b \in pAp\ ;\ S(b)w = 0\} = M \cap pAp.$$

Furthermore $\mathscr{F}$ is (pAp)-equivariant. As the norm $|||\cdot|||$ on $p\mathscr{A}p/M$ is equivalent to the norm $\|S(\cdot)w\|$, the mapping $\mathscr{F}$ can be extended to an isomorphism of the (pAp)-modules $Y^{|||} = \overline{(pAp/M)}^{|||} \subset X^{|||}$ and Y. On the other hand, left multiplication with p is continuous on Ap/M for the norm $|||\cdot|||$. Hence $L(p)$ is a well-defined bounded projection on $X^{|||}$. Furthermore it is easy to see that $L(p)X^{|||}$ is the closure of pAp/M in $X^{|||}$.

Let us show that $X^{|||}$ is cyclic. Let $\xi \in X^{|||}$ and let $\epsilon > 0$. There exists $\dot{x} = x + M \in Ap/M$ such that $|||\xi - \dot{x}||| < \epsilon$.

The factorization theorem in [29] tells us that $Ap = A \cdot pAp$, hence

$$Ap/M = L(A)(pAp)/M.$$

Therefore there exists $a \in A$ and $b \in pAp$, such that $\dot{x} = L(a)\dot{b}$. Let us take $\omega \in L(p)X^{|||}$, such that $\mathscr{F}(\omega) = w$. Then $\mathscr{F}(L(b)\omega) = S(b)w$ and so $L(b)\omega = b + M$. Hence $L(ab)\omega = \dot{x}$ and $|||\xi - L(ab)\omega||| < \epsilon$, implying $(L, X^{|||})$ is cyclic. □

Remark 8.3.11 Proposition 8.3.8 implies that for every extension norm $|||\cdot|||$ on Ap/M there exists a bounded linear mapping $i_{X^{|||}} : X^{|||} \to X^{\min} = X^{\|\cdot\|_{\min}}$, which is the identity on Ap/M. This allows us to give the following definition

Definition 8.3.12 Any norm $|||\cdot|||$ on Ap/M such that the previous remark's map $i_{X^{|||}}$ is injective will be called a *topologically simple extension norm* (TS for short).

Proposition 8.3.13 *The kernel $\mathscr{N}$ of $i_{X^{|||}}$ is the set of all $\xi \in X^{|||}$ such that $L(pc)\xi = 0$ for every $c \in A$.*

Proof Let $\xi \in \mathscr{N}$ and $c \in A$. Then as $i_{X^{|||}}$ is A-equivariant, we have that

$$i_{X^{|||}}(L(pc)(\xi)) = L(pc)(i_{X^{|||}}\xi) = 0.$$

Since $|||\cdot||| \cong \|\cdot\|_{\min}$ on pAp/M, then $i_{\mathscr{Y}^{|||}}$ is injective on $L(p)X^{|||} = \overline{pAp/M}^{|||} \simeq Y$ and $L(pc)\xi = 0$.

Conversely, if $L(pc)\xi = 0$ for every c, then $i_{X^{|||}}(L(pc)\xi) = 0$ and thus

$$\sup_{\|c\|_A \le 1} \|L(pc)(i_{X^{|||}}\xi)\|_{\min} = 0.$$

Let us show that

$$\sup_{\|c\|_A \le 1} \|L(pc)\eta\|_{\min} \ge \frac{1}{C}\|\eta\|_{\min} \tag{8.3.1}$$

for every $\eta \in X^{\min}$. Indeed, for $x \in A$, we have that

$$\sup_{\|c\|_A \le 1} \|L(pc)\dot{x}\|_{\min} = \sup_{\|c\|_A \le 1} \sup_{\|d\|_A \le 1} \|S(pdpcx)w\|_Y$$

$$\ge \frac{1}{C} \lim_j \sup_{\|c\|_A \le 1} \|S(pa_j pcx)w\|_Y = \frac{1}{C} \sup_{\|c\|_A \le 1} \|S(pcx)w\|_Y = \frac{1}{C}\|\dot{x}\|_{\min}.$$

On the other hand, by (8.3.5) we have

$$\sup_{\|c\|_A \le 1} \|L(pc)\eta\|_{\min} \le K\|\eta\|_{\min}$$

for every $\eta \in X^{\min}$. Hence (8.3.1) is valid for every η in $X^{\min}$ and so, if $\xi \in X^{|||}$ satisfies $L(pc)\xi = 0$ for every c, then $\|i_{\mathscr{Y}^{|||}}(\xi)\|_{\min} = 0$ and $\xi \in \mathscr{N}$. □

Example 8.3.14

(a) The norm $\|\cdot\|_{\min}$ on $X^{\min}$ is evidently TS.
(b) Suppose that $\mathscr{A}pA$ is dense in A. Then every extension norm $|||\cdot|||$ is a TS extension norm.

    Indeed if $\|i_{X^{|||}}\xi\|_{\min} = 0$ for some $\xi \in X^{|||}$, i.e., $L(pa)\xi = 0$ for every $a \in \mathscr{A}$, then by the previous proposition $\{0\} = (L(A)L(pA)\xi)^- = (L(A)\xi)^-$ and so $\xi = 0$.

Proposition 8.3.15 *Let $|||\cdot|||$ be an extension norm on Ap/M. The A-module $(L, X^{|||})$ is irreducible if and only if $|||\cdot|||$ is TS.*

Proof If $|||\cdot|||$ is not TS, then the closed $L(A)$-invariant subspace

$$\mathscr{N} = \{\xi \in X^{|||}; \quad L(pA)\xi = \{0\}\}$$

is not $\{0\}$ by (8.3.13). Hence $X^{|||}$ is not topologically simple.

Let us suppose that $|||\cdot|||$ is TS, and let us show that $(L, X^{|||})$ is topologically simple. Let $\xi, \eta \in X^{|||}$, $\eta \ne 0$. Let $\epsilon > 0$. Since $X^{|||}$ has a cyclic vector ω contained in $L(p)X^{\|} = Y^{\|}$, there exists $a \in A$ such that $|||\xi - L(a)\omega||| < \epsilon$. As the vector η is different from 0 and the norm $|||\cdot|||$ is TS, there exists $c \in A$ such that $Y^{\|} \ni L(pc)\eta \neq 0$. The (pAp)-module $Y^{\|}$ is topologically simple. Hence there exists $b \in pAp$ such that $|||L(b)L(pc)\eta - \omega||| < \frac{\epsilon}{C_{\|}\|a\|_{\mathbb{I}_{\mathscr{A}}}+1}$. Let $d = abpc \in A$. Then

$$|||L(d)\eta - \xi||| \leq |||\xi - L(a)\omega||| + |||L(a)\omega - L(d)\eta|||$$

$$\leq \epsilon + C_{\|}\|a\|_{\mathbb{I}_{\mathscr{A}}}|||\omega - L(bpc)\eta||| < 2\epsilon.$$

Hence $X^{\|}$ is topologically simple. □

Remark 8.3.16 Take an extension norm $|||\cdot|||$ on Ap/M. Let us suppose now that the closed subspace

$$\mathscr{N} = \{\xi \in X^{\|}; \quad \|i_{X^{\|}}\xi\|_{\min} = 0\}$$

is different from 0. Then on the quotient space

$$\mathscr{Y}^{\|} = X^{\|}/\mathscr{N}$$

the quotient norm $|||\xi|||' = \inf_{\eta \in \mathscr{N}} |||\xi + \eta|||$ becomes a TS norm. Indeed, the mapping $i_{\mathscr{Y}^{\|}}$ applied to $\mathscr{Y}^{\|}$ is injective by construction and it is easy to see that the properties of (8.3.6) hold for $|||\cdot|||'$.

Remark 8.3.17 All TS extensions of the (pAp) module (S, Y) can be realized as Banach submodules $X^{\|} = \overline{Ap/M}^{\|}$ of the universal module $(L, X^{\min})$.

Indeed, the canonical mapping

$$(Ap/M, |||\cdot|||) \to (Ap/M, \|\cdot\|_{\min})$$

is bounded and A-equivariant and can be extended to an injective intertwining operator of $X^{\|}$ onto a dense subspace of $X^{\min}$, which contains $\mathscr{A}p/M$.

8.4 Restricting and Extending Ideals

8.4.1 Definitions

Definition 8.4.1 Let H be a closed normal subgroup of the locally compact group G. Suppose that there exists a closed subset $S = S^{-1}$ in G containing the identity e such that the mapping $S \times H \to G$; $(s, h) \to s \cdot h$ is a homeomorphism. We say that S is a *complement* for H. Then S is homeomorphic to the locally compact group G/H.

We obtain in this way the two coordinate functions $\sigma : G \to S$ and $\iota : G \to H$ defined by the identification $G = S \cdot H$, $g = \sigma(g) \cdot \iota(g)$.

We can provide S with the multiplication coming from the group G/H:

$$s \cdot s' := \sigma(ss'), s, s' \in S,$$

and then put on it the Haar measure $ds = d\dot{g}$ of the group G/H. We then have

$$\int_G \varphi dg = \int_S \int_H \varphi(sh) dh ds, \varphi \in C_c(G).$$

Recall that the function

$$\tilde{\varphi}(g) := \int_H \varphi(gh) dh, \varphi \in C_c(G),$$

is contained in $C_c(G/H)$ (see [70]). Let

$$D : L^1(G) \to L^1(S, L^1(H))$$
$$D(F)(s)(h) := F(sh), h \in H, s \in S.$$

Then D is a bijective linear isometry between the two Banach spaces $L^1(G)$ and $L^1(S, L^1(H))$. Similarly, let

$$D^* : L^\infty(S, L^\infty(H)) \to L^\infty(G)$$
$$D^*(\varphi)(sh) := \varphi(s)(h), s \in S, h \in H.$$

Then D^* is a bijective linear isometry between the two Banach spaces $L^\infty(S, L^1(H))$ and $L^\infty(G)$.

Definition 8.4.2 Let G be a locally compact group and let H be a normal closed subgroup of G admitting a complement. Let I be a closed ideal of the algebra $L^1(G)$.

1. We say that I is $L^\infty(G/H)$-invariant if for every $\varphi \in L^\infty(G/H)$ and every $F \in I$ the function $\varphi \cdot F$ is also contained in I.
2. Define the *restriction* $r_H(I) = r(I)$ of I to H by

$$r(I) := \overline{\{F_{|H};\ F \in I, s \mapsto F(s) \in L^1(H) \text{ is continuous}\}}.$$

3. Let J be a closed G-invariant ideal of the algebra $L^1(H)$. Let

$$e(J) := \overline{\text{span}\{\alpha * f;\ f \in J, \alpha \in C_c(G)\}}^{L^1(G)}$$

be the *extension* of J to G.

Lemma 8.4.3 *Let $\alpha \in C_c(G)$, $F \in L^1(G)$. The mapping*

$$s \to L^1(H),\ \ s \to \alpha * F(s)$$

is continuous.

Proof Indeed we have for $u \in G$ that

$$\begin{aligned}\alpha * F(u) &= \int_G \alpha(v)F(v^{-1}u)dv \\ &= \int_G \alpha(uv^{-1})\Delta_G(v^{-1})F(v)dv \\ &= \int_S\int_H \alpha(uk^{-1}t^{-1})\Delta_G(t^{-1}k^{-1})F(tk)dkdt.\end{aligned}$$

Then

$$\begin{aligned}\|\alpha * F(s) - \alpha * F(s')\|_1 &= \int_H |\int_S\int_H \Big(\alpha(shk^{-1}t^{-1}) - \alpha(s'hk^{-1}t^{-1})\Big)\Delta_G(t^{-1}k^{-1})F(tk)dkdt|dh \\ &\le \int_H\int_S(\int_H |\alpha(shk^{-1}t^{-1}) - \alpha(s'hk^{-1}t^{-1})|dh)\Delta_G(t^{-1}k^{-1})|F(tk)|dkdt \\ &= \int_H\int_S(\int_H |\alpha(st^{-1}h) - \alpha(s't^{-1}h)|dh)|F(tk)|dkdt \\ &= \int_S(\int_H |\alpha(st^{-1}h) - \alpha(s't^{-1}h)|dh)|F(t)|_1dt \\ &\le \int_S \||\widetilde{\alpha}|(st^{-1}) - |\widetilde{\alpha}|(s't^{-1})\||F(t)|_1dt.\end{aligned}$$

Now, since the function $|\widetilde{\alpha}|$ is uniformly continuous, the lemma follows. □

Remark 8.4.4 Lemma 8.4.3 shows that for every closed ideal I in $L^1(G)$ the closed subspace $r(I)$ is well defined. Since I is translation-invariant we obtain a closed G-invariant ideal of $L^1(H)$.

It is also clear that for every closed G-invariant ideal J in $L^1(H)$ the extension $e(J)$ is a closed ideal of $L^1(G)$ since $e(J)$ is translation-invariant by definition.

8.4.2 Description of Extended and Restricted Ideals

Proposition 8.4.5 *Let J be a G-invariant closed ideal in $L^1(H)$. Then $e(J)$ is $L^\infty(G/H)$-invariant.*

Proof Indeed for any $\varphi \in L^\infty(G/H)$, $\alpha \in L^1(G)$ and $f \in L^1(H)$ we have that

$$\begin{aligned}\varphi(g)(\alpha * f)(g) &= \varphi(g)\int_H \alpha(gh)f(h^{-1})dh \\ &= \int_H \varphi(gh)\alpha(gh)f(h^{-1})dh \\ &= (\varphi\alpha) * f(g), g \in G.\end{aligned}$$

Hence

$$\varphi(\alpha * f) = (\varphi\alpha) * f.$$

Therefore $e(J)$ is $L^\infty(G/H)$-invariant. □

Proposition 8.4.6 *Let I be a closed $L^\infty(G/H)$-invariant ideal of $L^1(G)$. Then*

$$I = e(r(I)).$$

Proof Take $\varphi \in I^\perp \subset L^\infty(G)$ and $F \in I$ such that the mapping $s \to F(s) \in L^1(H)$ is continuous. Then for any $\psi \in L^\infty(G/H)$ we have that

$$\begin{aligned}0 &= \int_G \varphi(g)\psi(g)F(g)dg \\ &= \int_S\int_H \varphi(sh)\psi(s)F(sh)dhds \\ &= \int_S \psi(s)\int_H \varphi(sh)F(sh)dhds.\end{aligned}$$

This shows that

$$\int_H \varphi(gh)F(gh)dh = 0 \tag{8.4.1}$$

for all $g \in G$. For $F \in I$ and $\alpha, \beta \in C_c(G)$ we have that

$$\begin{aligned}\alpha * \beta * F(g) &= \int_S\int_H \Delta_G(t^{-1})\Delta_G(h)^{-1}\alpha(gh^{-1}t^{-1})\beta * F(th)dhdt \\ &= \int_S (r(t)\alpha) * (\beta * F)(t)(g)dt, g \in G,\end{aligned}$$

where

$$r(u)F(v) := \Delta_G(u^{-1})F(vu^{-1}), F \in L^1(G), v \in G.$$

Now $(r(t)\alpha) * ((\beta * F)(t)) \in e(r(I))$ for every $t \in S$ and $\int_S \|\beta * F(t)\|_{1,H} dt = \|\beta * F\|_{1,G} < \infty$. Therefore $\alpha * \beta * F \in e(r(I))$, whence $F \in e(r(I))$, since $e(r(I))$ is closed. Therefore $I \subset e(r(I))$.

Let now $\varphi \in I^{\perp} \subset L^{\infty}(G)$ and let $F \in I$ such that the mapping $s \to F(s)$ is continuous. Then

$$\begin{aligned} \langle \varphi, \alpha * (F_{|H}) \rangle &= \int_G \int_H \Delta_G(h^{-1}) \varphi(u) \alpha(uh^{-1}) F(h) dh du \\ &= \int_G \alpha(u) \int_G \varphi(uh) F(h) dh du \\ &\overset{(8.4.1)}{=} \int_G \alpha(u) \int_S 0 \, du \\ &= 0. \end{aligned}$$

Therefore $\alpha * (F_{|H})$ is contained in I for every $\alpha \in C_c(G)$, $F \in I$, and so

$$e(r(I)) \subset I.$$

□

Proposition 8.4.7 *Let I be an $L^{\infty}(G/H)$-invariant closed ideal in $L^1(G)$. Then*

$$I = \{F \in L^1(G);\ F(s) \in r(I),\ s\text{-almost everywhere}\}$$

Furthermore, the Banach spaces $L^1(G)/I$ and $L^1(S, L^1(H)/r(I))$ are isometrically isomorphic.

Proof Let

$$I' := \{F \in L^1(G);\ F(s) \in r(I),\ s\text{-almost everywhere}\}$$

Then it is clear that I' is a closed subspace of $L^1(G)$ and that I' is translation-invariant.

We have seen in the proof of Proposition 8.4.6 (use (8.4.1)) that

$$I^{\perp} = \{\varphi \in L^{\infty}(G);\ \varphi(s) \in r(I)^{\perp}, s\text{-almost everywhere}\}. \tag{8.4.2}$$

This shows that $I' \subset I$ and $I \subset I'$.

Consider the mapping

$$D_I : L^1(G) \to L^1(S, L^1(H)/r(I));\ D_I(F)(s) := F(s) \bmod r(I), s \in S.$$

Then $D_I(I) = \{0\}$ and so D_I factorizes through I. This gives us an injective mapping

$$\tilde{D_I} : L^1(G)/I \to L^1(S, L^1(H)/r(I)).$$

Relation (8.4.2) tells us that the mapping $\tilde{D_I}$ is in fact isometric. Obviously the image of D_I is dense in $L^1(S, L^1(H)/r(I))$. Hence $\tilde{D_I}$ is also surjective. □

8.5 Polynomial Growth and Functional Calculus

8.5.1 *Definitions and Elementary Properties*

Definition 8.5.1 Let G be a locally compact group. Let K be a bounded measurable subset of G. We denote by $|K|$ the Haar measure of K.

For a subset S of G and $n \in \mathbb{N}$, let

$$S^n := \{s_1 \cdot \cdots \cdot s_n;\ s_i \in S, i = 1, \cdots, n\}.$$

We say that G has *polynomial growth* if, for any compact subset K of G, we have a polynomial function $p = p_K : \mathbb{Z} \to \mathbb{R}_+$ such that

$$|K^n| \le p(n), n \in \mathbb{N}.$$

Remark 8.5.2 If G is compactly generated and of polynomial growth, then there exist $d \in \mathbb{N}$, and for every compact subset K of G a constant $C_K > 0$, such that

$$|K^n| \le C_K n^d, n \in \mathbb{N}.$$

Indeed, if we fix a compact symmetric neighbourhood U of G, which generates G, then for every compact subset K of G, we have that

$$K \subset U^N$$

for some $N = N_K \in \mathbb{N}$. There exists $d \in \mathbb{N}$ and a constant $C > 0$ such that

$$|U^n| \le Cn^d, n \in \mathbb{N}.$$

Therefore

$$K^n \subset (U^N)^n \Rightarrow |K^n| \le |U^{N \cdot d}| \le C(n \cdot N)^d = C_K n^d, n \in \mathbb{N} \text{ (where } C_K := CN^d).$$

Definition 8.5.3 Let G be a locally compact, compactly generated group. We call *growth* of G the smallest $d \in \mathbb{N}$ such that

$$|K|^n \leq C_K n^d, n \in \mathbb{N}$$

for any compact subset K in G.

Example 8.5.4

- Let $d \in \mathbb{N}^*$. Then the growth of the group $(\mathbb{R}^d, +)$ is obviously equal to d.
- Let $(N = \exp \mathfrak{n}, \cdot)$ be a connected and simply connected nilpotent Lie group. Then N is compactly generated by the compact subset $U = \exp B_1$, where B_1 is the unit ball $B_1 := \{X \in \mathbb{N}; \|X\| \leq 1\}$ (for some norm $\| \|$ of $\mathfrak{n}$). It follows from the proof of Proposition 6.2.5 in [70] that there exist a constant $C > 0$ and an $m \in \mathbb{N}$ such that

$$\| \log x \| \leq C j^m, x \in U^j, j \in \mathbb{N}.$$

Therefore

$$|U^n| \leq |B_{Cn^m}| \leq C_1 n^{\dim G \cdot m}, n \in \mathbb{N}.$$

- A group G with polynomial growth is unimodular. Indeed, if there exists $x \in G$ such that $\Delta_G(x) > 1$, then take any compact neighbourhood U of e containing x. But now $U \cdot x^n \subset U^{n+1}, n \in \mathbb{N}$, and so

$$\begin{aligned}
\Delta_G(x)^n \cdot |U| &= \Delta_G(x)^n \cdot \int_G 1_U(t) dt \\
&= \int_G 1_U(tx^{-n}) dt \\
&= \int_G 1_{U \cdot x^n}(t) dt \\
&= |U \cdot x^n| \\
&\leq |U^{n+1}| \\
&\leq C(n+1)^d
\end{aligned}$$

for some $C > 0, d \in \mathbb{N}$. Therefore $\Delta_G(x)$ cannot be different from 1.

Definition 8.5.5 Let G be a locally compact unimodular group. A *weight* ω on G is a Borel function

$$\omega : G \longrightarrow [C, +\infty[$$

(for some $C > 0$) such that

$$\omega(st) \leq \omega(s)\omega(t) \quad \forall s, t \in G.$$

All weights are assumed to be symmetric, i.e.,

$$\omega(s^{-1}) = \omega(s) \quad \forall s \in G.$$

Moreover, we shall always require the function ω to be bounded on compact sets. We say that two weights ω, ω' on G are *equivalent* if there exist two constants $0 < c < C$ such that

$$c\,\omega(g) \leq \omega'(g) \leq C\omega(g), \; g \in G.$$

Let us show that we can replace any measurable weight ω by an equivalent continuous weight.

Let $\alpha \in C_c(G)$ with values in $\mathbb{R}_{\geq 0}$ be such that

$$\alpha(s^{-1}) = \alpha(s), s \in G, \text{ and } \int_G \alpha(s)ds = 1.$$

Let $\omega' := \alpha * \omega * \alpha$, i.e.,

$$\omega'(s) = \int_G \int_G \alpha(u)\omega(u^{-1}sv)\alpha(v^{-1})dudv, s \in G.$$

Then we see that

$$\begin{aligned} \omega'(s) &\leq \int_G \int_G \alpha(u)\omega(u^{-1})\omega(s)\omega(v)\alpha(v^{-1})dudv \\ &= C_1\omega(s) \\ &\text{where } C_1 = \int_G \alpha(u)\omega(u)du \int_G \omega(v)\alpha(v)dv \end{aligned}$$

and

$$\omega'(s) \geq C \int_G \int_G \alpha(u)\alpha(v^{-1})dudv = C,$$

where $C > 0$, such that $\omega(s) \geq C, s \in G$. Furthermore

$$\begin{aligned} \omega(s) &= \int_G \int_G \alpha(u)\alpha(u^{-1})\omega(s)dudv \\ &= \int_G \int_G \alpha(u)\omega(uu^{-1}svv^{-1})\alpha(u^{-1})dudv \end{aligned}$$

$$\leq \int_G \int_G \alpha(u)\omega(u)\omega(u^{-1}sv)\omega(v^{-1})\alpha(u^{-1})dudv$$
$$\leq C_2 \int_G \int_G \alpha(u)\omega(u^{-1}sv)\alpha(u^{-1})dudv$$
$$= C_2\omega'(s),$$

where $C_2 = \sup_{u,v\in \mathrm{supp}(\alpha)} \omega(u)\omega(v)$. Therefore

$$\begin{aligned}\omega'(ss') &\leq C_1\omega((ss'))\\ &\leq C_1\omega(s)\omega(s')\\ &\leq C_1C_2^2\omega'(s)\omega'(s'), s, s' \in G.\end{aligned}$$

We replace now ω' by $C_3\omega'$ where $C_3 := C_1C_2^2$. Then this new function ω' satisfies the relationships

$$\omega'(ss') \leq \omega'(s)\omega'(s'), s, s' \in G,$$

and

$$\omega'(s) \geq C_3C, s \in G.$$

It is easy to see that ω' is also a continuous function. Since α is uniformly continuous and compactly supported, we can fix a compact neighbourhood U of the identity e, and then for every $\varepsilon > 0$ we have a compact neighbourhood $U_\varepsilon \subset U$ of e such that for any $s, t \in U_\varepsilon$ we have that

$$|\alpha(su) - \alpha(tu)| \leq \varepsilon 1_K(u), u \in G,$$

where $K := U \cdot \mathrm{supp}(\alpha)$.

Hence for any $s, t \in U_\varepsilon, x \in G$, we have that

$$|(|\omega'(sx) - \omega'(tx)) \leq C_3|\int_G \int_G (\alpha(sxu)\omega(uv)\alpha(v^{-1}) - \alpha(txu)\omega(uv)\alpha(v^{-1})dudv|$$
$$\leq C_3 \int_G \int_G |\alpha(sxu) - \alpha(txu)|\omega(u)\omega(v)\alpha(v)dudv|$$
$$\leq C_3\varepsilon \int_G \int_G 1_K(xu)\omega(u)\omega(v)\alpha(v)dudv$$
$$= C_4(x)\varepsilon$$

for the new constant

$$C_4(x) := C_3 \int_G \int_G 1_K(xu)\omega(u)\omega(v)\alpha(v)dudv \leq C_3|K| \sup_{u\in x^{-1}K} \omega(u) \int_G \alpha(v)\omega(v)dv.$$

Definition 8.5.6 Let G be a compactly generated locally compact group of polynomial growth. Let $U = U^{-1}$ be a compact, symmetric generating neighbourhood of the identity e. We define the fundamental weight ω_U of G by

$$|x| = |x|_U = \omega_U(x) = 1 + \tau_U(x),\, x \in G, \text{ where } \tau_U(x) = \min\{n \in \mathbb{N};\, x \in U^n\}.$$

We say that a given weight ω on G is *polynomial* if it is dominated by a power of ω_U. This means that there exist $C > 0$ and $N \in \mathbb{N}$ such that

$$\omega(x) \leq C\omega_U^N(x),\, x \in G.$$

Let us define

$$s(n) = s_n = \sup_{x\in U^n} \omega(x) \quad \text{for } n \in \mathbb{N}^*,\, s(0) = 1.$$

Then s is an increasing function from $\mathbb{N}$ to $\mathbb{R}$.

Furthermore, let us consider the function

$$\tilde{\omega}(x) = \sup_{y\in U^n} \omega(y) = s(|x|).$$

This function is also a polynomial weight on G, that depends only on $|x|$, since $\omega_U = \widetilde{\omega_U}$. It satisfies $\omega \leq \tilde{\omega}$ and is an increasing function of $|x|$. For a weight ω on G and $p \in [1, \infty[$, we let as before

$$L^p(G, \omega) := \{f : G \to \mathbb{C};\, f \text{ measurable}, \|f\|_{\omega,p}^p = \int_G |f(t)|^p \omega(t)dt < \infty\}$$

Then $L^1(G, \omega)$ is an involutive Banach algebra (contained in $L^1(G)$).

It is clear that equivalent weights ω, ω', give the same space $L^p(G, \omega) = L^p(G, \omega')$.

By changing to an equivalent Banach algebra norm on $L^1(G, \omega)$, we may assume, without loss of generality, that the weight ω is continuous.

We denote by $L\mathscr{P}(G)$ the involutive Fréchet algebra

$$L\mathscr{P}(G) := \bigcap_{\omega \text{ polynomial}} L^1(G, \omega) \cap L^2(G)$$

$$= \{f \in L^2(G);\, \int_G |f(t)|\omega_U(t)^N dt < \infty,\, N \in \mathbb{N}\}.$$

Since the convolution algebra $C_c(G)$ of continuous functions on G with compact support is contained in $L\mathscr{P}(G)$, it follows that $L\mathscr{P}(G)$ is dense in $L^1(G,\omega)$ for every weight ω on G.

8.5.2 Principles of Functional Calculus

Let G be a compactly generated, locally compact group with polynomial growth and let ω be a polynomial weight on G. One aim of this subsection is the construction of a functional calculus for self-adjoint functions $f = f^*$ that are contained in $L\mathscr{P}(G)$, along the lines of [48] and [91]. To do this, let us define

$$u(nf) = \sum_{k=1}^{+\infty} \frac{n^k f^{*k}}{k!} = e^{inf} - \delta_e, \quad n \in \mathbb{Z},$$

where δ_e is the Dirac measure. We have then the estimate:

$$\|u(nf)\|_\omega \leq e^{|n|\|f\|_\omega}, n \in \mathbb{Z},$$

and hence $u(nf) \in L^1(G,\omega)$, $n \in \mathbb{Z}$. But in order to construct a functional calculus, we shall need a much better bound for $\|u(nf)\|_\omega$. This bound will be given in Sect. 8.5.3.

We shall then look for non-zero 2π-periodic maps $\varphi : \mathbb{R} \to \mathbb{R}$ with $\varphi(0) = 0$ and such that

$$\varphi\{f\} = \sum_{n\in\mathbb{Z}} u(nf)\hat{\varphi}(n)$$

converges in $L^1(G,\omega)$. As a matter of fact we shall even require the convergence of

$$\sum_{n\in\mathbb{Z}} \|u(nf)\|_\omega |\hat{\varphi}(n)|.$$

Here $\hat{\varphi}$ denotes the Fourier transform of φ on the interval $[0, 2\pi]$ defined by

$$\hat{\varphi}(n) = \frac{1}{2\pi}\int_0^{2\pi} f(x)e^{-inx}\,dx.$$

8.5.3 Estimate for $\|u(nf)\|_\omega$

Let $U = U^{-1}$ be a compact generating neighbourhood of the identity element e in G. As G has polynomial growth, there exist constants $Q \in \mathbb{N}$ and $C > 0$ such that the Haar measure of U^n satisfies

$$|U^n| \le Cn^Q.$$

We have the following result:

Theorem 8.5.7 *Let $f = f^* : G \to \mathbb{C}$ be in $L\mathscr{P}(G)$. Let $r : \mathbb{N} \to \mathbb{N}$ be an arbitrary strictly increasing function. We then have the following bounds: there exist positive constants K_1, K_2 such that*

$$\|u(nf)\|_\omega \le K_1(1+|n|)(1+|n|r(|n|))^{\frac{Q}{2}} s(|n|r(|n|))e^{K_2(\frac{|n|}{s(r(|n|))})},$$

for all $n \in \mathbb{Z}$.

Proof See Sect. 8.5.5. □

Remark 8.5.8 We observe that for polynomial weights the ω-norms of $u(nf)$ are polynomially bounded in n. Hence if $\varphi : \mathbb{R}/\mathbb{T} \to \mathbb{C}$ is smooth then $\varphi\{f\}$ converges for every self-adjoint element in $L\mathscr{P}(G)$.

8.5.4 Properties of Functional Calculus

Functional calculus has the following properties.

For any self-adjoint function $f = f^* \in L\mathscr{P}(G)$, for any smooth functions φ and ψ,

$$(\varphi+\psi)\{f\} = \varphi\{f\} + \psi\{f\}, (\lambda\varphi)\{f\} = \lambda(\varphi\{f\}), \quad \forall \lambda \in \mathbb{R}.$$

If χ is any character of the Abelian Banach $*$-algebra generated by f, then

$$\chi(\varphi\{f\}) = \varphi(\chi(f)).$$

If π is any $*$-representation of $L^1(G,\omega)$ on a Hilbert space H, then

$$\pi(\varphi\{f\}) = \varphi(\pi(f))$$

where $\varphi(\pi(f))$ is the operator obtained by the functional calculus of Hermitian operators.

If φ and ψ are real-valued and smooth, and $f = f^* \in L\mathcal{P}(G)$, then $\varphi\{f\} \in L\mathcal{P}(G)$ is self-adjoint and so

$$\psi\{\varphi\{f\}\} = \psi \circ \varphi\{f\}.$$

Furthermore for two smooth functions φ, ψ on $\mathbb{R}/\mathbb{T}$ and $f = f^* \in L\mathcal{P}(G)$ we have that

$$\varphi\{f\} * \psi\{f\} = \varphi \cdot \psi\{f\}.$$

To see this, one may apply the left regular representation to $(\varphi \cdot \psi)\{f\}$ and $\psi \circ \varphi\{f\}$.

Remark 8.5.9 Instead of the "discrete" definition $\varphi\{f\} = \sum_{n\in\mathbb{Z}} u(nf)\hat{\varphi}(n)$, one may also use a "continuous" definition given by $\varphi\{f\} = \frac{1}{2\pi}\int_{\mathbb{R}} u(\lambda f)\hat{\varphi}(\lambda)d\lambda$ for a real-valued $\mathscr{C}^\infty$-function φ with compact support, such that $\varphi(0) = 0$ and such that $\hat{\varphi}$ has an appropriate decay. In fact, there exists an equivalent result of Mandelbrojt for the decay of $\hat{\varphi}(\lambda)$. The properties of functional calculus remain of course the same.

8.5.5 Computation of the Bound Used in Functional Calculus

Let G be a compactly generated locally compact group of polynomial growth and let $U = U^{-1}$ be a compact neighbourhood of the identity e. Let ω be a polynomial weight on G. For our estimates we may assume that $\omega = \tilde{\omega}$. In particular, $\omega(x) \geq \omega(y)$ for any $y \in U^n$, $x \in G \setminus U^n$, $n \in \mathbb{N}$. The methods we use are an adaptation of those in [91].

Let us use the following decomposition of $u(f)$:

$$u(f) = g + k$$

with

$$g = g_n := u(f) \cdot \chi_{U^{r(n)}} \quad \text{and} \quad k = k_n := u(f) \cdot \chi_{(U^{r(n)})^c},$$

where χ_A stands for the characteristic function of the set A. Then

$$\begin{aligned}\|k\|_1 = & \int_{(U^{r(n)})^c} |u(f)(x)|dx \leq \frac{1}{\inf_{x\in(U^{r(n)})^c}\omega(x)}\|u(f)\|_\omega \\ & \leq \frac{1}{\sup_{x\in U^{r(n)}}\omega(x)}\|u(f)\|_\omega = \frac{1}{s(r(n))}\|u(f)\|_\omega.\end{aligned}$$

Consider the map

$$(u(f)+\delta_e): L^2(G) \longrightarrow L^2(G): F \longmapsto (u(f)+\delta_e)*F = e^{if}*F = u(f)*F+F.$$

This is a unitary operator on $L^2(G)$ and hence $\|u(f)+\delta_e\|_{L^2\to L^2}=1$. Similarly, $\|u(nf)+\delta_e\|_{L^2\to L^2}=1$ and so

$$\|(u(f)+\delta_e)*F\|_2 = \|e^{if}*F\|_2 \le \|F\|_2,\ \ \|(u(nf)+\delta_e)*F\|_2 = \|e^{inf}*F\|_2 \le \|F\|_2.$$

for $n\in\mathbb{Z}$. Finally

$$\|g+\delta_e\|_{L^2\to L^2} = \|u(f)+\delta_e-k\|_{L^2\to L^2} \le 1+\|k\|_{L^2\to L^2} \le 1+\|k\|_1 \le 1+\frac{1}{s(r(n))}\|u(f)\|_\omega$$

and

$$\|(g+\delta_e)*F\|_2 \le (1+\frac{\|u(f)\|_\omega}{s(r(n))})\|F\|_2.$$

Let $n\in\mathbb{N}$, let us make the following estimates (using the methods of [91]):

$$\|(\delta_e+u(f))^{*n}\|_\omega = \|\big((\delta_e+g)+k\big)^{*n}\|_\omega \le \sum_{m=0}^{n}\sum_{\underline{a},\underline{b}}\|(\delta_e+g)^{a_1}*k^{b_1}*\cdots*(\delta_e+g)^{a_n}*k^{b_n}\|_\omega,$$

where $\underline{a}=(a_1,\ldots,a_n)\in\{0,1\}^n, \underline{b}=(b_1,\ldots,b_n)\in\{0,1\}^n$, $|\underline{a}|=m, |\underline{b}|=n-m$, $a_i+b_i=1$, $(\delta_e+g)^{a_j}=\delta_e$ if $a_j=0$ and $k^{b_j}=\delta_e$ if $b_j=0$. Fix $\underline{a}$ and $\underline{b}$ as above.

Assume that $\underline{a}\neq 0$. Then

$$\begin{aligned}
&\|(\delta_e+g)^{a_1}*k^{b_1}*\cdots*(\delta_e+g)^{a_n}*k^{b_n}\|_\omega\\
&=\int_G\Big|\int_G\cdots\int_G\big((\delta_e+g)^{a_1}*\delta_{s_1}^{b_1}*\cdots*(\delta_e+g)^{a_n}*\delta_{s_n}^{b_n}\big)(s)\\
&\cdot k(s_1)^{b_1}\cdots k(s_n)^{b_n}ds_1\cdots ds_n\Big|\omega(s)ds\\
&\le\int_G\cdots\int_G\|(\delta_e+g)^{a_1}*\delta_{s_1}^{b_1}*\cdots*(\delta_e+g)^{a_n}*\delta_{s_n}^{b_n}\|_\omega\\
&\cdot|k(s_1)^{b_1}\cdots k(s_n)^{b_n}|ds_1\cdots ds_n,
\end{aligned}$$

where the convention is such that there is no integral in ds_ℓ if $b_\ell=0$. One can easily check that

$$\begin{aligned}
&\|(\delta_e+g)^{a_1}*\delta_{s_1}^{b_1}*\cdots*(\delta_e+g)^{a_n}*\delta_{s_n}^{b_n}\|_\omega\\
&\le\sum_{a_j=1}\|(\delta_e+g)^{a_1}*\delta_{s_1}^{b_1}*\cdots*(\delta_e+g)^{a_{j-1}}*\delta_{s_{j-1}}^{b_{j-1}}*g*\delta_{s_j}^{b_j}*\cdots*\delta_{s_n}^{b_n}\|_\omega+
\end{aligned}$$

$$+ \|(\delta_e + g)^0 * \delta_{s_1}^1 * \cdots * (\delta_e + g)^0 * \delta_{s_n}^1\|_\omega$$

$$\leq \sum_{a_j=1} \|(\delta_e + g)^{a_1} * \delta_{s_1}^{b_1} * \cdots * (\delta_e + g)^{a_{j-1}} * \delta_{s_{j-1}}^{b_{j-1}} * g * \delta_{s_j}^{b_j} * \cdots * \delta_{s_n}^{b_n}\|_\omega + \prod_{b_i=1} \omega(s_i)$$

Let us write $B(n) = U^n$. Then for $\underline{a} \neq 0$, the function

$$(\delta_e + g)^{a_1} * \delta_{s_1}^{b_1} * \cdots * (\delta_e + g)^{a_{j-1}} * \delta_{s_{j-1}}^{b_{j-1}} * g * \delta_{s_j}^{b_j} * \cdots * \delta_{s_n}^{b_n}$$

is supported in $B(nr(n) + |s_1|^{b_1} + \cdots + |s_n|^{b_n})$ with the convention that $|s_\ell|^{b_\ell} = 0$ if $b_\ell = 0$. Moreover

$$\begin{aligned}
&\left|B(nr(n) + |s_1|^{b_1} + \cdots + |s_n|^{b_n})\right| \\
&\leq C \cdot \left(nr(n) + |s_1|^{b_1} + \cdots + |s_n|^{b_n}\right)^Q \\
&\leq C \cdot (1 + nr(n) + |s_1|^{b_1} + \cdots + |s_n|^{b_n})^Q \leq C \cdot (1 + nr(n))^Q \prod_{b_i=1} (1 + |s_i|)^Q.
\end{aligned}$$

If h is any function with compact support in $B(r) = U^r$, then

$$\begin{aligned}
\|h\|_\omega &= \int_{U^r} |h(x)|\omega(x)dx \\
&\leq \|h\|_2 \|\omega 1_{U^r}\|_2 \\
&\leq s(r)\|h\|_2 |B(r)|^{\frac{1}{2}}.
\end{aligned}$$

In addition

$$\|g\|_2 \leq \|u(f)\|_2$$

since $g = u(f) \cdot \chi_{U^{r(n)}}$. Finally

$$\begin{aligned}
&\|(\delta_e + g)^{a_1} * \delta_{s_1}^{b_1} * \cdots * (\delta_e + g)^{a_n} * \delta_{s_n}^{b_n}\|_\omega \\
&\leq \sum_{a_j=1} \left(1 + \frac{\|u(f)\|_\omega}{s(r(n))}\right)^{a_1 + \cdots a_{j-1}} \cdot \|u(f)\|_2 \cdot \left|B(nr(n) + |s_1|^{b_1} + \cdots + |s_n|^{b_n})\right|^{\frac{1}{2}} \cdot \\
&\quad \cdot s(nr(n) + |s_1|^{b_1} + \cdots + |s_n|^{b_n}) + \prod_{b_i=1} \omega(s_i) \\
&\leq C^{\frac{1}{2}} \|u(f)\|_2 \sum_{a_j=1} \left(1 + \frac{\|u(f)\|_\omega}{s(r(n))}\right)^{a_1 + \cdots + a_{j-1}} (1 + nr(n))^{\frac{Q}{2}} \prod_{b_i=1} (1 + |s_i|)^{\frac{Q}{2}} \cdot e \\
&\quad \cdot s(nr(n) + |s_1|^{b_1} + \cdots + |s_n|^{b_n}) + \prod_{b_i=1} \omega(s_i)
\end{aligned}$$

$$\leq C^{\frac{1}{2}} \|u(f)\|_2 \cdot n(1+nr(n))^{\frac{Q}{2}} \cdot s(nr(n)) \cdot \big(1 + \frac{\|u(f)\|_\omega}{s(r(n))}\big)^n$$

$$\cdot \prod_{b_i=1} \big((1+|s_i|)^{\frac{Q}{2}} \cdot \omega(s_i)\big) + \prod_{b_i=1} \omega(s_i)$$

$$\leq C^{\frac{1}{2}} \|u(f)\|_2 \cdot n(1+nr(n))^{\frac{Q}{2}} \cdot s(nr(n)) \cdot \Big(\big(1 + \frac{\|u(f)\|_\omega}{s(r(n))}\big)^{s(r(n))}\Big)^{\frac{n}{s(r(n))}}$$

$$\cdot \prod_{b_i=1} \big((1+|s_i|)^{\frac{Q}{2}} \cdot \omega(s_i)\big) + \prod_{b_i=1} \omega(s_i)$$

$$\leq C^{\frac{1}{2}} \|u(f)\|_2 \cdot n(1+nr(n))^{\frac{Q}{2}} \cdot s(nr(n)) \cdot e^{\|u(f)\|_\omega \cdot (\frac{n}{s(r(n))})}$$

$$\cdot \prod_{b_i=1} \big((1+|s_i|)^{\frac{Q}{2}} \cdot \omega(s_i)\big) + \prod_{b_i=1} \omega(s_i)$$

as $s(|s_i|) = \tilde{\omega}(s_i) = \omega(s_i)$ by assumption. For $\underline{a} \neq 0$ we have hence

$$\begin{aligned} &\|(\delta_e + g)^{a_1} * k^{b_1} * \cdots * (\delta_e + g)^{a_n} * k^{b_n}\|_\omega \\ &\leq C^{\frac{1}{2}} \|u(f)\|_2 \cdot n(1+nr(n))^{\frac{Q}{2}} \cdot s(nr(n)) \cdot e^{\|u(f)\|_\omega \cdot (\frac{n}{s(r(n))})} \\ &\cdot \prod_{b_i=1} \Big(\int_G |k(s_i)|(1+|s_i|)^{\frac{Q}{2}} \omega(s_i) ds_i\Big) + \prod_{b_i=1} \int_G |k(s_i)|\omega(s_i) ds_i. \end{aligned}$$

Let us put

$$\omega_1(x) = (1+|x|)^{\frac{Q}{2}} \omega^2(x).$$

Obviously, ω_1 is again a weight on G and

$$\begin{aligned} \int_G |k(s_i)|(1+|s_i|)^{\frac{Q}{2}} \omega(s_i) ds_i &= \int_{(U^{r(n)})^c} |k(s_i)|(1+|s_i|)^{\frac{Q}{2}} \omega(s_i) ds_i \\ &\leq \frac{1}{\inf_{x \in (U^{r(n)})^c} \omega(x)} \int_{(u^{r(n)})^c} |k(s_i)|(1+|s_i|)^{\frac{Q}{2}} \omega^2(s_i) ds_i \\ &\leq \frac{1}{\sup_{x \in U^{r(n)}} \omega(x)} \int_G |u(f)(x)|(1+|x|)^{\frac{Q}{2}} \omega^2(x) dx \\ &\leq \frac{1}{s(r(n))} \|u(f)\|_{\omega_1} \end{aligned}$$

and

$$\int_G |k(s_i)|\omega(s_i) ds_i \leq \int_G |k(s_i)|(1+|s_i|)^{\frac{Q}{2}} \omega^2(s_i) ds_i \leq \frac{1}{s(r(n))} \|u(f)\|_{\omega_1}.$$

Hence for $\underline{a} \neq 0$,

$$\|(\delta_e+g)^{a_1} * k^{b_1} * \cdots * (\delta_e+g)^{a_n} * k^{b_n}\|_\omega$$
$$\leq \left(C^{\frac{1}{2}}\|u(f)\|_2 \cdot n(1+nr(n))^{\frac{Q}{2}} s(nr(n)) \cdot e^{\|u(f)\|_\omega \cdot (\frac{n}{s(r(n))})} + 1\right) \cdot \left(\frac{\|u(f)\|_{\omega_1}}{s(r(n))}\right)^{b_1+\cdots+b_n}$$
$$\leq (C^{\frac{1}{2}}\|u(f)\|_2+1)(1+n)(1+nr(n))^{\frac{Q}{2}} s(nr(n)) \cdot e^{\|u(f)\|_\omega \cdot (\frac{n}{s(r(n))})} \cdot \left(\frac{\|u(f)\|_{\omega_1}}{s(r(n))}\right)^{b_1+\cdots+b_n}$$
$$= C_1(1+n)(1+nr(n))^{\frac{Q}{2}} s(nr(n)) \cdot e^{\|u(f)\|_\omega \cdot (\frac{n}{s(r(n))})} \cdot \left(\frac{\|u(f)\|_{\omega_1}}{s(r(n))}\right)^{b_1+\cdots+b_n},$$

where the constant C_1 is given by

$$C_1 = (C^{\frac{1}{2}}\|u(f)\|_2 + 1).$$

For $\underline{a} = 0$, $b_1 + \cdots + b_n = n$ we have that

$$\|(\delta_e+g)^{a_1} * k^{b_1} * \cdots * (\delta_e+g)^{a_n} * k^{b_n}\|_\omega = \|k^{*n}\|_\omega \leq \|k\|_\omega^n$$

with

$$\|k\|_\omega \leq \frac{1}{s(r(n))}\int_G |u(f)(x)|\omega^2(x)dx \leq \frac{\|u(f)\|_{\omega_1}}{s(r(n))}.$$

Finally we get the following bounds:

$$\|(\delta_e+u(f))^{*n}\|_\omega \leq \sum_{m=0}^{n}\sum_{\underline{a},\underline{b}} \|(\delta_e+g)^{a_1} * k^{b_1} * \cdots * (\delta_e+g)^{a_n} * k^{b_n}\|_\omega$$
$$\leq C_1(1+n)(1+nr(n))^{\frac{Q}{2}} s(nr(n)) \cdot e^{\|u(f)\|_\omega \cdot (\frac{n}{s(r(n))})} \cdot \sum_{m=0}^{n}\sum_{\underline{a},\underline{b}} 1 \cdot \left(\frac{\|u(f)\|_{\omega_1}}{s(r(n))}\right)^{b_1+\cdots+b_n}$$
$$= C_1(1+n)(1+nr(n))^{\frac{Q}{2}} s(nr(n)) \cdot e^{\|u(f)\|_\omega \cdot (\frac{n}{s(r(n))})} \cdot \left(1+\frac{\|u(f)\|_{\omega_1}}{s(r(n))}\right)^n$$
$$\leq C_1(1+n)(1+nr(n))^{\frac{Q}{2}} s(nr(n)) \cdot e^{\|u(f)\|_\omega \cdot (\frac{n}{s(r(n))})} \cdot e^{\|u(f)\|_{\omega_1} \cdot (\frac{n}{s(r(n))})}$$
$$\leq C_1(1+n)(1+nr(n))^{\frac{Q}{2}} s(nr(n)) \cdot e^{2\|u(f)\|_{\omega_1} \cdot (\frac{n}{s(r(n))})}$$

So, in the end,

$$\|u(nf)\|_\omega \leq \|\delta_e + u(nf)\|_\omega + 1$$
$$= \|(\delta_e+u(f))^{*n}\|_\omega + 1 \leq 2C_1(1+n)(1+nr(n))^{\frac{Q}{2}} s(nr(n)) \cdot e^{2\|u(f)\|_{\omega_1} \cdot (\frac{n}{s(r(n))})}.$$

The previous bounds hold for $n \in \mathbb{N}$. If we want a bound valid for $n \in \mathbb{Z}$, we first put $nf = (-n)(-f)$ if $n < 0$. We replace $\|u(f)\|_2$ by $\max(\|u(f)\|_2, \|u(-f)\|_2)$ in the constant C_1, $\|u(f)\|_1$ by $\max(\|u(f)\|_1, \|u(-f)\|_1)$ in the constant C_2 and also set $C_3 = 2\max(\|u(f)\|_{\omega_1}, \|u(-f)\|_{\omega_1})$. We then have

$$\|u(nf)\|_\omega \leq 2C_1(1+|n|)(1+|n|r(|n|))^{\frac{Q}{2}} s(|n|r(|n|)) \cdot e^{C_3(\frac{|n|}{s(r(|n|))})}, \quad n \in \mathbb{Z}. \tag{8.5.1}$$

If D is an element of the enveloping algebra of $\mathfrak{g}$ then we can use estimate (8.5.1) for our f and for $n \in \mathbb{N}$, and obtain

$$\begin{aligned}\|D * u(nf)\|_\omega &= \|(D * u(f)) * u((n-1)f)\|_\omega \\ &\leq C_{D,f}(1+|n|)^{d_\omega}, n \in \mathbb{Z},\end{aligned} \tag{8.5.2}$$

for every polynomial weight ω on G and for some constants $C_{D,f}, d_\omega > 0$.

8.6 Simple Modules of $L^1(G)$ for Nilpotent Lie Groups

In this section G is a locally compact nilpotent group. We want to show that every simple $L^1(G)$-module can be made unitary.

Lemma 8.6.1 *Let A be an involutive Banach algebra. Let I be a maximal modular left ideal in A. Suppose that there exists a positive linear functional $0 \neq p$ on A such that $\langle p, I\rangle = \{0\}$. Then the simple A-module A/I can be made unitary.*

Proof Let $u \in A$ be a modular unit for I. Since $a - au \in I$ for any $a \in A$ it follows that

$$p(a) = p(au), a \in A.$$

Since p is positive, we have that [29]

$$|p(b^*a)| \leq p(b^*b)^{1/2} p(a^*a)^{1/2}, a, b \in A.$$

As p is different from 0,

$$p(u^*u) \neq 0. \tag{8.6.1}$$

We define on A/I the inner product

$$\langle a + I, b + I\rangle := p(b^*a).$$

Let

$$I_p := \{a \in A;\ p(b^*a) = 0,\ \forall b \in A\}.$$

Then I_p is a left ideal containing I. Since $u \notin I_p$ by (8.6.1) and since I is maximal it follows that $I_p = I$.

Let $\mathscr{H}$ be the completion of A/I for the Hilbert norm

$$\|a + I\|_{\mathscr{H}} := \langle a + I, a + I\rangle^{1/2}.$$

We obtain in this way a unitary representation $(\pi, \mathscr{H})$ of A. Indeed, by Lemma 37.6 in [29] we have that

$$p(b^*a^*ab) \le \|a\|^2 p(b^*b), a, b \in A,$$

which says that

$$\|a \cdot (b + I)\| \le \|a\|\|b + I\|_{\mathscr{H}}, a, b \in A.$$

Let now

$$\Phi(a + I) := a + I \in \mathscr{H}.$$

Then the linear mapping Φ is injective and intertwines the simple module $(l, A/I)$ with the A-module $(\pi, \mathscr{H})$. □

Definition 8.6.2 Recall that the group G acts on $L^1(G)$ by left translations, ${}_xf(y) := f(x^{-1}y)$, and that $\|{}_xf\|_1 = \|f\|_1$. Denote by S the space of bounded, positive Hermitian sesquilinear forms on $L^1(G)$. Then the group acts on S by ${}_xB(f, g) = B({}_xf, {}_xg)$, i.e., ${}_{(xy)}B = {}_y({}_xB))$. For $f, g \in L^1(G)$ we have

$$|{}_xB(f, g)|] \le \|B\|\ \|f\|_1\|g\|_1 \quad \text{for all } x \in G$$

resulting in the estimate

$$\begin{aligned} B(f * g, f * g) &= \int_G f(x)B({}_xg, f * g)\,dx \\ &\le \int_G |f(x)|^{\frac{1}{2}}|f(x)|^{\frac{1}{2}}B({}_xg, {}_xg)^{\frac{1}{2}}B(f * g, f * g)^{\frac{1}{2}}\,dx \\ &\le \Big(\int_G |f(x)|\,dx\Big)^{\frac{1}{2}}\Big(\int_G |f(x)|B({}_xg, {}_xg)\,dx\Big)^{\frac{1}{2}}B(f * g, f * g)^{\frac{1}{2}}. \end{aligned}$$

Hence

$$B(f * g, f * g) \leq \|f\|_1 \int_G |f(x)|\ {}_x B(g, g)\ dx. \tag{8.6.2}$$

We use the following notations from [110]. Let F be a subspace of $L^1(G)$ and $H \subset G$ a subgroup. Define $S_F^H \subset S$ by

$$S_F^H = \{B \in S \mid {}_h B = B\ , \forall h \in H \text{ and } B(F, f) = 0\ , \forall f \in L^1(G)\},$$

$$S_F = \{B \in S \mid B(F, f) = 0\ , \forall f \in L^1(G)\}.$$

Let I again be a closed modular left ideal of $L^1(G)$ with modular unit u.

Now, if $S_I^G \neq 0$, any bilinear form $B \neq 0$ in S_I^G defines a positive linear functional p on $L^1(G)$ vanishing on I. Indeed, for $f, g, h \in L^1(G)$ we have that

$$\begin{aligned} B(f * g, h) &= \int_G f(s) B({}_s g, h) ds & (8.6.3)\\ &= \int_G f(s) B(g, {}_{s^{-1}} h) ds \\ &= \int_G f(s^{-1}) B(g, {}_s h) ds \\ &= B(g, f^* * h) & (8.6.4) \end{aligned}$$

As seen above, we have that

$$B(f, g) = B(f, g * u),\ f, g \in L^1(G),$$

since $B(L^1(G), I) = \{0\}$. Define the linear function p on $L^1(G)$ by

$$p(f) := B(f, u).$$

Then for any $f \in L^1(G)$ we have that $|p(f)| = |B(f, u)| \leq \|B\| \|f\|_1 \|u\|_1$, and by (8.6.3)

$$\begin{aligned} p(f^* * f) &= B(f^* * f, u) \\ &= B(f, f * u) \\ &= B(f, f) \geq 0,\ f \in L^1(G). \end{aligned}$$

Hence p is a positive linear functional on $L^1(G)$ vanishing on I. Hence any simple module (T, X) can be made unitary if and only if $S_I^G \neq \{0\}$ for every maximal proper modular left ideal $I \subset L^1(G)$.

So we must show that

$$S_I^G \neq \{0\}.$$

Since I is closed, the Hahn-Banach theorem guarantees the existence of a continuous linear functional $q \neq 0$ on $L^1(G)$ vanishing on I. Then $(f, g) \mapsto q(f)\overline{q(g)}$ is in S_I. Hence

$$S_I = \{B \in S \mid B(I, f) = 0 \,, \forall f \in L^1(G)\} \neq \{0\}.$$

Lemma 8.6.3 *Let G be locally compact group of polynomial growth. Let H and N be closed normal subgroups of G such that $N \subset H$ and such that H/N is central in G/N. Let I be a proper closed modular left ideal in $L^1(G)$ with modular right unit u. Then:*

$$S_I^N \neq \{0\} \Longrightarrow S_I^H \neq \{0\}.$$

Proof Assume that $S_I^N \neq \{0\}$ and choose a non-zero linear form $B \in S_I^N$. Let $K \subset H/N$ be compact and $\varepsilon > 0$ arbitrary. We shall show the existence of $\tilde{B} \in S_I^N$ such that $\tilde{B}(u, u) \geq \frac{1}{2}$,

$$\|\tilde{B}\| = \sup_{\|f\|_1 = \|g\|_1} |\tilde{B}(f, g)| \leq 1$$

and

$$|\, {}_k\tilde{B}(f, f) - \tilde{B}(f, f)| \leq \varepsilon \|f\|_1^2 \tag{8.6.5}$$

for all $f \in L^1(G)$ and all $k \in K$. Let $U = U^{-1}$ be a compact neighbourhood of the identity $\dot{e}$ in H/N. Let $V = \bigcup_{k=1}^{+\infty} U^k$ be the group generated by U. Then V is open in H/N. We use the function $\rho = \rho_\varepsilon$, introduced by Jenkins [92], defined on V by

$$\rho(s) = \rho_\varepsilon(s) = (1 + \varepsilon)^{-k} \quad \text{if } s \in U^{k+1} \setminus U^k.$$

It is easy to check that

$$|\rho(st) - \rho(t)| \leq \varepsilon\rho(t) \quad \text{and} \quad |\rho(ts) - \rho(t)| \leq \varepsilon\rho(t)$$

for all $s \in U$ and all $t \in V$. We use ρ to define $B' \in S_I^N$ by

$$B'(f, g) = \int_V \rho(\dot{v}) \,_{\dot{v}^{-1}}B(f, g) d\dot{v}$$

for all $f, g \in L^1(G)$, where $d\dot{v}$ is the Haar measure on H/N. To show that this integral is well defined and convergent, we first notice that G/N acts on S_I^N. In fact, for $v \in G$ and $n \in N$

$$_{vn}B = \,_{vnv^{-1}\cdot v}B = \,_{v}(\,_{vnv^{-1}}B) = \,_{v}B.$$

The polynomial growth of the Abelian group H/N tells us that $|U^k| \le Ak^D$ for some $D \in \mathbb{N}$ and some $A > 0$. We obtain the estimate:

$$\begin{aligned} |B'(f, g)| &\le \sum_{k=0}^{\infty} \int_{U^{k+1} \setminus U^k} |\rho(\dot{v})| |\,_{\dot{v}^{-1}}B(f, g)| d\dot{v} \\ &\le CA\|f\|_1\|g\|_1(1+\varepsilon) \sum_{k=0}^{\infty} (k+1)^D (1+\varepsilon)^{-(k+1)}. \end{aligned}$$

Obviously, the last series converges. It is easy to check that B' is non-zero and contained in S_I^N, and that

$$0 < b = \sup_{x \in G} \,_{x}B'(u, u) < \infty.$$

Now choose $y \in G$ such that

$$_{y}B'(u, u) \ge \frac{1}{2} b$$

and define $\tilde{B} \in S_I^N$ by

$$\tilde{B} = \frac{1}{b} \cdot_{y} B'.$$

Then by definition $\tilde{B}(u, u) \ge \frac{1}{2}$ and by (8.6.2) we have

$$\begin{aligned} |\tilde{B}(f, g)| &\le \tilde{B}(f, f)^{\frac{1}{2}} \tilde{B}(g, g)^{\frac{1}{2}} \\ &\le \|f\|_1\|g\|_1 \sup_{x \in G} \,_{x}\tilde{B}(u, u) \\ &\le \|f\|_1\|g\|_1 \cdot \frac{1}{b} \cdot \sup_{x \in G} \left[B'(\,_{yx}u, \,_{yx}u)\right] \\ &= \|f\|_1\|g\|_1. \end{aligned}$$

This proves that $\|\tilde{B}\| \leq 1$. To show the continuity property (8.6.5) of $\tilde{B}$, we write the estimate, for all $u \in K$ and $f \in L^1(G)$,

$$
\begin{aligned}
|{}_u\tilde{B}(f, f) - \tilde{B}(f, f)| &= \frac{1}{b} \cdot |B'({}_{yu}f, {}_{yu}f) - B'({}_yf, {}_yf)| \\
&= \frac{1}{b} \cdot \left| \int_V \rho(\dot{v})\big[{}_{\dot{v}^{-1}}B({}_{yu}f, {}_{yu}f) - {}_{\dot{v}^{-1}}B({}_yf, {}_yf)\big] d\dot{v} \right| \\
&\leq \frac{1}{b} \cdot \int_V |\rho((yuy^{-1}v)^{\cdot}) - \rho(\dot{v})| \, {}_{(v^{-1}y)^{\cdot}}B(f, f) d\dot{v}.
\end{aligned}
$$

As $(yuy^{-1})^{\cdot} = u^{-1} \in U$ and $\dot{v} \in V$,

$$
\begin{aligned}
|{}_u\tilde{B}(f, f) - \tilde{B}(f, f)| &\leq \frac{\varepsilon}{b} \cdot \int_V \rho(\dot{v}) \, {}_{\dot{v}^{-1}}B({}_yf, {}_yf) d\dot{v} \\
&= \frac{\varepsilon}{b} \cdot B'({}_yf, {}_yf) \\
&= \varepsilon \tilde{B}(f, f) \\
&\leq \varepsilon \|f\|_1^2.
\end{aligned}
$$

For $K \subset H/N$ compact and for $\varepsilon > 0$, let

$$
A_{K,\varepsilon} = \{B \in S_I^N \mid \|B\| \leq 1, \; B(u, u) \geq \frac{1}{2}, \qquad |{}_kB(f, f) - B(f, f)| \leq \varepsilon \|f\|_1^2, \; \forall f \in L^1(G), \forall k \in K\}.
$$

Then $\tilde{B} \in A_{K,\varepsilon}$ and $A_{K,\varepsilon} \neq \emptyset$. The intersection of finitely many sets $A_{K,\varepsilon}$ is non-empty and each $A_{K,\varepsilon}$ is weak $*$-closed in

$$
\left(S_I^N\right)_1 = S_I^N \cap \{B \in S \mid \|B\| \leq 1\}.
$$

Since $(S_I^N)_1$ is compact in the weak $*$-topology, there exists $B_1 \in \bigcap\{A_{K,\varepsilon} \mid K$ compact, $\varepsilon > 0\}$. Then $B_1(u, u) \geq \frac{1}{2}$ and so $B_1 \neq 0$. Moreover $\|B_1\| \leq 1$ and ${}_hB_1(f, f) = B_1(f, f)$ for all $f \in L^1(G)$ and all $h \in H$. By the polarization identity, ${}_hB_1(f, g) = B_1(f, g)$ for all $f, g \in L^1(G)$ and all $h \in H$. Hence $B_1 \in S_I^H$ and $S_I^H \neq \{0\}$. □

Theorem 8.6.4 *Let G be a locally compact nilpotent group. Then every simple $L^1(G)$ module (T, X) can be made unitary. There exists an irreducible unitary representation $(\pi, \mathcal{H})$ of G such that (T, X) is equivalent to a submodule of $(\pi, \mathcal{H})$.*

Proof There exists a finite sequence $N_0 = \{e\} \subset N_1 \subset \cdots \subset N_r = G$ of closed normal subgroups such that for every $j = 1, \cdots, r$, the quotient subgroup N_j/N_{j-1} is central in G/N_{j-1}. Let I be a maximal modular left ideal in $L^1(G)$. We apply Lemma 8.6.2 inductively to the groups G/N_{j-1}. This shows that $S_I^G \neq \emptyset$ and thus the simple module $(l, L^1(G)/I)$ can be made unitary. We can choose in the compact convex set S_I^G an extremal point B which gives us an irreducible unitary representation of $L^1(G)$ (and of G). □

Definition 8.6.5 Let A be an involutive Banach algebra. We say that A is *symmetric*, or *Hermitian*, if every simple A-module can be made unitary.

Theorem 8.6.6 *Let A be an involutive Banach algebra. Then A is symmetric if and only if for every self-adjoint element $a \in A$*

$$\sigma_A(a) \subset \mathbb{R}.$$

Proof See [29]. □

Definition 8.6.7 Let A be a Banach algebra.

1. We denote by $Prim(A)$ the collection of primitive ideals of A.
2. If G is a locally compact group then we write $Simple(G)$ instead of $Simple(L^1(G))$ and $Prim(G)$ instead of $Prim(L^1(G))$.
3. Let A be a Banach algebra. We denote by $Max(A)$ the collection of maximal closed ideals of A.
4. Let A be an involutive Banach algebra. Let $\hat{A}$ be the family of equivalence classes of irreducible unitary representations of A. We denote by $Prim^*(A)$ the space

$$Prim^*(A) := \{\ker(\pi); (\pi, \mathcal{H}_\pi) \in \hat{A}\}$$

Lemma 8.6.8 *Let A be an involutive Banach algebra. Let $(\pi, \mathcal{H})$ and $(\pi', \mathcal{H}')$ be two irreducible unitary representations of A. If $\mathcal{H}^{fin}$ and $\mathcal{H}'^{fin}$ are different from $\{0\}$ and if $\ker(\pi) \subset \ker(\pi')$, then π and π' are unitarily equivalent.*

Proof Choose an element $a = a^* \in A$ such that $\pi(a) = P_\xi$ is the one-dimensional projection onto the line $\mathbb{C}\xi \subset \mathcal{H}$, for some $\xi \in \mathcal{H}$ of length 1. Then for every $b \in A$ we have that

$$\pi(aba) = P_\xi \circ \pi(b) \circ P_\xi = \langle \pi(b)\xi, \xi\rangle P_\xi.$$

Hence since $\ker(\pi) \subset \ker(\pi')$, we have that

$$\begin{aligned} \pi(aba - \langle \pi(b)\xi, \xi\rangle a) &= 0 \\ &\Rightarrow \pi'(aba - \langle \pi(b)\xi, \xi\rangle a) = 0 \\ &\Rightarrow \pi'(aba) = \langle \pi(b)\xi, \xi\rangle \pi'(a). \end{aligned} \tag{8.6.6}$$

Furthermore

$$\begin{aligned}\pi(a^2-a) &= 0 \\ &\Rightarrow \pi'(a^2-a)=0 \\ &\Rightarrow \pi'(a) \text{ is a self-adjoint projection } \neq 0.\end{aligned}$$

Choose now a unit $\xi' \in \mathcal{H}$ contained in the image of $\pi'(a)$. Define the positive linear functional c_ξ^π of A by

$$c_\xi^\pi(b) := \langle \pi(b)\xi, \xi\rangle, b \in A.$$

Then we have for any $b \in A$ that

$$\begin{aligned}\langle \pi'(b)\xi', \xi'\rangle &= \langle \pi'(b)\pi'(a)\xi', \pi'(a)\xi'\rangle \\ &= \langle \pi'(aba)\xi', \xi'\rangle \\ &\overset{\text{by (8.6.6)}}{=} \langle\langle \pi(b)\xi, \xi\rangle\pi'(a)\xi', \pi'(a)\xi'\rangle \\ &= \langle \pi(b)\xi, \xi\rangle.\end{aligned}$$

Hence the two positive linear functionals c_ξ^π and $c_{\xi'}^{\pi'}$ coincide, which means that the two irreducible unitary representations $(\pi, \mathcal{H})$ and $(\pi', \mathcal{H}')$ are equivalent. □

Theorem 8.6.9 *Let $G = \exp \mathfrak{g}$ be a simply connected nilpotent Lie group. Then $Simple(G)$ is in one-to-one correspondence with the coadjoint orbit space $\mathfrak{g}^*/G$ and with $Prim(G) = Prim^*(G)$.*

Proof We know that the spectrum $\hat{G}$ of G is in 1-1 correspondence with the orbit space $\mathfrak{g}^*/G$, see [70]. Furthermore for every $(\pi, \mathcal{H}_\pi) \in \hat{G}$ the finite-rank submodule $\mathcal{H}_\pi^{fin}$ is not trivial. Hence for every $\ell \in \mathfrak{g}^*$ and every polarization $\mathfrak{p}$ at ℓ the module $(\pi_{\ell,\pi}, \mathcal{H}_{\ell,\pi})$ contains the unique simple submodule $\mathcal{H}_{\ell,\mathfrak{p}}^{fin}$. If now ℓ' is contained in the G-orbit of ℓ and $\mathfrak{p}'$ is a polarization at ℓ', then the representations $\pi_{\ell,\mathfrak{p}}$ and $\pi_{\ell',\mathfrak{p}'}$ are unitarily equivalent and the unitary intertwining operator maps $\mathcal{H}_{\ell,\mathfrak{p}}^{fin}$ onto $\mathcal{H}_{\ell',\mathfrak{p}'}^{fin}$. Hence the corresponding two simple modules are equivalent. Every simple module (T, X) of $L^1(G)$ can be made unitary, so equivalent to a simple submodule of an irreducible unitary representation, and hence to the simple $L^1(G)$-submodule $\mathcal{H}_{\ell,\mathfrak{p}}^{fin}$ for some $\ell \in \mathfrak{g}^*$ and some polarization $\mathfrak{p}$ at ℓ. But then $\ker(T) = \ker(\pi_{\ell,\mathfrak{p}})$. Therefore if two simple modules are equivalent, the kernels of the corresponding unitary representations $\pi_{\ell,\mathfrak{p}}$ and $\pi_{\ell',\mathfrak{p}'}$ in $L^1(G)$ coincide. This implies that $(\pi_{\ell,\mathfrak{p}})$ is unitarily equivalent to $(\pi_{\ell',\mathfrak{p}'})$. The Kirillov mapping induces a bijection of $\mathfrak{g}^*/G$ onto $Simple(G)$. We also see that the spaces $Simple(G)$ and $Prim(G)$ are in 1-1 correspondence. □

8.7 Fell's Topology on $Prim(G)$ and the Wiener Property

Definition 8.7.1 Let A be an involutive Banach algebra. We say that A has the *Wiener property* if every proper closed ideal of A is contained in the kernel of a unitary representation $(\pi, \mathcal{H})$. If so we may always assume that π is irreducible.

Definition 8.7.2

1. Take $I \subset A$. We call "hull of I" the subset

$$h(I) := \{J \in Prim(A);\ J \supset I\}$$

2. We say that a subset S of $Prim(A)$ is a "hull" if there exists a subset I in A such that $S = h(I)$.
3. For a subset S of $Prim(A)$ we define the kernel ker(S) of S in A to be the ideal of A

$$ker(S) := \bigcap_{J \in S} J$$

4. Let A be an involutive Banach algebra.

   Let $Irr^*(A)$ be the collection of irreducible unitary representation $(\pi, \mathcal{H}_\pi)$ of A. Denote by $\widehat{A}$ the *spectrum* of A, i.e., the family of all equivalence classes of irreducible unitary representations of A.

   The C^*-algebra $C^*(A)$ of A is the completion of A in the C^*-norm $\| \|_*$ of A:

$$\|a\|_* := \sup_{\pi \in Irr(A)} \|\pi(a)\|_{\text{op}},\ a \in A.$$

   We let $Prim^*(A)$ be the family

$$Prim^*(A) := \{\ker_{C^*(A)}\pi;\ \pi \in Irr^*(A) = Irr^*(C^*(A))\}.$$

   The canonical projection $P_{\widehat{A}} : \widehat{A} \to Prim^*(A)$ is defined by

$$P_{\widehat{A}}([\pi]) := \ker_{C^*(A)}\pi,\ (\pi, \mathcal{H}_\pi) \in Irr^*(A).$$

Theorem 8.7.3 *Let A be a Banach algebra. The collection*

$$\mathcal{C} := \{S \subset Prim(A);\ S \text{ is a hull}\}$$

can be regarded as the family of closed subsets in a topology on $Prim(A)$.

Proof Let $S = h(I)$ and $S' = h(I')$ in $\mathscr{C}$. Let us show that $S \cup S' = h(I \cap J)$. Let $(T, X) \in Simple(A)$ be such that $J := ker(T) \in S$. Then evidently

$$\begin{aligned} T(I) = \{0\} &\Rightarrow T(I \cap I') = \{0\} \\ &\Rightarrow J \in h(I \cap I'). \end{aligned}$$

Hence $S \cap S' \subset h(I \cap I')$. Suppose now that $J \in h(I \cap I')$. Then $J \supset I \cdot I'$. Since J is a prime ideal in A it follows that $J \supset I$ or $J \supset I'$. Therefore $J \in S \cup S'$.

Let now $(S_\ell = h(I_\ell))_{\in L}$ be a collection of hulls. Then

$$\begin{aligned} \bigcap_{l \in L} S_\ell &= \bigcap_{l \in L} h(I_\ell) \\ &= h(\sum_{l \in L} I_\ell). \end{aligned}$$

Furthermore

$$Prim(A) = h(\{0\}), \emptyset = h(A).$$

Hence the family $\mathscr{C}$ satisfies the axioms of closed sets of a topology. □

Lemma 8.7.4 *Let A be a Banach algebra. Let $S \subset Prim(G)$ be closed. Then*

$$S = h(ker(S)).$$

Proof Since S is closed there exists an ideal I in A such that

$$S = h(I).$$

For any $i \in I$ and $J \in S$, we have that $i \in J$. Therefore $i \in \cap_{J \in S} J = ker(S)$. Hence

$$\begin{aligned} I \subset \ker(S) &\Rightarrow h(\ker(S)) \subset h(I) \\ &\Rightarrow S \subset h(ker S) \subset h(I) = S \\ &\Rightarrow S = h(\ker(S)). \end{aligned}$$

□

Definition 8.7.5 Let A be a Banach algebra. The topology defined by the family $\mathscr{C} \subset Prim(A)$ of Theorem 8.7.3 is called the *Fell topology*.

Let A be an involutive Banach algebra. We also equip $Prim^*(A)$ with the Fell topology. The dual topology on $\widehat{A}$ is determined by its open subsets $P_{\widehat{A}}^{-1}(V)$, for V open in $Prim^*(A)$.

Definition 8.7.6 Let A be an involutive Banach algebra. We denote by

$$P(A) = \{c^{\pi}_{\xi}; (\pi, \mathcal{H}_{\pi}) \in \widehat{A}, \xi \in \mathcal{H}_{\pi}, \|\xi\| = 1\}$$

the set of *states* of A. On $P(A)$ we put the weak topology. A net $(c_i)_{i \in I}$ in $P(A)$ converges to an element $c \in P(A)$ if and only if for any $a \in A$

$$\lim_{i \to \infty} \langle c_i, a\rangle = \langle c, a, \rangle.$$

For an irreducible unitary representation $(\pi, \mathcal{H}_{\pi})$, let $P^{\pi}(A)$ denote the states of A associated with π:

$$P^{\pi}(A) := \{c^{\pi}_{\xi}; \xi \in \mathcal{H}_{\pi}, \|\xi\| = 1\}.$$

Theorem 8.7.7 *Let A be an involutive Banach algebra, let $(\pi_i)_{i \in I}$ be a net in $\widehat{A}$ and let $\pi \in \widehat{A}$. Let $J_i := ker(\pi_i), i \in I$, $J := ker(\pi)$. Then the net $(J_i)_{i \in I}$ converges to J in the Fell topology (and the net (π_i) converges to π in $\widehat{A}$) if and only for every unit $\xi \in \mathcal{H}_{\pi}$ there exists a net $(\xi_i \in \mathcal{H}_{\pi_i})_{i \in I}$ such that the coefficients $c_i = c^{\pi_i}_{\xi_i}$ weakly converge to the coefficient $c = c^{\pi}_{\xi}$.*

Proof See [49, Theorem 3.4.4]. □

Lemma 8.7.8 *Let A be a Banach algebra. Let S be closed subset of $Prim(A)$. Suppose there exist $a \neq 0, b \in A$ such that*

$$b \cdot a = a, b \in ker(S).$$

Then a is contained in every ideal I of A with $h(I) \subset S$.

Proof Suppose that there exists an ideal I of A such that $h(I) \subset S$ and $a \notin I$. We define the left ideal L of A by

$$L := I + \{c - c \cdot b; c \in A\}.$$

Then L admits b as modular right unit. We see that $b \notin L$, because otherwise for some $i \in I$ and $c \in A$ we would have

$$\begin{aligned} b &= i + c - c \cdot b \\ &\Rightarrow \\ b \cdot a &= i \cdot a + c \cdot a - c \cdot b \cdot a \\ &\Rightarrow \\ a &= i \cdot a + c \cdot a - c \cdot a \\ &= i \cdot a \in I. \end{aligned}$$

This contradiction tells us that L is a proper (modular) left ideal and so there exists a maximal left ideal M containing L. Define the simple module (T, X) by

$$(T, X) := (l, A/M).$$

Then

$$\ker(T) = \{j \in A; j \cdot A \subset M\}.$$

Since M is a modular left ideal with modular unit $e = b$, we have for every $j \in A$ that $j - je = j - jb \in L \subset M$. Hence for $j \in \ker(T)$, it follows that $jb \in M$ and so

$$j = j - jb + jb \in L + M \subset M.$$

Hence

$$\ker(T) = \{m \in M; m \cdot A \subset M\}.$$

Since $I \subset L \subset M$ it follows that $I \cdot A \subset I \subset L \subset M$ and so

$$I \subset \ker(T).$$

Therefore $\ker(T) \in h(I) \subset S$. Now b is contained in $ker(S)$ and so $b \in \ker(T) \subset M$. This implies that $A = M$, since b is a modular unit for M and so M cannot be maximal. Therefore $a \in I$. □

Theorem 8.7.9 *Let $G = \exp \mathfrak{g}$ be a simply connected nilpotent Lie group. Let S be a closed subset of $Prim(G)$. Then there exists a unique two-sided ideal $j(S)$ in $L^1(G)$ such that*

$$h(j(S)) = S$$

and such that $j(S)$ is contained in every ideal I' of $L^1(G)$ whose hull is contained in S.

Proof Let S be a closed subset of $Prim(G)$ and let $J = \ker((T, X)) \in Prim(G) \setminus S$. Therefore, by Lemma 8.7.4, we have that $\ker(T) \not\supset \ker(S)$. Hence there exists $f \in ker(S)$ such that $f \notin \ker(T)$. It follows from Theorem 8.6.9 that we can take $(\pi, \mathcal{H}_\pi) \in \hat{G}$ such that $\ker(T) = \ker(\pi)$. Hence we can assume that $f = f^*$, $\|f\|_1 = 1$ and $\|\pi(f)\|_{\text{op}} = c > 0$. Let now $\frac{c}{10} > \varepsilon > 0$. Since $L\mathscr{P}(G)$ is dense in $L^1(G)$, there exists $h = h^* \in L\mathscr{P}(G)$ such that

$$\|f - h\|_1 < \varepsilon^2.$$

This implies that

$$\begin{aligned}\pi - 1 > \frac{11c}{9} + \varepsilon \ &\geq \ c + \varepsilon \\ &\geq \ c + \varepsilon^2 \\ &\geq \ \|\pi(f)\|_{\text{op}} + \|\pi(f-h)\|_{\text{op}} \\ &\geq \ \|\pi(h)\|_{\text{op}} \\ &\geq \ c - \varepsilon > \frac{9c}{10} > \varepsilon\end{aligned}$$

and

$$0 \ \leq \ \|\sigma(h)\|_{\text{op}} \leq \varepsilon^2, \text{ for every } (\sigma, \mathscr{H}_\sigma) \in \hat{G} \text{ with } \ker(\sigma) \in S.$$

By Theorem 8.5.7 there exist a constant $C > 0$ (depending on h) and an integer $d \in \mathbb{N}^*$ (depending on the growth of G) such that

$$\|e(nh)\|_1 \leq Cn^d, n \in \mathbb{Z}. \tag{8.7.1}$$

Take $\mu \in C^\infty(\mathbb{T})$ such that

$$\mu(t) = t^{d+2}, |t| \leq \frac{11}{9}c + \varepsilon, \mu = 0 \text{ on } [\pi - 1, \pi] \cup [-\pi, -\pi + 1,].$$

Then

$$\mu\{h\} = h^{d+2}.$$

There exists a constant $R > 0$ such that for every $\varepsilon > 0$ we have a smooth function φ on $\mathbb{R}/2\pi\mathbb{Z}$ such that φ vanishes on $[-\varepsilon, \varepsilon]$ and

$$|\varphi^{(j)}(t) - \mu^{(j)}(t)| \leq R\varepsilon, t \in \mathbb{R}, j = 0, \cdots, d+1. \tag{8.7.2}$$

Then we have that

$$\begin{aligned}\|h^{d+2} - \varphi\{h\}\|_1 &= \|\mu\{h\} - \varphi\{h\}\|_1 \\ &= \|(\mu - \varphi)\{h\}\|_1 \\ &\leq \sum_{n\in\mathbb{Z}} \|u(nh)\|_1 |\widehat{(\mu - \varphi)}(n)| \\ &\leq \sum_{n\in\mathbb{N}} C|n|^d \frac{R\varepsilon^2}{|n+1|^{d+2}}\end{aligned}$$

$$\leq RC\varepsilon^2 (\sum_{n \in \mathbb{Z}} \frac{1}{|n|^2 + 1})$$
$$\leq C'\varepsilon^2 \tag{8.7.3}$$

for some new constant $C' > 0$ independent of ε. We see that for ε small enough we have that

$$\|\pi(\varphi\{h\})\|_{\text{op}} > 0$$

(since $\mu(\|\pi(h)\|_{\text{op}}) > 0$), and for any $(\sigma, \mathcal{H}_\sigma) \in \hat{G}$ with $\ker(\sigma) \in S$ it follows that

$$\|\sigma(\varphi\{h\})\|_{\text{op}} = \|\varphi\{\sigma(h)\}\|_{\text{op}} \leq \sup_{|t| \leq C'\varepsilon^2} |\varphi(t)| \leq \sup_{|t| \leq \varepsilon} |\varphi(t)| = 0.$$

This shows that $\varphi\{h\} \in \ker(S)$. Choose now a function $\psi \in C^\infty(\mathbb{R}/2\pi\mathbb{Z})$ such that $\psi = 0$ on the interval $[-\frac{\varepsilon}{2}, \frac{\varepsilon}{2}]$, and $\psi \cdot \varphi = \varphi$. For ε small enough we have $\psi\{h\} \in \ker(S)$ (in analogy to $\varphi\{h\}$). Furthermore, the relation $\psi \cdot \varphi = \varphi$ implies that

$$\psi\{h\} * \varphi\{h\} = \varphi\{h\}.$$

We see in this way that for every $J \in Prim(G) \setminus S$ there exist $b_J, a_J \in L^1(G)$ such that $b_J \in \ker(S)$, $a_J \notin J$ and $b_J * a_J = a_J$. Hence by Lemma 8.7.8 the ideal $j(S)$ generated by the set $\{a_J; J \in Prim(G) \setminus S\}$ is contained in very ideal I of $L^1(G)$ with hull $h(I) \supset S$, and the hull of $j(S)$ is equal to S. Evidently $j(S)$ is unique under these properties. □

Theorem 8.7.10 *Let $G = \exp \mathfrak{g}$ be a simply connected nilpotent Lie group. Let S be a closed subset of $Prim(G)$. Then the minimal ideal $j(S)$ with hull S of Theorem 8.7.9 is generated by Schwartz functions.*

Proof From the proof of Theorem 8.7.9 we know $j(S)$ is generated by functions of the form $\varphi\{h\}$, where $\varphi \in C_c^\infty(\mathbb{R})$ vanishes on neighbourhoods of 0 and where $h = h^* \in C_c(G)$. We can of course always take h in $C_c^\infty(G)$, since these functions are dense in $L^1(G)$. Now it follows from Theorem 8.7.9 that $\|e^{inh} - \delta\|_\omega \leq C_\omega(1 + |n|)^{d_\omega}, n \in \mathbb{Z}$, for some integer $d_\omega \in \mathbb{N}$ depending on the polynomial weight ω. Therefore

$$\|\varphi\{h\}\|_\omega \leq \sum_{n \in \mathbb{Z}} |\hat{\varphi}(n)| C_\omega (1 + |n|)^{d_\omega}$$
$$< \infty.$$

A similar estimate holds by (8.5.2) for the norm $\| \|_\omega$ of the function $D * \varphi\{h\}$, for every D in the enveloping algebra of $\mathfrak{g}$. □

Theorem 8.7.11 *Let G be a simply connected nilpotent Lie group. Then $L^1(G)$ has the Wiener property.*

Proof Let I be a closed ideal of $L^1(G)$ with empty hull. Relation (8.7.3) tells us that there is a $d \in \mathbb{N}^*$ such that for every $h = h^* \in C_c(G)$, there exists a sequence $(\psi_k) \subset C^\infty(\mathbb{R}/\mathbb{Z})$ of maps all vanishing on a neighbourhood of 0 and such that

$$\lim_{k\to\infty} \|h^{d+2} - \psi_k\{h\}\|_1 = 0$$

Lemma 8.7.8 tells us that $\psi_k\{h\} \in j(\emptyset)$ and so $\psi_k\{h\} \in I$ for every k. Therefore $h^{d+2} \in \overline{I} = I$. Choose a decreasing sequence $(U_k)_k$ of compact symmetric neighbourhoods of e in G such that $\bigcap_{k\in\mathbb{N}} U_k = \{e\}$. Choose also for any k an element $h_k = h_k^* \in C_c(G)$ such that $\mathrm{supp}(h_k) \subset U_k$, $h_k(s) \geq 0, s \in G$, and $\int_G h_k(s)ds = 1, k \in \mathbb{N}$. Then the sequence $(h_k)_k$ is a bounded approximate identity in $L^1(G)$ and therefore

$$\lim_{k\to\infty} \|h_k^{d+2} * f - f\|_1 = 0$$

for every $f \in L^1(G)$. This shows that $f \in \overline{I} = I$ and so $I = L^1(G)$. □

Theorem 8.7.12 *Let G be a simply connected nilpotent Lie group. Then*

$$Prim(G) = Prim^*(G) = Max(G).$$

Proof We already know from Theorem 8.6.9 that $Prim(G) = Prim^*(G)$. Let $I \in Max(G)$. The Wiener property of $L^1(G)$ implies that there exists $(\pi, \mathcal{H}_\pi) \in \hat{G}$ such that $\ker(\pi) \supset I$. Hence $I = \ker(\pi)$ by the maximality of I. If $\ker(\pi)$ is not maximal for some $(\pi, \mathcal{H}_\pi) \in \hat{G}$, then there exists a proper closed ideal $J \in L^1(G)$ containing $\ker(\pi)$. The Wiener property implies again that $\ker(\pi) \subset J \subset \ker(\pi')$ for some $(\pi', \mathcal{H}_{\pi'}) \in \hat{G}$. But then π and π' are equivalent by Lemma 8.6.8 and so $J = \ker(\pi)$. This shows that $Prim^*(G) = Max(G)$. □

8.8 Variable Groups

Definition 8.8.1 A *variable* locally compact group $(VLCG)$ is a triple

$$\Gamma = (M \times G, \mu, \iota)$$

where

- $M \times G$ is the topological product of the locally compact Hausdorff spaces M and G;

- The maps

$$\mu : M \times G \times G \to M \times G$$
$$\iota : M \times G \to M \times G$$

are continuous and such that

$$(x, y) \to x \underset{m}{\cdot} y := \mu(m, x, y), \quad x \to x_m^{-1} := \iota(m, x)$$

defines a topological group structure G_m on G, for every $m \in M$.
- There exists an element e in G such that e is the identity in G_m for every $m \in M$.
- There exist a positive Radon measure dx on G and a continuous function

$$\beta : M \times G \to]0, \infty[; \ (m, x) \to \beta_m(x)$$

such that $\beta(m, e) = 1$ and $dx_m := \beta_m(x)dx$ is a Haar measure on G_m, for every $m \in M$.

Definition 8.8.2 An involutive variable Banach algebra (VBA) is a pair

$$\mathbb{A} = (M \times A, \mu, *)$$

where M is a locally compact topological space, A a Banach space with norm $\|a\|$ and

$$\mu : M \times A \times A \to M \times A$$

a continuous mapping such that $(a, b) \to \mu(m, a, b) =: a \underset{m}{\cdot} b$ defines an associative Banach-algebra multiplication on A with

$$\|a \underset{m}{\cdot} b\| \leq \|a\|\|b\|, m \in M, a, b \in A,$$

and every $m \in M$ defines an isometric involution $\overset{*}{m}$ on A_m.

Let Γ be a $LCVG$. Then Γ defines the family of the group algebras $\{L^1(G_m)\}_{m \in M}$ with Haar measures $d_m x = \beta_m(x)dx$.

Let A_∞ be the Banach space $C_\infty(M, L^1(G, dx))$ of continuous mappings $F : M \to L^1(G, dx)$ vanishing at infinity and such that

$$\|F\|_\infty := \sup_{m \in M} \|\varphi(m)\|_1 < \infty.$$

Define on A_∞ the multiplication $*$:

$$F * F'(m)(x) := \beta(m)(x)\Big((\beta(m)^{-1}F(m)) \underset{m}{*} (\beta(m)^{-1}F'(m))\Big)(x), x \in G, m \in M,$$

and the involution

$$F^*(m)(x) := \Delta_{G_m}(x)^{-1}\overline{F(m)(x^{-1})_m}, \, x \in G, m \in M, F, F' \in A_\infty.$$

It is easy to check that

$$L^1(\Gamma) := (\overline{C_c(M, L^1(G, dx))}^{\|\|_\infty}, *,^*)$$

is an involutive Banach algebra.

Denote by $C^*(\Gamma)$ the universal C^*-algebra of $L^1(\Gamma)$.

The commutative Banach algebra $C_\infty(M)$ acts by multiplication on $L^1(\Gamma)$, and for $\psi \in C_\infty(M), \varphi, \varphi' \in L^1(\Gamma)$ we have

$$\psi \cdot (F * F') = (\psi \cdot F) * F' = F * (\psi \cdot F').$$

This means that $C_\infty(M)$ is contained in the center of the multiplicative algebra of $L^1(\Gamma)$.

Let $m \in M$ and let

$$P_m : L^1(\Gamma) \to L^1(G_m)$$

be the linear mapping defined by

$$P_m(F) := F(m), \, F \in L^1(\Gamma).$$

Then P_m is a surjective involutive Banach-algebra homomorphism from $L^1(\Gamma)$ onto $L^1(G_m)$.

Proposition 8.8.3 *The Banach algebra $L^1(\Gamma)$ admits bounded approximate units.*

Proof For every locally compact group G the family $\mathcal{F}(G)$ of continuous non-negative self-adjoint functions f supported on a symmetric neighbourhood of the identity, and satisfying $\int_G f(x)dx = 1$, form an approximate identity of $L^1(G)$ bounded by 1. Therefore it is easy to show that the family

$$\mathcal{F}(\Gamma) = \{\psi \cdot \beta^{-1} \otimes f | f : G \to \mathbb{R}_+ \text{ continuous, supported on a compact nhbd of } e,$$
$$\int_G f(x)dx = 1, \; \psi \in C_\infty(M), \|\psi\|_\infty = 1, \psi = 1 \text{ on a compact subset of } M\}$$

determines a left approximate unit of $L^1(\Gamma)$ bounded by 1. □

Proposition 8.8.4 *The unitary spectrum $\widehat{L^1(\Gamma)}$ of the algebra $L^1(\Gamma)$ is the disjoint union $\bigcup_{m\in M} \widehat{G_m}$.*

Proof For every $m \in M$ and $\pi \in \hat{G_m}$ we obtain an irreducible unitary representation $\pi_m \in \hat{\Gamma}$ through the formula

$$\pi_m(F) := \pi(P_m(F)), \; F \in L^1(\Gamma).$$

Since every $\psi \in C_\infty(M)$ defines a central multiplier of $L^1(\Gamma)$, we can extend any irreducible unitary representation $(\pi, \mathscr{H})$ of $L^1(\Gamma)$ to a unitary representation $\overline{\pi}$ of the C^*-algebra $C_\infty(M)$. Then we have that $\overline{\pi}(\psi) \circ \pi(F) = \pi(F) \circ \overline{\pi}(\psi)$ for every $\psi \in C_\infty(M)$, $F \in L^1(\Gamma)$. Since π is irreducible, we have by Schur's lemma that $\overline{\pi}(\psi)$ is a multiple of the identity. Hence

$$\overline{\pi}(\psi) = \chi_\pi(\psi)\mathbb{I}_{\mathscr{H}}, \; \psi \in C_\infty(M),$$

for some character χ of the algebra $C_\infty(M)$. Hence there exists $m_0 \in M$ such that $\chi(\psi) = \psi(m_0)$, $\psi \in C_\infty(M)$. We can write every $F \in L^1(\Gamma)$ as $F = \varphi \cdot F'$ for some $F' \in L^1(\Gamma)$, $\varphi \in C_\infty(M)$. Indeed it suffices to write

$$\varphi(m) = \sqrt{\|F(m)\|_1}, \; m \in M.$$

Now, if that $F(m_0) = 0$ for some $F \in L^1(\Gamma)$, then by writing $F = \varphi \cdot F'$ we see that $\varphi(m_0) = 0$. Hence

$$\pi(F) = \overline{\pi}(\varphi) \circ \pi(F') = \varphi(m_0)\pi(F') = 0.$$

Hence $\ker(P_{m_0}) \subset \ker(\pi)$ and so π defines an irreducible unitary representation of $L^1(\Gamma)/\ker(P_{m_0}) \simeq L^1(G_{m_0})$. Therefore we can write $\pi \simeq \pi'_{m_0}$ for some element $\pi' \in \hat{G_{m_0}}$.

□

Definition 8.8.5 A *variable Lie group VLG* is a locally compact variable group $\Gamma = (M \times G, \mu, \iota)$ where, in addition to the $VLCG$ properties, we also request the following: G is a smooth real n-manifold; the map $\mu(m, x, y)$ admits second partial derivatives in (x, y) with continuous derivatives in (m, x, y); $\iota(m, x)$ admits second partial derivatives in x and the derivatives are continuous in (m, x).

The variable Lie group $\Gamma = M \times G$ admits a variable Lie algebra $\mathfrak{g}_M = M \times \mathfrak{g}$, where $\mathfrak{g}$ is the tangent space at e, which gives us the Lie bracket $[,]_m$ on $\mathfrak{g}$ for any $m \in M$. This latter is clearly continuous in $m \in M$, since the multiplication in $\Gamma = G_M$ is of class C^2.

Let $\mu : M \to G$ be a continuous mapping and let $a_\mu : M \times \mathfrak{g}^* \to M \times \mathfrak{g}^*$ be defined by

$$a_\mu(m, \ell) := (m, \mathrm{Ad}^*_m(\mu(m))\ell).$$

Then a_μ is obviously a homeomorphism of $M \times \mathfrak{g}^*$ since it is continuous with continuous inverse. Let $\mathfrak{g}^*$ be the algebraic dual of $\mathfrak{g}$.

We say that a subset $S \subset M \times \mathfrak{g}^*$ is Γ-*invariant* if for every continuous mapping $\mu : M \to G$ we have that

$$\mu(S) := a_\mu(S) \subset S.$$

The *orbit topology* on $M \times \mathfrak{g}^*$ is given by the collection of G-invariant open subsets of $M \times \mathfrak{g}^*$.

We denote by

$$M \times \mathfrak{g}^*/G$$

the space of G-orbits in $M \times \mathfrak{g}^*$:

$$M \times \mathfrak{g}^*/G = \{(m, \mathrm{Ad}_m^*(G)\ell); m \in M, \ell \in \mathfrak{g}^*\}$$

We put on $M \times \mathfrak{g}^*/G$ the quotient topology: denote by $P_{M\times\mathfrak{g}^*/G}$ the canonical projection $M \times \mathfrak{g}^* \to M \times \mathfrak{g}^*/G$ given by

$$P_{M\times\mathfrak{g}^*/G}(m, \ell) := (m, \mathrm{Ad}_m^*(G)\ell).$$

Then a subset S of $M \times \mathfrak{g}^*/G$ is open in this topology if and only if $P^{-1}_{M\times\mathfrak{g}^*/G}(S)$ is open in $M \times \mathfrak{g}^*$. Of course this subset $P^{-1}_{M\times\mathfrak{g}^*/G}(S)$ is automatically G-invariant. On the other hand any open G-invariant subset S of $M \times \mathfrak{g}^*$ is mapped under $P_{M\times\mathfrak{g}^*/G}$ onto an open subset of the orbit space $M \times \mathfrak{g}^*/G$.

Lemma 8.8.6 *Let $\Gamma = M \times G$ be variable Lie group. A sequence $((m_k, \Omega_{m_k}))_k$ in $M \times \mathfrak{g}^*$ converges in the orbit topology to an orbit (m, Ω_m) if and only if $m = \lim_{k\to\infty} m_k$ and for every $\ell \in \Omega$ there exists a sequence $((m_k, \ell_k)) \in \{m_k, \Omega_{m_k}\}$ such that and $\lim_{k\to\infty} \ell_k = \ell$ in $\mathfrak{g}^*$.*

Proof Suppose that $((m_k, \Omega_k))_k$ converges in the orbit topology to (m, Ω). Then it is clear that $\lim_{k\to\infty} m_k = m$, since for any open subset $U \subset M$ the set $U \times \mathfrak{g}^*$ is open and G-invariant. Choose for (m, ℓ) in Ω an open neighbourhood U of m and an open neighbourhood V of $\ell \in \mathfrak{g}^*$. Let

$$\tilde{V} := \bigcup_{\mu:M\to G \text{ continuous}} a_\mu(U \times V).$$

Then $\tilde{V}$ is open and G-invariant in $M \times \mathfrak{g}^*$ and therefore $\Omega_k \subset \tilde{V}$ for k large enough. This shows that for k large enough, for any $(m_k, l'_k) \in \Omega_k$ there exists some $g_k \in G$ and a $v_k \in V$ such that

$$(m_k, \ell'_k) = (m_k, \mathrm{Ad}_{m_k}(g_k)v_k)$$

Hence $(m_k, \ell_k) := (m_k, \mathrm{Ad}^*_{m_k}(g_k^{-1})\ell'_k)$ is contained in $U \times (V \cap \Omega_{m_k})$ for k large enough and so $\lim_{k\to\infty} \ell_k = \ell$.

The statement's necessary implication is evident. □

Example 8.8.7 Let $\mathfrak{n}$ be a finite-dimensional real vector space and let $\mathscr{X} = \{X_1, \cdots, X_n\}$ be a basis of $\mathfrak{n}$. Let M be a locally compact topological space and suppose that for every $m \in M$ we have a Lie bracket on $\mathfrak{n}$

$$[,]_m : \mathfrak{n} \to \mathfrak{n};\ [X_i, X_j]_m = \sum_{k=1}^{n} \nu_k^{m,i,j} X_k,\ i, j \in \{1, \cdots, n\}, m \in M,$$

with continuous real functions $\nu_k^{m,i,j}$ in $m \in M$. We assume furthermore that the Lie algebras $\mathfrak{n}_m$, $m \in M$, are all nilpotent and that the basis $\mathscr{X}$ is a Jordan-Hölder basis for all m. Let now $N := \mathfrak{n}$ as a topological space, and define a group multiplication $\cdot_m$ on N using the Campbell-Baker-Hausdorff product. Then obviously $\mathbb{N} = (M \times N, \mu, \iota)$ is a VLG.

In the sequel the following situation will appear. Let $N = \exp \mathfrak{n}$ be a simply connected, connected nilpotent Lie group. Let $\dim N > d > 0$ and let $(H_k = \exp \mathfrak{h}_k)_{k\in\mathbb{N}}$, $\dim H_k = d$, be a sequence of closed connected subgroups of N converging to the subgroup $H^\infty = \exp \mathfrak{h}^\infty$ in the Grassmann topology. This means that for every $k \in \mathbb{N}$ we find a Jordan-Hölder basis $\mathscr{X}_k = \{X_1^k, \cdots, X_d^k\}$ of $\mathfrak{h}_k$ and a basis $\mathscr{X}^\infty = \{X_1^\infty, \cdots, X_d^\infty\}$ of $\mathfrak{h}^\infty$ such that $\lim_{k\to\infty} X_j^k = X_j^\infty$ for $j = 1, \cdots, d$.

We take now $N = \mathfrak{n} = \mathbb{R}^d$ with the canonical basis $\mathscr{E} = \{E_1, \cdots, E_d\}$. Let $M := \mathbb{N} \cup \{\infty\}$ with the endpoint topology, and define for $m \in M$ the Lie bracket $[,]_m$ on $\mathfrak{n}$ by

$$[E_i, E_j]_m := \sum_{l=1}^{d} \nu_\ell^{m,i,j} E_\ell,$$

where the numbers $\nu_\ell^{m,i,j}$ are coming from the Lie brackets in the subalgebras $\mathfrak{h}_m$, $m \in M$:

$$[X_i^m, X_j^m] = \sum_{l=1}^{d} \nu_\ell^{m,i,j} X_\ell^m.$$

It is evident now that we obtain a VLG structure on $M \times N$ in this way.

Definition 8.8.8 A variable nilpotent Lie algebra $\mathfrak{n}$ is a pair $\varphi\mathfrak{n} = (M \times \mathfrak{n}, \beta)$ where

- M is a locally compact topological space and $\mathfrak{n}$ is a finite-dimensional real vector space with a basis $\mathscr{X} = \{X_1, \cdots, X_n\}$;

- the continuous mapping $\beta : M \times \mathfrak{n} \times \mathfrak{n} \to \mathfrak{n}$ defines for every $m \in M$ a nilpotent Lie bracket $[,]_m$ on $\mathfrak{n}$ such that $\mathscr{X}$ is Jordan-Hölder basis of $\mathfrak{n}_m$ for any m.

Define now for every $m \in M$ the group multiplication $\underset{m}{\cdot}$ on $N = \mathfrak{n}$ by using the Campbell-Baker-Hausdorff multiplication on $\mathfrak{n}_m$. Then the triple $\mathscr{N} = (M \times N, \mu, \iota)$, where

$$\mu(m, x, y) := x \underset{m}{\cdot} y,$$
$$\iota(m, x) := -x, x \in N, m \in M, x, y \in N$$

defines a VLG, which we call variable nilpotent Lie group $(NVLG)$.

8.8.1 Kirillov's Conjecture for Nilpotent Lie Groups

Definition 8.8.9 Let A be an involutive Banach algebra with bounded approximate units. We denote by $P(A)$ the set of positive linear functionals of A of norm ≤ 1. The subset $P(A)$ of A' is weakly closed, convex and compact in the weak topology. By $EP(A) \subset P(A)$ we denote the extremal points of $P(A)$. The Alaoglu theorem tells us that $P(A)$ is the closure in A' of all convex combinations of $EP(A)$. Furthermore, by Dixmier [49], an element $\varphi \in A'$ is contained in $EP(A)$ if and only if there exists an irreducible unitary representation $(\pi, \mathscr{H})$ of A such that φ is the coefficient c_ξ^π for some $\xi \in \mathscr{H}$ of length 1. This means that

$$\langle \varphi, a \rangle = \langle \pi(a)\xi, \xi \rangle_{\mathscr{H}}, a \in A.$$

If we consider the Banach algebra $A = L^1(\Gamma)$ of a $VLCG$ $\Gamma = (M \times G)$, then we have seen in Proposition 8.8.4 that

$$EP(L^1(\Gamma)) = \bigcup_{m \in M} \bigcup_{\substack{\pi \in \widehat{G_m} \\ \xi \in \mathscr{H}_\pi, \|\xi\|=1}} c_\xi^\pi$$

To simplify notations, we write

$$\pi^m$$

for a representation $\pi \in \widehat{G_m}$, $m \in M$.

Denote by $EP(\Gamma)$ the states of Γ, i.e., the set of coefficients c_ξ^π, where $(\pi, \mathscr{H}_\pi)$ is any irreducible unitary representation of Γ and ξ any unit element of $\mathscr{H}_\pi$. Then

$$EP(\Gamma) = EP(L^1(\Gamma)).$$

Lemma 8.8.10 *Let $(\pi, \mathcal{H})$ be a unitary representation of a locally compact group G. Take a unit vector $\xi \in \mathcal{H}$ and let*

$$\psi = c_{\xi}^{\pi}.$$

Then for any $u \in G$ we have that

$$|\psi(u^{-1} \cdot x) - \psi(x)| \leq \sqrt{2}|1 - \operatorname{Re} \psi(u)|, x \in G.$$

Proof Indeed we have

$$\begin{aligned}
|\psi(u^{-1} \cdot x) - \psi(x)| &= |\langle \pi(u^{-1} \cdot x)\xi, \xi\rangle - \langle \pi(x)\xi, \xi\rangle| \\
&= |\langle \pi(x)\xi, \pi(u)\xi\rangle - \langle \pi(x)\xi, \xi\rangle| \\
&= |\langle \pi(x)\xi, \pi(u)\xi - \xi\rangle| \\
&\underset{\text{Cauchy-Schwartz}}{\leq} \|\pi(x)\xi\| \|\pi(u)\xi - \xi\| \\
&= \sqrt{\|\pi(u)\xi\|^2 + \|\xi\|^2 - 2 \operatorname{Re} \langle \pi(u)\xi, \xi\rangle} \\
&= \sqrt{2}|1 - \operatorname{Re} \psi(u)|.
\end{aligned}$$

□

Theorem 8.8.11 *Let $\Gamma = M \times G$ be a second countable variable group. Let $(\pi_k^{m_k})_k \in \mathbb{N}$ be a sequence in $\widehat{\Gamma}$ and let $\pi^m \in \widehat{\Gamma}$. Then the sequence of coefficients $(\varphi_k = c_{\xi_k}^{\pi_k^{m_k}})_k$ in $EP(\Gamma)$ weakly converges to $\varphi = c_{\xi}^{\pi^m} \in EP(\Gamma)$ if and only if $\lim_{k\to\infty} m_k = m$ and the functions $x \to c_{\xi_k}^{\pi_k^{m_k}}(x), k \in \mathbb{N}$, converge to $x \to c_{\xi}^{\pi^m}(x)$ uniformly on compact set in G.*

Proof Suppose that $\lim_{k\to\infty} m_k = m$ and that the functions φ_k converge uniformly on compact sets to the function φ. Let $F \in C_c(M \times G)$. Then there exist compact subsets $K \subset G, L \subset M$ such that

$$F(m, x) = 0, \text{ whenever } x \notin K \text{ or } m \notin L.$$

The function

$$\psi_k(x) := \beta_{m_k}(x) F(m_k, x)\varphi_k(x), x \in G, k \in \overline{\mathbb{N}}, x \in G,$$

are bounded by the function $\|\beta_{|L\times K}\|_\infty \|F\|_\infty 1_K$ and

$$\lim_{k\to\infty} \psi_k(x) = \psi_\infty(x), x \in G.$$

Therefore Lebesgue's theorem tells us that

$$\lim_{k\to\infty}\int_G \psi_k(x)F(m_k,x)dx = \int_G \psi_\infty(x)F(m_\infty,x)dx.$$

This means that

$$\lim_{k\to\infty}\langle \pi_k^{m_k}(F)\xi_k,\xi_k\rangle = \langle \pi^m(F)\xi,\xi\rangle.$$

Suppose now that the sequence of coefficients $(\varphi_k = c_{\xi_k}^{\pi_k^{m_k}})_k$ in $P(\Gamma)$ converges weakly to $\varphi = c_\xi^{\pi^m}$ in $P(\Gamma)$.

Suppose that the sequence $(m_k)_k$ does not converge to m_∞. Then we can assume that there exists a neighbourhood U of m in M such that $m_k \notin U$ for infinitely many indices k. Passing to a subsequence, we can assume that $m_k \notin U$ for every $k \in \mathbb{N}$. Take a function

$$F(m,x) = \rho(m)f(x), x \in G, m \in M,$$

where $\rho \in C_c(M)$ has support in U, $\rho(m) = 1$ and $f \in C_c(G)$ with $\pi^m(f)\xi \neq 0$. Then

$$\begin{aligned} 0 &= \rho(m_k)\langle \pi_k^{m_k}(f)\xi_k,\xi_k\rangle \\ &= \langle \varphi_k, F\rangle \\ &\overset{k\to\infty}{\to} \langle \varphi, F\rangle \\ &= \langle \pi^m(f)\xi,\xi\rangle \\ &\neq 0. \end{aligned}$$

This shows that $\lim_{k\to\infty} m_k = m$.

For simplicity of notation we identify now m_k with k for $k \in \mathbb{N}\cup\infty$.

Let now $F \in C_c(\Gamma)$. Let us show first that the functions

$$\psi_k(x) := F * \varphi_k(m_k,x), x \in G,$$

converge uniformly on compact sets to the function

$$\psi(x) := F * \varphi(m,x), x \in G.$$

Let K be a compact subset of G, let $L = \{m_k; k \in \mathbb{N}\} \cup \{m\}$ and let $\varepsilon > 0$. We choose a compact neighbourhood V of e such that

$$\|l(v)F - F\|_1 \leq \varepsilon, v \in V,$$

which means that

$$\sup_{m\in M}\int_G |F(m,u)-F(m,v^{-1}\underset{m}{\cdot} u)|\beta_m(u)du \le \varepsilon, v\in V. \tag{8.8.1}$$

There exists a finite family $\{x_j; j=1,\cdots,N\}\subset G$ such that

$$K\subset\bigcup_{j=1}^N x_j \underset{m}{\cdot} V, m\in L.$$

Since $(\varphi_k)_k$ converges weakly to φ there exists $k_0\in\mathbb{N}$ such that

$$|\langle\varphi_k, l(x_j)F\rangle-\langle\varphi, l(x_j)F\rangle|\le\varepsilon, j=1,\cdots,N, k\ge k_0. \tag{8.8.2}$$

Hence for $k\ge k_0$, $x=x_jv$ for some $1\le j\le N$, and some $v\in V$, we have by (8.8.1) and (8.8.2) that

$$\begin{aligned}|F*\varphi_k(m_k,x)-F*\varphi(m,x)| &= |l(v)F*\varphi_k(m_k,x_j)-l(v)F*\varphi(m,x_j)|\\ &\le 2\varepsilon+|F*\varphi_k(m_k,x_j)-F*\varphi(m,x_j)|\\ &= 2\varepsilon+|\langle\varphi_k, l(x_j)F\rangle-\langle\varphi, l(x_j)F\rangle|\\ &\le 3\varepsilon, j=1,\cdots,N, k\ge k_0.\end{aligned}$$

Choose now a compact neighbourhood U of e such that

$$|\varphi(m,u)-1|\le\varepsilon, u\in U. \tag{8.8.3}$$

We take a function F in $C_c(\Gamma)$ such that $\mathrm{supp}(F)\subset M\times U$, such that $F\ge 0$ and such that $\|F(m_k)\|_1=1, k\in\mathbb{N}$.

Then for U small enough,

$$|F*\varphi(m,x)-\varphi(m,x)|\le\varepsilon, x\in G.$$

Because of weak convergence, we can assume that for $k\ge k_0$, we have that

$$\begin{aligned}\varepsilon &\ge |\langle\varphi, F\rangle-\langle\varphi_k, F\rangle|\\ &= |\int_G\varphi(m,u)F(m,u)\beta_m(u)du-\int_G\varphi_k(m_k,u)F(m_k,u)\beta_{m_k}(u)du|.\end{aligned}$$

Hence since $F\in C_c(\Gamma)$, for k large enough, we see that

$$2\varepsilon\ge|\int_G\varphi(m,u)F(m_k,u)\beta_{m_k}(u)du-\int_G\varphi_k(m_k,u)F(m_k,u)\beta_{m_k}(u)du|. \tag{8.8.4}$$

Now, since $\mathrm{supp}(F) \subset M \times U$ and $\varphi(m, u) = 1$ modulo ε on U, it follows that for k large enough,

$$
\begin{aligned}
&|\varphi_k(m_k, x) - F * \varphi(m_k, x)| \\
&= |\varphi_k(m_k, x) \int_G F(m_k, u)\beta_{m_k}(u)du - \int_G F(m_k, u)\varphi_k(m_k, u^{-1} \underset{m_k}{\cdot} x)\beta_{m_k}(u)du| \\
&= |\int_G F(m_k, u)(\varphi_k(m_k, x) - \varphi_k(m_k, u^{-1} \underset{m_k}{\cdot} x))\beta_{m_k}(u)du| \\
&\leq |\int_G F(m_k, u)\,\mathrm{Re}\,(1 - \varphi_k(u))\beta_{m_k}(u)du| \\
&\leq \varepsilon + |\int_G F(m_k, u)\,\mathrm{Re}\,(\varphi(m, u) - \varphi_k(m_k, u))\beta_{m_k}(u)du| \\
&\leq 3\varepsilon \text{ (by (8.8.4))}
\end{aligned}
$$

for any $x \in K$. Finally we obtain the estimate

$$
\begin{aligned}
|\varphi_k(m_k, x) - \varphi(m, x)| &\leq |\varphi_k(m_k, x) - F * \varphi_k(m_k, x)| + |F * \varphi_k(m_k, x) - F * \varphi(m, x)| \\
&\quad + |F * \varphi(m, x) - \varphi(m, x)| \\
&\leq 3\varepsilon + 3\varepsilon + \varepsilon = 7\varepsilon,
\end{aligned}
$$

which is valid for large enough k and any $x \in K$. □

8.8.2 Coefficients of Monomial Representations

Let G be a locally compact group, let H be a closed subgroup of G and $\chi : H \to \mathbb{T}$ a unitary character of H. We take the spaces of functions

$$
\begin{aligned}
\mathscr{E}(G/H, \chi) := \{\xi : G \to \mathbb{C};\ &\xi(gh) = \Delta_{G,H}^{1/2}(h)\overline{\chi(h)}(\xi(g)), \quad \forall g \in G, h \in H, \\
&\xi \text{ is continuous with compact support modulo } H\}
\end{aligned}
$$

and

$$
\begin{aligned}
\mathscr{E}(G/H) := \{\xi : G \to \mathbb{C};\ &\xi(gh) = \Delta_{G,H}(h)(\xi(g)), \quad \forall g \in G, h \in H, \\
&\xi \text{ is continuous with compact support modulo } H\}
\end{aligned}
$$

We consider also the canonical surjective mapping

$$
R_{G/H,\chi} : C_c(G) \to \mathscr{E}(G/H, \chi)
$$

defined by

$$R_{G/H,\chi}(\varphi)(g) = \int_H \varphi(gh)\Delta_{G,H}^{1/2}(h)\chi(h)dh,\ g \in G, \varphi \in C_c(G).$$

and the linear surjection

$$R_{G/H} : C_c(G) \to \mathscr{E}(G/H)$$

defined by

$$R_{G/H}(\varphi)(g) = \int_H \varphi(gh)\Delta_{G,H}(h)dh,\ g \in G, \varphi \in C_c(G).$$

We denote as earlier by τ_χ the monomial representation $\tau_\chi = \mathrm{ind}_H^G \chi$ of the group G.

Proposition 8.8.12 *Let $\varphi, \psi \in C_c(G)$ and $x \in G$. Let $\tilde{\psi} := R_{G/H,\chi}(\psi), \tilde{\varphi} := R_{G/H,\chi}(\varphi)$. Then we have for any $x \in G$ that*

$$c^{\tau_\chi}_{\tilde{\varphi},\tilde{\psi}}(x) = \int_G \tilde{\varphi}(x^{-1}u)\overline{\psi}(u)du.$$

Proof Let $\varphi, \psi \in C_c(G)$ and $x \in G$. Let $\tilde{\psi} := R_{G/H,\chi}(\psi), \tilde{\varphi} := R_{G/H,\chi}(\varphi)$. Then we have:

$$\begin{aligned}
c^{\tau_\chi}_{\tilde{\varphi},\tilde{\psi}}(x) &= \oint_{G/H} \tilde{\varphi}(x^{-1}u)\overline{\tilde{\psi}(u)}d\dot{u} \\
&= \oint_{G/H}\int_H\int_H \varphi(x^{-1}uh)\Delta_{G,H}^{1/2}(h)\chi(h)dh\overline{\psi}(uk)\Delta_{G,H}^{1/2}(k)\chi(k^{-1})dhdkd\dot{u} \\
&\overset{h\to kh}{=} \oint_{G/H}\int_H\int_H \varphi(x^{-1}ukh)\Delta_{G,H}^{1/2}(kh)\chi(kh)dh\overline{\psi}(uk)\Delta_{G,H}^{1/2}(k)\chi(k^{-1})dhdkd\dot{u} \\
&= \oint_{G/H}\int_H\int_H \varphi(x^{-1}ukh)\Delta_{G,H}^{1/2}(h)\chi(h)dh\overline{\psi}(uk)\Delta_{G,H}(k)dhdkd\dot{u} \\
&= \oint_{G/H}\int_H \tilde{\varphi}(x^{-1}uk)\overline{\psi}(uk)\Delta_{G,H}(k)dkd\dot{u} \\
&= \int_G \tilde{\varphi}(x^{-1}u)\overline{\psi}(u)du.
\end{aligned}$$

□

Remark 8.8.13 Let $G = \exp\mathfrak{g}$ be a connected, simply connected nilpotent Lie group. Kirillov's orbit picture tells us that the spectrum $\widehat{G}$ of G is in one-to-one correspondence with the coadjoint orbit space $\mathfrak{g}^*/G$. Therefore, using Proposition 8.8.4, one has that the spectrum of $L^1(\Gamma)$ is in one-to-one correspondence with the disjoint union $\dot{\bigcup}_{m\in M}\mathfrak{g}^*/G_m = M \times \mathfrak{g}^*/G$.

Theorem 8.8.14 *Let $\Gamma = M \times G$ be a simply connected nilpotent variable group. Then the spectrum $\widehat{\Gamma}$ is homeomorphic to the orbit space $M \times \mathfrak{g}^*/G$.*

Proof We shall use the following two results from the representation theory of nilpotent Lie groups.

Theorem 8.8.15 *Let $G = \exp\mathfrak{g}$ be a connected, simply connected nilpotent Lie group.*

1. *Let $H = \exp\mathfrak{h}$ be a closed normal subgroup of G. Let $\ell \in \mathfrak{g}^*$ and let $\Omega := \mathrm{Ad}^*(G)\ell$. Then*

$$\ker_{C^*(H)}(\pi_{\ell|H}) = \bigcap_{p\in\Omega_{|\mathfrak{h}}} \ker_{C^*(H)}(\pi_p)$$

2. *Let $q \in \mathfrak{h}^*$. Let $\tau := \mathrm{ind}_H^G\pi_q$. Then $\ker_{C^*(G)}\tau = \bigcap_{\ell\in\mathfrak{g}^*,\ell_{|\mathfrak{h}}=q} \ker_{C^*(G)}(\pi_\ell)$.*

Fix an inner product $\langle , \rangle$ on $\mathfrak{g}$. Let $(m_k)_{k\in\mathbb{N}}$ be a convergent sequence in M with limit m, and let $(\mathfrak{h}_k)_k$ be a sequence of subspaces of fixed dimension d in $\mathfrak{g}$ such that $\mathfrak{h}_k$ is an ideal of $\mathfrak{g}_{m_k}$, $k \in \mathbb{N}$. We pick for any k an orthonormal Jordan-Hölder basis $\mathscr{B}_k = (X_1^k, \cdots, X_n^k)$ of $\mathfrak{g}$, which passes through $\mathfrak{h}_k$. Using an appropriate subsequence, we can assume that $\lim_{k\to\infty} X_j^k = X_j$ exists for every j and so we obtain a limit ideal $\mathfrak{h} = \lim_{k\to\infty} \mathfrak{n}_k$ of $\mathfrak{g}_m$ and a limit Jordan-Hölder basis of $\mathfrak{g}_m$ passing through $\mathfrak{h}$.

Using the techniques developed in Example 8.8.7, we can always assume that $M = \overline{N} := \mathbb{N} \cup \{\infty\}$ and $\mathfrak{n}_k = \mathfrak{n}$ for every $k \in \mathbb{N}$. This gives us the variable nilpotent Lie group $\mathbb{H} = M \times N$.

Suppose now that a sequence (m_k, ℓ_k) converges in $M \times \mathfrak{g}^*$ to (m, ℓ).

Let $\mathfrak{p}_k$ be a polarization in $\mathfrak{g}_{m_k}$ at ℓ_k for every k. Passing to a subsequence, we can assume that $\dim\mathfrak{p}_k = d$ is fixed for every k that $\lim_{k\to\infty}\mathfrak{p}_k = \mathfrak{p}$ exists in $\mathfrak{g}$. If $d = \dim\mathfrak{g}$ then the characters χ_{ℓ_k} of the groups G_{m_k} converge uniformly on compact sets to the character χ_ℓ of G_m. So we can assume that $d < \dim G$.

Since $\langle \ell_k, [\mathfrak{p}_k, \mathfrak{p}_k]\rangle = \{0\}$ for any k, it follows that $\langle \ell, [\mathfrak{p}, \mathfrak{p}]\rangle = \{0\}$. Let $P_k := \exp\mathfrak{p}_k = \mathfrak{p}_k$, $k \in \mathbb{N}$, $P = \exp\mathfrak{p} = \mathfrak{p}$.

We fix now a function $\varphi \in C_c(G)$ such that $\tilde{\varphi} = R_{G/P,\chi_\ell}(\varphi) \neq 0$. Let $\tilde{\varphi_k} := R_{G/P_k,\chi_{\ell_k}}$, i.e.,

$$\tilde{\varphi_k}(x) = \int_{P_k} \varphi(x \underset{m_k}{\cdot} p_k)\chi_{\ell_k}(p_k)dp_k.$$

Here dp_k is the Lebesgue measure on the subspace $\mathfrak{p}_k$ determined by the inner product chosen on $\mathfrak{g}$. This means that we choose for every k an orthonormal Jordan-Hölder basis $\mathscr{Y}_k = \{Y_1^k, \cdots, Y_d^k\}$ of $\mathfrak{p}_k$, which gives us the measure

$$\int_{P_k} f(\sum_{j=1}^{d} y_j Y_j^k) dy_1 \cdots dy_d,\ f \in C_c(G),$$

on P_k. We can assume (if necessary passing to a subsequence) that $\lim_{k\to\infty} Y_j^k = Y_j, j = 1, \cdots, d$. This gives us an orthonormal Jordan-Hölder basis of $\mathfrak{p}$.

Since $\lim_{k\to\infty} P_k = P$ and $\lim_{k\to\infty} \ell_k = \ell$, it is easy to see that $\|\tilde{\varphi}_k\|_\infty \leq C, k \in \mathbb{N}$ for some $C > 0$ and that the functions $\tilde{\varphi_k}$ converge uniformly on compact sets to the function $\tilde{\varphi}$. Let

$$\tau_{\ell,P} := \operatorname{ind}_P^{G_m} \chi_\ell.$$

Using Proposition 8.8.12, we see that the coefficients $c_k := c_{\tilde{\varphi}_k}^{\pi_{\ell_k,P_k}}$ converge uniformly on compact sets to the coefficient $c = c_{\tilde{\varphi}}^{\tau_{\ell,P}}$. This implies that the coefficients c_k also converge weakly on $L^1(\Gamma)$, hence on $C^*(\Gamma)$ too, to the coefficient c. This shows that

$$\bigcap_{j\geq k} \ker_{C^*(\Gamma)} \pi_{\ell_j} \subset \ker_{C^*(\Gamma)} \tau_{\ell,P}, k \in \mathbb{N}.$$

By Theorem 8.8.15, we then have that

$$\bigcap_{j\geq k} \ker_{C^*(\Gamma)} \pi_{\ell_j} \subset \ker_{C^*(\Gamma)} \pi_\ell, k \in \mathbb{N}.$$

Hence π_ℓ is contained in the closure of the sets $S_k := \{\pi_{\ell_j}, j \geq k\}$. Therefore π_ℓ is contained in the limit set of the sequence $(\pi_k)_k$.

For the other direction we use an induction argument based on the dimension of G. Suppose we have a sequence $(\pi_k)_k$ in $\widehat{\Gamma}$ which converges to some $\pi \in \widehat{\Gamma}$. We write $\pi_k = \operatorname{ind}_{P_k}^{G_{m_k}} \chi_{\ell_k}$ for some ℓ_k in the Kirillov orbit $\Omega_k^{m_k}$ of π_k and some polarization $P_k = \exp \mathfrak{p}_k$ at ℓ_k. Let also $\pi = \operatorname{ind}_P^{G_m} \chi_\ell$. We know already, by Theorem 8.8.11, that the sequence $(m_k)_k \subset M$ converges to m.

We can assume (passing to a subsequence) that $\dim \mathfrak{p}_k = d$ for some $d \leq \dim G$ for all $k \in \mathbb{N}$. If $d = \dim G$, then obviously π_k is the character χ_{ℓ_k} for any k and for every $X \in \mathfrak{g}$ we then have the identity

$$X \underset{m_k}{*} \chi_{\ell_k} = \langle \ell_k, X \rangle \chi_{\ell_k}.$$

This shows that $\lim_{k\to\infty} \ell_k = \ell$.

Hence we can assume that $d < n$.

Take for every k an ideal $\mathfrak{n}_k$ of codimension 1 in $\mathfrak{g}_k$ containing $\mathfrak{p}_k$. We can assume that $M = \overline{\mathbb{N}}$, the one-point compactification of $\mathbb{N}$, and that $\mathfrak{n}_k = \mathfrak{n}$, $k \in M$. Then

$$N = N_k = \mathfrak{n} = \exp \mathfrak{n}$$

is a closed normal subgroup of G_k for any $k \in M$. Since π_k converges to π, by Theorem 8.7.7, for any coefficient $c \neq 0$ of the representation π there exists for every $k \in \mathbb{N}$ a coefficient c_k of π_k such that c is the weak limit of the sequence $(c_k)_k$. But then the functions c_k converge uniformly on compact sets to the function c by Theorem 8.8.11. This implies that for any $k \in M$, we have that $\bigcap_{j \geq k} \ker_{C^*(\mathbb{H})}(\pi_{j|H}) \subset \ker_{C^*(\mathbb{H})}(\pi_{|H})$. Let $p \in (\Omega_\ell)_{|\mathfrak{h}} \subset \mathfrak{h}^*$. By Theorem 8.8.15, we have that $\pi_p \in \overline{\{\pi_{p_k};\ p_k \in \Omega_{k|\mathfrak{h}} \subset \mathfrak{h}^*\}}$. We use now the induction hypothesis for the $VNLG$ $(M \times \mathfrak{n})$. There exists a sequence $(p_k \in \Omega_{k|\mathfrak{n}})_k$ such that p_k converges in $\mathfrak{n}^*$ to $p = \ell_{|\mathfrak{n}}$. But since the subsets $\ell_k + \mathfrak{p}_k^\perp$ are contained in the G_k-orbits Ω_k ($\mathfrak{p}_k$ being a polarization at ℓ_k), we know that $\omega_{\ell_k} + \mathfrak{n}^\perp = \Omega_k$, $k \in \mathbb{N}$. Therefore we find for every $k \in \mathbb{N}$ and element $l'_k \in \Omega_k$ such that $\lim_{k\to\infty} l'_k = \ell$. □

8.9 $\mathscr{D}$-prime Ideals in the Schwartz Algebra of a Nilpotent Lie Group

8.9.1 Exponential Actions

Let $N = \exp \mathfrak{n}$ be a connected, simply connected nilpotent Lie group.

Definition 8.9.1 We say that a group of automorphisms $\mathscr{D}$ of N is *exponential* if

(i) $\mathscr{D} = \exp \delta$ is an exponential solvable Lie group with Lie algebra δ, which acts continuously on N (and on $\mathfrak{n}$),
(ii) the eigenvalues λ of any $D \in \delta$ on $\mathfrak{n}_{\mathbb{C}}$ are of the form $\lambda = a + ib$ with $a \neq 0$
(iii) $\mathrm{ad}(\mathfrak{n}) \subset \delta$.

The group $\mathscr{D}$ acts on functions and representations of N in the following way. For an integrable function $f : N \to \mathbb{C}$ we write

$${}^d f(x) = \delta(d) f(d^{-1}(x)), \quad x \in N, d \in \mathscr{D},$$

where δ denotes the modular function of $\mathscr{D}$, i.e., $\delta(d)$ is the unique positive number such that

$$\delta(d) \int_N f(x)dx = \int_N {}^d f(x)dx, \quad f \in L^1(N), d \in \mathscr{D}.$$

For a representation (T, X) of N on a vector space X, we write $({}^dT, X)$ $(d \in \mathscr{D})$ for the representation

$$ {}^dT(x) = T(d(x)), \quad x \in N, $$

on X.

Let $\mathscr{S}(N)$ be the Schwartz algebra of N, i.e., the Fréchet space

$$ \mathscr{S}(N) = \{f : N \to \mathbb{C};\ f \circ \exp \in \mathscr{S}(\mathfrak{n})\}. $$

Here $\mathscr{S}(\mathfrak{n})$ denotes the usual space of smooth fast decaying functions on $\mathfrak{n}$. Let us put the convolution product on $\mathscr{S}(N)$. Then $\mathscr{S}(N)$ is a Fréchet algebra with approximate units. The space of primitive ideals $\mathrm{Prim}(\mathscr{S}(N))$ of $\mathscr{S}(N)$ is homeomorphic to $\hat{N}$ of N, hence by Kirillov to the coadjoint orbit space $\mathfrak{n}^*/N$ of N (for more details see [114, 117]).

The group $\mathscr{D}$ leaves the Schwartz space $\mathscr{S}(N)$ invariant, and the mapping

$$ \mathscr{D} \times \mathscr{S}(N) \to \mathscr{S}(N) : (d, f) \mapsto {}^df $$

is continuous (see [116]).

Definition 8.9.2 We say that a subspace W of $\mathscr{S}(N)$ is $\|\cdot\|$-closed if $\|\cdot\|$ is a continuous norm on $\mathscr{S}(N)$ for which W is closed in $\mathscr{S}(N)$.

Let $J \trianglelefteq L^1(G)$ be a closed ideal. Then $I := J \cap \mathscr{S}(N)$ is $\|\|_1$-closed. Furthermore, if $I \trianglelefteq \mathscr{S}(N)$ is a $\|\|_1$-closed ideal in $\mathscr{S}(N)$, then $J := \overline{I}$ is a closed ideal in $L^1(G)$ and $I = J \cap \mathscr{S}(N)$.

Definition 8.9.3 Let $\tau \in \hat{N}$ be an irreducible unitary representation of N. We define the ideal $I_{\mathscr{D}(\tau)}$ of $\mathscr{S}(N)$ by

$$ I_{\mathscr{D}(\tau)} = \{f \in \mathscr{S}(N);\ {}^d\tau(f) = 0, \quad \forall d \in \mathscr{D}\}. $$

The subspace $I_{\mathscr{D}(\tau)}$ is closed for the norm $\|\cdot\|_1$ in $\mathscr{S}(N)$ and it is Δ-invariant.

Let I_1, I_2 be two Δ-invariant ideals of $\mathscr{S}(N)$ such that $I_1 * I_2 \subset I_{\mathscr{D}(\tau)}$. Then either I_1 or I_2 is contained in $I_{\mathscr{D}(\tau)}$. Indeed, if $I_1 \not\subset I_{\mathscr{D}(\tau)}$, then $I_1 \not\subset \ker({}^d\tau)$ for a certain $d \in \Delta$. But since $I_1 * I_2 \subset \ker({}^d\tau)$ and as $\ker({}^d\tau)$ is a prime ideal, we see that $I_2 \subset \ker({}^d\tau)$ and so $I_2 \subset \bigcap_{d' \in \Delta} \ker({}^{d'd}\tau) = I_{\mathscr{D}(\tau)}$, because I_2 is Δ-invariant.

Definition 8.9.4 Let I be an ideal in $\mathscr{S}(N)$. We say that I is Δ-*prime*, if I is Δ-invariant and if for every pair I_1, I_2 of D-invariant in $\mathscr{S}(N)$ such that $I_1 * I_2 \subset I$, then either $I_1 \subset I$ or $I_2 \subset I$.

Example 8.9.5 Let G be an exponential solvable Lie group. Let N be a closed connected normal subgroup of G. Let (T, X) be a bounded irreducible (or topologically simple) representation of $L^1(G)$ on a Banach space X. Then by Fujiwara and Ludwig [70] there exists a bounded continuous representation (T, X) of G such that

$$T(f) = \int_G f(g)T(g)dg, \ f \in L^1(G).$$

Hence we can restrict the corresponding representation of G to N. Let I be the kernel of $T|_N$ in $\mathscr{S}(N)$. Then I is $\|\cdot\|_1$-closed and Δ-prime, where $\Delta = \mathrm{Ad}(G)_{|\mathfrak{n}}$. Indeed, if I_1, I_2 are two Δ-invariant ideals of $\mathscr{S}(N)$ such that $I_1 * I_2 \subset I$, then we have that

$$\begin{aligned}(L^1(G) * I_1 * L^1(G)) * (L^1(G) * I_2 * L^1(G)) &\subset L^1(G) * \overline{L^1(G) * I_1 * I_2 * L^1(G)} * L^1(G) \\ &\subset \ker_{L^1(G)} T,\end{aligned}$$

where $\overline{A}$ denotes the closure of A in $L^1(G)$. Since $\ker_{L^1(G)} T$ is a prime ideal, and T is irreducible, we see that $L^1(G) * I_1 * L^1(G) \subset \ker_{L^1(G)} T$ or $L^1(G) * I_2 * L^1(G) \subset \ker_{L^1(G)} T$. Hence either $I_1 \subset I$ or $I_2 \subset I$.

Theorem 8.9.6 *Let Δ be an exponential solvable Lie group of automorphisms of the simply connected, connected nilpotent Lie group N. Then every non-trivial Δ-prime ideal in $\mathscr{S}(N)$ which is closed for some norm is the kernel of a Δ-orbit in $\hat{N}$. That is, there exists $\tau \in \hat{N}$ such that*

$$I = \bigcap_{d \in \Delta} \ker_{\mathscr{S}(N)} d \cdot \tau.$$

8.9.2 *Proof of Theorem 8.9.6*

We need some lemmas and definitions.

Definition 8.9.7 For a subset A of $\mathscr{S}(N)$ we denote by $h(A)$ the hull of A in $\hat{N}$:

$$h(A) = \{\tau \in \hat{N}; \tau(A) = \{0\}\}.$$

The subset $h(A)$ of $\hat{N}$ is closed in $\hat{N}$, if we put on $\hat{N}$ the coarsest topology making the canonical surjection $\tau \mapsto \ker(\tau)$ of $\hat{N}$ onto $\mathrm{Prim}(\mathscr{S}(N))$ continuous (see [49]).

Definition 8.9.8 The kernel $\ker(C)$ of a closed subset C of $\hat{N}$ is by definition the (closed) ideal

$$\ker(C) = \bigcap_{\tau \in C} \ker(\tau) \subset \mathscr{S}(N).$$

Definition 8.9.9 Let C be a closed subset in $\hat{N}$. We recall that there exists a minimal ideal $j(C)$ in $L^1(G)$ with hull C, which is generated by Schwartz functions (see

Theorem 8.5.7). We denote the minimal ideal in $\mathscr{S}(N)$ associated to C by $j(C)$. In other terms, $j(C)$ is the unique ideal in $\mathscr{S}(N)$ with hull h equal to C and which is contained in every ideal J with $h(J) = C$. (See [111, 114] for more details.)

Lemma 8.9.10 *Every Δ-prime ideal I of $\mathscr{S}(N)$ which is closed for the norm $\| \|_1$ is the kernel of its hull $h(I) = C$.*

Proof Since I is Δ-invariant, its hull C is also Δ-invariant, together with ker(C). Let us show that I contains $(\ker(C))^M$ for a certain $M \in \mathbb{N}$. As a consequence, $I = \ker(C)$, since I is Δ-prime. To show that $(\ker(C))^M \subset I$, we recall that $j(C)$ is generated as an ideal by the elements $\varphi(f)$, $f = f^* \in \ker(C) \subset \mathscr{S}(N)$, $\varphi \in C_0^\infty(\mathbb{R})$, where $C_0^\infty(\mathbb{R})$ is the space of smooth, compactly supported maps $\varphi : \mathbb{R} \to \mathbb{C}$ that vanish around 0. Here $f^*(x) = \overline{f(x^{-1})}$, $x \in N$ (see [114]).

The symbol $\varphi(f)$ denotes the integral

$$\varphi(f) = \frac{1}{2\pi} \int_{\mathbb{R}} \hat{\varphi}(\lambda)(e(i\lambda f))d\lambda,$$

and

$$e(i\lambda f) = \sum_{n>0} \frac{(i\lambda f)^{n(*)}}{n!}.$$

We have seen in Theorem 8.5.7 that for every continuous norm $\| \cdot \|$ on $\mathscr{S}(N)$, there exist $m \in \mathbb{N}$ and, for all $f = f^* \in \mathscr{S}(N)$, a constant $C = C(f, \| \cdot \|)$ such that

$$\|e(i\lambda f)\| \leq C(1 + |\lambda|)^m, \quad \lambda \in \mathbb{R}.$$

We take $f = f^* \in \ker(C)$. Let us show that $f^{m+3} \in I$. Let ϕ be a smooth function with compact support such that $\phi(t) = t^{m+3}$ on an interval containing $[-\|f\|_1, \|f\|_1]$. Then

$$\phi(f) = f^{m+3}.$$

Let $\epsilon > 0$ and choose an element $\varphi = \varphi_\epsilon$ in $C_0^\infty(\mathbb{R})$ such that

$$|\varphi^{(k)}(t) - \phi^{(k)}(t)| \leq \epsilon, \quad t \in \mathbb{R}, k = 0, \cdots, m+2.$$

Then

$$\|f^{m+3} - \varphi(f)\| = \|\phi(f) - \varphi(f)\| \leq \frac{1}{2\pi} \int_{\mathbb{R}} |\hat{\phi}(\lambda) - \hat{\varphi}(\lambda)| \|e(i\lambda f)\| d\lambda \leq C'\epsilon,$$

for a certain constant C' independent of ϵ. Hence $f^{m+3} \in \overline{j(C)}^{\|\cdot\|} \subset \overline{I}^{\|\cdot\|} = I$, since $\varphi_\epsilon(f) \in j(C) \subset I$ for every ϵ. Finally let us show that $(\ker(C))^{2^{m+3}-1} \subset I$. We have, for a family of self-adjoint elements $f_1, \cdots, f_{m+3}$ in $\ker(C)$, for any $s_1, \cdots, s_{m+3} \in \mathbb{R}$, that $I \ni (\sum_{j=1}^{m+3} s_j f_j)^{m+3} = (\prod_{i=1}^{m+3} s_i)(\sum_{s\in S_{m+3}}(\prod_{i=1}^{m+3} f_{\sigma(i)}))$ + expressions of the form $s_1^{n_1} \cdots s_{m+3}^{n_{m+3}}$ with $n_j > 1$ for at least one j. Therefore

$$\sum_{\sigma\in S_{m+3}} (\prod_{i=1}^{m+3} f_{\sigma(i)}) \in I. \tag{8.9.1}$$

Here S_{m+3} denotes the permutation group on $m+3$ elements. We can write every $f \in \ker(C)$ in a unique way in the form $f = g + ih$, where g, h are self-adjoint elements of $\ker(C)$. Then taking first g, then h as f_1 in (8.9.1) and adding the first expression to i times the second expression, we see that

$$\sum_{\sigma\in S_{m+3}} (\prod_{i=1}^{m+3} f_{\sigma(i)}) \in I,$$

as soon as f_1 is in $\ker(C)$ and $f_2, \cdots, f_{m+3}$ are self-adjoint elements in $\ker(C)$. Finally, repeating this argument, we see that

$$\sum_{s\in S_{m+3}} (\prod_{i=1}^{m+3} f_{\sigma(i)}) \in I,$$

for every $f_1, \cdots, f_{m+3} \in \ker(C)$ and in particular $f^{m+3} \in I$ for every $f \in \ker(C)$. The theorem of Nagata-Higman [133, Proposition 4.4.10] implies that $(\ker(C))^{2^{m+3}-1} \subset I$. □

Definition 8.9.11 Let C be a Δ-invariant subset of $\hat{N}$. We say that C is Δ-prime whenever it cannot be written as union of two closed Δ-invariant subsets C_1, C_2 (unless $C_1 = C$ or $C_2 = C$).

Lemma 8.9.12 *Let C be a Δ-invariant closed subset of $\hat{N}$. Suppose that $\ker(C) \subset \mathcal{S}(N)$ is Δ-prime. Then C is Δ-prime.*

Proof Let C_1, C_2 be two closed Δ-invariant subsets of C such that $C = C_1 \cup C_2$. Let $I_j = \ker(C_j) \subset \mathcal{S}(N), j = 1, 2$. Then $I_1 * I_2 \subset \ker(C_1) \cup \ker(C_2) = \ker(C_1 \cup C_2 = C)$. As $\ker(C)$ is Δ-prime, we have that $\ker(C_j) \subset \ker(C)$ for $j = 1$ or $j = 2$. But then either $C \subset C_1$ or $C \subset C_2$. □

Lemma 8.9.13 *Let C be a closed Δ-prime subset of $\hat{N}$. Then C is the closure of a Δ-orbit in $\hat{N}$.*

Proof For $f \in \mathcal{S}(N)$, let $\|f\| = \sup_{\tau\in C} \|\tau(f)\|_{op}$ be the C^*-semi-norm of f associated to C. We know that $\hat{N}$, hence also the closed subset C, is a Baire space

[49]. Choose a dense sequence $(f_n)_{n\in\mathbb{N}}$ in $\mathscr{S}(N)$ and take for $n \in \mathbb{N}$ the open subset

$$U_n = \{\tau \in C; \quad \|\tau(f_n)\|_{op} > \frac{1}{2}\|f_n\|\}.$$

Let $\mathbb{N}' = \{n \in \mathbb{N}; \quad U_n \neq \emptyset\}$. Then $V_n = \bigcup_{x\in\mathscr{D}} {}^x U_n$, $n \in \mathbb{N}'$, is an open $\mathscr{D}$-invariant subset of C.

As C is $\mathscr{D}$-prime, V_n is dense in C. The Baire property of C tells us that $V_\infty = \bigcap_{n\in\mathbb{N}'} V_n$ is also dense in C. Let $\tau \in V_\infty$. Let us show that the $\mathscr{D}$-orbit of τ is dense in C. If not, there exists an open subset W of C such that $({}^{\mathscr{D}}\tau) \cap W = \emptyset$. We take $\rho \in W$ and $g = g^* \in \mathscr{S}(N)$ such that $\tau'(g) = 0$ for all $\tau' \in ({}^{\mathscr{D}}\tau)^-$ and such that $\|\rho(g)\|_{op} = 2$. Replacing if necessary g by $\phi(g)$ for a certain ϕ, we can suppose furthermore that $\|g\|$ is also equal to 2.

There exists $n \in \mathbb{N}$ such that $\|f_n - g\| < \frac{1}{4}\|f_n\|$. Hence $\|\rho(f_n)\|_{op} \geq \frac{3}{4}\|f_n\|$. Therefore $\rho \in U_n$ (and so $n \in \mathbb{N}'$) and we have that $\|\tau'(f_n)\|_{op} < \frac{1}{4}\|f_n\|$ for every $\tau' \in {}^{\mathscr{D}}\tau$ (since $\tau'(g) = 0$). This tells us that $\tau \notin V_n$, which contradicts the fact that $\tau \in V_\infty$.

□

End of the Proof of the Theorem we have seen in (8.9.10) that I is the kernel of its hull C. Hence C is Δ-prime by Lemma 8.9.12 and, by Lemma 8.9.13, C is the closure of a Δ-orbit. Finally

$$I = \ker(C) = \ker({}^{\mathscr{D}}\tau)$$

for a certain $\tau \in \hat{N}$.

□

Example 8.9.14 Let $N = \mathbb{R}$, $c \in \mathbb{R}$ and let $J = \{f \in \mathscr{S}(\mathbb{R}); \hat{f}(c) = \int_{\mathbb{R}} f(t)e^{-2\pi itc}dt = 0\}$. Let

$$I = \{f \in \mathscr{S}(\mathbb{R}); \quad (\frac{d}{dt})^k \hat{f}(c) = 0, \quad \forall k \in \mathbb{N}\}.$$

Then I is a closed ideal of $\mathscr{S}(\mathbb{R})$, but it is not closed for any norm. In fact, I is the closure of the minimal ideal $j(\{c\})$ and $I \neq \ker(\{c\})$. It is however easy to see that I is prime. In fact, $I_1 * I_2 \subset I$ for two ideals I_1, I_2 of $\mathscr{S}(\mathbb{R})$, and if $I_j \not\subset I$ for $j = 1, 2$, then there exists $f_j \in I_j$ such that $(\frac{d}{dt})^{n_j}\hat{f}_j(c) \neq 0$ and $(\frac{d}{dt})^i\hat{f}_j(c) = 0, 0 \leq i < n_j$ for a certain $n_j \in \mathbb{N}$, $j = 1, 2$. But then $(\frac{d}{dt})^{n_1+n_2}(f_1 * f_2)\hat{}(c) = (\frac{d}{dt})^{n_1+n_2}(\hat{f}_1\hat{f}_2)(c) \neq 0$ by the Leibniz rule, against $f_1 * f_2 \in I$.

The closed prime ideal I of $\mathscr{S}(\mathbb{R})$, which is not closed for any norm, is not a kernel either. Therefore the condition of being "closed for some norm" is necessary in Theorem 8.9.6.

Remark 8.9.15 It is easy to see that the trivial ideal $\{0\}$ is not a prime ideal. Indeed, let Z be the center of N. We can take $\mu \neq 0$, $\nu \neq 0$ in $\mathscr{S}(Z)$ such that $\nu * \mu = 0$. Then the convolution product of the ideals $I_1 = \mathscr{S}(N) * \mu$ and $I_2 = \mathscr{S}(N) * \nu$ is $\{0\}$, but evidently $I_1 \neq \{0\}$ and $I_2 \neq \{0\}$.

8.10 A Retract Theorem

Let $G = \exp \mathfrak{g}$ be a connected, simply connected exponential Lie group. We take $\ell \in \mathfrak{g}^*$, a Pukanszky polarization $P = \exp \mathfrak{p}$ at ℓ and we form the representation $\pi_\ell = \pi_{\ell,\mathfrak{p}} = \mathrm{ind}_P^G \chi_\ell$. Let $F \in L^1(G)$. We analyze the operator $\pi_\ell(F)$. Let as usual $d\dot{x}$ denote the (unique) linear, G-invariant, positive form $\oint_{G/P} \varphi(x) d\dot{x}$ on the vector space introduced in Sect. 1.2.2. The Hilbert space of the representation $\pi_{\ell,P}$ is the space of functions $L^2(G/P, \ell)$, where for $1 \le p < \infty$ we define

$$L^2(G/P, \ell) := \{\xi : G \to \mathbb{C}; \xi \text{ measurable},$$
$$\xi(gp) = \chi_\ell(p^{-1})(\frac{\Delta_H(p)}{\Delta_G(p)})^{1/2}\xi(g), g \in G, p \in P, \oint_{G/P} |\xi(u)|^p d\dot{u} < \infty\}$$

and

$$\mathscr{E}(G/P, \ell) = \{\xi \in L^2(G/P, \ell); \xi \text{ is continuous with compact support modulo } P\}.$$

Then

$$\begin{aligned}
\pi_\ell(F)\xi(g) &= \int_G F(u)\xi(u^{-1}g)du \\
&= \int_G \Delta_G(u)^{-1} F(gu^{-1})\xi(u)du \\
&= \oint_{G/P} \int_P \Delta_G(u)^{-1}\Delta_G(p)^{-1} F(gp^{-1}u^{-1})\Delta_{G,P}(p)\xi(up)dpd\dot{u} \\
&= \oint_{G/P} \int_P \Delta_G(u)^{-1}\Delta_G(p)^{-1}\chi_\ell(p^{-1})\Delta_{G,P}(p)^{1/2} F(gp^{-1}u^{-1})dp\xi(u)d\dot{u} \\
&= \oint_{G/P} \int_P \Delta_G(u)^{-1}\Delta_G(p)\chi_\ell(p)\Delta_{P,G}(p)^{1/2}\Delta_P(p^{-1})F(gpu^{-1})dp\xi(u)d\dot{u} \\
&= \oint_{G/P} \Delta_G^{-1}(u) \int_P \Delta_{G,P}^{1/2}(p)\chi_\ell(p)F(gpu^{-1})\xi(u)dpd\dot{u}.
\end{aligned}$$

Therefore the operator $\pi_\ell(F)$ is a kernel operator with kernel function

$$F_{\ell,P}(g,u) = F_\ell(g,u) = \Delta_G(u)^{-1} \int_P \Delta_{G,P}(p)^{1/2}(p)\chi_\ell(p)F(gpu^{-1})dp, \quad g, u \in G.$$

Definition 8.10.1 Let $G = \exp \mathfrak{g}$ be an exponential Lie group and let $\mathfrak{n} \subset \mathfrak{g}$ be a nilpotent ideal containing $[\mathfrak{g}, \mathfrak{g}]$. Let Λ be the set of roots of $\mathfrak{g}$. An element $X \in \mathfrak{g}$ is called regular if $\lambda(X) \neq 0$ for every $\lambda \in \Lambda$. Put

$$\mathfrak{g}_0 = \mathfrak{g}_0^X := \{U \in \mathfrak{g}; (\mathrm{ad}X)^m u = 0 \text{ for some m in } \mathbb{N}\}$$

for a regular X. Then $\mathfrak{g}_0^X$ is a nilpotent Lie subalgebra of $\mathfrak{g}$ and

$$\mathfrak{g} = \mathfrak{g}_0^X + [\mathfrak{g}, \mathfrak{g}].$$

Definition 8.10.2 Let $\mathfrak{n}$ be a nilpotent Lie subalgebra of $\mathfrak{g}$ containing the ideal $[\mathfrak{g}, \mathfrak{g}]$ of $\mathfrak{g}$. For instance we can take for $\mathfrak{n}$ the nilradical of $\mathfrak{g}$.

Let $\ell \in \mathfrak{g}^*$. We take an $(\mathrm{ad}(\mathfrak{g}(\ell_{|\mathfrak{n}})))$-invariant polarization $\mathfrak{p}_0 \subset \mathfrak{n}$ at $\ell_{|\mathfrak{n}}$. For instance a Vergne polarization built from an $\mathrm{ad}(\mathfrak{g})$ Jordan-Hölder sequence of $\mathfrak{n}$. Let σ be a subspace of $\mathfrak{g}(\ell)$ such that

$$\mathfrak{g}(\ell) + \mathfrak{n} = \sigma \oplus \mathfrak{n}$$

and let $\varphi(\ell) = \varphi \subset \mathfrak{g}$ be a subspace of $\mathfrak{g}(\ell_{|\mathfrak{n}})$ such that

$$\mathfrak{g}(\ell_{|\mathfrak{n}}) = \varphi(\ell) \oplus (\mathfrak{g}(\ell) + \mathfrak{n}).$$

We take the skew form B_ℓ on φ defined by

$$B_\ell(Y, Y') := \langle \ell, [Y, Y'] \rangle$$

and fix a maximal isotropic subspace $\mathfrak{y}$ of φ. Then the subspace

$$\mathfrak{p} := \mathfrak{y} + \mathfrak{g}(\ell) + \mathfrak{p}_0$$

is a Pukanszky polarization at ℓ (called adapted to $\mathfrak{n}$). Let $P := \exp \mathfrak{p}$. Obviously

$$\langle \ell, [\mathfrak{p}, \mathfrak{p}] \rangle = \{0\}$$

by construction and it is easy to check that $Ad^*(P)\ell = \ell + \mathfrak{p}^\perp$, since $\mathrm{Ad}^*(G(\ell_{|\mathfrak{n}})\ell) = \ell + (\mathfrak{g}(\ell) + \mathfrak{n})^\perp$.

We can identify then the Hilbert space $L^2(G/P, \ell)$ with $L^2(\mathbb{R}^{\dim \mathfrak{g}/\mathfrak{p}})$, and for every $F \in L^1(G)$ the operator $\pi_{\ell,\mathfrak{p}}(F)$ is then defined on $\mathbb{R}^{\dim \mathfrak{g}/\mathfrak{p}} \times \mathbb{R}^{\dim \mathfrak{g}/\mathfrak{p}}$.

Definition 8.10.3 We denote by $\mathscr{P}_m$ the space of polynomial functions on $\mathbb{R}^m$ and by D_m the space of differential operators on $\mathbb{R}^m$ with polynomial coefficients (also denoted by $\mathscr{P}_V$ if V is a finite-dimensional real vector space). For a subspace V of $\mathbb{R}^m$ denote by E_V the subalgebra of maps on $\mathbb{R}^m$ generated by the functions $x \to e^{\langle q,x \rangle}, q \in V^*$.

We also take an exponential basis $\mathscr{E}$ of $\mathfrak{g}$ modulo $\mathfrak{p}$ containing a Malcev basis $\mathscr{E}_0 = \{X_1^0, \cdots, X_e^0\}$ of $\mathfrak{n}$ modulo $\mathfrak{p}_0$. We refine this basis $\mathscr{E}$ by choosing a basis $\mathscr{T} = \mathscr{T}(\ell) = \{T_1, \cdots, T_t\}$ modulo $\mathfrak{g}(\ell_{|\mathfrak{n}}) + \mathfrak{n}$ of $\mathfrak{g}$ contained in $\mathfrak{g}_0$ and a basis of

$\mathscr{X} = \{X_1, \cdots, X_d\}$ of $\varphi(\ell)$ modulo $\mathfrak{y}$. Then we can identify G/P with $\mathbb{R}^t \times \mathbb{R}^d \times \mathbb{R}^e$ by using the diffeomorphism

$$E_{G/P} = E_{G/P}^{\mathscr{T}, \mathscr{X}, \mathscr{E}_0} : \mathbb{R}^{t+d+e} \to G/P$$

$$E_{G/P}(t, x, x_0) = \exp(t_1 T_1) \cdots \exp(t_\ell T_\ell) \exp(x_1 X_1) \cdots \exp(x_d X_d) \exp(x_1^0 X_1^0) \cdots \exp(x_e^0 X_e^0) \cdot P.$$

Definition 8.10.4 Let $G = \exp \mathfrak{g}$ be an exponential group, $\mathfrak{n}$ a nilpotent ideal of $\mathfrak{g}$ containing $[\mathfrak{g}, \mathfrak{g}]$, take $\ell \in \mathfrak{g}^*$ and a polarization $\mathfrak{p}$ at ℓ adapted to $\mathfrak{n}$ as per Definition 8.10.2. We define for $a \in \mathbb{N}$ the space $E\mathscr{S}(a, G/P, \ell)$ as space of smooth functions $K : \mathbb{R}^a \times G \times G \to \mathbb{C}$, which have the following properties:

(a) $F(\overline{u}, gp, g'p' = \overline{\chi_\ell(p)}\chi_\ell(p')F(\overline{u}, g, g'), \overline{u} \in \mathbb{R}^a, g, g' \in G, p, p' \in P.$
(b) if $\tilde{F}$ denotes the function $\tilde{F}(\overline{u}, E_G/P(t, x, x_0), E_{G/P}(t', x', x_0')$ on $\mathbb{R}^a \times \mathbb{R}^{t+d+e} \times \mathbb{R}^{t+d+e}$, then

$$\int_{\mathbb{R}^a \times \mathbb{R}^{t+d+e} \times \mathbb{R}^{t+d+e}} |\eta(\overline{x}) D \cdot \hat{\tilde{K}}(\overline{u}, t, x + x', x_0, t', x', x_0')| d\overline{u} dt dt' dx dx' < \infty$$

for every $D \in D_{\mathbb{R}^a \times \mathscr{T} \times \mathscr{X} \times \mathfrak{n} \times \mathscr{T} \times \mathscr{X} \times \mathfrak{n}}$ and every $\eta \in E_{\mathscr{X} + \mathscr{T}}$. Furthermore the symbol $\hat{\tilde{K}}$ denotes the partial Fourier transform of $\tilde{K}$ along $\mathscr{X}$:

$$\hat{\tilde{K}}(\overline{u}, t, x, x_0, t', x', x_0') = \int_{\mathscr{X}} \tilde{K}(\overline{u}, t, x, x_0, t', w', x_0') e^{i\langle x', w \rangle} dw.$$

Lemma 8.10.5 *For every $\infty > p \geq 1$, the space $E\mathscr{S}(a, G/P, \ell)$ is dense in $L^p(\mathbb{R}^a \times G/P, \ell)$.*

Proof Let $q \in \mathscr{P}_{d+d}$ and let

$$\psi_q(x, x') = q(x, x') e^{-\|x\|^2 - \|x'\|^2}, x = \sum_1^d x_i X_i, \|x\| = \sum_{i=1}^d x_i^2.$$

We denote by $\mathscr{A}$ the function space generated by the functions

$$K(\overline{u}, \exp t \exp x \exp x^0, \exp t' \exp x' \exp x_0') = \varphi(\overline{u}, t, x_0, t', x_0') \psi_q(x, x'),$$

where $\varphi \in C_c^\infty(\mathbb{R}^a \times \mathbb{R}^t \times \mathbb{R}^d \times \mathbb{R}^e \times \mathbb{R}^t \times \mathbb{R}^d \times \mathbb{R}^e)$ and $q \in \mathscr{P}_{d+d}$. This space is obviously contained in $E\mathscr{S}(a, G/P, \ell)$ since the f_q are finite linear combinations of Hermite functions on $\mathbb{R}^d \times \mathbb{R}^d$. Furthermore, Hermite functions generate the Schwartz space $\mathscr{S}(\mathbb{R}^d \times \mathbb{R}^d)$, so $\mathscr{A}$ is dense in $L^2(\mathbb{R}^a \times G/P, \ell)$. □

Definition 8.10.6 Let $\mathfrak{g}_0$ be any nilpotent Lie subalgebra of $\mathfrak{g}$ such that $\mathfrak{g} = \mathfrak{g}_0 + \mathfrak{n}$. Let $B \subset \mathfrak{g}_0$ be a complementary subspace of $\mathfrak{g}$ modulo $\mathfrak{n}$ contained in $\mathfrak{g}_0$. We say that a function $F : \mathbb{R}^a \times G$ is in $E\mathscr{S}(a, G, N)$ if F is smooth and if

$$\int_{\mathbb{R}^a \times G} |\varphi DF(\overline{u}, g)| dg d\overline{u} < \infty$$

for every $\varphi \in E_B$ (where $B = \text{span}\, \mathscr{B}$) and $D \in D_B + D_{\mathfrak{n}}$.

Lemma 8.10.7 *Let $G = \exp \mathfrak{g}$ be an exponential group and $\mathfrak{g}$ a nilpotent ideal containing $[\mathfrak{g}, \mathfrak{g}]$. For any $X \in \mathfrak{g}$ and $U \in \mathfrak{n}$ we have that*

$$R(X, U) := \log(\exp(-X) \exp(X + U)) \in \mathfrak{n}.$$

If $\{c_1, \cdots, c_m\}$ is a Jordan-Hölder basis of $\mathfrak{n}$ and if we write $R(X, U) = \sum_{i=1}^m R_j(X, U) c_i$, then the functions $(X, U) \to R_j(X, U)$ are analytic for any fixed polynomial X in U. Furthermore, for every partial derivative D of R_j there exist $C > 0$ and $\alpha \in \mathbb{N}$ such that

$$|DR_j(X, U)| \le ce^{c|X|}(1 + \|U\|)^{\alpha}, \ (X, U) \in \mathfrak{g} \times \mathfrak{n}.$$

Proof We can assume that $\mathfrak{g}$ is a triangular subalgebra of $\mathfrak{gl}(M, \mathbb{C})$ for some $M \in \mathbb{N}$ and that every $U \in \mathfrak{n}$ is a nilpotent matrix. Furthermore exp is the usual exponential mapping from $\mathfrak{gl}(M, \mathbb{C})$ to $Gl(M, \mathbb{C})$, and $G = \exp \mathfrak{g}$ is closed in $Gl(M, \mathbb{C})$ (see [89, XVIII 3. Theorem 3.1]). Hence $N = \exp \mathfrak{n}$ is unipotent and $\log : N \to \mathfrak{n}$ is a polynomial map. For any character λ of $\mathfrak{g}$ we have that

$$e^{\lambda}(\exp(R(X, U))) = e^{\lambda}(\exp(-X) \exp(X + U)) = e^{-\lambda(X)} e^{\lambda(X+U)} = 1$$

for every $X \in \mathfrak{g}$ and $U \in \mathfrak{n}$. Hence $R(X, U)$ is contained in $[\mathfrak{g}, \mathfrak{g}] \subset \mathfrak{n}$. Now $\exp(-X) = (\exp X)^{-1}$ is rational in the coefficients of $\exp X$ and the denominator of the matrix $\exp X$ has the form $e^{\varphi(X)}$ for some $\varphi \in \mathfrak{g}^*$. As $R(X, U)$ is a polynomial map in the coefficients of the unipotent matrix $\exp(R(X, U))$, to prove the lemma it is enough to study the map $(X, U) \to \exp(X + U)$.

Since $(\mathfrak{g} \cdot \mathfrak{n})^M = \{0\}$, we have that $(X + U)^n = \sum X^{n_1} U^{m_2} \ldots X^{n_r} U^{m_r}$, where the sum ranges over integers $n_1, m_1, \cdots, n_r, m_r$ which satisfy $\sum n_j + \sum m_j = n$ and $\sum m_j \le M$. Hence for fixed X

$$\exp(X + U) = \sum_{j=0}^{\infty} \frac{(X + U)^j}{j!}$$

is polynomial of degree $\le M$ in U, and the coefficients of $\exp X + U$ and its derivatives are bounded by finite sums of the type

$$(1 + \|U\|^{\beta})(e^{\|X\|} \|X\|^j + \text{ constant}).$$

□

Let us state without proof the following fact:

Theorem 8.10.8

(a) The Fréchet space $E\mathscr{S}(a, G, N)$ is an algebra under convolution

$$F * F'(\overline{u}, x) = \int_G F(\overline{u}, y)F'(\overline{u}, y^{-1}x)dy.$$

(b) $E\mathscr{S}(a, G, N)$ is invariant under translations by elements of G.

8.10.1 Smooth Kernels

In this subsection we prove that $E\mathscr{S}(a, G/P, \ell)$ is contained in the image of $E\mathscr{S}(a, G, N)$ under the map $F \to F_{\ell,P}$.

Theorem 8.10.9 *Let $G = \exp \mathfrak{g}$ be an exponential Lie group, let $\mathfrak{n}$ be a nilpotent ideal of $\mathfrak{g}$ containing $[\mathfrak{g}, \mathfrak{g}]$. Take $\ell \in \mathfrak{g}^*$ and a polarization $\mathfrak{p}$ at ℓ adapted to $\mathfrak{n}$. Then for every $a \in \mathbb{N}$ and $K \in E\mathscr{S}(a, G/P, \ell)$ there exists an $F = F_K \in E\mathscr{S}(a, G, N)$ such that*

(a) the operator $\pi_{\ell,P}(F(\overline{u}))$ admits the function $K(\overline{u})$ as kernel function for $\overline{u} \in \mathbb{R}^a$, $a \in \mathbb{N}$.
Furthermore
(b) if for some $\overline{u} \in \mathbb{R}^a \in \mathbb{N}$ we have that $K(\overline{u}) = 0$, then $F(\overline{u}) = 0$.
(c) If for some continuous function $q : \mathbb{R}^a \to \mathbb{C}$ one has $q \cdot K \in E\mathscr{S}(a, G/P, \ell)$ then $q \cdot F \in E\mathscr{S}(a, G, N)$.

The mapping $K \to F_K$ can be chosen to be linear and continuous for the Fréchet topologies on $E\mathscr{S}(a, G/P, \ell)$ and on $E\mathscr{S}(a, G, N)$. We call this mapping a retract and denote it by $R_{a,G/P,\ell}$.

Proof We proceed by induction on the dimension of $\mathfrak{g}$. If there exists an ideal $\mathfrak{a}$ of $\mathfrak{g}$ contained in $\mathfrak{n} \cap \ker(\ell)$, then the subgroup $A = \exp \mathfrak{a}$ is contained in the kernel of π_ℓ and we can consider the group $\tilde{G} = G/A$.

Hence we can assume that ℓ does not vanish on any non-trivial ideal of $\mathfrak{g}$ contained in $\mathfrak{n}$. Furthermore we can suppose that the center $\mathfrak{z}$ of $\mathfrak{g}$ is contained in $\mathfrak{n}$, since $\mathfrak{z}$ is anyway contained in $\mathfrak{p}$.

(1) The first case to study is that of a non-central ideal $\mathfrak{a} = \mathbb{R}Y$ of dimension 1 in $\mathfrak{n}$. Since now $\langle \ell, Y \rangle \neq 0$, we can assume that $\langle \ell, Y \rangle = 1$. There exists a non-trivial character $\alpha : \mathfrak{g} \to \mathbb{R}$ of $\mathfrak{g}$ such that

$$[U, Y] = \alpha(U)Y, \ U \in \mathfrak{g}.$$

Then Y is contained in the center of $\mathfrak{n}$ and hence in $\mathfrak{p}_0 \subset \mathfrak{p}$.

Choose a $T \in \mathfrak{g}$ with the property

$$\beta(T) = 1, \text{ that is, } [T, Y] = Y.$$

Let $\mathfrak{g}_1 := \ker(\beta)$, $G_1 := \exp(\mathfrak{g}_1)$, $\ell_1 := \ell_{|\mathfrak{g}_1}$. Then $\pi_{\ell,\mathfrak{p}} = \mathrm{ind}_{G_1}^{G} \pi_{\ell_1,P}$. Take $K \in E\mathcal{S}(a, G/P, \ell)$. Define the function K_1 on $\mathbb{R}^{a+2} \times G_1 \times G_1$ by

$$K_1(\overline{u}, s, t, x, y) = K(\overline{u}, \exp(sT)x, \exp(tT)y), \ x, y \in G_1, s, t \in \mathbb{R}, \overline{u} \in \mathbb{R}^a.$$

Let $\delta(t) := \log(\Delta_G(\exp(tT))e^t)$ and define

$$K_2(\overline{u}, s, t, x, y) := \delta^{-1}(t) F_1(\overline{u}, s, \log t, x, y), t > 0, \overline{u} \in \mathbb{R}^a, x, y \in G_1, s, t \in \mathbb{R},$$

and

$$K_2(\overline{u}, s, t, x, y) = 0 \text{ for } t \leq 0.$$

It follows that $F_2 \in E\mathcal{S}(a + 2, G_1/P, \ell_1)$. By the induction hypothesis for G_1, ℓ_1, we have the retract $R_1 := R_{a+2,G_1/P,\ell_1}$ with all the properties of the Theorem. Let $F_1 = R_1(K_2)$. Let

$$F_1'(\overline{u}, s, t, g_1) := \int_{\mathbb{R}} F_1(\overline{u}, s, t, g_1 \exp rY) e^{ir} dr, \overline{u} \in \mathbb{R}^a, s, t \in \mathbb{R}, g_1 \in G_1.$$

As $K_2(-, -, t, -, -) = 0$ for $t \leq 0$, we have that $F_1'(-, t, -) = 0$ for $t \leq 0$. For $g_1 \in G_1$ let $g_1^\alpha := \exp(\alpha T) g_1 \exp(-\alpha T)$. Define the function $F := R_{a,G/P,\ell}(K)$ by

$$F(\overline{u}, \exp sTg_1) = \int_0^\infty F_1'(\overline{u}, s + \log(v), v, g_1^{-\log(v)}) dv, \overline{u} \in \mathbb{R}^a, s \in \mathbb{R}, g_1 \in G_1.$$

In particular we then have that

$$F(\overline{u}, \exp sTg_1 \exp v_1 Y) = -\int_0^\infty \frac{1}{v_1^2} F_1'(\overline{u}, s - \log(v), \frac{1}{v}, g_1^{-\log(v)}) e^{iv_1 v} dv$$

for $\overline{u} \in \mathbb{R}^a$, $s \in \mathbb{R}$, $g_1 \in G_1$. It follows from this expression that F is contained in $E\mathcal{S}(a, G, N)$ and that F satisfies the conditions (b), (c) of the Theorem.

Let us show that F also satisfies condition (a). Let $\xi \in L^2(G/P, \ell)$. Then for almost all $s \in \mathbb{R}$ the function $g_1 \to \xi(\exp(sT)g_1) = \xi(s)(g_1)$ is contained in $L^2(G_1/P, \ell_1)$. Hence we may write

$$\begin{aligned} &\pi_{\ell,P}(F)\xi(\exp sTh_1) \\ &= \int_{\mathbb{R}\times G_1} F'(\overline{u}, \exp tTg_1)\pi_{\ell_1,P} g_1^{t-s})(\xi(s-t)(h_1) dg_1 dt \end{aligned}$$

$$= \int_{\mathbb{R}\times G_1} \Delta_G(\exp(s-t)T)F'(\overline{u}, \exp(s-t)Tg_1^t)\pi_{\ell_1,P}(g_1)\xi(t)(h_1)dg_1dt$$

$$= \int_{\mathbb{R}\times G_1} \Delta_G(\exp(s-t)T)\int_0^\infty F_1(\overline{u}, s-t+\log(v), v, g_1^{t-\log(v)})dv\pi_{\ell_1,P}(g_1)\xi(t)(h_1)dg_1dt$$

$$= \int_{\mathbb{R}\times G_1/\exp\mathbb{R}Y\times\mathbb{R}} e^t\Delta_G(\exp(s-t)T)\cdot$$

$$\cdot\int_0^\infty -\frac{1}{v^2}F_1(\overline{u}, s-\log(v), \frac{e^t}{v}, g_1^{\log(v)})e^{ivr}dve^{-ir}dr\pi_{\ell_1,P}(g_1)\xi(t)(h_1)dg_1dt$$

$$= \int_{\mathbb{R}\times G_1/\exp\mathbb{R}Y\times\mathbb{R}} e^t\Delta_G(\exp tT)\int_0^\infty -\frac{1}{v^2}F_1(\overline{u}, s-\log(v), \frac{e^t}{v}, g_1^{\log(v)})$$

$$\cdot e^{ivr}dve^{-ir}dr\pi_{\ell_1,P}(h_1)\xi(t)(h_1)dg_1dt$$

$$= \int_{\mathbb{R}\times G_1/\exp\mathbb{R}Y} F_1(\overline{u}, s, e^t, g_1)\pi_{\ell_1,P}(g_1)\xi(t)(g_1)dg_1dt$$

by the Fourier inversion theorem (omitting all constants)

$$= \int_{\mathbb{R}\times G_1} e^t\Delta_G(\exp tT)F_1(\overline{u}, s, e^t, g_1)\pi_{\ell_1,P}(g_1)\xi(t)(g_1)dg_1dt$$

$$= \int_{\mathbb{R}\times G_1/P} e^t\Delta_G(\exp tT)K_2(\overline{u}, s, e^t, h_1, g_1)\pi_{\ell_1,P}(g_1)\xi(t)(h_1)d\dot{g}_1dt$$

$$= \int_{G/P} K(\overline{u}, \exp sTh_1, \exp tTg_1)\xi(\exp tT(g_1)d\dot{g}_1dt.$$

Therefore $F(\overline{u}, -, -)_{\ell,P} = K(\overline{u}, -, -)$ for all $\overline{u} \in \mathbb{R}^a$.

(2) In the next case there exists a minimal non-central ideal $\mathfrak{a}$ of $\mathfrak{g}$ of dimension 2 in $\mathfrak{n}$. Then we have that $\mathfrak{a} = \text{span}\{Y_1, Y_1\}$ and there exist $\theta \in \mathbb{R}\setminus\{0\}$ and a non-trivial real character β of $\mathfrak{g}$ such that

$$[U, Y_1 + iY_2] = (1+i\theta)\beta(U)(Y_1+iY_2), U \in \mathfrak{g}.$$

This case is similar to the previous case (although slightly more involved).

(3) In this case there exists a two-dimensional ideal $\mathfrak{a} = \text{span}\{Y, Z\} \subset \mathfrak{n}$, where Z is contained in the center of $\mathfrak{g}$ and

$$[U, Y] = \alpha(U)Y + \beta(U)Z, U \in \mathfrak{g},$$

for a real character α of $\mathfrak{g}$ and a linear functional β of $\mathfrak{g}$. We can assume that $\langle \ell, Y\rangle = 0$, $\langle \ell, Z\rangle = 1$. Furthermore, if α and β are linearly independent, then we have $T \in \mathfrak{g}$, $X \in \mathfrak{g}$, with $\langle \alpha, T\rangle = 1$, $\langle \beta, T\rangle = 0$ and $\langle \alpha, X\rangle = 0$, $\langle \beta, X\rangle = 1$. We can then replace the vector X with the vector $-[T, X] \in [\mathfrak{g}, \mathfrak{g}] \subset \mathfrak{n}$ and then $X \in \mathfrak{n}$. If $\alpha = 0$, then the kernel $\mathfrak{g}_1$ of β is an ideal in $\mathfrak{g}$. Let $G_1 = \exp\mathfrak{g}_1$ and $A := \exp\mathfrak{a}$. The subspace $\mathfrak{g}_1$ is a subalgebra of $\mathfrak{g}$, $\mathfrak{p} \subset \mathfrak{g}_1$ and

$$G = \exp(\mathbb{R}X)\cdot G_1 \simeq \mathbb{R}\times\exp(\mathbb{R}T)\times G_2/A\times A$$

as topological products. Let $\mathfrak{g}_2 := \ker(\alpha) \cap \ker(\beta)$, $G_2 := \exp \mathfrak{g}_2$. The subspace $\mathfrak{g}_2$ of $\mathfrak{g}$ is an ideal. Let $\ell_1 := \ell_{|\mathfrak{g}_1}$. We have that

$$\pi_{\ell,\mathfrak{p}} = \operatorname{ind}_{G_1}^{G} \pi_1,$$

where

$$\pi_1 := \pi_{\ell_1,\mathfrak{p}}.$$

Define for $r, s, t \in \mathbb{R}$ the element $Q(r, s, t) \in G$ by

$$Q(r, s, t) := \exp(-sX) \exp(-tT) \exp(rX) \exp((s - e^t r)X) \exp(tT) \in N \cap G_2.$$

The mapping Q is analytic and its partial derivatives do not grow faster than $e^{ct}(1 + r^2 + s^2)^M$ at infinity (for some $c > 0$ and some $M \in \mathbb{N}$ depending on the derivatives of Q). For $K \in E\mathscr{S}(a, G/P, \ell)$ define K_1 on $\mathbb{R}^{a+2} \times G_1 \times G_1$ by

$$K_1(\overline{u}, s, t, x, y) := F(\overline{u}, \exp(sX) \cdot x, , \exp(tX) \cdot y), \overline{u} \in \mathbb{R}^a, s, t \in \mathbb{R}, x, y \in G_1.$$

Then $K_1 \in E\mathscr{S}(a + 2, G_1/P, \ell_1)$. Let $F_1 := R_{a+2,G_1/P,\ell_1}(K_1) \in E\mathscr{S}(a + 2, G_1, N \cap G_1)$ and define $F' : \mathbb{R}^{a+2} \times G_1 \to \mathbb{C}$ by

$$F'(\overline{u}, s, t, x \exp rY + r'Z) := \int_{\mathbb{R}^2} F_1(\overline{u}, s, t, x \cdot \exp(rY + r'Z))e^{-ir'} dr dr'$$

for $(\overline{x}, s, t) \in \mathbb{R}^{a+2}, x \in G_1$. Let $q \in C_c^\infty(\mathbb{R})$ with $\hat{q}(1) = 1$. We can now define the function $F = R_{a,G/P,\ell}(K)$ by

$$F(\overline{u}, \exp(tT) \exp(sX) w \exp(u_2 Y) \exp(u_1 Z))$$

$$:= e^{-t/2} q(u_1) \int_{\mathbb{R}} F'\Big(\overline{u}, e^{-t}(s + u), u, \exp(tT) \cdot ([Q(t, s, e^{-t}(u + s))^{-u}]^{-1}) \cdot (w^{-u})\Big) e^{iu_2 u} du$$

for $\overline{u} \in \mathbb{R}^a$, $s, t \in \mathbb{R}$, $w \in G_2/A$, $u_1, u_2 \in \mathbb{R}$. Here for an element $w \in G_2$ and $u \in \mathbb{R}$ the w^u of G_2 is defined by $w^u = \exp(uX) w \exp(-uX)$. It follows from this expression that F satisfies conditions b) and c) of the Theorem.

As $\pi_{\ell,\mathfrak{p}} = \operatorname{ind}_{G_1}^{G} \pi_{\ell_1,\mathfrak{p}}$ we can write for $\xi \in L^2(G/P, \ell)$

$$\xi(\exp sX \cdot v) = \xi(s)(w), s \in \mathbb{R}, w \in G_1,$$

and for almost all $s \in \mathbb{R}$ the function $\xi(s)$ is contained in $L^2(G_1/P, \ell_1)$. For $\overline{u} \in \mathbb{R}^a$ we then have

$$
\begin{aligned}
&\pi_{\ell,\mathfrak{p}}(F(\overline{u}, -)(\xi)(\exp(s'X) \cdot v) \\
&= \int_{\mathbb{R}^2 \times G_2/A \times A} F(\overline{u}, \exp tT \exp(-s + e^t s')X \cdot Q(t, -s + e^t s', s')w^s \exp rY \exp r's')e^{i(sr - r')} \\
&\cdot \pi_1(\exp tTw)\xi(s)(v)dtdsdwdrdr' \\
&= \int_{\mathbb{R}^2 \times G_2/A} \int_{\mathbb{R}} \int_{\mathbb{R}} F'\Big(\overline{u}, e^{-t}(-s + e^t s' + u), u \exp tT([Q(t, -s + e^t s', e^t(u - s + e^{-t}s'))]^{-u})^{-1} \\
&\cdot ([Q(t, -s + e^t s')]^{-u})^{-1}\Big)e^{-iur} du e^{isr} dr \pi_1(\exp tTw)\xi(s)(v)dtdwds \\
&\cdot \pi_1(\exp tTw)\xi(s)(v)dtdsdwdrdr' \\
&\text{(using Fourier's inversion formula)} \\
&= \int_{\mathbb{R} \times G_1/A} F'(\overline{u}, s', s, \exp tTw)\pi_1(\exp tTw)\xi(s)(v)dsdwdt \\
&= \int_{\mathbb{R}} \pi_1(F_1)(\overline{u}, s', s)\xi(s)(v)ds \\
&= \oint_{G/P} K(\overline{u}, \exp s'X \cdot v, r)\xi(r)d\dot{r}.
\end{aligned}
$$

Hence $\pi_{\ell,\mathfrak{p}}(F(\overline{u}))$ admits $K(\overline{u})$ as kernel function ($\overline{u} \in \mathbb{R}^a$).

(4) We will not treat the case where the center $\mathfrak{z}$ of $\mathfrak{g}$ is one-dimensional and contained in $\mathfrak{n}$ and where we have a non-central minimal ideal $\mathfrak{z} \subset \alpha$ of $\mathfrak{g}$ contained in $\mathfrak{n}$ of dimension 3. This is similar to the previous case (only more complicated to write down).

(5) We are now in the situation where $\mathfrak{n}$ contains no non-central minimal ideal of $\mathfrak{g}$, i.e., $\mathfrak{n} \subset \mathfrak{z}$. Then $\mathfrak{n}$ must be one-dimensional and so $\mathfrak{n} = \mathbb{R}Z$. In particular either $[\mathfrak{g}, \mathfrak{g}] = \{0\}$ or $[\mathfrak{g}, \mathfrak{g}] = \mathfrak{z}$.

The case of an Abelian algebra $\mathfrak{g}$ is trivial.

If $[\mathfrak{g}, \mathfrak{g}] = \mathbb{R}Z$ then $\mathfrak{g} = \mathfrak{h}_n \oplus \mathfrak{z}_1$, where $\mathfrak{h}_n$ is the Heisenberg algebra of dimension $2n + 1$ and $\mathfrak{z}_1$ is a central subspace. We then have that $\mathfrak{n} = \mathfrak{z}_1 \oplus \mathbb{R}Z \oplus \mathfrak{y}$, where $\mathfrak{y}$ is an n-dimensional Abelian subalgebra of $\mathfrak{h}_n$. We can assume that $\mathfrak{z}_1 + \mathfrak{y}$ is contained in the kernel of ℓ. We write the group G as topological product

$$G = \mathscr{X} \cdot \mathfrak{p}.$$

Choose a function $q \in C_c^\infty(\mathfrak{z}_1 \oplus \mathbb{R}Z)$ such that

$$\hat{q}(Z^*) = 1.$$

For any kernel function $K \in E\mathcal{S}(a, G/P, \ell)$, let

$$F(\overline{u}, X \cdot Z_1 \cdot Y \cdot (zZ)) = R_{a,G/P,\ell}(K)(\overline{u}, X \cdot Z_1 \cdot Y \cdot (zZ))$$
$$= q(Z_1 + zZ) \int_{\mathcal{X}} K(\overline{u}, X + U, U) e^{i\langle Y,U\rangle} dU,$$
$$X \in \mathcal{X}, Y \in \mathfrak{y}, Z_1 \in \mathfrak{z}_1, z \in \mathbb{R}.$$

One checks that F satisfies all the conditions of the theorem. □

Corollary 8.10.10 *Let $G = \exp \mathfrak{g}$ be an exponential Lie group, $\ell \in \mathfrak{g}^*$ and $\mathfrak{p}$ a Pukanszky polarization at ℓ. Then there exists $F \in L^1(G)$ such that $\pi_{\ell,\mathfrak{p}}(F)$ is a rank-one operator.*

Proof Choose a smooth element $0 \neq \xi \in L^2(G/P, \ell)$ such that the function $K(x, y) := \xi(x)\overline{\xi(y)}, x, y \in G$ is contained in the space of kernels $E\mathcal{S}(0, G/P, \ell)$. Then by the previous theorem there exists $F \in E\mathcal{S}(G, [G, G])$ such that the operator $\pi_{\ell,\mathfrak{p}}(F)$ admits K as kernel function. This shows that $\pi_{\ell,\mathfrak{p}}(F)$ is the projection onto the one-dimensional subspace $\mathbb{C}\xi$ of $L^2(G/P, \ell)$. □

If we apply Lemma 8.6.8 we obtain

Corollary 8.10.11 *Let $G = \exp \mathfrak{g}$ be an exponential Lie group. The canonical mapping $\widehat{G} \to Prim(G), [\pi] \to ker(\pi)$ is a bijection.*

8.10.2 A Retract Theorem for Exponential Orbits in a Nilpotent Lie Group's Spectrum

Let $G = \exp \mathfrak{g}$ be an exponential group, let $\mathfrak{n} \supset [\mathfrak{g}, \mathfrak{g}]$ be a nilpotent ideal of $\mathfrak{g}$ and let $N := \exp \mathfrak{n}$.

Fix an irreducible representation $(\pi, \mathcal{H})$ of N. Let $\Omega := G \cdot \pi \subset \widehat{N}$ be the G-orbit of $\pi \in \widehat{N}$. Since G is exponential we know that Ω is locally closed in $\widehat{N}$. Let $O = \mathrm{Ad}^*(N)\ell = K(\pi) \subset \mathfrak{n}^*$ be the Kirillov orbit of π and let

$$G_\pi = G_O := \{d \in G; d \cdot O = O\} \tag{8.10.1}$$

be the stabilizer of O (and of π). Then the G-orbit $\Omega = G \cdot 0$ of O in $\mathfrak{n}^*/N$ is homeomorphic to the quotient space G/G_0. Fix $\ell_1 = \ell_{|\mathfrak{n}} \in O$ and let $\mathfrak{g}(\ell_{|\mathfrak{n}}) = \{X \in \mathfrak{g}; \langle \ell, [X, \mathfrak{n}]\rangle = \{0\}\}$ and let $G(\ell_{|\mathfrak{n}}) = \{d \in G; d \cdot \ell_1 = \ell_1\} = \exp \mathfrak{g}(\ell_1)$ be the stabilizer of ℓ_1 in G. Then since N is normal in G, we have that

$$G_O = G(\ell_{|\mathfrak{n}})N.$$

Indeed, if $s \in G_O$, then $s \cdot \ell_{|\mathfrak{n}} \in O$. Hence there exists $n \in N$ such that $n \cdot \ell_1 = s \cdot \ell_1$. Therefore $n^{-1}s \in G(\ell_1) \Rightarrow s \in NG(\ell_1) = G(\ell_1)N$. If $x = sn \in G(\ell_1)N$ then for any $n' \in N$ we have that

$$\begin{aligned} x \cdot n' \cdot \ell_1 &= s \cdot n' \cdot s^{-1} \cdot s \cdot \ell_1 \\ &= s \cdot n' \cdot s^{-1} \cdot \ell_1 \\ &\in N \cdot \ell_1 = O. \end{aligned}$$

Hence $x \in G_O$. Let $\mathfrak{g}_O$ be the Lie algebra of the group G_O. We see from (8.10.1) that

$$\mathfrak{g}_O = \mathfrak{g}(\ell_1) + \mathfrak{n}.$$

Now the quotient group $G/G_O = G/G(\ell)N$ is homeomorphic to $\mathfrak{g}/(\mathfrak{g}(\ell_{|\mathfrak{n}}) + \mathfrak{n}) \simeq \mathbb{R}^r$ for some $r \in \mathbb{N}$. Choose a basis $\{S_1, \cdots, S_r\}$ of $\mathfrak{g}$ modulo $\mathfrak{g}_O$ and let $\mathscr{S} :=$ span $\{S_1, \cdots, S_r\}$. Let

$$S := \exp \mathscr{S}.$$

Then S is a closed subset of G and G is the topological product of S and G_O:

$$G = S \cdot G_O. \tag{8.10.2}$$

The subset

$$\Omega := S \cdot O \subset \mathfrak{n}^*/N$$

is locally closed in $\mathfrak{n}^*/N$ and homeomorphic to $S \simeq \mathbb{R}^r$.

We define now the space $\mathscr{S}(\mathbb{R}^N \times \widehat{\Omega})$ of *Schwartz kernels of* Ω. We realize the representation $\pi_{\ell_1,\mathfrak{p}}$ of N as

$$\pi_{\ell_1,\mathfrak{p}} = \mathrm{ind}_P^N \chi_{\ell_1}$$

for a certain polarization $\mathfrak{p} \subset \mathfrak{n}$ at ℓ_1 and $P = \exp \mathfrak{p}$.

Then every $g \in G$ the conjugate representation

$$\pi_{g\cdot(\ell,\mathfrak{p}))}(n) := \pi_{\ell_1,\mathfrak{p}}(g^{-1} n g), n \in \mathbb{N},$$

of N is equivalent to the monomial representation

$$\mathrm{ind}_{g\cdot P}^N \chi_{g\cdot\ell_1} = \pi_{g\cdot\ell_1, g\cdot P},$$

where $g \cdot P := g P g^{-1}$ and $\langle g \cdot \ell_1, X\rangle = \langle \ell, \mathrm{Ad}(g^{-1})X\rangle$, $X \in \mathfrak{n}$.

We fix a Malcev basis $\{X_1, \cdots, X_d\}$ of $\mathfrak{n}$ modulo $\mathfrak{p}$. Let

$$\mathscr{X} := \exp(\mathbb{R}X_1)\cdots\exp(\mathbb{R}X_d),\ E_{\mathscr{X}} : \mathbb{R}^a \to G,$$
$$E_{\mathscr{X}}(x_1, \cdots, x_a) := \exp(x_1X_1)\cdots\exp(x_aX_d).$$

Then

$$N = \mathscr{X} \cdot P$$

as topological product. Let us write down explicitly the operator $\pi^g(f)$ for $f \in L^1(N)$, $g \in G$. Let $\xi \in \mathscr{H}_\pi = L^2(G/P, \chi_\ell)$. Let us put

$$g \cdot n := gng^{-1},\ g \in G, n \in N.$$

Then

$$\begin{aligned}
\pi_{g\cdot\ell, g\cdot P}(f)\xi(u) &= \int_N f(n)\xi((g^{-1}\cdot n)^{-1}u)dn\\
&= \Delta_G^{-1}(g)\int_N f(g\cdot n)\xi(n^{-1}u)dn\\
&= \Delta_G^{-1}(g)\int_N f(g\cdot(un^{-1}))\xi(n)dn\\
&= \Delta_G^{-1}(g)\int_{N/P}\int_P f(g\cdot(up^{-1}n^{-1}))\xi(np)dpd\dot{n}\\
&= \Delta_G^{-1}(g)\int_{N/P}\int_P f(g\cdot(upn^{-1}))\chi_\ell(p)dp\xi(n)d\dot{n}
\end{aligned}$$

Hence the operator $\pi_{g\cdot\ell,g\cdot P}(f)$ is determined by its kernel function

$$k_f(g\cdot\ell, g\cdot P)(u,n) = \int_P f((g\cdot u)(g\cdot p)(g\cdot v)^{-1})\chi_{g\cdot\ell}(g\cdot p)dp,\ u, v \in N.$$

Definition 8.10.12

$$E\mathscr{S}(S, a, N/P, \ell) := \Big\{K : S\times\mathbb{R}^a\times N\times N \to \mathbb{C};$$

$$K(s, \overline{u}, gp, g'p') = \chi_{s\cdot\ell}(p^{-1})\chi_{s\cdot\ell}(p')K(s, \overline{u}, n, n'), n, n' \in N,\ p, p' \in P, \overline{u} \in \mathbb{R}^a, s \in S,$$

$$\text{and } K\circ(E_{\mathscr{Y}}\otimes E_{\mathscr{Y}}) \in E\mathscr{S}(S\times\mathbb{R}^a, \mathscr{S}(\mathbb{R}^d\times\mathbb{R}^d)),\ \text{compactly supported in } S\Big\}.$$

If $a = 0$, we just write $E\mathscr{S}(S, N/P, \ell)$.

These Schwartz kernels are just the kernel functions of smooth operator fields defined on $S \times \mathbb{R}^a$, vanishing exponentially on $\mathbb{R}^a \times \mathscr{S}$ and with compact support in S.

Theorem 8.10.13 *Let $G = \exp \mathfrak{g}$ be an exponential solvable Lie group and $N = \exp \mathfrak{n}$ a nilpotent ideal of $\mathfrak{g}$ containing $[\mathfrak{g}, \mathfrak{g}]$. Take $\ell \in \mathfrak{n}^*$ and let $\mathfrak{p}$ be a polarization at ℓ. Let $S \subset G$ be a cross-section of the quotient $G/G(O)$ as in (8.10.2). There exists for every $a \in \mathbb{N}$ a continuous linear mapping, which we call a* retract,

$$R = R_{S,N/P,\ell} : E\mathscr{S}(S, a, N/P, \ell) \mapsto \mathscr{S}(\mathbb{R}^a, N)$$

such that for every $s \in S, \overline{u} \in \mathbb{R}^a$ and $K \in E\mathscr{S}(S, a, N/P, \ell)$ the (smooth operator)

$$\pi_{s\cdot(\ell_{|\mathfrak{n}},\mathfrak{p})}(R(K(s, \overline{u})))$$

admits for every $s \in S$ the function $K(s, \overline{u})$ as kernel function.

Proof We apply Theorem 8.10.9. Let $\mathfrak{h} := \mathscr{S} \oplus \mathfrak{n} \subset \mathfrak{g}$ and $H := \exp \mathfrak{h}$. Define $\ell' := \ell_{|\mathfrak{h}}$. Let $\mathfrak{p} \subset \mathfrak{n}$ be a Vergne polarization at $\ell_{|\mathfrak{n}}$ for an $\mathfrak{h}$-Jordan-Hölder sequence of $\mathfrak{n}$. Then $\mathfrak{p}$ is also a Pukanszky polarization at ℓ', since $\mathfrak{h}(\ell_{|\mathfrak{n}}) = \mathfrak{n}(\ell_{|\mathfrak{n}})$. We use the retract $R_{a,H/P,\ell'}$. Let $K \in E\mathscr{S}(S, a, N/P, \ell)$ from Theorem 8.10.9. We define the kernel function $K' \in E\mathscr{S}(a, H/P, \ell)$ by

$$K'(\overline{u}, s \cdot n, t \cdot m) = \psi(t) K(s, \overline{u}, n, m), s, t \in S, \overline{u} \in \mathbb{R}^a, n, m \in N,$$

where $\psi \in C_c^\infty(S)$ equals 1 on the support of the map $s \to K(s, -)$. Let now $F' = R_{a,H/P,\ell_1}(K)$. We define

$$F(\overline{u}, n) := F'(\overline{u}, n), \overline{u} \in \mathbb{R}^a, n \in N,$$

which means that we restrict the function F' to $\mathbb{R}^a \times N$. Then $F \in E\mathscr{S}(a, N)$. For $s \in S$ and $\xi \in \mathscr{E}(H/P, \ell_1)$ we have that $\xi_{|N}$ is contained in $\mathscr{E}(N/P, \ell)$ and then for $s \in S$ we have that

$$\begin{aligned}
\pi_{s\cdot(\ell_{|\mathfrak{n}},\mathfrak{p})}(F(\overline{u}))\xi(m) &= \int_N F(\overline{u}, n)\xi(s^{-1} \cdot n^{-1} \cdot s \cdot m) dn \\
&= \Delta_G(s)^{-1} \int_{N/P} \Big(\int_P F(\overline{u}, s \cdot mpn^{-1}s^{-1}) \chi_\ell(p) dp \Big) \xi(n) d\dot{n} \\
&= \int_{N/P} \Big(\int_P R_{a,H/P,\ell}(K')(\overline{u}, s \cdot mpn^{-1}s^{-1}) \chi_\ell(p) dp \Big) \xi(n) d\dot{n}
\end{aligned}$$

$$= \int_{N/P} K'(\overline{u})(s \cdot m, s \cdot n)\xi(n)d\dot{n}$$

$$= \int_{N/P} K(s, \overline{u}, m, n)\xi(n)d\dot{n}.$$

Therefore $K(s, \overline{u}, -)$ is the kernel function of the representation $\pi_{s\cdot(\ell_{|\mathfrak{n}}, \mathfrak{p})}(F)$, $s \in S$.

□

8.10.3 An Application

In this subsection $G = \exp \mathfrak{g}$ is an exponential Lie group and $\mathfrak{n} \supset [\mathfrak{g}, \mathfrak{g}]$ a nilpotent ideal of $\mathfrak{g}$. We shall prove that every proper, closed, G-prime ideal of $L^1(N)$ is the kernel of a G-orbit.

Definition 8.10.14 Let $\ell \in \mathfrak{n}^*$, $\ell = \ell_{|\mathfrak{n}}$, $O := N \cdot \ell$ and $\Omega := G \cdot \ell_{|\mathfrak{n}}$. We write $G_O = G(\ell) \cdot N$ as in Sect. 8.10.2 and take a subspace $\sigma \subset \mathfrak{g}$ such that $\sigma \oplus (\mathfrak{g}(\ell) + \mathfrak{n}) = \mathfrak{g}$. Let $S := \exp \sigma$. We have that

$$\Omega \simeq S \simeq \sigma \simeq G/G_O \simeq \mathbb{R}^{G/G_O}$$

as topological spaces.

Definition 8.10.15 Let $K \subset \Omega$ be a compact subset. We denote by $I_K(\Omega)$ the ideal in the algebra $\mathscr{S}(N)$ defined by

$$I_K(\Omega) := \{F \in \mathscr{S}(N); \overline{\{s \in |\pi_{s\cdot\ell}(F) \neq 0\}} \subset K\},$$

and put

$$I_\Omega := \{F \in \mathscr{S}(N); \overline{\{s \in |\pi_{s\cdot\ell}(F) \neq 0\}} \text{ is compact}\}.$$

Then $I_\Omega = \bigcup_{\Omega \supset K \text{ compact}} I_K(\Omega)$ is a G-invariant ideal of $\mathscr{S}(N)$. Obviously $I_\Omega \not\subset \ker(\pi_\ell)$.

Proposition 8.10.16 *With the notations of Definition 8.10.15 we have that*

$$\overline{I_\Omega}^{L^1(N)} * \overline{I_\Omega}^{L^1(N)} * ker(\Omega) * \overline{I_\Omega}^{L^1(N)} * \overline{I_\Omega}^{L^1(N)} \subset \overline{ker(\Omega) \cap \mathscr{S}(N)}^{L^1(N)}.$$

Here $ker(\Omega) = \ker_{L^1(N)}(\Omega) = \{F \in L^1(N); \pi_{s\cdot\ell}(F) = 0; s \in S\}$.

Proof Let us denote by a_K the operator on $L^2(N/P, \ell)$ determined by its kernel function K, for any $K \in E\mathscr{S}(S, N/P, \ell)$. By the Hahn-Banach theorem, it is enough to show that every $\varphi \in L^\infty(N)$ which vanishes on $\ker(\Omega) \cap \mathcal{S}(N)$ is 0 on the subspace

$$I_K(\Omega) * I_K(\Omega) * \ker(\Omega) * I_K(\Omega) * I_K(\Omega)$$

for every compact subset $K \subset \Omega$.

Let $K \subset S$ be a compact set and let $F_1, \cdots, F_4 \in I_K(\Omega)$.

For a function $0 \leq \rho \in C_c^\infty(S)$ and a function $0 \leq \tau \in C_c^\infty(N)$ such that $\int_S \psi(s)ds = 1$ and $\int_N \tau(n)dn = 1$ and any $F \in L^1(N)$ we define the element $F^{\rho,\tau}$ in $L^1(N)$ by

$$F^{\rho,\tau} := \int_S \int_N \rho(t)\tau(n) F^{t\cdot n} dnds$$

and the element $\varphi^{\rho,\tau} \in L^1(N)$ by

$$\langle \varphi^{\rho,\tau}, F\rangle := \langle \varphi, F^{\rho,\tau}\rangle,$$

where

$$F^s(n) := \Delta_G(s^{-1})F(s \cdot n \cdot s^{-1}), n \in N, s \in S.$$

We remark that for every $s \in S$ we have the relation

$$\begin{aligned}(F^{\rho,\tau})^s &= \int_S \int_N \rho(t)\tau(n)(F^{t\cdot n})^s dndt \\ &= \int_S \int_N \rho(t)\tau(n) F^{s\cdot t\cdot n} dndt \\ &= \int_S \int_N \rho(t-s)\tau(q(t,s)\cdot n) F^{t\cdot n} dndt,\end{aligned}$$

where $q : S \times S \to N$ is a polynomial map (since $S \subset G_0$ and G_0 is nilpotent group).

We show that there exists a constant $C = C_{\rho,\tau} > 0$ such that

$$|\langle \varphi^{\rho,\tau}, F_1 * F_2 * F * F_3 * F_4\rangle| \leq C \sup_{s\in S} \|\pi_{s\cdot\ell}(F)\|_{\text{op}}, F \in L^1(N). \quad (8.10.3)$$

First we define a linear form μ on the Fréchet space $E\mathcal{S}(S, N/P, \ell)$ by the relation

$$\langle \mu, K\rangle := \langle \varphi, R_{S,\mathrm{N}/P,\ell_{|\mathrm{n}}}(K)\rangle.$$

This linear form is continuous by Theorem 8.10.13. Since φ vanishes on $\ker(\Omega) \cap \mathscr{S}(N)$ we have the relation

$$\langle \varphi, F \rangle = \langle \mu, K \rangle$$

whenever $F \in \mathscr{S}(N)$, $K \in E\mathscr{S}(S, N/P, \ell)$ and $\pi_{s\cdot\ell}(F) = a_K(s, -, -)$, $s \in S$.

Through our coexponential basis $\mathscr{Y}$ the Fréchet space $E\mathscr{S}_K(S, N/P, \ell) := \{K \in E\mathscr{S}(S, N/P, \ell); K(s, -) = 0 \text{ for } s \notin K\}$ can be identified with the Fréchet space $C^\infty(K, S(\mathbb{R}^d \times \mathbb{R}^d))$ where $d := \dim \mathfrak{n}/\mathfrak{p}$. Using [148] we can write $\langle \mu, K \rangle$ as

$$\langle \mu, K \rangle = \sum_{|\alpha|+|\beta|+|\gamma|\le m} \int_{K\times\mathbb{R}^d\times\mathbb{R}^d} \tau_{\alpha,\beta,\gamma}(s, x, y)(\frac{\partial}{\partial s})^\alpha(\frac{\partial}{\partial x})^\beta(\frac{\partial}{\partial y})^\gamma K(s, x, y)dsdxdy$$

where the functions $\tau_{\alpha,\beta,\tau}$ are continuous and of slow growth in x, y.

Denote now by $F_j^\ell(\cdot, -, -)$, $j = 1, \cdots, 4$, the kernel functions of the operators $\pi_{s\cdot\ell}(F_j)$ and by $K^\ell(s, -, -)$ the kernel function of the operator $\pi_{s\cdot\ell}(F)$ for $F \in \mathscr{S}(N)$. This gives us elements of $E\mathscr{S}(S, N/P, \ell)$ by composing the fields corresponding to the F_j with the field coming from $F \in \mathscr{S}(N)$.

We obtain then the following identity

$$\begin{aligned}
&\langle \varphi^{\rho,\tau}, F_1 * F_2 * F * F_3 * F_4 \rangle \\
&= \langle \varphi, (F_1 * F_2 * F * F_3 * F_4)^{\rho,\tau} \rangle \\
&= \langle \mu_{s,x,y}, \int_S \int_N \rho(t-s)\tau(q(t,s)\cdot n)(F_1 * F_2 * F * F_3 * F_4)^{t\cdot n})^\ell(t, x, y)dtdn \\
&= \sum_{|\alpha|+|\beta|+|\gamma|\le m} \int_S \int_N \int_{K\times\mathbb{R}^d\times\mathbb{R}^d} \tau_{\alpha,\beta,\gamma}(s, x, y)(\frac{\partial}{\partial s})^\alpha(\frac{\partial}{\partial x})^\beta(\frac{\partial}{\partial y})^\gamma \\
&\quad \rho(t-s)\tau(q(t,s)\cdot n)(F_1 * F_2 * F * F_3 * F_4)^n)^\ell(t, x, y)dtdndsdxdy \\
&= \sum_{|\alpha|+|\beta|+|\gamma|\le m} \int_S \int_N \int_{K\times\mathbb{R}^d\times\mathbb{R}^d} \tau_{\alpha,\beta,\gamma}(s, x, y)(\\
&\quad (\frac{\partial}{\partial s})^\alpha(\rho(t-s)\tau(q(t,s)\cdot n)) \int_{\mathbb{R}^d}\int_{\mathbb{R}^d} ((\frac{\partial}{\partial x})^\beta(((F_1^n) * (F_2^n))^\ell(t, x, x'))(F^n)^\ell(t, x', x'') \\
&\quad ((\frac{\partial}{\partial y})^\gamma((F_3^n) * (F_4)^n))^\ell(t, x'', y))dx'dx''dtdndsdxdy.
\end{aligned}$$

Since the functions $(F_j^{t\cdot n})^\ell$ are zero outside K and the function ρ has compact support, the integral over S is in fact an integral over a new compact subset K_1 of S. For $s \in K$, $t \in \mathbb{R}$, $n \in N$, $x, y \in N/P$ we may estimate the function

$$\psi_{\alpha,\beta,\gamma}(s, t, x, y) := |\tau_{\alpha,\beta,\gamma}(s, x, y)(\frac{\partial}{\partial s})^\alpha(\rho(t-s)\tau(q(t,s)\cdot n))|$$

by

$$\psi_{\alpha,\beta,\gamma}(s,t,n,x,y) \leq \psi(s)\psi(t)\psi_2(n)(1+\|x\|)^R(1+\|y\|)^R$$

for some $C > 0$, $R > 0$, $\psi_1 = C1_{[-R,R]} \in C_c(S, \mathbb{R}_+)$, $\psi_2 \in C_c(N)$ independent of $|\alpha|+|\beta|+|\gamma| \leq m$. Furthermore, we have that

$$(\frac{\partial}{\partial x})^\beta(((F_1^n)*(F_2^n))^\ell(x,x')) = \int_{\mathbb{R}^d} \frac{\partial}{\partial x}^\beta((F_1^n)^\ell(t,x,y'))((F_2^n))^\ell(t,y',x'))dy'$$

and

$$(\frac{\partial}{\partial y})^\gamma(((F_3^n)*(F_4^n))^\ell(x'',y)) = \int_{\mathbb{R}^d}(F_3^n)^\ell(t,x'',y')(\frac{\partial}{\partial y})^\gamma((F_4^n))^\ell(t,y',y))dy'$$

Let $a_{\beta,t,n}$ (resp. $a_{\gamma,t,n}$) be the operator on $L^2(N/P,\ell)$ defined by its kernel function $(\frac{\partial}{\partial x})^\beta(((F_1^n)^\ell(t,x,y)$ (resp. by $(\frac{\partial}{\partial y})^\gamma((F_4^n))^\ell(t,x,y)))$. Then we have that

$$\begin{aligned}
&|\langle\varphi^{\rho,\tau}, F_1*F_2*F*F_3*F_4\rangle| \\
&\leq C \sum_{|\alpha|+|\beta|+|\gamma|\leq m} \int_{K_1\times\mathbb{R}^d\times\mathbb{R}^d}\int_{K_1\times N} \psi_2(n)(1+\|x\|)^R(1+\|y\|)^R|(\frac{\partial}{\partial x})^\beta((F_1^n)^\ell(t,x,y'))| \\
&|\int_{\mathbb{R}^d}\int_{\mathbb{R}^d}((F_2^n))^\ell(t,y',x'))(F^n)^\ell(t,x',x'')(F_3^n)^\ell(t,x'',y'')dx'dx''| \\
&|(\frac{\partial}{\partial y})^\gamma((F_4^n))^\ell(t,y'',y))|dy'dy''dtdndsdxdy.
\end{aligned}$$

Hence using Cauchy-Schwartz inequality we obtain the estimate

$$\begin{aligned}
&|\langle\varphi^{\rho,\tau}, F_1*F_2*F*F_3*F_4\rangle \\
&\leq C \sum_{\substack{|\alpha|+|\beta|+|\gamma| \\ \leq m}} (\int_{K_1\times N\times\mathbb{R}^{2d}} \psi_2(n)|\int_{\mathbb{R}^d}\int_{\mathbb{R}^d}((F_2^n))^\ell(t,y',x'))(F^n)^\ell(t,x',x'')(F_3^n)^\ell(t,x'',y'')dx'dx''|^2 \\
&dtdndy'dy'')^{1/2}\cdot\int_{K_1\times\mathbb{R}^d\times\mathbb{R}^d}(1+\|x\|)^R(1+\|y\|)^R \\
&(\int_{K_1\times N\times\mathbb{R}^{2d}}\psi_2(n)|(\frac{\partial}{\partial x})^\beta((F_1^n)^\ell(t,x,y'))|^2|(\frac{\partial}{\partial y})^\gamma((F_4^n))^\ell(t,y'',y))|^2dy'dy''dtdn)^{1/2} \\
&dsdxdy \\
&= C \sum_{\substack{|\alpha|+|\beta|+|\gamma| \\ \leq m}} (\int_{K_1\times N}\psi_2(n)\|a_{\alpha,t,n}\circ\pi_\ell^t(F)\circ a_{\gamma,t,n}\|_{H.S}^2 dtdn)^{1/2} \\
&\int_{\mathbb{R}^d}\int_{\mathbb{R}^d}((F_2^n))^\ell(t,y',x'))(F^n)^\ell(t,x',x'')(F_3^n)^\ell(t,x'',y'')dx'dx''|^2dtdndy'dy'')^{1/2}
\end{aligned}$$

$$\cdot \int_{K_1\times\mathbb{R}^d\times\mathbb{R}^d} (1+\|x\|)^R(1+\|y\|)^R$$

$$(\int_{K_1\times N\times\mathbb{R}^{2d}} \psi_2(n)|(\frac{\partial}{\partial x})^\beta((F_1^n)^\ell(t,x,y'))|^2|(\frac{\partial}{\partial y})^\gamma((F_4^n))^\ell(t,y'',y))|^2 dy'dy''dtdn)^{1/2}$$

$$dsdxdy$$

$$\leq C \sup_{t\in S}\int_N \psi(n)\|\pi_\ell^t(F^n)\|_{\mathrm{op}}dn$$

for a new constant $C > O$. This inequality remains now true for every $F \in L^1(N)$. If now $F \in \ker(\Omega)$, then

$$\langle\varphi, \int_{S\times N}\rho(t)\tau(n)F^{t\cdot n}dtdn\rangle = 0$$

for every $\rho \in C_c(S)$, $\tau \in C_c(N)$. Finally

$$\langle\varphi, F\rangle = 0$$

for every $F \in \ker(\Omega)$. □

Proposition 8.10.17 *Let $N = \exp\mathfrak{n}$ be a simply connected nilpotent Lie group. Let I be a closed ideal in $L^1(N)$. Denote by $h(I)$ the hull of I in $\hat{N} \simeq Prim(N)$. Then*

$$h(I) = h(I\cap\mathscr{S}(N)).$$

Proof We have that

$$\begin{array}{rcl} h(I) & \subset & h(I\cap\mathscr{S}(N)) \\ & \overset{(\text{Theorem 8.7.10})}{=} & h(j(h(I))) \\ & = & h(I). \end{array}$$

□

Theorem 8.10.18 *Let $G = \exp\mathfrak{g}$ be an exponential group. Let $\mathfrak{n} \supset [\mathfrak{g},\mathfrak{g}]$ be a nilpotent ideal of $\mathfrak{g}$ and let $N = \exp\mathfrak{n}$. Proper, closed G-prime ideals in $L^1(N)$ coincide with the kernels ker(Ω) of G-orbits in $\mathfrak{n}^*$. Moreover the map*

$$\Omega \to ker((\Omega))$$

is a bijection between the set $\mathfrak{n}^*/G$ of G-orbits in $\mathfrak{n}^*$ and the set $Prime^G(N)$ of proper, closed G-prime ideals of $L^1(N)$.

Proof Let $I = \ker(\Omega) \subset L^1(N)$. Let us show that I is G-prime. Let J_1, J_2 be two G-invariant ideals in $L^1(N)$ such that $J_1 * J_2 \subset I$. Then if $\ell_1 \in \mathfrak{g}^*$ is any extension of ℓ to $\mathfrak{g}$, we have that $\ker_{L^1(N)}(\pi_{\ell_1|N}) = \ker(\Omega)$ and so

$$(L^1(G) * J_1 * L^1(G)) * (L^1(G) * I * L^1(G)) * (L^1(G) * J_1 * L^1(G)) \subset$$
$$\subset L^1(G) * I * L^1(G) \subset \ker(\pi_{\ell_1})$$

since I is G-invariant. Now $\ker_{L^1(G)}(\pi_{\ell_1})$ is a prime ideal, hence it follows that either $L^1(G) * J_1 * L^1(G)$ or $L^1(G) * J_2 * L^1(G)$ is contained in $\ker_{L^1(G)}(\pi_{\ell_1})$. Therefore either J_1 or J_2 is contained in I. So I is G-prime.

Suppose now that I is a proper closed G-prime ideal in $L^1(N)$. Then $I_{\mathscr{S}} := I \cap \mathscr{S}(N)$ is a proper G-invariant ideal in $\mathscr{S}(N)$ that is closed for $\| \|_1$. Let us show that $I_{\mathscr{S}}$ is also G-prime. Let J_1, J_2 be two G-invariant ideals in $\mathscr{S}(N)$ such that $J_1 * J_2 \subset I_{\mathscr{S}}$. Then of course

$$\overline{J_1}^{L^1(N)} * \overline{J_2}^{L^1(N)} \subset \overline{I_{\mathscr{S}}}^{L^1(N)} \subset I$$

and so, as I is G-prime, we have that $\overline{J_1}^{L^1(N)} \subset I$ or $\overline{J_2}^{L^1(N)} \subset I$, which implies that either J_1 or J_2 is contained in I_S. By Theorem 8.9.6 we have that $I_{\mathscr{S}} = \ker(\Omega) \cap \mathscr{S}(N))$ for some G-orbit $\Omega \subset \mathfrak{n}^*$. Now by Proposition 8.10.17 we have that

$$h(I) = h(I \cap \mathscr{S}(N)) = h(\ker(\Omega) \cap \mathscr{S}(N)) = h(\ker(\Omega)) = \overline{\Omega}$$

and so $I \subset \ker(\Omega)$. With the notations of Proposition 8.10.16 we have that

$$\overline{I_\Omega}^{L^1(N)} * \overline{I_\Omega}^{L^1(N)} * \ker(\Omega) * \overline{I_\Omega}^{L^1(N)} * \overline{I_\Omega}^{L^1(N)} \subset \overline{\ker(\Omega) \cap \mathscr{S}(N)}^{L^1(N)} \subset \overline{\ker(\Omega \cap \mathscr{S}(N))}^{L^1(N)}$$
$$= \overline{I \cap \mathscr{S}(N)}^{L^1(N)} \subset I.$$

As $I_\Omega \not\subset \ker(\Omega)$ and $I \subset \ker(\Omega)$ we have that $I_\Omega \not\subset I$. Therefore $\ker(\Omega) \subset I$ as I is G-prime. So $I = \ker(\Omega)$.

The map $\Omega \to \ker(\Omega)$ is clearly surjective. Let us prove that it is also injective. Suppose that Ω_1 and Ω_2 are two distinct orbits such that $\ker(\Omega_1) = \ker(\Omega_2)$. Then $\Omega_1 \cap \Omega_2 = \emptyset$. As

$$\overline{\Omega_1} = h(\ker(\Omega_1)) = h(\ker(\Omega_2)) = \overline{\Omega_2}$$

we have that $\Omega_1 \subset \overline{\Omega_2} \setminus \Omega_2$ and $\Omega_2 \subset \overline{\Omega_1} \setminus \Omega_1$. As every G-orbit in $\mathfrak{n}^*$ is open in its closure (G is exponential), we have that $\Omega_1 \subset \overline{\Omega_1} \subset \overline{\Omega_2} \setminus \Omega_2 \subset \overline{\Omega_2} \subset \overline{\Omega_1} \setminus \Omega_1$. This contradiction shows that $\Omega_1 = \Omega_2$. □

8.11 Bounded Irreducible Representations of G

8.11.1 G-prime Ideals

Let us recall some definitions.

Definition 8.11.1 Let G be a locally compact group. We recall that a bounded G-module (or representation) (T, X) is a homomorphism

$$T : G \to Gl(X)$$

of G into the group $Gl(X)$ of invertible bounded endomorphisms of the Banach space $(X, \| \|)$ such that

$$C = \sup_{s \in G} \|T(s)\|_{\text{op}} < \infty$$

and all mappings

$$t_x : G \to X, \quad t_x(s) = T(s)x, \quad s \in G \quad (x \in X),$$

are continuous.

Let us mention that if G is in particular amenable, then every bounded G-module is equivalent to an isometric one. There exists on X a new norm $\| \|'$, equivalent to $\| \|$, such that $\|T(s)x\|' = \|x\|'$ for all $x \in X$ and every $s \in G$ (see [137, Prop. 17.5] in the Hilbert case).

Indeed it suffices to take a left-invariant averaging M on G, and consider for $x \in X$ the continuous bounded function $no_x(g) := \|T(g)x\|$, $g \in G$, then define the new norm $\| \|'$ by

$$\|x\|' := \langle M, no_x \rangle.$$

Definition 8.11.2 We say that the module (T, X) is *irreducible or topologically simple* if any G-invariant closed subspace of X is trivial, i.e., either $\{0\}$ or X itself.

We know from Theorem 2.3.3 in [70] that a bounded G-module (T, X) on a Banach space X can be integrated to a bounded $L^1(G)$ module:

$$T(f) = \int_G f(s)T(s)ds, \quad f \in L^1(G),$$

where ds denotes the Haar measure of G. Then

$$\|T(f)x\|_X \leq C\|f\|_1\|x\|_X, \quad x \in X, f \in L^1(G).$$

Conversely, every bounded non-degenerate (meaning $T(L^1(G))X$ is dense in X) $L^1(G)$-Banach module (T, X) comes from the integration of a bounded representation of G. In particular, a bounded Banach G-module (T, X) is irreducible if and only if the corresponding $L^1(G)$-module is irreducible.

A particularly interesting class of modules is the class of simple, or algebraically irreducible, $L^1(G)$-modules. To describe them the orbit space $\mathfrak{g}^*/G$ is no longer sufficient. Every simple $L^1(G)$-module is associated to linear forms $\ell + i\nu \in \mathfrak{g}^*_{\mathbb{C}}$ for some $\nu \in [\mathfrak{g}, \mathfrak{g}]^{\perp}$. We develop this theory in Sect. 8.15.3.

Lemma 8.11.3 *Let (T, X) be an irreducible Banach module of $L^1(G)$. Then the closed ideal $\ker_{L^1(N)} T$ of $L^1(N)$ is G-prime.*

Proof Let J_1, J_2 be two G-invariant ideals in $L^1(N)$ such that $J_1 * J_2 \subset \ker_{L^1(N)} T$. We must show that either J_1 or $J_2 \subset \ker_{L^1(N)} T$. But since the ideals J_1 and J_2 of $L^1(N)$ are G-invariant, it follows that

$$T(L^1(G) * J_1 * L^1(G)) * J_2 * L^1(G) = \{0\}.$$

Hence

$$(L^1(G) * J_1 * L^1(G)) * (L^1(G) * J_2 * L^1(G)) \subset \ker_{L^1(G)} T.$$

The irreducibility of the representation T tells us that $\ker_{L^1(G)} T$ is a prime ideal in $L^1(G)$. Therefore either $L^1(G) * J_1 * L^1(G) \subset \ker_{L^1(G)} T$ or $L^1(G) * J_2 * L^1(G) \subset \ker_{L^1(G)} T$. Hence $T(J_1) = \{0\}$ or $T(J_2) = \{0\}$. □

Let $G = \exp \mathfrak{g}$ be as always a solvable exponential Lie group and let $\mathfrak{n} \supset [\mathfrak{g}, \mathfrak{g}]$ be a nilpotent ideal. Take $\ell \in \mathfrak{g}^*$ and let $\mathfrak{g}(\ell)$ be the stabilizer of ℓ in $\mathfrak{g}$, and $\mathfrak{g}(\ell_{|\mathfrak{n}})$ the stabilizer of $\ell_{|\mathfrak{n}}$ in $\mathfrak{g}$. We choose a $\mathfrak{g}(\ell_{|\mathfrak{n}})$-invariant polarization in $\mathfrak{n}$, a polarization $\mathfrak{p}^0$ of $\ell_{|\mathfrak{g}(\ell_{|\mathfrak{n}})}$ and we let $\mathfrak{p} := \mathfrak{p}^0 + \mathfrak{p}_0$. Then $\mathfrak{p}$ is a Pukanszky polarization at ℓ.

Take an ideal $\mathfrak{h}$ of $\mathfrak{g}$ containing $\mathfrak{n}$ and such that $\mathfrak{g} = \mathfrak{g}(\ell) + \mathfrak{h}$ and $\mathfrak{g}(\ell) \cap \mathfrak{h} \subset \mathfrak{n}$. Let $H := \exp \mathfrak{h}$.

Denote by $\pi_{\ell,\mathfrak{p}} = \pi_\ell = \pi$ the unitary irreducible representation of G associated to l and $\mathfrak{p}$ and let $\gamma = (\pi_\ell)|_H$. This representation is irreducible and can be identified with the representation $\pi_{l_{|\mathfrak{h}}}$ of H associated to the restriction of l to $\mathfrak{h}$.

Let $\tau := \pi_{\ell_{|\mathfrak{n}}} \in \widehat{N}$. By Theorem 8.10.18, we know that $\ker_{L^1(N)} T$ is the kernel of a G-orbit in $\widehat{N}$. We choose ℓ such that τ is an element of this G-orbit.

Let σ be a subspace of $\mathfrak{g}(\ell)$ such that $\mathfrak{g} = \sigma \oplus \mathfrak{h}$. Let $S = \exp \sigma$. Then $G = G(\ell)H = SH$.

We can identify S with the linear subspace σ via the exponential mapping. This gives the structure of an Abelian group on S by letting

$$s + t := \exp(\log s + \log), s, t \in S. \tag{8.11.1}$$

In particular we have that

$$S^{-1} = (\exp \sigma)^{-1} = \exp(-\sigma) = S,$$
$$s^{-1} = -s, s \in S.$$

This group $(S, +)$ is also isomorphic to the Abelian group $G(\ell)/G(\ell) \cap N$.

Definition 8.11.4 Let

$$\mathscr{J}_0 = \{f \in L^1(H);\; \gamma(f) \text{ has finite rank }\}$$

and

$$\mathscr{J} = \overline{\mathscr{J}_0}^{L^1(H)}.$$

Evidently $\mathscr{J}_0$ and $\mathscr{J}$ are ideals in $L^1(H)$ and $\mathscr{J}_0/\ker(\gamma)$ is the minimal ideal of $L^1(H)/\ker(\gamma)$.

Lemma 8.11.5 *Let $\mathscr{J}_0$ and $\mathscr{J}$ be the ideals defined in (8.11.4). Let χ be a character of H such that $\chi|_N \equiv 1$. Then $\mathscr{J}_0$ and $\mathscr{J}$ are invariant under the multiplication by χ.*

Proof As $\gamma(\chi \cdot f) = \gamma \otimes \chi(f)$ and $\gamma \otimes \chi$ and γ are unitarily equivalent by (8.11.8), we see that if $\gamma(f)$ has finite rank, then so does $\gamma(\chi \cdot f)$. □

Theorem 8.11.6 *Let $G = \exp \mathfrak{g}$ be an exponential Lie group, let $N = \exp \mathfrak{n}$ be a closed nilpotent subgroup of G containing $[G, G]$ and let (T, X) be an irreducible bounded Banach space representation of G. Let $l \in \mathfrak{g}^*, \tau := \pi_{l|\mathfrak{n}}$ such that $ker_{L^1(N)} T_{|N} = \bigcap_{g \in G} ker_{L^1(N)}(g \cdot \tau)$. Choose a subspace σ of $\mathfrak{g}(\ell)$ such that $\mathfrak{g}(\ell) = \mathfrak{g}(\ell) \cap \mathfrak{n} \oplus \sigma$ and a subspace $\mathfrak{h}$ of G containing $\mathfrak{n}$ such that $\mathfrak{g} = \sigma \oplus \mathfrak{h}$. Let $H = \exp \mathfrak{h}$, $\gamma := \pi_{\ell|H}$. Then*

$$ker_{L^1(H)}(\gamma) = (L^1(H) * \ker({}^G\tau))^- = \ker_{L^1(H)} T_{|H}.$$

Proof Since for every $\chi \in N^\perp \subset \hat{H}$ the representation $\gamma \otimes \chi$ is equivalent to γ, we have that $\chi \ker(\gamma) = \ker(\gamma)$. This implies that $L^\infty(H/N)\ker(\gamma) = \ker(\gamma)$. Therefore, by Proposition 8.4.6,

$$\ker_{L^1(H)}\gamma = (L^1(H) * \ker_{L^1(N)}\gamma)^-.$$

On the other hand, it is well known [96, 5. Theorem 9], that $\pi_m|_N$ is weakly equivalent to ${}^G\pi_q = {}^G\tau$, i.e., $\ker_{L^1(N)}\gamma = \ker_{L^1(N)}({}^G\tau)$. Hence

$$\ker_{L^1(H)}\gamma = (L^1(H) * \ker_{L^1(N)}({}^G\tau))^-.$$

By (8.10.18) we know

$$(L^1(H) * \ker_{L^1(N)}(^G\tau))^- = (L^1(H) * \ker_{L^1(N)} T)^- \subset \ker_{L^1(H)} T.$$

Hence $\ker_{L^1(H)}(\gamma) \subset \ker_{L^1(H)}(T)$. Let us suppose that there exists $f \in \ker_{L^1(H)} T$ such that $\gamma(f) \neq 0$. Then there exists a $\nu \in ES(\gamma)$ such that $\gamma(f)\nu \neq 0$. Take as before $\lambda \in ES(\gamma)$ and $p_\lambda \in L^1(H)$ such that $\gamma(p_\lambda) = P_\lambda$. Then

$$P_\lambda = \frac{1}{\|\gamma(f)\nu\|^2} P_{\lambda,\gamma(f)\nu} P_{\gamma(f)\nu,\lambda} = \frac{1}{\|\gamma(f)\nu\|^2} P_{\lambda,\nu} \circ \gamma(f^*) \circ \gamma(f) \circ P^*_{\lambda,\nu}.$$

Furthermore, as $\lambda, \nu \in ES(\gamma)$, there exists $f_0 \in \mathscr{J}_0$ such that $\gamma(f_0) = P_{\lambda,\nu}$. Hence

$$p_\lambda = \frac{1}{\|\gamma(f)\nu\|^2} f_0 * f^* * f * f_0^* \ \mathrm{mod} \ker_{L^1(H)}(\gamma).$$

Because $\ker_{L^1(H)}(\gamma) \subset \ker_{L^1(H)} T$ and $f \in \ker_{L^1(H)} T$, the same goes for p_λ. Since $\mathscr{J}_0$ is minimal modulo $\ker(\gamma)$ and since $p_\lambda \in \mathscr{J}_0 \cap \ker_{L^1(H)} T$, we conclude that $\mathscr{J}_0$ and $\mathscr{J}$ are contained in $\ker_{L^1(H)} T$. But $\mathscr{J}$ is invariant by multiplication with $\chi \in N^\perp$. Hence $\mathscr{J} = (L^1(H) * \mathscr{I})^-$ by Proposition 8.4.6, where $\mathscr{I}$ is an H-invariant ideal of $L^1(N)$. As $\mathscr{J} \subset \ker_{L^1(H)}(T)$, we have that $\mathscr{I} \subset \ker_{L^1(N)}(T) = \ker_{L^1(N)}(^G\tau)$ and so $\mathscr{J} = (L^1(H) * \mathscr{I})^- \subset (L^1(H) * \ker_{L^1(N)}(^G\tau))^- = \ker_{L^1(H)}(\gamma)$, which is absurd. Hence $\ker_{L^1(H)}(\gamma) = \ker_{L^1(H)} T$. □

Remark 8.11.7 This line of reasoning shows in particular that $T(p_\lambda) \neq 0$, and so $Y := T(p_\lambda)X \neq \{0\}$.

8.11.2 The Representation $\pi^\gamma = \mathrm{ind}_H^G \gamma$

Take $\ell \in \mathfrak{g}^*$ and let $\pi = \pi_\ell \in \hat{G}$ be the element of $\hat{G}$ associated to $\ell \in \mathfrak{g}^*$. Let $\gamma = \pi|_H$. Since $G = G(\ell) \cdot H$, we know that γ is irreducible and that $\gamma \simeq \pi_m \in \hat{H}$, where $m = l|_{\mathfrak{h}} \in \mathfrak{h}^*$ (see [96] or [70]).

Remark 8.11.8 For every character $\chi = e^{-ir(\log(\cdot))}$ of H such that $\chi|_N \equiv 1$ i.e., $\chi \in N^\perp$ and $r \in \mathfrak{n}^\perp$ (in $\mathfrak{h}^*$), the representations π_{m+r} and $\gamma \otimes \chi$ are unitarily equivalent. Indeed, it is well known that for any connected Lie group $D = \exp \delta$, for every $f \in \delta^*$, for any ideal $\mathfrak{i} \supset [\delta, \delta]$, we have that

$$\mathrm{Ad}^*(D(f|_{\mathfrak{i}}))f = f + (\delta(f) + \mathfrak{i})^\perp.$$

Since in our case $\mathfrak{h}(m) = \mathfrak{h}(\ell) \subset \mathfrak{n}$ (since $\mathfrak{h}(\ell) \subset \mathfrak{n}$), because $\mathfrak{h}(\ell)$ is contained in $\mathfrak{g}(\ell) \cap \mathfrak{h}$, we deduce that $\mathrm{Ad}^*(H(q))m = m + \mathfrak{n}^\perp$. Hence $\gamma \otimes \chi$ is unitarily equivalent to γ, because $\gamma \otimes \chi$ is associated to the linear form $m + r \in \mathrm{Ad}^*(H)m = \mathrm{Ad}^*(G)m$.

Define now the unitary representation π^γ of G by

$$\pi^\gamma := \mathrm{ind}_H^G \gamma.$$

Since H is a normal subgroup of G, we can identify the Hilbert space $\mathcal{H}_{\pi^\gamma}$ of π^γ with $L^2(G/H, \mathcal{H}_\gamma) \simeq L^2(S, \mathcal{H}_\gamma)$, where as before $\sigma \subset \mathfrak{g}(\ell)$, $\mathfrak{g} = \sigma \oplus \mathfrak{h}$ and $G = S \cdot H$, $S = \exp \sigma$.

For $F \in L^1(G)$, $\varphi \in L^2(F, \mathcal{H}_\gamma)$ let us compute the element $\pi^\gamma(F)(\varphi)$ of $\mathcal{H}_{\pi^\gamma}$. We have for $s \in S$ that

$$\begin{aligned}
\pi^\gamma(F)(\varphi)(s) &= \int_G F(g)\varphi(g^{-1}s)dg \\
&= \int_G \Delta_G(g^{-1})F(sg^{-1})\varphi(g)dg \\
&= \int_S\int_H \Delta_G(h^{-1}t^{-1})F(sh^{-1}t^{-1})\varphi(th)dhdt \\
&= \int_S\int_H \Delta_G(h^{-1}t^{-1})F(sh^{-1}t^{-1})\gamma(h^{-1})(\varphi(t))dhdt \\
&= \int_S\int_H \Delta_G(t^{-1})F((s-t)c(s,-t)(tht^{-1}))\gamma(h)(\varphi(t))dhdt \\
&= \int_S \gamma(d(s,t)) \circ \gamma(F^t(s-t))(\varphi(t))dt \\
&= \int_S \gamma(t^{-1}) \circ \gamma(c(s,-t))^{-1} \circ \gamma(F(s-t))(\gamma(t)(\varphi)(t))dt.
\end{aligned} \tag{8.11.2}$$

Here F^t denotes the mapping of $S \to L^1(H)$ defined by

$$F^t(s)(h) := \Delta_G(t^{-1})F(s \cdot tht^{-1}), h \in H, s \in S, t \in S,$$

and

$$d : G \times G \to N; \quad d(s,t) := t^{-1}(c(s,-t))^{-1}t.$$

is a smooth mapping. It follows from this formula that

Proposition 8.11.9

$$\begin{aligned}
\ker_{L^1(G)}(\pi^\gamma) &= \{F \in L^1(G); \gamma(F(s)) = 0, \ s \textit{ almost everywhere}\}, \\
&\quad \{F \in L^1(G); T(F(s)) = 0, \ s \textit{ almost everywhere}\} \\
&= \overline{L^1(G) * ker_{L^1(H)}(T)} = \overline{L^1(G) * ker_{L^1(N)}(T)} \\
&\subset ker_{L^1(G)}(T).
\end{aligned}$$

Proof It suffices to remark that $\gamma^s \equiv \gamma$ for any $s \in S \subset G(\ell)$. □

So in order to characterize the irreducible $L^1(G)$-modules with kernel $\ker_{L^1(G)}(T)$, it suffices to study the representations of the algebra

$$\begin{aligned} A &= L^1(G)/\big(L^1(G) * \ker_{L^1(H)}({}^G\tau)\big)^- \\ &= L^1(G)/\ker_{L^1(G)}(\pi^\gamma). \end{aligned}$$

8.11.3 An Example: Representations on Mixed L^p-spaces

Definitions

1. Let as always $G = \exp \mathfrak{g}$ be an exponential solvable Lie group. The nilradical of $\mathfrak{g}$ will be denoted by $\mathfrak{n}$ and $N = \exp \mathfrak{n}$ will be the corresponding subgroup of G. Take $\ell \in \mathfrak{g}^*$ and write $q = \ell|_{\mathfrak{n}} \in \mathfrak{n}^*$. We recall the following definitions:

$$\begin{aligned} \mathfrak{g}(\ell) &:= \{X \in \mathfrak{g} \mid \ < \ell, [X, \mathfrak{g}] > \equiv 0\}, \\ \mathfrak{g}(q) &:= \{X \in \mathfrak{g} \mid \ < q, [X, \mathfrak{n}] > = < \ell, [X, \mathfrak{n}] > \equiv 0\}, \\ \mathfrak{n}(q) &= \{X \in \mathfrak{n} \mid \ < q, [X, \mathfrak{n}] > = < \ell, [X, \mathfrak{n}] > \equiv 0\} = \mathfrak{g}(q) \cap \mathfrak{n}. \end{aligned}$$

Then we decompose the Lie algebra as follows:

$$\begin{aligned} \mathfrak{g}(\ell) + \mathfrak{n} &= \sigma \oplus \mathfrak{n} \text{ with } \sigma \subset \mathfrak{g}(\ell) \subset \mathfrak{g}(q), \\ \mathfrak{g}(q) + \mathfrak{n} &= \mathfrak{w} \oplus (\mathfrak{g}(\ell) + \mathfrak{n}) = \mathfrak{w} \oplus \sigma \oplus \mathfrak{n} \text{ with } \mathfrak{w} \subset \mathfrak{g}(q), \\ \mathfrak{g} &= \varphi \oplus (\mathfrak{g}(q) + \mathfrak{n}) \\ &= \varphi \oplus \mathfrak{w} \oplus \sigma \oplus \mathfrak{n} \\ \mathfrak{h} &:= \varphi + \mathfrak{w} + \mathfrak{n}. \end{aligned}$$

2. Now we choose $\mathscr{Y} \subset \mathfrak{w} \subset \mathfrak{g}(q)$ a maximal ℓ -isotropic subspace of $\mathfrak{w}$, i.e., a maximal subspace of $\mathfrak{w}$ such that $< \ell, [\mathscr{Y}, \mathscr{Y}] > \equiv 0$. Then there exist a subspace $\mathscr{X}$ in $\mathfrak{w}$ and bases $\{X_1, \cdots, X_c\}$ of $\mathscr{X}$, $\{Y_1, \cdots, Y_c\}$ of $\mathscr{Y}$ such that

$$\mathfrak{w} = \mathscr{X} \oplus \mathscr{Y}, < \ell, [X_i, X_j] > = 0, < \ell, [Y_i, Y_j] > = 0, < \ell, [X_i, Y_j] > = \delta_{ij}.$$

This means $\mathscr{X}$ is a dual to $\mathscr{Y}$ with respect to ℓ. This is possible because

$$\{Z \in \mathfrak{w} \mid \ < \ell, [Z, \mathfrak{w}] > \equiv 0\} = \{0\}.$$

As a matter of fact, $\mathfrak{w} \oplus \mathfrak{n}(q)$ modulo $\ker(q|_{\mathfrak{n}(q)})$ is a Heisenberg algebra. We write $S = \exp \sigma$, $\mathscr{V} = \exp \varphi$, $\mathscr{W} = \exp \mathfrak{w}$, $\mathscr{X} = \exp \mathscr{X}$, $\mathscr{Y} = \exp \mathscr{Y}$, $H := \exp \mathfrak{h}$.

3. First let us choose a $\mathfrak{g}(q)$-invariant polarization $\mathfrak{p}^0$ of q in $\mathfrak{n}$ (for example a Vergne polarization). Then $\mathfrak{p} = \mathscr{Y} \oplus \mathfrak{p}^0 \oplus \sigma$ is a Pukanszky polarization of ℓ in $\mathfrak{g}$ and $\mathfrak{p}^1 := \mathfrak{p}^0 + \mathscr{Y}$ is a Pukanszky polarization at $\ell_{|\mathfrak{h}}$. Moreover $\mathfrak{p}^0 = \mathfrak{p} \cap \mathfrak{n}$, $\mathfrak{p}^1 = \mathfrak{p} \cap \mathfrak{h}$. For the rest of this section we shall stick to these polarizations. We write

$$P^0 := \exp(\mathfrak{p}^0),$$
$$P := \exp \mathfrak{p},$$
$$P^1 := \exp \mathscr{Y} + \mathfrak{p}^0 = P \cap H.$$

4. Let

$$\mathfrak{n} = \mathfrak{n}_0 \supset \mathfrak{n}_1 \supset \cdots \supset \mathfrak{n}_k \supset \mathfrak{n}_{k+1} = \{0\}$$

be a Jordan-Hölder sequence for the action of $\mathfrak{g}$ on $\mathfrak{n}$. Let

$$I = \{i \mid \mathfrak{p}^0 + \mathfrak{n}_i \neq \mathfrak{p}^0 + \mathfrak{n}_{i+1}, i = 0, \cdots, k\} = \{i_j \mid 1 \leq j \leq m, 0 \leq i_1 \leq \cdots \leq i_m \leq k\}.$$

We write $\mathfrak{p}_j = \mathfrak{p}^0 + \mathfrak{n}_{i_j}$, for $j = 1, \cdots m$. Obviously $\mathfrak{p}_{m+1} = \mathfrak{p}^0$ and $\mathfrak{p}_1 = \mathfrak{n}$. For each $j \in \{1, \cdots, m\}$ we choose a subspace $\varphi_j \subset \mathfrak{n}_{i_j} \subset \mathfrak{p}_j$ such that

$$\varphi_j \oplus \mathfrak{p}_{j+1} = \mathfrak{p}_j.$$

Then

$$\sum_{j=1}^{m}{}^{\oplus} \varphi_j \oplus \mathfrak{p}^0 = \mathfrak{n} \tag{8.11.3}$$

and

$$\Phi : \sum_{j=1}^{m}{}^{\oplus} \varphi_j \longrightarrow N/P^0 \simeq V_1 + \cdots + V_m \equiv (V_1, \cdots, V_m) \longmapsto \exp(V_1) \cdots \exp(V_m) \cdot P^0$$

is a diffeomorphism.

Representations on Mixed L^p-spaces

Let us write

$$\tilde{\mathfrak{n}} = \sum_{j=1}^{m}{}^{\oplus} \varphi_j, \; \mathscr{V}_j = \exp \varphi_j \text{ and } \tilde{N} = \prod_{j=1}^{m} \mathscr{V}_j = \prod_{j=1}^{m} \exp \varphi_j.$$

Consider the following decomposition of G:

$$G = \mathscr{V} \cdot \mathscr{X} \cdot \tilde{N} \cdot P = S \cdot \mathscr{V} \cdot \tilde{N} \cdot P_1.$$

Take $\overline{p} = (p_1, \cdots, p_m) \in [1, \infty]^m$.
Define the real character $\Delta^{\overline{p}}$ of G by

$$\Delta^{\frac{1}{\overline{p}}}(g) = e^{\sum_{j=1}^m \frac{1}{p_j} \operatorname{tr} \operatorname{ad}_{\mathfrak{p}_j/\mathfrak{p}_{j+1}}(\log g)} = e^{\sum_{j=1}^m \frac{1}{p_j} \operatorname{tr}\lambda_j(\log g)}, \tag{8.11.4}$$

$$\delta^{\overline{p}}(X) := \sum_{j=1}^m \frac{1}{p_j} \operatorname{tr}\lambda_j(X),\ X \in \mathfrak{g},$$

where we use the notation $\lambda_j(\cdot) = \operatorname{ad}_{\mathfrak{p}_j/\mathfrak{p}_{j+1}}(\cdot)$.
There is a relationship between the Haar measures on G, P and the measure on $G/P \equiv \mathscr{V} \cdot \mathscr{X} \cdot \tilde{N} = \mathscr{V} \cdot \mathscr{X} \cdot \prod_{j=1}^m \mathscr{V}_j$: if the Lie algebra $\mathfrak{g}$ is decomposed by

$$\mathfrak{g} = \varphi \oplus \mathscr{X} \oplus \mathscr{Y} \oplus \sigma \oplus (\tilde{\mathfrak{n}} \oplus \mathfrak{p}^0),$$

we get a Haar measure on G by

$$\int_G f(g)dg$$
$$= \int_\varphi \int_{\mathscr{X}} \int_{\mathscr{Y}} \int_\sigma \int_{\varphi_1} \cdots \int_{\varphi_m} \int_{P^0} f\big(\exp V \cdot \exp X \cdot \exp Y \cdot \exp U \cdot \exp V_1 \cdots \exp V_m \cdot p^0\big)$$
$$dp^0 dV_1 \cdots dV_m dU dY dX dV,$$

where we use the Haar measure on P^0 and the Lebesgue measures on $\varphi, \mathscr{X}, \mathscr{Y}, \sigma, \varphi_j$. The Haar measure on P is given by

$$\int_P f(p)dp = \int_{\mathscr{Y}} \int_\sigma \int_{P^0} f\big(\exp Y \cdot \exp U \cdot p^0\big)dp^0 dU dY.$$

Using the fact that

$$f(\exp V \cdot \exp X \cdot \exp Y \cdot \exp U \cdot \exp V_1 \cdots \exp V_m \cdot p^0)$$
$$= f(\exp V \cdot \exp X \cdot [\exp Y \cdot \exp U \cdot \exp V_1 \cdots \exp V_m \cdot \exp U^{-1} \cdot \exp Y^{-1}] \cdot$$
$$\cdot \exp Y \cdot \exp U \cdot p^0),$$

and that

$$\Delta(\exp V \cdot \exp X \cdot \exp V_1 \cdots \exp V_m \cdot \exp Y \cdot \exp U \cdot p^0) = \Delta(\exp Y \cdot \exp U),$$

we get

$$\int_G f(g)dg$$
$$= \int_{\varphi}\int_{P_a(G\cdot\Gamma_f)}\int_{\varphi_1}\cdots\int_{\varphi_m}\int_{\mathscr{Y}}\int_{\sigma}\int_{P^0} f(\exp V\cdot\exp X\cdot\exp V_1\cdots\exp V_m\cdot$$
$$\cdot\exp Y\cdot\exp U\cdot p^0)\Delta^{-1}(\exp Y\cdot\exp U)dp^0dUdYdV_m\cdots dV_1dXdV$$
$$= \int_{G/P}\int_P f(gp)\Delta^{-1}(gp)dpd\dot{g}.$$

Here we used the notation

$$\int_{G/P} f(g)d\dot{g} = \int_{\varphi}\int_{P_a(G\cdot\Gamma_f)}\int_{\varphi_1}\cdots\int_{\varphi_m} f(\exp V\cdot\exp X\cdot\exp V_1\cdots\exp V_m)dV_m\cdots dV_1dXdV.$$

The representation space $L^{\overline{p}}(G/P,\ell)$ is then defined to be the completion, for the norm $\|\cdot\|_{\overline{p}}$ given below, of the space $\mathscr{E}^{\overline{p}}(G/P,\ell)$ of continuous maps

$$\xi:\mathscr{V}\cdot\mathscr{X}\cdot\tilde{N}\cdot P\longrightarrow\mathbb{C}$$

with compact support mod P such that

$$\xi(x\cdot p) = \overline{\chi_\ell(p)}\Delta^{\frac{1}{\overline{p}}}(p)\xi(x),\quad \forall x\in G,\forall p\in P,$$

and

$$\|\xi\|_{\overline{p}} = \Big(\int_{\mathscr{V}}\int_{\mathscr{X}}\Big(\Big(\int_{\mathscr{V}_1}\Big(\cdots\Big(\int_{\mathscr{V}_m}|\xi(vxv_1\cdots v_m)|^{p_m}dv_m\Big)^{\frac{1}{p_m}}\cdots\Big)^{p_1}dv_1\Big)^{\frac{1}{p_1}}\Big)^2dxdv\Big)^{\frac{1}{2}} < \infty.$$

The above measures are the Lebesgue measures on φ, $\mathscr{X}$, $\varphi_1,\cdots,\varphi_m$. If $p_j=\infty$, then $\left(\int_{\mathscr{V}_j}|\cdots|^{p_j}dv_j\right)^{\frac{1}{p_j}}$ is replaced by the corresponding sup-norm. On the space $L^{\overline{p}}(G/P,\ell)$ the group G acts by left translations, which gives a representation by isometric operators.

We denote this representation by

$$\pi_{\ell,\overline{p}} = \pi_{\ell,\mathfrak{p},\overline{p}} = \mathrm{ind}_P^G\chi_{\ell,\overline{p}}.$$

The representation $\pi_\ell = \pi_{\ell,\mathfrak{p},\overline{2}}$ is the usual induced unitary representation $\pi_{\ell,\mathfrak{p}}$.

8.11.4 The Spaces $\mathscr{E}S$

Spaces of Kernel Functions

Let $\ell \in \mathfrak{g}^*$ and let the polarization $\mathfrak{p}$ at ℓ be chosen as in (Section "Definitions"). Let $\mathscr{B}_1 = \{A_1, \ldots, A_j\}$ be a coexponential basis for $\mathfrak{p}_0$ in $\mathfrak{n}$, which has for instance been chosen in the subspaces φ_j. Let $\mathscr{B}_2 = \{B_1, \ldots, B_k\}$ be a coexponential basis for $\mathfrak{n} + \mathfrak{p}$ in $\mathfrak{g}$. Then $\mathscr{B} = \mathscr{B}_1 \cup \mathscr{B}_2$ is a coexponential basis for $\mathfrak{p}$ in $\mathfrak{g}$. Given a function F on $G/P \times G/P$, we define a function $\tilde{F}$ on $(\mathbb{R}^k \times \mathbb{R}^j) \times (\mathbb{R}^k \times \mathbb{R}^j)$ by

$$\begin{aligned}&\tilde{F}(b_1, \ldots, b_k, a_1, \ldots a_j; b_1', \ldots, b_k', a_1', \ldots a_j')\\ := {} & F(\exp(b_1 B_1) \cdots \exp(b_k B_k) \exp(a_1 A_1) \cdots \exp(a_j A_j); \exp(b_1' B_1)\\ &\cdots \exp(b_k' B_k) \exp(a_1' A_1) \cdots \exp(a_j' A_j)).\end{aligned}$$

We proceed similarly for a function defined on G/P. This allows us to give the following definition (see also Section "Definitions"):

Definition 8.11.10

(a) The space $\mathscr{E}S(G/P \times G/P, \ell, \overline{p})$ is the space of C^∞-functions $F : G \times G \to \mathbb{C}$ such that

1.
$$F(xs, x's') = \Delta^{\frac{1}{p}}(ss')\overline{\chi_\ell(s)}\chi_\ell(s')F(x, x'), \quad \forall x, x' \in G, \forall s, s' \in P$$

2.
$$\|F\|_{\partial,\alpha,\alpha',R,R'} = \sup_{a,a' \in \mathbb{R}^j, b,b' \in \mathbb{R}^k} \left(e^{\alpha|b|} e^{\alpha'|b'|} |R(a)R'(a')\partial_a \partial_b \partial_{a'} \partial_{b'} \tilde{F}(b, a; b', a')|\right) < \infty$$

for all $\alpha, \alpha' \geq 0$, for all polynomials R and R', for all derivation operators ∂, if $|b|$ and $|b'|$ denote the Euclidean norm on $\mathbb{R}^k$.

3. The same conditions as in (2) are required for all partial Fourier transforms of $\tilde{F}$ in the variables b and b'.

(b) The space $\mathscr{E}S(G/P, \ell, \overline{p})$ is defined similarly. It is the space of C^∞-functions $\xi : G \to \mathbb{C}$ such that

1.
$$\xi(xs) = \overline{\chi_\ell(s)}\Delta^{\frac{1}{p}(s)}\xi(x), \quad \forall x \in G, \forall s \in P$$

2.

$$\|\xi\|_{\partial,\alpha,R} = \sup_{a\in\mathbb{R}^j, b\in\mathbb{R}^k} \left(e^{\alpha|b|}|R(a)\partial_a\partial_b\tilde{\xi}(b,a)|\right) < \infty$$

for every $\alpha \geq 0$, for every polynomial R, for all derivation operators ∂.

3. The same conditions as in (2) are required for all partial Fourier transforms of $\tilde{\xi}$ in the variable b.

Remark 8.11.11 The previous spaces are independent of the coexponential bases' choice. They also contain functions that are real analytic and may therefore be extended to functions with complex variables (see Theorem 8.10.13).

A Retract

Let $\mathscr{B}_3 = \{C_1, \ldots, C_i\}$ be a coexponential basis for $\mathfrak{n}$ in $\mathfrak{g}$. We may choose the elements of $\mathscr{B}_3$ in a nilpotent subalgebra $\mathfrak{q}$ of $\mathfrak{g}$ such that $\mathfrak{g} = \mathfrak{q} + \mathfrak{n}$. Let $\mathscr{B}_4 = \{D_1, \ldots, D_g\}$ be a Jordan-Hölder basis for $\mathfrak{n}$. For a function f defined on G, we define $\tilde{f}$ on $\mathbb{R}^i \times \mathbb{R}^g$ as previously. We then define:

Definition 8.11.12 The space $\mathscr{E}S(G)$ is the space of all C^∞-functions $f : G \to \mathbb{C}$ such that

$$\|f\|_{\partial,\alpha,R} = \sup_{c\in\mathbb{R}^i, d\in\mathbb{R}^g} \left(e^{\alpha|c|}|R(d)\partial_c\partial_d\tilde{f}(c,d)|\right) < \infty$$

for every $\alpha \geq 0$, for every polynomial R, for all derivation operators ∂, where $|c|$ denotes the Euclidean norm on $\mathbb{R}^i$.

Remark 8.11.13 The space $\mathscr{E}S(G)$ is independent of the choice of the bases. It is dense in $L^1(G)$. Similarly for $\mathscr{E}S(G/P,\ell)$ and $L^{\overline{p}}(G/P,\ell)$.

It is easy to check that

$$\Delta^{\frac{1}{2}-\frac{1}{p}}\mathscr{E}\mathscr{S}(G/P,\ell,\overline{p}) = \mathscr{E}\mathscr{S}(G/P,\ell,\overline{2}), \overline{p} \in [1,\infty]^m. \qquad (8.11.5)$$

Proposition 8.11.14 *For every $F \in \mathscr{E}\mathscr{S}(G), \xi \in \mathscr{E}\mathscr{S}(G/P,\ell,\overline{p})$ we have that*

$$\pi_{\ell_|,\overline{p}}(F)(\xi) = \Delta^{\frac{1}{p}-\frac{1}{2}}(\pi_{\ell,\overline{2}}((\Delta^{\frac{1}{2}-\frac{1}{p}}F))(\Delta^{\frac{1}{2}-\frac{1}{p}}\xi).$$

Proof We have for $\eta \in \mathscr{E}\mathscr{S}(G/P,\ell,\overline{2}), \xi = \Delta^{\frac{1}{p}-\frac{1}{2}}\eta \in \mathscr{E}\mathscr{S}(G/P,\ell,\overline{p})$ and $u \in G$ that

$$\begin{aligned}\pi_{\ell,\overline{p}}(F)\xi(u) &= \int_G F(h)\xi(h^{-1}u)dh \\ &= \int_G \Delta_G(h)^{-1}F(uh^{-1})\xi(h)dh\end{aligned}$$

$$= \oint_{G/P}\int_P \Delta_G(h)^{-1}\Delta_G(p)^{-1}\Delta_{H,G}(p)^{-1}F(up^{-1}h^{-1})\xi(hp)dpd\dot{h}$$

$$= \oint_{G/P}\Delta_G(h)^{-1}\int_{P^1}\Delta_H(p^{-1})^{-1}\chi_\ell(p^{-1})\Delta^{\frac{1}{p}}(p)F(up^{-1}h^{-1})dp\xi(h)d\dot{h}$$

$$= \oint_{G/P}\Delta_G(h)^{-1}\int_P \chi_\ell(p)\Delta^{\frac{1}{p}}(p^{-1})F(uph^{-1})dp\xi(h)d\dot{h}$$

$$= \oint_{G/P}\Delta_G(h)^{-1}\int_P \chi_\ell(p)\Delta^{\frac{1}{p}}(p^{-1})F(uph^{-1})dp\Delta^{\frac{1}{p}-\frac{1}{2}}(h)\eta(h)d\dot{h} \tag{8.11.6}$$

$$= \Delta^{\frac{1}{p}-\frac{1}{2}}(u)\oint_{G/P}\Delta_G(h)^{-1}\int_P \chi_\ell(p)\Delta^{\frac{1}{2}}(p^{-1})(\Delta^{\frac{1}{2}-\frac{1}{p}}F)(uph^{-1})\eta(h)dpd\dot{h}$$

$$= \Delta^{\frac{1}{p}-\frac{1}{2}}(u)(\pi_{\ell,\overline{2}}((\Delta^{\frac{1}{2}-\frac{1}{p}}F))(\Delta^{\frac{1}{2}-\frac{1}{p}}\xi)(u).$$

□

An immediate consequence of Proposition 8.11.14 is the irreducibility of the representation $\pi_{\ell_{|\mathfrak{h}},\overline{p}}$.

Proposition 8.11.15 *For any $\overline{p} \in [1,\infty]^m$ the representation $\pi_{\ell,\overline{p}}$ is irreducible.*

Proof Let $0 \neq \xi \in \mathscr{E}S(G/P,\ell,\overline{p})$, $\xi' \in \mathscr{E}\mathscr{S}(G/P,\ell_{|\mathfrak{h}},\overline{q})$, $\frac{1}{\overline{p}}+\frac{1}{\overline{q}} = \overline{1}$. By the previous Proposition 8.11.14 and the retract Theorem 8.10.9 there exists $p_{\xi,\xi'} \in \mathscr{E}\mathscr{S}(G)$ such that $\pi_{\ell,\overline{p}}(p_{\xi,\xi'}) = P_{\xi,\xi'} \in B(L^{\overline{p}}(G/P,\ell)$.

Let $0 \neq \eta \in L^{\overline{p}}(\ell,\mathfrak{p},\overline{p})$ be arbitrary. We shall show that η is a cyclic vector and hence that $\pi_{\ell,\mathfrak{p},\overline{p}}$ is irreducible. Let $\xi \in \mathscr{E}S(G/P,\ell,\overline{p})$ be arbitrary. Choose $\xi' \in \mathscr{E}S(G/P,\ell,\overline{q})$ such that $<\eta,\xi'>\neq 0$. There is $F \in \mathscr{E}S(G)$ such that

$$\pi_{\ell,\mathfrak{p},\overline{p}}(F)(\eta) = <\eta,\xi'>\xi.$$

Hence

$$\xi = \pi_{\ell,\mathfrak{p},\overline{p}}\Big(\frac{1}{<\eta,\xi'>}F\Big)(\eta) \Rightarrow \xi \in \pi_{\ell,\mathfrak{p},\overline{p}}\big(L^1(G)\big)(\eta) \subset L^{\overline{p}}(G/P,\ell)$$

and so

$$\mathscr{E}S(G/P,\ell,\overline{p}) \subset \pi_{\ell,\mathfrak{p},\overline{p}}\big(L^1(G)\big)\eta \subset L^{\overline{p}}(G/P,\ell).$$

As $\mathscr{E}S(G/P,\ell,\overline{p})$ is dense in $L^{\overline{p}}(G/P,\ell)$, this proves that η is a cyclic vector. □

8.12 Using Projections in $L^1(G)/\ker_{L^1(G)}(\pi^\gamma)$

Let as before $G = \exp \mathfrak{g}$ be an exponential solvable Lie group with Lie algebra $\mathfrak{g}$. Let $\mathfrak{n}$ be the nilradical of $\mathfrak{g}$ and $N = \exp \mathfrak{n}$. Let (T, X) be an irreducible representation of G (and of $L^1(G)$) on a Banach space X with $\ker_{L^1(G)} T$ in $L^1(G)$. We know that $\ker_{L^1(N)}(T) = \ker_{L^1(N)}(\tau^G)$ for $\tau = \pi_{\ell_{|\mathfrak{n}}} \in \widehat{N}$ for some $\ell \in \mathfrak{g}^*$.

Let $\mathfrak{h}$ be an ideal of $\mathfrak{g}$ containing a nilpotent ideal $\mathfrak{n}$ of $\mathfrak{g}$ that contains $[\mathfrak{g}, \mathfrak{g}]$, and such that

$$\mathfrak{h} + \mathfrak{g}(\ell) = \mathfrak{g}, \mathfrak{g}(\ell) \cap \mathfrak{h} \subset \mathfrak{n}.$$

Let $\gamma = \pi_{\ell_{|\mathfrak{h}}} \subset \widehat{H}$ and $\pi^\gamma = \mathrm{ind}_H^G \gamma$. We have seen in Proposition 8.11.9 that $\ker_{L^1(G)}(\pi)^\tau \subset \ker_{L^1(G)}(T)$. Hence we have only to study the irreducible representations of the algebra

$$L^1(G)/(L^1(G) * \ker_{L^1(N)} T)^- = L^1(G)/\ker_{L^1(G)}(\pi^\tau).$$

This will be done by studying the irreducible modules of the quotient subalgebra $p_\lambda * L^1(G/\ker_{L^1(G)}(\pi^\gamma)) * p_\lambda$, where $p_\lambda \in L^1(H)$ is such that $\gamma(p_\lambda)$ is a one-dimensional projection on the Hilbert space of γ. We choose now any extension $\tilde{\gamma} = \gamma$ of γ to the whole group G, for instance $\tilde{\gamma} = \pi_{\ell,\mathfrak{p}}$.

We write $\gamma(s)\eta = s \cdot \eta, \eta \in L^2(G/P, \ell), s \in S$. Let $\lambda \in \mathscr{E}\mathscr{S}(G/P, \ell)$. Let $\sigma \subset \mathfrak{g}(\ell)$ such that $\mathfrak{g} = \sigma \oplus \mathfrak{h}$ and let $S = \exp \sigma$.

We consider here only the case:

$$\sigma \neq \{0\}$$

The trivial case $\sigma = \{0\}$ will be mentioned in Theorem 8.14.1.

There exists by the retract theorem a smooth mapping $S \to L^1(H), s \to p_{s^{-1}\cdot\lambda,\lambda}$ such that

$$\gamma(p_{s^{-1}\cdot\lambda,\lambda}) = P_{s^{-1}\cdot\lambda,\lambda}, s \in S.$$

Let $p_\lambda := p_{0\cdot\lambda,\lambda}$. Then p_λ is self-adjoint and idempotent modulo $\ker_{L^1(H)}(\gamma)$. Define the subalgebra $L^1(G)_\lambda$ by

$$L^1(G)_\lambda := p_\lambda * L^1(G) * p_\lambda.$$

We shall analyze the algebra $p_\lambda * L^1(G) * p_\lambda/\ker_{L^1(G)}(\pi^\tau)$.

Let us first compute $\pi^\gamma(p_\lambda)$:

$$\begin{aligned}
\pi^\lambda(p_\lambda)(\varphi)(s) &= \int_H p_\lambda(h)\pi^\gamma(h)(\varphi)(s)dh \\
&= \int_H p_\lambda(h)\varphi(h^{-1}s)dh \\
&= \int_H p_\lambda(h)\gamma(s^{-1}hs)(\varphi)(s)dh \\
&= \gamma(s^{-1}) \circ \int_H p_\lambda(h)\gamma(h)\gamma(s)(\varphi)(s))dh \\
&= \gamma(s^{-1}) \circ P_\lambda(\gamma(s)(\varphi)(s)) \\
&= \langle\varphi(s), s^{-1}\cdot\lambda\rangle s^{-1}\cdot\lambda \\
&= P_{s^{-1}\cdot\lambda}(\varphi(s)), \varphi \in L^2(S, \mathcal{H}_\gamma), s \in S.
\end{aligned}$$

For $F \in L^1(G)$, using (8.11.2) we compute

$$\begin{aligned}
&\pi^\gamma(p_\lambda * F * p_\lambda)(\varphi)(s) \\
&= \pi^\gamma(p_\lambda)(\int_S \gamma(t)^{-1} \circ \gamma(c(s,-t))^{-1} \circ \gamma(F(s-t)) \circ \gamma(t)(\pi^\gamma(p_\lambda)(\varphi)(t)))dt) \\
&= \pi^\gamma(p_\lambda)(\int_S \langle\varphi(t), t^{-1}\cdot\lambda\rangle\gamma(t)^{-1} \circ \gamma(c(s,-t))^{-1} \circ \gamma(F(s-t))(\lambda)dt) \\
&= (\int_S \langle\varphi(t), t^{-1}\cdot\lambda\rangle\langle\gamma(t)^{-1} \circ \gamma(c(s,-t))^{-1} \circ \gamma(F(s-t))(\lambda), s^{-1}\cdot\lambda\rangle dt))\ s^{-1}\cdot\lambda \\
&= (\int_S \langle\varphi(t), t^{-1}\cdot\lambda\rangle\langle\gamma(F(s-t))(\lambda), \gamma(c(s,-t)) \circ \gamma(t)\gamma(s^{-1})\lambda\rangle dt)\ s^{-1}\cdot\lambda \\
&= (\int_S \langle\varphi(t), t^{-1}\cdot\lambda\rangle\langle\gamma(F(s-t))(\lambda), (t-s)\cdot\lambda\rangle dt)\ s^{-1}\cdot\lambda.
\end{aligned}$$

Let

$$q_F(t) := \langle\gamma(F(t))(\lambda), \lambda\rangle, t \in S.$$

Then

$$\pi^\gamma(p_\lambda * F * p_\lambda)(\varphi)(s) = (\int_S \langle\varphi(t), t^{-1}\cdot\lambda\rangle q_F(s-t)dt)\ s^{-1}\cdot\lambda \tag{8.12.1}$$

$$= \int_S q_F(s-t)P_{s^{-1}\cdot\lambda,t^{-1}\lambda}(\varphi(t))dt. \tag{8.12.2}$$

We see that

$$|q_F(t)| \leq \|F(t)\|_{L^1(H)/\ker_{L^1(H)}(\gamma)}, t \in S.$$

Let now $q : S \to \mathbb{C}$ be such that

$$\int_S |q(t)| \|p_{s^{-1}\cdot\lambda,\lambda}\|_1 dt < \infty.$$

We define $F_q : G \to \mathbb{C}$ by

$$F_q(sh) = q(s)p_{s^{-1}\cdot\lambda,\lambda}(h), \; h \in H, s \in S.$$

Then $F_q \in L^1(G)$. We have that

$$\begin{aligned}
\pi^\gamma(F_q)(\varphi)(s) &= \int_S \gamma(t^{-1}) \circ \gamma(c(s,-t))^{-1} \circ \gamma(F(s-t))(\gamma(t)(\varphi)(t))dt \\
&= \int_S \gamma(t^{-1}) \circ \gamma(c(s,-t))^{-1} \circ q(s-t)P_{(t-s)\cdot\lambda,\lambda}(\gamma(t)(\varphi)(t))dt \\
&= \int_S q(s-t)\langle\varphi(t), t^{-1}\cdot\lambda\rangle\gamma(t^{-1}) \circ \gamma(c(s,-t))^{-1}((t-s)\cdot\lambda) \\
&\quad \circ P_{(t-s)\cdot\lambda,\lambda}(\gamma(t)(\varphi)(t))dt \\
&= (\int_S q(s-t)\langle\varphi(t), t^{-1}\cdot\lambda\rangle dt)s^{-1}\cdot\lambda \\
&= \int_S q(s-t)P_{s^{-1}\cdot\lambda,t^{-1}\lambda}(\varphi(t))dt.
\end{aligned}$$

Hence for $F \in L^1(G)_\lambda \cap \mathscr{E}\mathscr{S}(G)$ we have that

$$\pi^\lambda(F) = \pi^\lambda(F_{q_F}).$$

Now, for $F \in L^1(G)$ we have that

$$\begin{aligned}
F^*(s)(h) &= \Delta_G(s^{-1})\Delta_G(h^{-1})\overline{F}(h^{-1}s^{-1}) \\
&= \Delta_G(s^{-1})(F^s(-s))^*(h), \\
&\text{i.e., } F^*(s) = \Delta_G(s^{-1})(F^s(-s))^*, s \in S.
\end{aligned}$$

Furthermore

$$\begin{aligned}
F_q^*(s) &= \Delta_G(s^{-1})\overline{q(-s)}(p^s_{s\cdot\lambda,\lambda})^* \\
&= \overline{q(-s)}p_{s^{-1}\cdot\lambda,\lambda}, s \in S.
\end{aligned} \tag{8.12.3}$$

8.12.1 The Weight ω

Let as before $v(g) = p_{g^{-1}\cdot\lambda,\lambda}$, $g \in G$.

Definition 8.12.1 Let

$$\omega(g) := \|v(g)\|_{L^1(H)/\ker_{L^1(H)}(\gamma)}, \ g \in G.$$

For a function $f \in L^1(H)$ and $y \in G$ let

$$f^y(h) := \Delta_G(y)^{-1} f(yhy^{-1}), h \in H.$$

Then $f^y \in L^1(H)$ and

$$\|f^y\|_1 = \|f\|_1.$$

Proposition 8.12.2 *The function ω is a symmetric weight on G, constant modulo H.*

Proof Remark first that

$$\gamma(v_\lambda(x)) = \gamma(x)^{-1} \circ \gamma(p_\lambda), x \in G.$$

Then the identity

$$\begin{aligned}
\gamma(v_\lambda(xy)) &= \gamma(xy)^{-1} \circ \gamma(p_\lambda) \circ \gamma(p_\lambda) \\
&= \gamma(y)^{-1} \circ \gamma(x)^{-1} \circ \gamma(p_\lambda) \circ \gamma(y) \circ \gamma(y)^{-1} \circ \gamma(p_\lambda) \\
&= \gamma(y)^{-1} \circ \gamma(v_\gamma(x)) \circ \gamma(y) \circ \gamma(v_\lambda(y)) \\
&= \gamma(v_\lambda(x)^y *_{L^1(H)} v_\lambda(y))
\end{aligned}$$

implies that

$$v_\lambda(xy) = v_\lambda(x)^y *_{L^1(H)} v_\lambda(y) \bmod \ker(\gamma)$$

and so

$$\omega(xy) \le \omega(x)\omega(y).$$

As

$$\|\gamma(v_\lambda(x))\|_{\mathrm{op}} = 1,$$

we have that

$$\omega(x) = \inf_{r \in \ker_\gamma} \|v_\lambda(x) + r\|_{L^1(H)} \geq \inf_{r \in \ker_\gamma} \|\gamma(v_\lambda(x) + r)\|_{\mathrm{op}} = \|\gamma(v_\lambda(x))\|_{\mathrm{op}} = 1.$$

Finally the identity

$$\gamma\big((v_\lambda(x^{-1})^x)^*\big) = \gamma(v_\lambda(x))$$

implies that

$$\omega(x) = \omega(x^{-1})$$

and ω is a symmetric weight on G. Furthermore

$$\gamma(v_\lambda(xh)) = \gamma(_{h^{-1}}) \circ \gamma(v_\lambda(x)), \quad \forall h \in H$$

implies that

$$\omega(xh) = \omega(x) \quad \forall h \in H$$

and ω is constant modulo H. □

8.12.2 The Algebras $(p_\lambda * L^1(G) * p_\lambda)/ker_{L^1(G)}(\pi^\lambda)$ and $L^1(S, \omega)$

We have seen in Proposition 8.11.9 that $\ker_{L^1(G)}(\pi^\gamma)$ is the ideal of $L^1(G)$ given by

$$\ker_{L^1(G)}(\pi^\gamma) = \{F \in L^1(G);\ F(t) \in \ker_{L^1(H)}(\gamma) \text{ a.e.}\}.$$

We can thus identify the Banach space $L^1(G)/\ker_{L^1(G)}(\pi^\gamma)$ with the Banach space $L^1(S, L^1(H)/\ker(\gamma))$ using the morphism:

$$M^\gamma(F)(t) := F(t) \bmod \ker_{L^1(H)}(\gamma),\ t \in S.$$

It is easy to see that the linear mapping M^γ is isometric (use Sect. 8.4). It follows from the identification of the linear dual spaces $(L^1(G)/\ker_{L^1(G)}(\pi^\gamma))^*$ and $(L^1(S, L^1(H)/\ker(\gamma)))^*$ that the mapping M^γ is isometric. It is also easy to see that the image of M^γ contains the dense subspace $C_c(S, L^1(H)/\ker_{L^1(H)}(\gamma))$, and so M^γ is an isometric isomorphism of Banach spaces.

We shall prove now that the closed subalgebra $A_\lambda := (p_\lambda * L^1(G) * p_\lambda)/\ker_{L^1(G)}(\pi^\lambda)$ of $A^\gamma := L^1(G)/\ker_{L^1(G)}(\pi^\gamma)$ and the Abelian Banach algebra $L^1(\mathcal{S}, \omega)$ are isomorphic. The multiplication in the algebra $L^1(S, \omega)$ is the convolution defined by the Abelian group $(S, +)$.

Theorem 8.12.3 *The involutive Banach algebras $p * L^1(G) * p/ker_{L^1(G)}(\pi^\lambda)$ and $L^1(S, \omega)$ are isometrically isomorphic.*

Proof Using relations (8.12.2) and (8.12.3), we see that

$$p_\lambda * F * p_\lambda = F_{q_F} \bmod \ker_{L^1(G)}(\pi^\gamma) \tag{8.12.4}$$

for any $F \in \mathcal{E}\mathcal{S}(G)$. We define the mapping M_γ by

$$M_\gamma : A_\gamma \to L^1(S, \omega), \;\; M_\gamma(F) := q_F.$$

We then have that

$$\begin{aligned}
\|F\|_{L^1(G)/\ker_{L^1(G)}(\pi^\gamma)} &= \int_S \|F(t)\|_{L^1(H)/\ker_{L^1(H)}(\gamma)} dt \\
&= \int_S |q_F(t)| \|v_\lambda(t)\|_{L^1(H)/\ker_{L^1(H)}(\gamma)} dt \\
&= \int_S |q_F(t)\omega(t) dt \\
&= \|M_\gamma(F)\|_1.
\end{aligned}$$

Let now $q, r \in C_c(S)$. Then

$$\begin{aligned}
\pi^\gamma(F_q * F_r)(\varphi)(s) &= \pi^\gamma(F_q) \circ \pi^\gamma(F_r)(\varphi)(s) \\
&= \int_S q(s-t) P_{s^{-1}\cdot\lambda, t^{-1}\lambda}((\pi^\gamma(F_r)\varphi(t))) dt \\
&= (\int_S q(s-t) \langle(\pi^\gamma(F_r)\varphi(t)), t^{-1}\cdot\lambda\rangle dt) s^{-1}\cdot\lambda \\
&= \int_S q(s-t)(\int_S r(t-u)(\varphi(u)), u^{-1}\cdot\lambda\rangle du)\langle t^{-1}\cdot\lambda, t^{-1}\cdot\lambda\rangle dt) s^{-1}\cdot\lambda \\
&= \int_S\int_S q(s-t) r(t-u) dt \langle\varphi(u), u^{-1}\cdot\lambda\rangle du)) s^{-1}\cdot\lambda \\
&= \int_S q * r(s-u)\langle\varphi(u), u^{-1}\cdot\lambda\rangle du) s^{-1}\cdot\lambda \\
&= \pi^\gamma(F_{q*r})(\varphi)(s), \varphi \in L^2(S, \mathcal{H}_\gamma), s \in S.
\end{aligned}$$

Hence

$$F_q * F_r = F_{q*r} \text{ modulo } \ker_{L^1(G)}(\pi^\gamma).$$

This shows that M_γ is an isometric algebra homomorphism. Its image contains clearly the dense subspace $C_c(S)$ and so M_γ is surjective, since it is isometric. Now by (8.12.3) we have that

$$F_q^* = F_{q^*}.$$

Therefore the mapping M_γ is a $*$-isomorphism. □

Remark 8.12.4

(a) Distinct extensions ℓ of $\ell_{|\mathfrak{h}}$ to $\mathfrak{g}^*$ give representations γ_ℓ which differ only by a unitary character. Hence since $\gamma(v_\lambda(x)) = \gamma_\ell(x)^{-1} \circ \gamma(p_\lambda)$, $v(x) = v_\lambda(x) \bmod \ker(\gamma)$ and $\omega(x) = \|v(x)\|_{L^1(H)/\ker_{L^1(H)}(\gamma)}$, two distinct extensions ℓ give us the same weight ω.

(b) Let us recall that $p_\lambda \in L^1(H)$ is such that

$$\gamma(p_\lambda) = P_{\lambda,\lambda} \text{ with } \lambda \in ES(\gamma) \text{ and } \langle\lambda, \lambda\rangle = 1.$$

Let $p_\mu \in L^1(H)$ such that

$$\gamma(p_\mu) = P_{\mu,\mu} \text{ with } \mu \in ES(\gamma) \text{ and } \langle\mu, \mu\rangle = 1.$$

Then the algebras $L^1(G)_\lambda/\ker_{L^1(G)}(\pi^\tau)$ and $L^1(G)_\mu/\ker_{L^1(G)}(\pi^\tau)$ are $*$-isomorphic. Indeed, since $\lambda, \mu \in ES(\gamma)$, there exists $r \in L^1(H)$ such that $\gamma(r) = P_{\lambda,\mu}$. It follows then that the mapping

$$L^1(G)_\lambda/\ker_{L^1(G)}(\pi^\tau) \mapsto L^1(G)_\lambda/\ker_{L^1(G)}(\pi^\tau),$$
$$F \bmod \ker_{L^1(G)}(\pi^\tau) \mapsto (r^* * F * r) \bmod \ker_{L^1(G)}(\pi^\tau)$$

is a $*$-isomorphism for the algebras.

8.12.3 Conclusion: Two Problems to Solve

Let (T, X) be a bounded irreducible representation of G. Then the orbit ${}^G\tau \subset \hat{N}$, $\tau \in \hat{N}$ such that $\ker_{L^1(N)} T = \ker({}^G\tau)$ is uniquely determined by (T, X). One then forms the subgroup $H \subset G$ and $\gamma \in \hat{H}$ such that $\ker_{L^1(H)} T = \ker(\gamma)$. This shows that $(L^1(G) * \ker(\gamma))^- \subset \ker_{L^1(G)} T$. Secondly, it implies that to characterize completely the representation (T, X), it suffices to study the irreducible representations of the algebra $L^1(G)/\ker_{L^1(G)}(\gamma) \simeq L^1(S, L^1(H)\ker_{L^1(H)}(\gamma))$ and of its Abelian subalgebra

$$L^1(G)_\lambda/\ker_{L^1(G)}(\pi^\tau) \simeq L^1(S, \omega).$$

The last quotient does not depend on H, up to an isometric isomorphism. Since the subalgebra $L^1(G)_\lambda/\ker_{L^1(G)}(\pi^\tau)$ is (up to an isomorphism) independent of the choice of p_λ, with $\mathfrak{h}$ fixed once and for all, we have to solve the two following problems:

(i) study the irreducible representations $L^1(S, \omega)$
(ii) extend the irreducible representations of the subalgebra $L^1(G)_\lambda/\ker_{L^1(G)}(\pi^\tau)$ to the algebra $L^1(G)/\ker_{L^1(G)}(\pi^\gamma))$.

The first problem will be addressed in Sect. 8.13, the second one in Sect. 8.14.

8.13 Irreducible Representations of $L^1(S, \omega)$

8.13.1 Characters and Other Examples of Irreducible Representations

Let $\omega : S \simeq \mathbb{R}^n \to [1, \infty[$ be a measurable symmetric weight defined on $\mathbb{R}^n$ which is bounded on bounded sets. Call $\mathscr{B}_\omega = \mathscr{B} = L^1(\mathbb{R}^n, \omega)$ the involutive Banach algebra of functions defined on $\mathbb{R}^n$ which are integrable in the measure $\omega(x)dx$, where dx is the Lebesgue measure. The multiplication on B is the convolution product of $L^1(\mathbb{R}^n)$.

Let us suppose that this weight ω is exponential, i.e., there exists a non-trivial semi-norm r on $\mathbb{R}^n$ such that $r(x) \leq \log\omega(x)$ for all $x \in \mathbb{R}^n$. Since the weight ω is bounded by a constant $C \geq 1$ on the unit ball of $\mathbb{R}^n$ we know that for every $x \in \mathbb{R}^n$, $\|x\| \geq 1$, and for every a $s \in \mathbb{N}$ such that $\|x\| \leq s \leq 2\|x\|$,

$$\omega(x) = \omega(s(\frac{x}{s})) \leq \omega(\frac{x}{s})^s \leq C^s \leq e^{d\|x\|},$$

where $d = 2\log(C)$.

The characters of $\mathscr{B} = L^1(\mathbb{R}^n, \omega)$ are then determined by the continuous homomorphisms

$$\chi_\alpha(x) = e^{-i\alpha(x)}, x \in \mathbb{R}^n,$$

where α is a complex linear form of $\mathbb{R}^n$ such that $|Im\alpha(x)| \leq \log(\omega(x))$ for any x.

The algebra $\mathscr{B}$ has many irreducible representations of infinite dimension. Indeed, let Y be a Banach space of infinite dimension admitting a bounded linear operator u which has no non-trivial closed u-invariant subspaces (see [19, Chapter XIV]). We take a real linear form η of $\mathbb{R}^n$ different from 0 such that

$$|\eta(x)| \leq (1/\|u\|\text{op})r(x); \quad \forall x \in \mathbb{R}^n.$$

Then for any $f \in \mathscr{B}$ the linear operator

$$T(f) = \int_{\mathbb{R}^n} f(x)e^{\eta(x)u}dx$$

is bounded, for

$$\|T(f)\|\text{op} \leq \int_{\mathbb{R}^n} |f(x)|e^{|\eta(x)|\|u\|\text{op}}dx \leq \int_{\mathbb{R}^n} |f(x)|\omega(x)dx = \|f\|_\omega.$$

We obtain a bounded representation (T, Y) on Y. Let us show that this representation is topologically simple. Since $\mathscr{B}$ admits a bounded approximate unit, it suffices to show that the corresponding representation of $\mathbb{R}^n$ defined by

$$T(x) = e^{\eta(x)u} = \sum_{j=0}^{\infty} \frac{(\eta(x)u)^j}{j!}, \quad \forall x \in \mathbb{R}^n$$

is topologically simple. Let $\mathscr{Z}$ be a closed T-invariant subspace of Y. Take $x \in \mathbb{R}^n$ such that $\eta(x) = 1$. For every $z \in \mathscr{Z}$ we have that

$$T(sx)z = \sum_{j=0}^{\infty} \frac{(su)^j}{j!}(z) \in \mathscr{Z}; \quad \forall s \in \mathbb{R}.$$

Differentiating at $s = 0$ we see that $u(z) \in \mathscr{Z}$. Hence $\mathscr{Z}$ is u-invariant. Since u has no non-trivial closed invariant subspaces, we know that $\mathscr{Z} = \{0\}$ or $\mathscr{Z} = Y$, and Y is an irreducible $\mathscr{B}$-module.

Hence there exists many irreducible $L^1(\mathbb{R}^n, \omega)$-modules. Their characterization is an open problem. However, Schur's lemma tells us that the simple are one-dimensional and given by characters, since the algebra $L^1(\mathbb{R}^n, \omega)$ is commutative.

8.13.2 Estimating the Weight ω

Preparations

Poguntke found an estimate for the weight ω (see [139]). To describe this estimate, let $\mathfrak{m}(\ell) = \mathfrak{g}(\ell_{|\mathfrak{n}}) + \mathfrak{n}$, where ℓ is a linear functional of $\mathfrak{g}$ and $\mathfrak{n}$ is the nilradical of $\mathfrak{g}$. Let σ be a subspace of $\mathfrak{g}(\ell)$ such that $\mathfrak{g}(\ell) = \sigma \oplus (\mathfrak{n} \cap \mathfrak{g}(\ell))$. Let us take a polarization $\mathfrak{p}_0$ at $q := \ell|_{\mathfrak{n}}$, which is $G(q) = \exp \mathfrak{g}(q)$-invariant. Let $\mathfrak{p}_1 \subset \mathfrak{g}(q)$ be a polarization at $\ell_{|\mathfrak{n}}$) and let

$$\mathfrak{p} := \mathfrak{p}_1 + \mathfrak{p}_0.$$

Then $\mathfrak{p}$ is a Pukanszky polarization at ℓ. Choose a Jordan-Hölder sequence

$$\mathfrak{n} = \mathfrak{n}_0 \supset \mathfrak{n}_1 \supset \cdots \supset \mathfrak{n}_{m+1} = \{0\}$$

of $\mathfrak{n}$ for the action of $\mathfrak{m}(\ell)$. Let

$$I = \{j \in \{0, \cdots, m\};\ \mathfrak{p}_0 + \mathfrak{n}_{j+1} \neq \mathfrak{p}_0 + \mathfrak{n}_j\}.$$

Let us write $I = \{j_1, \cdots, j_k\}$ and put $\mathfrak{p}_i = \mathfrak{p}_0 + \mathfrak{n}_{j_i}, i \in \{1, \cdots, k\}$. We can choose subspaces $\varphi_i, i = 1, \cdots, k$ such that $\mathfrak{p}_i = \varphi_i \oplus \mathfrak{p}_{i+1}, i = 1, \cdots, k$. For every i the dimension d_i of φ_i is 1 or 2. The $\mathfrak{g}(q)$-modules $\mathfrak{p}_i/\mathfrak{p}_{i+1}$ are irreducible. Let us write $\lambda_i(T) = \mathrm{ad}_{\mathfrak{p}_i/\mathfrak{p}_{i+1}}(T), T \in \mathfrak{g}(q)$. Poguntke has shown that every character χ of the algebra $L^1(\mathbb{R}^n, \omega) \simeq p_\lambda * L^1(S, L^1(H)/\ker\gamma) * p_\lambda$ is of the form

$$\chi(h) = \int_{\mathbb{R}^n} h(t)\chi(t)dt, \quad h \in L^1(\mathbb{R}^n, \omega)$$

where

$$\chi(t) = \chi_\nu(t) = e^{\nu(T)}, \quad t \simeq T \in \mathbb{R}^n \simeq \exp\sigma$$

and where ν is a complex linear form on σ such that

$$1/2 \sum_0^k |\operatorname{tr}(\lambda_i)(T)| \geq |\mathbb{R}(\nu(T))|, \quad T \in \sigma. \tag{8.13.1}$$

Since we always have that $|\chi_\nu(h)| \leq \|h\|_\omega$ for every $h \in L^1(\mathbb{R}^n, \omega)$, we see that $\omega(t) \geq |\chi_\nu(t)|$ almost everywhere in t. Since the previous relations must hold for every linear form ν such that (8.13.1) is satisfied, we must have

$$\omega(\exp T) \geq e^{1/2 \sum_0^k |\operatorname{tr}(\lambda_i)(T)|}, \quad T \in \sigma.$$

Hence we see that the weight is of exponential growth in at least one direction of $S = \exp\sigma$ precisely when $\lambda_i \neq 0$ for at least one i. Let us recall that Boidol has shown in [26] that this condition is equivalent to the following. Let $\mathfrak{m}(\ell)^k = [\,m(\ell), \mathfrak{m}(\ell)^{k-1}], k = 1, 2 \cdots$ be the lower central series of $\mathfrak{m}(\ell)$. Then one of the λ_i is different from 0 if and only if $\ell(\mathfrak{m}(\ell)^\infty) \neq \{0\}$, where $\mathfrak{m}(\ell)^\infty = \bigcap_{k \in \mathbb{N}} \mathfrak{m}(\ell)^k$.

The Estimates

Let us fix $x = \exp X \in S \subset G(\ell)$ and let us study the growth of $\omega(\exp tX)$ for $t \in \mathbb{R}$ and X fixed. Hence in the following arguments the constants may depend on the choice of X, but not on t. Take λ, p_λ, v_λ, v as in Sect. 8.12. Recall that

$$\omega(\exp tX) = \|v(\exp tX)\|_{L^1(H)/\ker(\gamma)} \text{ with } v(\exp tX) = v_\lambda(\exp tX) \bmod \ker(\gamma).$$

We shall make the computations for a special choice of λ (a Gaussian function). This is possible because different choices of λ give equivalent weights.

Put $\sigma(g) = e^{\frac{1}{2}\sum_{j=1}^{m} |\operatorname{tr} \lambda_j(\log g)|}$ for $g \in G(\ell)$, where $\lambda_j(\cdot) = \mathrm{ad}_{\mathfrak{p}_j/\mathfrak{p}_{j+1}}(\cdot)$. We are going to see that there are constants C and C' such that

$$\omega(\exp tX) \leq C' \cdot (1 + |t|)^C \cdot e^{\frac{|t|}{2}\sum_{j=1}^{m} |\operatorname{tr} \lambda_j(X)|} = C' \cdot (1 + |t|)^C \cdot \sigma(\exp tX).$$

Let

$$D = \mathrm{ad}_{\mathfrak{h}} X : \mathfrak{h} \to \mathfrak{h}$$

and let for $X \in \sigma \subset \mathfrak{g}(\ell)$:

$$D = D_s + D_n, \quad [D_s, D_n] = 0$$

be the Jordan decomposition of D into a semisimple part D_s and a nilpotent part D_n. Let us write $\mathrm{Exp}(tD_i) = \sum_{k=0}^{\infty} \frac{t^k}{k!} D_i^k$, for $i = s$ or n, and $\exp tD_i$, $i = s$ or n, for the corresponding automorphism of H such that

$$\exp tD_i \circ \exp = \exp \circ \mathrm{Exp}(tD_i), \quad i = s \text{ or } n.$$

Hence

$$\exp tD_s \cdot \exp tD_n = \exp tD_n \cdot \exp tD_s$$

is equal to the inner automorphism corresponding to $\exp tX$. Let $\gamma \equiv \pi_{\ell,\mathfrak{p}}$ be as above:

$$\Big(\gamma(\exp tX)\xi\Big)(y) = \Delta^{-\frac{1}{2}}(\exp tX)\chi_\ell(\exp tX)\xi(\exp -tX\, y \exp tX)$$

for $\xi \in \mathfrak{H}_\gamma \equiv L^2(H/P \cap H, \ell_{|\mathfrak{h}})$. Here Δ is the real character of $G(\ell)$ defined by

$$\Delta(s) := e^{\operatorname{tr}(\mathrm{ad}_{\mathfrak{n}/\mathfrak{p}_0}(\log s))}, \; s \in G(\ell).$$

Then we define γ^s and γ^n on $\mathfrak{H}_\gamma$ by

$$\left(\gamma^n(\exp tX)\xi\right)(y) = \xi(\exp -tD_n)(y))$$

$$\left(\gamma^s(\exp tX)\xi\right)(y) = \Delta^{-\frac{1}{2}}(\exp tX\chi_\ell(\exp tX)\xi(\exp -tD_s(y)).$$

With these definitions

$$\gamma(\exp tX) = \gamma^n(\exp tX)\gamma^s(\exp tX) = \gamma^s(\exp tX)\gamma^n(\exp tX).$$

Moreover, $\exp tD_n$ induces an action on $L^1(H)$ and $L^1(H)/\ker(\gamma)$ by

$$f^{\exp(tD_n)}(h) = f(\exp tD_n h), \forall h \in H.$$

Take $\lambda \in \mathscr{E}\mathscr{S}(H/P \cap H, \ell_{|\mathfrak{h}})$ (to be fixed more precisely later). By (8.10) there are $b_t, c_t \in L^1(H)/\ker(\gamma)$ such that

$$\gamma(b_t) = <\cdot, \lambda> \gamma^s(\exp tX)^{-1}\lambda = P_{\gamma^s(\exp tX)^{-1}\lambda,\lambda},$$

$$\gamma(c_t) = <\cdot, \lambda> \gamma^n(\exp tX)^{-1}\lambda = P_{\gamma^n(\exp tX)^{-1}\lambda,\lambda}, \quad t \in \mathbb{R}.$$

Proposition 8.13.1 *We have*

$$v(\exp tX) = b_t^{\exp(tD_n)} * c_{t\,ker(L^1(H))}(\gamma).$$

Proof To check this, just apply the representation γ to both sides. □

In order to compute $\omega(\exp tX)$, we then have to determine $\|b_t^{\exp(tD_n)}\|_{L^1(H)/\ker\gamma}$ and $\|c_t\|_{L^1(H)/\ker\gamma}$. To do this, we choose λ to be a Gaussian function in the following way. Notice first that

$$\mathfrak{h}/\mathfrak{p} \simeq \mathfrak{h}/\mathfrak{h} \cap \mathfrak{p} + \mathfrak{n} \oplus n/\mathfrak{p}_0.$$

As $\mathfrak{h}$ and n are ideals, they are both $\mathrm{ad}_\mathfrak{h} X$- and D_s-invariant. Moreover $D_s(\mathfrak{h}) \subset \mathfrak{n}$. By the semisimplicity of D_s there exist vectors $W_1, \cdots, W_k$ such that $D_s W_i = 0$ for all i and $\mathfrak{h} = \sum_{i=1}^k \mathbb{R}W_i \oplus (\mathfrak{p} + n)$. Let us choose the vectors W_i so that $|W_i| = 1$, where $|W_i|$ stands for the Euclidean norm of W_i. The vector subspaces $\mathfrak{v}_j, j = 1, \cdots, m$, in the decomposition of (8.11.3) may also be chosen to be D_s-invariant. As

$$\mathfrak{h} = \sum_{i=1}^k \mathbb{R}W_i \oplus \sum_{j=1}^{m}{}^{\oplus} \mathfrak{v}_j \oplus \mathfrak{p},$$

we may then define $\lambda \in \mathscr{E}\mathscr{S}(H/P, \ell_{|\mathfrak{h}})$ by

$$\lambda(\exp t_1 W_1 \cdots \exp t_k W_k \cdot \exp V_1 \cdots \exp V_m \cdot p_1) = e^{-\sum_{j=1}^m |V_j|^2 - \sum_{i=1}^k |t_i|^2} \cdot \overline{\chi_\ell(p)} \Delta_{P,H}^{1/2}(p)$$

$$V_j \in \mathfrak{v}_j, p \in P,$$

where $|V_j|$ stands for the Euclidean norms of the corresponding vectors. An easy computation shows that

$$< \gamma^s(\exp tX)\lambda, \lambda > = \Delta^{-\frac{1}{2}}(\exp tX)\chi_\ell(\exp tX)(\frac{2\pi}{\sqrt{2}})^k \cdot \prod_{j=1}^m \int_{\mathfrak{v}_j} e^{-|\operatorname{Exp}(-tD_s)V_j|^2 - |V_j|^2} dV_j$$

$$= \Delta^{-\frac{1}{2}}(\exp tX)\chi_\ell(\exp tX)(\frac{2\pi}{\sqrt{2}})^k \cdot \prod_{j=1}^m \frac{2\pi}{\det_{\mathfrak{v}_j}(E_s^* E_s + I)^{\frac{1}{2}}}$$

with $E_s = \operatorname{Exp}(-tD_s)$. Finally,

$$| < \gamma^s(\exp tX\lambda, \lambda >^{-1} | \leq C \cdot e^{\sum_{i=1}^m |\frac{t}{2}||\operatorname{tr}\lambda_i(X)|} = C \cdot \sigma(\exp tX).$$

Let us now define the action of $\exp tD_s$ on $L^1(H)/\ker(\gamma)$ by

$$f^t(h) = f^{\exp(tD_s)}(h) = \det\Big(\operatorname{Exp}(tD_s)\Big) f(\exp tD_s(h)) = e^{\operatorname{tr}(tD_s)} f(\exp tD_s(h)).$$

This action is isometric and

$$\gamma(f^t) = \gamma^s(\exp tX)^{-1} \circ \gamma(f) \circ \gamma^s(\exp tX).$$

It is then easy to check that

$$b_t =< \gamma^s(\exp tX)\lambda, \lambda >^{-1} p_\lambda^t * p_\lambda \bmod \ker(\gamma)$$

and hence that

$$\|b_t\|_{L^1(H)/\ker_{L^1(H)}(\gamma)} \leq C \cdot \|p_\lambda\|^2_{L^1(H)/\ker_{L^1(H)}(\gamma)} \sigma(\exp tX) = \beta \cdot \sigma(\exp tX).$$

Lemma 8.13.2 *The norm of the function c_t is bounded by*

$$\|c_t\|_{L^1(H)/ker_{L^1(H)}(\gamma)} \leq C' \cdot (1 + |t|)^C$$

for some constants C and C'.

Proof Recall that $\gamma(c_t) = P_{\gamma^n(\exp tX)^{-1}\lambda, \lambda}$ has kernel

$$F_t(x, y) = \lambda(\exp tD_n x)\overline{\lambda(y)}.$$

Hence

$$c_t = R(F_t)\ker_{L^1(H)}(\gamma)$$

where R is the retract of Theorem 2.7.3. The result follows then from the continuity of the retract, from the nilpotency of the action and from the formula of a nilpotent orbit with respect to a Jordan-Hölder basis. □

Proposition 8.13.3 *There exist constants $C > 0$, $C' > 0$ such that for every $X \in \mathfrak{s} \subset \mathfrak{g}(\ell)$ of length one, the weight ω satisfies*

$$\omega(\exp tX) \leq C'(1+|t|)^C \sigma(\exp tX) = C'(1+|t|)^C \prod_{i=1}^{m} e^{\frac{|t|}{2}|\, tr\lambda_i(X)|}, t \in \mathbb{R}.$$

Proof This result is evident because the weight ω is continuous on S, because of the previous computations and due to

$$\omega(\exp tX) = \|v(\exp tX)\|_{L^1(H)/\ker_\gamma} = \|b_t^{\exp(tD_n)} * c_t\|_{L^1(H)/\ker_\gamma}.$$

□

Notice that in the previous proposition the functions $\omega(\exp tX)$ and $\sigma(\exp tX)$ are in fact constant on the cosets of $G(\ell)$ mod $G(\ell) \cap N$, and may therefore be considered as functions on the quotient $G(\ell)/G(\ell) \cap N$.

Corollary 8.13.4 *Let χ be a character of the Banach algebra $L^1(G(\ell)/G(\ell) \cap N, \omega) \equiv L^1(S, \omega) \equiv L^1(\mathbb{R}^n, \omega)$. Then*

$$|\chi(\exp tX)| \leq \prod_{i=1}^{m} e^{\frac{|t|}{2}|\, tr\lambda_i(X)|} = \sigma(\exp tX)$$

for all $t \in \mathbb{R}$, $X \in \mathfrak{s} \equiv \mathfrak{g}(\ell)/\mathfrak{g}(\ell) \cap \mathfrak{n}$.

Proof As χ is automatically continuous on $L^1(S, \omega)$, there exists a constant $K > 0$ such that $|\chi(\exp tX)| \leq K \cdot \omega(\exp tX)$, for all $t \in \mathbb{R}$, $X \in U$. As a matter of fact, we may take $K = 1$, since we can replace $\exp tX$ by $\exp ntX$, $n \in \mathbb{N}^*$. Let us write $\chi(\exp tX) = e^{\rho(tX)}$ with ρ a complex linear form on $\sigma \equiv \mathbb{R}^n$. Then

$$|\chi(\exp tX)| = e^{\mathfrak{Re}\rho(tX)}, \quad \forall t \in \mathbb{R}, \forall X \in \sigma.$$

Assume that there is $X_0 \in \sigma$ such that $\mathfrak{Re}\rho(X_0) > 0$ (otherwise, change X_0 to $-X_0$) and such that

$$|\chi(\exp X_0)| = e^{\mathfrak{Re}\rho(X_0)} = e^{|\mathfrak{Re}\rho(X_0)|} > \prod_{i=1}^{m} e^{\frac{1}{2}|\, \mathrm{tr}\lambda_i(X_0)|}.$$

Hence

$$\prod_{i=1}^{m} e^{\frac{|t|}{2}|\mathrm{tr}\lambda_i(X_0)|} < e^{|t||\Re e\rho(X_0)|} = |\chi(\exp |t|X_0)| \le \omega(\exp |t|X_0) \le C'(1+|t|)^C \prod_{i=1}^{m} e^{\frac{|t|}{2}|\,\mathrm{tr}\lambda_i(X_0)|}$$

and so

$$1 < e^{|t|(|\Re e\rho(X_0)| - \frac{1}{2}\sum_{i=1}^{m} |\,\mathrm{tr}\lambda_i(X_0)|)} \le C'(1+|t|)^C$$

for all $t \in \mathbb{R}^*$. As this is impossible,

$$|\chi(\exp X)| \le \prod_{i=1}^{m} e^{\frac{1}{2}|tr\lambda_i(X)|} = \sigma(\exp tX),$$

for all $X \in \sigma$. □

8.14 Classifications of Bounded Irreducible G-modules: The Main Theorem

We have proved the following theorem.

Theorem 8.14.1 *Let G be an exponential solvable Lie group. Every $T \in L^1(G)^{\hat{}top}$ is characterized by a triple $({}^Gq, S, \|\!| \cdot \|\!|)$, where: $q \in \mathfrak{n}^*$, ${}^Gq$ is the orbit of q in $\mathfrak{n}^*$, with $S \in L^1(\mathbb{R}^n, \omega)^{\hat{}top}$, and $\|\!| \cdot \|\!|$ is a TS extension norm.*

The G-orbit ${}^Gq$ is obtained in the following way. Let $N = \exp\mathfrak{n}$ be a closed connected subgroup of G which contains $[G, G]$. There exists $q \in \mathfrak{n}^*$ such that $\ker_{L^1(N)}T = ker_{L^1(N)}({}^G\tau)$, $\tau = \pi_q \in \hat{N}$. Let $\ell \in \mathfrak{g}^*$ be such that $\ell_{|\mathfrak{n}} = q$. Choose a subalgebra $\mathfrak{h}$ of $\mathfrak{g}$ containing $[\mathfrak{g}, \mathfrak{g}]$ such that $\mathfrak{g}(\ell) \cap \mathfrak{h} \subset \mathfrak{n}$.

Suppose first that $\mathfrak{h} \neq \mathfrak{g}$. Let $n := \dim(\mathfrak{g}/\mathfrak{n})$. The weight ω on the group $(\mathbb{R}^n, +)$ has polynomial or exponential growth, and is determined by an idempotent element p acting on the algebra $L^1(G)/ker_{L^1(G)}(\pi^\tau)$, ($\pi^\tau = \mathrm{ind}_N^G\tau$) so that

$$p * \left(L^1(G)/L^1(G)/ker_{L^1(G)}(\pi^\tau)\right) * p \cong L^1(\mathbb{R}^n, \omega) \cong L^1(G(\ell)/(G(\ell) \cap N), \omega).$$

To find p, one constructs a unitary representation $\gamma \in \hat{H}$ such that $\ker \gamma = \ker_{L^1(H)}T$.

The previous constructions are independent of the idempotent element p, once we fix the subgroup H.

If $\mathfrak{h} = \mathfrak{g}$, i.e., $n = 0$, then the algebra $\left[p * L^1(G)/(L^1(G) * ker_{L^1(N)}T)^- * p\right]$ is one-dimensional and isomorphic to the Banach algebra $(\mathbb{C}, +, \cdot, | \cdot |)$.

Suppose now that $\mathfrak{h} \neq \mathfrak{g}$, i.e., $n > 0$.

The extension norm $|||\cdot|||$ *allows us to build irreducible* $\left[L^1(G)/(L^1(G) * \ker_{L^1(N)} T)^-\right]$*-modules starting with a topologically simple* $\left[p * L^1(G)/(L^1(G) * \ker_{L^1(N)} T)^- * p\right]$*-module* (S, Y) *as submodule of the universal module* X^{min}.

This determines all irreducible representations for a given kernel in $L^1(N)$. *Furthermore,*

$$T = ({}^G q, S, |||\cdot|||) \text{ and } T_1 = ({}^G q_1, S_1, |||\cdot|||_1)$$

are equivalent if and only if ${}^G q = {}^G q_1$, S *and* S_1 *are equivalent representations,* $|||\cdot|||$ *and* $|||\cdot|||_1$ *are equivalent extension norms.*

8.14.1 Relationships Between Kernels

Let G be an exponential solvable Lie group. The following relation between the kernel of an irreducible representation of $L^1(S, L^1(H)/\ker_{L^1(H)}(\gamma))$ and the kernel of its restriction to $p * L^1(S, L^1(H)/\ker_{L^1(H)}(\gamma)) * p$ is easy to obtain. In order to simplify notations, let us identify $L^1(G)$ and $L^1(S, L^1(H))$ and also the representation (T, X) of $L^1(G)$ and the corresponding representation of $L^1(S, L^1(H)/\ker_{L^1(H)}(\gamma))$.

Proposition 8.14.2 *Let* (T, X) *be an irreducible representation of* $L^1(G)$. *Let as before* p *be the idempotent element acting on* $L^1(G)/(L^1(G) * \ker_{L^1(N)} T)^-$. *Let* $Y = T(p)X$ *and* $S = T|_Y$. *Then*

$$ker_{L^1(G)}(T) = \{F \in L^1(G)\,;\ S(p * u * F * v * p) = 0 \quad \forall u, v \in L^1(G)\}$$

and $ker_{L^1(G)}(T)$ *is completely determined by the restriction* S *of* T.

Proof It suffices to apply the irreducibility of (T, X). □

Corollary 8.14.3 *Let* (T, X) *and* (T', X') *be two irreducible representations of* $L^1(G)$ *such that* $ker_{L^1(N)}(T) = ker_{L^1(N)}(T'))$. *For* $Y = T(p)X$ *and* $Y' = T(p)X'$ *let* (S, Y) *and* (S', Y') *be the irreducible representations of* $p * L^1(G) * p/ker_{L^1(G)}(T)$ *obtained by restriction. Then*

$$\ker S = \ker S' \iff ker_{L^1(G)}(T) = ker_{L^1(G)}(T').$$

Corollary 8.14.4 *Let* G *be a non-symmetric exponential Lie group. Then there exist many closed prime ideals of* $L^1(G)$ *which are not primitive.*

Proof Let T be an irreducible representation of $L^1(G)$, built as explained in the paper and associated to an exponential weight ω. Such a weight exists, since G is not symmetric [139]. Let $A = L^1(S, L^1(H)/\ker_{L^1(H)}(\gamma))$. If $T|_{pAp}$ is not a character, then its kernel is not of codimension 1. Such a T exists since ω is exponential (see Sect. 8.13.1). According to Proposition 8.2.26 and Schur's lemma, the kernel of the restriction of a simple module T' to pAp has codimension 1. Hence by (8.14.3), $\ker(T) \neq \ker(T')$. Then $\ker(T)$ is a non-primitive prime ideal. □

8.15 Characterization of Simple Modules of $L^1(G)$

8.15.1 A Family of Simple Modules

Let $G = \exp \mathfrak{g}$ be a connected, simply connected, solvable exponential Lie group. Let $l \in \mathfrak{g}^*$ and let $\mathfrak{p}$ be a suitable Pukanszky polarization for ℓ in $\mathfrak{g}$. For every $\overline{p} = (p_1, \ldots, p_m) \in [1, \infty]^m$ we have defined in Sect. 8.11.3 the representation $\pi_{\ell,\mathfrak{p},\overline{p}}$ by induction on an $L^{\overline{p}}$-space, where the norm $\| \cdot \|_{\overline{p}}$ was obtained by successive L^{p_j}-norms, with distinct p_j in different directions. By Proposition 8.11.15 this representations is irreducible. Therefore we have

Corollary 8.15.1 *Let $L^{\overline{p},fin}(G/P, \ell)$ be*

$$\text{span}\,\{\pi_{\ell,\mathfrak{p},\overline{p}}(F)\xi \mid \xi \in L^{(\overline{p}}G/P, \ell), F \in L^1(G) \text{ such that } \pi_{\ell,\mathfrak{p},\overline{p}}(F) \text{ has finite rank}\}.$$

Then $\pi^{fin}_{\ell,\mathfrak{p},\overline{p}} := \pi_{\ell,\mathfrak{p},\overline{p}}|_{L^{\overline{p},fin}(G/P,\ell)})$ is a simple representation of $L^1(G)$.

Proof We have seen in the proof of Proposition 8.11.15 that the G-invariant subspace is different from $\{0\}$. It suffices to apply Proposition 8.2.17. □

8.15.2 A Character

Let us determine $\ker_{L^1(N)}(\pi_{\ell,\mathfrak{p},\overline{p}})$. For $a \in L^1(N)$ and $\xi \in \mathscr{E}\mathscr{S}(G/P, \ell, \overline{p})$ we have for $u \in G, m \in N$ that

$$\begin{aligned}
\pi_{\ell,\overline{p}}(a)(\xi)(u) &= \int_N f(n)\xi(n^{-1}um)dn \\
&= \int_N \Delta_G(u^{-1}) f(umn^{-1}u^{-1})\xi(un)dn \\
&= \int_{N/P^0} \Delta_G(u^{-1}) f(ump^{0^{-1}}n^{-1}u^{-1})\xi(unp^0)dn \\
&= \int_{N/P^0} \Delta_G(u^{-1}) \int_{P^0} \chi_\ell(p^0) f(ump^0n^{-1}u^{-1})dp^0\xi(un)d\dot{n}.
\end{aligned}$$

Hence we see that

$$\ker_{L^1(N)}(\pi_{\ell,\mathfrak{p},\overline{p}}) = \{f \in L^1(N); \pi_{\mathrm{Ad}^*(g)(\ell_{|\mathfrak{n}})} = 0, \forall g \in G\}.$$

Let $(T, X) := (\pi^{fin}_{\ell,\mathfrak{p},\overline{p}}, L^{\overline{p},fin}(G/P, \ell, \overline{p}))$, $\tau = \pi_{\ell_{|\mathfrak{n}}} \in \hat{N}$, $\gamma = \pi_{\ell_{|\mathfrak{h}}} \in \hat{H}$ and $\pi^{\gamma} := \mathrm{ind}_H^G \gamma$. Choose $\lambda \in \mathcal{ES}(H/P^1, \ell_{|\mathfrak{h}}, \overline{p}) \simeq \mathcal{ES}(G/P, \ell, \overline{p})$ and choose as in Sect. 8.12, for every $s \in S$, an element

$$p_{s^{-1}\cdot\lambda,\lambda} \in L^1(H)$$

such that

$$\gamma(p_{s\cdot\lambda,\lambda}) = P_{s^{-1}\cdot\lambda,\lambda}.$$

We need to compute

$$\pi_{\ell,\mathfrak{p},\overline{p}}(F)$$

for any F in the commutative subalgebra $L^1(G)_\lambda = p_\lambda * L^1(G) * p_\lambda \bmod \ker_{L^1(G)}(\pi^\lambda)$. In order to do that we choose a special polarization $\mathfrak{p}$ at ℓ. From Sect. 8.11.3 we have

$$\begin{aligned} \mathfrak{g}(\ell) + \mathfrak{n} &= \sigma \oplus \mathfrak{n} \text{ with } \sigma \subset \mathfrak{g}(\ell) \subset \mathfrak{g}(\ell_{|\mathfrak{n}}), \\ \mathfrak{g}(\ell_{|\mathfrak{n}}) + \mathfrak{n} &= \mathfrak{w} \oplus (\mathfrak{g}(\ell) + \mathfrak{n}) = \mathfrak{w} \oplus \sigma \oplus \mathfrak{n} \text{ with } \mathfrak{w} \subset \mathfrak{g}(\ell_{|\mathfrak{n}}). \end{aligned}$$

In that section we took a maximal isotropic subspace $\mathcal{Y}$ in $\mathfrak{w} \subset \mathfrak{g}(\ell_{|\mathfrak{n}})$ and then

$$\mathfrak{p} = \sigma + \mathcal{Y} + \mathfrak{p}^0.$$

Now choose $\mathcal{Y}$ in the kernel of the character $\delta_{\overline{p},\overline{2}}$ of $\mathfrak{g}$ defined by

$$\delta_{\overline{p},\overline{2}}(X) = \sum_{j=1}^{m} \left(\frac{1}{p_j} - \frac{1}{2}\right) \operatorname{tr} \operatorname{ad}_{\mathfrak{p}_j/\mathfrak{p}_{j+1}}(X) = \sum_{j=1}^{m} \left(\frac{1}{p_j} - \frac{1}{2}\right) \operatorname{tr}\lambda_j(X), \quad X \in \mathfrak{g}. \tag{8.15.1}$$

In order to do that it suffices to take a maximal isotropic subspace in $\ker((\delta_{\overline{p},\overline{2}})) \cap \mathfrak{w}$ and if necessary intersect it with the subspace $\{Y \in \mathfrak{g}; \langle \ell, [X_1, Y] \rangle = 0\}$ for some fixed $X_1 \in \mathfrak{w} \setminus \ker((\delta_{\overline{p},\overline{2}}))$. Such a polarization $\mathfrak{p}$ is then contained in the kernel of the character $\delta_{\overline{p},\overline{2}}$. Hence for any $p \in P$ we have that

$$\Delta^{\frac{1}{\overline{p}}}(h) = \Delta^{\frac{1}{\overline{2}}}(p).$$

This means that

$$\mathscr{E}\mathscr{S}(G/P, \ell, \overline{p}) = \mathscr{E}\mathscr{S}(G/P, \ell, \overline{2}).$$

In particular for a function $f \in \mathscr{E}\mathscr{S}(H)$ and $\xi \in \mathscr{E}\mathscr{S}(H/P^1, \ell_{|\mathfrak{h}}, \overline{p}) \simeq \mathscr{E}\mathscr{S}(G/P, \ell, \mathfrak{p}, \overline{p})$ we have by (8.11.6) that

$$\begin{aligned}
\pi_{\ell_{|\mathfrak{h}}, \overline{p}}(f)\xi(u) &= \oint_{H/P^1} \Delta_G(h)^{-1} \int_{P^1} \chi_\ell(p)\Delta^{\frac{1}{p}}(p^{-1}) f(uph^{-1}) dp \xi(h) d\dot{h} \\
&= \oint_{H/P^1} \Delta_G(h)^{-1} \int_{P^1} \chi_\ell(p)\Delta^{\frac{1}{2}}(p^{-1}) f(uph^{-1}) dp \xi(h) d\dot{h} \\
&= \pi_{\ell_{|\mathfrak{h}}, \overline{2}}(f)\xi(u).
\end{aligned}$$

Hence

$$\pi_{\ell_{|\mathfrak{h}}, \overline{p}}(p_{s^{-1}\lambda, \lambda})\xi = \langle \xi, \lambda \rangle s^{-1} \cdot \lambda, \xi \in L^{\overline{2}}(G/P, \ell).$$

Since for $F \in p_\lambda * L^1(G) * p_\lambda / \ker_{L^1(G)}(\pi^\lambda)$, it follows from (8.12.4) that

$$F(s) = q_F(s) p_{s^{-1}\cdot\lambda, \lambda}, s \in S,$$

and so for $u \in H$:

$$\begin{aligned}
\pi_{\ell, \overline{p}}(F)\xi(u) &= \int_G F(h)\xi(h^{-1}u) dh \\
&= \int_S \int_H F(sh)\chi_\ell(s)\Delta^{\frac{1}{p}}(s^{-1})\xi(h^{-1}s^{-1}us) \\
&= \int_S \int_H F(s(s^{-1}us)h)\chi_\ell(s)\Delta^{\frac{1}{p}}(s^{-1})\xi(h^{-1}) dh ds \\
&= \int_S \chi_\ell(s)\Delta^{\frac{1}{p}}(s^{-1}) \int_H \Delta_G(h^{-1}) F(s(s^{-1}us)h^{-1})\xi(h) dh ds \\
&= \int_S \chi_\ell(s) q_H(s)\Delta^{\frac{1}{p}}(s^{-1}) \int_H \Delta_G(h^{-1}) p_{s^{-1}\cdot\lambda, \lambda}(s^{-1}us)h^{-1})\xi(h) dh ds \\
&= \int_S \chi_\ell(s) q_H(s)\Delta^{\frac{1}{p}}(s^{-1}) \oint_{H/P^1} \Delta_G(h^{-1}) \\
&\quad \int_{P^1} \chi_\ell(p_1)\Delta^{\frac{1}{2}}(p_1) p_{s^{-1}\cdot\lambda, \lambda}(s^{-1}us) p_1 h^{-1}) dp_1 \xi(h) d\dot{h} ds \\
&= \int_S \chi_\ell(s) q_H(s)\Delta^{\frac{1}{p}}(s^{-1}) \oint_{H/P^1} \Delta_G(h^{-1})
\end{aligned}$$

$$\int_{P^1} \chi_\ell(p_1)\Delta^{\frac{1}{2}}(p_1)p_{s^{-1}\cdot\lambda,\lambda}(s^{-1}us)p_1h^{-1})dp_1\xi(h)d\dot{h}ds$$
$$= \int_S \chi_\ell(s)q_H(s)\Delta^{\frac{1}{p}}(s^{-1})\Delta^{\frac{1}{2}}(s)ds\langle\xi,\lambda\rangle.$$

Hence

$$\pi_{\ell,\mathfrak{p},\overline{p}}(F) = \Big(\int_S \chi_\ell(s)q_H(s)\Delta^{\frac{1}{2}-\frac{1}{p}}(s)q_F(s)ds\Big)P_{\lambda,\lambda}. \tag{8.15.2}$$

8.15.3 Analysis of Simple $L^1(G)$-modules

Let now (T,X) be any simple $L^1(G)$ module. We know by Theorem 8.10.18 that there exists $\ell \in \mathfrak{g}^*$ such that $\ker_{L^1(N)}(T)$ is the G-prime ideal $\ker_{L^1(N)}(\tau^G)$ for $\tau = \pi_{\ell_{|\mathfrak{n}}} \in \hat{N}$. We take as before $\mathfrak{h} \supset \mathfrak{n}$ such that $\mathfrak{g}(\ell)+\mathfrak{h} = \mathfrak{g}$, $\mathfrak{h}\cap\mathfrak{g}(\ell) \subset \mathfrak{n}$ and consider again the irreducible representation $\gamma = \pi_{\ell_{|\mathfrak{n}},\mathfrak{p}^1}$ of H and the representation $\pi^\gamma = \mathrm{ind}_H^G\tau$ of G. Let $\lambda \in \mathscr{E}\mathscr{S}(H/P^1,\ell,\overline{2})$.

We can again restrict our attention to the case $\mathfrak{h} \neq \mathfrak{g}$.

Let $L^1(G)_\lambda = p_\lambda * L^1(G) * p_\lambda/\ker_{L^1(G)}(\pi^\lambda)$. This Banach algebra is isomorphic to the weighted function algebra $L^1(\mathscr{S},\omega)$, where $\omega(s) = \|p_{s^{-1}\cdot\lambda,\alpha} \bmod \ker_{L^1(H)}(\gamma)\|_1$, $s \in S$ (see Theorem 8.12.3). Since T is simple, the module (T,X) is determined by a character χ_T of $L^1(S,\omega)$ (Proposition 8.2.27). This character χ_T is then given by a continuous function $q_T : S \to \mathbb{C}$ which satisfies the relations

$$q_T(s+s') = q_T(s)q_T(s'),\ s,s' \in S,$$
$$|q_T(s)| \le \omega(s),\ s \in S.$$

We have computed this character $q_{\ell,\overline{p}}$ for the simple module $\pi^{fin}_{\ell,\mathfrak{p},\overline{p}}$ in Sect. 8.15.2. Let

$$\delta_{c+i\delta_{\overline{p}}}(U) = \langle c,U\rangle + i\delta_{\overline{p}}(U),\ U \in \sigma, \tag{8.15.3}$$

where $c := \ell_{|\sigma}$ and $\delta_{\overline{p}} := \delta_{|\sigma}^{\frac{1}{p}-\frac{1}{2}}$. Then

$$\chi_{\pi^{fin}_{\ell,\mathfrak{p},\overline{p}}}(s) = e^{-i\delta_{c+i\delta_{\overline{p}}}(\log(s))},\ s \in S. \tag{8.15.4}$$

We have seen in Corollary 8.13.4 that any character χ of the algebra $L^1(S,\omega)$, which is automatically a linear functional bounded by 1, satisfies the inequality

$$|\langle\chi, f\rangle| \leq \int_S \prod_{i=1}^{m} e^{\frac{1}{2}|\operatorname{tr}\lambda_i(X)|} |f(\exp X)| dX,$$

i.e.,

$$|\chi(s)| \leq e^{\frac{1}{2}|\operatorname{tr}\lambda_i(\log(s))|}, s \in S.$$

Furthermore, by (8.15.4) the weight ω also satisfies

$$\omega(s) \geq \chi_{\pi^{fin}_{\ell,\mathfrak{p},\overline{p}}}(s), s \in S, \ell \in \mathfrak{g}^*, \overline{p} \in [1,\infty]^m,$$

$$\Rightarrow$$

$$\omega(s) \geq \prod_{i=1}^{m} e^{\frac{1}{2}|\operatorname{tr}\lambda_i(\log(s))|}, s \in S. \tag{8.15.5}$$

Proposition 8.15.2 *Define the following subsets of* $(\mathbb{R}^n)^* \equiv \sigma^*$*:*

$$S_1 = \{\sum_{i=1}^{m}(\frac{1}{2} - \frac{1}{p_i})\operatorname{tr}\lambda_i(\cdot) \mid 1 \leq p_i \leq \infty\} = \{\sum_{i=1}^{m} C_i \operatorname{tr}\lambda_i(\cdot) \mid |C_i| \leq \frac{1}{2}, C_i \in \mathbb{R}\}$$

and

$$S_2 = \{\rho \in \sigma^* \mid |\rho(X)| \leq \sum_{i=1}^{m} \frac{1}{2}|\operatorname{tr}\lambda_i(X)|, \forall X \in \sigma \equiv \mathbb{R}^n\}.$$

Then $S_1 = S_2$.

Proof Notice first that the linear form $\nu(\cdot) = \sum_{i=1}^{m}(\frac{1}{2} - \frac{1}{p_i})\operatorname{tr}\lambda_i(\cdot)$ of $\mathfrak{g}(\ell)$ is constant on cosets modulo $\mathfrak{g}(\ell) \cap \mathfrak{n}$, and may hence be considered as a linear form on $\mathfrak{g}(\ell)/\mathfrak{g}(\ell) \cap \mathfrak{n}$. The sets S_1 and S_2 are closed convex subsets of $(\mathbb{R}^n)^*$ such that $S_1 \subset S_2$. Assume there exists $\rho \in S_2 \backslash S_1$. By the Hahn-Banach theorem there are $X_0 \in \mathbb{R}^n \equiv \sigma$ and $\alpha \in \mathbb{R}$ such that

$$s_1(X_0) < \alpha < \rho(X_0), \quad \forall s_1 \in S_1.$$

Let us then choose $s_1 \in S_1$ by

$$s_1(X) = \sum_{i=1}^{m} \frac{1}{2}\varepsilon_i \operatorname{tr}\lambda_i(X), \quad \forall X \in \sigma,$$

where $\varepsilon_i = 1$ if $\mathrm{tr}\lambda_i(X_0) \geq 0$ and $\varepsilon_i = -1$ if $\mathrm{tr}\lambda_i(X_0) < 0$. Hence

$$\sum_{i=1}^{m} \frac{1}{2}|\mathrm{tr}\lambda_i(X_0)| = s_1(X_0) < \rho(X_0),$$

which contradicts the fact that $\rho \in S_2$. □

Corollary 8.15.3 *Let χ be a continuous character of $L^1(S, \omega)$. Then there exist a multi-index $\overline{p} = (p_1, \ldots, p_m)$ and $c \in \sigma^*$ such that*

$$\chi(\exp X) = e^{-i\langle c, X\rangle} \cdot e^{\sum_{j=1}^{m}(\frac{1}{2}-\frac{1}{p_j})tr(\lambda_j(X))}, \quad \forall X \in \sigma.$$

In other words, every character χ of $L^1(S, \omega)$ is of the form

$$\chi(\exp X) = \chi_c(\exp X)e^{\nu(X)},$$

where $c \in \sigma^*$ and $\nu \in \sigma^*$ is such that $|\nu(X)| \leq \frac{1}{2}\sum_{j=1}^{m} |\mathrm{tr}\lambda_j(X)|$.

Proof We may write $\chi(\exp X) = e^{-i\rho_1(X)} \cdot e^{\rho_2(X)}$ with $\rho_1, \rho_2 \in (\mathbb{R}^n)^* \equiv \sigma^*$. By Proposition 8.15.2 we have $\rho_2 \in S_2 = S_1$ and hence there is a multi-index $\overline{p} = (p_1, \ldots, p_m)$, $p_i \in [1, \infty]$ for all i, such that

$$|\chi(\exp X)| = e^{\rho_2(X)} = e^{\sum_{i=1}^{m}(\frac{1}{2}-\frac{1}{p_i})\,\mathrm{tr}\lambda_i(X)}, \quad \forall X \in \sigma.$$

We may then choose $c \in \sigma*$ such that $c = \rho_1$ on σ. □

We can finally determine $Simple(L^1(G))$!

Theorem 8.15.4 *Let (T, X) be a simple $L^1(G)$-module. Then there exist $\ell \in \mathfrak{g}^*$, a polarization $\mathfrak{p}$ for ℓ in $\mathfrak{g}$ and a multi-index $\overline{p} \in [1, \infty]^m$ such that (T, X) is equivalent to the simple module $(\pi^{fin}_{\ell,\mathfrak{p},\overline{p}}, L^{\overline{p},fin}(G/P, \ell))$.*

Proof It suffices to apply Proposition 8.2.26 and Corollary 8.15.3. □

8.15.4 Equivalence of Two Simple $L^1(G)$-modules

Recall that the linear form $\nu(\cdot) = \sum_{j=1}^{m}(\frac{1}{2} - \frac{1}{p_j})\,\mathrm{tr}\,\mathrm{ad}_{\mathfrak{p}_j/\mathfrak{p}_{j+1}}(\cdot)$ may be viewed as a function on $\mathfrak{g}(\ell)/\mathfrak{g}(\ell) \cap \mathfrak{n}$. Moreover the characteristic inequality of such a linear form, i.e.,

$$|\nu(X)| \leq \sum_{j=1}^{m} \frac{1}{2}|\,\mathrm{tr}\,\mathrm{ad}_{\mathfrak{p}_j/\mathfrak{p}_{j+1}}(X)| = \frac{1}{2}\sum_{j=1}^{m} |\,\mathrm{tr}\lambda_j(X)|, \quad \forall X \in \mathfrak{g}(\ell),$$

is independent of the choice of the Jordan-Hölder sequence and independent of the choice of the polarizations $\mathfrak{p}^0$ and $\mathfrak{p}$, for a given ℓ satisfying $q = \ell|_{\mathfrak{n}}$. In fact, if $\mathfrak{p}^0$ is an arbitrary $\mathfrak{g}(q)$-invariant polarization of q in $\mathfrak{n}$, then there is a non-degenerate pairing between $\mathfrak{n}/\mathfrak{p}^0$ and $\mathfrak{p}^0/\mathfrak{n}(q)$ given by the bilinear form $B(X, Y) =< q, [X, Y] >$, $X \in \mathfrak{n}/\mathfrak{p}^0$, $Y \in \mathfrak{p}^0/\mathfrak{n}(q)$ (see [139]). Let us write $\mathfrak{p}_j^{\perp} = \{X \in \mathfrak{n} \mid < q, [X, \mathfrak{p}_j] >\equiv 0\}$. Then

$$\mathfrak{n} = \mathfrak{p}_1 \supset \mathfrak{p}_2 \supset \cdots \supset \mathfrak{p}_{m+1} = \mathfrak{p}^0 = \mathfrak{p}^{0\perp} = \mathfrak{p}_{m+1}^{\perp} \supset \cdots \supset \mathfrak{p}_2^{\perp} \supset \mathfrak{p}_1^{\perp} = \mathfrak{n}(q)$$

is a Jordan-Hölder sequence for the action of $\mathfrak{g}(q)$ on $\mathfrak{n}/\mathfrak{n}(q)$. As, for every $T \in \mathfrak{g}(\ell)$,

$$(\mathrm{ad}_{\mathfrak{p}_j/\mathfrak{p}_{j+1}} T)^* = -\mathrm{ad}_{\,\mathfrak{p}_{j+1}^{\perp}/\mathfrak{p}_j^{\perp}} T,$$

we get the equality

$$\sum_{j=1}^{m} \frac{1}{2} |\mathrm{tr}\ \mathrm{ad}_{\mathfrak{p}_j/\mathfrak{p}_{j+1}}(T)| = \frac{1}{4}\Big(\sum_{j=1}^{m} |\mathrm{tr}\ \mathrm{ad}_{\mathfrak{p}_j/\mathfrak{p}_{j+1}}(T)| + \sum_{j=1}^{m} |\mathrm{tr}\ \mathrm{ad}_{\mathfrak{p}_{j+1}^{\perp}/\mathfrak{p}_j^{\perp}}(T)|\Big).$$

This last sum is computed over all $\mathfrak{g}(q)$-irreducible sub-quotients of $\mathfrak{n}/\mathfrak{n}(q)$ and is independent of the choice of the $\mathfrak{g}(q)$-invariant polarization $\mathfrak{p}_0$. Hence the function $\frac{1}{2}\sum_{j=1}^{m} |\mathrm{tr}\lambda_j(\cdot)| = \frac{1}{2}\sum_{j=1}^{m} |\mathrm{tr}\ \mathrm{ad}_{\mathfrak{p}_j/\mathfrak{p}_{j+1}}(\cdot)|$ is independent of the Jordan-Hölder sequence and of the choice of the polarization.

In the following propositions we shall study the equivalence of two representations of the form $\pi^0_{\ell,\mathfrak{p},\overline{p}}$. The results of (8.2.26) give us first:

Proposition 8.15.5 *Let $\overline{p}, \overline{q} \in [1, \infty]^m$ be two multi-indices. Then*

$$\pi^{fin}_{\ell,\mathfrak{p},\overline{q}}, \simeq \pi^{fin}_{\ell,\mathfrak{p},\overline{p}}$$

if and only if

$$\sum_{i=1}^{m}(\frac{1}{2} - \frac{1}{q_i})\mathrm{tr}\lambda_i(\cdot) = \sum_{i=1}^{m}(\frac{1}{2} - \frac{1}{p_i})\mathrm{tr}\lambda_i(\cdot) = \nu(\cdot)$$

on σ and hence on $\mathfrak{g}(\ell)$, i.e., if the corresponding linear forms $\nu \in (\mathfrak{g}(\ell)/\mathfrak{g}(\ell) \cap \mathfrak{n})^*$ are the same. We shall hence write $\pi^{fin}_{\ell,\mathfrak{p},\nu}$ instead of $\pi^{fin}_{\ell,\mathfrak{p},\overline{p}}$.

Proposition 8.15.6 *Up to equivalence the representation $\pi^{fin}_{\ell,\mathfrak{p},\nu}$ is independent of the choice of the polarization $\mathfrak{p}$ (chosen so that $\mathfrak{p} \cap \mathfrak{n}$ is a $\mathfrak{g}(q)$-invariant polarization of $q = \ell|_{\mathfrak{n}}$ in $\mathfrak{n}$). Hence we shall simply write $\pi^{fin}_{\ell,\nu}$ instead of $\pi^{fin}_{\ell,\mathfrak{p},\nu}$.*

Proof Take $\pi^{fin}_{\ell,\mathfrak{p},\nu}$ and $\pi^{fin}_{\ell,\mathfrak{p}',\nu}$ for two different choices $\mathfrak{p}$ and $\mathfrak{p}'$ of the polarization. As $\ker_{L^1(N)}\pi^{fin}_{\ell,\mathfrak{p},\nu} = \ker_{L^1(N)}\pi^{fin}_{\ell,\mathfrak{p}',\nu}$, we can make the same study for both representations, with the same choice of H, l, p, S, ω. We get the same algebras $L^1(G/H, L^1(H)/\ker(\gamma))$ and $p * L^1(G/H, L^1(H)/\ker(\gamma)) * p \equiv L^1(G(\ell)/G(\ell) \cap N, \omega)$ as well as the same character χ on $p * L^1(G/H, L^1(H)/\ker(\gamma)) * p$. Thus the construction leads to the same maximal modular left ideals M and hence to the same simple module $(L, L^1(G/H, L^1(H)/\ker(\gamma))/M)$ which is equivalent to both $\pi^{fin}_{\ell,\mathfrak{p},\nu}$ and $\pi^{fin}_{\ell,\mathfrak{p}',\nu}$. □

Proposition 8.15.7 *The representations $\pi^{fin}_{{}^g\ell,{}^g\nu}$ and $\pi^{fin}_{\ell,\nu}$ are equivalent for every $g \in G$.*

Proof Changing $(\ell, \mathfrak{p}, \nu)$ into $({}^g l, {}^g\mathfrak{p}, {}^g\nu)$, with ${}^g\nu \in (\mathfrak{g}({}^g\ell)/\mathfrak{g}({}^g\ell) \cap \mathfrak{n})^*$, gives an equivalent representation. The intertwining operator is obtained by conjugation. □

Proposition 8.15.8 *The representations $\pi^{fin}_{\ell,\nu}$ and $\pi^{fin}_{\ell',\nu'}$ are equivalent if and only if $\ell' = {}^g\ell = \mathrm{Ad}^*(g)(\ell)$ and $\nu' = {}^g\nu$ for some $g \in G$.*

Proof The "only if" part of the proof is given by (8.15.7). Assume now that $\pi^{fin}_{\ell,\nu}$ and $\pi^{fin}_{\ell',\nu'}$ are equivalent. Then $\ker_{L^1(N)}\pi^{fin}_{\ell,\nu} = \ker_{L^1(N)}\pi^{fin}_{\ell',\nu'}$ and hence $\ell'|_\mathfrak{n} \in G \cdot (l|_\mathfrak{n})$. Take $g \in G$ such that $\ell'|_\mathfrak{n} = {}^g(l|_\mathfrak{n})$ and replace ℓ by $\ell'' = {}^g l$ and ν by $\nu'' = {}^g\nu$. By (5.9.4)

$$\pi^{fin}_{\ell',\nu'} \simeq \pi^{fin}_{\ell,\nu} \simeq \pi^{fin}_{{}^g\ell,{}^g\nu} = \pi^{fin}_{\ell'',\nu''}$$

where ℓ' and ℓ'' have the same restriction q to $\mathfrak{n}$. Hence in the analysis of $\pi^{fin}_{\ell',\nu'}$ and $\pi^{fin}_{\ell'',\nu''}$ we may use the same subgroup H. As $\ell''|_\mathfrak{n} = \ell'|_\mathfrak{n}$, we know then that there exists $h \in G$ such that ${}^h\ell''|_\mathfrak{h} = \ell'|_\mathfrak{h}$, since $l + (\mathfrak{g}(\ell) + \mathfrak{n})^\perp \subset G \cdot l$. If we replace ℓ'' by ${}^h\ell'' = {}^{(hg)}l = \ell'''$ and ν'' by ${}^h\nu'' = {}^{(hg)}\nu = \nu'''$, we get $\ell'|_\mathfrak{h} = \ell'''|_\mathfrak{h}$ and $\pi^{fin}_{\ell',\nu'} \simeq \pi^{fin}_{\ell''',\nu'''} \simeq \pi^{fin}_{\ell,\nu}$ by (5.9.4). Since $\ell'|_\mathfrak{h} = \ell'''|_\mathfrak{h}$ we may, in the analysis of $\pi^{fin}_{\ell''',\nu'''}$ and $\pi^{fin}_{\ell',\nu'}$, use the same representation γ and the same projection p. Then $\pi^{fin}_{\ell',\nu'} \equiv \pi^{fin}_{\ell''',\nu'''}$ if and only if the corresponding characters on $L^1(G(\ell')/G(\ell') \cap N, \omega)$ satisfy $\chi_{\ell'-\ell'_0}(\exp X)\cdot e^{\nu'(X)} = \chi_{\ell'''-\ell'_0}(\exp X)\cdot e^{\nu'''(X)}$ for every $X \in \mathfrak{g}(\ell')$ and for some fixed $\ell'_0 \in \mathfrak{g}^*$ such that $\ell'_0|_\mathfrak{h} = \ell'|_\mathfrak{h}$. Equivalently, $\nu' = \nu'''$ on $\mathfrak{g}(\ell')/\mathfrak{g}(\ell')\cap\mathfrak{n}$ and $\ell' - \ell''' \in (\mathfrak{g}(\ell') + \mathfrak{h})^\perp = \mathfrak{g}^\perp = \{0\}$, i.e., $\ell' = \ell''' = {}^{(hg)}l$. Hence ℓ and ℓ' are in the same G-orbit ($\ell' = {}^{(hg)}l$) and we have the same relation $\nu' = {}^{(hg)}\nu$ for the corresponding linear forms ν and ν'. □

Let us write $(\tilde{\mathfrak{g}})^*$ for the collection of pairs (l, ν) with $l \in \mathfrak{g}^*$, $\nu \in (\mathfrak{g}(\ell)/\mathfrak{g}(\ell)\cap\mathfrak{n})^*$ such that $|\nu(X)| \leq \frac{1}{2}\sum_{j=1}^m |\operatorname{tr}\lambda_j(X)|$, $\forall X \in \mathfrak{g}(\ell)$. The group G acts on $\tilde{\mathfrak{g}}^*$ by conjugation. Let $\tilde{\mathfrak{g}}^*/G$ be the set of equivalence classes for this action.

We then get our final theorem about simple modules:

Theorem 8.15.9 *The mapping*

$$\tilde{K} : \tilde{\mathfrak{g}^*}/G \to Simple(G),$$
$$[\ell + i\nu_{\overline{p}}] \to [\pi^{fin}_{\ell,\overline{p}}]$$

is a bijection.

8.15.5 Symmetric Group Algebras

Let $G = \exp \mathfrak{g}$ be an exponential group. We want to know under which conditions the involutive Banach algebra $L^1(G)$ is symmetric (see Definition 8.6.5).

Let $\mathfrak{n}$ be a nilpotent ideal of $\mathfrak{g}$ which contains $[\mathfrak{g}, \mathfrak{g}]$.

Definition 8.15.10 Let $\ell \in \mathfrak{g}^*$. Let

$$\mathfrak{m}(\ell) := \mathfrak{g}(\ell) + \mathfrak{n}.$$

We can now define *Boidol*'s criterium for $*$-regularity (see [26]). We say that the group G satisfies Boidol's criterium if

$$\langle \ell, \mathfrak{m}(\ell)^\infty \rangle = \{0\}$$

for every $\ell \in \mathfrak{g}^*$.

Theorem 8.15.11 *Let* $G = \exp \mathfrak{g}$ *be an exponential solvable Lie group. Then the algebra* $L^1(G)$ *is symmetric if and only if* G *satisfies Boidol's criterium.*

Proof Suppose that Boidol's criterium holds. Let $\ell \in \mathfrak{g}^*$. Since $\mathfrak{m}(\ell)$ is an ideal of $\mathfrak{g}$, so is $\mathfrak{m}(\ell)^\infty$. Now this ideal is contained in the kernel of ℓ and so $\mathfrak{m}(\ell)^\infty$ is a subspace of every polarization $\mathfrak{p}$ at ℓ. This implies that $M(\ell)^\infty = \exp \mathfrak{m}(\ell)^\infty$ is contained in the kernel of every representation $\pi_{\ell,\mathfrak{p},\overline{p}}$. We can therefore suppose that $\mathfrak{m}(\ell)^\infty = \{0\}$. But then $\mathrm{ad}(X)$ is nilpotent for every $X \in \mathfrak{g}(\ell)$, hence $\lambda(X) = 0$ for every root λ of $\mathfrak{g}$. Hence by Proposition 8.15.5, every simple module of $L^1(G)$ is equivalent to a submodule of a unitary one. Therefore G is symmetric.

Suppose now that Boidol's criterium is not satisfied. Let $\ell \in \mathfrak{g}^*$ such that $\langle \ell, \mathfrak{m}(\ell)^\infty \rangle \neq \{0\}$. Let $\mathfrak{p}^0 \subset \mathfrak{n}$ be a $\mathfrak{g}(\ell)$-invariant polarization at $\ell_{|\mathfrak{n}}$. Suppose that $\mathrm{ad}(T)$ is nilpotent on $\mathfrak{n}/\mathfrak{p}^0$ for every $T \in \mathfrak{g}(\ell)$. Then $(\mathfrak{g}(\ell) + \mathfrak{p}^0)^\infty = (\mathfrak{g}(\ell) + \mathfrak{n})^\infty = \mathfrak{m}(\ell)^\infty$. But then

$$\begin{aligned} \langle \ell, \mathfrak{m}(\ell)^\infty \rangle &\subset \langle \ell, [\mathfrak{g}(\ell) + \mathfrak{p}^{fin}], [\mathfrak{g}(\ell) + \mathfrak{p}^{fin}] \rangle \\ &= \{0\}. \end{aligned}$$

This contradiction tells us that there exists $T \in \mathfrak{g}(\ell)$ such that $\mathrm{ad}(T)$ is not nilpotent on $\mathfrak{n}/\mathfrak{p}^0$. Hence we have a root λ of $\mathfrak{g}$ and an $X \in \mathfrak{n}_{\mathbb{C}} \setminus \mathfrak{p}^0_{\mathbb{C}}$ such that $\lambda(T) \neq 0$ and such that $\mathrm{ad}(T)X = \lambda(T)X$ modulo $p^0_{\mathbb{C}}$. Again by Proposition 8.15.5, the algebra $L^1(G)$ admits a simple module which cannot be made unitary, meaning $L^1(G)$ is not symmetric.

□

8.15.6 Final Remarks

As we pointed out in the introduction, the simple $L^1(G)$-modules for a solvable exponential Lie group are essentially obtained in the same way as in the case of nilpotent groups, except that one has to generalize the induced representations. This is no longer true for topologically irreducible representations, as shown in [117]. Two major differences exist. Usually there are a lot of extensions of a topologically irreducible representation of the subalgebra $p * L^1(G/H, L^1(H)/\ker(\gamma)) * p$ to a topologically irreducible representation of the algebra $L^1(G/H, L^1(H)/\ker(\gamma))$, whereas this extension is unique in the algebraic case. These different extensions are characterized by different extension norms, hence by topological issues. But the main difference arises from the irreducible representations of $L^1(S, \omega)$. These representations coincide with the characters in the algebraic case. In the topological case there are many irreducible infinite-dimensional representations of $L^1(S, \omega)$ if the weight ω is exponential, which happens if and only if the group G is non-symmetric. The corresponding representations of $L^1(G)$ are fundamentally other than induced representations. The construction of such representations is related to the invariant subspace problem.

Bibliography

1. J. Abdennadher, J. Ludwig, Disintegrating tensor representations of nilpotent Lie groups. Trans. Amer. Math. Soc. **361**(2), 819–848 (2009)
2. D. Arnal, H. Fujiwara, J. Ludwig, Opérateurs d'entrelacement pour les groupes de Lie exponentiels. Am. J. Math. **118**(4), 839–878 (1996)
3. L. Auslander, B. Kostant,Polarizations and unitary representations of solvable Lie groups. Invent. Math. **14**, 255–354 (1971)
4. A. Baklouti, Nouvelle désintégration lisse de $L^2(G)$ pour les groupes résolubles exponentiels, J. Lie Theory **8**, 25–50 (1998)
5. A. Baklouti, Harmonic analysis on invariant differential operators on nilpotent homogeneous spaces. Russian J. Math. Phys. **6**, 125–136 (1999)
6. A. Baklouti, H. Fujiwara, Opérateurs différentiels associés à certaines représentations unitaires des groupes de Lie résolubles exponentiels. Compositio Math. **139**, 29–65 (2003)
7. A. Baklouti, H. Fujiwara, Commutativité des opérateurs différentiels sur l'espace des représentations restreintes d'un groupe de Lie nilpotent. J. Math. Pures Appl. **83**, 137–161 (2004)
8. A. Baklouti, H. Hamrouni, On the down-up representations of exponential solvable Lie groups. Russ. J. Math. Phys. **8**, 422–432 (2001)
9. A. Baklouti, J. Ludwig, Désintégration des représentations monomiales des groupes de Lie nilpotents. J. Lie Theory **9**, 157–191 (1999)
10. A. Baklouti, J. Ludwig, Invariant differential operators on certain nilpotent homogeneous spaces. Monatsh. Math. **134**, 19–37 (2001)
11. A. Baklouti, J. Ludwig, Entrelacement des restrictions des représentations unitaires des groupes de Lie nilpotents. Ann. Inst. Fourier **51**, 395–429 (2001)
12. A. Baklouti, A. Ghorbel, H. Hamrouni, Sur les représentations mixtes des groupes de Lie résolubles exponentiels. Publ. Matemàtiques **46**, 179–199 (2002)
13. A. Baklouti, H. Fujiwara, J. Ludwig, Analysis of restrictions of unitary representations of a nilpotent Lie group. Bull. Sci. Math. **129**, 187–209 (2005)
14. A. Baklouti, H. Hamrouni, F. Khlif, Analysis of some monomial representations of exponential solvable Lie groups. Russ. J. Math. Phys. **13**, 363–379 (2006)
15. A. Baklouti, H. Hamrouni, The multiplicity function of mixed representations on completely solvable Lie groups. Tokyo J. Math. **30**, 41–55 (2007)
16. A. Baklouti, H. Fujiwara, J. Ludwig, A variant of the Frobenius reciprocity for restricted representations on nilpotent Lie groups, in *Infinite Dimensional Harmonic Analysis IV* (World Scientific, Singapore, 2008), pp.13–31

A. Baklouti et al., *Representation Theory of Solvable Lie Groups and Related Topics*, Springer Monographs in Mathematics,
https://doi.org/10.1007/978-3-030-82044-2

17. A. Baklouti, H. Fujiwara, J. Ludwig, Intertwining operators of irreducible representations for exponential solvable Lie groups. Forum. Math. **27**(4), 2231–2257 (2015)
18. A. Baklouti, H. Fujiwara, J. Ludwig, The polynomial conjecture for restrictions of some nilpotent groups representations. J. Lie Theory **29**, 311–341 (2019)
19. B. Beauzamy, *Introduction to Operator Theory and Invariant Subspaces,* (North-Holland Mathematical Library, North-Holland, 1988)
20. R. Benedetti, J.J. Risler, *Real Algebraic and Semi-Algebraic Sets* (Hermann, Paris, 2010)
21. Y. Benoist, Espaces symétriques exponentiels, Thèse de 3e cycle, Univ. Paris VII, 1983
22. Y. Benoist, Analyse harmonique sur les espaces symétriques nilpotents. J. Func. Anal. **59**, 211–253 (1984)
23. Y. Benoist, Multiplicité un pour les espaces symétriques exponentiels. Mém. Soc. Math. France **15**, 1–37 (1984)
24. P. Bernat, et al., *Représentations des groupes de Lie résolubles* (Dunod, Paris, 1972)
25. J. Bochnak, M. Coste, M.F. Roy, *Géométrie algébrique réelle* (Springer, Berlin, 1998)
26. J. Boidol, $*$-regularity of exponential Lie groups. Invent. Math. **56**, 231–238 (1980)
27. J. Boidol, H. Leptin, J. Schürman, D. Vahle, Räume primitiver Ideale von Gruppenalgebren. Math. Ann. **236**, 1–13 (1978)
28. P. Bonnet, Transformation de Fourier des distributions de type positive sur un groupe de Lie unimodulaire. J. Funct. Analy. **55**, 220–246 (1984)
29. F.F. Bonsall, J. Duncan, *Complete Normed Algebras* (Springer,, Berlin, 1973)
30. N. Bourbaki, *Intégration* (Hermann, Paris, 1967)
31. J. Brezin, Harmonic analysis on nilmanifolds. Trans. Amer. Math. Soc. **150**, 611–618 (1970)
32. J. Brezin, *Harmonic Analysis on Compact Solvmanifolds*. Springer Lecture Notes, vol. 602 (Springer, Berlin, 1977)
33. N. Chabot-Chevalarias, Formules de Plancherel pour des groupes de Lie résolubles exponentiels, thèse Poitiers, 2001
34. C. Y. Chao, Uncountably many nonisomorphic nilpotent Lie algebras. Proc. Amer. Math. Soc. **13**, 903–906 (1962)
35. L. Corwin, F.P. Greenleaf, Character formulas and spectra of compact nilmanifolds. J. Func. Analy. **21**, 123–154 (1976)
36. L. Corwin, F.P. Greenleaf, Intertwining operators for representations induced from uniform subgroups. Acta Math. **136**, 275–301 (1976)
37. L. Corwin, F.P. Greenleaf, Integral formulas with distribution kernels for irreducible projections in L^2 of a nilmanifold. J. Func. Analy. **23**, 255–284 (1976)
38. L. Corwin, F.P. Greenleaf, Spectrum and multiplicities for restrictions of unitary representations in nilpotent Lie groups. Pacific J. Math. **135**, 233–267 (1988)
39. L. Corwin, F.P. Greenleaf, *Representations of Nilpotent Lie Groups and Their Applications, Part I: Basic Theory and Examples* (Cambridge University Press, Cambridge, 1990)
40. L. Corwin, F.P. Greenleaf, Commutativity of invariant differential operators on nilpotent homogeneous spaces with finite multiplicity. Comm. Pure Appl. Math. **45**, 681–748 (1992)
41. L. Corwin, F.P. Greenleaf, Spectral decomposition of invariant differential operators on certain nilpotent homogeneous spaces. J. Funct. Anal. **108**, 374–426 (1992)
42. L. Corwin, F.P. Greenleaf, R. Penney, A canonical formula for the distribution kernels of primary projections in L^2 of a nilmanifold. Commun. Pure Appl. Math. **30**, 355–372 (1977)
43. L. Corwin, F.P. Greenleaf, R. Penney, A general character formula for irreducible projections on L^2 of a nilmanifold. Math. Ann. **225**, 21–32 (1977)
44. L. Corwin, F.P. Greenleaf, G. Grélaud, Direct integral decompositions and multiplicities for induced representations of nilpotent Lie groups. Trans. Amer. Math. Soc. **304**, 549–583 (1987)
45. B.N. Currey, The structure of the space of coadjoint orbits on an exponential solvable Lie group. J. Trans. Am. Math. Soc. **332**(1), 241–269 (1992)
46. B.N. Currey, Smooth decomposition of finite multiplicity monomial representation for a class of completely solvable homogeneous spaces. Pacific. J. Math. **170**, 429–460 (1995)

47. B.N. Currey, R. Penney, The structure of the space of coadjoint orbits of a completely solvable Lie groups. Michigan Math. J. **36**, 309–320 (1989)
48. J. Dixmier, Opérateurs de rang fini dans les representations unitaires. Inst. Hautes Etudes Sci. Publi. Math. **6**, 305–317 (1960)
49. J. Dixmier, *Les C^*-Algèbres and leurs Representations* (Gauthiers-Villard, Paris, 1969)
50. J. Dixmier, *Algèbre Enveloppantes* (Gauthier-Villars, Paris, 1974)
51. J. Dixmier, P. Malliavin, Factorizations de fonctions et de vecteurs indéfiniment differentiables. Bull. Sci. Math. **102**, 305–330 (1978)
52. R.S. Doran, J.M.G. Fell, *Representations of $*$- Algebras, Locally Compact Groups and Banach $*$-Algebraic Bundles*, vol. 2 (Academic, Cambridge, 1988)
53. M. Duflo, Sur les extensions des représentations irréductibles des groupes de Lie nilpotents. Ann. Sci. ENS **5**, 71–120 (1972)
54. M. Duflo, *Open Problems in Representation Theory of Lie Groups*, ed. by T. Oshima, Katata in Japan (1986), pp. 1–5
55. M. Duflo, M. Rais, Sur l'analyse harmonique sur les groupes de Lie résolubles. Ann. Sci. Ec. Norm. Supér., IV. Sér. **9**, 107–144 (1976)
56. P. Eberlein, Geometry of 2-step nilpotent Lie groups, in *Modern Dynamical Systems and Applications* (Cambridge University Press, Cambridge, 2004), pp. 67–101
57. S.M. Elhacen, C. Molitor-Braun, Etude de l'algèbre $L^1_\omega(G)$ avec G groupe de Lie nilpotent et ω poids polynomial. Travaux mathématiques, Publications du Centre Universitaire de Luxembourg **X**, 77–94 (1998)
58. S.M. Elhacen, J. Ludwig, C. Molitor-Braun Characterization of the simple L1(G)-modules for exponential Lie groups. Pacific J. Math. **212**, 133–156 (2003)
59. J. Fox, On the spectra of compact nilmanifolds. Trans. Amer. Math. Soc. **290**, 577–583 (1985)
60. H. Fujiwara, Certain opérateurs d'entrelacement pour des groupes de Lie résolubles exponentiels et leur applications. Mem. Fac. Sci. Kyushu Univ. Ser. A, **36**, 13–72 (1982)
61. H. Fujiwara, Polarisations réelles et représentations associées d'un groupe de Lie résoluble. J. Func. Anal. **60**, 102–125 (1985)
62. H. Fujiwara, Représentations monomiales des groupes de Lie nilpotents. Pacific J. Math. **127**, 329–351 (1987)
63. H. Fujiwara, Représentations monomiales des groupes de Lie résolubles exponentiels, in *The Orbit Method in Representation Theory, Proceedings of a Conference in Copenhagen*, ed. by M. Duflo, N.V. Pedersen, M. Vergne (Birkhäser, Boston, 1990), pp. 61–84
64. H. Fujiwara, Sur les restrictions des représentations unitaires des groupes de Lie résolubles exponentiels. Inv. Math. **104**, 647–654 (1991)
65. H. Fujiwara, *La formule de Plancherel pour les représentations monomiales des groupes de Lie nilpotents* (World Scientific Publishing, River Edge, 1992), pp. 140–150
66. H. Fujiwara, Sur la conjecture de Corwin-Greenleaf. J. Lie Theory **7**, 121–146 (1997)
67. H. Fujiwara, Analyse harmonique pour certaines représentations induites d'un groupe de Lie nilpotent. J. Math. Soc. Jpn. **50**, 753–766 (1998)
68. H. Fujiwara, Correction à "Analyse harmonique pour certaines représentations induites d'un groupe de Lie nilpotent" J. Math. Soc. Jpn. **52**, 483–483 (2000)
69. H. Fujiwara, *Une réciprocité de Frobenius, in Infinite Dimensional Harmonic Analysis III*, ed. by H. Heyer, et al. (World Scientific, Singapore, 2005), pp. 17–35
70. H. Fujiwara, J. Ludwig, *Harmonic Analysis on Exponential Solvable Lie Groups* (Springer Japan, 2014)
71. H. Fujiwara, S. Yamagami, Certaines représentations monomiales d'un groupe de Lie résoluble exponentiel. Adv. St. Pure Math. **14**, 153–190 (1988)
72. H. Fujiwara, G. Lion, B. Magneron, Opérateurs d'entrelacement et calcul d'obstruction sur des groupes de Lie résolubles, in *Non Commutative Harmonic Analysis and Lie Groups* Lecture Notes in Mathematics, vol. 880 (Springer, Berlin, 1981), pp. 102–137
73. H. Fujiwara, G. Lion, B. Magneron, Algèbres de fonctions associées aux représentations monomiales des groupes de Lie nilpotents. Prèpublication de l'Universitè Paris-Nord, 13, 2002–2 (2002)

74. H. Fujiwara, G. Lion, B. Magneron, S.Mehdi, Commutativity criterion for certain algebras of invariant differential operators on nilpotent homogeneous spaces. Math. Ann. **327**, 513–544 (2003)
75. I.M. Gelfand, M.I. Graev, I.I. Pyatetskii-Shapiro, *Representation Theory and Automorphic Functions* (W. B. Saunders Company, Philadelphia, 1969)
76. A. Ghorbel, The Bonnet-Plancherel formula for monomial representations for classes of completely solvable Lie groups. Trav. Math. **XV**, 45–68 (2004)
77. A. Ghorbel, H. Hamrouni, Discrete co-compact subgroups of the five dimensional connected and simply connected nilpotent Lie groups. SIGMA **5**, 020, 17 (2009)
78. A. Ghorbel, H. Hamrouni, J. Ludwig, M. Selmi, The Bonnet-Plancherel formula for normal monomial representations of exponential solvable Lie groups. Forum Math. **17**, 823–849 (2005)
79. A. Ghorbel, H. Hamrouni, J. Ludwig, Harmonic analysis on two and three-step nilmanifolds. Bull. Sci. Math. **138**, 887–911 (2014)
80. C.S. Gordon, The Laplace spectra versus the length spectra of Riemannian manifolds. Contemp. Math. **51**, 63–79 (1986)
81. C.S. Gordon, E.N. Wilson, Isospectral deformations of compact solvmanifolds. J. Diff. Geom. **19**, 241–256 (1984)
82. C.S. Gordon, E.N. Wilson, The spectrum of the Laplacian on Riemannian Heisenberg manifolds. Mich. Math. J. **33**, 253–271 (1986)
83. R. Gornet, Equivalence of quasi-regular representations of two and three-step nilpotent Lie groups. J. Funct. Anal. **119**, 121–137 (1994)
84. G. Grélaud, Désintégration des représentations induites des groupes de Lie résolubles exponentiels, Thèse de 3e cycle, Univ. de Poitiers, 1973
85. G. Grélaud, La formule de Plancherel pour les espaces homogénes des groupes de Heisenberg. J. Reine Angew. Math. **398**, 92–100 (1989)
86. G. Grélaud, La formule de Plancherel pour les espaces homogénes des groupes de Lie nilpotents. J. Reine Angew. Math. **398**, 92–100 (1989)
87. W. Hauenschild, J. Ludwig, The injection and the projection theorem for spectral sets. Monatsh. Math. **92**, 167–177 (1981)
88. M.E. Herrera, Integration on a semianalytic set. Bull. Soc. Math. France **94**, 141–180 (1966)
89. G. Hochschild, *The Structure of Lie Groups* (Holden-Day, San Francisco, 1965), ix+230 pp.
90. L. Hörmander, *The Analysis of Linear Partial Differential Operators II*. Grundlehren der Mathematischen Wissenschaften, vol. 257 (Springer, Berlin, 1983)
91. A. Hulanicki, A functional calculus for Rockland operators on nilpotent Lie groups. Studia Math. **78**, 253–266 (1984)
92. J.W. Jenkins, A fixed point theorem for exponentially bounded groups. J. Funct. Analy. **22**(4), 346–353 (1976)
93. A.A. Kirillov, Unitary representations of nilpotent Lie groups. Russian Math. Surv. **17**, 57–110 (1962)
94. S. Kobayashi, K. Nomizu, *Foundations of Differential Geometry I* (Interscience Publishers, a Division of Wiley, New York-London, 1963), XI, 329 p.
95. H. Leptin, Ideal theory in group algebras of locally compact groups. Invent. Math. **31**, 259–278 (1976)
96. H. Leptin, J. Ludwig, *Unitary Representation Theory of Exponential Lie Groups* (W. De Gruyter, Berlin, 1994)
97. G. Lion, *Intégrales d'entrelacement sur des groupes de Lie nilpotents et indices de Maslov*. Lecture Notes in Mathematics, vol. 587 (Springer, Berlin, 1977), pp. 160–176
98. G. Lion, Indice de Maslov et représentation de Weil. Pub. Math. Univ. Paris VII **2**, 45–79 (1978)
99. R. Lipsman, Orbital parameters for induced and restricted representations. Trans. Amer. Math. Soc. **313**, 433–473 (1989)
100. R. Lipsman, Induced representations of completely solvable Lie groups. Ann. Scuola Norm. Sup. Pisa **17**, 127–164 (1990)

101. R. Lipsman, The Penney-Fujiwara Plancherel formula for symmetric spaces, in *The orbit method in representation theory. Proceedings of a Conference Held in Copenhagen, August to September 1988*, ed. by M. Duflo, N.V. Pedersen, M. Vergne (Birkhauser, Basel, 1990), pp. 135–145
102. R. Lipsman, Restriction representations of completely solvable Lie groups. Can. J. Math. **42**, 790–824 (1990)
103. R. Lipsman, The Penney-Fujiwara Plancherel formula for Abelian symmetric spaces and completely solvable homogeneous spaces. Pacific J. Math. **151**, 265–295 (1991)
104. R. Lipsman, The multiplicity function on exponential and completely solvable homogeneous spaces. Geom. Dedicata **39**, 155–161 (1991)
105. R. Lipsman, The Penney-Fujiwara Plancherel formula for homogeneous spaces, in *Representation Theory of Lie Groups and Lie Algebras, Proceedings of Fuji-Kawaguchiko Conference* (World-Scientific, Singapore, 1992), pp. 120–139
106. R. Lipsman, Representations of exponential solvable Lie groups induced from maximal subgroups. Michigan Math. J. **40**, 299–320 (1993)
107. R. Lipsman, The up-down formula for nil-homogeneous spaces. Annali di Matematica Pura ed Applicata, (IV) **CLXVI**, 291–300 (1994)
108. R. Lipsman, The Plancherel formula for homogeneous spaces with exponential spectrum. J. Fur Reine und Ang. Math. **500**, 49–63 (1998)
109. S. Lojasiewicz, Triangulations of semianalytic sets. Ann. Sc. Norm. Sup. Pisa **18**, 449–474 (1964)
110. J. Ludwig, A class of symmetric and a class of wiener group algebras. J. Funct. Analy. **31**, 187–194 (1979)
111. J. Ludwig, Polynomial growth and ideals in group algebras. Manuscr. Math. **30**, 215–221 (1980)
112. J. Ludwig, Irreducible representations of exponential solvable Lie groups and operators with smooth kernels. J. Reine Angew. Math. **339**, 1–26 (1983)
113. J. Ludwig, On primary ideals in the group algebra of a Nilpotent Lie group. Math. Ann. **262**, 287–304 (1983)
114. J. Ludwig, *Minimal C^*-Dense Ideals and Algebraically Irreducible Representations of the Schwartz-Algebra of a Nilpotent Lie Group*. Harmonic Analysis (Springer, Berlin, 1987), pp. 209–217
115. J. Ludwig, Dual topology of diamond groups. J. Reine Angew. Math. **467**, 67–87 (1995)
116. J. Ludwig, C. Molitor-Braun, L'algèbre de Schwartz d'un groupe de Lie nilpotent. Trav. Math. **VII**, 25–67 (1995)
117. J. Ludwig, C. Molitor-Braun, Exponential actions, orbits and their kernels. Bull. Austral. Math. Soc. **57**, 497–513 (1998)
118. J. Ludwig, D. Müller, Sub-Laplacians of holomorphic L^p-type on rank one AN-groups and related solvable groups. J. Funct. Analy. **170**, 366–427 (2000)
119. J. Ludwig, H. Zahir, On the nilpotent $\star$- Fourier transform. Lett. Math. Phys. **30**, 23–34 (1994)
120. J. Ludwig, G. Rosenbaum, J. Samuel,The elements of bounded trace in the C^*-algebra of a nilpotent Lie group. Invent. Math. **83**, 167–190 (1986)
121. G.W. Mackey, Induced representations of locally compact groups II. Ann. Math. **58**, 193–221 (1953)
122. G.W. Mackey, Borel structure in groups and their duals. Trans. Amer. Math. Soc. **85**, 134–165 (1957)
123. G.W. Mackey, Unitary representations of group extensions, I. Acta Math. **99**, 265–311 (1958)
124. G.W. Mackey, *The Theory of Unitary Group Representations*. Chicago Lectures in Mathematics (The University of Chicago Press, Chicago, 1976) X, 372 p.
125. B. Magneron, Une extension de la notion d'indices de Maslov. C. R. Acad. Sci. Paris, série A **289**, 683 (1979)
126. A.I. Malcev On a class of homogeneous spaces. Amer. Math. Soc. Transl. **39**, 33 (1951)
127. Y. Matsushima, On the discrete subgroups and homogeneous spaces of nilpotent Lie groups. Nagoya Math. J. **2**, 95–110 (1951)

128. C. Molitor-Braun, Actions exponentielles and idéaux premiers, Thèse Metz, 1996
129. C. Molitor-Braun, Exponential actions and maximal $\mathscr{D}$- invariant ideals. Manuscr. Math. **96**, 23–35 (1998)
130. C.C. Moore, Decomposition of unitary representations defined by discrete subgroups of nilpotent Lie groups. Ann. Math. **82**, 146–182 (1965)
131. C.C. Moore, J. Wolf, Square integrable representations of nilpotent groups. Trans. Amer. Math. Soc. **185**, 445–462 (1973)
132. O.A. Nielsen, M. Rains, A Fourier transform for compact nilmanifolds with flat orbits. Math. Ann. **271**, 209–223 (1985)
133. T.W. Palmer, *Banach Algebras and the General Theory of $*$-Algebras, vol. I, Algebras and Banach Algebras*. Encyclopedia of Mathematics and Its Applications, vol. 49 (Cambridge University Press, Cambridge, 1994)
134. N. Pedersen, On the infinitesimal kernel of irreducible representations of nilpotent Lie groups. Bull. Soc. Math. France **112**, 423–467 (1984)
135. H. Pesce, Calcul du spectre d'une nilvariété de rang deux et applications. Trans. Amer. Math. Soc. **339**, 433–461 (1993)
136. R. Penney, Abstract Plancherel theorem and a Frobenius reciprocity theorem. J. Func. Anal. **18**, 177–190 (1975)
137. J.P. Pier, *Amenable Locally Compact Groups* (Wiley, New York, 1984)
138. D. Poguntke, Operators of finite rank in unitary representations of exponential Lie groups. Math. Ann. **259**, 371–383 (1982)
139. D. Poguntke, Algebraically irreducible representations of L^1-algebras of exponential Lie groups. Duke Math. J. **50**, 1077–1106 (1983)
140. N.S. Poulsen, On C^{∞}-vectors and intertwining bilinear forms for representations of Lie groups. J. Func. Anal. **9**, 87–120 (1972)
141. L. Pukanszky, On the theory of exponential groups. Trans. Amer. Math. Soc. **126**, 487–507 (1967)
142. L. Pukanszky, *Leçon sur les représentations des groupes* (Dunod, Paris 1967)
143. L. Pukanszky, On the unitary representations of exponential groups. J. Funct. Anal. **2**, 73–113 (1968)
144. L. Pukanszky, Unitary representations of solvable Lie groups. Ann. Sci. ENS **4**, 457–608 (1971)
145. M.S. Raghunathan, *Discrete subgroups of Lie Groups*. Ergebnisse der Mathematik, Bd. vol. 68 (Springer, New York, 1972)
146. L. Richardson, Decomposition of the L^2 space of a general compact nilmanifolds. Amer. J. Math. **93**, 173–190 (1971)
147. J.J. Risler, Some aspects of complexity in real algebraic geometry. J. Symb. Computat. **5**, 109–119 (1988)
148. L. Schwartz, *La Théorie des distributions* (Hermann, Paris, 1973)
149. W. Soergel, An irreducible not admissible Banach representation of $SL(2, \mathbb{R})$. Proc. Amer. Math. Soc. **104**, 1322–1324 (1988)
150. M. Vergne, Constructions de sous-algèbres subordonnées à un élément du dual d'une algèbre de Lie. C. R. Acad. Sci. Paris **270**, 704–707 (1970)
151. M. Vergne, Étude de certaines représentations d'un groupe de Lie résoluble exponentiel. Ann. Sci. ENS **3**, 353–384 (1970)
152. G. Warner, *Harmonic Analysis on Semi-Simple Lie Groups I* (Springer, Berlin, 1972)

Index

A. Baklouti et al., *Representation Theory of Solvable Lie Groups and Related Topics*, Springer Monographs in Mathematics,
https://doi.org/10.1007/978-3-030-82044-2